SECOND EDITION

# WORLD REGIONAL GEOGRAPHY

## THE NEW GLOBAL ORDER

MICHAEL BRADSHAW

Boston Burr Ridge, IL Dubuque, IA Madison, WI New York San Francisco St. Louis
Bangkok Bogotá Caracas Lisbon London Madrid
Mexico City Milan New Delhi Seoul Singapore Sydney Taipei Toronto

*McGraw-Hill Higher Education*

*A Division of The **McGraw-Hill** Companies*

WORLD REGIONAL GEOGRAPHY: THE NEW GLOBAL ORDER,
SECOND EDITION

 This book is printed on recycled, acid-free paper containing 10% postconsumer waste.

1 2 3 4 5 6 7 8 9 0 QPD/QPD 0 9 8 7 6 5 4 3 2 1 0

ISBN 0–697–38514–0

Vice president and editorial director: *Kevin T. Kane*
Publisher: *Edward E. Bartell*
Sponsoring editor: *Daryl Bruflodt*
Developmental editor: *Lu Ann Weiss*
Marketing manager: *Lisa L. Gottschalk*
Senior project manager: *Marilyn Rothenberger*
Senior production supervisor: *Mary E. Haas*
Coordinator of freelance design: *Rick Noel*
Photo research coordinator: *John C. Leland*
Supplement coordinator: *Stacy A. Patch*
Compositor: *Precision Graphics*
Typeface: *10/12 Berkeley Medium*
Printer: *Quebecor Printing Book Group/Dubuque, IA*

Freelance designer: *Maureen McCutcheon*
Cover photograph: *Dhaka-Bangladesh © Mufty Munir/AFP*

**Library of Congress Cataloging-in-Publication Data**

Bradshaw, Michael J. (Michael John), 1935–
World regional geography : the new global order / Michael Bradshaw. — 2nd ed.
p. cm.
Includes bibliographical references and index.
ISBN 0–697–38514–0
1. Geography. I. Title.
G116.B73 2000
910—dc21 98–49408
CIP

www.mhhe.com

*To Valerie*

# BRIEF CONTENTS

# CONTENTS

Chapter Three

## AFRICA SOUTH OF THE SAHARA



Chapter Four

## NORTHERN AFRICA AND SOUTHWESTERN ASIA

**Chapter Five**

## SOUTHERN ASIA



**Chapter Six**

## EASTERN ASIA

**Chapter Seven**

## EUROPE



**Chapter Eight**

## COMMONWEALTH OF INDEPENDENT STATES

Chapter Nine

## ANGLO AMERICA



Chapter Ten

## LATIN AMERICA

**Chapter Eleven**

## SOUTH PACIFIC

# PREFACE AND LEARNING SYSTEM IDEAS AND RESPONSES

*This is a new type of preface. It combines:*

- *an introduction to the ideas that stimulated the writing of the book;*
- *the ways in which the author responded to reviewers across the United States; and*
- *connections to ways in which instructors and students may use the book's resources in the learning process.*
- *connections to ways in which instructors and students may use the book's resources in the learning process.*

New features in the second edition, often suggested by reviewers, are highlighted in bold type. As we enter a new century, a new global order is emerging. Travel and communications extend our experience of the world, while international trade and political agreements become more wide-ranging in significance. Knowledge of world geography—and of how local places relate to the global context—is increasingly important for living in our world.

What problems face the Israeli-Palestinian negotiators? Why are some West Indian islands vulnerable to volcanic eruptions? What made the African countries fall behind the rest of the world's economic and political development? Why did the United States ignore Africa until recently, and why is it now showing interest? What differences are the political changes in Russia making to the lives of people in different parts of that country? Will China become a leader among world countries? What are the real problems behind conflicts in Northern Ireland, Bosnia, Kosovo, or Chiapas? Why did the "Asian Miracle" turn sour? Can the European Union overcome national differences among its members—let alone admit former Soviet bloc countries? How are Islamic extremists affecting the governments of their own countries? These issues get headlines in newspapers and on TV news. Each has a geographic component, affecting a particular country or region within that country. Many issues have wider implications for the surrounding countries and may be worldwide.

This text provides a geographic account of the world for the new century. Geographic studies not only inform as to where places are in relation to each other. They evaluate current movements of people from place to place and shifts in the distribution of economic activities. Geographers examine such features in the contexts of continuing cultural characteristics and environmental conditions.

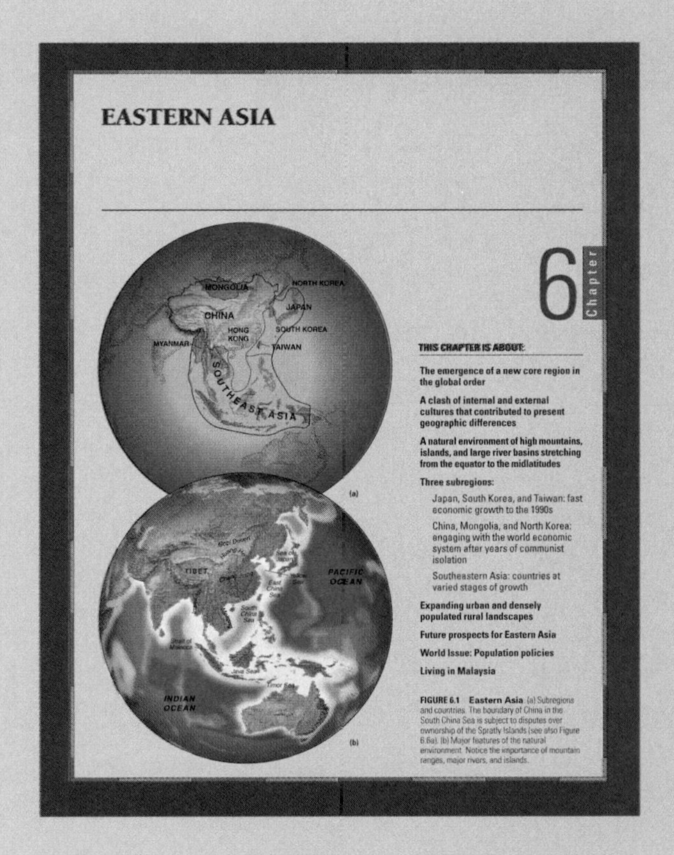

EASTERN ASIA

6 Chapter

THIS CHAPTER IS ABOUT:

**The emergence of a new core region in the global order**

**A clash of internal and external cultures that contributed to present geographic differences**

**A natural environment of high mountains, islands, and large river basins stretching from the equator to the midlatitudes**

**Three subregions:**

Japan, South Korea, and Taiwan: fast economic growth to the 1990s

China, Mongolia, and North Korea: engaging with the world economic system after years of communist isolation

Southeastern Asia: countries at varied stages of growth

**Expanding urban and densely populated rural landscapes**

**Future prospects for Eastern Asia**

**World Issue: Population policies**

**Living in Malaysia**



**FIGURE 6.1 Eastern Asia** (a) Subregions and countries. The boundary of China in the South China Sea is subject to disputes over ownership of the Spratly Islands (see also Figure 6.6a). (b) Major features of the natural environment. Notice the importance of mountain ranges, major rivers, and islands.

## Global Dimension

The global dimension is growing, but does not—and may never—reduce the significance of the regional dimension. Since 1991, world political relationships are no longer inevitably divided by the Cold War. And yet the end of that conflict did not lead to worldwide political leadership from the United Nations or the United States. A single economic system now operates around the world, and yet regional variations may give advantages to some countries. The globalizing of popular culture is likewise gaining ground wherever radio, TV, and the Internet penetrate, but it cannot destroy the influence of older cultural factors. There is a growing sense of environmental oneness, but differences of concern and ability or willingness to deal with issues remain. **At the outset, Chapter 1 now emphasizes the impacts of global forces on local activities—a major focus throughout the text.**

## Regional Dimension

Today regions at various levels operate within the global context. They contribute their own resources and character to the overall picture and are the basic element of geographic studies.

Geographers focus on human and physical elements of our world. **Chapter 2 is devoted to describing the general geographic principles** that underlie regional studies. They include the significance of political divisions, population distributions and dynamics, economic processes, cultural impacts, and the influence of natural environments. Geographers examine the world using a variety of tools that have been enhanced by modern technology. Maps combined with satellite images and statistical data become geographic information systems; book-based information is extended by Internet websites. As geographic studies become more aware of the global, technology has made it possible to gain a greater comprehension of the daunting task of understanding such a large and complex world.

Each of Chapters 3 through 11 contains two boxes that emphasize the significance of both global and regional elements. The *World Issue* boxes focus on issues that affect the region concerned but also have worldwide implications. The *Living in . . .* boxes focus on day-to-day issues in a specific country through the eye of an individual who has lived there; they give a more personalized view of local conditions. **A *Living in Russia* box is added to Chapter 8.**

**WEBSITES: Southern Asia** ***New!***

Further sites are available through the Dorling Kindersley Atlas online facility.

*http://www.virtualbangladesh.com/* is an excellent site. Take the *"Grand Tour"* and access satellite views and maps. What have you learned about the country that affects its place in the wider world?

*http://www.dailystarnews.com* is a Bangladeshi newspaper site, giving news highlights and fuller information. What events happened over the last few days that have an influence on the political or economic environment of the country?

*http://www.indiagov.org*

*http://www.pak.gov.pk/*

The government sites of India and Pakistan. Find out reasons given for hostilities between the two countries. Look up the sections on *"Economy—Industry"* (India) and *"Economy—Energy"* (Pakistan).

*http://www.lk/* The Sri Lankan site. Access *"Economy"* and *"Tea Board"* to find out how one of this country's major crops is marketed.

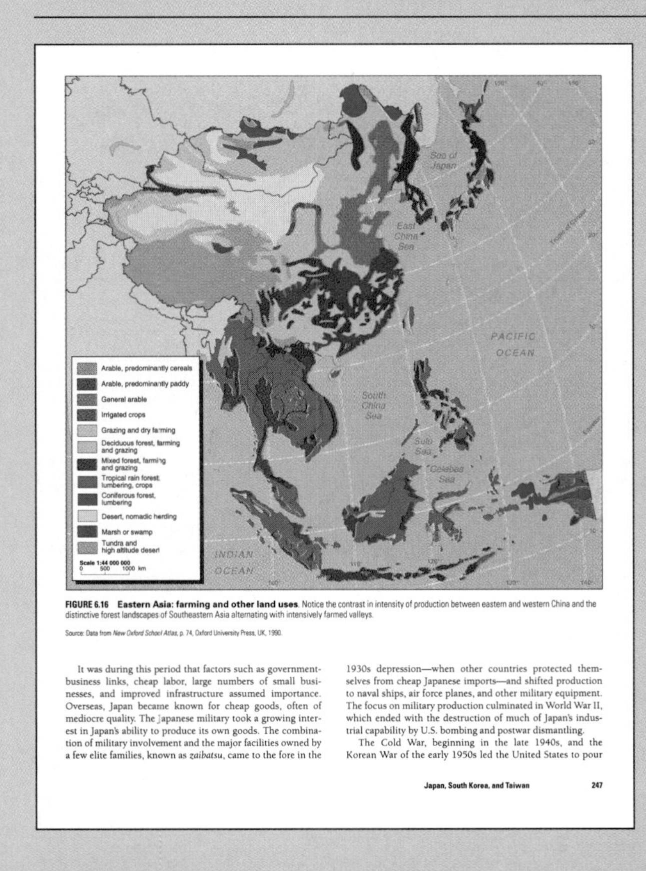


**FIGURE 6.16 Eastern Asia: farming and other land uses.** Notice the contrast in intensity of production between eastern and western China and the distinctive forest landscapes of Southeastern Asia alternating with intensively farmed valleys.

Source: Data from *New Oxford School Atlas*, p. 74, Oxford University Press, UK, 1990.

It was during this period that factors such as government-business links, cheap labor, large numbers of small businesses, and improved infrastructure assumed importance. Overseas, Japan became known for cheap goods, often of mediocre quality. The Japanese military took a growing interest in Japan's ability to produce its own goods. The combination of military involvement and the major facilities owned by a few elite families, known as *zaibatsu*, came to the fore in the 1930s depression—when other countries protected themselves from cheap Japanese imports—and shifted production to naval ships, air force planes, and other military equipment. The focus on military production culminated in World War II, which ended with the destruction of much of Japan's industrial capability by U.S. bombing and postwar dismantling.

The Cold War, beginning in the late 1940s, and the Korean War of the early 1950s led the United States to pour

Japan, South Korea, and Taiwan 247

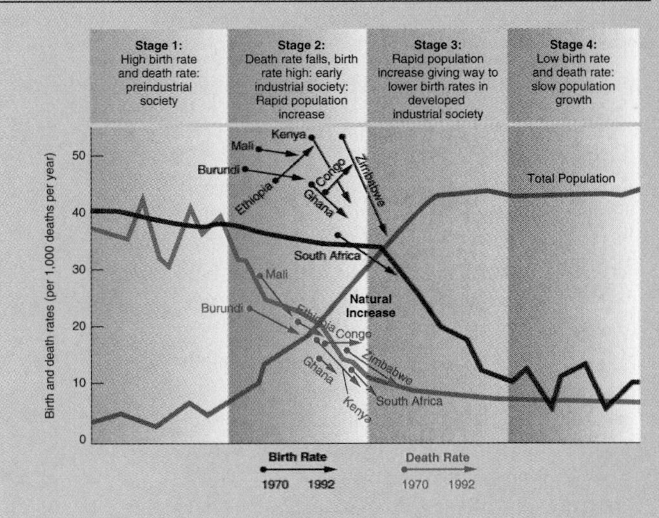


## Comparisons

World regions are the subject of most of this book—Chapters 3 through 11. The global outlook of regional geographic studies, based around contrasts in human development makes comparisons inevitable. Such comparisons further the understanding of problems faced and of ways in which those problems are met in other regions.

## Learning System

A textbook is designed to help students make the most of their course. A number of features in this text facilitate study:

- Each regional chapter follows a pattern that makes it straightforward to compare one region with another. Chapters begin by covering regionwide features such as the place that region holds in the world, the major common cultural historical factors that formed the region's character, and the natural environmental backdrop. The common sections are reinforced in the maps and diagrams, providing the same pattern in each chapter.
- Each major region is divided into subregions, and studies of these focus on the internal variety of countries and smaller regions, the characteristics of the people, and the level of economic development.
- At the end of each chapter two regionwide sections summarize the geographic character of a region through its landscapes and assess the region's future prospects.
- The first two pages of each chapter now combine two globes showing the region, its subregions, and its physical features. There is an overview of chapter contents and a diagram that highlights the relationships of countries in the region to the rest of the world.
- The text is written in a style that is clear but does not insult by being too simple.
- Much has been written for the new edition, paragraphs are shorter, there are more sideheads, and some sections (like this one) are bulleted.
- Some five or so times in a chapter the information is summarized in RECAP sections that have been

**RECAP 7C: Western Europe**

Western Europe includes three of the four largest countries by population in the region (France, Germany, United Kingdom) and the Low Countries (Belgium, the Netherlands). It initiated the 1700s industrial revolution and dominated the 1800s world, and in combination, its countries are still the wealthiest and largest affluent market in the world. The Republic of Ireland is part of the less affluent western margins of this subregion.

The population of Western Europe is static in overall numbers and aging. Decreasing numbers of people in the working age groups support increasing numbers of retired people. In the 1960s and 1970s, immigration was encouraged to bring in guest workers.

Agriculture was modernized so that the subregion that was a net importer is now a net exporter of food. The manufacturing industries modernized in type and levels of productivity. Service industries, especially financial and tourism aspects, are the main employers, contributing increasingly to the continuing affluence of the subregion.

7C.1 Which five features of each give France, Germany, and the United Kingdom their geographic distinctiveness?

7C.2 What level of success do you think was achieved by regional policies in the countries of Western Europe?

7C.3 Why do so many difficulties confront the movement toward peace in Northern Ireland?

**Key Terms:**

| | | |
|---|---|---|
| intensification (agriculture) | specialization (agriculture) | agribusiness |
| concentration (agriculture) | extensification (agriculture) | producer services |

**Key Places:** *(cities in italics)*

| | | |
|---|---|---|
| Belgium | *Düsseldorf* | United Kingdom |
| *Brussels* | *Cologne* | *London* |
| France | *Hamburg* | *Manchester* |
| *Paris* | *Munich* | *Leeds* |
| *Lyon* | *Stuttgart* | *Birmingham* |
| *Marseilles* | *Frankfurt* | *Glasgow* |
| Germany | Luxembourg | Ireland, Republic of |
| *Berlin* | The Netherlands | *Dublin* |
| *Essen* | *Rotterdam* | |
| | *Amsterdam* | |

The approach of European governments to economic growth remains very different from that of governments in Eastern Asia. While European governments subsidize and protect their industries, the Eastern Asian governments protect some industries but help companies to identify new areas of technology and market growth and encourage them to develop the capability of competing in those markets on a global scale. It has been suggested that Europe is wealthy enough to control its own future within "Fortress Europe" of the EU, surrounded by tariff walls and with internal laws that guarantee its labor force cozy conditions. It is unlikely that this attitude will enable Europe to maintain its place in the forefront of the affluent and growing economies of the world.

The European Union is negotiating, however, with the Mercosul trading group in Southern South America (see Chapter 10) and the NAFTA group (see Chapter 9). Such links suggest a prospect of increasing integration in world trade, rather than a continuation of "Fortress Europe."

While such prospects are considered, countries in Eastern and Balkan Europe that were impoverished under Soviet bloc conditions wait in line to join the European Union, seek opportunities for trade with the rest of Europe, and try to attract finances and expertise into their own economies. Germany is in the forefront of developing economic links with countries to its east that are likely to form areas of economic expansion.

**CHAPTER REVIEW QUESTIONS**

1. Which of the following countries was an original member of the European Economic Community in 1957? (a) The United Kingdom (b) Denmark (c) Spain (d) France (e) All of the above were original members of the EEC.
2. In 1997, NATO invited which former communist country(ies) to apply for membership in 1999? (a) Ukraine (b) Poland (c) Czech Republic (d) Hungary (e) b, c, and d
3. Transportation in Europe during the early periods of colonialization and industrialization was eased by (a) an integrated road network (b) an abundance of navigable major rivers and sea ports (c) comprehensive railroad networks (d) early airline links (e) all of the above
4. Which of the following is a reason for the pollution problem in the Mediterranean Sea? (a) Many surrounding countries discharge their sewage into the sea (b) Oil and other pollutants leak from ships passing through the sea (c) The relatively closed basin limits the amount of seawater coming in and out to dilute the pollutants (d) There are many urban-industrial centers on, or close to, the Mediterranean coast (e) All of the above
5. Europe has some of the highest population densities in the world. Which subregion of Europe has the highest population density? (a) Alpine (b) Western Europe (c) Eastern Europe (d) Balkan Europe (e) Mediterranean Europe
6. Which sector of the economy employs the largest proportion of the workforce in most Western European countries? (a) Mining (b) Agriculture (c) Manufacturing (d) Service sector (e) Military
7. Which of the following European countries are monolingual (i.e. have only one official language)? (a) Belgium (b) France (c) Switzerland (d) Serbia-Montenegro (e) All of the above
8. Which of the following European countries were part of the Soviet Union before 1991? (a) Estonia (b) Czech Republic (c) Slovenia (d) Poland (e) Hungary
9. Which of the former Yugoslav republics is the wealthiest and has the closest ties with the stronger European economies to the north and west? (a) Macedonia (b) Serbia-Montenegro (c) Croatia (d) Slovenia (e) Bosnia-Herzegovina
10. Cities with medieval cores and strict protections on building preservation are characteristic of which subregional urban landscape? (a) Northern European cities (b) Swiss cities (c) Balkan cities (d) Eastern European cities (e) All of the above
11. The modern world economic system was established in Europe.
    True / False
12. The European Union (EU) arose from attempts to reclaim lands invaded by the Soviet Union at the end of World War II.
    True / False
13. From 1750 to 1900, northwest Europe took a central role in the technological innovations that led to fresh expansions of industrial activity in products and new places.
    True / False
14. Countries which are invited to apply for EU membership in the near future include Russia, Turkey, Ukraine, and Switzerland.
    True / False
15. The climatic pattern most prevalent in Europe, and extending from northern Norway to northern Italy, is the Mediterranean climate.
    True / False
16. Europe is the world's major center of tourism.
    True / False

Chapter Review Questions 353

expanded to include critical theory questions (some suggested by reviewers) about that region, and a list of Key Terms and Key Places.

- Websites provide a huge potential set of connections to materials that extend and update available information. Websites are introduced in Chapter 2 and listings close each regional chapter. There is a special website for this book (http://www.mhhe.com/earthsci/geography/bradshaw) that includes updated examples and a bulletin board.
- The illustrations—photos, maps, and diagrams—are designed to work with the text and to provide a visual impression of each region. The legends draw your attention to details, sometimes by asking questions about the content. Where diagrams, such

as the demographic transition diagram, are repeated in each chapter, they are explained in Chapter 2.

- NASA space shuttle photos are a special feature of the new edition (see Figures 3.30, 4.5, 5.10, 6.21).
- Maps are designed to be appropriate to the course level. There are new maps of population distribution and natural vegetation/land uses in each regional chapter (see Figures 3.12, 3.17).
- At the end of each chapter are 30 sample test questions that will help students review the chapter.
- The Reference Section at the end of the book contains glossaries of the Key Terms and Key Places, together with additional information on map projections, a list of further reading and CD-ROMs that can be consulted, and an index. The new data tables have been designed to provide information on each country.


## ACKNOWLEDGMENTS

This book has become more complex to produce as it has included new ideas, resources, and techniques. The list of acknowledgments grows.

Many contributions were made by reviewers of the first edition and during the preparation of drafts of the second edition. Reviews were of a high caliber and many of their comments and ideas were incorporated in the new edition. The author thanks them profusely.

At McGraw-Hill, Marilyn Rothenberger and her team managed the complex process of producing the text from drafts with skill and pleasant efficiency. A special word of thanks must go to my two editors—Daryl Bruflodt and Lu Ann Weiss. It has been stimulating to work with them and to bring together complementary ideas for the new edition. Their readiness to discuss points and make suggestions for innovations that the author could not refuse, and their continued availability to maintain progress on such a complex project were major contributions to the new edition.

The author thanks all those who have contributed to the book. He, however, takes responsibility for the final product.

Michael Bradshaw
Canterbury, England

### End of Chapter Review Question Contributions

Elizabeth Leppman
*St. Cloud State University*

Max Lu
*Kansas State University*

Raoul Miller
*University of Minnesota–Duluth*

Eric Prout
*Louisiana State University*

### Book Web site Materials

Barry Mowell
*Broward Community College*

### Focus Group Members

Sherrie Hajek
*Florida State University*

Elizabeth J. Leppman
*Miami University*

Max Lu
*Kansas State University*

Raoul Miller
*University of Minnesota-Duluth*

Phillip Thiuri
*William Patterson University of New Jersey*

Robert Voeks
*California State University*

## Reviewers

Robert M. Barclay
*University of Sioux Falls*

Roger A. Becker
*St. Louis Community College, Meramec Campus*

Reuben H. Brooks, Ph.D.
*Tennessee State University*

Michael Camille
*Northeast University of Louisiana*

Craig S. Campbell
*Youngstown State University*

James Doerner
*University of Northern Colorado*

Paul J. O'Farrell
*Middle Tennessee State University*

Paul B. Fredric
*University of Maine at Farmington*

James W. Harris
*U.S. Air Force Academy*

Marcia M. Holstrom
*San Jose State University*

Dr. Richard R. Kroll
*Kean University*

David E. Kromm
*Kansas State University*

Dr. Hilary Lambert Hopper
*American Geographical Society*

Elizabeth J. Leppman
*St. Cloud State University*

Max Lu
*Kansas State University*

Debra L. Lundberg
*University of Kentucky*

Raoul Miller
*University of Minnesota-Duluth*

Robert D. Mitchell
*University of Maryland*

Barry Mowell
*Broward Community College*

Pasquale A. Pellegrini
*Ohio State University*

Michael P. Peterson
*University of Nebraska at Omaha*

Eric Prout
*Louisiana State University*

Lynn A. Rosenvall
*University of Calgary*

Daniel A. Selwa
*Coastal Carolina University*

Charles L. Smith, Ph.D.
*Blinn College*

Joseph Spinelli
*Bowling Green State University*

Joseph Stoltman
*Western Michigan University*

Richard Walasek
*University of Wisconsin-Parkside*

David Zurick
*Eastern Kentucky University*

## First Edition Reviewers

John Anderson
*University of Louisville*

Jerry Aschermann
*Missouri Western State College*

Joseph M. Ashley
*Montana State University*

Paul Butt
*University of Central Arkansas*

David Iyegha
*Alabama State University*

Tarek Joseph
*Lansing Community College*

Ezekiel Kalipeni
*University of Illinois at Urbana–Champaign*

Richard Kujawa
*St. Michael's College*

Paul E. Phillips
*Fort Hays State University*

Jeff Roet
*San Jacinto College–South*

Harry J. Schaleman
*University of Southern Florida–St Petersburg*

Burl Self
*Southwest Missouri State University*

Roger Selya
*University of Connecticut*

Ronald V. Shaklee
*Youngstown State University*

Gerald Showalter
*Ball State University*

J. Andrew Slack
*Ball State University*

Bruce W. Smith
*Bowling Green State University*

Royse Smith
*Community College of Southern Nevada*

Richard Walasek
*University of Wisconsin–Parkside*

Robert Wales
*University of Southern Mississippi*

## Supplements

McGraw-Hill Geosciences provides a series of products that can be linked with this text for your course, such as:

- Nystrom Desk Atlas
- Rand McNally Atlas of World Geography
- Student Atlas of World Geography/Allen
- Base Map Collection
- Annual Editions: Geography 00/99
- Transparencies
- Slide Set
- GeoScience Visual Resource Library CD-ROM
- GeoScience Videotape Library
- Eyewitness World Atlas CD-ROM / Dorling Kindersley
- New Millennium CD-ROM / Rand McNally

**World Regional Geography Web site including:**

- Instructor's Manual and Test Item File
- PowerPoint Lecture Outlines
- Student Study Guide
- Quizzing
- Over 300 World-wide Regional Links
- Embassy Resource Links
- Link to Dushkin Online
- News Update

http://www.mhhe.com//earthsci/geography/bradshaw2e

# A NEW GLOBAL ORDER AND GEOGRAPHY

Chapter 1

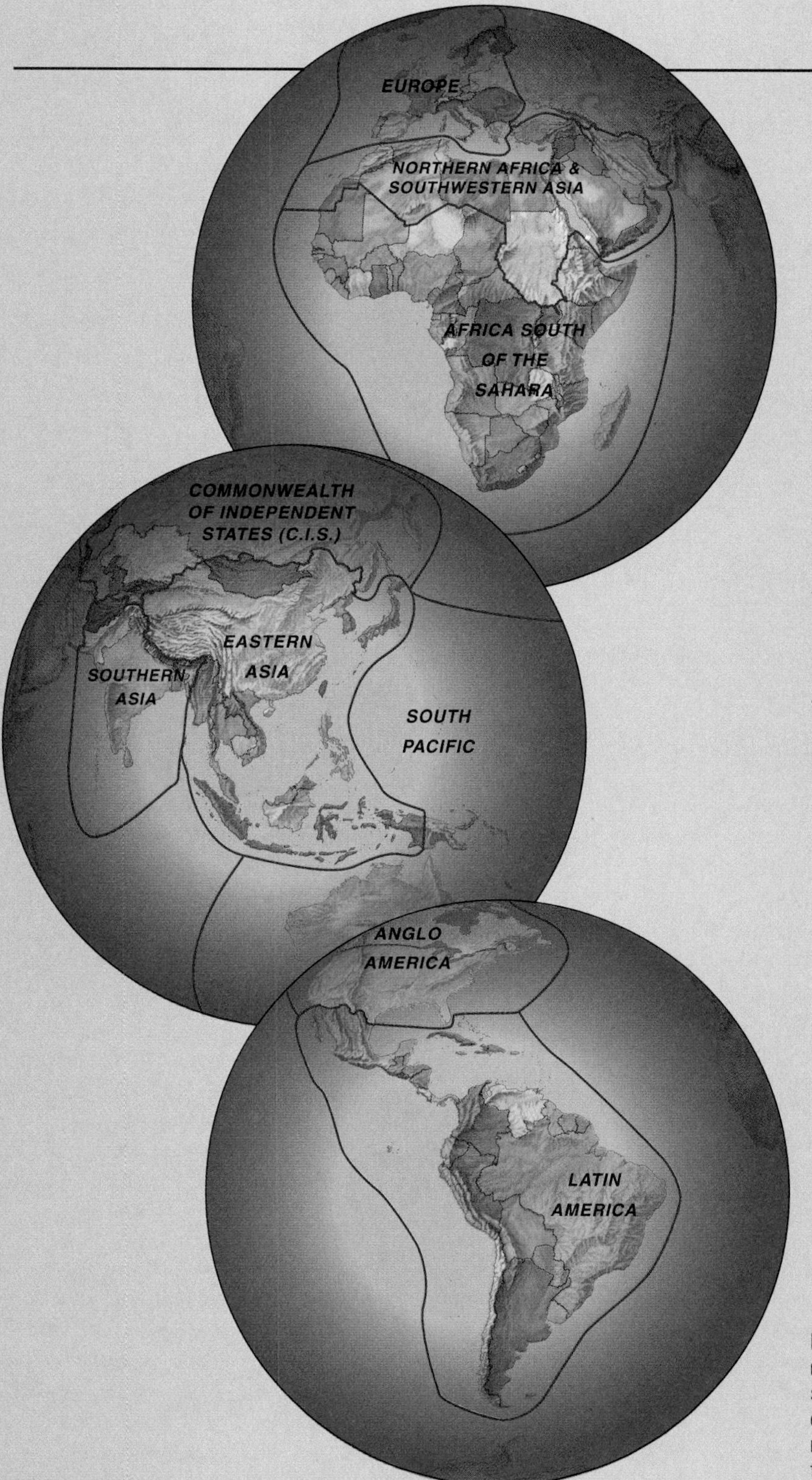


**THIS CHAPTER IS ABOUT:**

**What comprises a new global order**

**The importance of geography in understanding a new global order**

**The special nature of regional geography**

**How major world regions are chosen: the basis for the rest of the book**

**How world regions developed through history**

**FIGURE 1.1 Global view of Earth and the major world regions.** Views of Earth from space led to a wider global consciousness and concern for the planet. The globes show the major world regions that are the subject of this book.

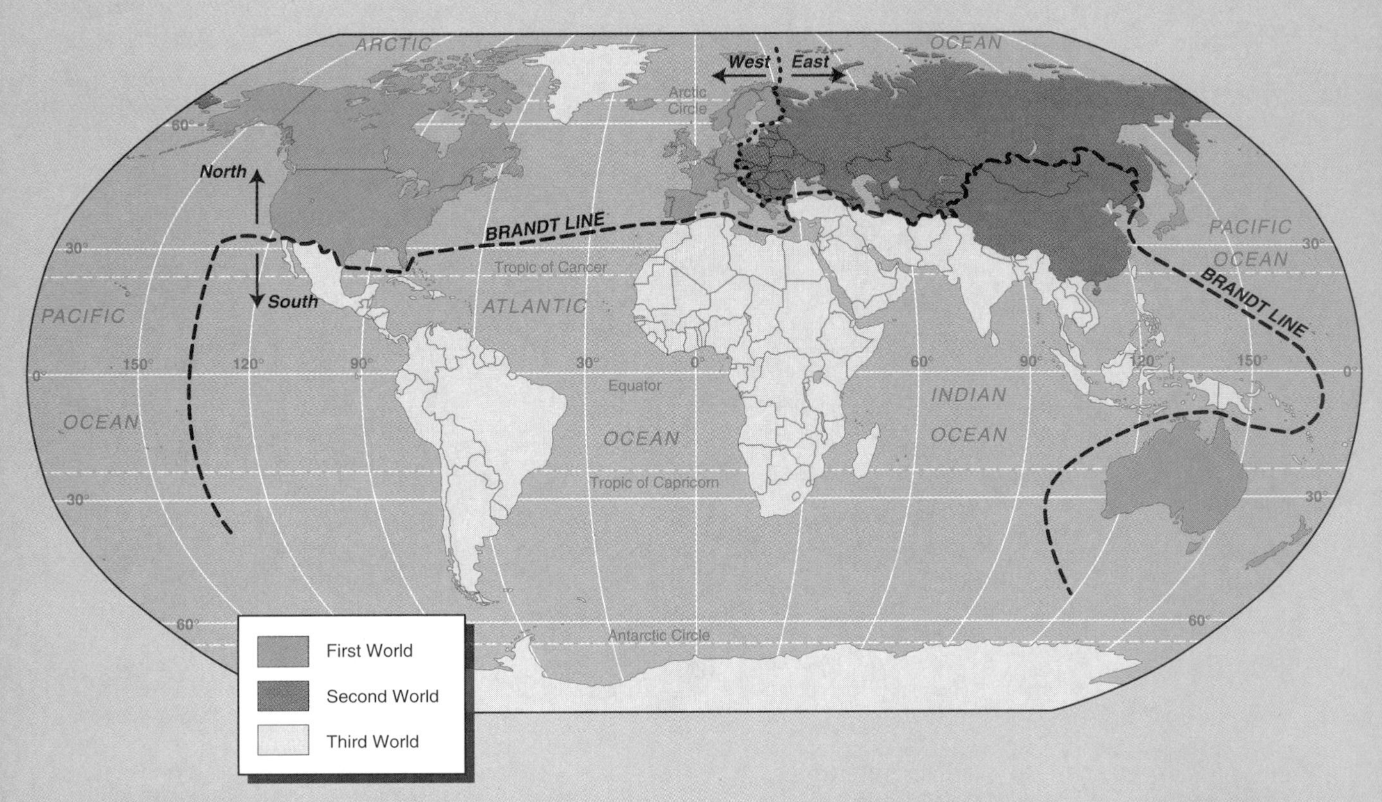


**FIGURE 1.2 Old global order of the three worlds.** The First World comprised the affluent countries of the West, dominated by the United States and western Europe. The Second World included the Soviet Union, China, and other communist countries. The Third World made up the rest—the world's poor countries. The West-East division marked a Cold War boundary of potential conflict. The Brandt line, proposed in the early 1970s by Willy Brandt, chancellor of West Germany, separated the rich North from the poor South.

## FACETS OF A NEW GLOBAL ORDER

From the 1960s, space vehicle views of Earth (Figure 1.1) brought a deeper consciousness of our planet's unity. They emphasized the finite limits of the land areas and materials on "spaceship Earth." On the eve of a new millennium, all parts of the world are interlinked increasingly by political, economic, and cultural ties and environmental concerns. Within this new global order, however, local, regional, and national areas retain their distinct characters, or geographies.

Any text that sets out to involve the whole world needs central themes to organize and select what will be included. It also needs a set of basic principles to provide a basis for comparison among the various parts of the whole. The main theme of this text, "a new global order," challenges geographers to evaluate its meaning and how it affects different parts of the world. This book sets out to explore the importance of regional geographic variations within wider global contexts. It begins with a discussion of what makes a new global order and how geography—particularly regional geography—is relevant to its study (this chapter). This is followed by a summary of the basic geographic principles that contribute to regional geographic analysis (Chapter 2). The remaining chapters (3–11) are each concerned with a major world region.

### Global Political Order

Politics is about power and the ways in which countries are governed. During the 1990s, a new global emphasis grew out of major political events in the late 1980s. A single superpower—the United States—displaced the old divisive geopolitical order of the Cold War.

#### Cold War

In the Cold War, which lasted from 1950 to 1990, two rival groups of countries advocated different political systems (Figure 1.2). One group was led by the United States—the **First World**, or *free market, countries*—and the other by the Soviet Union—the **Second World**, or *planned economy, countries*. They competed for

**FIGURE 1.3 Third World city: midday street scene in Bangalore, southern India.** Compare this view of a main shopping street in Bangalore with a shopping area or main street with which you are familiar. What ways do people use to get about in each place?



influence among countries in the **Third World** (Figure 1.3). The First World countries were mainly in western Europe and Anglo America. They had democratically elected governments that placed importance on the rights of the individual. The Second World countries attempted to implement the communist ideals of Marx and Lenin, in which the needs of the state came before those of the individual. In the Soviet Union and its subordinate neighbors in eastern Europe, and in countries such as China, central governments controlled most aspects of political, economic, and cultural life.

The Cold War competition between two rival political systems resulted in huge investments on both sides in weapons of mass destruction. Wars of a more or less local nature erupted every few years, mostly in the Third World. New military technology made wholesale slaughter possible. The nightmare of a "hot war," in which the leaders of the two groups fought with nuclear weapons, was, however, averted. The wars in Korea in the early 1950s and in Vietnam in the 1960s and 1970s were the main conflicts, but civil wars such as those in Angola, Mozambique, and Malaysia were instigated and supported by the two sides. Some countries in the Third World took sides, as when Fidel Castro aligned Cuba with the Soviet Union after 1959. In the late 1970s, Egypt linked itself with the United States, obtaining economic support in exchange for moves toward reconciliation with Israel. This phase of world history resulted in countries becoming parts of political blocs.

## After the Cold War

The 1991 breakup of the Soviet bloc Second World countries cost them political influence. The United States became the sole world political, military, and economic superpower. When it invaded Kuwait in 1990, Iraq found itself up against U.S. military forces and those of other countries drawn into the conflict by their links to the United States. Iraq also had to face its Arab neighbors, who supported the U.S. forces in the absence of intervention from the weakened Russian remnant of the former Soviet bloc. Although the 1990s showed that the United States is not able by itself to maintain a role as policeman to the world, countries had to take note of American wishes since there were no longer two superpowers to play off against each other. The greater interest in peace between the Arab countries and Israel in Southwest Asia and the end of apartheid in South Africa owe much to the shift from the Cold War to a new global order.

The full political implications of a new global order were not worked out by the late 1990s but will become clearer as they affect world political events into the next century. Challenges to the political leadership of the United States may come from a revived Russia that retains pretensions to world power, from the growth of Chinese aspirations, or from a grouping of Muslim countries. Throughout most of the 1900s, the view prevailed that the rest of the world should become Westernized. It is now becoming understood that Western politics must share the global stage with other countries to form the world's future.

### New Global Disorder

The world is not politically united, and is unlikely to be so. It remains a dangerous place of rivalries within and among countries. The armaments industry makes its wares widely available. New sources of conflict arise from racism and ethnic or religious antagonisms. Some have suggested that the term "New Global *Dis*order" is more appropriate for world politics in the 1990s, bringing into question whether or not world events are dominated by a state of confused or unruly behavior. If global political order is unlikely, it is necessary to examine the economic, cultural, and environmental dimensions of a new global order.

## Global Economic Order

After the end of the Cold War global order, economic events became a major factor in growing worldwide linkages and exchanges. The breakup of the Soviet bloc not only left the United States as the single world political superpower, it also enabled the capitalist economic system—of which the United States and multinational corporations are the most powerful exponents—to dominate the world. Politics and economics work closely together.

### Economic Systems

An **economic system** is a way in which goods and services are produced, distributed, and consumed. The capitalist economic system emphasizes the private or corporate organization of business and investment, together with the determination of prices, production, and distribution of goods by market forces. Governments intervene in capitalist economies mainly to regulate terms of trade.

The alternative economic system to the capitalist system during the Cold War was that of state-run, socialist economies exemplified by the former Soviet bloc and China. They failed to generate the high levels of wealth, consumer goods, and tourism options that became a mark of most people living in Western countries. Although socialist countries often had universal medical care, comprehensive educational systems, and strong military defense, they did little to reduce the numbers of poor and had few people with middle and upper incomes. The inability of the socialist economic system to provide consumer goods had much to do with the failure of the related communist political system.

### Global Economic System

The capitalist system now operates in virtually all parts of the world and is regarded as the new global economic system. The few countries, such as the People's Republic of China, that attempt to maintain central state control of a planned economy work within the capitalist system and are being drawn increasingly into it. Russia and the other countries that were part of the former Soviet system are in transition from their planned economies to market-related systems.

The few exceptions to this trend are in the remotest parts of the world, such as the most isolated parts of the Amazon rain forest and Papua New Guinea, where a few tribes still live outside the world economy. In parts of Africa, people are returning to hunting and gathering modes of livelihood after failing to become part of the world economy.

During the 1990s, the United States and other Western countries promoted the geographic expansion of capitalist activities, thus strengthening their own economies. They encouraged former Soviet bloc countries to become part of the world economic system, and discouraged countries, such as Iran and Iraq, from finding alternatives. The end of the Cold War spread capitalist power to many countries inside and outside the former Soviet bloc.

As more of the world becomes involved in this economic order, complaints arise that rich countries get richer—along with rich people within those countries—and poor countries (and poor people) generally get poorer. Even though the world's poorest countries as a whole experienced relative economic growth in the 1990s, those in Africa did not; neither did large groups of people in Asia and Latin America. Poorer countries in the Third World that received financial assistance from several sources in the Cold War era can now obtain aid only from a limited group of countries. New loans from the World Bank and the International Monetary Fund—both headquartered in the United States—are often based on agreements requiring stringent actions to open a country's internal economy for external investment and foreign goods and to reduce its bureaucracies.

### Core and Periphery

The capitalist world economic system, like the political order, has a geographic dimension (Figure 1.4). It reflects the relative success of countries in the system. The richer countries are termed the **core** countries. Core countries have a wide range of products and services, use advanced technology, and enjoy relatively high wages. They play major roles in world trade by importing raw materials from poorer countries and exporting manufactured goods and services back to them. Core countries establish favorable balances of trade with poorer countries and build up reserves of capital that can be invested in their own or other countries. The core countries mainly invest in each other's economies, and their

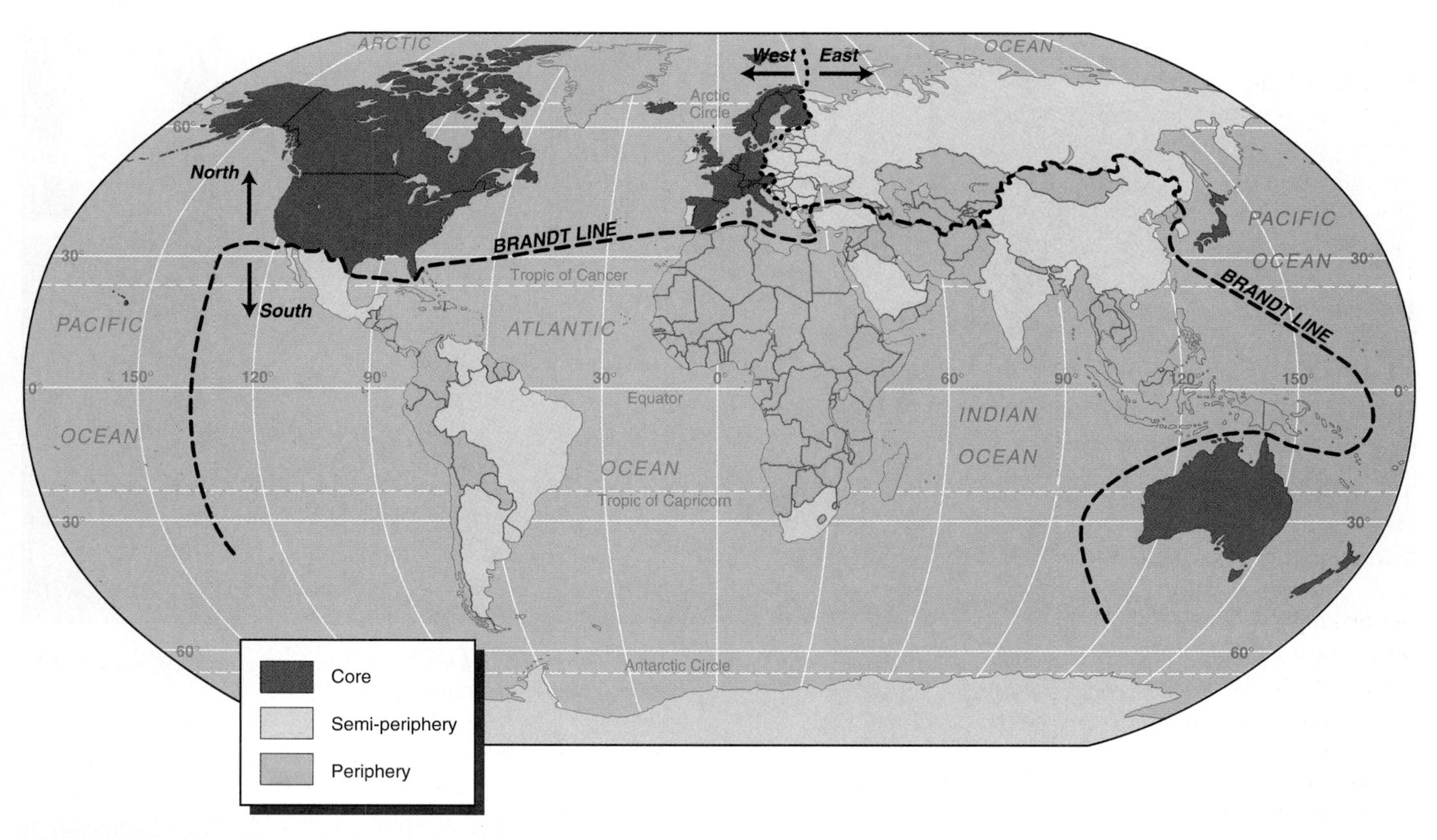


**FIGURE 1.4 New global economic order: core and periphery.** The core countries are the rich countries of the world that dominate world trade: they have a wider range of products, use advanced technology, pay high wages, and consume massive amounts of resources. The countries of the periphery are poor, have a limited range of products, and often depend on trade with the richer countries. Countries in the semiperiphery are moving up or down in this order. Discuss changes you might make to the country placings.

most valuable trade is with each other. The levels of such intracore investment and trade multiplied several times in the 1980s and 1990s.

The poorer countries belong to the **periphery** of the world economic system. Peripheral countries have narrow ranges of products, less advanced technology, and lower wages. They play a secondary role in world trade. Peripheral countries commonly depend on core countries for purchasing their exports, supplying their imports, and providing capital. Some peripheral countries have a part of their economy that is involved deeply in world trade by supplying raw materials to core countries. Many areas of peripheral countries, however, have subsistence-type economies based on local needs and production. The core-periphery relationship is generally one of core countries exploiting dependent peripheral countries. By the late 1990s, agencies such as the United Nations and World Bank were more optimistic about the future of poorer countries as a whole—although some of the poorest continue to have little prospect of making life better for their peoples.

Transition between core and periphery is possible. Some countries move from dominant to dominated status, or vice versa; countries in the process of moving between the two extremes are grouped as the **semiperiphery.** Countries in the semiperiphery retain dependent relationships with core countries but have peripheral countries dependent on them. At present newly industrializing countries (NICs), such as South Korea, Mexico, and Argentina, are clearly part of the semiperiphery, moving toward the core. Former Soviet bloc countries, including Russia and its neighbors, are either in the periphery or the semiperiphery.

## Impacts of Economic Change

As with the shifts in global political order, the outcome of changes within the world economic order are not fully apparent in the late 1990s. The late 1980s and early 1990s were bad years for most countries in the peripheral group because of recession in the core countries on which they depended. Core countries switched capital investments away from peripheral countries and toward other core countries. Despite this, several countries, including Mexico, Malaysia, and Thailand, entered the semiperiphery with an upward momentum, although it is not certain that any will join the core group. Such countries were marked by a greater participation in the world economy as they relaxed duties on imports and produced more goods for export. The late 1990s, however, brought problems for many of these countries when banking systems failed and currencies were devalued. Although the global economic system was widely established by the 1990s, local and regional variations remained and appeared to be an integral part of the system.

**FIGURE 1.5 City center, Bangalore, southern India.** What aspects of transportation types might help to define an Indian technological culture, as compared with your own? Remember that India has the technology to launch satellites, produce computer software, and explode nuclear bombs.

© Michael Bradshaw

## Global Cultural Order

Some evidence suggests a cultural trend toward "One World." Cultural characteristics at national and local levels, however, modulate the influences of global political and economic pressures, producing cultural variety. The **culture** of a group of people living in a particular part of the world is based on the ideas, beliefs, and practices they hold in common and pass on from one generation to the next. It is demonstrated in their language and religion, the ways in which they do things socially, and the design of items they make (Figure 1.5). Different cultures have distinctive approaches to family life, the roles of women, the structure and decoration of houses, and the values they place on communal and individual actions.

### One World

The One World idea is based on extending the Western urban-industrial culture to the rest of the world in the wake of the world economic system. This process is often referred to as "Westernization," or "Europeanization." It is met in the so-called "Cocacola-ization" of eating and drinking habits, the worldwide extension of Western popular music, and the global presence of the same personal computers, cars, and other consumer goods.

### Cultural Differences

Although the One World trend is powerful, many local cultural features derived from past events and constructions still demonstrate lasting qualities. People can conclude which part of the world they are in from clues that inform about local habits. For example, the housing styles and shop signs in Japan, Australia, Pakistan, Brazil, Romania, or Kenya can be told apart. Awareness is widespread through the increasingly global medium of television that distinctive cultural traits, economic contrasts, political systems, and environmental conditions mark out different parts of the world.

During the 1990s, cultural factors often replaced political ideologies in generating local conflicts. Historic cultural differences existed among the dominating Christian Orthodox Serbs, the Roman Catholic Croats and Slovenes, and the Muslims in the former Yugoslavia. They led to declarations of independence as Yugoslavia broke apart in 1991, leaving the Serbs alone in a remnant of that country. Determined to reassert their former dominance, the Serbs instigated years of civil war, first in Bosnia, then in Kosovo in the late 1990s.

## Global Issues in the Natural Environment

The world's **natural environment** is the world as it might be without human modifications. It includes the mountains and rivers, atmosphere and oceans, plants, animals, and soils. It provides contrasts at the regional and local levels. The variations from tropical forest to desert (Figure 1.6) and from mountain peak to flat plain and polar ice cap are determined by atmospheric and interior Earth activities that are often too powerful for humans to modify.

### People Modify Nature

Human numbers and technologic abilities continue to increase rapidly, however, and have rising impacts on natural

(a)

(b)

**FIGURE 1.6 Contrasting natural environments.** (a) Tropical island in the Caribbean with tree-covered hills and vigorous plant growth resulting from high temperatures and plenty of rain. (b) Desert in northern Africa, covered by sand dunes except where water is available at a palm-fringed oasis. What other contrasts in natural environments have you experienced?



environments. People modify all landscapes in which they live. The effect is greater where those activities are more intense or the environment is more fragile, as where farming or urban activities occur on the edge of a desert. It may be less where fewer people live in a more resilient environment, such as a well-watered plain.

For thousands of years the human modification of landscapes was the means of expanding food production. From around 3000 B.C., river plains in dry countries were managed for irrigation farming, but they became subject to erosion, waterlogging, and salinization. Vast tracts of forest were converted to farmland in medieval Europe and Asia, and when settlement moved westward across the United States in the 1800s. By the early 1900s, it was widely understood that some farming practices destroyed soils that had taken hundreds or thousands of years to develop. Industrial practices fouled rivers and the atmosphere. In the later 1900s, governments in the wealthier countries took measures to restrict soil erosion and air and water pollution. In poorer countries, such steps are seldom seen as necessary or affordable. The core countries even exported some of the polluting industries that could not meet local regulations to peripheral countries, a development highlighted by the chemical industry disaster at Bhopal, India, in 1984.

## Global Environmental Issues

People throughout the world are becoming increasingly aware of the linkages among different aspects of the natural environment, the human impacts on the natural environment, and the need to initiate policies that will sustain significant environmental resources into the future. The environment is likely to become a major issue of regional and international political tensions in the new century.

A major issue facing the new global order is providing for the needs of the increasing numbers of people in the world from the same stock of natural resources. **Natural resources** are those materials stored in the natural environment, such as metallic minerals, stone, water, timber, fish, and soils, that human societies use to build and maintain their economic and cultural systems. In the next 25 years (to A.D. 2025), the world population is likely to increase from 6 billion to 9 billion. The 4.6 billion in peripheral countries will rise to 7.5 billion, while the populations of core countries will increase slowly or stop growing. This will place additional pressures on the natural environments of peripheral countries by increasing the extraction of minerals and the usage of water and by placing more demands on soils. The core countries may be able to afford to preserve their own surroundings through costly conservation measures.

Although the world's resources of water, soils, timber, fish, and minerals are limited, people have continually expanded their usage by discovering new sources or inventing new technologies to manage what is there. So far, as population has expanded, discoveries of new resources or technologies have made it possible for all to be fed and provided with the necessities of life. Many people in the world live in extreme poverty, and some die of starvation or lack basic facilities. This is more a problem of politics, economics, and culture than of environmental resource availability or technical ability to produce sufficient food. Distribution through the world economic system is often hampered by political restrictions on trade. The paradox of setting aside farmland in the United States and Europe while people in other countries starve highlights a dilemma that human systems have not yet solved.

**RECAP 1A: A New Global Order**

The new global order has political, economic, cultural, and environmental dimensions. These dimensions are interdependent. The global economic system led the way, but culture and environment assumed greater significance in the 1990s. Although a new global order is developing, local and regional differences are maintained.

1A.1 What do you think are the main elements of a new global order? Support your answer with examples of recent events.

**Key Terms:** Make sure you understand the meaning of the following (see definitions in the Glossary in the Reference Section):

| | | |
|---|---|---|
| First World | core | natural environment |
| Second World | periphery | natural resources |
| Third World | semiperiphery | |
| economic system | culture | |

# GEOGRAPHY AND A NEW GLOBAL ORDER

**Geography** is the study of how human beings live in varied ways on different parts of Earth's surface. Geographic studies relate the political, economic, cultural, and environmental aspects of a new global order to countries around the world and to regions within countries. The features and products of each region complement others. For example, people living in small towns in coastal China work in new factories built since the 1980s to make goods that will be sold in America, often financed by money from Japan or Taiwan. Yet they still depend on local farms for their food and are subject to communist governmental controls and to a cultural heritage from the more distant past. These characteristics—global, national, and local, with inputs from past and present—give character to each part of Earth's surface. They form the basis of geographic studies.

## A World of Differences

Although linked and interdependent in many ways, the world is marked by differences and inequalities among its parts and by rapid changes that affect people's lives. So much has been made of a global order focused on worldwide politics and economics determining local events that some have forecast "the end of geography"—implying that differences between one part of Earth's surface and another may no longer be significant. Such a view is premature; the reverse is true. Geography—differences among places—still matters.

Multinational corporations with their worldwide networks have to adapt their manufacturing and marketing strategies to local conditions. The 1980s concept of the world car was abandoned because people in different countries demanded special variants. Manufacturing industries still locate in concentrated areas, such as Silicon Valley, California (microelectronics), or Dalton, Georgia, USA (carpets). Service industries focus specialist activities on major cities such as London, England (currency trading), Hartford, Connecticut (insurance), Chicago (futures trading), or Los Angeles (entertainment). There is no likelihood that locations of people, jobs, or demands for goods and services will disperse evenly around the world. Country governments or distinctive local cultures set local conditions and priorities. This will not change.

## Nature of Geography

Geography begins with the study of places and their locations within physical and human environmental contexts. When we say we go to visit a place, it might be an individual building (ski center), small town (Freeport, Maine), large city (New York), rural area (Midwest), another state, or another country. Places may be points on a map or areas of several thousand square kilometers. Each place has its own character, defined by what it looks like and what people do there and how they feel about it. Each place is linked to other places by transportation and telecommunications networks.

A place's location is defined by reference to its position on Earth's surface, or **absolute location**, often recorded as its latitude and longitude, or by its level of interactions with other places, or **relative location**. The increasing availability of rapid transportation facilities and the "global information highway" bring people into easier contact with each other, making them relatively—but not absolutely—closer. This process helps a new global order penetrate more aspects of local life.

Geographers study places on Earth's surface as the environments and spaces in which humans live. Studies of the surrounding environment with its natural and human conditions at a place join with studies of the links among different places. Geographers sometimes separate **physical geography** as the study of nonhuman processes and environments across Earth's surface (e.g., distribution of climate variety, plant ecology, or soil formation, or the location of mountain building or river action). **Human geography** is the study of the distribution of people and their activities (e.g., economies, cultures, politics, urban changes). Although it helps to study aspects of geography separately, such divisions may obscure interactions between human activities and the natural environment. This text seeks to present geography as a unified field of study in which the understandings of physical and human geography work together.

The two basic geographic concepts of place and location are combined in three main approaches to geographic studies.

- **Regional geography** evaluates the differences among places and is based on recognizing the uniqueness of some places and the features that several places may have in common. It provides an informed approach to assessing the roles of global processes and their impacts on people in different places. A **region** is an area of Earth's surface with similar physical and human characteristics that distinguish it from other regions and cause it to interact with other regions in specific ways (Figure 1.7a). Regional geography involves the description of the characteristics within each region that give rise to distinctive landscapes. It places each region and its location in a national and global context. The other approaches contribute to regional geography.
- The second approach is called **spatial analysis**, highlighting the relationships among places that are based

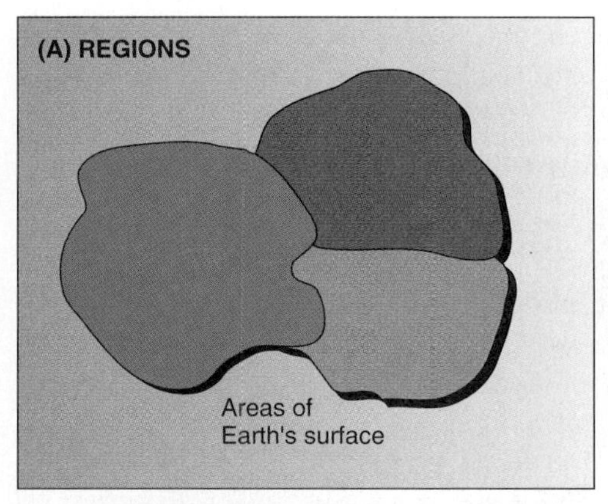


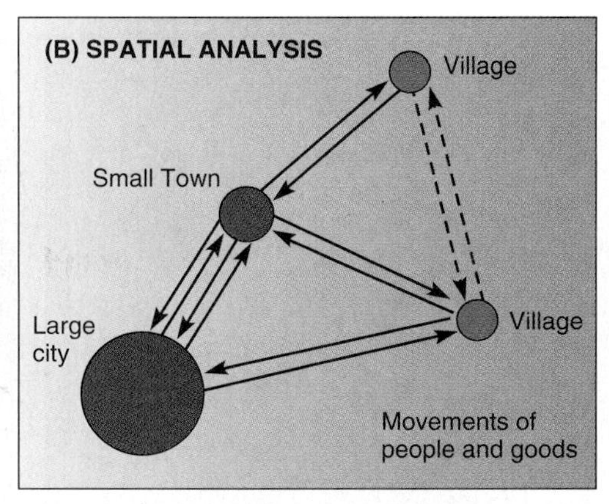


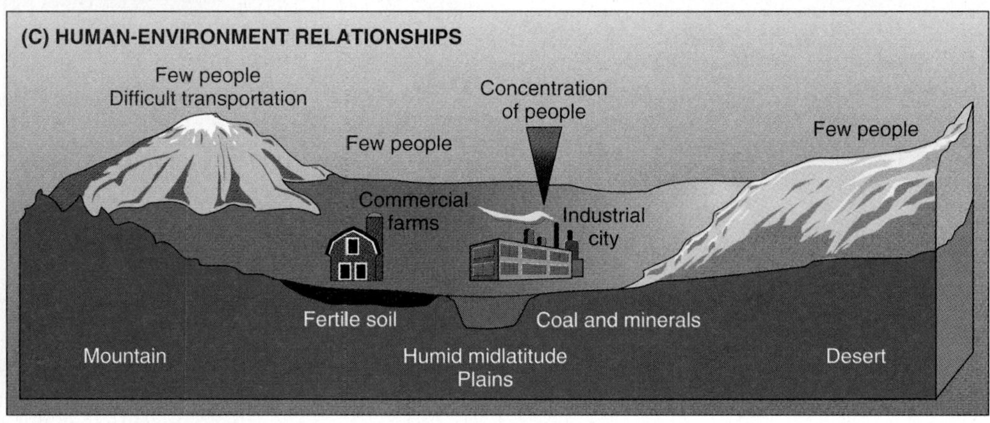


**FIGURE 1.7 Geographers study how people use Earth's surface.** (a) By dividing the surface into regions. (b) By analyzing spatial relationships among places. (c) By evaluating human-environment relationships.

on linkages over the space of Earth's surface (Figure 1.7b). The character and location of places are often considered in terms of geometrical points, lines, and areas. Statistical links among places add to the mathematical basis of spatial analysis. Studies often focus on a particular aspect of geographic significance, such as economic or population changes, or on geographic relationships among different sectors of towns or rural areas. Spatial analysis helps to assess linkages among regions.

- The third approach adopted by geographers is that of investigating **human-environment relationships** (Figure 1.7c). The study of the interactions between human activities and the natural environment broadened from one earlier in the 1900s that was concerned mainly with the impacts of climate, mountain and lowland relief, and soil types on human affairs. More recent studies assess the impacts of human activities on the environment. A landscape is seen by geographers as summarizing the outcomes of interactions between people and natural environments over time. Comparisons among town-scapes and rural landscapes help to define regional characteristics.

## REGIONS IN A NEW GLOBAL ORDER

This text's central theme of "a new global order" makes demands on geographic studies. Global processes and tendencies in the post-Cold War world interact in different ways with local areas, countries, and major world regions. Contrasts among countries in the core, periphery, and semiperiphery are linked to the changing global context. Geographers carry out regional studies to unravel some of the connections and differences. Regions provide comparable units in relation to scale and function.

### Regional Scales

Regions within the overall global realm are of varied scales from local to continental. In this text, we identify three regional scales. At each scale human and physical geographic features combine to produce distinctive regions.

- **Major world regions** encompass several countries and occupy a major part or all of a continent. Nine such regions are identified, and their human geography is considered through distinctive subregions within each major region in Chapters 3 through 11.
- **Countries** are self-governing political entities within major world regions, and their borders provide the boundaries of major world regions and subregions. Very large countries, such as Russia, Canada, China, the United States of America, Brazil, and India form subregions.
- **Local regions** are subdivisions of countries that may vary in size from an urban neighborhood to an area of distinctive farm products or physical features. There is not

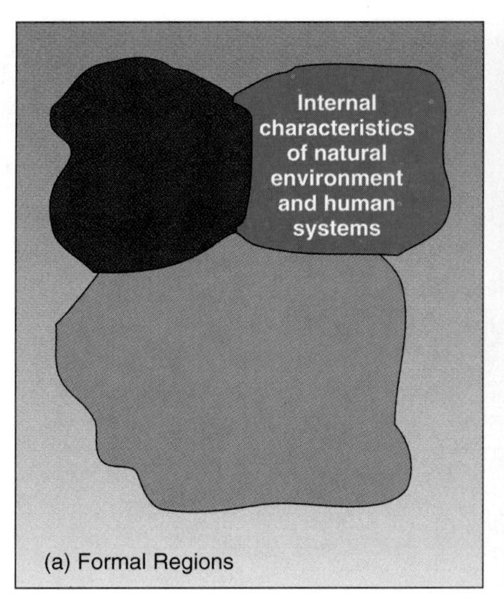


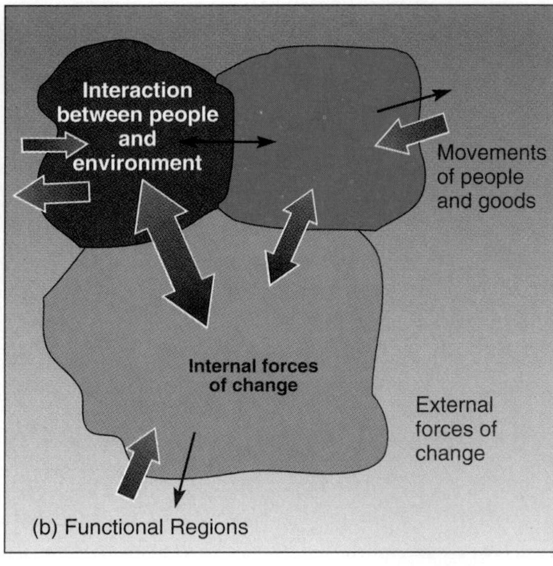


**FIGURE 1.8 Formal and functional regions.** (a) Formal regions reflect a static understanding of regions, each one seen as separate and different from others. (b) Functional regions are a dynamic concept that involves exchanges of raw materials, products, people, and information, and considers the relative influences of regions on each other. This understanding is more appropriate to world regional geography.

room in a text of this length to give detailed analyses of all local regions; some are mentioned to exemplify points and illustrate their existence.

Boundaries of regions may be major physical features, such as coastlines, mountain ranges, and rivers, political borders between countries, provinces, or states, or cultural borders between lands occupied by different peoples.

The ever-changing world and its regions can be studied in terms of *multiple geographies*. In these, a hierarchy of multiple spaces is linked to multiple time periods. The hierarchy of regional scales from local to country, major world region and global scales is linked with timescales from short-term events through medium-term economic cycles to longer-term changes. Many events that change the nature of regions occur over short time spans at a local level but take longer to influence a whole country or wider area. The link between geography and history is a vital part of understanding our world.

## Formal and Functional Regions

### Formal Regions

**Formal regions** are distinctive parts of Earth's surface that have characteristics determined by a combination of physical and human geographic features. Each region is unique and unconnected to the rest of the world (Figure 1.8a). The definition of a region may be based on such criteria as the arrangement and type of physical features, descriptions of local landscapes, or economic characteristics. Businesses and governmental bodies use formal regional divisions in defining their marketing and administrative areas.

### Functional Regions

**Functional regions** are dynamic geographic entities that interact with other regions in national and global geographic patterns (Figure 1.8b). The growth of worldwide economic and communications systems results in regions changing their roles and relative locations within the systems. Some become political, economic, or cultural cores, while others become parts of peripheries. Interactions between cores and peripheries result in continuing changes as regions grow or decline over time. Each phase in this process leaves behind features in the landscapes and built environments of regions. Seen from a viewpoint of rapid changes in the late 1900s, the complex mosaic of geographic regions has five characteristics that influence the past, present, and possible future character of each region.

- **People create regions.** The people who live in a region—not the natural environment—have a dominating role in determining its characteristics. People act in different ways in similar natural environments and regard each type of natural environment as having different potentials for development. For example, the margins of the Sahara Desert of northern Africa remained home to nomadic tribes into the 1900s, while the deserts of Pakistan in Southern Asia were reclaimed for commercial agriculture by British engineers from the late 1800s (see Chapters 4 and 5). The expansion of Los Angeles in southern California was another different—American urban—response to a desert environment.

  Furthermore, similar patterns of human organization are applied across different environments. Within the United States, for instance, similar economic, social, and political systems occur across the arid, humid, subtropical, and temperate environments within the country's borders.

  The primary importance of people in fashioning regional character demonstrates that the relative importance of regions is determined less by criteria such as the possession of natural resources and more by human actions at critical phases of history. People create their own images of their country's role, portraying their idea of its importance and relating other parts of the world to it. For example, the patriotism and political postures engendered within the United Kingdom during the period of the British Empire continue to give that country a greater world political significance and cultural impact than its economic performance would justify.

- **Regions shape people's activities.** The established characteristics of a region influence people's lives. Each region is an environment for human activity. People in a region not only mold its characteristics but also live and work within limits imposed by the economic, social, and political systems and the natural environments that operate there. People living in very dry or cold regions have different options than those in humid midlatitudes. The lives of Amazon tribal peoples reflect the local availability of plants, animals, and fish for food, as well as tribal custom and the penetration of external influences. People living in New York are constrained by the way in which it is built and how the different neighborhoods have developed.

  There is thus a two-way process at work. While people are the main forces creating the distinctive features of regions, they are themselves affected by the regional characteristics that others established. Regional characteristics are perpetuated through people developing cultural traits in a place.

- **People remake regions.** A third set of interactions alters regions over time at varying rates when internal or external forces impose stresses on established regional characteristics. An example of changes generated internally over a long period of time was when people in medieval Europe cut down forest to expand farmland. They built market towns that became the basis for craft industries and trade (see Chapter 7).

  By contrast, external market conditions may rapidly alter the basic pattern of a region's products once it is incorporated in the world economic system. For example, during the 1950s, farmers in northern Nigeria largely abandoned the traditional subsistence economies that evolved over centuries to cope with climatic fluctuations on the Sahara margins and began to grow commercial crops such as peanuts and cotton for export (see Chapter 3). When this was successful in generating income for purchasing manufactured goods, farmers ceased their former practices of growing their own food, keeping a few cows, and maintaining woodland as a source of firewood. They turned all their land over to peanuts or cotton. Education and health both improved as the sale of crops generated income to supply the need for healthy, educated workers. The economy, social structures, and political influences in this part of Nigeria changed along with the land uses and communal practices. Beginning in the 1970s, however, low world prices for peanuts, combined with more frequent droughts, brought poverty and environmental damage that disrupted the new economy. There was no backup system to provide food during the decline of the single crop on which people now depended. Living standards fell, and many families were forced to leave the land and migrate to overcrowded cities.

  Within a country, particular groups or individuals may have such an influence that they determine the course and pace of regional changes. The former Soviet Union arose out of the revolution of 1917 that brought communists to power through leaders such as Vladimir Lenin. During the next 70 years, successors such as Josef Stalin changed the way in which the country's economy functioned. In 1991, the ultimate failure of the system Lenin and Stalin created left behind changing patterns of regions within the former Soviet Union and the neighboring countries that had been added to the Soviet bloc. Mikhail Gorbachev in the late 1980s and Boris Yeltsin in the 1990s had important roles in facilitating these changes.

  Changes in regions generated by the people who live there or by external forces have impacts on other regions. The poverty of northern Nigeria led to people moving into urban areas, which had effects on both urban areas in other parts of the country and the abandoned farmland. The breakup of the Soviet bloc from 1991 led to a major restructuring of regional geographies from eastern Europe to Central Asia within a few years.

- **Regions interact with other regions.** No region is an isolated entity. Each region interacts with other regions through processes that encourage political, economic, and cultural exchanges on a global scale. Coal miners in Appalachia and farmers in the American Midwest see their jobs and incomes affected by world market prices for fuels and cereal crops (see Chapter 9). European countries lacking mineral resources increasingly supply their manufacturing industries from African sources. From 1950 to 1990, the countries of Eastern and Balkan Europe were tied to the Soviet bloc, which tried to isolate itself economically from the world economic system. When the Soviet Union broke up in 1991, the other countries suddenly had to reorient their economies toward the global economic system (see Chapters 7 and 8). Even the world's remotest regions, such as the upper Amazon River basin and interior Borneo, are being increasingly affected by external economic influences.

  Interactions among regions may result in conflicts instead of positive linkages. The presence of several cultural groups within a country, each having territorial claims, often means that one group becomes dominant and others are discriminated against. The Bosnian conflict stemmed from differences among Serbs, Croats, and Muslims (Figure 1.9). Each group had supporters outside Bosnia that supplied them with arms and so kept the conflict going. Conflicts over the ownership and allocation of water resources, as along the Nile River Valley, or over the sources of pollution, as in northwestern Europe, lead to political tensions among countries and regions. Regional differences thus become sources of separation instead of integration.

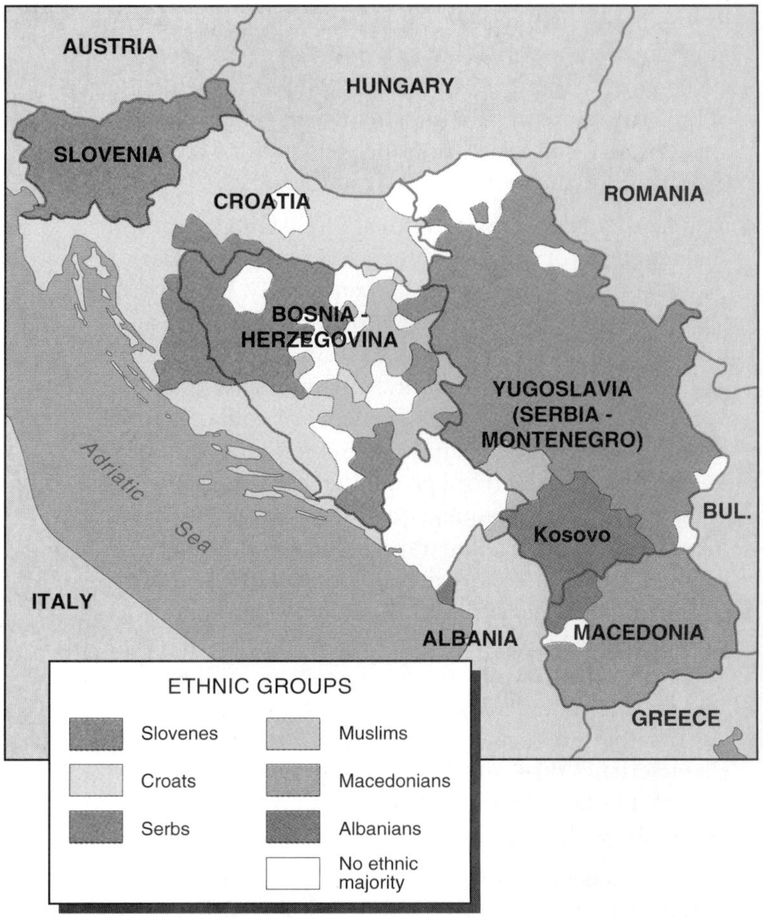

**FIGURE 1.9 Importance of cultural differences.** The former country of Yugoslavia in the Balkans of Europe. After 1991, Yugoslavia broke into five independent countries, although Serbia-Montenegro continues to call itself "Yugoslavia." Slovenia, Croatia, Bosnia, and Macedonia maintain their independence with difficulty, and the Albanian southern part of Serbia (Kosovo) became an area of civil conflict in the late 1990s. The colors show areas where particular ethnic groups comprise over 50 percent of the population. Which country has the greatest potential for ethnic conflict?

- **Regions are used by those in power.** Regional characteristics are not always the outcome of the chance interactions among peoples or of the influence of environmental, economic, or social processes. Regional character often reflects the deliberate actions of powerful governments. Regions were manipulated to change people's lives in the former Soviet Union, where rural areas were industrialized on a major scale, even north of the Arctic Circle. Groups of people were also moved from their traditional homes to other parts of the country for state security reasons. In other countries, attempts to modernize regions that lag behind in economic development may destroy local characteristics of culture and discourage local enterprise. In the United States, federal government attempts to develop the economically lagging region of Appalachia were criticized because the modernization introduced was accompanied by the decline of distinctive local cultures (see Chapter 9).

  Political power is also exercised against foreign countries. Despite the ending of the era when major powers held territory as colonies, the strategic interests of powerful countries—as defined by the countries themselves—maintain strong influences in some parts of the world. This was particularly noticeable during the Cold War from 1950 to 1990, but it still exists. The United States, for example, views the oil-producing Persian Gulf countries as important to its own economic well-being as major suppliers of energy. It went to war without hesitation in 1991 to free the oil-producing country of Kuwait after Iraq's invasion. It long resisted, however, involvement to stem the advances of invading Serbs in the Muslim and Croat sectors of Bosnia.

  Some regions have continuing strategic roles because of their positions at narrow throughways on ocean trade routes that may be termed **global choke points:** examples include the Suez and Panama Canals, Strait of Hormuz (entrance to the Persian Gulf), Dardenelles and Bosporus (entrance to the Black Sea in Turkey), and Strait of Malacca (near Singapore). Such strategic places are of importance to core countries for maintaining access to their markets and raw material sources. They provide bargaining points for the countries in which they are situated.

## MAJOR WORLD REGIONS

Dynamic regional geography provides a strong basis for understanding the regional differences and changing circumstances within the global order. Such regional geography highlights inequalities among regions and evaluates attempts

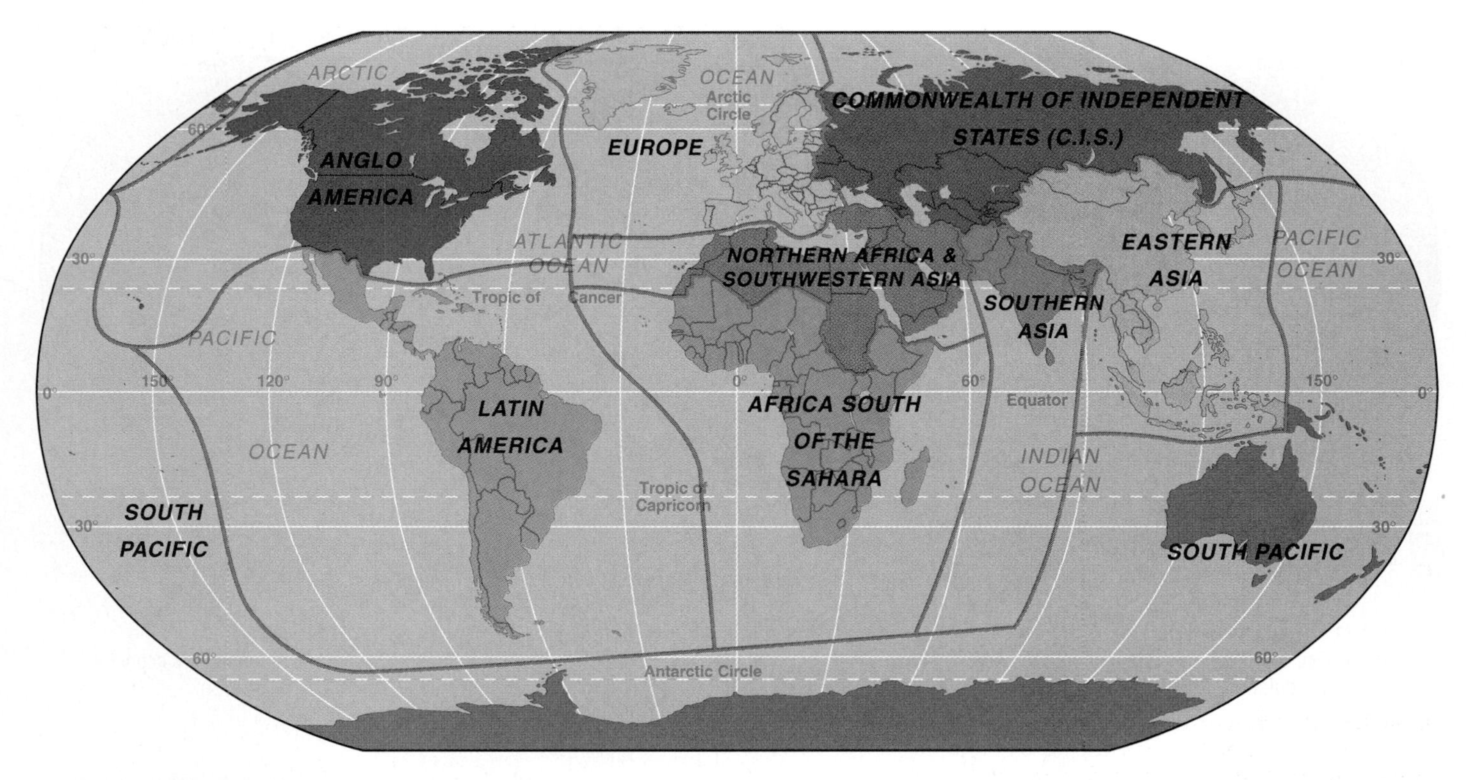


**FIGURE 1.10 Major world regions based mainly on cultural characteristics.** These regions form the subjects of Chapters 3 through 11. World maps in Chapter 2 are divided on this basis so that comparisons may be made.

to reduce them. It provides opportunities for assessing the potential impacts on places of the changing global political order, the expanding world economic system, the trends that may reduce cultural variety, and the pressures on the environment.

In this text, the globe is divided into nine major world regions and a chapter is devoted to each. Countries that have many features in common are grouped into regions (Figure 1.10) and subregions. The criteria for the major groupings are cultural, economic, political, and physical—in that order. Groupings of countries with common cultural and historic experiences produce the following major regional divisions. These divisions are not established for all time. Shifts that are currently in progress led to changes in the eastern boundary of the "Western Europe" major region in the first edition (1997). In this edition, the countries of eastern Europe that set their sights westward by applying to join the European Union are grouped together as "Europe" with those already in the EU. The major region previously titled "Eastern Europe, Balkans, and the Former Soviet Union" becomes the "Commonwealth of Independent States," incorporating the countries of the former Soviet Union, minus the Baltic states.

**Africa South of the Sahara** has traditional cultures based on small ethnic groups. Its religions have environmental, or animistic, bases with Islamic or Christian overlays resulting from trade and colonization.

**Northern Africa and Southwestern Asia** has the twin uniting cultural features of Islamic religion and Arabic language and a natural environment that provides much oil but limited water resources.

**Southern Asia** is characterized by the intermingling of and clashes among Hindu, Islamic, and Buddhist religious cultures and British colonial influence.

**Eastern Asia** was the scene of historic civilizations in China, from which cultural influences permeated the rest of the region. It had less European colonial intervention than other world regions.

**Europe** is the home of mainly Catholic and Protestant Christian religious groups, and of capitalism and modern industrial technology. It is increasingly defined as those countries that are present or potential members of the European Union, including those in Eastern Europe and the Balkans.

**Commonwealth of Independent States** stretches from easternmost Europe across northern Asia. Russia, the largest and still dominant country, was the scene of long-term clashes between the Orthodox Christian and Islamic faiths, overlaid and subdued for most of the 1900s by the communist political culture of the Soviet Union.

**Anglo America** is dominated by the culture brought by settlers beginning in the A.D. 1500s, at first mainly from northern and western Europe. Later people came from

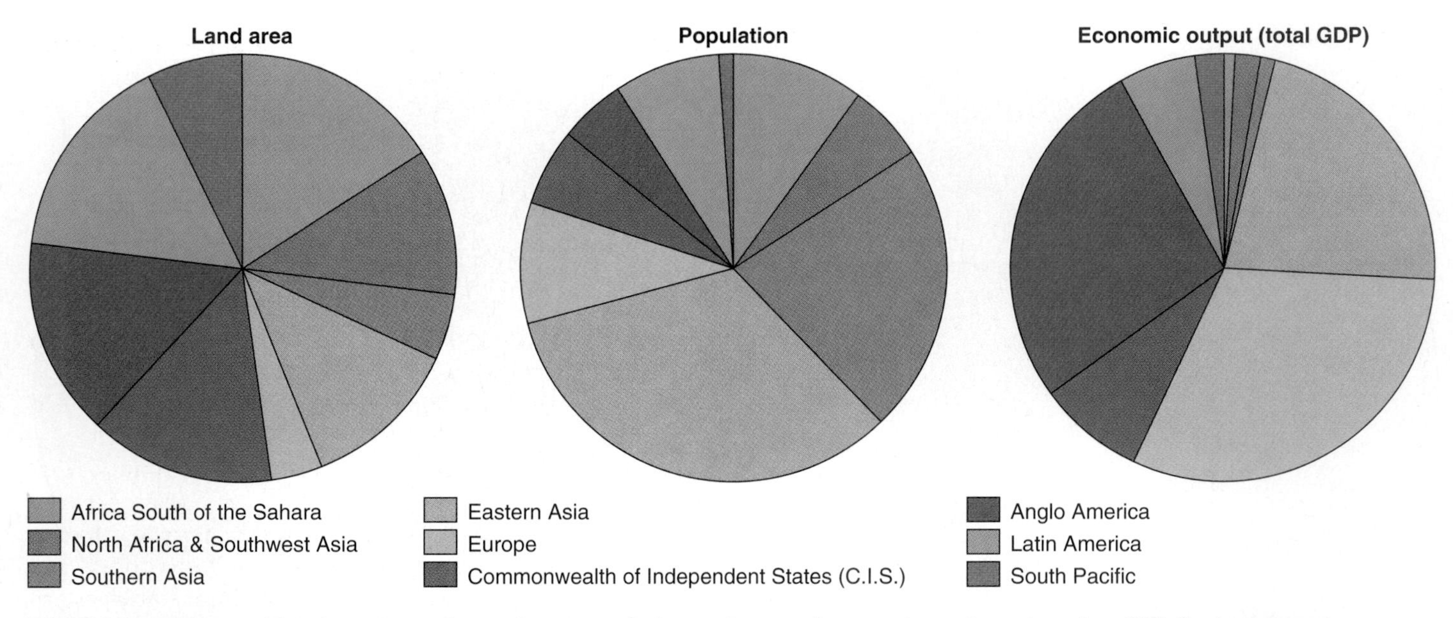


**FIGURE 1.11 Major world regions: comparisons of area, population, and economic output** (gross domestic product, GDP). Pie charts show the variations. Which major regions have more of the world's land area than its population? Which major regions have a higher proportion of the world's economic output than its population?

southern and eastern Europe and eventually from most parts of the world. It comprises the United States and Canada.

**Latin America** is dominated by the Roman Catholic culture and languages brought by settlers mainly from Iberia (Portugal and Spain), who interacted with native peoples and their cultures.

**South Pacific** includes former European colonies, of which the largest and wealthiest are Australia and New Zealand. Distant from the home countries, they are integrating trade with their close neighbors in Eastern Asia. Antarctica, the uninhabited continent, is included in this region.

The major world regions reflect the world economy core-periphery relationships (see Figure 1.4). Europe and Anglo America were established as the world core regions by the early 1900s and continued to grow economically at the expense of the rest of the world in the periphery (Figure 1.11). During the 1900s, some countries in the periphery, mainly those in Eastern Asia and Latin America, improved their economic positions and rose into the semiperiphery (e.g., South Korea, Taiwan, Hong Kong, Singapore, Brazil, Argentina, Mexico, Thailand, and Malaysia). Japan joined the core countries to form a new core that interacts with and encourages the growing economies of Eastern Asia. The former Soviet Union and its linked bloc of socialist countries appeared to be challenging the wealthiest core countries up to the 1980s, but their political breakup exposed an economic position that was no better than that of many peripheral countries. South Africa is likely to establish itself in a higher group than other African countries, while the People's Republic of China may become a leading economy in the new century after A.D. 2000.

Political boundaries between countries are used to define the extents of each major region and subregion. Geographers do not agree in detail upon any division of world regions. Other texts draw some boundaries differently. For example, Mauritania (an Arab country) is included with Western Africa rather than with North Africa because of its closer historic ties to the former; Sudan, another country that straddles the Muslim and southern African cultures, is included with Egypt because of their common reliance on Nile River waters. Myanmar (Burma), which dictators make an isolated country, was included in Southern Asia in the first edition because of historic colonial ties, but its 1997 membership of the Association of South East Asian Nations (ASEAN) relocated it in Eastern Asia. The boundary between Eastern Asia and the South Pacific cuts New Guinea in half, since the western half of the island (Irian Jaya) is part of Indonesia, while Papua New Guinea has a culture and economy akin to that of the South Pacific islands.

**RECAP 1B: Geography and Regions**

Geographers study the differences between places and interactions among them. They take account of political, economic, cultural, and environmental factors and their influences on particular sections of Earth's surface, or regions.

Regions are areas of Earth's surface with distinctive characteristics that emerge from people's interactions with the natural environment and human institutions. Regional character and boundaries change over time. Regions interact with other regions.

1B.1 How would you define *geography* and *regional geography?*

1B.2 What do you think are the major differences among the nine major world regions identified here? See if your perceptions change as you go through this course.

**Key Terms:**

| | | |
|---|---|---|
| geography | regional geography | country |
| absolute location | region | local region |
| relative location | spatial analysis | formal region |
| physical geography | human-environment relationships | functional region |
| human geography | major world region | global choke point |

**Key Places:**

| | | |
|---|---|---|
| Africa South of the Sahara | Eastern Asia | Anglo America |
| Northern Africa and Southwestern Asia | Europe | Latin America |
| Southern Asia | Commonwealth of Independent States | South Pacific |

## GEOGRAPHIC GROWTH OF A GLOBAL ORDER

The growth and expansion of a global order can be traced from the earliest city-based civilizations, through the development of unified countries and expanding trade links across the globe, to the current position that incorporates the whole world. An understanding of historic events and their influences is basic to the study of today's world regions.

Periods of wealth, based on expanding political power and extensive trade areas, alternated with periods of empire breakup, unsettled conditions, and reduced trade. Each left an inheritance of traditions, buildings, and land patterns that later groups of people incorporated in their own developments. Each prosperous stage expanded the global area in which wealth accumulated and which engaged in the trading of high-value goods—the area subject to growing **world systems**.

For much of the world's history, core and periphery can be recognized over limited areas of the globe, with the remaining territory, or hinterland, remaining outside the main trading areas. Making some huge generalizations and simplifications, world history can be summarized in five major phases.

### Phase I. Before World Systems: Hunting and Farming

Around 10,000 years ago (8000 B.C.), humans formed small hunting and gathering bands, the numbers in each and the distribution of which depended on the natural productivity of the local ecosystems. Productivity could be raised in places, as where human-set fire extended natural grassland into forested areas, increasing the numbers of larger, meat-providing animals.

Settled farming began in Southwest Asia, around 8000 B.C., at first located in the subhumid hilly areas between mountains and plains. Major changes in human patterns began with the first agriculture, which was based on selected species of plants and animals that varied from southwestern Asia (wheat, barley), to China (millet, rice), the Americas (corn, squashes, beans, potatoes, peppers, tomatoes, cotton), and Africa (sorghum, yams). Animals were herded near cultivated fields, providing meat, wool, and milk, and pulling carts and carrying loads. Some of the domestic animals, especially the horse and camel, gave mobility to nomadic herders. Villages grew into the first small towns, such as Jericho, that produced pottery, textiles, and metal goods. Farming based on early irrigation techniques spread from the surrounding hills to the Tigris-Euphrates River valleys around 5000 B.C.

The economic, political, and technical changes that followed spread to other areas with similar natural environments—the lower Nile River valley (modern Egypt) and the Indus River valley (modern Pakistan). China and southeast Asia were also important centers for the domestication of plants and animals and settled farming. Europe was late on the scene, however, and settled farming did not reach northwestern Europe until 4000 B.C. Although the Americas were isolated from these developments, recorded plant cultivation there began around 7000 B.C. By 1500 B.C., corn formed the basic crop of settled villages and some larger towns in the Americas with craft industries and local trades.

Different parts of the world developed at different rates, but the geographic scale of interaction in this phase was dominantly local. Wider trading links began when areas of surplus farm production exchanged food and craft goods for building materials.

### Phase II. First World System: City Civilization and the Bronze Age, 2500 to 1000 B.C.

The irrigated lowlands of Mesopotamia—drained by the Tigris and Euphrates Rivers—became the central core and periphery of a world system that eventually stretched from

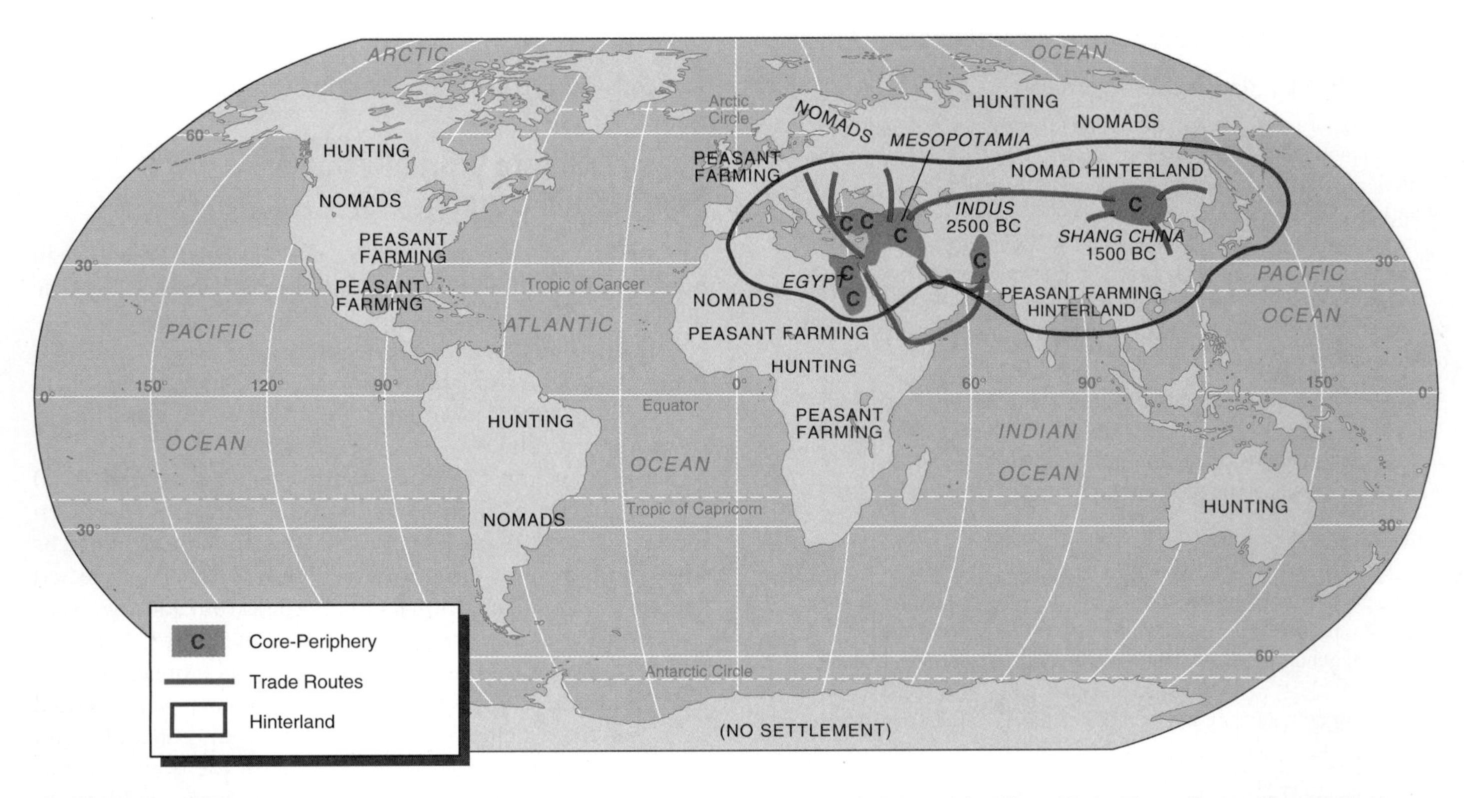


**FIGURE 1.12 First world system, about 2500 to 1000 B.C.** The first civilizations of southwest Asia and the Nile and Indus River valleys, with somewhat later developments in China. The core-periphery areas were the main centers of world trade and development, linking with each other along sea and overland routes. The near hinterland supplied some raw materials, but most of the world lay beyond in an unconnected outer hinterland.

the Mediterranean Sea to China (Figure 1.12). During this phase, rival city-based states in Mesopotamia, such as Ur, Akkad, and Sumer, controlled local irrigation farming. At times the processes of wealth accumulation and trade broke down under the onslaught of peoples from the hinterland. From 2000 to 1000 B.C., more powerful groupings, including the Babylonians, the Hittites, and the Assyrians, took control of this central region of the world system.

The main achievements of this phase included the first writing—to record government activities and trade; the extension of arts in warlike human likenesses, animal paintings, and statues; the foundations of mathematics; technological innovations in metal-working, pottery, and construction materials, and the invention of the wheel; religious beliefs linked to mythical prehistories; and a codification of laws, including those giving rights to women. Mesopotamia became the center of a trading network based on exporting its agricultural and craft products and importing timber, metal ores, stone, and gemstones from the local periphery. Longer routes dealt in luxuries, with connections to Egypt, Persia, and the Indus River valley, and, after 1500 B.C., to China. Slavery provided the main source of labor—often supplied by men and women captured in war and moved to work in the core area cities.

Other cores imitated developments in Mesopotamia, sometimes extending human achievements, but Mesopotamia remained central to trade links. Irrigation farming formed the basis of wealth accumulation in Egypt, which incorporated many of the Mesopotamian developments in its own distinctive kingdoms that controlled the lower Nile River valley and built huge memorials in the pyramids. Three periods of Egyptian kingdoms were separated by invasions of outsiders, including the Hyksos, who captured the northern delta region around 1750 B.C.

The island of Crete flourished as the center of the Minoan civilization from 1900 to 1600 B.C. Less is known about the civilization in the Indus River valley around Harappa and Mohenjo-Daro, but its influence extended widely and its products have been found in Mesopotamian sites. It ended around 1550 B.C. after invasions from the north.

In China, a confederation of cities and states was gradually drawn together, but it was not until the Shang dynasty (1523–1028 B.C.) that there was wider control and an opening of trade relations to the north and west. Most of the world remained outside the immediate hinterlands, with economies based on hunting, fishing, and some farming, and was only occasionally caught up in the affairs of the core regions of wealth accumulation.

## Phase III. Second World System: Expansion in the Classical Civilizations of the Iron Age, 1000 B.C. to A.D. 600

"Classical" periods that formed the basis of modern world systems occurred in Persia, Greece, Rome, India, and China during this phase, when several major world universal religions

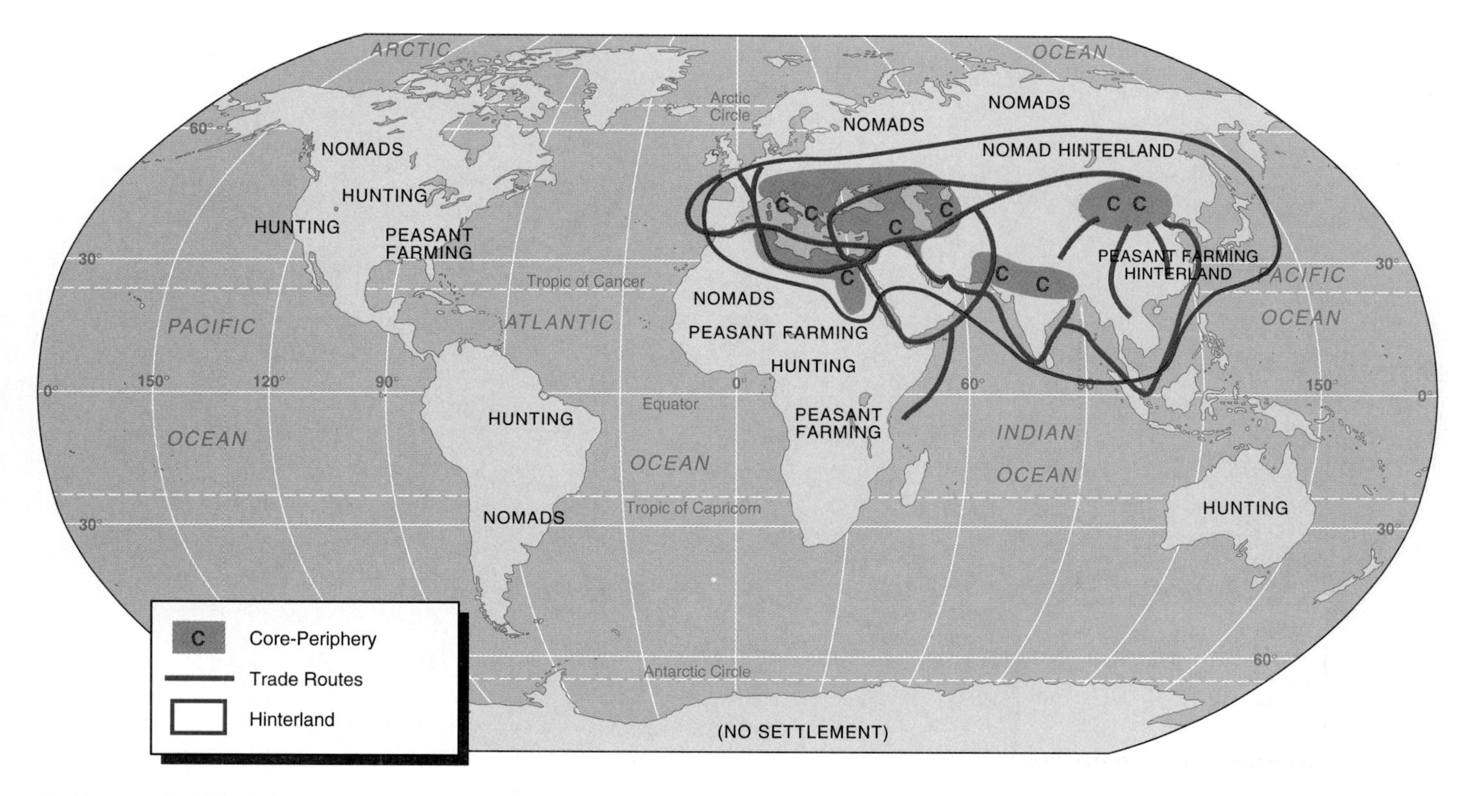


**FIGURE 1.13 Second world system, about 1000 B.C. to A.D. 600.** During this phase, the Greek and Roman empires expanded westward, the central Persian empire and northern India provided new cores, and the Chinese empires became organized and expansionist. By A.D. 200, there was a considerable east-west trade in luxury goods and a major growth of trade within the core-periphery zones.

and cultural systems originated. Confucius, Zoroaster, Buddha, the greatest Hebrew prophets and Greek philosophers, Jesus Christ, and Mohammed all lived between the 500s B.C. and the A.D. 600s. The Greek pantheon of gods (taken up by the Romans), the Celtic druidical religion, Scandinavian gods under Wotan, and the Egyptian Isis failed to attain the same prominence. Only in Hinduism did more ancient traditions and polytheism remain.

The geographic coverage of political and commercial activities expanded, spreading from the centers established in Phase II and shifting outward from Mesopotamia—which lost its central role (Figure 1.13). In the west, Roman legions incorporated much of northern Africa, modern Spain, France, and England into its empire. New controlling interests took over Egypt, Persia, and India. In China, the Zhou, Qin, and Han dynasties expanded the empire, opening the route through Central Asia to Persia and Rome by A.D. 200.

Economic and cultural contacts from one end of Eurasia to the other had never been so close. Luxury goods from China moved west by camel caravans through central Asia, a trade fostered by the Kushan (Afghan) and Parthian empires in the center. Other trade went by ship through the Red Sea and Persian Gulf to and from the Indian Ocean. Slavery remained the main form of labor in the fields, workshops, and domestic realms of life.

India emerged from a phase of complex movements of people and changes after the invasion of Aryan peoples, becoming a region of settled farmers concentrated along the Ganges River valley. After 800 B.C., the Aryans penetrated into southern India and Ceylon, with their iron implements enabling the clearance of woodland. Seafaring ties to southwestern Asia increased. Many of the subsequent features of Indian religious culture were enshrined in Sanskrit literature during this phase, including the emphasis on sacrifice and the notion of caste as a basis for social relationships. Hinduism combined a complex mixture of ancient deities, right-living based on vegetarianism, human sacrifice, and asceticism. Buddhism began as an attempt to reform Hinduism in the 500s and 400s B.C., later spreading to Ceylon, Burma, and China. It repudiated the caste system and the idea of a central god and claimed deliverance from suffering through the annihilation of desire. By A.D. 100, Indian merchants established links with southwestern Asia and Egypt, exporting precious stones, indigo, and silk yarn on their way to Rome. By the end of this phase, the Gupta dynasty ruled the whole of northern India.

China was for long affected by internal disorder as emperors became figureheads without power and a hundred feudal states fought for power. Yet this was also a time when Chinese culture was consolidated and spread into surrounding lands. The collapse of the previous social order gave rise to the teachings of Kong Fuzi (Confucianism), with its strongly ethical and organizational emphasis. The more mystical Taoism (the "way" of Lao Zi) recognized a different approach to harmony with the universe through quietude. Warring states within China continued to vie for power, with the scale of destruction increasing as iron weapons

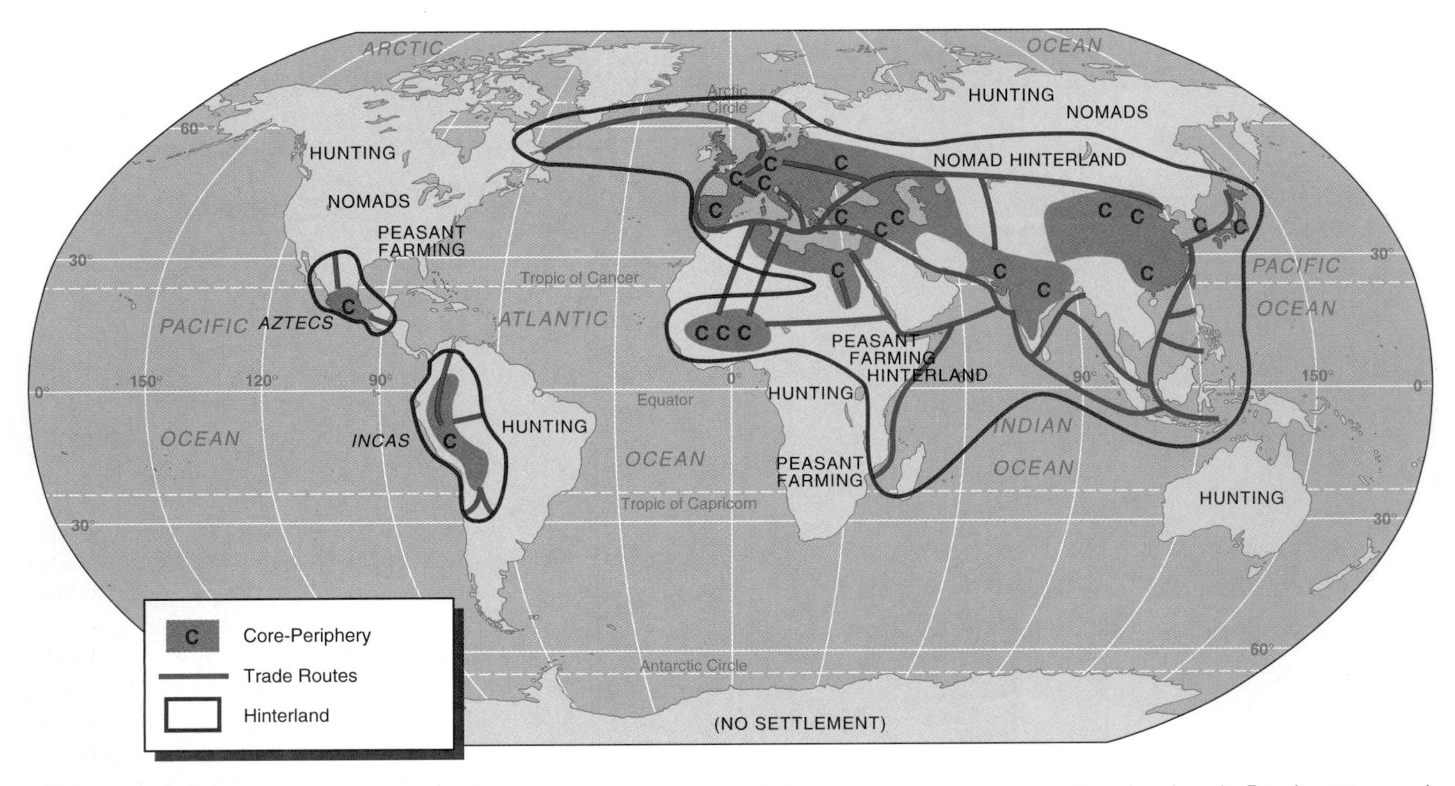


**FIGURE 1.14 Third world system, about A.D. 600 to 1450.** This period began with disruptions in Europe and included invasions from the Eurasian steppes and Viking homelands. It also saw the development of new empires outside the previous world systems, including the Americas. Some regions experienced important periods of political consolidation and wealth accumulation.

came into use alongside the implements that brought new land into cultivation. China combined a rising population, innovative technology, and expanding trade. The Qin controlled the northwest and west by 300 B.C. and all China by 221 B.C. The Han, under whom the Chinese empire became the most prosperous in the world, succeeded the Qin. This was the time when the Roman Empire reached its greatest extent in the west. By A.D. 100, Buddhism entered China, linking to both Confucianism and Taoism. The combination of a court weakened by factions and child emperors, incursions by nomadic Xiongnu from the north, and poor harvests, however, led to a breakdown and division of the empire. This ended phase III.

## Phase IV. The Third World System: Medieval Times, A.D. 600 to 1450

Although the term *medieval* is applied mainly to Europe, it signifies a middle phase in the development of world systems between the classical and modern eras. It involved interruptions to wealth accumulation and trade. The classical world that set the tone for much of succeeding world history and culture was followed by "Dark Ages" with invasions from central Asia of core areas in Rome, Greece, northern China, Persia, and India. Western Europe remained disorganized and backward as the result of continuous invasions from the east.

New empires arose in Byzantium (centered on Constantinople), Persia, northern India, China, Indo-China, Korea, and Japan. The Muslim expansion from the A.D. 600s out of southwestern Asia to Persia and North Africa—one of the most important events in history—was followed by Arab traders penetrating across the Sahara in northern Africa and taking a major role in the Mediterranean Sea and Indian Ocean. In western Africa, the kingdoms of Ghana, Mali, and Songhai followed each other. In the Americas, the Mayas, Aztecs, and Incas established military-economic control over large areas (Figure 1.14).

### Mass Migrations

Huge movements of peoples were set off by invading hordes from the steppe grasslands of central Asia. Mounted archers with light armaments were mobile and effective, attacking crucial centers of power. As they moved out of central Asia, they often pushed other groups ahead of them into southern China, India, and northern Europe. They spread Turkish languages westward but were seldom able to govern settled peoples for long. In the westward movements caused by such pressures during the 300s to 500s, German and Slavic tribes moved on to former Roman Empire lands, often pushed westward from present-day Ukraine by Goths and Huns. Further disruption and resettlement followed the Magyar invasions of Hungary in the late 800s. The Great Wall of China, however, proved an effective defense frontier for the Chinese. Other nomad groups followed throughout Phase IV, with the most significant and final invasions in the 1200s and 1300s by the Mongols, who conquered and ruled China for 200 years. In India, Muslims gained political control of the north by the 1200s, setting up the Sultanate of Delhi, which lasted until European conquests.

From the late 700s to the 1100s, the Vikings of northern Europe set out by sea to raid and settle lands from Iceland, Greenland, and the British Isles in the west, through France and northern Europe and down the Volga River valley in present-day Russia—where they traded furs and slaves with Muslims.

### New Geographic Patterns

Toward the end of this phase, the future cultural and territorial patterns of European and other powers began to emerge. The Christian church, previously confined to the Roman Empire, sent missions to northern Europe, resulting in the conversion of most of the region by A.D. 1000. With Roman power diminished by invasions, the eastern (Orthodox) church centered in Constantinople spread its influence northward and eastward. Muslim advances, however, swamped the church in southwestern Asia, India, and northern Africa. Muslim armies conquered the Balkan peninsula, southern Italy, and Iberia (Spain and Portugal), as well as northern Africa.

China again became the world's most prosperous empire with extensive sea power. Hangzhou was the world's greatest city, with 1.5 million inhabitants. China's visual arts, literature, philosophy, science, technology (first printing), and education were in advance of the rest of the world. Rich merchants and a growing middle class emerged through the increasing numbers of trained state officials, bankers, and independent peasants. The basics of Chinese culture (language and political organization) were adopted by surrounding states in Korea, Japan, Manchuria, and Yunnan (southern China). Southeastern Asia became a world crossroads between India and China, later influenced by Muslim trading and military missions.

In northern Europe, the Franks of central Germany established control over modern France and Germany by A.D. 800, while the Saxons moved into England and fought off the Vikings. Slavic states emerged in eastern Europe, including the first Russian state in the mid-800s. Throughout Europe, slavery gave way to a feudal economic system in which local lords held land from a king or prince. Peasants, or serfs, had to provide the lord with farm labor or military service for part of the year. In Russia, such serfdom continued until the 1800s.

### Further Changes

By the 1100s, Europe was climbing out of political and economic dislocation. The population rose from around 30 million in A.D. 1000 to 42 million some 150 years later. The increases were concentrated mostly in France, Germany, and England. New lands were opened for farming by cutting woodland and draining marshes. In western Europe, the feudal system was breaking down by the 1400s as serfs changed into freeholders or wage workers. Peasant farmers gained more independence, and towns expanded with craft industries and markets, some of which had international significance.

Toward the end of phase IV, the attacks of Genghis Khan and his Mongol hordes, the expansion of the Ottoman Turks, the Black Death, and a worsening climate resulted in economic recession and population decimation (up to one-third in some parts of western Europe). Trade across Eurasia was once again reduced.

## Phase V. Fourth World System: Modern Times, Capitalism, Industrialization, and Socialism, A.D. 1450 to the Present

This phase, building on previous phases, resulted in most of the detailed geographic differences among places at the end of the 1900s. Some historians restrict a discussion of world systems and global orders to this phase, when the growth of capitalism paralleled the centralization of economic and political power in western Europe. The new form of wealth accumulation spread outward—through trade and colonization—to the Americas, Africa, the South Pacific, and parts of Asia. By the late 1800s, much of the world was ruled from European countries and incorporated in dependent trade relations as part of a major expansion of the world system. Although the United States rose to become the world's most productive industrial country during the 1800s, it remained mostly aloof from overseas involvements (except for Cuba, the Philippines, and other former Spanish colonies that it took over at the end of the 1800s). Following the disruptions of two world wars, decolonization after 1945, and the Cold War, the 1990s initiated a new global order by involving virtually the entire globe in the capitalist world economic system.

**Capitalism** is a mode of production in which goods and services are produced and sold by private individuals, corporations, or governments in competitive markets. Most of the means of production (e.g., land, machinery) are owned by a few people or corporations. The producers, or workers, own neither the means of production nor the products: they sell their labor. Capitalism is linked to notions of individual freedom—at least for those who own the means of production—together with private property ownership and limited government interference in decisions about production, distribution, and consumption. This type of capitalism is often termed "free-market." Capitalism's emphases on individual enterprise, industrial corporations owned by individuals, and making profits encourage its own expansion. Such emphases also lead to uneven levels of wealth over time and geographic space. When governments, rather than individuals or corporations, control production and distribution, as, for example in the Soviet Union up to 1991, the socialist ideals result in a centrally planned economic system that treats capitalism as an opposed system. The outcome is termed **state capitalism** because of the way socialist states function in an essentially capitalist world.

Capitalism, now the dominant world economic system of wealth accumulation, developed when technologies of

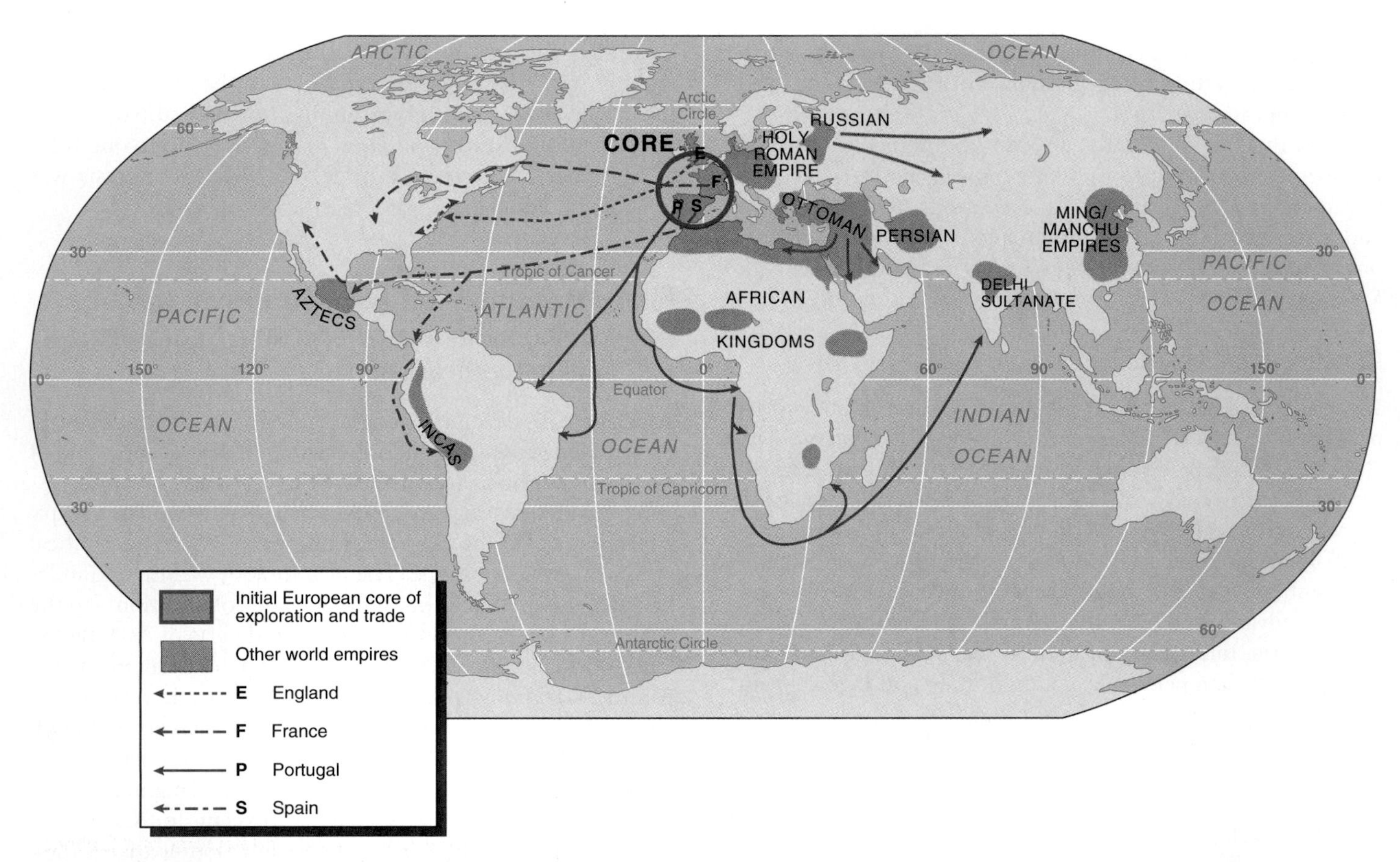


**FIGURE 1.15 Fourth world system, part 1, about A.D. 1450 to 1750.** This phase was dominated by the capitalist means of wealth accumulation. In the first period, about A.D. 1450 to 1750, mercantile capitalism led to the exploration from Europe of new lands, along with some colonization. Other empires existed outside the European realm and traded with each other (see also Figures 1.16 and 1.20).

production and communication expanded the geographic scope of exchange and control from the 1400s until today. At first, in the period A.D. 1450 to 1750 (Figure 1.15), **mercantile capitalism** grew out of European feudalism as merchants based in independent market towns and ports traded local products, such as wool and cloth, for spices and dried fish from overseas. Primary products of the fishing, timber, mining, and farming industries were the main goods distributed through mercantile capitalism. Owners of capital invested it in land and farming improvements, such as the drainage of wetlands in Holland and England, or in trading expeditions to the new colonies. Some regions specialized in producing cloth, at first in people's homes with merchants collecting and selling the output.

Such trade resulted in the growth of linkages among market towns within European countries and led to the overseas colonization of new territories to provide sources of new products. European countries, beginning with Spain and Portugual in the 1500s but giving way to French, Dutch, and British merchants and military in the 1600s, invested in exploration and trade with Africa, Asia, and the Americas. Western Europe became the core of the new capitalist economic system, while its colonies and trading stations were its periphery. Other parts of the world remained in the hinterland of isolated subsistence or feudal systems until the 1800s.

From A.D. 1750 until the early 1900s, the first industrial revolutions set off growth in factory-based manufacturing industries, bringing a new impetus to the capitalist economic system and hastening its world dominance (Figure 1.16). The concentration of manufacturing production in factories involved high levels of investment in buildings and machinery. Production in factories added significant value to the raw materials and components and brought good returns on the investments. Manufacturing also required investment in transportation facilities linking the factories to their sources of raw materials, component producers, and markets. New social groups emerged out of the production processes, from the owners to the managers, foremen, and workers. New distinctive patterns of towns, new functional areas within towns, transportation links, and regional wealth inequalities marked this phase of **industrial capitalism** (Figure 1.17).

Industrial capitalism experienced alternating periods of expansion and stagnation from the mid-1700s to the present

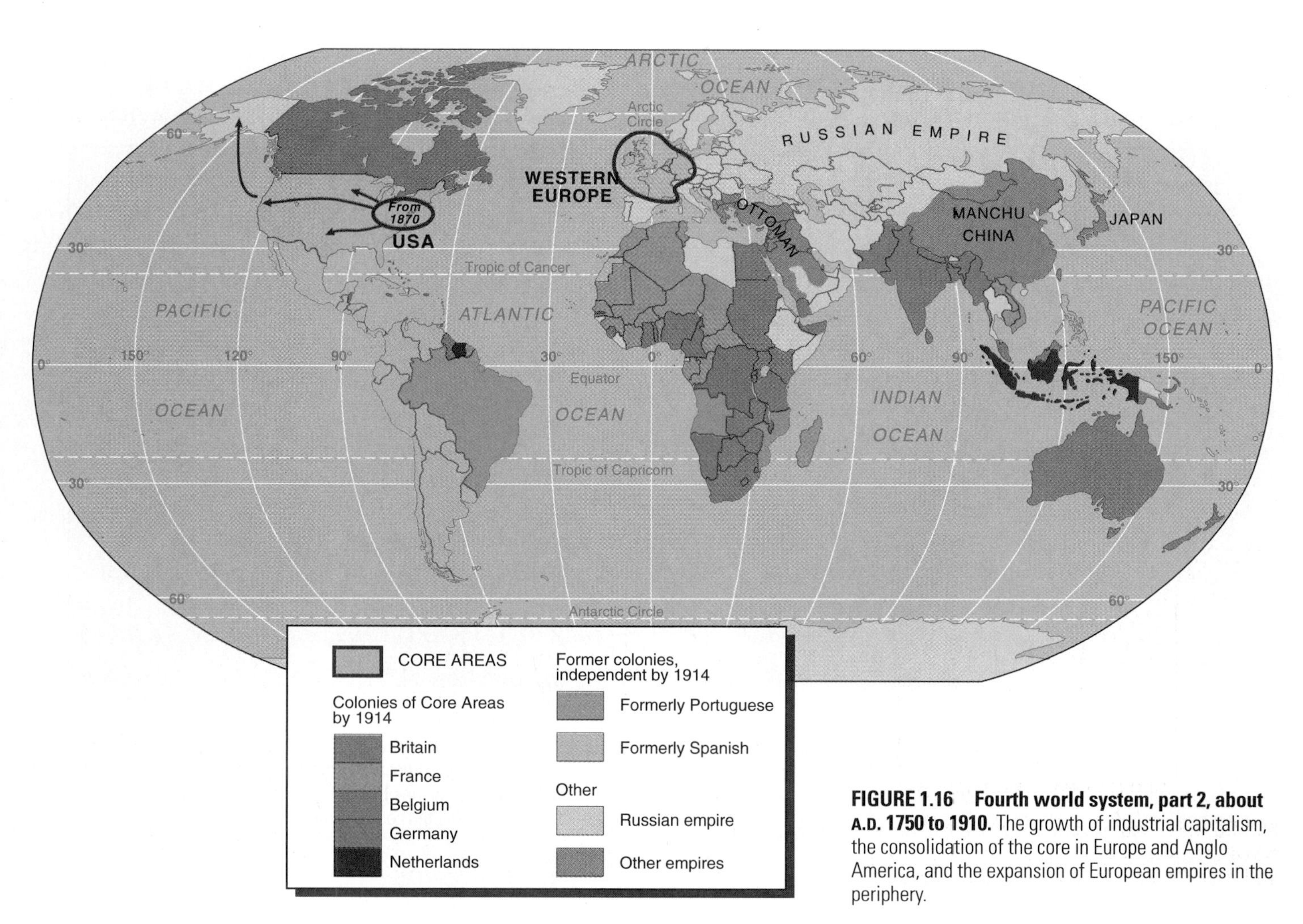


**FIGURE 1.16 Fourth world system, part 2, about A.D. 1750 to 1910.** The growth of industrial capitalism, the consolidation of the core in Europe and Anglo America, and the expansion of European empires in the periphery.

**FIGURE 1.17 Industrialization in the 1800s.** A textile mill in North Grovnerdale, Connecticut. Built in 1872 beside a canal, its four floors enabled steam power based on burning coal to be used more efficiently.

© James Marshall

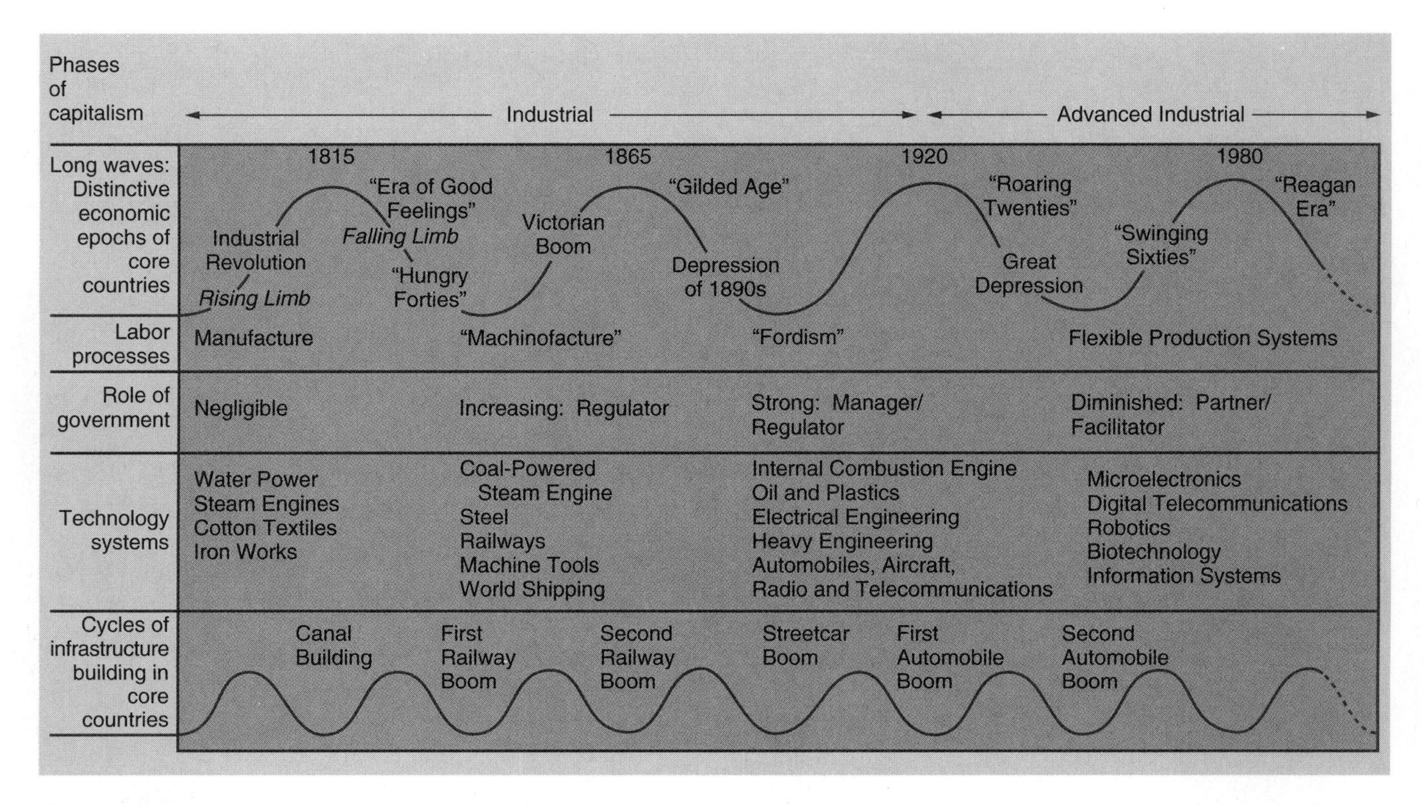


**FIGURE 1.18 Fourth world system: economic fluctuations.** This summary mainly reflects experiences in the United States. The rising limb of each long wave was accompanied by the development and use of new technologies of manufacture and transportation that enlarged productive capacity. Core-periphery relations expanded into new world areas. During the falling limb, costs to manufacturers (such as the price of raw materials from peripheral countries) increased, and there was less money for new investments. Suggest how these fluctuations affected countries in the world periphery.



(Figure 1.18). During periods of expansion, new technologies of production and transportation caused a series of further industrial revolutions and spread the influence of capitalism. Cities expanded as more factories and homes were built to accommodate workers moving from the countryside to work in factory jobs. Overseas exploration opened new lands to the commercial production of goods that entered world trade. During periods of stagnation, technological innovation slowed, while surpluses of labor, productive capacity, and capital resulted in economic recession. Places still dominated by subsistence and feudal economic systems were reoriented as the world economic system advanced further in each phase of expansion.

Four major periods of industrial revolution, followed by slower economic expansion, mark the world economic system since 1750.

- **First industrial revolution, 1750 to 1850.** The first factories were in Britain, powered by water mills and then steam engines. They processed agricultural and mine outputs. The early manufacturing industries included iron working, cotton textiles, leather goods, and pottery. This was a period of competition among family-owned manufacturing firms, particularly in western Europe and the United States. Transportation was mainly by water around the coasts, up rivers, and along specially constructed canals linking the manufacturing areas with ports, raw material sources, and markets. After boom conditions up to the 1820s, further economic growth proved slow in the 1830s and 1840s as production capacity caught up with demand.

  Territorial expansion overseas accompanied the growth phase. Britain's booming economy in the early 1800s, and its world naval dominance following the defeat of Napoleonic France in 1815, led to it establishing political control over India (see Chapter 5) and Australia (see Chapter 11), followed by commercial dominance in the newly independent Latin American countries (see Chapter 10). Other European countries followed this example. The United States (Chapter 9) began its industrial growth in the early 1800s. It accomplished a remarkable expansion of territory as it pushed out its borders by incorporating the western two-thirds of its conterminous lands between 1803 and 1853 and adding Alaska in 1867.

- **Second industrial revolution, 1850 to 1910.** Manufacturing production techniques spread from Britain to several parts of northern Europe by the 1850s. During the late 1800s, new technologies based on steel and

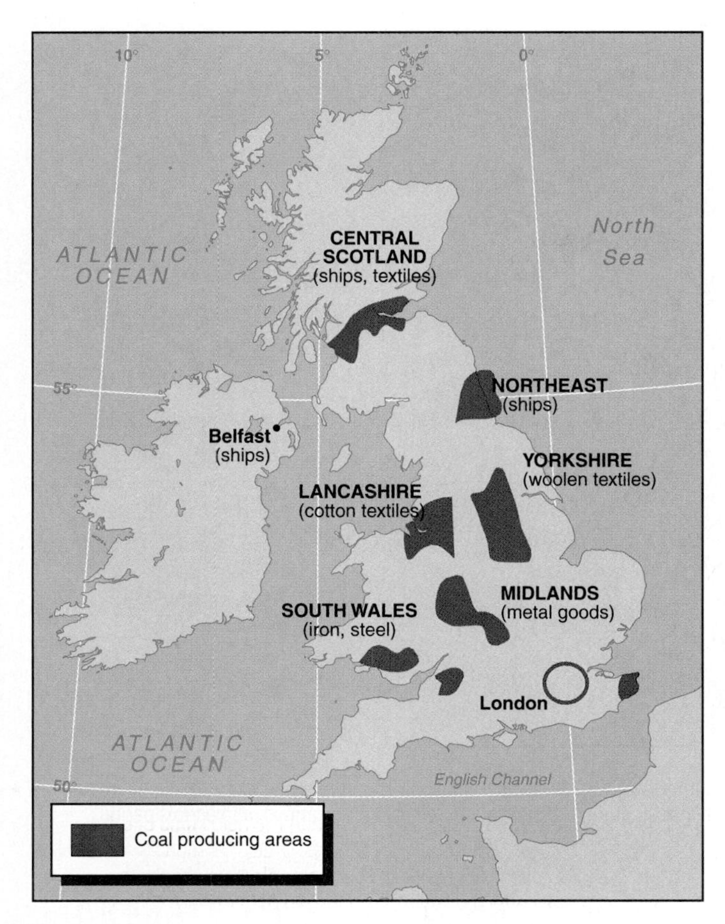


**FIGURE 1.19 Industrialization in the 1800s and coalfield locations.** Areas of Britain that produced coal were industrialized and urbanized, growing rapidly in wealth and population. They suffered setbacks in the 1900s as their products competed at a disadvantage with new areas in other parts of the world.

chemical products, steam power, and railroad transportation networks instigated another industrial revolution and a massive expansion of manufacturing in both scale of production and geographic extent. Coal became the main energy source and a major raw material in steel production; many coalfield areas became centers of industrial expansion (Figure 1.19). Germany and the United States rivaled Britain in output as Britain concentrated on expanding its empire into Africa and failed to continue its lead into new technologies. Once again, an initial spurt of industrial production in the 1860s and 1870s gave way to slower growth in the 1880s and 1890s.

During the late 1800s, manufacturing industry underwent organizational changes that developed further capacity for growth. Family firms struggled in times of recession, and many sold out to growing corporations. For example, Andrew Carnegie, after a rapid rise in the U.S. railroad industry, understood the potential of the new steel technology and established a steel mill near Pittsburgh in the early 1870s; he later bought out other steelmakers in a process of horizontal integration. During the 1880s, he took control of all aspects of production from mining coal and iron ore to making steel and selling steel rails and other steel products—a process of vertical integration. By 1900, Carnegie Steel had 20,000 employees and was the world's largest industrial corporation. In the following year, it joined other U.S. steelmakers to form the US Steel Corporation with 168,000 employees—a totally new scale of manufacturing activity and control. The growth of corporation size and power enabled one or a few firms to control prices and wages. This trend sparked resistance to unfettered corporate control as labor became organized and the U.S. Congress passed antitrust legislation to regulate price-fixing abuses.

- **Third industrial revolution, 1910 to 1950.** In the early 1900s, another industrial revolution brought further technological innovations and geographic expansion of capitalist enterprise. There was more economic growth in the world's expanding cores of Western Europe and the United States (Figure 1.20). The energy dominance of coal and steam was challenged by the use of oil, natural gas, and electricity. Roads, cars, trucks, pipelines, and airplanes challenged the transportation dominance of ocean liners, cargo ships, and railroads. Product ranges diversified into electrical consumer goods, cars and trucks, aircraft, plastics, radios, and telecommunications.

Production methods in the core manufacturing industries changed to the mass production on assembly lines that had been pioneered by Henry Ford in car production—and hence termed "Fordist." Scientific management and mass consumption in the richer countries backed such methods. Corporate ownership replaced the family ownership of firms, with U.S. corporations beginning to transfer their cost advantages in car and consumer goods manufacture to new factories in Europe.

During the early 1900s, large new areas of agricultural and mineral production opened in southwestern Asia, South America, Australia, and parts of Africa, thus expanding the resource base and increasing competition among peripheral resource providers. Costs were reduced for core-country manufacturers. The core countries produced nearly all the manufactured goods and sold their products in reciprocal trade to peripheral countries after having benefited from the added value in manufacturing. Core countries gained further wealth and often invested in urban expansion through the building of suburbs. Public transportation and private cars provided access to suburban housing. More people could afford the new mass-produced consumer products.

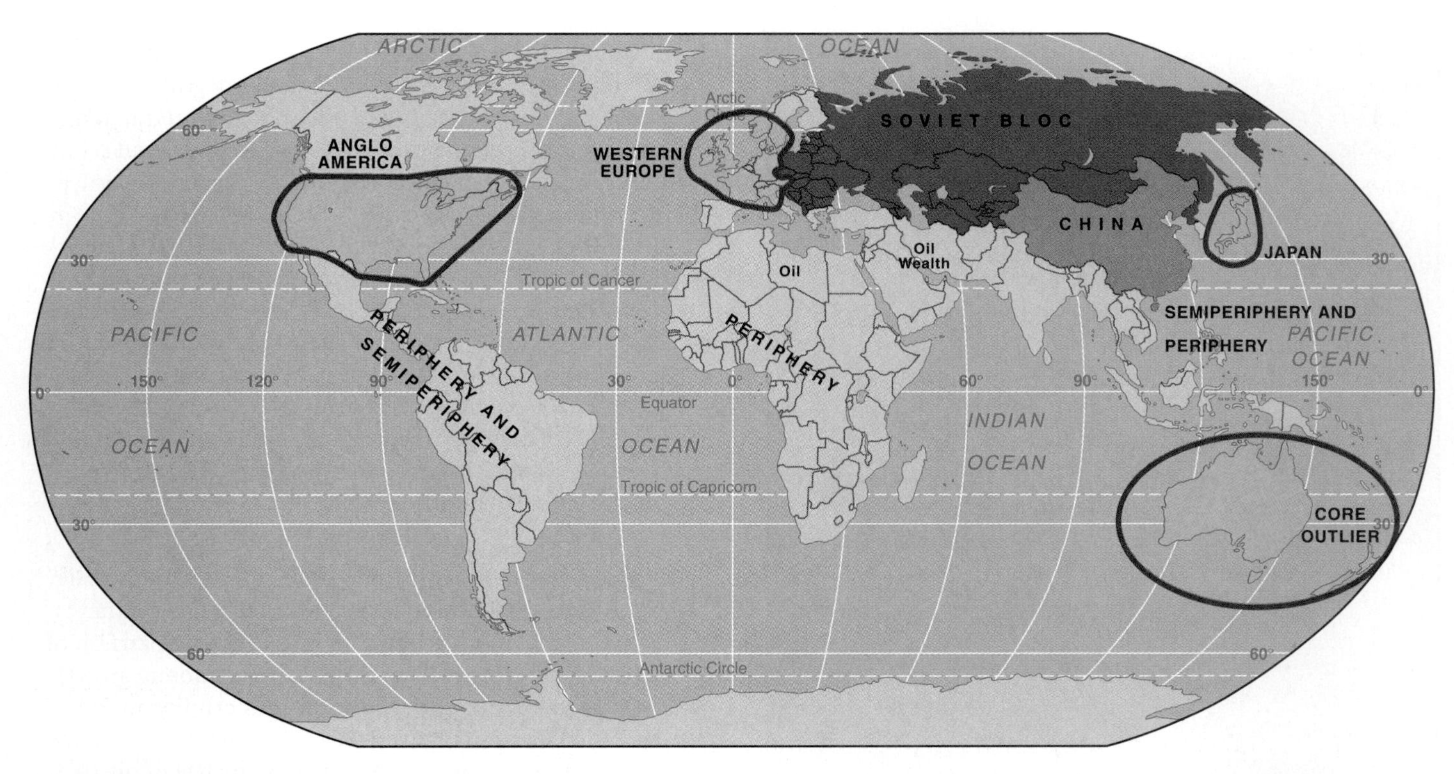


**FIGURE 1.20 Fourth world system, part 3, about A.D. 1910 to present.** State capitalism, advanced industrial capitalism, two world wars, independence of former European colonies, expansion of world trade, and technologic developments in manufacture, transportation, and communications marked this period.

Up to 1950, the world economic core remained in western Europe and Anglo America, while most of Africa and southwestern Asia were drawn into the colonial linkages that dominated the rest of the world outside of Latin America, China, and Japan. Japan went through its own period of industrialization but had limited participation in the expanding world economic system before 1950. Ex-colonial or largely uncolonized areas entered the periphery as their commercial relations with the core countries grew. The gap between core and periphery increased.

The years of core country economic growth ended temporarily in the economic depression of the 1930s, followed by World War II in the 1940s. Government and trade unions became more involved in the regulation and mediation of industrial affairs during this period, while governments achieved further prominence through organizing the production of war materials.

- **Fourth industrial revolution, 1950 on.** World War II was followed by a new period of technological diffusion, often based on products developed during the war—jet engines, electronics, and aerospace—together with expanded ranges of consumer goods and increasingly packaged foods. This led to further urban growth in core countries to accommodate workers in the expanding manufacturing and service industries.

The wider world economic order experienced political changes, such as the independence of most of the former core-country colonies, and the eruption of the Cold War. Some of the newly independent countries sought economic links with, and assistance from their former colonizers, such as Britain and France. Others looked to alternative sources of markets, guidance, and support from the United States or the Soviet bloc.

Large corporations, beginning with those in the United States and extending to their counterparts in western Europe and Japan, controlled production and distribution in many countries. Such multinational corporations began to maneuver governments and labor interests by siting new production plants in countries where labor costs were low.

Technological advances further mechanized and automated farm and factory production in the core countries. Such processes required less semiskilled labor in many industries and led to redundancies. The service sector grew in significance as additional jobs were created in expanded health, education, professional, business, recreation, and especially government services. This trend began in core countries.

Manufacturing employment shifted to newly industrializing countries, mainly in Eastern Asia and Latin America (South Korea, Taiwan, Hong Kong, Singapore, Malaysia, Thailand, Brazil, and Mexico). These countries competed with the core countries of the West, and a new core began to form in Eastern Asia, centered on Japan.

In the 1970s, world economic growth faltered as raw material, and especially energy, prices rose. New prospects for economic growth emerged in the 1980s. The development of microelectronics, biotechnology, robotics, and information systems may signal the start of another period of industrial revolution and world economic growth at the beginning of the new century. Another view of the current situation, however, is that the post-Fordist, or postmodern, period is one of uncertainty and confusion resulting from the global extension of the world economic system.

**RECAP 1C: Growing Global Order**

World history records how civilizations spread their geographic impact through core and peripheral regions. Periods of wealth accumulation and expansion of empires and trade alternated with periods of disruption. During the 1900s, the world's core regions expanded their periphery to include the whole world in the capitalist economy.

1C.1 Construct a chart to summarize the main features of the expanding global order, using the hierarchy of timescales and the geographic differentiation of local, regional, countrywide, and global.

1C.2 How far is a knowledge of the growth of the global order essential to our understanding of modern world geography?

**Key Terms:**
world system
capitalism
state capitalism
mercantile capitalism
industrial capitalism

# CHAPTER REVIEW QUESTIONS

1. Before 1990, the Second World consisted of countries that (a) placed individual rights as the first priority of government (b) tried to implement the communist principles of Marx and Lenin (c) had democratic governments (d) were newly independent
2. A group's culture includes (a) ideas (b) beliefs (c) gender roles (d) all of the above
3. To a geographer, a landscape is (a) pretty scenery, suitable for an artist to paint (b) planting flowers and shrubbery in front of one's house (c) the result of interaction between people and environment over time (d) the design of a new park
4. Which of the following is *not* an example of a global choke point? (a) Strait of Malacca near Singapore (b) Davis Strait in northern Canada (c) Panama Canal in Central America (d) Strait of Hormuz at the entrance to the Persian Gulf
5. Settled farming began (a) in Egypt about 3000 B.C. (b) in the Tigris-Euphrates Valley around 5000 B.C. (c) in Africa around A.D. 100 (d) in hill lands of Southwest Asia around 8000 B.C.
6. The core of the first world system was in (a) Mesopotamia (b) Egypt (c) China (d) Greece
7. Which of the following was *not* a characteristic of the Second Industrial Revolution? (a) use of electricity (b) importance of coal (c) growth of railroad networks (d) development of large industrial corporations through vertical and horizontal integration
8. Which of the following countries is in the "North" according to the "Brandt Line" but is in the periphery in the New World Order? (a) Russia (b) China (c) Kazakstan (d) Mexico
9. The major world region with the largest share of population is (a) Europe (b) Commonwealth of Independent States (c) Anglo-America (d) Eastern Asia
10. Which of the following is *not* a characteristic of countries with socialist economic systems? (a) state-directed central economic planning (b) health care and education for a large portion of the population (c) widespread ownership of consumer goods (d) most of the population in the lower-income group
11. After 1990, Russia was the sole remaining superpower. True / False
12. The spread of Western food and drink, popular culture, and consumer goods has completely obliterated cultural differences in the world.
    True / False
13. Because of an increasingly interconnected world, in which global forces impact local events, geography no longer matters.
    True / False
14. "Classical" periods that formed the basis of modern world systems occurred in Persia, Greece, Rome, India, and China between 1000 B.C. and A.D. 600.
    True / False

15. The rise of Islam and expansion of the Muslims took place in a period considered "backward" in Europe. True / False
16. The first phase of the Industrial Revolution began in the United States.
    True / False
17. In the Third Industrial Revolution, more parts of the world were drawn into the capitalist industrial system, and the gap between core and peripheral countries narrowed.
    True / False
18. In the Fourth Industrial Revolution, production and distribution of many goods began to be controlled by large multinational corporations.
    True / False
19. During the Cold War, countries of the First World competed with countries of the Second World for influence over countries of the Third World.
    True / False
20. The economic system that emphasizes private or corporate organization of business and investment is the ____________________ system.
21. Poor countries that have a narrow range of products, less advanced technology, and poorly paid workers are said to be in the ____________________.
22. Materials stored in the natural environment that human societies use in their cultural and economic systems are ____________________.
23. The first form of capitalism, developing from the 1450s and involving European merchants trading local products for fish and spices overseas, was ____________________ capitalism.
24. The use of assembly lines and scientific management in manufacturing is called ____________________.
25. As an example of the power of core countries, the World Bank and the International Monetary Fund are headquartered in ____________________.
26. Countries that are in the process of moving between being in the core and being in the periphery are called ____________________.
27. The world as it would be without human modification constitutes the ____________________.
28. The location of a place as defined by its position on the Earth's surface, usually in terms of latitude and longitude, is its ____________________.
29. Distinctive parts of the Earth's surface with characteristics determined by a combination of physical and human geographic features are ____________________ regions.
30. When the government, rather than individuals or corporations, control production and distribution in an essentially capitalist world, the result is ____________________ capitalism.

# BASICS OF WORLD REGIONAL GEOGRAPHY

Chapter 2

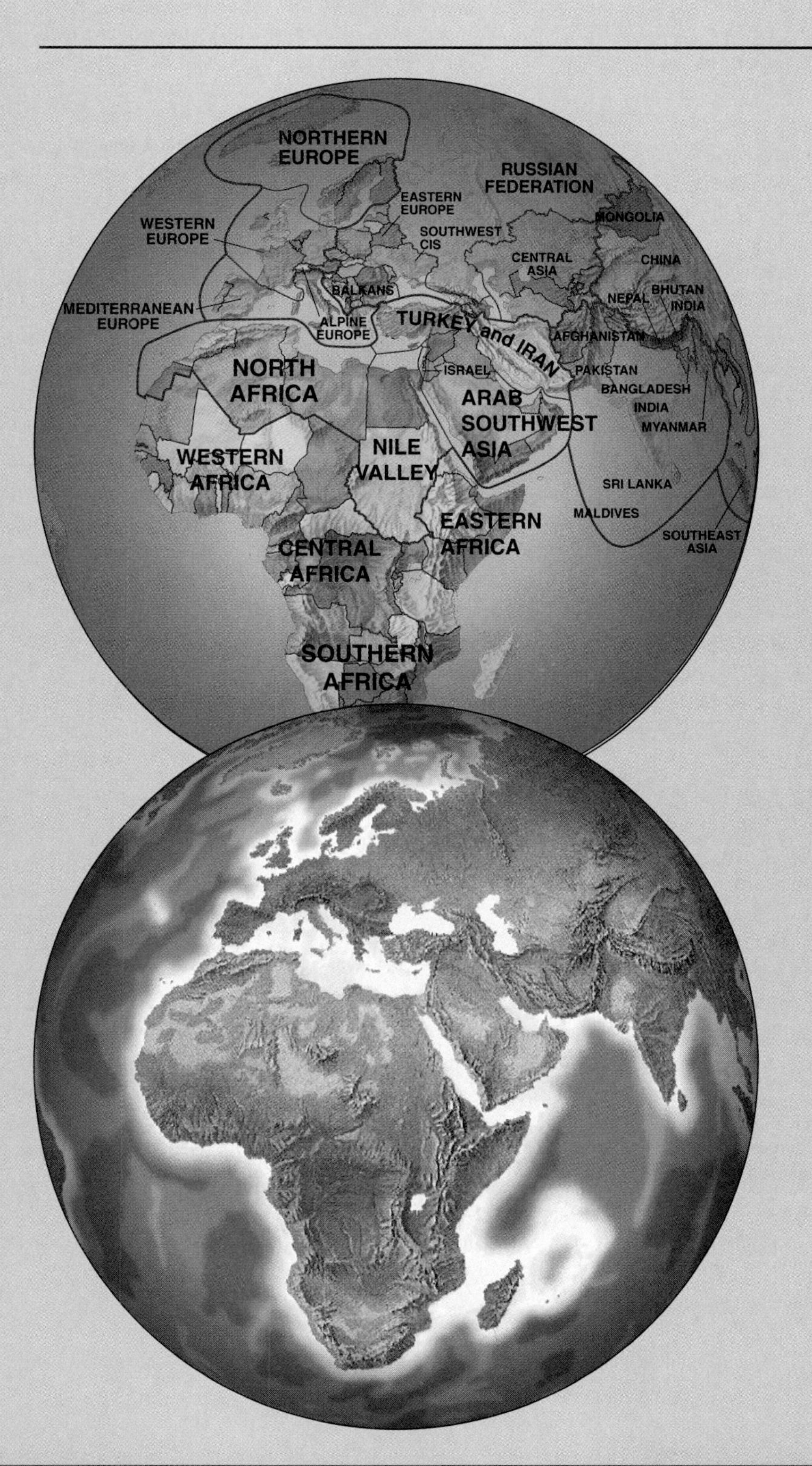


## THIS CHAPTER IS ABOUT:

**How geographers study regions**

**The fundamental importance of place and location**

**Basics of population geography: people matter**

**Basics of political geography: countries and governments**

**Basics of economic geography: wealth, poverty, and development**

**Basics of cultural geography: lifestyle differences**

**Basics of physical geography: environmental variations**

**Factors that decide how many people Earth can support**

**How landscapes provide a summary of regional geographies**

**FIGURE 2.1 Worlds of geographic study.** The globes compare aspects of human (country boundaries) and physical geography.

Geographers attempt to understand the formation of a place's characteristics and location through studying the interactions of a wide variety of factors. Such factors include those identified in Chapter 1 as major aspects of a new global order—political, economic, and cultural processes and the workings of the natural environment. This chapter introduces a range of geographic concepts that are important in regional analysis (Figure 2.1) and describes the sources of data that inform them—many of which are accessible from daily or annual updates through the World Wide Web. A selection of data for each country is included in Reference Section tables.

In investigating a particular topic, geographers first define their goals, then collect relevant information, analyze it in various ways, and attempt to explain the result. This process often provides deeper understandings that can be applied in communal planning or individual decision making. Geographers increasingly go beyond this point, and many get involved in the subject of their studies in such areas as housing markets or conditions in poor countries. In this book, we use the results of many geographic studies. Training as a geographer leads to a variety of potential careers and job opportunities, as set out in the Reference Section (Jobs for Geographers, page 581).

**WEBSITES: Introduction**

A good introduction to using the World Wide Web is found on the Public Broadcasting System site at http://www.pbs.org. Under the heading "Select a PBS Online Web site," choose "Understanding and Using the Internet." Try this site and check out one of its special current presentations. There are thousands of websites, and it is easy to become confused and discouraged in using them. Throughout the text, we suggest a few long-term sites that are worth exploring and visiting frequently.

## Information Sources

Geographers collect their information about places and about human characteristics and actions from a variety of sources. They use the results of censuses that are published by each country at varying time intervals. The most useful censuses come mainly from the richer countries that have sophisticated data-collecting systems. In many poorer countries, the censuses are not so frequent, regular, or accurate, and some countries publish little of the information they collect. International agencies—such as the United Nations and the World Bank—and the Population Reference Bureau in Washington, D.C., collect data published by individual countries and point out the levels of accuracy and acceptability of the information. Such international sources are often more convenient than collecting data from each country.

Geographers use maps and satellite images produced by government mapping agencies around the world. Maps are a specifically geographic way of presenting information, since they show locations of places and the spatial relationships between them. Other sources of information used by geographers include written reports, such as those from other geographers, the United Nations, World Bank, governments, independent think tanks, and a range of current news items. Such varied sources of data and maps are combined into geographic information systems (GIS) that are used increasingly by business and government agencies.

Geographers also collect information by carrying out fieldwork. This involves visits to the places being studied to make observations of landscapes and to interview people about the ways by which they make a living or how their culture affects the geography of the region or country. Such observations may be compared to the statistical record. Overall, geographers use a rich variety of source materials.

## Using Information

Once the information is gathered, geographers use cartographic, graphical, and statistical techniques to analyze it. They may make their own maps or produce overlays on published maps; they may use a range of visual diagrammatic means such as graphs; and they may compare their data with other data and assess the significance of their findings through a range of statistical means. Geographic analysis is usually carried out with reference to a specific objective. For instance, regional geographers seek answers to such questions as: What is unique about this region? What links this region to others in its country? How is this region changing? Why is this region richer or poorer than others? Answering such questions helps us to understand the inequalities in the world, the influences on each region, and the changes that are continually taking place.

## Understanding Other People and Places

Trying to understand how people living in other countries think and feel about their own region and country and its place in the wider world can be carried out through such means as field experience or by becoming involved in a specific issue affecting people in a region. Geographers take up emotive issues and get involved in politics at local or national level, studying how processes act in society from the inside. Appropriate techniques of observation and reporting make it possible to express and collate individual impressions. For example, when I carried out a study of the impact of U.S. federal government programs in Appalachia in the early 1980s, I not only mapped and analyzed statistical data at the county level, but also interviewed officials of the local development districts concerning their work and their views of its impacts. This provided a partial, but new, aspect to studies of the region (see Bradshaw, 1992).

## This Chapter

The rest of this chapter introduces ideas in the major areas of geographic analysis that contribute to regional studies. Major concepts in each area of study are reviewed so that reference may be made to them in the later study of each major region.

- It is important to be able to locate places with respect to the rest of the world and to understand how geographers represent information on maps and use sources of information gathered by aerial and satellite means.
- The basics of population geography include factors influencing the distribution of people and the processes contributing to changes in population (demography).
- The basics of political geography include the ways in which countries and international groupings function.
- The basics of economic geography include a consideration of different sectors of production and levels of development within the global economic system.
- The basics of cultural geography include studies of religious affiliation, languages, and some social factors such as race, class, and gender.
- The basics of physical geography include a consideration of the distribution of climatic environments, ecosystems, and soils, and the formation and location of surface features such as continents and mountain systems. Human actions not only influence the functioning of processes in the natural environment but also define aspects of natural environments as resources or hazards, depending on human interactions with them.
- Finally, there are two sections that bring together aspects of the above list. First, there is the question as to how many people Earth can support, relating population growth to natural resources, political, economic, and cultural conditions. Second, there is the idea of landscape as a summary of geographic character in a region.

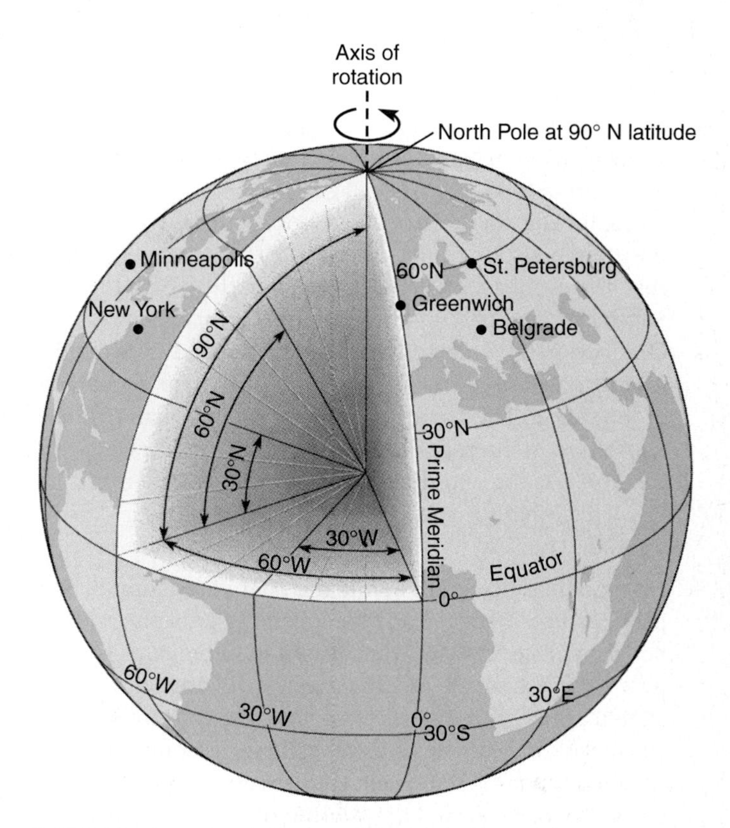


**FIGURE 2.2 Location: latitude and longitude.** The coordinate system used for locating places on Earth's surface, and the network of parallels of latitude and meridians of longitude. Give the latitude and longitude of each place marked on the globe.



# SENSE OF PLACE: LOCATION, LOCATION, LOCATION

## Location of Places

When you buy a house or set up a commercial facility, realtors emphasize the importance of location. Basic geographic concepts include location and the related characteristics of direction, distance, and scale.

- **Location** is the precise place on Earth's surface where someone lives or an event occurs. Latitude and longitude form the framework of the international reference system that pinpoints a place's absolute location (Figure 2.2). **Latitude** describes how far north or south of the equator a place is, measured in degrees. The equator is at 0° of latitude. The North Pole is at 90° N and the South Pole at 90° S. A circle joining places of the same latitude at Earth's surface is called a **parallel of latitude.** The distance covered by one degree of latitude is approximately 110 km (69 mi.) on Earth's surface. To pinpoint locations more accurately, each degree is divided into 60 minutes (written: 42′) and each minute into 60 seconds (written: 26″). The precise location of Meades Ranch, Kansas—the control point for a survey of the United States—is 39°13′26.686″ N.

Each place on Earth is given a precise east-west position by its longitude. **Longitude** measures position east or west of a half circle drawn from the North to the South Pole and passing through the former Royal Observatory at Greenwich, London, England. Such lines joining places of the same longitude are called **meridians of longitude.** The position of the prime meridian (0°) was chosen by an international conference in 1884, when London was the world's most powerful decision-making city. The longitude of Meades Ranch, Kansas, is 98°32′30.506″ W.

- **Distance** and **direction** define the relative location of one place with reference to another. Although distance between places is usually measured in kilometers or miles, travel time or travel cost may be of greater significance. Time-distance and cost-distance are often substituted for measured distance in geographic studies. The increasing cost and time of distance between places gives rise to the idea of the **friction of distance**, since there is likely to be less interaction between people across the distances where costs are higher or journey time is longer. Geographers give directions by the points of the compass.
- **Scale** defines whether one is studying a smaller or a larger area. Horizontal ground distance is related to map distance by a scale, usually quoted as a fraction (e.g., 1/10,000) or ratio (e.g., 1:10,000) in which 1 unit on the map represents 10,000 of the same units on the ground. For example, on a map with a scale of 1:10,000, 1 cm represents 10,000 cm, or 100 m, on the ground.

  Map scales vary with the size of the area to be mapped and the purpose of the map. *Small-scale maps* usually show areas at ratios of 1:250,000 or smaller (e.g., 1:1 million). They provide a small amount of detail about extensive areas (Figure 2.3). The world maps used in this text (see, e.g., Figure 1.2 or 2.10) are examples of small-scale maps in which the scale along the equator is approximately 1:120 million. *Large-scale maps,* usually with ratios less than 1:10,000 and greater than 1:250,000 (e.g. 1:50,000), contain more details, as in town maps. Not everything can be drawn to scale on maps, so roads, rivers, buildings, and other features are replaced by symbols.

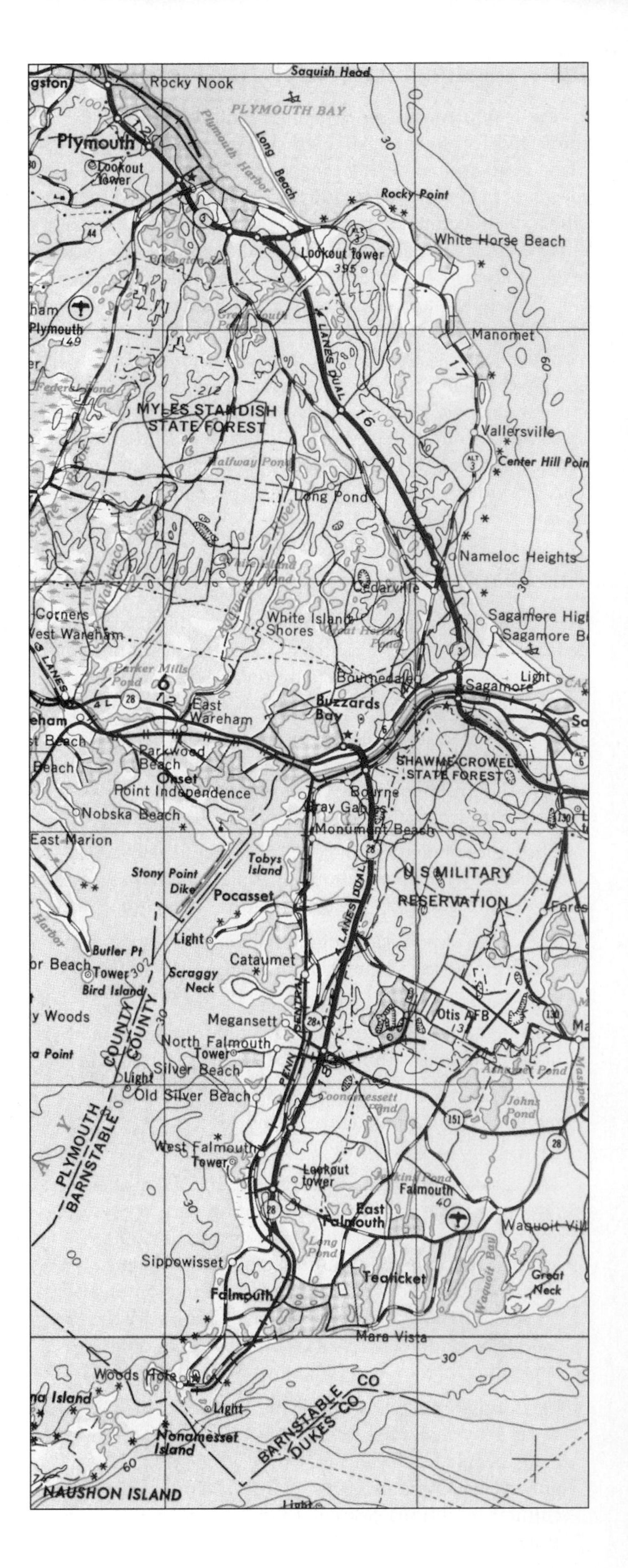


**FIGURE 2.3 Topographic map: location, distance, direction, and scale.** A 1:250,000 map of part of eastern Massachusetts. What other information does the map provide? In what direction does Route 3 travel on leaving Plymouth toward Sagamore, Cape Cod?

Source: U.S. Geological Survey

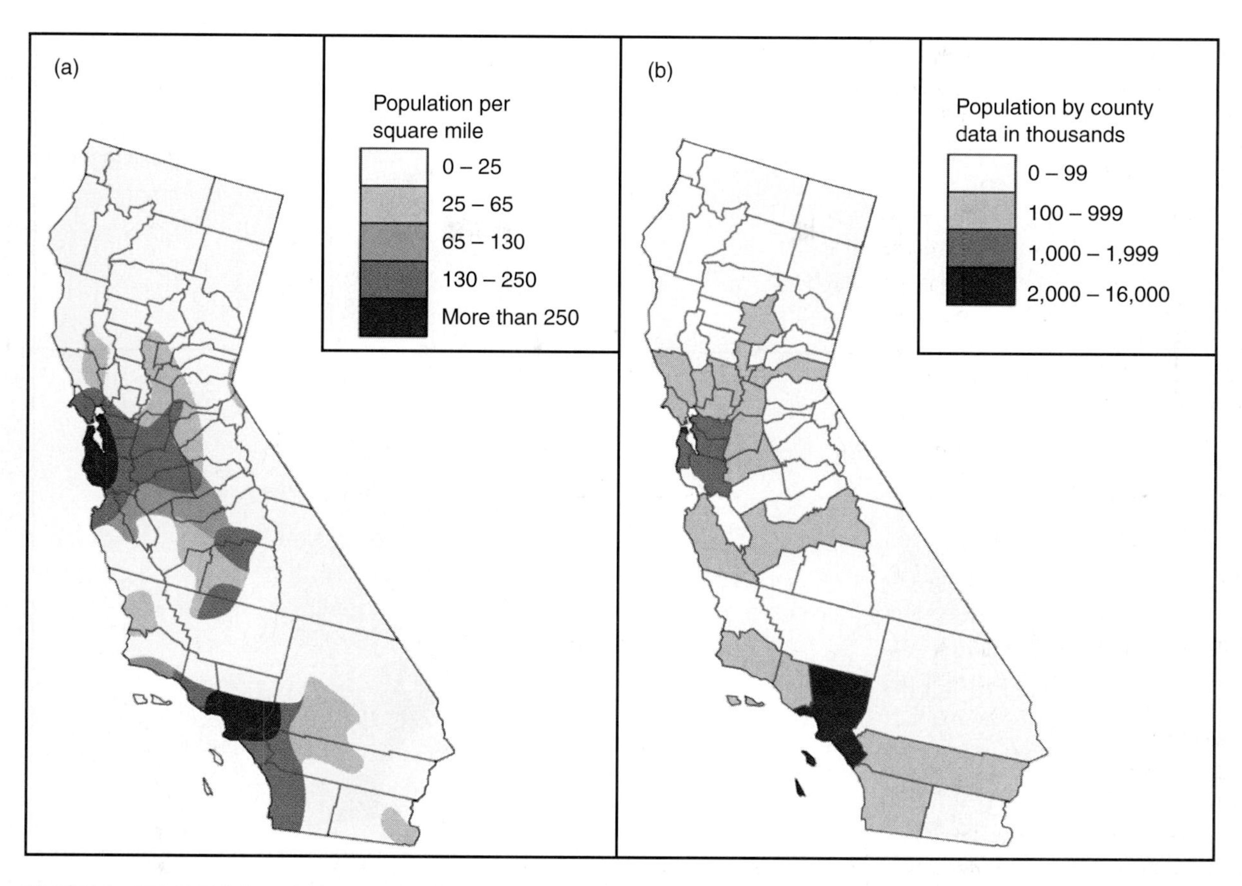


**FIGURE 2.4 (A) AND (B) Thematic maps.** What does each tell us about California? (a) Isoline map, where lines join places of equal population density. (b) Choropleth map, where the population totals for each county are grouped and the county areas shaded by category.

## Maps

Geographers often use maps to present information about location, direction, distance, and other characteristics of areas. **Maps** are relatively small pieces of paper that represent much larger areas of Earth's surface. Representing Earth's sphere on flat paper is a problem that troubled cartographers for generations. They projected the framework of parallels of latitude and meridians of longitude onto flat paper in various ways. Some of the solutions, known as map projections, are described in the Reference Section (Map Projections, page 582).

Location maps present information about where places are, the distances between them, and the directions from one to another. Thematic maps present information about such features as population totals, densities, or distributions. Thematic maps often use **isolines**, which join together points having the same value. For instance, contours join places of the same height above sea level and are used to show the relief (rise and fall) of a landscape. Isolines are also used to show such information as equal distances, costs of travel from a place, or population densities (Figure 2.4a, see Figure 2.10). *Choropleth maps* (Figure 2.4b) are used where units of area such as counties or countries are shaded. The values for the total number of areas are grouped in class intervals and a shade assigned to each group. *Dot maps* and *graduated circle maps* also show distributions of people or product output (Figure 2.4c,d). In dot maps, each dot has a specific value (e.g., 5,000 people), while graduated circles are drawn in proportion to the total population of a given area.

## Geographic Information Systems (GIS)

Modern maps are important sources of information, but ancient maps were also greatly valued as summaries of the available knowledge about the world. Even before locational systems and accurately surveyed details were considered essential features of maps, they were widely used.

**FIGURE 2.4 (C) AND (D) Thematic maps.** What does each tell us about California? (c) Graduated circles, where the area of a circle represents the total population of a county. (d) Dot map, in which each dot is equal to the same number of people.

The *Mappa Mundi* ("World Map") produced in England in A.D. 1289 (Figure 2.5) has north at the left, coastal boundaries that are scarcely identifiable, and details of mythological creatures and Biblical places, as well as names of the major European cities. It was regarded as one of the most important information systems available to scholars at the time.

## Components of a GIS

Today geographers often use the power of computers to combine maps and the statistical analysis of data that is tied to geographic locations, producing **geographic information systems** (GIS). The results are often displayed as maps and associated statistical values. A series of overlay maps, such as those shown in Figure 2.6, can be extracted individually or in combinations. If geographers wish, for example, to identify neighborhoods in a city in the United States that are experiencing the economic and social upgrading process known as gentrification (see Chapter 9), they might use a base map that shows census tracts—the smallest division for which the U.S. Bureau of Census reports a wide range of data. They will then produce maps and graphs showing the distribution of higher-income groups, tracts where income increased markedly between censuses, and those where there is a predominance of single professional people or couples without children. Such analyses make it possible to identify localities for detailed

**WEBSITES: Maps and GIS**

http://www.usgs.gov/research/gis/title.html

http://www.esri.com/library/gis/abtgis/what_gis.html

Both these sites contain "What Is GIS?" features, including examples of the use of this technology. Follow through one of them to find out more about GIS. The U.S. Geological Survey produces topographic maps for the United States and has a site on place-names. It also has a site devoted to LANDSAT images. ESRI Corporation is the main producer of mapping software.

http://earth.jsc.nasa.gov

This site provides a selection of space shuttle views that can be downloaded, together with short descriptions of each view. You can select by topic and area, leading to thumbnail views of those that meet selection criteria. Try accessing a view for your home region.

**FIGURE 2.5 An ancient GIS: Mappa Mundi.** This map, on exhibition in Hereford Cathedral, England, is the only surviving complete world map of its time. Drawn in A.D. 1289, it includes Europe, Africa, and Asia, summarizing information about the known world (no Americas). It puts north on the left, and its religious origin is shown by having Jerusalem at the center and paradise at the top. The dark area in the lower center is the Mediterranean Sea; the British Isles are at the bottom left; and the Red Sea (top right) is colored red. This map began a process that led to later political and business maps and to the military electronic maps of today.

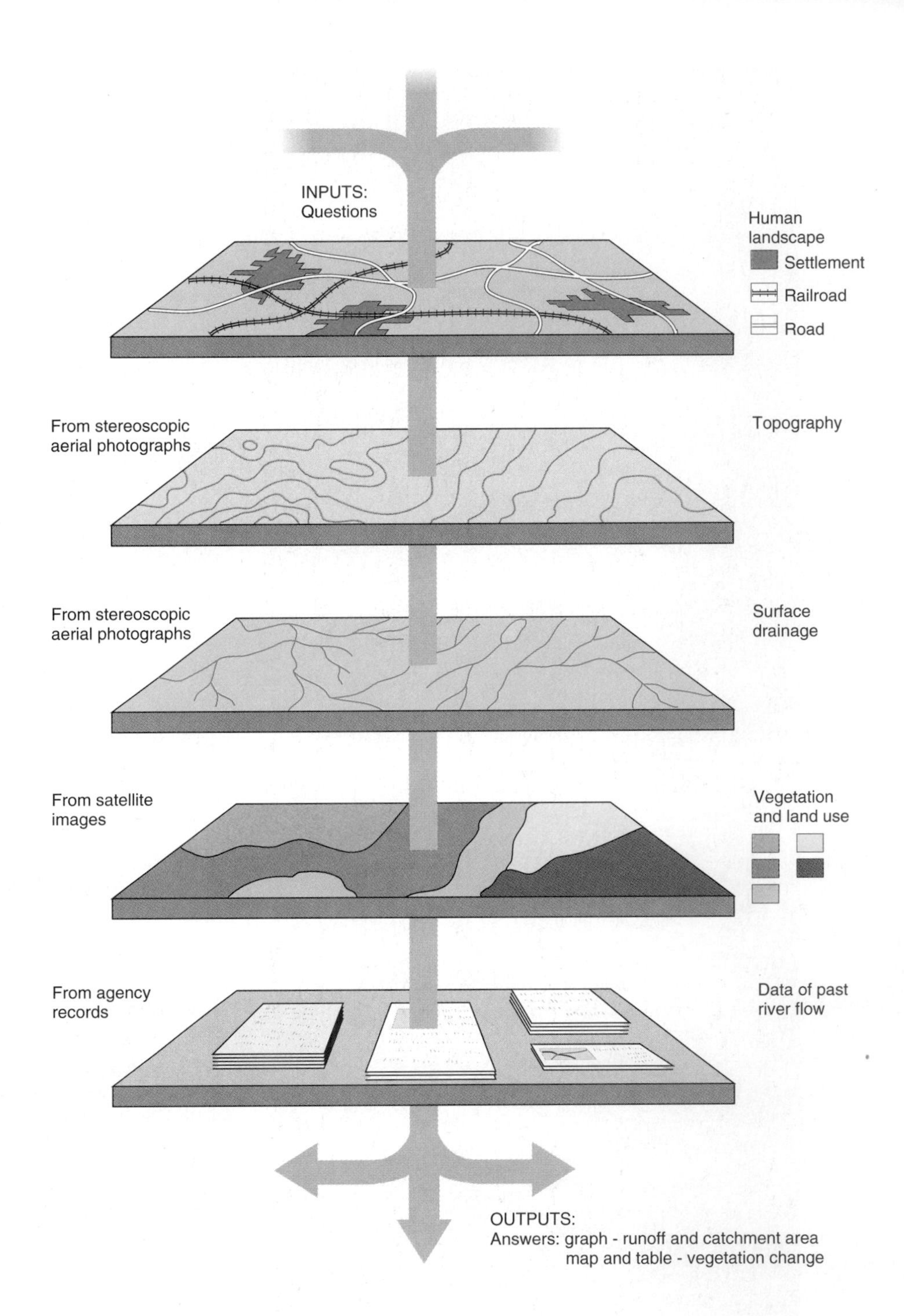


**FIGURE 2.6 Geographic information system.** The layers of information in this example could be used to monitor a river system that feeds a reservoir. Outputs from the system might include graphs of seasonal stream flow divided by drainage basin area and dominant vegetation cover type, and maps of changing land use.

fieldwork. The census tracts are too large and the data is not sufficiently detailed, however, to enable geographers to identify gentrifying areas without linked field-based inquiries.

## Aerial and Satellite Views of Earth

**Remote sensing**—a set of techniques that uses aerial photographs or satellite images to gather information about the land uses of Earth's surface—is increasingly part of a geographer's analytic tools. Aerial photographs and satellite images often figure in geographic information systems. They provide a cost-efficient means of surveying an area, and government agencies depend on them to make maps.

Most remote sensing systems use cameras and scanners that record the amount of solar energy reflected toward them from Earth's surface. Artificially generated radiation is also used in radar systems. Different objects can be identified because they reflect the energy in distinctive ways. Human eyes detect only a small part of the sun's radiation, termed visible light, but remote sensing systems are sensitive to a

(a)

(b)

**FIGURE 2.7 Aerial photographs.** (a) Black and white photographs distinguish among different light reflectances of ground surface land uses. (b) False color adds information sensed by infrared radiation (red colors), assisting in the interpretation of a variety of land uses, such as crops or grass in fields and buildings in urban areas

(a) & (b) NASA

wider range of radiated waves. They provide a view of Earth's surface that enables scientists to make maps and detect the nature of surface features from different crops to mineral deposits.

Aerial photographs can be black and white or color and record visible and infrared reflected energy (Figure 2.7). True color photographs record the landscape in the colors seen by the pilot. False-color photographs drop some of the visible spectrum and replace it with infrared. In such false-color photographs, healthy vegetation shows up as bright red. Aerial photographs are often taken with a 60 percent overlap between adjacent photos. Such overlapping photographs can be placed beneath an arrangement of mirrors called a stereoscope to recreate the three-dimensional landscape seen from the plane. Measurements of height from the 3D view make it possible to plot contours from such photographs. Space shuttle astronauts take photographs of Earth with handheld cameras (Figure 2.8).

Satellite images are from electronic scanners on board satellites in fixed orbits around Earth. Images of a whole hemisphere are obtained from satellites positioned some 36,000 km (22,500 mi.) above the equator. Their orbital speed matches Earth's rotation, making the satellite appear stationary (a geostationary orbit). Images from such satellites are of small scale and are used mostly for weather observations. Satellites at lower altitudes of 900 km (550 mi.) above the surface do not keep up with Earth's rotation, and their orbits cross the poles every 12 hours. Such satellite images show more detail and are particularly useful in monitoring surface land uses.

A major problem affecting the wider use of satellite images is the barrier formed by clouds. Much of Earth's surface is obscured, especially in the more humid parts of the world where most people live. Systems using radar technology penetrate cloud cover, giving clear views of the surface during day or night. Radar systems do not depend on the sun's illumination but send out and collect their own radiation at wavelengths that "see through" clouds.

A close look at a satellite image shows it is composed of a grid of small squares, or pixels ("picture elements"). The ground area represented by a pixel determines the amount of detail on the image. The best pixel size achieved in civilian systems is 10 m by 10 m from the French SPOT sensor. The digital nature of the data used to construct each image means that computers can store, analyze, and develop such images, making them an integral part of geographic information systems.

(a)

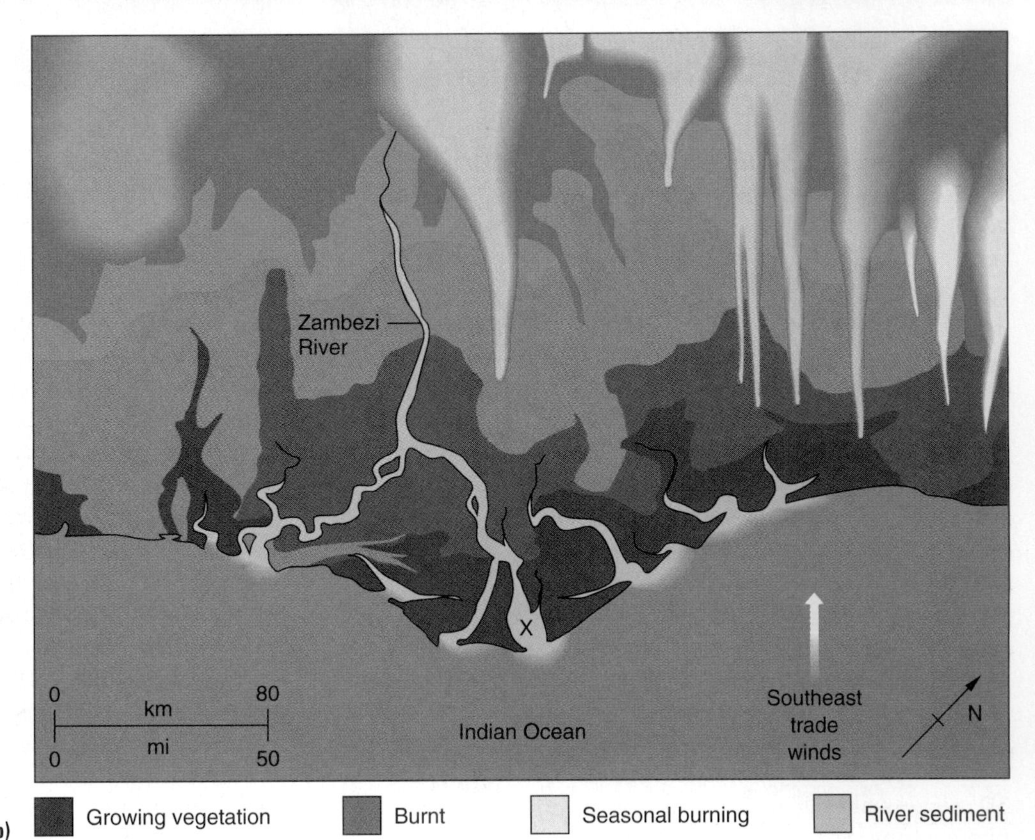


(b)

**FIGURE 2.8 Space shuttle photo: the coast of Mozambique, Southern Africa.** (a) The false-color reds are from the infrared radiation of growing vegetation, picking out wetlands near the coast. Red colors are absent from areas of seasonal burning. The smoke from burning areas shows the constancy of the trade winds blowing from the Indian Ocean. The map (b) provides details and scale. Space shuttle photos feature in each regional chapter of this text.

(a) Nasa, Michael Helfert

## Distributions, Density, and Diffusion

Maps like those in Figure 2.4 show **distributions** of events and phenomena, such as where earthquakes occur or where computers are made. Distributions have patterns that may be regular (in lines or clusters) or random. Distributions also show **densities**—the frequency of a phenomenon in a unit of land area. For instance, population density (see Figure 2.10) is measured in numbers of people per square kilometer (or mile).

> **RECAP 2A: Sense of Place**
>
> Basic geographic concepts include location (defined by latitude and longitude), direction (points of the compass), distance (actual in kilometers, time, cost), distribution, density, and diffusion.
>
> Maps are basic tools of geographers for representing details of places and their relationships to each other. Geographic information systems merge a range of geographic data including maps, statistics, and aerial or satellite images.
>
> 2A.1 When would it be best to use thematic maps with isolines, choropleths, proportional symbols, or dots? Find other examples of these maps.
>
> **Key Terms:**
>
> | | | |
> |---|---|---|
> | location | direction | map |
> | latitude | friction of distance | isoline |
> | parallel of latitude | scale | geographic information system (GIS) |
> | longitude | distribution | |
> | meridian of longitude | density | |
> | distance | diffusion | remote sensing |

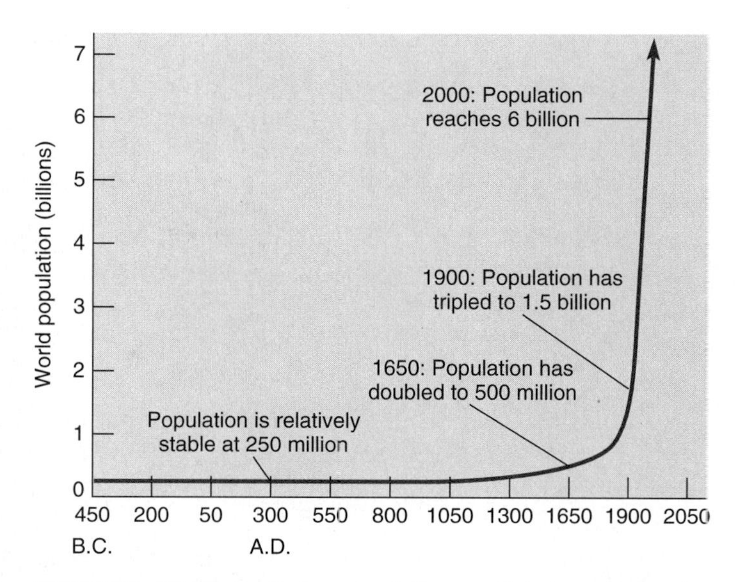


**FIGURE 2.9 World population growth.** For most of the human occupation of Earth, population growth was slow compared to the last 300 years. The population took 1,300 years to double from 250 to 500 million, then doubled again in 200 years. In the 1900s, world population quadrupled.



Distributions and densities change as people, ideas, or items that people use move from place to place. This process is known as **diffusion**. When people move in large numbers, they diffuse from older to new areas and relocate there. For instance, the rural black population of the southern United States diffused into local cities and then northward to major cities such as Washington, D.C., New York, and Chicago. Large numbers of African Americans moved in these directions between 1920 and 1960 (see Figure 9.30). Since then, the diffusion of African Americans has slowed and partly reversed its direction. From the 1940s to the 1980s, both black and white Americans diffused from the Northeast toward the West Coast.

Diffusion is not only characteristic of people movements. New technologies, fashions in clothes, or music diffuse by expansion, a process aided by easier communications. Diseases spread by contagion in person-to-person contacts. Barriers such as international borders and language differences hamper diffusion, although both are proving less resistant to change as transportation, communications, and political agreements reduce their significance.

## WORLD POPULATION GEOGRAPHY: PEOPLE MATTER

In mid-1998, there were estimated to be over 5.9 billion people living in the world. Projections suggest that by 2010 the total will be 6.9 billion, rising to 8 billion by 2025. After thousands of years in which world population rose steadily to around 0.25 billion 2,000 years ago and 0.5 billion in A.D. 1600, it increased rapidly from the start of the industrial revolution in the 1700s (Figure 2.9). In 1900, the world population reached 1.9 billion and then multiplied four times during the following century. Questions asked by regional geographers include: How is the world population distributed? How do the processes of population change distinguish among regions? And, how many people can Earth support—as a whole and in each region? The last of these will be discussed at the end of this chapter.

### Population Distribution

The world's highest densities of population occur in Europe, Southern Asia, Eastern Asia, and the eastern United States (Figure 2.10). Population densities in other world regions are low, with peaks in limited, mainly coastal areas around industrial port cities. Large expanses of the world's lands have very low population densities, often linked to economies based on low-productivity farming or to the more difficult natural environments of desert, ice sheet, high mountain, and areas with very long winters.

The number of people in a region and their density—or people per square kilometer—depend on the natural resources available, the historic use of those resources, and the type of economy and technology that is dominant. For example, the traditional world hinterland way of life still adopted by many Native American tribes in the Amazon basin of South America and by peoples in the cold northern lands of North America and Eurasia always supported low densities of population. Higher densities resulted from settled and more intensive agriculture, particularly in the rice lands of Eastern Asia. Both here and in the richer countries of western Europe and Anglo America, high-density rural populations dependent on farming were thinned in the 1900s by the mechanization of agriculture and migration to growing

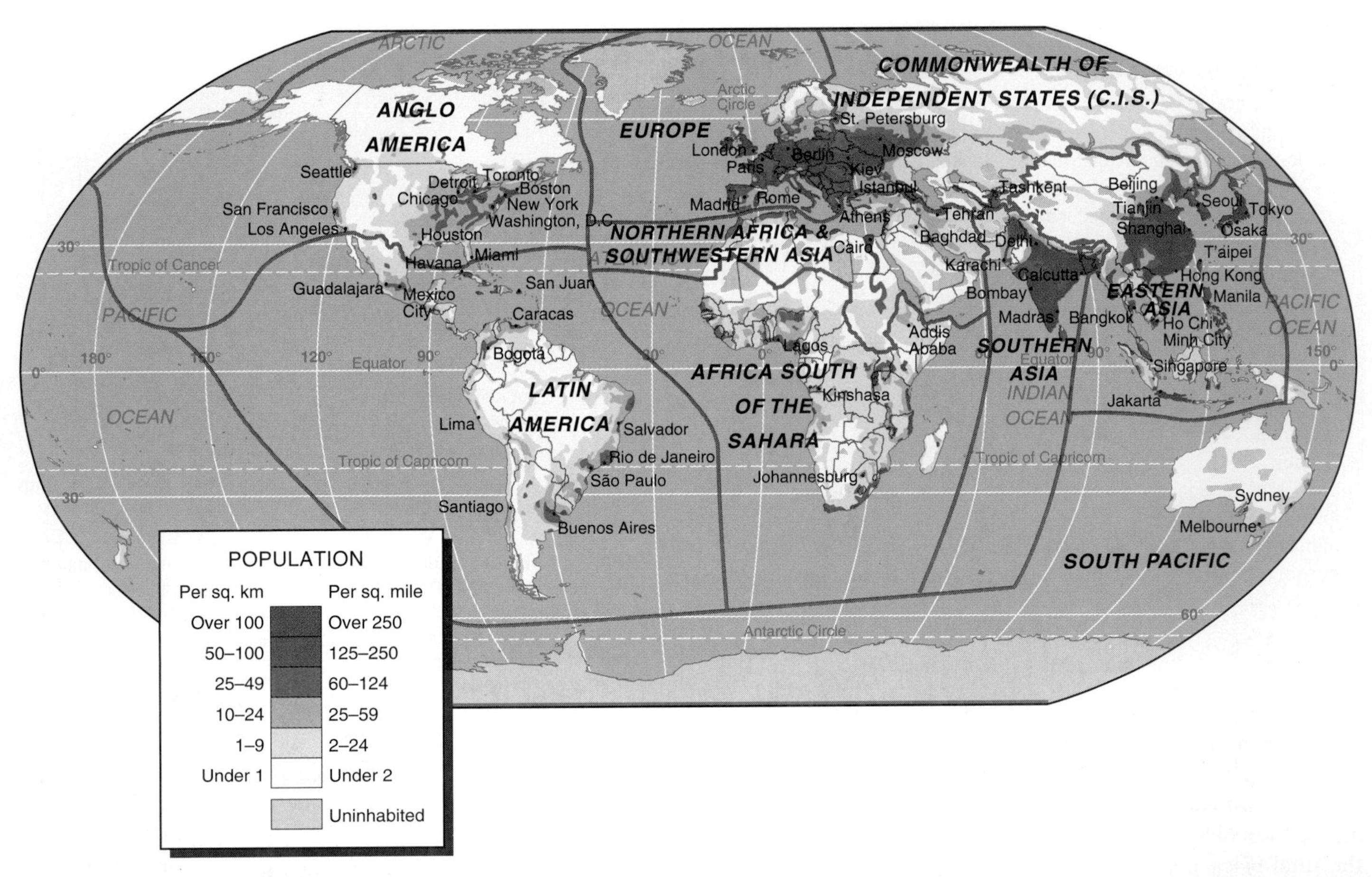


**FIGURE 2.10 World population distribution.** Which major regions have the highest and lowest densities of population? As you read through this chapter, look for evidence that might explain the differences.

urban-industrial areas. Most current population growth occurs in large metropolitan areas.

## Demography

**Demography** is the study of human populations in terms of their numbers, density, growth or decline, and their movements from place to place. Changes in the population of a region are determined by whether births exceed deaths, immigration exceeds emigration, and the overall balance between natural change and migration change.

- Trends in the **birth rate** (births per 1,000 inhabitants in a year) are often related to the **total fertility rate** (average number of births per woman in her lifetime) to work out the likely future rate of births. Total fertility rates of 6 to 7 are typical of many poorer countries, while richer countries have rates of 2 or below.
- The **death rate** (deaths per 1,000 inhabitants in a year) is often broken down into age groups. **Infant mortality** (deaths per 1,000 live births in the first year of life) and child mortality (deaths per 1,000 live births in the first five years of life) are examples. Infant mortality rates are below 10 in richer countries but still above 100 in many poor countries, including Haiti and several in Africa and Asia.
- The combination of birth rates and death rates defines the **natural population change**—the rate of increase or decrease of the total population. The composition of a country's population is often shown in an age-sex diagram, also termed "population pyramid" (Figure 2.11), the shape of which provides strong indications about the population's recent history and potential future.
- **Migration** is the long-term movement of people in and out of a place. If immigration to a country or region exceeds emigration, numbers of people there increase. Immigration is a major cause of population increase in the United States at the present time, since natural increase is slow. The U.S. population grew by 2.6 million between July 1989 and July 1990; of that gain, 1.5 million (58%) were legal immigrants (i.e., this total does not include refugees or illegal aliens, who added more). The United States does not report emigration but assumes a net annual loss of around 0.2 million. Net immigration was thus at least half of U.S. population growth.

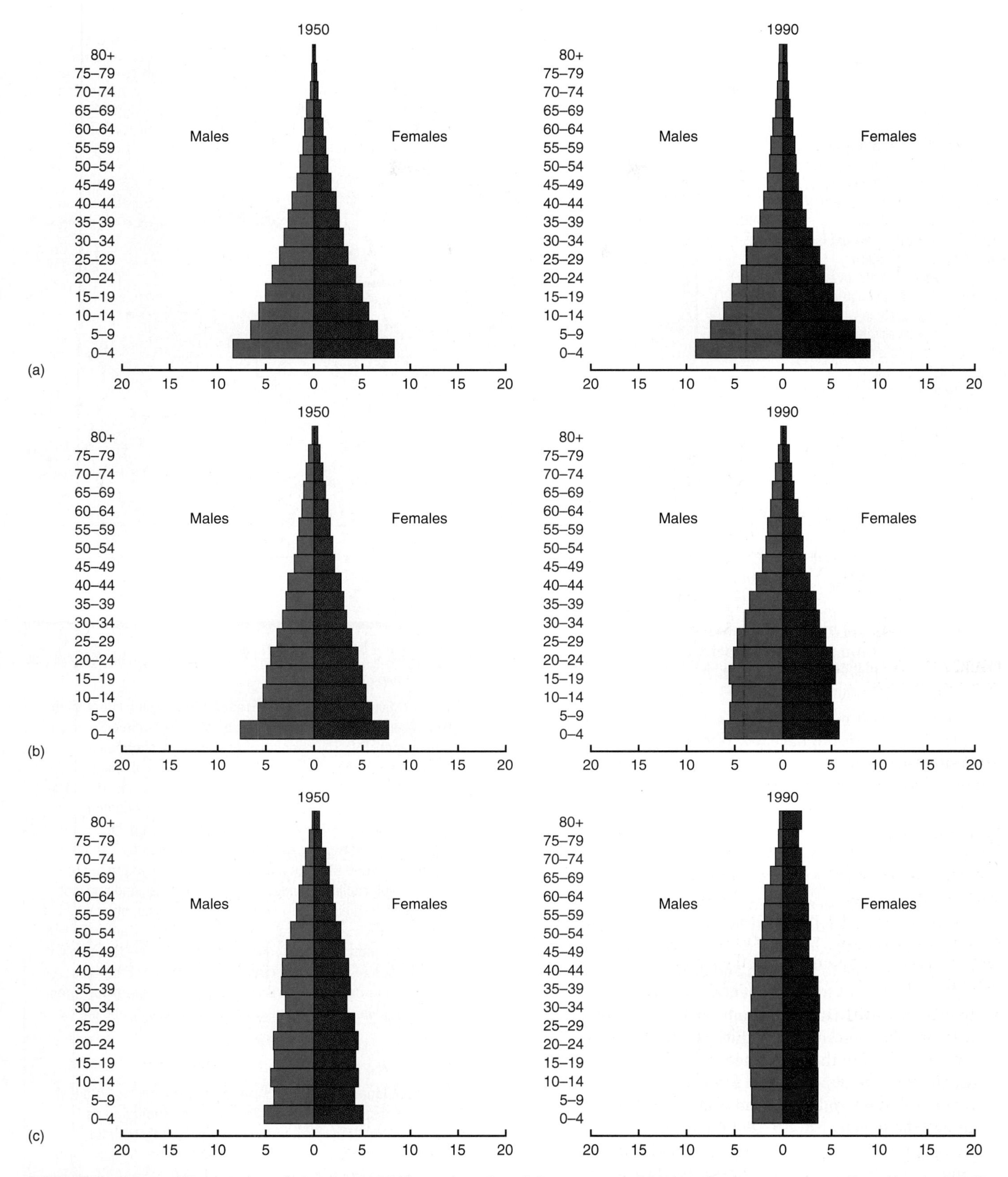


**FIGURE 2.11 Age-sex diagrams (population pyramids).** The numbers of people in a country are divided into five-year age groups, male and female, up to 80 years; those over 80 are included in a single group for each sex. This gives the sequence of bars. Each age group is represented as a percentage of the whole, making the diagrams comparable over time and from country to country. The shape of the diagrams indicates important characteristics of the population. The three examples here show: (a) countries with high birth rates in both 1950 and 1990 have two pyramid-shaped graphs; (b) countries that reduced birth rates from 1950 to 1990 have one pyramid and one evenly distributed graph; (c) countries that had low birth and death rates in both 1950 and 1990 have two evenly distributed graphs.

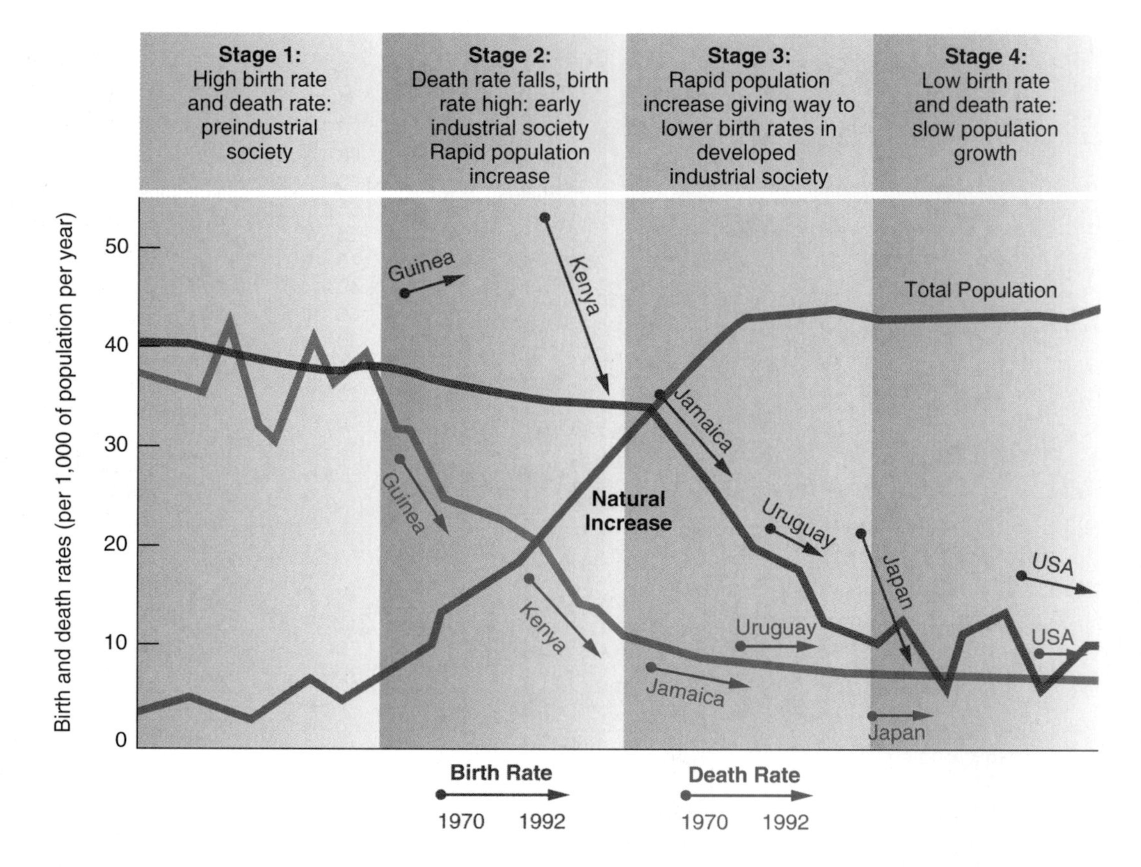


**FIGURE 2.12 Demographic transition.** Population change is plotted, from high birth and death rates (Stage 1) to low birth and death rates (Stage 4). All countries were in Stage 1 until the 1800s, when Western countries began to move toward Stage 4. Today many countries are in Stages 2 and 3, in which population increases rapidly as death rates fall, but birth rates remain high. Each stage is linked to a phase of economic growth. Similar diagrams are drawn for each world region in this text. Notice where countries depart from this general progression. It has been suggested that there should be a Stage 5, in which birth rates and death rates are both low, but the death rate is greater than the birth rate and the population declines.

- When natural and migration changes are combined, an overall population *increase* of 1 percent per year means that the population will double in 70 years; one of 2 percent means it will double in 35 years; and one of 3 percent means it will double in 23 years. It is common for richer countries to have low rates of overall population increase, below 0.5 percent, while poorer countries have rates of increase around 2 to 3 percent. Rapid increases of population make it difficult for countries to plan for providing jobs or housing, especially if the rate of population growth exceeds the rate of economic growth.
- As countries progress economically, they generally pass through a process of **demographic transition**, in which the high birth and death rates typical of poorer countries give way over time to low birth and death rates (Figure 2.12). At both ends of the sequence, there is slow population growth. In the second stage, medical science and improved food supplies reduce death rates but keep birth rates high. In the third stage, birth rates fall, reducing the rate of population growth. The core countries of western Europe and Anglo America passed through the sequence in the 1800s and early 1900s. The rapid population increases during the middle phases of demographic transition place stresses on the resources available to feed the extra mouths. Although rates of population increase in the world are slowing at present, doubt remains as to whether some countries will emerge from the middle stage of demographic transition, how long emergence will take, and whether this process applies to all countries. At the same time, countries in the former Soviet Union and some European countries have declining populations as death rates exceed birth rates (see Chapters 7 and 8).

**WEBSITES: Population Change**

http://www.popnet.org

This site gives access to a wide range of population information sources and is organized by the Population Reference Bureau in Washington, D.C. The opening page lists some alternative approaches to the information—through listings of other websites of the world's countries supplying up-to-date data. The main headings are: "Organizational Sources" (by country), "Selected Topics" (demographic statistics, economics, education, environment, gender, policy, reproductive health), "Clickable World Map" (major regions), and "Keyword Search." One valuable facility provides the ability to devise and print out population pyramids (age-sex graphs) for each country for several years, including future projections. Try this: enter the site homepage, then choose "Organizational Sources—Government Organizations—U.S. Census Bureau International Programs Center—International Data Base—Population Pyramids." Then you can choose a country, years, and size of graph, and print it out.

http://www.prb.org

This Population Reference Bureau site provides up-to-date information and data on population issues, including its annual world data sheet and text of its monthly bulletin, *Population Today.*

## Population Policies

In the late 1990s, variations in directions and rates of population change are major world issues causing differences in regional geography and highlighting potential future crises for regions. The main consensus of the 1994 International Conference on World Population and Development in Cairo, Egypt, was that population growth should be contained. Although opinions differed about how many people the world and its resources can support, it was agreed that the present rates of growth of over 2 percent per year in many countries are too high.

In the 1950s, governments first identified population growth as a major issue facing the world. Initially, contraceptive techniques were thought to be the key to controlling population growth, so they were made widely available in poorer countries. Then the demographic transition process suggested that reduced birth rates accompany economic growth, so policies in the 1960s and 1970s were based on the "development is the best contraceptive" slogan. In some countries, however, large families continue to be seen as necessary for survival or as a means of gaining status within the culture. By contrast, parts of India and Kenya had major falls in fertility rates with little or no economic growth. The demographic transition process may reflect the experience of western Europe and Anglo America in the 1800s. By A.D. 2000, it is not yet fulfilled beyond the middle stage in much of the rest of the world.

Many factors are important in achieving population stabilization, including education and the improvement of women's status—both points that were emphasized at the 1994 world population conference. A greater role for women in deciding on the numbers of children they will have is still resisted strongly in some countries because of ingrained cultural beliefs.

The causes of population growth (or decline) are complex and closely related to cultural expectations. In considering overall planning policies, countries need to relate their population growth rate to what can be sustained by available resources of food and other basic needs. Some countries are experiencing a surprising level of success in population control. For example, Bangladesh, a very poor and overpopulated country in which women have a very low status, is experiencing rapid falls in birth rates. Its family planning policies provide families with access to a variety of choices that they had not previously considered.

# WORLD POLITICAL GEOGRAPHY: COUNTRIES AND GOVERNMENTS

**Political geography** is the study of how governments influence the human geography of the world and its regions. Self-governing countries are the basic political units: within its borders, a government has political control, or sovereignty, over the country's inhabitants. The world is divided into around 200 self-governing countries, each of which is recognized by other countries. Country governments have powers to promote and protect their peoples in world affairs, provide public services, and encourage economic and social development internally. They often have systems of regional and local government that carry out some of the governmental responsibilities at different geographic levels. Country governments may also join with other country governments in mutual trading or defense agreements. In world regional geography, countries provide the main units of study.

## Nations and Nationalism

A country as defined above is not always the same as a nation. A nation is an "imagined community" having common cultural features, usually linked to a specific area of land. The cultural features may be language, religion, or other characteristic with a historic background. In the country known as the United Kingdom, for example, the English, Scottish, Welsh, and Irish consider themselves distinctive nations and often enter separate teams in world sporting events. The notion of "tribe" is close to that of "nation" (see Chapter 3), but many African countries contain several tribes, while their modern boundaries, imposed by colonizers, often cut across tribal territories. When countries contain more than one nation, scope exists for political tensions to arise.

Nationalism—the desire to combine cultural and territorial features—became basic to the formation of countries in Europe from around A.D. 1800. It gained significance as the rise of capitalism turned allegiances away from overriding feudal principles or ecclesiastical loyalties. The increased levels of communication made possible through printed books and newspapers, the telegraph, telecommunications, radio, and television bolstered nationalism. From the 1800s, universal education supported nationalist themes through selective views of history that glorified the national experience. European nations, such as Germany, emerged as countries in the 1800s, when a group of smaller states united under the linguistic banner. In the 1900s, Germany used the idea of uniting separated German-speaking minorities in other countries as an excuse for talk of national supremacy. This twice led to world wars, when Germany discovered limits to the expansionism its neighboring countries would tolerate.

## Governments

Governments range from those elected democratically to those dominated by dictators. Some countries have a unitary governmental structure, administering all parts from the center for all aspects of government. Other countries have a federal government structure in which they divide the authority for various activities between a central government and partitions called states or provinces. In the United States, for example, the federal government has responsibility for

> **WEBSITE: Governments**
>
> http://www.odci.gov/cia/publications/factbook/index.html
>
> The CIA *World Factbook* is produced for U.S. government officials. It includes data for every country, including information on the type of government. Some of this information is also available on the CD-ROM sources listed in this text's Reference Section.

external relations, defense, and interstate relations, while the states have authority for education, local roads, and physical planning. Both the federal and state governments may levy income tax.

Governments differ in the direct influence they impose on internal affairs. Dictatorships and governments where state capitalism is the dominant economic system both intervene much more than democratic governments in economic activities from production to distribution. During the 1900s, however, all governments increased their level of intervention in regulating and promoting economic activity.

Government functions are concentrated in capital cities, where the head of state lives and the parliament and government offices are situated. Many capital cities are the largest cities in the country, as, for example, London (United Kingdom), Tokyo (Japan), and Nairobi (Kenya). Especially in federal countries, new capital cities were often built to avoid jealousies among established cities. Washington, D.C., Brasilia (Brazil) (Figure 2.13), Abuja (Nigeria), and Canberra (Australia) are examples.

## Country Groupings

Countries make agreements with other countries, primarily to foster common defense and trade interests. Many defense agreements were generated during the Cold War period, but some continue. The North Atlantic Treaty Organization (NATO) linked Anglo America and western Europe in a common response to a perceived military threat from the Soviet bloc. Since 1991, it extended its agreements to some of the former Soviet bloc countries, a move resisted by Russia. The Association of Southeast Asian Nations (ASEAN) was formed to oppose the extension of communism in that region but became a forum for economic cooperation (see Chapter 6).

Governments have a major influence on the conduct of world trade. They may encourage their people to export goods and may control certain imports by charging taxes, or tariffs, on them. Making imports of food or manufactures more expensive by charging tariffs gives home firms and farmers cost advantages. During the early 1990s, there was a widespread political will among countries to free world trade from barriers. The **General Agreement on Tariffs and Trade** (GATT) was established in 1948 to encourage countries to lower tariffs. After 1993, the **World Trade Organization** (WTO) took over the role of preventing discrimination among trading partners.

(a)

(b)

(c)

**FIGURE 2.13 Brasilia, the planned capital city of Brazil.** (a) The central avenue lined by public buildings. (b) The cathedral. (c) Unplanned satellite town of Paranoa that sprang up to hold over a quarter of a million people, including the enterprising shopkeeper on the main road.

All photos © Michael Bradshaw

Most progress on liberalizing trade has been made at the regional level in **free-trade areas**, the members of which have common tariff rates on imports. Opinions differ as to whether the regional free-trade area agreements encourage

**RECAP 2B: People and Political Geography**

The population distribution of a country depends on historic factors that encouraged local concentrations. The present trend is toward metropolitan center expansion. Populations change according to the balance among births, deaths, and migration. Demographic transition suggests that populations move from high birth and death rates toward low birth and death rates as economic development occurs. Population policies are based on factors that stimulate population growth or decline.

Countries are self-governing political units having recognized borders with other countries; nations are culturally united groups of people occupying or claiming a specific territory. Political systems affect the ways in which countries are governed. In the 1900s, governments of most countries took an increasing role in guiding their economies.

2B.1 How would you define: *birth rate, total fertility rate, death rate, infant mortality, immigration, emigration?*

2B.2 Why may some countries not follow the demographic transition pattern?

2B.3 Compare government types in several different countries (e.g., United States, United Kingdom, Russia, Japan, Cuba). (Get data about this from the CIA *World Factbook* website.)

**Key Terms:**

| | | |
|---|---|---|
| demography | infant mortality | political geography |
| birth rate | natural population change | General Agreement on Tariffs and Trade |
| total fertility rate | migration | World Trade Organization |
| death rate | demographic transition | free-trade area |

members to further lower their tariffs or whether they discriminate increasingly between members and nonmembers. The largest trading group at present is the European Union (EU, see Chapter 7). The North American Free Trade Area (NAFTA, see Chapter 9) may extend to much of Latin America by A.D. 2005 (see Chapter 10). Even more ambitious is the prospect of the Asia-Pacific Economic Cooperation Forum (APEC, see Chapters 6 and 11) that has pledged itself to free internal trade among its richer countries by A.D. 2010 and among other countries by A.D. 2020. The European Union is considering extensions into Southwest Asia and Eastern Europe (see Chapters 4, 7, and 8), and it has been suggested that Anglo America and the EU could form links. Such groupings with regional jurisdiction are considered in each chapter of this text.

## WORLD ECONOMIC GEOGRAPHY: WEALTH, POVERTY, AND DEVELOPMENT

**Economics**—the study of the production, distribution, and consumption of goods and resources—has strong geographic expressions. Producing and consuming areas are often separated and linked by transportation routes.

Economic systems are linked closely to political systems. During the Cold War, Soviet and Chinese communism linked political ideology to the economics of state capitalism. Many Western-style democratic governments also gave the state extended responsibilities in running major industries, such as coal, steel, railways, airlines, and utilities. After the 1991 breakup of the Soviet economic bloc, capitalism became the world economic system. State-owned enterprises in many countries were privatized. Capital and labor became increasingly mobile. Multinational corporations that already operated in several countries spread the location of innovation and, especially, production facilities. Producers and markets were integrated more closely across continents.

Governments of all countries now find themselves having to work within this global economic order. Countries have different mixes of economic sectors, relate differently to the global economy, and are at different levels of economic development.

### Economic Sectors

The range of economic activities is divided into sectors that include specific industries or business activities. An **industry** is a group of business activities, such as the farming industry, manufacturing industry, or tourist industry. Economic sectors create wealth in different ways.

- The **primary sector** of an economy involves what is produced from natural sources. The mining industry extracts minerals from rocks, the fishing industry catches fish from the ocean and fresh waters, the timber industry cuts trees from forests, and the farming industry cultivates crops and rears animals by using the soil nutrients, water, solar energy, and genetic stock of plants and animals. Although some products such as fruit and fish can be distributed and consumed as they are, most primary products require further processing and so become raw materials for the secondary sector. The primary sector is the dominant source of wealth creation in many countries of the world's periphery.
- The **secondary sector** of an economy changes the raw materials from the primary sector into useful products and by so doing increases their value. Manufacturing industries produce finished and partly finished goods. For instance, a chewing gum factory takes in and mixes the gum, processed flavors, sugar, and chalk, and packages the final products for sale. An engineering factory may, however, take in and shape metal that has already been refined from minerals and then send the resultant tubes to another factory to be made into hydraulic systems that are assembled elsewhere in the braking system of a jet aircraft. At each stage of manufacturing, human labor and machinery make a more useful and valuable product. The construction industry is another secondary sector industry. It uses bricks, tiles, timber, and steel girders to make houses, factories, and office blocks. The growth of the secondary sector was the economic engine of development in many countries of the world's core and semiperiphery.

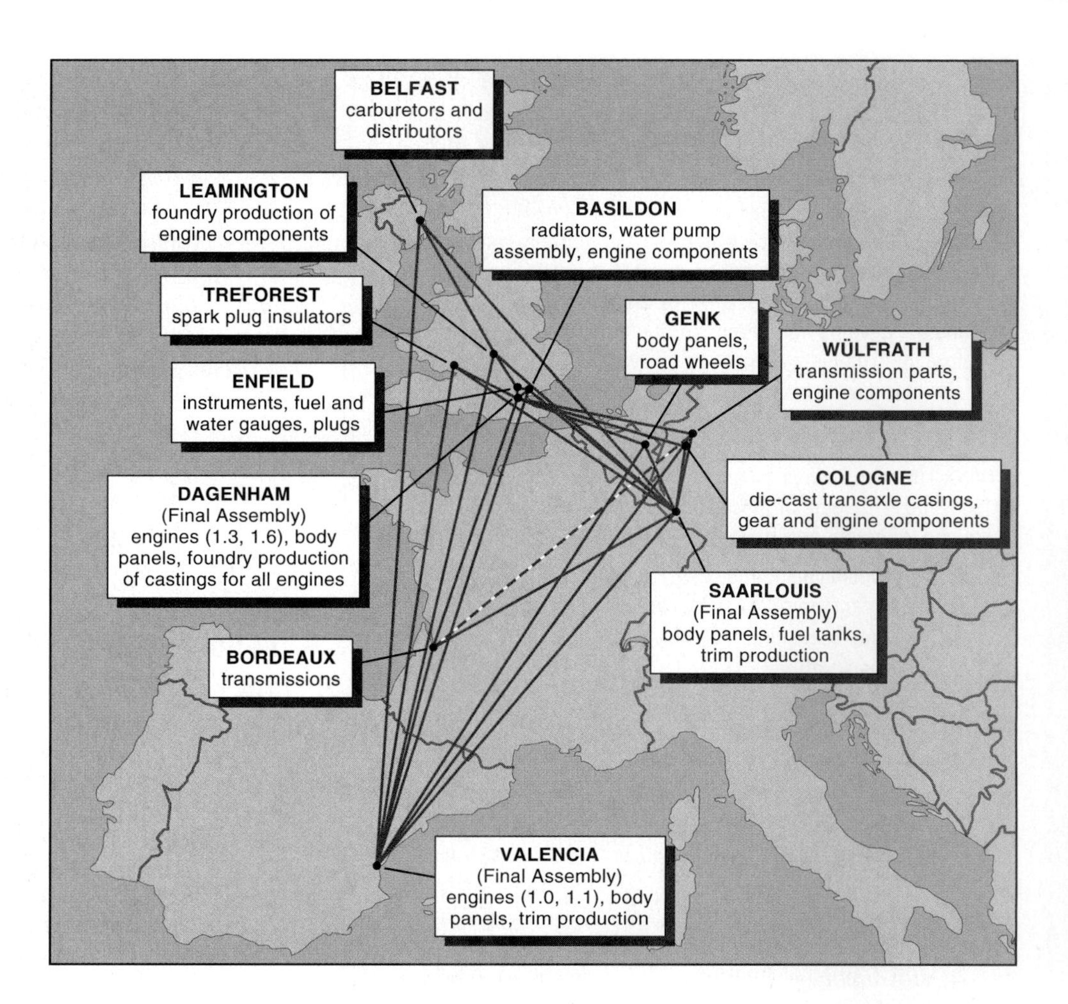


**FIGURE 2.14 Multinational corporation's international linkages.** The European spread of factories include makers of parts and final assembly locations for the Ford Fiesta in the late 1980s.

From Peter Dicken, *Global Shift.* Copyright © 1992 Guilford Press, New York, NY. Reprinted by permission.

- The **tertiary sector** of an economy is concerned with the distribution and consumption of goods and services, and is often called the service sector. This sector includes distribution (wholesale and retail), business and professional service, transportation, tourism, and entertainment industries, as well as government. Such industries make goods and services available where and when they are required at different locations from the primary extraction of raw materials or secondary manufacture. Value may also be added during these tertiary sector processes, since, for example, people will pay more for a product purchased with others in a supermarket than if bought on its own from a farm. The farmer's or bottler's income from a gallon of milk is a fraction of what it sells for in a supermarket.
- A **quaternary sector** is sometimes identified separately from the tertiary sector. It specializes in the assembly, transmission, and processing of information and in controlling businesses by such means. This sector includes professions, finance industries, education, media, and government and, with the tertiary sector, employs an increasing proportion of the work force and generates new forms of wealth, mainly in core countries.

## Global Economy

After the 1960s, the capitalist system expanded its hold. Global-scale economic activity became increasingly common as a result of multinational corporation development, greater ease of transportation and communications, and major changes in worldwide financial control. Although these factors reduced the power of individual countries to make some decisions, each country maintained a distinctive role within the world economic system.

### Multinational Corporations

**Multinational corporations** are those that make goods or provide services in several countries but direct operations from a headquarters in one country. A corporation establishes factories and offices in other countries to avoid tariff barriers and to take advantage of cheaper land, labor, and energy. For example, many car manufacturers spread the manufacture of automobile components across several countries to ensure continuity of supplies during labor strikes, react to local needs, and take advantage of lower production costs (Figure 2.14).

Based mainly at first on manufacturing industries in the United States, multinational corporations with European, Japanese, and South Korean headquarters are now significant.

By the 1990s, movements of goods among countries by multinational corporations accounted for 40 percent of world trade. Thousands of multinational corporations and their foreign affiliates produce a range of products under such brands as Coca Cola, PepsiCo, Ford, General Motors, Volkswagen, Exxon, Shell, Toyota, Sharp, Samsung, Kellogg, Nestlé, Hyundai, and IBM. Clothing, electronics goods and computers, cars and trucks, food and drink, and oil derivatives are the major products. The 600 largest multinational corporations contribute a quarter of the value added by world manufacturing.

The expansion of multinational corporations was eased by improved air transportation, limited-access highways, and telecommunications that brought people and firms closer in both time- and cost-distance. International airlines, courier services, and multinational car rental firms make it possible for people and high-value goods to get from almost any part of the world to another within 48 hours. Faxes, electronic mail, and video conferences ease instant personal communication and information transfer.

Multinational corporations in service industries spread around the world from the 1970s, following the development of rapid air travel and telecommunications. By the mid-1990s, some 40 percent of direct inward foreign investment to countries was directed at services rather than manufacturing. The services favored by foreign investment include tourism and travel (hotels, airlines, and car rentals), information (data processing, software, and telecommunications), business (accountancy, advertising, market research, and public relations), and financial (banking and insurance). Many of the manufacturing multinational corporations developed involvements in the service sector. For example, the Ford Motor Company has an increasingly profitable financial loan company, and General Motors issues one of the top credit cards.

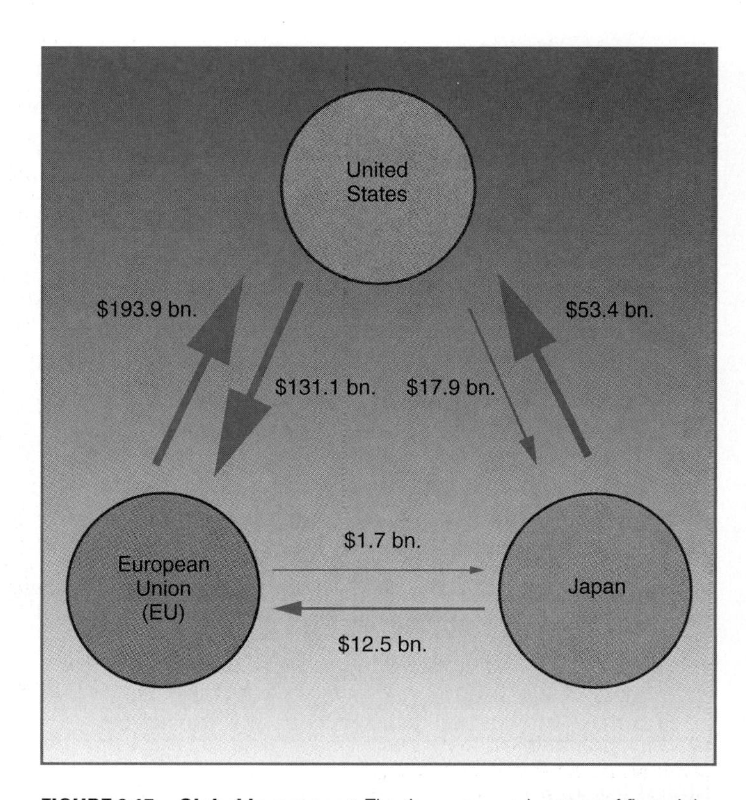


**FIGURE 2.15 Global investment.** The three-cornered pattern of financial flows in the late 1980s, concentrating on three world core areas. Previously, much investment went to countries in the periphery, and that began to increase again in the 1990s.

Adapted from P. Knox and J. Agnew, *Geography of the World Academy,* 1989. Copyright © 1989 Edward Arnold Publishers, Ltd., London.

## Financial Services

Global financial services expanded rapidly in the late 1900s, as both a result of the internationalization of economic activity and an enabler of it. A series of events contributed to this process, starting with easier movements of money after 1970.

- The system of fixed exchange rates among different currencies broke down and exchange rates floated, subject to continuous adjustment. This encouraged more frequent and freer flows from one currency to another.
- U.S. dollars began to accumulate in foreign banks. The rate of dollar accumulation rose sharply after the oil crisis of the early 1970s, when the price of oil (paid in U.S. dollars) went up and the oil-producing countries had massive surpluses of income. There was a further rise from the early 1980s, when the United States ran a huge budget deficit that was financed by borrowing and repayment in dollars.

  The banks that held dollars created new financial markets, using some of the money for loans to peripheral countries. Between 1980 and the 1990s, international lending by the 24 wealthiest countries, members of the Organization for Economic and Commercial Development (OECD), rose by 30 times. Foreign exchange turnover per day quadrupled between the mid-1980s and mid-1990s. Countries such as Brazil used the loans in the early 1980s to develop their infrastructure of roads, dams, and power stations. In the harder times of the late 1980s, however, interest rates rose and made it difficult for many debtor countries to repay the loans.
- A significant change in the direction of money flows occurred in the 1980s. In the 1970s, direct foreign investment was mainly from the United States and other core countries toward the peripheral countries. During the 1980s, however, foreign investment from western Europe and Japan into the United States increased, and a triangular pattern was established between the United States, Japan, and western Europe (Figure 2.15). The United States was the chief recipient, taking 40 percent of global direct foreign investment in the 1980s. Japan invested heavily in both the United States and western Europe. The flow of foreign capital to peripheral countries slowed to a trickle as the debt crisis mounted and nearly all financial resources were shared among core countries—which also increased their wealth by retrieving loan funds or interest from the indebted peripheral countries. Other countries, such as Taiwan and South Korea, generated their own capital by forcing their peoples to save.

- In the early 1990s, huge financial surpluses accumulated in Japan, Taiwan, Hong Kong, and South Korea. They were reinvested mainly in other Eastern Asian countries outside the core (see Chapter 6). Banks and other financial houses expanded to service the global money explosion. American banks started this trend, but others in Britain, France, Germany, Arab countries, and Japan soon followed. Financial markets proliferated, dealing in both equities (stocks and shares) and forward contracts based on expectations of commodity production and world prices. A growing infrastructure developed to service the financial markets, based on trading around the clock in New York, London, Tokyo, and Hong Kong. All these trends increased the number, speed, and global penetration of such transactions.
- In the late 1990s, confidence in the system was shattered by great numbers of bad loans taken out with Japanese and South Korean banks in particular. This caused a crisis in many Asian countries, which had to devalue their currencies, slowing the pace of economic growth.

### Global Information Services

Information services also became available worldwide, with rapid growth of Internet telephone-computer services in the 1990s. Just as some multinational corporations moved manufacturing to places with cheap labor costs, so others moved information handling to such places.

In 1983, American Airlines established Caribbean Data Services in Bridgetown, Barbados, to process its tickets and boarding passes. It is now the largest single employer in Barbados, and the documents processed increased from 38 million in 1983 to over 200 million in the mid-1990s. In Montego Bay, Jamaica, over 3,500 employees in the Jamaica Digiport Interstate Communication System link clients in Canada, the United States, and the United Kingdom. The Bangalore area of southern India became a major world center for computer software development by multinational corporations such as Apple, IBM, Intel, Oracle, and Texas Instruments. U.S. insurance companies process claims in Ireland, and U.S. manufacturers conduct research and development in countries such as Hungary and Thailand. These trends follow basic capitalist tenets of locating economic activities at low-cost places to achieve greater profits.

### Global Cities

Urban areas are major nodes in the world economy. The increasing speed of analysis and financial transfers made possible by information technology and telecommunications encourages the concentration of corporate headquarters. The internationalization of business services, accountancy, and advertising causes corporate activities to concentrate in the central business districts of the core's largest cities. There, they have access to office space, business services, telecommunications, government offices, international airports, and each other. Improved communications now give them control of large areas and access to other large world cities.

Several layers of world cities are emerging. At the top are three truly *global cities:* New York, Tokyo (Japan), and London (United Kingdom) (Figure 2.16). They contain the greatest number of multinational corporation and major bank head offices and the three largest stock exchanges. They account for most financial dealings on the world scale. At the second level are *zonal centers,* such as Paris (France), Frankfurt (Germany), Singapore, Sao Paulo (Brazil), and Los Angeles, which also contain many head offices but relate to specific zones of the world rather than the whole globe. The third level is of *regional centers,* which contain many corporate offices and foreign financial outlets but are not essential links in the flows of international finance.

The largest global centers are becoming as tuned to the world economy as they are to their internal country concerns. New York, for instance, is at the center of a metropolitan region that has 18 million people and an economic product greater than whole countries such as Canada, Brazil, or China. Its role in international business increased after 1970 despite some loss of corporate headquarters. The remaining businesses receive 40 percent of their revenues from foreign sources (compared to 20 percent in other U.S. cities). The number of foreign banks rose from 47 in 1970 to over 200 in the 1990s. New York has over half of all U.S. law firms with overseas business, and its airports service major international links. In the 1980s, the expanding class of managers and professionals working in the well-paid international sector led to the growth of conspicuous consumption, generating the "yuppie" image. Such factors are reflected in the special landscapes of world cities with their clusters of high-rise office blocks.

## Poverty and Development

Economic differences throughout the world lead to a focus on the contrast between those peoples and countries that are in the core and are relatively wealthy, and those that are in the periphery and do not enjoy such wealth or opportunities. **Poverty** is not merely a lack of income but a deprivation of minimally acceptable material requirements, health and education, and access to jobs. Poverty is the absence of basic capabilities that enable people to function fully as human beings—including income, clothing, and shelter, together with the expectation of a long life and full participation in community affairs. Poverty may include those who are not very income-poor but are kept out of the mainstream of society by disablement, ethnic or racial segregation, or destructive behaviors associated with alcohol or drug abuse.

### Get a Life

The term **development** implies efforts that improve one's lot. The poor countries of the world see their hope in terms of development toward the levels of well-being in the richer countries. We often use the terms *less-developed countries* and *more-developed countries* to indicate both these differences

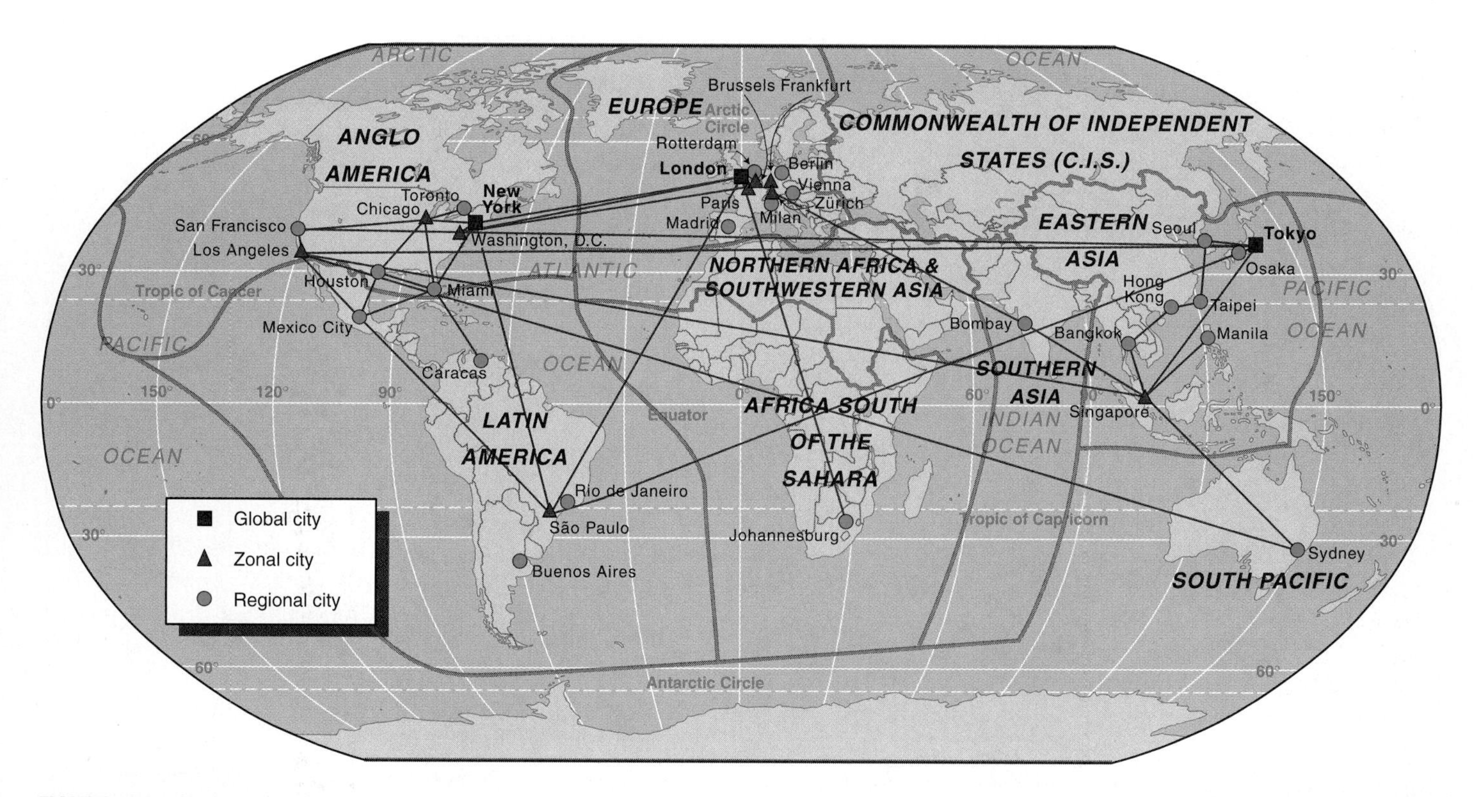


**FIGURE 2.16 Global cities.** A hierarchy of cities based on the amount of international financial trading, corporate headquarters of multinational corporations, and the growth of communications and media activities. Which major world regions have many or few of these global cities?

Adapted from P. Knox and J. Agnew, *Geography of the World Academy,* 1989. Copyright © 1989 Edward Arnold Publishers, Ltd., London.

and the fact that all countries are still developing: the process does not stop and is relative. When first used, the term *development* linked to economic growth and was measured by increases of income. It now has a broader meaning that includes cultural, political, and environmental aspects of living. People—rather than economic systems—are central, and economic growth is regarded as a means to the end of enabling people to enlarge their capabilities to the full and enjoy the richness of being human. The aim of **sustainable human development** is to plan development at a rate that does not deplete basic resources for the future.

The inequalities of economic and social well-being in the world and the gap between the poorest countries, such as Mozambique and Somalia, and the richest, such as the United States and Switzerland, have caused people to ask how poorer countries can catch up with the richer countries. That question leads to three further questions. What measures are used to compare economic and social well-being? What explains the differences between countries? How do poor countries narrow the gap?

## Measures of Wealth, Poverty, and Economic and Human Development

Several indicators are used, reflecting the range of meanings of the above terms.

- One vivid indicator of wealth differences among countries is the ownership of consumer goods such as televisions, cars, telephones, and personal computers (Figure 2.17).

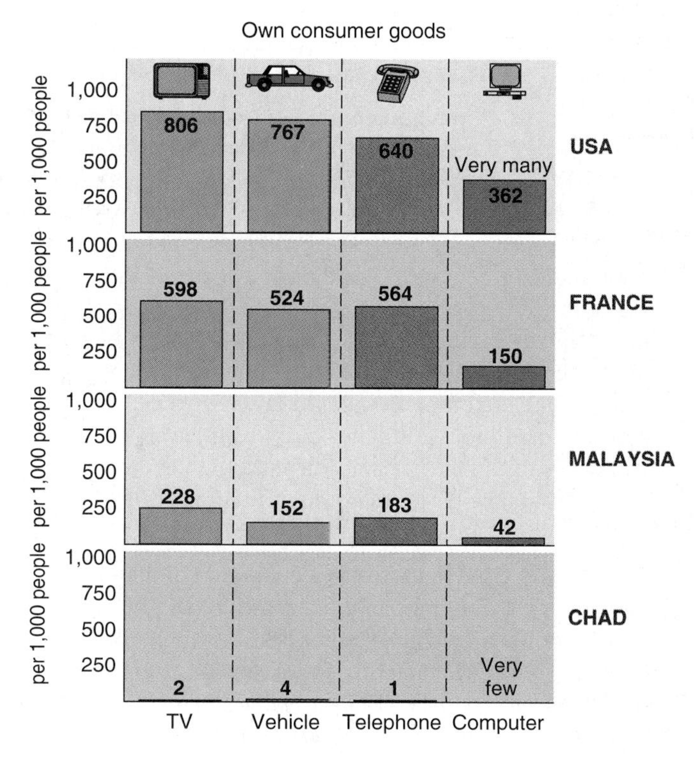


**FIGURE 2.17 Ownership of consumer goods.** Numbers of goods per 1,000 people in specific countries. Contrasts in affluence occur among core countries (United States, France), countries in the semiperiphery (Malaysia), and periphery countries (Chad). Similar diagrams occur in each chapter, with data for all countries in the Reference Section.

Source: Data (for 1996) from *World Development Indicators.* World Bank, 1998.

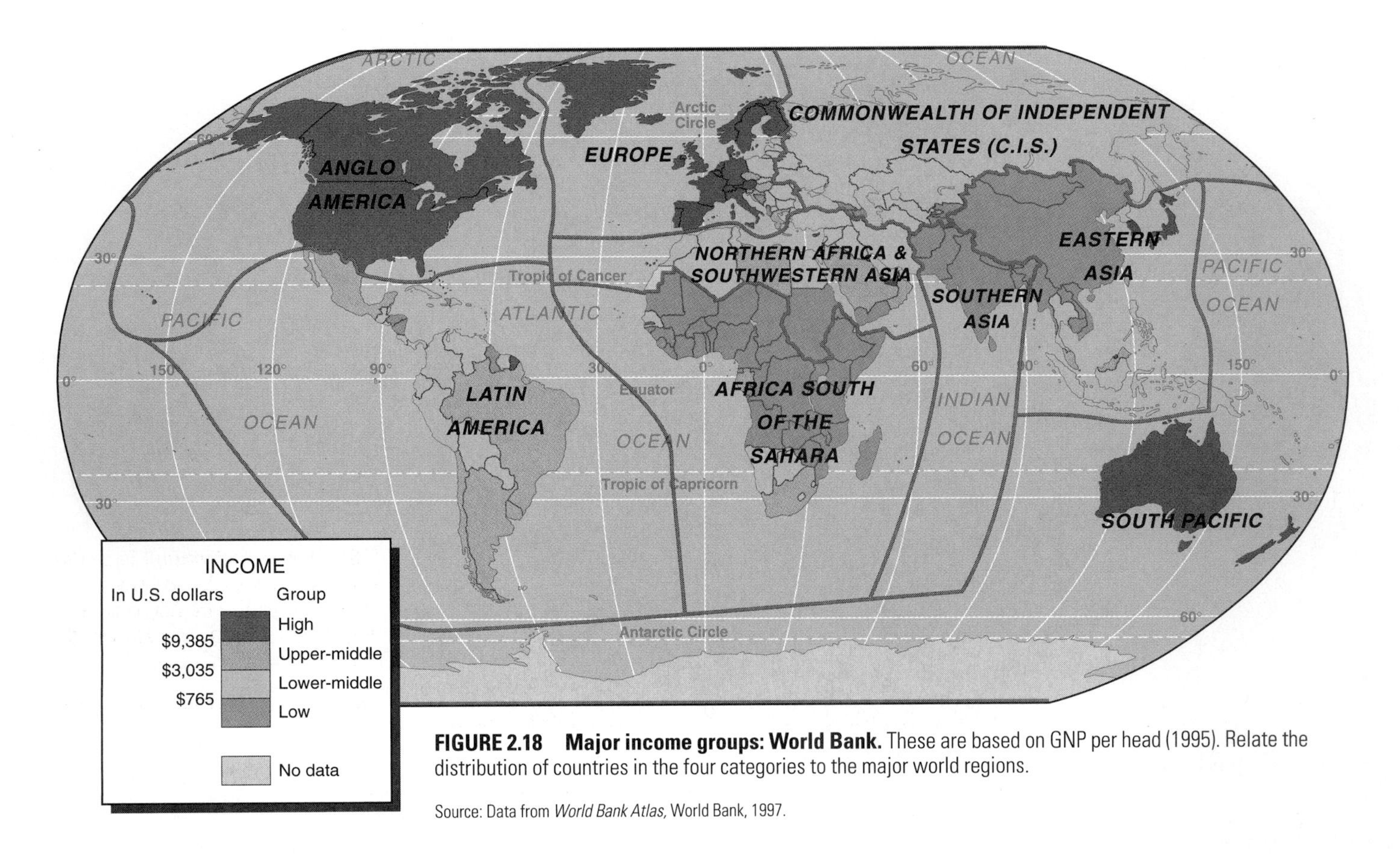


**FIGURE 2.18 Major income groups: World Bank.** These are based on GNP per head (1995). Relate the distribution of countries in the four categories to the major world regions.

Source: Data from *World Bank Atlas*, World Bank, 1997.

This indicator highlights the availability of income to spend on items that poorer people regard as luxuries, but wealthier people see as a vital part of their lives.

- Economic development is more commonly measured by two statistics of income that are widely reported. **Gross domestic product** (GDP) is the total value of goods and services produced within a country in a year. **Gross national product** (GNP) is the GDP plus income from labor and capital working or invested abroad less deductions for payments to those who live abroad. GDP and GNP per capita are averaged on the number of people and the country's total annual income: they do not indicate personal incomes. Based on GNP per capita, the World Bank divides countries into four income groups: low, lower-middle, upper-middle, and high (Figure 2.18). The "champagne glass" diagram of Figure 2.19 shows how unevenly the world's GNP is distributed.
- All GDP and GNP figures are calculated in national currencies and for comparison are converted to U.S. dollars at official exchange rates. Because such exchange rates may not reflect actual costs of living within a country, *purchasing power parity* (PPP) estimates of GNP and GDP are now published. Countries with high incomes and high living costs have a lower PPP estimate than the exchange rate GDP or GNP, while poorer countries often have higher estimates. For example, Luxembourg topped the 1996 GNP list at $45,360 per head, but its PPP GDP was $24,371; Zimbabwe in Southern Africa had a GNP of $610 and a PPP GDP of $1,737.
- The United Nations' **human development index** (HDI) provides a broader estimate of development. This is a composite statistic calculated from information about life expectancy, educational attainment, and health, as well as income. Data that contribute to HDI such as life expectancy, adult literacy, infant mortality, and per capita GNP are widely reported and can be compared over periods of time. Countries that invest in education and health care, such as the affluent countries, together with others such as Costa Rica, Sri Lanka, and Russia, provide a better quality of life for their people and are higher on this list than on the GDP or GNP lists. By comparison, many countries around the Persian Gulf with a high income from oil are high on the income per capita list but low on the HDI list. HDI figures are included in the data tables at the end of each chapter.

As the emphasis in thinking about development shifted toward the needs of the poorest peoples, the *United Nations Human Development Report for 1997* introduced a **human poverty index** (HPI). Linked to the HDI, it pointed out levels of personal deprivation among the least favored groups in society. It is based on three indicators that strike a balance between individual material deprivation and lack of public provisions, and between relevant data and available data.

- The percentage of people expected to die before age 40 (Figure 2.20), indicating vulnerability to early death and hence a restriction on human capability development.

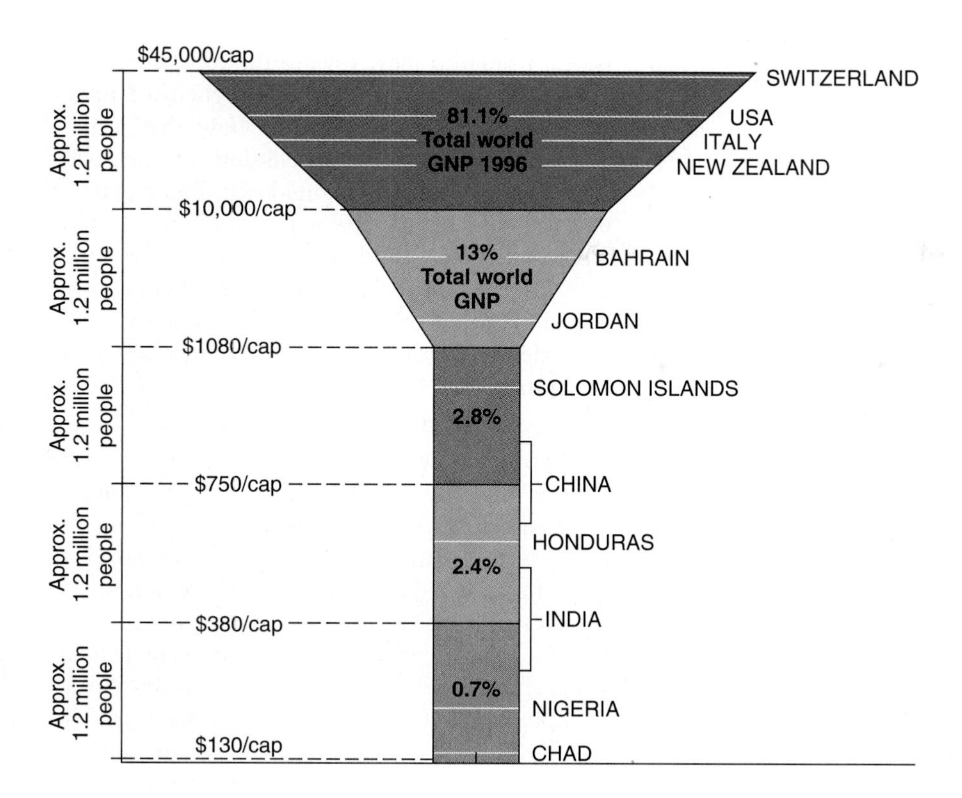


**FIGURE 2.19 Uneven distribution of world wealth.** The 1996 world population was divided into five groups with equal numbers and world GNP output divided among the five groups. What striking information does this diagram provide? Similar diagrams occur at the start of each regional chapter (3–11).

Source: Data (for 1996) From *World Development Indicators,* World Bank, 1998.

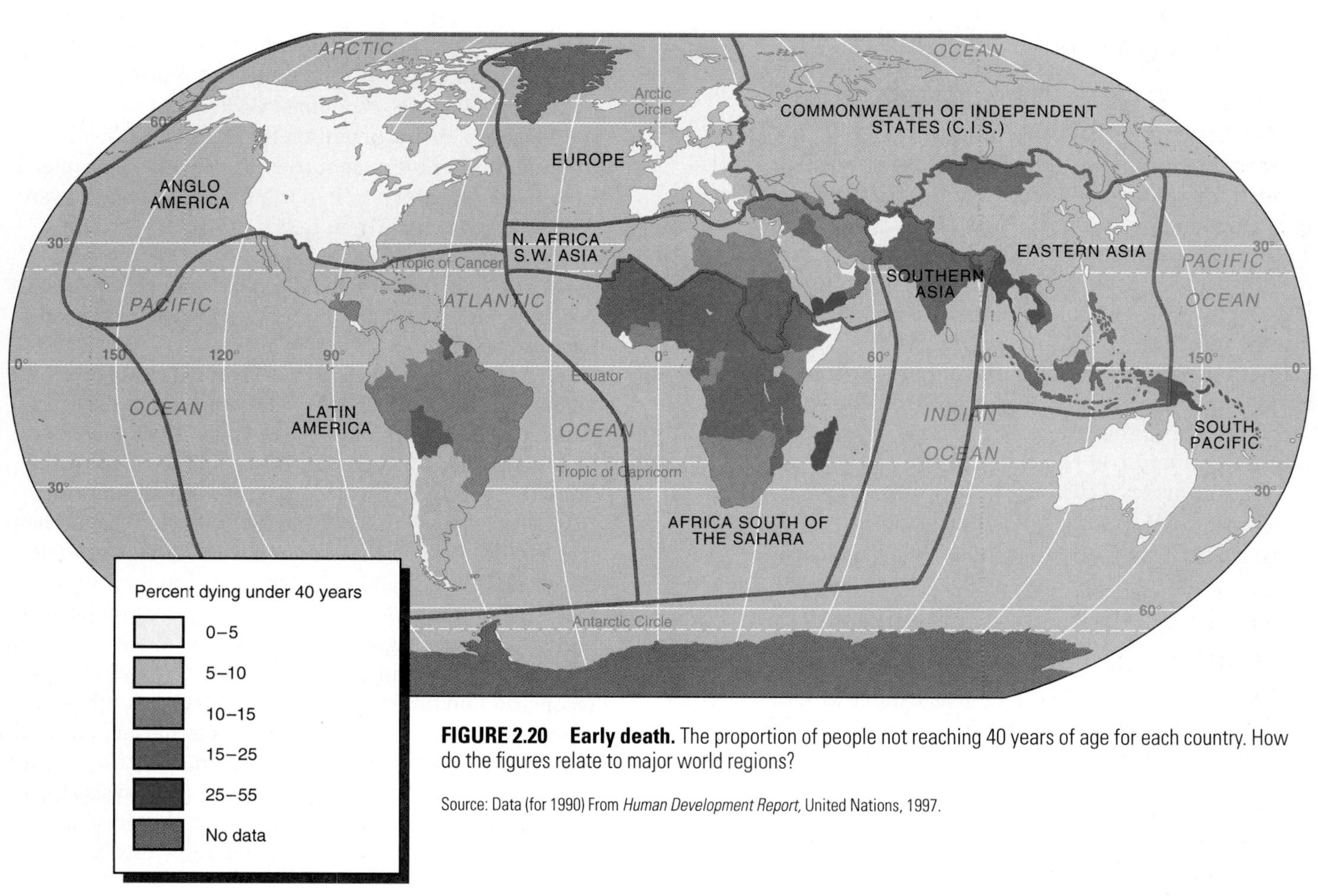


**FIGURE 2.20 Early death.** The proportion of people not reaching 40 years of age for each country. How do the figures relate to major world regions?

Source: Data (for 1990) From *Human Development Report,* United Nations, 1997.

**WEBSITES AND DATA SOURCES: Development**

The recommended CD-ROMS (see the Reference Section) and the following websites contain a range of economic, social, and demographic data and research project reports that are linked to development. The media resources describe relevant recent events. Explore some.

http://www.worldbank.org (World Bank data and research studies)

http://www.un.org (United Nations data and reports)

http://www.wto.org (World Trade Organization reports on trade issues)

http://www.economist.com (*The Economist* weekly articles and surveys)

http://www.nytimes.com/specials (Current issues analysis)

http://www.washingtonpost.com (Country studies including online discussions)

http://www.pbs.org (Current issues in special presentations)

http://www.newsweek.com (Special articles and links to Encyclopedia Britannica)

- The percentage of illiterate adults (restricting entry to better jobs and fuller involvement in community life).
- A composite of three variables that prevent a decent standard of living (percent of people with access to health services and safe water, and percent of malnourished children under age 5).

The HPI indicates the proportions of a population affected by such deprivations: 10 percent of the population is affected in such countries as Cuba, Chile, Costa Rica, Trinidad, and Tobago (all Latin America), and Singapore (Eastern Asia); over 50 percent of the population is affected in such countries as Niger, Sierra Leone, Burkina Faso, Ethiopia, Mali, and Mozambique (all Africa South of the Sahara), and Cambodia (Eastern Asia). There are also contrasts within countries: northeastern Brazil has a 46 percent HPI, the southeast has 14 percent; in China, the coastal regions have 18 percent and the remote interior 44 percent.

The HPI makes a distinction between income poverty, as shown by GNP or GDP, and human poverty (HPI). Some countries are better at reducing income poverty than human poverty: Côte d'Ivoire and Egypt both have 20 percent income poverty but over 35 percent human poverty. Other countries, such as China, Costa Rica, Kenya, Peru, Philippines, and Zimbabwe, have lower human poverty than income poverty. Countries with a very low HPI ranking also have a very low HDI. Overall, human poverty (as defined by the HPI) affects 25 percent of the population in poor countries, while income poverty affects 33 percent. Human poverty is most widespread in Africa South of the Sahara and Southern Asia, where it affects 40 percent of the population.

## Processes of Economic and Human Development

If *development* describes the processes that increase a country's economic and social well-being and elevate its world system position from periphery toward the core, it has to be admitted that there is no simple or comprehensive understanding of what the processes are or how they can be applied worldwide. The above discussion of measuring poverty and development highlighted the complexity of analysis required. This complexity is increased by variations in political systems. In the Cold War, the Western countries' emphasis on modernization that would bring other countries to a stage equal to their wealth was combated by Soviet bloc alternatives. Other distinctive approaches emerged in Eastern Asia and Latin America.

- **Western views: modernization.** The Western core-country attitudes toward the rest of the world derive from visions in the 1800s of a "manifest destiny" to improve government, order, and commercial prosperity, and to take Christianity to what were seen as the less civilized parts of the world. The sense of mission and leadership led to geopolitical interventions such as the British naval control of ocean trade and the extension of British, French, Belgian, Dutch, and German empires. The United States' war against Spain and subsequent occupations of Cuba, Dominican Republic, and Haiti followed in the early 1900s. Modernization in non-Western countries was encouraged by investments in education, health care, and sanitation; judicial reform; improved agriculture; and the bringing of core-country scientific advances to bear on common problems.

  In the 1950s, U.S. economist Walter Rostow defined the process of development on the basis of the historic experience of Western countries, as detailed in Chapter 1. In his view, "take-off" from traditional economies toward the modern occurred from the late 1700s as Western countries manufactured goods and sold them abroad. Personal wealth increased more rapidly than population numbers, along with urban living and better education, health, and ownership of consumer goods.

  The sequence of stages through primary, secondary, and tertiary economic sectors that occurred in the countries of western Europe and Anglo America in the 1800s and 1900s was seen by Rostow as a formula for **modernization** that was recommended to the countries that became newly independent after World War II. The recipe proved partly effective. A few countries negotiated the stages—some in 30 years, compared to the 150 years taken by Western countries—but others did not. Japan's industrialization began in the late 1800s, and it rose to join the most affluent countries after post–World War II reconstruction funded by the United States. South Korea took a shorter time, helped by huge internal and external investments following the wars on its territory in the early 1950s. Thailand, Taiwan, Singapore, and Malaysia also rose rapidly, but these Asian examples followed somewhat different paths in the late 1900s than the Western countries in the 1800s.

  Other countries that attempted to apply the modernization wisdom were less successful, while some found it difficult to start. The newly independent countries after 1945 had to work within a global system in which they were the periphery. The world's core countries and their multinational corporations remained the dominant political influences in markets and set trading terms. Furthermore, local cultural norms interacted with the modernization process, encouraging or discouraging its progress. The societies of Japan and South Korea, organized around agreed-upon goals and strong government, adapted more easily to industrialization than, for example, African countries with their multiplicity of small ethnic group interests (see Chapters 6 and 3, respectively).

- **Western views: structural adjustment.** In the later 1980s, new approaches, termed **structural adjustment**, were designed by the core-based financial lenders to reduce and reschedule debts that had accumulated in that decade. They mandated better government and greater involvement in world markets before poor countries could be supported by further aid or loans. This demanded basic changes in economic and political structures, including the balancing of budgets, privatization of government corporations, dismantling of tariffs to encourage free trade, and more democratic constitutions. Countries were invited to make industries globally competitive and **export-oriented.**

  During World War II, many peripheral countries became isolated from their core-country sources of manufactured goods and began to industrialize by protecting their internal factories against external competition. This process, known as **import substitution**, removed their need to import some products. In many cases, such protection resulted in uncompetitive and poorly managed nationalized industries with high costs arising from too many employees. They could not sell their goods on world markets. Reducing the protection against foreign products under structural adjustment led to local markets being flooded by cheaper (and often better) foreign goods and the collapse of sections of local manufacturing. A publicized example occurred in Mexico during 1994 and 1995 (see Chapter 10). The instability introduced by the exposure to world markets threatened poor countries' economies. Furthermore, many African countries had such a small industrial base and home market that attempts to implement aspects of structural adjustment only led to rising levels of unemployment and political unrest.

- **Non-Western views: dependency.** During the 1960s and 1970s, the countries of the Soviet bloc and China denounced Western approaches by claiming that the capitalist world economic system itself makes some places more underdeveloped than others. In short, the prosperity of the core countries is based on transfers of wealth from the peripheral countries, making the latter dependent on the core. Such ideas are summarized in **dependency theory.** More development in one place necessarily involves less elsewhere. Core countries grow at the expense of peripheral countries; metropolitan regions grow at the expense of rural areas.

  Dependency theory, however, while providing a valid critique of some aspects of modernization, offers few hints as to how peripheral countries may attain core status if they are locked into dependency. The expectation of working class revolutions breaking out in the capitalist world never happened, while the authoritarian embodiment of communist ideas and the collapse of the Soviet Union in 1991 made the alternative of state capitalism unattractive and extremely unlikely. Where it was applied, as in Cuba and Vietnam, it placed countries at a disadvantage in the world economic system.

  Much of the dependency arose out of the earlier political colonialism, particularly in Africa and Southern Asia. Modern economic colonialism has similar results of dependency relationships and stark differences between rich and poor countries by exploiting cheap labor in peripheral countries. Whereas the division of labor inherent in factory production processes led to the class differences inside Western countries, a **new international division of labor** resulted from the growth of multinational corporations, which established production units in peripheral countries. Multinational corporations take advantage of cheap labor and land around the world through technical developments such as the containerized transportation of goods and easy communications links that reduce other production costs. For instance, 90 percent of the labor force in the electronics industries of Eastern Asia are females age 15 to 25 years who work twice the hours of British workers for 40 percent of the wages.

- **Non-Western views: marginalization.** While the Western and Soviet bloc views were being imposed on the rest of the world, new insights arose from the experiences of peripheral countries. In Latin America, it was observed that the social and cultural shifts associated with the economic changes of modernization left behind groups of people who remained marginal to the process of improving economic well-being.

  The squatter settlements and shantytowns of major cities in peripheral countries (Figure 2.21) provided a cultural poverty trap for rural-to-urban migrants from which they and their children could not escape. Although this theory of **marginalization** led to widespread rejection of the idea of an automatic progression in development through economic sectors, it did not suggest an alternative prescription for success. Later experience showed that, for many, the shantytowns provided a transition into the urban economy, beginning with people earning money outside the regular wages or salaries of the formal economic sector (see Chapter 10).

- **Non-Western views: dual sectors.** Events in Indonesia led to the recognition of **dual sectors** in a country's

(a)

(b)

(c)

**FIGURE 2.21 Shantytowns (favelas) in Sao Paulo, Brazil.** (a) Favela built on unused land between inner-urban developments and by a main road, close to a supermarket. (b) Favela in a valley among apartment blocks and suburban houses. (c) Team from a local high school helps to install concrete sewers in a favela along a ditch that previously carried off wastewater after rains.

All photos © Michael Bradshaw.

economy and expanded the concept of marginalization (see Chapter 6). Traditional rural areas formed a growing contrast in well-being with urban-industrial areas; traditional Asian ways of life conflicted with the Western lifestyles that many migrants to towns wished to imitate.

This understanding highlighted how large and essential sectors of the economy in many countries were based on occupations outside the formal economy, not reflected in measures such as GDP or numbers of workers or taxpayers. Activities from unlicensed street trading to small-scale manufacturing and squatter housing met the basic needs of large numbers of people at little cost to the authorities. Governments began to implicitly support such activities by providing building materials for shanty-type housing and adding infrastructure such as electricity supplies, water, and sewerage. The **informal sector** of the economy in many poorer countries is thus not merely a transition to the formal but is often closely related and subordinate to it. For example, pedicabs in Indonesia are driven by people outside the formal employment sector but are often owned by low-paid teachers and government officials (formal sector) who rent them out to supplement their main income. Some informal activities, such as scavenging trash, are not linked to the formal sector, although those engaged in such "recycling" activities may earn more than factory workers. Since the 1980s, aid agencies and institutions such as the World Bank have recognized informal activities in their development strategies, seeing them as a means of encouraging self-assistance.

- **Non-Western views: East Asian "tigers."** The countries making the greatest steps to increasing their GNP included several in Eastern Asia (see Chapter 6), such as Japan, and countries that are called the Asian "tigers," including South Korea, Taiwan, Hong Kong, and Singapore. In these, strong single-party governments kept democracy to a minimum and focused investment in protected growth areas of manufacturing for export sales. Their populations were paid low wages to achieve high levels of domestic investment but were promised future rewards that began to come once huge trading surpluses were achieved. In short, they followed some, but not all, of the goals and means of structural adjustment. In the late 1990s, bad debts and poor banking systems often dominated by political rather than economic goals led to a major economic downturn in these countries.

## Varied Outcomes of Development

To date, economic growth has been uneven throughout the world. It has seldom been to the benefit of the poor; 97 of 166 countries had lower real per capita income in 1990 than in 1970. Although some commentators judge world economic growth to have generally bad outcomes, and others are too optimistic about its potential, no human development occurs without strong economic growth. The *United Nations Human Development Report for 1997* concluded that the reduction of poverty can be achieved only by accelerating economic growth and aiming that growth at the poorer sections of society. During the 1990s, of the major world regions, only Eastern Asia (growing at 12% per year from 1990 to 1994) and Southern Asia (5.1%) grew faster than the 3 percent per year estimated to be the figure required to reduce poverty by half within 10 years. By contrast, Africa South of the Sahara declined (–2.4%), along with the Arab countries (–4.5%) and the former Soviet bloc countries (–9.1%), while Latin America slowed (–1.3%).

Priorities identified for encouraging growth that helps poorer groups of people include establishing focused programs with targets for reducing the numbers of the poor; raising the productivity of small-scale farming; prioritizing small-scale enterprises, including those in the informal sector; promoting labor-intensive industries; and expanding human capabilities through education and health programs.

In the 1990s, more fundamental questions were asked about the development process. Currently, voices are calling for opposition to development as it has been understood and applied by the core countries and their organizations. Such voices stress the need to understand local conditions and culture in order to redefine what people can cope with, expect, or need. Some of the themes include the promotion of indigenous grassroots interests, the greater involvement of female interests, and concerns for the natural environment. For example, the clashes between indigenous groups and multinational mining corporations in Brazil (see Chapter 10) and Papua New Guinea (see Chapter 11) illustrate the current dominance of the world economic system over local interests. There is a need for broader agreement on the objectives of development in the light of the concerns of those living in peripheral parts of the world.

In this text, the economic development of each major region and subregion is a central consideration. By beginning with the study of Africa South of the Sahara and other countries in the world periphery, it may be possible to raise an appreciation of concerns and needs from the local standpoint as well as from external Western perspectives.

### Civil Wars, Disease, and Environmental Stress

Major global forces in the late 1900s worked against the prospects for economic growth and human development. One-third of countries with declining income and development indicators had major armed conflicts within their borders that affected the poorest civilians; 110 million landmines remained undetonated in 68 countries—a deterrent to reinstating farmland. Millions of refugees left poor countries and migrated to other poor countries, deflecting additional resources for their upkeep. Economic sanctions on renegade countries affected the poor in Haiti, Cuba, and Iraq.

The HIV/AIDS pandemic brought impoverishment to Southern Africa in particular, where 14 million of the world's 23 million cases lived: although Botswana and Zimbabwe made economic advances from the 1980s, life expectancies fell. Malaria and diarrhea still kill more, but AIDS is linked to poverty, loss of schooling, increasing numbers of orphans, and the spread of new infections.

**WEBSITES: Cultural and Social Data**

The CD-ROM sources listed in the Reference Section provide data on language, religion, race, gender, and other social matters. Websites of interest include:

http://www.un.org

http://www.undp.org

The United Nations and U.N. Development Program sites contain databases and reports on social development, human rights, humanitarian affairs, international law, peace and society, and sustainable human development.

http://www.lonelyplanet.com

This travel site provides accounts and photos of travels in different parts of the world, including cultural and social data.

Environmental degradation increased in lands where 50 percent of the poorest world population live among fragile ecosystems, often isolated from world trade routes, including the southern margins of the Sahara and the margins of the Himalayan Mountains. The fact that rapid population increase adds to the environmental stresses in these lands highlights another continuing problem affecting growth and development—that of rising populations in poor countries.

## CULTURAL GEOGRAPHY AND WORLD REGIONS

The culture of a group of people:

- includes the code of acquired beliefs, passed on from generation to generation, that supplements intuitive behavior. It produces a pattern of meanings, copied and updated, through which people communicate and develop attitudes to life: language and religion are two expressions of such codes.
- is a system of social relations, exemplified in race, class, and gender.

Cultural geography studies the distribution of such phenomena. Regional geography uses these concepts, particularly in historical explanations of regional differences, and in understanding the form and personality of a landscape as the outcome of a succession of cultural inputs over time.

### Language

**Languages** are spoken means of communication among people. Languages provide a shared identity for a cultural group and grow out of historic experiences and traditions. Minority groups, such as the Québecois in Canada, the Welsh in Britain, and the Hispanics in the United States are often keen to preserve their language as a means of maintaining their

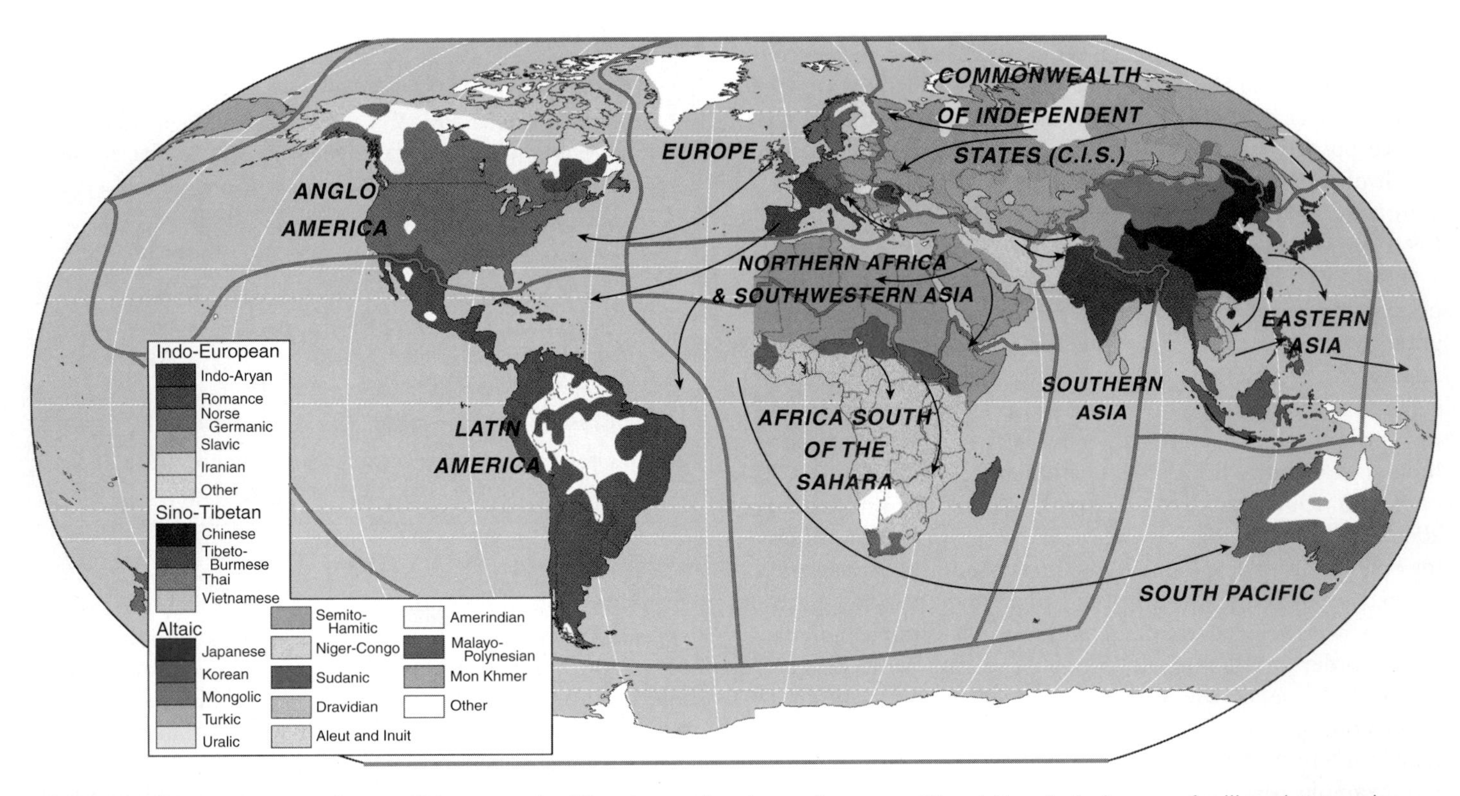


**FIGURE 2.22 Cultural geography: world language families.** Arrows show how major groups diffused. How do the language families relate to major world regions?

cultural identity. Language is an important factor in geographic diversity.

Thousands of languages are spoken around the world, many by small groups of people in isolated environments such as the Borneo hills and the Amazon River basin. Twelve dominant languages each have over 100 million people speaking them (Figure 2.22). Six of these (English, French, Spanish, Russian, Arabic, and Mandarin Chinese) are official languages of the United Nations. The six were chosen at the end of World War II by the victorious allies, who established the United Nations and recognized the growing importance of Arabic.

The present geographic distribution of languages is the result of major language families developing in specific regions and diffusing to other regions by conquest or trade. Such groupings of languages are very general, each containing hundreds of linked languages and dialects.

The Indo-European family includes most of the languages of the Indian subcontinent, the Slavic languages of eastern Europe and Russia, and the languages of southern, western, and northern Europe. These languages, especially Spanish, Portuguese, English, French, and Dutch, spread worldwide with colonization from around A.D. 1450, dominating the Americas and South Pacific and becoming important in many parts of Africa and Asia. English is increasingly important in communications among scientists throughout the world, in international air traffic control, and in computer software. It remains a common language in ex-colonies of Britain where there is rivalry among local languages.

Of the other language families, the Sino-Tibetan family includes the languages of China and the southeast Asian land area. The Altaic family includes the languages of Central Asia that diffused westward to Hungary and Finland and eastward to Japan and Korea during medieval conquests. The Semitic-Hamitic (Arabic) family of languages dominates northern Africa and the Arabian peninsula, while the Niger-Congo family includes many languages spoken in Africa South of the Sahara. The Malayo-Polynesian family occurs through Malaysia, Indonesia, the Philippines, and the South Pacific. Smaller family groups include those of the pre-European populations of the Americas (Aleut, Inuit, Amerind) and Cambodia (Mon Khmer).

## Religions

Religion generates strong group loyalties. It is reflected in regional emphases, such as social and legal practices, and in visual features such as building designs (Figure 2.23). Each **religion** is an organized system of values and practices involving faith in and worship of a divine being or beings. Religions are significant in transferring cultural values and practices from one generation to the next.

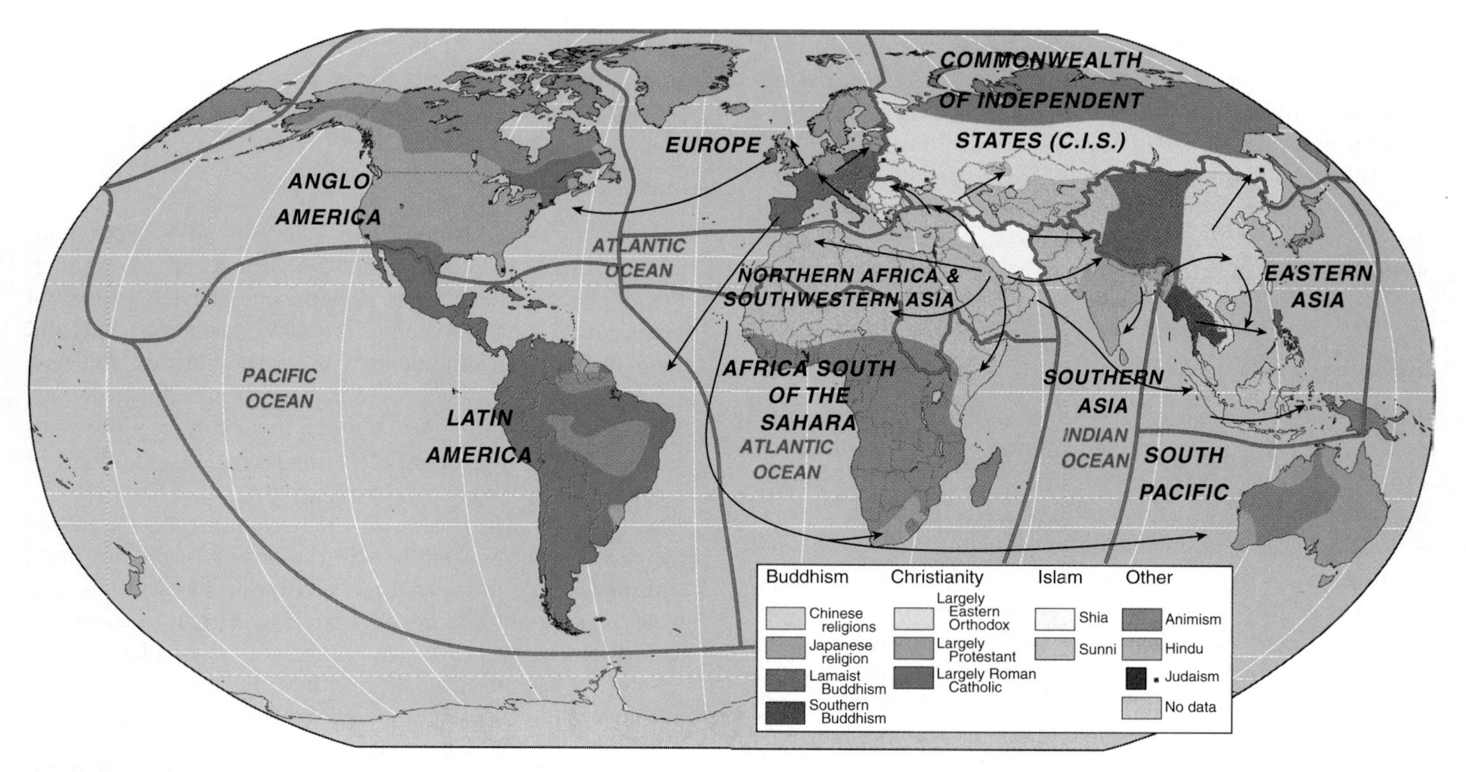


**FIGURE 2.23 Cultural geography: world religions.** Many religions are practiced, but only Christianity, Islam, Buddhism, and Hinduism have large numbers of adherents. Islam, Christianity, and Buddhism diffused through trade, mission, or conquest. The Jewish religion is spread around the world, mainly in cities. Older and local religious practices remain in small areas, mainly in peripheral countries. Is there a link between religion and major world regions?

## Major World Religions

The religions claiming the largest numbers of adherents are Christianity, Islam, Hinduism, and Buddhism. Judaism has a smaller number of adherents but widespread influence, especially in Western countries. Christianity, Islam, and Buddhism are religions that can be joined by anyone in any country, and they often seek to extend their membership; Hinduism and Judaism are more closely tied to family and region and do not encourage conversions.

Branches of major religions are linked to specific world regions. In Christianity, for instance, the split between Roman Catholicism and the Orthodox churches set apart the cultures of western and eastern Europe. The growth of Protestant churches in northern Europe after A.D. 1500 resulted in further divisions as well as renewed Christian missionary expansion by all the groups.

Today some 60 percent of world Christians are Roman Catholics, 25 percent belong to Protestant groups, and 10 percent are Orthodox. Christianity has some 1.5 billion adherents and is the main religion of Europe, the Americas, South Pacific, and much of Africa South of the Sahara. Islam is dominant throughout the Arab world of Northern Africa and Southwestern Asia and extends eastward to Central Asia and through Pakistan and Bangladesh to Indonesia (the most populous Muslim country). Buddhism, often combined with older local religious elements, is the main religion of Eastern Asia, while Hinduism is the predominant religion of Southern Asia.

## Religion and Society

Religious adherence often determines responses to issues affecting society. Roman Catholics and Muslims, for instance, resist population policies aimed at reducing births. Most religions demonstrate a care for the natural environment, although the practice of societies has often not matched these beliefs. Since religious beliefs are often held deeply, religious differences between people, when linked to inequalities of wealth and power, may result in conflicts. Such conflicts often arise after being fanned by those in power for their own ends. Christians versus Jews, Muslims versus Hindus, Christians versus Muslims, and Muslims versus Jews underlie some of the basic human conflicts. The conflicts extend to Catholic versus Orthodox (former Yugoslavia) or Protestant (Northern Ireland) Christians, and to Shia versus Sunni Muslims.

While religious beliefs help to explain many cultural features of present world geography, it has been suggested that economic globalization will be accompanied by an increasing secularization based on scientific explanations of phenomena, thereby reducing religious influences. Evidence exists that attendance at religious ceremonies and observance of religious rites is decreasing. Science does not, however, provide a system of values or guides to the meaning of life. Today, both Western and non-Western societies question the significance of scientific and rational approaches as prescriptions for living. It is likely that religion will continue to influence culture, perhaps in new ways. The resurgence of Islamic militancy in some countries is one sign that religious differences may once again become significant after the Cold War period of largely political conflicts.

## Race, Class, and Gender

Status—and the power to carry out or stifle changes in society—varies with culture. In many cultures, status is inherited through the family, as in the caste groups of India or the aristocracies of Europe. In other cultures, status may be achieved by creating wealth, political allegiance, media prominence, or sporting performance. While many of the world's richer countries increasingly base status on achievement, some people have a greater chance of achieving higher status than others by virtue of their birth into wealthy families, their upbringing, or their education. Throughout the world, race, class, and gender influence access to position in society.

### Racial Variety

**Race** is a biologic condition shown in skin, hair, and eye colors, height, and body shape. It is often confused with *ethnicity,* which is based on nonbiological cultural differences. Although no scientific evidence exists that race is related to intelligence or ability, racial differences result in varied levels of social and economic opportunity in all parts of the world. Most African Americans in the United States still have less status than those of European origin. In South Africa, the minority white population operated an apartheid policy until 1993 that separated whites, blacks, and other peoples into segregated neighborhoods. Even in Brazil, where European, African, and Asian peoples mix freely, most rich people are of European origin and most people of African or Native American origin are poor.

### Class Distinctions

**Class** arises from a stratification of society that is imposed on the basis of religious, economic, or social criteria. Wealth, education, or birth are common bases for class distinctions in all countries. In Britain, the royal family is linked to a hierarchy of dukes, earls, lords, and knights. In India, the caste system and, increasingly, education define position in society. In communist countries, the expectation of a classless society is contradicted by the status and privileges given to members of the Communist Party. Most Americans identify general class differences on the basis of possessions and appearances. That basis produces a black underclass, a lower middle class of black and white factory and shop workers, a mainly white upper middle class of managers and professionals, and a group of exceptionally rich financiers, property owners, and sports and media stars (both white and black).

### Gender Inequalities

**Gender**—the cultural implications of one's sex—is also responsible for differences and inequalities within and among societies. Males dominate most societies and have a history of denying full rights to women. Illiteracy among women is still much higher than among men in most of Africa and Asia because more men than women have been educated. Few European countries allowed women to vote in elections until the 1900s, with Switzerland delaying it until 1971 and one of its cantons until 1990.

Women, even in the world's richer countries, commonly receive lower wages than men for the same job and constitute a minority of doctors, engineers, corporate executives, and elected politicians. Some jobs such as nursing, secretarial work, elementary school teaching, and shop clerking are regarded as "women's work" and have lower status. Women commonly take a major share of home management and child care, which affects their career prospects. Marriage breakups usually leave women at a financial and social disadvantage.

The *United Nations Human Development Report* for 1995 included a **gender-related development index** (GDI) for the first time. The GDI focuses on the same criteria as the HDI but reflects inequalities between men and women. While 175 countries were reported for real GDP (PPP$) per capita and HDI values, only 148 were reported for GDI in the 1997 report (reflecting figures for 1994). The data are included in the Reference Section tables.

Some aspects of female inequality, such as illiteracy in poorer countries, are scarcely being tackled but have major implications for future population-resource issues in these countries. Until women are able to control how many children they have and can read birth control leaflets, it is unlikely that great progress will be made in reducing population growth.

Despite being undervalued and disadvantaged in most societies, women play major roles in the expanding world economy. During the 1970s and 1980s, the proportion of women in manufacturing jobs in the United States and Canada grew as employment opportunities shifted in their favor. For example, female employment in the electronics industry expanded but men became redundant when "male" jobs were taken over by machine tools. Expanding numbers of jobs in services such as retailing, health care, teaching, banking, and tourism further increased female employment. In countries of the world's periphery, women comprise around 80 percent of the export-oriented labor force in electronics, apparel, and textile industries. Yet while the core countries and Eastern Asia experienced greater household prosperity and female involvement in the labor force, women's employment prospects in Africa and Latin America deteriorated as the result of measures imposed to solve the debt crises in those countries.

## Cultures and Regions

Cultural differences make the world's major regions and smaller regions within them distinctive and thus have strong geographic impacts. As the Chapter 1 section "Geographic Growth of a Global Order" showed, major cultural differences can be traced back to early human history and the development of agricultural technology, religion, and language in small regions with particular environmental conditions. Such regions are termed **cultural hearths** and were centers from which techniques, useful materials, and social norms diffused to other regions.

Current world geography includes the imprint of many newer cultural overlays on older ones. For example, there are still remnants of the Mayan and Aztec empires, as well as of Spanish colonization, in Mexico. The Mayan remains in Yucatan are being revealed from beneath the forest, while modern urban-industrial buildings are replacing Aztec and Spanish relics around Mexico City. In Southwest Asia, the Muslim culture removed most pre-Muslim relicts, but some, such as Crusader castles, still exist. In communist China, the palaces of former imperial dignitaries occupy large areas of Beijing, while the new cities of the eastern coastlands impose modern urban-industrial features on the rural landscape that evolved over centuries.

### Cultural Fault Lines

Just as cultural hearths identify areas that had formative influences on cultures in wider areas, cultural "fault lines" exist between cultural regions. Long-lasting tensions along these lines lead to conflict in the absence of other constraints. For example, Bosnia in the former Yugoslavia is where the three conflicting cultural interests of Roman Catholic, Orthodox, and Muslim people meet. After the 1991 breakdown of the Yugoslav communist government, Bosnia became the focus of a renewal of the tensions that had ignited World War I.

### Social Development

The growing worldwide realization that cultural factors are important in people's well-being and development led to a World Summit for Social Development, held in Copenhagen, Denmark, in March 1995. It brought together world leaders pledged to supporting human development, focusing on the need to alleviate poverty and invest in women. It called on major institutions of economic change to protect vulnerable people from the effects of rapid change and requested core countries to devote 20 percent of their aid budgets to social development programs, including health and education. Commitments to such measures, however, were not as strong as had been hoped by some but may provide a framework for further measures.

**RECAP 2C: Economic and Cultural Geography**

World economic activities are divided into primary, secondary, tertiary, and quaternary sectors. Economic systems, based on modes of production, are dominated worldwide today by capitalism.

Development focuses on achieving sustainable systems in relation to economics, environments, and culture. Levels of development are measured by GDP, GNP, HDI, GDI, and HPI. Understanding the "how" of development arose from Western or Soviet views of the world, but local factors are now taken more seriously.

Cultural characteristics with important geographic impacts include language, religion, race, class, and gender.

2C.1 How would you define primary, secondary, tertiary, and quaternary industries, listing examples of each?

2C.2 What are the main ideas that contribute to an understanding of development?

2C.3 In what ways do languages and religions define cultural groups?

2C.4 How are gender issues determined by cultural expectation? How may present discrimination be changed?

**Key Terms:**

economics
industry
primary sector
secondary sector
tertiary sector
quaternary sector
multinational corporation
poverty
development
sustainable human development
gross domestic product (GDP)
gross national product (GNP)
human development index (HDI)
human poverty index (HPI)
modernization
structural adjustment
export-oriented industries
import substitution
dependency theory
new international division of labor
marginalization
dual sectors
informal sector
language
religion
race
class
gender
gender-related development index (GDI)
cultural hearth

## PHYSICAL GEOGRAPHY AND WORLD REGIONS: ENVIRONMENTAL VARIATIONS

Physical geography is the study of natural environments and their world distribution. In world regional geography, the outcomes of interactions between human activities and natural environments help to define the nature of each region. Major physical features, such as oceans, seas, mountain ranges, or rivers, often form boundaries between countries. The length of growing season, amount of water available, soil types, and mineral-bearing rocks all influence the type and location of human activities and the level of development possible in each region.

Environmental issues focus around the possibility of human activities resulting in a lowering of quality and productivity

(a)

(b)

**FIGURE 2.24 Contrasting natural environments.** (a) The Canadian Rockies: high mountains raised in movements powered by Earth's internal energy and then sculpted by glaciers. Such environments are favorites of outdoor recreation buffs. (b) Bonneville Salt Flats, Utah. The area was covered by the expanded Great Salt Lake during a period of wetter climate several thousands of years ago. Evaporation of the water during drier conditions left behind the white salt crust. The salt provides a raw material for chemical industries, while its solid, smooth surface is used for world land-speed record attempts.

(a) © Norman Benton/Peter Arnold, Inc.; (b) © Michael Bradshaw

among natural resources. Lower environmental quality often goes with a less sustainable prospect for the future. Each chapter in this text contains a section on the natural environment of the major world region being considered, together with specific concerns about the management of the natural environment there.

Earth's natural environment is a dynamic system of interacting parts—the solid earth, the atmosphere and oceans, plants, animals, and soils—that produce regional differences from high forest-clothed mountains (Figure 2.24a) to dried-up desert areas (Figure 2.24b). Natural processes include:

- the workings of the atmosphere that produce weather and long-term climate changes;
- the workings of Earth's interior that cause huge sections of the crust to collide with each other, producing mountain systems;
- the effects of rain, glacier ice, wind, and the ocean that shape the continental surfaces to produce individual landforms such as hills, valleys, and beaches—the stages for human activities; and
- the actions of living organisms in responding to and modifying the local climate, landforms, and soils.

Natural features and processes not only impinge on human activities, but human beings also increasingly intervene in and influence the ways in which natural processes function.

## Climatic Environments

The **climate** of a place is determined by the transfers of heat and moisture in the linked atmosphere-ocean system that is powered by energy from the sun. Solar energy is filtered as it passes through the atmosphere, so that it is mainly visible light rays that reach Earth's surface. Absorption of these rays causes rock, soil, and ocean water to be heated and to radiate heat rays upward, raising the temperature of the lower atmosphere. The gases and liquids composing the atmosphere and oceans circulate, distributing the heat and moisture to places that have little.

### Heat Transfers

Because the sun is more directly overhead for more of the year at the tropical regions, its heating impact on the atmosphere is greatest there (Figure 2.25). Tropical areas have an excess of incoming energy over that lost back to space (Figure 2.26). The polar regions, however, have a deficit of energy. In winter, they have several months of almost complete darkness, losing energy to space without any coming in. The tropical excess and polar deficit are compensated by flows of air and ocean water between the two regions. Tropical oceans become huge reservoirs of heat that is moved poleward by ocean currents. This heat is transferred into the midlatitude atmosphere. The cooled air and waters return to the tropics, where they are reheated.

**WEBSITES: Weather and Climate**

The following websites provide information on current weather conditions, some worldwide, including weather satellite images:

http://www.ems.psu.edu/wx/usstats/uswxstats.html (Penn State University)

http://www.AccuWeather.com

http://www.intellicast.com

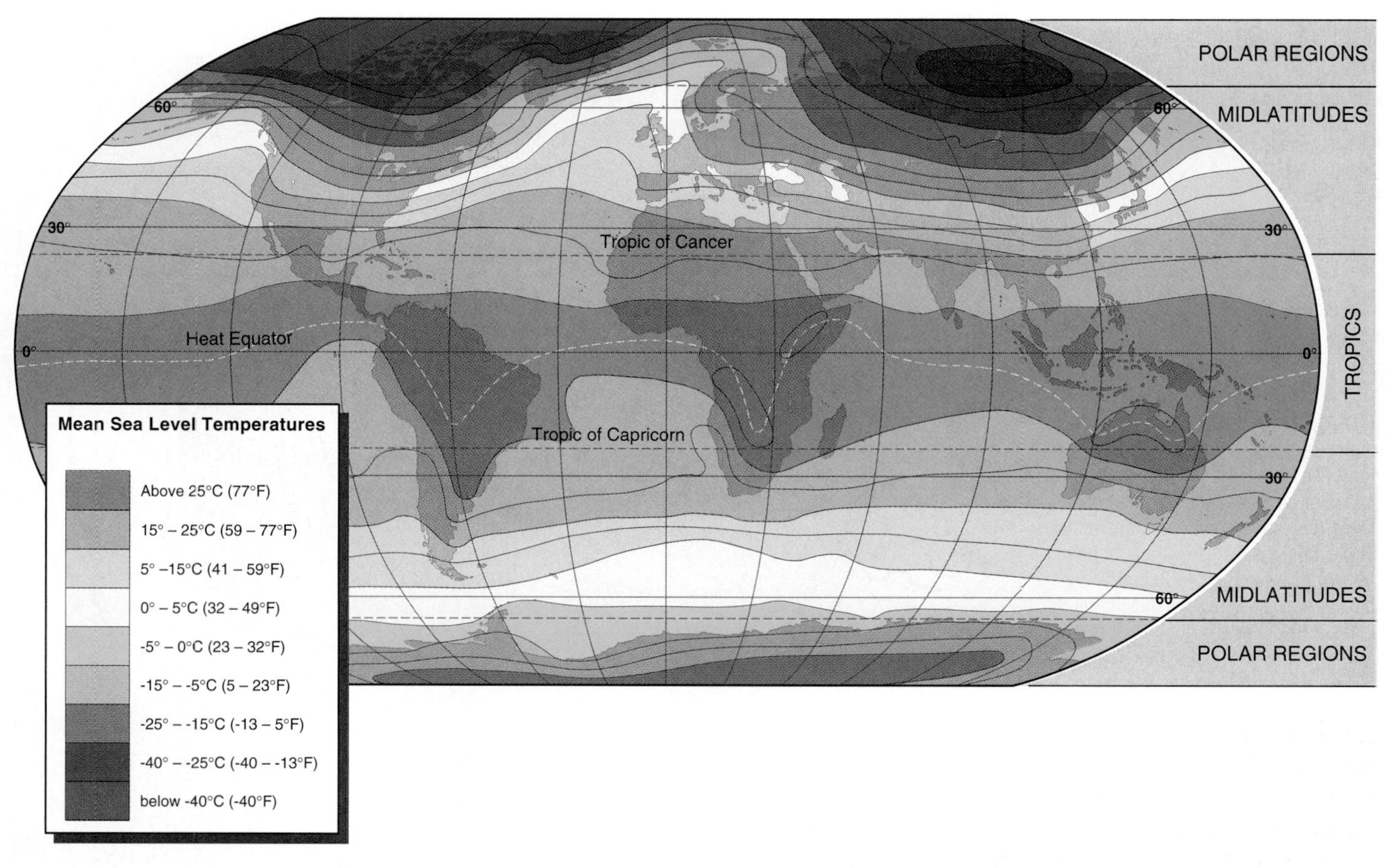


(a)

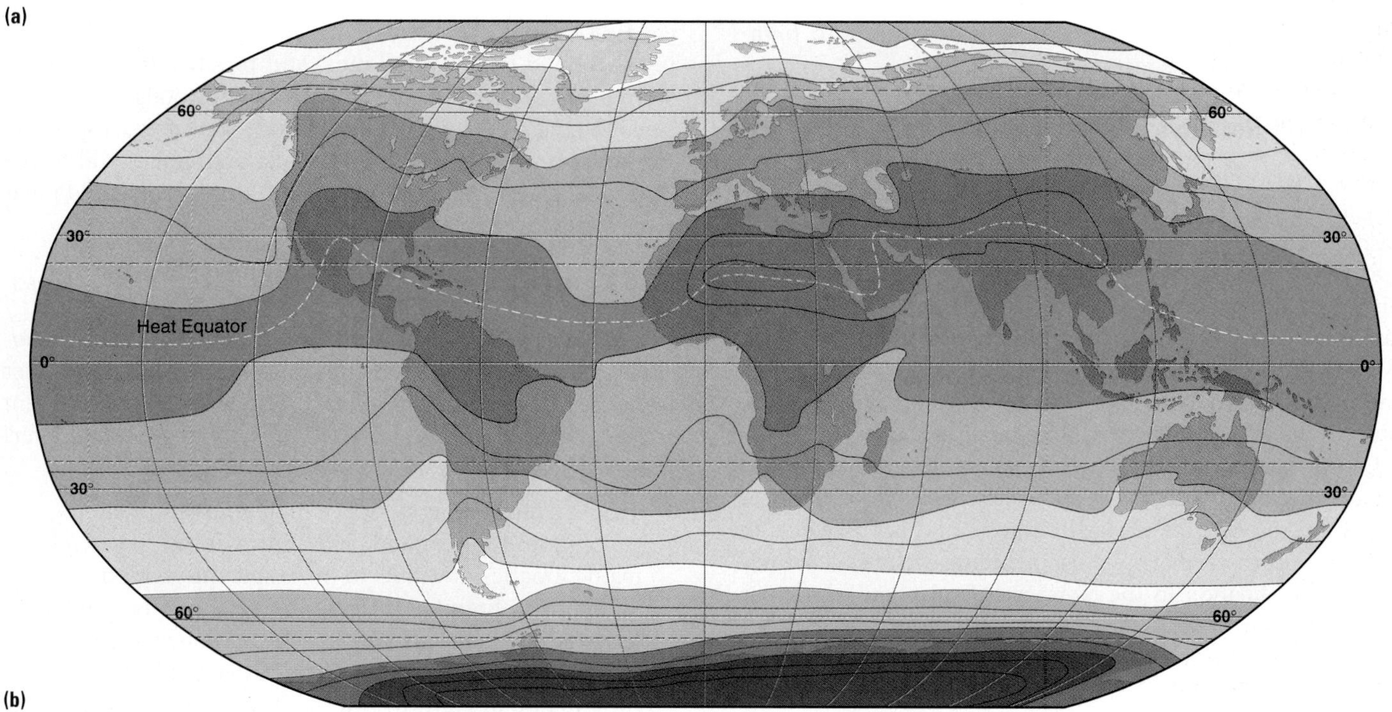


(b)

**FIGURE 2.25 Global temperatures in air near the ground.** Isotherms (joining places of equal temperature) for (a) January and (b) July. The heat equator connects points of highest temperature at each meridian of longitude. Compare the Northern and Southern Hemispheres at these two periods in terms of the extent of very cold temperatures and the position of the warmest band of temperatures. What effect on air temperatures do the oceans have?

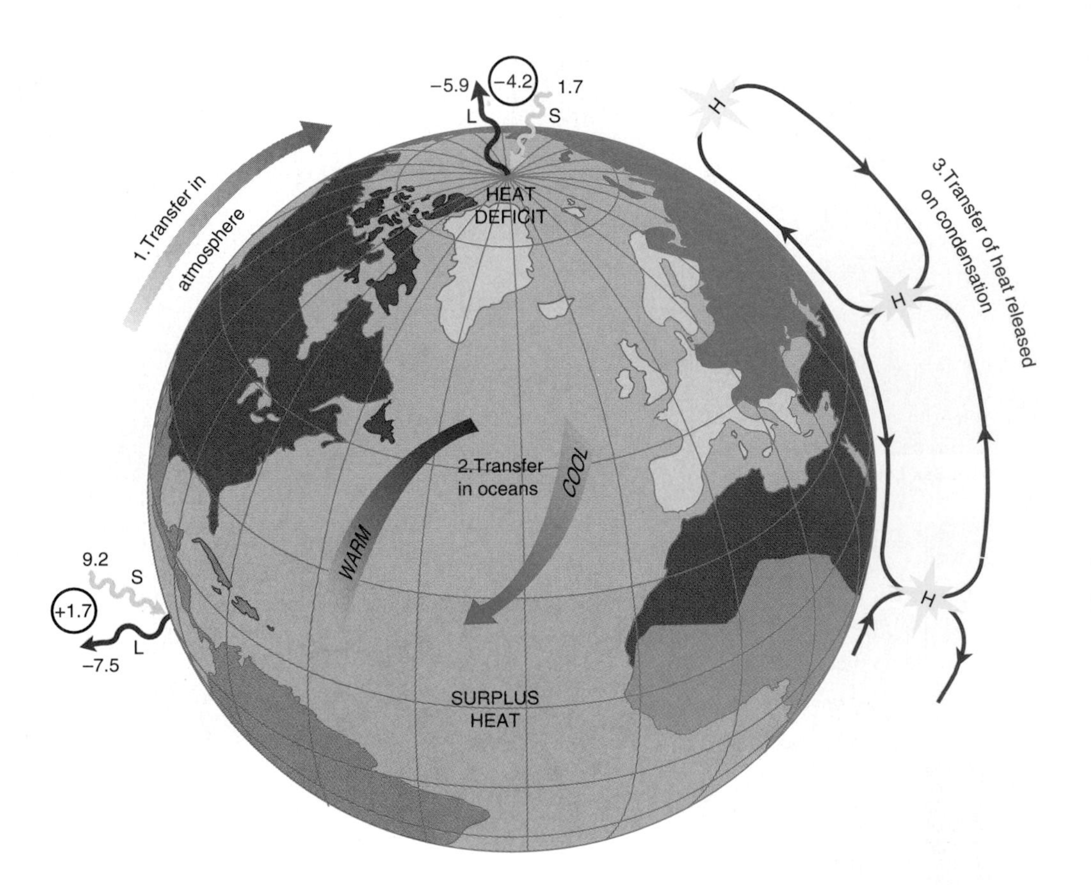


**FIGURE 2.26 Global heat transfers.** In the tropics, heat losses by longwave radiation from Earth are less than incoming shortwave solar radiation; at the poles, the losses exceed the gains. Three main mechanisms transfer heat from low toward high latitudes, balancing the excess and deficit: winds in the atmosphere, ocean currents, and movements of humid air (heat is trapped during evaporation and released when clouds condense). Flows of warm air and water toward the poles are paralleled by flows of cold air and water toward the equator.

This system makes human habitation possible into high-latitude regions near coasts.

## Earth's Daily Rotation and Annual Solar Orbit

Earth rotates on its axis once a day, traveling around the sun once a year. The former creates day and night, and the latter seasonal changes. Because of Earth's axial tilt, the sun is directly overhead at the Tropic of Cancer (Northern Hemisphere) at noon between June 19 to 23 and the Tropic of Capricorn (Southern Hemisphere) between December 19 to 23. Overhead sun brings summers of warmer weather and long polar days to each hemisphere, while the winter hemisphere has cooler weather and long polar nights.

Earth's rotation affects air and water movements—winds and ocean currents—across the surface. The effect increases away from the equator toward the poles and bends winds to form circulating weather systems, including cyclones (counterclockwise circulation in the Northern Hemisphere, clockwise in the Southern Hemisphere) and anticyclones (clockwise circulation in the Northern Hemisphere, counterclockwise in the Southern Hemisphere).

## Water Transfers

Oceans are major sources of water that evaporates into the atmosphere, condenses into clouds, and gives rise to rain and snow. Parts of the world with the highest rainfalls lie near the equator (Figure 2.27), where humid airstreams collide, forcing the air to rise, form clouds, and produce frequent rainstorms. Another zone of high rainfall totals is on the ocean-facing west coasts of midlatitude continents, especially where high mountains (as in Canada and Chile) add to the lifting caused by the meeting of warm tropical and cold polar air. Areas between have little, or seasonal, rain.

## World Climatic Environments

The receipt and redistribution of solar energy, the atmospheric and oceanic movements, and the circulation of water from oceans to continents all vary in their effects on different parts of the world, giving rise to weather systems that differentiate climatic zones, or environments (Figure 2.28).

- **Tropical climates** experience high temperatures throughout the year and have short winters, if any. The main climatic variations in the tropics are seasonal differences of rainfall. Places close to the equator often have rain in all months of the year, although a few months may be drier than the rest. Places farther from the equator, but still within the tropics, have a marked alternation of dry and wet seasons. Eventually the dry season becomes so long that annual water shortages occur, and the climate becomes arid. The main tropical climates thus have a north-south distribution, from the equatorial climate with rain at all seasons, through the

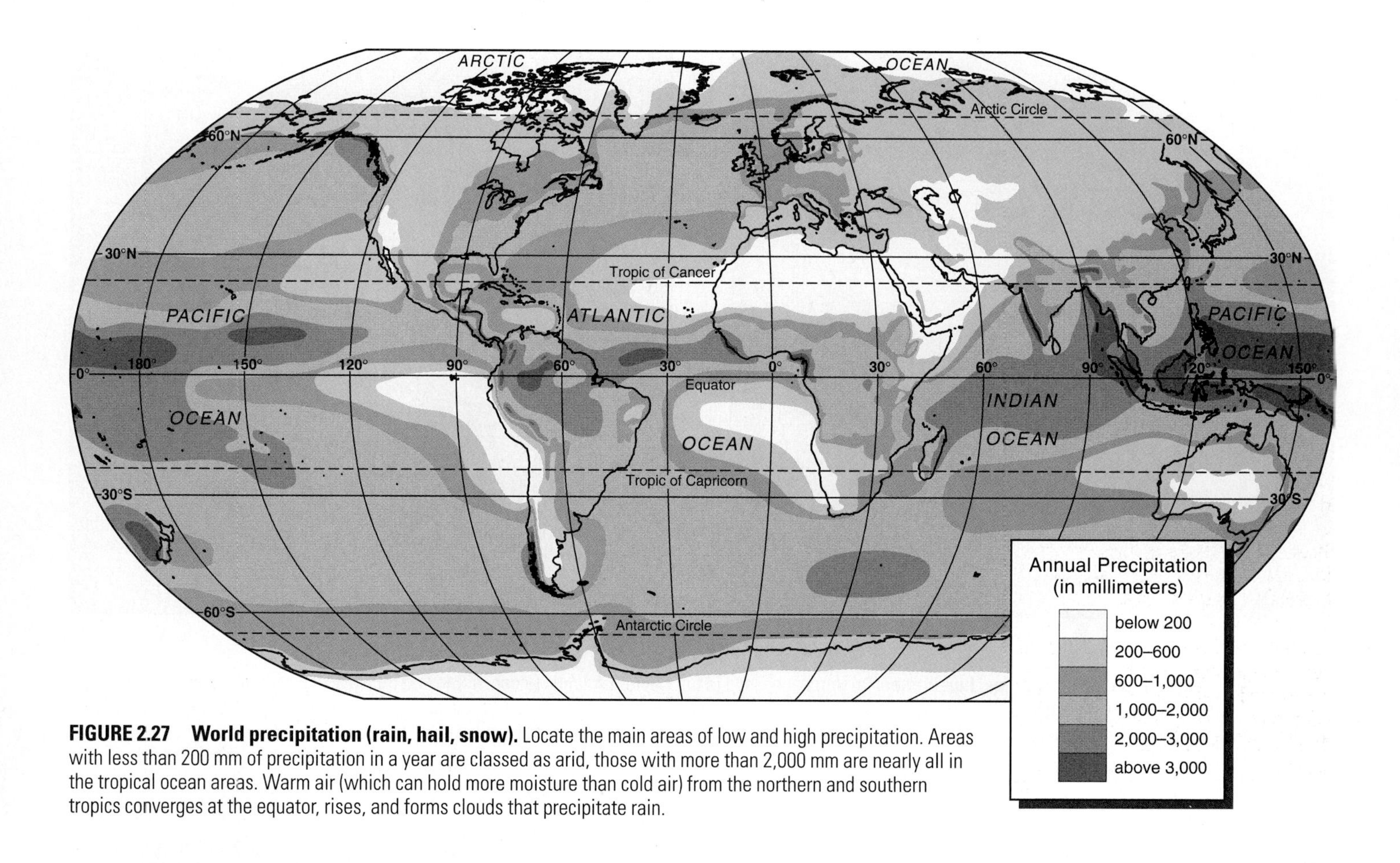


**FIGURE 2.27 World precipitation (rain, hail, snow).** Locate the main areas of low and high precipitation. Areas with less than 200 mm of precipitation in a year are classed as arid, those with more than 2,000 mm are nearly all in the tropical ocean areas. Warm air (which can hold more moisture than cold air) from the northern and southern tropics converges at the equator, rises, and forms clouds that precipitate rain.

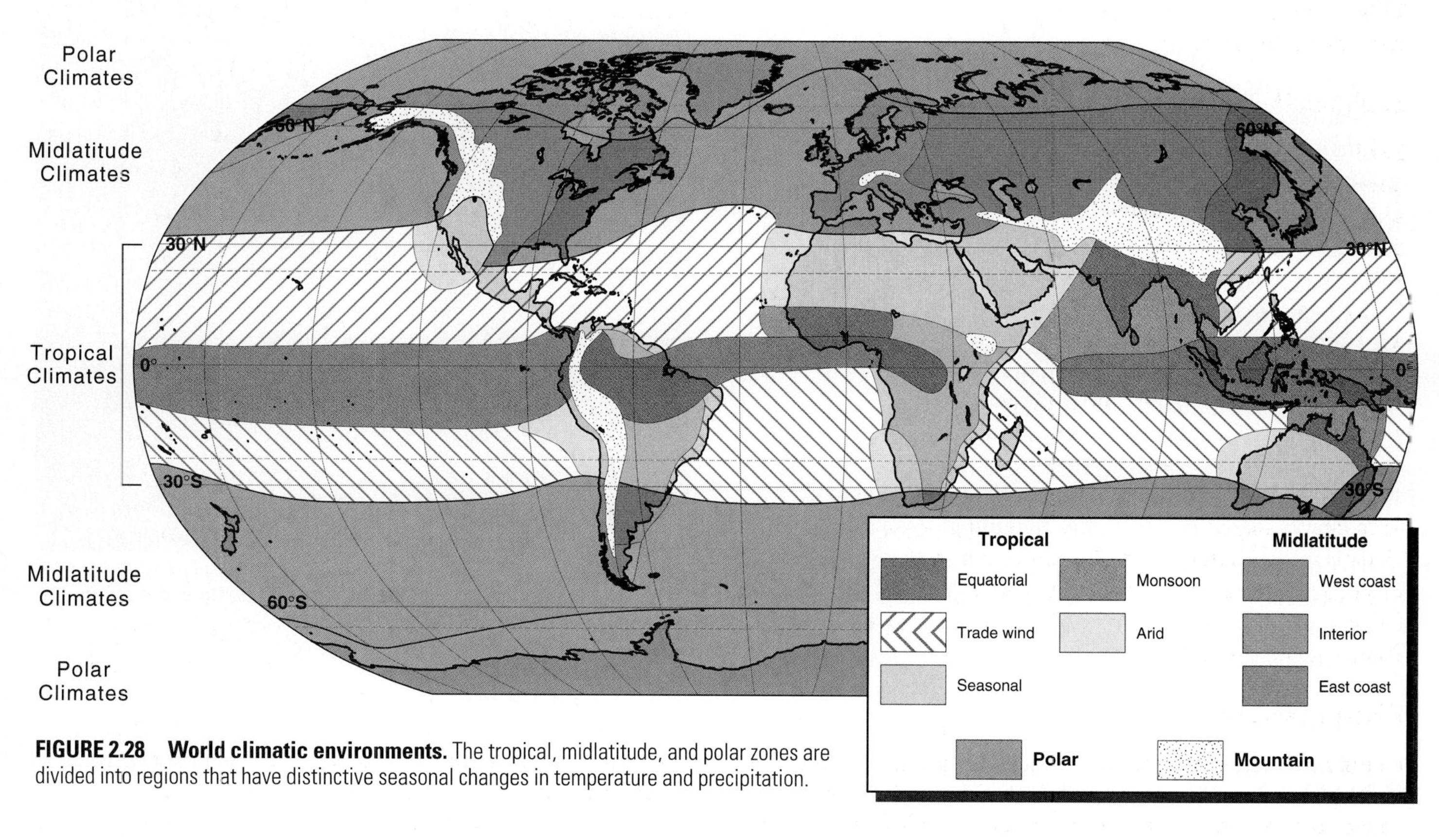


**FIGURE 2.28 World climatic environments.** The tropical, midlatitude, and polar zones are divided into regions that have distinctive seasonal changes in temperature and precipitation.

(a)

(b)

**FIGURE 2.29 Tropical weather systems: hurricanes.** (a) Hurricane Elena in the western Atlantic Ocean, as viewed from a space shuttle. Several hundred kilometers across, it has the typical eye of descending air, surrounded by strong winds and swirls of cloud. Hurricanes form over tropical oceans where the water temperature and evaporation rates are high. (b) Damage caused by Hurricane Hugo when it hit the South Carolina coast in 1989.

(a) NASA, Michael Helfert; (b) S. J. Williams, U.S. Geological Survey

wet-dry seasonal climates, to the very dry climates. The seasonal wet-dry contrasts are greatest in the monsoon climates of the Indian subcontinent and Western Africa. Tropical climatic environments are marked by distinctive weather systems, including tropical cyclones (also called hurricanes and typhoons), which may cause loss of life and property (Figure 2.29). The frequent rains near the equator come from massive thundercloud developments (Figure 2.30).

- Climates of **midlatitude lands** have marked seasonal temperature contrasts. Such contrasts are greatest in the centers of the North American and Eurasian continents and least where winds blowing from the oceans bring winter warmth and summer coolness to the west-facing coasts of Europe and Anglo America. Most precipitation falls on hills and mountains facing west near these coasts and declines inland to the point where it may become insufficient to support vegetation or agriculture. Midlatitude climates have a west-east distribution, with mild and moist west coasts, dry continental interiors, and east coasts with summer rains. Midlatitude weather systems are subject to the greater degree of turning in the atmosphere in these latitudes and include frontal cyclones—which bring rain and high winds—and calmer anticyclones.
- **Polar climates** are extremely cold throughout the year. Although expeditions to explore Antarctica and the Arctic Ocean take advantage of the local summer, temperatures remain low and the exploration season is mainly when sunlight is available. In winter, there is almost total darkness for several months in polar regions. Polar regions have few weather systems of their own but are dominated by dense, cold air that sinks and flows outward

**FIGURE 2.30 Tropical weather systems: thunderstorm groupings over central Africa.** Seen from a space shuttle, several cloud tops amalgamate in fibrous masses of ice that spread outward and downwind.

NASA

**WEBSITES: Shaping Earth's Surface and Natural Environments**

http://www.usgs.gov
http://daac.gsfc.nasa.gov
The U.S. Geological Survey site contains information on current environmental issues and education resources on hazards, resources, ecosystems, and global change. Sample some of the fact sheets provided. The NASA Goddard DAAC site includes a series of space views of landforms.

to deepen the winters of midlatitude regions. Midlatitude cyclones sometimes invade polar regions, bringing high winds and precipitation.

### Climate Change

Climates change over time—small amounts over shorter periods and larger over longer periods. The "Little Ice Age," from approximately A.D. 1430 to 1850, resulted in upland glaciers advancing several kilometers down valleys and cultivation retreating from higher areas in midlatitude countries. A more intensive freeze, which caused huge ice sheets to dominate the northern parts of North America and Europe and sea levels around the world to fall, lasted nearly 100,000 years. It ended around 10,000 years ago with a period of warming, ice melting, and a sea surface rise to its present level of around 100 m (300 ft). After the last ice advances ended, fluctuations of climate brought the warmest conditions around 5,000 years ago. Periods of warmer conditions coincided approximately with each expansion of world human systems described in Chapter 1. Periods of cooler world climates occurred during the periods of disruption.

A present concern is that the carbon gases humans pour into the atmosphere in using coal, oil, and gasoline will add to the gases that trap solar energy in the lower atmosphere and lead to climatic changes through **global warming.** This may cause ice sheet margins to melt and the ocean level to rise a meter or so within the next 50 years, putting the low-lying coasts of coral islands, wetlands, and port areas at risk. The precise trends and impacts are difficult to predict and illustrate a point about natural environments. Events are not only too complex for us to predict with great accuracy but also are totally outside the realm of human ability to alter them.

## Shaping Earth's Surface

Earth's surface is 71 percent covered by ocean water and only 29 percent occupied by the continents on which people live. The natural environment of continental surfaces forms the height and slope of the land, or its relief. The differences among mountains, valleys, and plains are caused by the interaction of internal Earth-building forces with external Earth-molding influences resulting from weather and the action of the sea (Figure 2.31).

**FIGURE 2.31 Internal and external Earth forces mold the landscape.** The gash across the Carrizo Plain in southern California was caused by the San Andreas fault splitting a section of Earth's crustal rocks. Rain and rivers then etched out the line of weakness. Further movement along the fault shifted and offset the line of stream valleys on either side. The San Andreas fault occurs where two plates move against each other, with the western side moving northward.



### Earth Interior Forces

Earth is a multilayered planet with a hot molten core. Earth's internal heat provides the energy that forces large blocks of surface rock, thousands of kilometers across and 100 km (65 mi.) thick, known as **tectonic plates**, to move and crash into each other (Figure 2.32a). The heat from the interior is sufficient to drive the movements of the six major and seven minor plates (Figure 2.32b).

The plate movements produce the major features of Earth's surface, including mountain systems and continents, and cause the opening and closing of ocean basins. The rock that forms the main part of the plate solidifies from the molten state where it erupts along fissures opened up as plates move apart. Such gashes mostly occur in ocean basins, along ocean

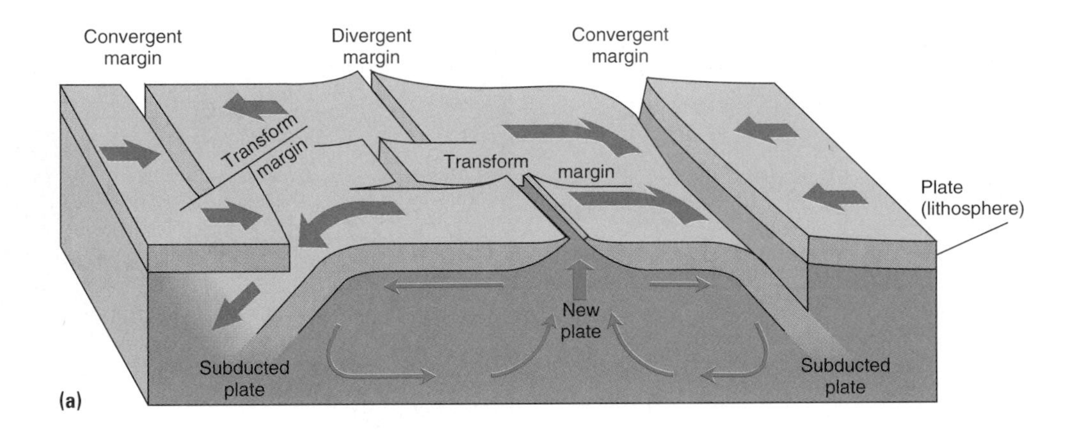


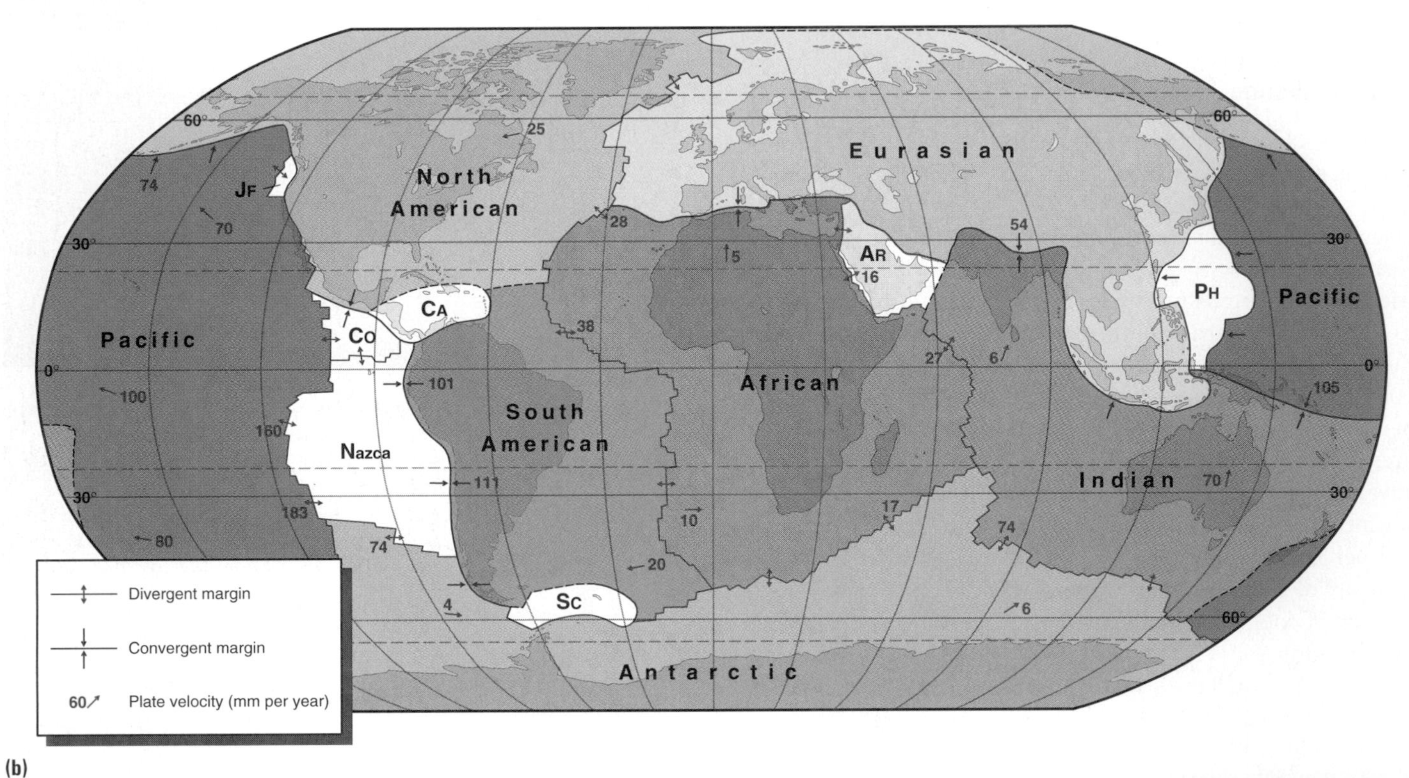


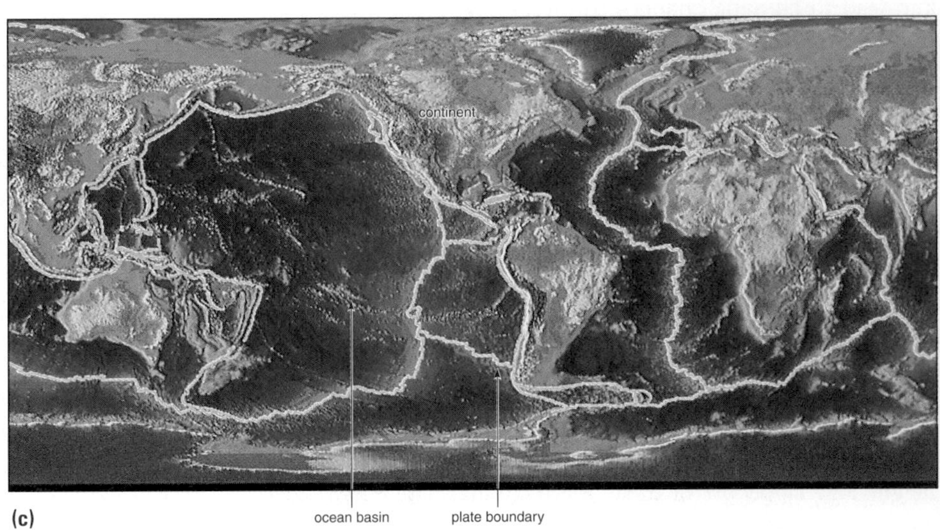


**FIGURE 2.32 Plate tectonics, continents, ocean basins, and mountain ranges.** (a) The main features of plates and plate margins. Red arrows show plate movement directions, blue arrows depict possible internal Earth convection currents driving movements in the top 100 km of Earth's interior. (b) World map of major and minor plates. The minor plates include Nazca, Cocos (Co), Caribbean (Ca), Juan de Fuca (Jf), Arabian (Ar), Philippine (Ph), and Scotia (Sc). (c) Plate boundaries, ocean basins, and continents. Which boundaries are convergent and divergent, and what features are each associated with?

Sources: (a) and (b) Plate velocity data from NASA; (c) © NOAA

ridges (Figure 2.32c). Plates also collide with each other, forcing one side upward to form mountain systems. The plate that is forced beneath the raised plate is said to be subducted. As the subducted solid rock melts under high temperatures of burial, the molten rock under pressure rises toward the surface, creating volcanic mountains there. Earthquakes and volcanic outbursts of molten rock from beneath Earth's surface are concentrated along plate boundaries, signaling the effects of plates colliding and moving apart.

## Surface Changes

Once land emerges above sea level, it is attacked by the atmosphere and ocean, which mold the details of surface relief. Changes in the temperature and chemical composition of the atmosphere, together with water from rain and snowmelt, react with the rocks exposed at the surface and break them into fragments. Small particles are acted on by chemicals and, in some cases, dissolved in water. Such changes are called **weathering.** The broken rock material forms the mineral basis for soil. Some of it moves downslope under the influence of gravity. Such movement may be rapid in slides, flows of mud, or avalanches, or slow in local heaving and downslope creep of the surface. The mobile fraction of this broken rock material often falls into rivers or on to glaciers and is moved toward the ocean.

The concentrated flows of water or ice and rock particles in rivers and glaciers gouge valleys in the rocks—a process called **erosion** (Figure 2.33). When the flows reach a lake or the ocean, they stop and the rock particles drop to the lake or ocean floor in the process of **deposition.** Wind blows fine rock particles (dust-size) long distances, while sand-sized particles are moved around deserts or across beaches to form dunes. Along the coast, sea waves and tides fashion distinctive eroded cliff and deposited beach features, often moving the rock particles supplied by river, glacier, or wind.

## Physical Landscape

In a particular region, the relief features of the land surface are determined by the combination of these internal and

(a)

(b)

(c)

**FIGURE 2.33 Eroding the land: rivers, glaciers, and the sea.** (a) Stream in the White Mountains of New Hampshire with boulders up to a meter across, used by the stream to wear away its channel bed. (b) Deeply crevassed Meldenhall Glacier eroded a deep valley as the ice spilled downward from the Juneau ice field, Tongass National Forest, Alaska. (c) Ocean waves attack the cliffs at Shore Acres State Park, Oregon.

(a) © Tim Hauf/Visuals Unlimited; (b) © Kim Heacox/Peter Arnold, Inc.; (c) © Scott Blackman/Tom Stack & Associates

(a)

(b)

**FIGURE 2.34 Landscape evolution: two examples explained.** (a) Western Scotland. The rocks in this northwestern part of Britain were formed over 400 million years ago and then crushed and raised into mountain ranges by colliding plates as an earlier Atlantic Ocean closed. Rivers wore down the mountains. The Atlantic Ocean opened again some 50 million years ago, raising this area with others along its margins. Within the last million years, the Ice Age covered the higher parts with thick ice that moved outward, forming deep valleys such as the one in this photo. On melting, the ice poured water back into the oceans, raising its level and drowning the lower part of this valley as a sea loch. Rivers eroded the hills, depositing deltas along the edge of the deep water. (b) Zimbabwe, Southern Africa. The contrast between the rocky hills and the flat plain results from a position far from any plate margin and millions of years of intense chemical decomposition in a tropic environment of seasonal rains. Rock particles weathered from the rocky hills fall to the plain, where rivers at the start of the wet season carry off dry surface particles.

(a) © Michael Bradshaw; (b) © Paul Bradshaw

external forces, as the two examples in Figure 2.34 show. The landscape outcome depends on whether the region contains a plate margin, which climatic elements fashioned the surface, and how long the forces operated. Changes of climate are important in altering the surface forces at work, such as rivers, glaciers, or wind. For example, eastern Canada was covered by an ice sheet only 20,000 years ago but is now the domain of rivers.

## Plants, Animals, and Soils in Ecosystems

Living organisms are a unique feature of Earth: we know of no other planet where life exists. Plants and animals are sustained by a combination of energy from the sun's rays, water circulating from oceans to the continents and back again, and nutrient chemicals in the soils. The nutrients are made available to living organisms by the atmospheric processes that break up surface rocks, by the actions of other organisms, and by the water flowing through them. Most plants have the ability to capture and store the sun's energy in chemical form, combining with the mineral nutrients drawn from soil to produce the foods that animals require.

### Ecosystems and Biomes

Plants and animals live in communities in which they share the physical characteristics of heat, light, water availability, and nutrients. The total environment of such linked communities and physical conditions is known as an **ecosystem.** Ecosystems exist at all geographic scales but for the purposes of this text are discussed in relation to the largest scale, or **biome**. There are five main types of biome—forest, grassland, desert, polar, and ocean. They reflect the major climatic environments, although the natural vegetation and soils of many biome areas have been greatly modified by human activities.

- **Forest biomes** exist in both tropical and midlatitude climatic environments. Tropical forest types range from the dense rain forest of equatorial regions, which is characterized by the crowding of great numbers of plant and animal species, to more open forests with fewer species in areas having long dry seasons. Midlatitude forests include the deciduous forests, once typical of the Eastern United States in which the leaves are shed in winter, and the evergreen forests typical of Canada and the Pacific coast of North America.

(a)

(b)

**FIGURE 2.35 World biome types: savanna grassland and desert.** (a) Tropical savanna grassland, Eastern Africa. The grasses grow in areas of moderate rainfall, supporting large numbers of grazing animals, such as zebras and gazelles. These become food for lions and other carnivores at the end of the food chain. (b) Desert vegetation, southwestern USA, including teddy bear cactus, which survives by storing water in its tissues beneath a tough outer skin that both reduces transpiration and deters animals from eating it. This biome supports few large animals.



- **Grassland biomes** occur where longer dry seasons, or annual burning by humans, restrict tree growth. In the tropics, savanna grasslands are characteristic of much of Africa and South America (Figure 2.35a). Midlatitude grasslands occurred in the prairies of Anglo America, the steppes of southern Russia, the pampas of Argentina, and the veldt of Southern Africa. These environments are now dominated by grain farming and grazing.

  With its low shrub vegetation, the Arctic tundra does not qualify as grassland but has some similar characteristics of that form and supports grazing animals. Tundra is present where long cold seasons and strong winds prevent tree growth.
- **Desert biomes** occur in lands dominated by aridity (where evaporation exceeds water supply). Deserts are characterized by little or no vegetation. Any plants that survive in such conditions will have special means for storing water, as in the cactus (Figure 2.35b). Tropical deserts include the zone extending from the Sahara in northern Africa through southwestern Asia to Pakistan. Other tropical deserts occur in central Australia, southwestern North America, and coastal Peru and northern Chile. Midlatitude deserts occur in Central Asia, Patagonia (southern Argentina, South America), and in some areas of western North America.
- **Polar biomes.** In polar lands, ice cover, combined with low inputs of solar energy, provide little sustenance for living organisms. Coastal areas of Antarctica, however, support a variety of animals through the offshore marine resources.
- **Ocean biomes** contrast with those on the land, but the differences in water temperature between tropical, midlatitude, and polar zones and between surface and deep waters, together with the differences in nutrient availability, produce zones of more or less abundant life. Microscopic plankton plants form the basis of living systems in the oceans, and they are often most profuse where cold water containing nutrients rises to the surface, as along the west coasts of Africa and North and South America. Such differences affect the productivity of fisheries, with smaller numbers of a wide range of different species occurring in the tropics and greater numbers of fewer species in midlatitude oceans.

## Soils

Soils are a particularly significant aspect of the natural environment for humans. Soil fertility governs whether a region will produce good crops or support livestock farming. **Soils** form as broken rock matter interacts with weather, plants, and animals. The original rock materials supply or withhold nutrients. Water from rain and snowmelt makes any nutrients present available to plants; decaying plant and animal matter releases the nutrients back to the soil in mineral form. In the United States, for example, the soils of the eastern half developed beneath forest in which the annual leaf fall returned nutrients to fertile brown soils. Farther north the pine needles of evergreen forests made the soils difficult to

(a)

(b)

**FIGURE 2.36 Soil types: podzol and black earth.** (a) A podzol in France formed by rain removing nutrients from sandy material. Such soils, with their ash-gray layer, are common beneath northern forests in Anglo America and Eurasia. Their low nutrient content makes them difficult to cultivate. (b) A black prairie soil in Texas. The dark surface layers have a high organic and nutrient content, contrasting with the lower white section rich in calcium carbonate. These are among the most fertile soils for farming.

(a) Hari Eswaran, USDA; (b) USDA

cultivate (Figure 2.36a). In the South, rainwater and higher temperatures increased the rates of chemical reactions and removed most nutrients. In the Midwest, development beneath prairie grassland (Figure 2.36b) produced the black soils that have proved the most fertile and resilient of all. Farther west, the presence of mountains and arid areas resulted in the formation of few well developed soils.

## Human Impacts

Natural environments operate largely outside human controls, being powered by energy from the sun or Earth's interior. Early in human history, livelihoods depended on the local characteristics of weather, rocks and minerals, landforms, water supply, vegetation, animals, and soils. Even before the development of more effective farming and manufacturing technologies, human activities made major changes in the workings of the natural environment. Burning of vegetation in dry seasons around forest margins encouraged the extension of grassland and the increase of grazing animals—such as the bison in North America that provided meat and hides for Native Americans (see Chapter 9).

### Farming and Erosion

The first farmers used lighter soils where there was not much vegetation to clear. Around 1000 B.C., the use of iron implements made it possible to fell trees on a larger scale and extend farmland into heavy clay soil areas. Both phases of woodland clearance led to soil erosion and changes in the species composition of animals and plants. In Asia and elsewhere, the construction of terraces extended the area that could be cultivated to steep slopes at the expense of forest cover.

### Industrial Revolution

The industrial revolutions after A.D. 1750 resulted in a greater intensity and scale of human intervention in natural environments (Figure 2.37). The rate of soil erosion from deforested slopes and plowed fields increased, extending to areas of the world that were opened up in the 1800s and 1900s by settlers from the core countries with the aim of producing commercial crops for sale to the core. Factories poured their wastes into the atmosphere and rivers. Pollution became a health problem in many industrial cities. The building of cities caused rain to run off the surface more rapidly, adding to the numbers of floods downstream.

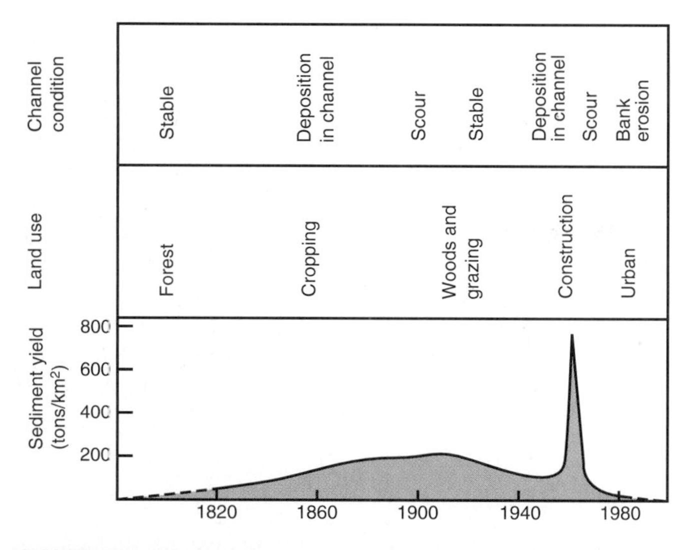


**FIGURE 2.37 Human impacts on the environment.** The example of a watershed near Washington, D.C., that has been subject to land-use changes over the last 200 years. Describe how clearance for farming, land abandonment, and urban expansion affected the sediment flow and hence changed rates of erosion by the rivers of the area.

Reprinted from "A Cycle of Sedimentation and Erosion in Urban River Channels" by M. G. Wolman from Geografiska Annaler, 1967, 49A:385-395 by permission of Scandinavian University Press.

Increasing demand for fish by a rising world population led to overfishing of the oceans; increasing demand for timber and land led to the cutting of tropical rain forest.

### Atmospheric Pollution

Further potential problems built up as the materials entering the atmosphere from industrial processes increased. The carbon gases from burning coal and gasoline exceeded the amounts that natural systems could absorb. In the 1980s, fears arose that the increasing atmospheric levels of carbon gases would encourage greater heating of the air near the ground around the globe to cause global warming.

The growth of the "ozone hole" over Antarctica is a different and potentially greater problem for continuing human existence on Earth. Ozone is a gas that is concentrated in the upper atmosphere above the sector in which weather movements occur. The gas forms a protective shield, intercepting incoming ultraviolet rays from the sun that would burn up plants and animals. In the 1980s, it became clear that chlorine gases, including the human-made chlorofluorocarbons used in refrigeration systems, gradually permeate upward and destroy ozone by chemical reactions. The reactions are most intense during the polar winters over Antarctica: the October measurements (at the end of Antarctic winter) show the greatest depth of ozone depletion—the hole. Elsewhere around the globe the ozone layer is thinning slowly. Governments have agreed to end the use of the ozone-depleting gases, but the effects will last for several decades.

While global warming and ozone depletion may have worldwide effects, acid rain is a regional problem that affects areas downwind of major urban-industrial areas, especially in core countries. Sulfur and nitrogen gases from power stations and vehicle exhausts react with sunlight in the atmosphere and return to the ground as acids. Soils and lakes that are close to pollution sources and are already somewhat low in plant nutrients suffer first. The phenomenon is a growing menace around new industrial areas in peripheral countries.

### Desertification

**Desertification** is the destruction of the productive capacity of an area of land. It occurs as deserts, such as the Sahara, expand (see Chapter 3) but also in humid areas where vegetation is removed or the soil is stripped away. Environmental contamination, as when mining wastes foul the land or nuclear or other toxic wastes leak, also makes land unusable for many years.

## Resources and Hazards

Natural resources and hazards are distributed unevenly around the world and have important influences on regional geography. They unite aspects of physical geography and cultural perceptions of what is useful or presents difficulties.

### Natural Resources

The natural environment provides human societies with materials that they identify as resources and use to maintain their living systems and built environment. Resources valuable to one society are not always rated highly by others. For example, Stone Age peoples used flint and other hard rocks that flaked with sharp edges to make tools and weapons, but such rocks have few uses today. The clay mineral bauxite was ignored until it was found that refining it produced the strong, lightweight metal, aluminum. The recognition and availability of resources is basic to economic development and regional differences.

Natural resources include minerals in the rocks, fertile soils, and water. **Renewable resources** are those that are replaced by natural processes faster than they are used. The best example is solar energy that provides a constant stream of light and heat to Earth. Water is also considered to be a renewable resource since it is recycled from ocean to atmosphere and back to the ground and oceans. Many countries use less than 20 percent of the water falling on their territory, returning much of that to natural systems after use. All renewable resources are, however, ultimately finite in quantity and quality. This fact limits irrigation-based development in arid countries, for example, along the Nile River valley and in the western United States.

Other resources are **nonrenewable:** once extracted, there is no more left. Nonrenewable resources include the fuels and metallic minerals available in rocks. Technological advances, however, assist in finding new sources, extracting sources that were once thought to be uneconomic, and recycling metal products. Such technologies extend the lifetime usefulness of nonrenewable resources.

### Natural Hazards

The natural environment poses difficulties and challenges for human settlement in **hazards** such as volcanic eruptions, earthquakes, hurricanes and other storms, river and coastal floods, and coastal erosion. Hazards interrupt human activities to a greater or lesser extent, although they seldom deter humans from settling or developing a region if its resources are attractive. For example, people are drawn to living in California or the major cities of Japan, despite the likely occurrence of earthquakes. As major earthquakes in 1994 and 1995 showed, it is impossible to predict precisely the location or timing of such events or to construct highways or buildings that resist the largest shocks. Similarly, people continue to live and work in areas liable to hurricane damage or river flooding. In areas such as the Mississippi River valley and the lower Rhine River valley in the Netherlands, protection walls have been built to designs intended to cope with all but the worst floods. The highest levels of flooding, as in 1993 along the Mississippi and in 1995 along the lower Rhine, may overtop these walls.

Hazards cause loss of life and destruction of property, but the costs of protection against hazards are also high. Most protection is provided in core countries, where hazards cause the least loss of life and the greatest damage to property. Many peripheral countries have few resources available to construct protective measures against natural hazards and often suffer major losses of life. Comparisons between the few people killed when hurricanes attack the United States and the high number killed in the hurricane affecting Central America in 1998 illustrate this point.

## HOW MANY PEOPLE CAN EARTH SUPPORT?

Concerns about whether Earth and its resources can continue to support its growing populations have been expressed for many years. In the early 1800s, English economist Thomas Malthus predicted that world population would rise faster than the agricultural production needed to feed it. He lived at a time of population increase that had not been previously experienced but before industrial processes raised agricultural productivity. Since his time, environmentalists and scientists have often placed hard upper limits on the numbers of people Earth can support, while population experts and economists have been more optimistic in view of technological, political, and cultural changes.

Estimates of Earth's carrying capacity vary because they make different assumptions. Recent estimates focus on natural resources (soil type and climatic conditions) and production systems (impacts of technology, power, capital, infrastructure). An estimate in the late 1600s of 13.4 billion people is similar to several in the 1990s, although 1994 estimates ranged from 3 billion to over 44 billion.

Questions need to be asked in assessing the level of Earth's population that might outstrip the resources available and people's ability to manage and distribute them. These questions arise from elements covered earlier in this chapter.

- **Political questions.**

  What sort of political system might resolve internal and external conflicts among countries?

  Will organized violence continue to waste human lives and resources?

  How will domestic and international trade arrangements work out?

- **Economic questions.**

  What level of well-being is expected? Are estimates based on the continuation of a small proportion of rich people and many poor or on increasing numbers of rich people?

  Will more subsistence farmers (who make fewer claims on the natural environment) move toward urban-industrial areas?

  What levels of technology are to be used in growing food, manufacturing goods, and providing services?

- **Cultural questions.**

  How will average family sizes change?

  What support will be provided for young and old?

  How far are people wedded to current habits?

  Are people willing to adopt ways of life that include vegetarian diets, cycling to work, and spending more tax money on schools and health care?

- **Questions about natural environments.**

  Are people willing to trade off producing just enough food to feed the population in order to maintain a clean and wholesome environment with conserved wilderness areas?

  How much natural hazard risk can people accept?

  How long will any prediction last, given the uncertainties over the usage of such resources as water and fish stocks?

For example, recent crop failure and famine in Sudan, northern Africa, resulted partly from drought (natural environ-

**RECAP 2D: Natural Environments of Human Activities**

Natural environments have climatic, rock, landform, plant, animal, and soil characteristics. Human activities modify the operation of natural processes. Natural processes provide resources recognized by human cultures and present hazards to human activities.

The number of people that Earth might support depends on complex political, economic, cultural, and environmental factors. No firm estimate can be made of the number of people Earth can support and at what level

2D.1 How do (a) natural environments affect human activities and (b) human activities affect natural environments?

2D.2 Do natural environments change?

2D.3 Is there a relationship between population numbers and quality of life?

**Key Terms:**

| | | |
|---|---|---|
| climate | erosion | polar biomes |
| tropical climates | deposition | ocean biomes |
| midlatitude climates | ecosystem | soils |
| polar climates | biome | desertification |
| global warming | forest biomes | renewable resource |
| tectonic plate | grassland biomes | nonrenewable resource |
| weathering | desert biomes | hazard |

ment), partly from inadequate investment in water supply facilities and transportation facilities (economics), and partly from civil war preventing international relief workers getting food supplies into what the government considered to be a rebel area (politics, culture). Thousands starved to death, but this was not a matter simply of carrying capacity.

Choices for the future have been summarized in three basic approaches. The "bigger pie" school proposes expanding production through technological advances. The "fewer forks" school emphasizes environmental considerations by slowing, stopping, or reversing population growth. The "better manners" school highlights cultural values through improving the terms on which people interact in political and economic interchanges. Our choices—those of people throughout the world—in what we eat and wear, in our lifestyles, and in our treatment of the environment, will determine the future of humankind on Earth and how well this planet can provide for that future.

## WORLD REGIONAL LANDSCAPES

The characteristic regional geographies of different places throughout the world produce distinctive local landscapes. **Landscape** is defined in various ways. For many people, a landscape is scenery—the visual appearance of a stretch of countryside or town (townscape) that the eye can take in in one view. Artists mostly treat landscapes as an arrangement of rural scenes in a specific composition. Tourists find particular landscapes inviting and go there or repelling and avoid them. For geographers, a landscape may be the typical appearance of a regional environment, such as that of the Mississippi River delta or Florida, including landforms, ecologic relationships, land use, cultural elements such as buildings, and other human works.

In all these uses of the term, there is an element of overview and combining different elements, particularly those of environmental and human significance. Ultimately, landscapes are cultural interpretations of interactions between the natural environment and human history. Landscapes can also have meanings in terms of the economic, political, or cultural systems that produce them or in terms of their effects on the people who live in them. At present, urban landscapes are expanding in all parts of the world, while populations in rural landscapes are declining in proportion. Urban landscapes are the outcomes of secondary, tertiary, and quaternary economic activities, while primary sector activities are more common in rural areas.

In this text on world regional geography, each chapter ends with accounts of a selection of the distinctive and contrasting landscapes typical of the major region being described. These sections seek to integrate and summarize the combination of features described in the individual subregional accounts.

### Urban Landscapes

Urban landscapes are dominated by buildings and transportation routes, although physical relief is also important in providing character and channeling development. Rome is built on seven hills, and New York's landscapes are greatly affected by its central site on Manhattan Island. Special industrial, commercial, government, and residential localities can be identified as distinctive landscapes within urban areas. Historic legacies of buildings reflect the manner and timescale of urban growth, changing economic and social status of the inhabitants, and cultural values (Figure 2.38). Over time, urban landscapes change as older buildings are surrounded or replaced by new ones, a process that depends on the strength of new demands as against conservation movements that attempt to preserve older buildings. Urban landscapes expanded rapidly throughout the 1900s, first in core countries and more recently in peripheral countries, reflecting the spread and dominance of the world economic system through essentially urban functions in manufacturing and financial services

### Rural Landscapes

Rural landscapes depend to a greater extent on coverage of the ground by agricultural land uses, forests, wetlands, and wilderness areas and so contrast with the complex built environments of urban areas (Figure 2.39). Mining and tourist activities are other common features of rural areas,

(a)

(b)

**FIGURE 2.38 Urban landscapes.** Two European landscapes that demonstrate how historic legacies become parts of subsequent urban expansion. (a) In Athens, Greece, the Parthenon Temple and Acropolis hill leave a monument from the 400s B.C., the focus of tourism today, surrounded by the modern city. (b) In Nimes, southern France, the Roman amphitheatre is at the heart of densely packed medieval city buildings.

(a) © James P. Blair/National Geographic Image Collection; (b) © Hubertus Kanus/Photo Researchers, Inc.

although increasingly people in core countries who live in rural areas work in urban areas rather than in rural occupations. Urban areas make intensive use of land by crowding houses, offices, and factories together, leading to high densities of people and high prices of land. Rural areas have more extensive uses of land and lower land prices. Rural landscapes include small village settlements, individual homesteads, occasional factories, the large homes of wealthy people set in extensive estates, and dormitory housing developments. With the expansion of urban areas and urban employment opportunities in the 1900s, many rural areas within reach of major cities retain a rural appearance while being reoriented to city political, economic, and cultural opportunities.

Rural landscapes reflect a variety of economic systems and modes of production. A few regions still exist in which isolated, subsistence cultures determine the landscape through their land uses. Hunting and gathering cultures, largely restricted to the tropical rain forests of interior Borneo and Brazil, together with the Arctic margins, make few changes in the natural environment. Subsistence farmers develop long-term strategies that sustain production,

(a)

(b)

**FIGURE 2.39 Rural landscapes.** What features do the two areas have in common, and what is unique to each? (a) Rice fields in Guangdong province, near the Xi Jiang delta, southern China. (b) Wheat fields near Chadron, Nebraska, USA.

including the building of terraces for rice cultivation in Asia. Commercial agricultural uses, growing crops on family farms, plantations, or communal farms, and the keeping of livestock at varied intensities and densities dominate most of the rural world. Forested and mining landscapes add to the rural variety, while fishing landscapes consist of small coastal ports.

**RECAP 2E: Landscapes**

Landscapes represent a summary of human influences in regions, demonstrated in the built urban environment and in rural landscapes that include farms, forests, and mineral extraction sites and wilderness areas.

2E.1 What are the typical features of urban and rural landscape types known to you?

**Key Term:** landscape

# CHAPTER REVIEW QUESTIONS

1. Which of these is a source of data for geographic studies? (a) Maps (b) Censuses (c) Field studies (d) All of the above
2. Which of the following is the *smallest* scale? (a) 1:10,000 (b) 1:100,000 (c) 1:1,000,000 (d) 1:10,000,000
3. Which of the following is not among the areas of highest population density? (a) Eastern United States (b) California (c) Eastern Asia (d) Europe
4. Independent country governments have the power to do all of the following except (a) promote and protect their peoples in world affairs (b) provide public services (c) encourage economic and social development internally (d) insist that other countries do what they say
5. The organization that took over supervision of world trade in 1993 is the (a) North Atlantic Treaty Organization (NATO) (b) Association of Southeast Asian Nations (ASEAN) (c) Universal Postal Union (UPU) (d) World Trade Organization (WTO)
6. Which of the following is an example of a primary industry? (a) a flour mill (b) your university (c) a wheat farm (d) your local video store
7. Global warming (a) would be the first major climate change in history (b) may be the result of increased burning of coal, oil, and gasoline in recent times (c) is caused by the earth and sun moving closer together in space (d) is of no concern to people
8. Major mountain systems are created at places (a) where tectonic plates collide (b) where tectonic plates slip past each other (c) in the middle of tectonic plates
9. Thomas Malthus wrote that (a) the Industrial Revolution would provide food for the expanding population (b) population would grow faster than food supply (c) food supply was growing faster than population (d) governments should institute birth-control programs
10. All of the following have increased the globalization of the world's economy except (a) a world government (b) multinational corporations (c) global financial services (d) increased world transportation and communications
11. In order to maintain scientific objectivity, geographers should never get involved in political issues of concern to the people of the region they are studying.
    True / False
12. Satellites in geostationary orbits are primarily used to monitor weather.
    True / False
13. The 1994 International Conference on World Population and Development in Cairo, Egypt, concluded that population growth was occurring too slowly.
    True / False
14. Education and the status of women play large roles in the success of policies to limit population growth rates.
    True / False
15. A nation and a country are the same thing.
    True / False
16. The construction industry is a tertiary activity.
    True / False
17. The quaternary sector deals with assembling and disseminating information.
    True / False
18. The 1980s and 1990s have seen increased globalization of financial and data services.
    True / False
19. According to dependency theory, core countries grow rich by taking wealth from peripheral countries, which become dependent on them.
    True / False

20. The largest-scale ecosystem is a biome.
True / False

21. Distance north and south of the equator is measured by ________________.

22. Distance east and west of a line drawn through the Royal Observatory in Greenwich, England, is measured by ________________.

23. A circle joining places of the same latitude is called a ________________.

24. A line joining places of the same longitude is called a ________________.

25. The effect of increasing the cost and time of distance between two places is ________________.

26. Lines on a map that connect points of equal value are called ________________.

27. The process of the spread of people, ideas, or items from place to place is called ________________.

28. The total value of goods and services produced in a country in a year is its __________.

29. The biological differences among human beings constitute __________, while ________ is based on non-biological cultural differences.

30. The cultural implications of one's sex are referred to as ________________.

# AFRICA SOUTH OF THE SAHARA

Chapter 3

**THIS CHAPTER IS ABOUT:**

**Africa in the new world order**

**The range of African cultures**

**Africa's tropical natural environments**

**Four subregions of Africa South of the Sahara:**

- Central Africa: greatest poverty at the heart of the continent
- Western Africa: economic growth slowed
- Eastern Africa: prospects disappointing
- Southern Africa: potential for improvement

**Traditional, colonial, and modern urban and rural landscapes**

**Africa's future prospects**

**World Issue: Famine in Africa**

**Living in Zimbabwe**

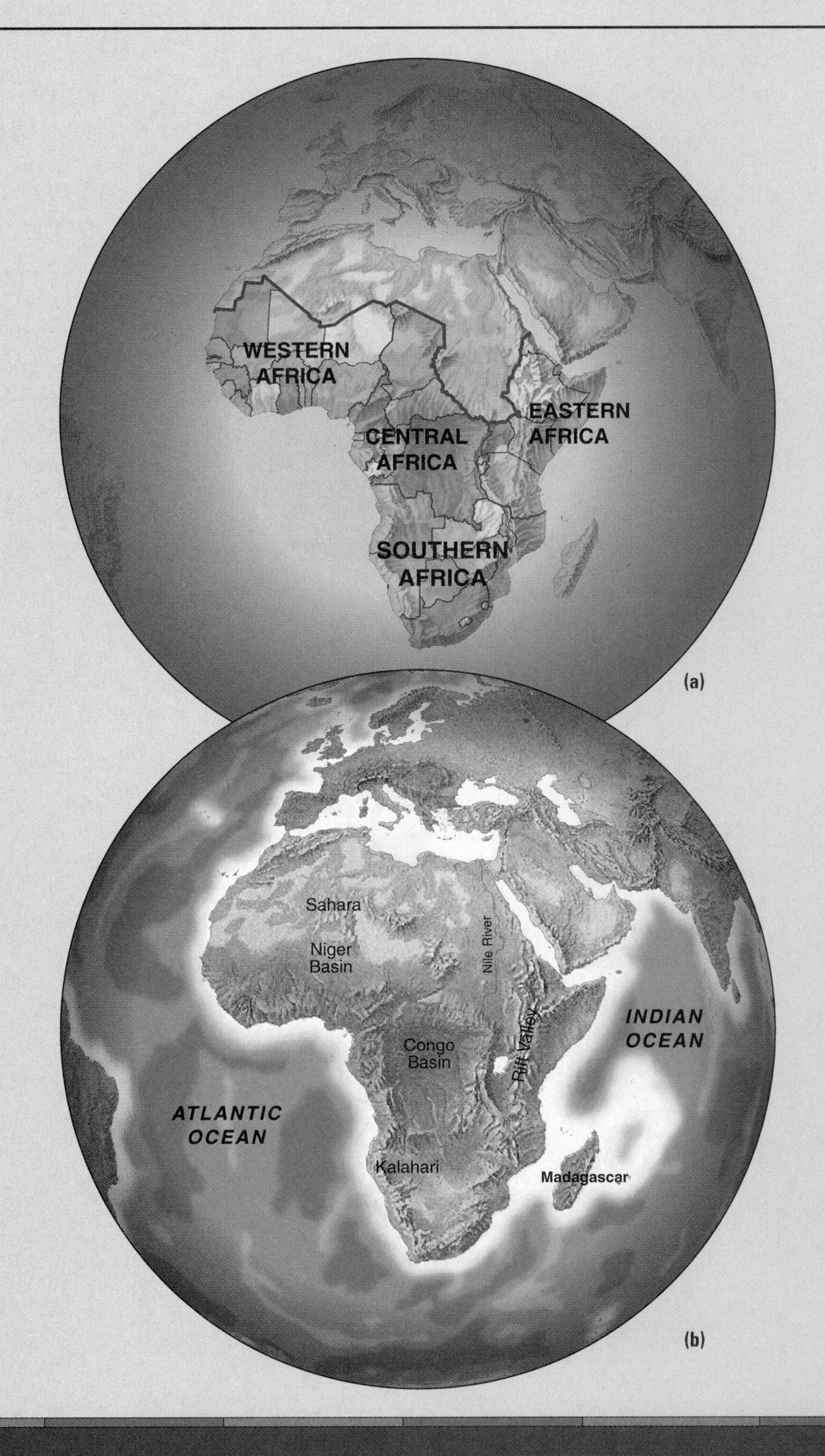


**FIGURE 3.1 Africa South of the Sahara: global features.** (a) Subregions and the countries in each: further details are shown on the maps accompanying each subregion later in the chapter. (b) The major natural environments, showing the contrast between the drier areas to the north and south (browns) and the deep green of the tropical rain forest in the Congo River basin.

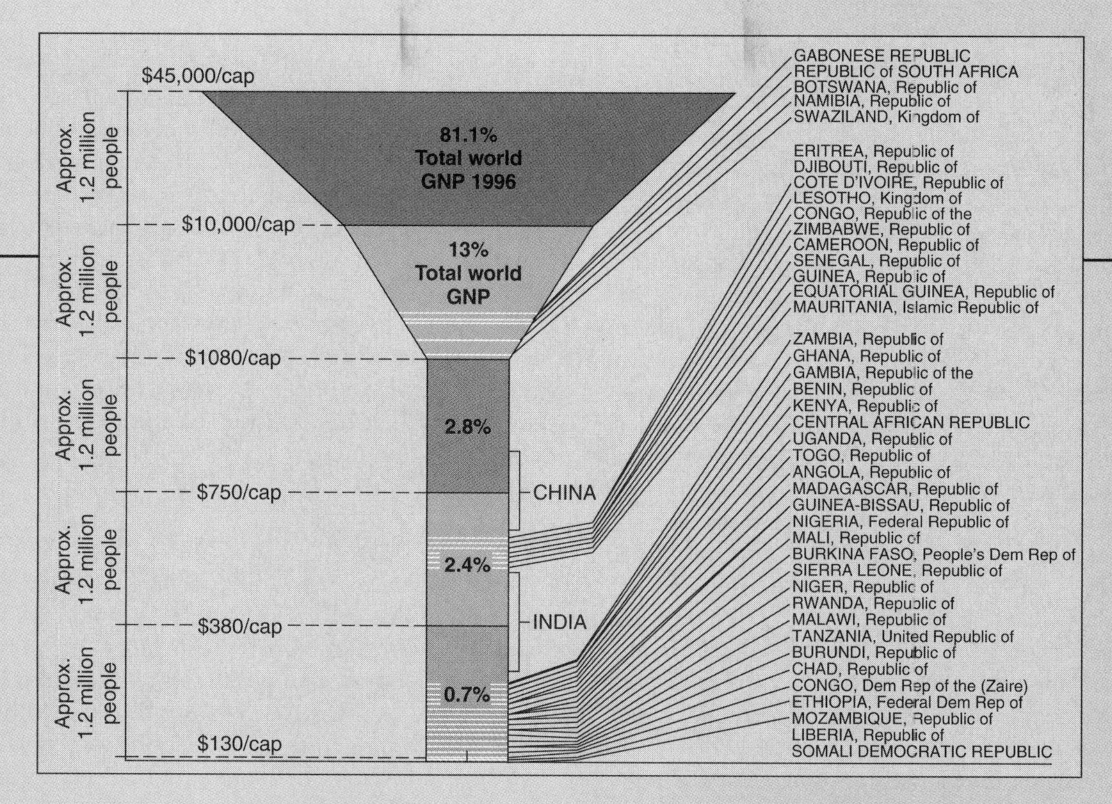


**FIGURE 3.2 Countries of Africa South of the Sahara on the world scale of GNP.** None occurs in the richest-fifth group. The countries farther up the scale have income from oil or metallic mineral sales. The lowly positions of Central African and Eastern African countries contrast with the Southern African countries, while the countries of Western Africa are spaced throughout the list.

Source: Data (for 1996) from *World Development Indicators,* World Bank, 1998.

## AT THE EXTREME PERIPHERY

Africa South of the Sahara (Figure 3.1), once the cradle of the human species, is today the poorest of the nine major world regions. In 1998, it had nearly 10 percent of the world's population, but its 580 million people contributed only 1 percent of the worlds' GNP—down from 2.4 percent in 1980 and 1.8 percent in 1965 (Figure 3.2). The region is the least integrated into the world economic system and the least able to take advantage of prospects for increasing its status within that system. Many of its countries depend on external aid. It is the most peripheral part of the world economy.

African countries are not just the world's poorest; they also fell farther behind the rest of the world in the 1980s and 1990s. Countries in Asia and Latin America that were poorer than several African countries up to the 1970s are now better off. While Africa South of the Sahara receives around 3 percent of total foreign investment to poorer countries, Asian countries receive 40 percent.

Politically, many African countries experience difficulties in finding their place in a wider world. The continent contains the most desperate and politically the most explosive mix of conflicts at a time of rapid population growth, slow economic development, poor environmental management, disrupted social relationships, and inadequate human welfare programs. Many African countries face civil wars and famine.

And yet, Africa contains a richness of cultural life that makes major contributions, for example, to world art and sports. Past contributions to American musical styles and present innovations in popular music, together with striking artistic creations, make links into the everyday lives of people around the world.

African sports teams are admired for the dedication of the Kenyan long-distance runners and the exuberance of national soccer teams. Such achievements, together with involvements at the highest level in United Nations administration, ridicule suggestions that African people cannot be expected to catch up with the rest of the world.

During the 1990s, a note of hope crept into the crisis warnings that were made about Africa South of the Sahara for over 40 years. The end of the Cold War reduced superpower competition and aid, forcing African countries to adopt more open politics and different economic management. After years of the region being ignored by the world's rich countries as having little strategic significance, the United States moved closer to it through new policies set out in 1997 and followed up by President Bill Clinton's tour in 1998. He confirmed that aid would be available to countries adopting democratic procedures and improving their human rights records. It was widely agreed among richer countries that Africa's economic and political collapse should be avoided. More favorable political environments and the need of richer countries for Africa's natural resources prompted such interest.

Negative trends also emerged in this decade. The HIV/AIDS pandemic hit the region, especially in Eastern and Southern African countries; few countries improved their economic performance in the new environment; new civil wars erupted; and the two countries that could be leaders in the region—Nigeria and South Africa—stagnated economically or were more concerned with resolving internal problems. It is unlikely that African countries will see much improvement until well into the new century. They will remain volatile, with many political ups and downs.

Such poverty, economic shrinking, and internal disorder—alongside the positive contributions of Africa—challenge world citizens, as a priority, to examine the background, causes, and possible futures of this region. These are reasons for studying Africa before other parts of the world since they place the geography of other regions in a context. As was indicated in Chapter 2 (see page 50), there are no simple answers to the question of how countries improve the quality of life of their peoples. Each part of the world has to be viewed in its natural, cultural, demographic, economic, and political environments.

Africa South of the Sahara has distinctive boundaries. The northern boundary is drawn along country borders through the **Sahara**, while the remaining boundaries are ocean coasts. Not only does the desert form a physical break that interrupts continuous settlement, but it is also an economic and cultural break between the Arab-speaking Muslim countries along the Mediterranean Sea and the rest of Africa. There are two exceptions to this rule. Although Mauritania is an Arab-speaking country, it is included in Africa South of the Sahara because it lacks a Mediterranean coast and has historic and economic links to Western Africa. Sudan is included in Chapter 4 because of its historic ties to Egypt and their common dependence on the Nile River waters. Data for this region is found in the Reference Section, pages 586–610.

## AFRICAN CULTURES

The distinctive cultural geography of Africa South of the Sahara reflects the region's ethnic and religious makeup and its history of population movements. In Africa, ethnicity is expressed through tribal loyalties.

A **tribe** is a group of people marked out by ethnic characteristics such as a common cultural heritage including a common language (Figure 3.3a). Each tribe is linked to a specific land area that is often smaller than the modern African countries (Figure 3.3b). The boundaries between tribal areas are not, however, always clear-cut. Many changed in the past as the result of migrations stemming from push factors, such as the medieval Muslim invasions and slave trade, or pull factors, such as a perception of better lands elsewhere. The terms *tribe* and *tribalism,* however, are now less favored because of their implications of primitive feuding. *Ethnicity* replaces such terms.

### Tribes and Kingdoms

Within African tribes, family ties are particularly important. The extended family of several generations and distant cousins offering support to each other provides strength at times of economic and political stress. Marriages are generally outside the family unit but are often within the same tribe, establishing further connections over wide areas. Today links remain strong between rural and urban branches of a family. When people move from rural to urban areas, they frequently live with relatives until they find their own niche. As well as family ties, tribal links are strong among craft specialists and age-group organizations, and are fostered by the work of religious shamans. Tribal and family links often occur across country borders, making it difficult to police movements of refugees, drugs, gold, or diamonds.

Before the colonial era, the more powerful African tribes often formed kingdoms controlling extensive geographic areas. Trade patterns involving metals, ivory, cattle, slaves, and craft goods were important in the economies of the West African kingdoms of Ghana and Mali. The Great Zimbabwe ruins in Southern Africa also testify to widespread trade (Figure 3.4). There was, however, a great variety of tribal structures. Tribes lacking a dominant leader operated in small clans based on subsistence agriculture or livestock herding.

For most Africans today tribal loyalties remain strong, predating the colonial acquisition of African territory by European countries during the late 1800s. Such tribal loyalties are often greater than loyalties to the country in which they live, since colonial rulers established the countries and their borders. The internal politics of many African countries continue to be based on tribal loyalties and rivalries. Conflicts arise between tribal customs and new ways resulting from education, exposure to world media, and living in urban rather than rural areas. Such conflicts may inflame tribal loyalties and result in fighting or may cause people to abandon their tribal links.

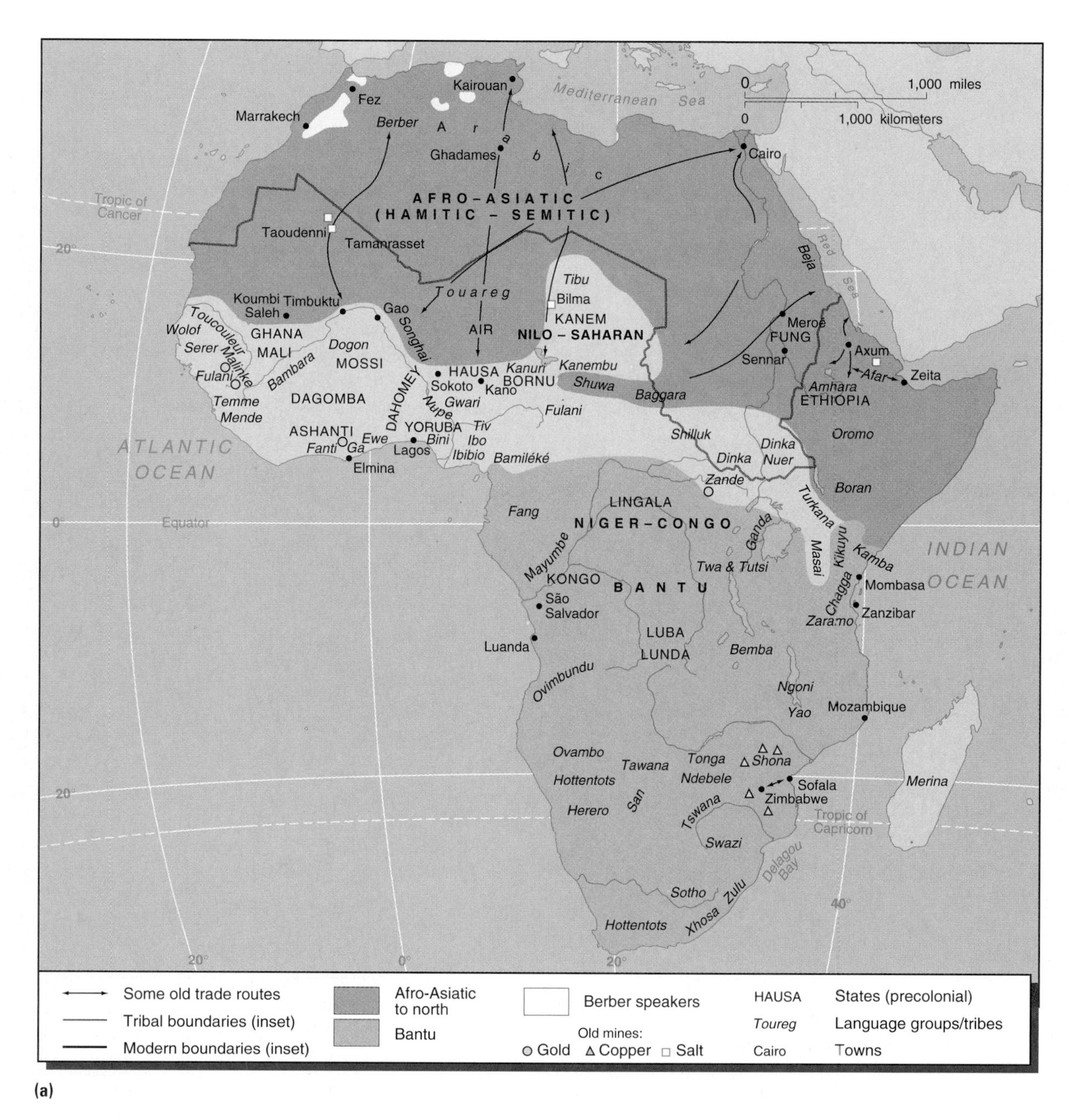


(a)

**FIGURE 3.3 Africa South of the Sahara: cultural features.** (a) Tribes, languages, and precolonial kingdoms. In the zone between that dominated by the Afro-Asiatic groups of northern Africa and that dominated by the Niger-Congo peoples to the south, a diversity of languages exists.

## Religious Influences

African religious beliefs are often linked to tribal loyalties. Before the Muslim invasions and European trading and colonial influences, indigenous African religions were based on local tribal conceptions of deities, often linked to their experience of the natural environment. Such religions are termed *animist* and are still important to many Africans.

In medieval times, Muslim traders and warlords broke through the physical barrier of the Sahara and brought their religion to areas south of the desert. Whole tribes were converted to Islam. Today the northernmost countries in Western, Central, and Eastern Africa have Muslim majorities or substantial minorities. The Muslim push from the north displaced many African tribes southward. Some moved into the equatorial forests, expelling pygmy groups, while, farther south, migrating tribes caused the smaller and less organized groups of peoples living in southernmost Africa to move into harsher environments.

The Christian influence in Africa is long-standing but was limited in area until the 1800s. The Coptic church in Ethiopia

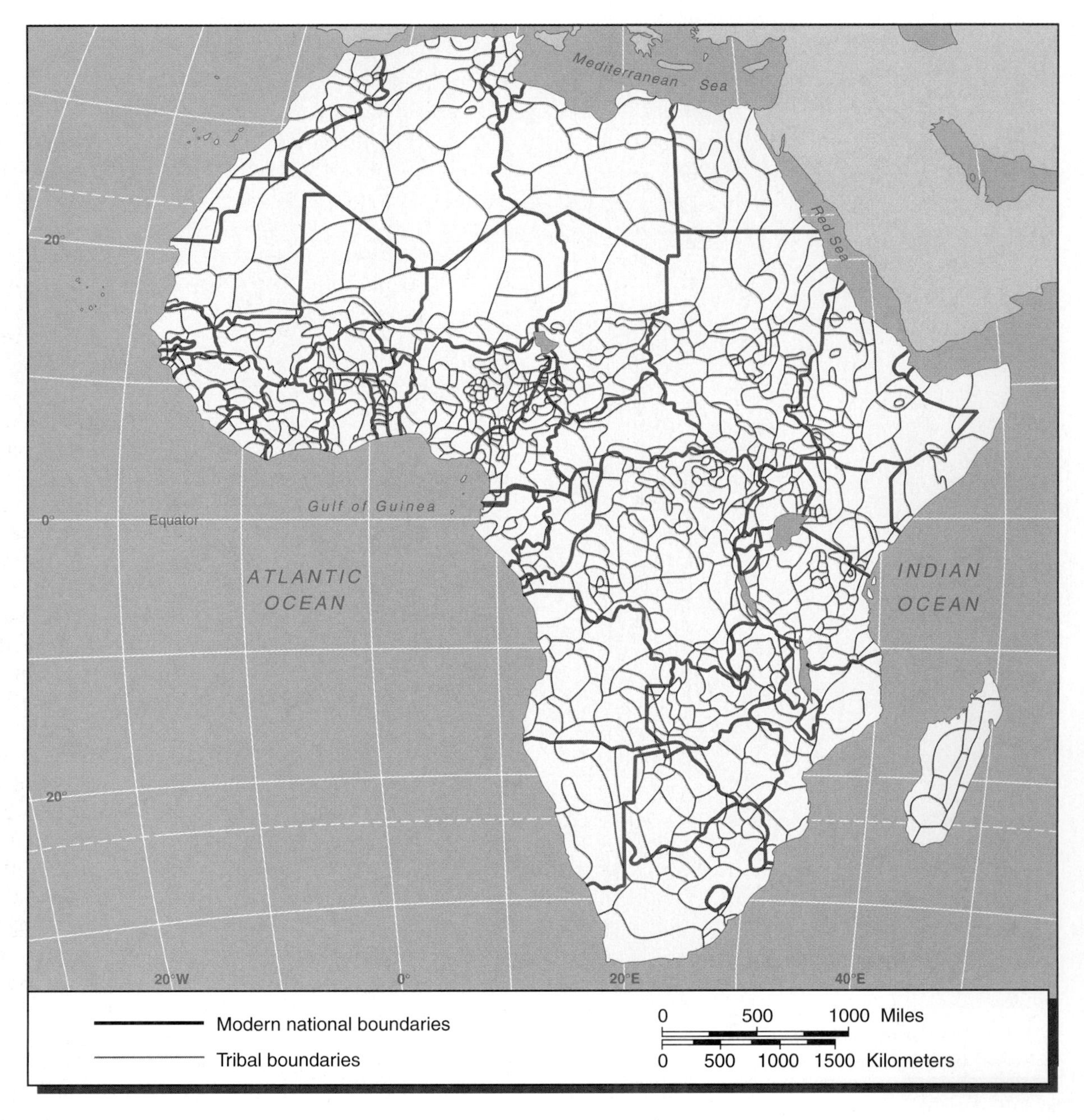


**(b)** The relationship of colonial-imposed boundaries of the modern countries to tribal areas.

Source: (b) From *World Regional Geography: A Question of Place,* by Paul W. English and James A. Miller, 3rd edition. Copyright © 1989. Adapted by permission of John Wiley & Sons, Inc.

traces its heritage to the first century A.D. It was only when Europeans set up coastal missions from the 1400s and penetrated the interior during the 1800s that Christianity extended its influence to the whole region. Roman Catholic priests came with Portuguese, Spanish, French, and Belgian traders and colonists, while Protestants came with British, Dutch, and Germans. Most of the Christian influence is in the area south of the Muslim countries, and tensions among Muslim, Christian, and traditional religious beliefs remain important in African tribes and countries that have significant Muslim and Christian groups.

## Colonial Impacts

Occupation by colonial powers often lasted less than 100 years, from the late 1800s to the mid-1900s. In this relatively short period, European colonial powers tried to impose their ways on Africans. Colonists often came for short periods merely to extract the mineral wealth or to establish tropical crop plantations that produced food and raw materials for their industries at home. Africans remained a largely rural people in which extended family relationships and tribal customs continued to rule. Other colonists remained longer and

**FIGURE 3.4 Great Zimbabwe, Zimbabwe, Southern Africa.** The ruins record the existence of a major trading center from around A.D. 1100. It was based on a sophisticated political organization and economy before European exploration and colonization. Each layer of rock in the walls is approximately 7 to 10 cm (3–4 in.) high.

© Michael Bradshaw

had greater impacts on local lives. Christian mission schools and hospitals provided a source of Western education and health care. The British, and to some extent the French, encouraged European farmers and miners to settle permanently. They invested more than other colonial powers in building roads and railroads, schools, and hospitals, and in setting up a civil service, even though such facilities were aimed mainly at improving the flow of raw materials to home industries.

Some tribes had strong leaders who either worked with the colonial powers or fought against them; in either case, such strong leaders prompted moves toward independence and took over government when colonial powers left. Tribes with weaker leadership acquiesced in the colonial period and give way to other tribes following independence. Colonial governors commonly used and enhanced rivalries between tribes to achieve their ends.

## Independence Outcomes and Prospects

When the colonial powers left at independence, some countries were ruled by people from a dominant tribe. They often generated the hatred of other tribes by attempts to repress them. For example, the minority Tutsi tribe in Burundi and Rwanda assumed repressive dictatorial powers but was itself subject to atrocities as the majority Hutus took power in Rwanda or rebelled in Burundi. Other countries were more evenly split by tribal interests, and civil war became common. For example, the 1960s civil war in Nigeria stemmed from tribal differences and rivalries that began before but were intensified during colonial occupation. The tensions between the Muslim north and the rest of Nigeria continued to be significant at the end of the 1900s as military rulers from the north of Nigeria maintained their hold on political power within the country.

Although many newly independent African countries disliked their colonial borders, the alternative of renegotiating their political areas raised worse prospects. Within a few years of the main round of independence, the Organisation of African States (OAS) agreed to keep the established borders. Some country amalgamations were tried among former French colonies in Western African and former British colonies in Eastern and Southern Africa but none persisted.

The future of many African countries depends on reconciling racial, tribal, and religious differences. The first free elections in the Republic of South Africa after apartheid ended in 1994 were threatened by Zulu tribal groups claiming special rights, as well as by the most conservative white people. To date these tensions are being resolved politically. The many examples of intertribal rivalries leading to strife and economic breakdown in other African countries provided the new South African government with examples to avoid and a stimulus for working together.

In the 1990s, intertribal cooperation occurred in experiments that tried to persuade peoples of different tribes and religious affiliations to work together. Examples include the attempts at regionalization of government in Ethiopia along ethnic lines, the multitribal governments in Uganda and Zimbabwe, and the government of national unity in the Republic of South Africa. More examples of tribal rivalries and relationships will be examined as specific cases are studied in the rest of this chapter.

At the same time that tribal rivalries remain significant, the future of both tribal groups and social customs in Africa are under threat as rural ways and traditional economic patterns give way to urban living and commercial links into the world economic system. Urban areas are places where many people loosen their ties with traditional cultural ways or change their religious allegiance. In particular, westernized modes of life, often symbolized by the material cultures of clothing, food, owning cars and domestic gadgets, and leisure activities, have become the goals of many Africans.

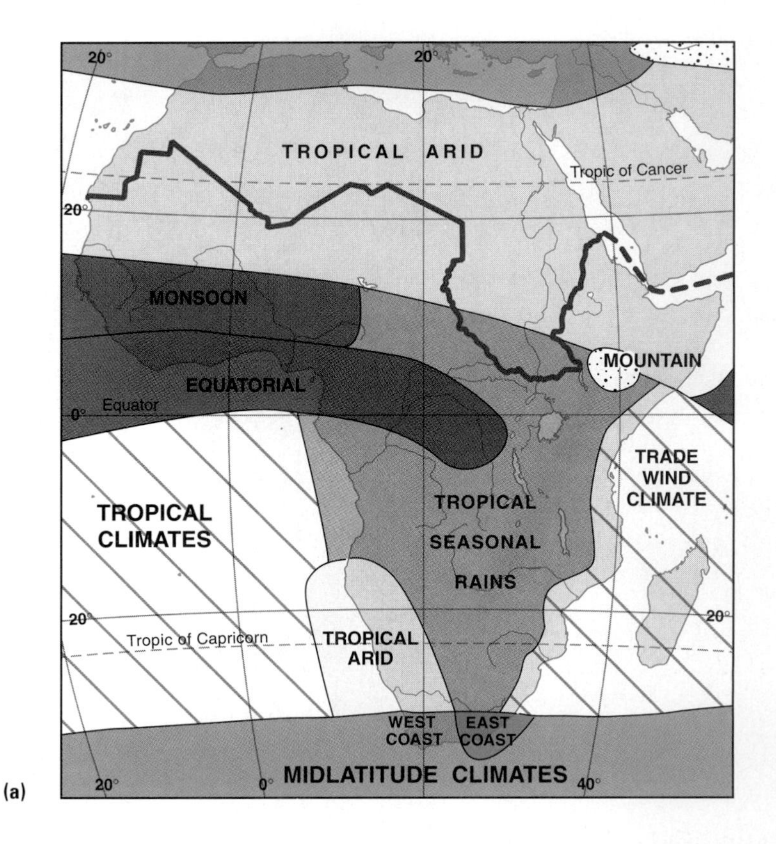


(a)

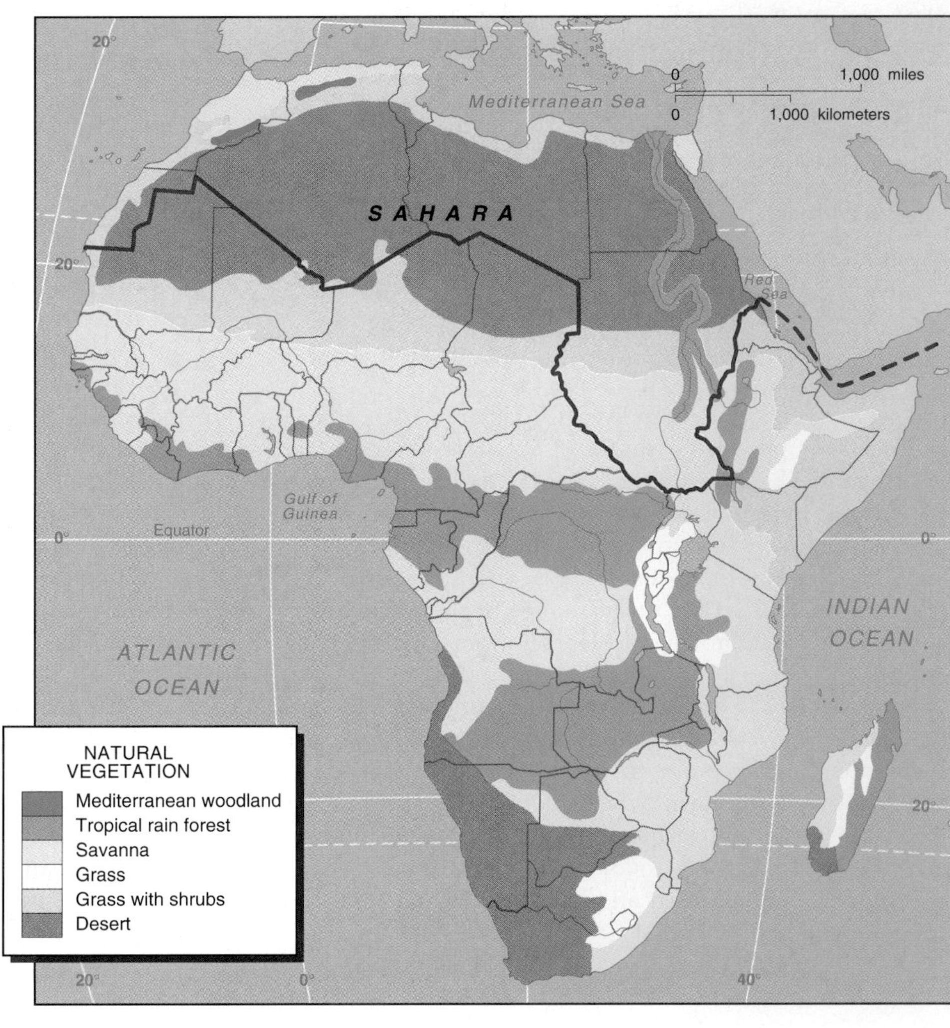


(b)

**FIGURE 3.5 Africa South of the Sahara: climatic environments and natural vegetation.** (a) Climatic environments. Note the transition from equatorial to arid conditions north and south of the equator. (b) Natural vegetation. How far does the natural vegetation follow the pattern of climates?

## NATURAL ENVIRONMENT

The natural environments of Africa South of the Sahara have generally been perceived by Westerners as negative. Images of the impenetrable tropical rain forest, the difficult communications, poor soils, and the prevalence of diseases delayed external colonial interest in this region until the late 1800s. Such views ignored significant positive aspects of the natural environment. Many large mineral deposits and areas with farming potential remain largely unexploited, and the climate, fauna, flora, soils, and water resources provide potential over much of the continent. Today revised perceptions and fuller exploration of the natural environment and resources are vital considerations for the peoples of countries trying to catch up with the rest of the world.

### Tropical Climates

The climates of Africa South of the Sahara—apart from the extreme south—are tropical in character, ranging from the tropical arid environment of the Sahara through seasonal tropical to equatorial climatic environments (Figure 3.5a). The southern tip of the continent has warm temperate climates.

The **equatorial climatic environment** dominates the basin of the **Congo (or Zaire) River** in Central Africa (Figure 3.6). Temperatures remain high all year and rain comes in all months. The high rainfall causes the Congo River to carry the second highest volume of water in the world to the oceans (after the Amazon in South America). Although Africa has a large area close to the equator, the extent of the equatorial climate is restricted by the presence of high plateaus in Eastern Africa. The plateau areas near the equator have lower rainfalls than the Congo River basin and may suffer droughts outside of the rainy periods in April and November. In Western Africa, the coastal areas have a **monsoon climatic environment** with contrasting wet and dry seasons. Flows of moist air from the Atlantic Ocean to the south of Western Africa bring heavy rains, concentrated on the coastal areas from July to September. The moist airflows give way in the rest of the year to dry northerly harmattan winds from the Sahara.

Most of Africa South of the Sahara has a **tropical seasonal climatic environment** that receives rains in the season when the sun is high. Such rains are not as heavy as those in the monsoon areas of Western Africa. They are most reliable close to the equatorial climatic areas, but all areas are subject to variations from year to year. The dry seasons increase in length and the rains become more variable as distance increases from the equatorial climatic areas toward the tropical arid areas. The **tropical arid climatic environments** of the Sahara, Kalahari, and Namib deserts have little rain at any season. The little that falls is rapidly returned to the atmosphere by high rates of evaporation.

Winters become cooler toward the southern tip of Africa in the small zone of midlatitude climates. Winter cold in southwestern Africa comes from a combination of heat loss through clear skies in the dry climatic environment and the cooling of air above the cold ocean current along the Atlantic coast.

**FIGURE 3.6 Southern Africa: weather patterns.** A space shuttle photo shows a clear distinction between the cloudy (rainy) zones along the equator in the north and near the southern coasts. Moist air rises near the equator, condensing into clouds and spreading northward and southward at higher levels in the atmosphere. The air then descends at around 30 degrees of latitude, drying out and preventing cloud formation over the deserts. To the south of the continent, the swirls of midlatitude weather systems and frontal clouds over the ocean move from west to east.

NASA

### Changing Climates

The present arrangement of climatic environments in Africa is part of a sequence of **changing climates.** Over periods of a few decades, the southern boundary of the Sahara shifts north and south by up to 150 km (100 mi.). This affects a broad zone of very low rainfall, known as the Sahel, in which annual amounts are critical to plant life and human activities. Drier periods cause the grasses covering old sand dunes to die and expose the sands to movement by winds. Wetter periods cause the grasses to extend their coverage.

Over longer periods it is clear that the Sahara boundary shifted greater distances. During the last 5,000 years, the Sahara has extended and become drier. Evidence for the Sahara's smaller extent before that includes the dry lake and

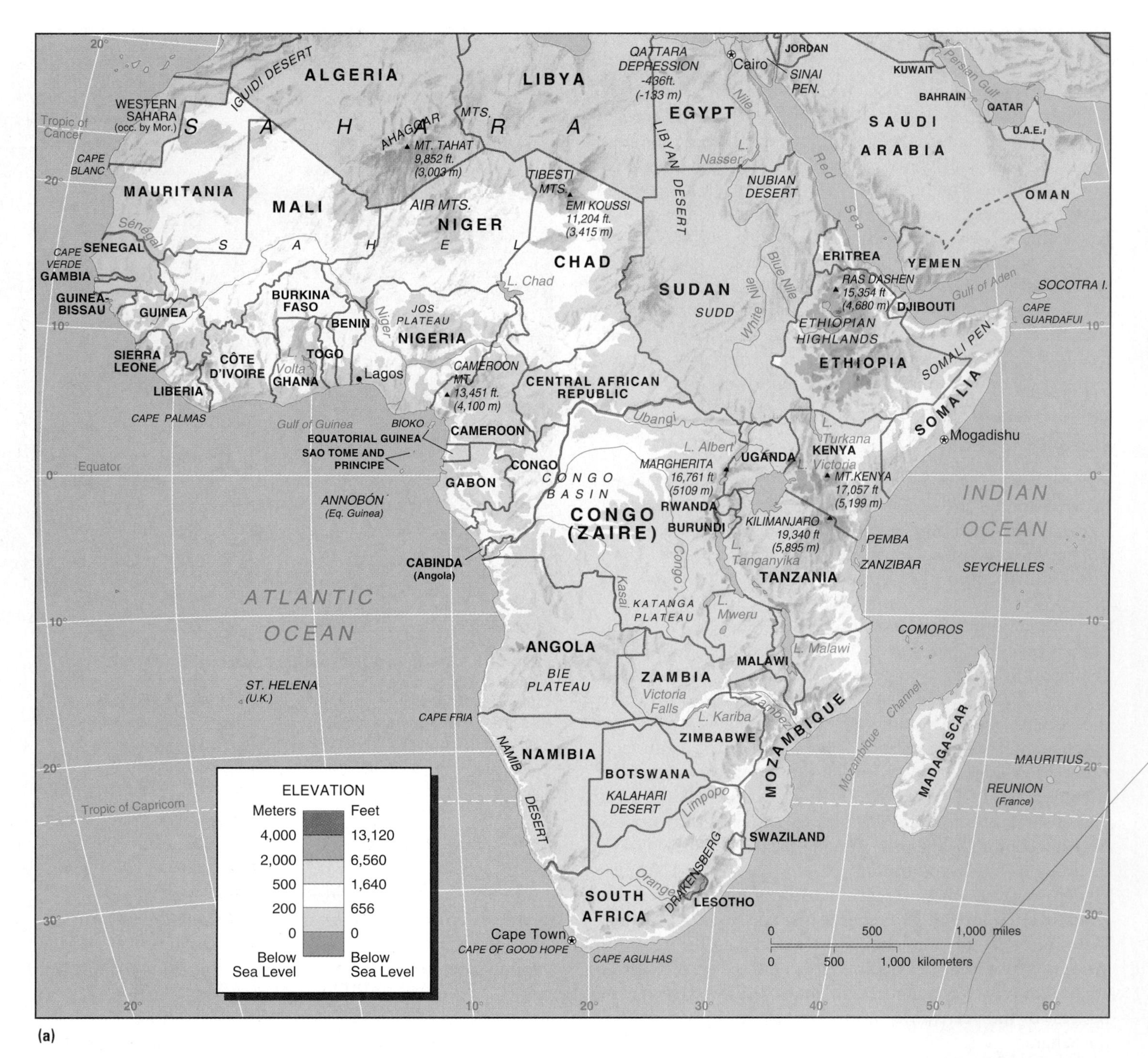


(a)

**FIGURE 3.7 Africa South of the Sahara: major geologic features.** (a) Map of relief (elevation) and rivers. Identify the highest points, which are mainly of volcanic origin. (b) Africa and tectonic plates: what is the significance of divergent margins (arrows moving apart) and convergent margins (arrows moving together)? (c) Africa at the center of the former southern continent of Gondwanaland. In the last 200 million years, the continents have moved apart as oceans opened along constructive plate margins.

riverbeds around Lake Chad. Remains of human communities that once inhabited the now arid areas are common. The desiccation forced large numbers of people southward and into the Nile River Valley around 3000 B.C., giving the Pharaohs greater power (see Chapters 1 and 4). The margins of the Namib and Kalahari deserts in Southern Africa experienced similar fluctuations, but the numbers of people living there were fewer than along the southern edges of the Sahara and, therefore, left fewer relict features of human occupation.

As the deserts retreated or expanded, the areas affected by seasonal or equatorial climates occupied more or less land

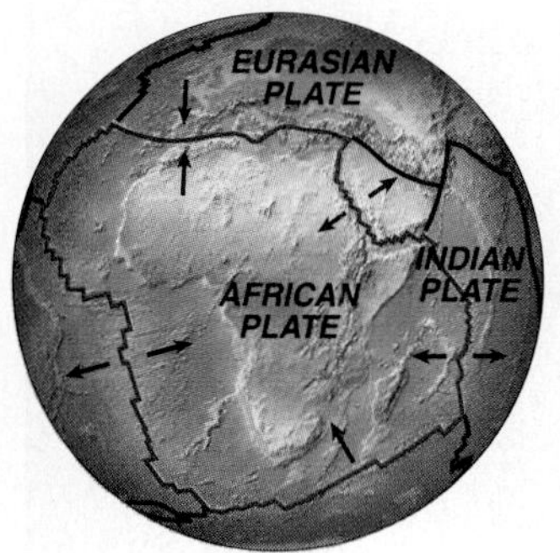


(b)

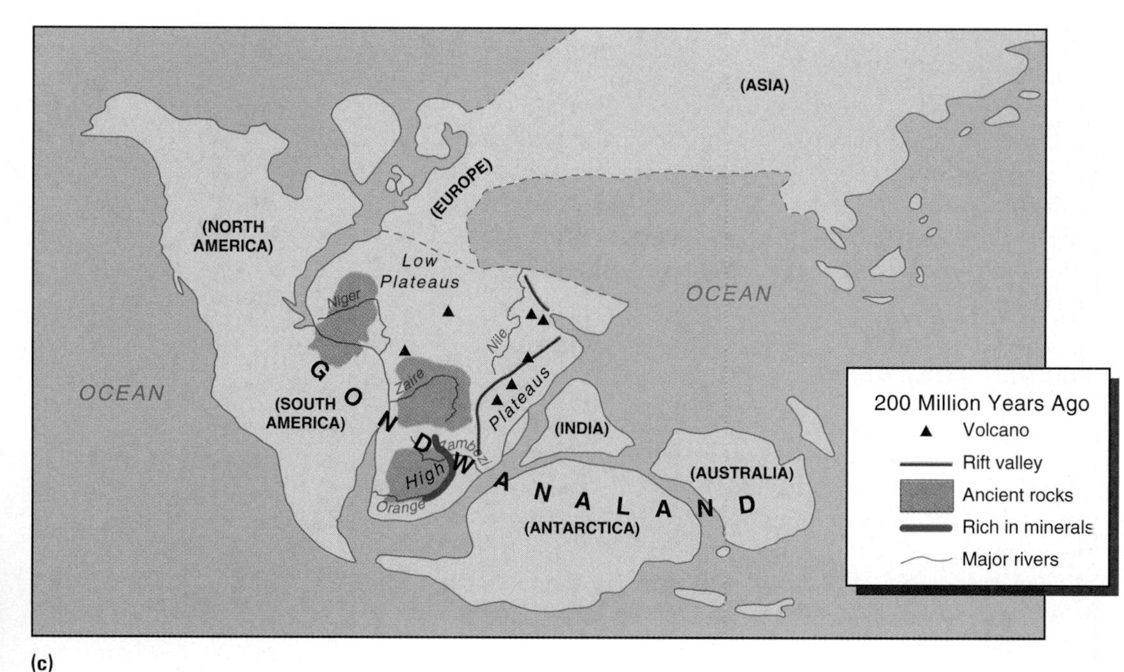


(c)

respectively. Such shifts affected patterns of human occupation, including the extent of tribal lands, but little evidence exists concerning specific changes.

Uncertainty over the climatic future fuels debates about whether the drying of the continent is a natural or a human-induced phenomenon. While humans can do little about the natural changes, it might be possible to modify their activities and so reduce the impacts of change. For example, where the expansion of desert conditions resulted from overgrazing or the removal of vegetation for cultivation, some African countries are now implementing tree-planting programs to reverse the trend toward desertification.

## Ancient Rocks, Plateaus, and Rifts

African landscapes are dominated by extensive plateau surfaces on ancient rocks (Figure 3.7a). **Plateaus** are elevated areas with relatively flat tops. In places, more recent sedimentary layers that give variety to the scenery cover the very ancient rocks. The major rivers such as the Niger, Nile, Congo, Zambezi, and Orange flow across the plateaus. Navigation along them is made difficult in places by seasonal variations of flow related to rainfall inputs or by waterfalls and rapids where the rivers descend from one plateau level to the next or from a plateau level to the coast. Where possible, rivers such as the Congo and the Niger are much used for internal transportation.

The main reason for Africa's monotonous plateau landscapes is that most of the continent is not crossed or bounded by tectonic plate margins (Figure 3.7b). Only in the extreme north (outside the present region) does the convergent margin (see Figure 2.32) along the Mediterranean Sea bound the continent. Some 200 million years ago, Africa was at the center of a huge worldwide continent (Figure 3.7c). The world continent then broke apart, and its fragments moved to form the present continents, separated from Africa by divergent plate margins. This maintained Africa as a continent of ancient rocks and relatively small geologic changes over very long periods.

Geologic activity is more evident in some parts of the continent. Around 50 million years ago, Africa crashed into the Eurasian plate, forming the Atlas Mountains in the far northwest and the Alpine mountain system across southern Europe. More recently, the rift valleys that cross Eastern Africa from north to south and the associated groups of volcanic mountains are linked to the divergent plate margin activity along the line of the Red Sea. **Rift valleys** lie along two zones where Earth's crust has arched, pulled apart, and broken, with the central part collapsing. They may signal the beginnings of a new plate margin and the opening of a new ocean. So much is suggested by the links from the East African rift valleys across Ethiopia into the southern end of the Red Sea, which geologists identify as a new ocean that is already widening. The rift valleys are marked by deep, elongated lakes along their length and by volcanic activity along their margins. The highest mountains in Africa include **Mount Kilimanjaro** (5,895 m, 19,340 ft.) and Mount Kenya in Eastern Africa, which are extinct volcanoes close to the rift valley. The **Ethiopian Highlands** are partly built of lava flows, also linked to rift valley volcanic activity. The **Cameroon Highlands** on the Nigeria-Cameroon border include active volcanoes that are distant from plate boundaries and rift valleys and result from local melting of deeply buried rocks.

## Ancient Landscapes

The very stable plateaus of Africa South of the Sahara were molded by the action of running water and wind on the underlying rocks. The surfaces developed slowly over long periods of time marked by little geologic movement and the predominance of seasonal tropical climates. The distinctive African landscapes of plateaus carved by planation contrast with the deep valleys incised by powerful rivers or glaciers in midlatitudes. The planation results from a combination of factors. Consistent high temperatures and the availability of water in deep soils for much of the year lead to rapid rock breakdown by chemical action. During the dry season, the grasses and small shrubs die and the soil surface dries so that the particles become detached. At the start of the wet season, the combination of bare, loose soil and water carries fine sand and clay across the landscape and into the rivers. This process lowers the landscape as a whole and leaves individual hills with bare rocky sides, or **inselbergs**, as prominent landforms (Figure 3.8). Similar rocky slopes characterize the steps from one plateau level to another.

**FIGURE 3.8** A **kopje,** a type of inselberg occurring commonly in Zimbabwe and other parts of southern Africa. The rocky outcrops have steep sides and a sharp angle with the sandy material at their base. The steepness is maintained by chemical weathering where rock and soil meet.



Although climate change affected much of Africa, long-term changes did not greatly alter the set of processes producing the extensive plateaus. The most recent glacial phases elsewhere in the world produced shifts in the climate zones in Africa but did not replace the dominant landform processes and left few glacial landforms, except on the highest mountains. The landscapes of the desert areas were probably formed under more seasonal rainfall conditions during periods of greater humidity. Increasing aridity led to the desiccation of soil and drying of the rivers. The wind then blew away the finest soil components and concentrated the sandy fractions into large dune seas. The landscapes of the equatorial regions were also subject to more extensive seasonal conditions as the climate zones shrank toward the equator during glacial advances.

## Forests, Savannas, and Deserts

The natural vegetation of Africa South of the Sahara follows closely the patterns established by the climatic regime (see Figure 3.5b). Equatorial climatic areas are covered by dense **tropical rain forest** containing a huge variety of tree and other plant species and of birds and insect species (Figure 3.9). The seasonal climate areas between the deserts and forests are characterized by **savanna grasslands**, in which there is varying amounts of tree cover. The savanna grasslands are noted for their large plant-eating animals such as elephants, giraffes, zebras, and a variety of deerlike forms (see Figure 2.35a), together with their predators, the lions, leopards, and wild dogs. Arid areas produce **deserts** with little vegetation growth and a few small animals.

While climate type and the distribution of natural vegetation are closely linked, there are several anomalies, especially where savanna interrupts tropical rain forest in humid areas. Some of the anomalies are explained by changing climate or by poor soils on which trees will not grow. Others result from human activities. For centuries, the annual burning of grasses by Africans prevented trees from establishing themselves and expanded the grassed savanna area—at the expense of woodland areas and savanna with more trees. This in turn encouraged the expansion of herds of large grazing and browsing animals that provided food and hides for the peoples in the region.

(a)

(b)

(c)

(d)

**FIGURE 3.9 Africa South of the Sahara contrasting ecologies.**
(a) Sossusulei, Namibia. The barren surface, sparse vegetation, and sand dunes of desert. (b) Near Bandiagara, Mali. The Sahel zone where people gather around a well amid signs of increasing drought in dead and truncated trees and uncultivated fields. (c) Masai Mara Game Reserve, Kenya. Grazing wildebeest, gazelle, and zebra in savanna grassland with scattered trees. (d) Congo (Zaire). Small village by a river in the heart of tropical rain forest—an ecosystem based on plentiful rainfall.

(a) © Betty K. Bruce/Earth Scene; (b) © Michael A. Dwyer/Stock, Boston; (c) © Michele Burgess/Stock, Boston; (d) © Torleif Svensson/The Stock Market

The majority of soils in Africa, especially in the tropics, are poor in nutrients as the result of rapid chemical weathering and removal of the nutrients by water flowing through the soils. They are widely workable for agriculture, however, if care is taken to cope with the high clay and iron contents that can turn exposed soil into rock-hard laterite in the dry season. Areas around some of the volcanoes are more fertile because of the ejection of rocks rich in the nutrients that support plants.

## Resources

Africa has a wealth of natural resources, the economic implications of which are considered in each of the subregional studies. The ancient rocks of the African continent contain some of the world's most significant mineral deposits, including iron, copper, uranium, gold, diamonds, and bauxite (the ore of aluminum). Many African countries depend for much of their foreign income on such minerals, although the resources are distributed unevenly and much exploration still remains to be done. Large deposits, such as the iron ores of Equatorial Guinea, remain unused at present because of the lack of internal transportation.

Coal is mined in some countries of Southern Africa. Up to the 1990s, known deposits of oil and natural gas were limited to a few land areas of coastal subsidence and sediment accumulation in areas such as the Niger River Delta. The lack of fossil fuels proved a problem for many African countries trying to industrialize, and high world market prices for energy restricted their economic growth. Exploitation of fuel resources was determined mainly by the demands of former colonial powers or multinational mining corporations, for whom the production of fuels for local industry was not an important factor. In the 1990s, evolving offshore oil exploration technology identified major oil fields along the Atlantic coasts of Africa.

Angola, in particular, is becoming a major world producer as most of the world's major oil corporations invest there.

The tropical climates make it possible to grow a variety of distinctive subsistence crops, including yams, rice, cassava, maize, millet, and sorghum. Commercial crops such as bananas, cocoa, coffee, tea, palm oil, rubber, cotton, tropical fruits, and peanuts are exported to the core countries in midlatitude areas. Forest resources were extensive in the equatorial countries, and plentiful fish resources occur in the major rivers and in the areas of cold ocean currents off northwestern and southwestern Africa. Both forests and fisheries are experiencing increased rates of exploitation for export commodities.

Water resources range from the plentiful amounts in the Congo River basin to many areas where they are poor or uncertain because of seasonal rainfall, frequent droughts, and continuous aridity. The major rivers with their rapids and waterfalls provide potential for generating electricity, but this potential is only partially developed.

Africa also possesses natural resources that form the basis for tourism. Such resources include sunshine, coastal beaches, and large numbers of animals now protected in the national parks of countries such as Kenya, Tanzania, Zimbabwe, Botswana, Namibia, and the Republic of South Africa. Tourist interests also center on scenic wonders such as the **Victoria Falls** on the Zambia-Zimbabwe border, and historic sites such as the former slave markets of Western Africa. Many opportunities for tourist development, however, await political stability and the development of transportation facilities.

## Environmental Problems

Environmental problems, identified by early European settlers, continue to be significant. The tropical African environments are associated with poor soils, diseases, and drought.

### Soil Quality Losses

Many tropical soils lack nutrients or are difficult to work with simple tools. The decline in soil quality following agriculture is one of the most widespread problems in Africa. For example, removal of the forest cover in the Ethiopian Highlands led to rapid erosion of the soils, forcing people to move elsewhere and impose further pressures on land resources in already crowded areas. In semiarid areas, the overgrazing of slow-growing vegetation and the removal of woody plants for firewood expose soils to erosion by water and wind—processes that may extend desert areas, particularly along the southern margin of the Sahara, in the Sahel region. Rapidly declining yields on commercial farms require expensive inputs of fertilizer to maintain productivity.

### Tropical Diseases

Despite massive efforts and advances in health care, tropical **diseases** remain among the major environmental problems of Africa for people and food production. Farming losses resulting from cattle and plant diseases fell after 1970, especially outside the Congo River basin. Crops and livestock introduced from other parts of the world, however, are often susceptible to African diseases and pests, and may bring their own. The use of pesticides and livestock disease treatments usually means reliance on multinational drug companies and their costly products. Some countries in Africa are beginning to increase their use of local remedies.

Great strides in immunology and antibiotics allow recovery from many diseases such as cholera, various fevers, and even malaria, lowering human death rates. New deadly viruses, such as the Ebola that affected Congo (Zaire) in 1995, keep appearing. Other major diseases are still killers on a larger scale than the Ebola virus. Sleeping sickness, for example, kills 200,000 people a year in Congo (Zaire). Although a cure for sleeping sickness is known, few people get treated for it. The high cost of continuing treatments for diseases such as sleeping sickness and malaria and the continual risk of infection from waterborne diseases where water supply is untreated cause death rates to remain higher than those in other parts of the world. In many parts of Africa, a lack of sufficient food makes people less resistant to disease.

### HIV/AIDS Pandemic

By 1998, AIDS was listed by the World Health Organization as the fifth main cause of global deaths and expected to rise to third by 2005. Although reduced in Europe and Anglo America by expensive triple-drug therapy, the disease diffused rapidly through Africa from the 1960s, achieving pandemic (affecting large numbers of people over large areas) status (Figure 3.10). In the 1990s, HIV/AIDS also spread rapidly through the former Soviet bloc countries and China as they opened frontiers to international trade and tourism. Life expectancies in Botswana and Zimbabwe that rose to 60 years by 1990 fell by 10 years in the following decade. Other African countries experienced smaller but important effects. Some, like Kenya, resisted public admission of the disease's prevalence to protect their international tourist industry. Throughout the continent, the lives of young, often skilled workers were shattered.

HIV/AIDS emerged in Africa, hopping over the species barrier (like Ebola from other species) from chimpanzees around 1950 and diffused into the seven (HIV-1, A to G) strains known today around the world. The 1960 Congo civil war provided the means of diffusion through refugees and armies roaming the country. While Ebola virus acts rapidly and its contagion is short-lived, HIV is a long-term condition and so spreads widely. In the West it was first limited to homosexuals but was always a heterosexual problem elsewhere and affected wider sections of the population. Furthermore, drug research focuses on the Western form, HIV-1B, and the marketed drugs are too expensive for poor countries.

A 1998 international conference highlighted possible measures for controlling the spread of AIDS in poorer countries. In the absence of long-term vaccination, well-broadcast

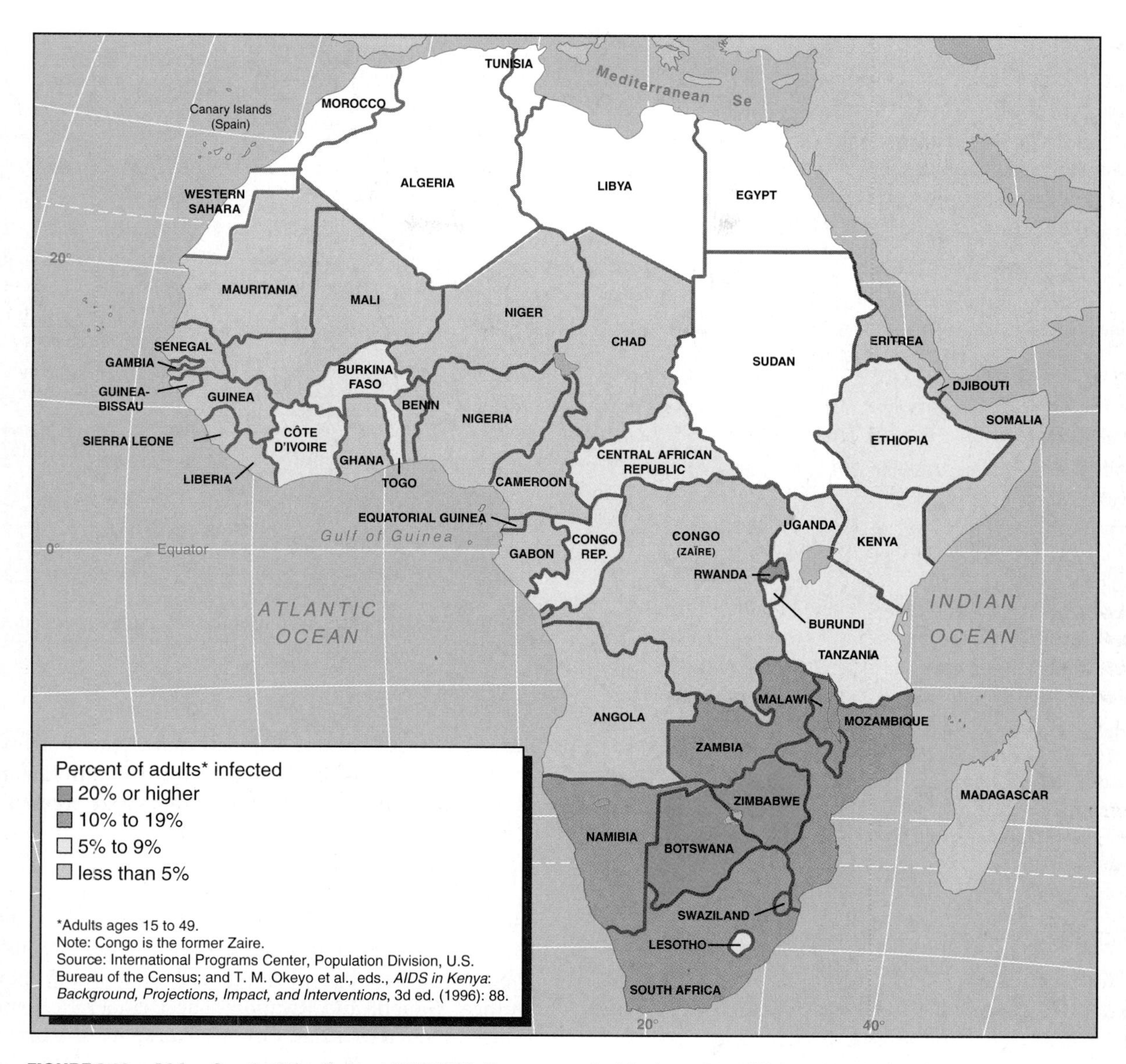


**FIGURE 3.10 Africa South of the Sahara: HIV/AIDS.** The percent of adults (age 15 to 49) infected by HIV/AIDS. What factors might account for the higher prevalence in Southern Africa?

Source: Data from T. J. Goliber, *Population Health in Sub Saharan Africa,* Population Reference Bureau, 1997.

public health measures are having some effect. In the face of HIV being spread by sexual activity and drug addicts reusing needles, moralistic "better behavior" policies failed in comparison with those that recognized human frailty by providing cheap or free condoms and needles. In Senegal, religious groups were involved in a program that included sex education in schools, the "social marketing" (at low prices) of condoms and needles, and a focus on at-risk groups (prostitutes and young men in the army). HIV infections remained below 2 percent, compared to over 13 percent in neighboring Côte d'Ivoire. In Tanzania, treatment of other sexually transmitted diseases, such as syphilis and gonorrhea, reduced HIV by 40 percent, while counseling of at-risk groups to improve behavior also had success. All these, however, required early action, open discussion of sensitive issues, and repeated targeting of vulnerable groups. Where HIV already has a major hold, it is transmitted to the next generation at birth, blood banks are suspect, and the reuse of medical equipment in poor countries makes HIV/AIDS a wider problem that is not confined to sexually promiscuous groups.

## Drought

Shortages of water through longer dry seasons or periods of dry years appear to be increasing in the seasonal rainy areas of Africa. Such shortages lower crop productivity and hydroelectric project efficiency. National resources have to be diverted into buying food to prevent famine and fuels to generate electricity.

**RECAP 3A: Africa's Culture and Environment**

The countries of Africa South of the Sahara are very poor compared to the rest of the world and are part of the periphery of the world economy. Ethnic cultures and the impacts of colonialism are major influences on the internal politics of many African countries today.

The natural environment is dominated by tropical climates that produce frequent droughts outside the rainy Congo River basin; by plateau, rift valley, and river basin landscapes; and by vegetation that ranges from tropical rain forest through savanna grasslands to desert. Africa's natural environments provide major mineral resources. Tropical climates attract tourists and enable tropical crops to be grown for sale in rich midlatitude countries. Soil quality in Africa is often poor, and diseases continue to reduce the productivity of people and farms.

3A.1 What evidence places Africa South of the Sahara in the world's periphery?

3A.2 How do the histories of Native Americans and African ethnic groups compare in their relations with European colonizers?

3A.3 How are climate, natural vegetation, and landforms linked in Africa South of the Sahara?

3A.4 What would be the response in the United States if the African levels of HIV/AIDS infection were repeated there?

**Key Terms:**

tribe
equatorial climatic environment
monsoon climatic environment
tropical seasonal climatic environment
tropical arid climatic environment
changing climate
plateau
rift valley
inselberg
tropical rain forest
savanna grassland
desert
disease

**Key Places:**

Sahara
Congo (or Zaire) River
Mount Kilimanjaro
Ethiopian Highlands
Cameroon Highlands
Victoria Falls

## CENTRAL AFRICA

Although **Central Africa** is at the heart of the continent, its largely equatorial climate, dense rain forest environment, and difficult access from world trade routes caused it to be perceived by Europeans as the most uninhabitable part of Africa. It was colonized late, mainly by French and Belgians, at the end of the 1800s (Figure 3.11). Much of it remains remote and isolated from the world economy, and the subregion includes some of the poorest countries in Africa, despite potentially rich natural resources (see Figure 3.2). Although it possesses a central position within Africa, this subregion is unlikely to act as a hub of development for the continent.

The difficulties of entry to the Congo River basin past the rapids at its mouth continue to restrict economic development by adding costs to the inland transportation of exports and imports. Moreover, civil disorder dismembers government, education, and trade. Recent civil strife in Rwanda and Burundi, Chad's continuing battles with its neighbors, the chaos that continues in Congo (Zaire), and internal unrest in other countries of the region prevent economic progress.

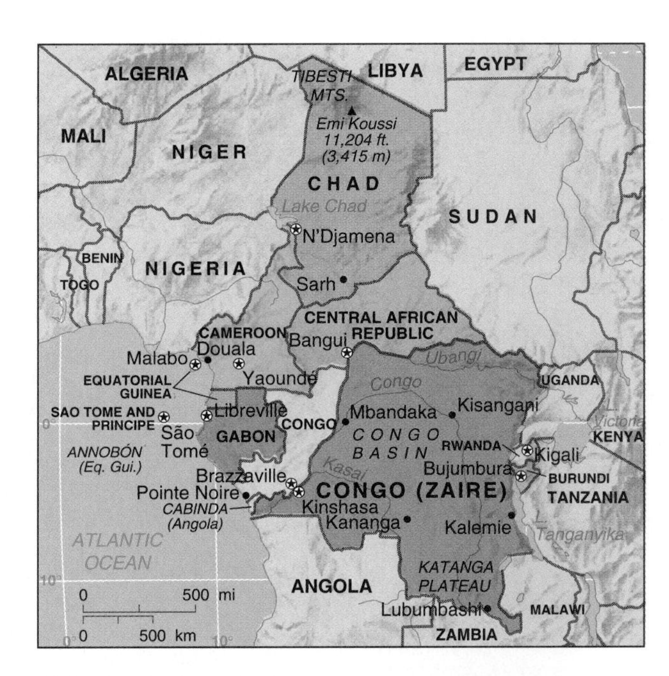


**FIGURE 3.11 Central Africa: main features.** The map shows the countries included in the subregion, the major cities, and rivers.

## Countries

The countries of Central Africa are **Burundi, Cameroon, Central African Republic, Chad, Congo (Republic of), Democratic Republic of Congo** (formerly Zaire), **Equatorial Guinea, Gabon,** and **Rwanda.** The Democratic Republic of Congo is referred to as Congo (Zaire) in this text. All are extremely poor, and most people live at a bare subsistence level. Congo (Zaire) is by far the largest country, covering about 40 percent of the area and having over half the total population. Despite its huge mineral resources, Congo (Zaire) is now one of the poorest countries in the world following the collapse of internal order in the 1980s and 1990s. Robbery of national wealth under President Sese Seko Mobutu led to a new description of his government as "kleptocratic." The 1997 takeover by President Laurent Kabila faces immense difficulties.

Apart from Congo (Zaire), the other countries are of small to moderate size in land area and have populations of only a few million people. Burundi, Rwanda, and Equatorial Guinea have tiny areas. Burundi, Rwanda, Chad, and the Central African Republic are landlocked, produce few commercial goods, and have no easy communications through the coastal countries on either side. Countries with such small and poor populations find it difficult to develop their own manufacturing industries because there are virtually no home markets and they cannot make and market products at prices that are competitive in world markets. Chad, Burundi, Rwanda, Equatorial Guinea, and Congo (Zaire) are among the world's poorest countries (see Figure 3.2). The coastal countries of Gabon, Congo, and Cameroon have higher incomes than the other countries and positive trade balances. These countries get most of their wealth from relatively small exports of offshore oil. Their low HDI rankings, however, indicate that health and education facilities are not much better than those of the poorer countries.

## People

### Population Distribution

The population distribution in Africa South of the Sahara (Figure 3.12) is marked by large areas of very low densities (under 10 people per km$^2$). Most of Central Africa has such low densities and densities of over 100 per km$^2$ occur only in major urban areas and in Burundi and Rwanda.

The populations of the nine Central African countries are still largely rural, although rapid urban growth poses problems. As yet, no country has as many as half of its population in urban areas and most have between 30 and 45 percent. Civil strife, however, leads to more people migrating into towns for perceived safety. Other reasons for urbanization, such as industrial development, have less relevance in Central Africa, although the growing cities are important centers of the informal economy.

The largest urban areas are often former colonial trading or governmental centers such as **Kinshasa** (Congo [Zaire], former Leopoldville), **Yaoundé** (Cameroon), **N'Djamena** (Chad), and **Brazzaville** (Congo). **Libreville** (Gabon) was set up for freed

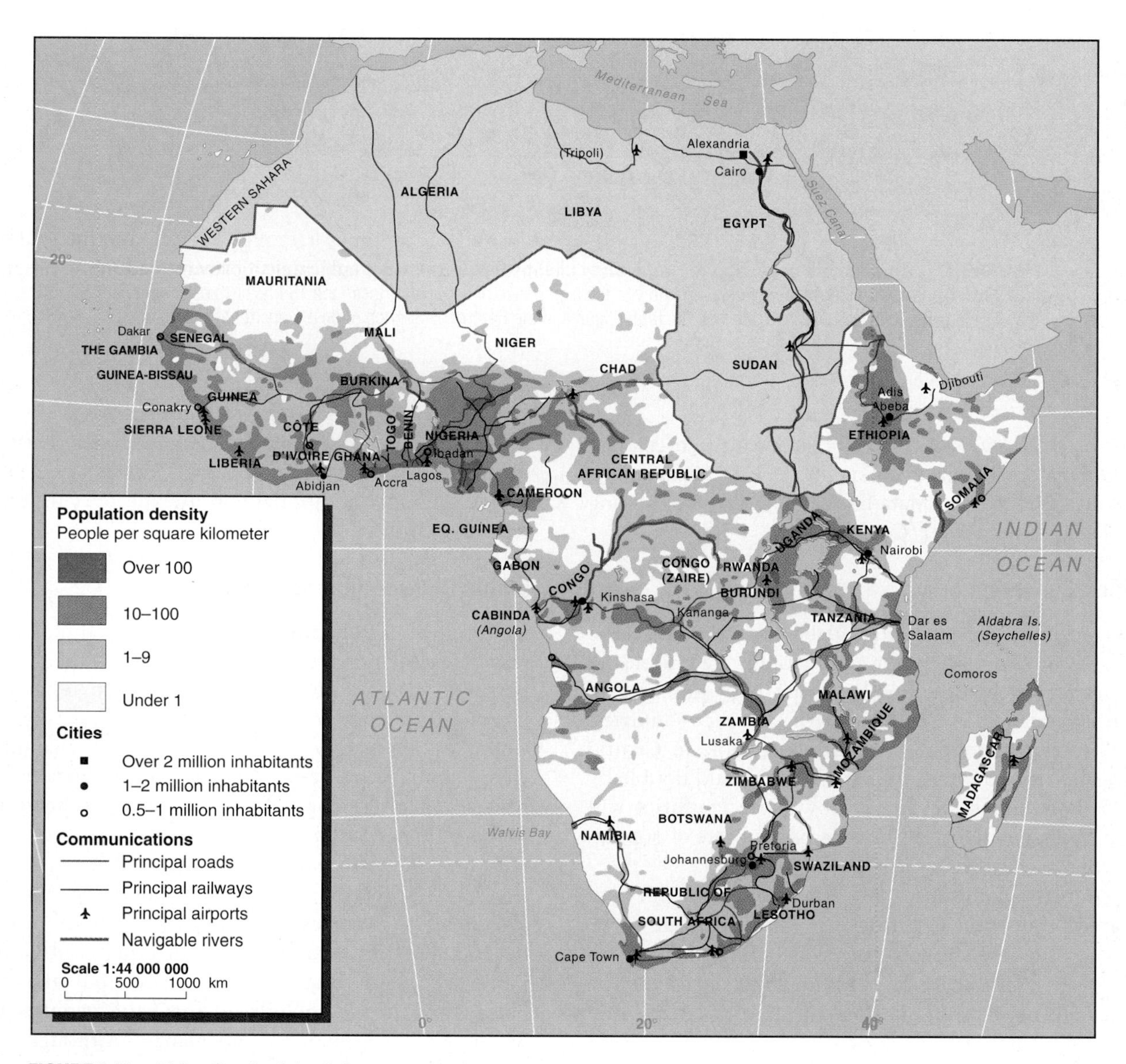


**FIGURE 3.12 Africa South of the Sahara: population distribution.** What is the relationship between the population densities and the natural environments (see Figure 3.5)?

Source: Data from *New Oxford School Atlas*, p. 91, Oxford University Press, UK, 1990.

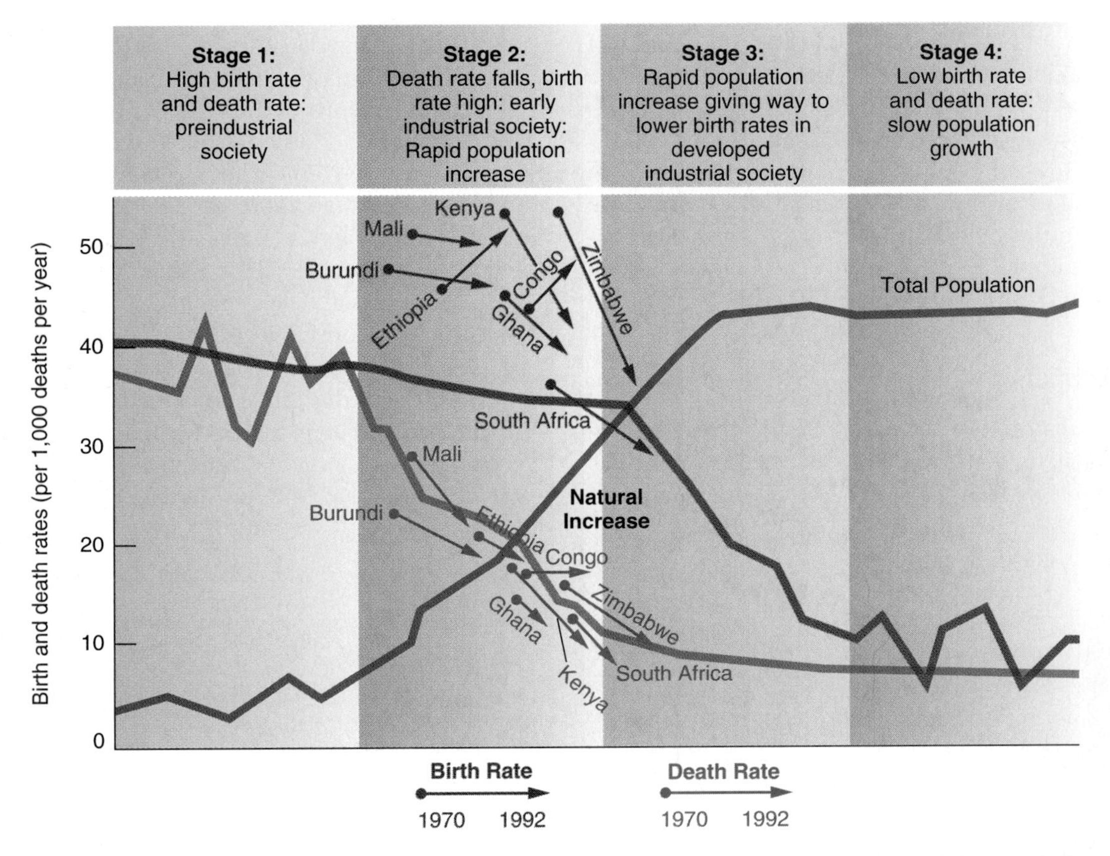


**FIGURE 3.13 Africa South of the Sahara: population change linked to demographic transition between 1970 and 1992.** Although the death rates fell comparably to the experience of core countries in the 1800s, birth rates remained very high and some even increased. Look up the latest figures in the Reference Section or on a website: are there any changes?

slaves in the early 1800s and later became a French colonial town. Other major cities are ports such as **Douala** (Cameroon), **Pointe Noire** (Congo), and **Matadi** (Congo [Zaire]), or mining centers such as **Lubumbashi** (Congo [Zaire]).

## Population Dynamics

Although accurate data are unavailable for some of these countries, it is clear that their populations are increasing rapidly despite such local events as the high numbers of deaths in Rwanda in 1994 and civil strife in other countries. The 1998 total population of 90 million people in Central Africa was up from 54 million in 1980 and could double by A.D. 2025. It is impossible to see how these countries will feed such a massive increase since the present rate of population growth exceeds economic growth.

Most countries conform to the falling death rate curve of demographic transition (Figure 3.13), but birth rates continue high, with few showing signs of falling. Some birth rates increased over the period between 1970 and 1998. The high rates of population increase in Central Africa are related to high total fertility rates at a time of falling death rates. The age-sex graphs (Figure 3.14) show that new births and young age groups dominate populations. Life expectancy in 1998 ranged from 43 to 58 years, up from 35 to 42 years in 1960. Flows of refugees, such as those from Rwanda and Burundi into Congo (Zaire), Tanzania, and Uganda, impose additional and unexpected burdens on the receiving countries.

## Ethnic Distributions and Tensions

Culturally, Central Africa is a mixture of ethnic groups and religious affiliations. Chad in the north has the greatest proportion (50%) of Muslims. The Christian religion has an increased presence southward and Islam less (Figure 3.15). Christians are mainly Roman Catholics in the countries that were colonized by France or Belgium. Indigenous religious beliefs remain strong, often influencing Roman Catholic practices in particular.

Some countries have a dominant tribe, but most countries have a diversity of tribal allegiances (see Figure 3.3b). The smaller countries, such as Equatorial Guinea, tend to be dominated by a single tribe. In Equatorial Guinea, the Fang are the largest mainland tribe, but the country's boundaries do not include all the Fang peoples, many of whom live in Cameroon and Gabon. Although the Fang form the largest group in Gabon, a pact among the smaller tribes keeps the Fang out of government. In Chad, the tensions between the Muslim northerners, backed by Libyan

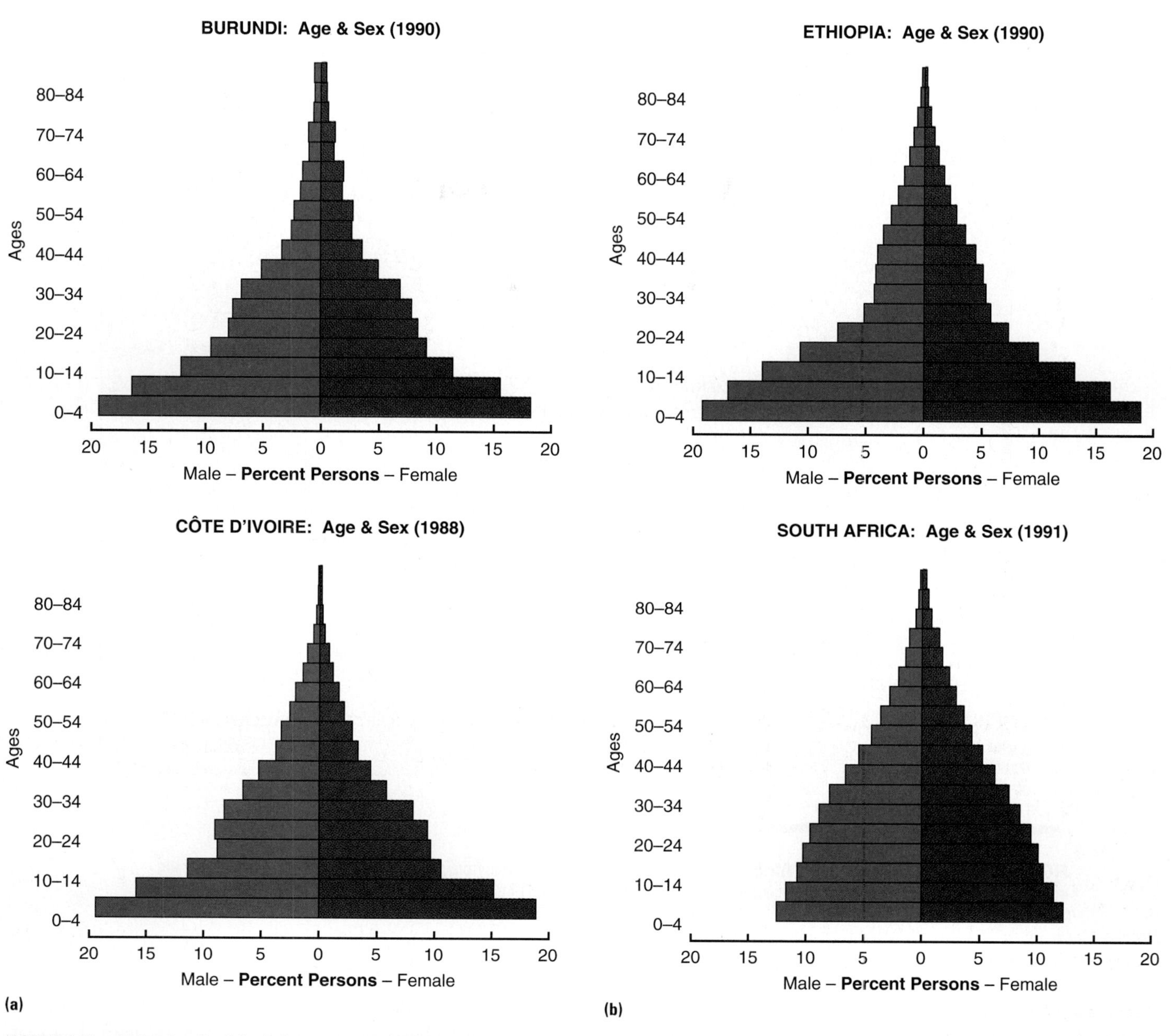


**FIGURE 3.14 Africa South of the Sahara: age-sex diagrams for countries.** Explain the differences between the South African pattern and that of the other countries shown. What might be the effect of such population structures on the size of families, numbers of young or old dependents, and numbers of new job seekers?

Source: Data from *Demographic Yearbook.* New York: United Nations, 1994.

forces, and Christian southerners erupted in open warfare in the 1980s and early 1990s. Congo is another tribally conscious country, in which the Bakongo (48% of total) inhabit the north and the Mbochi (12%) dominate the more prosperous south.

In Cameroon, the north-south enmity between Muslims and non-Muslims is reduced by the great diversity of tribal groups. Over 200 are recognized and none are dominant. The main differences in Cameroon society are between French- and British-speaking groups. In Congo (Zaire), Bantu groups make up 80 percent of the total population, although there are some 200 ethnic groups and 12 major language groupings. Diversity does not, however, preclude tensions in Congo (Zaire). The original inhabitants were pygmies who were largely displaced by other tribes, and with the breakdown of central government control in the 1990s, tribal wars erupted in some of the provinces.

Originally populated by the Twa pygmies, the agricultural **Hutu tribe** moved to the Rwanda-Burundi area from the Congo River basin in the A.D. 1300s. They later came under the feudal rule of the cattle-herding **Tutsi tribe**, who invaded from the north in the following century. Roman Catholic missionaries reached the area in the 1800s, and German colonists took control at the end of the century, linking the area to German East Africa. After World War I, Belgium took over as colonial power, redirecting links westward.

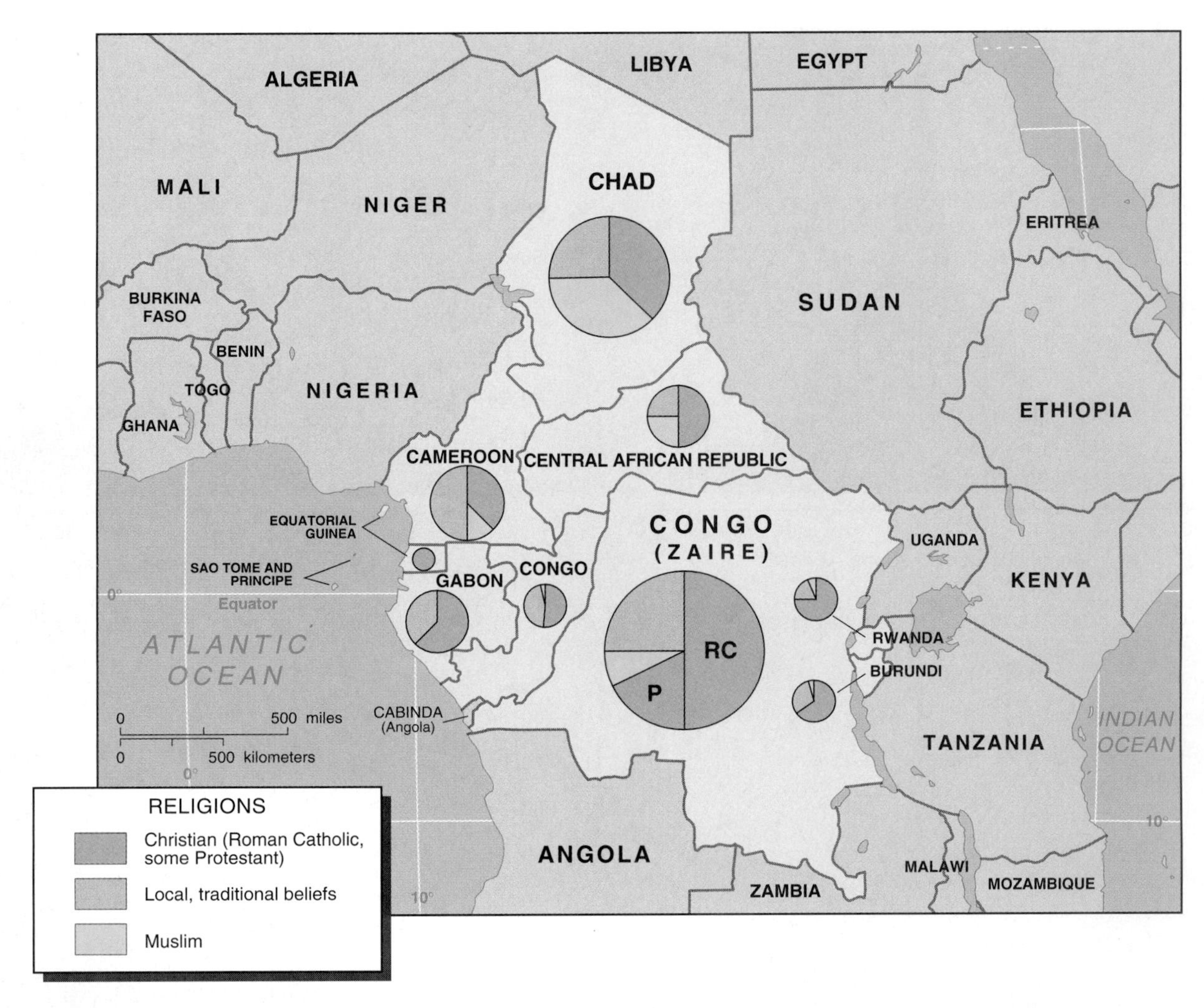


**FIGURE 3.15 Religions in Central Africa.** The proportion of Muslims declines from north to south. Traditional beliefs remain important. The sizes of the pie charts are proportional to the total population of each country.

Tensions between the Hutu and Tutsi came to a head after independence in 1962. The majority Hutus took control in Rwanda, but the minority Tutsis remained in charge in Burundi, from which Hutus fled as refugees to Rwanda. In 1994, the deaths of the presidents of the two countries in a missile-caused air crash set off tribal war in Rwanda. The Hutus massacred Tutsis and the Tutsis responded with armed rebellion, eventually taking over the government. It is estimated that up to 2.5 million people, one-third of the pre-1994 population of Rwanda, were killed or fled to Zaire and other surrounding countries.

Tutsi-related tribes in Uganda and eastern Congo (Zaire) became important in their own countries, aided by the former Rwandan Tutsi soldiers. In late 1996, the former rebel Laurent Kabila led Tutsi-dominated groups in eastern Congo (Zaire) to a conquest of the whole of Congo (Zaire) within six months. There was little opposition from the underpaid and poorly equipped Zairean army. Kabila entered Kinshasa to acclaim ousted President Mobutu's corrupt and incompetent regime, and changed the country's name. He gained support from the United States and the European Union, but then faced major problems in rehabilitating the country, since central government control had broken down into a series of regional baronies whose leaders paid Mobutu for legitimizing their power. By 1998, Kabila lost most of his allies and his discarded Tutsi-led invasion force attempted to overthrow his regime. Other African countries took sides, threatening an escalation of war in the subregion.

### External Impacts on People

The cultural makeup of the Central African countries has to be seen in the context of Arab influences followed by European exploration and colonization. After several centuries of the slave trade—which began among African groups and was developed by Arab traders before European ships redirected the trade to Atlantic coasts and the Americas—the African population was depleted and African kingdoms and their economies were disrupted. European explorers passing through the subregion during the mid-1800s returned reports to their governments about resources that could be significant for the industrializing countries of western Europe. Territory was claimed in an indiscriminate and competitive rush following the 1884 Congress of Berlin, when Africa was divided by the main powers.

In 1910, France established French Equatorial Africa from a group of annexed territories. Belgium laid claim to the large territory that it named Belgian Congo and is now the country of Congo (Zaire). While the Belgians did little apart from developing and extracting the wealth from the copper mines in the southern part of its main colony, the French invested more in education and a developing civil service. When independence came in the 1960s, the French retained links and supported local currencies, while Belgium handed over power reluctantly and then severed ties, apart from some trading links. France never left Central and Western Africa and in 1998 still propped up governments, maintaining military bases in seven countries. Although there were moves to replace its patronage system with more normal ties, France returned to its imperial attitudes in the late 1990s, not endearing itself to African countries.

After independence, most countries in Central Africa had single-party constitutions, but in the 1990s some moved to multiparty systems and held democratic elections. Unfortunately, this move frequently awakened ethnic rivalries that had been subdued under single-party, often repressive, rule.

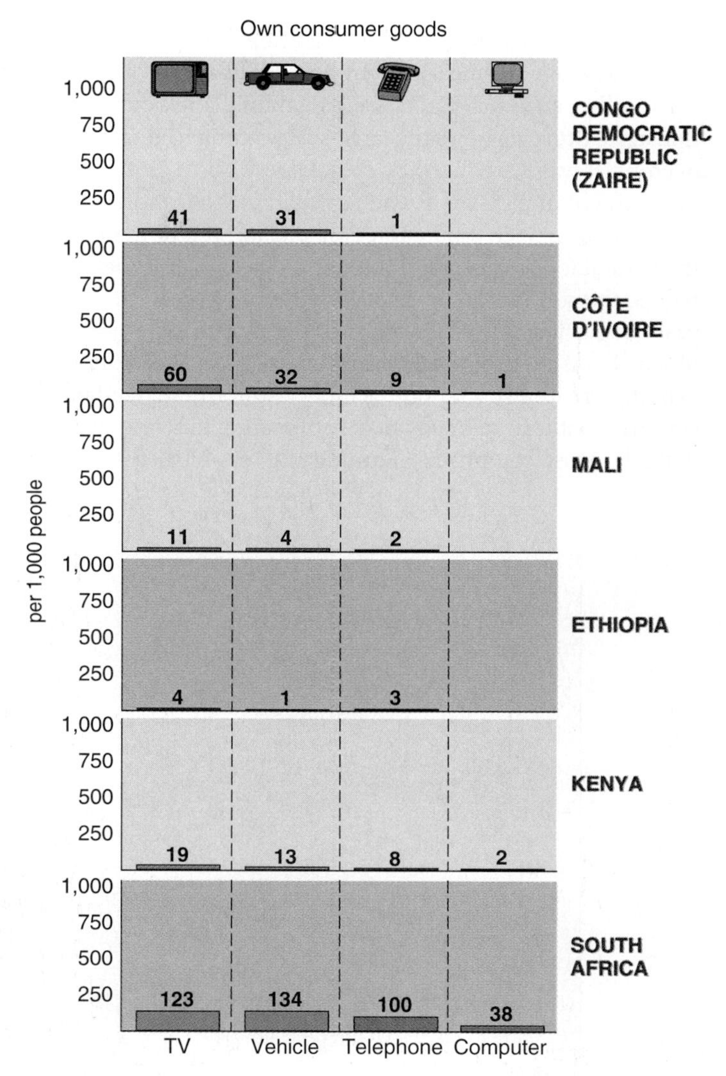


**FIGURE 3.16 Africa South of the Sahara: consumer goods ownership.** It is very low in African countries, apart from South Africa: compare levels with those in other world regions. Which goods are regarded as most important?

Source: Data (for 1996) from *World Development Indicators*, World Bank, 1998.

## Economic Development

The combination of physical isolation, ethnic tensions, repressive colonial history, and rapid population growth holds back the economic development of Central Africa. Although many countries are endowed with mineral deposits and tropical timber resources that are in demand by richer countries, infrastructure remains poor, and few people inside or outside the new countries have ideas as to what policies might help to overcome their lag in the world economic system. The Central African countries experience slow economic development and now produce a smaller proportion of total African output than they did in the mid-1960s. The low levels of possession of consumer goods in Congo (Zaire) (Figure 3.16) show that only a tiny group of people in these countries can afford such items. In real terms, incomes in all the larger countries fell from 1980. Only Equatorial Guinea and Gabon have rising middle incomes. HPIs are 30 percent or higher, rising to 50 percent. Despite improvements in education and health since the 1960s, incomes have scarcely risen at all, and levels of all measures of life in Central African countries remain low. Several countries depend almost totally on external aid since their own products earn little income from world markets.

### Dominant Agriculture

Agriculture remains the main economic involvement of over 70 percent of the populations of these countries, although only a small part of the surface is cultivated (Figure 3.17). For example, much of Congo (Zaire) has potential for cultivation but only 3 percent is used. Most farming is for subsistence, using traditional methods to grow tropical root crops, cereals, fruits, and vegetables. There are still peoples who continue traditional practices of gathering forest products or of shifting agriculture (Figure 3.18). Cattle farming in Central Africa is severely restricted by diseases, particularly those borne by the tsetse fly. Modern protection for animals is available, but expensive.

Commercial farming is poorly developed in Central Africa apart from some **tropical tree crops** such as cacao (for chocolate), coffee, and rubber that are grown on plantations. **Plantation agriculture** involves the large-scale commercial cultivation of crops. It was introduced by colonial powers to produce large quantities of tropical foods and raw materials cheaply for European markets. Such plantation crops relate well to the form of the natural forest vegetation but not its variety of species. They reduce the soil fertility over 30 to 40 years instead of 5 to 10 years under field crops. Farmers in the drier northern regions of Chad and the Central African Republic produce cotton for export, but transportation facilities from these landlocked countries without railroads make competing in world markets difficult. Most commercial farming was established by European planters in the colonial era and its continuation is often under threat. Although some countries maintained and extended such production after independence in the 1960s, Zairian nationalization in the 1970s resulted in a decline of commercial farming within its borders.

Despite the dominance of farming occupations, most Central African countries find it difficult to feed their populations. Food production is often affected by policies such as overvaluing their currency (and so making exports costly to others) or maintaining low prices for urban populations (and so making incomes low for farmers). The combination discourages farmers from increasing yields but encourages food imports that add to overseas debts. In Congo (Zaire), the European Union funded a project for developing commercial

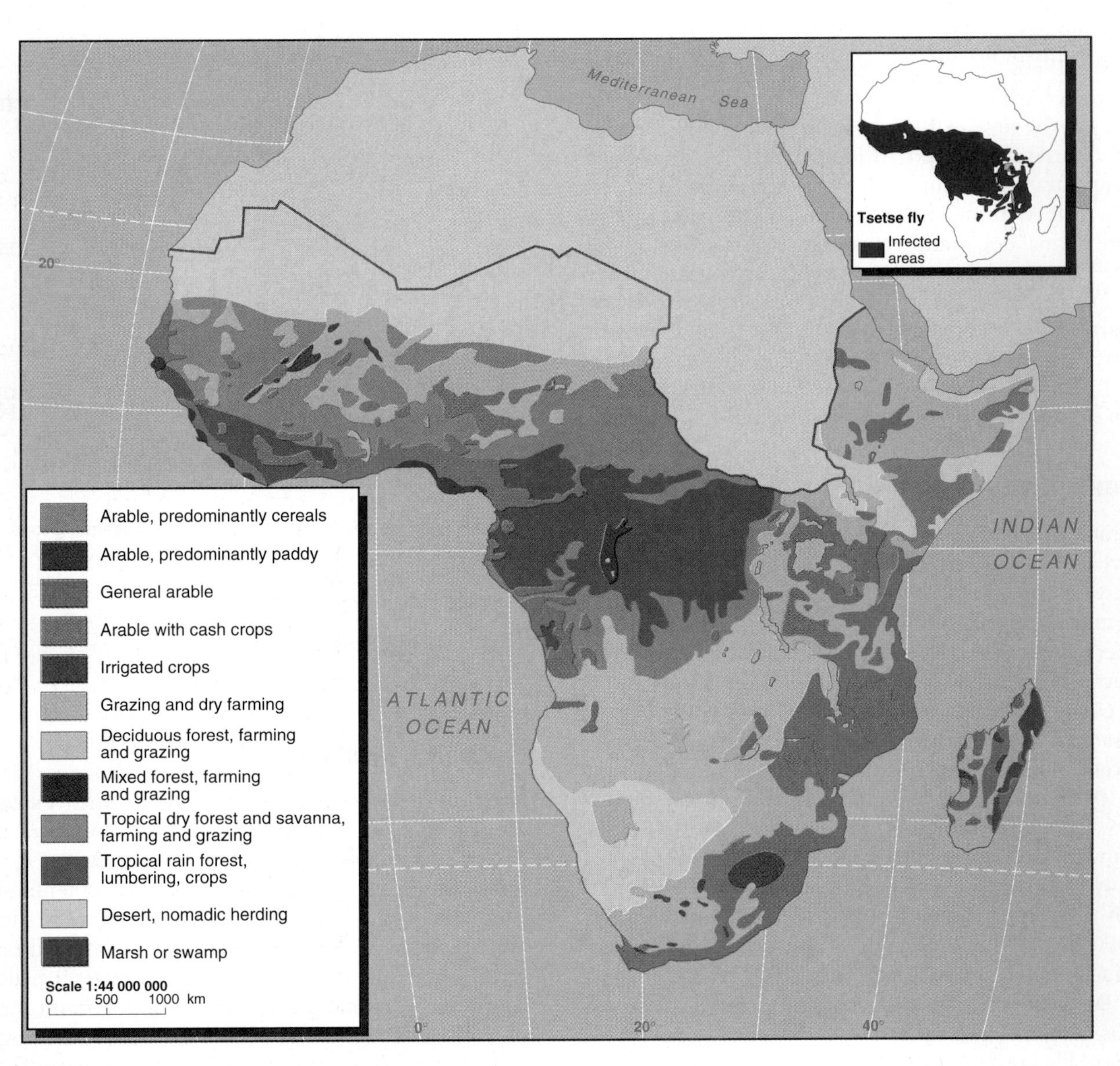


**FIGURE 3.17 Africa South of the Sahara: agricultural land use.** Note the distribution of intensive, commercial uses and of unusable land or low-intensity uses. The insert showing the area affected by tsetse flies is an area where cattle rearing is difficult.

Source: Data from *New Oxford School Atlas,* p. 90, Oxford University Press, UK, 1990.

farming around Kinshasa in order to supply the local market with food, but it had only a modest impact because of the political upheavals since the late 1980s.

## Forestry, Fishing, and Mining

Timber products from the tropical rain forest, including mahogany and ebony wood, are of increasing importance to the exports of Cameroon and Congo, while Gabon produces a softwood that is used in plywood. Deep-sea fishing is increasing in the coastal countries such as Cameroon and Equatorial Guinea. Chad obtains fish from its lake.

Mining brings in most foreign exchange to the countries of Central Africa. It could bring more if transport facilities, world market demand, and investment were available. Transportation costs are high and foreign investors are few at present. Southern Congo (Zaire) is one of the world's major copper mining regions and the country mines diamonds, cobalt, and some oil as well, although its exports fell because of civil disorder. Gabon's relatively high per capita income results from its oil and manganese production, while oil makes up 90 percent of Congo's export revenues. Such a high proportion emphasizes the low value of other exports. Cameroon, Chad, and Equatorial Guinea also have oil deposits awaiting further investment before exploitation. Gabon possesses the world's largest unexploited iron ore deposit. Until the 1990s, mining companies wishing to invest in Central Africa found that the combination of poor infrastructure, dislike of foreign investors, and local corruption discouraged attempts to develop new mines. By the late 1990s, however, the countries that were not disrupted by civil war were making conditions easier for foreign investors. Congo (Zaire) remained a problem. Contracts with international mining corporations negotiated by Kabila's men before he gained power were canceled in 1998 in favor of possible better deals, adding to uncertainty.

## Underdeveloped Manufacturing

Manufacturing industries are little developed in Central African countries. No country has more than 15 percent of its low GDP base contributed by manufacturing, and many have less than 10 percent. Internal markets are tiny because the countries have small populations and few people who can afford to purchase consumer products. Most factories are engaged in processing local mine, forest, and farm products or in making bulky products that do not stand transport (e.g., cement, bottled drinks). Cotton textiles are made in Chad from the local crop.

Energy to power manufacturing is insufficient. Of the Central African countries, only Gabon generates more than 250 kWh (kilowatt hours) of electricity per capita each year, compared to 12,000 kWh in the United States; Burundi, the Central African Republic, and Chad generate under 30 kWh per head. A possible source of power for manufacturing and other forms of economic development is available in the huge water resources of the Congo River basin, estimated to contain one-sixth of the world's hydroelectricity potential. Already the Inga Dam, downstream from Kinshasa, has harnessed some of this power, but the electricity from Inga is carried overland southward to the Shaba mines in southern Congo (Zaire) without intermediate allocations of electricity to local uses.

**FIGURE 3.18 Central Africa: traditional farming in tropical rain forest** Some 85 km south of Kinshasa (Congo [Zaire]), the rain forest is being cleared by crude "slash and burn" methods. Some of the smaller wood is being carried home for fuel. Crops will be grown for a few years until the soil nutrients are exhausted, then the land will be left to return to forest, often in degraded form



## Transportation Is Vital

Prospects for economic development rest on improving transportation links, but so far Gabon is the only country to invest some of its oil revenues in constructing all-weather roads and a railway into the interior. Elsewhere, the extensive navigable waterways of the Congo River system are used for inland transport but have no direct outlet to the sea. The railways on either bank linking Kinshasa and Brazzaville with ocean ports are narrow gauge, limited in capacity, and unconnected to each other. The road from Kinshasa to the port of Matadi below the rapids, on which trucks once took 5 hours to cover the 350 km (220 mi.), was in such poor repair in the mid-1990s that trucks needed up to five days for the trip. Mineral production in the southern Shaba Province of Congo (Zaire) has no railroad outlet northward. At first mineral exports were routed through Angola to Lobito or to Beira in Mozambique, but local wars cut these connections. Most exports now take the long and expensive route through Zambia and Zimbabwe to ports in the Republic of South Africa. A planned Pan-African Highway from Lagos in Nigeria through Cameroon, Congo, and Congo

(Zaire) to Mombasa in Kenya would have opened up many parts of Central Africa, but it suffered from a lack of investment and political squabbles. The project was shelved in the early 1990s but is being considered again.

## Informal Economy

In the absence of developments in major sectors of the economy, the informal sector is extensive. People who are not supported by subsistence agriculture or do not have permanent paid jobs earn a living in various ways. Occupations range from individuals selling goods in the towns to smuggling gold, diamonds, ivory, or drugs. The informal sector, by definition, is difficult to keep account of or to tax, but its activities are often maintained by illegal payments to civil servants. The informal sector has grown to be such a major feature of the economy that many countries depend on it and even encourage it.

## Problems of the Periphery

The countries of Central Africa play a small part in the world economy. This subregion demonstrates the problems of countries at the outer margins of the world economic system's periphery. With two-thirds of the population engaged in subsistence agriculture, it is only the primary products of the cacao and coffee plantations, tropical forests, and mines that are in demand by world markets. Poor inland transport and increasing numbers of mainly poor people hold back prospects of economic or human development. At present, there is little scope for entrepreneurs to establish businesses that have more than local significance. The world's bankers do not invest in the building of new infrastructure such as ports, roads, or water resources in countries where returns are likely to be small, come slowly, or focus mainly on elites. The early effects of building such projects are often negative for the majority of people and their environments. At the same time, however, the promise of joining the world's wider economy may discourage efforts to become involved in smaller-scale projects linked to local and sustainable development.

## French Links

One of the major external economic factors affecting many of the countries of both Central Africa and Western Africa is that they were once French colonies and retained financial links with France after independence (Figure 3.19). Instituted in 1948, the **Communauté Financière Africaine** (CFA) franc shared by 14 African countries, including Chad, Central African Republic, Congo, Equatorial Guinea, Gabon, and Cameroon, continued after independence. Its value was tied to the French franc and guaranteed by the Bank of France. This link held currencies at artificially high exchange rates but allowed French companies to retain dominant positions in local contracting. The high exchange rates, however, made it difficult for French-speaking African countries to export goods, used up their foreign exchange reserves on imports, and made them increasingly dependent on France. In January 1994, the French government, in line with other economic moves encouraged by the World Bank, devalued the CFA franc by 50 percent. Initially, this brought hardship to people in the countries that depended on France, especially by raising further the cost of imports. For its part, France wrote off the debts of the poorer countries and the World Bank made increased grants available. Foreign investors began to show some interest in these countries, but activity was slow because of the other factors that retard development there.

## Attempts at Structural Adjustment

The debt crisis of the late 1980s resulted in most countries in Central Africa having high debts with the world's richer countries. The World Bank and International Monetary Fund stipulated that in the future, countries applying for loans should reorganize their economies to encourage realistic exchange rates, reduce inflation, balance their internal budgets by reducing government economic intervention, and open their markets and economic activities to external competition and investment. This approach of structural adjustment (see Chapter 2, page 51) appears to have had few positive results so far in Africa and none in Central Africa. In Cameroon, a once-growing economy shrank and income per capita was halved between 1985 and the mid-1990s. Civil servants struck against pay cuts and antigovernment feelings rose. The World Bank and the foreign aid donors who use World Bank

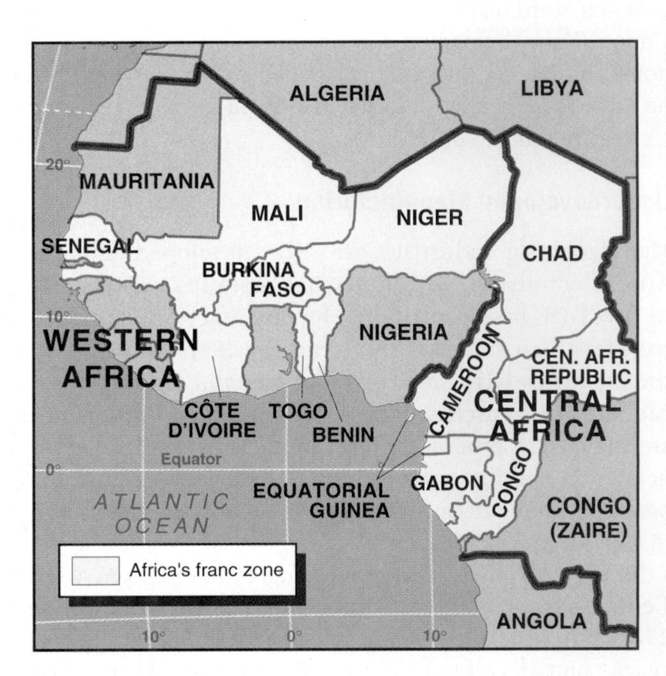


**FIGURE 3.19 French franc (CFA) area of Africa.** These countries, formerly French colonies, were supported by French loans and their currency was linked to the French franc. The French government devalued the CFA franc by 50 percent in 1994, causing major changes in economic policies in the countries.

criteria saw little progress toward change and classified Cameroon as ineligible for assistance, along with the Central African Republic, Congo, and Gabon. By the late 1990s, richer world countries realized that too rapid implementation of structural adjustment measures would have negative effects on Central African economies. New approaches included the encouragement of countries committed to improving their economic management and sustainable human development.

### Internal Politics

Many of the problems of Central African countries stem from poor or bad political leadership. The paternalistic colonial powers discouraged democracy and were generally replaced by one-party governments, often with dictatorial presidents. This type of government often achieved a degree of national unity by force but generally had a limited vision of sustainable human development. Large sums of scarce public funds were spent on bringing the countries under military control for long periods.

Congo (Zaire) in particular was plagued by poor government and internal disruption. President Mobutu was rumored to be the richest man in Africa despite the fact that his country had declined toward being one of the poorest. In the 1980s and 1990s, civil strife left Congo (Zaire) as an uncoordinated country of local power bases operating without reference to the central government. In other Central African countries, small groups of prosperous people maintained their elitist base and resisted reforming influences. Military and political leaders siphoned much of the wealth created by mineral exports or made available by international funding and aid agencies into personal accounts. Tribal and religious rivalries erupted into wars that destroyed the infrastructure built with foreign aid funds, deterred new foreign investors, and devolved political and economic activities to the local level.

### Summary

Central Africa illustrates the continuing influence of the unequal relationships of external links that began in colonial times and are still paramount in core-periphery relationships. Fluctuating income from overseas markets for raw materials is produced by commercial agriculture, logging, and mining. Loans or grants for development and infrastructure construction projects relate to the likely growth of trade with overseas markets rather than a broader consideration of a country's internal needs or furthering human development. Multinational corporations decide which oil deposits should be exploited and which metal ores should be mined. For long, the CFA franc and other French influences kept many Central African countries dependent on France and its technical advice and expertise. In light of such external controls, it is perhaps not surprising that governments by single parties and dictatorial leaders should be susceptible to personal aggrandizement. Central Africa remains peripheral to the world economy and operates largely outside it, effectively cutting it off from opportunities to raise levels of well-being.

## WESTERN AFRICA

**Western Africa** is defined by the Atlantic Ocean to the south and west, and by the Sahara on the north (Figure 3.20). It is the most populous subregion of Africa South of the Sahara, having over 200 million people—nearly 40 percent of the total.

The countries of Western Africa have more links into the world economic system than those of Central Africa. Being nearer to Europe and having a longer coastline, the area was penetrated from the 1400s by slave traders and those looking for gold or ivory. During the 1800s, the British and French established plantations for producing cocoa and palm oil, together with some mining output, and the area was incorporated as a peripheral area within the world economy. Twelve of the 15 countries had roads and railroads built to connect ports with inland plantations and mines, facilitating raw material exports. The growth of an educated colonial civil service also served the ends of getting export products to the colonial powers' markets.

At independence Nigeria, Ghana, and Côte d'Ivoire were among the most economically advanced African countries and had per capita incomes rivaling Asian countries that are now among world leaders in economic growth. Since independence, which was granted between 1957 and 1964, internal rivalries, changing governments, smuggling, high rates of population increase, and shifting world markets have worked against these countries, many of which are now just as poor as those in Central Africa (see Figure 3.2). Their indicators of sustainable human development have shown modest increases, but population growth continues to outstrip economic growth. These countries remain poor.

### Countries

The countries of Western Africa with coastlines are: **Benin**, **Côte d'Ivoire (Ivory Coast)**, **The Gambia**, **Ghana**, **Guinea**, **Guinea-Bissau**, **Liberia**, **Mauritania**, **Nigeria**, **Senegal**, **Sierra Leone**, and **Togo**. **Burkina Faso**, **Mali**, and **Niger** have no coasts and are in isolated interior positions without direct access to world markets by ocean routes. The different geographical positions and sizes of the countries, the multitude of tribal groups, and the variety of environmental and accessibility conditions create political and economic differences and tensions within Western Africa.

The countries with south-facing coastlines (Côte d'Ivoire to Nigeria) contain a greater range of productive environments, ranging from coastal tropical forest to inland savannas, than those with west-facing coastlines (Liberia to Mauritania), which have savanna or desert environments. They

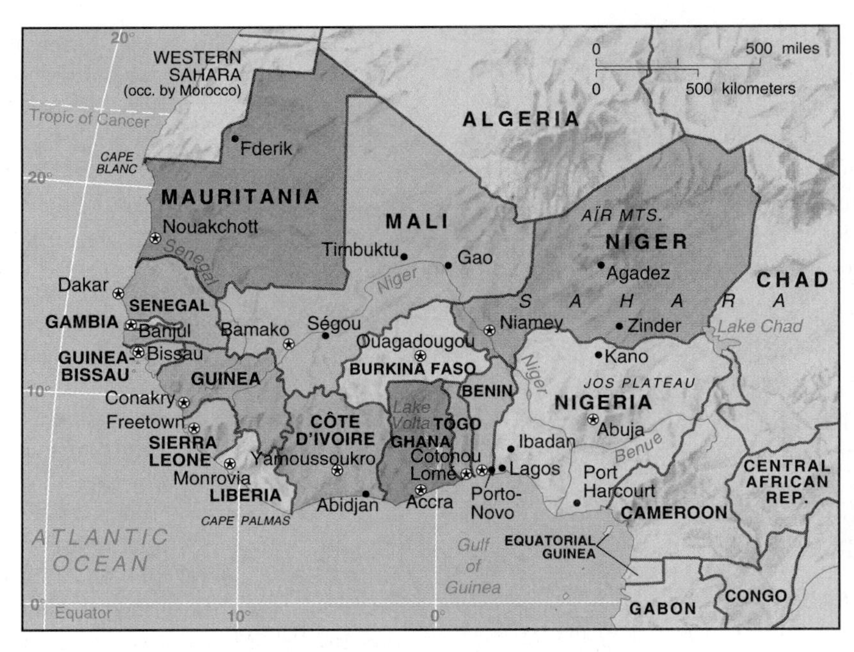


**FIGURE 3.20 Western Africa: main features.** The map shows the countries included in the subregion, the major cities, and rivers.

combine more varied natural conditions for producing tropical crops with easier access to world markets and are generally more diversified economically. The ports established by colonial powers, despite a lack of natural harbors, remain the largest cities in many of these countries. The climatic conditions of the inland countries bring frequent water shortages, but governments—both colonial and independent—have invested less in the development of the inland areas.

Nigeria is by far the largest and most populous country with over 100 million people, half the total of Western Africa and over twice the population of the next largest African country. All the other countries are small in population and some are tiny. Only Côte d'Ivoire (16 million in 1998) and Ghana (19 million) have totals above 10 million, while six of the 15 countries have 5 million or fewer people. The former British colonies of Nigeria, Ghana, The Gambia, and Sierra Leone retained the boundaries that were imposed by Britain after independence. Most of the remaining countries were part of French West Africa but split into smaller units after independence while retaining strong links to France through the CFA (see Figure 3.19). Guinea-Bissau was a Portuguese trading port with a tiny hinterland.

## People

### Population Distribution and Dynamics

Western Africa is the most populous part of Africa (see Figure 3.12), and densities exceed 100 people per km$^2$ along most of the coast from Nigeria to Côte d'Ivoire. The northernmost parts have densities of less than one person per km$^2$ on the Saharan margins. The total population is increasing rapidly. The 1998 total of almost 230 million was up from just over 150 million in 1980 and could reach 400 million by A.D. 2025. The rapid increase is the result of some of the world's highest total fertility rates of over 5 in all countries except for Liberia (3). Although death rates continue to fall, birth rates scarcely change (see Figure 3.13). The age-sex graph for Côte d'Ivoire (see Figure 3.14) shows the results of high birth rates in very large numbers of children and young people. As death rates fell, life expectancies increased from 35 to 50 years in 1960 to 35 (Sierra Leone) to 60 (Togo) years in 1998 but remain low by world standards.

Urbanization is occurring rapidly, although no country in Western Africa outside the deserts of Mauritania has as much as 45 percent of its population living in towns, and some are still below 25 percent. Compared with Central Africa, the urbanization in Western Africa still partly reflects perceptions of safety but is more related to job opportunities and better health and education provision in towns. The major cities nearly all began as ports or port-related colonial developments. **Lagos** (Nigeria), **Abidjan** (Côte d'Ivoire), **Dakar** (Senegal), **Accra** (Ghana), **Freetown** (Sierra Leone), **Monrovia** (Liberia), and **Conakry** (Guinea) are of this type. Inland centers such as **Ouagadougou** (Burkina Faso) and **Bamako** (Mali) began as ancient trade centers at the southern end of trans-Saharan routes and became modern capital cities of the interior countries.

### Precolonial Events

Western Africa has a distinctive cultural history. In medieval times, Muslim groups forged trade routes across the Sahara and established powerful empires on the southern edge of the desert. This forced major migrations of African tribes toward the south and east. Many of the Muslim empires, some with names repeated in modern countries (Ghana, Mali)—although the borders are different—were centered in the Niger River valley and extended their influence southward to the coast. Muslim African groups, such as the Hausa and Fulani, dominated the region and resisted the colonial intentions of European countries. From the time of the first Portuguese traders in the mid-1400s to the late 1800s, the African kings limited intrusion to trading agreements for slaves, ivory, and gold. The slaves from Western and Central Africa went to the Caribbean and North America. In the early 1800s, returning slaves from Anglo America were provided with land in Liberia and the port of Freetown.

### Colonial Competition and Long-term Influences

In the later 1800s, however, Britain, France, and Germany raced each other to colonize Africa. Their 1884 agreement gave Britain precedence in Nigeria, the Gold Coast (now Ghana), Sierra Leone, and The Gambia, while France had its influence acknowledged in most of the rest apart from German Togoland, Portuguese Guinea (Guinea-Bissau), and Liberia.

After a series of military engagements against local rulers, France established its French West Africa province with a capital at Dakar. Britain extended its rule northward through Nigeria and Ghana, necessitating military action against the Ashanti in Ghana during the 1800s and against the northern kings in Nigeria in the early 1900s. During World War I, the Germans in Togoland surrendered and most of that colony joined French West Africa. Following the colonial takeovers, European countries and companies developed plantations for tropical tree crops such as palm oil, cacao, and rubber, while mining concessions were also allocated and exploited.

The combination of historic African cultural features and the overlay of European trading and colonial influences left important cultural features. English and French remain the official and commercial languages of former colonies. The Muslim influences remain most significant in the northern parts of Western Africa: Mauritania, Burkina Faso, Mali, Niger, Senegal, The Gambia, and Guinea are dominantly Muslim countries in which indigenous religious elements have been largely eradicated. The countries with south-facing coasts have Muslim populations in their northern parts and mixtures of Christian and indigenous beliefs in the southern. Within these countries, religious rivalries and memories of domination by the northern kingdoms that consigned many southern tribes to slavery often heighten the tribal rivalries. Such sensitivities remain evident in Nigeria, where the military government is largely comprised of northerners. Even within the Muslim parts of the subregion, racial and ethnic differences raise conflicts, as between the Maurs and black Africans in Mauritania. In Liberia, the differences are between the local tribal groups and the resettled ex-slaves, whose descendants make up only 5 percent of the modern families but who ruled as an elite with little concern for others until 1980. Since 1990, there has been continuous civil war in Liberia, and also in Sierra Leone since 1995, and in Guinea-Bissau since 1998.

### Independence

Ghana became the first independent ex-colony in 1957. Most others in Western Africa were independent by 1960. Their subsequent histories differed, but many have had periods of single-party or military government and little experience of fully democratic government. Most countries have preoccupations with internal ethnic conflicts and long droughts in the Sahel region of the interior. They continue to develop the early footholds in the world economy that began with their former colonial linkages. For a time, Guinea linked to the former Soviet bloc but gained little from this. Côte d'Ivoire experienced the longest uninterrupted period of continuing business-centered government and economic progress after independence, lasting into the 1990s.

Attempts to merge some of the small countries failed. At independence, Mali and Senegal formed a single country, but Senegal soon withdrew from the arrangement. A 1990 attempt to join ex-French Senegal with ex-British Gambia ended in reversion to the present borders with economic links retained.

## Economic Development

The newly independent countries of Western Africa began with much greater hopes of an improving future than those in Central Africa. To date, few have managed to advance very far along the road of sustainable human development. The combination of restricted economic base, rapid population growth, drought hazards, political conflict and mismanagement, and especially fluctuating world market conditions got in the way. The low levels of possession of consumer goods, though somewhat better than in Central African countries, reflect this outcome (see Figure 3.16). Almost all countries in Western Africa experienced declining real income from 1980, and all were in the lowest group of world countries in 1996 (under $1500 GDP per capita PPP). Burkino Faso, Mali, Niger, and Sierra Leone have the highest HPIs in Africa, between 55 and 66 (Niger) percent.

### Primacy of Agriculture

Agriculture remains the main source of employment (50–90%) and income (25–50% of GDP) for most countries (see Figure 3.17). For the majority of people, this means subsistence—based on growing crops in the south but increasingly on herding livestock toward the north. Commercial agriculture was developed in the coastal countries and especially in Nigeria, Ghana, Côte d'Ivoire, and Senegal. Nigeria, Ghana, and Côte d'Ivoire are major world producers of a range of commercial crops including palm oil, cacao, rubber, tropical fruits, rice, and coffee (Figure 3.21). Benin and Togo produce smaller amounts. Senegal, The Gambia, Nigeria, and Mali export peanuts that are grown in the drier areas. The interior countries have few commercial farm products apart from their livestock, which provided meat for the coastal countries until the major droughts of the 1970s decimated the herds.

Through the 1980s, Côte d'Ivoire diversified its established products such as cocoa (of which it is the world's leading producer) into bananas, pineapples, and rubber. Côte d'Ivoire is now the second largest producer of rubber in Africa after Nigeria. In the 1990s, the government privatized the largest rubber plantations. Growers benefited from the CFA devaluation, with exported crops earning up to twice the previous price. Liberia became a major producer of rubber following the establishment of plantations by the Firestone Rubber Company in 1926, although civil strife disrupts the output and trade.

**FIGURE 3.21 Western Africa: plantation agriculture.** Cacao trees in a small plantation near Oyo, Nigeria. The pods contain beans from which chocolate and cocoa are made. They grow directly from the trunk of the tree.



Ghana was less successful in its farming policies until the 1990s. Its agriculture sector grew slowly compared to that of Nigeria in the late 1980s and early 1990s. While Ghana retained its major crop marketing boards, which set quotas and took a percentage of income, Nigeria abolished its crop boards, bringing its farmers greater income. Nigeria invested more in crop research and farm management education and has a better rural infrastructure enabling farmers to get their produce to market more easily. Cocoa, which has been through a long period of low world prices, is more important to Ghana's economy than to Nigeria's. Ghana's structural adjustment policies did not affect agriculture until the late 1980s, whereas Nigeria, which took little interest in structural adjustment, implemented agricultural reforms quickly from the mid-1980s.

Shifts of agricultural development policy led first to large externally funded projects and later to smaller-scale projects that involve the local farmers more centrally. In the 1970s, countries that had neglected the development of food production found themselves paying for expensive food imports. They attempted to invest income from oil, mineral, and crop sales to make up the shortfall by producing more food at home. Large-scale water projects in the drier parts of northern Nigeria, Senegal, and The Gambia were established to increase the production of wheat, maize, and rice for local and national markets. Unfortunately, poor management of human and natural resources in these projects resulted in crop yields that were lower than expected, costs that were higher, and disaffected farmers who were given little latitude for their own decision-making. Subsequent development projects in the 1980s shifted emphasis to focus on the need to involve local communities. Funds were used to sink boreholes for water, worked by hand pumps, and to provide rural infrastructure, advice, and credit. It was soon learned that a sudden and total shift to commercial crop production using high-priced new strains of crops and heavy applications of fertilizer was unsustainable environmentally and financially. Better projects combined commercial and subsistence crops, as well as row and tree crops with some pastoral farming. The large-scale projects were examples of unsustainable development and were followed by more sustainable development that supported local people as well as produced crops for sale in world markets.

### Forestry, Fishing, and Mining

Forest products remain significant in Côte d'Ivoire and Liberia. In many countries of Western Africa, however, the former coastal forests were largely cut and replaced by tree-crop plantations. In Nigeria, attempts to maintain the woodland set aside as national forest reserves are continually being challenged by farmers and timber companies.

Fishing is of growing significance for a number of coastal countries, particularly those close to the cold Canaries' current such as Mauritania, Senegal, and Guinea-Bissau. Lake fish and river fish are important foods in the interior countries such as Niger and Mali.

Western Africa does not contain the mineral riches of Southern Africa, but mining provides significant contributions to income for a number of countries. Metallic minerals are important in Ghana, where production in the gold-mining industry increased in the 1990s following major new investments and where bauxite mining linked to local smelting of aluminum powered by the Volta Dam is a major output. Iron ore is mined in large quantities in Mauritania and Liberia. Nigeria produces tin and Niger has major uranium deposits. Phosphate deposits are mined in Senegal and Togo and others exist in Mali. Guinea possesses one-third of the world's reserves of high-grade bauxite and is the world's second producer. Further development of its mineral resources requires investment capital to improve infrastructure. Guinea also has major gold and diamond reserves and iron mines near the Liberian border, where production was halted by civil war.

The main mineral developments are in the oil and natural gas fields of southern Nigeria (Figure 3.22) and in smaller quantities in Ghana and Côte d'Ivoire. Nigeria is now a major world producer of oil and oil products. Oil was discovered in Nigeria in 1956 and production expanded to 3.4 million tons

**FIGURE 3.22 Western Africa: Nigerian oil refinery.** Africa's main oil field is located in the tropical rain forest of the Niger River delta. Notice the storage tanks, the refinery area on the left, and the stack flaring off natural gas. Can you imagine why local people have objected to the oil field developments on environmental grounds?

in 1962 and a maximum of 114.2 million tons in 1979 before falling back to 73 million tons in the mid-1980s. Oil continues to provide over 90 percent of Nigerian exports. The oil income led to grandiose infrastructure projects funded by loans that turned into debts and to a reduced enthusiasm for economic diversification and for selling Nigeria's other products in world markets. Despite the great hopes placed on using its oil income to boost its economy, living standards in Nigeria are now lower than before the oil boom of the 1970s.

The use of energy resources as a basis for local development is beginning with the building of thermal power stations in Ghana and Côte d'Ivoire, using the natural gas that was previously flared off from Nigerian oil wells. Plans to build a gas pipeline along the coast await agreement on whether it should be on land (Nigeria's preference) or offshore along the seabed (Ghana's preference).

Western Africa supplies world markets with a wide range of products from farm, forest, mines, and ocean waters. Some countries diversified their output and so are not as dependent on one product as they were. Yet all these products remain sensitive to world markets and exchange rates. After the 1970s, during which the prices of oil, metallic minerals, and agricultural exports remained high, the 1980s were disastrous for many countries in Western Africa as world prices fell. Crops are affected by weather as well as market fluctuations. From the 1970s, droughts affected peanut and cattle production in particular. Furthermore, former western European buyers of palm oil and peanuts as sources of vegetable oil began to produce their own oil crops, causing a decline in demand for palm oil and peanuts.

## Manufacturing

Manufacturing has limited development in Western Africa, remaining only 5 to 15 percent of GDP. Most countries have import-substitution industries producing food and drink products, construction materials, and other low-price or high-bulk goods for local markets. Export crops and minerals are sometimes processed locally in oil-refining, textile, fertilizer, alumina, rubber, or fishmeal factories. In Nigeria and Côte d'Ivoire, there are car assembly factories. The lowest proportions of manufacturing industry occur in the landlocked inland countries. As a result of the devaluation of the CFA franc, some local manufacturing industries developed in Western Africa. One of the most obvious is the manufacture in Côte d'Ivoire of the highly colored printed "pagne" cloth used in women's dresses across Western Africa—after years of importing it from Europe.

## Services

The service sector of the economy in Western African countries grows slowly, although government employment is a major factor in many countries. Structural adjustment counsels the reduction of civil service bureaucracies, but this is causing pain to many countries because of the importance of family links into such employment. Tourism, a service industry that is part of the development strategy of many poor countries, was little developed in Western Africa, apart from servicing small groups of travelers crossing the Sahara, until numbers of African Americans began seeking their family roots in the area.

## Development Policies

The countries of Western Africa appeared as stronger candidates for structural adjustment in the 1980s than the countries of Central Africa. Ghana is more committed to the idea than most. In 1965, Ghana and Nigeria combined produced 65 percent of Western Africa's GDP because of their established markets for tropical crops in Britain. Ghana then declined and was replaced by Côte d'Ivoire in many of its markets as the latter country invested in its production and marketing facilities. The military rulers of Ghana kept payments to cocoa farmers low and many farmers ceased production. After 1983, Ghana restructured its economy with a view to encouraging exporters. This policy is moderately successful in attracting investors to expand mineral production and in placing the marketable agricultural products on a better footing. The growing prosperity in Ghana led to urban households buying more domestic electrical goods and therefore placed extra demand on electricity supplies. Unfortunately, the increased demand coincided with dry conditions that lowered the lake levels at the main source of hydroelectricity. This led to proposals for building new hydroelectric facilities and thermal power plants.

Nigeria's military government began to implement structural adjustment in the mid-1980s but by 1990 abandoned market reform, snubbed external creditors, and returned to the state control of economic policy. By 1995, loss of confidence internally and externally forced the Nigerian government to return to policies of opening the economy, even though this went against the vested interests of powerful groups inside the country. The 1995 budget included liberalizing Nigeria's exchange rate, reducing its budget deficit, allowing foreign investment, and making oil revenues subject to central control. Many still doubt Nigeria's will and ability to implement such reforms as its infrastructure decays.

## Political Management

One of the continuing problems for the countries of Western Africa, as in Central Africa, is their political leadership. The colonial period left a varied legacy of specialized economies and tribally based organizations. Initial constitutions were often soon converted into one-party systems or were overridden by dictators and military rule. The result was a narrow vision for the future in which political favors formed the basis of internal politics on an almost feudal basis. Short-term political goals, such as providing cheap food for urban populations and building grandiose projects, appeared to be of more importance than more efficient and longer-term management of the economy. Although many countries changed to multiparty constitutions in the 1990s, evidence for their implementation is not convincing, and the adoption of structural adjustment measures is often attempted with a minimum of enthusiasm.

**RECAP 3B: Central and Western Africa**

Central Africa is isolated from the world economy, and its countries' economies and human resources are poorly developed. There is some production and export of tropical crops, timber, and minerals. Most people depend on subsistence agriculture. Political disruption is common.

Western Africa is more involved in the world economy through its tropical tree crops and mineral exports. Its Atlantic coast provided connections to European traders and colonizers. Since independence around 1960, development has been slow, and the impact of Nigeria's oil industry is less positive than might have been expected.

3B.1 Which five geographic characteristics would you choose to distinguish Central Africa and Western Africa?

3B.2 Why is there a lack of engagement with the world economic system in Central Africa?

3B.3 What factors explain why some African countries have poorly developed manufacturing sectors?

3B.4 How important are tropical tree crops to Western Africa?

**Key Terms:**

Hutu tribe
Tutsi tribe
tropical tree crops
plantation agriculture
Communauté Financière Africaine

**Key Places:**

Central Africa
Burundi
Cameroon
*Yaoundé*
*Douala*
Central African Republic
Chad
*N'Djamena*
Congo (Republic)
*Brazzaville*
*Pointe Noire*
Congo (Democratic Republic, formerly Zaire)
*Kinshasa*
*Matadi*
*Lubumbashi*
Equatorial Guinea
Gabon
*Libreville*
Rwanda
Western Africa
Benin
Côte d'Ivoire
*Abidjan*
The Gambia
Ghana
*Accra*
Guinea
*Conakry*
Guinea-Bissau
Liberia
*Monrovia*
Mauritania
Nigeria
*Lagos*
Senegal
*Dakar*
Sierra Leone
*Freetown*
Togo
Burkina Faso
*Ouagadougou*
Mali
*Bamako*
Niger

## EASTERN AFRICA

**Eastern Africa** (Figure 3.23) has environmental, cultural, economic, and political contrasts with Central and Western Africa. Its landscapes are of high plateaus and towering volcanic peaks cut by rift valleys containing lakes. Its climates are generally water-deficient. Its economies are marked by a lack of mineral resources and a greater dependence on farm products. Its ethnic history includes a mixture of African influences, followed by peoples moving in from Arabia, while colonial development came from Britain and Italy rather than from France and Britain. Ethiopia was the only African country south of the Sahara that was never colonized from Europe, since the Italian invasion in 1936 lasted only five years. Whereas Western Africa was involved in the early years of the growing world economic system, Eastern Africa gained the interest of colonial powers mainly after the opening of the Suez Canal in 1869, when it achieved strategic significance and was brought closer to European markets through the new routeway.

**FIGURE 3.23 Eastern Africa: main features.** The map shows the countries included in the subregion, the major cities, and rivers.

Apart from Kenya, the countries of Eastern Africa have not developed their inland areas by linking them to major ports as in Western Africa. Ethiopia no longer has a coastline, the small countries of Eritrea and Djibouti have unproductive inland areas, and Somalia and Tanzania have not developed the hinterlands of their ports. The result of continuing internal conflicts, lack of natural resources, rising populations, and fluctuating world markets is that all these countries remain poor (see Figure 3.2). Ethiopia, Somalia, and Tanzania are among the poorest eight in Africa.

### Countries

The countries of Eastern Africa are **Djibouti, Eritrea, Ethiopia, Kenya, Somalia, Tanzania,** and **Uganda.** The physical diversity is among the greatest in Africa, including plateaus, volcanoes (Figure 3.24), rift valleys, desert, savanna, and tropical rain forest. Although they share location and physical characteristics, these countries are a diverse group.

Ethiopia, Tanzania, Kenya, and Uganda have some of the largest populations among the countries of Africa South of the Sahara, while Djibouti and Eritrea have very small populations. Ethiopia was ruled, until the 1974 communist revolution, by a monarchy established in the A.D. 100s, and then the revolution was overturned in 1991. Other countries are the creation of European colonialism from the late 1800s. Eritrea gained its independence only in 1993 after 30 years of fighting Ethiopia, but border fighting flared again in 1998.

The countries vary greatly in the levels of internal disruption, but all have suffered from civil and international wars in

**FIGURE 3.24 Eastern Africa: Mount Kenya.** A space shuttle view shows the remnant of former glaciers near the peak (white), the open forest of the upper slopes, the rain forest on the lower slopes, and the surrounding savanna that has been largely converted to farmland.

NASA

the second half of the 1900s. The Eritrean fight for independence left much of the new country with little working infrastructure. Ethiopia not only experienced two major revolutions within 20 years but also had wars on its territory against Eritrean guerrillas and Somali groups and has faced several periods of famine. Civil wars and harsh dictatorships from the 1960s to the late 1980s destroyed Uganda's prosperity. Somalia initiated attacks on Ethiopia in the late 1970s and reaped the outcome of having to accommodate over a million refugees. It then faced drought and civil wars in the early 1990s.

Kenya and Tanzania have, however, been affected less by civil strife since independence. Tanzania successfully resisted attacks from Uganda in 1978 to 1980 and is now trying to cope with refugees from Rwanda and Burundi. After prolonged preindependence guerrilla warfare against the British, Kenya experienced few further conflicts and used its political stability to develop links in the world economic system. It attracted investments and regional headquarter facilities for United Nations and multinational corporations, and has a strong tourist industry. Tanzania did not engage with the world economic system to the same degree as Kenya and attempted to develop its largely internal economy on socialist lines until the early 1990s.

## People

### Population Distribution and Dynamics

Most people in Eastern Africa—with population densities of 10 and over per km$^2$ (see Figure 3.12)—live in the better-watered upland areas of Ethiopia, Kenya, and Uganda. Few inhabit the semiarid areas where nomadic herding supports the people.

The population of Eastern Africa, like that of Central and Western Africa, is growing rapidly and probably even faster than the other two regions. The population of Eastern Africa rose from 91 million in 1980 to over 150 million in 1998 and could reach 250 million by A.D. 2025. In 1998, total fertility rates varied from 4 (Kenya) to 7 (Ethiopia). With death rates falling slowly (see Figure 3.13), population increase is rapid. Like most other African countries, the youthful element remains high (see Figure 3.14). Life expectancy rose from 35 to 40 years in 1960 to 40 to 55 years in 1998. In Uganda, however, it remained around 40 because of HIV/AIDS.

Apart from Djibouti and Eritrea, where almost all the people are urban, the other countries remain largely rural with fewer than 25 percent living in towns. The largest cities are **Addis Ababa** (Ethiopia) and **Nairobi** (Kenya), **Dar Es Salaam** (the current capital of Tanzania), **Kampala** (Uganda), Djibouti, and **Mombasa** (the main port of Kenya). They are growing fast because of migration from rural areas and natural population growth and contain extensive shantytown areas, contrasting with the impressive office towers in Nairobi.

### Ethiopian Cultural History

The cultural backgrounds of the countries in Eastern Africa divide them into two groups. The northern group centers on Ethiopia and reflects a history of attempts to expand or displace the Ethiopian monarchy. Established in the A.D. 100s by the Sheba peoples who claimed ancestry from King Solomon and Queen Sheba, the monarchs and people converted to the Coptic Church. Muslims failed to dislodge the monarchy on many occasions but succeeded in establishing coastal settlements that are now the countries of Eritrea, Djibouti, and Somalia. These mostly became part of the Ottoman Empire in the 1500s. Djibouti was the lone French toehold in Eastern Africa, becoming a colony in 1888 until it gained independence in 1977. Britain established its presence at Aden in Yemen to guard the approaches to the Suez Canal but also took over Egyptian-occupied towns in the territory that became British Somaliland. Italy colonized Eritrea and the Indian Ocean coast of present Somalia in the late 1800s. At that time, Italy's attempts to invade Ethiopia were repulsed, but Ethiopia succumbed to Italian invasion in 1936 for just five years until Britain restored the Ethiopian emperor. After World War II, the United Nations decided that Eritrea should be federated with Ethiopia and that British and former Italian Somaliland should be joined in the new country of Somalia. Eritrea fought Ethiopian and Soviet Union forces until it gained independence in 1993; the former British Somaliland declared its independence from Somalia in 1991 but is not recognized by other countries.

### Southern Coasts and Plateaus

The southern group of countries were affected less by the migrations and conflicts around the southern Red Sea area, although ports were established by Arabs as part of their

Indian Ocean trading system that was dominated by the sultans of Oman and Muscat until the late 1800s. Inland this region was a major crossroads of African peoples moving southward and eastward. Cattle herders from the Nile River valley lived alongside farming Bantu peoples migrating from Western and Central Africa. These people's lives were organized in small clans rather than the large powerful tribal kingdoms in Western Africa.

When European colonists arrived, they met little organized resistance to settlement. Britain got involved initially to end the slave trade based on Zanzibar Island by the sultans of Oman and Muscat (Figure 3.25), who also controlled trade in cloves and palm oil. In 1886, Britain annexed Kenya and Uganda and in 1902 built a railroad from Mombasa to Lake Victoria in Uganda. This encouraged British settlers to farm the fertile Kenya highland area and lands around Lake Victoria. German colonists began to settle German East Africa for coffee and tea production, but after World War I it became the British protectorate as Tanganyika. The British did little to develop Tanganyika and did not encourage white settlement there. These three countries gained their independence in the early 1960s following guerrilla warfare in Kenya. Tanganyika and Zanzibar united to form Tanzania in 1964, although there is still a Zanzibar independence movement. Zanzibar was one of the richest countries in Africa before it became part of Tanzania.

## Economic Development

Nearly all the countries of Eastern Africa are very poor. Most of them made limited progress in human development aspects such as education and health care in the period 1960 to 1992. Kenya stands out as the exception to this. Possession of consumer goods remains low (see Figure 3.16). All countries in Eastern Africa had very low incomes up to 1996, and the northern three countries had high HPIs of over 50 percent. Kenya, Tanzania, and Uganda had HPIs of less than 40 percent (Kenya, 26%).

**FIGURE 3.25 Eastern Africa: Zanzibar.** Zanzibar was a center of Arab trade along the coasts of Eastern Africa and across the Indian Ocean before Europeans arrived. The view from the top of the old Arab fort takes in part of the old city. Now that Zanzibar is part of Tanzania, many in the city campaign for its independence to be renewed.

### Agriculture Is Central

In the general absence of mineral resources and the slow development of manufacturing and service industries, most countries in Eastern Africa depend on agriculture as an economic base (see Figure 3.17). They earn foreign exchange by exports of farm products, but this is difficult for some countries with arid climates and poor soils. Many countries depend on a restricted range of products, which makes them susceptible to changes in world market prices. A boom in beverage crop prices in the early 1980s was caused by a shortfall in world production, but as production rose again prices fell. Coffee makes up 90 percent of Ethiopian exports; coffee, tea, and tobacco are 97 percent of Uganda's exports; coffee, tea, sisal, cotton, cashews, and cloves are 85 percent of Tanzania's exports. Even in Kenya, with its greater economic diversity, coffee, tea, sisal, fruit, and vegetables provide around 50 percent of exports. The arid countries in the north and in the adjacent parts of Ethiopia are not able to grow crops and rely on exports of livestock. Somalia's livestock exports make up 65 percent of its total exports, augmented by bananas grown in the south. Few of these dry countries balance their imports by exports, and they rely on external aid.

Most people in Eastern Africa gain a livelihood by subsistence farming or nomadic herding, but this is subject to the vagaries of weather. Long droughts have been a feature of the region since the 1970s, leading to major famines in Ethiopia in the 1970s and 1980s, made worse by civil war that prevented supplies reaching the famine areas. In the famine of 1983 to 1985 that was publicized worldwide, over half a million people died of starvation or diseases contracted at the feeding sites. This famine was related partly to soil erosion in the formerly fertile highlands area of Ethiopia, which led to migrations of people northward into the zone of drier conditions and civil war (see "World Issue: Famine in Africa," page 112).

Kenya developed the most diversified farming community. Following independence, large areas of the highlands around Nairobi that had been farmed by Europeans were taken over by the government and the land was redistributed as smallholdings. Kenyans were encouraged to grow both commercial export crops, such as coffee and tea, and food crops for the home market. Kenya provides some comfort for those who see more hope than alarm in Africa. For example, the Machakos district east of Nairobi was a region of periodic drought that suffered famine and soil erosion in the colonial period. After independence and land reform, smallholders invested in high-value crops, earning money to support this by off-farm jobs. Better tools and crop varieties were introduced and a second crop grown. Former badlands were transformed into terraced hills and fenced fields. Good road links to Nairobi markets made it possible to get fruit and vegetables there. By 1990, the area had 1.4 million people and output had risen 15 times with yields per hectare (2.47 acres) increasing 10 times.

Tanzania took a different approach from the Kenyan emphasis on small, commercial family farms as its government applied socialist principles. It tried to regroup the scattered farmers into communal farms and centralized villages, and guaranteed low prices for food across the country. Farmers lost interest in increasing productivity, and the Tanzanian economy failed to grow. Although it also exports some of its crops, its total exports are less than half those of Kenya. Since the late 1980s, however, efforts have been made to improve farm management, including using more fertilizers. Experimental maize crops improved yields from 2 to 3 kilograms per hectare in 1988 to 18 to 20 kg in 1994. Lack of storage, the high cost of fertilizer imports, and poor roads are problems in the way of further advances.

### Few Mineral, Forest, or Fish Resources

Other primary products are few in Eastern Africa. There is some offshore fishing and lake fishing, but the waters do not have great fish stocks. There are few mineral products apart from some salts from dried lakes in Ethiopia and Kenya, a little oil in Kenya, and a small output of diamonds and gold from Tanzania. Tanzania has iron ore and coal reserves in the south of the country, but transportation links are poor. Many Tanzanian resources remain isolated from world markets. When the Chinese built the TanZam railroad between 1968 and 1975 to connect Zambian copper mines to the Tanzanian port of Dar es Salaam, they took no account of potential development opportunities within Tanzania.

### Variable Manufacturing Development

Manufacturing makes up between 5 and 13 percent of GDP in Eastern African countries. Although several countries gave emphasis to manufacturing in their early development plans, only Kenya achieved much. In addition to processing its crops, Kenya has oil refineries and medium-sized consumer goods factories. It is a growing regional center of production and distribution. The other countries do not have the home demand for manufactured products or the foreign exchange needed to buy machinery and spares. Most countries have small-scale enterprises that produce goods for local markets, based on processing local crops and manufacturing foods and drinks. Ethiopia has metal and chemical industries that were nationalized in the communist era, but such large-scale industries contrast with the small-scale craft manufactures of other goods.

### Services and Tourism

For most countries the service economy is poorly developed apart from a government bureaucracy. Once again, Kenya is ahead of the others. Its relatively stable political environment attracted offices of the United Nations and the regional and continental headquarters of multinational corporations to Nairobi. Kenya also made progress in its tourist industry, attracting 0.75 million visitors in 1996. In years when coffee and tea suffer poor prices, tourism is Kenya's main source of foreign exchange. People from the core countries visit Kenya mainly to see the many animals in the national parks (Figure 3.26). Hotels and transportation facilities are developed to support this industry.

Tourism provides a major potential for earning foreign currency through attracting core-country visitors to the unrivaled scenic grandeur and wildlife. The commitment of African governments to conservation in designated game and national parks resulted in Africa having a higher proportion of such land uses than any other continent. Unfortunately, the governments have little money to spend on the prevention of poaching, while the sentimental insistence of people in core countries on total preservation of wildlife makes realistic policies difficult to devise and manage. Elephant numbers in Africa are around 700,000—down from several million in the 1800s and reduced by poaching in the countries of Eastern Africa. In Zimbabwe, Botswana, and South Africa, however, where poaching is controlled by policies that involve local people in management, elephant numbers increased from around 4,000 when Europeans arrived to around 70,000 today—twice the numbers the ecosystems can support. High annual culls are necessary if the parks and local agriculture are not to suffer and a range of wildlife are not to be endangered. Yet concerns about the situation of elephants in Kenya were publicized and led to a worldwide ban on ivory sales. The ivory accumulating in Southern African warehouses could have funded famine relief in the early 1990s or new projects that would employ more people in managing the wildlife.

Although many of the staff of the parks and hotels are poorly paid, tourism brings a welcome injection of foreign currency. Like other products, however, tourism is subject to fluctuations according to cycles of prosperity and recession in the core countries. The hotels in Nairobi have a twofold clientele of tourists and business people. Other countries, especially Tanzania, also have tourist industries, but the egalitarian Tanzanians resisted building large luxury hotels and their tourist industry remained small until the government encouraged a major expansion of park resorts in the mid-1990s. It had 0.3 million visitors in 1996. Warfare in the other countries deters most visitors.

## Political Changes

The 1990s saw improvements in the political situations of several countries in Eastern Africa. The independence of Eritrea appeared to end the civil strife affecting Ethiopia. The new Ethiopian government that took over in 1991 is wrestling more realistically with the latter country's economic problems, although its attempts to work toward a federal country based on tribal loyalties is not easy to implement. After 20 years of civil war and repressive regimes, Uganda returned to peaceful conditions. Somalia, however, descended into bitter civil war among several armed factions. Even the might of the U.S. military could not impose peace on that country. At the end of the 1990s, however, Kenya faced potential strife and disruption that focused on the future of political control, while Ethiopia and Eritrea reopened hostilities.

Few of the Eastern African countries apart from Kenya exhibit advantages for competing in the world economy.

**FIGURE 3.26 Eastern Africa: Masai Mara National Park, Kenya.** Tourists view a lion from the safety of their bus. Tourism based on such viewings is a major source of foreign income in Kenya and other countries in Africa, but its methods are often criticized by those who wish to preserve the animals.

Even in Kenya, the continued economic growth has had to contend with poor weather that reduced crop yields, population pressure on land, a shortage of urban employment, and political strife.

## SOUTHERN AFRICA

**Southern Africa** (Figure 3.27) has the greatest potential for leading the rest of Africa into sustainable human development. It contains countries that have some of the best records of economic progress in the continent based on mineral wealth, diversified agriculture, and some manufacturing. Over a third of Africa's rail mileage is in this subregion. Such potential rests on a fragile immediate past. For most of the second half of the 1900s, civil wars devastated Angola and Mozambique. The Republic of South Africa was isolated over a crucial period by its apartheid policy and resulting sanctions, and countries such as Malawi and Zambia remained among Africa's poorest (see Figure 3.2).

The future of this region depends very much on the Republic of South Africa after its rejection of apartheid policies, its democratic elections of a truly national government in 1994, and its restoration to the world economic system following decades of isolation and sanctions. South Africa has one-third of Southern Africa's population but produces three-fourths of its GDP and is regarded as the "engine" that could drive the economy of Southern Africa. Its own outputs and exports of minerals, farm products, and manufactures are much greater than those of the other countries in the subregion, which have many economic ties with it. Many families derive considerable income from the remittances of their men who work in South Africa.

The Republic of South Africa's presence dominated the subregion through the apartheid years, largely as a perceived adversary. Countries such as Botswana, Malawi, and Zimbabwe made public pronouncements against apartheid but maintained economic relations with South Africa. Lesotho and Swaziland, virtually encircled by South Africa, maintained political relations. The South African military supported guerrilla groups in the civil wars in Angola and Mozambique to destabilize its potentially antagonistic neighbors. Namibia was occupied by South African troops until 1990, when the United Nations finally got its way and made Namibia independent. The deal included the removal of Cuban forces from Angola in 1988.

When the other countries in Southern Africa banded together in 1980 to form the Southern Africa Development Coordination Conference (SADCC), their first purpose was collective opposition to South African apartheid. In 1992, the name was changed to **Southern Africa Development Conference** (SADC) and the emphasis moved toward providing economic arrangements to reduce their dependence on South Africa. Since the abandonment of apartheid, SADC welcomed the 1994 membership of South Africa to strengthen mutual linkages in trade and to help control traffic in illegal drugs and arms. The other countries of Southern Africa are pleased that South Africa is putting regional cooperation at the head of its foreign policy agenda.

Southern Africa stands at the southern margin of the tropical climatic environments and includes warm midlatitude conditions (Figure 3.28). The Republic of South Africa has midlatitude climates with winter rains at the Cape and summer rains along the southeastern coasts. This made it particularly attractive to European colonial settlement. The northern

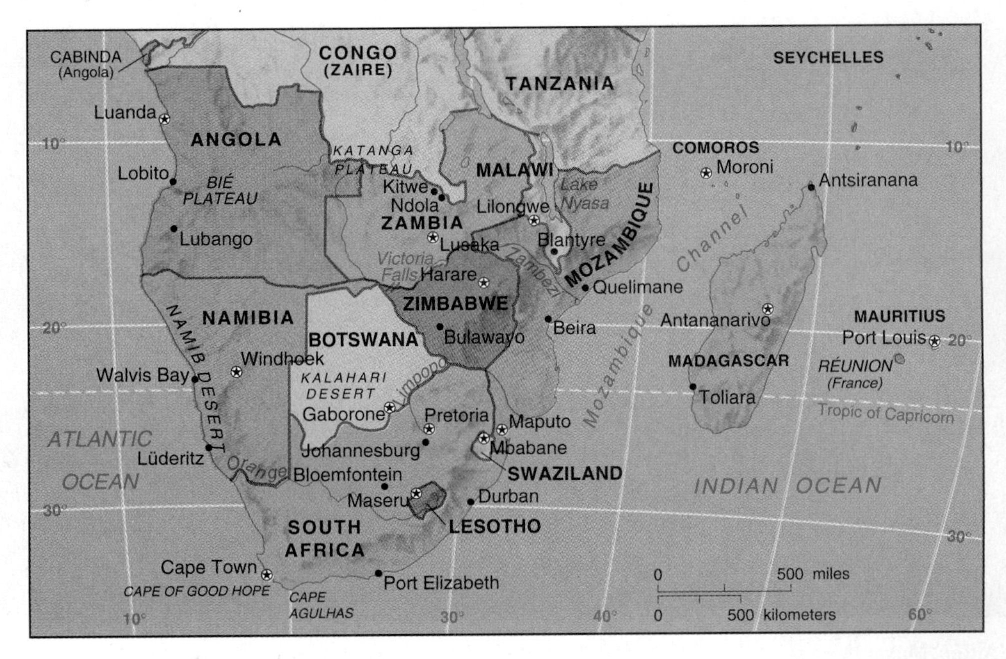


**FIGURE 3.27 Southern Africa: main features.** The map shows the countries included in the subregion, the major cities, and rivers.

**(a)**

**(b)**

**FIGURE 3.28 Southern Africa: contrasting environments.** (a) An early morning view of Cape Town and Table Mountain, South Africa. The site provided the foundation for one of Africa's major ports. The development of the area is constrained by the landforms. It is the midlatitude southernmost tip of Africa, with a climate marked by dry summers and cool, wet winters. (b) Desert near Usakos, Namibia: an area largely empty of people. Describe the barrenness.

WORLD ISSUE

## FAMINE IN AFRICA

Some of the points at issue include:

How is famine defined?

What causes famine?

Can famine be prevented?

Famine is a severe shortage of food occurring over a wide area and causing deaths by starvation and by diseases that take advantage of a reduced ability to fight them following malnutrition. Famines occurred in various parts of the world throughout human history, but in the last 20 years, most have occurred in Africa South of the Sahara. Despite huge injections of aid from other parts of the world, many countries remain in danger of famine from year to year. It is likely that Africa will continue to be the hungry continent into the new century as its population growth outstrips economic growth. Famine and chronic malnutrition will continue to be major problems.

Famine results from shocks to natural and human systems that disrupt food production or distribution. Natural shocks include drought, river flooding, insect plagues, and plant diseases. Human shocks include wars, civil conflict, widespread poverty, inefficient food distribution, and population growth that exceeds the ability of a country or region to feed the extra mouths. It is often not so much the immediate result of such shocks but the failure of human systems to cope with them that causes famines. For example, recent African famines have occurred on the semiarid margins of deserts where droughts that are part of climate changes have had greater famine impacts than before. The reasons for these greater impacts include pressures from increasing populations, civil wars that disrupt transportation systems, and the abandonment of long-established systems of cultivation and livestock keeping that made allowances for coping with drought.

The much publicized famines in Ethiopia and Somalia occurred when stresses resulting from drought were made worse by the difficulties of getting food to populations cut off from supplies by civil war. By mid-1992, up to 4.5 million people in Somalia faced starvation. Baidoa and Bardera, two small towns west of the capital Mogadishu, lie in some of the country's best farmland. After refugees from other areas moved in, famine struck both. At Baidoa, where the population was swollen to 60,000, over 100 died each day; in Bardera, 30 to 40 died each day. Neither town had community kitchens, and the issue of dry food was irregular. Farther inland, Belet Uan had to rely on daily flights sponsored by the Red Cross to bring in food. Distribution was slowed by the stipulation that all food be distributed through Mogadishu instead of being landed directly at points along the coast. Half the food distributed by truck was lost to looters.

Famines in Zimbabwe and southern Sudan occurred in 1991 and 1992, when the drought impacts were increased by the arrival of hundreds of thousands of refugees from war in Mozambique or famine in Ethiopia, respectively. Famine in Mali and Niger occurred in the 1970s when livestock herds had to be slaughtered because the grass had dried up rapidly when overgrazing was followed by drought.

Famines get emergency attention from governments and from international aid agencies. Governments can usually cope with the onset of famine if they have distribution systems that deal equally with urban and rural areas. Famines are more prevalent in rural areas where provision is uneven, where transportation is poor, and where it is difficult for the people to migrate rapidly into the better provisioned urban areas. Rural areas are also at a disadvantage since many people there are undernourished compared to those in towns. Poor nourishment gives famine conditions a start as the young and old succumb rapidly to starvation and killing diseases. One of the best means of using limited resources to control famine in southern Zimbabwe in 1991–1992 was the provision of lunches for elementary school children by aid agencies. This policy kept most children attending school and ensured a good level of nutrition. In previous famines, children were neglected at home and many died. Agencies also provided seed for planting maize in the following year, leading to a more rapid return to economic health in rural areas.

International agencies bring food and medical aid to emergency situations. They are limited in what they can do by response time and access to affected areas. It takes time to assemble staff and purchase, ship, and deliver emergency food to the needy country. Once there, internal transportation is often so poor that only limited quantities can be shipped rapidly. Such emergency food often brings wheat flour and milk powder not normally consumed by local people, who may need a cultural shift to continue consuming such food.

It is clear that better systems are required to cope with the threat of famine. Early-warning systems are being imple-

parts of this subregion have seasonal tropical climates with summer rains decreasing toward the arid region of Namibia and western Botswana. Most of the subregion is formed by a series of plateaus drained by major rivers such as the Zambezi, Limpopo, and Orange-Vaal system and cut by the southernmost extension of the East African rift valleys.

### Countries

The countries of Southern Africa are: **Angola**, **Botswana**, **Lesotho**, **Madagascar**, **Malawi**, **Mozambique**, **Namibia**, **South Africa**, **Swaziland**, **Zambia**, and **Zimbabwe**. They range from Africa's richest (South Africa) to its poorest

(a)

(b)

**Images of famine.** (a) Women and children at a Red Cross feeding center near Rama, Ethiopia, in 1986. Disease and death often spreads rapidly in the centers. (b) Starving people in the Wamma camp, Kismayo, Somalia, in 1992.

Steve Raymer/National Geographic Image Collection

© Norbert Schiller/The Image Works

mented in some areas by monitoring nutrition levels. Improvements in education, population control, political stability, and infrastructure in the poor countries of Africa are basic to reducing the threat of famine in the long term. For some years to come, it will be necessary to improve short-term aid delivery.

Mass starvation through famine is, however, not the main problem of hungry Africa. Inadequate food supplies affect all parts of the region long term. Commonly seasonal, and particularly affecting women and children, malnutrition is at crisis levels, linked to urban poverty and poor harvests in rural areas. Malnutrition forms a basis for short-term dramatic famine episodes. The causes are complex and wide-ranging, from the decline in environmental quality, the greater value placed on inexpensive food for urban areas and commercial crop production, and the lack of rural-urban transportation facilities, to the growing population creating demands that exceed supplies.

### TOPICS FOR DEBATE

1. Famine is the result of failures in human systems rather than an "act of God."
2. Parts of the world outside Africa are liable to suffer from famine in the next decade.
3. Famine was common through human history and is a normal part of human experience.

(Mozambique and Malawi) and include a few that have intermediate economic conditions, such as Botswana, Namibia, and Zimbabwe. They include the war-wrecked countries of Angola and Mozambique and countries that have had more stable governments such as Botswana, South Africa, Zambia, and Zimbabwe. "More stable" is a relative term, since Zimbabwe went through a period of guerrilla and open warfare in the late 1970s and South Africa experienced considerable unrest during the apartheid years.

The presence or absence of mineral resources is one of the major distinctions within Southern Africa. Countries such as Angola, Botswana, Namibia, South Africa, Zambia, and Zimbabwe are well endowed, while Malawi, Mozambique, Lesotho, and Swaziland have few exploited minerals. The large island of Madagascar has a unique history and people but its future could be increasingly linked to Southern Africa.

## People

### Population Distribution and Dynamics

In Southern Africa, the population is concentrated around the coasts, along major inland routeways, and in the urban growth areas of the Republic of South Africa (see Figure 3.12). Almost half the area has fewer than one person per $km^2$ because of the extent of semiarid and arid environments.

The population of Southern Africa is set to rise rapidly—as it is in the rest of Africa South of the Sahara. Total fertility rates range from 7 in Angola to 3 or 4 in Botswana, Lesotho, South Africa, and Zimbabwe. Birth rates fell in all countries since 1970 (see Figure 3.13), but high proportions of younger age groups dominate the population structures (see Figure 3.14). Although the Zimbabwe population grew faster than the African average of over 3 percent in the mid-1980s, its 1998 increase was 1.5 percent, a rate that could be soon outstripped by economic progress. At first, Zimbabweans were reluctant to take up birth control when it was seen as a white colonialist plot to reduce black population growth, but a rural-based family planning network made it popular. A male motivation campaign helped to reduce the beating of wives discovered using contraceptive methods and made men aware of the need to control HIV infection by using condoms. By the late 1990s, HIV/AIDS was worst in Zimbabwe and Botswana, increasing death rates and lowering life expectancy and natural increase.

Life expectancy is around 60 years in South Africa but below 45 years in nearly all the other countries. The population of Southern Africa, which grew from 81 million in 1980 to 110 million in 1998, may rise to 180 million by A.D. 2025—a slower rise than expected in the early 1990s.

South Africa is the only country in Africa South of the Sahara with a sizeable nonblack population. In 1998, 75 percent of the population was black, 14 percent white, 3 percent Asian, and 8 percent mixed race. These differences are reflected in life expectancy (72 years for whites, 64 years for blacks) and in total fertility (1.8 for white women and 5.2 for black women).

### Urbanization

Zambia and Zimbabwe are over 60 percent urban because of their mining towns and urban industries, respectively. The Republic of South Africa is 50 percent urban, but other countries are less than 25 percent urbanized. The process of urbanization is thus under way but still in its early stages in most countries. The lack of manufacturing in countries outside South Africa, Zambia, and Zimbabwe gives their towns less drawing power, although the major cities have rapid growths of shantytowns. The South African proportion living in towns might be even higher without the apartheid policies that forcibly sent black Africans to their tribal homelands. It is likely that urban growth will be more rapid now that freedom of movement is allowed.

South Africa has most of the large cities in Southern Africa, including **Cape Town, Johannesburg, Durban, Pretoria,** and **Port Elizabeth. Maputo** (Mozambique) and **Luanda** (Angola) are also very large cities that have grown mainly as a result of migration from rural areas to national capitals during the civil wars in those countries. **Harare** and **Bulawayo** are the largest cities in Zimbabwe, while **Lusaka** (Zambia) and **Antananarivo** (Madagascar) are capital cities with large government bureaucracies.

### Waves of Settlement

The cultural environment of Southern Africa is varied, reflecting the last two centuries of history that witnessed major shifts of peoples of all colors. For a long time, low densities of the San and Khoikhoi (Hottentot) peoples occupied the area. They were gradually displaced by the advance of agricultural and cattle-herding Bantu tribes from the north during and after the A.D. 600s. Kongo tribespeople moved into northern Angola first, but it was not until the 1700s that Bantu peoples reached the present South Africa. The building of Great Zimbabwe (see Figure 3.4) as the center of trade in gold, ivory, and cattle occurred from the A.D. 700s. The Mutapa Empire was based there from the 1300s. By the early 1800s, the Bantu tribes in southernmost Africa—the Swazis, Zulus, Xhosas, and Sothos—were fighting each other for territory and building strong kingdoms.

### South Africa

Europeans came to settle, increasingly contesting the ownership of land with the Bantu tribes and each other. The Dutch first settled the Cape area after A.D. 1652, forcing the indigenous Khoikhoi northward. They built Cape Town as a major port on the route to the East Indies and established settlers on cattle farms. Separation from the Netherlands gave these settlers a distinctive name (Boers) and Dutch-based language (Afrikaans). In 1814, the British purchased the Cape Colony from the Dutch. New colonists arrived demanding the use of English, the end of slavery, and the protection of the Khoikhoi. The Boers undertook the "Great Trek" northward to the Orange and Vaal river valleys, where they established the Orange Free State and Transvaal. The Boers displaced the Ndebele tribe, who moved north of the Limpopo River, and the Zulus, who moved south into Natal, creating tensions with the local tribes. The British then extended their colonies east of the Cape into Natal, where there were wars against the Xhosas and Zulus.

Although the Boers declared a South African Republic in their new lands, the discovery of diamonds and gold there from the 1860s and the occupation of South West Africa (modern Namibia) by the Germans in 1884 caused Britain to take action. It annexed Bechuanaland (modern Botswana) to block German-Boer links and extended protectorates to Basutoland (modern Lesotho) and Swaziland after threats from Boer attacks. In 1899, the Boers declared war on the British, who had been instigating local conflicts. Although the British soon took the major centers, a costly and inconclusive guerrilla war followed.

After the Boer War, the four colonies—Cape, Natal, Orange Free State, and Transvaal—joined to form the Union of South Africa as a self-governing dominion within the British Empire in 1910. Afrikaans speakers led the governments while attempting to keep a degree of unity among the whites until 1948, when the apartheid policy was imposed. Apartheid laws and linked informal measures increased separation of the white, black, and mixed races. Special housing areas were designated for each group, and each black person was assigned to a homeland area. From the 1960s, while policies of racial separation became less significant in other countries, including the United States, the Republic of South Africa stepped up its efforts to continue and extend the separation. This largely isolated the country from the world political and economic systems.

The combination of internal resistance to apartheid by leaders of the African National Congress (ANC), such as Nelson Mandela, and external isolation and sanctions led to the relaxation of the apartheid policy in the 1990s, signaled by the release of Mandela from long-term imprisonment. Democratic elections in 1994 resulted in a "Government of National Unity"—effectively, the first by the majority racial group in South Africa—with Mandela as its first president. South Africa continues to struggle with the issues raised by apartheid, including the geographic expressions of townships, homeland policies, and separate racial schooling. The dual communities of wealthy whites and poor blacks have tolerated each other despite a slow economic transition. The Reconciliation Commission dealt with many sensitive issues in a positive manner. The end of Mandela's political presence in 1998 will be a major test for the future of this experiment.

By the late 1990s, South Africa—which, along with Brazil, has the greatest rich-poor extremes in the world—was experiencing economic development difficulties as managers and professionals left. They moved out of the country for apparently better opportunities and living quality, especially less crime, in Australia, Canada, the United Kingdom, and the United States. This "white flight" reduced other employment opportunities in South Africa, although some potential leavers stayed for the sunny, mild climate and lower costs of living.

South Africa also faces problems of internal management, since the provinces set up by the 1994 constitution have some local control but no financial resources. Although the idea helped to obtain agreement to the post-apartheid arrangements, it may be changed if the 1999 elections return a large ANC majority.

### Portuguese Colonies

Although events in South Africa are particularly important for understanding the cultural, economic, and political environments of Southern Africa as a whole, the colonization processes in other countries were also significant. The Portuguese were the first European colonists, reaching northern Angola in 1483 and establishing relations with the local Kongo ruler. They visited the Mozambique coast in 1498 and took over rule in the following century from the Arab traders who had built coastal cities as centers for their Indian Ocean trade. Both Angola and Mozambique were devastated by the slave trade, which continued through the 1800s. Over 3 million slaves were sent across the Atlantic to Brazil from Angola alone.

Portugal did not encourage settlement by Europeans in its colonies until the early 1900s, when there was a push toward greater economic exploitation led by European technology, finances, and administrators. There was little attempt to provide schooling for the African population, and any dissension was brutally repressed. Mineral resources in Angola and plantation crops in Mozambique were exploited. As European settlement increased, nationalism gained strength and rival groups carried on guerrilla warfare in both countries until a revolution in Portugal led to the independence of Angola and Mozambique being granted in 1975. From that date until the 1990s, the rival guerrilla groups fought each other, with one side often backed by South Africa. The Mozambique civil war ended in 1989. In Angola, a treaty was signed in 1994, but UNITA forces keep inland bases and their leader does not emerge to take part in national politics.

### British Colonies

Britain developed the swathe of land between Angola and Mozambique, sometimes in response to cries of help from tribes who perceived the threat of Portuguese colonial repression to be greater than that of British administration and economic development. Tribes in Malawi (1891) and Zambia (1889) requested British protection, while the Ndebele tribe in western Zimbabwe signed contracts with Cecil Rhodes for a British mining company to exploit the minerals on their land (1888). Railroads built from South Africa into this area shipped out the minerals. Britain extended protection to these new colonies, and settlers came to take up the good farming land, particularly in what was Southern Rhodesia (modern Zimbabwe).

### Madagascar

Peoples from the East Indies and Africa occupied Madagascar, but the Portuguese, French, and English all attempted to colonize the island. They found it difficult to overcome the forces of the powerful rulers in hilly and forested terrain. By the end of the 1800s, the French established a colony, but internal dissent gradually rose until independence was granted in 1960.

### Late Independence

Independence came relatively late to many of the other countries in Southern Africa, apart from South Africa, which became a self-governing country with loose ties to Britain in 1910. South Africa declared itself to be a republic without ties to Britain in 1961. While most countries in the rest of Africa became independent around 1960, Angola and Mozambique had to wait until 1975 and Zimbabwe until 1979. The white settlers in Zimbabwe first tried to dominate the former countries of Nyasaland (now Malawi) and Northern and Southern Rhodesia (now Zambia and Zimbabwe) by establishing a united Federation of Rhodesia in 1953, but this was resisted by black Africans and dissolved in 1963. In 1965, the whites in Southern Rhodesia unilaterally declared independence, but Britain and other countries did not accept this move. Years of internal strife and international sanctions delayed full independence until 1979.

Namibia became a protectorate under South Africa in 1920. Efforts by the United Nations to cut this tie when apartheid became South African policy after 1948 were defeated by the South African refusal to give up the territory until 1990, when Namibia became independent. Of the other countries, Zambia and Malawi achieved independence on the breakup of the Rhodesian federation in 1964, Botswana and Lesotho in 1966, and Swaziland in 1968.

## Economic Development

Southern Africa has around 20 percent of the population of Africa South of the Sahara but produces over 50 percent of the total GDP. This is largely because of the economic dominance of the Republic of South Africa, but some other countries in the subregion also have growing economies. Economic growth and human development in the subregion have not paralleled that in the rest of the world but are better than in the rest of Africa. Botswana, Namibia, South Africa, and Swaziland have real incomes that place them in the World Bank's middle-income group. Madagascar, Malawi, Mozambique, and Zambia remain very poor, with HPIs of over 45 percent (Zambia, 35%). Outside of South Africa, however, economic development is related largely to the presence of minerals and the absence of civil war.

### Agricultural Diversity

Agriculture is the economic mainstay in Lesotho, Madagascar, Malawi, Mozambique, and Swaziland, where over 85 percent of the population gains a living from subsistence farming (see Figure 3.17). Maize is a common staple food, grown on thousands of small farms in Southern Africa.

Farming is also significant in the more diversified economies of South Africa and Zimbabwe, where commercial farming related to world markets is expanding. South Africa made itself self-sufficient in food during the years of international sanctions against its apartheid policies. It produces a variety of temperate grains, vegetables, and fruits, together with sugar, cotton, and livestock products. Commercial crops provide nearly 40 percent of Zimbabwe's exports by value. Although its tobacco production is declining, commercial farmers are diversifying into vegetables and even flowers for European markets.

In Malawi, tobacco, tea, and coffee make up 70 percent of the country's exports, and in Madagascar, coffee, sugar, vanilla, cloves, and cacao also make up 70 percent of the exports. In Zambia, the potential for commercial farming was neglected in the years of high copper prices, and only one-fifth of its good arable land is utilized. The opening of Zambian frontiers to trade with Zimbabwe in the 1990s led to the flooding of Zambian markets with cheaper Zimbabwe food. In Mozambique, agricultural production declined to less than three-fourths of the 1980 level during the civil war and food imports were necessary—although by the late 1990s there were signs of a rapid return to self-sufficiency in food.

Land ownership is a major issue in the parts of Southern Africa where white settlers took land. Newly independent governments have to weigh the political advantages of returning that land to black Africans against the loss of income from commercial farms (see "Living in Zimbabwe," page 120). In Zimbabwe, most commercial farms are efficient and provide valuable export crops. The government took over some commercial farmland for redistribution. Evidence exists that peasant farmers in Zimbabwe can be more productive than large commercial farms because they use the land more intensively, but there are also cases where the new smaller landholdings were quickly affected by soil erosion and abandoned. In the early 1990s, droughts affected the southern part of Zimbabwe where the communal areas and small farms are mostly situated but had less impact on the commercial farms in the center and north. This emphasized the differences between the two types of landholding and created pressure for political action to convert more large commercial farms to smallholdings. The commercial farms gained further advantages as marketing controls were dropped as part of structural adjustment and small farmers lost various forms of government assistance. In the late 1990s, the Zimbabwean government continued to threaten action to appropriate the white-run commercial farms.

### Mining Wealth

Mining dominates the economies of Angola, Botswana, Namibia, South Africa, Zambia, and Zimbabwe, involving them deeply in the world economic system (Figure 3.29). The possession of such resources is not always a recipe for prosperity; it needs to be linked to stable political conditions, adequate transportation facilities, and good management of the national economy if multinational corporation investors are to be attracted. Multinational corporations do not always return wealth they gain to local communities. Mining often has negative environmental impacts, including air and water pollution and desertification of local areas.

The history of mining in Southern Africa illustrates the manner of European colonization. In the early 1900s, groups

**FIGURE 3.29 Southern Africa: mining.** The Okiep Copper Mining Company at Nababiep, northwestern Cape Province, South Africa. This is typical of many mines in the subregion: the piles of waste after the ore is separated from its containing rock, the pithead buildings, the smelter with its tall stack; location in a rural area with the surrounding vegetation killed by gases from the smelter.



such as that under the British entrepreneur Cecil Rhodes bought or appropriated lands that contained diamonds or gold. Previous black African or Afrikaans landowners were dispossessed, or paid small sums of compensation, at times with British military backing. Railroads linked mining centers around Johannesburg to the Cape ports and then extended northward into new mining and farming areas named after Rhodes—the Rhodesias. Rhodes and the company he controlled, De Beers, achieved a monopoly of the world diamond trade, laying a foundation for further mineral explorations.

South Africa is the world's top producer of platinum and one of the major producers of gold, diamonds, and several strategic metal ores. Platinum is of growing importance because of its use in aerospace construction and pollution control (catalytic converters) in core countries. South Africa has 40 percent of the world's gold reserves and 68 percent of the platinum reserves. Together with minerals that are important in special steels, such as chrome, manganese, and vanadium, mining products, often refined in South Africa, make up two-thirds of the country's exports. South Africa's mining industry, however, is subject to increasing challenges. It mined 70 percent of the world's gold in 1970 but by 1995 this was down to 27 percent as Australia, Canada, and the United States increased their shares, together with new mines in other African countries such as Mali and Ghana. As the world gold price fell, South Africa's costs of deep mining went from some of the world's lowest to its highest. The major South African corporations are now investing in gold mines in other parts of Africa, rather than in their own mines.

Namibia is the world's fourth largest exporter of nonfuel minerals and one of the major producers of uranium, together with diamonds, zinc, and copper. In Botswana, diamonds make up 80 percent of exports, and the country became wealthy by exploiting these resources since the late 1970s.

Zambia was a major world producer of copper, sharing a rich ore field with southern Zaire, but the need to restructure its copper industry lies at the heart of the Zambian future. Nationalization in Zambia led to a halved mineral output. In

**FIGURE 3.30 Southern Africa: Namibian desert.** The red sands of the desert dunes are a distinctive feature of this arid area, made drier by the offshore cold Benguela current that flows northward toward Angola in the distance. The clouds provide some rain to the higher inland areas, the cold ocean current is a rich fishing ground, and some of the world's richest diamond mines are found in the foreground.

NASA

the 1980s, income was further affected by falling copper prices. A lack of investment to replace exhausted mines, combined with heavy debts, prevented the Zambian mines from making a profit in the 1990s. In world markets, it is up against Chilean mines that produce more copper with half the workers (see Chapter 10). By the late 1990s, plans to sell the mines to foreign corporations expecting to redevelop them were well advanced.

Zimbabwe's mineral output, making up 40 percent of its exports, includes coal, gold, chrome, and nickel in a wide belt along a major igneous intrusion between Bulawayo and Harare. Output is increasing after some years of standstill, and a new platinum mine, opened in 1994, is the largest foreign investment in Zimbabwe for 25 years. It is expected to make Zimbabwe the world's second largest producer of platinum after South Africa. In Swaziland, the iron ores have been mined out and asbestos has ceased to be in demand.

Angola has oil fields along its northern coast, together with diamond mines inland. Although its northern land-based oil installations were destroyed during the civil war, oil companies now recognize the deepwater (over 1000 m deep) offshore resources as of major world significance. By 1996, annual oil sales of $4.5 billion made up 80 percent of Angola's revenues as nine new oil fields came on stream. Production could triple within 20 years. Angola was once the world's fourth largest diamond producer and could be higher than that when its huge deposits are exploited. For the moment, that is held up by the continued UNITA occupation of the interior diamond lands (and the support UNITA receives from annual illegal diamond sales of $500 million).

### Timber and Fishing

Of the other primary products, timber is of little significance, apart from locally, and most countries in Southern Africa have to import their needs. Fishing is likely to increase in importance in Namibia (Figure 3.30), following the 1992 start to effective policing of its territorial waters. Angola and South Africa also have access to the productive cold current off southwestern Africa.

### Manufacturing Contrasts

South Africa also dominates manufacturing industry in Southern Africa, together with the linked power utility and transportation systems. As soon as mining began a century ago, there was a need for engineering services and chemical supplies, since the ores had to be refined near the mines to reduce the bulk for transportation. During World War II, sources of machinery and other manufactured goods in Europe were cut off and South Africa diversified its manufacturing base into food products, textiles, clothing, armaments, and motor vehicle assembly in a process of import substitution.

International sanctions resulting from its apartheid policy caused South Africa to increase its manufacturing base further, raise its coal output from 50 million tons in 1970 to 200 million tons by 1990, and develop technologies for using coal as a source of chemicals. The main coal-mining area is on the high veld east of Pretoria, where coal-burning plants generate 80 percent of South Africa's electricity and are responsible for acid rain downwind. The great majority of South Africa's manufacturing is concentrated in the urban-industrial area around Johannesburg and around the ports of Cape Town and Durban. During the period of apartheid policy and the establishment of tribal homelands, factories were built just outside the homeland areas for access to cheap labor.

Six large South African corporations run almost totally by white people dominate mining and manufacturing. The new South African government has a commitment to enable black people to enter the white-controlled business world throughout the country. At present, most black businesses, apart from a major brewery and an insurance company, are small and in trade rather than manufacturing. The restoration of South Africa's access to world markets following the end of apartheid and the free elections of 1994 exposed its manufacturing industries to competition from cheaper overseas products. It is hoping to export some of its own products, especially armaments, which have been in demand from other countries and previously were exported in a clandestine manner. Unemployment remains a great threat to political stability in the country as the business sector expands cautiously and the government spends money on training but not on creating many new jobs.

As South African markets open to foreign manufactures, the large local business conglomerates and continuing political uncertainty often deter foreign investment. There is considerable capital within South Africa for investing in local and wider industrialization and infrastructure in Southern Africa, but much goes abroad. Some foreign investment has occurred, largely because multinational corporations bought back companies they pulled out from in the 1980s. New investment was slow to start, but by 1994 companies such as Levi Strauss, Nestlé, and some South Korean electronics manufacturers established factories.

After some years of progress, the late 1990s raised questions about the future of the Republic of South Africa's (RSA) economy. Pessimists predicted it will slide into economic collapse, lawlessness, ethnic warfare, and corruption; optimists retort that South Africa has already refuted many such worries and is building a prosperous and equitable economy by removing sources of instability. Both overstate their case. Economic growth continues but at a slower-than-desired rate to carry out necessary reforms; drought could wreck the harvests that underpin the stable economy; gold and other metal prices on world markets are low. Although the economy expands, jobs do not at a time when 20 to 30 percent of adults have no formal employment. Education is not at the forefront of reforms and many schools have declined in quality. Crime levels remain high, and the future as President Mandela retires is uncertain in terms of maintaining current policies and of the treatment of white people. Yet economic growth continues, and there has been no need to resort to external aid. Some benefits, such as domestic water and electricity supplies, are now available to areas of poorer housing; a growing black middle class enjoys the fruits of affluence; and the African National Congress-led government seems to be managing the economy well.

Zimbabwe has the only other development of diversified manufacturing in Southern Africa. Industrialization began before the fight for independence, when cotton textile plants (Kadoma) and an iron and steel mill (Kwe Kwe) were established. The country possesses good power sources from its own coalfields at Hwange and hydroelectricity at the Kariba Dam, although more than 10 years of drought in the 1990s lowered the Kariba lake level to the point where only a small fraction of the potential hydroelectricity could be generated. During the period of unilaterally declared independence, factories were built to produce goods that were kept out by sanctions. Manufacturing grew from 5 percent of GDP in 1965 to 26 percent in the 1990s. With the reopening of its South African trade and easier access to world trade, Zimbabwe now faces the need to move from import-substitution industries to compete with other world manufacturers.

Among the other countries, agricultural products are processed in Malawi and Swaziland and minerals are refined in Namibia and Zambia. Portugal did not encourage manufacturing in Angola or Mozambique until the 1960s, but by independence Angola had vehicle assembly and chemical factories, while Mozambique produced steel and textiles using hydroelectricity from Cabora Bassa Dam on the Zambezi River. After independence, skilled managers and technicians departed, some sabotaging the factories, and further decline occurred during the civil wars. In the mid-1990s, it was estimated that Mozambique industries operated at less than half their capacity.

### Services

Service industries, from shops and banks to government jobs and tourist facilities, are increasing in importance throughout Southern Africa, but especially in South Africa and Zimbabwe. Tourism is a growing feature of the economies in Botswana, South Africa, and Zimbabwe. In 1996, South Africa received around 5 million tourists, Zimbabwe 1.5 million, and Botswana 0.7 million. In Zimbabwe, tourism is the fastest growing sector of the economy, and the country attracts more tourists to its side of the Victoria Falls than Zambia does to the north. As with Kenya, there is a special interest in the national parks where large animals are protected and able to live in something like natural conditions.

# ZIMBABWE



Visitors arriving for the first time in Zimbabwe's capital, Harare, are often surprised at how very western the city is, with its shiny glass skyscrapers, wide avenues, and leafy suburbs. The more adventurous who travel to the rural communal lands, where most of the country's population lives, may be equally surprised at just how like the Africa of story books life appears, with villages of small, round thatched huts, cattle pens, the sound of drums at night, and perhaps an occasional passing game animal.

Such a brief tour of Zimbabwe may leave the impression of two quite separate countries in one—an untouched piece of Africa inlaid with a rather incongruous modern infrastructure. Yet this would be a misapprehension. Speak to any black businessperson or academic in Harare and you will probably find that they still regard the family village as their home. Alternatively, head out into the bush in search of the remotest village you can find. You will almost certainly discover there a shop selling ice-cold beer, soap, cookies, and bread, with perhaps a record player in the corner, belting out Township Jive music at full volume. These apparently separate worlds are, despite appearances, deeply interwoven into the whole that is Zimbabwe. For most Zimbabweans the bright lights of the cities are no more than a few hours' bumpy bus ride away, and, if they have the money, they can purchase many of the gadgets available to Americans. The livelihoods of the majority, however, are still mainly dependent on subsistence agriculture.

Gutu Mission is situated in communal lands in the middle of the country (see map) and has a history that stretches back a hundred years or so to around the time the country was founded by the British diamond tycoon Cecil Rhodes. He and his motley band of pioneers ruthlessly claimed for themselves a chunk of Africa that would become the Crown Colony of Southern Rhodesia and later Zimbabwe. Since that time, the country experienced development through the 1900s. The mission played its part, educating children from the surrounding villages—including Simon Mazenda, one of the country's vice presidents—in Shakespeare and Newton's Laws of Motion and providing the modern facilities of a hospital. Yet only a stone's throw from the school's well-equipped science labs, people live almost exactly as they did when the first German missionary set up camp in Gutu over a century ago. Despite the dramatic turns of history over the subsequent years, the pressing issues for most people remain: Will it rain this year? Will my cattle survive? Will I be able to feed my family?

Sadza, a stiff maize porridge, is the staple food of Zimbabwe. And very good it is, too, with a little rich meat stew or vegetable relish. During the hot, rainy season that lasts, with luck, from November to February, the communal lands turn green and maize grows everywhere. In a good year, families will have plenty to eat and will make enough money by selling what they have left over to the government for resale or export. They can pay the school fees for their children and perhaps buy a few luxuries, such as new clothes or even a radio.

During the cooler dry season, the sandy soils of the communal lands are left bare, turning the area into what resembles a desert. This is the time of hunger, when people make do by living on stored grain and what they can afford to buy with their savings. In the 1991–1992 rainy season, barely a drop fell in Gutu. Countrywide, crop production fell to a fraction of what it would normally have been. The government was forced to spend scarce foreign currency on importing maize from the United States.

Drought has been the trend over the last decade or so. And this, combined with increasing population pressure, is putting a noticeable strain on natural resources. There are fewer and fewer trees and less grazing land. People in Gutu say that a few years ago everywhere around the mission was covered in thick bush, which was home to a host of wild animals, including antelopes and leopards. These are now gone, and you would be lucky to spot a few baboons through the thin covering of trees that still exists on the more inaccessible rocky outcrops.

More frequently now, people are inclined to look with envy toward the large, profitable, mainly white-owned commercial farms. These farms, established by settlers during 90 years of white rule, tend to be situated on the richer soils and are worked intensively by armies of black laborers. The inequitable distribution of land in white-ruled Rhodesia was one of the main grievances that fueled the Bush War of the 1960s and 1970s that finally led to an independent Zimbabwe in 1980. Yet even now, the division of land is little different to what it was before the war began.

President Robert Mugabe, who was a rebel leader during the war, has yet to find a solution to the problem of land distribution. Around 4,500 white farmers own over half of Zimbabwe's cropland, while 8 million black peasant farmers crowd into the rest. Since independence, the United Kingdom paid for purchases of land at market prices on which 62,000 African families were resettled. If Mugabe turns over more large areas of commercial farmland to peasant farmers, he runs the risk of badly disrupting the economy. These farms produce tobacco, cotton, meat and dairy products, grain, horticultural produce, and game, forming a pillar of the Zimbabwean economy. Most of these areas of farming, particularly on a large scale, are alien to traditional African agriculture, and it is difficult to foresee any successful wholesale takeover. Alternatively, if the president does nothing to address the problem, he faces the accusation of letting down his own people.

Attempts at land redistribution have been, and are being, made. A few miles from Gutu lie several old commercial farms, abandoned by whites around the time of independence. Half the white population fled across the border to South Africa or away from Africa, probably in fear of what a new black, Marxist-inspired government might do. The abandoned farms were divided into resettlement schemes

**(a) Map of Zimbabwe**

**(b) Urban Zimbabwe: central Harare, capital city.** Jacaranda trees in bloom in Harare Gardens, with African Unity Square, the Episcopalian Cathedral, and new office and hotel buildings beyond.



for families of ex-combatants of the war and other landless people. One immediate result of this policy was that former farm laborers became squatters overnight. In the long term, however, these schemes produced very little. In fact, many of the people who now live on these schemes work on the commercial farms that still exist nearby. One school teacher, an ex-combatant who fought under Mugabe, lamented one night after a few beers about how one of these resettled farms, which had been a major milk producer for the region before independence, now produced nothing.

Recent government policy is moving toward the more heavy-handed approach of forcibly buying underutilized commercial farmland at a predetermined price and using it for further resettlement. Predictably, there has been much disquiet among the farm owners and potential investors about the unfairness of this policy and the likelihood of its failure. In a populist bid during 1997, President Mugabe threatened to take half of the commercial farmland without payment to the present owners. Requests to the U.K. government for help in making payments were refused on the grounds that the Zimbabwe proposals did not guarantee that subsequent land distribution would be open and would benefit the poor. Mugabe's threat coincided with falling tobacco and mineral export income and his promises to pay further sums to civil war veterans, and led to a financial crisis. Such a massive takeover of white commercial farms would affect laborers employed on the farms and could reduce exports by one-third.

The land issue is not the only huge challenge facing Zimbabwe. There is also the need to restructure the economy, the mounting scourge of HIV/AIDS, unemployment, population explosion, and environmental destruction. But by other African standards, Zimbabwe has still got a lot going for it. It has what must be one of the best educated populations in Africa, strong agricultural, mining, and tourist industries, a good infrastructure, and the enduring bedrock of a vibrant African culture, which has produced music and art that have become famous throughout the world. Indeed, Zimbabwe has been held up by many as a model for the new Southern Africa. Whether it continues to develop into the next century or falters badly seems to depend critically now on two things: wise political leadership and the progress of the economic giant to the south.

**RECAP 3C: Eastern and Southern Africa**

Eastern Africa received cultural influences from Arabia as well as interior Africa and was colonized mainly by Italy and Britain after the opening of the Suez Canal. Kenya has experienced most economic development and has a diversified economy based on agriculture, tourism, and service industries. Ethiopia and its surrounding countries have been held back by drought and civil war. The latter also affected Uganda. Eastern Africa has few mineral resources.

Southern Africa contains poor countries such as Angola, Malawi, and Mozambique but also the most affluent African country (South Africa) and some that are diversifying their economies (Zimbabwe) or have mineral wealth (Botswana, Namibia). The newly democratic South Africa sees its role as being involved in the wider development of Southern Africa.

3C.1 Which five geographic characteristics would you choose to distinguish between Eastern and Southern Africa?

3C.2 How did the precolonial cultures and the impacts of colonial occupations by different European countries affect the present geographies of the subregions of Africa South of the Sahara?

3C.3 What are the past, present, and likely future impacts of the Republic of South Africa on other African countries?

**Key Term:** Southern Africa Development Conference (SADC)

**Key Places:**

| | | |
|---|---|---|
| Eastern Africa | Uganda | Namibia |
| Djibouti | *Kampala* | South Africa |
| Eritrea | Southern Africa | *Cape Town* |
| Ethiopia | Angola | *Johannesburg* |
| *Addis Ababa* | *Luanda* | *Durban* |
| Kenya | Botswana | *Pretoria* |
| *Nairobi* | Lesotho | *Port Elizabeth* |
| *Mombasa* | Madagascar | Swaziland |
| Somalia | *Antananarivo* | Zambia |
| Tanzania | Malawi | *Lusaka* |
| *Dar Es Salaam* | Mozambique | Zimbabwe |
| | *Maputo* | *Harare* |
| | | *Bulawayo* |

## Outlook

In the late 1990s, grounds for optimism in the future of this subregion included:

- Infrastructure destroyed during civil wars in Angola and Mozambique is being reconstructed. In the latter, transmission lines take power from the Cabora Bassa Dam on the Zambezi River across 1,400 km of land containing few people to reconnect South African users.
- Port facilities at Maputo, Beira, and Nacala in Mozambique were expanded, together with the repair of railroad links into Zimbabwe and Malawi. The line from Lobito (Angola) into the southern mining area of Congo (Zaire) is being rebuilt.
- Privatization of copper mines in Zambia took place under partnerships with South African, Canadian, and U.S. companies.
- The development of new tourist facilities in Malawi and Mozambique testified to increasing security within those countries and greater openness to the rest of the world.
- The development of major industrial zones along new roads between Johannesburg (RSA) and Maputo (Mozambique) and between Windhoek (Namibia) and southern Angola testifies to increasing confidence by external investors and to the creation of regional marketing and development strategies.

Southern Africa possesses the potential for further growth, but world recessions, local droughts, and civil wars have held back the area and may occur again. South Africa is in the middle of a transition toward involvement in the world economy. If this can be accomplished with minimum civil unrest at home, it may be able to lead the other nations of Southern Africa forward. In recent years, some conflicts have been solved, including the granting of independence to Namibia and the end of fighting in Mozambique and (officially at least) in Angola. The democratization of South Africa leaves Angola as the one major area of potential unrest—a country that could also enjoy growing prosperity with the peaceful exploitation of its resources and the rebuilding of its rail links into southern Congo (Zaire). The HIV/AIDS epidemic, however, threatens the future of this subregion (see Figure 3.10) as death rates rise and life expectancy falls.

# LANDSCAPES IN AFRICA SOUTH OF THE SAHARA

## Urban Landscapes

The urban landscapes of Africa South of the Sahara reflect the historic cultural, economic, and political changes that molded the continent's geography. Towns are becoming increasingly important features of African landscapes as people move to them for higher incomes, better education and health facilities, safety, or better access to formal and informal job opportunities. People in major towns are more involved in the political life of their country and tend to set fashions in clothes, music, and food. Expanding towns are growing markets for food and consumer goods. The current rapid growth of urban areas leads to extensive tracts of new housing, shantytowns, and commercial and industrial landscapes. Many cities, however, still retain evidence of past functions although the uses of buildings may have changed. Accra, Ghana (Figure 3.31), and Kano, Nigeria (Figure 3.32) illustrate many of the features typical of African cities.

### Precolonial Cities

The oldest urban landscapes occur mainly in Western and Eastern Africa and are unusual from Congo (Zaire) southward. They include the trading centers of Timbuktu, Sokoto, and Kano at the southern end of trade routes across the

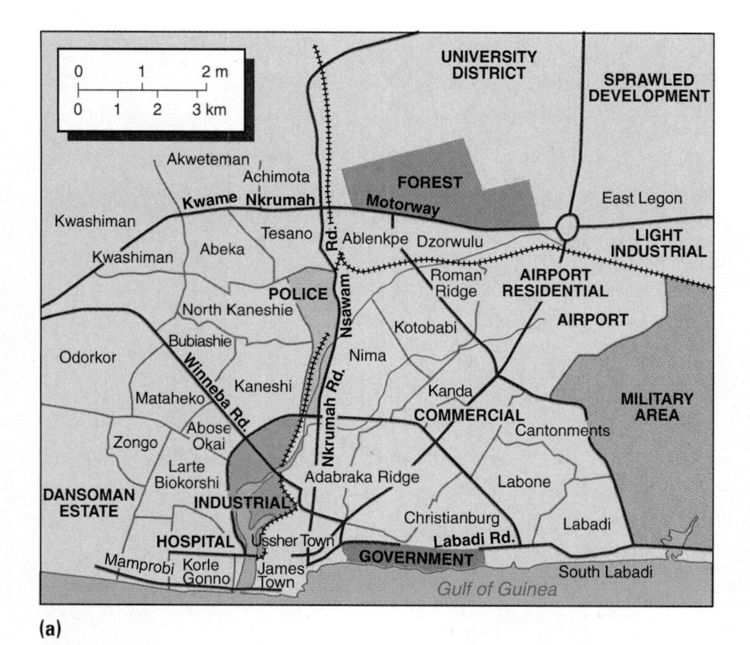


(a)

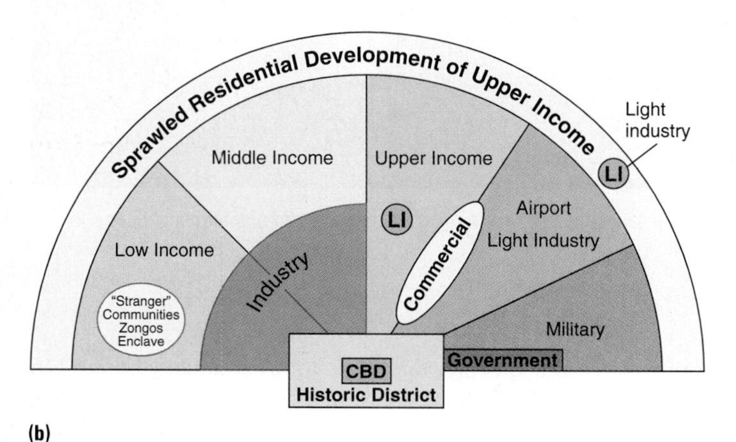


(b)

**FIGURE 3.31 African urban landscapes: Accra, Ghana.** (a) The port city of Accra, Ghana, expanded from a village for colonial trade, with port facilities along the lower river, flanked by government offices to the east and poor housing to the west. Account for locations of military, airport, and university districts. (b) A generalized pattern based on Accra that is repeated (with variations) in other Western African port cities.

3.31a Source: Data from S. Aryeetey-Attoh, *Geography of Sub-Saharan Africa,* Fig. 7.4, pp. 192–193, Prentice-Hall, 1997.

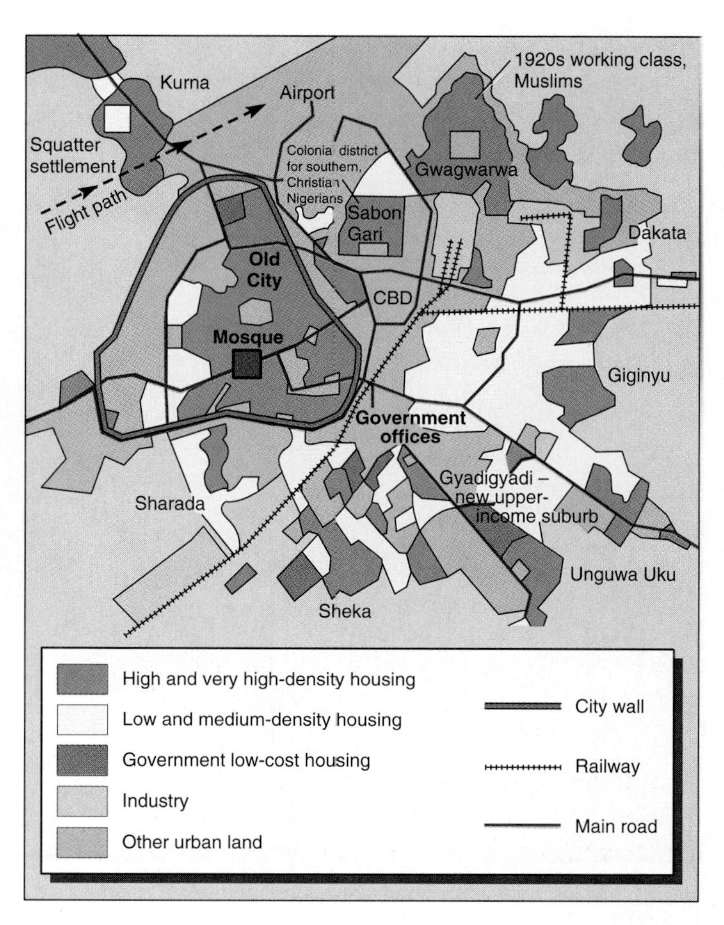


**FIGURE 3.32 African urban landscapes: Kano, Nigeria.** The inland center of Kano, northern Nigeria, reflecting over 1,000 years of change. The old city was built in the early A.D. 1100s and enlarged several times, with a central market linking trade routes northward across the Sahara and southward to the coast. The colonial period disrupted former trade patterns, but the railroad made Kano the major center in northern Nigeria with exports of cotton and peanuts, and led to the establishment of government offices, industrial and residential areas outside the walls. Since independence, manufacturing, education, and civil service jobs have increased, giving rise to the new residential areas ranging from high-income to poor squatter shantytowns.

Source: Data from R. Stock, *Africa South of the Sahara,* Vignette 13.1, p. 199, and Fig. 13.5 p. 201, Guilford Press, 1995.

Sahara. These towns are still dominated by craft workshops rather than modern industry (Figure 3.33). In Eastern Africa, the old Arabian ports extend from the Red Sea to Mozambique, including Mogadishu, Malindi, Mombasa, Zanzibar (see Figure 3.25), and Sofala (Mozambique), and contain historic buildings reflecting the pre-European era. Towns such as Kumasi (Ghana), Kampala (Uganda), and Addis Ababa (Ethiopia) also contain pre-European buildings, including palaces and churches. One particularly distinctive urban area that preserves a pre-European landscape in southwestern Nigeria is that of Ibadan, the center of the Yoruba culture, linked to a series of surrounding towns. Ibadan itself has a central palace and nearby market at the focus of streets radiating toward other towns. Fortifications were added later to resist Muslim attacks. The modern populations of these older towns remain much more ethnically unified than those in many of the newer towns. Islamic cities along the Sahel zone (e.g., Kano, Sokoto, N'Djamena) have central markets, mosques, citadels, and public baths.

## Colonial Cities

European settlement brought new types of urban centers to new locations, often leaving precolonial cities to decay as relics of the past. Colonial administrations established towns that sometimes used old city cores but were often built nearby to avoid being too close to centers of local opposition or disease. Many were ports linking the new colony to the home country

**FIGURE 3.33 African urban landscapes: precolonial.** The old buildings of Timbuktu, Mali, founded in the A.D. 1000s at the center of a former African kingdom and on trans-Saharan trade routes. Compare this urban landscape with others known to you.



for both administrative and commercial purposes. Major inland towns were linked to the ports by railroads. Brazzaville (Congo) and Kinshasa (Congo [Zaire]) provide a contrast on opposite banks of the Congo River. Brazzaville became the chief city of French Equatorial Africa where French diplomats and an educated African elite built large houses on quiet avenues. Its population grew after 1960 independence, from around 100,000 to around 600,000 today, and it remained a fairly sedate capital city. Kinshasa reflects grandiose Belgian purposes—continued in the Mobutu era—with its wide boulevards and post-independence construction of prestigious tall office blocks. It grew from around 100,000 people at independence to over 4 million people, with huge sprawling shantytowns. Other new towns grew as mining settlements, such as Johannesburg in South Africa, Enugu in Nigeria, and the series of towns that developed around the copper mines in southern Congo (Zaire) and northern Zambia. Nairobi in Kenya began at a place where railroad workshops were established along the route from Mombasa to Lake Victoria.

The colonial towns were marked by government offices surrounded by the large homes built for European civil servants, often at a distance from the commercial sectors or native townships. The commercial sector was often located near the docks or railroad station. Even though the people occupying them have changed, the former homes of white settlers in Harare, Zimbabwe, can be easily located on one side of the main shopping and business center and close to ornamental gardens. On the other side of the shopping center are industries and the bus station, and beyond that a series of townships in which the poorer African people live. The newer towns contain a great variety of peoples from the different parts of their country and beyond. Eastern African countries have Asian shopkeepers, and many who were expelled in the early days of independence, particularly from Uganda, returned in the 1990s.

## New Cities

There are also totally new cities, built since independence to express the national need for establishing a center of government without the associations of the colonial era. These towns, including Abuja (Nigeria) and Dodoma (Tanzania), often contain imposing buildings and wide streets, designed as a whole and often with grandiose intentions. In many cases, they were planned at a time when the country had expectations of income to be able to complete the construction but are often half built with people living in shantytowns.

**Shantytowns** are a feature of African large towns (Figure 3.34)—as they are of growing cities in many developing countries—and their inhabitants are often associated with the informal economy. The towns are unplanned, constructed of any materials that come to hand, from packing cases to builders' blocks and corrugated iron, and basic in

**FIGURE 3.34 African urban landscapes: shantytown.** Cape Town, South Africa. Comment on the building materials, size of buildings, arrangement of buildings, and security.



their services. Some condemn families to a hopeless future of poverty. In many cases, however, people move into them on arriving from a rural area but eventually find better accommodations. Governments may supply utilities and build schools, hospitals, and roads to integrate shantytowns with the rest of a large urban area.

Many African urban landscapes carry the scars of civil war. Massawa, the port of Eritrea, was heavily bombed in the fight for independence and is now being rebuilt. Cities in Liberia, Angola, Congo (Zaire), and Mozambique also show evidence of prolonged fighting, destruction, and abandonment.

## Rural Landscapes

Rural landscapes include cultivated landscapes, areas of little-disturbed natural vegetation, mining areas, and human settlements in villages and small towns. Some of the most familiar African landscapes are those in the national parks where large animals roam across savanna grassland (see Figure 2.35a). These landscapes preserve elements that are close to those that existed before intensive human intervention. Africa also contains extensive areas of tropical rain forest and desert, although most of the rain forest has been cut or modified.

The cultivated landscapes of Africa South of the Sahara include those of both commercial and subsistence farming (Figure 3.35). Commercial farming covers relatively small areas. Tree crop and bush plantations for palm oil, cacao, coffee, and tea give greater regularity to the landscape with their focus on single species growing to sizes that promote the best productivity. Field crops such as peanuts, cotton, tobacco, and vegetables produce landscapes similar to those producing such crops elsewhere in the world.

Subsistence farming reflects older farming methods used on small plots of land, often only partly cleared. Most villages grow a variety of crops and keep a few animals. In Central and Western Africa, root crops as well as grains are grown, but in Eastern and Southern Africa, maize has become the staple food crop.

Mining leaves local scars—deep pits or high spoil heaps—on the landscape. Some concentrations are more extensive, as along the Zambia and Congo (Zaire) border or around Johannesburg. Mines may be at a distance from major towns and have only a local village for the workers (see Figure 3.29).

Villages and small towns are a growing feature of the African landscape. The villages of mud huts roofed with thatched leaves or grasses are giving way to those built of concrete blocks and roofed with corrugated iron. Some

(a)

(b)

**FIGURE 3.35 African rural landscapes: commercial and subsistence farming.** (a) The Hex River valley, Cape Province, South Africa: commercial vineyards with large cultivated areas and a wine product that is sold overseas. (b) Terraced farmlands in the Shona area of the Eastern Highlands of Zimbabwe. Contrast the two landscapes.

villages have mushroomed into small service centers with rapidly built shops and market stalls, where food and small consumer goods are sold and buses from the surrounding rural area focus their routes. Other small towns, built once for colonial administration, are being encouraged by governments to take on new urban functions in rural areas. Both Tanzania and Zimbabwe have policies to develop small rural towns as service centers and so reduce pressures on the largest cities. Modern facilities such as electricity and piped water are now available in such towns, and people like the combination of a small business to run and their own nearby farm to live on. These towns signal the growing mobility of people, the importance of commercial exchange, and increasing African involvement in the world economic system.

## FUTURE PROSPECTS FOR AFRICA SOUTH OF THE SAHARA

Clearly the immediate prospects of people living in Africa South of the Sahara are poor. In most countries incomes are low, primary products—the source of most export income—are subject to world market fluctuations, and internal conflicts destroy any temporary gains that are made. One estimate holds that it will take 40 years for African countries outside of South Africa to climb back to their incomes of the mid-1970s. After the bleak years of the 1980s, some indicators show that African GDP began to rise in the 1990s. No account of Africa's geography would be complete, however, without an analysis of what caused the continent to lag behind the rest of the world and what might provide a better way ahead.

### Legacy

Africa South of the Sahara is caught in the worst core-periphery relationship, in which the core countries buy Africa's minerals and commercial crops at relatively low prices and sell back their own manufactured goods at high prices. Few African countries have their own manufacturing industries that can compete in world markets. This situation can be traced to past history and, perhaps strangely, to Africa's rich endowment of minerals.

#### Tribal Kingdoms and Slavery

African peoples developed a mosaic of tribal kingdoms that occasionally extended into wider empires or broke down into local clans. Major migrations across the continent continued into the 1800s. Medieval invasions of Muslim peoples from Northern Africa set off migrations of Bantu tribes from areas south of the Sahara eastward and southward. Arab traders on the east and Portuguese and other European traders on the west then depopulated areas through the slave trade, often assisted by more powerful African kings.

#### European Colonization

During the late 1800s, the European colonial countries divided Africa among themselves. The interior was settled selectively, where minerals or potential commercial farming areas were discovered. The colonial economy placed a reliance on primary products in demand by the core countries. Valuable minerals such as gold and diamonds, and other materials, such as ivory, were the initial draw to economic exploitation and the construction of transportation routes; attraction to other minerals followed as world market needs and available transportation dictated. Large untapped mineral resources, however, still await shifts in world market prices. Commercial crops, such as the tropical products of palm oil, cacao, coffee, tea, sugar, cotton, and peanuts, were encouraged, civil services organized, and railroads and ports built to export the produce efficiently.

The colonial powers imposed this economic system but ruled out manufacturing, apart from local needs, in case it should compete with their home products. Some colonial powers, such as France and Britain, were more paternalistic and provided basic education and health care. Others, such as Portugal, Belgium, and Germany (before 1914), were more repressive and provided little infrastructure outside the mining and plantation areas. Moreover, colonial powers often governed their territories by inflaming tribal rivalries or by generating so much hostility that rival guerrilla groups vied for a leading role in the push toward independence. Both sowed the seeds of conflict in newly independent countries. Such actions consigned African colonies to the world periphery.

#### Independence and Economic Colonialism

After independence, the African countries struggled with a limited economic role based on primary products in a world where people in urban-industrial countries made major leaps in well-being. Multinational mining companies and makers of coffee, tea, and chocolate products imposed their control of world markets in what is called **economic colonialism.** For example, major aluminum manufacturers encouraged external funding for the Volta River project in Ghana that generated hydroelectricity to refine bauxite, the ore of aluminum. Multinational corporations often maintained low world prices for raw materials by opening up new areas of production as growing markets absorbed that from established areas. Although raw material prices soared in the 1970s (minerals) and in the early 1980s (beverages), these booms were unusual. They often tempted the producer countries by tempting them into taking out loans on the strength of the high prices; when the prices fell, those countries faced debts they could not repay.

The Cold War from 1950 to 1990 resulted in political rivalries between the First and Second Worlds, worked out in civil wars that destroyed Third World countries. Internally, single-party governments often held back local enterprise but were supported by core Western countries on the grounds that they were stable and opposed to communism. After the end of the Vietnam War in the mid-1970s, Africa became the main Cold

War battlefield. When the Cold War ceased in the early 1990s, the authoritarian brand of socialism adopted by many African countries was discredited and the powers that had engaged in the Cold War viewed Africa as without strategic interest. In 1995, 47 African countries attracted only 3 percent of the foreign investment flowing to developing countries, while fewer countries in Latin America attracted 20 percent and Eastern Asian countries about 60 percent. This led many countries to depend on external aid.

### Poor Government and Internal Strife

The combination of unequal positions in world markets, external interference in their politics and economics, and internal tribal rivalries resulted in unstable governments and a trend to dictatorships in many countries as a response to the instability. Some countries were destabilized by the external activities of South Africa, which maintained a policy of supporting guerrilla groups in any of its neighbors that might help those antagonistic to apartheid within South Africa. When Angola drew in Soviet and Cuban forces to help its internal struggles, South Africa supported the UNITA opposition. Ethiopia became a battlefield as Soviet forces supported its communist government against rebels from Somalia or those fighting for Eritrean independence. French troops fought the Libyans in Chad. The catalogue of civil wars in African countries is long and continues, although they were reduced for a while after the end of the Cold War. Instability in government is also caused by the lack of experience of many African leaders, the military background of many, and the corruption that is part of so many administrations.

The 1990s saw a trend toward greater democratization in African countries. Many moved from single-party to multiparty constitutions and held open elections. Some evidence exists that political repression and corruption are decreasing. It is now conceivable that entrenched leaders may be removed by ballots instead of bullets. The full implementation of the new constitutions is often hesitant, people are easily discouraged if new governments do not produce economic growth, and tribal rivalries have surfaced in open conflict.

### Population Growth

In addition to such problems, African countries have to cope with high rates of population growth. In the period from 1980 to 1998, the total population of Africa South of the Sahara rose from around 380 million to 580 million. The rate of increase equaled or exceeded rates of economic growth and kept African countries in a continuous struggle to provide sufficient food and jobs for their growing populations. The total population could be around 1,000 million by A.D. 2025, adding urgency to reduce population growth or instigate more economic growth to cope with the enlarged and rapidly urbanizing populations. Efforts to encourage smaller families through family planning have had moderate success because of cultural pressures that regard large families as a high priority and give little control in these matters to women. Death rates in most African countries plummeted as modern medicine was applied, but birth rates in many African countries fall slowly if at all. Population control is one major key to Africa's future. Sadly, the most likely curb on population growth is the HIV/AIDS pandemic, which is increasing death rates in many African countries.

### Global Warming

As if such problems were not enough, world scientists now agree that global warming is happening. Previous doubts have receded. The consequences for Africa are likely to be worst in the southern part of the continent, the subregion with greatest prospects of a better future. Droughts, which got worse after 1980, are likely to be more frequent over the next 100 years. Such changes would affect both farming and water supplies for industry.

## Disappointments

African countries are well aware of the problems facing them, but disappointments arose from the lack of progress made since independence. In many countries economic well-being has declined.

### Failure of Inappropriate Policies

Some of the disappointments result from a history of applying policies that have not worked. At the time of independence for most countries in the 1950s and 1960s, it was received wisdom that countries could grow economically by following the history of the richer core countries in moving from the primary sector into manufacturing and from there into services. Both sides in the Cold War advised similar policies. Soviet Union experts were contemptuous of an unskilled peasantry, urging the building of a manufacturing base, even if it caused the destruction of other sectors.

Many African countries attempted this route, but none achieved a major breakthrough. External funding was used in major power projects such as the Volta River hydroelectric project in Ghana and the Kariba project on the borders of Zambia and Zimbabwe. Such external funding was often tied to the benefits to be gained by specific multinational corporations and ignored local concerns about drowning good farmland.

One of the wider outcomes was the neglect of farming, affecting both food production for the home market and commercial farm products for export. Countries such as Nigeria and Ghana, which had been among the leading African countries economically at independence, allowed farming to decline until the 1980s. Worsening farm prospects led many to migrate to large towns, where food was cheap. The falling local production was substituted by imports of temperate grains from the excess of core countries, establishing new tastes. At the time when countries wished to substitute local manufactures for costly imports, they had to substitute imported food for local produce. The imbalance of imports over exports was little affected. Further difficulties arose as increased droughts in the 1970s led to famine in the Sahel area and Ethiopia (see "World Issue: Famine in Africa," page 112), including the death of livestock herds as well as people.

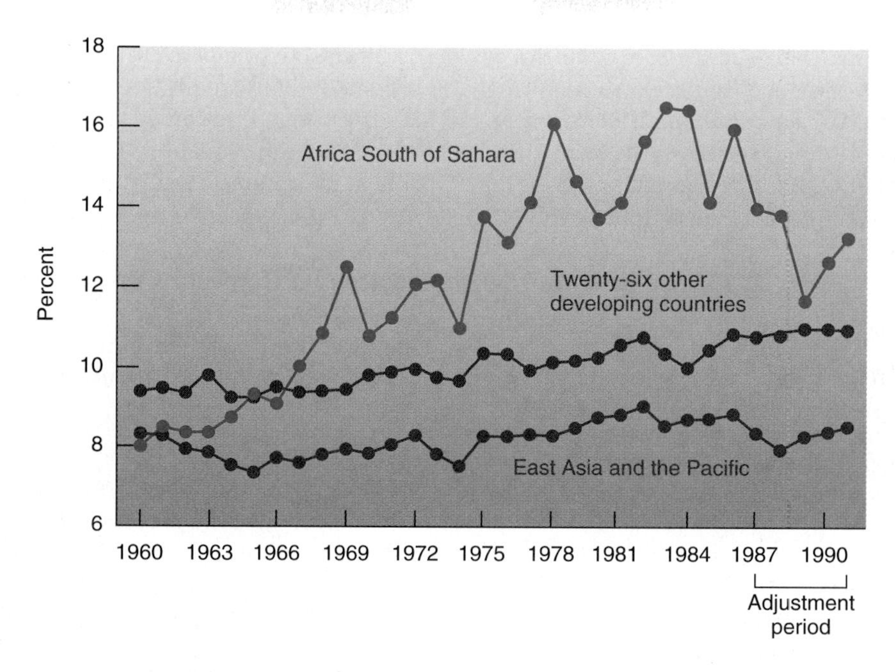


**FIGURE 3.36 Expensive governments in 25 countries in Africa South of the Sahara.** African governments spent more on goods and services than others in the world periphery. Although there was initial response to economic adjustment from 1987, expenditure soon rose again.

Source: Data from World Bank

## False Economies of Large-Scale Projects

By the 1970s, it was clear that African countries needed to invest more in agriculture to provide food for the home market and increase farm product exports. The first response was to invest funds earned in the growing receipts from rising mineral prices in major agricultural projects. Large areas of northern Nigeria, for example, were cleared for extensions of peanut growing. Although they had some initial success, such projects often failed when world market prices fell. Inappropriate management methods took responsibility away from local farmers, treating them as laborers. Subsequently, projects became community-based and relied more on supplying local markets and involving local farmers in decision making.

By the 1990s, the rate of population increase in Africa South of the Sahara was still around 3 percent, while that of agricultural production growth was only 2 percent. Per capita food production dropped by 12 percent between 1961 and 1995 in contrast to the Green Revolution increases in Asia (see Chapters 5 and 6). Food imports and food aid had to be increased to meet the shortfall, yet many Africans still had insufficient diets. Countries such as Nigeria, Kenya, Tanzania, and Benin are now increasing agricultural production faster than their rates of population increase, but all African countries have to take care that an increased intensity of production does not result in losses of soil quality or erosion.

## Currency Exchange Rates

Many African countries suffered by maintaining artificially high exchange rates for their currency or by not allowing their currencies to adjust to world markets. This was perceived as a matter of prestige and was encouraged by the CFA system in former French colonies. High exchange rates kept imports relatively cheap (thus favoring the products of core countries and discouraging home producers) but gave home products high prices in export markets. African countries became less competitive as other parts of the world grew faster and competed in world markets for primary products through cycles of boom and bust.

## Slow Results from Structural Adjustment

In the 1980s, the World Bank and the International Monetary Fund—the two major lending institutions for developing countries—established new guidelines for structural adjustment. This was a response to the low rates of success achieved by previous loans and the large proportions that were absorbed by high exchange rates and the cost of internal government bureaucracies (Figure 3.36). To date, these policies have not been successful in African countries, although Ghana, Tanzania, Burkino Faso, Nigeria, and Zimbabwe have come closest to following their precepts. Even these countries have not been great successes in terms of improving incomes or the quality of life for most of their inhabitants.

Some of the problems with these policies were discussed earlier in this chapter. Perhaps most African countries are at a stage of development where structural adjustment either is not yet appropriate or needs careful encouragement. At present, African countries seem to lose out whether they decide to follow structural adjustment policies or not. If they do not adopt them, they lose access to funds from the World Bank, International Monetary Fund, and aid agencies. If they do, they often alienate their people by adopting stringent policies.

Moreover, after the end of the Cold War, the core countries of the First World trimmed their aid budgets, so reducing another source of funds to African countries. If demands on the reduced aid budgets are too great, it is likely that Africa will be marginalized in the world of development loans and aid. Whatever is attempted, the outcome is disappointing.

## Basic Needs

Many African countries are still at the stage where they need major improvements in basic human resources and infrastructure before population control, economic development, and political democracy can be effective.

### Education

All countries have shown increases in vital educational achievements since independence (Figure 3.37). By the 1990s, countries such as Botswana, Cameroon, Kenya, South Africa, Zambia, and Zimbabwe had virtually total enrollment, male and female, in elementary schools, up from 50 percent in 1965. A second group of countries including Benin, Central African Republic, Ghana, Malawi, Nigeria, Tanzania, and Zaire had increased male and female elementary education from 50 to 75 percent and from 40 to 60 percent, respectively. Countries such as Burkina Faso, Ethiopia, Guinea, Mali, and Niger still have under 30 percent of children in elementary school, and female education lags behind male. Burundi, Chad, and Mauritania made major strides by increasing primary education from 10 to 20 percent in 1965 to 50 to 70 percent in the 1990s.

Increasing numbers of Africans also have opportunities to become fully literate in secondary school and to earn higher academic qualifications. It is often disappointing to many who gain higher qualifications that there are few jobs available in their home country. African doctors, lawyers, and airline pilots are increasing in numbers, but many find their employment abroad in core countries. Perhaps there should be a greater push to provide elementary education for all rather than put increasing funds into higher education at this stage. Other educational and health improvements await developments in infrastructure.

### Infrastructure

All African countries lack infrastructure. This means that they do not have the necessary roads or other forms of internal transportation, lack clean water supplies and adequate sanitation, have shortages of electricity and telecommunications, and need the development of further public works such as irrigation canals. Such facilities are imperative if countries are going to be an integral part of the world economic system—or even feed their own populations: 40 percent of crops are lost through poor transportation and storage. Lack of such facilities places many parts of Africa outside that system. All forms of infrastructure are expensive to provide and maintain, but former assumptions that the home government should provide (and thus control) essential services have been replaced in many African countries in the 1990s by privatization and the encouragement of foreign investment and management to develop these facilities. This is a further key to Africa's future.

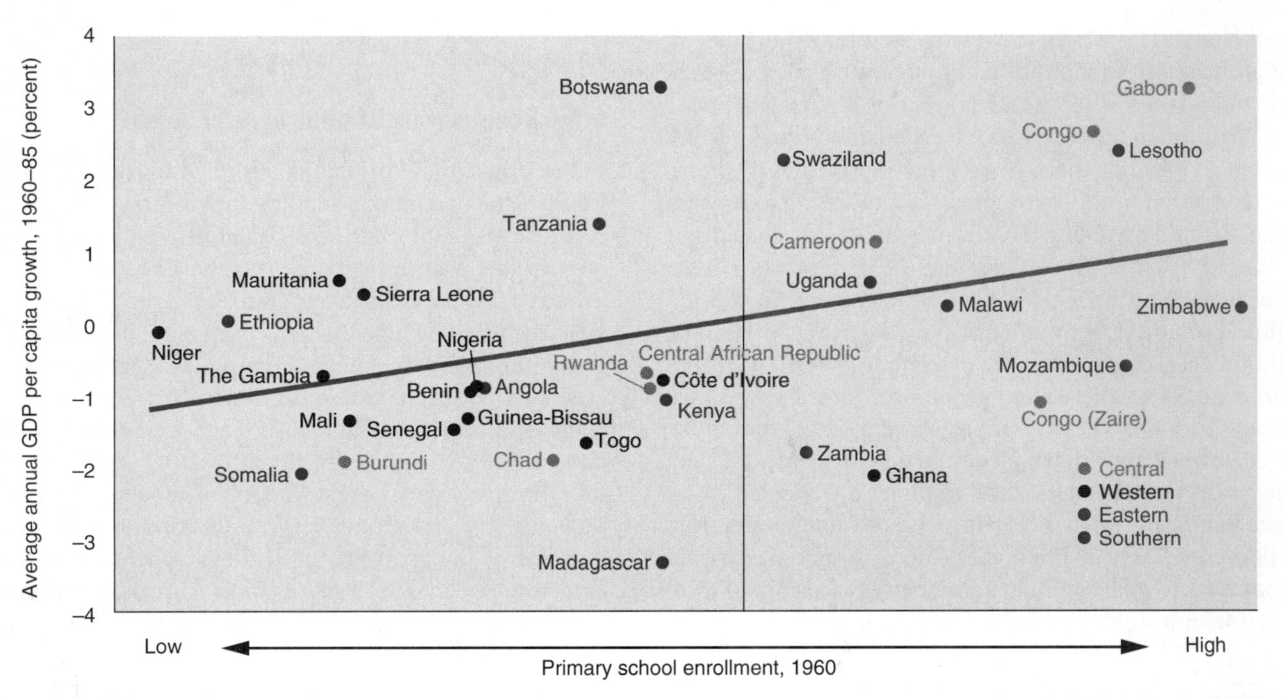


**FIGURE 3.37 Comparison between primary school enrollments and economic growth.** In Africa South of the Sahara, countries with higher primary school enrollments generally experienced higher economic growth. What other factors affected countries that are below the thick line?

Source: Data from World Bank

**RECAP 3D: Landscapes and African Futures**

Urban landscapes in Africa South of the Sahara include elements of precolonial, colonial, industrial, commercial, and shantytowns. Rural landscapes include cultivated areas, national parks, forests, deserts, mining excavations and waste heaps, and villages and small towns.

The future of Africa South of the Sahara depends on major improvements in human resources and infrastructure, leading to population control, economic development, and political democracy. This is a daunting list of major changes to be faced by people still very involved in traditional tribal cultures.

3D.1 Analyze photos of African urban and rural landscapes in terms of cultural, economic, political, and environmental influences.

3D.2 What are the economic and political problems facing African countries, and what possible solutions are suggested for them?

3D.3 Can you identify examples of sustainable development in African countries?

**Key Terms:** shantytown economic colonialism

## Internal or External Impetus

It remains to be seen whether Africa's growth can be generated from inside the continent or whether external leadership, loans, and aid will continue to be needed or imposed. To date, few attempts have been made to seriously reconcile the needs of the African people with the demands of the world economy. In early 1995, Nelson Mandela, president of the Republic of South Africa, spoke to a conference on oil and minerals in Johannesburg. He made a plea that African countries, often well endowed with mineral resources, should work together. "Africa needs to reclaim its minerals by way of indigenization, by developing our own institutions, by enabling the African entrepreneur to come to the fore," he said. He called for more production of basic construction materials to reduce import costs—and so increase capacity for new infrastructure. He also pledged that South Africa would overlook short-term national interests to encourage regional cooperation in mineral development, the building of an electricity grid across Southern Africa, and the renewal and coordination of railroad and road networks. The revival of South Africa's credibility in the rest of Africa appears threatening to some local interests but could be an important key to progress. By the end of the 1990s, the Republic of South Africa had made important contributions to this process in Southern Africa. Elsewhere, new African leaders were determined to keep control of internal affairs while entering investment partnerships with foreign funders—at times overseeing competition among American, European, and Asian bidders for projects.

**WEBSITES: Africa South of the Sahara**

Select countries from each subregion and find out more about each, or update your knowledge, from the CIA *World Factbook* site: http://www.cia.gov/cia/publications/factbook/index.html

Access http://www.economist.com to see what articles are written about African countries this week. Is there evidence of changing politics or economies?

http://hyperion.advanced.org/16645 *The Living Africa* site with sections on people, land, wildlife, and national parks. Try "The Land" and "Congo River."

The CD-ROM Dorling-Kindersley *Eyewitness World Atlas* and Rand McNally *New Millennium World Atlas* include online facilities to access information from most countries around the world. The following are some of the sites they recommend.

- The *Central Africa Watch* site provides a range of newspaper and other reports on Central African countries, broadly defined (including Niger, Angola, etc). Seek out information that updates this chapter: http://www.marekinc.com/index.html
- http://www.kenyaweb.com is the Kenya government site. Access the *Economy* section and compare districts in different parts of the country. Access the *Tourism* section and compare the facilities of state parks.
- http://www.ghana.gov.gh/ is the Ghana government site. What business opportunities are advertised?
- http:www.inc.co.za/ is the Republic of South Africa *Independent Online* digest of news that is updated daily. What is the latest state of political and economic affairs?
- http://www.anc.org.za is the African National Congress site, giving the history of its efforts to change the political life of South Africa. How important has been the life and work of Nelson Mandela?

# CHAPTER REVIEW QUESTIONS

1. The fault system in East Africa, which gave rise to mountain peaks and lakes, is the: (a) Tell (b) Rift Valley (c) Great Escarpment (d) Great Divide
2. All of the following characterize Africa South of the Sahara, *except* (a) The region has nearly 10 percent of the world's population, but contributes only 1 percent of the world's GDP (b) Many of the countries face civil wars and famine (c) Most countries have Islamic religious majorities (d) The end of the Cold War has forced many countries to adopt open politics and different economic management
3. The landform term that best describes the African continent as a whole is (a) plateau (b) coastal plain (c) mountain-and-valley flatland (d) deserts.

4. Eritrea is (a) the name of a lake in the Rift Valley area in Africa (b) the name of a new country in Africa (c) the name of a tribe in Africa South of the Sahara (d) the name of a mountain in Africa
5. In which vegetation type do Africa's many large plant-eating animals and their predators live? (a) Tropical rain forest (b) Savanna grasslands (c) Deserts (d) Sahara
6. According to this chapter, whether Zimbabwe continues to develop or falter, may depend on two things (a) economic restructuring and population control (b) land redistribution and population control (c) wise political leadership and the progress of the economic giant to the south (d) democracy and population control
7. Residents of shantytowns in Africa are (a) often in poverty (b) associated with the informal economy (c) usually new migrants from rural areas who may eventually find better accommodations (d) isolated from the rest of the large urban area because governments usually ignore their infrastructure needs
8. The European colonial powers in Africa established (a) slash and burn agriculture (b) large scale plantation agriculture (c) informal economy (d) a communal growth policy
9. Which of the following statements is not true about Central Africa? (a) The rate of population growth exceeds economic growth (b) The proportion of people living in urban areas is less than 50 percent (c) Agriculture remains the main economic involvement of over 70 percent of the population (d) Since independence, most countries have developed an extensive railroad system
10. Which of the following is not among the major environmental problems in Africa South of the Sahara? (a) The decline in soil quality due to soil erosion (b) Tropical diseases (c) Earthquakes (d) AIDS
11. The land area linked to a tribe is often smaller than the modern African countries.
    True/False
12. African countries are poor because they lack the mineral resources needed to develop manufacturing industry.
    True/False
13. Hatred between tribes, stemming from repressive acts toward each other, is the major cause of ethnic conflicts and civil wars in many countries of the region.
    True/False
14. Most of the area in Africa South of the Sahara is influenced by a tropical climate, which is mostly warm and humid year round.
    True/False
15. In many parts of Africa, a lack of sufficient food makes people vulnerable to disease.
    True/False
16. Many countries in Africa South of the Sahara have experienced a sharp decline in birth rates and death rates since their independence.
    True/False
17. Few African countries have a manufacturing industry that can compete in world markets.
    True/False
18. Most people in Africa depend on subsistence farming.
    True/False
19. South Africa is the only country in Africa South of the Sahara with a sizable, non-black population.
    True/False
20. In the 1990s, some African countries have shown a trend toward greater democratization.
    True/False
21. A group of people, marked out by ethnic characteristics such as a common culture, are called a ____________.
22. The two major ethnic groups in Rwanda and Burundi are ______________ and ______________.
23. ______________ was the only African country, south of the Sahara, that was never colonized by Europe.
24. The apartheid policy was practiced in ____________ until the early 1990s.
25. The Central African country, Democratic Republic of Congo, was once called ______________.
26. The most affluent African country is ____________.
27. ANC stands for ______________.
28. The highest mountain in Africa is ______________.
29. The country from which Cuban forces were removed in 1988 is ______________.
30. The Bush War of the 1960s and 1970s took place in ______________.

# NORTHERN AFRICA AND SOUTHWESTERN ASIA

Chapter 4

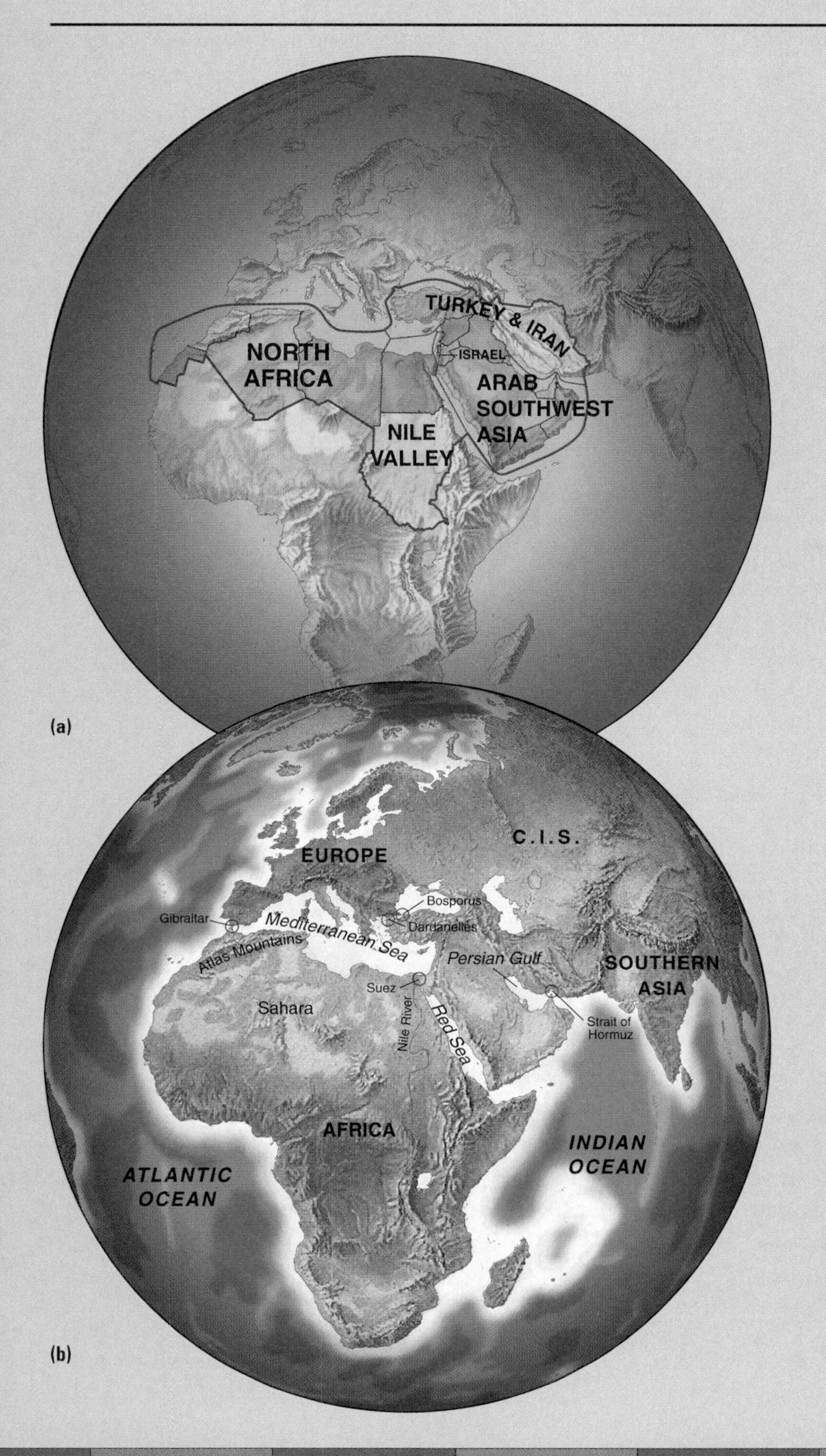


**THIS CHAPTER IS ABOUT:**

**The region's geographic location where Africa, Asia, and Europe meet**

**The prime significance of religion**

**The arid natural environment**

**The geography of five subregions:**

- North Africa: between the Sahara and Europe
- Nile River valley: dependent on water flows
- Arab Southwest Asia: heartland of Islam
- Israel, Gaza, and the West Bank: tiny irritant enclave
- Turkey and Iran: contrasting largely Muslim countries

**How urban and rural landscapes reflect changing geographies**

**Future prospects for the region**

**World Issue: Islamic revivalism**

**Changing politics and life in Syria**

**FIGURE 4.1 Northern Africa and Southwestern Asia: the location of this region "in the middle" of four other major world regions.** (a) The subregions. (b) Natural environments, dominated by deserts. The red circles identify major strategic points on ocean routes.

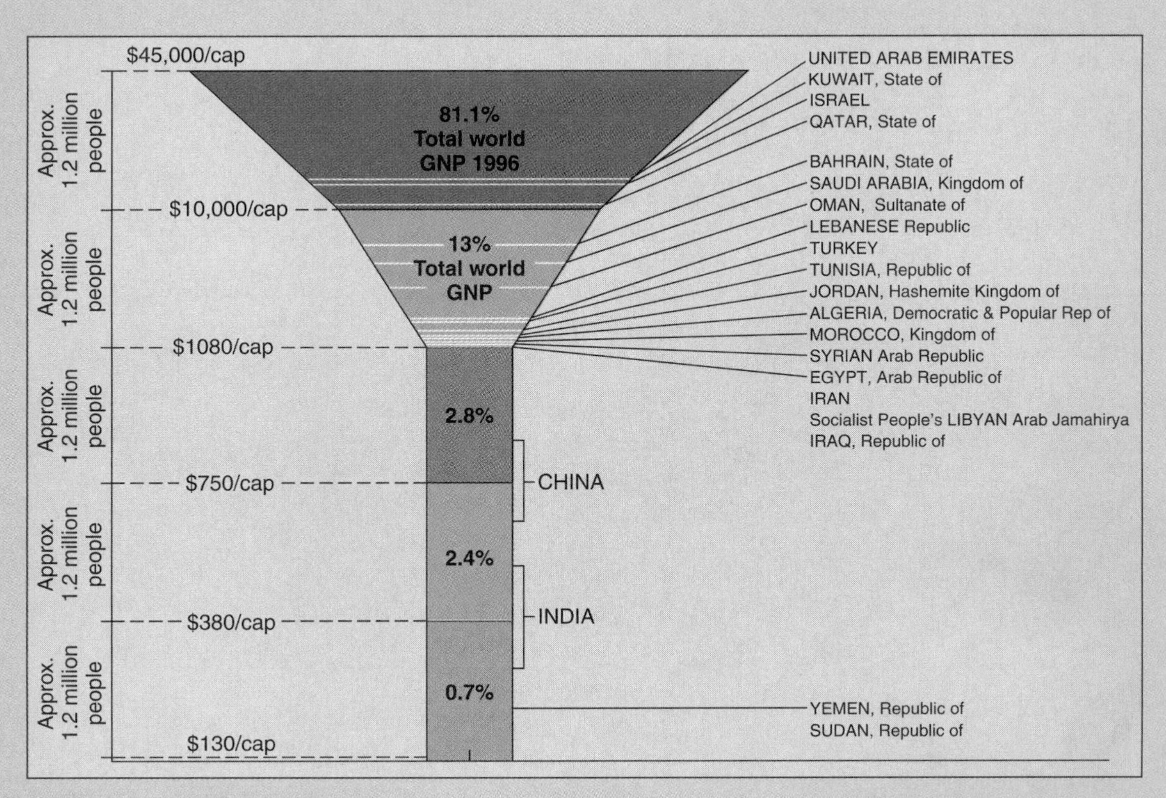


**FIGURE 4.2 Northern Africa and Southwestern Asia: relation to world wealth distribution.** The countries of the region and their position on the scale of income levels. Which countries are in the richest one-fifth and which are in the poorer groups?

Source: Data (for 1996) from *World Development Indicators*, World Bank, 1998

## IN THE MIDDLE

The countries of northern Africa fronting the Mediterranean Sea and those of southwestern Asia lie geographically among several continents and power blocs (Figure 4.1a). They are at one of the world's great physical and human junctions (Figure 4.1b). In the geologic past, continental masses that now form Africa and Arabia rammed into the southern edges of Europe and Asia, raising mountains along the line of collision. Throughout human history, the region has been a battleground for external influences. It was also a source of religious, political, and technical cultures that diffused outward.

During the 1900s, the main changes in the human geography of this region resulted from external global influences impinging on the established cultures of the region. The dominance of the Islamic religion, to which 93 percent of the people belong, led to reactions against such influences, often focused on the 1948 establishment of the Jewish state of Israel in the midst of Islamic countries.

The main intrusions from outside the region came with the realization that the rocks of parts of this region contain around two-thirds of the world's known oil reserves. After World War I, the exploitation of oil resources from its lands gave the region new strategic significance within the global economy and political system, wrenching countries out of traditional ways and isolation. The whole region remains sensitive to both international relationships and internal changes. The 1991 Gulf War, which freed Kuwait from Iraqi invaders, highlighted jealousies within this region that are linked to geographic inequalities of resource and income distributions (Figure 4.2).

Interactions among the countries of Northern Africa and Southwestern Asia and surrounding regions in the rest of Africa, Europe, and Asia occurred throughout history and continue today. In this respect, Northern Africa and Southwestern Asia remain "in the middle" of global forces that influence its people's ways of life. It is also "in the middle" in terms of core-periphery. Apart from Israel, which is a unique case but approaches core-country conditions, other countries in the region are in the semiperiphery (dependent on core-country markets for oil but with others dependent on them), with a few in the periphery. The area that includes Southwest Asia and Egypt is also often called the "Middle East," a term that originated in a European view of Asia as consisting of the Near East (Turkey and Palestine), the Middle East (Gulf states), and the Far East (Pakistan and beyond). "Middle East" is not used in this text, since "southwestern Asia" is more appropriate in the global context.

The region contains many global choke points within the world economy's trade routes that link the core and peripheral areas across the strategic hub of Northern Africa and Southwestern Asia. They range from the **Strait of Gibraltar** in the west to the **Dardenelles** and **Bosporus** (Turkey), the **Suez Canal** (Egypt), and the **Strait of Hormuz** (Iran) in the east. Additionally, the major world oil users in the core countries of Anglo America, Europe, and Japan are concerned by the fact that the lands bordering the **Persian Gulf** contain most of the world's oil reserves.

As the world economic system increasingly involved this region in the 1900s, Islamic religious ideas were revived. After attempting to resist the external influences of the world economic system and Western culture through political means, such as nationalizing oil resources, many in the region tried to replace such influences with alternative perspectives based on Islamic beliefs. Such views spread through the actions of militant groups, causing increasing agitation not only within the countries of Northern Africa and Southwestern Asia but also in the countries of surrounding regions. Major tensions in the region exist among Islamic activists and more conservative groups: there is no single Islamic way.

The human geography of Northern Africa and Southwestern Asia is unified by being "in the middle" culturally (Islamic religion), economically (oil), and politically (resistance to the West). The region also has a distinctive physical environment that is dominated by arid climatic conditions. Most people adapt to the harsh environment by living on desert margins where water is available. The Mediterranean coastlands of North Africa, the Nile and Tigris-Euphrates River valleys, and the uplands of Turkey and Iran provide the region's most favorable environments for human settlement and food production. The scarce water resources, however, are under pressure from the present population, and it is difficult to understand how a rapid increase of population can be accommodated without huge gains in productivity.

Today the Northern African and Southwestern Asian countries occupy nearly 12 percent of the world's land and have 6 percent of the total population. This is one of the smallest major world regions but also one of the most influential—often because it is at the center of both strategic economic interest and political instability. Its status may become even more significant as its population increases rapidly—from 390 million people in 1998 to over 600 million by A.D. 2025. Its oil reserves are likely to give it increasing political and economic power in a world of rising demand and declining reserves.

The boundaries of this region are coastlines or are drawn through the Sahara in the south and around Turkey and Iran in the north and east. Sudan is included in the region because of its strong links with Egypt along the Nile River valley. Mauritania (Chapter 3), the largely Muslim countries of Central Asia (Chapter 8), and those that are part of Southern and Eastern Asia (Chapters 5 and 6) are discussed elsewhere. Data for this region are found in the Reference Section tables (pages 586–610).

Five subregions of Northern Africa and Southwestern Asia (see Figure 3.1a) are recognized. The countries of North Africa, west of Egypt, have desert interiors and a Mediterranean outlook that links them northward to Europe. The Nile River valley is the major geographic feature linking Egypt and Sudan. The region's heartland of Arab Southwest Asia extends from Syria in the northwest to Oman in the east. The small but distinctive grouping of Israel, Gaza, and the West Bank are a continuing focus of tensions within the region. Iran and Turkey are two of the largest countries in the area, forming the northern and eastern borders and having mostly mountainous terrains.

## RELIGIONS AND LANGUAGES

Southwest Asia and the Nile River valley formed two of the world's early cultural heartlands (see Figures 1.12 and 1.13). Even in times of slow and limited transportation, the region at the hub of early world systems had constant movements of people and changing control, witnessing some of the first developments in farming and urbanization. Such early achievements diffused to the surrounding continents along with the religious beliefs, languages, legal systems, and political systems that originated in this region.

### Religions

Religion is the most basic and influential of the cultural elements that give character to Northern Africa and Southwestern Asia. Islam is dominant today, but other world religions that preceded Islam trace their origins to the region. The early civilizations of Mesopotamia and Egypt gave rise to religions having many gods linked to natural phenomena. The people often idolized a human emperor as one of such a pantheon. After approximately 1000 B.C., religions with many gods gave way to religions based on a single god. Monotheistic religions diffused from their hearths in Southwest Asia to Europe, Africa, and Asia.

#### Mazdaism, Judaism, and Christianity

"Mazdaism" worships Ahura Mazda ("wise lord"), as taught by the Persian Zoroaster. It remained important in Persia until the Islamic expansion of medieval times.

**Judaism** worships Yahweh, the creator and lawgiver. Beginning in Palestine some 3000 years B.C., Jews spread through southwestern Asia and into Europe, especially after the Roman armies destroyed Jerusalem in A.D. 70 and dispersed them.

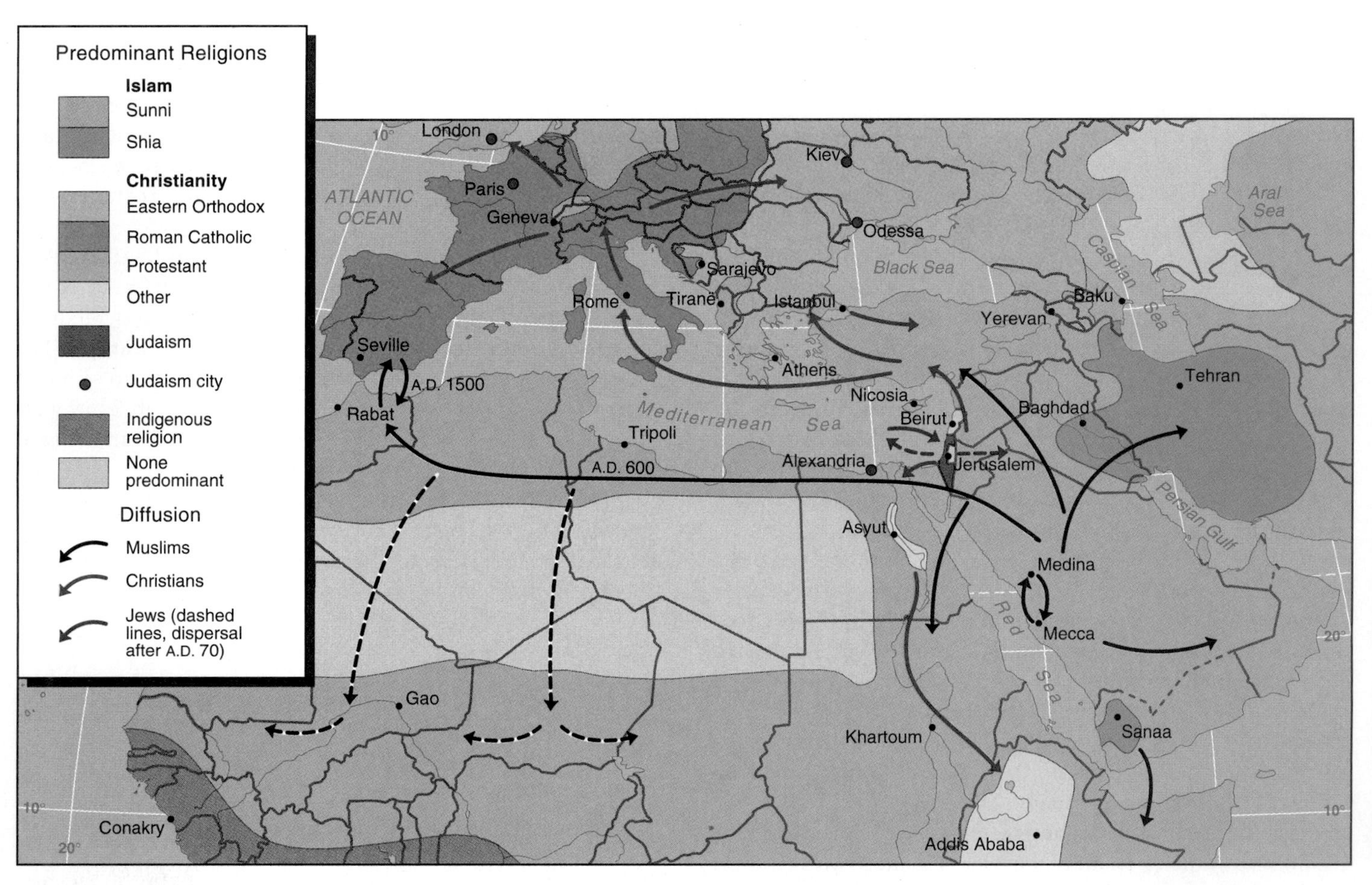


**FIGURE 4.3 Major religions and movements of diffusion within and from Northern Africa and Southwestern Asia.** Muslim Arabs conquered the area shown in green in the first century after Muhammad emerged as a new religious leader around A.D. 600. Christianity had become the religion of the Roman Empire, spreading through Europe. Jews were dispersed at the fall of Jerusalem in A.D. 70, and the nation of Israel was not formed again until 1948.

This began a process of religious diffusion (see Figure 2.21) that continued for nearly two thousand years.

**Christianity** stemmed from Jewish elements in Palestine and gave a fresh impetus to monotheism. It became the official religion of the Roman Empire in the A.D. 300s. Many ancient religions waned, leaving relics of their influence in local shrines. Christian churches spread across southwestern Asia, and into Africa and Europe. Controversies over how much Jesus Christ was man and how much God resulted in divisions among the churches of Northern Africa and Southwestern Asia. A later division occurred between the eastern (Orthodox) and western (Catholic) groups of churches in Europe. The eastern church was for long centered in Constantinople (modern Istanbul, Turkey), the focal point of the Byzantine Empire until overcome by Arabs and Turks.

## Islam

**Islam** ("the reconciliation to God") arose in Arabia during the A.D. 600s. **Muslims** are "those who are reconciled to God," or followers of Islam. After the death of the prophet Muhammad, the empire based on his teachings and ruled by Arab caliphs spread rapidly westward through northern Africa to Spain and eastward into central Asia (Figure 4.3). The Arab invasions were followed by conversion of local peoples to become Muslims, and Islam became the region's dominant religion. Muslim mosques are distinctive features of towns (Figure 4.4) throughout Northern Africa and Southwestern Asia.

Early divisions within Islam continue to be significant today. The main, or orthodox, Muslim group is the **Sunni Muslims**, or Sunnites, who base their way of life on the Koran holy book, supplemented by traditions, and have given temporal power to a succession of leaders descended from the caliphs. While many Sunni Muslims are conservative and take on Western ways of life, some groups—such as the rulers of Saudi Arabia and the Taleban faction in Afghanistan—impose rigid behavior on their people.

Other Muslims also follow the Koran but dispute the caliph succession accepted by Sunnites. They look to Ali, a cousin and son-in-law of Muhammad, whom murder prevented from becoming a caliph. Supporters of Ali as the only *Imam*—authoritative interpreter of the Koran—are known as "partisans of Ali," "Shi'at 'Ali," "**Shia Muslims**," or Shiites,

**FIGURE 4.4 Islam and architecture.** Mosques are a common feature of urban landscapes in Muslim countries. The Mohammed Ali Mosque towers over the citadel in Cairo, Egypt.



and look to the return of a related Imam. Shia Muslims comprise large proportions of the population in a few countries, including 90 percent of the Iranian population. The Shia branch of Islam produced a new vigorous sect in the 1800s that forms the basis of current Iranian isolationism and wider Muslim dissident behavior (see "World Issue: Islamic Revivalism," page 138).

Gender inequalities are a continuing feature of Muslim countries. Saudi Arabia boycotted the 1995 fourth world conference on women in Beijing on the grounds it was un-Islamic. In Iran it is a criminal offence for women not to wear a headscarf. There is currently, however, considerable debate in Muslim countries over what the Koran means on these matters. Muslim-dominated countries such as Turkey, Pakistan, and Bangladesh have elected women leaders in recent years. Women's education is increasing more rapidly in Arab countries than elsewhere, starting from very low levels but doubling female literacy rates from 1970 to the 1990s. Arab countries still exhibit high levels of gender inequality, as the GDI rankings show. There are also differences among countries on this matter. While Saudi Arabia excludes women from all public activity, Tunisia promoted women's rights after its 1957 independence.

## Languages

Northern Africa and Southwestern Asia demonstrate how language diffusion (see Figure 2.22) followed conquest as military groups, such as Greeks, Romans, and Arabs, imposed their language on others. Other diffusions occurred as the use of a particular language spread throughout a population as a result of trading or religious use.

- **Arabic** is the main written language of almost all countries in the region. Its origins are in a type of language that spread from the Arabian Peninsula with the Islamic conquests in the A.D. 600s. Arabic has a special place in all Islamic countries because it is the only language allowed in the Koran holy book and Muslim prayers. "Arabs," who were once defined as living in the Arabian Peninsula, now include all those using Arabic as their first language; there are over 180 million Arabic speakers in Northern Africa and Southwestern Asia. Spoken Arabic is widely used in trade, but there are major differences among dialects that evolved from Morocco in the far west to Egypt, Syria, Iraq, and Arabia in the east. Radio and television are now reducing the significance of those dialects.

# ISLAMIC REVIVALISM

The issues connected with this topic include:

Is Islamic revivalism different from Islamic fundamentalism?

Is Islamic revivalism good or bad for the countries in which it is active—or for other countries?

Does Islamic revivalism have links with Pan-Arabism?

"Islamic fundamentalism" is a widely used but poor description for a radical movement that is largely confined to minority sects of Muslims. The term implies that all Muslims are part of a movement that takes them back to their roots and that its only application might be in terms of the attempts of some Muslims to revive their faith and its influence by radical, even terrorist, means. The truth is that, although Muslims agree in general that their religion should affect their daily lives, most Muslims separate religious observance from commercial and political activities and the need to deal with non-Islamic countries. Confusion also arises from activities of radical groups that do not have a religious basis.

Islamic fundamentalism in the late 1900s was mainly linked to the resurgence of the Shiite sect and its insistence on religious participation in politics and economic management. Fundamentalism requires Islamic states to separate themselves from the "satanic" influences of the Western world. "Islamic revivalism" is a better term since it seeks to reestablish Islam as a world religion and political force. Shiites are the majority and governing sect only in Iran and the ruling minority group in Syria's dictatorship. They form majorities in some Gulf States and minorities elsewhere in Northern Africa and Southwestern Asia—where they are a source of radical agitation and are often persecuted by the governments. The persecution of the Shiites in southern Iraq, the suppression of Islamic radicals in Algeria, Libya, and Syria, and the failure of the Welfare Party in Turkey's 1991 elections suggest it is unlikely that radical Shiite Islam will take political and economic control of Northern Africa and Southwestern Asia.

The Shiites are not the only revivalist groups within Islam. The Wahhabi sect that motivated the Saudis to extend their area of influence in the 1800s and the Egyptian Muslim Brotherhood that began in 1928 were two Sunnite groups that had major influences. Today the Taleban faction in Afghanistan has a similar basis.

A basic issue facing Muslims is the extent to which their countries become absorbed into the world economy and whether such involvement offends the basic tenets of their faith. This question increased in importance during the later 1900s, when the region's vital oil supplies placed it in a strategic global position. From 1950 to 1990, Northern Africa and Southwestern Asia were a focus for Cold War antagonisms that raised desires to find an alternative way of life avoiding the excesses of state capitalism, market capitalism, and Western consumerism and artistic tastes. The best known expression of these desires occurred when Muslim leaders in Iran took control of their country in the 1979 revolution. They remain in power in Iran. Although Islamic law became the official basis for managing the country in 1983, with over a hundred offenses carrying the death penalty, its implementation was not complete because of differences within the ruling elite of clerics and laymen. In 1997, a moderate was elected president of Iran, with a 70 percent vote, reflecting the population's distaste for strict Islamist and anti-Western policies.

Sudan is another country where radical Islamic views dominate the government, and links with Iran are strong. The resultant policy of a religious war, or *jihad,* against the southern Christian provinces of Sudan led to economic disaster for Sudan. Despite support for radical movements in other Muslim countries and the popular appeal of the radical message, no other country has yet been taken over by the radical groups.

One reason for this lack of success by militant Islam is the fact that kings or dictators govern most Muslim countries and allow a limited democratic element. Radical elements of all types are suppressed if they appear to provide opposition. Rulers such as Qaddafi in Libya and Assad in Syria keep several hundred militants in prison. In most Arab countries the radical Islamic groups are illegal and suppressed. Radicals made their presence felt in Algeria, Egypt, Morocco, Tunisia, and Saudi Arabia and remain a significant irritant to the governments of those countries. In Algeria, the army stepped in to avoid the possibility of a radical-led government being elected in 1991 but was plagued by guerrilla warfare that killed tens of thousands through the 1990s. In Jordan, the radicals lost their parliament seats when Jordan began negotiating with Israel after the Gulf War. In Lebanon, the radical Muslims vied with other Muslim and non-Muslim groups in the seemingly insoluble civil strife until the fighting ceased in 1990.

Where radical Islamic groups are allowed to take part in democratic elections, as in Turkey, they did not fare well until the mid-1990s. The constitutional separation of religion and politics in Turkey since 1928 made it difficult for religious groups to become political. The continuing weakness of the Turkish government in the mid-1990s enabled radical Muslim groups to increase their representation and become the largest group in government—although far from an overall majority. After a short period in power, however, the Islamic Welfare Party government was replaced by the military and declared illegal in 1998 despite continuing popular support.

Although all countries in northern Africa and southwestern Asia acknowledge Islam as the main or only state reli-

**The Koran,** Islamic holy book, written in Arabic. Arabic reads from right to left.

**Ayatollah Khomeini,** an Iranian religious and political leader who led a revolution after the collapse of the Shah's rule in 1979. He is seen here at prayer with his followers. After his death in 1989, a somewhat more liberal regime took over in Iran.



gion, different reactions to radical Islamic movements demonstrate that a broad alliance of Arab and Muslim states is unlikely despite their common dislike of Israel. Other interests of the rulers or countries prevent this. Several attempts have been made at Pan-Arab cooperation, including that by President Gamal Nasser of Egypt linking with Syria in 1956, and the anti-Israel and anti-American groupings up to 1990. The Gulf War of 1991 made it clear that Northern Africa and Southwestern Asia comprise nations with varied interests and little trade with each other. Their aims fit more closely with economic growth within the world economy than with a desire to see Islam predominant throughout the world at all costs.

As Islamic revivalism became less militant in the late 1990s, the December 1997 session of the Organization of the Islamic Conference in Tehran, Iran, chaired by that country's new moderate president for the next three years, attracted representatives from all Arab countries. It highlighted the growing split within Iran between the hard-line Islamic religious leaders and the forces that elected the president. The latter contradicted plans for Muslims to unite against Israel and the West, urging the conference to avoid rejection of the West and calling for a "dialogue of civilizations."

### TOPICS FOR DEBATE

1. "Islamic fundamentalism" is a poor term that does not describe the situation in this region.
2. Islam is producing an alternative to the world economic system.
3. The strict Islamic policies of Iran are better than liberal Western economic and cultural ways.

- **Berber** languages are African-origin languages spoken by 40 percent of the peoples of Morocco, 30 percent of Algerians, and smaller groups in surrounding countries. Berber languages, which have no written form, survived the Arabic onslaught in places such as the hilly areas of the Atlas Mountains. Most Berbers—who are similar physically to Arabs in the same area—now speak or write Arabic as well.
- **Persian** came from a different language family than Arabic. It was brought by invaders from Eurasia around 3000 B.C.—the Aryan origin being reflected in "Iran." These invaders included the Medes, Parthians, and Scythians who formed the basis of successive Persian empires. After the Arab invasions in the A.D. 600s, Arab words were introduced and the Persian language was written in an Arabic script. Modern Persian is also known as Farsi. Only 50 percent of Iranians speak Persian today, but it is the official language and linked to the Shiite Islamic beliefs of the leaders.
- **Kurdish** is a language akin to Persian kept alive by a group of tribes in the hilly lands of southeastern Turkey, northern Iraq, western Iran, and parts of Syria. The numbers of Kurdish speakers are unknown because the language is at the heart of demands for a separate state: those who do not want the state talk in terms of 10 million people, while the Kurds inflate their numbers to over 20 million.
- **Turkish** is part of the Altaic family of languages (see Figure 2.22), which came to Central Asia in the A.D. 700s and from there spread to Southwest Asia. Turks form the majority of the population of Turkey—a country that emphasizes nationalism and decrees the use of the language. It uses Roman script rather than Arabic. Another group of Turks, the Azeri (or Azerbaijanis), still use Arabic script and make up minority groups in Iran as well as being the main peoples of Azerbaijan (see Chapter 8). Attempts by the former Soviet Union to unite its Azeri Turks with others failed because many are Shiites who look to a home in Iran. Additional groups of people with Turkish affinities live in the Central Asian republics that were part of the former Soviet Union (see Chapter 8). Proposals to form a united Turkish-speaking country from these various elements have not received wide support.
- **Hebrew** is the official language of Israel. It is related to Arabic and was the language of the Jewish scriptures and religious services. The movement devoted to creating a Jewish state from the 1800s, known as Zionism, revived the use of Hebrew, modernizing it for secular usage in order to provide a common religious and secular element among Jews from many countries.

## Arab and Islamic Political Movements

In the mid- and late 1900s, the dominance of Arabic language and Islamic religion in Northern Africa and Southwestern Asia led to post-independence movements that sought to unite Arab countries in a single nation of increased world influence. None succeeded in this aim.

### Arab League and Palestine Liberation Organization

The **Arab League** was created in 1945 to encourage the united action of Arab countries. Its 7 founding members were the only independent Arab states at the time and increased to 21 as more countries became independent. They eventually included the **Palestine Liberation Organization** (PLO), a secular, left-wing political organization that focuses its demands on reestablishing the country of Palestine.

The linking of Arab countries in the western Mediterranean with those in the eastern Mediterranean and southwestern Asia, and the ostracizing of Egypt after its 1979 accord with Israel, led to the transfer of the Arab League headquarters to Tunis. Disunity over the Israel-Palestine issue and over tensions among countries with different resource bases weakened the Arab League and its ability to foster unity among Arab countries in the 1980s and 1990s. Religious right-wing organizations, such as Hezbouallah and Islamic Jihad, competed with the secular PLO as the voice of Muslims in the "holy war" (jihad) against Israel. Smaller in numbers and international influence, they engaged in terrorism to achieve their goals and resisted the growing accord between the PLO and Israel.

Some attempts were made to link Arab countries even more closely. Egypt and Syria joined under the leadership of Colonel Jamal Nasser with the intent of persuading other countries to commit their futures to a single **Pan-Arab state**. This combination lasted only three years, from 1958 to 1961, before breaking up. The Arab League continues to provide a forum united by opposition to the existence of Israel. Little has been heard of the Arab League since the Gulf War, when one Arab country (Iraq) attacked another (Kuwait). In the 1990s, there is renewed emphasis on national rather than Pan-Arab identity.

### Organization of the Islamic Conference

The **Organization of the Islamic Conference** (OIC) was set up in 1970 by foreign ministers of Muslim countries. It has 45 members, extending to Muslim countries outside Northern Africa and Southwestern Asia such as Pakistan and Indonesia. It gained an increasing voice in world councils but is more successful in advancing individual member countries' interests than in defining and pursuing a common agenda. Although closer economic integration was agreed upon in principle in the late 1970s, most Muslim countries still carry out the great majority of their trade with the world core countries in Europe, Anglo America, and Japan. One of OIC's most important affiliates is the Islamic Development Bank that opened in 1975 and is dedicated to economic development among OIC members. The December 1997 meeting of the OIC in Tehran (Iran) was unusually well attended by all members and focused on pragmatic world trading issues rather than on Islamic religious policies.

**FIGURE 4.5 Arid lands: parts of Egypt, Israel, Jordan, and Saudi Arabia.** A space shuttle view taken over the northern Red Sea across desert lands to the Mediterranean Sea. On the left, the delta mouth of the Nile River and the Suez Canal; in the center, the Sinai Peninsula; on the right, the Gulf of Aqaba leading northward to the Dead Sea.

NASA

## NATURAL ENVIRONMENTS

The strongly unifying Islamic-Arabic culture of Northern Africa and Southwestern Asia is paralleled by common features of the natural environment, especially the region's dry climates, water shortages, and oil resources. The geographic inequalities of the distribution of water and oil resources within the region go a long way toward explaining economic and political differences among countries within a fairly uniform cultural environment.

### Dry Climates and Desert Vegetation

Arid climatic environments, in which evaporation rates are greater than precipitation, dominate virtually the whole region (Figure 4.5). Most places have some rain, but it falls irregularly and with increasing uncertainty as aridity increases. The region boasts the world's highest recorded shade temperature (58° C, 136° F at Azzizia, Libya), but the lack of cloud cover makes nights cold, and frosts are possible in winter.

Some of the marginal areas receive more rain. The coasts of North Africa and the eastern Mediterranean have a Mediterranean climate with relatively rainy winters caused by the seasonal southward shift of the frontal weather systems that bring precipitation to Europe. The summers are largely without rain and river channels dry up. The mountains of Turkey and Iran force air to rise and precipitate rain and snow. The winter snowfall there provides a source of meltwater that feeds the Tigris-Euphrates River system in spring.

The combination of coastal locations and almost guaranteed sunshine makes the Mediterranean area attractive to tourists from the colder and cloudier countries of northern Europe. The combination of winter rains, early springs, and summer sunshine gives the Mediterranean areas advantages in growing citrus fruits, olives, grapes, and early vegetables for the urban-industrial regions of northern Europe.

Water resources are precious. Water comes from melting snows on the nearby mountains in the Maghreb (northern Morocco, Algeria, and Tunisia), Turkey, and Iran, from low and variable amounts of rainfall, from underground stores, or from external sources such as the equatorial and Ethiopian rains that feed the Nile River. The historic extension of irrigated land in lowland areas, using these water resources, brought increased economic activity to Egypt, Iraq, and Syria but often led to the abandonment of hilly areas.

The growth of the oil industry and associated urbanization had major impacts on water resources. Groundwater extraction increased and costly **desalination plants** were built to provide fresh water for coastal settlements. A large proportion of the land is without fresh water, however, and, having little potential for economic development, is left as desert.

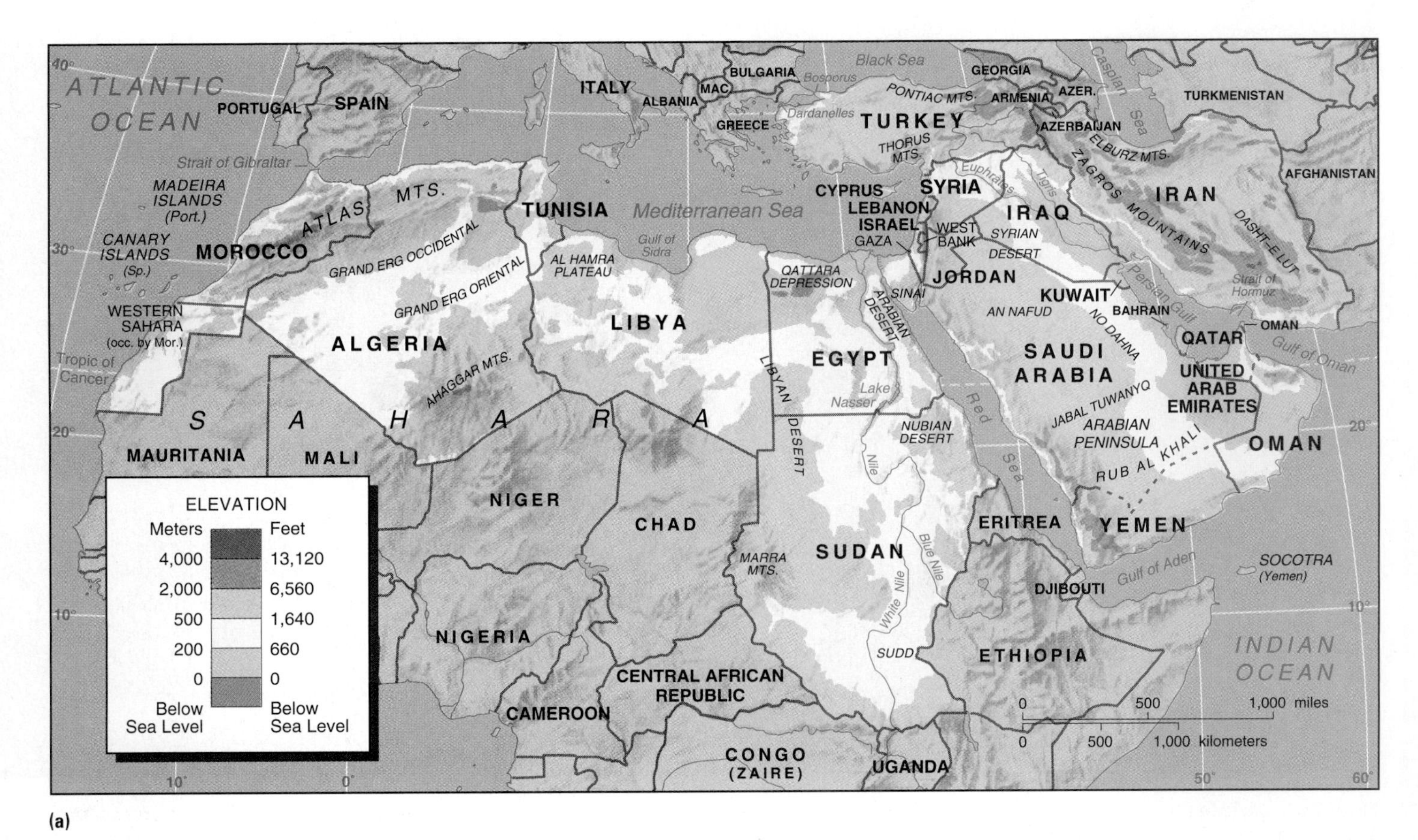


(a)

**FIGURE 4.6 The physical features of Northern Africa and Southwestern Asia.** (a) As well as being arid, the region is mountainous, especially along the northern margins. Locate the main river basins. (b) The plate margins of this region coincide with mountain ranges and seas.

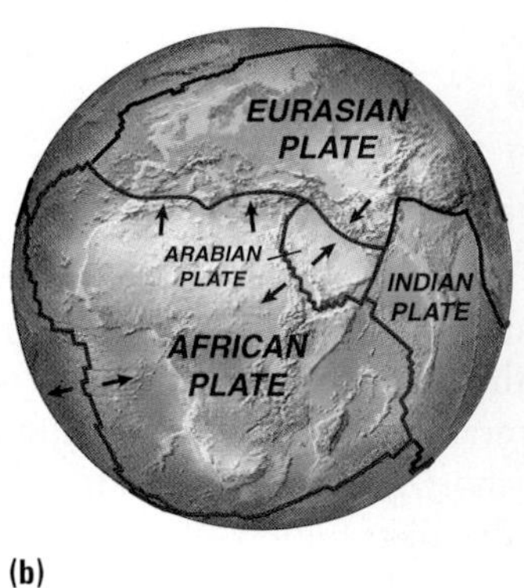


(b)

The arid climatic environments are associated with drought-resisting desert plants that form a partial ground cover. This cover thickens and becomes denser on uplands having higher precipitation totals and lower rates of evaporation. Large areas of desert, however, have no vegetation and are gravel-strewn, rocky, or sand-covered. Mediterranean climate areas once had a natural vegetation of pines and shrubs before being largely cleared for timber and agriculture or modified by grazing. The uplands of Turkey and Iran have grassland and woodland vegetation as the altitude reduces temperatures and evaporation levels and makes the low-to-moderate precipitation more effective. Soils are poor and undeveloped through most of the region: the best occur in the Mediterranean climate rainy areas and along the valley floors of rivers where annual floods deposit mud and so replenish the nutrients.

Climate change left its mark on the region and affected human settlement. The present aridity is one stage in a sequence of fluctuations between more and less arid conditions. Cave paintings and dried lakes suggest that the climate was more humid and cattle herders more widespread a few thousand years ago. The stocks of underground water that accumulated in such humid phases provide much of the water being used today, but current precipitation levels do not replenish them after use. The present phase of desertification began some 5,000 years ago, forcing many into the watered valleys of the Nile and Tigris-Euphrates Rivers. The desert margins continue to fluctuate, retreating in a series of wetter years and advancing in drier years or as human actions removed vegetation and lowered water levels in the soil and rocks.

## Clashing Plates

Contrasting landscapes from high mountain ranges and extensive plateaus to wide river valleys stand out where an arid climate supports little vegetation. It is common to see bare rock rather than grassed or wooded slopes. Such landform contrasts are influenced by the location of tectonic plate boundaries (Figure 4.6). The African and Eurasian continents and the Arabian Peninsula are the tectonic plates of this region. They comprise ancient rocks that were worn down over many millions of years. Such rigid sections of Earth's crust underlie the extensive plateaus of northern Africa and Arabia. The ancient rocks are often covered by flat layers of sedimentary rocks, such as the limestones and sandstones forming the plateaus on either side of the Nile River valley.

The highest mountains are the Atlas in North Africa and the Taurus, Zagros, and Elburz ranges in Turkey and Iran. The **Atlas Mountains** rise to nearly 4,000 m (12,500 ft.) and the Elburz to over 5,500 m (17,000 ft.). Both are parts of the **young folded mountain ranges** that extend from North Africa and Southern and Alpine Europe, through southwestern Asia, to the Himalayas in Southern Asia. They were formed as the African and Arabian plates collided with the Eurasian plate. Rocks that formed in the seas filling the basins between the plates were caught up in the collision and raised to form the mountains. Volcanic eruptions along this active belt led to the injection of mineralized fluids that solidified as veins in the rocks. The compression and folding of thick sedimentary rocks also provided many places where oil and natural gas could be trapped. The Persian Gulf area became the environment for the world's largest concentration of these fuels.

Steep slopes and mountain ranges, rising to around 2,500 m (8,000 ft.), occur along the lands bordering the **Red Sea**. The tectonic plates here are pulling apart to form a new ocean. Rising molten rock beneath the surface first raised the crustal rocks into a dome that cracked to form a rift valley. Then the sides pulled away from the center and the Red Sea opened. Whereas the **Mediterranean Sea** and the Tigris-Euphrates River valley occupy the site of a closing ocean, the Red Sea is considered to be an opening ocean. Volcanic activity occurs along its margins. At its northern end, the rift splits in two around the Sinai Peninsula (see Figure 4.5). The eastern branch forms the valley into which the **Jordan River** flows from the uplands around the **Sea of Galilee** toward the **Dead Sea** some 400 m (1,312 ft.) below sea level.

## Major River Valleys

Rivers are unusual in the arid climatic environment of Northern Africa and Southwestern Asia. Most flow from water sources in the surrounding mountains outside the region.

### Tigris-Euphrates Rivers

The **Tigris-Euphrates River system** supplies water to southeastern Turkey, Syria, and Iraq. Snows falling on the high mountain ranges of Turkey and Iran melt in spring, causing the Tigris and Euphrates to flood. The floods gouge sediment from the upland areas, depositing the coarser materials in giant fans where they leave the mountains and the finer sand and mud across the Syrian and Iraqi lowlands.

### Nile River

The **Nile River system**, supplying water to Sudan and Egypt, begins in the rainy equatorial area around Lake Victoria and the seasonal rainy area of the Ethiopian Highlands. A narrow zone a few kilometers wide allows cultivation and settlement as the river flows across Sudan and Egypt through a valley carved into the surrounding plateaus. Away from the Nile River on either side and northward of central Sudan is arid land that has little prospect of development apart from the river waters.

Two major Nile River tributaries join near Khartoum (Figure 4.7). The White Nile River, flowing from Lake Victoria on the equator, has tributaries that are fed by rains throughout the year and supply a fairly constant flow of water. On reaching southern Sudan, the White Nile River flows through the Sudd swamp in which it loses nearly half its water by evaporation. The Blue Nile River has its source in the Ethiopian Highlands where there is a marked summer maximum rainfall: this flow is more important because it is less subject to evaporation and produces the annual Nile River flood in Sudan and Egypt in September.

This pattern may vary. In 1962 unusually heavy rains led to a massive rise in the level of Lake Victoria, supporting continually high flows in the White Nile River during years when Ethiopian droughts caused the Blue Nile River to contribute much less than normal. By the 1980s, however, the flows from Lake Victoria returned to normal, but the Blue Nile River continued to be low. The lower flows from both major sources coincided with increasing demands on the water in the source countries of Uganda, Tanzania, Kenya, and Ethiopia and began to threaten the long-term availability of water to Egypt and Sudan. By 1988, the expectation that the Aswan High Dam would ensure Egypt's water needs for 50 years was proved wrong, and discharges below Aswan had to be reduced.

Egypt and Sudan now seek to enlarge the scope of their agreement for the use of Nile River waters by involving Ethiopia—the source of 80 percent of the long-term Nile River flow—and other upper basin countries, such as Tanzania, that wish to extract from Lake Victoria. Sudan seeks to renegotiate its allocation so that it may develop water projects in its southern region. The outcome is certain to reduce the water available to both Egypt and Sudan.

The competition for use of Nile River waters demonstrates that political and social factors are as important as environmental events in the economic development of natural resources. Action on such projects is delayed by unforeseen events. Political upheavals in Ethiopia and Sudan halted development in those countries, allowing Egypt to have maximum use of Nile River water in the 1970s and 1980s. At present, international finance is not available for major water projects in the upper basin countries to meet the rising needs there. Such needs, however, must eventually increase pressure for greater water extractions before the Nile River reaches Sudan. At the same time, population growth in the lower basin countries adds pressures for making extra water available there.

## Oil Resources

The huge oil and gas production from Saudi Arabia and the Gulf states amounted to 22 percent of rising world oil production from the 1960s to the 1990s. Despite the addition of new producers in Latin America, Asia, and Africa, total

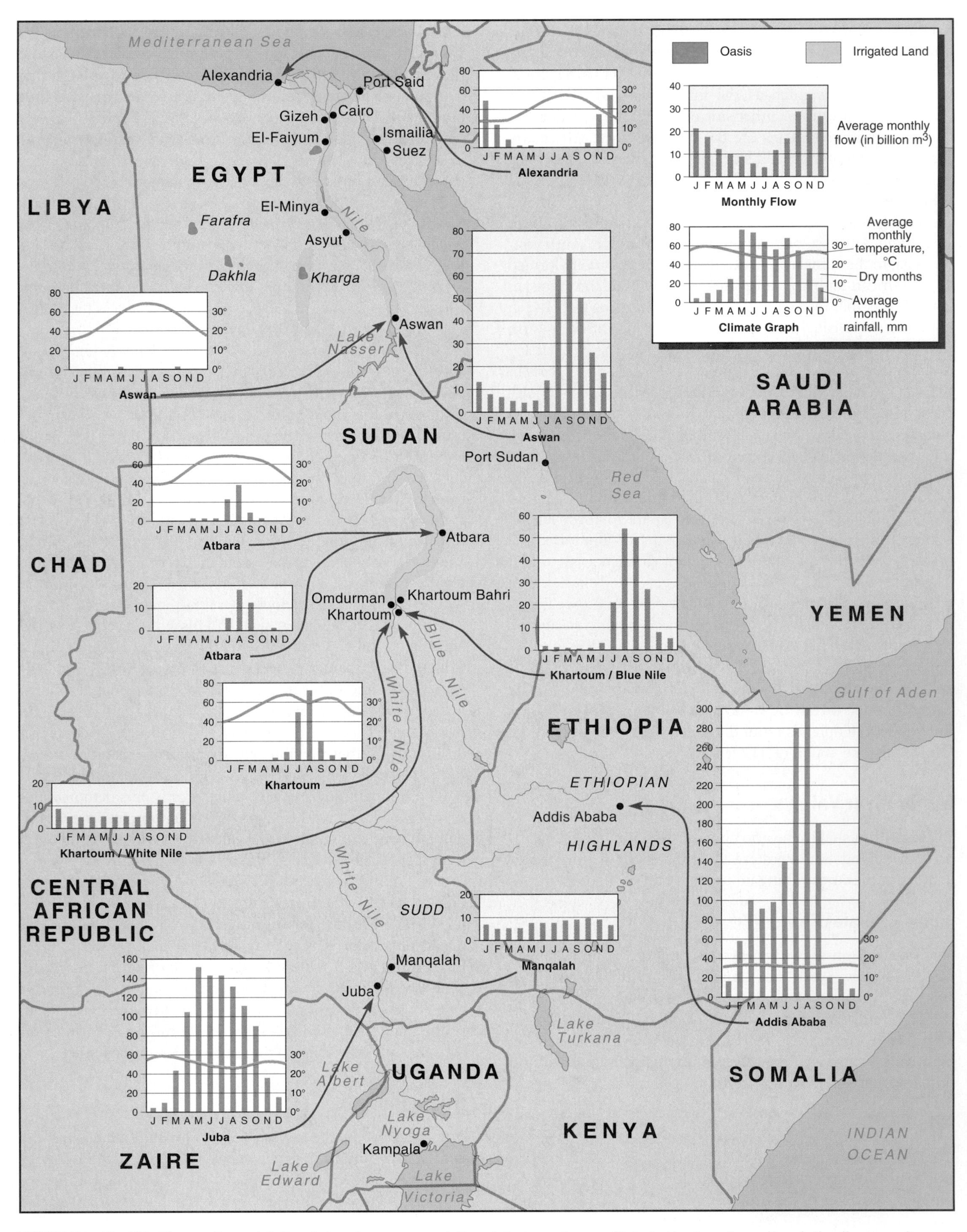


**FIGURE 4.7 The Nile River valley: rainfall and river flow compared.** The relatively small areas of irrigated land are shown. How do the climate graphs relate to the river flow at the places shown?

known reserves around the Persian Gulf increased their proportion of world reserves (Figure 4.8). Continued discoveries boosted them from 61 (1965) to 65 (1998) percent of total world reserves (from 215 billion barrels to 560 billion barrels). These figures do not include Iran, Libya, Egypt, or Algeria, which are also major oil and natural gas producers in this region.

Iran has 10 percent of world oil reserves and experienced rapid oil income growth from the 1960s, reaching 40 percent of the value of national production in the oil boom of the 1970s. Much of the additional wealth was at first spent on importing foreign goods rather than on investing in domestic agricultural and industrial production. Algeria is so massively indebted as a result of loans raised on the strength of 1970s oil income that it now has difficulty attracting foreign investment. Libya, however, has smaller debts and, with a smaller population, managed to adjust its economy to the ups and downs of oil income.

The rapid growth of oil income and its outlay on construction works required more labor than the often small oil-producing

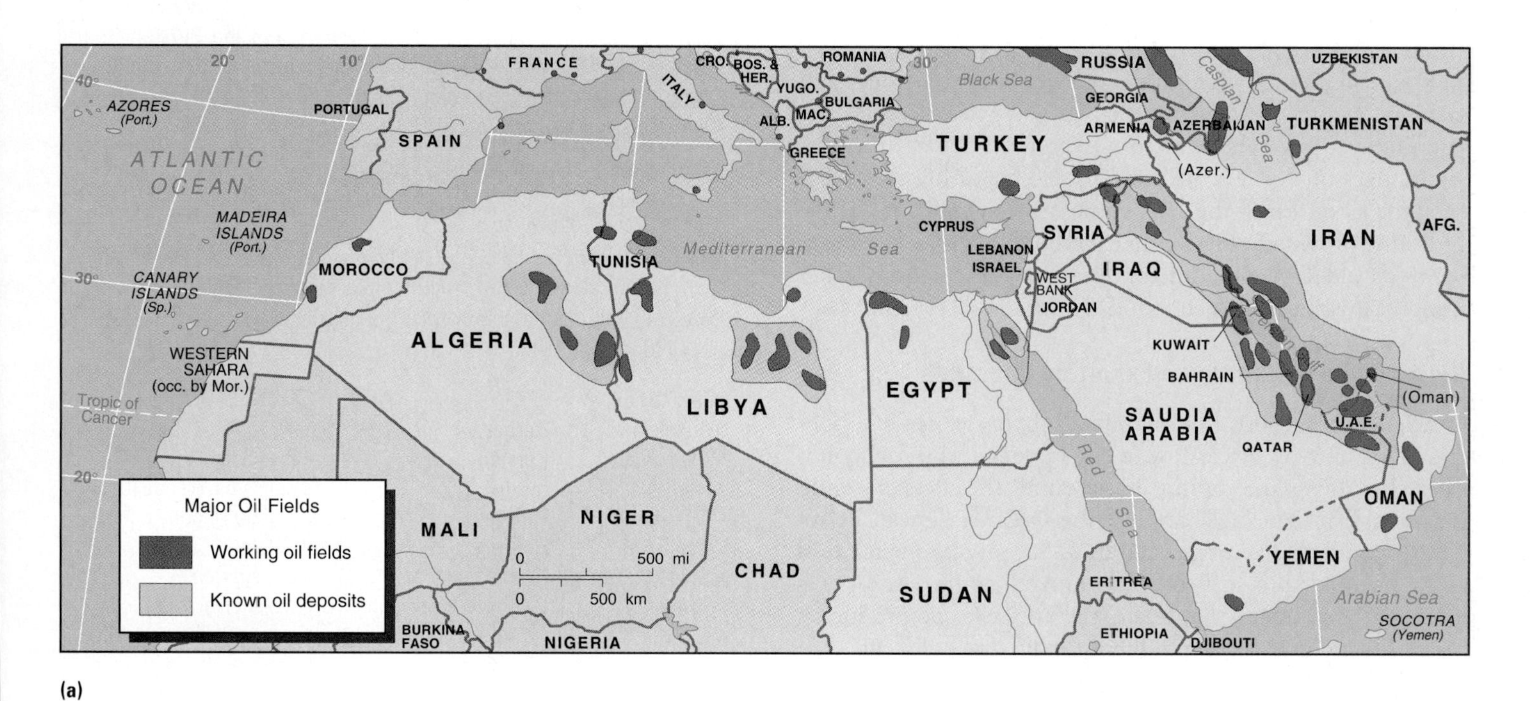


(a)

ESTIMATED PROVEN RESERVES OF OIL
end 1997, billion barrels

RESERVES BY REGION, %

Southwest Asia — North America — Europe — Latin America — Africa — Asia-Pacific

Years of reserves remaining

|  | Years of reserves remaining |
|---|---|
| Indonesia | 9 |
| OMAN | 16 |
| Angola | 20 |
| Canada | 9 |
| Azerbaijan | >100 |
| Kazakhstan | 43 |
| ALGERIA | 19 |
| Norway | 9 |
| Nigeria | 20 |
| China | 21 |
| LIBYA | 56 |
| United States | 10 |
| Mexico | 34 |
| Russia | 22 |
| Venezuela | 60 |
| IRAN | 69 |
| KUWAIT | >100 |
| UAE | >100 |
| IRAQ | >100 |
| SAUDI ARABIA | 80 |

Saudi Arabia: 262

(b)

**FIGURE 4.8 Oil resources in Northern Africa and Southwestern Asia.** (a) Map of major oil fields. (b) World oil resources at the end of 1997 (billion barrels), showing the continuing significance of this region. The share, 65 percent of world reserves, is a little greater than 10 years ago, and reserves in this region are being used more slowly, while many countries in core regions have only 10 to 20 years of reserves. Note which countries in the region have major oil reserves and which have little or no oil.

Source: Data from *The Economist,* 1997.

countries possessed. Instead of developing skills internally, they imported labor from other parts of the region and from Southern Asia. Immigrant labor soon formed large proportions of the population in Kuwait, Qatar, United Arab Emirates, and Saudi Arabia. The reconstruction of Kuwait after the 1991 Gulf War also depended on immigrant labor.

### Oil-Poor Countries

Not all the countries of Northern Africa and Southwestern Asia, however, are major oil producers. Morocco, Sudan, Turkey, Israel, and Jordan are not. Tunisia, Syria, and Yemen produce and export small amounts. Countries that import oil face the burden of purchasing oil whatever its price, and many went into debt during the 1970s high prices. At the same time they often received an economic boost from the remittances of workers attracted by the major oil producers' need for labor. From the 1970s, the oil-rich countries of the Gulf also provided aid to other Arab countries and causes. Both the aid and demand for foreign workers from other Arab countries were much reduced after the 1991 Gulf War.

### Organization of Petroleum Exporting Countries

From the early 1900s, international oil companies had policies of keeping oil prices low to consumers in the core countries. In 1960, the producers around the Persian Gulf, together with Venezuela, formed the **Organization of Petroleum Exporting Countries (OPEC)**. Seventy-five percent of OPEC membership still comes from Arab countries. OPEC's main purpose is to coordinate the interests of producing countries and gain higher and more stable prices for oil.

In 1968, Saudi Arabia, Kuwait, and Libya formed the Organization of Arab Petroleum Exporting Countries (OAPEC) to counterbalance Arab League pressures to use oil as a weapon in the struggle against Israel. Libya left after its 1969 revolution, but by 1972, all other Arab oil producers were members. OAPEC carefully avoided involvement in pricing or production decisions, which were left to OPEC, although OAPEC was accused of causing the 1973 rise in oil prices. OAPEC coordinated many Arab oil projects and energy industry developments, but the fall in oil prices in the 1980s and 1990s starved OAPEC of the necessary funds, and it lost impetus.

The successful oil embargo during the Arab-Israeli War in 1973 established OPEC for a few years as the controlling factor in world oil distribution. Prices rose fourfold. Revenues of the oil producing countries boomed in the mid- and late 1970s, and they invested their money in new roads, hospitals, government buildings, airports, and military hardware. During this period, countries such as Saudi Arabia, Kuwait, United Arab Emirates, and Iraq gained over 90 percent of their export earnings from oil sales.

### World Oil Price Fluctuations

As world oil prices fell in the 1980s, many of the oil producing countries borrowed heavily to maintain the levels of internal investment that had been possible in the 1970s. The continuing purchase of armaments placed additional pressure on reduced oil incomes.

Kuwait, a country with a very small population, had less need or opportunity to invest its huge oil wealth internally, and by the late 1980s, its overseas investments generated more income than oil sales. Such income was the major source of Kuwaiti funds for several years after the destruction of its oil facilities in the Gulf War of 1991. Iraq's oil production was also reduced by military action during the Gulf War, and its sales were decreased by subsequent international sanctions.

Although the 1980s and 1990s saw falling oil prices, costing, for example, Saudi Arabia $2.5 billion per year for every $1 fall in the price per barrel, the future could be much brighter. The massive gulf oil reserves contrast with the relatively small amounts being used up rapidly elsewhere in the world. Unless alternative energy sources are developed economically, the Gulf could again set world prices.

**RECAP 4A: Islam, Water, and Oil**

The countries of Northern Africa and Southwestern Asia have a strategic geographic position between Europe and the rest of Africa and Asia. The region was subject to clashes of tectonic plates in geologic time and of people and ideas throughout human history.

The region has strategic significance greater than the numbers of people might suggest because of its oil resources and location at control points on world sea routes. The Arab language and Islamic religion are important to all the countries, although Turkey and Iran have their own language and Israel stands apart from the rest.

Oil production provides a basis for economic development in Algeria, Libya, Saudi Arabia, Iraq, Kuwait, and the Persian Gulf states, but the geographic outcomes vary from country to country. An arid environment and water shortages are common in these countries, influencing where people live.

4A.1 How do language and religion contribute to the culture of countries? Use examples from Northern Africa and Southwestern Asia.

4A.2 What are the advantages and disadvantages of dependence on oil production by countries in this region?

4A.3 To what extent does water act as a limit to economic development in the region?

**Key Terms:**

| | | |
|---|---|---|
| Judaism | Berber | Pan-Arab state |
| Christianity | Persian | Organization of the Islamic Conference (OIC) |
| Islam | Kurdish | desalination plant |
| Muslims | Turkish | young folded mountain ranges |
| Sunni Muslims | Hebrew | Organization of Petroleum Exporting Countries (OPEC) |
| Shia Muslims | Arab League | salinization |
| Arabic | Palestine Liberation Organization (PLO) | |

**Key Places:**

| | | |
|---|---|---|
| Strait of Gibraltar | Persian Gulf | Sea of Galilee |
| Dardanelles | Atlas Mountains | Dead Sea |
| Bosporus | Red Sea | Tigris-Euphrates River system |
| Suez Canal | Mediterranean Sea | Nile River system |
| Strait of Hormuz | Jordan River | |

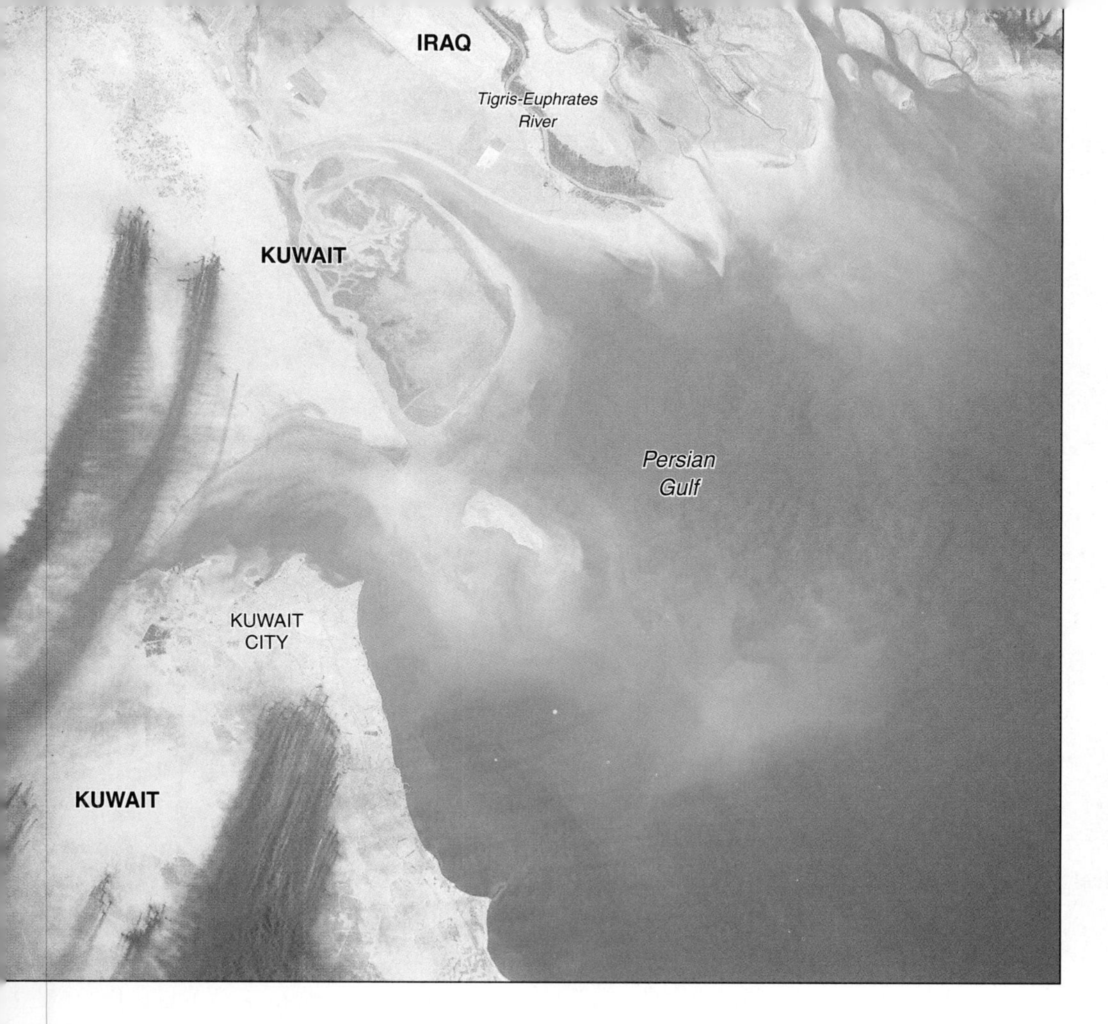


**FIGURE 4.9 Kuwait after the Gulf War.** An April 1991 space shuttle photograph of Kuwait City and the surrounding areas, showing a number of blazing oil wells. The soot from the smoke soon fell to the ground, staining the sand and small amounts entered global air circulation.

NASA

## Environmental Problems

Arid environments are particularly fragile and susceptible to pollution. The oil industry is a major polluter throughout the world, and its concentration around the Persian Gulf brings many problems. It pollutes the atmosphere of areas around oil wells and seepages, especially where unwanted gases are flared off. Additional fogs occur and respiratory complaints are made worse. It pollutes the waters near ocean terminals because of leakages. Even before the Gulf War, the Persian Gulf had lost most of its plant and animal life as a result of pollution since the first oil production there earlier in the 1900s. The huge oil slick that was released at the end of the Gulf War caused less damage than it might have if the Persian Gulf had not already been so polluted. The burning oil wells that were set on fire by retreating Iraqis also added large quantities of carbon gases to the atmosphere, the full impact of which remains unknown (Figure 4.9).

The margins of arid areas are often affected by soil erosion after vegetation is removed from dry soils on moderate slopes. Such soil quality degradation occurred in North Africa following the extension of grain farming in Roman times. Irrigation farming in arid areas requires good management if the high rates of evaporation are not to leave so much salt in the soils that crop productivity is reduced or ended. This process is known as **salinization.** Under careful management, a balance should be maintained between applying sufficient water to flush the salts downward and adding too much water that waterlogs the soil. In ancient irrigation schemes, much of the land in the Tigris-Euphrates River lowlands became unusable because such management was not maintained. Modern usage led to the loss of farmland for similar reasons. Poor management of irrigation in Iraq rendered 60 percent of the watered land too saline for further use.

# NORTH AFRICA

**North Africa** is the westernmost sector of the Arab world (Figure 4.10). It is closest to Europe and many of its past political ties and present economic links are northward to Spain, France, and Italy. Algeria, Morocco, and Tunisia, in particular, retain close ties with France and have strong links to markets in Europe for selling products, buying goods, and sending emigrant labor. Libya reflects past links by selling

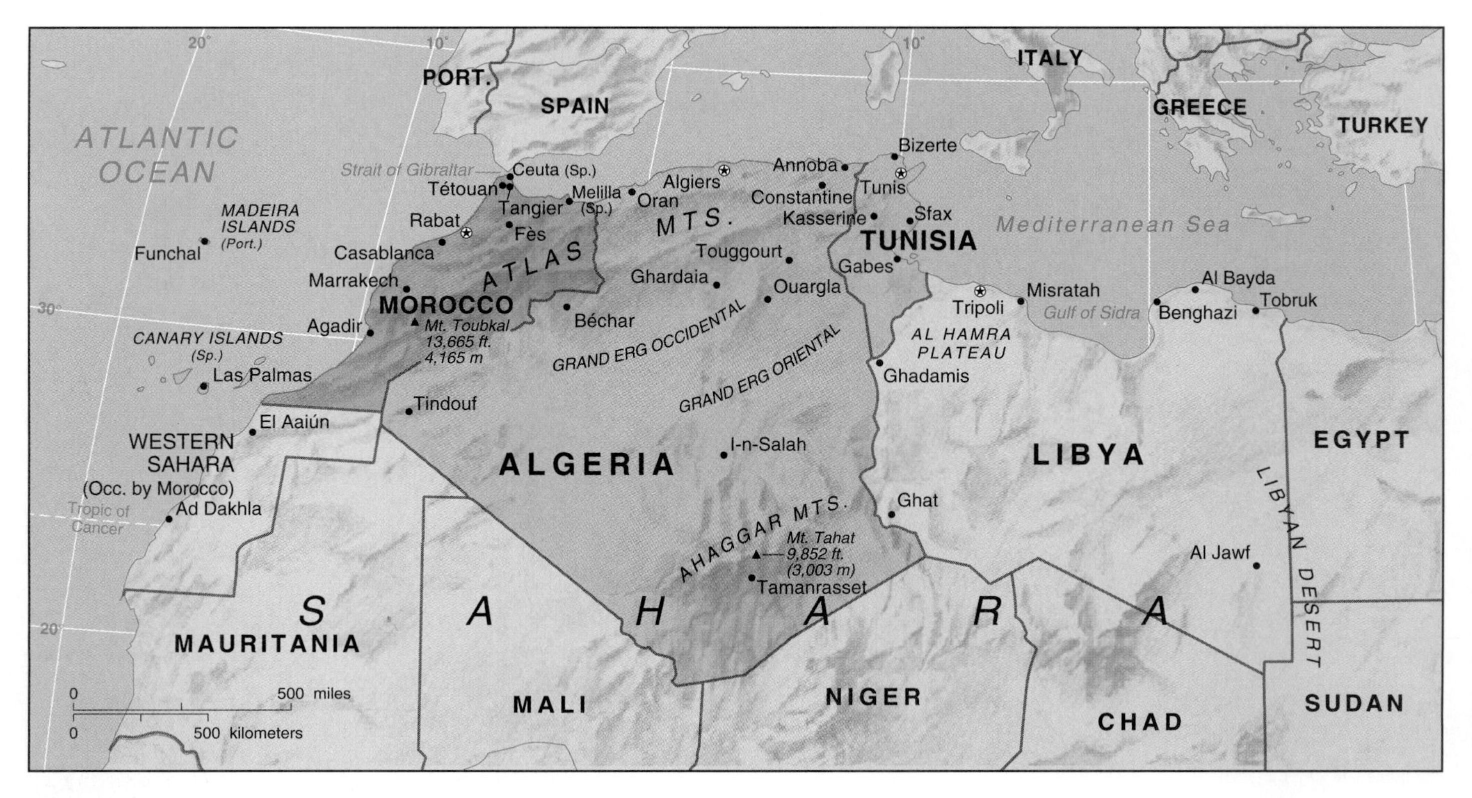


**FIGURE 4.10 North Africa: countries, physical features, and main cities.**

much of its oil and gas to Italy. The development of the European Union and its 1986 extension to include Spain, Portugal, and Greece—which compete for similar agricultural produce markets—placed stress on these relationships.

The North African countries, although a considerable distance from Southwest Asia, share an adherence to Islam as an almost exclusive religion and Arabic as the official language. In 1979, this link with other Muslim Arab countries was strengthened by the move of the Arab League headquarters from Cairo (Egypt) to Tunis.

North African countries exist in a harsh, largely arid and often mountainous environment that restricts agriculture and most human settlement to a small percentage of their territory. This environment will bring increasing problems of water supply if the rapid population increases continue. Algeria, Libya, and, to a smaller extent, Tunisia are now oil producers, and much of the income from this source has been invested in broadening the economic base into manufacturing. The oil income and more diversified economies place these countries higher in the world wealth rankings than other African countries (see Figure 4.2).

## Countries

The four countries of North Africa are **Algeria**, **Libya**, **Morocco** (with Western Sahara), and **Tunisia**. Algeria and Morocco had around 30 million people each in 1998, Libya had nearly 6 million, and Tunisia nearly 10 million. Each has contrasting natural environments that include cultivated coastal areas in the north, desert interiors, and the high Atlas Mountain ranges with their interior plateaus. Over 80 percent of Algeria's and Libya's territory is desert, but Morocco and Tunisia do not extend so far into the arid environment. The northern parts of Morocco, Algeria, and Tunisia are dominated by the Atlas Mountains (Figure 4.11), an upland area that is known collectively as the **Maghreb.**

Algeria has had the most tumultuous recent history of the countries in North Africa. Independence was won at the cost of a war between its nationalists and France in the 1950s. Since then, Algeria has been ruled by democratically elected governments with socialist policies and at times by the military. The curtailment of elections in 1992 and the army takeover of government led to a civil war with the dispossessed Islamic militants. By the late 1990s, terrorist activity had devastated Algeria's economy and people.

Morocco and Tunisia, which border Algeria, view the self-destruction in Algeria with fear and take steps to avoid involvement. Morocco has political stability under its moderate king, who gained international Muslim credibility following his mediating role in Arab issues and the construction of a massive new mosque. Tunisia is modernizing under democratic rule. Libya remains under the strong direction of Colonel Muammar al Qaddafi, who seized power in 1969 and runs the country as a military republic. The former Spanish Sahara was annexed by Morocco in 1976, but groups within the arid area known as Western Sahara maintain their claims

**FIGURE 4.11 North Africa: the Atlas Mountain environment.** A Berber village in the Atlas Mountains, with desert vegetation (cacti, thorn bushes) in the foreground, olive trees and arable land around the village, and trees planted on the hillside to prevent flash floods by slowing the passage of rainwater into the gullies.



to independence. Negotiations about this issue continue between the United Nations and Morocco.

## People

### Population Distribution and Dynamics

The people of North Africa live mainly along the coasts: the harsh inland environments are home to few (Figure 4.12). The populations of all the North African countries continue to grow rapidly, although total fertility rates fell from levels of 7 in 1965 to under 5 in 1998. Morocco's and Tunisia's fertility fell to 3.3, but Libya's remained high at 6.3. Tunisia's lower population growth resulted from policies instigated soon after independence, including the forbidding of polygamy, setting a minimum age for marriage, and instituting a successful family planning program. Morocco set up a program to empower women, including family planning, maternal, and child services. Algeria began its program later. The demographic transition diagram (Figure 4.13) shows how the birth rates fell sharply from 1970 to 1992; the age-sex diagrams show the preponderance of young people in the populations (Figure 4.14). One advantage of this phase of population growth is that there are few older dependents, but these will increase in coming years since life expectancies are now over 65 years in all but Libya (62). Despite the reduction of fertility in the most populous countries, the total population of these five countries rose from just under 50 million in 1980 to 73 million in 1998 and could be over 115 million by A.D. 2025.

### Urban Population Growth

Much of the growth of population in North African countries occurs in urban areas, which now contain nearly half the total population and over three-fourths in Libya. The decline of employment opportunities in the rural areas has been only partly compensated by the growth of urban employment in factories and offices. The largest cities include **Algiers** (Algeria), **Casablanca** (Morocco), **Tripoli** (Libya), and **Tunis** (Tunisia). Other cities with around a million people include **Oran** (Algeria), **Benghazi** (Libya), and **Marrakech** and **Fez** (Morocco).

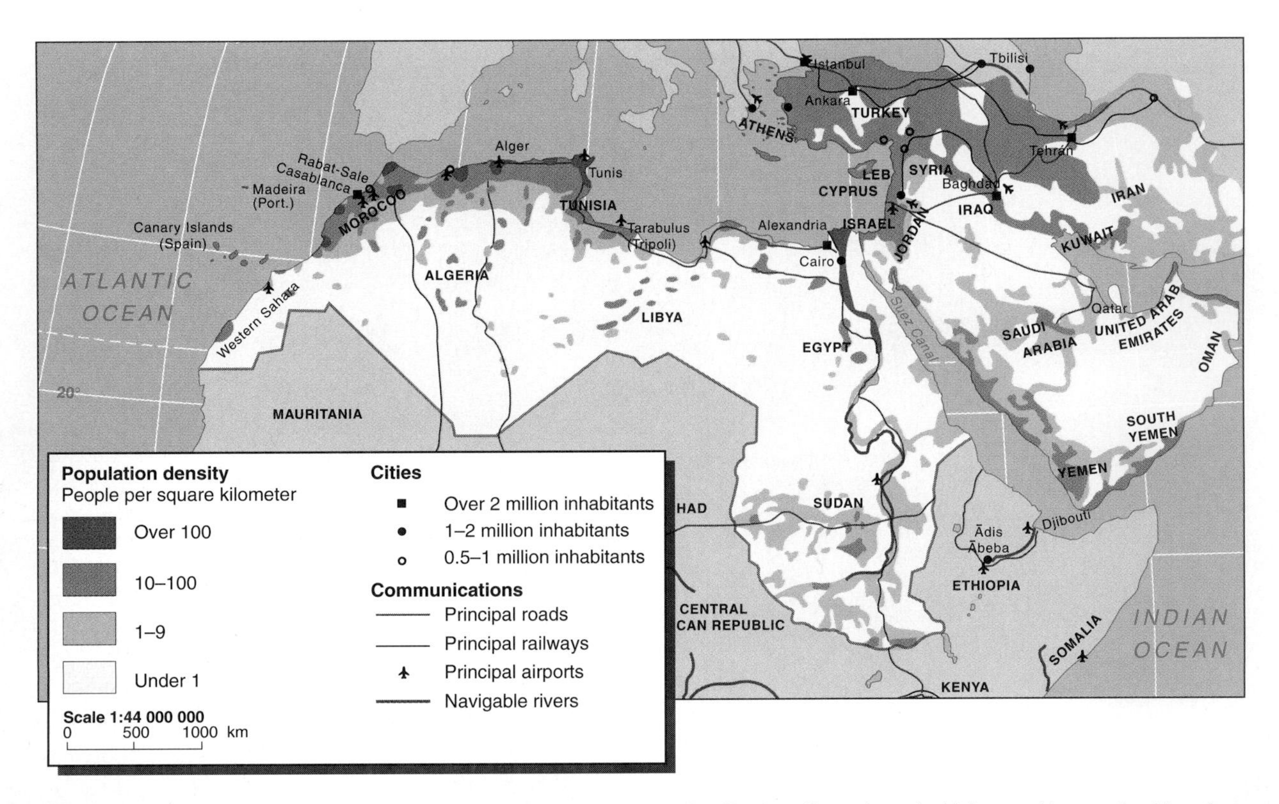


**FIGURE 4.12 Northern Africa and Southwestern Asia: population distribution.** Note where the highest and lowest densities of population occur in each subregion.

Source: Data from *New Oxford School Atlas*, p. 75 & p. 91, Oxford University Press, UK, 1990.

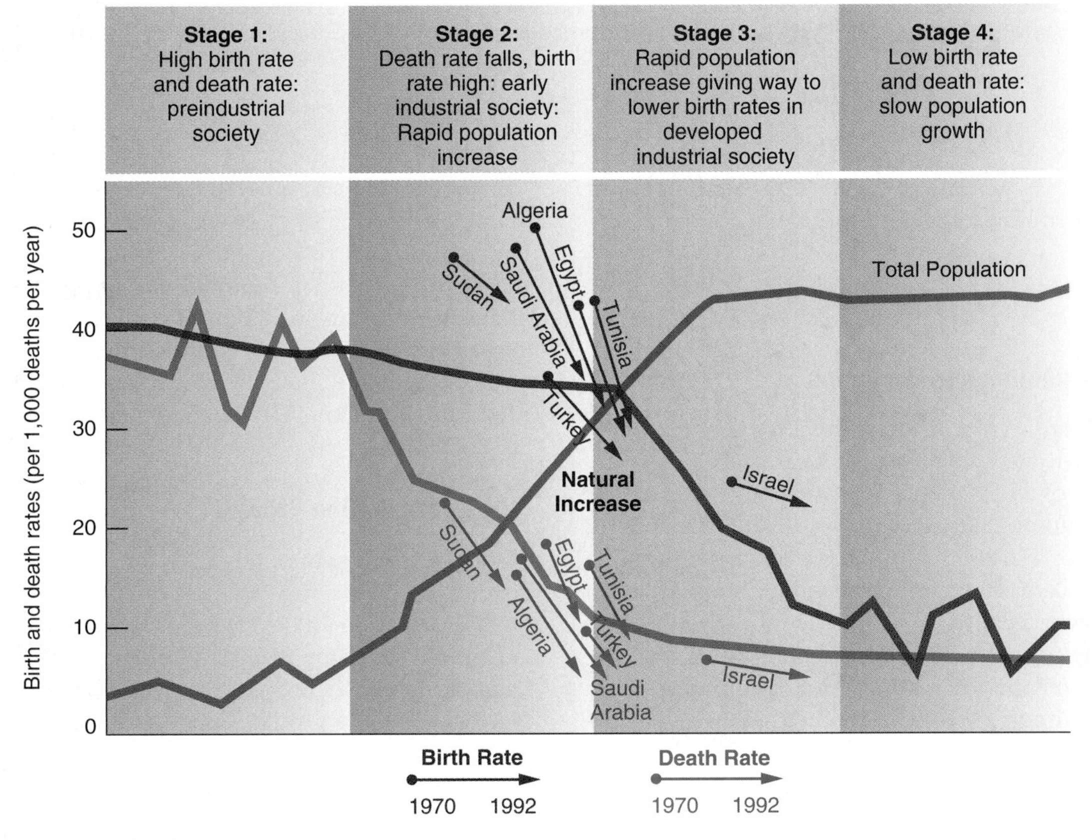


**FIGURE 4.13 Northern Africa and Southwestern Asia: demographic transition.** Birth and death rates in selected countries. Compare the experiences of these countries with those in other parts of Africa.

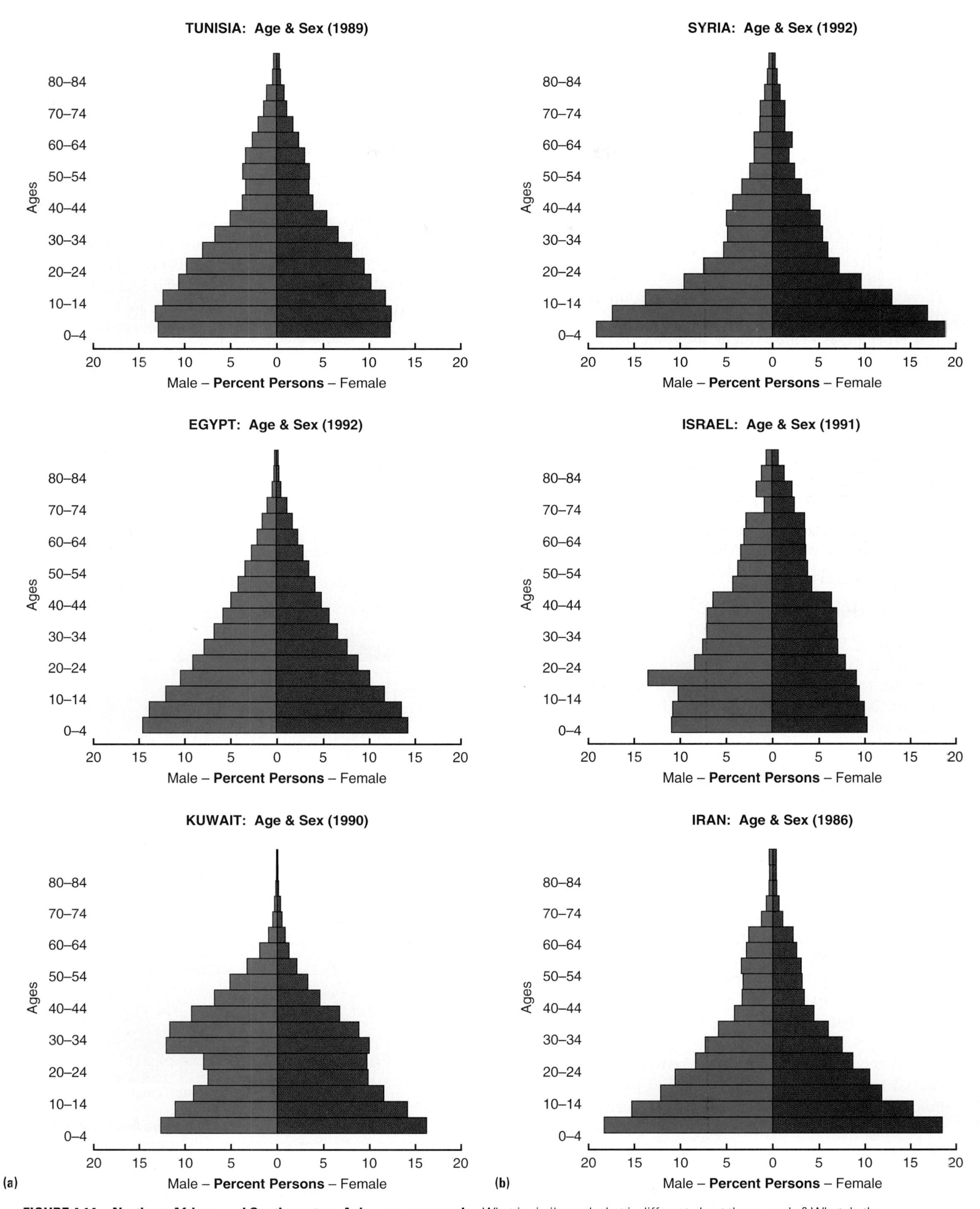


**FIGURE 4.14 Northern African and Southwestern Asia: age-sex graphs.** What is similar and what is different about these graphs? What do the differences show?

Source: Data from **Demographic Yearbook,** United Nations, 1994.

## Employment and Migration

The current rapid population growth creates problems for the education systems and employment prospects. Only in Libya is there are shortage of labor and need for immigrants. The other countries have more entrants to the labor force than there are jobs. Algeria and Morocco have more than 20 percent unemployed. Shortages of skilled labor continue despite intensive education programs to provide primary school literacy and to redress the male-female differences of opportunity. Female literacy, although improving, lags behind that of men: for example, in Morocco in 1994, 55 percent of adult men and 28 percent of women could read. There is a growing university-educated group in each country, but they find few employment outlets there.

In the 1970s, employment at home led many North Africans to migrate to France and other European countries, where low-paying jobs were available, but there have been few vacancies since. There are around 1.5 million workers from North Africa in European countries, especially France and Spain. The remittances of money sent home are important additions to local income. When those who went to Europe to work for a number of years return, they often reflect new priorities by setting up a shop or business and buying a good house. In the 1980s, migrant workers from Morocco and Tunisia also worked in Libya and in the Persian Gulf, but these opportunities fell away after the Gulf War.

## Population History

The cultural backgrounds of the countries in North Africa are similar. In the 800s B.C., the coast was settled by traders from the eastern Mediterranean who established the kingdom of Carthage. This was conquered by Rome, then overrun by Vandals in the A.D. 400s but retaken by the Christian Byzantine king. Arabs invaded in the A.D. 600s, converting the local Berber tribes to Islam and imposing new ways of life. The Muslim invasion of Spain followed, lasting until the expulsion of the Moors around A.D. 1500, when over a million were forced to move back to North Africa.

From the 1200s, pirate groups controlled most of the port cities, making the western Mediterranean dangerous for merchant shipping. The coast of this subregion became known as the Barbary Coast, and the pirate activities contributed to the decline of Mediterranean ports in Southern Europe. Even after the Ottoman Empire took control of this area in the 1500s, the pirates continued their activities. During the 1800s, European countries and the United States stopped the pirate attacks and France annexed territory in Algeria and later in Tunisia and Morocco.

In Algeria and Tunisia, French settlers established commercial farms that grew citrus fruit and produced wine for European markets. The settlers also assumed political and administrative control of the colonies. After World War II, nationalist groups, especially in Algeria, fought for and obtained independence—in 1956 in Morocco and Tunisia and 1962 in Algeria. Libya remained poor, a mainly desert area of little economic or political outside interest until Italy occupied it in 1911. After World War II, Italy was replaced by a British-French protectorate until independence was achieved in 1952. Then oil was discovered.

# Economic Development

Despite the gradual fall in the rate of population growth, the populations of countries in North Africa continue to increase almost as fast as economic growth. Major problems face all four countries in their attempts to generate further economic and human development. Ownership of consumer goods is higher than in the African countries south of the Sahara but is still modest (Figure 4.15). Morocco and Tunisia are rising middle-income countries. Algeria's income fell in the 1990s because of civil strife. Libya does not report income, but its oil reserves and a relatively small total population put it into the upper-

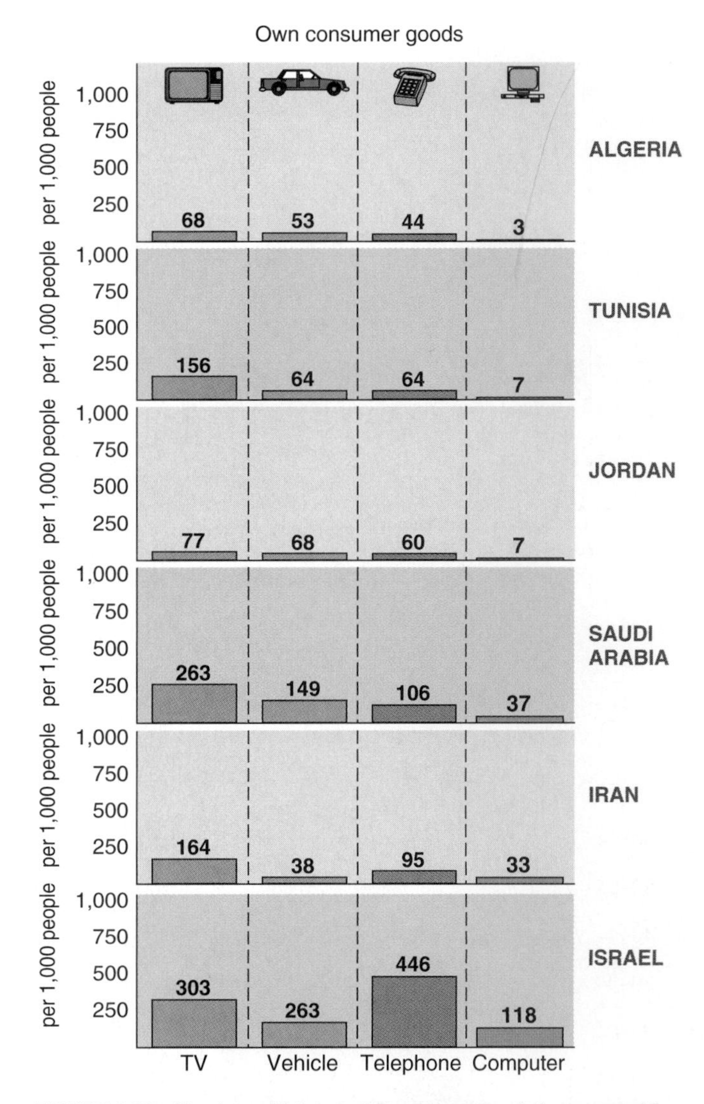


**FIGURE 4.15 Northern Africa and Southwestern Asia: ownership of consumer goods.** Compare these levels with those of other countries in Africa.

Source: Data (for 1996) from *World Development Indicators,* World Bank, 1998.

middle income group of countries. In spite of U.N. sanctions, Libya's oil and gas are sold to Europe, and the income is used to subsidize housing, basic foods, oil products, education, and health care. This subsidy makes up for low wages.

## Colonial Influences and Farming

Many of the problems facing North Africa stem from the type of economy established by colonial countries. The colonial system involved land appropriation for European settlers, who used irrigation water for intensive commercial farming that was tied to markets in Europe. Manufacturing and oil exploration were not encouraged. Export crops, such as citrus fruits and olive oil, continue to be produced on large holdings of over 50 hectares (124 acres). Although disrupted and pushed off the best land, the traditional farming sector remains a major employer, growing cereals on holdings of 5 to 10 hectares. Yet North African countries still need to import up to half their food needs, with Libya up to three-fourths. The growth of private, as opposed to collectivized, agriculture in the 1980s led to increases in the commercial production of vegetables, fruit, and dairy and poultry products.

Although a few Italian settlers took up land in Libya before 1945, the main development of farming there was since 1969, when rotary-sprinkler irrigation was introduced along the coast and at a few inland sites to grow cereals. Oil revenues funded this investment, and the production costs are high while yields are not. Libyan farming is thus uneconomic, and farmers are upset when priority for water use is given to urban inhabitants.

## Morocco's Farming Economy

Only Morocco has as much as half of its population still dependent on agriculture; none of the other countries has more than a third (Figure 4.16). This reflects the arid environments of the subregion, as well as the growing returns from oil production, manufacturing, and service industries in the other countries. Morocco's farmers use irrigation water to produce citrus fruits, vegetables such as tomatoes and potatoes, and cut flowers for European markets.

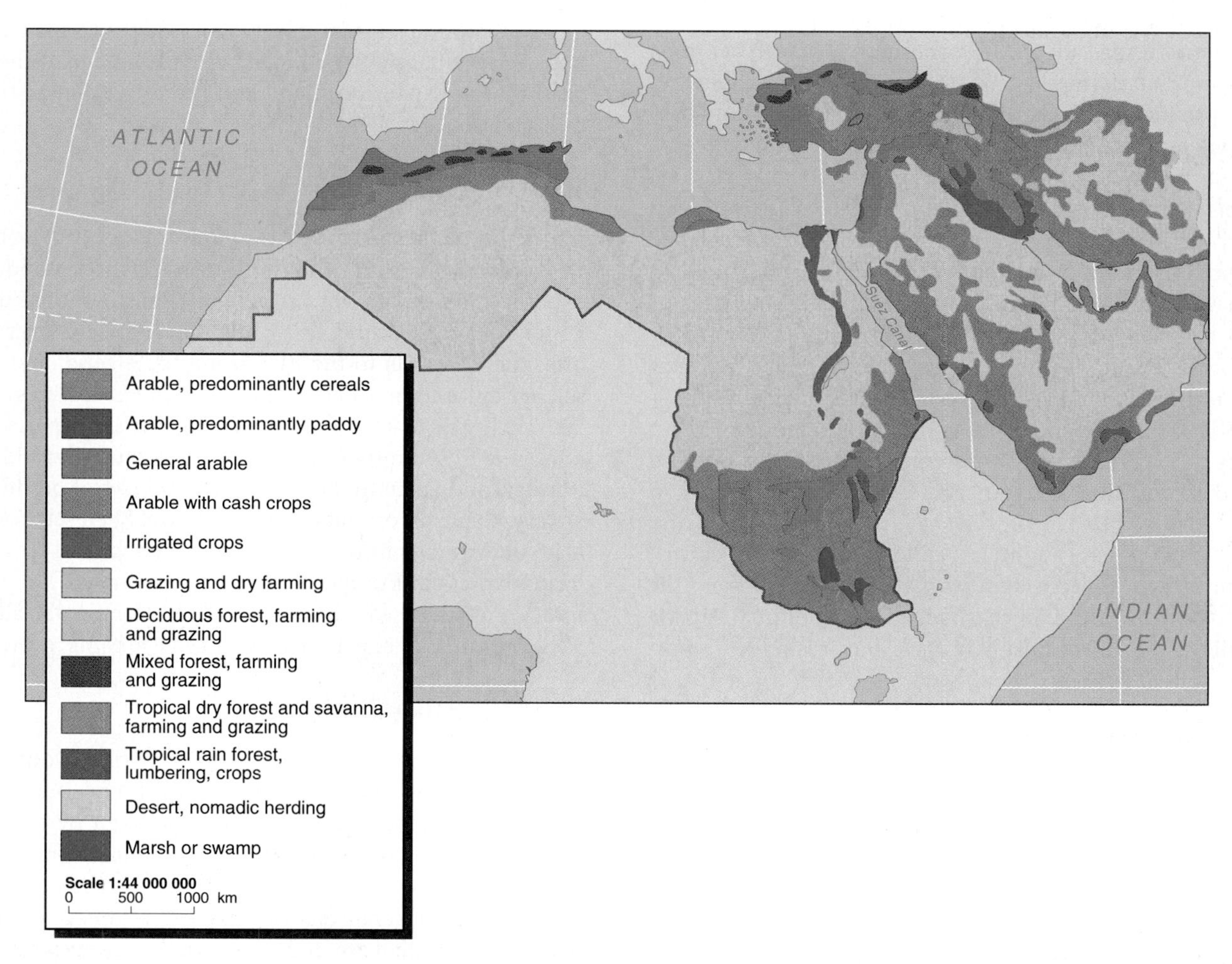


**FIGURE 4.16 Northern Africa and Southwestern Asia: agricultural land use.** Note the extent of desert and other dry-farming uses.

Source: Data from *New Oxford School Atlas,* p. 74 & p. 90, Oxford University Press, UK, 1990.

**FIGURE 4.17 North Africa: oil refinery at Skikda, Algeria.** Oil refineries take up huge areas of land, here necessitating the reclamation from the sea. They include ocean terminal facilities, storage tanks, and the complex pipework of the processing plant.



Farming in Morocco uses most of the water and land available. At present, 85 percent of water used in Morocco is taken by farmers, but this is expected to fall as domestic and industrial demands increase. The shift toward urban uses is bringing crisis to farms, especially in drier southern Morocco at a time when Moroccan farmers need to invest further capital to improve productivity and quality. It may be necessary for farmers to switch to more valuable crops or those that require less water, but the markets for such crops are not obvious. Hashish (cannabis) remains a major—though illegal—source of hard currency for the country. Although some efforts are being made with European Union help to eradicate the crop, it is difficult to do so in its main growing area of the Rif Mountains, a virtually autonomous region ruled by Spain until 1957.

Morocco's economy also rests on other primary products from mining and fishing. Its exports of phosphate are significant, and its fishing industry has grown to the point where it provides 15 percent of Moroccan exports. At present, Morocco is contesting the management of fishing grounds off western Africa with the European Union and particularly Spain. The main fish caught are squid (for export to Japan), tuna, and hake. Morocco continues to export cork from the bark of oak trees in the northern area of the country. Algeria and Tunis also export phosphate rock, but it is a less important component of their overall exports.

## Oil and Gas

Oil and natural gas are the chief mining products of North Africa and dominate the economies and exports of Algeria and Libya. It was only after independence that these countries discovered their oil reserves or were able to exploit them. Libya, with its small population, suddenly gained great riches that were nationalized after an initial phase of development by multinational oil companies headquartered in Europe and the United States. Algeria and Libya are major world producers. Pipelines bring the oil and natural gas from interior locations to coastal ports and refineries (Figure 4.17). Up to the late 1990s, Islamic militants in Algeria had not disrupted the oil and gas flows or discouraged foreign investment in the expansion of production, but the pipelines remained vulnerable.

## Expanding Manufacturing and Services

Manufacturing is growing in all countries and now contributes 20 (Morocco) to 30 (Algeria) percent of GDP. Morocco's manufacturing sector is the least developed, but it has a substantial craft industries sector. All countries have basic food processing, constructional materials, and textile industries located in and around major cities. Apart from Morocco, all countries have oil-refining and petro-chemical industries. Algeria also has light industries, including the manufacture of electrical components; Libya makes steel and aluminum; Tunisia has a small steel industry. North African countries have growing government bureaucracies and tertiary-sector occupations in education and health care. In Morocco and Tunisia, tourism is a major source of income, based on their sunshine, coastal locations, historic and cultural features, and stable political environments. Tunisia had nearly 4 million visitors in 1996 and Morocco over 2.5 million. Terrorist activity in Algeria after 1992 ended the envisaged tourist expansion there, and there were fewer than 0.7 million visitors in 1996.

## Trade with Europe

The trade of the North African countries is tied to Europe, both the source and end of over half their imports and exports. Algeria and Libya supply one-fourth of the European Union's oil and natural gas, while Morocco sells over half its phosphate exports to the EU. Morocco, Algeria, and Tunisia sell agricultural products and textiles to EU markets, but the 1992 EU rules made it more difficult for these exports to penetrate the EU countries. This is partly because the addition of Portugal and Spain to the EU led to surpluses of products that are also the main outputs of North Africa. Such setbacks came at an unfortunate time when tourist revenues and remittances from workers in Europe were unlikely to increase and, therefore, compensate for the lost income. As a result, all North African countries experienced reduced economic growth.

## Economic Policy Changes

The economic problems of the North African countries are similar to many other developing countries but have not always been helped by the major role taken by the state in trying to mastermind economic development. The setting of five-year plans was often wastefully bureaucratic and inflexible at a time of rapid change. Only the oil-exporting countries receive adequate foreign currency to pay for major developmental projects. Algeria, however, borrowed so much on top of this income that its debt servicing now takes 75 percent of the export earnings. Tunisia's smaller oil income and problems of drought and locust attacks caused a financial crisis when its oil revenues decreased in the 1980s. The collapse of world phosphate prices and the costs of conflict over Western Sahara upset Morocco's overambitious planning in the late 1970s, leading to a massive increase of its debts.

In the 1980s, the International Monetary Fund required Tunisia, Algeria, and Morocco to carry out measures that would cut internal budget deficits by reducing the public sector of the economy and encourage trade in export products. The resultant unpopular measures of structural adjustment, such as reduced civil service employment and the removal of subsidies on basic foods (flour, oil, and sugar), led to riots.

In the 1990s, North African countries attempted to privatize large sections of their economies. Tunisia is farthest ahead and even has its own stock exchange. Morocco is following with unparalleled sales of state holdings that reversed the previous policy of "Moroccanization" up to 1989. A new balance of ownership between foreign investors, Moroccan interests, and employees affects the oil distribution, sugar refining, and even utility industries.

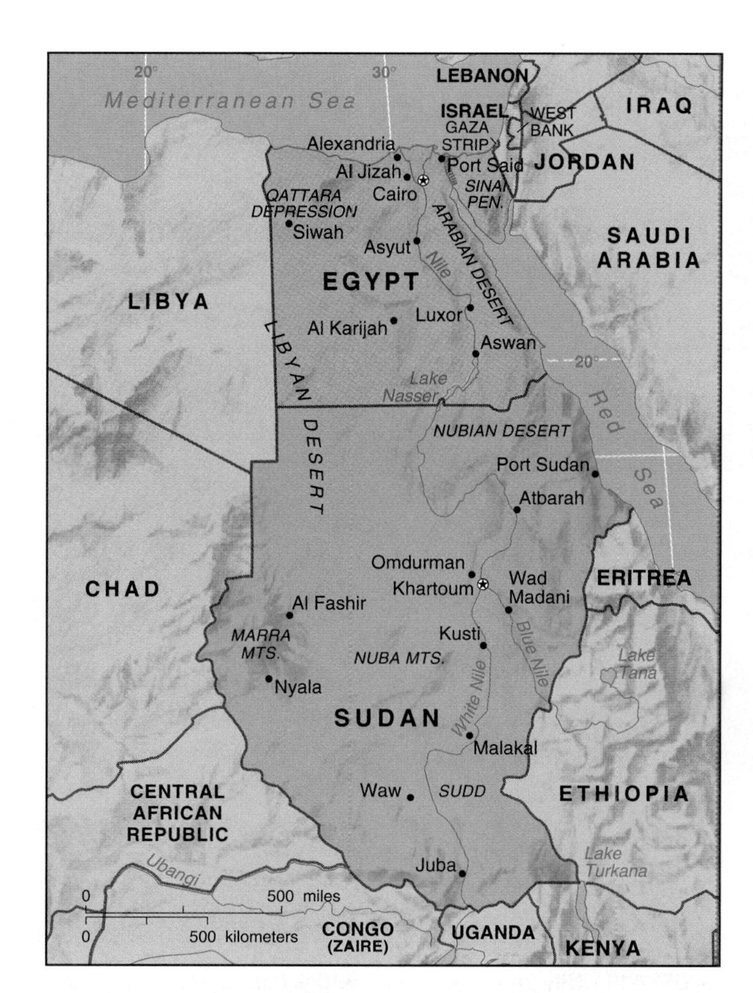


**FIGURE 4.18 Nile River valley: countries, physical features, and main cities.**

## NILE RIVER VALLEY

The Nile River provides a life-giving water supply to Egypt and Sudan that has maintained a human presence in the dry eastern Sahara since the early days of farming and civilization. The modern histories and economies of the two countries are connected by this flow of water (see Figure 4.7). The 1959 **Nile Waters Agreement** shared the use of Nile River waters between Egypt and Sudan, with Sudan receiving less than 30 percent of the annual flow. These two countries in the lower valley will continue to be linked as they contest the use of the water with countries in the upper Nile River watershed such as Ethiopia, Uganda, and Tanzania.

Although the Nile River waters provide a basis for living in an arid climatic environment, the pressures imposed by increasing populations have made it difficult for Egypt and especially Sudan to modernize and diversify their economies. Both are poor countries (see Figure 4.2) with major balance of payments problems. Neither produces sufficient export income to pay for the needs of their people. Both rely on attracting aid from core countries and the major Arab oil producers, but both sources reduced their contributions in the 1990s.

### Countries

The countries of the **Nile River valley—Egypt** and **Sudan** (Figure 4.18)—share the river waters and problems of developing their economies. Their geographic positions, political environments, and products provide a contrast. Egypt is the largest Arab country in population (66 million in 1998) and one of the foremost in international prestige. Egypt retains control over the global choke point of the Suez Canal, which has been widened to accommodate increasingly large shipping vessels. It also controls the Sinai Peninsula and the land route from Africa into Southwest Asia.

The year 1956 marked a turning point in Egypt's modern history. After centuries of dependence on the annual Nile River flood—for food and, more recently, the growing of cotton for export to the British textile industry—Egypt became a socialist state with new priorities for farming and manufacturing focused on its own internal needs. The civil service expanded to implement the centralized direction of civil affairs. As issues of national importance became paramount, the first major goal was to make the water supply secure to support farming expansion and industrial power supplies. The 1959 agreement to share the Nile River waters with Sudan—rather than a basin-wide management system—led to the building of the Aswan High Dam. On completion of the dam in 1970, Lake Nasser behind the dam stored three times Egypt's annual water usage (Figure 4.19).

President Nasser adopted an assertive attitude externally, leading the Arab world. By getting the Soviet Union to help

**FIGURE 4.19 Nile River valley: Aswan High Dam, Egypt.** An earlier dam was built on the site in 1902 and extended in 1934. The massive modern dam was completed in 1970, causing the relocation of 90,000 people and the drowning of archaeologic sites.



build the Aswan High Dam, he also exploited Cold War politics. Despite a growing population, Egypt moved forward under Nasser's leadership and grew economically at a more rapid pace than the rest of Africa in the 1960s and 1970s. Nasser died in 1970, and his successors shifted foreign policy in the late 1970s to accommodate Israel and so receive more support from the United States. This brought isolation from the rest of the Arab world but generous aid from the United States. Today, Egypt has a diverse economy based on farming but with substantial manufacturing and tourist industries; its total GDP doubled between 1980 and 1995. It maintains political stability despite terrorist activity. Egypt's main problem remains the continuing increase in its already large population, but measures to reduce this are beginning to be successful.

Sudan is the largest country in Northern Africa and Southwestern Asia and the poorest. It has half the population of Egypt but less than a third of its annual production. After Sudan achieved independence in 1956, the seasonal rains and fertile soils of the southern region gave hope for economic potential. The new leaders began to develop areas that could sustain rain-fed agriculture along the Nile River valley and on sandy areas in the west of the country. The full potential of such developments was not realized—partly because of droughts and partly because of political instability as the Muslim leaders of the national government tried to impose their language and religion on Christian and traditional religious communities in the south. In the late 1980s, the total irrigated area declined, remittances from workers overseas fell away, and war broke out in the south. Over a million refugees from Ethiopia's famines compounded these problems.

Sudan's people are more rural than those of Egypt and more dependent on agriculture. With its low income, narrow economic base, continuing high rates of population increase, low life expectancy, and civil war, Sudan resembles other African countries more than those of northern Africa. Sudan, like Egypt, remains dependent on the Nile River waters for its dominantly agricultural economy. Sudan is also part of the southern Saharan rim with conditions somewhat similar to those of its neighbors in Chad and Ethiopia.

## People

### Population Distribution and Dynamics

The populations of Egypt and Sudan are concentrated along the Nile River valley (see Figure 4.12). Both Egypt and Sudan

**FIGURE 4.20 Nile River valley: Cairo, Egypt.** A busy street with traffic congestion and many people on foot. Is there much difference between this view and the center of a town familiar to you?



have rapidly expanding populations within this limited natural environment. The 1998 total of 94 million people in the two countries was up from 60 million in 1980 and may reach over 160 million by 2025. These totals, and the potential in the large groups of young people for continuing high rates of population increase as they reach child-bearing age (see Figure 4.13), have to be considered in the context of the pressures they will place on only 5 percent of the land that is cultivated in each of these largely agricultural countries.

Egypt is making progress in reducing rates of population growth (see Figure 4.14). In the late 1970s, worries over its ability to support the projected population increases led to mass media and government support for family planning, backed by U.S. aid. The effect has been to reduce total fertility rates from over 5 to under 4 and to under 3 in cities. More women are stopping at small families. Further progress depends on raising the status of women in Egyptian society. At present, only 37 percent of women, compared to 63 percent of men over 15 years old, can read and write. In Sudan, overall population growth remains high at nearly 3 percent per year. Little is done to reduce this high rate of increase.

Egypt is an increasingly urbanized country, with almost half its population now living in towns and cities. **Cairo** has expanded to the foot of the pyramids and its population, including Giza, is around 11 million (Figure 4.20). **Alexandria**, the main Egyptian port, has nearly 4 million people. Sudan has only 20 percent of its population in towns, of which **Khartoum**, the capital, is by far the largest.

## Population History

The cultural environments of Egypt and Sudan are related as much to the history of the Nile River valley as to the more recent Arab and colonial phases. It is not fully understood how the Egyptian civilization came to dominate the lower Nile River valley, but a united empire emerged around 3200 B.C., based on centers at Memphis in the north and Thebes in the south of modern Egypt. The Nubian kingdom in the area of modern Sudan became part of this Egyptian empire, based on slave labor and the annual Nile River floods that renewed the soil. The empire history was complex but ended when Alexander the Great conquered it in 332 B.C., and was followed by Roman control up to the A.D. 500s. During this period, the Christian Coptic church spread through the area and southward into Ethiopia.

In the A.D. 600s, Arabs conquered Egypt but did not fully control the area of modern Sudan until the 1500s. In the interim, Sudan had a series of independent kingdoms, some of which were Christian. Although it was an Arab kingdom, Egypt suffered invasions from Turks and Mongols before becoming part of the Ottoman Empire in the 1500s. Egypt's position at a crossroads of international trade brought prosperity to its markets, followed in the 1800s by British encouragement of the Nile River valley farmers to grow cotton for British textile mills. The opening of the Suez Canal in 1869 added to Egypt's strategic value. The British occupied Egypt until the latter gained its independence in 1952. After decades of British discouragement of Egyptian manufacturing, the new government made industrial development its major priority.

When it became a Muslim country in the 1500s, Sudan extended its territory southward although many of the tribes in the southern area retained their Christian or traditional beliefs. In the 1800s, it was attacked by Egyptian forces but kept its independence until the British joined in the 1899 invasion. After this, Sudan came under the joint control of Britain and Egypt, with British engineers constructing irrigation works in the Gezira area south of the junction between the White Nile and Blue Nile Rivers. Sudan gained its independence in 1956, despite Egypt's claims to keep control. The tensions resulting from the period of Egyptian occupation of Sudan and fears of future Egyptian expansion remain in the minds of many Sudanese and affect negotiations over the use of Nile River water.

## Economic Development

Egypt has been part of the world economy since the late 1800s and has a rising lower-middle income. Sudan is poor and still has a tenuous hold on external economic links because of its narrow economic base, isolation from major transportation routes, and internal breakdown of security. The two countries face different problems in trying to improve their low world ratings in HDI, GDI, and GNP. Low incomes are reflected in low ownership of consumer goods (see Figure 4.15).

### Economic Diversification in Egypt

Egypt's agriculture moved from a concentration on cotton exports at the beginning of the 1900s to increasing food production during the 1950s (see Figure 4.16). It is still not self-sufficient in food production. After independence, the land holdings were controlled in size to allow more people to share Egypt's main resource. The Agricultural Reform of 1952 limited holdings to 80 hectares, but after a series of further reductions over the years, the largest were reduced to 20 hectares (49 acres) by 1969. The cultivated area rose steadily until the 1970s, and further expansion took place outside the Nile River valley in the early 1980s at a time when Egypt was able to use more than its agreed-upon share of water. Productivity increased as the application of fertilizers and the double-cropping of some areas made it possible to harvest two crops within a year. Egypt remains a major world exporter of cotton and also produces sugar, rice, vegetables, and fruit. After 1987, however, some land had to be withdrawn from cultivation because of subsidence near the Nile Delta coast along with salinization, waterlogging, and shortages of water. Such withdrawals highlighted the limits of Egypt's land resources.

The main hopes for Egyptian development based on the generation of electricity at Aswan and its potential to supply power for manufacturing are not fully realized. The manufacturing base of iron and steel and chemicals production, together with the assembly of cars, food processing, and bus, textile, tire, and television manufacturing is more diverse than that of other African countries. The largely nationalized industries are overregulated, however, and in the 1990s needed modernization and considerable investment to maintain their importance to the country. Egypt's armament production received a boost in the Iraq-Iran War (1980–1988) but was less busy in the 1990s.

Other aspects of Egypt's economy that added income up to the 1980s suffered a decline in the early 1990s. Remittances from 2 million Egyptian workers in the Persian Gulf equaled agriculture and manufacturing as sources of income for a few years in the 1980s, but over a million workers with their families returned to Egypt during the Gulf War. Tourism based around Egypt's historic treasures, such as the pyramids and temples along the Nile River, formed another growth area until Islamic fundamentalists began to attack foreigners in the 1990s, causing a major slump in 1993. Although 1995 bookings returned to 1992 levels, and there were nearly 4 million visitors in 1996, renewed terrorist action in 1997 was a further setback.

The ending of the Cold War and Israeli-Arab conflict will also probably lead to a reduction in massive aid from the United States—which amounted to as much as $10 billion, or 10 percent of Egyptian GDP, at its height. The Suez Canal revenues, static at just below $2 billion per year, and the low level of oil exports do not compensate for these downturns of income. By switching the focus of its external relationships from the former Soviet Union to the United States, Egypt was the forerunner of many Arab countries in assessing where its best interests lay—a pragmatic rather than an ideologic judgment. During the 1970s and 1980s, this switch brought economic benefits to Egypt, but the future is less certain.

Within Egypt, the Nile River valley south of Cairo, designated "Upper Egypt," experienced less development than the northern part of the country despite grandiose plans. Incomes in Upper Egypt are half those of the rest of the country, and it has the poorest measures of infant mortality, adult literacy, health services, schools, and unemployment. Many Egyptians who worked abroad came from Upper Egypt, and those who returned at the time of the Gulf War came back to a penniless area. Much of the civil unrest in Egypt, and nearly all the terrorist killings, occur there.

### Sudan's Economic Plight

Sudan's position is worse. It never had much support from the United States or Soviet Union, and the little it received from other Arab sponsors fell away after the Gulf War. Remittances from overseas workers also declined in the 1990s. The attempts to forcibly integrate the southern part of Sudan in an Islamic state led to 15 years of civil war that destroyed production in the south and diverted northern resources to military expenditure.

**RECAP 4B: North Africa and Nile River Valley**

North Africa has close ties with the former colonial powers in Europe, particularly France, and continues to supply energy needs but faces economic difficulties as the European Union expands to include countries that market similar products.

The Nile River valley countries of Egypt and Sudan still rely heavily on the river's water, but the supply cannot be increased to match the population growth. Economic diversification in Egypt scarcely keeps income above population growth rates. Civil war slows development in Sudan.

4B.1 Which five geographic characteristics do you think make each of the North African countries distinctive?

4B.2 Outline the causes of the annual Nile River flood. Why should Egyptians be pessimistic about their future water supply?

4B.3 What are the arguments for and against import substitution and export-based manufactures in the economies of developing countries?

**Key Term:** Nile Waters Agreement

**Key Places:**

North Africa
Algeria
*Algiers*
*Oran*
Libya
*Tripoli*
*Benghazi*

Morocco (with Western Sahara)
*Casablanca*
*Marrakech*
*Fez*
Tunisia
*Tunis*
Maghreb

Nile River valley
Egypt
*Cairo*
*Alexandria*
Sudan
*Khartoum*

Around 1 million people have died in the south, and 1.5 million have been displaced. After losing its influence in the early 1990s because of its brutal attacks on local peoples, the Sudan People's Liberation Army (SPLA) developed a friendlier political wing in 1994 to win over its people. It attracted external military support and by 1998 was besieging Juba, the main city of southern Sudan. But people still died of hunger and disease.

Sudan's main products are from farming. Cotton provided 50 percent of Sudan's exports in better times, when the southern part of the country produced livestock products for export. Although Sudan is normally able to feed its population, civil war, drought, and the pressures of refugees from Ethiopia combined to cause famine in the early 1990s. Sudan has no major mineral products, and its manufacturing sector is at an early stage of development. Textile factories, paper mills, and an oil refinery, all based around Khartoum, contrast with small-scale production for local needs in the rest of the country and together make up a small part of the country's total product.

Price declines in major staples such as wheat and livestock feed, resulting from competition between the European Union countries and the United States, affected both Egypt and Sudan. Imports became cheaper but discouraged home producers, who could not compete with the low prices on world markets. Furthermore, the main trading partners of both countries in Anglo America, Europe, and Eastern Asia now buy cotton from Egypt and cotton and sugar from Sudan at decreasing world prices. The focus of both countries on developing commercial agriculture for export crops provides a fragile future. Markets are fiercely competitive and the returns on new projects decline as competitors follow suit. A small encouragement was the worldwide 1990s switch from artificial fiber textiles to cotton goods, thus increasing the demand and price for cotton—at least until other areas raised their production. Egypt could face a better future if regional trade with its neighbors, including Israel, develops, but the political will for such enterprise is lacking.

## ARAB SOUTHWEST ASIA

**Arab Southwest Asia** is the heart of the Arab World, comprising the Arabian Peninsula and the Fertile Crescent from the Tigris-Euphrates River basin to the Lebanon coast (Figure 4.21). Despite small numbers of people, the countries of this subregion play a major part in world affairs because of their oil wealth and opposition to the Arab-Israeli peace process.

The world centers of Islamic religion and Muslim pilgrimage are located in the Arabian Peninsula at Mecca (Makkah), where Muhammad was born in A.D. 570, and at Medina, which became his power base after he was thrown out of Mecca. This subregion has a greater uniformity of religious affiliation and language than other parts of the region. The main ethnic distinctions are between groups holding contrasting Islamic beliefs—the Sunni and Shia Muslims.

### Oil Wealth

This subregion contains a large proportion of the world's known oil reserves (see Figure 4.8). This gives it a huge strategic significance within the world economic and political systems. Sales of oil and oil products brought great wealth to some countries in the 1970s, when oil prices quadrupled.

Although the oil reserves are being used up slowly enough to be compensated by new discoveries, the 1990s were not easy for even the wealthiest oil states. After building huge financial reserves when oil prices rose during the 1970s, the oil-rich countries of Arab Southwest Asia took on major commitments to increase military strength, build internal infrastructure, and provide welfare services. They made donations to the oil-poor Arab countries. Paying for the Gulf War and subsequent continuing military expenditure at a time of low oil prices drained the financial reserves, forcing cutbacks in expenditure by the mid-1990s. Such cutbacks caused their inhabitants to question the quality of leadership in the oil-rich countries, particularly where hereditary princes continue to maintain extravagant lifestyles and remain unanswerable to their people. The poorer countries now receive fewer donations from the rich and miss the remittances from their workers who had worked in the oil-rich countries but returned home during the Gulf War.

By the late 1990s, faced again with possible punitive actions against Iraq, many of the Gulf states began to draw

**FIGURE 4.21 Arab Southwest Asia and Israel: countries, physical features, and cities.** The Fertile Crescent stretches from the Mediterranean coastlands through Syria and Iraq to the Persian Gulf.

back from support for proposed U.S. actions. This grew out of a desire to demonstrate their independence in the face of public sympathy toward the impoverished Iraqis. Many continued to feel that Iraq might still be a counterweight to Iran.

## Israel Problem

The presence of the country of Israel in the midst of strongly Muslim countries is one of the major political issues for countries united by religious concerns that are antagonistic toward Judaism. Since 1948, there have been three major conflicts between Israel and the surrounding countries. Despite defeat in those conflicts and continued support for Israel from the United States and other Western countries, most Arab countries claim they are engaged in a holy war against Israel. Such views leave little room for compromise. During the mid-1990s, recognition of Israel spread from Egypt to Jordan, but no further as the Israelis hardened their approach from compromise to confrontation after 1997. Peace in and around Israel remains fragile and subject to other future changes in Southwest Asia.

## Countries

The Arab Southwest Asia countries are **Bahrain, Iraq, Jordan, Kuwait, Lebanon, Oman, Qatar, Saudi Arabia, Syria, United Arab Emirates** (UAE), and **Yemen.** They are differentiated by natural resources of oil and water, emphases within Islam, and forms of government. The differences in natural resources and economic management produce a wide range of economic status—from countries that are little better off than India and China to those that rival the wealthiest core countries.

## Oil or No Oil

Saudi Arabia, Kuwait, United Arab Emirates, Bahrain, Qatar, Oman, and Yemen produce oil, but water is scarce. Lebanon and Jordan have some water but little or no oil; Iraq and Syria have both oil and water. These differences cause tensions among countries that share a Muslim faith and Arabic language. Although efforts were made to redistribute some of the oil wealth in the 1970s and early 1980s, the donor countries found that low oil prices and indebtedness forced them to reduce such support in the late 1980s and 1990s. Tensions remain between the donor countries (Gulf States) and those that receive aid (Jordan, Lebanon, Syria, and Yemen). Related problems arise from the lack of skilled or unskilled labor in the oil-rich Gulf states and from the other countries' reliance on remittances sent by their migrant workers. In the Gulf States, over half the workforce is expatriate. Many of the migrants returned home during the Gulf War, placing further burdens on their home countries.

In the future, as populations continue to increase rapidly, the differences in water resource distribution may become more vital than the possession of oil reserves. Many countries are already close to the maximum use of their water resources. The Tigris-Euphrates River system is the major surface-water source in the subregion but is increasingly the basis of tension among Turkey, where the headwaters rise, and the downstream countries of Syria and Iraq.

### Sunni and Shia Muslims

Differences also arise between the rival Muslim groups from the more conservative Sunnites to the more radical Shiites (see "World Issue: Islamic Revivalism," page 138). Most countries in this subregion have Sunnite majorities or governments but fear the growth of Shiite groups and the aggressive external policies of Shiite-dominated Iran. Countries such as Bahrain, Iraq, Lebanon, and Oman have Sunnite leadership but Shiite majorities that have often been repressed or persecuted. A further complication is the outcome of the radical policies of the Sunnite Muslim leader Muhammad ibn 'Abd al-Wahhabi, who sought in the 1800s to restore Islamic basics and energies. Wahhabi principles were taken up by the Saud family, using aggression in the early 1900s to extend their kingdom into the modern country of Saudi Arabia. In the process they conquered provinces now claimed by Yemen.

### Governments

Southwest Asian countries gained independence in different ways from different colonial overlords and now operate a range of political systems. The area remained under the gradually disintegrating Ottoman Empire until World War I, after which Syria and Lebanon became French Protectorates while the rest of the area was under British protection. The French encouraged republican governments. British protection guaranteed the survival of local kingdoms, including Jordan, Iraq (until 1958), and the small emirates along the Persian Gulf, seven of which united as the United Arab Emirates (UAE) in the 1970s. It also led to the increasing exploitation of local oil resources by British and American oil companies. The British had begun to extract Iranian oil in 1901, and such activities spread to Iraq in 1931 and Saudi Arabia in 1938.

Internal conflicts spring up from time to time, including those in Oman from 1970 to 1975 and in Yemen in the 1980s. The latter resulted in the merger of North Yemen with Yemen People's Democratic Republic in 1990, but 1994 violence between these two parts of Yemen demonstrated that important differences remain. Internal conflicts among tribal groups or Islamic sects in such countries as Syria and Iraq led to dictatorship rule and the fostering of socialist nationalism, repressing internal conflicts. In the Gulf states, the tribal rulers maintain their leadership positions but are gingerly introducing democratic measures.

## People

### Population Distribution and Dynamics

The desert environment causes the population of this subregion to cluster along the coasts and in the Tigris-Euphrates River valley (see Figure 4.12). The 11 countries of the subregion have relatively small populations. In 1998, Iraq and Saudi Arabia had 20 million people each, while Syria and Yemen had 16 million. The rest had 5 million or fewer.

As in the rest of the major world region that stretches from North Africa to Iran, population growth in Arab Southwest Asia is rapid from a low base earlier in the 1900s. In 1970 the total population was around 30 million, growing to nearly 50 million by 1980 and 90 million by 1998. The estimate for 2025 is 180 million people in these countries of limited non-oil resources. By then, oil resources may be dwindling and water will be in even shorter supply.

The population growth is maintained by some of the world's highest total fertility rates of between 6 and 7.5 children per female (Oman, Saudi Arabia, Yemen). In most of the small and rich Gulf states—Bahrain, Kuwait, Qatar, and UAE—fertility is under 5. Lebanon's fertility rate is 2.3—after a long period of civil war. Kuwait's population declined during and after the Gulf War. High birth rates combined with falling death rates result in rapid population growth (see Figure 4.13), large proportions of young people (see Figure 4.14), and life expectancies of over 65 years in most countries. In Yemen, life expectancy remains low at 58 years, and in Iraq it fell to the fifties after years of war and deprivation.

In many Southwest Asian countries, migration is a way of life that has major impacts on population change. Much of the high population growth in the oil states reflects high rates of immigration. Foreign labor made up nearly 30 percent of the total labor force in the Gulf oil states in the mid-1980s. Although the origins of foreign labor shifted from other Arab countries to Southern Asia in the 1990s, this proportion remains high. Figure 4.14 shows how the presence of male foreign workers affects the age and sex structure of Gulf States such as Kuwait. Jordan, Lebanon, and Yemen exported many workers and depended on their remittances. They were all affected adversely by the reduction in demand for migrant workers in the Gulf and changes in their origins after the 1991 Gulf War. In 1995, estimates put the number of migrant workers at around 1.6 million, with over 90 percent coming from India, Pakistan, Bangladesh, and Egypt. Much of the migration is short-term labor migration, but there are also increasing numbers of refugees escaping civil war in Lebanon and Iraq, and moving out of Israeli-held territories.

**FIGURE 4.22 Arab Southwest Asia: Baghdad, Iraq.** Modern bridge over the Tigris River, with view looking from city center to new buildings on the west bank of the river.



### Highly Urbanized

Arab Southwest Asia is highly urbanized and, apart from Jordan, Oman, Syria, and Yemen, all countries have over 70 percent of their populations living in towns. This high level reflects the changes brought by the oil industry, the lack of farming land in many countries, and the development of manufacturing and government employment. Between 1950 and 1998, Kuwait and Qatar moved from 50 percent to over 90 percent urban, while Saudi Arabia went from 10 to 67 percent urban. The largest cities are all capitals (see Figure 4.2), including **Baghdad** (Iraq) (Figure 4.22), **Amman** (Jordan), **Beirut** (Lebanon), **Riyadh** (Saudi Arabia), **Damascus** (Syria), **Abu Dhabi** (UAE), and **Sanaa** (Yemen). Other large cities include the port of **Basra** and the oil center of **Mosul** in Iraq; **Irbid** in Jordan; the cities of **Jeddah, Mecca,** and **Medina** in Saudi Arabia; **Aleppo** (Halab) in Syria; and **Aden,** the port and commercial center of Yemen.

### Population History

The subregion includes the earliest major world center of civilization in the Mesopotamian region of the Tigris-Euphrates River valley that is now largely Iraq. The subsequent cultural history of the subregion reflected its position "in the middle" between major intercontinental influences. It was the seat of several polytheistic religions but then witnessed the origins of the monotheistic Jewish, Christian, and Islamic religions. It suffered from clashes among external powers such as Egypt, Persia, Greece, and Rome and became the source from which Arab dominance and Islam expanded in the A.D. 600s. It was the focus for conflicts between Arabs and Christian Crusaders, and between Arabs and the Mongols from Central Asia in the 1100s. Many of these clashes resulted in destruction and impoverishment of previous economies.

In the 1500s, the subregion was conquered by Ottoman Turks, who retained overall control of the local tribal arrangements, through which they ruled until the early 1900s. The Ottomans, however, did little after the 1700s to encourage modernization or involvement in the expanding world economic system. The subregion became isolated from wider contacts. Some of the more isolated tribes became largely independent under their own kings or sultans before 1914. The British and French built on upwelling Arab nationalism to expel the Turks during World War I. Although the Arab and Muslim influences remain dominant, animosities generated by the establishment of the country of Israel, clashes with significant Christian minorities in Lebanon and Syria, and feelings of hatred toward Turks and European colonial powers remain significant factors in understanding internal tensions.

## Economic Development

Arab Southwest Asia has a major economic contrast between countries with high oil revenues and those that produce little or no oil. The different levels of income are reflected in the ownership of consumer goods (see Figure 4.15).

The richer countries bordering the Persian Gulf, including Bahrain, Kuwait, Oman, Qatar, Saudi Arabia, and UAE, sit on huge oil reserves, the exploitation of which brought high incomes per head of the population in the 1970s and early 1980s but then declined as oil prices fell. They have small total populations and rely on immigrant labor. Their high world ranks in GNP per capita fell somewhat. Their HDI and GDI ranks are much lower, indicating that the benefits of oil wealth have not had a great impact on education and health care. Women have a low status in many of these countries.

The poorer countries of Lebanon, Jordan, and especially Yemen have much lower per capita GNP, HDI, and GDI ranks, more available labor, and some capacity for agriculture. Iraq and Syria are the exceptions to this opposing rich-poor duality, since up to the mid-1980s, they both had oil revenues, significant water resources and agriculture, and large labor forces. Since the Iran-Iraq War, Gulf War, and subsequent U.N. sanctions, however, Iraq's economy has been virtually destroyed.

### Changing Impacts on Oil-Producing Countries

Before the exploitation of oil, the lands of Arab Southwest Asia were the scene of low-intensity farming wherever water was available, nomadic herding where water was scarce and vegetation sparse, and extensive empty desert areas. A few crops, such as dates and citrus fruits, were exported from the better watered places in southern Iraq and along the Mediterranean coasts. Such activities are still found in parts of the subregion, but the increasing output of oil and gas from the oil states transformed the economies of many countries, particularly since 1965.

By 1980, the oil-rich Gulf countries assumed a strategic place in the world economy that gave them increased political significance. The income from oil was partly used to generate internal changes such as industrialization; agricultural intensification; the provision of roads, airports, health services, and education; and increased living standards. These changes managed only part of the shift from a dependence on oil revenues to a pattern of sustainable development based on diversified economies. A **diversified economy** is one in which manufactured (secondary sector) goods are more important than primary products and a variety of manufactured products is joined by a growing service (tertiary) sector. Crude oil and oil products still make up between 80 and 95 percent of exports from the oil countries apart from Syria, where the figure is 40 percent. The great challenge faced by the oil-rich countries is to make the move to broader diversification in the face of falling oil prices, or before the oil runs out. In 1995, Syria had only 8 years of recoverable reserves.

From 1965 to 1990, the oil-producing countries greatly increased their proportions of world GDP, although there were changes of emphasis from country to country. In 1965, Saudi Arabia's GDP was $2 billion, but it rose to $156 billion in 1980, when oil prices reached their highest point, before falling to $125 billion in 1995. Kuwait's GDP rose from $2.1 billion in 1965 to $28 billion in 1990, before the Gulf War. Iraq, once the leading oil producer in this group, spent much of its resources on the 1980s war with Iran and then suffered massive destruction and further loss of life in the 1991 Gulf War. Iraq is now subject to sanctions that prevent exports and imports apart from food and medical supplies. As of 1994, Iraq could have sold some oil on world markets but did not choose to do so until 1998.

Meanwhile, by the late 1990s, Kuwait largely restored the oil production that was wiped out in the 1991 Iraq invasion. The 742 oil wells torched by the retreating Iraqi armies were repaired rapidly to restore pre-war levels of oil output. The costs of reconstruction and funding the Gulf War coalition, however, reduced Kuwait's overseas assets from around $100 billion to around $35 billion. Kuwait increased its oil output and reconsidered the employment and social welfare conditions of its people and the dominance of the ruling Al-Sabah family. During the 1980s and 1990s, some smaller states, such as Bahrain, Qatar, and Oman, began to experience falling oil production, while the UAE had a dramatic rise in output. Saudi Arabia took over the position of leading producer from Iraq and Kuwait, with major growth in its economy.

## Diversification in Oil-Producing Countries

The oil-exporting countries spend a large proportion of their income on food imports, a need that grew as the population has increased. Despite major investments in home-based irrigated agriculture, costs are very high and production is uneconomic compared to world prices.

Other aspects of economic diversification include the development of manufacturing. This began with the building of oil-refining capacity to replace the direct export of crude oil to European refineries. Petrochemical and other "downstream" activities based on the materials produced from oil followed. Bahrain and UAE used the local energy source as a basis for building aluminum smelters, Oman built a copper smelter, and Qatar and Saudi Arabia produce steel. Several countries also manufacture construction materials, including cement, to provide the materials for upgrading roads and building new housing and factories. Further diversification into food industries and consumer goods industries increased the proportion of GDP from manufacturing. The tertiary sector has also become important in many of these countries where banking and government employment are increasing rapidly.

In Saudi Arabia, new industrial towns are being developed at Al Jubayl on the Gulf coast and at Yanbu on the Red Sea. Al Jubayl is halfway toward its planned population of around 300,000 and has major industrial zones with airport, port, and highway linkages. It will cover an area the size of built-up Greater London or Atlanta.

The Gulf oil states formed the **Gulf Cooperation Council** in 1981 in the context of the Iran-Iraqi war. It excluded Iraq and focused on common economic and political interests. Saudi Arabia dominates the organization, which was effective in bringing together the countries to resist the 1990 Iraqi invasion of Kuwait.

## Impacts of Declining Oil Income

By the mid-1990s, years of budget deficits made these countries realize that prosperity was over, and they cut government spending. Reform is needed of their taxation structures and the lavish spending by wealthy elites that still relate to the high oil prices and incomes of the 1970s rather than the low oil prices of the 1980s and 1900s. These countries suffer several common problems that stem from their continued overspending through the period of falling oil prices.

They spend a third of their income on defense and internal security. Gasoline, electricity, water, and telephone charges are highly subsidized but are set to rise because the subsidies that support them cannot be maintained. The government by wealthy elites slows development by cutting out competition and fresh ideas while maintaining ancient legal systems and concentrating capital in the hands of a few. Education focuses on prestigious university provision rather than good basic systems that will produce much needed skilled labor and technicians. Having been used to possessing so much wealth of their own, they expected to be self-sufficient, and so lack any strategy to attract foreign and domestic investment to provide for the postoil era. A mid-1990s World Bank report on Oman, for example, stated that there could be social and economic disruption unless the economy is restructured. The other Gulf states, including Saudi Arabia, face similar issues that are linked to growing demands by their people for more democracy and information.

Iraq could have been the leading country in the area through political and economic influence, but its resort to war against Iran in the 1980s and its invasion of Kuwait in

LIVING IN . . .

## SYRIA

Syria continues to reflect its cultural past, but its people have experienced many changes since 1960. Most Syrians are members of the Sunni sect of Islam, a politically moderate group. The past for most people involved a tradition of family life and trade. Several related married couples shared living accommodations, with the women preparing the meals and caring for the home. Some families still recognize several thousand members who may be spread through southwestern Asia. The Syrians, along with the Lebanese, are the businesspeople of southwestern Asia, and Syrian families often have relatives in Lebanon, Palestine, and Jordan. Many Syrians regard these countries as "Greater Syria."

Syria became independent from French control in 1947. At independence, Syria felt itself much reduced in size from what it might have been. When the British and French took responsibility for parts of the former Ottoman Empire in 1919 at the end of World War I, they divided the territory into colonies or protectorates. These became the newly independent countries after World War II—with Israel carved out of Palestine in 1948.

Changes to the traditional way of life in Syria began in 1958 when President Nasser proclaimed the union of Syria and Egypt as the United Arab Republic. People in Syria saw this as part of a grand strategy to bring the Arab peoples into a single state. They thought that, with its strong agriculture, growing industry, and strategic location, Syria could become a major force in the Arab world. The next few years were a boom time for Syria, since it remained open to world trade and ideas, while Egypt adopted socialist policies that were increasingly restrictive. Egyptian businesspeople moved into Damascus to set up new enterprises since private businesses were prohibited in Egypt.

That phase lasted a mere three years, however, before the socialist ideas took hold in Syria. The Baath party, composed of people with socialist sympathies from both Muslim and Christian groups, was taken over by the Alawin, a Shiite minority group from the mountainous region of northern Syria. This group had a strong representation in the military and soon dominated the government. The 1960s were an uncertain time, with new Presidents every year (or week) or so, but in 1969, President Hafiz Assad, who had first been in military intelligence and then Defense Minister, became president. He is still there. Although a minority group within Syria, the Alawin were brought into all aspects of government, and especially the military, so that control was complete. Many formerly rural Shiite folk were encouraged to move into the expanding Damascus to balance the previously dominant Sunnite peoples.

Up to the late 1980s, the socialist policies of the Baath party, underwritten financially by agreements with Soviet Russia, had major effects on the Syrian economy. High tariffs were introduced to protect farmers and the currency from external competition, in a bid to establish self-sufficiency. Syria's oases produce a wide variety of vegetables and fruits that are the basis of its interesting and varied food. The manufacturing industries—mainly textiles based on locally grown cotton—were nationalized, and other manufactured goods were produced for internal consumption, including transport equipment and military goods.

The land reforms took land from previous landowners and distributed it to small farmers. Unable to make them profitable, the small farmers abandoned the land and migrated into the towns. New owners, including government and military officials, bought up the abandoned land and introduced new farming technologies from the former Soviet Union in attempts to develop commercial agribusiness. Many failed through poor management. The factories also became less productive. The lack of enterprise permitted in the controlled socialist society also reduced the national aspirations of a traditionally enterprising trading people, so that they took to various forms of corruption. The twin virtues of freedom and justice, held dear by many Syrians, were lost.

Another result of these changes was the rapid growth of Damascus to a city of around 5 million people (including the surrounding towns). The new suburbs were home to many that moved into the city to take advantage of government jobs in offices or the military. Military barracks were distributed through the suburbs.

Further changes came in the late 1980s as external trading restrictions were lifted in response to Syria's own economic performance. Syrians were encouraged to export goods and were then allowed to import a proportion of other goods without tariffs. This initiated more private enterprise as small firms grew in response. Private ownership of busi-

1990 impoverished its population and devastated its resources and infrastructure. Iraq's GDP is now around 10 percent of what it was in 1980. Although it has the largest area of well-watered farmland in Southwest Asia, Iraq paid little attention to making this commercially intensive, preferring to rely on oil income to pay for imports and military expansion. Most farmers are peasants who focus only on subsistence. The U.N. sanctions imposed after the Gulf War broke Iraq's overseas trade based on oil income. Consequently, food has to be rationed in the towns, and decreasing food rations result in high levels of malnutrition, although home food production is now encouraged. The Iraqi govern-

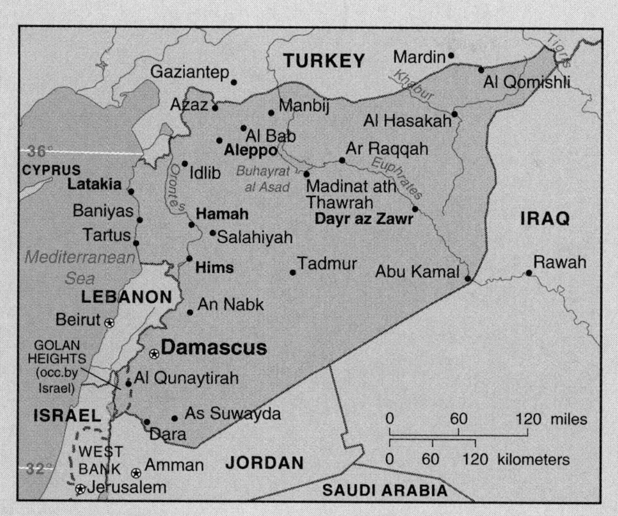


**(a) Map of Syria.**

**(b) Damascus:** view across suburbs to the multistoried buildings of downtown at the foot of a hill.



nesses doubled from 35 to 70 percent of the total, and economic growth reached 10 percent in 1992. These new policies paralleled the decline of Soviet support, the new demands of the World Bank for internal enterprise as opposed to government-run projects, and the need for Syria to have a stronger economy in the face of negotiations with Israel for a peaceful southwestern Asia. Economic strength was seen as a precondition of negotiating power—perhaps a healthier bargaining stance than one based merely on military strength and confrontation. President Assad was behind the new policies—a clever and astute politician fitting the system to changing circumstances. Although disliked by many of the people in the Syrian majority group, Assad is generally respected and preferred to the previous uncertainties.

By the late 1990s, however, economic growth returned to around 3 percent per year, and the Syrian government sought a balance between public and private ownership. President Assad was preoccupied with establishing the country's wider international relationships, continuing to act in a pragmatic manner in the face of huge debts to Russia, branding by the United States as a supporter of terrorists, and continuing opposition to the Arab-Israeli peace process. Like many Arab countries, Syria is torn between opening up to foreign investment and Western culture or pursuing an isolationist, socialist, and possibly Islamic path.

ment continues to harry the Shiite groups in the southern part of the country, and around 650,000 Shias have fled to Iran. Although some disenchanted Iraqis flee northward to join the Kurdish forces in their zone protected by the United Nations, disagreements among Kurdish leaders and a loyal Iraqi army prevent effective revolts.

## Countries with Little or No Oil

Syria has small oil deposits in its eastern area that are its major export at present. Failure to find new sources will lead to the end of oil income soon after A.D. 2000. Syria also has lands watered by the Euphrates River on which its farmers

(a)

(b)

**FIGURE 4.23 Arab Southwest Asia: Beirut, Lebanon.** (a) Devastation of the city during the civil war in the 1980s. (b) The world's largest rebuilding project: work being carried out by the Solidère Corporation in central Beirut, late 1997.



grow cotton, cereal grains, and fruit; however, it is not self-sufficient in food. After centuries of soil erosion on its northern hills, new projects are reseeding the slopes to expand the local sheep farming. Syria industrialized since the 1960s, at first based on import substitution policies. In the 1990s, Syrian markets were opened to foreign products and private enterprise and exports were encouraged (see "Living in Syria," pages 164–165). As with Iraq, however, even the best farms and factories are not highly productive.

Jordan and Yemen continue to have small economies with limited prospects of expansion. When Israel appropriated the West Bank area in 1967, Jordan lost half its agricultural land and much of its water and labor. Yemen's 1986 crisis led to uniting the two parts of the present country in 1990. The discovery of oil in the north gave hopes of an improving future, but the 1994 conflict showed that major difficulties still stand in the way of concerted efforts to improve the economy. Yemen remains the poorest country in the subregion. Before the civil war that began in 1975 tore the country apart, Lebanon was an agricultural country, but most of its income was generated in its capital, Beirut, the commercial center for much of the region. Many Beirut financiers fled abroad during the civil war.

These countries have many problems typical of peripheral nations, although aid during the Cold War, remittances of workers in the oil-producing countries, and subsidies from those countries helped to reduce financial deficits up to the 1980s. Lebanon and Yemen could be self-sufficient in food from irrigated and rain-fed farming, but manufacturing provides no more than 12 percent of GDP in any of these countries. In the later 1990s, the city of Beirut was being rebuilt (Figure 4.23), and much of the capital that went abroad during the civil war was returning. Farming production was being restored in the coastal plain (fruits, vegetables, tobacco) and the Bekaa valley (grains).

Jordan has high debts that led to austerity measures and to social unrest. Exports of phosphate, its main source of income, are affected by fluctuating world prices. Attempts to reestablish manufacturing industries around Amman when Israel took the West Bank were not successful. Aid income from oil countries supported Jordan as a "frontline state" adjoining Israel. Jordan also received aid from Iraq in exchange for providing access in the Gulf War. After the Gulf War, external earnings collapsed for a time as the Iraqi source stopped, other Arab countries withdrew their aid, overseas workers returned home, and a flood of refugees arrived from Iraq.

### Prospects for Regional Interchanges

The economies of all Arab Southwest Asian countries are linked closely to the world core countries' demands for oil and other products. Although an Arab Common Market was established in 1965, only Iraq, Syria, Jordan, and Egypt are full members, and the countries of Arab Southwest Asia have little trade with each other. They remain vulnerable to price instability in world oil markets, and some face European Union protectionism against agricultural imports. In the 1990s, increases of trade with Eastern Asia offset oil exports and stimulated hopes that markets there might develop further, reducing dependency on Europe and Anglo America. Egypt, Israel, Jordan, and Syria are at present considering new forms of economic cooperation with each other. Iraq is now isolated by United Nations sanctions but will make a major mark on the world economy once it begins to sell its oil even in the modest amounts allowed under the sanctions.

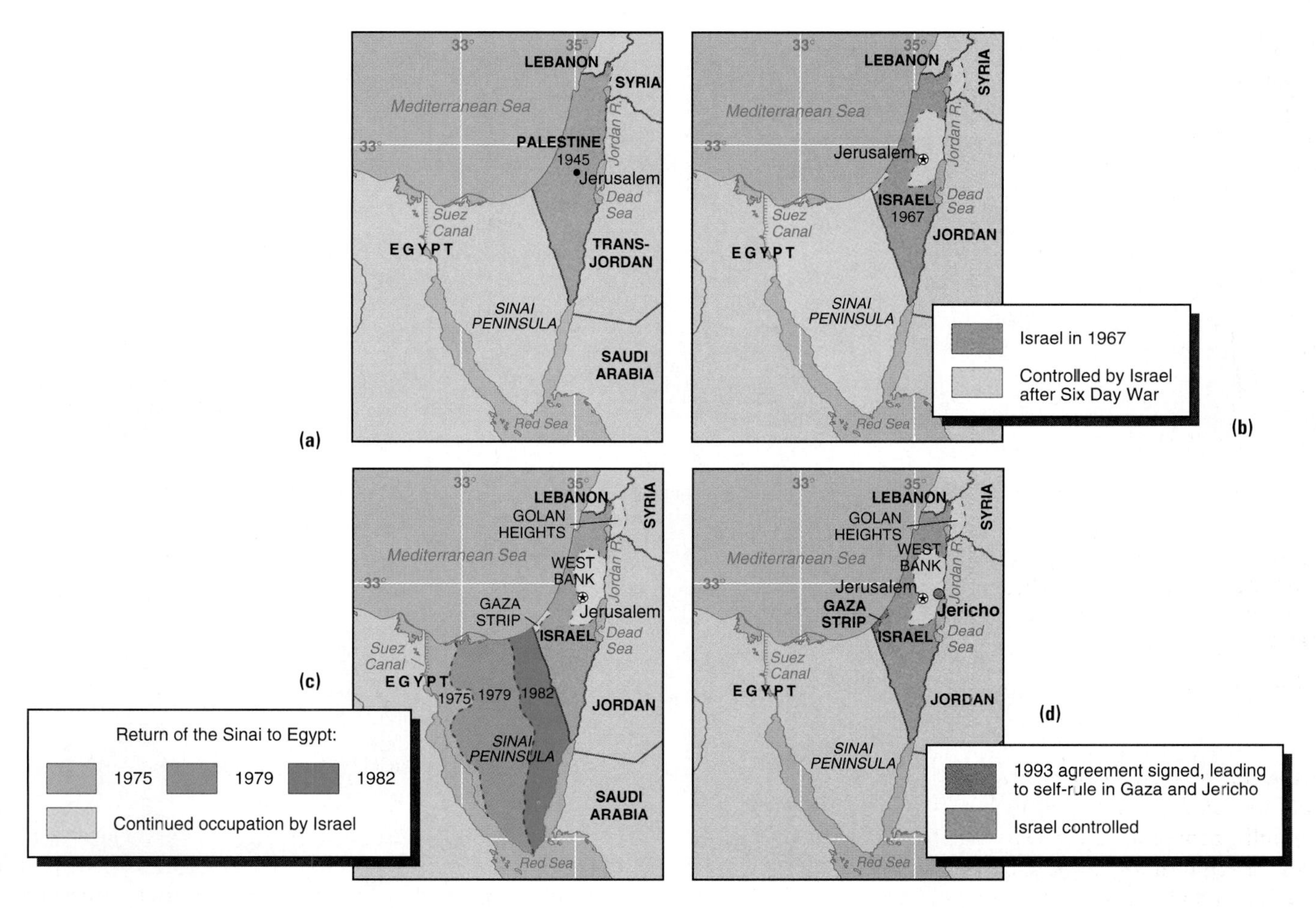


**FIGURE 4.24 Israel: changing boundaries.** (a) The preindependence protectorate of Palestine. (b) Attacks on Israeli territory provoked retaliation and occupation of surrounding lands to provide a defensible perimeter. (c) Land occupied outside the 1948 Israeli boundaries was gradually returned as a result of peace accords. (d) Agreements in the 1990s led to limited self-rule in small areas, but no further handing back of land.

## ISRAEL, GAZA, AND THE WEST BANK

Israel is a major anomaly in Southwest Asia. It is a unique example of a country created for a particular ethnic group despite opposition from those living in and around it. Subsequent territorial disputes, terrorism, and the creation of refugee groups have major geographic impacts and implications. In 1998, Israel celebrated 50 years of existence, growing prosperity, democratic government, and the bringing of one-third of the world's 15 million Jews to a separate homeland.

### Origins and History

A homeland for Jews was established in 1922 by the League of Nations, allowing the British to carve an area for the purpose from within its protectorate of Palestine—although few Jews had lived there after their dispersion by the Romans in the first century A.D. As Jews fled Nazi Germany and Europe before 1940 and moved to Palestine in larger numbers after World War II, pressure mounted for an independent Israel. The United Nations partitioned Palestine in 1947 as a prelude to Israeli independence, but Arab protests led to civil war and then an international conflict. The Arab countries could not recapture the land claimed by Israel, which became an independent country in 1948, recognized by the United Nations. Some 600,000 Palestinian refugees moved out, mainly into Jordan and Lebanon.

Arab countries refused to accept the existence of Israel and attacked it unsuccessfully in 1956, 1967, and 1973. During the 1967 war, Israel extended its territory into the Gaza strip and Sinai, across the West Bank area of Jordan, and to the Golan Heights (Figure 4.24). Israel subsequently refused to give up those lands for security reasons—at least until surrounding countries accepted the existence of Israel and renounced military activity against it. In 1994, Palestinian authorities were given limited jurisdiction in the first two of these territories. Israeli settlements in these areas (Figure 4.25) were maintained, however, and their populations expanded. The lack of interest by Syria in acknowledging Israel's existence caused Israel to hold on to the Golan Heights. Progress toward the return of land and the development of a Palestinian country halted after 1996 changes in the Israeli government, which resisted further devolvement of power and transfers of land to Palestinians. It continued to

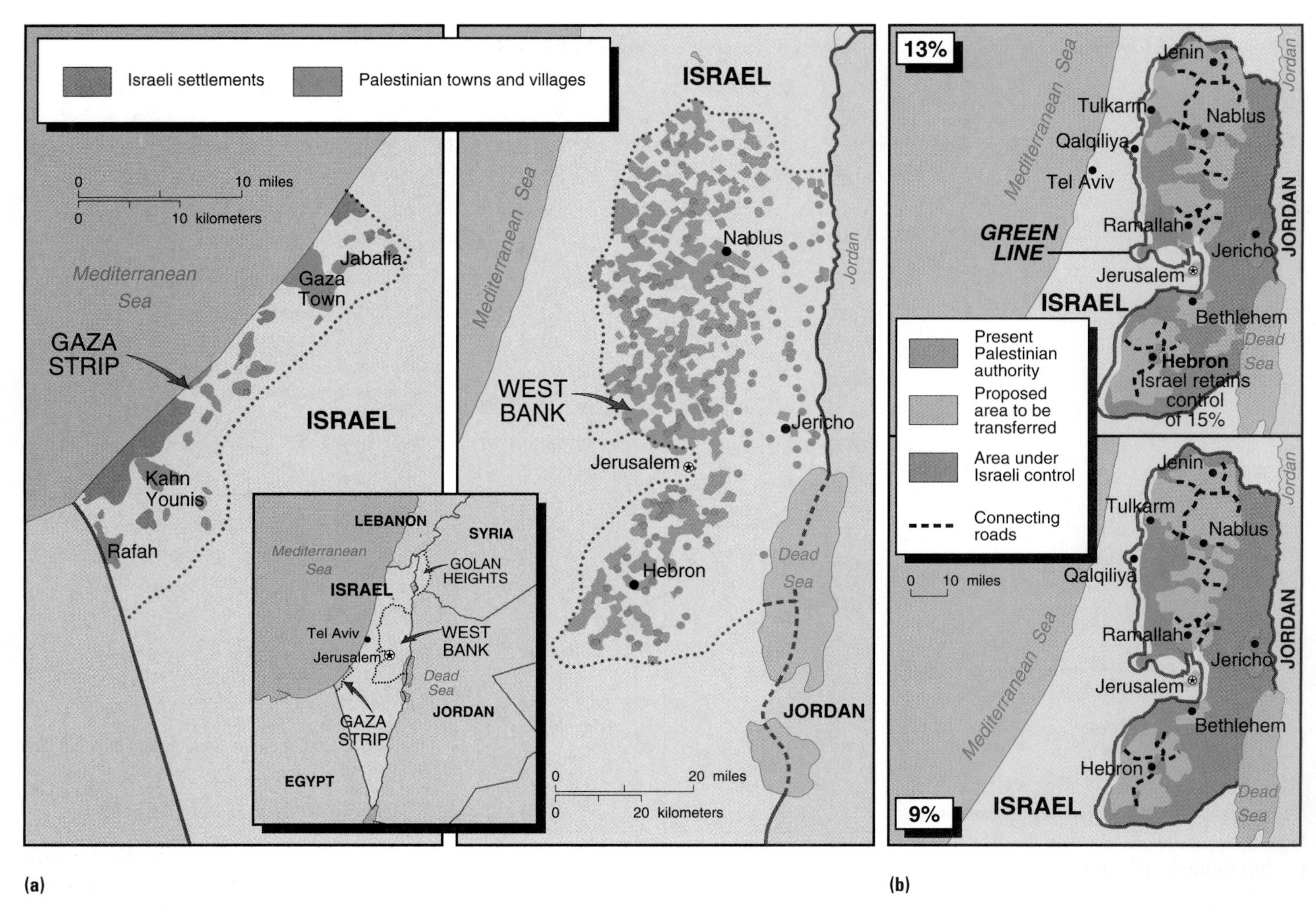


(a)

(b)

**FIGURE 4.25 Israel: the continuing presence of Israelis in Gaza and the West Bank.** (a) Many Israeli settlements continue to expand despite statements that the land will be given back to Palestinians. By mid-1995, four types of areas were recognized as a basis for West Bank negotiations: the largest towns; 460 Palestinian villages; sparsely populated rural areas; and 144 Israeli settlements and military areas. (b) 1998 proposals to transfer more land to Palestinian authorities: the 9 percent proposed by Israel affected few Jewish settlements; the 13 percent was proposed by the United States and reluctantly supported by Palestinians (who wanted more).

(a) Copyright © 1994 The Economist Newspaper Group, Inc. Reprinted with permission. Further reproduction prohibited. www.economist.com. (b) Source: Data from *The Times*, London, 1998.

build Israeli settlements in the West Bank against pressure from both friendly and unfriendly countries.

## Limited Resources

**Israel, Gaza,** and the **West Bank** occupy a small area of land lying between the Mediterranean coast and the Jordan valley (see Figure 4.21). The terrain includes a coastal plain and mostly hilly land with lower areas around Lake Tiberias and along the Jordan River to the Dead Sea (see Figures 4.5 and 4.6). Although Israel receives winter rain, the total rainfall is low and summer drought brings water shortages. Great efforts have been made to supply water to dry areas and to manage the environment efficiently. Despite such careful management, Israel relies on external sources for 25 percent of its water. Internal groundwater sources are now fully used. Brackish (slightly saline) water is used for some farming but can damage the soils if too much is used over many seasons. Urban water supply relies increasingly on coastal desalination plants.

The shortage of water affects politics. The extension of Israeli jurisdiction across the West Bank and to the Golan Heights was partly for reasons of defense and partly to guarantee access to water. Peace negotiations with Jordan included agreement on the use of the Jordan River basin waters, which became central in the negotiations over the future of the West Bank. Drilling for water in the West Bank area when it is controlled by Palestinians will take the water from the aquifer that supplies Israel's coastal cities. The water question remains a sensitive aspect of Israeli-Palestine peace negotiations. One initiative concerned a proposal to build a "peace" water pipeline from Turkey to Israel and Jordan, which are both short of water, but Saudi Arabia, which might have helped to provide funding for the project, is against assisting Israel.

### Continuing Stresses

Jerusalem presents another major problem for Israeli-Palestine negotiations. Since Israel took East Jerusalem in the 1967 war, it continued to confiscate mainly Arab land in the occupied area and made Arabs second-class citizens by denying property rights. This is part of a policy to confirm Jerusalem as capital of a Jewish state—which Israelis claim but others do not accept. Muslims regard Jerusalem as one of their holy cities and resist the Israeli takeover.

Israel established its political presence in the region, managed its unpromising environment to make it more productive than other countries in the region, and enhanced the quality of life of its people. And yet, many of its citizens found it difficult to celebrate wholeheartedly the 50th anniversary in 1998. They are still surrounded by political enemies and frequently reminded of this by terrorist acts. Internal tensions often bring greater concerns than the external enemies. Strong disagreements exist over whether the new territories gained in 1967 should be traded for peaceful relations or kept as part of Israel; how relationships among nation, state, land, and religion should be resolved in a conflict between religious and secular interests; how the move from a rural focus that made the desert bloom to a high-tech, urbanized country should be managed; and how the original melting pot required by national cohesiveness in the face of enemies can give way to the divisive ethnic interests of Russians, Moroccans, Western, and Asian Jews. These separate issues intertwine in practice to generate a pluralistic but polarized society. The dilemmas are real and evenly balanced, but they get in the way of rationalizing and coping with the question of the Palestinian citizens of Israel and their lands.

## People

### Population Distribution and Dynamics

The Israeli population has a higher density than that of other countries in the region. It is concentrated along the coastal plain, but political considerations result in settlements close to frontiers being supported (see Figure 4.12).

Israel's population growth has been irregular, involving periods when rapid immigration of Jews from other parts of the world added to natural population growth and periods when immigration was slow. In the early 1990s, over 1 million Russian Jews moved to Israel. The resident population of 6 million (1998) includes 18 percent non-Jewish, most of whom are Palestinian Arabs. Israel's population structure differs from the other Southwest Asian countries (see Figure 4.14) because it has a higher proportion of older people, who came as immigrants after 1945. Life expectation is almost 80 years. The Palestinians living in Gaza and West Bank have a more typical Arab-country population structure. Gaza has one of the highest total fertility rates in the world (over 7, compared to 2.9 in Israel as a whole in 1998).

### Ethnic Differences

Israel's population is composed of several elements, some of whom trace their family existence in the country to before Israeli independence, while others are newly arrived from Russia; some have adopted Western lifestyles and family patterns, while others follow Muslim cultural patterns. Israel's annual population growth rate fell from 2.8 percent in the period 1965 to 1980 to 1.8 percent in the 1980s, but rose again in the 1990s to 3.3 percent because of the Russian Jewish immigration. At times, immigration had a greater influence on population increase than the birth rate or death rate.

The Palestinian element within Israel has some rights, including voting, but is always suspected of acting against the state. As soon as terrorist or external threat occurs, Palestinians are confined to their living areas. Internal Jewish conflicts slow movements toward giving Palestinians more autonomy and freedom of movement. Israel holds onto territory along the Jordan River.

### Rural versus Urban Emphases

Jewish Israel means more to its inhabitants than most other countries do to theirs. Israel is a symbol to many Jews of the reestablishment of a religious national way of life after centuries of dispersal; to other Jews, it is a distinctive secular ethnic community. Many Jews place the making of a religious or ideological statement above demonstrating economic growth for its own sake.

Much of the initial thrust of settlement was in rural areas, where **kibbutzim**, in which the land is communally owned and decisions are taken collectively, provided a spiritual, agricultural, and social basis for the economy and tied families to the land (Figure 4.26). More recently, development of the rural areas has been through cooperative *moshav* farms and smallholdings (*moshawa*). Many Israelis, however, moved into the growing towns and left farming to Palestinians. Less than 5 percent of the Israeli labor force now works on kibbutzim or moshav farms.

Israel changed its goals as increasing numbers of immigrants came from urban areas in other parts of the world and as the development of sophisticated manufacturing and commercial functions became more town-based. In 1998, 90 percent of the Israeli population lived in towns. **Tel Aviv–Jaffa** is the largest urban area, while **Jerusalem** and **Haifa** are other large cities.

Israel has higher educational, health, and income levels than the Arab countries. It is becoming more outward-looking as it attempts to overcome the isolation imposed by its neighbors. In mid-1995 the minister of transport put forward plans to give Israel a "role as a trade and transport hub of a peaceful Middle East." Israel plans to invest heavily in domestic and wider regional transportation links, including better roads and urban rail networks. It is already engaged in the expansion of its airport and port facilities. The tensions arising from a preoccupation with heavy defense spending and the need to come to terms with Arab neighbors—instead of relying on aid

**FIGURE 4.26 Israel: communal farming.** Harvesting eggplant on Kibbutz Ein Gedi, located between the desert hills and the Dead Sea. Bananas are in the background.



from the United States, European countries, and Jews in other parts of the world—are causing divisions within Israeli society that add to those imposed by new waves of immigrants and the presence of Palestinians within Israel.

## Economic Development

Israel's economy places its people in the same league as countries of Southern Europe such as Spain. Ownership of consumer goods is high (see Figure 4.15). The steady growth and diversification of the economy up to 1980 gave way to slower growth and economic crisis in the mid-1980s. In 1985, Israel had a large budget deficit and its currency reserves were being depleted rapidly. Actions taken with the help of the United States put the country in a better position and reduced foreign debts to manageable proportions. There is still much to be achieved in the realm of privatization, which is resisted because Israel remains a highly centralized and heavily taxed country—maintained in that outlook by workers and managers fearing job losses, an influential group of politicians, and a large bureaucracy.

### Diversified, High-Tech Economy

Israel's economy has a diversified structure in which the agricultural sector, using intensive reclamation and irrigation farming methods, produces fruits, vegetables, and flowers for export to Europe but now constitutes only 5 percent of total GDP. Israel is self-sufficient in food, apart from grains.

Israel possesses economic advantages in its well-educated workforce and access to foreign aid and investments. The need to provide a sophisticated defense capability prepared a generation of engineers for work in Israel's high-technology industries. Although the influx of Russian Jews in the early 1990s generated worries about rising unemployment, they brought mainly professional skills and caused a new boom in house building.

Manufactures make up 44 percent of exports and include diamonds, machinery, military equipment, and chemicals. In the 1990s, Israel became a major center and leader of high-technology development in manufacturing areas such as telecommunications, electronic printing, diagnostic imaging systems for medicine, and data communications. Such products accounted for 50 percent of industrial output in the later 1990s as against 15 percent in 1990. The high-technology industries are supported by the government, which is anxious to continue in this field after taking advantage of the peace dividend of its military-based innovations. Most of the manufacturing industries are based around the coastal cities such as Tel Aviv and Haifa. There are proposals to build industrial estates along the borders with Gaza and the West Bank so that cheap Arab labor may be attracted without these workers posing security worries by living on Israeli soil.

## Services and Environmental Management

Israel gains over half of its GDP from the services sector. The quality of education and health care provision places it on a par with western Europe. The financial sector will grow increasingly with the opening of its economy to privatization and foreign investment. Tourism is a major industry that attracted over 2 million visitors in 1996 and would grow further if visitors could combine visits to, for instance, Jerusalem, Damascus (Syria), Petra (Jordan), and the pyramids (Egypt).

Conscious of the need for careful management in the sensitive arid environment of Southwest Asia, Israel afforested large areas. The pressure on space within Israel, however, leads to a continual expansion of building on good agricultural land and potential forestland around Jerusalem and in border areas. The port and tourist facilities developed around Elath in the south resulted in environmental pressures on that area.

## Diversifying Trade Links

Israel is now actively pursuing free-trade policies and opening its borders to products from neighboring Jordan and countries farther afield in Asia and Latin America. If peace can be added to structural adjustment and a strong base of human skills, Israel could have a leading economic role in Southwest Asia and the eastern Mediterranean area. In fact, people in neighboring countries fear that Israeli technology and business acumen will place others in a subordinate role.

One of Israel's major concerns is to increase its trade with new areas in Asia, Latin America, and the Arab world. Progress in the last of these will be slow because Arabs resist buying goods made in Israel. At present, half of Israel's trade is with the European Union, and the proportion is rising. Israeli attempts to join the EU as a trading partner—with access to the free-trade area, rather than as a political partner—are moving forward more slowly than Israel would like. Israel is particularly keen for its agricultural and telecommunications products to gain equal entry alongside those of European countries. It argues that joining the European free-trade area group of countries would compensate for opening its home markets to Jordanian produce and the future return of the Golan Heights to Syria.

Israel is also engaged in talks with Egypt, Jordan, and Palestinians about establishing a regional trading bloc. Progress in this field is subject to removing the Arab economic boycott of Israel and to allaying Arab suspicions that Israeli expertise will dominate any such arrangement. The advantages of the coordinated improvement of infrastructure in the region and attraction to foreign investors may eventually overcome such opposing considerations.

## Poverty in Gaza and the West Bank

While Israel has a growing economy that places it ahead of its neighbors in development and lifestyles, the Palestinian areas of Gaza and the West Bank continue to be problem areas with poorer conditions for human development. Gaza is home to many professional people, but its infrastructure is crumbling, 40 percent of its population lives in refugee camps, and unemployment is at least 50 percent of the work force. Access to jobs in Israel and remittances from workers overseas are now greatly reduced. Many families depend on emergency food supplies, while lawlessness has increased as gangs of children aged under age 14 control many aspects of life. Policing by the Palestinians is minimal because it cannot be afforded. Israeli settlements have a separate existence, occupying one-fourth of the land.

In the West Bank, Palestinian farmers increasingly find themselves close to Israeli settlements that have been built among their villages since 1967. Tensions are high, but the Palestinian groups opposed to peace, such as Hamas, take no part in consultations and planning. Large sums of money pledged to support the Israeli-Palestinian agreements are slow in arriving or await fuller plans for their use so that corruption may be minimized. Some projects—such as the provision of water, transportation facilities, and school buildings—easily attract funds, while funds that can be used for paying teachers, health programs, power provision, and waste disposal are more difficult to obtain, thus threatening the successful management of the projects.

# TURKEY AND IRAN

Turkey and Iran occupy the northern and eastern margins of this region. They are so distinct from the Arab-language countries of Southwest Asia that they are considered separately (Figure 4.27). Although these countries are still overwhelmingly Islamic in religion, Arabic is replaced by the Turkish and Persian languages respectively. Both countries had 1998 populations of 64 million people—together comprising over one-third of the total population of Northern Africa and Southwestern Asia.

Both countries are powerful in southwestern Asia and the wider world. Turkey and Iran have crucial strategic positions between the southern boundary of the former Soviet Union and the Persian Gulf oil fields. Their control of two major choke points—the Bosporus and Dardanelles in Turkey linking the Black and Aegean Seas, and the Hormuz Strait at the entrance to the Persian Gulf—adds to their strategic interest for the world's major powers. The increasingly democratic nature of their governments, after years of military and religious dictatorships, contrasts with other countries of southwestern Asia and gives them greater potential for success within the world economic system.

Iran and Turkey share other characteristics. They are largely mountainous countries along the plate collision margin between Arabia and Asia. This makes them subject to frequent and devastating earthquakes. The mountains attract precipitation, much of which falls as winter snow. Meltwater in the spring and early summer feeds rivers flowing southward to the Persian Gulf and northward into the Black and Caspian Seas. Both countries have Kurdish minorities and

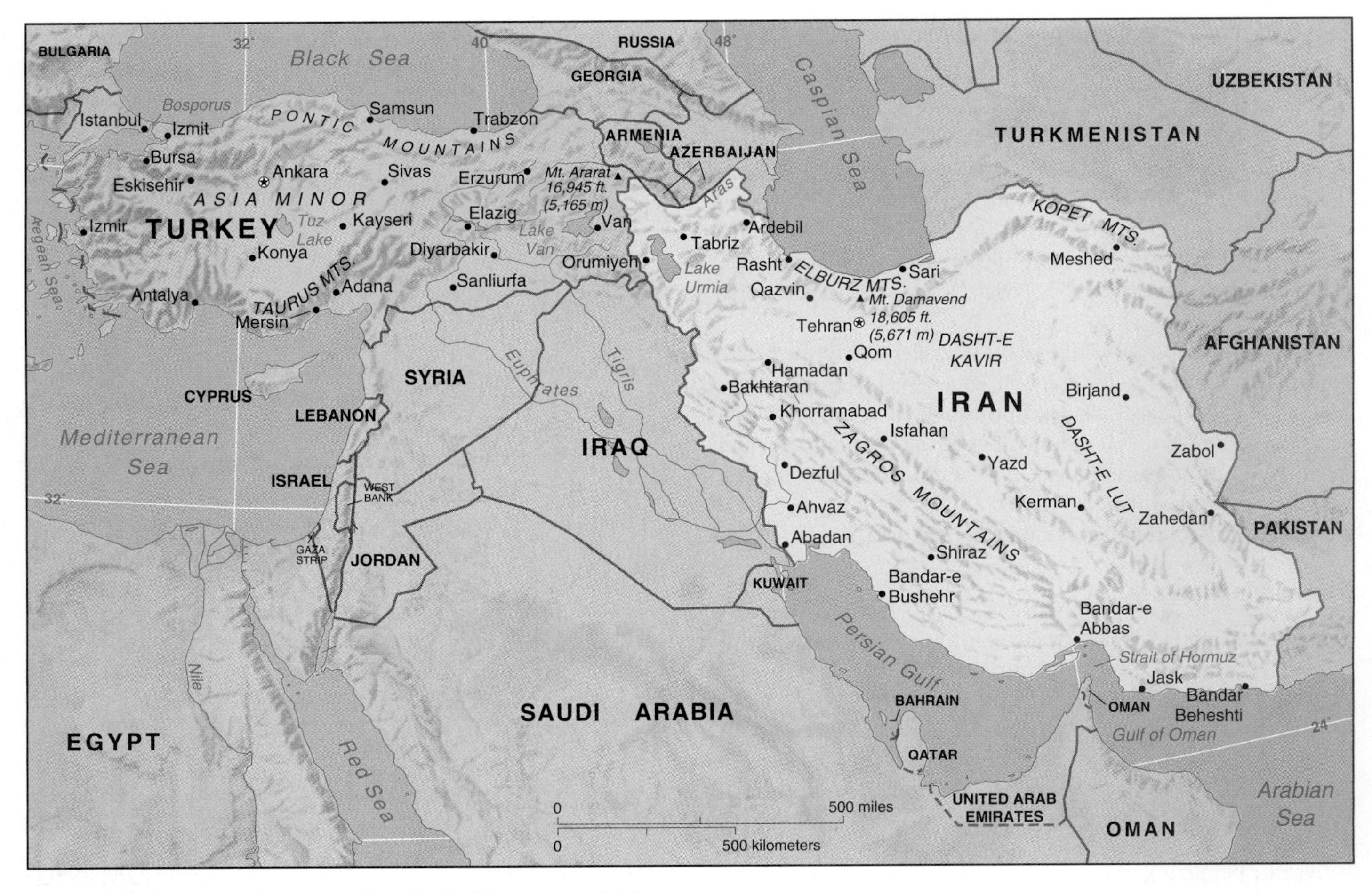


**FIGURE 4.27 Turkey and Iran: countries, physical features, and cities.**

must deal with the fight for Kurdish independence, a problem they share with Iraq and Syria. All these countries resist demands for a separate country of Kurdistan.

## Countries

Although **Iran** and **Turkey** have some similarities, they are long-term rivals with very different basic resources and approaches to the world economy. Iran's Muslims belong to the Shiite group, while Turkey's are mainly Sunnites.

In the 1900s, Iran was ruled by a leading family that kept it largely under military control but with a semblance of democracy. In 1979, the Shah fled, and Iran shifted to Islamic religious leadership and isolationism. Iran's oil wealth and control of the Strait of Hormuz (Figure 4.28) resulted in close involvements with the world economy but in less economic diversification at home; economic benefits have gotten through to fewer of the Iranian people than to ordinary Turks.

After the disastrous defeat of the Ottoman Empire in World War I, Turkey became a nationalist and secular republic, putting state before religion in questions of government. From the 1920s, Turkey modernized and became increasingly involved in the world economy through exports of crops and manufactured goods and a growing tourist industry. Its main natural resources are water and metallic minerals. It was only in the mid-1980s (Turkey) and 1990s (Iran) that democracy returned to both countries; it remains subject to faltering progress.

Iran and Turkey continue to be political as well as economic rivals. The rivalry is about to intensify. After World War II, Turkey identified with European interests, became a member of the North Atlantic Treaty Organization (NATO), and received support from the United States in resisting Soviet overtures. The control of the Bosporus made it possible to monitor Soviet ship movements from Turkey's southern flank of the Black Sea. U.S. Air Force bases in Turkey were close to the Soviet heartland. Iran avoided close links with the West, however, and developed agreements with the Soviet Union, especially after 1979. Following the breakup of the Soviet Union in 1991, rivalry between the two countries increased as both Iran and Turkey sought links with Muslim countries in the Caucasus and Central Asia. Turkey claims historic and language links with these new countries, but Iran is developing more practical links with the landlocked countries in trade and transportation (see Chapter 8).

**FIGURE 4.28 Iran and the Persian Gulf.** Space shuttle view of the Strait of Hormuz with the Zagros Mountains of Iran to the left (north) and the United Arab Emirates and Oman on the south side of the strait. All marine traffic into and out of the Gulf must pass through the strait—which has great strategic significance.

NASA

In the late 1990s, both Iran and Turkey shared the tensions between liberalizing Western influences and the influences that would take them toward control by Islamic-oriented political groups. As Iran began to emerge from such dominance, Turkey faced challenges to its long-term secular state principles (see "World Issue: Islamic Revivalism," page 138).

## People

### Population Distribution and Dynamics

The upland plateaus of Turkey and western Iran support higher densities of population than most of the region (Figure 4.12). Iran's population growth is more rapid than Turkey's, although both are troubled by the prospect of providing for projected increases. Iran caught up with Turkey from 1980 (Iran: 39 million; Turkey: 45 million) to 1998 (both around 64 million), and 2025 projections suggest over 92 million people for Iran and 88 million for Turkey.

Iran's total fertility rate (3) underlies annual natural increases of around 2 percent. In Turkey, the fertility rate is down to 2.6 and natural increase is 1.6 percent, but the legacy of past increases is a large population that needs rapid economic growth. Although birth rates are falling in both countries, death rates have also fallen (see Figure 4.13). The age-sex diagram for Iran (see Figure 4.14) shows the results of the recent high birth rates.

Both Iran and Turkey have varied ethnic compositions. Iran's population is just over half Persian, one-fourth Azeri (Shiite Turks who make up the majority in Azerbaijan), and about 7 percent Kurds. Turkey has around 80 percent Turks and 17 percent Kurds. There are also several small ethnic groups who remained in the area after it was conquered by the Turks.

### Urban Growth

Both countries have rapidly expanding urban areas, with people migrating to towns from the countryside. In Iran, such migration resulted from greater investments in urban jobs in the manufacturing and services sectors. The Turkish urban population increased from 25 percent of the total in 1945 to 50 percent in 1998. Urban jobs increased in Turkey, but many migrants to the largest cities are escaping the internal war against the Kurds in southeastern Turkey. **Tehran**, the capital, is the largest city in Iran. Other large Iranian cities include **Meshed** (Mashdad), **Isfahan**, **Tabriz**, and the port of **Abadan**. In Turkey, **Istanbul** is the largest city and has a population that is likely to grow to around 10 million by 2000. **Ankara**, the capital, was purpose-built after 1923 in the formerly "empty" interior. **Izmir**, **Adana**, and **Bursa** are other Turkish cities with 1 million people each.

### Historic Rivalries

Iran and Turkey are long-term rivals. Iran was the center of the wider Persian Empire conquered by the Greeks under

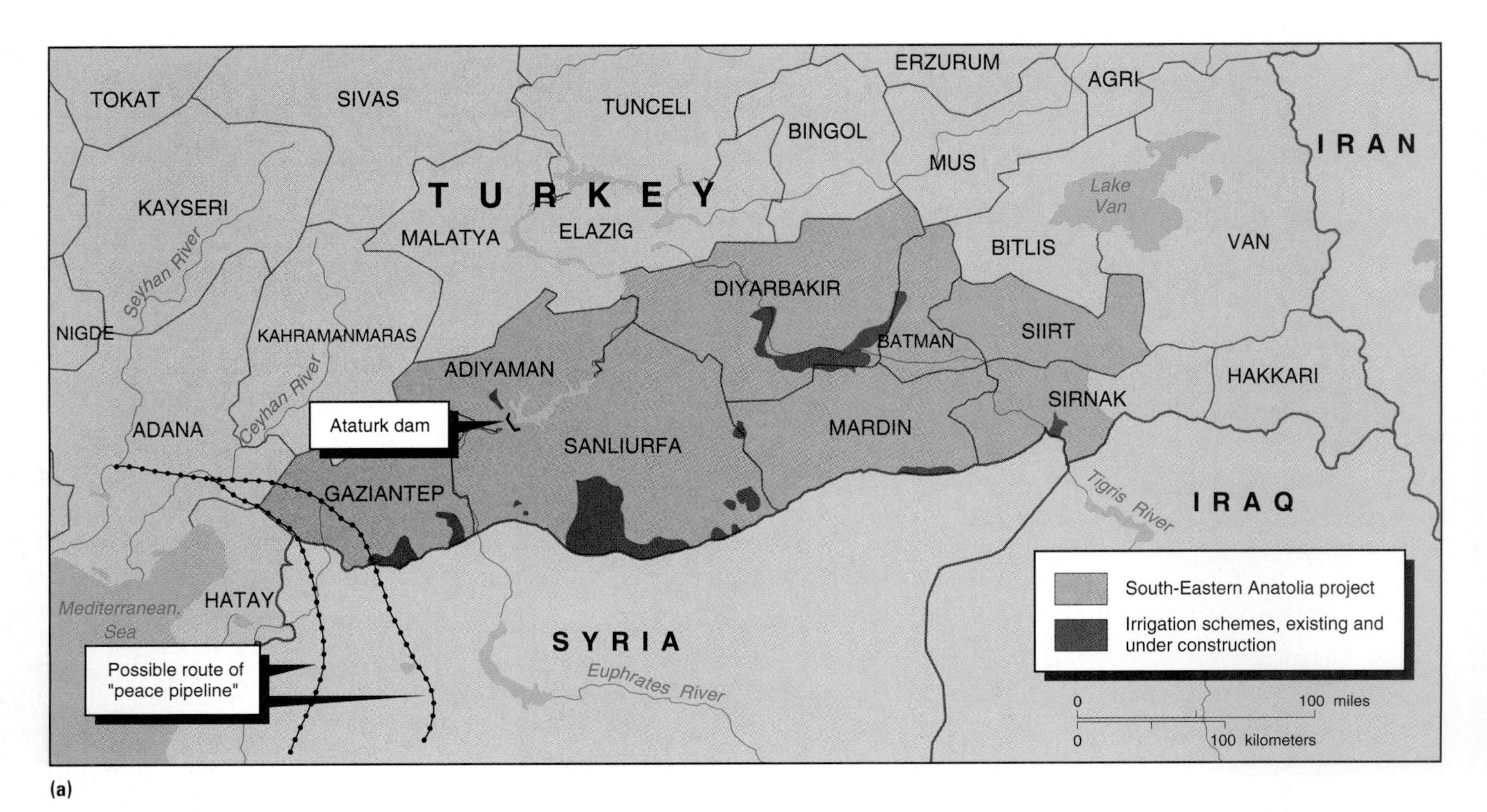


(a)

**FIGURE 4.29 Turkey: Southeast Anatolian water project.** (a) Storage of water behind huge dams, such as the Ataturk Dam, makes it possible to irrigate large new areas of land and provide hydroelectricity for local industry. Such uses of water, however, deprive Syria and Iraq of water they have been using from the Tigris and Euphrates Rivers. The "Peace Pipeline" was part of a project to supply water to Israel and its neighbors. (b) The Ataturk Dam on the Upper Euphrates River as it neared completion in 1992.



Alexander the Great. It resisted expansion by the Roman Empire and maintained links with central Asian peoples. In the Muslim conquest around A.D. 650, the new rulers took on Persian court manners as Zoroastrianism was replaced with Islam. In the A.D. 1000s, the Turks, followed by Mongols from Central Asia, overran the country—which then became a zone of conflict between Turks and Afghans and later between Russians and British as the country's significance rose in world geopolitics in the 1800s.

British discoveries of oil in Iran and its extraction from 1901 began the southwestern Asia oil era and increased the number of foreign influences that started to break down the previous isolation. This, in turn, generated internal nationalist resistance and the establishment of local leaders as shahs. During World War II, Britain and the Soviet Union occupied parts of Iran to guard its oil fields against German attack, but afterward Iran nationalized its oil industry and determined its own future.

Interior Turkey had its own civilization, developed under the Hittites around 1600 B.C. Later it was taken over in succession by Persians, Greeks, and Romans. It became Christianized as part of the eastern Roman empire of Byzantium, which extended around the eastern and southern Mediterranean. The Muslim eruptions during the A.D. 600s reduced Byzantine influence to the peninsula that is modern Turkey.

The Byzantine Empire lasted until Turks from Central Asia overran most of it after A.D. 1000, superimposing Islam on Christian traditions and establishing their own systems of government.

The Ottomans were leaders of the Turks fighting to take Constantinople, which did not fall until A.D. 1453. In the meantime, they bypassed this city and, with help from Arab Muslims, conquered much of southeastern Europe (see Chapter 7). In the 1300s, they reversed this direction and attacked the Arab world, bringing it under subjection by the late 1500s. Conquering other Muslims weakened Ottoman support but enabled them to exploit Mediterranean wealth through agents who were often local tribal leaders. By the late 1800s, this system began to break down, and a disastrous alliance with Germany in World War I led to defeat, destruction, famine, and the death of almost one-fourth of the Turkish population. This defeat led to the freeing and eventual independence of other countries in Northern Africa and Southwestern Asia and to the formation of the nationalist, secular state of modern Turkey.

## Economic Development

The two countries have different types of economic development based on oil income (Iran) and water resources

**(b)**



(Turkey). This produced different patterns of HDI, GDI, and GNP. While Iranian national income fluctuated after 1980 with world oil prices, Turkey's real income rose steadily to equal that of Iran by 1996. Ownership of consumer goods in Turkey is twice the rate as in Iran (see Figure 4.15).

## Oil or Water

Iran used its oil income to encourage urban-industrial, rather than agricultural, expansion, especially in the late 1970s period of high oil prices. The takeover of the country by religious leaders in 1979, and the war with Iraq in the 1980s, led to isolation from the world and the squandering of much of this wealth. In the 1990s, however, Iran began to look outward and restructure its economy and world links.

Turkey has little oil but invested heavily in the development of its water resources for more agricultural output and hydroelectricity. Industrial expansion increased from 1950 and especially in the 1970s and 1980s. The 1990s began with a period of political and economic confusion, partly due to the internal war against the Kurds that absorbed large resources of money and manpower. In the future, Turkey likely will gain income from proposals to construct oil pipelines giving Central Asia access to world markets without passing through Russia (see Chapter 8).

## Agricultural Contrasts

In agriculture, Turkey used mechanization and fertilizers to increase yields. Half the working population is engaged in farming, and the opening of new areas to irrigation in southeastern Turkey (Figure 4.29) is enabling the country to increase its production of cotton, soybeans, grains, fruits, and vegetables. The agricultural sector produces nearly 20 percent of Turkey's exports by value.

Although a third of Iran's workers are farmers, production is much less sophisticated than in Turkey, and Iran needs to import much of its food. Iran is also handicapped by large areas of arid land in the south and east of the country and has not invested as much as Turkey in the infrastructure needed to deliver irrigation water. Water projects in Iran are more concerned with urban-industrial requirements. Land reforms in Iran designed to please the peasant farming community are at odds with hopes of modernizing the sector and increasing productivity.

## Manufacturing Differences

Iran was the first major oil exporter in the Persian Gulf, but internal economic development was slow until the 1960s. Craft industries, such as carpet-making and weaving, remain signifi-

**FIGURE 4.30 Iran: manufacturing.** The Irana plastics factory, in which the machinery is basic and aging.



cant. Oil revenues, which now provide 90 percent of exports, were then used to fund public-sector manufacturing ventures such as oil refineries and petrochemical plants on the Gulf coast, the iron and steel works at Esfahan, and machine building at Arak. In the 1960s and 1970s, new roads were built to link the capital, Tehran, with the Gulf coast ports. Pipelines distributed oil and gas throughout the country. The private sector developed car assembly, textiles, leather goods, plastics (Figure 4.30), and other light industries around Tehran and other major cities. Small industrial estates were established in most medium-sized and larger towns. Rapid urban expansion drew people away from the rural areas, slowing agricultural expansion. Although Iran possesses resources of other minerals in its mountains, little attention has been devoted to exploring or extracting them. In 1995, the Iranian government began to involve international oil corporations in financing and developing new oil fields despite a continuing United States embargo of Iran.

Although it an oil importer, Turkey's manufacturing economy fared as well as that of Iran and is more diversified. The shortages arising from high oil prices in the 1970s caused Turkey to reform its economy. Economic growth was not smooth but affected the whole country, with industrial output increasing in the 1980s. Real GDP per capita rose by 50 percent between 1980 and 1998 while Iran's stayed the same. During the 1980–1988 Iraq-Iran War, Turkey supplied both nations with food and manufactures in exchange for oil.

The first Turkish manufacturing developed out of mining for chromite and copper in the mountain ranges. In the 1990s, gold was added to mineral exports. Import substitution industries, such as car assembly, textiles, and steel, were also established. The steel is made on the Black Sea coast near coal and iron ore deposits. During the 1980s, Turkey diversified into light industrial products, which now make up some 80 percent of exports.

#### Services and Tourism

Turkey has a more developed tertiary sector than Iran. Government employment is very important in the economy. International tourism grew and annually brings in over a billion dollars. In 1996, Turkey had 8 million visitors (up from 1 million in 1980). The rising income from tourism partly compensated for the falling level of remittances from 1.5 million Turkish workers in European and Arab countries—although they still contributed nearly half of foreign exchange income in the late 1980s. Turkey is the only country in Northern Africa and Southwestern Asia (outside of Israel) with such a diversified economy.

#### Prospects

While Turkey diversified its economy in the context of increasing democratic government and a slowing growth of population, Iran made slow progress in such diversification and human development. Iran also has to contend with a faster rise in population. Iran's oil revenues and scope for expanding agricultural production to feed this population give it a potentially prosperous future if suitable investments are made and peace is maintained. It may still be able to cope with the pressures on resources and emerge as a major power in southwestern Asia.

Environmental management has been a feature of Turkey's policies in the last 30 years, with controls on grazing and land reforms to offset soil erosion. The latter have been rather slow to take effect and have operated mainly in the western part of the country. In Iran, farm mechanization produced increased erosion rates, but in the 1960s and 1970s, the government instituted management controls for rangeland and forests.

## LANDSCAPES IN NORTHERN AFRICA AND SOUTHWESTERN ASIA

### Urban Landscapes

Urbanization was rapid in Northern Africa and Southwestern Asia after 1950. It is now one of the most urbanized world regions.

#### Ancient Urban Centers

Before this recent growth in urbanization, the region had a number of ancient but relatively small towns, often walled. The older towns survive as enclaves within newer cities, some much changed by clearing crowded buildings to make way for new highways.

High densities of homes and commercial buildings inside city walls marked the traditional towns of the region. They related to an agricultural and trading economy based on the products of the local region or to port activities connecting more widely. Within these towns was a rigid pattern of land

**FIGURE 4.31 North African urban landscapes: old town.** Small shops selling craft goods along narrow streets close to the marketplace in Fez, Morocco.



use in which prestigious craft workers, such as religious artisans, had premises close to a central mosque, castle, and square; those in lowlier occupations, such as leather workers, existed on the edges (Figure 4.31). Today, walls remain only in cities where tourist interests are considered, such as Fez and Marrakech (Morocco). In Kuwait, only the ruler's palace remains of the old town, while Iranians bulldozed new boulevards through their old towns. Elsewhere, old towns decayed into slums as the wealthier merchants moved to new suburbs and were replaced by poor rural immigrants. In the larger commercial centers, such as the Tehran (Iran) bazaar, imported manufactures and plastic goods replaced traditional craft goods.

### Modern Urban Expansion

New town sectors and whole new towns are also a feature of Northern African and Southwestern Asian urban landscapes (Figure 4.32). As traditional crafts gave way to manufacturing and service industries, the construction of factories, offices, and new residential suburbs expanded the built-up areas (Figure 4.33). Capital cities, especially in the oil-rich countries, built imposing government office blocks. New central business districts sprang up, often on new land near the old centers of towns.

New residential areas began with the European quarters built in the colonial era, especially in the former French colonies of North Africa. Wealthy Arab families now occupy such areas as economic segregation replaces racial segregation. In the Gulf states, wealthy Arabs often live in separate areas from immigrant workers. In many cases, expansion of demand for housing by poorer people was greater than public authorities could cope with in planning development or provision of utilities. Irregular patterns of roads and houses, together with shantytowns, are a feature of cities from Casablanca (Morocco) to Istanbul (Turkey). Some new towns have been built to more regular plans near major oil installations in Gulf states. For example, Al Jubayl and Yandu (Saudi Arabia) are based around oil-related manufacturing industries. In Israel, the growth of population because of immigration resulted in the rapid building of often shoddy settlements with minimal public buildings and service provision.

## Rural Landscapes

Rural landscapes in Northern Africa and Southwestern Asia range from open desert of rock and dune to highly intensive cultivated and irrigated areas where farmers have small plots of land (Figure 4.34). Irrigation is significant in Morocco, Egypt, Iraq, Israel, Syria, and Turkey but varies in the intensity of production and the patterns of land ownership. Local oases still using traditional methods of raising water from wells contrast with growing expanses of river-fed irrigation canals. Large feudal estates are still important in Turkey, but elsewhere land reform has redirected ownership toward peasant farmers. Other rural landscapes include those where the

**FIGURE 4.32 Arab Southwest Asia urban landscapes: Gulf waterfront.** Dubai, United Arab Emirates, where modern office and apartment blocks line Dubai Creek.



limits of water availability allow dry farming or livestock herding as the main occupations. In some desert areas, oil well pumps constitute a distinctive landscape.

## FUTURE PROSPECTS FOR NORTHERN AFRICA AND SOUTHWESTERN ASIA

The future of the countries of Northern Africa and Southwestern Asia depends on matters such as whether a peaceful environment will continue to allow their economies to develop, whether they will be able to take an increasing part in the world economic system, whether they can control population increase so that it is slower than economic growth, and whether tensions arising from the distribution of oil and water resources can be resolved. Signs, such as the mid-1990s increase of peaceful accords between Arab countries and Israel and the development of more diversified economies, were encouraging for the future, but others, such as rapid population increases, the terrorist actions in Algeria and Israel, the lack of thrust in providing for future development in the Gulf countries, and the hardening of Arab-Israeli negotiations in the late 1990s, are not.

### Cold War and Gulf War Impacts

The countries of Northern Africa, and particularly those of Southwestern Asia, are coming to terms with the outcomes of two major political events—the end of the Cold War and the impact of the Gulf War. These affected the region throughout the 1990s. The Cold War and the jousting for control or the sympathy of the oil producers between the United States and the Soviet Union often focused on local problems in this region. After 1990, this was followed by a period with the United States as the sole superpower. Countries in the region competed for U.S. favors that were often tied to moves contributing toward peace with Israel.

The Gulf War of 1991 temporarily ended Arab policies that rejected American and other Western influences. Many Arabs felt betrayed both by Saddam Hussein's invasion of another Arab country and by their own need to rely on outside help to redress the wrongs of this event. The war highlighted differences within the Arab world, since the Gulf oil states with a small proportion of the total Arab population could afford to pay for military assistance to defend themselves. Other Arab states watched, and some sent their own, inevitably modest, military support but remained conscious of their own per capita incomes that were a fraction of those in the oil countries. They gave small credit to the oil states for the financial aid they had distributed previously

**FIGURE 4.33 Nile River valley urban landscapes: new suburbs.** High-rise apartment blocks in the Maadi district of Cairo, Egypt.

(a)

**FIGURE 4.34 Northern Africa and Southwestern Asia: rural landscapes.** (a) Bedouin women taking goods to market near Foudouk al Aouerab, Tunisia. The cultivated valley behind them and the bare hillsides above are typical of North Africa and much of the wider region.

(b)

**FIGURE 4.34** (b) Sprinkler irrigation on a farm in the Negev Desert of southern Israel. Assess the importance of the presence or absence of water for landscapes in this region.



to other Arab countries. Against such an alliance, Iraq failed to mobilize the feelings of other Arabs to fight the rich Gulf states despite linking the actions to their hatred of Israel. This was a signal that the interests of individual countries remain more significant than any overriding Pan-Arabism. Such conclusions permeated the late 1995 Middle East and North African economic summit in Amman, Jordan. As well as links among themselves, however, countries in this region are looking to establish a partnership with the European Union and European-Mediterranean trade zone by A.D. 2010. Such agreements would bring greater trading opportunities, together with grants and loans for infrastructure and business modernization.

### Trends in the Postwar 1990s

At the end of the Gulf War, Saudi Arabia, the Gulf oil states, Egypt, and Syria assumed a new leadership in Northern Africa and Southwestern Asia, replacing Iraq and Iran for a while. Kuwait was preoccupied with rebuilding its oil industry and infrastructure, while sanctions and the narrow views of its leader, Saddam Hussein, prevented Iraq from restoring its devastated economy. The oil states turned to Southern Asia rather than other Arab countries for their labor sources. Jordan, Yemen, and Sudan, which sided with Iraq, all lost their aid from oil producers for some years and had to support the recognition of Israel to regain some external aid. Iran, which remained aloof in the Gulf War, began to build its technological capacity and sought better relations with Europe and Eastern Asia by relaxing its political-religious isolation. If Israel establishes peaceful relations with the other countries and the Cold War is not revived, there may be less reason for the United States to provide the external aid that has underpinned the finances of a number of countries in this region. The aid will have to be replaced by internal economic growth.

At the end of the 1990s, the incomes of oil-rich countries fell as world oil prices dropped by 40 percent below the already low prices of the 1980s. The poor countries improved their incomes slowly, helped by continuing aid from other countries. Inequalities among countries were less pronounced.

### Oil, Water, and Environmental Issues

By A.D. 2025—if not before—there may be new geographic differences and alignments that can only be guessed at now. The desert environment that supports the present population with difficulty may be home to twice the number of people—with little or no addition of resources. It is likely that countries worldwide will again be dependent on oil from this region as they use up the other sources that led to lower prices in the 1980s and 1990s. Alternatives to oil and natural gas are being developed too slowly to compete with the fossil fuels for many years. Higher oil prices would give an impetus to such developments.

Countries such as Egypt, which did quite well compared to other non-oil countries up to the 1980s, may increasingly feel the effect of the limits of water resources and cultivable land, together with the loss of aid income that came from its sympathetic approach to Israel. Iran, on the other hand, may have a more-than-doubled population in 2025 but has greater scope for extending its agricultural production and can continue to rely on oil income. It could become the dominant power in

**RECAP 4C: Arab Southwest Asia, Israel, Iran, Turkey, Landscapes, and Futures**

Arab Southwest Asia is the heartland of the Arab world but has internal divisions based on whether oil is produced; few countries have sufficient water for agriculture. Israel may face a future of less conflict with its neighbors but, like them, has problems of finding resources to handle an increasing population. Israel has a Western economy with strong human resources and technologic skills.

Turkey and Iran have more diversified economies than most countries in the region; Iran's oil income and reserves provide continuing potential, while Turkey's water resources are the basis for much of the country's development.

The future of this region will be decided by the world's needs for oil, the organization of water resources, the development of democracy, and the diversification of economic output.

4C.1 How do the economies of the oil producers and non-oil-producing countries in this region compare with each other in terms of income fluctuations and levels of diversification?

4C.2 What has been the impact of wars and civil strife on economic development in the countries of this region?

4C.3 How do Israel's geography and development contrast with those of the other countries of this region?

**Key Terms:**

| | | |
|---|---|---|
| diversified economy | Gulf Cooperation Council (GCC) | kibbutzim |

**Key Places:**

| | | |
|---|---|---|
| Arab Southwest Asia | Saudi Arabia | *Jerusalem* |
| Bahrain | *Riyadh* | *Haifa* |
| Iraq | *Jeddah* | Gaza |
| *Baghdad* | *Mecca* | West Bank |
| *Basra* | *Medina* | Iran |
| *Mosul* | Syria | *Tehran* |
| Jordan | *Damascus* | *Meshed* |
| *Amman* | *Aleppo* | *Isfahan* |
| *Irbid* | United Arab Emirates | *Tabriz* |
| Kuwait | (UAE) | *Abadan* |
| Lebanon | *Abu Dhabi* | Turkey |
| *Beirut* | Yemen | *Istanbul* |
| Oman | *Sanaa* | *Ankara* |
| Qatar | *Aden* | *Izmir* |
| | Israel | *Adana* |
| | *Tel Aviv–Jaffa* | *Bursa* |

Southwestern Asia. If Iraq escapes its present internal tyranny and external sanctions, it might increase its agricultural output and revive its oil production as a basis for becoming a major power. Iraqi farmers, however, will depend on the Turkish use of waters in the Euphrates and Tigris Rivers, and Iraq's portion is likely to decline to 40 percent of what it now receives. Water supply issues are likely to become increasing sources of tension.

**WEBSITES: Northern Africa and Southwestern Asia**

Further websites are available through the Dorling-Kindersley or Rand McNally CD-ROM atlas online facilities.

http://www.arab.net is a site of major importance for this region, since it covers all the Arab countries. For each country, there is an overview and material on history, geography, business, cultures, government, transport, and tourism. Try some of the following:

- Saudi Arabia. Select the "Business" area and the "Boom City of Yanbu."
- Egypt. Select "Business" and "Economy" to obtain an up-to-date view.
- Sudan. Select "History" and "Civil War" to assess its impact.
- Palestine. Select "History" and find out about "Zionism" and the history of Israel from an Arab point of view.
- Libya. Select "Culture" and assess the role of "Berber" tribes in an Arab world.

http://www.jpost.co.il/ is the *Jerusalem Post* site, with today's information on Israel and its place in the world. Compare views expressed with those on the arab.net site.

http://www.turkishdailynews.com/ is a Turkish newspaper site with today's news. Find if there is information about economic developments or the Kurdish people.

The future of the Northern African and Southwestern Asian countries also lies in the degree to which democracy and liberalization extend to their economies and internal politics. Moves in these directions are further legacies of the ending of the Cold War and the Gulf War, but they are likely to create internal, and possibly international, disruptions in this sensitive area that is so strategic for its oil production. That would affect the economic development of the countries involved.

A further question that might be asked at the end of this chapter on Northern Africa and Southwestern Asia is whether this region is in danger of being "overdeveloped" compared to Africa South of the Sahara (see Chapter 3). Many of the countries in Africa South of the Sahara have untapped mineral resources and large areas of uncultivated land where water is available, and can be considered to be "underdeveloped." The Northern African and Southwestern Asian countries by contrast have small cultivable areas that are already having difficulty in supporting their populations. Their expanding urban populations rely on oil wealth—either on their home territory or through aid and foreign worker remittances. The building of new cities in sensitive arid environments and the almost total reliance on a single product (oil) makes the future dependent on oil or water resources that will eventually run dry, with few alternatives.

## CHAPTER REVIEW QUESTIONS

1. The three most populous States in Northern Africa and Southwestern Asia are: (a) Israel, Jordan, Lebanon (b) Egypt, Kazakhstan, Georgia (c) Saudi Arabia, Iraq, Iran (d) Egypt, Iran, Turkey
2. The Northern Africa and Southwestern Asia region contains about ____ percent of the world's oil reserves. (a) 3 (b) 10 (c) 30 (d) 65 (e) 90
3. Which of the following statements is incorrect? (a) An overwhelming majority of Egypt's people live near (<100 km) the Nile River (b) The Blue Nile originates in Ethiopia (c) The largest Nile River control project is the Aswan High Dam (d) Cairo is located on the shores of Lake Nasser in Upper Egypt
4. Which of the following OPEC oil producers does not border the Persian Gulf? (a) Libya (b) Oman (c) Iran (d) Kuwait (e) Saudi Arabia
5. In August 1990, Iraq invaded which of its neighbors? (a) Jordan (b) Kuwait (c) Iran (d) Turkey (e) Syria
6. Jordan has absorbed large numbers of (a) Lebanese (b) Iraqis (c) Kurds (d) Palestinians (e) Syrians
7. Which of the following statements about Lebanon is false? (a) The country fell apart in 1975 when the civil war broke out (b) The population is about 25 percent Christian (c) The reformed Army now controls all of Lebanon's territory (d) Beirut has already been totally rebuilt
8. The modern state of Israel was created in: (a) 1919 (b) 1933 (c) 1948 (d) 1962 (e) 1975
9. What can be said of Turkey's capital? (a) Istanbul is located on the Bosporus (b) Ankara is located on the Anatolian Plateau (c) The name rhymes with Ataturk (d) all of the above
10. The capital of Iran is (a) named Tehran (b) at the foothills of the Elburz Mountains (c) the largest city (d) all of the above
11. Mesopotamia is renowned for its high number of domestications and innovations.
    True / False
12. Arabs are not allowed to be citizens in Israel.
    True / False
13. Israelis regard Jerusalem as their capital.
    True / False
14. The center of the Arabian Peninsula is uninhabited.
    True / False
15. Casablanca is more than a great movie, it is the largest city in Morocco.
    True / False
16. North Africa was a Roman province before Arabs conquered it.
    True / False
17. Khartoum, Sudan is located near the confluence of the White and Blue Niles.
    True / False
18. The small states in the Persian Gulf include Yemen.
    True / False
19. Sales of oil and oil products are central to all the countries in Northern Africa and Southwestern Asia.
    True / False
20. As regions evolve, Northern Africa and Southwestern Asia are likely to grow as they include more Islamic States.
    True / False
21. What is the main point on which Sunnite and Shiite Muslims differ?
22. What are the three holiest cities in Islam?
23. What are the three Western religions that originated in Northern Africa and Southwestern Asia?
24. How many speakers of Arabic are there?
25. Which country in Northern Africa and Southwestern Asia has open and free elections?
26. Where are U.S. troops stationed in this region?
27. What are the direct benefits of oil exporting to the people?
28. What is meant by the statement "Egypt is the gift of the Nile"?
29. Why is North Africa not in the same region as Nigeria?
30. What cultural factors link southwestern Asia to Europe?

# SOUTHERN ASIA

5 Chapter

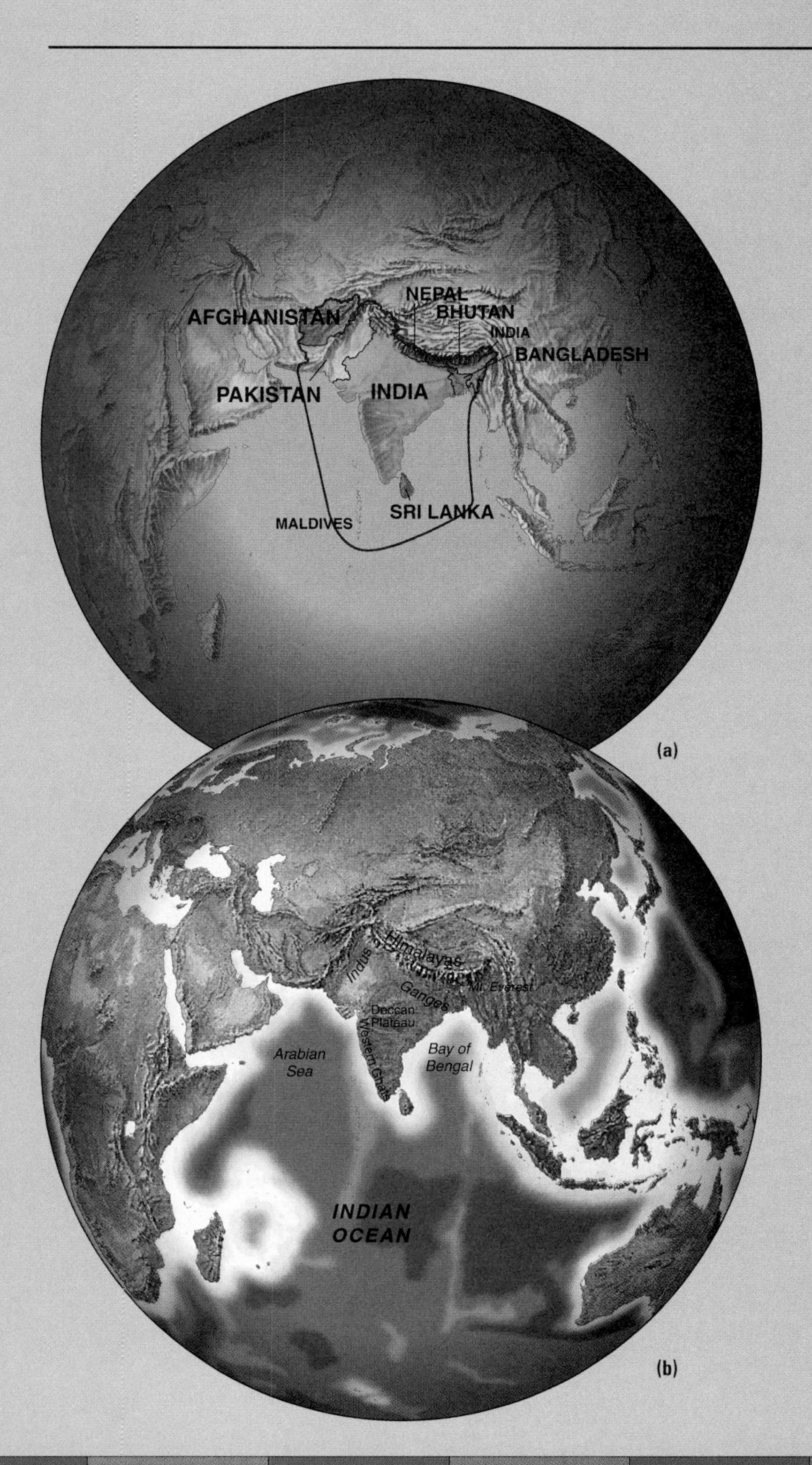


**THIS CHAPTER IS ABOUT:**

**Global significance of Southern Asia**

**Internal geographic differences resulting from Hindu, Buddhist, and Islamic religions and British colonialism**

**Contrasting monsoon and arid environments; the world's highest mountains and broad river plains**

**Three subregions:**

- India: world's second-most populous country
- Bangladesh and Pakistan: contrasting Muslim countries
- Mountain and island rim: isolation or position on trade routes

**Geographic aspects of landscapes in Southern Asia**

**Future prospects for Southern Asia**

**Living in Sri Lanka**

**World Issue: Economic growth and environmental problems**

**FIGURE 5.1 Southern Asia.** (a) The subregions and countries. (b) The natural environments, marked by the northern boundary formed by the Himalayan mountain wall, the arid northwest, the Indian peninsula, and the Indian Ocean.

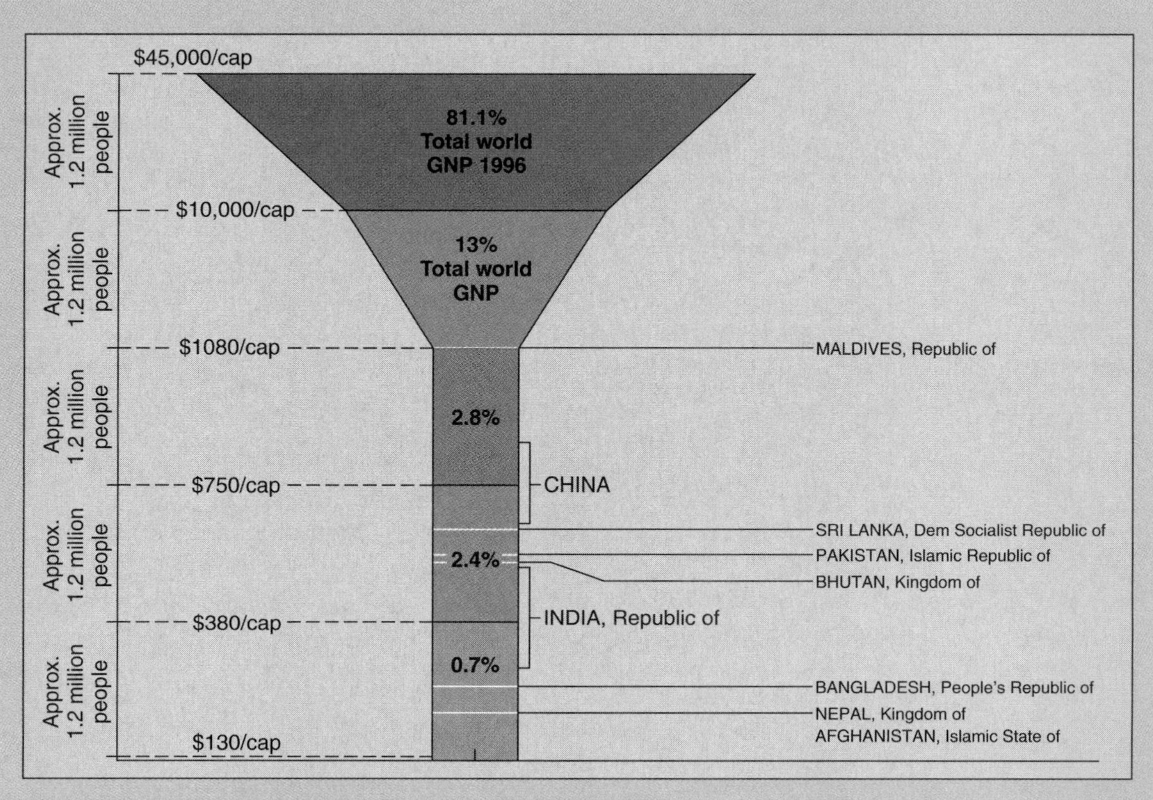


**FIGURE 5.2 Southern Asia and the world income range.** All countries in this region are poor, with most in the world's poorest one-fifth.

Source: Data (for 1996) from *World Development Indicators*, World Bank, 1998.

## PERIPHERY WITH DEFERRED PROMISE

Southern Asia comprises India and the surrounding countries, including Afghanistan, Pakistan, Bangladesh, Nepal, Bhutan, and Sri Lanka (Figure 5.1). The region occupies the **Indian subcontinent**—the part of Asia that consists of the Indian peninsula and is rimmed to the north by the Himalayan Mountain ranges. This is a region where extreme wealth confronts extreme poverty and where illiteracy exists alongside high levels of technical expertise. Some of the smaller countries are riven by civil war. In 1998, the animosities between India and Pakistan flared as both tested nuclear weapons.

Southern Asia is the world's second-poorest region after Africa South of the Sahara (Figure 5.2). In 1998, however, Southern Asia had a total population of over 1,300 million people, more than one-fifth of the world total population, compared to 580 million in Africa South of the Sahara. Southern Asia also has a higher density of population than other world regions and only a slightly larger percentage of world GDP than Africa South of the Sahara: 22 percent of the world's population live on only 4.3 percent of the land and produce just over 1 percent of world GDP. This region thus has the largest number of very poor people. It has major economic, social, and political problems, leaving little room for further population expansion unless modern technology can be used to increase the region's economic output—without the environmental damage that reduces resources for the future.

Although malnutrition continues to affect many people, no famines have occurred for several decades. Whereas there have been wars between India and Pakistan, they have not totally disrupted economic progress as in some of the African countries. Several countries have falling rates of population increase, and most countries of Southern Asia have made improvements in human development that are ahead of economic development. There are also major examples of economic growth, as in the farming region of northwest India and the high-tech region around Bangalore in southern India. Many parts of this region have a legacy of infrastructure that can form the basis of future development.

Such features suggest that Southern Asia may have greater promise for the future than most of Africa.

**FIGURE 5.3 Southern Asia: the Himalayan ranges.** A view westward along the ranges, taken by a space shuttle. The Tibetan Plateau lies to the north (right) of the snowcapped Himalayan peaks. The Ganges River valley is to the south.

NASA

Enthusiasm over the future promise has to be tempered by multiple current problems. In India alone, which has more than two-thirds of the total population of the region, fewer than half the adults can read, only 14 percent of the population has access to clean sanitation, and two-thirds of children under 5 years are malnourished. Even India's rich are poor by other standards: only 2 percent have household incomes of over $2,500 a year.

Increasing ethnic polarization is posing further problems to countries in the region that are beginning to see economic and human resource growth. Long-term civil wars in Afghanistan and Sri Lanka hold back economic development and threaten to destabilize other countries, while India and Pakistan continue to threaten each other militarily. Bangladesh has to fight floods from the rising Ganges River or ocean waters aroused by tropical cyclones.

Southern Asia is a region of technical and social contrasts. High technology that is used in building nuclear power plants, making nuclear bombs, supporting electronics and software research, making jet aircraft, and putting satellites into orbit occurs alongside abject urban poverty and subsistence farming that scarcely provides a living. The many university graduates contrast with low rates of adult literacy among the rest of the population. Feudal land ownership in Pakistan conflicts with a desire for greater democracy. As many people move toward secular modes of living, religious issues from bride-burning to pulling down the buildings of other religions remain public order issues in India.

Southern Asia is distinctive among world regions because none of the others has had women occupying so many of the top elected offices. In 1995, the prime ministers of Bangladesh, Pakistan, and Sri Lanka were women, and Indira Gandhi had been a long-term prime minister in India.

Whether the countries of Southern Asia fulfill the glimmers of promise opening to them will depend partly on how they relate their human and physical resources to wider markets and sources of investment within the world economic system. To external investors they offer the advantages of cheap labor and potentially large markets for goods produced, but have, until the 1990s, tended to isolate themselves from world markets.

After independence, many of the governments in the region tried to make progress on the basis of internal capabilities and resisted being drawn into wider world markets. Such isolation followed centuries in which the world's highest mountains (Figure 5.3) cut off the Indian subcontinent from direct land contact with surrounding countries. Colonial experiences led to reactions against foreign links. Reliance on internal sources of investment, however, held back the development of education and health care as well as transportation and power supply facilities that are basic in increasing the quality of life. In the 1990s, Southern Asia was slow to follow the economic growth in Eastern Asia as countries moved gradually from protected internal markets toward engagement with the world economic system.

Southern Asia is bounded by the Himalayan Mountains and Indian Ocean. Myanmar, which was included here in the first edition because of its British Empire connections, is now part of Eastern Asia (Chapter 6). Data for the region can be found in the Reference Section, page 586.

## DIVERSE CULTURES

Throughout history, Southern Asia was occupied by a succession of peoples who entered through narrow passes and initiated new cultures. Periods of invasion alternated with periods of assimilation and development of new cultures. The Hindu and Buddhist religions developed and flourished in the isolated conditions. Other peoples brought cultures from outside. Muslims dominated much of the region after the A.D. 1100s. The Hindu, Buddhist, and Islamic cultures all had immense impacts on the areas where they predominated. Their ideals affected behavior in ways that resulted in distinctive imprints on the landscapes of the region. Whereas Northern Africa and Southwestern Asia (Chapter 4) are marked by the single combination of the Arabic language and Islamic religion, Southern Asia has many languages and religions.

After the mid-1700s, the British Indian Empire, with its ocean links to the home country, became a major force. The impacts of the British on their Indian Empire were probably the world's most intense expression of European colonialism. A group of new countries emerged from the partitioning of that empire at independence in 1947.

### Precolonial Cultures

The earliest events in the evolution of the cultural traits in Southern Asia are not fully understood. By 3000 B.C., groups of dark-skinned Dravidian people (Figure 5.4) were forced out of the Indus Valley, where they had established irrigation farming and a major civilization that rivaled those in the Tigris-Euphrates and Nile Valleys (see Chapter 4). The Dravidians were driven southward into the Indian peninsula by the first of many invasions through the northwestern passes. This invasion brought lighter-colored and taller Indo-Aryan peoples to Southern Asia and resulted in a new society developing skills in arts and sciences.

#### Hindus and Buddhists

In this society, **Hinduism** crystallized around 1200 B.C., while **Buddhism** and Jainism were founded in the 500s B.C. A battle was waged for supremacy between Buddhism, which grew in significance by the 200s B.C., and Brahman Hinduism, which ousted Buddhism and irrevocably established the Hindu **caste system.** This caste system arose out of the differences between the Aryan and non-Aryan peoples. The Aryans were divided into priests (Brahmins), warriors, and people; the non-Aryans included cultivators, craftsmen, and the untouchables. In the 1990s, this caste system continued to impose rigid social divisions on Indian society, although the untouchables are now a major political constituency: one became prime minister. A major attraction of Buddhism was the rejection of the rigid caste system, allowing a greater degree of social mobility. The expelled Buddhists converted much of Sri Lanka, spreading their faith eastward through Asia.

**FIGURE 5.4 Southern Asia: Dravidian peoples.** Indian people engaged in a workshop industry in Bangalore, southern India. The dark skins often originated from the Dravidians—the people who occupied India before the invasions of lighter-skinned Aryan groups.

## Many Invasions

An invasion by the Greeks under Alexander the Great in 326 B.C., at the end of his conquests in western and central Asia, lasted for only two years but left behind distinctive effects on sculpture and science. Although Greek rule lasted such a short time, its influence continued in the subsequent empires. Further invasions and empires followed, and another phase of major cultural development occurred from the A.D. 300s, when the Gupta Empire brought peace, economic growth, and a new flowering of art, music, and literature. At the end of the 400s, Huns and later Turks invaded from the northwest. Although they were unable to establish lasting rule, they left Southern Asia with anarchic local government.

## Muslim Presence

The major Muslim invasion occurred in the 1100s, after which Muslim influences extended southward from the Indus and Ganges Plains. Cruel Mongol invaders in the 1300s alienated both Hindus and their fellow Muslims, and it was not until the 1500s that the Mughal (Mogul) dynasty extended Muslim beliefs to the rest of the region. The Mughals were Persian Turks who were mistaken for Mongols. Under the Mughals, India again experienced developments in such fields as architecture: the Taj Mahal and many examples of Muslim buildings that were built in the 1600s survive throughout the region (Figure 5.5).

## Mountain Isolation

While such changes affected the greater part of Southern Asia, some of the places at the margins of the region had linked but different histories. Tribes in the Himalayas remained isolated from the main flows of people and established their own ways of living. They were occasionally invaded from Tibet, and fortified monastic schools built in the A.D. 1600s mark Bhutan landscapes.

(a)

(b)

**FIGURE 5.5 Southern Asia: historic buildings in Delhi, India.** (a) Quwwatu'l Islam (Might of Islam) mosque inside Tomar fortress. The mosque formed the heart of a late 1100s city, and the buildings were constructed out of ruined Hindu temples. The buildings include a courtyard, cloisters, prayer hall, and the Qutb Minar, a giant prayer tower rising 72.5 m (230 ft.). (b) A 1500s mausoleum built by a Mughal emperor to commemorate the life of one of his ministers and set in a large, square-walled garden. Such buildings, similar to the Taj Mahal at Agra, which was built in the same period, reflect a period of great riches.

### Island Openness

Sri Lanka was inevitably much affected by its closeness to the Indian peninsula and openness to ocean-borne trade and conquest. The majority Sinhalese population was established by a movement of Buddhist peoples from northern India around 500 B.C. Although Buddhism scarcely survives in India, it thrived in Sri Lanka. These peoples developed a civilization based on irrigated rice agriculture in the north-center and southeast of the island, lasting from the First Century B.C. to the A.D. 1200s. During this period, Tamils from southern India established kingdoms in northern Sri Lanka, and around the A.D. 900s, Tamil mercenaries were brought in to help settle disputes over leadership among the Sinhalese. Tamils established a strong presence in the north of Sri Lanka around Jaffna. As they did so, Sinhalese peoples drifted southward, abandoning the irrigation systems and adopting rain-fed agriculture. They grew spices such as cinnamon. Muslim settlers controlled the overseas spice trade.

The growing Muslim control of Indian Ocean trade routes involved the Maldive Islands that form a large and widespread group in the south of the region. Chinese and Arab merchants set up trading posts for local fish, coconuts, and shells, and the islands became solidly Muslim by the A.D. 1100s.

## Colonial Impacts

In the early 1700s, the Mughal Empire broke up as internal factions caused political chaos. The region splintered into small kingdoms, often established by Muslim or Hindu adventurers, and into large independent states such as Hyderabad. This lack of cohesion made it easy for Europeans to take control of the increasing trade in cotton, cloth, rice, and opium.

### Trading Expansion

The European search for an alternative route to get Indian products to Europe without traveling through the Muslim countries of southwestern Asia brought the Portuguese explorer Vasco da Gama to India in 1498. The Portuguese and Dutch soon founded trading stations, such as Goa, around the coasts of Southern Asia.

The British followed, establishing their first trading post in 1612. During the 1600s, they ousted their Portuguese and Dutch rivals from most trading centers, set up their East India Company in the former Portuguese port of Bombay, built a new port at Madras, and began trading with the most populous area of Bengal around modern Calcutta. The French contested British control in the 1700s but were defeated, and left with only a few small trading ports after the Treaty of Paris in 1763—at which time the French also ceded their interest in what then became *Anglo* America (see Chapter 9). Over the next 100 years, the **British East India Company**, backed by the British Indian Army, increased its hold on Southern Asia, taking a particularly strong position in Bengal and the peninsula. Its area of influence was extended southward into Ceylon (modern Sri Lanka) in 1798, westward into Punjab in the mid-1800s, and eastward into Burma (modern Myanmar) in 1885.

### British Indian Empire

Following several uncoordinated revolts, a major mutiny of its Indian Sepoy troops in 1857 caused the British government to take full political control of Southern Asia. The East India Company was abolished and the **British Indian Empire** established. Control was never total because Britain did not conquer a number of the smaller princely states. It was not considered necessary to do so since those states were run peaceably by their rulers and in harmony with British policies. At the time of independence in 1947, 40 percent of Southern Asia, with almost one-fourth of the total population, remained under princely family governments. The 600 princely states ranged from Hyderabad with 14 million people in 1947, to Jathiamar, with only 200 (Figure 5.6).

During the 1800s, the period of British dominance in Southern Asia established a strong core-periphery relationship between the industrial "mother country" and its raw-material-producing colonies. The British saw their role as one of civilizing India by means of Western education, new technology, public works, and a new system of law. Seeing the advantage of redirecting India's cottage industries toward producing raw materials for British industries, they built a modern infrastructure including networks of railroads and irrigation canals. The process brought Indians into the periphery of the growing world economy after centuries of controlling their own destiny outside this system. British India's products played a crucial part in the home country's economy, since the export of Indian cotton to other countries compensated for negative trade balances that Britain had with those countries.

The British Indian Army, consisting of Indian soldiers and British officers, became a tool of imperial expansion into Afghanistan and Burma in the 1880s and into Tibet in 1903–1904. Nepal had a close relationship with Britain from the 1850s and many Nepalese served in the Indian Army.

British imperial rule had massive effects on the contemporary and future geography of the peninsula by defining the resources that could be developed, determining what was produced, altering patterns of land control, selecting areas for development and cities for growth, and controlling trade. Cotton was grown and land irrigated to produce it in the Indus and Upper Ganges River basins in order to balance the British budget. The cities of Calcutta, Bombay, Madras, and Delhi grew when the East India Company designated them as focuses where lines of overseas trade and internal communications met. Former communal land was reallocated to larger and smaller landowners, forming interest groups that were expected to support the colonial administration.

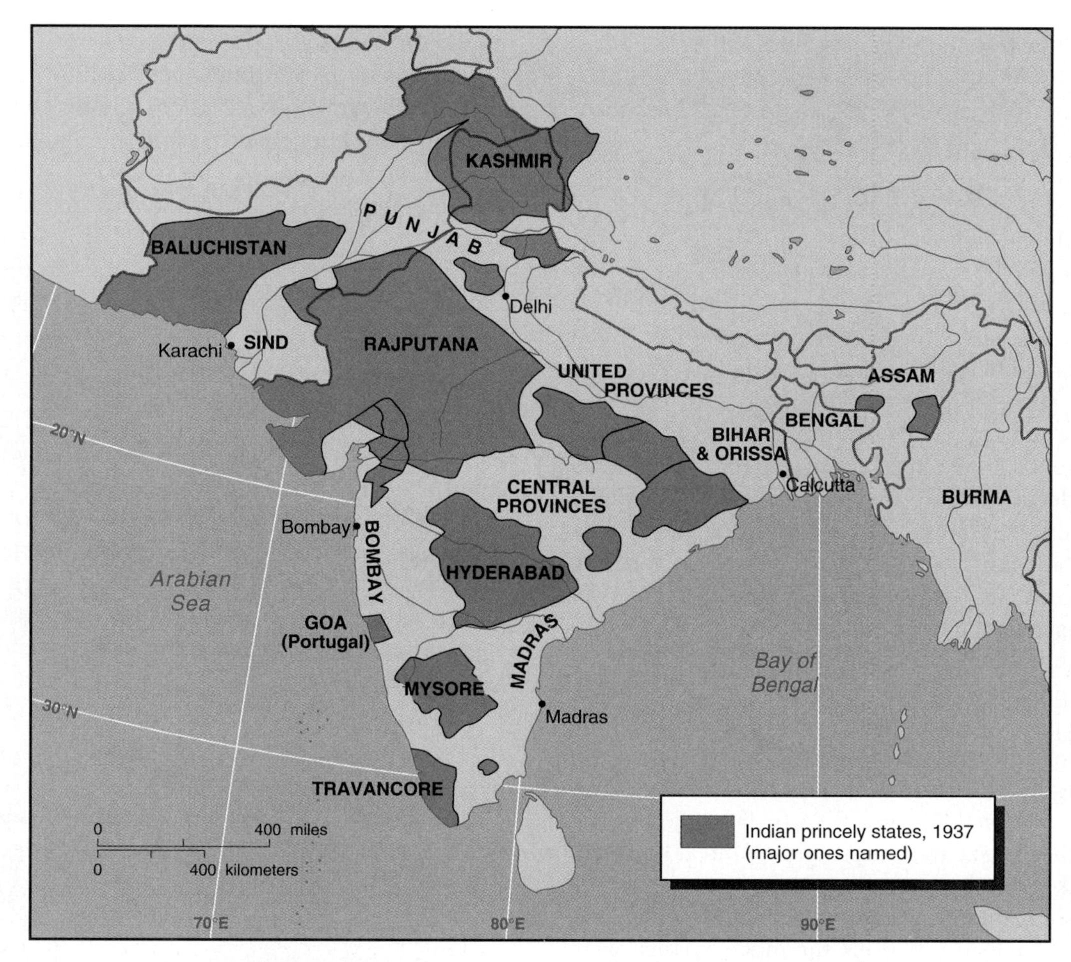


**FIGURE 5.6 Southern Asia: the British Indian Empire just prior to independence in 1947.** The princely states had working arrangements with Britain but were not ruled as part of the Indian Empire. At independence, they agreed to be part of the new countries of India and Pakistan (borders shown in gray) and subject to their governments. Kashmir is still disputed between India and Pakistan.

Source: From Francis Robinson (ed.), *The Cambridge Encyclopedia of India, Pakistan, Bangladesh, Sri Lanka, Nepal, Bhutan and the Maldives.* Copyright © 1989 Cambridge University Press. Reprinted with the permission of Cambridge University Press.

### Ceylon and the Maldives

Ceylon (modern Sri Lanka) and the Maldives were also much changed by the colonial experience. In Ceylon, the Portuguese and Dutch were the early traders and colonists, and the British did not take over until 1798. The Portuguese left a strong legacy of Roman Catholic missions; Portuguese remained a major trading language throughout Asia until the British imposed English in the 1800s. As on the mainland of India, the British reoriented the Ceylon economy to make it fit the needs of Britain. British companies set up plantations for growing tea, rubber, and coconuts, instead of spices. More Indian Tamil labor was brought in to work on the plantations, bringing experience of Indian plantation work but adding to local ethnic tensions. A new elite of local plantation owners was Western-educated and attempted to emulate the social life of the colonials. The emphasis on export crops, however, was accompanied by the neglect of traditional agriculture, leading to a decline in the rice crop. Although the British tried to rehabilitate the decayed ancient irrigation works, over half of Sri Lanka's rice was imported at the time of independence in 1948.

The Maldives became a British protectorate in 1887, an arrangement that lasted until 1965. Like the Indian Empire, Ceylon and the Maldives were incorporated as British colonies into the periphery of the world economic system that focused on the demands of the colonizing country rather than on local needs.

## Paths to Independence

### India and Pakistan

As British rule was extended during the late 1800s, Indian leaders became aware of a national identity arising from a combination of cultural background and personal interest. A mainly Hindu elite formed the **Indian National Congress** in

1885 and took an increasing part in politics. Muslim leaders at first thought it safer to build links with the colonial order, but the growth of the National Congress influence led them to form the **Muslim League** in 1906.

Although many events before 1914 showed that British rule would never be fully accepted in India, World War I revealed that colonial rule might end. During the war, the Indian Army fought on equal terms alongside Europeans, while many in India suffered as British and Indian resources were devoted to the battles in Europe. After the war, India became less vital to Britain's economic interests of trade, investment, and political empire. India could even impose a tariff on Lancashire cotton manufactures while it developed its own. In the 1920s, Mahatma Gandhi popularized political issues among a deeply religious people and welded together a coalition of interests. Gandhi's policy of nonviolent resistance brought the country to a standstill in the early 1920s. The British response was to widen political involvement at all levels with the stated intent of moving slowly toward self-government. Gandhi, however, wanted more rapid change.

By 1940, the push toward independence involved growing demands for a separate Muslim state because of fears by Muslims that they would be an underprivileged, even persecuted, minority if a single country covered the whole subcontinent. Britain and the National Congress Party favored a single country in Southern Asia and proposed a political structure based on a weak central government and strong provinces. They preferred such an arrangement to partitioning into separate countries. While negotiations for independence were proceeding, communal violence between Hindus and Muslims added pressure for partition.

The Muslim League requested a separate country to be called Pakistan, formed of provinces in which there was a majority of Muslims, including the whole of Punjab and Bengal. These two provinces, however, had Muslim majorities of only some 56 percent and large numbers of Hindus and Sikhs. The British government proposed dividing the provinces of Bengal and Punjab between India and Pakistan. This placed the great majority of territory and people in a new India and left the Muslims marginalized in two separate areas to the west and east of India. The pressures to hand over government rapidly at the end of World War II resulted in a grudging Muslim acceptance of the smaller East and West Pakistan. Despite wishing to eject the colonists, few were happy with the nature of the separate countries created. Muslims and Hindus caught in the wrong country were encouraged to move to avoid ethnic bloodshed. Some 12 million people were displaced, but over a million died in clashes. In 1971, the eastern part of Pakistan established its separate identity as Bangladesh following armed conflict with the Pakistan army.

### Ceylon Becomes Sri Lanka

The majority Buddhists in Ceylon pressured for independence early in the 1900s, and demands increased following the 1931 adoption of universal suffrage. The rural vote encouraged Buddhist revivalism but alerted the Tamil population to the need to take action to preserve their separate identity that had been protected in colonial times. Tamils living in northern Ceylon often had better schools and more interest in Western education than the Buddhist Sinhalese and so had become the main group of business people and civil servants. Ethnic differences and resentments surfaced before independence in 1948. Ceylon was renamed Sri Lanka in 1970. It suffers from civil war between the Tamils in the north and the central government, composed largely of Sinhalese.

### Resistance to Indian Dominance

The splitting of the former British Indian Empire into the large country of India and a group of smaller surrounding countries left fears of Indian dominance among the others. Pakistan and India went to war with each other over border disputes—especially the still-disputed ownership of Kashmir—but India established its military superiority. The degree of interaction between India and the other countries varies: it is great with the mountain states of Nepal and Bhutan, much less with Bangladesh and Sri Lanka, and least with Pakistan.

## NATURAL ENVIRONMENT

The natural environments of Southern Asia are marked by great contrasts between high mountains and broad plains, and between hot dry and hot wet climates. The alternation between summer rains and winter drought, known as the monsoons, is characteristic of much of this region. Long dry seasons, uncertain rainfall, and large areas of arid conditions make water basic and crucial in sustaining the livelihoods of the large population of the region, where some form of farming is still the most common occupation. Seven good (i.e., wet) monsoon seasons to the mid-1990s provided a foundation for improving economic prospects in India.

### Monsoon Climates

The monsoon climatic environment of Southern Asia brings heavy downpours of rain in summer to much of the Indian subcontinent but hardly any rain at other times of the year. Cut off from contact with Central Asia and its freezing winter air by the Himalayan Mountains, Southern Asia remains warm but dry in the winter. In this season of the dry monsoon, winds flow outward from high atmospheric pressure over Punjab in the northwest (Figure 5.7a). Only northern Sri Lanka and parts of southernmost India receive rain at this season. It comes from winds that blow out from the continent, over the Bay of Bengal, become moist by evaporation from the ocean surface, and turn toward the land.

As the land warms in early summer, temperatures in much of Southern Asia become unbearably hot until the wet monsoon breaks in June or July. At this time, the winds change direction, and southwesterly winds from the Indian Ocean bring moisture. Rain totals are highest on the western coastal

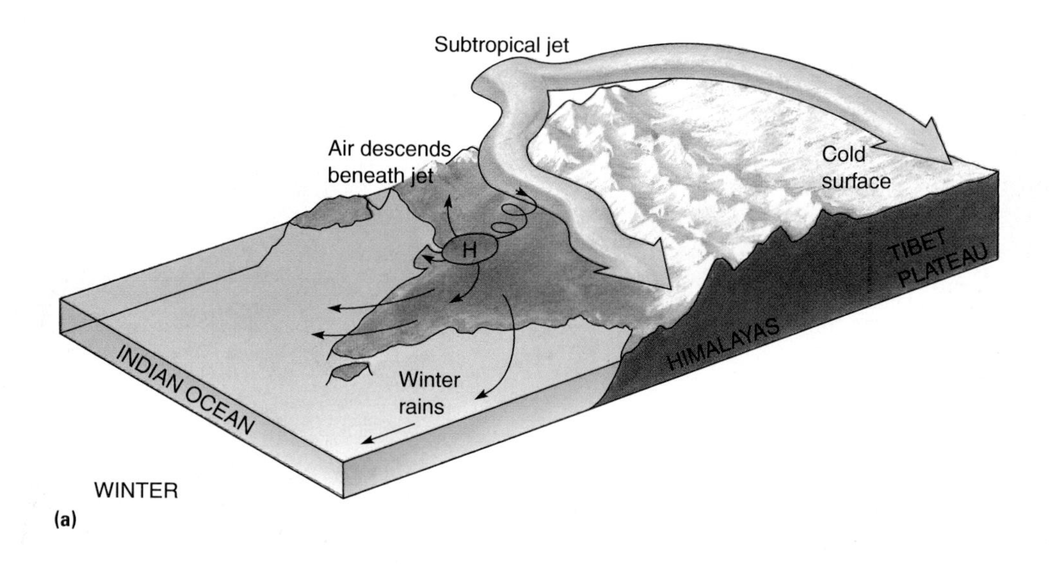


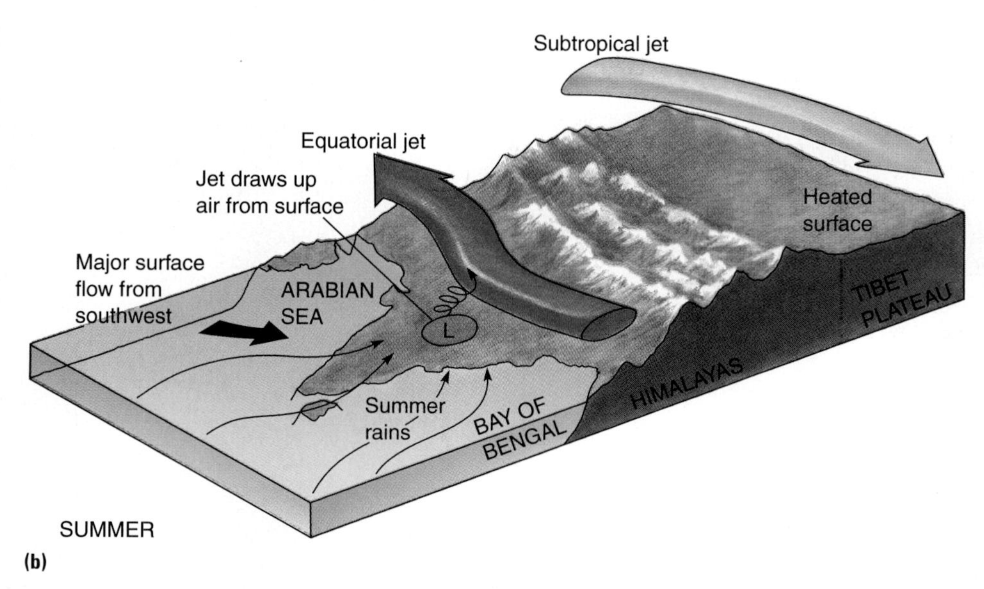


**FIGURE 5.7 Southern Asia: the monsoons of the Indian subcontinent.** (a) In winter, the high-level winds split around the Himalayan-Tibet highlands. The southern branch pumps air down toward the surface, causing high atmospheric pressure in the northwest of this region. Such high pressures lead to dry air at the surface and dry winds blowing outward. (b) In summer, the strong high-level wind shifts north of the mountains and is replaced by a weaker equatorial high-level flow. The heated continent causes air to rise, and this draws in very humid air from the Indian Ocean—the source of water for the monsoon rains.

mountains of the peninsula (Figure 5.7b). Precipitation (often as snow) also occurs on the high Himalayas to the north and on their continuations in Assam (northeastern India). Some of the world's largest rainfall totals, over 10,000 mm (400 in.) per year, fall on the Assam hills. Each year, a small number of tropical cyclones occur in the **Bay of Bengal** during the summer, bringing flooding and death to the Ganges-Brahmaputra delta in Bangladesh.

The eastern part of the peninsula and the Ganges Plains receive lower and more irregular rainfalls, varying sharply from year to year. Since most people live in regions of lower rainfall, the variability is of great human significance. Crop production often depends on water flow in the rivers that rise in the mountains and bring water to the plains.

The northwestern parts of Southern Asia remain dry throughout the year, forming one of the world's major arid regions. The main source of water in this part of the region is the rivers, such as the Indus and Ganges, carrying the melted snows of the Himalayas.

## World's Highest Mountains

The **Himalayan Mountains** include **Mount Everest** (8,848 m; 29,028 ft.), the world's highest mountain, and nearly 50 other peaks that rise over 7,500 m (25,000 ft.). They form a northern wall-like boundary to the region (Figure 5.8a). The mountain wall is the southern margin of the high and broad plateau of Tibet, forming a wide barrier to climatic

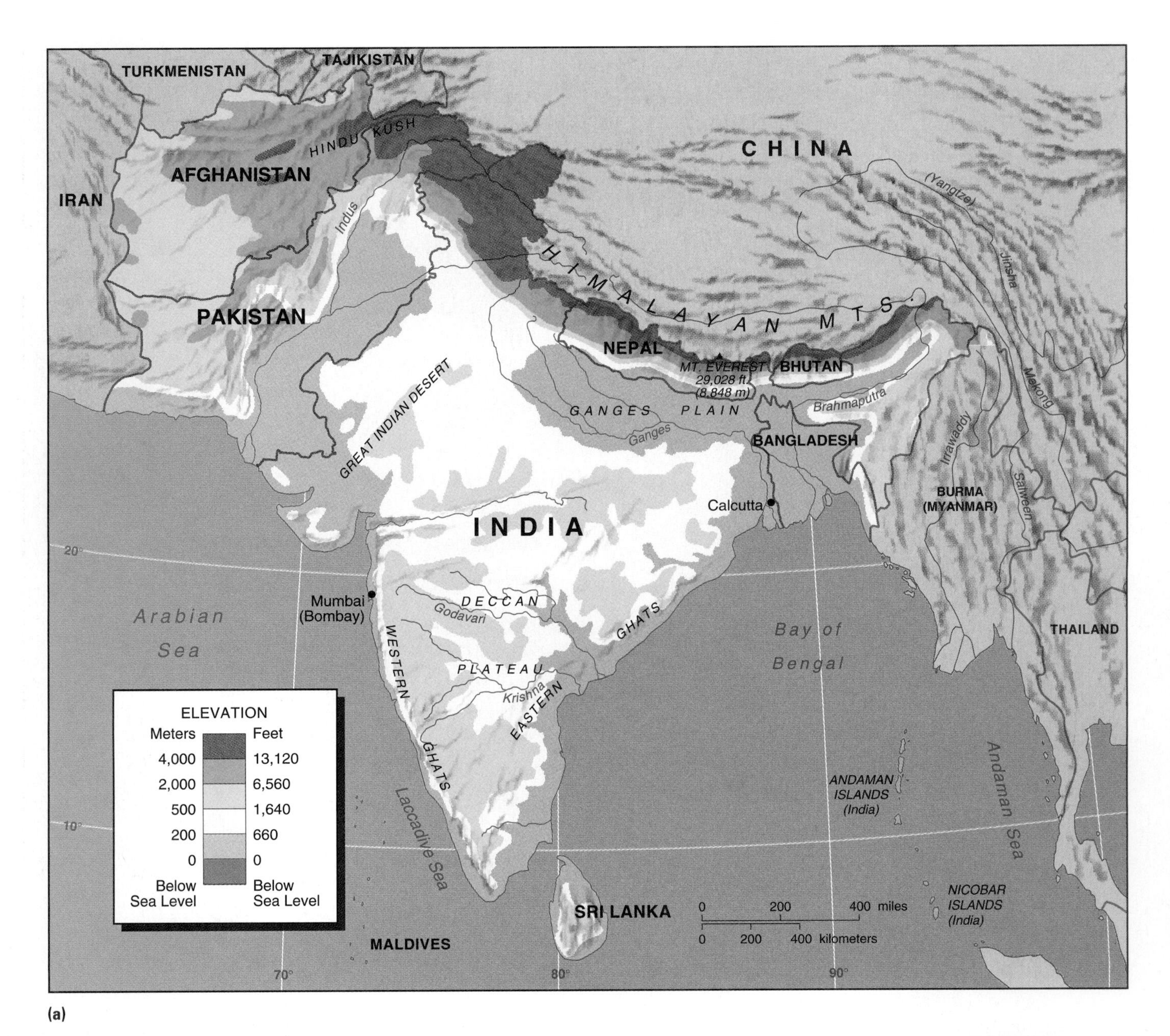


(a)

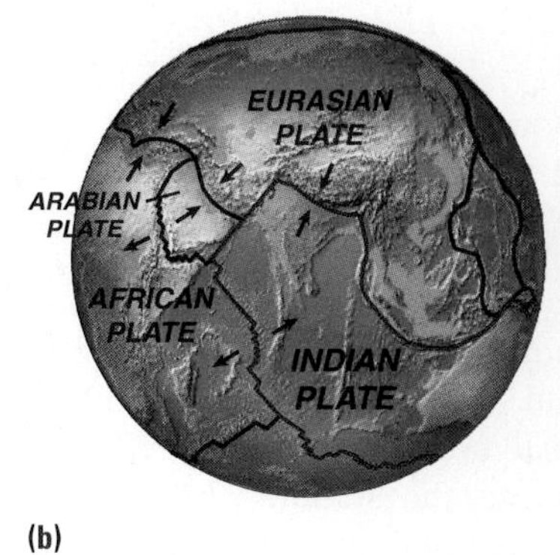


(b)

**FIGURE 5.8 Southern Asia: the major landforms of the Indian subcontinent.** (a) The Himalayas form the northern boundary and are separated from the hills and plateaus of peninsular India by the wide plains of the Indus, Ganges, and Brahmaputra Rivers. (b) The Indian plate crashed into the southern edge of the Eurasian plate, forming the Himalayan Mountains.

influences and incursions of all but the most aggressive peoples.

In the far northwest, the headwater streams of the Indus and Ganges River systems breach the Himalayan wall. Deep valleys lead to the passes through the Hindu Kush range, such as the Khyber, that give access to Afghanistan and Central Asia. In the past, these passes formed points of entry for Aryan and Central Asian groups seeking to take advantage of the fabled riches of the region. The passes continue to remain a focus of tensions between India and Pakistan, which vie for control of the Kashmir region. In the northeast, the ranges are lower, as they turn southward into Myanmar, and include breaks caused by the Brahmaputra River system that afford access from China.

The northern mountain wall of Southern Asia was formed when the continental plate fragment that is peninsular India crashed into the Asian continental plate (Figure 5.8b) and

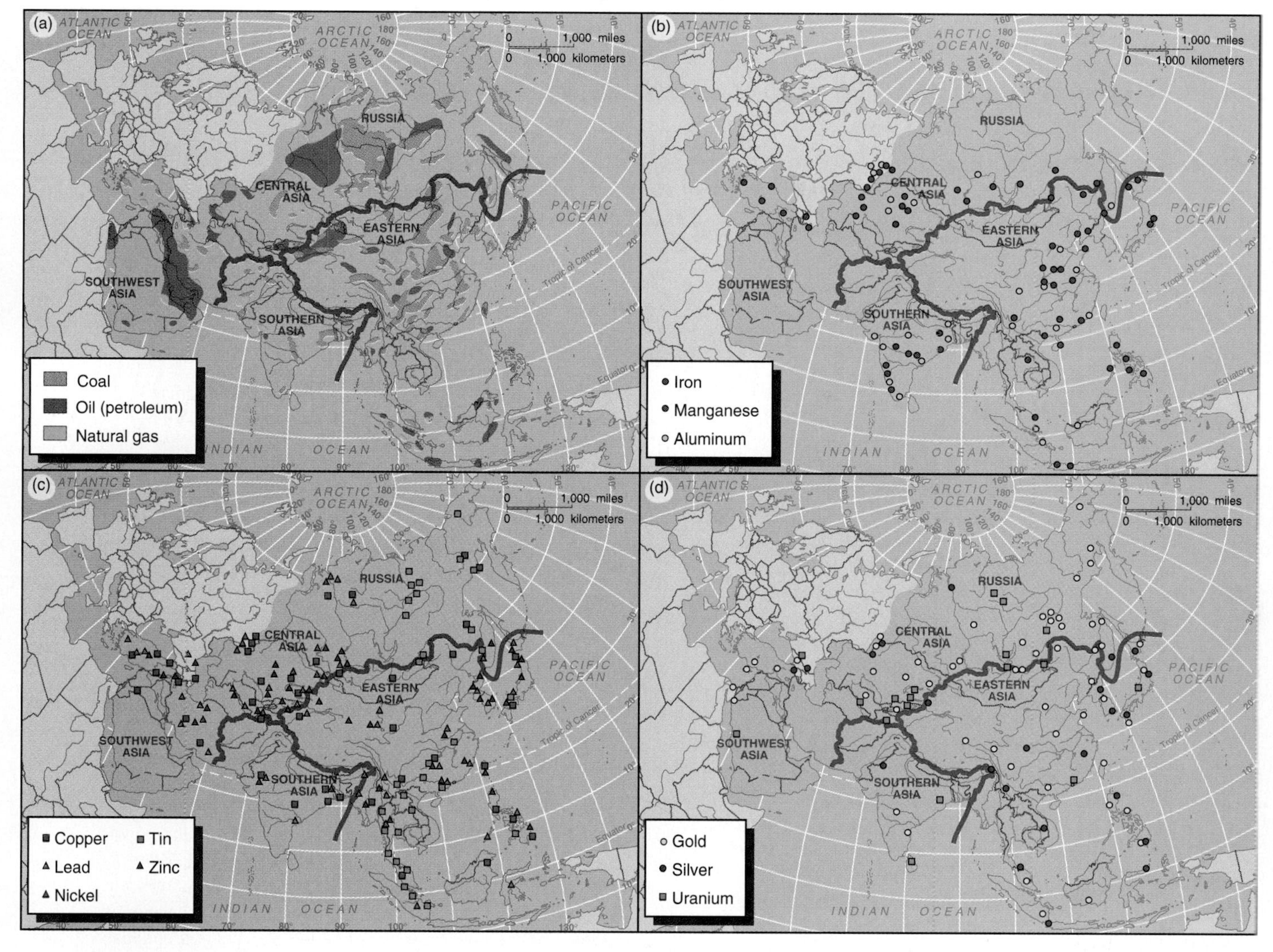


**FIGURE 5.9 Asia: mineral resources.** Compare the resources of Southern Asia with other parts of the continent. (a) Fuels: coal, oil (petroleum), natural gas. (b) Metals: iron, manganese ores, bauxite (aluminum ore). (c) Metals: copper. lead, nickel, tin, zinc. (d) Precious metals (gold, silver), uranium.

thrust itself beneath the main mass of Asian crustal rocks. The collision of two continent-carrying plates caused the thickening of the low-density crustal rocks to produce the highest mountains in the world. The process continues, signaled by earthquakes. The continuing thrust of peninsular India beneath Asia causes the mountains to rise at 6 cm per year. While they are rising, however, glaciers and rivers cut into them and wear them down. At present, there is an approximate balance between the rate of rise and the rate at which these mountains are being eroded. Many of the plentiful minerals of Asia (Figure 5.9) are related to such crustal activity.

## Peninsular Hills and Plateaus

Peninsular India is a continental plate fragment of very ancient rocks that was tilted with the highest points along the western coast—the **Western Ghats**—rising to just over 2,500 m (8,000 ft.). Much of the peninsula consists of plateaus sloping eastward to the Bay of Bengal—a relief pattern followed by many rivers. Along the east coast a broken line of hills, the **Eastern Ghats**, rises in places to 1,500 m (5,000 ft.).

As the continental plate fragment that forms the peninsula of Southern Asia broke away from the other southern continents over 150 million years ago (see Figure 3.7c), layers of volcanic lava poured out through cracks in Earth's crust. The lava flows are most prominent in the northwestern part of the peninsula in an area known as the **Deccan plateau.** A narrow coastal plain encircles the peninsula. It is widest where rivers such as the Krishna and Godavari drain off the plateau, creating deltas at their mouths on the eastern coast (Figure 5.10).

## Major River Basins

Between the Himalayas and the peninsular plateaus, a wide lowland zone is crossed by three of the world's major river systems—the **Indus, Ganges,** and **Brahmaputra Rivers**—after they flow down from their Himalayan sources. These plains are now densely populated and intensively farmed, with the

# ECONOMIC GROWTH AND ENVIRONMENTAL PROBLEMS

The issues that arise include:

How does economic growth affect the natural environment?

What are the governments in Southern Asia doing to prevent environmental damage?

Can economic growth in poor countries be achieved without environmental damage?

The countries of Southern Asia face major issues concerning the impact of economic growth on their environment. These issues affect the future of a region where the population is likely to increase to nearly 2 billion people by A.D. 2025. Increasing rates of economic growth in agriculture and manufacturing will be needed for even larger numbers of people but will have impacts on the environment that at present are merely being monitored by governments at state and central levels. The countries of the region appear to have little ability to prevent environmental degradation since they are concerned first with providing subsistence and jobs for their people.

The *Green Revolution* resulted in environmental impacts that varied from place to place. In the Indian states of Punjab and Haryana, production rose, prosperity increased, and the trend is to other forms of intensive farming. However, the revolution was based on applying more fertilizer and using more pesticides. Both are washed from fields into rivers. Both are manufactured in chemical factories that bring pollution of land and water to surrounding areas. If not well regulated, concentrations of toxic chemical manufactures may cause deaths, as they did at the Union Carbide pesticide plant at Bhopal in 1984, when leaking gases killed 3,000 people and seriously injured 50,000. This was a well-publicized, but not isolated, occurrence, and the death toll over a period of years has been estimated at ten times the initial level.

*Irrigation is basic to increasing the intensity of agricultural production,* especially in the drier northwestern parts of Southern Asia. The Indus tributaries and the Middle Ganges River valley have intricate systems of irrigation channels constructed initially by British engineers in the late 1800s to extend productive farming into arid areas and insulate farmers against the uncertainties of monsoon rainfall. Extensions continued after independence, so that the proportion of irrigated cropland in Southern Asia increased from 18 percent (132 million hectares) in 1960 to 32 percent (174 million hectares) in 1980. Nearly two-thirds of the irrigated cropland is in India.

Irrigation in Southern Asia is of many types. Large dams and canal systems, irrigating over 4,000 hectares (10,000 acres) each, are supplemented by medium-sized dams that have been built in planned and coordinated projects over the last 100 years. Smaller traditional earth dams, also known as tanks, were built hundreds of years ago by local communities and are most common in the peninsula. Over a thousand large or medium-sized projects were constructed after 1947, completing the irrigation potential of southern India. Most of the remaining sites involve grandiose construction plans for the upper Ganges and Indus in the Himalayas.

The environmental problems associated with irrigation occur in both the source and using areas. Upstream, the storage of water requires building dams and drowning deep valleys. In 1993, local activists, supported by foreign environmentalists, influenced the World Bank to withdraw its funding for the Narmada multipurpose project, which was due to cause the resettlement of 100,000 Indian people—often to their disadvantage. This action affected the funding of other irrigation and hydroelectric projects throughout the Himalayas, including those planned in Pakistan and Nepal. Even where dams are constructed in the mountains, they fill with sediment or may be subject to earthquake disaster. Returns from the sale of electricity generated in the highlands are often poor because of a lack of local demand and the high costs of transporting the electricity to distant cities.

Downstream in the using areas, poor management of water may feed it to fields where the terrain or subsurface conditions are unsuitable. Soils become waterlogged or saline. The increasing extraction of water for irrigation also alters river flow patterns; whereas the Ganges River flooded regularly, those floods are reduced, and dry season flow is much lower.

Wells already provide nearly half the irrigation water used and remain the main source of present and future irrigation growth. Tube wells drilled by modern rigs increased in numbers from 0.2 million in 1960 to over 4 million in 1990. Water is lifted from the wells by diesel or electric pumps, in contrast to the human or animal power used in older wells. Wells have environmental advantages over canal-fed irrigation because they use local water supplies, keep the water table low, and encourage downward water movement in the soil to avoid salinization. They are increasingly combined with canal irrigation so that local groundwater sources can be replenished.

*The growth of manufacturing and urbanization* leads to further environmental damage. Increasing damage of the Taj Mahal because of air pollution led to the closure of metal workshops in the surrounding area to reduce the sulfur gases that attack building stones, but it is only possible to take such action on a local scale. Another striking case of pollution is the effluent from the Kampur concentration of leather tanning works. There is no real control on this, and the heavily polluted waters enter the Ganges River close to some of the main Hindu ritual bathing sites. In recent years, the high cost of firewood for cremations led to an upsurge of body dumping in the Ganges. The faithful believe Mother Ganga will preserve them from harm, but health dangers from various sources increase.

There are many pressures on the Ganges River basin, where nearly 40 percent of the Indian population lives. The land is intensively cultivated and has growing numbers of factories and urban areas within its watershed. The Ganga Action Plan formulated and passed by the Indian government in the 1980s envisaged a cleanup of the river. A few projects were implemented, including some new sewage treatment plants, but lack of funding, continuing industrial malpractice,

(a) Air pollution from local industry causes the stonework of the Taj Mahal in Agra, India, to crumble, although local regulation reduced pollution to help save this monument of great tourist interest. (b) Water buffalo tethered beside a badly polluted river in northern India.



illegal passing of wastes into the river, and illiteracy brought the program few successes.

Urban areas in Southern Asia are increasingly subject to air pollution at levels that are not tolerated in Western cities. Some multinational corporations site "dirty" manufacturing facilities in the countries of Southern Asia because environmental restrictions are fewer than in the core countries. The increasing levels of vehicle use, domestic burning of wood and dung, and industrial emissions are monitored by government organizations, but little is done to reduce the rising levels of harmful gases and particles in the atmosphere. The climates of Southern Asia tend to encourage higher levels of pollution because the air is still for much of the year.

The expansion of growing cities in which so many poor people live brings other environmental problems, including the devastation of vegetation for firewood in surrounding areas and the gouging and quarrying of brick-pits. Easily made from clays, bricks are in demand for the construction of buildings and are pulverized to make road surfaces. The brick-pits form depressions that fill with water, become sites for disease-carrying insects, and make it difficult to develop the land further.

As the population of Southern Asia increases and resource usage grows in attempts to raise economic levels, environmental impacts will be greater. The end result will be the degradation of atmospheric and water quality. Government action so far is ineffective, since even those laws that were passed are seldom observed or enforced. It is difficult to add the costs of maintaining environmental quality to those of manufacturing goods cheaply enough to compete in world markets. The tension between current expedients and providing for continuing production into the future is a major problem that the poor countries in Southern Asia and in other world regions face increasingly as they adopt industrial practices and expand cities.

**TOPICS FOR DEBATE**

1. It is inevitable that economic and human development lead to environmental degradation.
2. You cannot expect people in Southern Asia to worry about the environment when they cannot see where their next meal is coming from.
3. The impact of economic growth on the environment is not just a problem for Southern Asia.

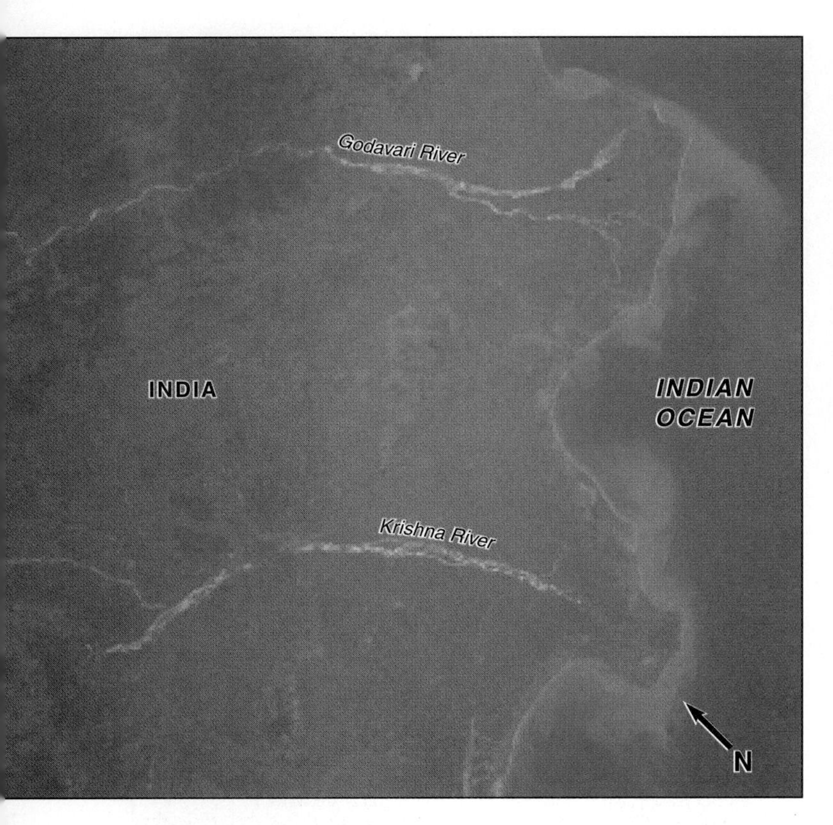


**FIGURE 5.10 Southern Asia: Godavari and Krishna River deltas.** Two sacred rivers of India, the Godavari (upper, north) and Krishna (lower, south), form deltas as they reach the Bay of Bengal to the east of India: locate on Figure 5.8a. Both rise in the Western Ghats and flow eastward. The 1450 km (900 mi.) long Godavari divides below the city of Rajahmundry to form a delta that was the site of some of the earliest European settlements in India. The Krishna is 1290 km (800 mi.) long. Both have large discharges of water in the wet monsoon season, but this space shuttle view shows them dry with many sand bars revealed in their braided winter courses.

NASA

river water being used for irrigation. The Ganges River has great religious significance for Hindus (Figure 5.11).

Rock particles and fragments worn from the Himalayas and carried by these rivers built deposits of **alluvium** up to 3,000 m (10,000 ft.) thick. The surface of the deposits is generally flat but marked by small-scale relief of a few meters or tens of meters where the rivers cut new channels in older material. Along the northern margins of this plain, large **alluvial fans** of gravel mark the junction of the steep mountain courses with the lowlands. The sudden change of gradient causes the rivers to drop the gravel and sand they can carry along their steeper-gradient mountain valleys. In the drier areas of the northwest, there are sections of **badlands topography** where occasional rainfall runoff cuts dense networks of steep-sided gullies into unvegetated alluvial deposits.

## Forests and Soils

Much of Southern Asia was originally forested, including the peninsula, northern river plains, and Himalayan foothills. After centuries of clearance by growing populations, the teak forests of southern India and the forests on the Himalayan

**FIGURE 5.11 Southern Asia: the sacred, life-giving Ganges River.** The river at Varanasi is supplied by Himalayan meltwaters and is liable to flooding. Pollution from agricultural chemicals, industrial wastes, and domestic wastes is an increasing problem in this river.



slopes are almost all that remain. These are still being reduced in size by increased cutting. Much of the former woodland has been cut to expand the cultivated area and for fuel. The northwest with its arid climate forms the **Thar Desert**.

The most fertile soils formed beneath the forests, in areas subject to annual flooding, or around the lava plateaus. Long utilization for farming, however, reduced the soil quality, while the drier areas experienced soil erosion, waterlogging, and salinization.

## Natural Resources

Southern Asia is rich in mineral resources (see Figure 5.9). The ancient rocks of the peninsula contain mineral ores including iron, while the newer rocks on top contain one of the world's largest coal reserves. Deposits of oil and natural gas occur in the thick sediments beneath the major valley areas and just offshore in India and Pakistan.

For most people in this region, the main natural resource is water. Early in history, the monsoon rains made it possible to support large populations at subsistence level. Irrigation and water storage techniques were developed. The Indus tributaries brought the water to its lower valley, where one of the world's earliest civilizations developed (see Chapter 1). The waters flowing northward from Sri Lanka's central hills also fostered a civilization based on irrigation. The Ganges and Brahmaputra brought economic life to the lowlands south of the Himalayas. Large hydroelectricity projects continue to harness the water power of the mountains.

In peninsular India, the monsoon rains are less certain in amount from year to year. Most of these rains fall on the Western Ghats, supplying short rivers that flow rapidly westward to the ocean and the headwaters of longer rivers draining eastward. These rivers flow with maximum discharge for only a week or so across low gradients and are not as powerful as those flowing from the Himalayas. Today the amount of monsoon rains still determines whether the region's economies continue to grow. In the many thousands of villages across Southern Asia, access to

**RECAP 5A: Southern Asia, the Wider World, and Cultural and Physical Variety**

Southern Asia is the world's second-poorest region after Africa South of the Sahara, but its much larger population means that it has more very poor people. Technical and political sophistication and existing infrastructure suggest prospects of better things.

Southern Asia occupies the Indian subcontinent, hemmed in by the Himalayan Mountains. Passes through the mountains provided routes for invaders. Incoming Muslims confronted the Hindu and Buddhist religions that originated in Southern Asia. The British Indian Empire was the world's most intense expression of modern colonialism. The region's economy was oriented to the needs of Britain until independence was achieved in 1947.

The physical environment of Southern Asia is marked by monsoon and arid climatic environments, the world's highest mountain system in the Himalayas, the wide plains of the Indus, Ganges, and Brahmaputra River basins, and the hills and plateaus of the peninsula. The dynamic physical environment produces earthquakes, summer river floods, and tropical cyclones. Human economic development makes use of the mineral and water resources and results in increasing modifications of the natural environment, including irrigation systems and pollution.

5A.1 How is Africa South of the Sahara the world's poorest region if Southern Asia has more poor people?

5A.2 Which major cultural elements affect the lives of people in Southern Asia today?

5A.3 How would you illustrate the statement that Southern Asia's natural environments are dynamic and full of contrasts?

**Key Terms:**

| | | |
|---|---|---|
| Hinduism | British Indian Empire | alluvium |
| Buddhism | Indian National Congress | alluvial fan |
| caste system | Muslim League | badlands topography |
| British East India Company | | |

**Key Places:**

| | | |
|---|---|---|
| Indian subcontinent | Western Ghats | Ganges River |
| Bay of Bengal | Eastern Ghats | Brahmaputra River |
| Himalayan Mountains | Deccan plateau | Thar Desert |
| Mount Everest | Indus River | |

water is a constant focus of efforts to manage the environment and provide communities with a continuing supply of food.

## Environmental Problems

People living in Southern Asia face a number of environmental hazards. Some are an integral part of the dynamic natural environment, and some are the outcomes of human clearance of land, population increase, and economic growth in the context of the expanding world system.

The dynamic natural environment is brought about partly by the clashing plates that produced the Himalayas and continue to set off earthquakes like the one in Hyderabad in 1993. Activity within the natural environment also results from the extremes of the monsoon climatic environment, particularly the high river runoff and flood levels in summer. Tropical cyclones developing over the Bay of Bengal bring disaster to the Ganges delta area, mainly in Bangladesh.

The combination of high mountains and heavy precipitation produces some of the most rapid physical landscape modifications in the world. People have always found it difficult to live with them. The melting Himalayan snows and the monsoon rains combine in flood flows in the Indus, Ganges, and Brahmaputra River systems. The strength of flow can be gauged from the fact that the Ganges and Brahmaputra sweep debris 3,000 km (nearly 2,000 mi.) out to sea in the Bay of Bengal. The flows in the Indus system bring water to the Thar Desert.

Human impacts modified the natural environment, at times increasing productivity and at others reducing it (see "World Issue: Economic Growth and Environmental Problems," page 194). Soil erosion following forest destruction and poor cultivation methods, pollution of the rivers and air by chemical factories, urban expansion, and the damming of rivers all changed and degraded the natural environment of Southern Asia.

# INDIA

The **Republic of India** dominates the geography of Southern Asia in area, population, and centrality (Figure 5.12). In 1998, it was home to 75 percent of the region's population and produced two-thirds of the total GNP. Its huge population was 990 million. When the British Indian Empire was partitioned at independence in 1947, India gained much of the best land, controlled most water resources, had most of the manufacturing industry, and included the largest cities in Southern Asia. The combination of size and productive potential continues to make India—firmly in the world periphery itself—a locally dominant presence for other countries in the region.

At independence, India's leaders hoped to establish a democracy among a multireligious and multilingual people that had never been unified and to end poverty in a return to precolonial "glory." The first prime minister, Jawaharlal Nehru, wanted to rid India of the rural feudalism that had been encouraged by Britain, allowing princes and rulers to take 20 percent of the national income while millions of poor without rights starved. Nehru wanted to reverse the deindustrialization that the British Raj had carried out by transferring wealth to the home country. His goals were to transfer land from the aristocracy to the peasants—which he did—and to make India an industrial power. Although trumpeted as a success with 3.5 percent GDP growth per year, the Indian approach was later seen as a failure in comparison with twice that rate realized in Eastern Asia. India neglected primary education for all in favor of higher and technical education for a few. Food production was a lower priority than the state industrial planning that resulted in overcapacity, uneconomic units, corruption, and a fall in productivity.

Nehru's daughter, Indira Gandhi, succeeded to power in 1966, holding office until 1977. Although her policies continued the move toward state capitalism, she reversed the neglect of agriculture by encouraging Green Revolution techniques supported by better distribution facilities. This ended the famines of the 1960s and dependence on U.S. food aid. Later,

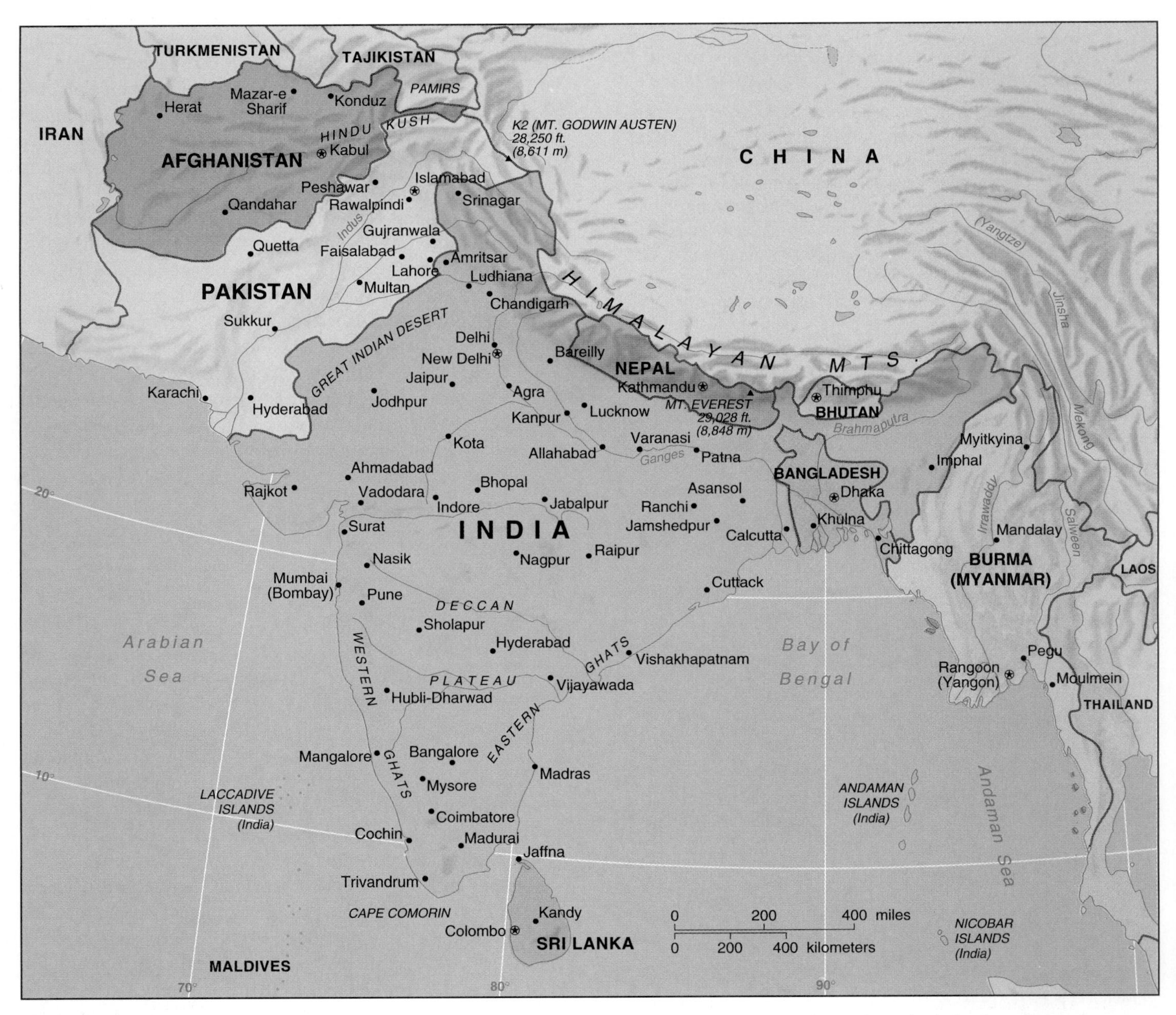


**FIGURE 5.12 Southern Asia: major geographic features, countries, and cities.** The Indian subregion dominates the region by its size and centrality; Bangladesh and Pakistan are two Muslim countries, created as one at independence but since broken apart; Afghanistan, Nepal, and Bhutan form the Northern Mountain Rim, while the Maldives and Sri Lanka are the Southern Island Rim. This is a region of huge, growing cities.

in the 1980s, after she was reelected prime minister, she and her successor, son Rajiv, began moves toward opening India to world economic forces, but the cost of internal subsidies and repeated elections took India into deep indebtedness. In 1991, reformist parties took over the government as a reluctant response to external insistence on dismantling the still-popular government subsidies and public sector involvement. At first, in the early 1990s, annual economic growth rose to 7 percent, but later it fell. Change came more slowly than hoped for by reformists, but by the late 1990s, more capital was being invested in revitalizing industry, new entrepreneurs emerged from all kinds of business sizes and social classes, rural development began to progress, and consumers saw benefits of competition in lower prices and greater choices of goods.

In 1998, the new government of the Bharatiya Janata Party, with a strong Hindu nationalist mandate, brought two surprises. First, in May, it detonated five nuclear weapon devices, causing horror around the world and drawing sanctions from major funding nations. Second, in June, its first budget might have introduced further reforms on the basis of internal euphoria resulting from the display of nuclear prowess, but it did not. It returned to the theme of *swadeshi* (self-reliance) that the Nehru government proclaimed in the 1950s. Import tariffs—already high but falling—were raised, public spending was increased (with defense by 14%), foreign investment was encouraged (but mainly by overseas Indians), and some "strategic" industries were exempted from privatization. This retreat from reform was in part a response to fears of multina-

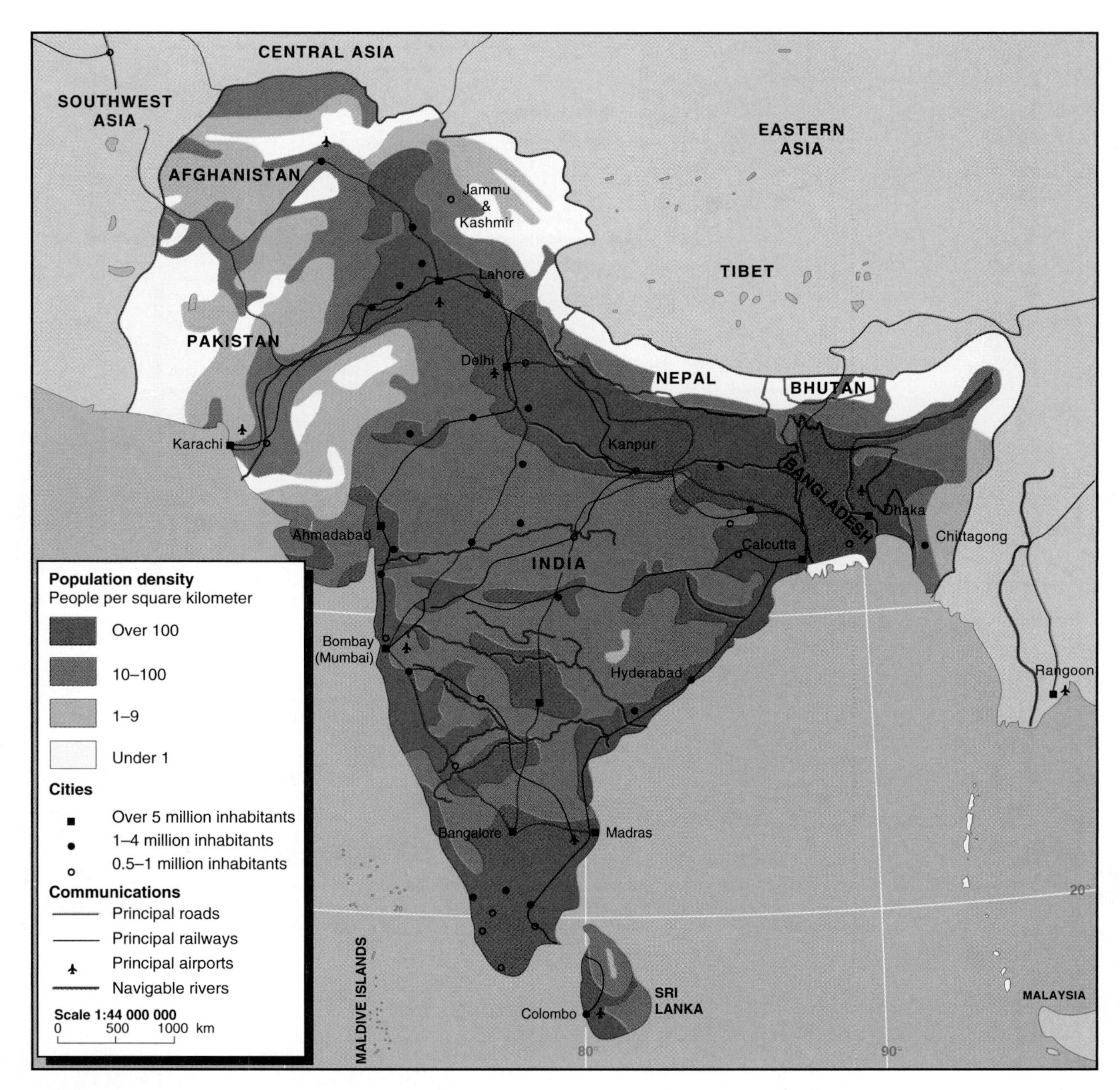


**FIGURE 5.13 Southern Asia: distribution of population.** Note the contrasts within Southern Asia, linking the areas of high population density to river valleys and of low population density to arid areas and mountains.

Source: Data from *New Oxford School Atlas*, p. 75, Oxford University Press, UK, 1990.

tional corporation dominance as India's businesses were opened to foreign competition. In part, it also reflected the shock waves of recession in Eastern Asia that made opening to the world economic system less attractive.

## Regions

The Republic of India can be divided into physical regions, into regions of population density, and into regions of economic and social prosperity.

- The physical regions (see Figure 5.8) include the Himalayan ranges in the north, the Ganges River plains, and the peninsular plateaus in the south. Across these contrasting landscapes, rainfall variations impose widely different conditions. They result in aridity and semiaridity in the northwest, higher rainfalls in the hilly Ghats of the peninsula and the foothills of the Himalayas and Assam, and lower and less certain rains in the eastern part of the peninsula and Ganges River plains.
- The physical contrasts are reflected in the distribution of population (Figure 5.13). The highest densities occur along the Ganges River plains and around the coastal plains of the peninsula. These areas of high density contrast markedly with the low population densities of the dry and mountainous areas. Large areas of the peninsula have moderate-to-high densities of rural population.

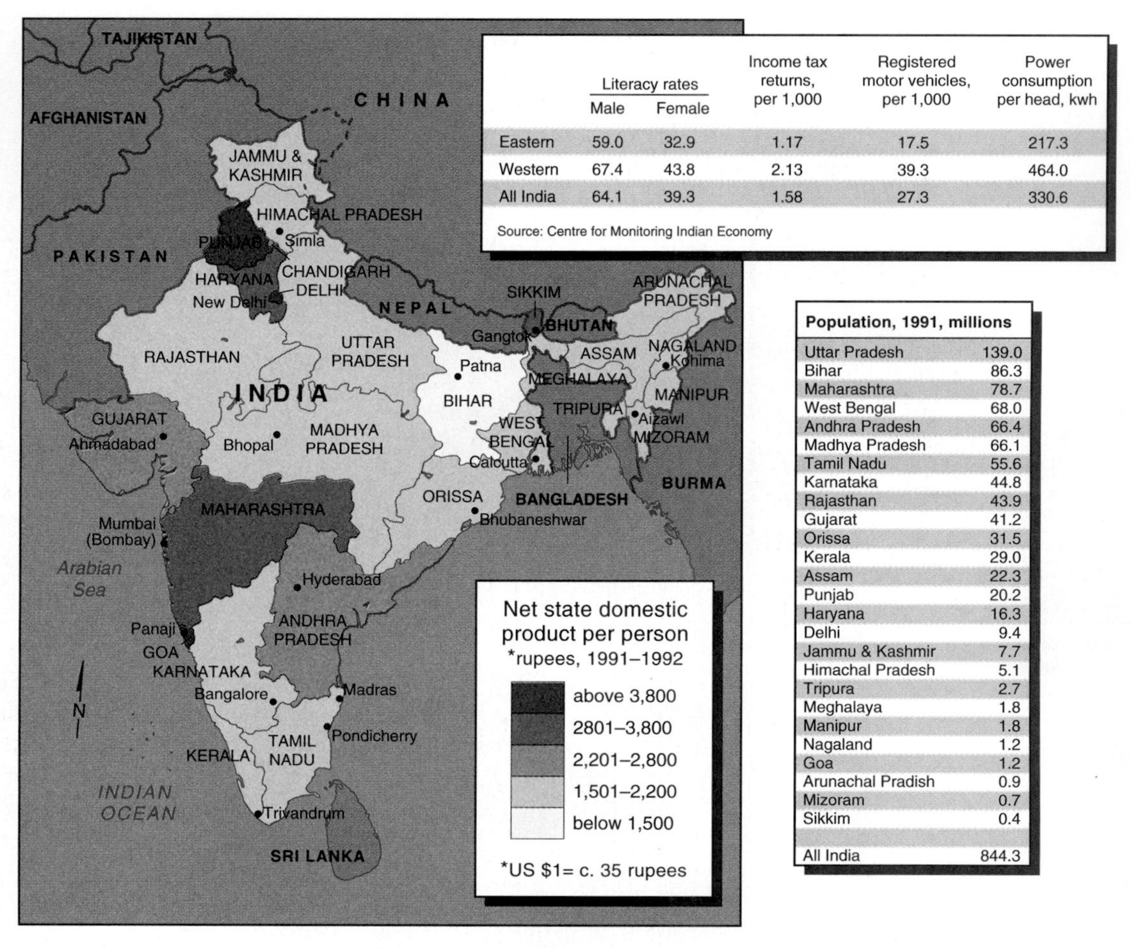


| | Literacy rates | | Income tax returns, per 1,000 | Registered motor vehicles, per 1,000 | Power consumption per head, kwh |
|---|---|---|---|---|---|
| | Male | Female | | | |
| Eastern | 59.0 | 32.9 | 1.17 | 17.5 | 217.3 |
| Western | 67.4 | 43.8 | 2.13 | 39.3 | 464.0 |
| All India | 64.1 | 39.3 | 1.58 | 27.3 | 330.6 |

Source: Centre for Monitoring Indian Economy

| Population, 1991, millions | |
|---|---|
| Uttar Pradesh | 139.0 |
| Bihar | 86.3 |
| Maharashtra | 78.7 |
| West Bengal | 68.0 |
| Andhra Pradesh | 66.4 |
| Madhya Pradesh | 66.1 |
| Tamil Nadu | 55.6 |
| Karnataka | 44.8 |
| Rajasthan | 43.9 |
| Gujarat | 41.2 |
| Orissa | 31.5 |
| Kerala | 29.0 |
| Assam | 22.3 |
| Punjab | 20.2 |
| Haryana | 16.3 |
| Delhi | 9.4 |
| Jammu & Kashmir | 7.7 |
| Himachal Pradesh | 5.1 |
| Tripura | 2.7 |
| Meghalaya | 1.8 |
| Manipur | 1.8 |
| Nagaland | 1.2 |
| Goa | 1.2 |
| Arunachal Pradish | 0.9 |
| Mizoram | 0.7 |
| Sikkim | 0.4 |
| All India | 844.3 |

**FIGURE 5.14 Southern Asia: east-west contrasts in India.** Relate the income figures shown on the map to the data on male-female literacy rates, motor vehicles, power consumption, and state population size.



- Economic and social well-being in India is increasingly a matter of east-west contrasts. This is a generalization with exceptions, but the western states of Punjab, Haryana, Gujarat, and Maharashtra experience faster economic development than the eastern states such as Bihar and Orissa (Figure 5.14). The greatest economic advances are in the irrigated and diversifying farming areas of Punjab and Haryana in the northwest, and in the industrializing states of Gujarat and Maharashtra in the center of the west around Bombay (Mumbai). Higher literacy rates, vehicles per head, and power consumption all demonstrate the west's advantages.

The northeastern corner of India, including Assam, was almost cut off from the rest of the country by partition in 1947. It contains a number of areas acquired by the British, including Sikkim, which became part of India in 1975. Many peoples in this part of India are not Hindu, and some have been influenced greatly by Christian missionaries. Assam has tea plantations and some oilfields. Tension between migrants from the rest of India and Bangladesh and local peoples results in clashes that force the Indian government to spend more on security in this area than elsewhere.

## People

India's population is the world's second largest after China's, and its rate of increase is faster than that of China. A sharp rise in population increase began after 1920 when the Indian Empire population numbered around 250 million people, but the British government did nothing to slow the rise. By 1961, India's population was 440 million—at a time of frequent famines. Estimates in the 1960s of growth to a total of 775 million by 1990 caused the Indian government to institute policies for reducing population increase.

### Impact of Birth Control Policies

Attempts to spread the use of birth control methods began in the 1950s but gained ground slowly among a largely illiterate people. From 1970 to the present, the proportion of Indian women using family planning increased from 15 to over 40 percent. The possibility of male sterilization in the 1970s, however, was widely rejected as economic suicide for families: the threat of its forced imposition was a major political issue in the 1977 elections that ousted Indira Gandhi's government until 1980. The greater use of family planning in the

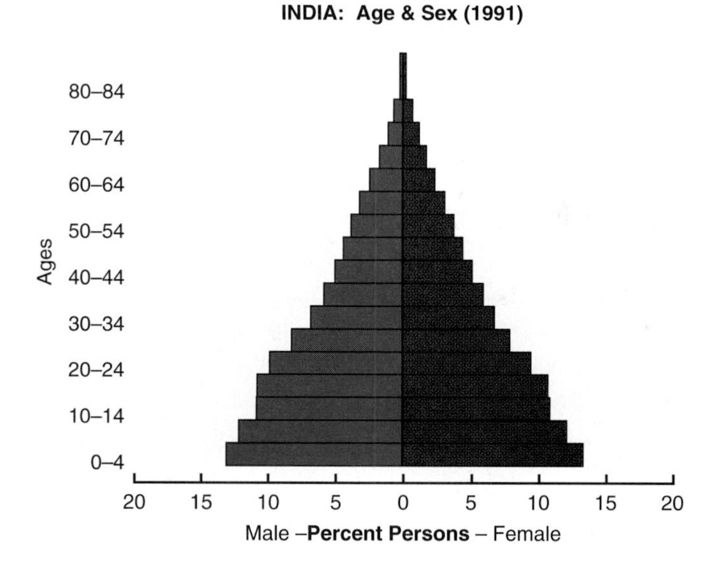


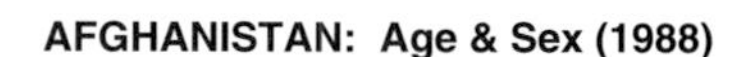


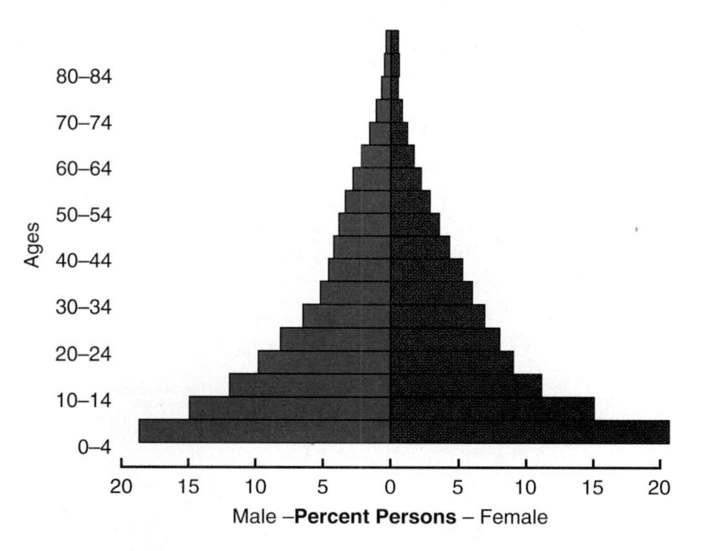


**FIGURE 5.15 Southern Asia: age-sex graphs.** The increasing numbers of young people will produce continuing rapid population growth when they reach childbearing age for many years to come—even if fertility declines. The smaller numbers in the 10–14 age group in India in 1991 suggest an impact of population policies in the late 1970s. (Age-sex graphs for most countries for current year, 2015, and 2050 can be printed out from an Internet website [see Chapter 2, page 40, for instructions]).

Source: Data from *Demographic Yearbook,* U.N., 1994.

1980s and 1990s was related to rising levels of literacy, increased urbanization, and improvements in the status of women. Hindu women of higher castes use family planning more than those among the poorer castes and less-educated Muslims. More women adopt family planning once they have given birth to a boy.

By 1998, India's population rose to 990 million despite the efforts to reduce population growth. Late 1990s' estimates put the population in 2025 at 1,440 million. Although famine is no longer a regular occurrence in India, a large proportion of the population is undernourished, and the struggle to provide food for the people continues to confront the government. India's population continues to grow because of the huge number of people in the child-producing ages (Figure 5.15) and the increased life expectancy. Even though total fertility rates fell to 3.4 by 1998, mortality rates fell more rapidly, and life expectancy rose to 59 years, as high as most countries in Southern Asia. A natural increase of just under 2 percent still adds over 1 million to the population each month. In the demographic transition process (Figure 5.16), India remains in the phase of rapid population increase.

Further slowing of the population increase could be difficult since many families, particularly in rural areas, still perceive a high risk of children dying. Death rates among those from birth to 5 years of age remain as high as 25 percent in many rural areas, where illiteracy is high. Large families of children are seen as important because they provide extra manual labor and parental care in later life. Cultural pressures, such as the great value that is placed on male succession, and the early marriage of girls that provides a longer childbearing period, contribute to continuing high birth rates, especially in rural areas. Moreover, 120 million Muslims disapproving of birth control live in India. The new urban middle class is the only group in which there are signs of the Indian population stabilizing around fertility rates of 2.3 children, but this group forms a small minority of the population.

## Mainly Rural Population

India's population remains largely rural. Despite rapid migration from rural to urban areas, only one-fourth of the total population lives in urban areas. Between 1960 and 1995, however, rural India went from 82 to 73 percent of the country's population and from 50 to under 30 percent of its GDP. Rural areas lag behind urban areas in education and health provision, and still have large proportions of the subsistence and low-paid farming lifestyles that are linked to long-term poverty. By 2025, it is projected that India's rural population will be down to 55 percent of the total, but this will comprise a greater number of people than at present (762 million instead of 630 million).

The future will see further increases in the proportions living in the urban places that provide most of India's economic opportunities. New jobs in factories and offices, better access to education and health services, and most opportunities in the informal sector of the economy occur in urban—especially large urban—areas (Figure 5.17a). Industrial zones, shantytowns (bustees) (Figure 5.17b), and middle-class suburbs expand with little planning control. The total population of urban areas is projected to rise from 250 million people in the mid-1990s to over 600 million by 2025. From 1980 to 1996, the number of cities with over a million people jumped from 12 to 30 (see Figure 5.12). The largest cities are becoming immense, especially the largest three. In 1996 United Nations estimates, **Bombay** had 15.7 million people, **Calcutta** 12.1 million, and **Delhi** 10.3 million. A projection for 2015 suggests that Bombay could become the world's

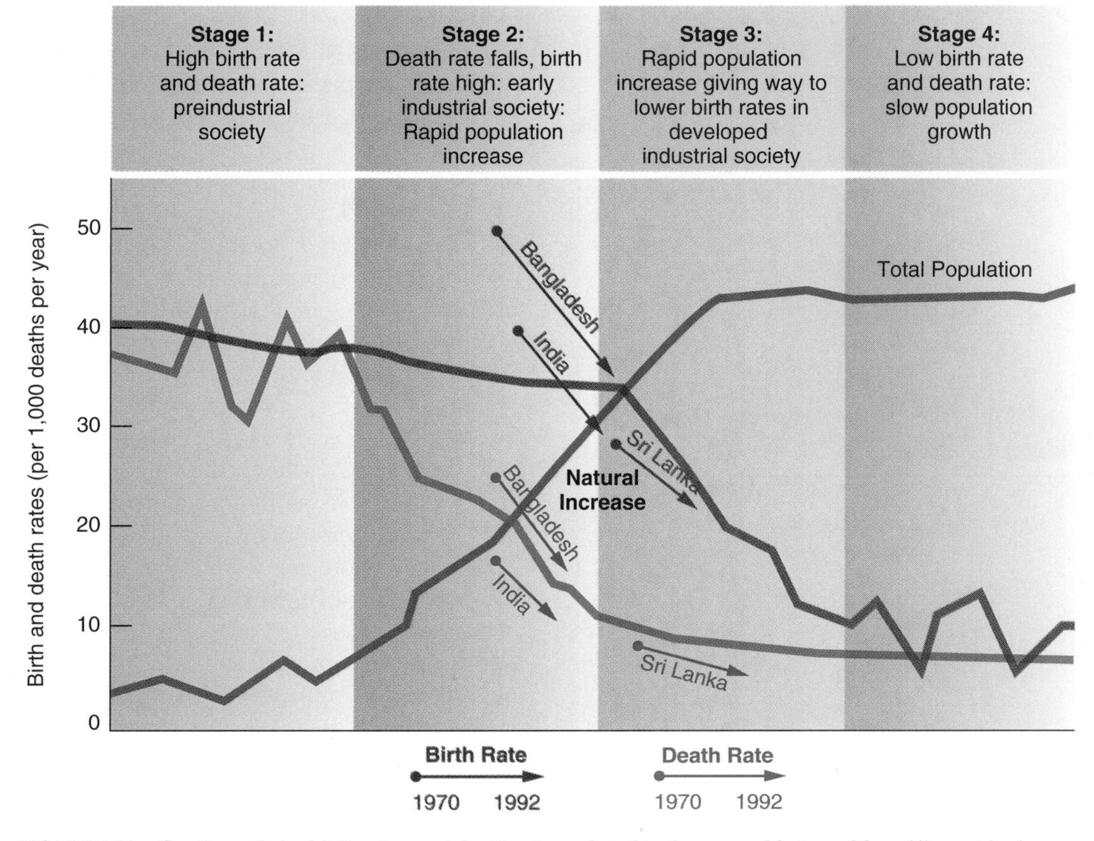


**FIGURE 5.16 Southern Asia: birth rates and death rates related to demographic transition.** All countries have a wide gap between falling birth rates and death rates. Sri Lanka is farthest ahead in reducing this gap. Compare with Africa South of the Sahara (see Figure 3.13).

(a)

**FIGURE 5.17 Southern Asia: urban landscapes in India.** (a) Former British colonial buildings in the center of Bangalore, southern India. People wait for buses as some enter an auto-rickshaw.

second-largest city with around 26 million people, with Calcutta and Delhi at around 17 million each. Other major cities include **Madras, Hyderabad, Bangalore,** and **Ahmadabad.**

### Religion, Language, and Race

India is a dominantly Hindu country with over 80 percent of the population having such an allegiance. It is also a country of minorities, with over 11 percent Muslims and just over 2 percent each for Christians and Sikhs. The proportions of Buddhists and Jains are under 1 percent. The religious diversity continues to raise contentious issues, as when militant Hindus pulled down a Muslim mosque in 1995.

There is even greater diversity in languages, with **Hindi** spoken by only 30 percent of the population and merely one of 15 official languages recognized by Indian states. The official languages vary by region across India, with Indo-Aryan languages strongest in the north and those of Dravidian origin in the south. In all, there are over 1,600 languages and dialects in India. English is the language of the legal system, is spoken by the educated elite, and is a common form of communication in commerce and national politics.

Ethnic variety is superimposed on racial variety. Nearly three-fourths of the population is of Indo-Aryan origin with lighter skins. Most of the rest is of the darker Dravidian group that occupied India before the invasions through the northwestern frontier passes and is still dominant in the southern peninsula. Small groups of east Asian peoples and tribal groupings, mainly in the northeast, make up the remainder.

The caste system, based originally on the differences between the Indo-Aryan and Dravidian groups, is still of great significance to Indian peoples. Although the Indian Congress Party makes much of its efforts to assist the poorest untouchables, special favors, such as reserving college and civil service places for them, tend to emphasize the caste contrasts while few in the lower castes see any improvement in their lot. Rapid urbanization blurs some of the caste divides, but protests of the poor castes increase as new political parties find a basis within this group for polarizing the Hindu population.

**(b)** Slum shantytown conditions in Bombay with dwellings made of sacking, plastic and metal sheeting, and boards of various types. Contrast the foreground dwellings with those in the distance.



### Indians Abroad

Around 15 million Indians live abroad, with the largest groups in the late 1990s in Nepal, South Africa, Malaysia, Sri Lanka, the Persian Gulf states, the United Kingdom, and the United States. When part of the British Empire, Indians provided a source of cheap labor and were moved to work on plantations and railroad construction in the Caribbean, Eastern and Southern Africa, Burma, Malaysia, and Pacific islands such as Fiji. Many stayed as shopkeepers and business people.

More recently, there have been flows to the oil-producing countries of the Persian Gulf and to the United Kingdom and United States to take up opportunities in education and the professions—a significant "brain drain." Indians are the fourth largest group of Asians in the United States, with over 400,000 immigrants from 1981 to 1994 and with 75 percent of those over 25 years old having a college degree, compared to less than 33 percent for all foreign-born immigrants to the United States. Many Indian expatriates send remittances home, estimated at over $3 billion per year in the mid-1990s, and are a source of investment in the Indian economy. Moreover, such emigrants make known the possibilities awaiting those able to escape the poverty in India and the value of having small families like others in their new countries.

## Economic Development

India has one of the largest economies outside the core countries, with areas of highly productive agriculture and growing manufactures. In the late 1990s, its total GDP was just behind that of the Russian Federation, Australia, or the Netherlands. Much of the country and many of its people, however, remain poor, tied to traditional low-productivity farming and owning few consumer goods (Figure 5.18).

Signs exist that parts of India's economy are expanding and diversifying, and the attempts to restructure it since 1991 resulted in faster growth, but India has not reached the growth rate, level of foreign investment, or reduction of poverty attained in Eastern Asia, deferring further growth. The economy is moving toward that of a more prosperous semiperipheral country, as demonstrated by changes between 1965 and 1995. Farming decreased its contribution to GDP from 44 to 29 percent, industry rose from 22 to 29 percent, and services rose from 34 to 41 percent. While India's economy grows and diversifies, however, many of its people remain poor, with a 1997 HPI value of 36.7 percent, and per capita income scarcely increases. There are demands for more spending on health and education to increase the human resource base.

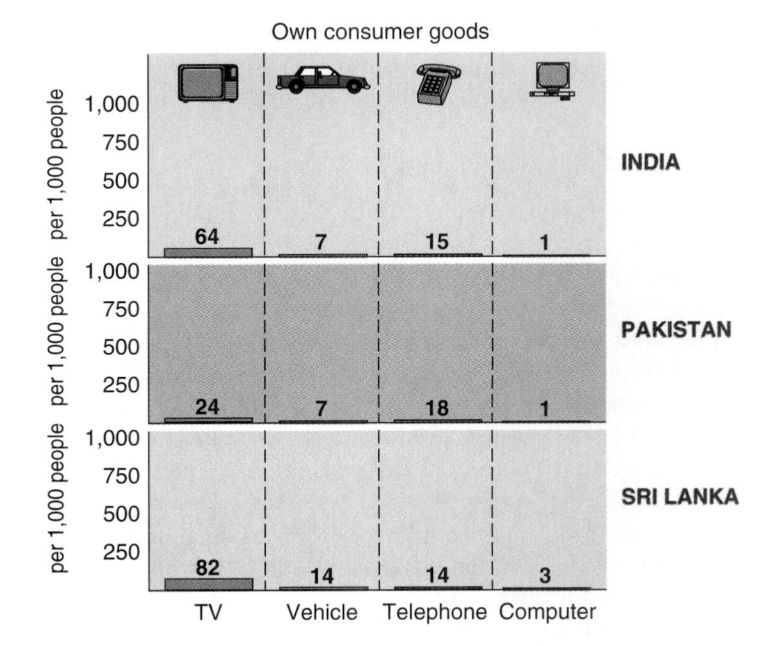


**FIGURE 5.18 Southern Asia: ownership of consumer goods.** How do these figures compare with other Asian countries?

Source: Data (for 1996) from *World Development Indicators,* World Bank, 1998.

## Protectionism

The country faces a continuing struggle between different views of the future. Its reliance on socialist principles and self-sufficiency arose from the teachings and actions of Mahatma Gandhi, who led the way to independence in the 1930s and 1940s. After the experience of more than two centuries of British rule and economic development related more to the needs of Britain than those of India, the country turned inward.

From its independence in 1947 up to 1991, India kept the world economy at bay, concentrating on providing its own food and developing industries to make the goods needed for internal consumption. It avoided close links with the core countries and multinational corporations. Trade with other countries was discouraged by import tariffs and export quotas. The central government and its bureaucrats attempted to control the economy, encouraging farmers to produce more and expanding publicly owned major industries such as iron and steel. This approach had its successes, and some mutually beneficial trade was established with Soviet bloc countries, mainly through political bargaining.

This policy led to dependence on limited internal funding for manufacturing enterprises and infrastructure support. Technology in textile, vehicle, and other manufacturing was outdated, and overmanning was encouraged. Transportation facilities were not upgraded to cope with increasing demands, while power shortages became common and led to regular outages of two to three hours per day. Education and health facilities were not geared to supporting economic growth in agriculture, industry, or services. Such policies did not prepare India for engagement with the world economic system.

## Role of States

India is a federation of 25 states and 7 union territories, presenting great diversity in geographic features, resources, socioeconomic profiles, and population size. Uttar Pradesh has 140 million people, and three other states have around 80 million, but 13 states have under 20 million each. Only 5 have per capita incomes above the national average—including the more progressive and reforming states such as Punjab, Gujurat, Maharashtra, and Haryana. The poorest states include Bihar, Uttar Pradesh, and West Bengal. Kerala is poor but has better social conditions than most.

The state role is a further factor that often holds back Indian economic development. In the 1990s, states gained major functions, particularly in the delivery of social programs, including health and education, but some states also became involved in foreign relations and defense. They face major funding problems, since they are responsible for 80 to 90 percent of public health, education, housing, urban development, power, and irrigation expenditures, and 50 to 60 percent of agriculture, public works, transportation, and communications expenditures. Local tax bases are low and poorly administered, however, so that while the states handle over half of all Indian government expenditures, they gather much less in revenue. Transfers from central government and borrowing make up the difference. The central government weights its payment transfers in favor of states with large populations and low per capita income. States thus have high, continuing budget deficits, holding back development. The need to finance increasing levels of current expenditures in new social programs reduces capital expenditures and creates infrastructure bottlenecks in the key modernization areas of transportation, power, and telecommunications. Although encouraging foreign investment in infrastructure is one way of dealing with this problem, some states in the late 1990s reverted to protectionist measures, canceling externally funded projects. Such actions in turn make the overall Indian market less attractive to foreign investors. In general, the states are slower to encourage changing economic attitudes than the central government.

## Economic Restructuring in the 1990s

In 1991, the Indian economy hit a crisis after much of its Soviet trade ended. Western countries persuaded India to restructure its economy by encouraging foreign trade and investment, and by producing more for export. The Indian government adopted policies to carry out the changes so that production and exports rose while foreign investment also increased dramatically. The balance of trade and foreign investment levels improved through the 1990s. Direct foreign investment in new facilities in India rose from $150 million in 1991 to $620 million in 1994. Foreign investment in company shares rose from $156 million to $4.1 billion in the same period, increasing India's foreign exchange reserves. Companies listed on the stock market increased from 5,560 in 1987 to 8,800 in 1996.

The new policies did not immediately generate the high rates of economic growth experienced in Eastern Asia and were not universally popular. Once the pent-up demand for foreign goods and the new export markets were satisfied, growth slowed in 1996—an election year and one in which infrastructure bottlenecks showed up. While business people and those with private capital profited from the new economic environment, the poor did not. Despite slow changes in state-owned enterprises, many civil servants began to lose their jobs or the control positions that enabled them to add to their basic income. The government was torn between the need to increase the pace of reform to attract outside capital and the need to retain votes by guarantees of job security to workers in state-owned enterprises and improvements in welfare programs. Some progress occurred: the targets achieved in the eighth Five-Year Plan (1992–1997) exceeded those in the seventh (1985–1990).

In 1997, the central government renewed its commitment to economic growth and involvement in the world economic system. Industry and services expanded more rapidly again, employment increased, and domestic investment in the private sector rose. In 1998, however, the government's budget appeared to return India to the self-reliant, isolationist policies of the 1950s.

Further economic modernization depends on India's ability to build the infrastructure to supply the power, telecommunications, and road transportation that will attract and keep productive businesses. The Indian government is aware of this need but is also conscious that it must make sure that the poor benefit from programs that emphasize basic education and health care. The ninth Five-Year Plan (1997–2001) envisaged increasing economic growth, mainly in industry, and focused on domestic investment and help for the poor. By the late 1990s, the plans to expand power generation were behind schedule and power shortages affected more areas. Plans to privatize telecommunications were held up by bureaucratic and legal squabbles. Although a Malaysian company offered to build 12,000 km (7,500 mi.) of modern highways for $25 billion, there were debates about toll payments and a shortage of funding to carry the project through.

## Farming and the Green Revolution

Indian farming output improved after independence, and the country now produces enough to support the basic food needs of its population. The vagaries of the monsoon, however, still determine the harvest: good years enable India to export food; bad years require imports. There are important regional variations in farm production (Figure 5.19). In most parts of the country, productivity is somewhat up, but some regions experienced little growth in output.

The Green Revolution had its greatest impact in Punjab and Haryana. Increases in wheat yields from the 1950s to the 1990s matched those in Western countries—up by 350 percent in Haryana and 200 percent in Punjab. Irrigated rice yields increased by 300 percent in Punjab. The so-called **Green Revolution** began in 1966 when high-yielding varieties of wheat and rice, developed at research institutes in Mexico and the Philippines, were introduced as part of Indian agricultural policy. From the early 1950s, Indian production of food grains almost quadrupled, from 50 million tons to over 190 million tons in 1996, making India self-sufficient in food. It became the world's third-largest exporter of rice in good monsoon years (after the United States and Thailand) and fourth-largest producer of wheat.

Benefits from the Green Revolution were geographically uneven. In Punjab and Haryana, the larger land holdings and the commercial drive of the Sikh and Jat farmers increased productivity through the large capital inputs required to purchase expensive seeds, fertilizers, pesticides, and tractors. The local infrastructure of available irrigation water, 90 percent rural electrification, 70 percent all-weather roads, and good bus and truck transportation provided a beneficial environment for the farmers' investments. The presence of many medium-sized towns and regulated grain markets further assisted the functioning of the new economic system that developed. Per capita income increased rapidly—but so did family size, reducing the economic impact of the increased grain harvest.

Elsewhere in India, the outcomes of the Green Revolution were less dramatic. Results were negligible in areas where irrigation water was not available, farm units were small, and infrastructure and access to markets at home and abroad were poor. Where the yields did not increase enough to match the increased costs, debt often put farmers out of business. Large numbers of landless laborers lost their jobs. Poor monsoon rains in the mid-1980s reduced the benefits for some areas. Bad management of irrigation water or fertilizers, leading to waterlogging or salinization of the soils, reduced output in others.

Even in Punjab and Haryana, diminishing returns from grain caused farmers to consider other ways of increasing their income. In the 1990s, some farmers diversified into new crops that produced higher incomes, including sales abroad, from the same land. Such crops include durum wheat (used in pasta), vegetables, fruit, and flowers. The crops are linked to local processing facilities for making pasta, french fries, tomato puree, and wines, much of which are exported. Nearly three-fourths of the exports are sold to the arid, oil-rich countries around the Persian Gulf. The main problems

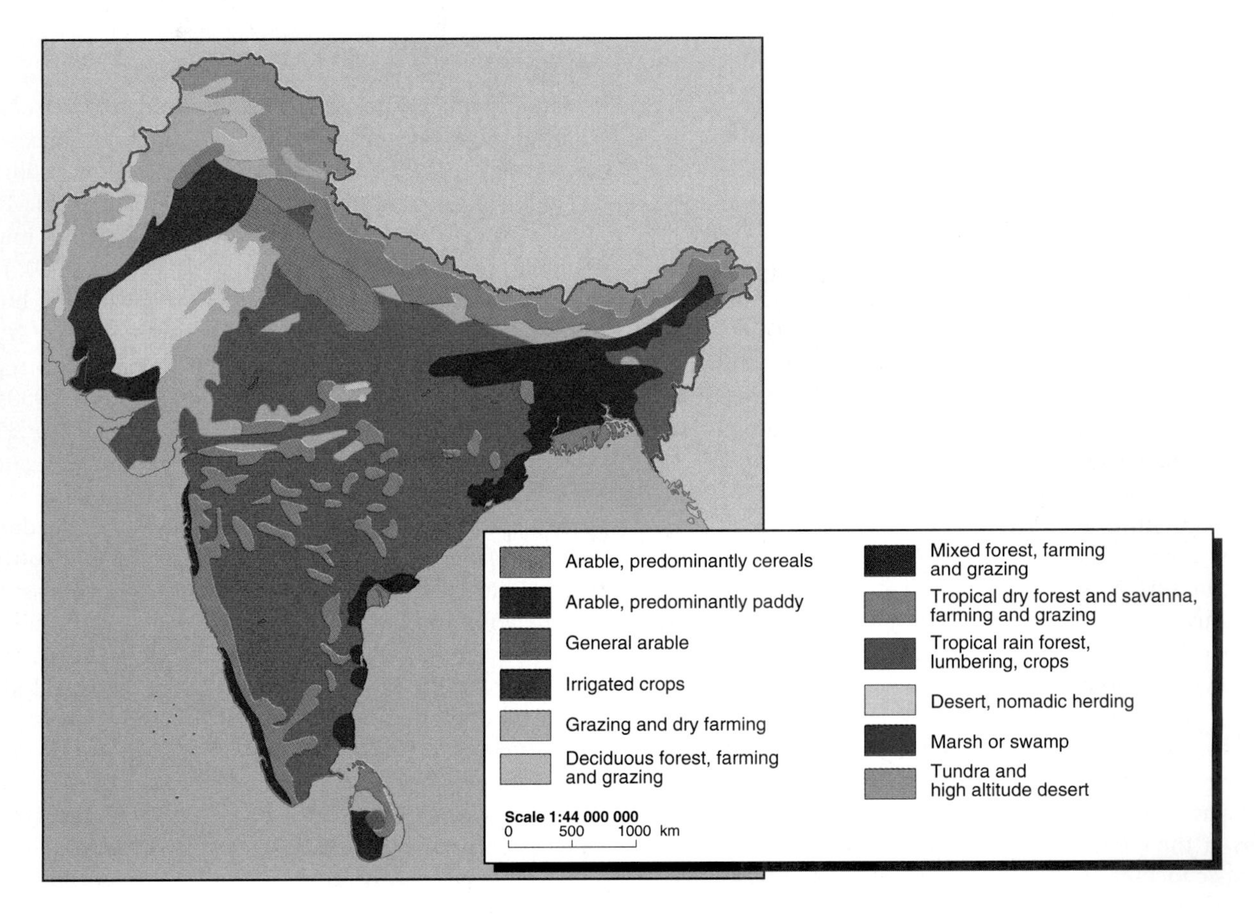


**FIGURE 5.19 Southern Asia: distribution of agricultural land uses.** Relate the main types and intensities of land use to climate and relief conditions.

Source: Data from *New Oxford School Atlas*, p. 94, Oxford University Press, UK, 1990.

facing producers of higher-value crops are the needs for investment capital, better transportation facilities, and storage. This area of India was at the center of a major expansion of India's agricultural exports, which quadrupled in value between 1990 to 1991 and 1994 to 1995.

## Other Farming Areas in India

Tea continues to be an important plantation crop in the Assam hills of northeastern India. Marketing is controlled by the multinational corporations that process, package, and distribute the product in core countries. This domination by external firms results in low prices for producers and low local wages.

The coastal deltas around the peninsula were drained for rice growing and cotton production. The most successful farming areas in southern India share conditions, such as electrification, good roads, and local markets, with Punjab and Haryana. Jute is a commercial fiber crop on the western margins of the Ganges-Brahmaputra Delta. Its growth there was developed after independence and the separation of the Calcutta factories from the Bangladesh jute-growing areas.

Farming in other parts of India is less intensive and commercial (Figure 5.19). The Ganges River plains east of Delhi retain traditional farming methods on, generally, small farms that provide subsistence. Large areas of the hilly peninsula have inadequate irrigation and grow coarse grains, such as millet and sorghum, that withstand dry conditions.

There are many landless laborers who find it increasingly difficult to make a living, especially since land reforms strengthened the middle peasantry and their ownership of land. People such as those in the northern part of Bihar state live in a poverty trap with few prospects of improvement. The lack of economic growth, poor infrastructure, rigid social structure, and little political drive inhibit the accumulation of capital and the development of wealth-making industries as an alternative to low-productivity farming.

One of the major features of India is the large numbers of livestock. The 200 million cattle are the largest number for any country in the world. Nearly all are used as working animals, since Hindus do not eat beef. A large proportion are water buffaloes, which work in the wetland environments of rice-growing areas (Figure 5.20).

## Mining

India has many mineral resources, particularly in the older rocks of the peninsula (see Figure 5.9). The most mineral-rich part of India is the Choto Nagpur Plateau of the poorer north-

**FIGURE 5.20 Southern Asia: rice paddy.** Fields in northern India. Intense summer rainfall and high river flows make it possible to grow rice in paddies. Water buffaloes and humans are the main forms of labor and energy used in cultivation, especially in the poorer areas of Bihar and Uttar Pradesh.

USDA

eastern peninsula. Plentiful iron ore and coal deposits were the basis of local steelmaking before independence. In the mid-1990s, India mined around 200 million tons of coal each year, becoming one of the world's leading producers. India also has the world's fifth-largest reserves of bauxite, the ore of aluminum, and has the advantage that the metal can be extracted from this ore at lower temperatures than from the ore in many other major deposits. The thick sediments of the Ganges lowlands cover deposits of natural gas and oil, and by the mid-1990s, India produced 75 percent of its oil needs, although output varied from year to year. India's rocks also contain other metal ores such as copper, gold, and manganese.

### Fishing and Forestry

India is one of the top 10 fishing countries in the world. Many coastal areas depend on income and food from the fishing industry. Most of the industry remains based on traditional methods of catching, with mainly unskilled labor and small boats.

India's forest resources were largely cut centuries ago, and many villages now find it difficult to produce sufficient firewood for their needs. There are still forests on the southern slopes of the Himalayas, from which softwood is cut, and on some of the hilly areas of the peninsula where hardwoods such as teak grow and are exploited. Overall, India imports 10 times as much timber as it exports.

### Small-Scale Manufacturing

India has a large and growing manufacturing sector. It has the most diverse range of products and organization among Southern Asian countries, ranging from small-scale craft industries to large-scale steelworks and modern high-tech facilities (Figure 5.21). There are distinctive regional emphases in manufactured products, attitudes toward industrialization, and entrepreneurial activities.

Much of India's manufacturing remains based on small-scale units. Small-scale industries in India employ 140 million people and produce 35 percent of all manufactured goods. Government encourages small producers in long-established craft industries, making brass metal goods, shoes, and carpets. Other small-scale industries have a more recent origin, being created following the closedown of large industries after union action. While foreign firms want to view India as a major emerging market for their own benefit, the Indian central government and some state governments wish to retain control over the type and rate of economic growth, and so favor small local businesses over investment by multinational corporations.

### Large-Scale Manufacturing

The Indian government developed a large-scale, heavy industrial sector of steel, chemical, and alumina works as part of its self-sufficiency and import-substitution policies after independence. The British had encouraged little local industry of this type during the colonial period, but the new Indian government saw it as basic. The main steelworks were built near the large iron and coal deposits in southern Bihar and Orissa. The first iron and steelworks in India, opened at Jamshedpur in 1914 to supply local needs, were expanded greatly after independence. The steel industry's output rose from 10 million tons in 1980 to over 20 million tons in 1997. Engineering industries are linked to the steel production.

Alumina works are distributed more widely across India close to bauxite deposits and hydroelectricity sources. Since the Indian economy was opened to foreign investment in 1991, the alumina industry received major investments from multinational aluminum corporations.

Car and vehicle assembly became major parts of the Indian manufacturing diversity. Up to the late 1980s, the state-owned factories producing Ambassador cars (based on a 1960s British model) and Tata trucks dominated production. In the 1990s, output and range of manufacturers increased with new auto and truck factories linked to major multinationals such as General Motors, Ford, Chrysler, Peugeot, Volvo, Hyundai, Mitsubishi, Daewoo, and, most recently, Volkswagen. The Indian Muarati firm has links to Suzuki and produces the most popular car.

(a)

**FIGURE 5.21 Southern Asia: Indian manufacturing contrasts.** High-tech industries exist alongside heavy industries, such as large-scale iron and steel and chemical works, and thousands of small-scale and craft industries. (a) The Union Carbide chemical plant and nearby housing in Bhopal.



### Textile Manufacturing

Textiles and garments comprise over one-fourth of India's exports, with sales doubling in the 1990s. Most of these exports go to the United States and European Union countries. Collaboration between Indian manufacturers and European and U.S. customers brought maker and market closer together. Although cotton remains the dominant textile fiber grown and manufactured in India, coarser textile fibers including coir (from coconuts) and jute are cheaper than cotton and are blended with cotton and synthetic fibers for use in denim cloth. Engineering textiles used in road construction and agricultural textiles that help to prevent soil erosion on slopes are other jute and coir products. The traditional process of manufacturing coir from coconuts polluted large areas of coastal lagoon during the retting (soaking) phase that lasted 10 months and left quantities of waste products that could not be disposed of. New processes involving concrete tanks and basic machinery for retting cut the time and improved the quality of both the fiber and the local environment.

### High-Tech Industries

Modern high-tech industry is increasing its presence in India. The Bangalore area in Karnataka state, southern India, where the state government has made foreign investment easy, attracted a range of multinational corporations such as 3M, AT&T, Digital, Ericsson, Hewlett-Packard, IBM, Motorola, and Texas Instruments. The area generated the first Indian hydroelectricity in 1902 and provided the location for many large public companies established as part of the Indian expansion in manufacturing after independence. The main new industries are electronics, computer engineering, software and services, telecommunications, aeronautics, and machine tools. Bangalore became India's first "electronic city" in the 1980s, and parallel developments are taking place at Mysore and Dharwad.

### Bombay: Center of Manufacturing

The main geographic concentration of manufacturing growth in India is around Bombay, in both its own state of Maharashtra and adjacent Gujurat. Bombay's population grew from 2 million in 1947 to over 15 million today. It handles one-fourth of India's trade in its port and 70 percent of its stock exchange transactions. It is the headquarters of nearly all of India's commercial banks. Maharashtra state accounts for over one-fourth of India's manufacturing output and is home to over half of India's top 100 companies. As well as having factories making a wide range of products from textiles to phar-

**(b)** Silk weaving for sari cloth, Kanchipuram.



maceuticals, Bombay is the center of India's very active film industry (often called "Bollywood"). Over half of the 500 to 600 titles produced annually are filmed in this area. Alongside the growth of new industries in the Bombay area, however, are many old textile mills with aging and inefficient machinery. They need capital to compete in world markets, but closing down and resiting in new areas requires government permission that comes very slowly.

Gujurat, to the north of Bombay, is the next most industrialized state. New investments in the 1990s almost equalled those in Maharashtra, with petroleum, petrochemicals, and pharmaceuticals making up over 60 percent of investment approvals from 1991 to 1994. The people of Gujarat have a reputation for business acumen, and many Indian business people in other countries come from this state. Ahmadabad was once the main textile manufacturing center in Asia, but textiles now compose only 10 percent of its manufacturing output, equal to engineering. The coastal zone between Ahmadabad and Bombay is marketed as the "Golden Corridor," and over 70 percent of its recent investment is in the southern sector close to Bombay.

### Northeastern India: The Calcutta Focus

The other main manufacturing region in India is in the northeast—in West Bengal around Calcutta and around Jamshedpur in southern Bihar, where the steel industry is important. Calcutta remains the headquarters of jute manufacturing, which is entering a revival after years of decline and union strangulation. At partition in 1947, the former province of Bengal was cut in two by the creation of East Pakistan (now Bangladesh). Calcutta's mills were separated from the main production areas of jute in the Ganges Delta. Although Indian farmers in West Bengal were encouraged to grow more jute, India still buys large quantities, especially of higher qualities, from Bangladesh. The Jamshedpur iron and steel area is an enclave within a poor rural and mining region. It imports many of its workers from other parts of India, and its products are shipped elsewhere for finishing. There is little investment in the infrastructure of the surrounding region, apart from small hotels and bars.

### Other Manufacturing Centers

Smaller centers of manufacturing exist throughout India, often based on the expansion of traditional craft industries. Three cities in the state of Uttar Pradesh illustrate the range and conditions of such industries.

- The Kanpur area specializes in tanned leather goods. Its small tanneries have good markets for their products but invest little in technology, training, health and safety, or environmental treatment of the toxic effluent. They pay low wages, employ children, and rely on unskilled, destitute labor for their "quality handmade" goods. For example, the Sultan Tanneries produce tanned leather goods mainly for export. Their workers take home one-tenth of the wages they might earn in the United States or Europe, working barefoot without protective clothing in an unfiltered atmosphere of toxic fumes and liquids to make safety boots for core countries.
- Small factories in Moradabad specialize in quality brass goods for export and sale in design shops in the core countries. Workers, including children on low wages, make them.
- Mirzapur is the center of an area that has a history of carpet-making going back to the A.D. 1500s, when the Mughal emperor brought in weavers from Persia. Many

modern factories were often set up by British owners in this century, but most closed following union militancy over the application of labor laws. Only basic processes such as dyeing wool and packing carpets are still carried on in factories. The 10,000 weavers work on 3,500 looms in homes and sheds in surrounding villages that rely totally on this income but face increasing competition in world markets from the cheaper products of Iran and China.

## Infrastructure Bottlenecks

Although Indian manufacturing industry grew remarkably after independence and even more rapidly since the 1991 opening of the country to international exchange, it is hampered by difficulties of infrastructure provision. Power shortages are likely to increase since the government's plans to increase electricity production are implemented slowly during the transition from a totally public service to increasing private investment. There is also a shift from the nuclear program to oil and gas-fired power plants.

Even greater problems face the expansion of telecommunications and the improvement of road transportation. These vital aspects of modern industrial expansion hold back a massive growth of Indian manufacturing.

## Service Sector Growth

The Indian service sector also grows rapidly. Government employment remains a major source of jobs for the well-educated middle class and higher castes. With reform, this source of jobs is likely to be reduced in favor of more jobs in the private sector. India is well supplied with some groups of professionals, from lawyers to accountants and economists.

The education and health sectors both need further development to deliver a good basic education to the whole population and to improve health care and public health levels. There is clearly scope for the expansion of employment in these areas, but it will need political will to shift investment from state-owned enterprises to the private ownership of industry so that government money can be channeled into physical infrastructure and human resources.

Events such as the 1994 outbreak of plague highlighted the problem of health in India, but it is more significant that 0.3 million children die of diarrhea each year and infant mortality remains high. Public health services do not work well. Delhi generates nearly 4,000 tons of trash each day but clears only 2,500 tons, leaving the rest on the streets. Of 1,800 tons of sewage, only two-thirds are collected.

Education is good in parts of India. Kerala, with its history of Christian missions and benevolent rulers, claims 90 percent literacy in contrast to the overall Indian figure of half that level. It is often the areas once ruled by Britain, and not the independent provinces, that have the worst record in education and where there is the greatest need for modern India to catch up. Education policy in India also needs to achieve a better balance between the basic levels of education that are still lacking and the oversupply of university graduates. The elementary, secondary, and higher education sectors divide the education budget into three almost equal parts, and higher education is subsidized for all who qualify. It is significant that a policy of primary education for all is recognized as one of the necessary conditions that led to faster economic growth in Eastern Asia (see Chapter 6).

India's tourist industry attracted over 2 million visitors in 1996. With its riches of historical buildings, mountains, and beaches, together with its exciting culture, India is a magnet to many from core countries. Visitor numbers could increase with improved infrastructure and public health.

## Reducing Poverty

When India became independent in 1947, its leaders blamed colonialism for the poverty of its people. The next 50 years saw major advances as famines were eliminated, literacy rates doubled, life expectancy rose from 33 to 59 years, infant mortality fell (165 per 1,000 live births in 1960 to 72 in 1996), income poverty was reduced, and overall income growth was substantial. Yet in the 1990s, the HPI is 36.7 percent, 60 million children under age 4 are undernourished, 50 percent of the population is illiterate, women remain at a disadvantage in society, and rural poverty is still about 40 percent of the population.

Economic growth over this period explains half of the reduction in income poverty. Other factors affecting income poverty are less clear. For example, the state of Kerala has one of the lowest incomes and slow economic growth but some of the best education and health-care statistics and an HPI of 15 percent, compared to one of 50 percent in another poor state, Bihar. Haryana has the fastest economic growth but treats

**RECAP 5B: India**

India has the world's second-largest population and dominates Southern Asia in land area, population, resources, and economic activity. India remains a largely rural country, although the major cities such as Bombay, Calcutta and Delhi are growing rapidly. Most of the people are Hindus, and still affected by the caste system. India's economy is characterized by low productivity in agriculture and manufacturing that produces goods for local markets, but the northwestern agricultural area and the west coast industrial area around Bombay are developing rapidly.

5B.1 What are the advantages and disadvantages of the Green Revolution in India?

5B.2 How do the poorer and richer areas of India contrast with each other? What factors are responsible for the geographic differences?

5B.3 Were India's nuclear tests in May 1998 really necessary? What would an Indian government minister say?

**Key Terms:**

Hindi

Green Revolution

**Key Places:**

India, Republic of
*Bombay*
*Calcutta*
*Delhi*
*Madras*
*Hyderabad*
*Bangalore*
*Ahmadabad*

women so badly that the ratio of females to males in the state is 86:100; in most states, there are more females than males.

Imbalances remain between rural areas, where increasing productivity is not matched by equitable distribution of the benefits arising from it, and urban areas, where there is the greatest emphasis on industry and state-owned enterprises. Smaller enterprises employing the poor pay very low wages. Education funding favors higher, rather than basic, education. Central bureaucracies reduce local involvement in decision making. The poor and disadvantaged see little change in their well-being; their children remain poorly fed and uneducated.

The central government plans for a major reduction in poverty by 2005—placing emphasis on human development priorities including health, education, safe water, and socially disadvantaged groups—together with pursuing favorable economic policies and increasing democratic participation. Optimism over the outcome of this plan is tempered by major cuts in public spending and the states choosing other goals as their priorities.

**FIGURE 5.22 Southern Asia: lower Indus River valley, Pakistan.** A space shuttle view of the Indus floodplain (dark area of irrigated fields) and the Makran Mountain Range in the foreground. Beyond the Indus is the Thar Desert and the (white) salt marsh of the Great Rann of Kutch in India. Pakistan's largest city and port, Karachi, lies at the western (near) end of the Indus floodplain.

NASA

## BANGLADESH AND PAKISTAN

Bangladesh and Pakistan share much of the same natural and cultural heritage as the Republic of India. Their central areas occupy lowland areas of Southern Asia along the Indus River valley and at the mouth of the Ganges and Brahmaputra Rivers (see Figures 5.1 and 5.12). They share a dominance of Islamic religious belief and were part of the British Indian Empire until 1947.

From independence in 1947 until 1971, West and East Pakistan were part of a single country, ending with a successful revolt in East Pakistan that led to the formation of the separate country of Bangladesh. At independence in 1947, the partitioning of the two former provinces—Punjab in the west and Bengal in the east—left both parts of the new Pakistan at a disadvantage in terms of natural resources and human relationships. India's territory included nearly all the mineral, agricultural land, and irrigation water resources, as well as strong military strategic positions and the largest cities. Further tensions developed during the initial transfers of people between India and Pakistan and the way in which India annexed undecided princely states on the borders. The wars with India that resulted and the civil war prior to Bangladesh's independence from the rest of Pakistan sapped the economies of both countries and required extensive reconstruction. Both countries continue to be subject to internal tensions between rich and poor and between those who migrated in at the time of independence and those who already lived there.

### Countries

The differences between **Bangladesh** and **Pakistan** outweigh the influence of their common Islamic faith. There are differences of physical environment—Pakistan is a country of arid lowlands and high mountains (Figure 5.22), whereas Bangladesh is low-lying apart from the small hilly eastern region inland of Chittagong and is well watered by rain and rivers. In Pakistan, the management of scarce water resources is a major problem, while in Bangladesh, flooding is more significant, with major flood disasters in 1974, 1988, 1991, and 1998.

The main differences, however, are in the people and in the political evolution of the two countries since their 1971 separation. Although they share a common Islamic faith, other ethnic factors proved stronger, while physical separation provided the opportunity for a breakup. Bangladesh is populated largely by Bengalis and Pakistan by a mixture of Punjabis, Sindhis, mountain Pashtans, and Muhajirs (exiles from Hindu India). When the people of East Pakistan became convinced that the Punjabi military controlling Pakistan was taking East Pakistan's resources to benefit West Pakistan, they revolted. With India's military help, they established Bangladesh as a separate country.

Bangladesh remains one of the world's poorest countries, and although Pakistan built on its established base of agricultural and manufacturing products, its people are little better off. Bangladesh had to rebuild after the civil war. It had to integrate Muslims expelled from Myanmar and Bengalis returning from India. Its attempts to nationalize the main economic functions were not a great success. Bangladesh needed huge quantities of aid from the world's core countries, and its internal politics involved military takeovers, coups, and assassinations. It was not until the 1990s that major changes to this pattern appeared possible under a settled democracy.

After Pakistan was defeated in its attempts to retain the link with Bangladesh in the 1971 civil war, its rulers took steps to work out an Islamic religious base in government and finances. Industry, financial, and educational institutions all became part of the public system, and some land reforms that had been begun were taken forward but not completed—leaving many long-term feudal landowners with great influence. The constitution was amended to include a basis of Islamic law. The 1979 occupation of Afghanistan by the Soviet Union added further pressures from the 3 million Afghan refugees who moved into areas of Pakistan close to the Afghanistan border. Subsequent civil war in Afghanistan led Pakistan to support the successful Taleban faction in the late 1990s. The association with the hard-line Islamic Taleban group, however, fed back into Pakistani politics with demands for increased Islamic influence backed by terrorist activity. Internally, the continued dominance of the government by an elite of feudal Punjabi landowners slowed economic modernization. As in Bangladesh, the 1990s saw renewed attempts to improve the country's economy and to integrate with the world economic system.

## People

### Population Distribution and Dynamics

The arid nature of Pakistan forces the main concentration of people into the Indus River valley (see Figure 5.13), leaving large areas in the mountains with few people. The small lowland country of Bangladesh has high population densities throughout.

Although Bangladesh's area is less than one-fifth of Pakistan's, its population was slightly higher until 1980 (88 to 83 million). By 1998, however, Pakistan's population exceeded that of Bangladesh (142 to 123 million), and estimates for 2025 suggest a further increase in the difference, with Pakistan having 260 million and Bangladesh just 166 million people. The greater rate of population increase in Pakistan is the result of a continuing high total fertility rate of around 6, compared to one of 3.3 in Bangladesh. Both have life expectancies in the upper fifties (see Figures 5.15 and 5.16).

Both countries confront increasing difficulties in feeding their populations. The massive increase in numbers of young people places pressures on their resources and ability to feed and find employment for their people. Bangladesh has greater problems of feeding its growing population, an issue that stimulated its government to institute effective family planning programs in the 1970s. This led to a significant reduction in fertility. By 1997, half the married women used contraception (8 percent in 1975). The loss of farming land and lack of farm jobs for adults caused moves into cities, where jobs require costly education, and reduced the number of births. Progress was made despite a continuing subordinate role for women, although new policies encourage their social and economic development. The shortage of jobs in Bangladesh resulted in many laborers seeking employment in the Persian Gulf countries and Malaysia in the 1980s and 1990s.

Pakistan's main lack is an educated work force. The combination of influences from Islamic religion, feudal landlords, and poor government keep Pakistan's school population low. Its enrollment rates in primary and secondary education are only 30 percent, compared to 70 percent in India and over 80 percent in Indonesia and China. In 1994, aid donors' demands led to new national programs for education, health, and population policies. Government funding shifted from elite university toward universal primary education.

### Urban and Rural Population

Both Bangladesh and Pakistan remain largely rural countries with one-third of the population living in the towns of Pakistan in 1998 and only 11 percent in Bangladesh. Nevertheless, the major cities struggle to cope with the inmigrations of rural people and the shantytowns and social tensions that the influx creates. **Karachi**, with over 10 million people and growing rapidly (see Figure 5.12), is the largest city in Pakistan and its major port and commercial center—although separated from the political power centers farther north. After over two years of military occupation, the army left the city at the end of 1994 and violence erupted as **Muhajirs**—the people who migrated into Pakistan from India after independence and who have been at a disadvantage since—demanded more jobs. Sunni Muslims campaigned to have Shiite groups declared non-Muslim. The violence was made worse by the glaring contrasts between rich and poor, the large numbers of unemployed teenagers, and the easy availability of guns following years of warfare in Afghanistan. **Hyderabad** is an industrial center near Karachi. Other major Pakistani cities are inland, where **Lahore** is the major center of Muslim culture and **Faisalabad** is the cotton industry center. The new capital of **Islamabad** is close to the Kashmir border, while **Peshawar** is the commercial center of the far north at the eastern end of the Khyber Pass into Afghanistan. Bangladesh, with a smaller urban population, has fewer large cities. **Dhaka**, the capital (see the front cover), has 9 million people. **Chittagong** is the main port and **Khulna** is a growing industrial center (see Figure 5.12).

### Ethnic Character

Bangladesh is the country of the Bengalis. **Bengali** is the language of 98 percent of the people, who have a racial makeup based on Indo-Aryan stock mixed with local tribal groups. The **Biharis** form a smaller group of non-Bengali Muslims who migrated to East Pakistan after independence in 1947 and speak **Urdu**.

As the site of the main historic entry routes of different groups of people to the Indian subcontinent, Pakistan has greater racial and ethnic diversity than Bangladesh. Although Urdu is the official language of the country, only 7 percent

**FIGURE 5.23 Southern Asia: jute harvest near Tangail, Bangladesh.** The fiber is stripped from the plant stalks using the plentiful water of the Ganges River delta area. What is jute used for?



speak it, while 64 percent speak **Punjabi** and 12 percent **Sindhi.** There are many smaller groups living in the border hills; the 1947 Muhajir immigrants still speak Hindi. Rivalries between groups and the elitism of Urdu-speaking feudal landowners and military officers continue to dominate social and political life in Pakistan.

## Economic Development

Bangladesh and Pakistan are among the world's poorest countries (see Figure 5.2) with low consumer goods ownership (see Figure 5.18). Pakistan's GDP total is around twice that of Bangladesh, but both have adverse trade balances and depend on aid donations. Pakistan's output reflects its greater range of products from both agriculture and manufacturing. Bangladesh has a narrower economic base, although it is attempting to widen it. While British governments invested heavily in building irrigation canals and railroads so that the Indus Valley could be developed as a source of cotton for British industry, they—and the Pakistan government from 1947 to 1971—largely ignored the Bangladesh economy. It is only since the 1970s that the Bangladesh government has tried to develop its highly populated area in the face of major environmental hazards. Little progress has been made: it takes huge efforts to maintain the same income levels while the population increases and factors contributing to productivity growth have such an uncertain basis.

### Prominence of Agriculture

Both are essentially agricultural countries. This makes the national economies sensitive to the weather and international market prices. Large areas are still dominated by traditional and feudal farming systems.

In Bangladesh, almost 75 percent of the labor force was engaged in farming in the 1990s. Their products made up 31 percent of GDP (down from 50% in 1980). Bangladesh grows over half of the jute that enters world trade (Figure 5.23), as well as rice, tea, and sugarcane. As in India, livestock are kept mainly as beasts of burden. Expansion of the farming area in Bangladesh occurred on new delta islands formed by deposition at the mouth of the Ganges River. Little of this land is above sea level, necessitating costly dikes and being vulnerable to ocean surges during tropical cyclones. Most deaths in recent storms occurred on these islands.

In Pakistan, agriculture occupies just under 60 percent of the labor force and its products accounted for 26 percent of GDP in the 1990s. Cotton remains the chief commercial crop, supported by government credits to farmers so that they will continue to produce what is now the raw material of the main manufacturing sector in the country. Continued poor crops because of heavy rains in the three years 1992 to 1994 threatened to undermine the country's major industry and caused farmers to question whether they should grow different crops. The greater use of virus-resistant strains, fertilizer, and tractors led to a record harvest in 1996, but further virus attacks in Punjab (where 80% of the cotton is grown) reduced the crop in 1997. The chief alternatives to cotton are wheat, rice, or sugarcane, although sugar suffers from low world prices, and some farmers are beginning to look to more profitable crops, such as vegetables and fruit.

Much of Pakistan's best farmland is owned by a small group of wealthy people who treat it more as a source of political power than as a basis for greater productivity. By supporting politicians, they have the patronage of many jobs in the government bureaucracy as well as being able to veto any measures that might change their dominant place in Pakistani society. Although Pakistan governments began to implement land reform by appropriating land and breaking it into smaller holdings, the feudal landlords continue to own more land than has been reallocated to peasant farmers.

### Mining and Forestry

Neither Bangladesh nor Pakistan developed other primary products to a great extent. Both have small reserves of oil and natural gas, although neither has exploited them fully because they resisted asking multinational corporations to carry out discovery and production until the 1990s. Bangladesh has no large stocks of other minerals. Pakistan's major chemical industry is built on deposits of gypsum, rock salt, and soda ash. Fishing is important in the coastal areas of both countries but poorly organized.

### Increasing Manufacturing and Service Industries

Manufacturing is of relatively low significance in Bangladesh, although there are signs that it will grow. The slow expansion of agricultural production at 2 to 3 percent per year cannot support the more rapidly growing population, and manufacturing growth needs to rise above the current 7 percent increase per year. The long-established jute industry requires restructuring, and the Bangladesh government is trying to privatize parts of it against the opposition of trade unions. Bangladesh's garment industry experienced major growth in the 1990s, jumping to become a major exporter. Bangladeshi apparel manufacturers work closely with their retailing customers in Europe, the United States, and Eastern Asia, although some countries are objecting to the use of child labor. Bangladesh also developed enterprise zones at Chittagong and Dhaka (Dacca), with others promised. These were taken up mainly by garment industries, but they hope to include high-tech industry.

Pakistan has a greater experience of manufacturing, which grew from 15 percent of GDP in 1965 to 25 percent in 1995. Stock market listings of private companies doubled from just under 4,000 in 1987 to almost 8,000 in 1996. In exports, primary products decreased from 48 to 11 percent of the total between 1975 and 1995, while manufactured goods rose from 52 to 89 percent. Textile manufacture is dominant, especially of cotton goods. Other industries include food processing, chemicals, and car assembly. Recent attempts to introduce cheaper tractors made in Poland and Belarus have not had great success because of the vested interests of tractor manufacturers within Pakistan.

Both countries lack many of the basic provisions that are needed to encourage more manufacturing industry to locate there. Power supplies are subject to shortages, although Pakistan hopes to double its power provision in the 1990s with the aid of private investment. It completed major hydroelectricity projects in the Himalayas, while a new thermal power station is being built with foreign capital and expertise at Hub near Karachi.

Service industries employ a growing proportion of the labor force in Pakistan and are expanding in Bangladesh: both countries produce around half of their GDP from such industries. Health, education, and financial services are growth areas. Neither country has a sizable tourist industry.

### Economic Restructuring

While Bangladesh adopted structural adjustment measures with enthusiasm, Pakistan remained reluctant to do so until the mid-1990s. In the early 1990s, Bangladesh reduced trade restrictions and encouraged external financing of garment factories, doubling its export of goods. During the 1996 elections, political unrest included strikes affecting half of the 1 million garment workers, but the subsequent economic reforms and new government soon returned people to work and the economy to growth. Privatization continues, but severe infrastructure (roads, power stations, communications) bottlenecks slow trade; the small level of domestic savings in such a poor country provide a meager basis for the major investments needed. Losses by state-owned enterprises are a heavy burden for the limited national budget to bear.

In Pakistan, the internal budget deficit and continuing unfavorable trade deficit at a time of unsettled political environment in the mid-1990s caused external bodies to refuse grant aid and discouraged external investment. The new government elected in 1997 gave signs that it would tackle problems that made its producers uncompetitive internationally and kept so many of its people poor and socially deprived. The growth in output and productivity in manufacturing increased jobs, even in rural areas, at a rate that equaled the rate of new entrants to the labor force, but such gains did not benefit the poorest groups of people. The detonation of nuclear weapon devices in 1998 was a response to the Indian demonstration of military power, but, although Pakistan was supported by Chinese technology in this field, it was a costly event, financially and politically.

### Poverty and Social Services

The United Nations Human Poverty Index is just under 50 percent for both countries. In Bangladesh, the problems faced are mainly those of overpopulation (very high densities with a low resource and infrastructure base). In Pakistan, they also include inequitable distribution of wealth, since the average per capita income is twice that of Bangladesh.

Bangladesh began family planning in the 1970s, reducing population growth, but still has poor education statistics: 40 percent of children are not at school; 24 percent of women and 48 percent of men are literate. Fewer than half the population has access to sanitation or legitimate electricity supplies and only 14 percent to garbage disposal.

Pakistan lagged behind other low-income countries in social conditions in the 1990s, with only 50 percent of primary-age children at school (31% of girls), high infant mortality (85 per 1,000 live births, compared to the average of 60 in all low-income countries), continuing high total fertility, and low literacy (23% for women, 49% for men). Following the failure of earlier efforts, the 1994–1998 Five-Year

Plan included a social action program that claimed major advances in the first three years. Seventeen thousand new primary schools, 750 buildings for shelterless schools, 200 new basic health units, and 50 new rural health centers were built. Ten thousand female health workers were recruited and trained. There was a small increase from 14 percent to 18 percent in contraceptive usage and some improvements in drinking water and sanitation. Yet the program suffered from slow bureaucratic responses and a lack of community participation, and only began to meet the needs. The numbers cited emphasize the immensity of the problem faced.

## MOUNTAIN AND ISLAND RIM

While Bangladesh and Pakistan form an inner periphery to India, the countries of the northern mountains, including Afghanistan, Nepal, and Bhutan, form a more isolated and restricted periphery (see Figure 5.12). The mountain countries are influenced not only by what happens in the rest of Southern Asia but also by interactions with Russia, Central Asia, and China to the north. The island countries of Sri Lanka and the Maldives have easier connections with the world economic system but experience other problems that restrict their economic development.

All these countries were involved with the British Indian Empire as it expanded into their territories in the 1800s. Afghanistan was the scene of three major wars in the mid-1800s, in which the British sought to prevent potential Russian intrusions. Britain recognized Afghanistan's independence in 1920. Nepal also suffered attacks by British forces in the 1800s but was allowed to remain independent while it had pro-British rulers who hired out Gurkha soldiers to the British. Bhutan was annexed by Britain in 1826 and given autonomy in 1907. Ceylon (now Sri Lanka) was taken by the British in 1795 and was a crown colony until its 1948 independence. The British years left marks on the economies, governments, infrastructures, and legal processes of these countries.

### Countries

The mountain countries are **Afghanistan**, **Bhutan**, and **Nepal**; the islands are the **Maldives** and **Sri Lanka.** Bhutan and the Maldives have tiny populations of fewer than 2 million people between them. Afghanistan, Nepal, and Sri Lanka each have 20 million to 25 million people. The mountain countries are among the world's poorest, while the Maldives and Sri Lanka have better living conditions. The varied histories of these countries reflect their distance from, or closeness to, world trade routes, internal feuding, and reluctance to get involved in the world economic system.

Afghanistan remains the most isolated and landlocked country. Since its independence in 1920, attempts to modernize the country's economy were held back by tensions with Pakistan over the northwest frontier lands that Pakistan swiftly annexed after its own independence in 1947. Afghanistan then linked more closely to the Soviet Union until 1979, when the latter invaded and occupied Afghanistan for 10 years. During the period of Soviet occupation, local warlords fought the Soviet armies and established their own military units. After the Soviet Union withdrew, the local warlords fought each other for control of the country. Most government systems broke down and poorly equipped schools opened and closed depending on the fighting; public health and health care are in such poor condition that epidemics are likely. In the late 1990s, the Taleban group conquered virtually all Afghanistan, although opposed fiercely by northern tribes who resisted the imposition of hard-line Islamic tenets. The rest of the country returned to some normality under the Taleban rule, but neighbors in Iran, Central Asia, and Pakistan feared the impacts of Taleban policies.

Nepal emerged from isolation as democracy was introduced in the 1980s, while alternative political parties were allowed in 1990. Its lowlying Terai Plain, where over half the population lives, gives way northward to hilly lands and the Himalayan Mountains. The upland areas are susceptible to earthquakes, landslides, and floods. Bhutan remains a tiny buffer state between India and China and has few cross-border contacts.

After independence, Sri Lanka's government by the dominant Sinhalese people instituted reforms that were disliked by the Tamil minority, leading to a still unresolved civil war (see "Living in Sri Lanka," pages 216–217). Following their independence, the Maldives went from a Muslim sultanate to a republic but called in Indian troops in 1988 to avert a coup by a group using Tamil mercenaries.

### People

#### Population Distribution and Dynamics

The populations of the mountain and island countries of Southern Asia are distributed in relation to the land terrain (see Figure 5.13). The mountainous states have valley-based settlements, while the islands have coastal concentrations.

There is a range of rates of population increase. Sri Lanka's population increases at only 1.3 percent per year. At the other end of the scale, Bhutan, the Maldives, and Afghanistan populations grow most rapidly at over 3 percent. In most countries the total fertility rates are declining but remain over 5 in Afghanistan (see Figure 5.16) and the Maldives. All the countries remain at a stage when the population increases rapidly and there are many young people (see Figure 5.16). For example, in Nepal, there is a strong family planning program, widely disseminated, but only 29 percent of women and 2 percent of men use birth control methods; fertility rates decline slowly. This is linked to female adult literacy rates, which were 3.7 percent in 1971 but rose to only 12.8 percent in the mid-1990s.

## SRI LANKA

Colombo, the largest city in Sri Lanka, is crowded and busy. People live in single-story houses or in apartments in buildings of four or five floors, but many live in self-built shanty houses in squatter settlements. The population grew too fast for the authorities to cope, and people built shanty houses on unused land. The shanty houses are without water or connection to sewage systems, but there are street taps. Transport in Colombo nearly copes with the great numbers of people moving in and out of the city center. Train services are run by the government, and there are privately owned bus and minibus companies.

Most people buy their food at markets, where the meat, fish, vegetables, and fruit are fresh each day. The Sri Lankan diet is similar to the Indian: rice is eaten with vegetables and meat or fish, more or less heavily spiced. The special feature of Sri Lankan food is the use of coconut milk and distinctive spices. The food sold in Colombo comes from all over Sri Lanka, brought by overnight truck services. Shops selling other goods often occur in clusters of similar businesses—a street of textile and clothing shops, followed by a street of jewelry or hardware shops. There are only a few department stores, but they are getting more popular.

The Sinhalese came from Bengal in northern India, settling in the southern and hilly part of Sri Lanka and becoming Buddhists by the 200s B.C. Ancestors of modern Tamils came from southern India after the A.D. 200s, settling the northern part of Sri Lanka, where the Hindu religion remains dominant. Muslim traders settled in the island and also speak the Tamil language. After A.D. 1795, the British colonized the island and missionaries set up Christian schools, especially in the north. Education became the major goal of the Tamil families in northern Sri Lanka. The dry climate there made commercial farming difficult in comparison to the south, where rain is plentiful. The greater number of schools in the north led to higher literacy among the Tamils and to their employment in the colonial bureaucracy.

Tensions between Sinhalese and Tamils developed after independence in February 1948. Families in Colombo were affected by communal riots in 1958, after which some Tamils returned north to Jaffna for a period. In 1960, the government elected by the majority Sinhalese increased the tensions by making Sinhalese the only official language and replacing Tamil civil servants with Sinhalese. Following objections by the Tamil community, their language was given equal official status in 1966. After 1970, university admission was made

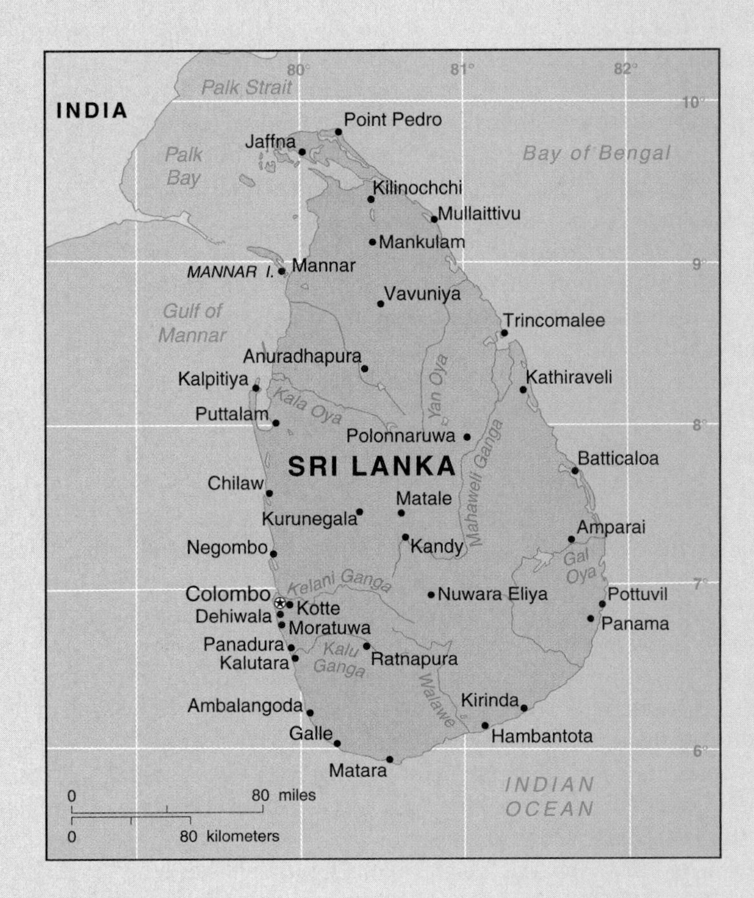


Map of Sri Lanka.

easier for Sinhalese on the grounds that they did not have access to such good schools as the Tamils.

Many Tamils who had left returned to Colombo. Until the 1980s, Tamils from the north of Sri Lanka ran many businesses in Colombo. Their homes were often in districts among the majority Sinhalese. Tamil families lived happily alongside friendly neighbors until political events stirred ethnic hatreds. When riots occurred in 1977 and 1983, large numbers of Tamil families were virtually forced to move to India.

The Tamils in the north continued to resist Sinhalese domination, setting up an independence movement and fighting for their future. This led to the north becoming virtually cut off from contact with the south. At present, Sri Lanka is mostly peaceful outside of the separated and cutoff north and northeast. Tamils living outside Sri Lanka can visit relations in the south of the country.

Sri Lanka remains poor in general and continues to go through difficult times. In 1970, for example, the international tea companies that had run the plantations in the hilly south were nationalized and left. The subsequent restructuring and transfer of ownership to smaller plantation units faced difficulties because of such events as the allocation of land in return for political favors, and resulted in lower production. The cut in export sales reduced the amount of hard currency coming into Sri Lanka. Since the mid-1980s, however, there has been a resurgence of business activities and international trade links, together with the privatization of many former state-owned monopolies.

The cost of living is high in relation to earnings, especially in Colombo. A house in a prime locality in Colombo costs more than one in an equivalent locality in an American city and has to be paid for in full, without loans—despite lower incomes. Many people travel to work from the suburbs where housing is somewhat cheaper. Despite the ethnic tensions, economic changes, and high costs of living in Colombo, living in Sri Lanka offers many benefits compared to life in neighboring countries in Southern Asia. Easy access to health care and education are major advantages of living in Sri Lanka. Government hospitals offer comprehensive service, and there are many private hospitals. For middle-class people, the pleasant environments of the coasts and the historic remains and cooler climate of the central hills that attract so many foreign tourists are close at hand.

The Buddhist Temple of the Tooth, Kandy, Sri Lanka, with its surrounding moat.

Only the Maldives and Sri Lanka have urbanization rates as high as 20 percent. **Kabul** (Afghanistan), **Colombo** (Sri Lanka), and **Kathmandu**, capital of Nepal, are the largest cities and contain extensive shantytowns.

### Varied Ethnic Histories

The mountain and island countries of Southern Asia have a variety of cultural backgrounds, ranging from Muslim dominance in the west (Afghanistan and Maldives), through Hindu (Nepal), to Buddhist in the east (Bhutan, Sri Lanka). Each country contains a variety of ethnic groups that make organized political life difficult because of their exclusive claims to dominance.

Afghanistan is 99 percent Muslim—84 percent Sunnite and 15 percent Shiite. Its diversity of ethnic groups results in only some 50 percent speaking the main language, an Afghan form of Persian. Pashtu is the language of the Pathan tribes in the north, some of whom also live in northwest Pakistan. There are groups of Central Asian peoples, such as the Tajiks and Uzbeks. The Maldives is another Muslim country, with that faith established by Arab traders from the A.D. 1100s among a diversity of ethnic groups from southern India, Sri Lanka, Arabia, and Africa associated with its wide trading links.

Nepal is the only Hindu country in the subregion. Over 90 percent of the people are of that persuasion, but Nepal also includes small proportions of Buddhists who have had a major impact on the local character of Hinduism. Like Afghanistan, however, Nepal has an ethnic diversity arising from peoples moving into the mountainous country from all directions—Tibetans, Indians, and many local hill tribes.

Bhutan is a Lamaistic Buddhist country like its neighbor, Tibet, but 35 percent of its population are Hindus from Nepal and India. The main group of people, the Bhote, drove out Indian peoples in the A.D. 800s and built impressive fortified monasteries in the 1500s.

Sri Lanka has a Buddhist majority, which is linked to the dominant **Sinhalese**-speaking group. Nearly 70 percent of Sri Lankans are Buddhists, 15 percent Hindus (mainly Tamil), and 8 percent each Muslims and Christians. The Muslims and Tamils are the main business groups, and since independence, both have been subject to repressions by the dominant Sinhalese. As part of the resulting civil conflict, the "Tamil Tigers" carry out terrorist activities throughout the country. The internal disruptions caused half a million Sri Lankans to move to southern India or to settle farther away in Europe or the United States.

### Quality of Life

The two island countries provide the best living conditions in the whole region, with the highest HDI and GDI ranks based on levels of education and health (infant mortality, life expectancy), and incomes. The Maldives and Sri Lanka have almost total adult literacy, contrasting with under 50 percent in the rest of the region. The civil war in Sri Lanka, however, holds back further economic development by deterring foreign investors and tourists. Public funds go into defense and social services expenditures and not into improving the infrastructure.

## Economic Development

The physical and political isolation of the mountain countries, together with internal conflicts, resulted in restricted economic development in the subregion. Agriculture is dominant in all the countries and most people still depend on subsistence farming and occasional paid jobs. The highland environments provide massive hydroelectric potential, but construction of the facilities is expensive, and the cost and environmental impacts slow projects. Long years of poor educational facilities are responsible for a lack of human skills and widespread unemployment.

### Agriculture Base

Agricultural output depends on the availability of level land, soil quality, water availability, and transportation infrastructure. In Nepal, 82 percent of workers are in agriculture, mainly on the Terai Plain along the Indian border. Sri Lanka (tea, rubber, and coconuts) and the Maldives (coconuts) have commercial plantations that sell products to world markets. These provide 35 percent of Sri Lanka's exports, down from over 90 percent in 1950 as a result of diversification in the economy since independence. Sri Lanka's adoption of new rice varieties in the late 1960s converted the major food deficit—which arose because of the colonial focus on export crops—into a virtual self-sufficiency in rice today. Noncommercial agriculture is based around the nomadic keeping of sheep in Afghanistan and the growing of rice and other cereals elsewhere. None of these countries is self-sufficient in food.

Despite international pressures, poppies remain an important crop in Afghanistan. That country's isolation, disrupted government, and lack of alternative commercial products makes it difficult to stop the production of opium and heroin and their entry to illegal drug traffic routeways. Although Afghanistan has been threatened with the withdrawal of aid because of its increasing production of these drugs, payments continue for fear of angering the warlords and as part of projects to persuade peasants to stop growing poppies.

### Mines, Forests, and Fisheries

Apart from natural gas in northern Afghanistan (which was formerly exported by pipeline to Russia), these countries are poor in mineral resources. Several produce gemstones. Sri Lanka has a major source of graphite.

The forests in Nepal, where there has been massive deforestation, and Sri Lanka are cut and used locally, mostly for firewood. Fishing is important to the economy of the Maldives, where tuna provide 60 percent of exports.

**FIGURE 5.24 Southern Asia: tourism in Nepal.** The peak of Annapurna in the Himalayan range rises behind the hotel in Kaski village. How would a trekking holiday in Nepal differ from one in the Rockies or Appalachians?



## Small-Scale Manufacturing

Only Sri Lanka produces more than 10 percent of its GDP by manufacturing. Much of the manufacturing that exists in the subregion is in small-scale factories or in a type of **cottage industry** based on the dispersal of manufacturing processes in homes. Local agricultural processing, together with the making of carpets and footwear, are most common. Nepal (Mahakali project) and Bhutan (Chukha) have encouraged external investment in dams for hydroelectricity and other uses. This resulted in additional local industry—making garments in Nepal and steel alloys in Bhutan. Sri Lanka has a more diversified manufacturing base than the other countries as the result of government policy since independence, including the production of steel, textiles, tires, and electrical equipment. By the 1990s, however, expansion placed pressure on existing transportation and telecommunications facilities, while power shortages reduced output and exports of garments and other textiles.

## Tourism

The mountain and coastal environments of these countries suggest tourist potential, but only the Maldives, Nepal, and Sri Lanka have developed tourist industries since the 1970s (Figure 5.24). Each has 0.3 million to 0.4 million visitors per year. Other countries are too isolated or unsafe for tourists. Bhutan keeps down the numbers of tourists entering to reduce the impact of foreign cultures on its people. In the Maldives, tourism spread to new islands, while older facilities were refurbished. By the late 1990s, the limits of expansion were being reached, signaled by coastal pollution, groundwater contamination, and the need to bring in labor to cope with the increasing demand. The tourist industry employed over 25 percent in the mid-1990s, but only 4 percent of the total labor force in 1985.

## Links to the World Economy

All the countries face major financial problems, importing more than they export and being dependent on overseas aid. During the Cold War, several obtained aid from both Soviet and Western sources, but aid amounts declined in the 1990s. The financial problems are typical of the whole of Southern Asia, including a lack of capital for investment in infrastructure such as transportation routes, telecommunications, and power generation.

Although these countries were very isolated economically up to the 1980s, they are now trying to enter the world economic system. For many, however, the combined difficulties of physical isolation, lack of infrastructure and skills, growing populations, and restricted diversification leaves them with few products in great demand. All the countries took slowly to structural adjustment measures, such as trade liberalization, privatization, and a reduction of government expenditures.

**RECAP 5C: Pakistan, Bangladesh, and Mountain and Island Rim**

Pakistan and Bangladesh were a single country from partition in 1947 until 1971. They remain largely agricultural, although Pakistan has a more diversified economy, with a long-established cotton textile industry.

Afghanistan, Nepal, and Bhutan are mountainous countries on the northern margins of Southern Asia. They have restricted involvement in the world economic system. Wars and political tensions have destroyed Afghanistan.

Sri Lanka and the Maldives produce plantation crops and fish for sale on world markets and have growing tourist industries. Sri Lanka's attempts to diversify its economy have been hampered by civil war.

5C.1 To what extent were Pakistan and Bangladesh incompatible as parts of a single country from 1947 to 1971?

5C.2 What different points of view might be held by the sides in the Sri Lankan civil war?

5C.3 Why do other countries object to funding development in Afghanistan?

**Key Terms:**

| | | |
|---|---|---|
| Muhajirs | Urdu | Sinhalese |
| Bengali | Punjabi | cottage industry |
| Biharis | Sindhi | |

**Key Places:**

| | | |
|---|---|---|
| Bangladesh | *Lahore* | Nepal |
| *Dhaka* | *Faisalabad* | *Kathmandu* |
| *Chittagong* | *Islamabad* | Maldives |
| *Khulna* | *Peshawar* | Sri Lanka |
| Pakistan | Afghanistan | *Colombo* |
| *Karachi* | *Kabul* | |
| *Hyderabad* | Bhutan | |

## LANDSCAPES IN SOUTHERN ASIA

### Urban Landscapes

The urban landscapes of Southern Asia reflect the waves of cultural influences that washed across the region. Many ordinary buildings were constructed of materials such as wood that do not have a long life, but most towns have more permanent religious, royal, or military buildings that have lasted for centuries (Figure 5.25). Buddhist buildings range from stupa temples in central India to the fortified monasteries of Bhutan. Hindu temples are often very ornate. Muslim mosques and tomb gardens (such as the Taj Mahal) also provide distinctive landscape elements. The many military rulers left behind fortified buildings.

High densities of low-rise housing with poor transportation access along winding alleys encroached by shops and artisans characterize older precolonial sections of towns in Southern Asia. Few buildings are more than two stories, except in Calcutta and Bombay, and there is no clear distinction of residential, commercial, and industrial functions. People live and work together in localities often determined by ethnic origins or type of trade. For long, the wealthy lived in the centers of such towns and the poorer people toward the outskirts.

The British rule from the 1700s to the 1900s added new aspects to towns in Southern Asia. The British commonly established a central market with administrative offices and a clock tower. British churches and civic buildings ranged from Gothic to neoclassical in style. New residential areas were built on European plans with wide streets and open places, separating business and residential functions. Such areas are now occupied by professional and business people. The more open style of town construction was adopted before and after independence in new government-dominated planned cities such as New Delhi and Islamabad.

Colonial urbanization brought fresh employment opportunities in administration and commerce, and after independence, manufacturing also grew in significance. Britain built and developed major ports such as Calcutta, Bombay, Colombo, Karachi, and Madras, initially as a means of estab-

**FIGURE 5.25 Southern Asia: urban landscapes in Delhi.** The main entrance to the Red Fort, constructed in the A.D. 1500s by the Mughal emperors. The massive fort walls enclose 6 ha (15 ac) of land, including several palaces and water gardens.

© Michael Bradshaw

(a)

**FIGURE 5.26 Southern Asia: rural landscapes.** (a) Lands around the market town of Pushkar, Rajasthan, western India: farmed plains contrast with dry Aravelli Mountain pastures. (b, next page) The mountain scenery of Tupsepdan Mountain in the Karakoram Range, northern Pakistan, with Shimshal village in the valley.



lishing trading security. The ports were then joined to inland centers by railroads. Where the railroad passed through a precolonial town, it grew in administrative and trading significance; where one was bypassed, it often declined.

The colonial expansion of cities brought in people from rural areas, but this flow was checked in the early 1900s by disease in the poorer districts. The British then made investments in infrastructure such as water supply, sewers, and roads.

Another feature of British imperialism was the construction of towns in the hills, known as hill stations, where British families went to escape the worst of the summer heat and humidity and the diseases that spread in that season. Simla, due north of Delhi, is typical of these hill stations, and is still accessed by a mountain railroad line built at great initial expense. The hill stations were built to remind the British colonials of home and often resemble small British towns in their architecture. They are now tourist centers.

After independence and partition, cross-border migrants moved into large cities and were accommodated in refugee camps that themselves became cities—Uhlanagar near Bombay now has 0.6 million people. The contrasts increased between the neglected districts of the poor and the well-kept suburbs and new cities. In some smaller cities, such as Bhopal, there was no real planned infrastructure, and even in major cities, funds for maintaining infrastructure were scarce. The extensive shantytowns in Southern Asian cities are landscapes that reflect dire poverty and lack of planning.

Modern cities have major industrial areas, including steelworks and alumina plants, chemical factories, textile mills, and factories assemble garments, vehicles, and electrical goods. Such factories are most common around Bombay in the Indian states of Gujarat and Maharashtra.

Urban landscapes in Southern Asia reflect their natural environments. The palm-lined coasts of southern Indian and Sri Lankan cities contrast with the starker lines of cities in the arid areas of northern Pakistan. The low-lying landscapes around Calcutta and Dhaka contrast with the mountains surrounding Kabul, Islamabad, and Kathmandu.

## Rural Landscapes

Rural landscapes in Southern Asia are distinguished by their physical environment and their intensity of use. The contrasts between the Himalayan ranges, their hilly margins, the Ganges River plains, and the peninsular plateaus make varied backdrops to human activities (Figure 5.26). The commercial farming of field crops such as wheat in Punjab and cotton in Gujarat create different landscapes from the plantation crops of tea and rubber in Sri Lanka. The large areas of unimproved traditional farming watered from local reservoirs known as tanks dominate much of the poor areas in the eastern parts of the Indian peninsula. Terraced mountain slopes, poppy fields, and nomadic livestock enterprises add further diversity to the rural landscapes of Southern Asia.

**(b)** © John Corbett/Ecoscene/Corbis

## FUTURE PROSPECTS FOR SOUTHERN ASIA

Southern Asia remains poor. The physical isolation of some countries, past histories of economic isolation, and extensive civil disorder suggest few prospects for rapid growth. Most countries now recognize the need to be part of the world economic system but approach the changes involved slowly. India and Pakistan—and Bangladesh and Sri Lanka to a lesser extent—are most likely to develop diversified economies. Sri Lanka and the Maldives have achieved the most in social service support for their peoples. In this region, however, any discussion of the future has to be linked to the growth of population and the likelihood of providing a continuing supply of food, let alone of increasing income per capita. A major priority of governments remains the slowing of population increase and the reduction of poverty.

Although India remains generally poor, areas in Punjab and Haryana states developed a more prosperous commercial farming base, while parts of Gujurat, Maharashtra, and the Bangalore area became centers of new manufacturing developments. The first of these was possible because the Green Revolution increases in productivity were linked to good local infrastructure and marketing facilities. The second area is based around the port facilities at Bombay and state administrations that welcomed new investments. Although in 1995 Maharashtra's new Hindu nationalist government first rejected and then renegotiated a contract for power plant construction, it has since encouraged further investments.

These examples of more developed areas may show the way forward to other parts of Southern Asia, but they may be unique, with limited lessons that can be applied elsewhere. They illustrate the need to modernize and develop more intensive production of agricultural and manufactured goods to accommodate the growing population and to export in exchange for foreign products. They also illustrate the dilemmas of governments, business people, and residents since inequalities are maintained and outcomes do not always confirm simplistic expectations.

Despite predictions that population expansion would slow as people became better off, the economically successful area of Indian Punjab has one of the highest fertility rates in India and one that has risen in the last 20 years; Bombay attracts 500 new families a day to its overcrowded shantytowns and streets, despite the terrible living conditions. Government funds spent on infrastructure to encourage further economic development have to be diverted from welfare programs in a vicious circle that is difficult to exit without hurting many people. Furthermore, the intensification of agriculture and the expansion of modern industry place

**RECAP 5D: Landscapes and Futures**

Urban and rural landscapes reflect the historic cultural contrasts of Southern Asia, the colonial developments, and attempts to modernize since independence.

Future prospects for Southern Asia depend on balancing population growth with food production and whether it is possible for the economy to grow without the hindrance of wars or isolationist policies.

5D.1 Which are the largest cities in India, Bangladesh, and Pakistan? What factors caused them to grow so big, and what problems do these cities face?

5D.2 What arguments would you make for and against the view that the policies of the British Indian Empire consigned Southern Asia to the world periphery?

5D.3 How would you illustrate the effects of postcolonial isolationism and import substitution policies on countries in Southern Asia? What problems are faced in orienting to the world economic system?

**WEBSITES: Southern Asia**

Further sites are available through the Dorling Kindersley Atlas online facility.

*http://www.virtualbangladesh.com/* is an excellent site. Take the *"Grand Tour"* and access satellite views and maps. What have you learned about the country that affects its place in the wider world?

*http://www.dailystarnews.com* is a Bangladeshi newspaper site, giving news highlights and fuller information. What events happened over the last few days that have an influence on the political or economic environment of the country?

*http://www.indiagov.org*

*http://www.pak.gov.pk/*

The government sites of India and Pakistan. Find out reasons given for hostilities between the two countries. Look up the sections on *"Economy—Industry"* (India) and *"Economy—Energy"* (Pakistan).

*http://www.lk/* The Sri Lankan site. Access *"Economy"* and *"Tea Board"* to find out how one of this country's major crops is marketed.

stresses on the natural environment (see "World Issue: Economic Growth and Environmental Problems," page 194).

In the 1990s, all the countries of Southern Asia took steps to become more integrated in the world economic system. Many of them started a long way behind the countries of Eastern Asia after decades of rejecting the Western economic system (see Chapter 6). Nearly all of them face major obstacles in opening their economies to world competition as a corollary of entering their goods in world markets. For Afghanistan, its physical isolation, history of war against the Soviet Union in the 1980s, continuing civil war in the 1990s, and reliance on income from opium and heroin repel other countries. Countries such as India and Sri Lanka have been slow to restructure their economies and wish to keep many of the state-owned facilities that have been a feature of their politically motivated isolation from world markets. Pakistan and Bangladesh are showing the greatest resolve in modifying their economies, but Pakistan is still hindered by the dominating position held by its feudal landowning elite, while Bangladesh lacks resources and continues to be affected by natural disasters that may kill a quarter of a million people in a single event.

The continuing wars and internal strife slow the efforts of these countries to catch up with their Eastern Asian neighbors. Afghanistan's civil disorder continues after 10 years of war against the Soviet Union in the 1980s. Sri Lanka's Tamil rebels continue to fight the rest of the country, and the government's major task is to reconcile the two sides. Within India, local eruptions based on union dissatisfaction with employers or on the increase of polarized tensions among Muslims, Hindus, and Sikhs show that full political stability is not yet achieved. Recent riots in Karachi, Pakistan's major commercial and industrial city, deterred foreign investors. The 1998 nuclear weapons testing by India and Pakistan can only slow their development by diverting funds away from infrastructure projects.

Many of the factors holding back economic and human development are similar to those affecting the countries of Africa South of the Sahara and those of Northern Africa and Southwestern Asia. Southern Asia shows signs of potential for moving forward but has the added difficulty of such a large population that will probably increase to nearly 2 billion, or one-fourth of the world total, by A.D. 2025.

To date, few attempts have been made to incorporate the countries of Southern Asia in wider trade agreements. In 1995, officials from Australia, India, South Africa, and Singapore met to consider setting up an International Forum of the Indian Ocean Rim. In the 1980s, such attempts to foster regional economic development and to free the Indian Ocean of big-power rivalry did not generate sufficient enthusiasm to result in an organization. The end of the Cold War and the re-entry of South Africa and India to the world economic system may bring new opportunities. The low volume of intraregional trade and the structure of the economies of the participant countries, however, may slow such cooperation. Trade among Indian Ocean countries is only around one-fifth of their total overseas trade, compared to three-fifths among the Asia-Pacific Economic Cooperation countries (see Chapters 6 and 11). There is also debate about which countries should belong and what issues should be discussed. Any trade agreement for the Indian Ocean that may benefit Southern Asia will develop slowly.

## CHAPTER REVIEW QUESTIONS

1. The "Partition" of Southern Asia occurred in (a) 1907 (b) 1927 (c) 1947 (d) 1967
2. The "Partition" resulted in which two entities? (a) India & Pakistan (b) India & China (c) China & Burma (d) Bangladesh & Afghanistan
3. Pakistan and Afghanistan are currently impacted by (a) Hindu nationalism (b) Marxist rebellion (c) Islamic fundamentalism (d) Christian evangelicalism
4. Mount Everest falls along the natural boundary between which two countries? (a) India & China (b) Nepal & China (c) Pakistan & China (d) India & Nepal
5. Buddhism dramatically departed from Hinduism in that it: (a) rejected the Caste system. (b) adopted materialism. (c) converted Christians. (d) moved out of India.
6. Which European colonial power established its rule over much of Southern Asia? (a) Britain (b) Dutch (c) France (d) Portugal
7. The Sepoy Rebellion dramatically altered colonial rule because the ________________. (a) French gained a foothold over all the major river deltas. (b) Portuguese lost Goa and their Oriental colonies. (c) Americans began to advocate independence. (d) East India Company rule reverted to direct Crown rule.
8. The correct monsoon sequence (season / climatic condition) is: (a) Summer / Wet; Winter / Dry (b) Summer / Dry; Winter / Wet (c) Summer / Cold; Winter / Warm (d) Summer / Mild; Winter / Wet
9. Which part of Southern Asia has been most contested between India and Pakistan? (a) Ghats (b) Assam (c) Tamil (d) Kashmir
10. In addition to Pakistan, which neighbor is India most concerned with? (a) Russia (b) China (c) Australia (d) Myanmar
11. At current population trends, India will become the most populous country.
    True / False
12. The prominent figure of Buddhism is Gandhi.
    True / False
13. Goa was a French colonial territory.
    True / False
14. The Dravidian people live mainly in southern India.
    True / False
15. The Khyber Pass is significant as an overland gateway to/from the Indian subcontinent.
    True / False
16. The Indus and Brahmaputra Rivers originate in Tibet (China).
    True / False
17. The majority of people on Sri Lanka are Tamil Hindus.
    True / False
18. India has aspirations of being a regional power in the Indian Ocean.
    True / False
19. Individuals from Southern Asia work in other parts of the world and send back money to their families.
    True / False
20. Deccan Plateau is the primary physical feature of Bengal.
    True / False
21. What legacies of British rule are present in India today?
22. By what criteria can India claim to be a (regional) "superpower?"
23. What two countries are in the insular southern part of the region?
24. What three countries are in the mountainous northern part of the region?
25. The culturally significant river for Hindus is the ________________.
26. At partition in 1947, the division of Bengal and Punjab gave ________________ great advantages.
27. India's three largest cities are __________, __________, and __________.
28. The Green Revolution led to greatly increased crops of ____________ and ____________ in ____________ (one Indian state).
29. List some of the common Caste system categories.
30. What does the federal system of government accomplish in India?

# EASTERN ASIA

Chapter 6

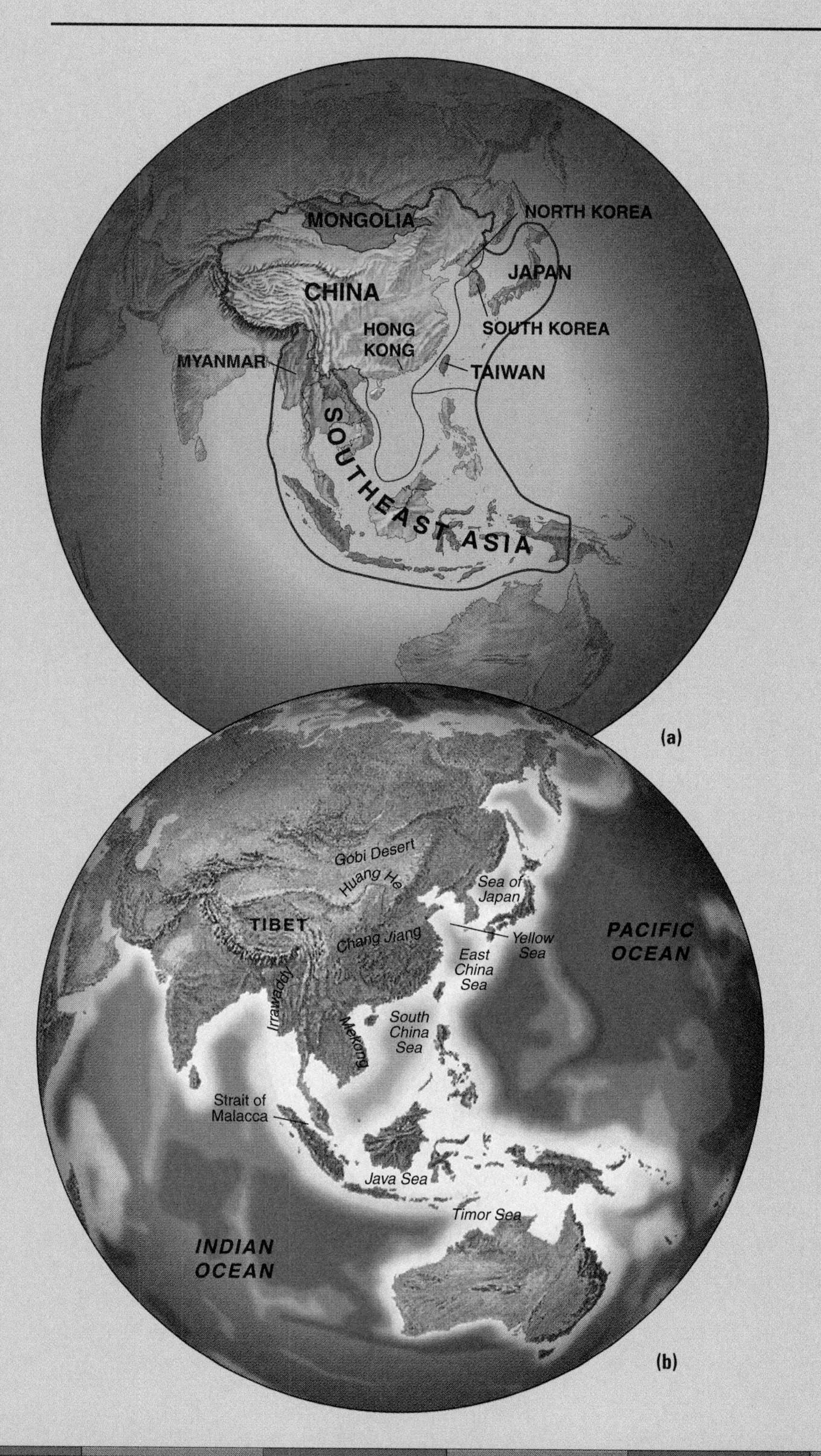


**THIS CHAPTER IS ABOUT:**

**The emergence of a new core region in the global order**

**A clash of internal and external cultures that contributed to present geographic differences**

**A natural environment of high mountains, islands, and large river basins stretching from the equator to the midlatitudes**

**Three subregions:**

Japan, South Korea, and Taiwan: fast economic growth to the 1990s

China, Mongolia, and North Korea: engaging with the world economic system after years of communist isolation

Southeastern Asia: countries at varied stages of growth

**Expanding urban and densely populated rural landscapes**

**Future prospects for Eastern Asia**

**World Issue: Population policies**

**Living in Malaysia**

**FIGURE 6.1 Eastern Asia.** (a) Subregions and countries. The boundary of China in the South China Sea is subject to disputes over ownership of the Spratly Islands (see also Figure 6.6a). (b) Major features of the natural environment. Notice the importance of mountain ranges, major rivers, and islands.

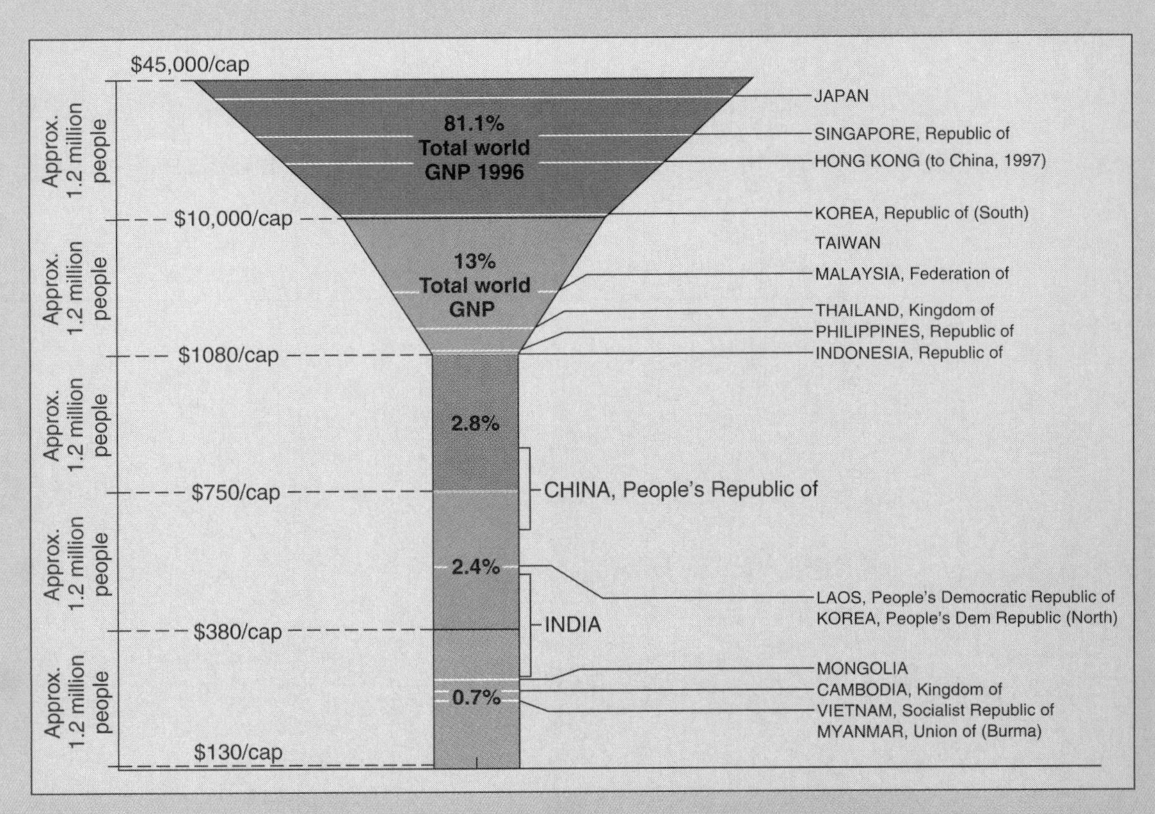


**FIGURE 6.2 Eastern Asia: national incomes per head in 1996.** Compare the range in this region with that of other major world regions. Which countries moved up or down in this diagram in the 1990s?

Source: Data (for 1996) from *World Development Indicators,* World Bank, 1998.

## FROM PERIPHERY TO CORE

In the 1980s and 1990s, many countries in Eastern Asia (Figure 6.1) shared the experience of having rapidly growing economies. They were drawn into the regional and worldwide financial and trading system, becoming a significant part of the global economy. Appearing to be generating a new core region, the experience was termed the "East Asian Miracle" by the World Bank in 1993. At the end of the 1990s, however, many of the leading Eastern Asian countries faltered, taking the region into what many forecast as a decade of stagnation to start the new century. Such an event would have major impacts on the rest of the world that is an integral part of the global economic system.

Eastern Asia has one-third of the world's population living on just over one-tenth of the land. Its richer countries cause it to produce one-fourth of the world's economic output, but there are still millions of extremely poor people in the region, with many countries remaining in the world's periphery (Figure 6.2). Its geographic variety is enhanced by some of the world's emptiest desert and mountain environments and some of the world's highest densities of urban and rural population.

Despite the contrasts of population densities, development, and natural environment, the region shares many common features. The historic rise of Chinese empires with their culture and rice-based diet, the 1800s intrusions of European colonists and traders, the mid-1900s Japanese invasions, and the later conflicts between communism and the world economic system are common experiences that affected the region's human geography.

"From periphery to core" reflects the range of current economic conditions, but it also indicates the dynamic progress of countries in Eastern Asia from the periphery, through the semiperiphery, to the core of the world economy. Japan began to industrialize in the late 1800s, developed manufacturing to the point where it could challenge the rest of the world in World

War II, and became one of the world's richest core countries in the later 1990s. South Korea, Singapore, Hong Kong (before it returned to China in 1997), and Taiwan—often called the "Asian Tigers" because of their aggressive economic progress—moved into the semiperiphery and almost into the core. Thailand and Malaysia are a little way behind and Indonesia and the Philippines farther back: this group is sometimes called the "Little Tigers."

Cambodia, Laos, Vietnam, Mongolia, Myanmar, and North Korea lag in the outer periphery, but most are changing slowly from their centrally planned socialist or military dictatorship base toward involvement in the world economic system. China, the largest country in Eastern Asia, is more difficult to place in the core-periphery spectrum, but since the mid-1970s, it has become increasingly involved economically with the rest of the world: its rate of economic growth in the late 1980s and mid-1990s was the world's highest.

Eastern Asia's increasing role in the world economy involves it in trade with other parts of the world that is less and less in primary products and more and more in manufactured goods and services. The countries in this region also have increasing amounts of intraregional trade—a major contrast with the countries in the previous regions studied in Africa and southwestern and Southern Asia, which have little trade among their constituent countries. The gradual evolution of the Asia-Pacific Economic Community (APEC) in the 1990s reflects the increasing prominence of wider trading links among Eastern Asian, Anglo American, and South Pacific countries.

Following rapid economic growth from the 1980s to the mid-1990s, Eastern Asian countries experienced economic slowdown in the late 1990s; some had major drops in output and currency exchange values. This was the result of a combination of factors triggered by falling prices due to the oversupply of electronics components but exposing overborrowing and leading to bank collapses. New investments inside and outside the region were cut, affecting all parts of the region. Since 44 percent of Japan's exports go to Asian countries (as opposed to 30% to the United States and 9% to Europe) and the country remains a major market for regional goods and tourism, its economy had a major recession. The events affected world confidence in the economies of emerging countries from Brazil to Russia; as a result, it is likely that core countries will again switch investment back to their own markets.

After a consideration of the contribution of historic cultural influences and natural environment to the geography of Eastern Asia, the rest of this chapter examines the human geography of the subregions. First, Japan is combined with South Korea and Taiwan—two other countries with growing economies that it occupied for the first half of the 1900s. Second, China—including Hong Kong since 1997—is considered with two of its immediate neighbors that have also been under communist governments, Mongolia and North Korea. Third, the countries in Southeastern Asia now include Myanmar (Burma) after it joined the Association of Southeast Asian Nations (ASEAN) in 1997. Data for each country in the region are listed in the Reference Section, pages 586–610.

## INTERNAL AND EXTERNAL CULTURAL INFLUENCES

Eastern Asian cultures arose out of the centrality of past Chinese civilizations and empires. A set of surrounding kingdoms emerged that were greatly influenced by Chinese customs but fought to establish and retain their own identities. Later, European and American colonialism had major impacts in the southern group of countries, including Malaysia, Indonesia, Indochina (modern Cambodia, Laos, and Vietnam), Myanmar (Burma), and the Philippines. Western traders backed by military action intruded China but it resisted colonial occupation. Thailand and Japan shunned external control but underwent modernization at their own pace, assisted by Western advisers. The outcomes of these historic influences produced the basis of current regional diversity.

### Chinese Empires

Chinese history was a sequence of empires expanding or retreating. Such fluctuations had impacts on much of the region. By 2000 B.C., a Chinese civilization developed in the Huang He (Yellow River) valley (Figure 6.3) based on agricultural and craft skills. It gained control of much of the lower Huang He and Chang Jiang (Yangtze River) basins, adding a separate center of ancient civilization to those in southwestern Asia, Egypt, and Pakistan (see Chapter 1).

The Chou dynasty (1122–256 B.C.) set the pattern for the future. Iron age technology produced economic growth based on irrigation and deeper plowing. Improving transportation resulted in a wider geographic trading area but also opened the wealthy lands to attacks from external groups. During the political and social upheavals of this period, three major sets of ideas had important impacts on Chinese culture. **Confucius** (Kong Fuzi), an administrator, called for more efficient and humane political and social institutions. He wished to weld the growing empire together on the basis of better trained and more able managers. **Taoism**, the teaching of Lao Zi, disdained the hierarchical system for organizing society that Confucius proposed, favoring a return to local, village-based communities with little government interference. The third set of ideas came from a group of leaders who wanted strict imperial laws to be imposed by a powerful state in response to the social disorder of the time.

The following Qin dynasty (after which China is named) imposed strict imperial laws around 220 B.C., welding the surrounding feudal states into a centralized and culturally

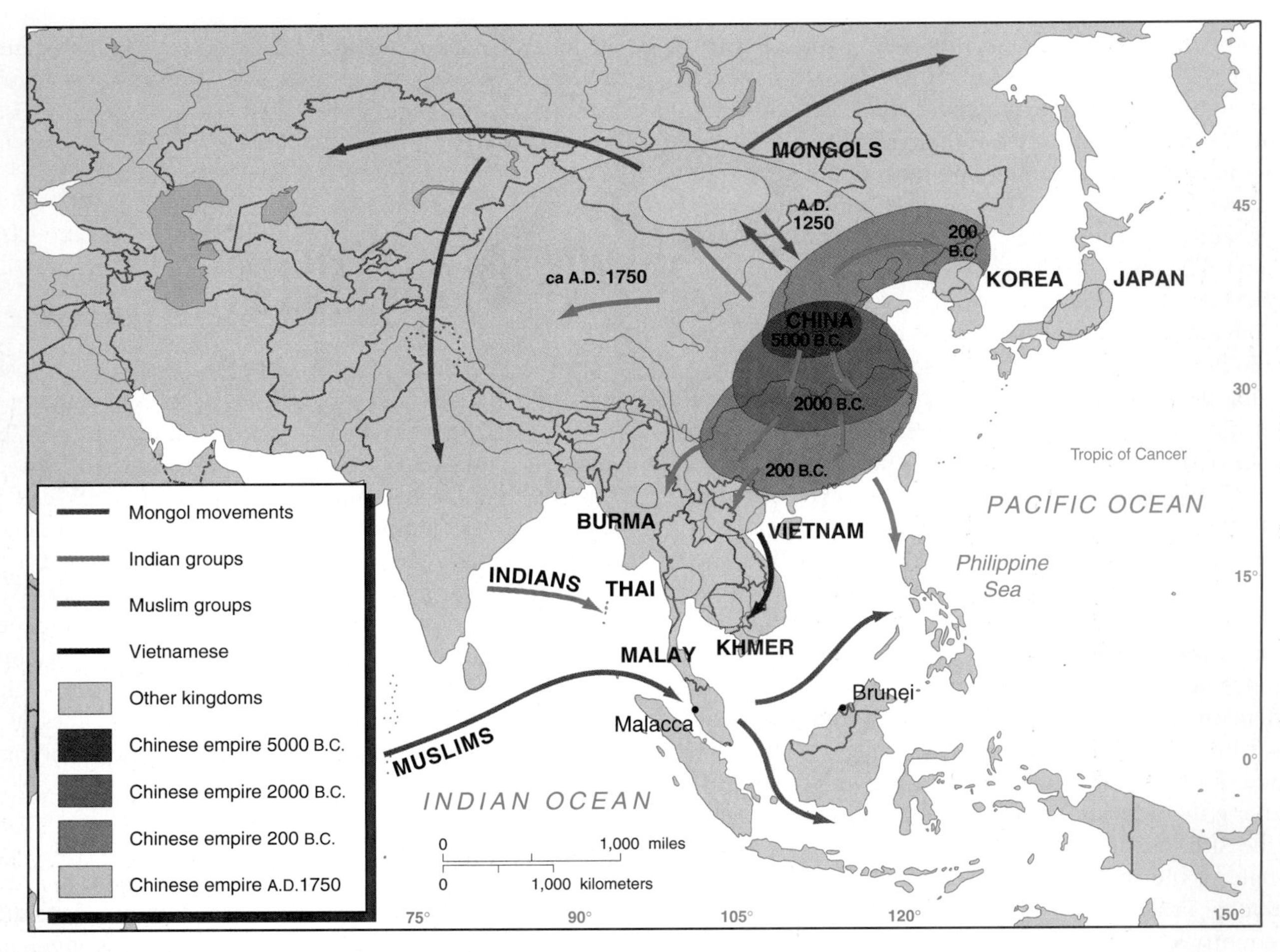


**FIGURE 6.3 Eastern Asia: the expansion of the Chinese empires and the location of adjacent kingdoms.** The Mongol kingdom was the only external one that expanded to take over China itself, but it lasted for only some 200 years. The green arrows show movements of Chinese peoples.

uniform empire with a standardized written script. It sited its capital at Xi'an. The empire extended southward to the Red River (modern Vietnam), westward to Lanzhou, and eastward into Korea. The Great Wall was completed (Figure 6.4), but the high cost of such expansion led to internal resentment. Confucian principles were adopted and appointments to public office were made on merit through a system of written exams.

In the early A.D. 600s, Buddhism reached China and flourished in a link with Taoist philosophies that produced a major period of literary and artistic brilliance, religious toleration, and foreign trade. A literate Chinese elite, however, regarded Buddhism as disruptive and returned to Confucianism. Over the next three centuries, the Tang dynasty spread Chinese influence again into Korea, northern Vietnam, and Manchuria.

In the 1200s, the Mongols invaded China from the north. Their initial impact drove the center of Chinese intellectual and economic development southward. Kublai Khan, the Mongol leader, established his capital in northern China at Beijing. He and his descendants ruled as Chinese emperors, controlling land and trade from Eastern Europe to eastern Siberia and into northern India (see Chapter 5). Chinese resentment at changes and taxes imposed on them, together with crop failures and famines in the 1300s, led to successful rebellions and the withdrawal of the Mongols from China in 1371.

Another period of Chinese expansion occurred under the Ming dynasty (1368–1644), during which the economy, literature, and education advanced. Chinese naval power ventured to India and Madagascar. The succeeding Qing (Manchu) dynasty ruled from 1644 into the 1800s and by the mid-1700s had expanded the Chinese realm to its farthest limits in Manchuria, Mongolia, Xinjiang, Tibet, Burma, and Taiwan.

## Other Kingdoms

While the Chinese empires expanded and retreated over the centuries, other peoples established kingdoms that formed the national basis of modern countries (see Figure 6.3). The most notable of these kingdoms developed in Japan, but others emerged in Mongolia, northern Korea, northern Vietnam, Cambodia, Thailand, and Burma. The Malay peninsula and offshore islands, which now form Malaysia, Indonesia, and the Philippines, were controlled by Muslim sultanates that

**FIGURE 6.4 Eastern Asia: Great Wall of China.** Built about 300 B.C., it is nearly 2,400 km (1,500 mi.) long. It was surprisingly effective in keeping out aggressors for many centuries.

© Iain Searle

were involved in the Indian Ocean trade. The Muslim sultanates were later brought together under British, Dutch, or Spanish colonial rule.

## Japan

In Japan, the Ainu people were among the earliest inhabitants but were gradually pushed northward by groups invading through Korea. Japanese traditional values were encapsulated in the principal of **Shinto**, a national religion, built on ancient myths and customs. Shinto became a political instrument of control, with emperor kings regarded as gods. Chinese influences increased, so that Chinese script was adopted in the A.D. 400s, and the orderly Confucian basis of Chinese government was copied. Buddhism reached Japan, becoming the official religion alongside Shinto in the 600s. Japanese emperors resided at Kyoto, retiring from public life in favor of the Fujiwara, a leading family of court nobles who ran the country. From the 1100s to the 1500s, feudal lords overrode the imperial administration, the armed shogun leaders acting over the heads of the powerless court. The shoguns dominated Japan and largely maintained its isolation until just before the last shogun resigned in 1867. The emperor then regained his position as head of government. The royal capital moved to Tokyo, which had become the shoguns' center.

## Khmer, Mongol, Thai, Burmese, and Vietnamese Empires

The Mongol expansion across Asia in the 1200s and 1300s was then contained. The Mongols retreated to their present territory, mainly within the country of Mongolia. During the Mongol advance, a group of Tai-speaking people moved from western to southern China and then into the western part of the Indochina peninsula, forming a unified political grouping by the 1300s, with its center at Ayutthaya. After conflicts with Burma and Cambodia, Thailand was never again occupied by other powers.

Mon and Khmer peoples moved into the Cambodian area, followed by the later arrival of the Vietnamese, Lao, Thai, and Burmese peoples in the territories that became their national centers. Indian influences brought Hindu and Buddhist religions to the area that became known as "Indochina," resulting in a distinctive architecture. The Khmer Empire waned in the 1400s, following attacks by Thais and Vietnamese.

Burma (modern Myanmar) became Buddhist following the expulsion of Buddhists from the Indian peninsula. Its main center was the Pagan kingdom based on Mandalay in a fertile, mountain-rimmed plain in northern Burma. Powerful tribes, including the Shan and Karen, lived in the surrounding hills and still resist Burmese attempts to draw them into the country's life. The Mongols ended the Pagan Empire, and when they left, Burma was a divided country until the 1700s, when efforts to promote Burmese expansion led to conflicts with the British in India and eventual colonization.

Northern Vietnam was the center of a kingdom by the first century B.C. The area was later conquered by various Chinese dynasties and took on Chinese methods of government and administration. The Vietnamese always fought back and retained their separate identity. In the A.D. 1400s, the Vietnamese not only fought off the Chinese but extended their lands southward across the older coastal kingdoms to take over some of the Khmer territories at the mouth of the Mekong River.

## Korean and Island Histories

A Korean kingdom was established first in what is now North Korea. The country was unified in the A.D. 600s. After the Mongol invasion and retreat, a Confucianist approach to government was adopted. Aggressive Japanese intentions were resisted until the late 1800s, when the newly strengthened Japanese forces overran Korea.

The island groups that now form Malaysia, Indonesia, and the Philippines had fewer and less powerful precolonial kingdoms and were subject to continuing overseas influences. The Indonesian islands experienced many phases of invasion and migration, bringing together over a hundred ethnic groups and languages. Groups of traders and invaders from India and China came to the islands before a kingdom based on rice-growing was established in Java in the A.D. 500s. Malays, Indonesians, and Chinese people invaded the Philippines. A Filipino culture emerged from such mixed influences around the A.D. 400s. Little is known of the Malay peninsula before the arrival of Muslim groups who occupied all these lands during the 1200s. Local sultanates were established at Malacca (now in Malaysia) and in Brunei to control the islands' trade. Malaysia and Indonesia remain dominantly Muslim in religion, with Indonesia being the world's largest Islamic state. Islam's presence in Eastern Asia is a major cultural factor from western China to the southern parts of Southeastern Asia.

## European Intrusions

The Portuguese led the European onslaught on the area at the start of the 1500s but were soon ousted by the Dutch, Spanish, French, and British. Although European colonists never occupied Thailand, China, Korea, or Japan, all countries in Eastern Asia were affected deeply by the arrival of traders, missionaries, and military forces (Figure 6.5). Their incorporation into the modern world economic system had begun.

### China Resists Colonization

During the 1800s, China was increasingly pressed by European countries to engage in trade as the world economic system expanded geographically. These interventions eroded Chinese power and the ability of its people to control their own destinies. Macao was the first trading port on the Chinese coast, established by the Portuguese in 1557, but China resisted other attempts to set up trade or political links. China never became the colony of a single European country, but the increasing external control of their economy affected the Chinese deeply. They believed that China was the only civilized country and all other peoples were barbarians who should rightly be bringing them tribute. Colonial status would have been unthinkable for the Chinese, but their main ports became subject to Western economic pressures, weakening the internal political structure as foreign trade conflicted with the self-sufficiency of the feudal system.

Foreign trade, mainly for Chinese tea, was at first paid for in silver and traded through the southern port of Guangzhou (Canton). Britain redressed the trade balance with opium from India. Local wars in the 1800s gave Britain, France, and the United States rights of trade and residence through specified ports. Hong Kong Island was ceded to Britain in 1842

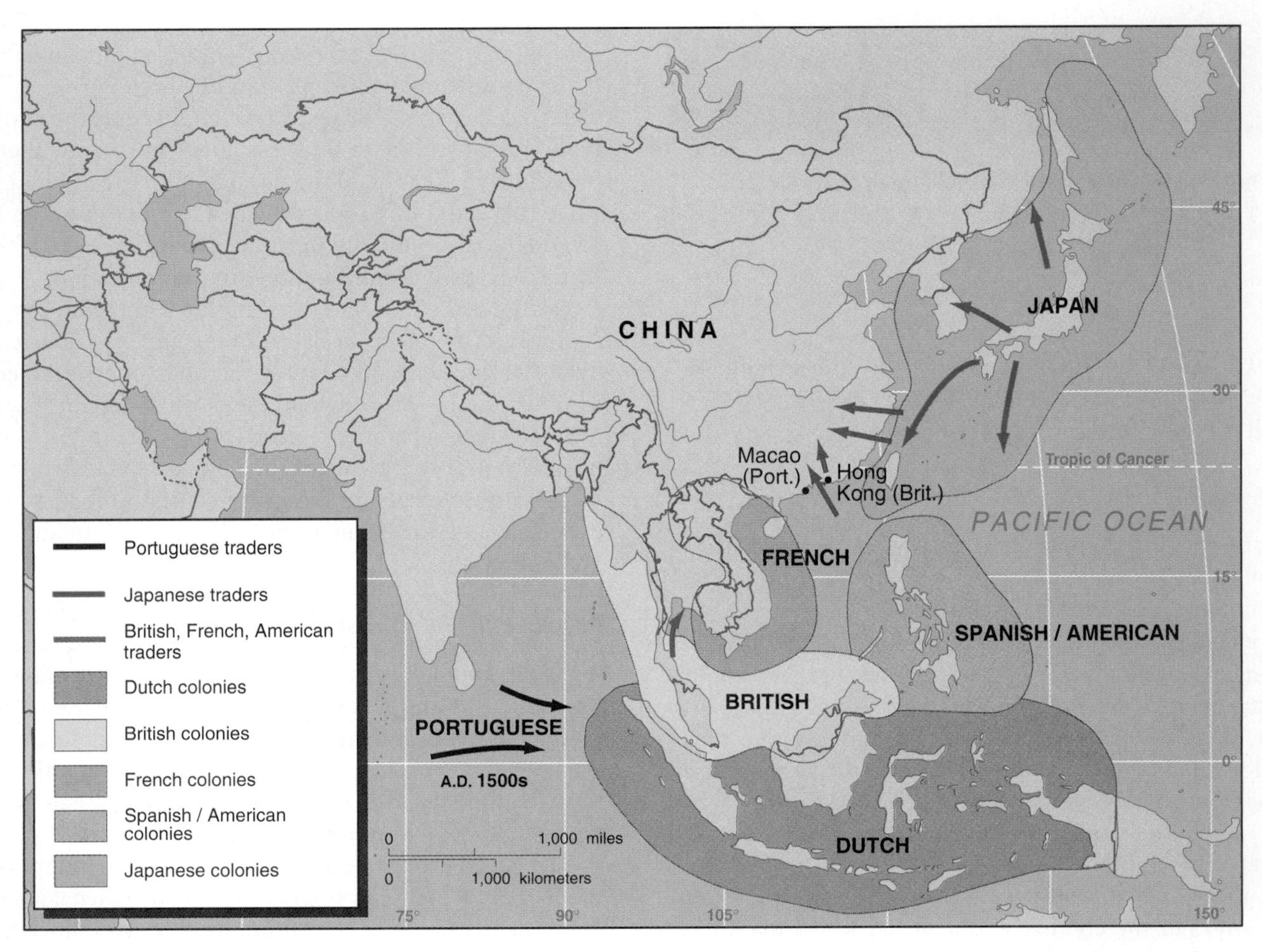


**FIGURE 6.5 Eastern Asia: colonization.** Portuguese, Dutch, British, French, Americans, and Japanese assumed control of areas easily accessed by sea.

and the Kowloon Peninsula in 1852. The New Territories, linking these to mainland China, were leased to Britain in 1898 for 99 years. The foreign powers did not allow the Chinese to tax imports—a source of income that might have provided capital for starting their own industries.

During the later 1800s, colonial powers moved into the areas surrounding China: Japan took the Ryukyu Islands, Taiwan, and Korea; France took Vietnam, Laos, and Cambodia; Britain took Burma; Russia took northern Manchuria until Japan dispossessed it in 1905; and the United States took the Philippines from Spain. In the 1920s a republic replaced the enfeebled Manchu dynasty. Sun Yat-sen's Kuomintang party then vied with the Chinese Communist Party that was established in 1921. Attempts to reunify China under a common government had to contend with local warlords, a communist rebellion, and Japanese attacks in the north. The Japanese invasion began with its 1933 expansion out of Manchuria, followed by a general declaration of war in 1937. It brought the Kuomintang and Chinese communists together in resistance. After World War II, they fought each other again, leading to the establishment of the communist People's Republic of China in 1949. The defeated Kuomintang leaders fled to Taiwan, where they established a new government, protected by U.S. forces.

## Japan Maintains Its Independence

After 1867, the new government of Japan began to imitate the Western world in rapid industrialization. Advice from the French was used in modernizing the army, British in the navy, Dutch in engineering, and the United States in the education system. At the end of the 1800s, new military power, backed by modernized industries, enabled Japan to take Taiwan and Korea and to win a war against Russia. Tensions with the United States first arose when America resisted the flow of Japanese immigrants into California. After invading China in the 1930s, the Japanese military-industrial war machine, invoking the past glories of the shogun leaders and overriding the civil government, occupied much of Southeastern Asia in the early 1940s. Its antagonism toward the United States came to a head with the Pearl Harbor bombing in December 1941. The Pacific war that followed was a major part of World War II and ended in the defeat of Japan. During that war, the rest of Eastern Asia was made wary of Japanese imperialism by the harsh manner in which the Japanese military occupied other countries.

## The Colonists

The Spanish pushed the Portuguese out of the Philippines in the late 1500s and established their capital at Manila on Luzon. Roman Catholic priests poured in, making rapid conversions to a faith that united the disparate groups of Filipinos in what became the only largely Christian country in Eastern Asia. Monastic orders took over large tracts of land and became wealthy. Spain retained its Philippine colonies until the end of the 1800s, when the United States placed them under its own jurisdiction. Local opposition to colonial rule increased during the 1930s and under Japanese occupation in World War II. The Philippine Republic was established in 1946, although the United States retained its military bases until the 1990s.

Java and the surrounding islands became the Dutch East Indies colony in 1799. The Dutch made Batavia (modern Jakarta) the trading headquarters of their Dutch East India Company. Farmers were organized to produce coffee, rubber, and other new crops for export, but neglect of local food production resulted in famines. Nationalist movements began in the early 1900s and were further motivated by harsh treatment from Japanese soldiers in World War II. Although the Dutch resisted granting independence after the war, it came in 1950.

Following the Portuguese annexation of Malacca in 1511, the Dutch took over until the British conquered the Malay peninsula in the 1700s. In 1819, the British built a new port at Singapore, and controlled the Malay States by diplomacy. The northern tip of Borneo (Sabah) and the northeastern coastlands (Brunei and Sarawak) were also British protectorates. Indians and Chinese were encouraged to migrate into these sparsely populated areas to work in the rubber plantations and tin mines. The Japanese took this area in World War II, but the British re-established their rule afterwards, fighting against communist guerrillas until independence was granted in 1957. Brunei decided to stay out of the Malaysian federation, and Singapore left in 1965.

The British colonized Burma from India, making it a province of British India in 1886. Developing rice growing in the Irrawaddy River delta, they built the port of Rangoon to export the rice. Attempts were made to unify the newly developed Lower Burma with the main Burmese population around Mandalay and with the hill tribes. Over a million Indians moved into Lower Burma, becoming the leaders of the commercial community. Burma separated from India in 1937 and gained independence in 1948. In 1989, the military government adopted the name "Union of Myanmar," when the name of the capital, Rangoon, was changed to Yangon.

The French occupied Indochina (Cambodia, Laos, and Vietnam) progressively during the 1800s and early 1900s. They built roads and railroads and encouraged manufacturing. During World War II, Nazi Germany occupied France and persuaded the French colonial government to allow in the Japanese forces. After the war, the French tried to reestablish control of the area, but communist groups forced them to leave the northern parts of Vietnam in 1954. The divided country with a communist North and free market South was subject to further warfare, in which South Vietnam was supported by the United States. When North Vietnam was victorious in 1975, it reunified the country. Laos went through much strife along its border with Vietnam, slowing economic development. Cambodia became independent in 1953 but suffered 30 years of civil war and invasions by the Vietnamese.

### Thailand

Thailand never became a colony but established commercial treaties with Britain in the 1800s, allowing increased British influence to encourage modernization. Following a short war with France in the 1890s, Thailand was forced to yield land in modern Cambodia to France and some of the Malay peninsula to Britain. In World War II, Thailand at first sided with Japan, allowing the latter's armies through to invade Burma and Malaya and regaining territories in Malaya and Cambodia. Although Thailand installed a new government in 1944 that wisely supported the victorious allies, it still had to return the disputed territories at the end of the war. The fighting in Vietnam in the 1960s and 1970s made it seem that Thailand could be the next "domino" to fall to the worldwide communist advance. The communist advance was stemmed at Thailand's borders, but the danger attracted considerable financial aid from the United States.

## NATURAL ENVIRONMENT

The natural environments of Eastern Asia add distinctive features to those imposed by cultural history. This is a region of earth activity and often violent weather, marked by earthquakes, volcanoes, and typhoons. Climatic environments vary from wet equatorial to arid midlatitude continental interior types, and there is a dominance of mountainous landscapes cut by major river basins. The large numbers of people crowding into the small proportions of well-watered lowland are affected by natural events, especially in rural areas, and have major impacts on the physical environment by modifying natural features and processes.

### Tropical and Midlatitude Climates

Eastern Asia has a greater north-south extent than Southern Asia, giving it a latitudinal range from south of the equator to over 50° N and a wide variety of climatic environments (see Figure 2.28). Malaysia, Indonesia, and the southern parts of the Philippines have hot and rainy weather all year in the equatorial climatic environment. Since islands and narrow peninsular environments surrounded by ocean occupy most of this area, strong land and sea breezes alternate by night and day. It is common for mornings to be sunny, but tall clouds build up in the afternoon and intensive thunderstorm rainfalls occur in the late afternoon and evening. Some places have over 300 days with thunder in a year.

The northern Philippines, Indochina mainland, Thailand, Myanmar, and southern China experience a monsoon climatic environment in which summer rains are brought by winds from the southwest. The subtropical environments are subject to typhoons that approach from the ocean most commonly in the late summer. Winters are cool and dry as winds blow outward from continental Asia.

Northern coastal China, the Koreas, and Japan have a similar regime of summer rains and drier winters but experience much colder winters caused by icy winds blowing from Siberia. This east coast midlatitude climatic environment is modified in Japan, where more precipitation falls in winter after the westerly winds pick up moisture on crossing the Sea of Japan. The western parts of China and the whole country of Mongolia experience arid conditions in the midlatitude continental interior climatic environment. Winters are extremely cold and little rain falls in summer. The western parts of China, including Tibet, are affected by high altitudes in the mountains and plateaus where there are even greater summer-winter and daily ranges of temperature.

### Mountains and Islands

Mountain systems and relatively small areas of lowland (Figure 6.6*a*) mark the relief of Eastern Asia. Over one-third of the region is occupied by the high mountains extending eastward from the Himalayas and Tibet. The volcanic islands of Indonesia, the Philippines, and Japan provide active landscapes with eruptions every decade or so. Such high relief, frequent earthquakes, and volcanic eruptions result from the clashing of three major and one minor tectonic plates (Figure 6.6*b*). Although these events suggest a difficult environment for human occupation, the region supports some of the highest densities of rural population in the world.

Convergent plate boundaries form much of the western and all the southern and eastern boundaries of the region. The northern margin of the Indian plate turns southward after following the line of the Himalayan Mountains along the northern boundary of Southern Asia. Where the Indian plate pushed into Asia from the west, it formed parallel north-south ranges in Myanmar, Thailand, and Malaya. The plate boundary then swings eastward so that the Indian Ocean floor is subducted beneath the southern margin of Eastern Asia. The islands of Indonesia are formed of volcanic material that was erupted just north of this plate boundary. Further groups of islands, from the Philippines to Japan, formed where the Pacific plate was subducted beneath the eastern margin of the Eurasian and Philippine plates. The outer ripples caused by these plate movements also affected the landscapes of interior China and Mongolia, lifting the Tibetan plateau and crumpling the rocks to form the ranges of mountains in the continent's interior. These movements continue to cause earthquakes.

### Major Rivers

The combination of high mountain systems and intense summer rains causes some of the world's most active rivers to flow across the continental areas of Eastern Asia. In many

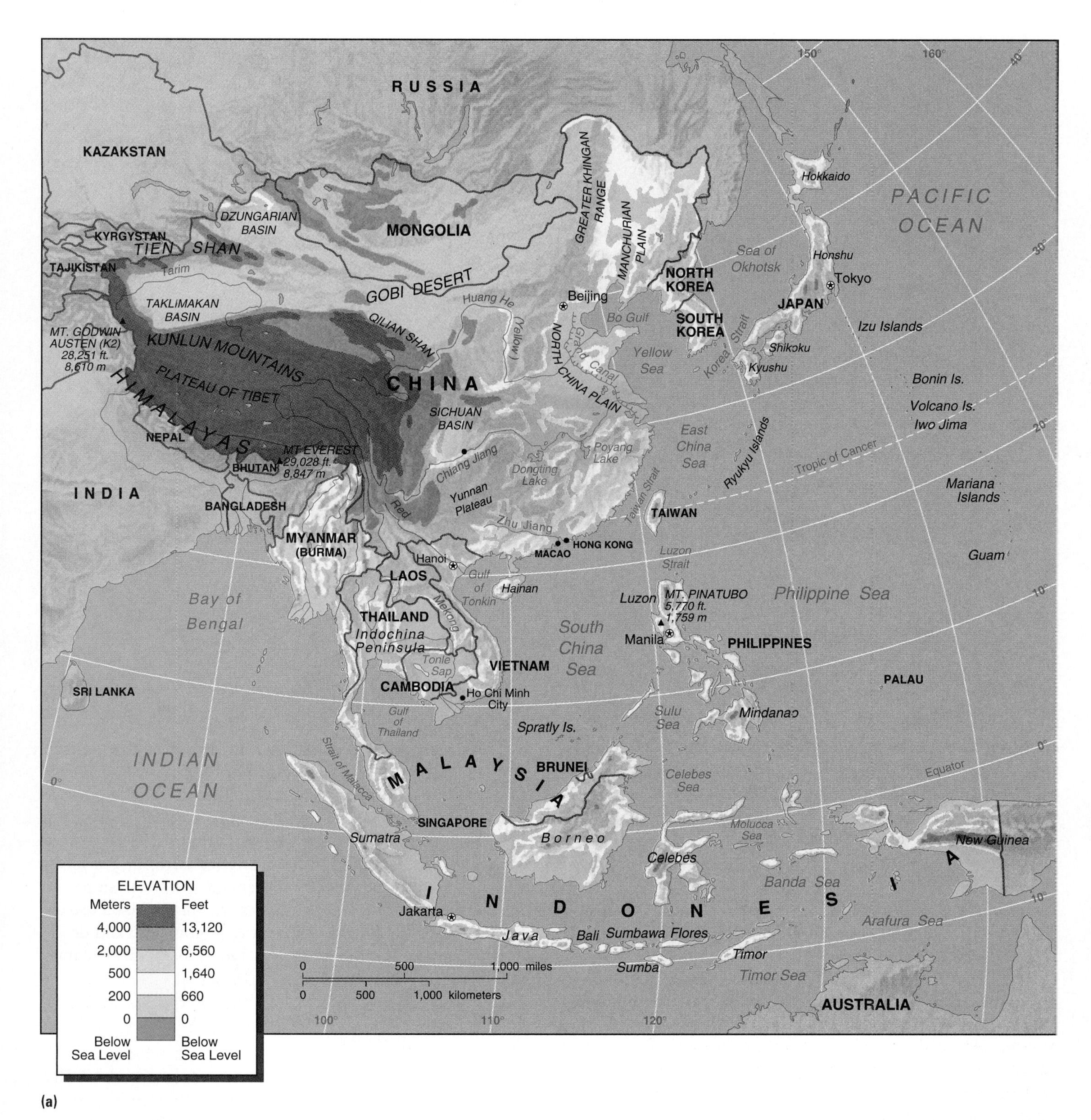


(a)

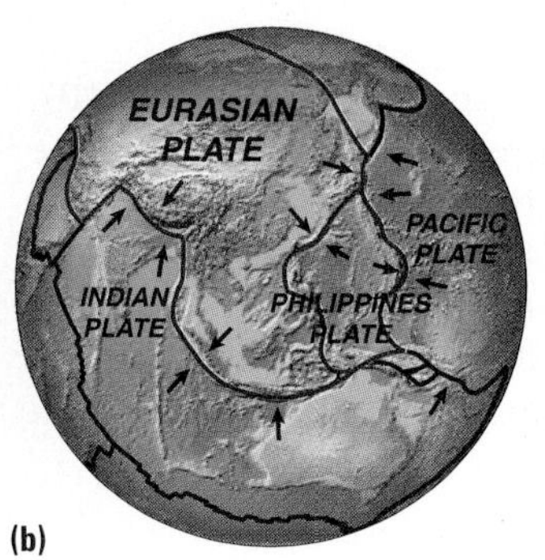


(b)

**FIGURE 6.6 Eastern Asia: landforms.** (a) The relief map shows how the region comprises islands, continental mountains, lowland plains, and major river basins. (b) Plate margins form two thirds of the region's boundaries.

(a)

**FIGURE 6.7 Eastern Asia: landscapes in southwest China.** (a) Limestone hills near Guilin, Guangxi province. To what extent has human activity affected this landscape? (b) The upper Mekong River (Lancang in China) carves a deep valley in the western part of Yunnan province that is accessible mainly by river.



parts of the region, human populations continue to be concentrated in the main river basin lowlands. The major rivers carry large quantities of water toward the sea, together with mud, sand, and rock fragments eroded from Earth's surface. Inland they carve deep valleys with steep slopes and boulder-strewn valleys (Figure 6.7). Toward the coast they drop their load of sand, silt, and clay to form wide plains and deltas extending out to sea. The tidal range and wave activity are both low around these coasts and, combined with large river-borne loads of silt and clay, cause deltas to be a common feature of the coastal landforms.

The major rivers include the **Irrawaddy** and **Mekong** that rise in the eastern Himalayan ranges of southern China. The Irrawaddy flows through Myanmar and the Mekong through the countries of Indochina to its delta in Vietnam. Thailand is drained by a series of rivers that join in the **Chao Phraya** lowlands near Bangkok. Northern Vietnam has the **Red River.** In China, the three major rivers are the longest in the region: from north to south they are the **Huang He** (Yellow), the **Chang Jiang** (Yangtze), and the **Zhu Jiang** (Pearl).

## Tropical Rain Forest to Midlatitude Desert

The range of climatic environments in Eastern Asia produced a variety of natural vegetation types, although human land uses replaced much of them. The tropical rain forest of the equatorial Indonesian and Malaysian islands and peninsulas contains many different tree species. The monsoon forests from Myanmar to Indochina, the Philippines, and southern China have fewer species and extensive areas are dominated by teak forest. Farther north, midlatitude deciduous and evergreen forests have largely been cleared from the lower and more cultivable areas but remain on steep slopes in northeast China, the Koreas, and Japan. Inland the increasing aridity causes the forests to give way to grassland and desert in northwest China and Mongolia. The **Gobi Desert** occupies much of Mongolia.

## Natural Resources

The major natural resources of Eastern Asia are the surface water flows, the fertile soils that floor the lowlands, and the minerals in the mountain rocks.

- The mineral resources of Eastern Asia (see Figure 5.10) include tin, iron, gold, and precious stones. Many of the tin deposits occur where active rivers eroded tin-bearing veins in the rocks and deposited this heavy mineral in concentrated alluvial layers lower down the valley or offshore. Coal deposits are widespread in China, while deposits of oil and natural gas occur in Indonesia and in

**(b)** © Michael Yamashita/Corbis

many offshore locations that are only just being explored.

- The water and fertile soils formed the material basis of abundant crops that supported civilizations and more localized kingdoms throughout the region at an early stage of history. The seasonal flows of water and the annually renewed alluvial soils made it possible to grow highly productive wet rice that fed large populations. The lowlands of Eastern Asia became some of the world's most densely populated agricultural areas.

The recent history of the Mekong River illustrates some of the resource potentials and political difficulties associated with modern water use in the region. It flows through or along the borders of six countries—China, Myanmar, Thailand, Laos, Cambodia, and Vietnam. It is one of the region's longest rivers, and its development potential includes hydroelectricity, irrigation water, and transportation. Improved flood control would also aid economic development of the lands along the Mekong's length.

Plans for an integrated and multipurpose development of the Mekong waters in the 1940s were abandoned during the subsequent civil wars in the area. Suspicions among the countries prevented further planning until an initial meeting of the four lower basin countries in 1994. An outline agreement was then signed to investigate the sharing of waters and their protection from environmental damage. Progress on this accord will be slow. For example, China and Myanmar did not sign the agreement at once; Cambodia and Vietnam are in dispute over transportation on the river; without consultation, Laos is adding to its two dams that generate electricity, a major foreign exchange earner; China is already building the large Manwan Dam on the upper waters; and Thailand is criticized for taking too much water from the river to irrigate its arid northeast. Furthermore, there are concerns over environmental impacts if the new dams hold back silt that now renews the fertility of soils downstream and if more regular flows in the river affect the freshwater fisheries in Tonle Sap, Cambodia. In 1995, Cambodia, Laos, Myanmar, Thailand, Vietnam, and China's Yunnan province signed an agreement for the future integrated development of the *Greater Mekong Subregion* based on a plan submitted by the Asian Development Bank. This adds formality to previous intentions but does not immediately overcome the difficulties of the cooperation among former enemies and potential economic competitors.

## Country Boundaries

Although most country borders in Eastern Asia follow physical features such as mountain ranges or rivers, they were established on the basis of cultural and colonial history. Boundary disputes remain central to the political concerns of many countries, including the disputes between Thailand and its neighbors and between China and Vietnam. Some countries act as **buffer states** between major powers, helping to reduce direct conflict between them. Mongolia in particular plays such a role along most of the boundary between China and Russia. The Koreas and Taiwan have experienced fluctuating dominance by China and Japan.

One of the remaining disputes that is likely to have wider impacts concerns the ownership of the **Spratly Islands** in the South China Sea (see Figure 6.6*a*, approximately 8° N, 114° E), where there are no physical boundaries. All the surrounding countries, including Taiwan, but mainly China and Vietnam, claim the islands. In the 1990s, ownership of the Spratly Islands became a significant issue. Oil was discovered in rocks beneath the surrounding seas at a time when Chinese offshore exploration farther north had not fully met forecasts. Both China and Vietnam issued licenses to international oil corporations for exploring adjacent sections of the area, raising tensions between them. The issue may draw in other interested parties, including Japan, which receives 70 percent of its oil imports after transportation through this sea and wishes to keep the sea lanes international.

## Environmental Problems

The presence of so many people in such environments results in a host of environmental issues. Some stem from the natural processes at work; others are caused by human interventions and attempts to wrest higher productivity from local resources.

### Natural Hazards

The combination of plate boundaries and monsoon rains not only provides resources but also presents hazards to human occupation. Volcanic and earthquake activity continues along the plate boundaries, with 1990s eruptions in the Philippines and major earthquakes in Japan. Such geologic activity also occurs some distance from the plate boundaries. In 1976, a massive earthquake hit the coal-mining area around Tangshan (near Tianjin) in northeastern China, killing 242,000 people. The area was closed to outsiders, but rehabilitation was carried out within a few years.

Flooding is a major hazard on the low-lying river plains in the monsoon climatic environment. Precipitation and high water flow in rivers are concentrated into the summer months. Chinese flood hazards attract publicity because of the large numbers of people involved when the Huang He or Chang Jiang flood, but this hazard also affects other countries in Eastern Asia. The 1981 and 1991 flooding by the Chang Jiang followed abnormal precipitation and was enhanced by rapid runoff increased after excessive deforestation in the upper parts of the river basin. The 1991 floods were the worst in China during the 1900s and affected both the middle basin and the provinces at the river's mouth. In the mid-1990s, the 180 million people in the Huang He basin began to suffer the opposite problem—too little water. The building of dams in headwater areas reduced flooding but also diverted waters from the lower basin to middle basin irrigation works. Successive years of drought from 1987 reduced flow in the lower basin to under half its previous level. Even in normal years, the river's waters do not meet the expanding demand.

### Human Impacts and Responses

Human impacts add to the natural hazards. Such impacts are especially significant in areas of high population density and modern industrialization. Deforestation resulted from expanding agriculture, cutting of forests for lumber, mining, and river damming for hydroelectricity and irrigation. Warfare in Vietnam involved defoliation by chemicals that left barren hillsides taking decades to recover. Rapid loss of forest often leads to soil erosion, downstream flooding, and silting, commonly destroying offshore fisheries and coral reefs as the silt is swept out to sea.

China experienced major problems of desertification as it tried to plow the formerly grassed semiarid areas of the northwest. From 1948 to 1978, soil erosion by wind and water removed over 700,000 hectares (1.7 million acres) per year from productive activities. It was decided to reduce that threat by planting a "Great Green Wall" of trees between 400 and 1,700 km (250–1,000 mi.) wide. The outcome of this program is not fully known, but dust-storm days in Beijing fell from 30 per year in the 1970s to 12 by the late 1980s, indicating a degree of success.

Forest planting was a feature of other parts of China in the 1970s and 1980s. The need for this was highlighted dramatically by the 1981 flooding in Sichuan that was made worse by poor land-use management, including excessive tree felling. Major plantings occur throughout the coastal region and as shelter belts in the intensive farming areas of the major rivers. By 1989, one-fourth of all administrative areas had at least a 10 percent forest cover and shelter belts. The Chinese focus on economic growth and efficiency since 1976, however, reduced funding for the planting program. The emphasis on private enterprise expands farm production but reduces controls. This produces soil erosion and salinization from the plowing of grassland in northeast China, where soil is exposed to runoff and evaporation.

### Diseases

The tropical environments of Southeastern Asia harbor many diseases. Although forest clearance, better sanitation, improved diets, and the wider availability of primary medical care have improved health since 1960, malaria, cholera, typhoid, and rabies remain serious. The expansion of water surfaces through dam building spread waterborne diseases to new areas. HIV/AIDS is an increasing threat in the major cities of Eastern Asia, especially in Bangkok (Thailand).

## Pollution

Pollution of air, water, and soils is a serious problem as industrial activities increase, more vehicles are used, and farming is made more productive by the use of fertilizers, pesticides, and mechanization. In large cities, the standards for concentrations of carbon monoxide, sulfur dioxide, suspended particles, and lead in the atmosphere are regularly exceeded. Five Chinese cities—Beijing, Shanghai, Shenyang, Xi'an, and Guangzhou—are among the 10 most polluted cities in the world. China's heavy dependence on coal for 80 percent of its energy needs makes it the world's largest coal user. Inefficient small factories and power plants produce one-third of the particulates and sulfur dioxide, while domestic users produce another one-third. Tokyo traffic police wearing smog masks became a symbol of deteriorating environmental quality in Japan, but legislation led to improvements. In China and other poorer countries, legislation is often not implemented. In 1997, the forest fires on Borneo and Sumatra—lit to clear forest for planting oil palms—covered much of Malaysia and Indonesia with a choking smog when the burning coincided with unseasonal drought.

Water pollution is particularly serious around the expanding major cities where raw sewage and untreated industrial effluent are common. The poisoning of fish with mercury killed over 1,000 Japanese people in the 1960s. Heavy metals polluted drinking and crop water around Shenyang in northeast China, leading to a surge of birth defects and a life expectancy of 10 years below the national average of 70 years. Although access to safe drinking water is widespread in China, water pollution from untreated urban wastes and rural fertilizers caused Shanghai to move its drinking water sources upstream in the mid-1990s. Polluted water and acid rain both reduce crop productivity and make it essential to cook all vegetables at high temperatures in a wok. At sea, the passage of around 40,000 supertankers through the Strait of Malacca each year is accompanied by oil spills.

**RECAP 6A: World Position, Cultural History, and Environments**

Eastern Asia has one-third of the world's population and produces one-fourth of the world's output. In the 1990s, the region contained countries that ranged from some of the world's poorest to its richest. Many countries at different levels of economic development experienced economic growth.

The Chinese culture dominated the region's history, but smaller kingdoms around the margins established separate national entities that formed the basis of modern countries. Muslim influences affected the southern and northwestern parts of the region. European colonizations had their greatest impacts in modern Myanmar, Malaysia, Indonesia, the Philippines, and Vietnam.

The physical environments of Eastern Asia are marked by monsoon seasonal contrasts of rainfall and drought, mountain systems, major river basins, and hazards such as typhoons, flooding, earthquakes, and volcanoes. Environmental pollution and degradation are increasing problems facing countries with high rates of economic development.

6A.1 To what extent can Eastern Asia be defined as the "Chinese realm"?

6A.2 Which cultural influences worked against the total Chinese domination of Eastern Asia?

6A.3 Does the range of natural environments in Eastern Asia contradict the idea of its character as a major world region?

**Key Terms:**

Confucius, Shinto, buffer state, Taoism

**Key Places:**

Irrawaddy River, Red River, Zhu Jiang, Mekong River, Huang He, Gobi Desert, Chao Phraya River, Chang Jiang, Spratly Islands

## Environmental Policies

Although many countries in Eastern Asia were late in adopting environmental policies, they are now busy setting aside wildlife sanctuaries, establishing forest management procedures, and considering the overall management of water resources. Some countries have passed environmental legislation to reduce pollution of air and water, but it is not easy to enforce the standards. In China, the government is raising coal prices to reduce the use of this inefficient fuel, but acid rain will become more extensive in the time it takes to implement the policy. The danger of companies corrupting government officials for short-term gains remains a problem that slows progress in environmental quality in many countries across the region.

Japan made major strides in environmental control since the worst years of the 1960s, and incidents of pollution declined. One of the major environmental problems Japan faces today is the disposal of trash, which increased by a third in the 1980s and 1990s, at a time when virtually all the dumping sites are filled. As a result, Japan is becoming a world leader in developing new technologies of recycling materials.

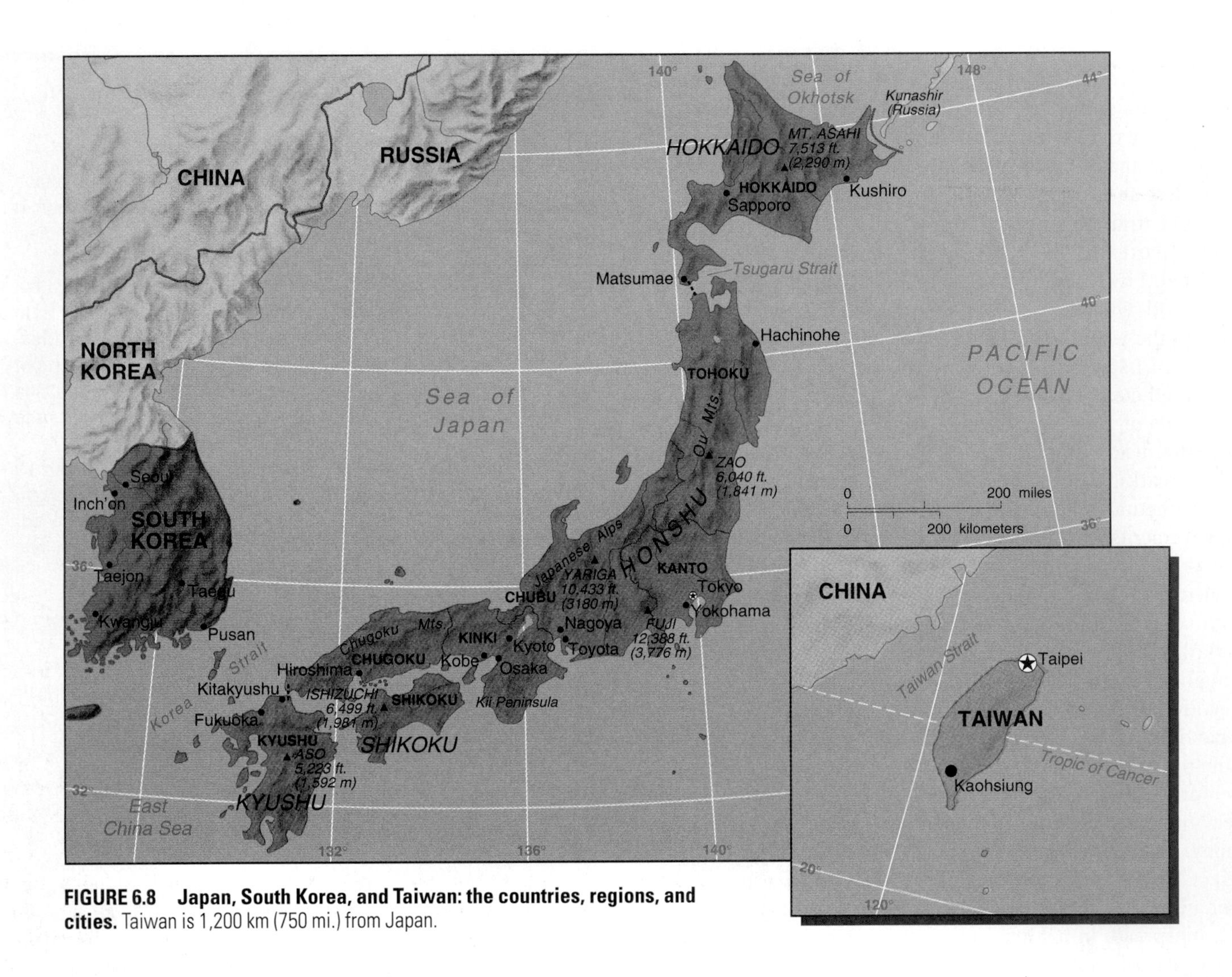


**FIGURE 6.8 Japan, South Korea, and Taiwan: the countries, regions, and cities.** Taiwan is 1,200 km (750 mi.) from Japan.

## JAPAN, SOUTH KOREA, AND TAIWAN

**Japan**, **South Korea**, and **Taiwan** (Figure 6.8) are among the leaders of economic growth in Eastern Asia—the others being the port cities of Hong Kong (part of China since 1997) and Singapore (in Southeastern Asia). Japan occupied both Korea and Taiwan (Formosa) in the early 1900s, and both followed its lead in economic development from the 1950s. They all received postwar assistance from the United States.

### Japan

Japan is the weathiest country in Eastern Asia and one of the richest in the world. Its industrial development from the late 1800s, its recovery from defeat and destruction in World War II, and its rapid economic growth since the 1950s make up the greatest economic success story of the 1900s. From 1965 to 1995, Japan's total GDP grew from 13 to 73 percent of the U.S. total and from 4.5 to 18 percent of the world total. This growth is reflected in the rising GDP per head and a HDI world ranking of $7^{th}$.

In the mid-1800s, Japan's leaders realized that its previous isolation placed it at an economic and political disadvantage compared to the expanding global influence of Western countries. As it modernized, however, Japan maintained its traditional culture that encouraged attitudes based on loyalty to the emperor, family, and workplace. This provided an alternative platform for competing in the world economic system. Although the rising costs of labor in Japan and the high value of the yen in the early 1990s placed Japanese manufacturers at an increasing disadvantage to those in other countries and overlending put stresses on its financial system, Japan should be able to maintain its new world position.

Japan's present government, economy, and constitution were put in place during the post-World War II U.S. occupation that lasted until 1952. Following postwar reconstruction, economic growth was rapid, encouraged and partly directed by the central government. Since the 1980s, Japanese investment has been vital to the growing economies of the whole region and particularly of Southeastern Asia. The strong yen led Japan's major corporations to invest vast sums in overseas manufacturing facilities that provided Japanese entry to new markets. It has been the most successful of the former kingdoms surrounding the Chinese hub but is being challenged again in Eastern Asia by the growing Chinese economy.

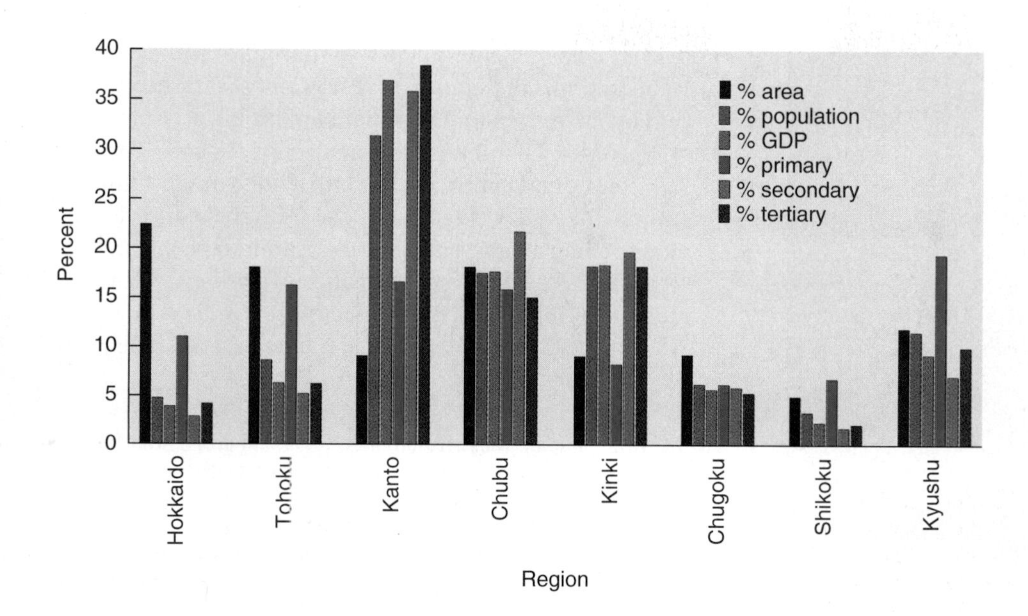


**FIGURE 6.9 Japan: regional differences of population and economy.** Note the relationships of area and population, the proportions of GDP, and the economic structure by sector—as percentages of each region's GDP.

## South Korea and Taiwan

South Korea and Taiwan grew economically in the later 1900s, assisted by U.S. protection, investments, and favored market status. This encouraged them to resist North Korean and Chinese expansion and made them good allies of the United States. Like Japan, they had governments that ranged from paternalistic democracies to dictatorships, promoting internal savings and exporting industries. As in Japan, such policies led to prosperity, large trade balances, and overseas investment by the 1980s and 1990s. It also led to more democratic governments in the 1990s. Also as in Japan, South Korea's financial sector, together with its mode of industrial organization in large conglomerate firms, or *chaebols,* threatened to destroy much of the country's economy and political stability in the late 1990s.

Taiwan maintains its independence although China sees it as a part of the mainland country. China's leadership is determined to follow the recent transfer of Hong Kong with that of Taiwan—extending the "one country, two systems" formula—but Taiwan watches events in Hong Kong before agreeing to become fully part of China again. Around half the Taiwanese population supports reunification with China. Taiwan is different from Hong Kong in that it is 170 km (100 mi.) offshore, has no built-in deadline date for return to China, and has a more mature democracy than Hong Kong and much of the rest of Eastern Asia. It has guarantees of Western support, especially from its strongest ally, the United States. Taiwan was less affected by the financial troubles of Eastern Asia in the late 1990s.

## Countries

Japan has more than three times the area of South Korea and nearly three times the population; South Korea is twice the size of Taiwan, with more than twice the population. The three countries are peopled by distinctive ethnic groups—Japanese, Koreans, and Taiwanese (mainly Chinese).

### Japan and Its Regions

Although Japan is a small country compared to China, it contains a considerable regional variety. The physical environments—from snowy Hokkaido in the north to subtropical Okinawa in the south—combine with distinctive contrasts in urbanization and manufacturing emphases. Japan consists of four main islands: Hokkaido, Honshu, Shikoko, and Kyushu (see Figure 6.8).

- The northernmost island of *Hokkaido* has over 20 percent of the land but only 5 percent of the Japanese population and produces 3.8 percent of total GDP (Figure 6.9). The island is hilly and, until recently, suffered from poor accessibility. A government development program, a direct rail link through the Seikan tunnel under the Tsugaru Strait, and better air services are helping to extend the growing Japanese economy into this island. Hokkaido's natural resource-based industries—fishing, coal mining, and forestry—were recently displaced in importance by farming based on new breeds of rice together with dairying and other crops linked to the increasing Japanese demand for western-style foods. Tourism is expanding, helped by holding the 1972 Winter Olympics at **Sapporo**—now Japan's fifth largest city. This region was affected badly by the late 1990s recession, with many bankruptcies and increasing unemployment.
- The main island of Honshu can be divided into three parts. *Northernmost Honshu* (Tohuko) also has a relatively sparse population for Japan—8 percent of the total population on 18 percent of the land—but its density is twice that of Hokkaido. Despite better transportation provision, the central mountains make crossings between the two coasts difficult. This region is Japan's leading rice producer with 25 percent of the total, much on reclaimed coastal lands. Fishing and forestry are also important. Since the 1960s, the Japanese government has had difficulties in trying to bring manufacturing to the area.

(a) NASA

- In *central Honshu*, the Tokyo region (Kanto) is the heart of Japan in terms of population concentration and economic activity: 31 percent of the population produce 37 percent of Japanese GDP on under 10 percent of the land. **Tokyo**, Kawasaki, and **Yokohama** fuse into a huge urban complex around Tokyo Bay (Figure 6.10) to form the world's largest metropolitan area. It had around 27 million people in 1996. Tokyo is the center of finance and other service industries, Yokohama is Japan's busiest port, and Kawasaki is a center of heavy industry. Oil refineries, steel mills,

**FIGURE 6.10 Japan: Tokyo-Yokohama-Kawasaki urban area.** (a) Space shuttle view of the harbor and the world's largest metropolis. The extensive port facilities reflect the country's involvement in trade. Although subject to earthquakes at a point where three major, active tectonic plates meet, most of the industrial building is on landfill and artificial islands: such materials are liable to liquefaction in an earthquake, and a major event could cause catastrophic damage. (b) Part of Tokyo harbor, showing detail of industrial building on landfill.

(b) © Michael Yamashita/Corbis

© Michael Yamashita/Corbis

**FIGURE 6.11 Japan: Kyoto.** Nijojo Castle and modern buildings in Japan's ancient capital. What is distinctively "Japanese" about this urban landscape?

chemical plants, and power stations line the waterfront and cover large sectors of reclaimed land. The rest of the plain around this megacity still grows much rice, but fruit, vegetables, poultry, and pigs are also produced for the nearby markets. Volcanic mountains around the northern and western margins of the bay area are used for winter sports facilities (with the 1998 Winter Olympic Games at Nagano), and their hot springs attract tourists.

West and north of the Tokyo complex, the land in Chubu is more mountainous. The northern coastal section facing the Japan Sea is a major rice-growing district. Mountains separate it from the main economic areas of the country, although highway and rail links are being built through these Japanese "Alps." The mountains include national parks, such as that around Mount Fuji, Japan's highest peak (3,776 m, 12,300 ft.), and have hot springs. Chubu leads the country in hydroelectricity generation. The Pacific coastal area is an industrial corridor of connected lowlands where former agricultural land has been urbanized by a line of cities along the "bullet train" (*shinkansen*) and major highway routes. **Nagoya** is Japan's third largest city and the center of a textile region that now has automaking (Toyota), oil refining, petrochemicals, and engineering industries. This region suffered least in the late 1990s recession.

- *Southern Honshu* contains Japan's second most prosperous region (Kinki province, often called Kansai). Three cities dominate the eastern part of this region: **Kyoto** is Japan's sixth largest city and was the capital from A.D. 794 to 1868 (Figure 6.11). It remains a cultural and tourist center and has craft industries such as silk, pottery, and traditional furniture. **Osaka**, the second largest city with 10 million, and **Kobe** are industrial ports on Osaka Bay. Following an early specialization in textile manufacturing, the area shifted into iron and steel, chemicals, and shipbuilding in the 1920s, adding oil refining, petrochemicals, automaking, and domestic appliances since World War II. Reliance on such industries resulted in this region becoming Japan's "rust belt" of old, declining industries, and it was hit very hard in the late-1990s recession. It needs to keep up with new trends. Much damage from the 1996 Kobe earthquake is not repaired.

  The western peninsula of southern Honshu (Chugoku) has coastal lowlands on either side of a mountainous spine, but the western coast has a harsher winter climate and poor links with the rest of Japan. The eastern coast facing the Inland Sea has a milder winter climate and responded economically to improved accessibility by train and road since 1945. Reclamation of land produced port facilities and heavy industry using imported raw materials. In 1995, the new Kansai airport was opened, built on reclaimed land south of Osaka.

- *Shikoko* is a mountainous, small island with only 5 percent of Japan's land. The southern part of the island has a warm, moist climate but is isolated by mountains, while the northern coasts facing the Inland Sea are irrigated for crops and industrialized. Major infrastructure investments in bridges now connect the island with the rest of Japan.

- *Kyushu* is the westernmost island and is connected by tunnel to Honshu. It was the historic point of entry to Japan for Buddhism, new writing script, and trade. Two-thirds of the island's population is concentrated in the northern region around Fukuoka and **Kitakyushu**, both having long-established metal, chemical, and shipbuilding industries. These are now reducing output and employees. Farming and fishing remain important in the economy. The rest of Kyushu is a rice-growing area. There is often a second crop of vegetables or rushes, while early crops are grown for the large cities farther north. Mandarin oranges, tobacco, and livestock are produced on the hills. In the 1980s, Kyushu attracted automaking and high-tech industries with expanding research and development facilities that eased the late-1990s problems. It has a growing tourist industry, although affected in the late 1990s by the decline in numbers of Korean shoppers.

- Offshore, the Ryukyu Islands including Okinawa extend 1,000 km (650 mi.) southward toward Taiwan. Many islands still have American military bases alongside the traditional fishing and agriculture, and there is a developing tourist industry. Pressure increases for the removal of the last group of U.S. military personnel on Japanese territory.

### South Korea and Taiwan

South Korea is a peninsula of mainland Asia that forms a hilly land with the highest ridges along the east coast. The main population and industrial facilities growth was in the northwest, including the capital, Seoul, and along the southern coast.

Taiwan is dominated by the north-south mountain ridge that has highest points just over 4,000 m (12,000 ft.), lying just east of center along the island's long axis. The western plains are the site of current urban expansion, while the uplands and east coast are less developed.

## People

### Population Distribution

All three countries are densely populated (Figure 6.12). Seventy percent of Japan's population is concentrated in the central southern coastlands from Tokyo to Osaka. South Korea's population has a major concentration around **Seoul** and **Taejon** in the northwest, with other large cities along the southern coast, including **Pusan**, **Taegu**, and **Ulsan**. In northern Taiwan, **Taipei** is the center of the country's main concentration of people.

### Population Growth

With few natural resources, people are central to past and future economic growth in these countries. Japan's total population increased from around 90 million in 1960 to 126 million in 1998. The combination of low fertility (1.4 in 1998) and population growth rates (natural increase of 0.2% in 1998) will cause the total to peak around 128 million in 2010 and fall to 120 million by 2025. Infant mortality at 4 per 1,000 live births is the world's lowest and life expectancy at 80 years is the highest.

In 1948, concern over providing for its large and rising population after World War II led Japan to pass its Eugenics Act, allowing abortions and instigating a series of family planning drives. As the birth rate fell over 50 years, from 34 to 10 per 1,000, the death rate fell from 14 to 7 per 1,000. Japan is in the final stage of demographic transition (Figure 6.13). The age-sex graph for Japan (Figure 6.14) has bulges related to more postwar and late-1960s births and reduced numbers of births between (as the economy grew) and since 1970 (when birth control and affluence combined).

Although Japan's population is not rising rapidly, changes in its structure and distribution have important economic and social implications. The population is aging. In 1970, 7 percent of the population was over age 65; by 1995, this proportion rose to over 14 percent. Traditional care of the elderly by eldest sons ended as smaller nuclear families and accommodation in small suburban houses became common. The shift to smaller households caused the total number of households to increase from 23 million in 1960, when they averaged 4.5 persons per household, to over 40 million in the 1990s, with 3 persons per household. The aging of the population affects pension provision, medical costs, and leisure provision for the retired, but it also increases firm's wage costs as more employees reach senior positions. Japanese people are now worried that there will not be enough labor for their industries or funds to support the increasing numbers of elderly people. Encouragements are being considered for couples to have more children.

South Korea and Taiwan both have low rates of population increase—under 1 percent in the mid-1990s—and are entering the final demographic transition stage. South Korea's population of 46 million in 1998 is projected to rise to 53 million by 2025, while Taiwan's 22 million could reach 25 million. Both will need to prepare for rapid increases in older dependent populations in the early years of the new century.

### Rural to Urban Populations

In the short period of 70 years after 1920, Japan changed from a largely rural country with some industrial towns housing one-fourth of the population to the present situation where three-fourths are city dwellers. After World War II, half of the Japanese population was still rural, but further industrialization brought more urbanization.

Up to 1940, Japan's developing industrialization caused gradual migration from rural locations to factory towns, especially into the growing metropolitan areas along the Pacific coast from Tokyo to Osaka. Between 1945 and the mid-1970s, such migration became more rapid as jobs in the manufacturing and service sectors multiplied. Growth continued in the Tokyo metropolitan area but extended to the industrializing area around Osaka and even to some towns on Hokkaido, southern Honshu, and Kyushu. The urbanized zone between Tokyo and Osaka is often known as the Tokaido **megalopolis.** The term *megalopolis* is used for a series of almost continuously welded metropolitan centers and was coined to describe the growing together of the U.S. cities between Boston and Washington, D.C. (see Chapter 9).

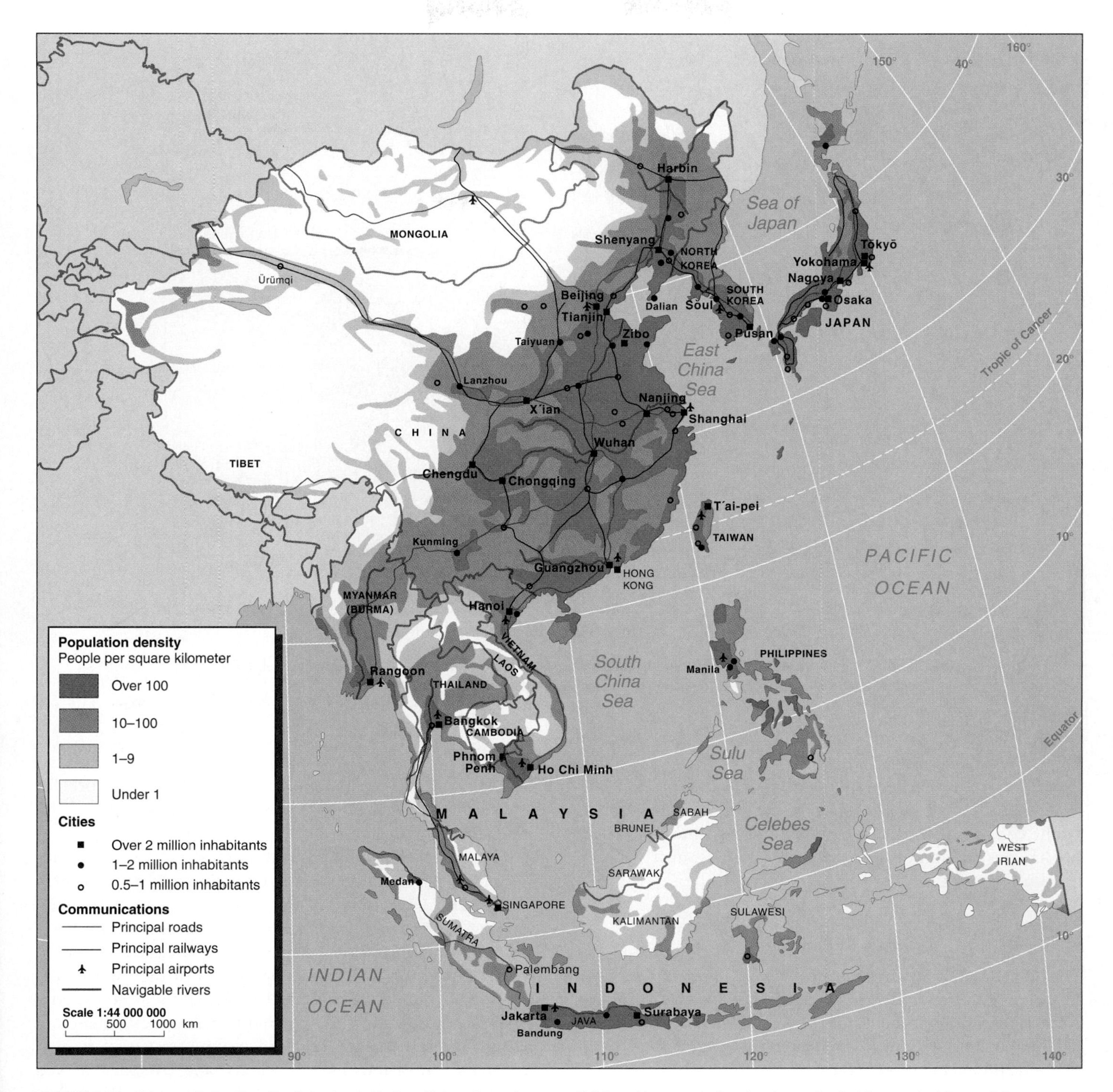


**FIGURE 6.12 Eastern Asia: distribution of population.** Relate the main areas of high and low population density to physical factors. Is this a satisfactory basis for explaining the differences?

Source: Data from *New Oxford School Atlas,* p. 75, Oxford University Press, UK, 1990.

From the mid-1970s, the main metropolitan areas grew more slowly, and there was a movement out of central cities catalyzed by congestion and helped by improving transportation. This is more a process of **suburbanization** than of **counterurbanization**; in the former, people move to the outskirts of existing cities, while in the latter, they move away from metropolitan centers to small towns in rural settings.

As urban population grew, the more remote areas in the mountains and along parts of the western coasts of Japan became depopulated. Such regions often have over 17 percent of their population over 65 years old and are assisted by government grants to help them retain some population.

In South Korea, urbanization and industrialization were even more rapid than in Japan. Urban population increased

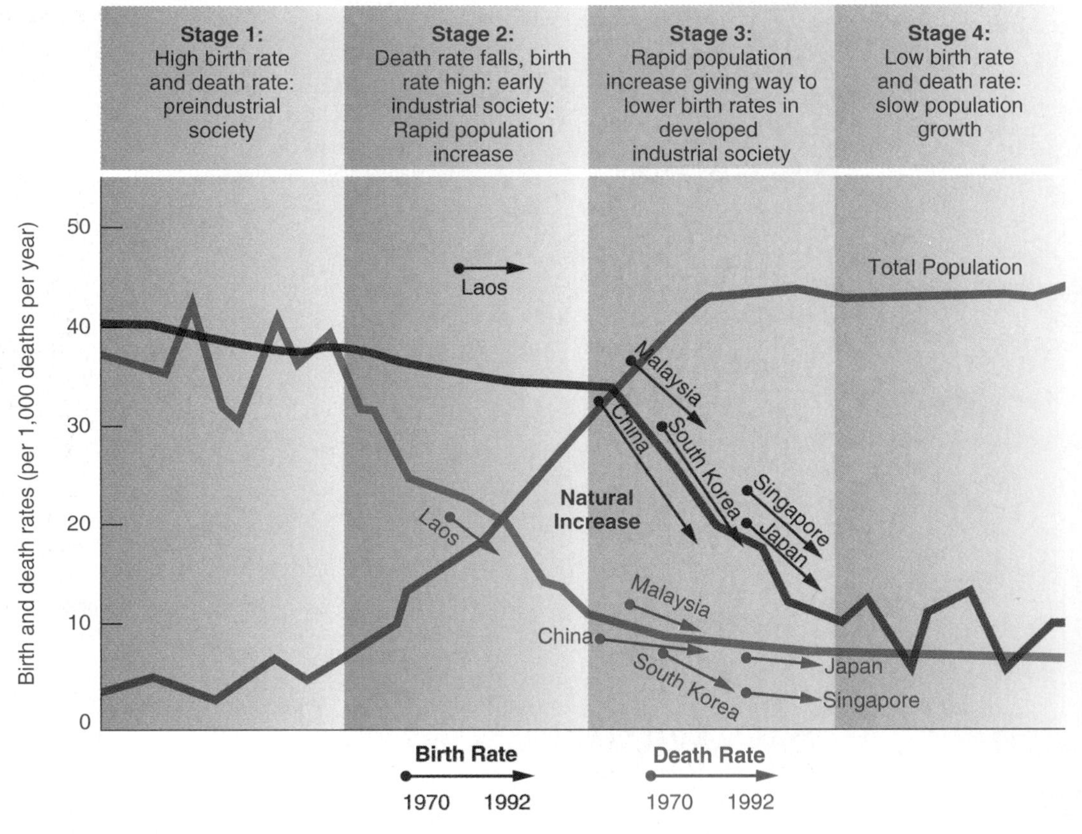


**FIGURE 6.13 Eastern Asia: demographic transition.** Compare the positions of Eastern Asian countries in this transition with each other and with countries in other major world regions.

from 25 percent in 1950 to 50 percent in 1980 and 80 percent in 1998. Taiwan also urbanized rapidly over the same period. Traffic congestion as the economy grew was almost matched by the construction of new roads and urban transit systems in both countries.

## Ethnicity

The Japanese, Korean, and Taiwanese populations are remarkably uniform in race and language, given the multicultural variations throughout the rest of Eastern Asia. Most Japanese people stem from an ancient mixture of Asian continental and Pacific island stocks. From approximately 1650 to 1850, the military shogun rulers isolated Japan from the European influx of traders. This policy had a formative and consolidating effect on its ethnicity and on its culture that brought together many elements of religion and philosophy, including the indigenous Shintoism, which is an animistic and pantheistic religion based on emperor worship and Japanese superiority. Buddhism, brought from Korea in the A.D. 600s, tolerates Shintoism and is followed by most of the population. Chinese Taoism and Confucianism are also common. Beliefs in any one of these are not seen as exclusive. Most agree on the subordinate status of women, although women's status is changing with their greater involvement in the labor force. Yet women still work for lower wages and few attain management positions or lifetime employment status.

Three small Japanese minority groups remain important. The Ainu of Hokkaido were overwhelmed by the Japanese occupation of the island in the 1800s but retain ethnic and language differences. On Honshu, the Burakumin, numbering around 1.2 million people, were outcasts allowed to work only in the lowest occupations until their emancipation in 1871, but it was not until 1969 that the Japanese government put funds into upgrading their settlements to nationwide standards. The third group consists of foreign nationals, three-fourths of whom are Koreans moved forcibly to Japan in the early 1900s. Until the 1980s, they found it difficult to gain Japanese citizenship and are still subject to discrimination.

South Korea has a dominantly Korean ethnic composition—a people who trace their ancestry to the Mongols but were much affected by Chinese culture and Japanese occupation. In Taiwan, 84 percent of its people claim to have had relatives in the country before 1949 but are similar ethnically to the immigrants from mainland China at that time. There are also some groups who were on the island before the Chinese arrived, mostly living in the less populated mountainous area.

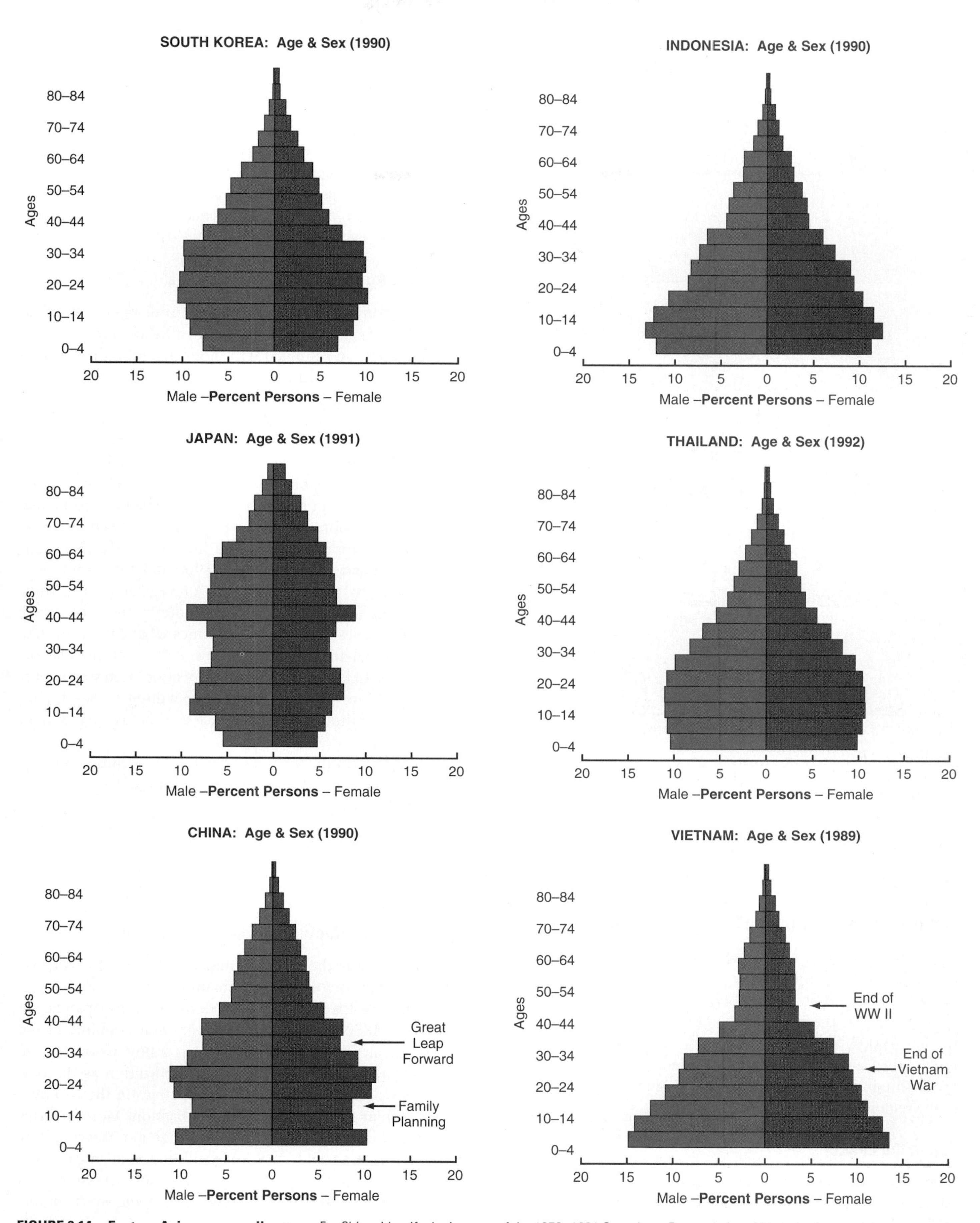


**FIGURE 6.14 Eastern Asia: age-sex diagrams.** For China, identify the impacts of the 1959–1961 Great Leap Forward, the 1966–1976 Cultural Revolution, and the one-child-per-family policy of the early 1980s. Compare the Japanese diagram with those of poorer countries in the region.

Source: Data from *Demographic Yearbook,* U.N. 1994.

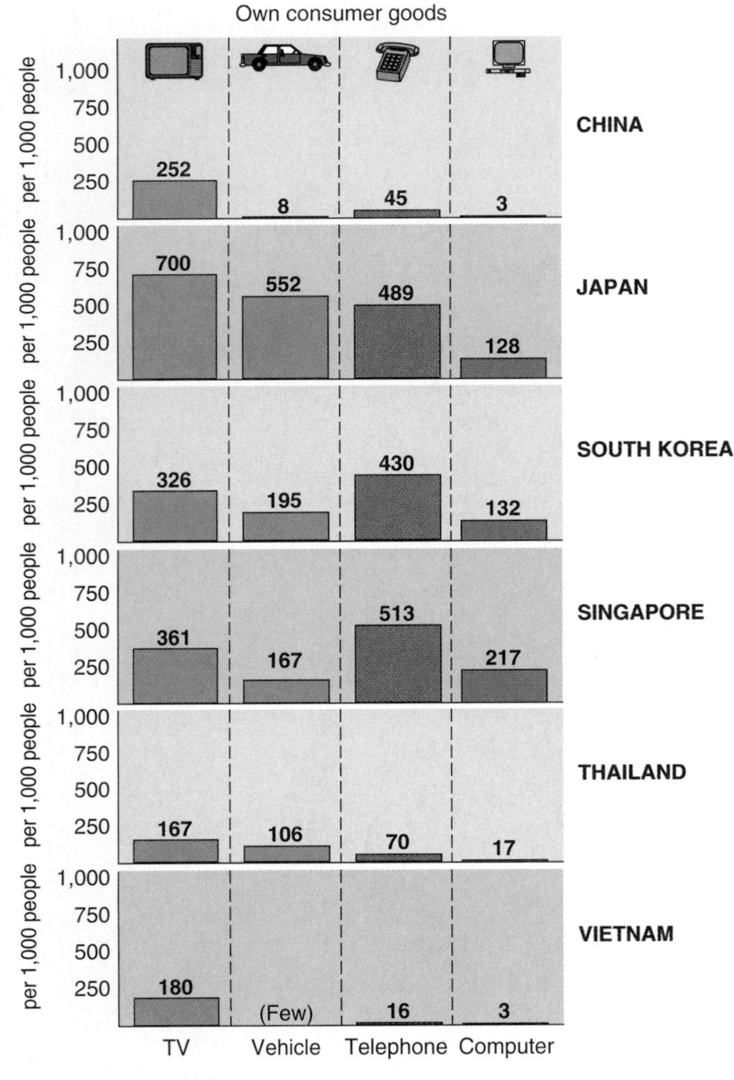


**FIGURE 6.15 Eastern Asia: ownership of consumer goods.** Account for the wide range of affluence among these countries.

Source: Data (for 1996) from *World Development Indicators,* World Bank, 1998.

## Economic Development

Japan, South Korea, and Taiwan show differences in size, population totals, and ethnicity. The greatest similarities are in terms of economic performance since the 1950s, for these countries lay at the heart of the "Eastern Asian Miracle" in the late 1900s. Yet, despite following somewhat similar formulas, they are each distinctive in detail. Consumer goods ownership in Japan and South Korea (Figure 6.15) stands out in the region.

### Japan the Leader

The Japanese economy experienced massive changes after World War II. From 1960 to the late 1990s, farming, fishing, and mining—the primary sector of the economy—declined from 33 to 2 percent of the work force, while the service sector rose from 38 to 60 percent. The American occupation of Japan after World War II resulted in democratization, including the female franchise and land reforms. The initial outcomes of postwar restructuring were hardship and uncertainty. It was not until the 1950s that American aid, substantial contracts during the Korean War (1951–1953), and a lack of need to invest government funds in defense (since it was forbidden) began to make a difference to the Japanese economy. Growth was rapid thereafter.

### Japanese Agriculture

Japanese farming is highly productive but concentrated in small areas of suitable land within the hilly and mountainous country, often intermingled with housing and industry (Figure 6.16). Urban and industrial users, who pay high prices, also compete for the land.

Agriculture remains a significant sector of the economy only because of government support. After World War II, the government bought rice at a price that supported the farmers and continued mechanization, and then sold it cheaply to the growing urban populations. Farming remained profitable, but overproduction led to a rice mountain that was too expensive to export as new tastes for Western foods caught on. Rice demand halved between 1960 and 1990, and more land was given to the cultivation of vegetables and fruit. Farm acreage fell, reducing the rice store by the 1980s. Rice still costs Japanese consumers six times what Americans pay and 13 times what Thais pay, because of the continued Japanese subsidies to farmers and tariff protection from world market competition. Japanese people are willing to pay higher prices because they believe their rice is of better quality than imported rice.

Although farm income was secure, new earning opportunities in towns attracted people from rural areas. Migration from rural to urban locations halved the rural population between 1960 and the 1990s, while only 12 percent of farmers were full-time without off-farm jobs after 1970. Farm size remained stable, however, because land increased in value and owners held on to their land as an investment.

### Japanese Manufactures and Modernization

Manufacturing is the basis of modern Japanese growth, but the complexity and diversity of manufactures did not happen overnight, or even since 1945. Modernization in Japan began in the mid-1880s, building on the economic stability of the shogun era and responding to external trading pressures. The restored Japanese monarchy saw modernization as the best way to resist colonialism and made changes in the administrative system to encourage industrialization. Most factories were built in the zone from Tokyo to Nagasaki. Heavy industries such as iron and steel, shipbuilding, and engineering gave a capability for further industrial expansion and brought overseas respect. Textiles became the most important of these early industries.

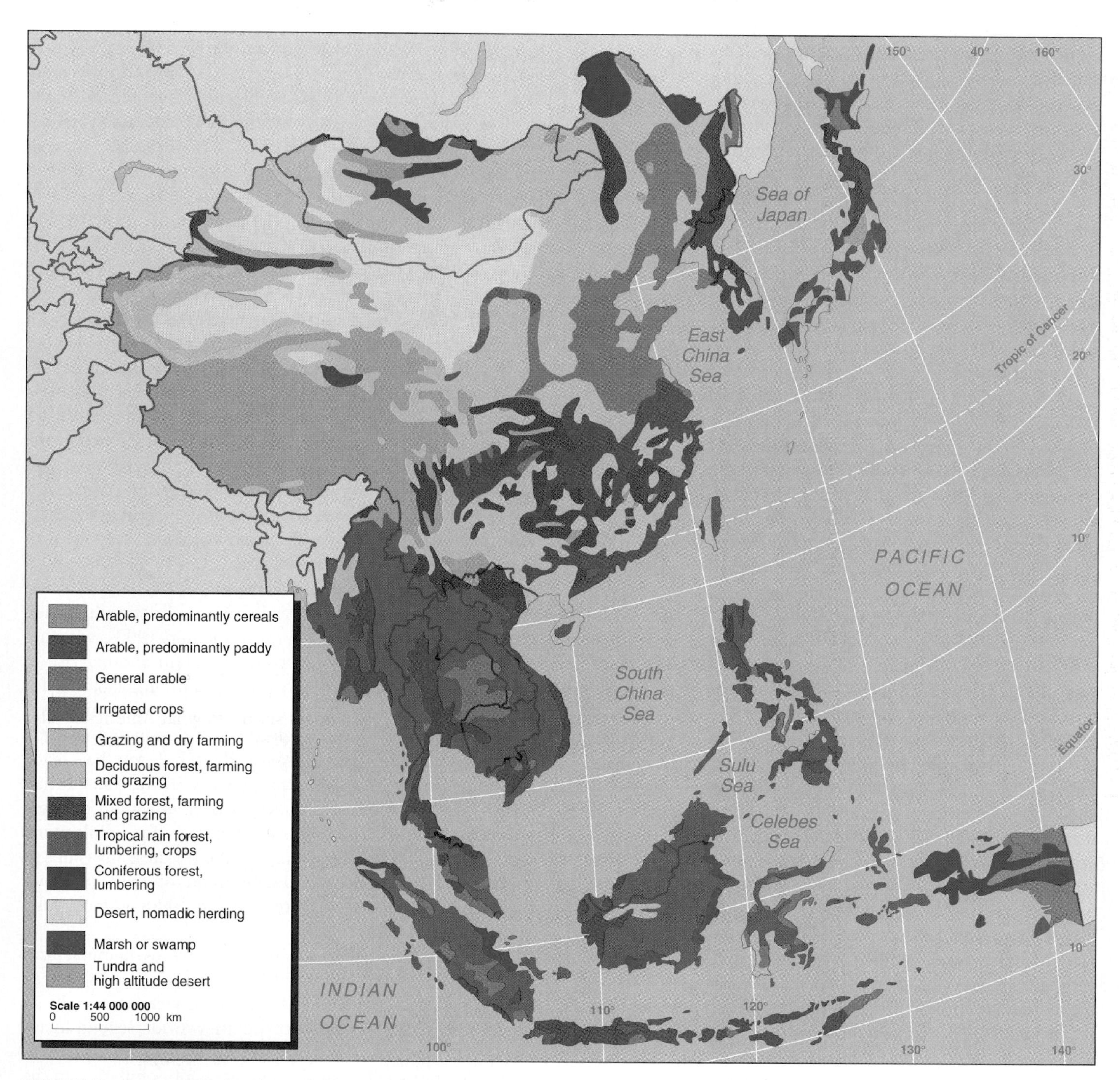


**FIGURE 6.16 Eastern Asia: farming and other land uses.** Notice the contrast in intensity of production between eastern and western China and the distinctive forest landscapes of Southeastern Asia alternating with intensively farmed valleys.

Source: Data from *New Oxford School Atlas*, p. 74, Oxford University Press, UK, 1990.

It was during this period that factors such as government-business links, cheap labor, large numbers of small businesses, and improved infrastructure assumed importance. Overseas, Japan became known for cheap goods, often of mediocre quality. The Japanese military took a growing interest in Japan's ability to produce its own goods. The combination of military involvement and the major facilities owned by a few elite families, known as *zaibatsu*, came to the fore in the 1930s depression—when other countries protected themselves from cheap Japanese imports—and shifted production to naval ships, air force planes, and other military equipment. The focus on military production culminated in World War II, which ended with the destruction of much of Japan's industrial capability by U.S. bombing and postwar dismantling.

The Cold War, beginning in the late 1940s, and the Korean War of the early 1950s led the United States to pour money into Japan to rebuild its industries and infrastructure. American markets opened to Japanese products. Although Japan lacks the mineral resources that were once considered necessary to a region's industrial growth, the period after World War II brought improved transportation technology that made it possible to transport bulk quantities of oil, coal, and iron ore from southwestern Asia and Australia at low prices. The cheapness of raw materials in world markets and its own low-cost, young labor force enabled Japanese manufacturers to undercut their rivals in the world's core countries and gain overseas markets. The Japanese government assisted and advised industry, with the **Ministry of International Trade and Industry** (MITI) managing export sales through a worldwide network of marketing intelligence-gathering offices.

Manufacturing industries expanded along the Pacific coast of Japan, often on reclaimed land. Following the rebuilding of heavy industries such as steelmaking and chemicals in the 1950s, Japan diversified into shipbuilding, automaking, and light manufactures that were demanded by consumers abroad. Japan made timely transportation innovations such as building very large oil tankers and bulk carriers for other commodities at a time when the Suez Canal was closed and western oil companies sought a cheaper means of transporting oil. During the 1960s, total output increased by 10 percent per year. At this stage, Japanese workers waited for gains in prosperity as their firms plowed profits back into new developments. It was not until the 1970s that Japan established a favorable trade balance.

In the 1970s, labor shortages and increased oil prices, together with a world shipping slump and reduced demand for some products, caused Japanese corporations and their supportive government to rethink policy. Japan was dependent on imported oil for 70 percent of its energy needs. A greater reliance on internal sources such as hydroelectricity and nuclear power became policy, while unprofitable coal mines were supported by government subsidy. Increased investment raised the output of such light industries as camera and household appliance manufacturers. International competitiveness was increased by applying high levels of technology to production, including electronics, robotics, and new materials. From being a country that took over technology and design to please Western markets, Japanese firms became initiators and built a quick response to changing markets. In the 1990s, after finding that the indiscriminate use of robotics and automation was more expensive than using it where most appropriate, Toyota moved to a system that involved the greater use of people skills in making cars and saw further increases in productivity.

### Japanese Responses to Changes

Japan turned from a country of producers to one of consumers. Its exporting was so successful that other countries had huge deficits of payments to Japan. In the 1980s, the value of the Japanese yen currency doubled, making Japan's exports more costly. Imports became cheaper, however, so that Japanese people's spending rose and produced a major boom in retailing and construction. As consumer demand rose, more luxury and prestige goods were sold and Japanese manufacturers moved into these fields, competing with imported goods.

By the 1980s, some of the older industries such as steel, shipbuilding, petrochemicals, and cement-making suffered overcapacity on world markets and competition from newly industrializing countries such as South Korea. The lack of Japanese competitiveness in these fields was made worse by the high-value yen. Many factories in Japan closed, leading to local social problems in the Osaka region in particular.

In the 1980s, increased investment overseas began with Japanese corporations buying out their mineral suppliers but later extended to their establishing Japanese factories in other countries. The rising value of the yen caused Japanese industries to build large productive units in other countries during the 1980s and 1990s. Such moves also helped in the penetration of markets that had previously placed quotas on the import of, for example, Japanese-made cars. After major investments in the United States and Britain, where there are now 14 Japanese auto plants, Japan switched investment to Southeastern Asia in the late 1980s.

The economic growth of Thailand, Malaysia, and other countries in that subregion received a major stimulus when Japanese investments poured in. In the mid-1990s, a further phase of foreign investment coincided with another rise in the yen. On this occasion, the policy of building factories for assembling Japanese cars or completing cheap-labor operations was broadened to include shifts of design and research facilities to Southeastern Asia. Although Japan has made some investments in China, particularly in the Manchurian port of Dallan, it remains cautious over uncertain Chinese government policies and poor local cooperation, while many Chinese remember the bad times of the Japanese invasions in the early 1940s. The worldwide trend of core countries investing beyond the core areas in the 1990s is most fully developed in Eastern Asia.

The huge trade surplus from exports of manufactures resulted in the rapid growth of Japan's retailing and financial sectors and of personal and business services, especially in the 1980s. In 1986, Tokyo became the world's second financial center after New York and in front of London. The additional functions led to further increases of population in the Tokyo region.

Japan became a place for tourists to visit but mainly remained a source of tourists visiting other parts of the world. In 1996, it attracted over 2 million visitors, while 17 million of its people vacationed abroad.

Japan thus passed through the economic sectors from feudally controlled agriculture to Western-type manufacturing and then to major service industries and investment abroad. In the later 1990s, Japan again faced a need to change. Its established systems of government-business linkages were undemocratic, and its banking systems came under pressure because of huge debts, particularly resulting from its investments in Southeastern Asia. While Western nations imitated

Japanese processes and were forced increasingly to compete with Japanese goods in world markets, the Japanese looked more closely again at Western ways, particularly in relation to financial control.

## South Korean Economy

South Korea suffered from a devastating war on its territory in the early 1950s, at a time when three-fourths of its people still gained a living from farming. In the subsequent land reform, the government took over large estates that had been run inefficiently and encouraged a new group of small landowners to use imported fertilizers for increasing productivity. As farm productivity rose, South Korea became largely self-sufficient in food.

When the Japanese ruled Korea, they built roads and railroads to develop the Korean economy for their benefit. After World War II and the Korean War (1951–1953), massive overseas aid and investment from the United States and Japan combined with strong government action, leading to economic growth and social changes. South Korea pursued the development of import substitution industries and, like Japan, forced its people to work for low wages while the economy improved.

From the early 1960s to the early 1990s, few other countries grew so fast. Exports rose from $33 million in 1960 to $130 billion in 1996. A country of subsistence farmers was transformed in a generation to the world's largest maker of large ships and memory chips, the fifth largest auto maker, and eleventh largest world economy—with greater output than the whole of Africa South of the Sahara. Life expectancy rose from 47 years in 1955 to over 70 years.

Iron and steel, shipbuilding (Figure 6.17), chemicals, automaking, and textiles continue to be important, although in the 1990s, there was a swing toward high-tech industry. South Korea is now a major world economy and trading nation, having a total GDP that is 66 percent of China's and 10 percent of Japan's. South Korea was a major investor in modern China in the early 1990s, taking advantage of cheap labor costs in the nearby Chinese province of Shandong. The initial rush of small companies is giving way to major South Korean corporations such as Daewoo, Samsung, and Hyundai.

The South Korean government supported the growth of larger companies that developed into huge conglomerates, the chaebol, such as Hyundai, Daewoo, LG (Lucky Goldstar), and Samsung. The chaebol became diverse in their products, starting up subsidiaries that had little chance of success, taking on massive debts, and providing major problems for the country. When Halla, the twelfth largest chaebol, became bankrupt in 1997 with debts of 20 times the company's assets, Hyundai, the largest, was found to have guaranteed around 15 percent of the debts because the founders of the two chaebol were brothers. Family loyalty—part of the Korean (and Confucian) culture—was exposed as being less successful in running large corporate businesses. The largest three chaebol each have over 100 businesses but get 80 percent of their revenues from 20 percent of them and all the profit from one or two. Thus, the chaebol make few profits yet crowd out smaller, innovative firms. The chaebol also control large parts of the South Korean financial markets, providing an environment for building up mutual debts that could result in massive financial ruin.

**FIGURE 6.17 South Korea: Ulsan City shipyard.** The Hyundai *chaebol* is a major builder of large ships. Note the space required for assembling the materials and the range of equipment used in construction.



In late 1997, indebtedness reached the point where the South Korean currency unit, the won, and its stock exchange, collapsed, leading to fears that the linked financial problems in Japan and Southeastern Asia might cause widespread financial failure and economic depression in the region and beyond. Devaluation of its currency raised South Korea's indebtedness to other countries. Increased bankruptcies and unemployment stirred militant trade unions into political action against the stringent conditions proposed by external loan providers.

## Taiwan Economy

Like Korea, Japan occupied Taiwan before 1945. The Japanese used the Taiwanese as slave labor to build roads and railroads, water projects, coal mines, and factories. They encouraged

**RECAP 6B: Japan, South Korea, and Taiwan**

Japan is a heavily populated, small, island-based country that has become one of the world's richest nations following a dramatic economic recovery since World War II.

Japan's agriculture remains highly protected from foreign competition. Its industrial economy was first based on importing raw materials and producing goods for export markets. It has moved from iron and steel manufacturing to cars and high-tech goods. In the 1990s, the rising value of the yen, caused by selling more exports than buying imports, led to competition in world markets from cheaper sources. This led to a slowing of Japan's economy and to major investments in other countries.

South Korea and Taiwan are two of the most economically advanced countries in Eastern Asia. Their growth is based on manufacturing for exports and strongly directive governments. South Korea's development was interrupted by financial disasters in the late 1990s.

6B.1 In tracing the history of economic growth in Japan from the mid-1800s to the present, how would you assess the impacts of the changes on Japan's human geography?

6B.2 What are the main regional differences within Japan?

**Key Terms:**

megalopolis
suburbanization
counterurbanization
Ministry of International Trade and Industry (MITI)

**Key Places:**

- Japan
  - *Sapporo*
  - *Tokyo*
  - *Yokohama*
  - *Nagoya*
  - *Kyoto*
  - *Osaka*
  - *Kobe*
  - *Kitakyushu*
- South Korea
  - *Seoul*
  - *Taejon*
  - *Pusan*
  - *Taegu*
  - *Ulsan*
- Taiwan
  - *Taipei*

farming and the growth of a local market for Japanese goods. In 1949, the Kuomintang forces escaping from China took control of Taiwan. The Kuomintang leaders brought their families to join several thousand military personnel who had been stationed on Taiwan to quell anti-Kuomintang riots. All opposition was doused and a repressive dictatorship imposed.

The new government instituted policies that led to very rapid industrialization, rural change, and urbanization. At first, agriculture provided the financial basis for industrial growth in Taiwan, which remained slow until the late 1960s. The increasing rice surplus gave way to more diversified farm products, including sugar, tea, vegetables, and fruits. The expansion of urban-based industrial jobs and buildings, plus the mechanization of farming, reduced the farm population from 6 million in 1965 to under 4 million in 1990, while an initial agricultural trade surplus gave way to increasing food imports. Agriculture's share of Taiwan's GDP fell from 27 percent in 1965 to around 5 percent in 1990.

Starting in the mid-1960s, export processing zones focused attention on manufacturing for export. By the 1980s, the emphasis switched to high-technology developments. Private enterprise increased its ownership of firms from 52 percent in 1960 to over 80 percent in the 1980s. The government retained control of banks, interest rates, and exchange rates to maintain a consistent financial environment. By the 1990s, Taiwan had one of the largest world trade surpluses and was able to export capital, particularly to China and Thailand.

The large financial surplus was a strength, and Taiwan was affected less when Japan, South Korea, and Southeastern Asia experienced major financial problems in the late 1990s. Taiwan was also helped by having mainly small corporations, more flexible conditions that enabled new firms to force others out of business, and higher productivity compared to the South Korean chaebol and Japanese conglomerates. Taiwan is using this strength to get diplomatic support for recognition by the United Nations and the International Monetary Fund (IMF). Taiwanese businesspeople take great care in making overseas investments at a time of widespread financial problems in the region.

## CHINA, MONGOLIA, AND NORTH KOREA

The **People's Republic of China, Mongolia,** and **North Korea** (People's Democratic Republic of Korea) (Figure 6.18) are linked by geography and recent history, although currently they operate independently. China came under communist government control in 1949. Korea north of the 38th parallel was occupied by the Soviet Union after 1945, leading to the creation of a separate communist state, and was supported by China in the 1951–1953 war against South Korea and the U.S.-led United Nations forces. Mongolia struggled to maintain its independence as a buffer state between China and the Soviet Union. Soviet military forces were withdrawn only in 1992. **Hong Kong** and **Macao**, previous colonies of the United Kingdom and Portugal, respectively, were returned to Chinese control in 1997 (Hong Kong) and at the end of 1999 (Macao).

### China, Mongolia, North Korea, and the World System

#### Chinese Perceptions of Its Role

China remains the geographic and strategic hub of Eastern Asia. A period of dynastic weakness and foreign intrusions in the 1800s was followed in the 1900s by conflicts over attempts to develop a republic, war with Japan, and isolation under communism until the mid-1970s. These events pushed China out of its accustomed dominant role in Eastern Asia for nearly two centuries. Japan in the north and European colonial powers in the south took over much of the historic Chinese realm by the mid-1900s, and Japan became a new core power within the world economic system in the late 1900s. From the late 1970s, however, China relaxed its isolationist communist ideology and began to bounce back. By the 1990s, it seemed likely to rejoin the ranks of the world's leading countries—as an integral part of the world economic system. After 20 years of economic growth, China in the late 1990s still grappled with challenges of two transitions: from a central-command econ-

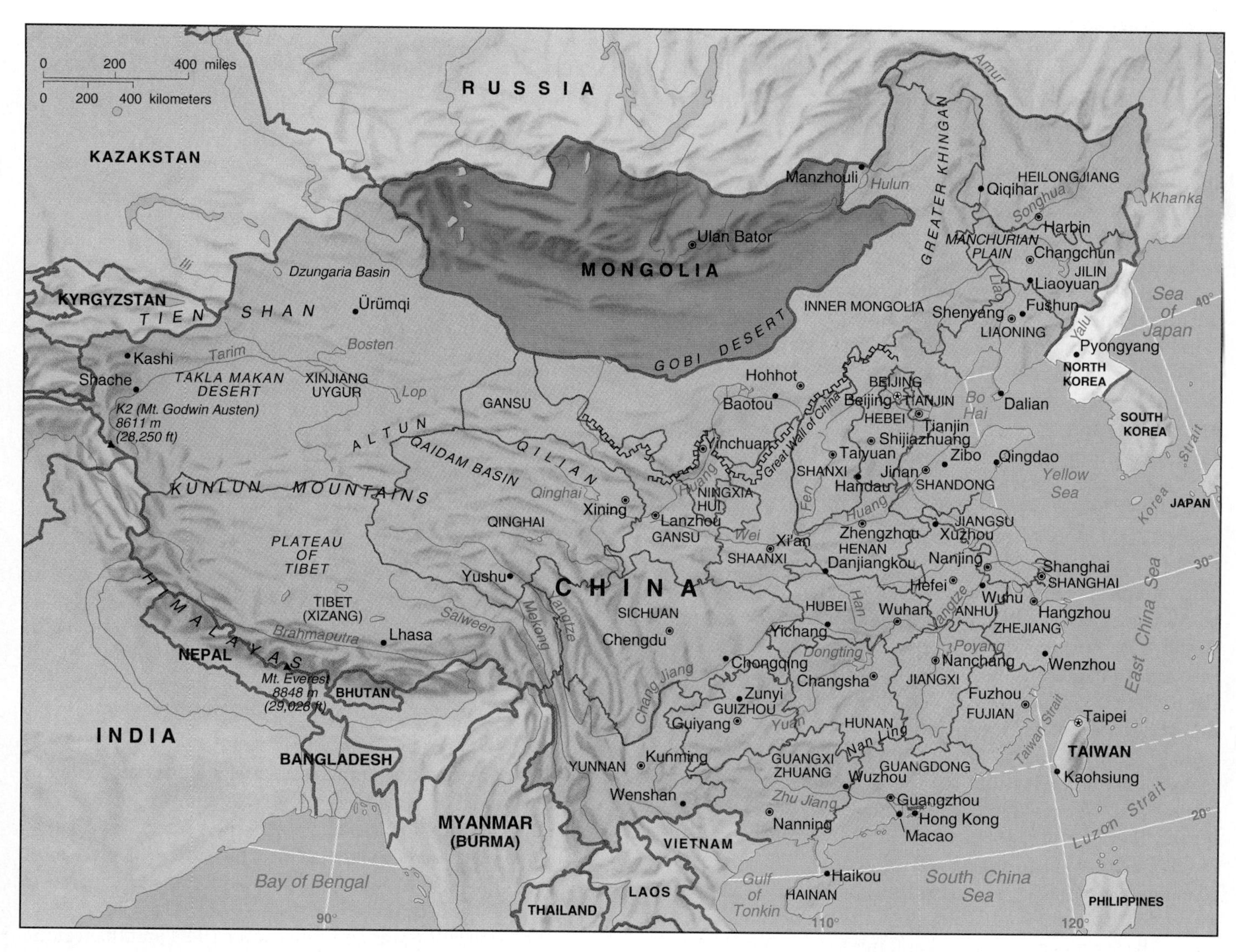


**FIGURE 6.18 The People's Republic of China, Mongolia, and North Korea: major geographic features, including towns and cities, major rivers, and provincial boundaries.** China has 23 provinces, four cities with provincial status (Beijing, Shanghai, Tianjin, and Chongqing), and five autonomous regions that have large national minorities (Tibet, Inner Mongolia, Guangxi, Ningxia, and Xinjiang).

omy to a unique socialist market-based economy, and from a rural-agricultural economy to an urban-industrial economy.

China continues to perceive its sphere of influence as that which it held during its period of greatest expansion in the Ming dynasty (see Figure 6.3). The surrounding countries remain wary of China's new economic and military strength. While other countries may see China as likely to play an increasing role in the world economically and politically in the new century, the Chinese themselves still view their country as poor and struggling to escape centuries of foreign intervention. The people are still largely rural, many are illiterate, and many have low incomes, although major strides have been made since the 1970s in reducing poverty. In the 1990s, the Chinese government established new programs to reduce poverty further. The Chinese protest when accused of aggressive military tendencies, claiming to have relatively small armed forces and a declining defense budget. At the same time, however, they expect much from increasing contacts with the rest of the world.

In 1998, the People's Republic of China had 22 percent of the world's population living on 6 percent of the world's land. Its very large population kept incomes low, despite having a total economy that was seventh in the world. Education and health characteristics and gender equality, however, are improving in major ways for a large proportion of the population.

## Ups and Downs under Communism

The proclamation of the People's Republic of China in 1949 brought a rule under which national policy was more unified. A set of false starts interrupted economic progress, however.

Mao Zedong, the leader and virtual dictator of China from 1949 to his death in 1976, relied initially on Soviet advice in setting up five-year plans based on the central control of large-scale industrialization and the collectivization of agriculture. Tensions among this policy, the traditional subsistence agriculture, and the previous feudal pattern of local control brought problems that slowed the pace of change.

The communist leaders attempted to force policies on the country by carrying out massive changes over short periods of time. The "**Great Leap Forward**" of the late 1950s attempted to increase the pace of industrialization by regimenting labor and setting up agricultural communes focused on manufacturing. It was a disaster. Both industrial and agricultural production dropped. Drought combined with economic collapse caused famine. Stung by failures and misunderstandings, the Soviet advisers went home. In the early 1960s, the most drastic policies were abandoned, leading to a few years of higher production and greater confidence in the new leaders. Another attempt to change the whole basis of Chinese society, the "**Cultural Revolution**" of 1966 to 1976, swept aside the short-lived gains in a massive persecution of the educated classes, technologists, and former leaders. This was partly an attempt by the aging Mao Zedong to reestablish his leadership, but it again slowed economic development.

After Mao Zedong died in 1976, economic policy was in the hands of Mao's former chief deputy, Deng Xiaoping. He reversed Mao's movement toward self-sufficiency and central control on the basis that China had to accumulate wealth before it could diffuse it to the poorer parts of the country. He returned property and decision making to peasant farmers, encouraged a more open economic policy that attracted investment from outside China, and generated the world's highest rate of economic growth from the mid-1980s to the mid-1990s. New industrial development was linked to world market demands. Deng Xiaoping died in 1997, following the rapid economic growth of many coastal regions of China and the acquisition of Hong Kong. His successors continued and further developed his policies. The economic openness to world markets and to foreign investment appeared to be leading to greater democratic involvement of the Chinese people until the Tiananmen Square massacre of 1989 demonstrated the communist regime's determination to hold on to centralized political power.

### Isolation of North Korea and Mongolia

North Korea and Mongolia remain isolated in a global sense—the former by political dogma and the latter by geography. It is not easy to obtain data for these countries. North Korea's economy appears to have collapsed, subjecting its people to famine. In Mongolia, the small population strives to eke out a living while attempting modernization in a semiarid environment.

## Countries

China is a large country—almost as big as Canada (world's second largest country) and a little larger than the United States of America (fourth largest)—and has the world's largest population. China's borders enclose a variety of regions based on differences in physical environments and on their histories of settlement, occupation, and external linkages. Mongolia is a large country with a small population of 2.4 million in 1998, while North Korea had 22 million people in less than one-tenth of Mongolia's area.

### Regions in China

Politically, China is divided into 23 provinces and other province-level units including four great cities (Shanghai, Tianjin, Beijing, and Chongqing) and five autonomous regions with large ethnic minorities (Inner Mongolia, Guangxi, Ningxia, Xinjiang, and Tibet) (see Figure 6.18). Physically, China is a country of contrasts between the extensive lowlands of the lower parts of the Huang He and Chang Jiang basins and the mountain ranges and high plateau of Tibet; between the subtropical monsoon southern coastlands, the midlatitude east coast climate of the northeast with long cold winters, and the arid inland areas; and between the former forests, grasslands, and deserts. Coal occurs mainly in the north and the highest production is in Shanxi. Petroleum is produced mainly in the northeast and far west, although offshore reserves are being explored and coming into production. Hydroelectricity is the main power source in the south, where nearly three-fourths of the national total is produced.

The combination of physical and human characteristics produces a broad regional division of China (Figure 6.19). The major contrasts are between the north and south, and between the coastal east and the inland west.

- The *northeast* is the former area of Manchuria. It has very cold winters contrasting with hot and humid summers, somewhat akin to New England in the United States. It was not much settled by Chinese until the early 1900s, after which the Japanese occupied it and developed it into one of the most industrialized regions of China. This industrialization was continued after 1949, and the region became the center of large-scale industry in China (Figure 6.20) that produces iron and steel products, machinery, and vehicles, and is supported by a dense railroad network. After 1949, agriculture was extended across the rich, black earth soils, where corn and soybeans are grown—but there is room for further development of these lands.
- The *north-central heartland* of China comprises the large area of plain into which the Huang He and Chang Jiang drain and is formed largely of river delta deposits. Nearly one-third of the Chinese people live in this region. Winter temperatures and rainfall totals both increase southward. After 1949, further drainage and irrigation works regulated water through this highly productive and densely populated area, which has an agricultural base of winter wheat. Its industrial cities include Beijing and Tianjin. Oil is produced near the mouth of the Huang He.
- The southernmost part of the Chinese coastal zone, the *maritime south*, is experiencing economic growth. Summer rainfall and mild winter temperatures increase southward. Crops range from winter wheat in the north to rice and tea farther south. Two crops of rice are harvested in the longer summers of the far south. Rocky coasts alternate with the

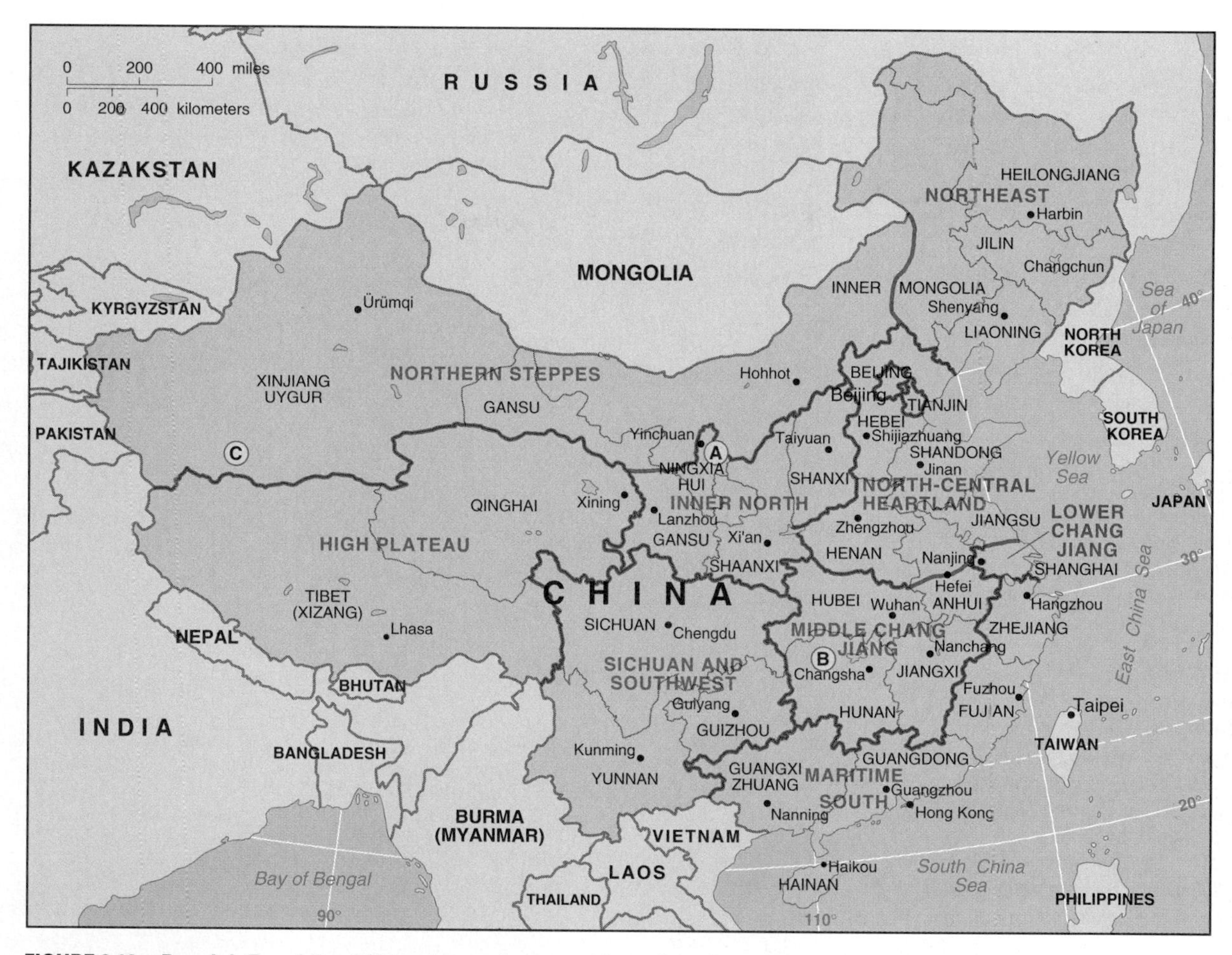


**FIGURE 6.19 People's Republic of China: internal geographic regions.** Provincial boundaries define the regions. Locate the three space shuttle views of Figure 6.21 at (A), (B), and (C).

**FIGURE 6.20 People's Republic of China: northeastern industrial region (Manchuria).** Workers cross railroad tracks at an Anshan steel mill, Liaoning province. Summarize the economic problems this part of China faces.

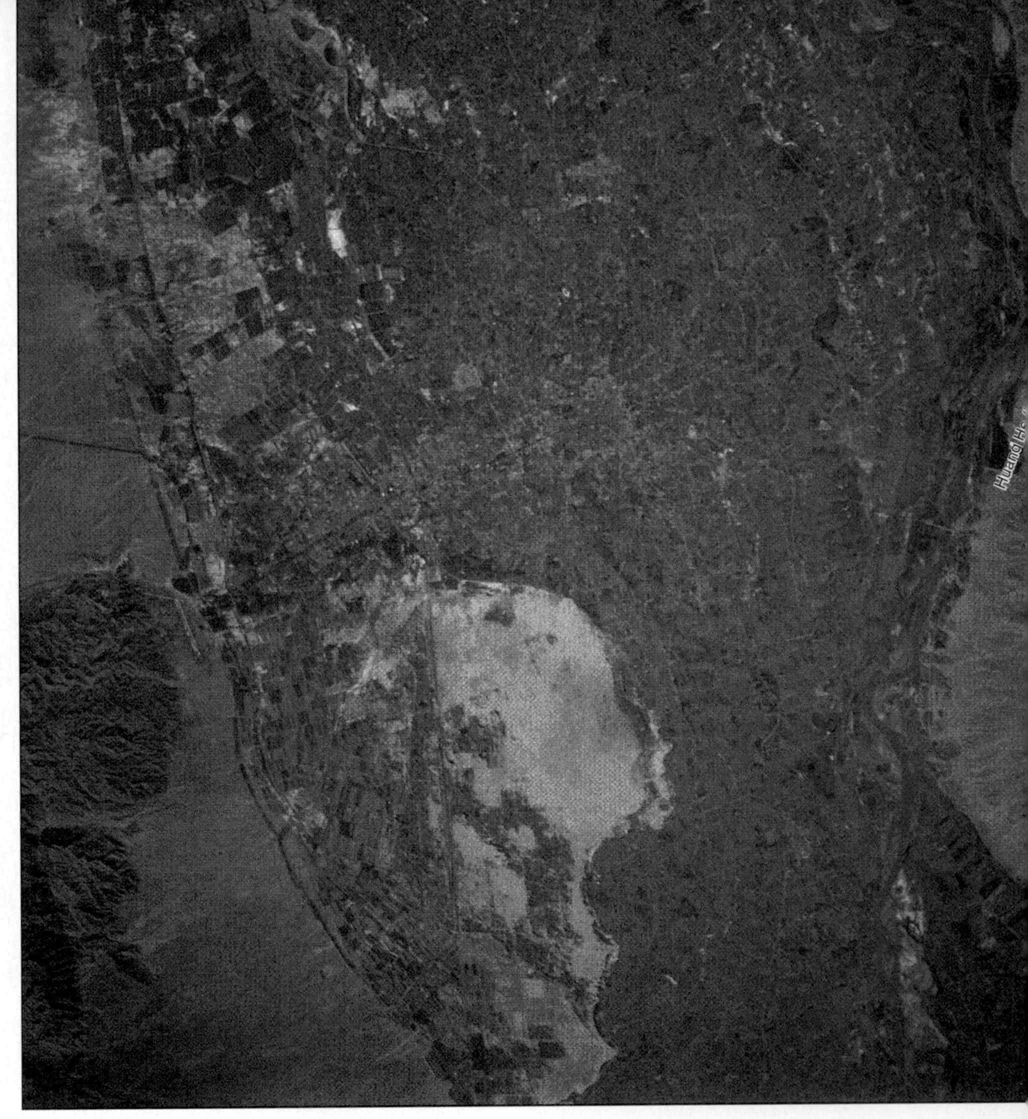


**(a)** NASA

**FIGURE 6.21 People's Republic of China: contrasting rural landscapes from space shuttles.** (a) Irrigated farmland that transformed lands around Yinchuan and Xincheng (center) in the arid northwest of China on the Huang He. The main crops are spring wheat, sorghum, corn, and sugar beets. Uncultivated lands are remnants of former desert. Two airport runways, roads, railroad, and irrigation canals can be identified. (b) Complex lake region 80 km (50 mi.) north of Changsha, Hunan province. The east-flowing Yuan and north-flowing Xiang Rivers merge and flow toward the Chang Jiang. This is a region of intensive paddy rice cultivation.
(c) Western China: the western Kunlun Mountains (7000 m, 20,000 ft.) and southern Tarim Basin. The belt of gravel at the mountain foot stores water from mountain snowmelt that is used in irrigated oases farther away on finer soils. These oases have been cultivated since ancient times, and one of the Europe-China Silk Road routes passed through the basin.

plains around Guangzhou at the mouth of the Zhu Jiang. Special economic zones established in this region since 1979 encourage foreign investment and technological innovation in manufacturing. The newly Chinese ports of Macao and Hong Kong, together with the offshore economy of Taiwan, provide links to the world economy.

- The *lower Chang Jiang lowlands*. Shanghai, at the northern end of the southern coastal belt, is the most advanced and prosperous city of a productive farming region at the mouth of the Chang Jiang. The region was newly industrialized in the 1990s after years of slow growth. The cities of the lower Chang Jiang region have both traditional manufacturing (textiles), post-1949 heavy industries (steel, shipbuilding, oil refining), and modern industries.
- Inland there are regions that are also well populated but do not have the ease of transportation or humid climates of the coast. The *inner north region* centered in the Shanxi provinces was an ancient center of Chinese civilization around Xi'an, but now forms the northwestern rim of Chinese culture. The loess soils of windblown origin are easy to work but also easy to erode, attracting major efforts to overcome soil losses since 1949 (Figure 6.21*a*).

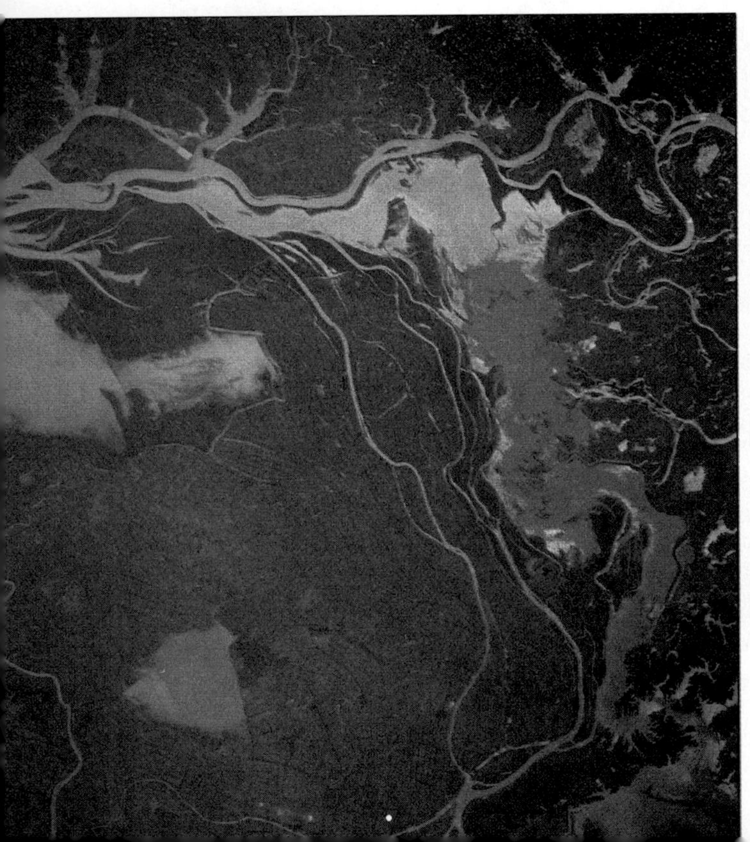

**(b)** NASA

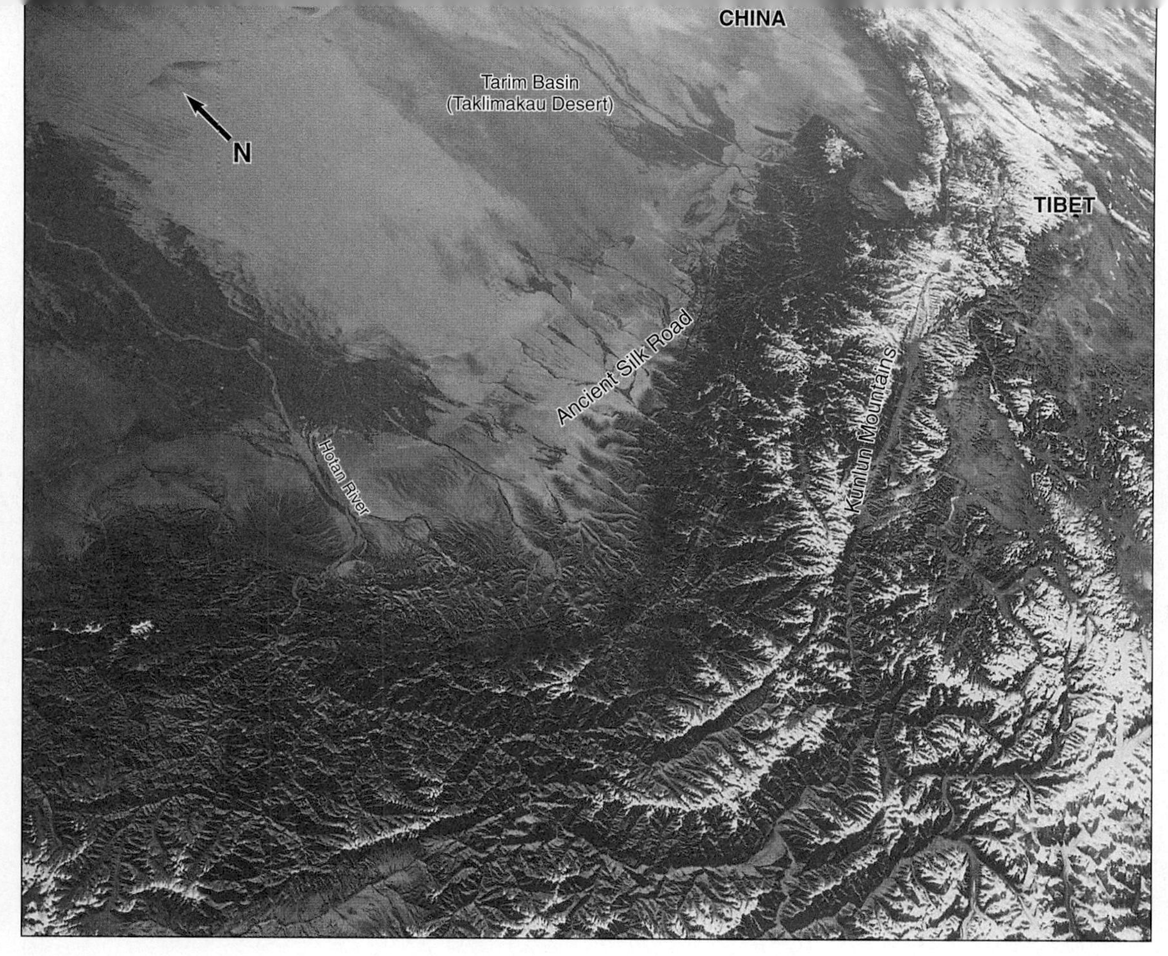


(c) NASA

Summers are hotter than on the coast, but rainfall diminishes westward. Spring wheat and millet are basic crops. Local coal mining and hydroelectric potential provide a basis for modern industry, but distance to markets and low levels of technology slow development.

- The *middle Chang Jiang basin* combines wider plains (Figure 6.21*b*) and hilly areas. It has cold winters and decreasing rainfall westward. Rice, winter wheat, and cotton are grown on the plains, and the cities are being industrialized, linked to local hydroelectricity potential and river transportation.
- The *Sichuan basin and Yunnan Mountains* in southwest China are less accessible, being cut off by gorges along the Chang Jiang (Figure 6.22). The highly cultivated rice-growing Sichuan plain contrasts with the surrounding mountains and plateaus. The isolation of this area, however, stimulated industrialization for local and national defense needs. Mineral resources are poorly exploited.
- The deep interior of China consists of inhospitable arid lands with very low densities of population. In the *northern steppes* from Inner Mongolia westward to Xinjiang, plateaus and encircling mountains (Figure 6.21*c*) separate grassy basins. The climate is dry with warm summers and very cold winters. Ethnic minorities such as Mongols, Uighurs, and Kazaks, rather than Chinese, live here. Livestock keeping is the main traditional occupation. The proximity to the Russian border led China to develop parts of this area, but there are extensive mineral deposits that await further exploitation.
- The *high plateau of Tibet and Qinghai*, which is mostly above 3,000 m (10,000 ft.), occupies the southern half of western China. The People's Republic of China reduced the region's isolation and introduced modern industry, but only 6 million people live in a huge area.

### Mongolia and North Korea

Mongolia is landlocked, a country of mountain ranges and grassy steppes akin to the northern steppes region of China. Although Mongols dominated the Asian world from the 1100s to the 1300s, the present remnant of that empire is poor, with few people. Much of its trade is through Russia.

North Korea remains an isolated outpost of centrally controlled state capitalism, wracked by internal economic collapse, dictatorial deception, and policies that attempt to gain advantages by threats rather than any accommodation with the world economic system. It is a hilly country and its main river, the Yalu, forms the border with China.

## People

Population distribution in China, Mongolia, and North Korea (see Figure 6.12) is marked by the contrast between

**FIGURE 6.22 People's Republic of China: hydroelectric dam.** The major reservoir is behind the dam built across the Chang Jiang in the Linjaxia Gorge; an overflow channel and building housing the electricity generators are on the left.



very low densities in the west and north and very high densities in the river valleys and coasts of the east. Mongolia has the former pattern and North Korea the latter. China's population increased from 583 million in 1953 to 1,243 million in 1998 and may grow to nearly 1,600 million by 2025. Hong Kong, part of China since 1997, had a 1998 population of around 7 million. Mongolia increased from 0.8 million in 1950 to 2.4 million in 1998, and could reach 3.3 million by 2025. North Korea's 1950 population of 9.5 million rose to 22 million in 1998 and may top 26 million by 2025.

## Chinese Population Dynamics

Population growth is a major factor affecting Chinese prospects for economic development. A large population was at first seen as an advantage by Mao Zedong, who attempted to mobilize the great numbers of Chinese people as a "cohesive force." When it was realized that about 20 percent of the country's annual income increase would be needed to merely keep up with the basic needs of the extra people, China made several attempts to restrict family size (see "World Issue: Population Policies," page 258–259).

Population growth rates fell from 2 percent per year on average between 1965 and 1980 to 1.4 percent in the 1980s and 1 percent in 1998. During the period 1965 to 1998, total fertility fell from 6.4 to 1.8. Even this major achievement merely slowed the continued growth of China's huge population, since infant mortality fell from 150 per 1,000 live births in 1960 to 31 in 1998; although fewer babies were born, more survived early childhood, while the number of people surviving to old age also increased.

China moved toward the later stages of demographic transition (see Figure 6.13), enjoying advanced levels of demographic development at an early stage of economic development. In the 1990 age-sex graph (see Figure 6.14), the lower numbers in the 30-to-34 age group reflect the impact of the major population disaster linked to the Great Leap Forward.

The large Chinese population and growth projections resulted in the acceptance of increasingly draconian family planning policies that markedly reduced births from the 1970s. Targets of two children per family in urban areas and three in rural were cut back to two throughout the country and eventually to one in the 1980s. Marriage ages rose. In the

**FIGURE 6.23 People's Republic of China: The Bund, Shanghai.** The British built these buildings along the Chang Jiang riverfront in the late 1800s. New building in Shanghai in the 1990s makes major changes in the city landscape.



1980s, the Family Planning Commission had strong powers, including imposing fines and forcing the sterilization of women who had more than two children. Such measures had an immediate effect in the numbers of children born in the late 1970s and early 1980s but were unpopular and were somewhat relaxed in the 1990s.

One projection shows that, with one child per family, China will reach a maximum population of 1,250 million in A.D. 2000 and will then decline to under 500 million by A.D. 2070. With two children, a maximum of over 1,500 million will be reached in A.D. 2050 and subsequent decline will be slow. With three children per family, the population will rise to over 4 billion by A.D. 2080. The policy of reduced families has so far gained greater hold in the eastern areas and cities than in the essentially rural west with its minority populations. China's ability to maintain small-family policies will be closely linked to continued economic growth.

The future growth of China's population is linked to potential political issues. One is the problem of caring for the elderly as they grow in numbers. Traditionally, they were looked after by extended families, but urbanization makes this difficult, while the decline in family size places pressure on smaller numbers of young people. In the face of small government pensions, Chinese families that can are increasing their savings.

### Urban versus Rural Population in China

China has been a country of cities for many centuries, but until the mid-1900s, most remained as market and administrative centers, often with walls that contained expansion. The exceptions were some port cities, of which Shanghai became the largest, growing because of external trade contacts from the late 1800s (Figure 6.23). The communist rule since 1949 alternately favored urbanization and movements back to the countryside. A debate continues inside China about encouraging the growth of large industrial cities or placing more resources in widespread small and medium-sized towns that are integrated with the surrounding rural area.

In the Mao Zedong era, the wish to encourage modernization through industrialization led to rapid urbanization from 1949 to 1960. Public rhetoric exhorted the Chinese to avoid the environmental dangers of urbanization and the formation of urban elites that had little interest in the rural population. In this period, the urban population more than doubled,

# POPULATION POLICIES

Some of the issues involved include:

What are the problems of increasing youthful populations and of stable aging populations?

How far can governments legislate for population size?

Is it better to slow overall population growth or to redistribute people to places with smaller populations?

In 1960, all the countries of Eastern Asia faced the problems of large and growing populations. For some this was seen as a matter of overpopulation; for others it was partly a problem of the uneven distribution of population. Governments responded with varied forms of family planning policies, and some resorted to wholesale movements of people from one part of the country to another. The best known examples are the family planning policies of Thailand and China and the land resettlement policies of Indonesia.

Until the 1960s, Thailand encouraged births and its population increased at more than 3 percent per year. A World Bank mission in 1958 advised a program of birth control, and this became national policy in 1970. The urban-focused National Family Planning Project became more oriented to rural needs in the Community Based Family Planning Service of 1974, a private, nonprofit organization. The approach adopted was to try to overcome both the ignorance about the mechanisms and value of birth control and the cultural constraints that favored large families. A traveling road show aimed to dispel tradition-based inhibitions. It included support for the policy by Buddhist monks who used religious texts and blessed contraceptive pills. Rapid implementation of the program occurred because there was no involvement of other government ministries and because farmers were encouraged to take part with the inducement of favorable bank loans. Thailand's population growth rate fell according to the targets set, from 3 percent in 1972 to 2.5 percent in 1976, 2.1 percent in 1981, 1.5 percent in 1986, and 1 percent in the 1990s.

China has the world's largest population, with 1,243 million in 1998. Its policy in the early 1990s was to hold the total at 1,250 million by A.D. 2000. The history of Chinese population change since 1950 is marked by major shifts in policies and attitudes. Although the Maoist government watchword of the early 1950s was "strength in numbers," the census of 1953 registered a total population of 583 million and the first birth control campaign was launched in 1956 in urban areas. Its impact was minimal, however, in comparison with the deaths caused by the Great Leap Forward beginning in 1958. This economic policy was designed to introduce industrialization, regimentation of labor, and communization at all levels. The particular policy of encouraging backyard iron smelting was not only a spectacular failure but took people away from food production. When bad climatic conditions

Family planning posters, Lat Yao, Nakhon Saionn, Thailand.



coincided with the main phase of this policy, 25 million to 30 million people died in one of the worst population disasters in the world's history.

Another program of birth control was launched by China in the early 1960s, based on delayed marriages and a wider distribution of birth control pills. It had scarcely time to become effective before the Cultural Revolution of 1966–1976 swept it aside. After this, the most sustained period of family planning was introduced, first extending earlier policies and then adding the single-child policy in the early 1980s. Couples pledging to have just one child received small financial benefits together with better educational and medical provision. The policy had some unfortunate outcomes such as the "granny police" who kept a close watch on families, forcing abortions and female infanticide in rural areas. In some country areas, male babies dominated the child population and millions of female babies remained unaccounted for. A relaxation of the policy allowed rural families to have a second child.

The most recent developments of Chinese family planning policy cover justifying the policy, guidance for bringing up healthy children, further measures of reward and punishment, and an extension to the minority ethnic groups that were previously exempt. Between 1982 and 1990, the Han Chinese (92% of the 1990 population) increased by 11 percent and the minorities by 35 percent. If the present policies continue, China's population numbers could peak in the early or middle part of the new century before decreasing.

Indonesia also had family planning policies to reduce family size but has placed more of its resources into redistributing its population. Sixty percent of Indonesia's population of 178 million in 1990 lived on 7 percent of the land in four largely metropolitan islands (Java, Madura, Bali, and Lombok). Rural population densities on these four islands are often above 1,000 per km$^2$, whereas they are often under 1 per km$^2$ on other islands such as Kalimantan (the Indonesian part of Borneo).

Large Chinese birth control poster at a busy intersection, showing proud, young parents with their one well-fed, well-clothed only child.



After 1950, when the transmigration program began, over 6.5 million people were resettled on the less populated islands. In the early years, practical problems included the selection processes that placed people of the wrong age groups and skills in inappropriate situations, poor site choices for resettlement, encouragement of unsuitable farming systems, and poor administration. These problems and their local impacts have been largely overcome in succeeding years. For example, cases of displaced tribal peoples or tropical rain forest destruction were often given sympathetic treatment. The lasting problem with this policy was that, despite its democratic rationale, it was regarded by many as an insidious form of political manipulation and social engineering. People introduced to outlying islands were seen as government supporters and importers of unpopular economic and social policies. In 1997, there were outbreaks of tribal groups murdering settlers as their forest habitats were cut and burned for the extension of palm oil plantings.

The Indonesian experience raises a further point: whether imbalances of population distribution can be redressed on a large scale. Few good tracts of agricultural land with lower densities of population remain in the islands, so there is little chance of accommodating major population movements in this way. Such considerations cause Indonesia to face the need for either reducing population growth or increasing productivity, or both.

**TOPICS FOR DEBATE**

1. Referring to Eastern Asian and other countries, should governments impose family planning policies on their people?
2. It makes sense to redistribute people in other countries as well as Indonesia.
3. Some countries do not see population expansion as a problem.

from 49 to 109 million, representing a growth from 9 to 16 percent of the total population. Part of this urbanization resulted from industrialization around previously administrative cities such as Xi'an and Nanjing. Another part came from the development of new inland centers such as Baotau, the iron and steel center in Inner Mongolia.

During the Great Leap Forward after 1958, the imposition of communes and the famine disaster of 1959–1961 resulted in further growth of urban centers as an unexpected result of the policy that aimed to focus on rural control. People moved to the towns because they perceived their chances of survival there as greater. Such migrations placed strains on the supply of food to increased numbers of town dwellers from fewer rural farmers.

From the 1960s to mid-1970s, the agricultural/nonagricultural basis of the registration system established in the 1950s was used to control urban population growth, along with a food rationing system and public security policing. During the Cultural Revolution, over 20 million young people were moved to rural areas, and almost as many people were moved there in other programs aimed at reeducating politically incorrect urbanites. In line with this policy, there was little investment in urban housing or transport infrastructure. Any capital available went into heavy industry. In the 1970s, the urban population reached a stable level of very slow increase. By 1978, the Chinese urban population made up just under 13 percent of the total, but living conditions in urban areas had declined.

The economic reforms of the late 1970s resulted in a further rapid rise of urban population numbers after 1980 as permits were no longer needed for living in cities or rural towns. From 1978 to 1998, the urban proportion rose from 13 to 30 percent of the Chinese population. Newly urbanizing areas that are still classified as rural, such as the dispersed, small-town urbanization of coastal provinces, the expanding Beijing-Tianjin area, and the corridors along the lower Huang He, Chiang Jiang, and Zhu Jiang, raise the total urban population to half the Chinese total.

After 1976, the premise of the Mao Zedong years—that production must come before consumption—was abandoned and investment shifted toward better living conditions and making consumer goods available. New industries recruited labor in the countryside on temporary contracts; such workers now make up as much as 10 percent of city inhabitants. Many new urbanites do not even have temporary residence certificates. The barrier between rural and urban residence is largely broken. Large urban industries that develop sophisticated technology link to simpler technology of rural production units in the surrounding countryside with lower labor costs. The rural-urban contrast in economic conditions and prospects, however, produces increasing income gaps within China between poverty-stricken groups and the new urban middle class.

Since 1978, increased investment in housing and urban infrastructure tripled the housing space built each year. Higher standards of housing, sewage services, transportation, and domestic water and gas supplies improved the lot of urban dwellers. Huge expanses of new apartment and condominium blocks encroached on farmland. The backlog of need in this area, however, is great, and it will take many years to complete basic service provision.

The largest cities in China in 1996 (see Figure 6.12) included **Beijing** with 11 million people, **Shanghai** with 14 million, and **Tianjin** with 10 million. **Shenyang**, **Wuhan**, **Changchan**, **Chengdu**, **Hangzhou**, **Harbin**, and **Guangzhou** had several million people each. It is likely that further growth of the major cities will take Beijing, Shanghai, and Tianjin to over 15 million people each by 2015.

### Hong Kong

Hong Kong comprises an island (centered around Victoria) and mainland peninsula (Kowloon) that were ceded to the British in the mid-1800s and a more extensive area of mainland, the New Territories, that was leased for 99 years in 1898. Hong Kong has an excellent natural harbor with deep-water access and extensive water frontages (Figure 6.24). The population of Hong Kong is almost totally Chinese.

The main urbanized area became overcrowded, so new towns, built in the New Territories from the 1950s, now accommodate around one-third of the total population. The high population density causes around half of government expenditure to be spent on roads and public transportation. The railroad route inland to Canton (Guangzhou) was upgraded by electrification, and its use has increased since the late 1970s. Two cross-harbor tunnels and an underground transit system were opened in the 1980s. In the 1990s, the Kai Tak international airport was built offshore north of Lantau Island with bridges to the mainland.

### Mongolian and North Korean Peoples

Mongolia has a dispersed, low-density population (see Figure 6.12), while North Korea has a population of much higher density, concentrated mainly around the west coast that faces China's political and economic core. In both countries, total fertility and natural increase rates fell in the late 1900s. Life expectancies are now in the 60s. Mongolia and North Korea are dominated by related Mongol and Korean peoples, respectively, although Mongolia also has 10 percent who are Chinese, Russians, or Kazaks.

Both countries have just over half of their populations in urban places as the result of industrialization and rural poverty. Mongolia's capital, **Ulan Bator**, houses nearly one-fourth of the country's population, including those living in large tented (yurt) areas. **Pyongyang** is the capital and largest city of North Korea.

### Chinese Ethnicity

The **Han Chinese** comprise 92 percent of the Chinese people. They are of similar racial type to most of the national minorities. The Han people derive their distinctive ethnicity from the emergence of the Chinese administrative culture in

**(a)** © Derek Allan/Travel Ink/Corbis



**(b)**

**FIGURE 6.24 People's Republic of China: Hong Kong.** (a) Hong Kong harbor seen from Victoria on the island and looking across to mainland Kowloon. (b) Space shuttle view of Hong Kong and adjacent southern China. The port and other features of the Hong Kong Special Zone are linked to the mainland by builtup transportation routes. The new city of Shenzhen (left center) is just outside the Hong Kong Special Zone.

NASA

the A.D. 200s and 300s. Most non-Han groups did not have their own writing scripts and adopted Han Chinese script. The Han culture was diffused by conquest and movements of people to administrate new areas. More recently, Han peoples moved into Manchuria after being excluded until the late 1800s and, since 1949, into western China.

Most of the minority peoples live along the northern border with Russia and Mongolia and in western China. Central Asian racial groups dominate the north and northwest, while Tibetans are the main group in the far west. Tibet, at the heart of Lamaistic Buddhism, was taken under Chinese control in 1950, but, following a 1959 rebellion, rapid conversion to Chinese economic, social, and political values occurred. Despite large subsidies, the material condition of Tibetans did not improve, and the Cultural Revolution was a disaster with further bloody repression. Xinjian province, once known as East Turkestan, is linked culturally to Central Asia (see Chapter 8) and was taken over by China in 1949. Since then, Han Chinese have been encouraged to settle, increasing from 5 to 38 percent of the 18 million people in the province. The local Uighurs resist with occasional terrorist acts. Local autonomy is unlikely, however, as China develops relations with Central Asian countries, extending its rail system to Kashgar and the border, and proposing a pipeline to take gas and oil into China.

From 1949, the Chinese government encouraged the more widespread use of Han language alongside the retention of minority languages. Minorities, such as the Mongols in the north, the Tibetans and Uighurs in the west, and the more dispersed Manchu and Muslim Hui, were given considerable jurisdiction over their affairs in autonomous regions and counties within the main provinces. Local resentment of the Chinese takeover of these lands causes major anti-Chinese demonstrations in Tibet and in Xinjiang that erupt from time to time. Khamba tribesmen continue to fight the Chinese in western Tibet.

### Overseas Chinese

Centuries of emigration made Chinese people—the overseas Chinese—prominent members of societies in other Eastern Asian countries. Although their numbers are relatively small—21 million people in Southeastern Asia, another 21 million in Taiwan, and most of the Singapore population—their influence is much greater. Of the four Asian Tigers, only South Korea does not have a majority Chinese population. All the Little Tigers have substantial Chinese populations.

Chinese capital and entrepreneurs dominate private-sector economies in Indonesia, Malaysia, Singapore, the Philippines, and Thailand. For example, Chinese comprise 3.5 percent of the Indonesian population but control around 75 percent of the listed companies. Most of the overseas Chinese in Eastern Asia come from the Shanghai or Guangdong areas and have strong family and clan links on which their business interests are based. Although some anti-Chinese discrimination continues in countries such as Indonesia, Malaysia, and Vietnam, their role in economic development makes the Chinese difficult to exclude. Overseas Chinese financiers, particularly those in Taiwan and formerly in Hong Kong, are major contributors of foreign investment to new developments within China. The **transnational Chinese economy** has become a major factor of economic growth in Eastern Asia and is increasingly linked to Chinese-based business activity in the western United States and Canada.

## Economic Development

Economic development in China, Mongolia, and North Korea is closely linked to government policies in socialist, or centrally controlled, countries. While China and Mongolia opened up their economies to the world economic system in the 1980s and 1990s, North Korea remained isolationist.

### China's Planned Economy, 1950 to 1976

After 1949, the Chinese seesaw of programs imposed new socioeconomic regimes and had different effects in the geographic regions. At the establishment of the People's Republic of China in 1949, the government's first aim was to rapidly turn a backward agricultural country into an advanced socialist industrial state. From being the world's most productive economy in the early 1800s, China saw little further growth in the rest of the 1800s, and its economy declined in the first half of the 1900s. Modeled on the example of the Soviet Union, China funneled two-thirds of available investment into manufacturing, where output rose rapidly, but hardly any into agriculture. This uneven investment planted the seeds of later disaster.

A second aim was to distribute economic activity more evenly through the country and so increase equality and enhance national defense. Previous industrial development was concentrated in the northeastern area (Manchuria) and in coastal ports such as Shanghai and Tianjin, which produced 70 percent of national output. The rapid fall of Manchuria to the Japanese in the 1930s and the vulnerability of Shanghai also played on Chinese minds. The policy of wider diffusion had an ideologic basis in which the Communist Party leadership wished to demonstrate that past locations and concentrations of industry were poor because they depended on the flawed precepts of the capitalist system. New locations had better economic foundations where factories were sited close to raw material sources. For strategic reasons production of military goods was moved away from the vulnerable coast to the "Third Front" (Figure 6.25).

From the mid-1950s, new manufacturing regions emerged based on local coal, hydroelectricity, oil deposits, or

**FIGURE 6.25 People's Republic of China: regional economic policy**. The Third Front of the Maoist era and the three economic belts established in the 1980s. The "golden coastline" policy focuses on special economic zones and open cities that have special advantages for economic development and attracting overseas investment.

strategic factors. The disruptions of the Great Leap Forward and the Cultural Revolution, however, slowed economic growth. The main centers of production continued to be in the northeast, which still had 60 percent of industrial output in 1976. Central and southern China developed slowly, mainly because of poor transportation linkages to established industrial centers. Overall, the locational policies of industrialization from 1950 to 1976 tended to maintain and enhance regional differences instead of reducing them. Shanghai's continuing rapid industrial and urban growth went against the party dogmas, both of widely distributing manufacturing activities and of integrating urban and rural areas.

## Collectivization and Communes

Despite the emphasis on manufacturing after 1949, China remained a largely agricultural country. As Figure 6.16 shows, intensive farming is concentrated along major valleys and coasts. In the arid and mountainous west and in many of the hillier parts of the east, low-productivity agricultural systems remain.

Little new investment went into agriculture, but major changes occurred in rural social structures. Before 1950, most Chinese were tenants or landless, working tiny parcels of land. Fewer than 10 percent of the population were the landlords or rich peasants who owned over 70 percent of the cultivated land. During the early and mid-1950s, central government control increased and individual decision making decreased.

- In 1952, 300 million landless peasants received their own plots of land, houses, implements, animals, free from debt.
- By 1954, over half had joined Mutual Aid Teams, each of up to 10 households, in which labor, tools, and work animals were pooled at seed time and harvest. Such teams often became permanent, grouped in larger Agricultural Producers' Cooperatives: by early 1956, over 90 percent of the rural population was part of cooperatives.
- Next, land was pooled under central management, with members of the cooperative receiving income based on their land and labor inputs. This led to "advanced" cooperatives in which the land became the property of the organization apart from tiny garden plots for each family. A typical advanced cooperative included approximately 150 households farming a total of about 400 acres of land. The process is known as **collectivization** and was completed by late 1956.
- Finally, communes were established as part of the Great Leap Forward. A **commune** combined agriculture, industry, trade, education, and the formation of a local militia. The commune leadership planned the agricultural year, other economic activities, and social functions. Food, clothing, education, housing, and

arrangements for weddings and funerals were guaranteed to members. By September 1958, some 26,425 communes included 98 percent of peasant households. The size of a typical commune was never finalized, but the total number of communes rose to 74,000 in 1965 and then fell to around 54,000 by the time they were dismantled in the 1980s. Although some communes in favored areas prospered, the potential economies of scale were seldom achieved. Their main failure was in motivating peasants, with whom punishments never had the same effects as incentives. The commune is now seen as an expensive experiment that 20 years of restructuring could not make work.

## Impacts of the Great Leap Forward and Cultural Revolution

The so-called Great Leap Forward of the late 1950s in China focused on heavy industry, such as iron and steel production, and attempted to integrate the rural and urban areas by setting up small-scale production units in both. Throughout much of China, community efforts devoted to backyard iron smelting caused neglect of farm work, resulting in the dreadful famine of 1960, while industrial production hardly rose. The discredited Great Leap Forward was replaced by efforts to restore food production when peasants responded to work incentives, state-supplied fertilizers, and the building of new irrigation and drainage projects that suggested a more secure future. Grain yields rose for a few years.

From 1966 to 1976, the Cultural Revolution provided another diversion from economic production with its protests against local authority encouraged by Mao Zedong's *Little Red Book*. Management structures at all levels of society and throughout the country were destroyed. Shifts of policy during this period kept Mao Zedong in power until his death but scarcely encouraged economic growth. An emphasis on staple crop production, for example, produced sufficient rice at the expense of a more diversified range of crops. By 1976, the great majority of the Chinese people lived in communes where production increased slowly if at all. Grain yield increases ceased.

## New Policies and New Growth

After 1976, major changes in outlook led to China entering the world economic system as a new approach to rural life was combined with rapid increases in manufactured exports and investment from foreign countries and corporations. Commune-based agriculture gave way to the **household responsibility system** in which families could decide which crops to grow. Collectively owned land was leased to households for up to 15 years. At first a local government initiative, it was supported by the central government after 1981 and became the hub of its reforms. By 1984, 99 percent of households participated. Small groups entered short-term contracts and leased land to achieve production quotas set by the government. Any surplus could be sold in local markets. The expanded crop output that resulted led to growing confidence, longer-term contracts, and lower quotas to be fulfilled before selling on the open market. The more profitable specialist crops, including industrial crops (cotton, oilseeds), fruit, vegetables, and livestock brought in better returns, threatening some of the traditional staples. Patterns of farm production moved toward diversification with more complex land use, such as the integrated planting of mulberry, sugarcane, and bananas around fishponds in the Zhu Jiang Delta.

While successful in terms of increasing production in many rural areas, the individualistic policy of household responsibility led to the neglect of some aspects of rural land use that communal working had encouraged, such as the long-term processes of planting woodland and maintaining road and irrigation canal infrastructure. In the 1990s, other problems facing agriculture included soil erosion and the extension of urban-industrial areas that removed good land from production.

## Chinese Farming in the 1990s

Agriculture remains predominant in the Chinese economy—the major employer, a source of food for the huge population, a supplier of industrial raw materials, and, it is hoped, a producer of surplus capital to be invested in industry. In short, and although not always recognized by the government, agriculture continues to determine the pace and level of the country's economic growth.

Today, China's agriculture attempts to feed 21.3 percent of the world's population from 7 percent of its arable area. Only 10 percent of China is cultivated, and attempts to extend this since 1949 failed. The proportion of farmland is slowly decreasing as cities and deserts expand. Increased productivity is now reaching a plateau.

Water is as important a constraint as the availability of land. While 64 percent of arable land is in northern China, that is the driest part with under 20 percent of the water; the 36 percent of arable land in southern China has 81 percent of the water but is menaced by floods. New projects to move water from south to north, including the South-North Transfer Project begun in 1995, will be successful only if they are linked to efficient management of the water.

In the 1990s, China increased its imports of grain and foods. These are likely to rise further as the Chinese population grows in numbers, if food prices are kept low and crop yields stagnate. In the long term, if China cannot grow enough food, the rest of the world could not make up the difference. Further emphasis on home production in the ninth Five-Year Plan encouraged grain growers to increase yields by using more potash fertilizer. Nitrogen and phosphate fertilizers do less for productivity but are available within China, while potash has to be imported. In the late 1990s,

China produced enough grain to store some. Other voices inside China, however, counsel growing more valuable agricultural products and importing more grain—with the prospect of the country returning to grain production in a crisis.

### Growth in Chinese Manufacturing

After 1978, new measures stimulated manufacturing growth. Efficiency became more important than equality, and self-sufficiency was replaced by an open-door policy to encourage foreign investment—what has been termed the "Great Leap Outward."

Rural industry boomed as agricultural productivity increased from the late 1970s. Commune industries were linked mainly to local needs in machinery workshops but did not produce consumer goods. A rise in savings in rural areas led to investment in new industries that had better returns than farming, used surplus labor not required on higher productivity farms, and created new markets among rural people. Farm product processing and subcontracting to urban manufacturers became the basis of new industries. Initially, this expansion was within collectively owned enterprises, rising from 22 to 36 percent of all Chinese industrial output by the late 1980s, but then gave way to individual and private enterprise as new driving forces in industrial expansion.

Government policy gradually relaxed to encourage increasing economic relationships with the outside world. At first, this was a response to the need for foreign exchange to buy grain and high-tech equipment. As foreign exchange balances improved, exports were stimulated, with exporters allowed to keep some of the foreign funds earned. Foreign direct investment was then encouraged.

These policies put pressure on state-owned enterprises that had formed the heart of Chinese industry up to 1978, often maintained by subsidies at a cost to central government. From the 1980s, reduced subsidies and the "management responsibility system" placed more decision making in such firms' hands in an equivalent to the household responsibility system. By the late 1990s, 95 percent of production was sold at market prices. State-owned enterprises, however, are less profitable than private firms, having greater legal requirements to provide job security and social benefits. Central government focuses its own investment on only 1 percent of existing state-owned enterprises, allowing others to be merged, taken over by workers, or become bankrupt.

China would like to generate its own multinational corporations with world-known brands. One such company is Qingdao Haier, centered in the city of Qingdao in Shandong province. From being a state firm producing shoddy home appliances, it became one selling 15 percent of the Chinese washing machines and 33 percent of the refrigerators. It built this position through a mixture of improved quality and listening to customer needs: when Sichuan peasants used a washing machine to rinse soil off potatoes, the machines were adapted to prevent clogging; a small-scale washing machine for a single change of clothing sells well in crowded urban centers. Haier took over seven other companies in two years, and others wish to join it. As Haier expands, however, it enters the very competitive production of TVs and pharmaceuticals, looking to overseas markets at a time when financial resources and qualified staff are in short supply. Such conditions face expanding Chinese enterprises.

State-owned enterprises remain at the heart of China's economic problems. Bad loans to failing state enterprises and grandiose socialist housing projects amount to around 30 percent of GDP, twice that of Southeastern Asia. In the late 1990s, bank crises due to overlending affected much of the rest of Eastern Asia, cutting the rapid economic growth rates in the region's countries. Part of the problem faced by other Eastern Asian countries was the expansion of Chinese output, causing oversupply to common markets. China may be forced to devalue its currency further to keep up its exports, leading other countries in the region to follow suit.

Much internal industry continues to be state-run, and large sections of the military-industrial complex—a major influence with over 600 factories that produced armaments and other military equipment—is being rapidly converted to consumer goods, shipbuilding, and aerospace, still under military ownership. China's economic development provides a complex and uncertain intermarriage of local enterprise, foreign multinational capital, state control, and largely uncontrolled military groups that operate their own facilities.

### Chinese Regional Economic Development

In the 1980s, the government divided China into three regions that government policies expected to have different types of development (Figure 6.25).

- The *eastern coastal region* includes active ports, the capital Beijing, and other economically developed areas. It is a zone one province deep along the eastern coast but contains considerable internal variety. It has the greatest potential for economic development and the easiest links to the world economic system. It is to specialize in export-oriented goods to motivate economic growth that will spread to the interior. Special concessions were made available to parts of the region, especially to the four **special economic zones** established in the southeast in 1979 and a fifth on the island of Hainan in 1988. Fourteen coastal ports were designated as open cities in 1984. Not only do these areas have advantages for access to foreign capital and technology, but they also offer cheap local labor and tax concessions for plants and equipment. Such measures are at the expense of the rest of the country

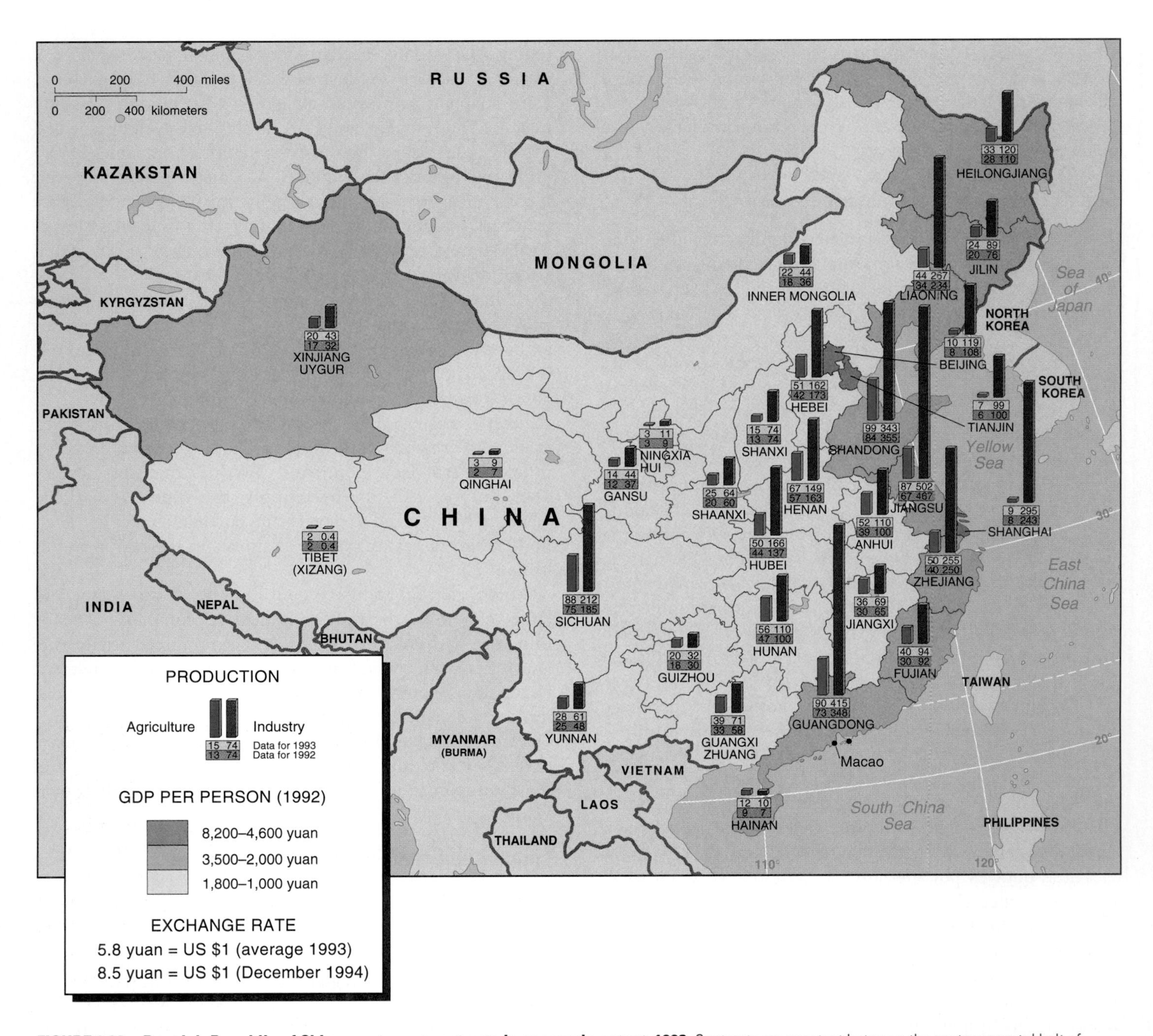


**FIGURE 6.26 People's Republic of China: east-west contrasts in economic output, 1993.** Contrasts are greatest between the eastern coastal belt of provinces and the inland areas. Compare the differences in agricultural and industrial output.

(Figure 6.26) and shifted much of the economic growth southward. Shanghai was not allowed to participate until 1991. It stagnated in the late 1980s but is booming now.

By 1990, 60 percent of over 12 million township small business enterprises were located in the southern coastal provinces—the new "sun belt" of China. The new industries included electronics, chemicals, and machine industries. A huge area of rapidly expanding production grew around Shenzhen in Guangdong province inland of Hong Kong (Figure 6.27). Facilities in this area include an oil refinery, nuclear power plant, airport, deepwater shipping berths, and inland superhighway links.

- The *central region* is an interior zone without coastal outlets but with natural resources of minerals, water, soils, and not-too-steep slopes that allow agriculture to occur. Such resources provide a basis for local industrialization. This region is to specialize in farming and energy production. Its main problem is the lower prices received for food and raw materials compared to

**FIGURE 6.27 People's Republic of China: Shenzhen, Guangdong province.** The downtown area of this city, just inland from Hong Kong, sprang up in the 1980s based on manufacturing goods for export. Many factories moved here from Hong Kong.



those paid for manufactured goods from the coastal zone factories. This tension has been partly relieved by the encouragement of growing interregional linkages along the Chang Jiang basin, which is being called "China's soaring dragon."

- The *western region* is generally sparsely populated with extensive arid and high mountain environments. It coincides with the farthest extent of the historic Chinese Empire. There is relatively little friction across external borders because so few people live near them. This region's future is concerned with animal husbandry and mineral exploitation.

## North versus South China

As economic growth took place in southern China, the old industrial area of Manchuria in the northeast was in trouble—becoming the country's "rust belt" (see Figure 6.20). The new China built its heavy industrial capacity in this area in the 1950s, and most factories and their equipment have not been updated since. Output of steel products continued, but they were often unmarketable on the basis of price or quality. Local coal mines (20% of China's total) and oil companies (33% of China's total) were not paid for their products by the industrial users, and in the mid-1990s, their workers were subsidized by the central government on half pay. Enterprise reform has not been extended to this area and would certainly lead to social unrest.

And yet, China's steel industry is the world leader with increasing production from new coastal and inland mills. The 1997 production was 105 million tons—just a little more than that of Japan or the United States and almost doubling in 10 years. The chemical industry is China's fourth largest after textiles, machinery, and metals, and China is the world's largest producer of fertilizers. Yet it only has 3 percent of total world chemicals production by value. The average Chinese person uses 5 percent of the plastics that an American uses. The Chinese government plans to build five huge petrochemical complexes, but the multinational corporations that have the funds and expertise are not getting involved quickly because of the large-scale and long-term nature of the investments in the context of the uncertainties of control and possible political instability.

Through the 1980s, the Shanghai area, like the northeast, was affected adversely by the policy to encourage growth in the south. While incomes in Guangdong and Fujian doubled, those in Shanghai saw no growth. In 1991, however, Shanghai was opened to foreign investment and became the center of intensive expansion. Work on the construction of bridges, better roads, river port and airport facilities, and an underground train system suggests that much of the city is being rebuilt.

The new economic zone in Pudong across the river from Shanghai already has 4,000 enterprises and 400,000 workers. Further developments are taking place inland of Shanghai around Suzhou, near the main inland expressway road. The Shanghai authorities initiated Suzhou New District in 1990 to move industry westward out of its central area. Then the Singapore government, in a publicized demonstration of inter-Chinese working, contracted to build the Suzhou Industrial Park and inland supercity facilities east of Suzhou. Competition between the two zones during the late 1990s and decline in foreign investment made rapid returns to Singapore unlikely.

Shanghai has the advantages of being China's main port and financial capital. It is also a major center of higher education and research and development. Shanghai is the center of China's main economic growth area in the 1990s—the Huang He Delta that now accounts for one-third of total Chinese output. Its products are mainly consumed internally instead of exported, and the output is based on internal Chinese investment.

Shanghai is the center of China's growing vehicle-making industries, often based on joint ventures with multinational corporations. Volkswagen of Germany made cars in Shanghai from 1985 and produces one-third of the national total, expecting to double its output in the late 1990s. Of the parts it uses, 85 percent are locally made. Peugeot and Citroen (France), Chrysler (U.S.), and Japanese companies also produce cars, trucks, and the minivans that are used as taxis. Other auto factories are located in the northeast and in southern coastal areas. In 1994, China stopped further joint ventures in car-making for three years to consolidate current developments, although several world manufacturers are keen to invest in the country. As with other countries in Eastern Asia, the multinational corporations press the Chinese government for greater openness to the world economy so that they can establish their own factories and offices inside a country but then encourage the government to protect their investment by excluding potential competitors.

### Chinese Power Policy

Although China ranks 3rd in the world for producing energy (1st in coal, 5th in oil, and 6th in hydroelectricity), it is still expensive to distribute electrical power over long distances. By 1994, the shortfall in power supplies was as high as 20 percent in the expanding southern areas. Oil finds are mainly distant from economic centers. Although offshore oil came onstream in the late 1990s, the results from the Yellow and East China Sea areas have been somewhat disappointing.

China has the world's greatest hydroelectricity potential, but even if it were all developed, it would supply just 6 percent of the country's needs. The building of the Three Gorges Project in the late 1990s illustrates the issues facing major hydroelectricity projects. The outcome should generate enough electricity to save the yearly burning of 45 million tons of high-sulfur coal and should help flood protection and navigation. The dam will create a lake 600 km (450 mi.) long below Chongqing. Over 1 million people will need to be resettled; farmers on good valley land will be moved to poor uplands; some industrial towns will be submerged. The Chinese have built many dams, but some have collapsed, and the buildup of silt in the slowed river upstream of the reservoir could submerge additional towns. But this project is Chinese government policy.

Coal reserves remain China's chief source of power. They occur mainly in the northern half of the country, resulting in heavy use of the rail network and massive investment in the railcars that carry it southward. Coal transportation takes up 60 percent of the railroad capacity, and plans are in hand to build coal slurry pipelines from interior mining areas to the coast. One of the positive outcomes of the Great Leap Forward and Cultural Revolution was the opening of small local mines run by villages in southern China; these now produce around half of the total output. China's mining economy has two aspects—one with large-scale, capital-intensive, and technologically advanced mines, and another with small-scale, labor-intensive, low-technology mines. The smaller rural mines help with chronic underemployment, but their operation often lacks environmental management.

One of the major outcomes of attracting more foreign investment will be to expand power provision. Discussions about the proportion of profits to be allowed held back such investments until late 1994, but agreement was then reached to construct the first of around 50 power stations that will use Chinese coal and foreign capital.

### Poverty in China

The slow economic growth to 1976 and more rapid growth since were paralleled by two phases of improvements in the living environments of Chinese people and reductions of the numbers of people in poverty. From 1945 to 1998, infant mortality declined from 200 to 31 per 1,000 live births, life expectancy doubled from 35 to 69 years, and adult illiteracy fell from 80 to 19 percent.

Rural life, however, remained harsh up to the late 1970s. In 1978, China estimated that 33 percent of its rural population (260 million people) lived below its defined poverty line. After 1976, the initial impacts of agricultural reforms doubled output and rural incomes, rural industries created over 100 million new jobs through the 1980s, and the incidence of rural poverty declined from 33 to 9 percent (from 260 million to 97 million people).

As the government's emphasis in the late 1980s shifted to the coastal enterprise centers, regional disparities widened, so that in the mid-1990s, the HPI for interior provinces was 44 percent, compared to 18 percent in the coastal provinces. Urban food subsidies rose to account for five times the funding of rural health, education, and relief programs.

The Chinese government acted again, instituting its 1994 "8–7 Poverty Reduction Program" (80 million still in poverty to be reduced over seven years). New antipoverty measures were established centrally and in the provinces, and funds were committed to investment in poor areas. During the 1990s, the numbers of poor fell to 60 million. Most of the remaining poor live in areas with few natural resources and poor infrastructure—where improvements will be more difficult.

**FIGURE 6.28 People's Republic of China: Beijing.** Bicycles are a common form of transportation. What are the signs of improving income?



## Future Prospects in China

During the early 1990s, with overall production continuing to rise at 7 to 10 percent per year, the ninth Five-Year Plan stated an intention to develop inland resources, particularly those in the least-developed western half of China, without slowing growth in the coastal zone. One aspect of this strategy to date is the construction of highway, energy, and water projects. These employ large numbers of people. Another aspect is the reopening of cross-border trade with Russia and the new countries of central Asia.

China's economic development in the 1900s took it from a feudal, largely subsistence economy after World War II to the building of heavy industries and the collectivization of agriculture. Despite, or because of, draconian measures to bring about such an economic revolution as quickly as possible, the results were achieved slowly. In 1978, China was poorer than South Korea or Taiwan in the 1960s, but rapid growth tripled its total GNP within 20 years. Rapid growth in the coastal regions, particularly inland of Hong Kong and around Shanghai and Beijing (Figure 6.28), brought problems of massive economic adjustment and regional imbalances as interior areas experienced declining income. Such tensions led to political pressures that were ruthlessly dealt with in Tiananmen Square in 1989. Yet economic growth continued with massive foreign direct investments in the early 1990s. Later in the decade, however, the combination of internal institutional and political barriers, together with oversupplied markets, led to poor or uncertain returns for many companies. This may ease as demand rises again, but foreign direct investment in China fell from $42 billion in 1996 to $30 billion in 1997.

Short-term problems include the need to reform or privatize state enterprises, make the banking system less shaky, and determine the correct value of its currency. In 1998, government policy moved to make such changes, including expecting state-owned enterprises to become competitive within three years and state banks to get rid of bad loans and become more commercial, privatizing housing, and shedding half the government workers. If such moves fail, political unrest could ensue as a result of rising unemployment. What remains to be seen are the further impacts of the latest changes on China's human geography and on its relations with other countries in Eastern Asia and the wider world. Certainly China has become a major force in the world economic system since the 1970s.

## Hong Kong in China

Hong Kong's harbor is the focus of its economy. Hong Kong is a port where goods received from inland China are sent to worldwide destinations, while others collected from the rest of the world are sent into China. These are the features of an **entrepôt**, like Singapore. From 1949, Hong Kong was China's contact with potentially hostile countries. As the Chinese economy opened up and expanded after 1976, particularly in the special economic zone around Shenzhen, Hong Kong's trade increased and it became the world's leading container port.

Before Hong Kong rejoined China in 1997, the narrow base of government decision making was centered in the governor appointed by Britain, an executive council, and a partly elected legislative council, which ensured support for extreme free-enterprise capitalism. Hong Kong became one of the world's major finance centers and the third largest gold market, accounting for up to 15 percent of the total.

Hong Kong's growth as a manufacturing center stemmed from the 1970s, and by 1990, manufacturing provided a third of all employment. Major expansion took place in the production of electronic goods and scientific equipment, but growth in the textiles and clothing industries slowed because of foreign competition and rising labor costs in Hong Kong. In the 1980s, the expansion of manufacturing just over the border in southern China, with its cheap land and labor costs, led to a decline of these activities in Hong Kong. By 1996, service industries accounted for 84 percent of Hong Kong GDP and 71 percent of employment. Hong Kong provides expertise in finance, trade, and public administration and has an important stock market. Its daily turnover of foreign exchange is fifth in the world. It has 1,300 accountancy firms and 3,000 management consultancy firms. It is the regional headquarters for 800 foreign companies.

Tourism became the second major industry with 11.7 million visitors in 1996, especially from Japan. Many new hotels catered to this influx of people that added an extra 10 percent to the population at peak times. After Hong Kong's transfer to China in 1997, tourism continued at a high rate at first, but declining economies in Eastern Asia (the source of 75% of Hong Kong tourists), and early 1998 reports of outbreaks of a flu spread by poultry cut the numbers of visitors.

The higher wages in factories and service industries make farming labor costs high as well. Even so, Hong Kong grows nearly a third of its food needs on the small area of nonurbanized land in the New Territories.

After the 1997 transfer of power to China, Hong Kong did not suffer further capital flight or brain drain but rather experienced an influx. The Basic Law, under which Hong Kong operates within China, guarantees retention of its involvement in a free-market system for 50 years and sees it as a complementary financial center to Shanghai. At the time of political transfer, Hong Kong was already deeply enmeshed with China in trade, investment, and personal contacts. In 1996, China accounted for 37 percent of Hong Kong's trade, and Hong Kong for 47 percent of China's trade. From the late 1970s, 57 percent of foreign direct investment in China was channeled through Hong Kong, and 90 percent of syndicated loans for the mainland were arranged there. China became a major investor in Hong Kong, with 50 mainland corporations ("red chips") having a stock exchange value of $24 billion located there. Over 20,000 trucks cross the Hong Kong-mainland border each day. Now part of China, Hong Kong has an edge over its rivals for dealings inside the growing country.

While such considerations suggested continuing stability and prosperity in Hong Kong, other signs were less optimistic, particularly in relation to Hong Kong's pride in its careful financial regulation. Shares in mainland red chip corporations were oversubscribed, borrowing to buy shares was heavy, and mainland corporations brought in corrupting measures that could destroy trust in Hong Kong's market. Hong Kong and Chinese regulations have different norms, with the latter subject to political pressures.

**FIGURE 6.29 North Korea: Pyongyang.** A typical communist urban landscape, including the National Monument and the 70-ton statue of Kim Il Sung on Mangu Hill. The statue dwarfs people paying homage to the country's former leader.



### Mongolia's Limited Resources

Mongolia is a small country with a strategic location in the heart of Asia but little economic potential. After it gained its independence in 1911, the Soviet Union modified Mongolia's Eastern Asian character, taking over the education system and replacing the Mongolian alphabet. Mongolia became dependent on Soviet aid and trade. The capital, Ulan Bator, is connected by rail to both China and Russia.

Many people still gain a living from herding livestock on the semiarid grasslands, and farmers responded to privatization and decontrol of meat prices in the early 1990s by massively increasing their herds. Cropland area declined as agricultural cooperatives were split and land abandoned since there was no equipment to cultivate it.

For the future, Mongolia is increasing its output of minerals such as copper, gold, and molybdenum and encouraging foreign investment to provide alternatives to the continuing Russian links. Private enterprises became the main source of economic growth in the 1990s, with rising production of gold, although a major fall in world copper prices and weaker prices for cashmere woolens slowed growth. Mongolia suffers from a limited range of resources and the problems of making the transition to a market economy.

### Economic Stagnation in North Korea

When the Japanese ruled the whole of Korea in the early 1900s, the northern part provided iron, coal, and raw materials for a chemical industry and had plenty of hydroelectricity potential. The southern part was largely agricultural. Since its creation in 1948, North Korea has been an inward-looking country. The personality cult of Kim Il Sung (Figure 6.29) lasted longer than that of Mao Zedong—until Kim's death in 1994. By that stage, however, North Korea had become poor and was getting poorer through a policy advertised as isolationist self-reliance. In fact, the country depended on aid and good prices from China and the Soviet Union. These arrangements ceased in 1989. Factories became idle and children lacked proper nutrition. By the mid-1990s, the GDP per head in North Korea was one-eighth of South Korea's.

North Korea faced two alternatives in the late 1990s. It could continue its path to greater poverty and eventual collapse. The next step could be reunification with South Korea—although that country viewed the prospect with unease after the experience of Germany incorporating its eastern area (see Chapter 7). Alternatively, it could follow a path of economic liberalization similar to that of other socialist countries in Eastern Asia, take offered aid in exchange for running down its nuclear program, and encourage investment from foreign countries. Some South Korean multinational corporations were keen to capitalize on low-wage labor in North Korea, and many South Korean families with relatives in North Korea wished to help them take advantage of better economic conditions.

**RECAP 6C: China, Mongolia, and North Korea**

China has over 20 percent of the world's population. In 1976, it emerged from isolation under a communist regime that had instigated a mixture of greater national unity, industrial development, collective farming, and disastrous policies that attempted to force people to fulfill state-set targets.

China's agriculture became more productive after 1976 but is reaching the limits of expansion; the country's industries are expanding in southern coastal China but elsewhere need new plant and infrastructure. China's economy is a mixture of state-controlled enterprises, military-industrial diversification, and foreign multinational investment in manufactured goods for export.

Mongolia and North Korea remain poor and isolated—the former because it is landlocked and the latter because of government policy. Since losing Soviet financial aid, their economies have slipped.

6C.1 How did Chinese government policies from 1949 to 1976 and from 1976 to the present differ? What impacts did changes have on China's agricultural, manufacturing, and urban geography?

6C.2 Why do the age-sex graphs for China and other countries in Eastern Asia show different patterns? What are the consequences of such population structures for education, labor forces, house-building policies, and the proportion of older people?

6C.3 What roles might Hong Kong play within China?

**Key Terms:**

Great Leap Forward
Cultural Revolution
Han Chinese
transnational Chinese economy
collectivization
commune
household responsibility system
special economic zones
entrepôt

**Key Places:**

China, People's Republic of
Mongolia
*Ulan Bator*
North Korea
*Pyongyang*
Hong Kong
Macao
*Beijing*
*Shanghai*
*Tianjin*
*Shenyang*
*Wuhan*
*Changchan*
*Chengdu*
*Guangzhou*
*Hangzhou*
*Harbin*

Following international concerns about North Korea's nuclear weapon capability, South Korea, Japan, and the United States agreed to provide nuclear power stations, but of the type that could not be used to make such weapons. The first was begun at Shupo in mid-1997. Until these power stations are functioning, North Korea receives 0.5 million tons of free fuel oil each year. The construction project involves 5,000 South Korean engineers.

Some of the problems facing North Korea are illustrated by its late-1995 promotion of external economic cooperation based on the Rajin-Sonbong free-trade zone founded in 1991 in the Tumen River valley close to the Russian border. The area, however, is rural, sparsely populated, and has little modern infrastructure such as telecommunications or an airport—although it does have rail and port facilities and power supplies. The South Korean government, however, prevented Samsung from establishing a plant there. Whatever happens, North Korea faces a difficult period. After years of a declining economy, the people of North Korea, duped by leadership hype or cowed by fear, do not appear likely to revolt.

## SOUTHEASTERN ASIA

**Southeastern Asia** includes countries (Figure 6.30) that were influenced in the past by Chinese expansionism, became colonies of European countries, and are today increasingly in the orbit of Japanese and Chinese economic activity. The former kingdoms of Burma, Thailand, Cambodia, Laos, and Vietnam, as well as the former sultanates and colonial territories of Malaysia, Indonesia, and the Philippines, became separate countries by the late 1900s.

With so many islands and extensive coastlines, the countries of Southeastern Asia, apart from landlocked Laos, were subject to many external influences. Peoples of Chinese origin, or those escaping Chinese expansion, dominated the early populating of the mainland countries of Thailand and Indochina. The Malay peoples dominated the populations of the island lands of Indonesia and the Philippines. Muslim influences began with trade in the East Indian Spice Islands, leading to a degree of political control by sultanates from the A.D. 1200s.

The Portuguese traders led a European tide of trade and colonization in the 1400s, but the Dutch took what is now Indonesia, the Spanish took the Philippines, the British took the Malay States, and the French took Indochina (now Cambodia, Laos, and Vietnam). These changes had economic and social impacts that are reflected in the modern countries. In particular, they restored a strong nationalism in the central parts of former kingdoms that led to resistance against colonial influences.

After World War II, the countries of Myanmar (1948), Indochina (1954), Malaysia (1957), Indonesia (1949), and the Philippines (1946) became independent of the former colonial powers. While Vietnam, Laos, and Cambodia became communist-run countries from the 1950s, Myanmar experienced military rule.

The private economies of Southeastern Asian countries are today largely controlled by minority groups of Chinese merchants and industrialists—or by Indians in Myanmar. Singapore has the highest living standard in Southeastern Asia and the highest proportion of Chinese people, around 70 percent. All these countries are increasing trade with others in the subregion, with China and Japan, and even with Europe and the United States.

### Countries

The countries of Southeastern Asia are **Brunei**, **Cambodia**, **Indonesia**, **Laos**, **Malaysia**, **Myanmar**, the **Philippines**, **Singapore**, **Thailand**, and **Vietnam**. In 1998, their populations

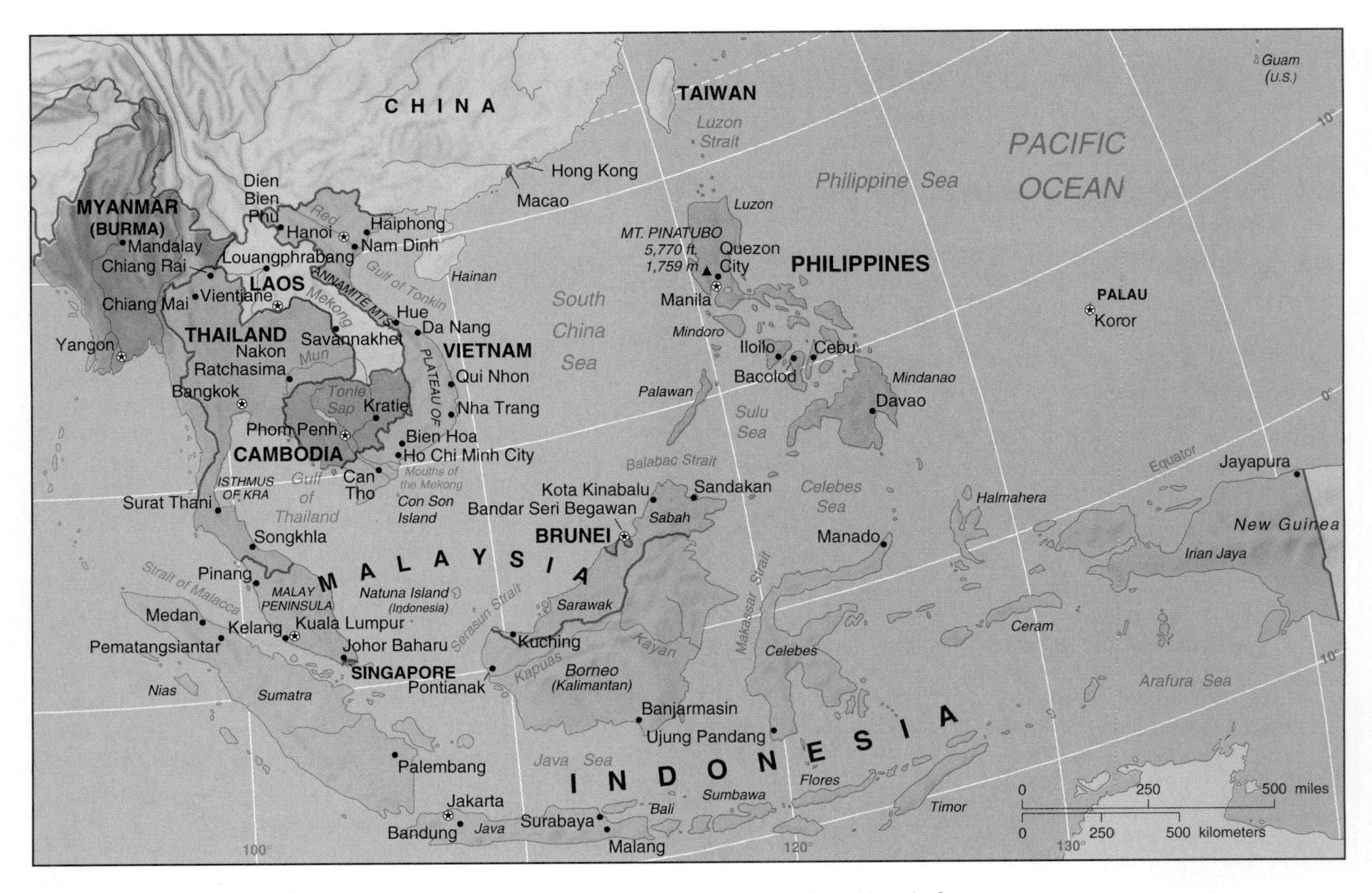


**FIGURE 6.30 Southeastern Asia: countries and major cities.** How important are ocean links in this region?

ranged in size from only 0.3 million people in Brunei to less than 6 million in Laos and Singapore, 11 million in Cambodia, 22 million in Malaysia, 50 to 80 million people in each of the large countries of Myanmar, the Philippines, Thailand, and Vietnam, and over 200 million people in Indonesia.

In economic development they range from the oil-rich country of Brunei and the trading center of Singapore—one of the Eastern Asian Tigers—through Malaysia, Thailand, Indonesia, and the Philippines (the Little Tigers), to the military dictatorship of Myanmar and the communist regimes of Cambodia, Laos, and Vietnam. Improvements in health, education, and income occurred over the period 1960 to 1996. These countries have relatively high levels of HDI and gender equality, as demonstrated by male and female adult literacy figures and GDI ranks. Malaysia made reducing poverty a major policy objective. Locally calculated as affecting 60 percent of the population in 1970, poverty fell to 21 percent by 1985 and 14 percent by the mid-1990s.

Internal variety within Southeastern Asia reflects the influences of colonialism; the physical division into continental and island countries, and the impacts of communist and free market policies. The communist advances in the 1950s and 1960s led the noncommunist countries (Indonesia, Malaysia, the Philippines, Singapore, and Thailand) to establish the **Association of Southeast Asian Nations** in 1967. It created a defensive alliance and a means of establishing peaceful cooperative links when the war in Vietnam and political chaos in Indonesia appeared to threaten the future of the region. Brunei joined in 1984.

After the United States withdrew from Vietnam in 1975, the noncommunist countries adopted a policy of peaceful coexistence but continued the polarization that was established after 1967 and enhanced after the 1978 invasion of Cambodia by Vietnam. ASEAN nations isolated Vietnam politically and economically, supporting rebel groups inside that country. Vietnam, Laos, and Cambodia were denied access to the World Bank, the Asian Development Bank, and other sources of external funding. After Vietnam withdrew from Cambodia in 1989 and adopted internal reforms in the 1990s, restrictions on trade were reduced and new business and technology-exchange links encouraged. Vietnam joined ASEAN in mid-1995.

Despite its isolation from most of the world because of its dictatorship government, poor human rights record, and involvement in drug trafficking, Myanmar was accepted into ASEAN with Laos in 1997. Cambodia had its membership application deferred because of a 1997 coup that appeared to return it to the days of internal repression.

## People

The distribution of population in Southeastern Asia (see Figure 6.12) is marked by high densities of people in the lowlands of major mainland river valleys—the Irrawaddy (Myanmar), Chao Phraya (Thailand), Mekong and Red (Vietnam)—and on the islands and peninsulas that were centers of colonial economic activity, southern Malaya, Singapore, Java, and the northern Philippines. Elsewhere, densities are much lower in the upland areas and on the less developed islands of Indonesia and the Philippines.

### Population Growth Slows

In 1965, the combined population of the countries of Southeastern Asia was 275 million, rising to 550 million in 1998, and is expected to top 700 million by 2025. In the period from 1965 to 1980, most countries had population growth rates averaging over 2 percent per year, with Thailand and the Philippines close to 3 percent. The exceptions were Cambodia and Laos, with growth rates close to 1 percent or less, reflecting the ravages of war. By 1998, annual natural population increase rates in most countries were falling. Thailand, Indonesia, Singapore, and Vietnam were down to almost 1 percent, while Malaysia, Myanmar, and the Philippines remained just above 2 percent. Cambodia, Laos, and Vietnam, however, rose to between 2 and 3 percent.

Slower population growth rates reflected falls in total fertility rates, all of which were over 6 in 1965. By 1998, only Laos and Cambodia were over 5, and the other countries under 4. Thailand's family planning policies reduced fertility in that country to 2 (see "World Issue: Population Policies," pages 258–259). There is progress in demographic transition (see Figure 6.13). The age-sex graphs (see Figure 6.14) demonstrate the impact of reduced numbers of births in Indonesia and Thailand. The Vietnamese population (those over 10 years old in 1975, over 25 in 1989) was affected by wartime losses, but after the end of hostilities in 1975, the population expanded slowly.

### Impacts of Rapid Urbanization

Urbanization is increasing in all the countries of Southeastern Asia, although the mainland countries of Cambodia, Laos, Thailand, and Vietnam still had under 20 percent of their populations living in urban areas in 1998. Indonesia had over 30 percent, and the Philippines and Malaysia a 40 percent urban population. The two smallest countries, Brunei and the city-state of Singapore, had 60 and 100 percent, respectively, living in towns and cities.

One of the common features of Southern Asian countries is the high proportion of urban populations living in the largest, or **primate**, city, and the rapid growth of these cities. In Thailand, **Bangkok** has over 20 times the population of the next largest city, Chiangmai. **Jakarta** (Indonesia), **Kuala Lumpur** (Malaysia), **Manila** (Philippines), and **Phnom Penh** (Cambodia) are not quite so dominant, but the pattern is similar. **Yangon**, the capital of Myanmar, and Mandalay form a dual focus in that country. In Vietnam, the capital, **Hanoi**, and its port, **Haiphong** in the north and **Ho Chi Minh City** (formerly Saigon) in the south also form a dual focus. Other large cities in Indonesia include **Surabaya** and **Bandung** on Java, **Medan** and Palembang on Sumatra, and **Ujung Padang** on Celebes.

The rapid growth of city populations changed the rural-urban balance and concentrated economic growth in large cities. Modern services such as hospitals and schools, business services, and transportation mainly occur in urban centers. Living in Bangkok, for instance, gives individuals and firms better access to such services than in the poverty-stricken northeast of the country. Inside cities, governments subsidize urban transportation and provide tax incentives for industrialists to locate factories. Manufacturing is also encouraged to a greater extent than agriculture—a further factor that leads to urban growth and higher incomes at the expense of rural areas. The negative side of growing cities—congestion and pollution—appears to have little effect on their expansion. Bangkok's traffic jams are known as the world's worst.

It is difficult for governments to expand services and facilities at the same rate as urban population growth. Government actions themselves tend to promote further city building, just as economic forces do. The uncontrolled growth of Jakarta in the 1970s caused the Indonesian authorities to issue housing permits only to those who could prove employment, but this policy only increased the growth of shantytowns.

### Urbanization under Communism

The communist countries adopted distinctive policies toward urbanization. The cities of Cambodia, such as Phnom Penh, were systematically depopulated in the late 1970s under the dictatorship of Pol Pot.

At its end, the Vietnam War focused on the American defense of Saigon, which was bolstered by U.S. aid and black-market profiteering. The frightened and war-torn from surrounding rural areas moved into the city during the war and into other coastal cities such as Da Nang. Saigon grew from 2.4 million people in 1964 to 4.5 million in 1975. When the communist Vietnamese government took over the former Saigon after the 1975 unification of North and South Vietnam, they renamed it Ho Chi Minh City.

A **deurbanization** policy was implemented to reduce the "unhealthy" overcrowding with insufficient water, housing, and power. The first target was to reduce the population of Ho Chi Minh City to three-fourths of a million as part of an attempt to move people back to rural areas to restore food production and to develop rural industries through local cooperatives. Families were directed back to their villages, and ethnic Chinese, who mainly lived in the city, were so persecuted that nearly three-fourths of a million left the country between 1977 and 1982, many fleeing by boat; these refugees became known as "boat people." Economic zones were set up in rural areas and over a third of a million people moved into them from the city. Eventually, it was realized that the initial target for Ho Chi Minh City's population was too low, and it was raised to 1.75 million. In the late 1980s, the policy

# MALAYSIA

Malaysia became noticeably more prosperous during and since the mid-1980s. More people moved into the towns, especially the capital, Kuala Lumpur, because of the jobs in government offices, businesses, and banks, the greater range of shopping available in supermarkets and chain stores, and the greater range of leisure opportunities from golf courses and swimming pools to badminton courts—the favorite sport. In the 1990s, Kuala Lumpur experienced a huge expansion of high-rise apartment and office construction, including investments by corporations moving out of Hong Kong. Housing is more expensive in Kuala Lumpur than in other parts of Malaysia, so many people live in suburbs outside the city and commute in by motorcycle, car, bus, train, and the light rail system. In 1995, Malaysia announced plans to build a new administrative capital called Putrajaya, 40 km (25 mi.) south of Kuala Lumpur by 2008. Life in the west coast cities still contrasts with life in rural villages and on the east coast, where people live in houses of wood with coconut-leaf or tin roofs instead of brick and concrete. Most Malaysian houses have stone floors to help cool the equatorial air.

Urban homes have refrigerators, TV, and video. TV programs include many from Australia and the United States, with a few British programs and local programs for Malay, Chinese, or Indian groups. There are radio channels for each of the main languages—Malay, Chinese, Hindi, and English. Most cooking is by gas, bought in cylinders delivered to the house. Water is metered, and most houses have showers rather than bathtubs.

Family ties remain strong, although family size decreased from around three or four children before 1980 to around two in the 1990s. Parents usually live with the oldest child, who looks after them when they finish earning. Government policy maintains a high birth rate in order to generate a large domestic market for manufactured goods—hoping to raise the present 20 million population to 70 million by 2100. Large families are encouraged in an attempt to slow the fertility decline linked to increasing affluence and use of contraceptives.

Schools are mostly subsidized by the government, although some are still run by churches; the Chinese and Indian communities run independent schools. Even at the subsidized schools, parents pay fees and buy uniforms and books. Poor families often cannot afford to send their children to school. For most children, schooling lasts from 7 to 17 years of age, with the possibility of another two years for university entrance exams. The school day lasts from 8 A.M. to 12:30 or 1 P.M., with a break during the hot part of the day. Any games take place after 4 P.M.

Most Malaysians use Western medicine, although some use herbal remedies. Visits to the doctor and hospital treatment all have to be paid for. Poor people get free medical care but may have to pay for prescriptions.

Malaysia has a democratic government and is a member of the Commonwealth (of former British colonies). Malaysia is a multiethnic society, with many people of mixed birth, such as an Indian father and Thai mother. Malays, who are mainly Muslims, retain control of Parliament and often regard

**Kuala Lumpur, Malaysia.** The city center, with the historic Sultan Abdul Samad building and gardens, backed by modern high-rise office blocks.

© David Ball/The Stock Market

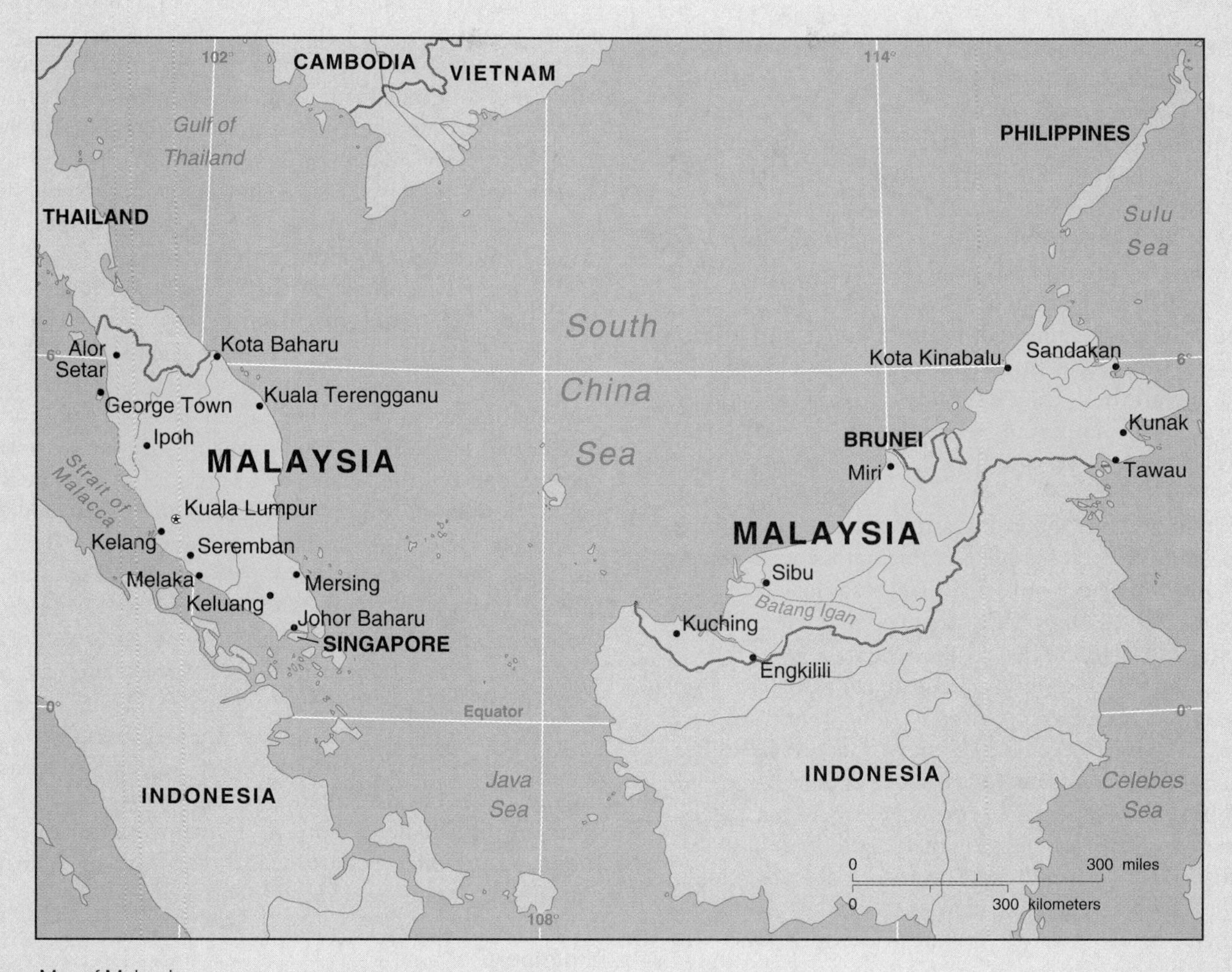


Map of Malaysia

themselves as first-class citizens and others as second-class citizens. The Chinese came to work in the tin mines during British rule, when Indians were also brought in for the rubber plantations. Chinese are now mainly middle class with commercial interests, while Hindu Indians include contrasts between poor laborers, owners of small businesses, and professional lawyers or doctors. Fearing that the Chinese and Indian groups would become too powerful, the Malaysian government passed laws that forced Chinese businesspeople to have Malay partners holding 50 percent control and enforced strict racial quotas in higher education, government jobs, and business ownership. Relationships among the different groups are not so tense as they were following government actions after race riots in the late 1960s and are helped by the increasing well-being of Malaysians.

From 1986 to 1995, the Malaysian labor force grew faster than the total population as foreign workers flowed in to take up increasing numbers of jobs. The economic growth exposed labor shortages, and the multiethnic nature of Malaysian society was further extended by bringing immigrant workers from Indonesia, Bangladesh, and the Philippines. More women are involved in the labor force for high-tech manufacturing, and this contributes to rising family incomes.

The orientation of Malaysian society changed with its economic focus. British influences remained strong until 1970, but since then, Japanese and U.S. firms established businesses in Malaysia that are seen as forming the foundation of recent economic growth. In addition to the manufactured products of Japanese and other international firms, palm oil, rubber, tin, and iron ore remain important exports, and oil comes from offshore wells in the South China Sea.

In 1997, Malaysians became aware of environmental quality loss as forest fires in Indonesian and Malaysian areas being cleared for oil palm plantings created intense smog over the country. A public health crisis followed in which schools closed. This event, repeated in early 1998, focused attention on lax environmental policies that encouraged timber cutting and removal and mining at the expense of silting estuaries, disappearing coral reefs, and depleted fisheries.

changed again, with a controlled but slower urban growth being allowed alongside a welcome for foreign capital investment in new industries.

### Ethnic Variety

Following centuries of invasions by land and sea and migrations instigated by Chinese expansion and colonial powers, the peoples of Southeastern Asia comprise very diverse groups. Chinese communities are significant in all countries, making up 7 percent overall. Overseas Indians are significant in Myanmar and Malaysia. Groups of indigenous peoples survive in the hilly and mountainous regions of the mainland and particularly in the uplands of Myanmar and the islands of Indonesia such as Kalimantan (Indonesian Borneo).

Religious diversity is marked. Islam dominates in Brunei, Malaysia, and Indonesia. Buddhism is the main religion of the mainland countries. The Philippines are Roman Catholic. Hinduism is locally important (e.g., in southern Myanmar, Malaysia, and Bali), as are animistic religions among indigenous tribes.

Such ethnic diversity often leads to tensions and conflict. In Malaysia, the 1969 race riots led to new policies to correct the economic "imbalance" in which Chinese owned most of the businesses; by 1990, non-Chinese people owned 30 percent of shares in Malaysian businesses. The policy was relaxed somewhat when Chinese investments that could have been made in Malaysia began to be placed in other countries. Such policies in Malaysia, however, tended to make racial differences into rigid and institutionalized social barriers.

Northern Myanmar remains a battlefield between the hill country peoples (Shan, Karen, and others) and the central government that tries to crush them. The genocidal eradication of hill tribes in Cambodia and Vietnam was a further example of ethnic hatreds being linked to political crusades. The Vietnamese hatred of the Chinese is exposed in occasional border skirmishes, in the expulsion of many Chinese businesspeople from southern Vietnam since 1975, and in competing claims for the Spratly Islands. In 1998, the overturn of the Suharto government in Indonesia more was accompanied by aggression toward the Chinese (often Christian) business community.

## Economic Development

The contrasts between the economies of the mainly noncommunist countries and the fewer communist countries in Southeastern Asia continue to have major geographic outcomes. In the late 1990s, all countries had the option of joining ASEAN and becoming part of the world economic system.

### ASEAN and Growth

The ASEAN grouping was initially a defensive alliance rather than a trading one, but the peace it established formed a basis for economic growth. From the late 1960s, the ASEAN countries experienced a pattern of rapid industrialization, stimulated by foreign investment and internal government encouragement. The economies of the communist countries and Myanmar, however, remained stagnant into the 1990s. From the late 1980s, they changed from isolationist policies to those that encouraged greater links to the world economy and joined ASEAN in the mid-1990s. To do so, they promoted greater economic freedom without modifying the political control, as in China.

Rapid economic growth in noncommunist ASEAN countries came to a halt in the late 1990s, when overlending by banks caused a major financial crisis. Thailand, followed by the Philippines, Malaysia, and Indonesia, was forced to devalue its currency in 1997. The problems centered on high levels of foreign borrowing, government budget deficits, major bank loans to glutted property markets, and slower-than-expected economic growth. Huge Japanese investments in the late 1980s and growth in exports from Southeastern Asia were linked to a high yen value against the U.S. dollar; by the late 1990s, however, the reverse was true, and Japanese slowed investment and technical transfer to Southeastern Asia. What appeared to be a problem for Southeastern Asia became one for the whole of Eastern Asia: too much economic growth, too soon, on too much unsecured credit.

Thailand had these problems to the greatest extent, but it also had the advantage of a new government that could impose austerity measures and distance itself from the past. Indonesia and Malaysia, ruled by long-term authoritarian governments under Suharto and Mahathir Mahomad, respectively, found it more difficult to admit fundamental mistakes in financial regulation. None of these governments found that the United States dashed to their aid, as in the past during the Cold War. The financial crisis led to a change of government in Thailand and to a major political crisis in Indonesia in which Suharto was forced to resign, leaving the country's economic and social life in chaos.

### Singapore

Singapore is the smallest country and has no natural resources but has become by far the richest in terms of GDP per capita. This is reflected in the comparisons of consumer goods ownership in Southeastern Asia (see Figure 6.15). Its main strengths are its strategic position on the Malacca Strait at one of the world's main transportation crossroads, its port facilities, and an enterprising and well-educated population. Like Hong Kong farther north, Singapore has long been an entrepôt port collecting goods from countries in its region and exchanging them on world markets. It continues to refine oil and reship the products, to collect rubber and other agricultural products from Malaysia and Indonesia and tranship them, and to import and distribute autos and machinery.

Singapore developed into a high-technology manufacturing country. The strong control exercised by the Singapore government redirected its economy from port-centered activities toward high-tech industry including information technology and automation. Singapore firms manufacture 40 percent of world computer disk-drives and include some of the leading makers of computer sound cards. As a transportation and communications hub, Singapore became the

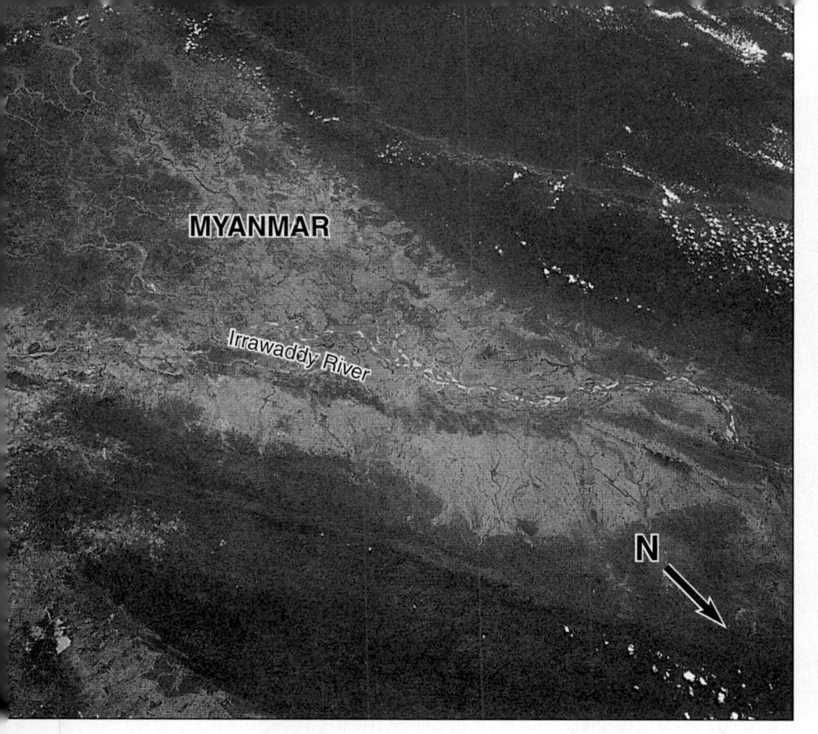


(a) NASA

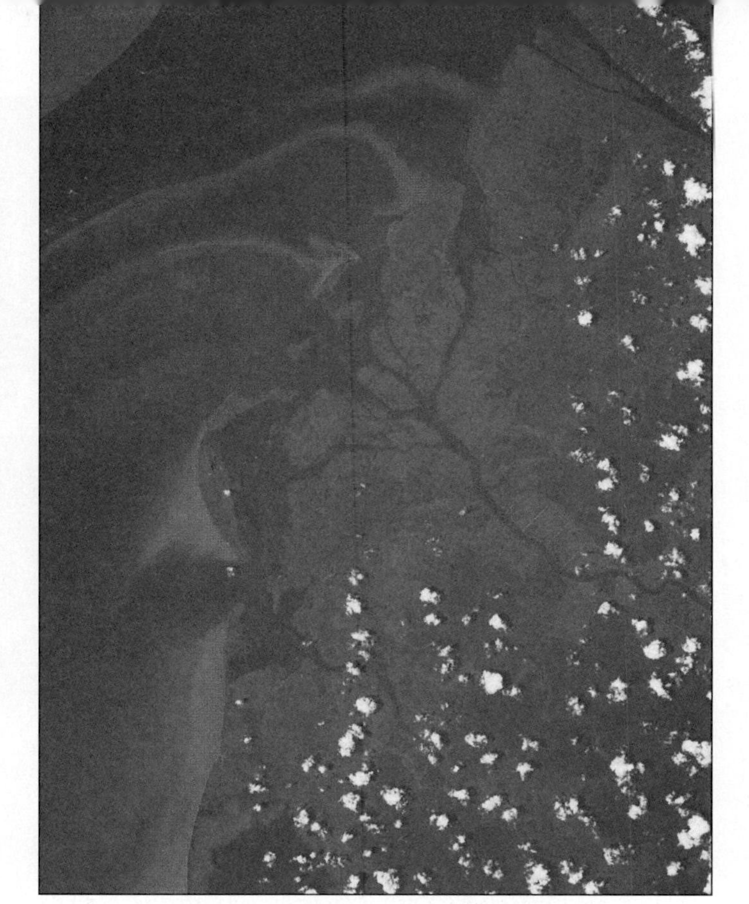

(b) NASA

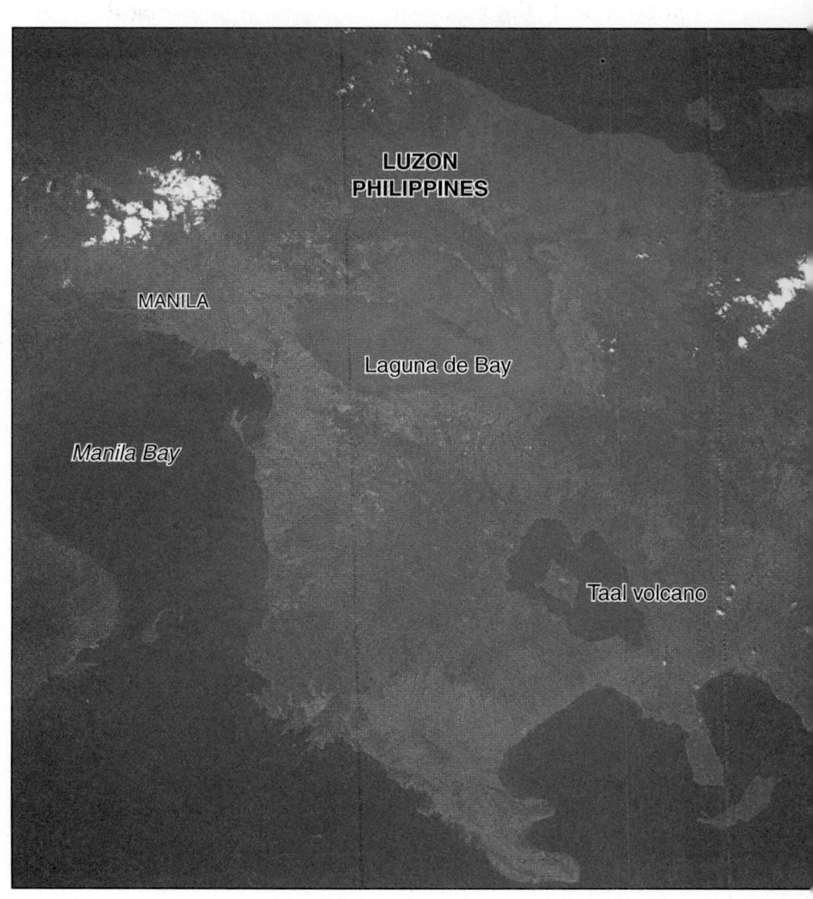


(c) NASA

**FIGURE 6.31 Southeastern Asia: variety of landscapes as viewed from space shuttles.** (a) Irrawaddy Delta, Myanmar, flanked by the forested Arakan and Pegu Mountains. The active river delivers large quantities of sand and silt in the wet monsoon season to form the delta; this view is in the dry season when the brightly colored sandbars stand out from the river. The area produces rice, sugarcane, beans, and other vegetables.
(b) Kalimantan, Indonesian Borneo. Resettlement of people from Java had impacts around the town of Pontianak. Shore currents carry southward sediment from the streams that enhance erosion on cleared forest. The light-green, deforested coastal area is distinguished from the darker green undisturbed patches of rain forest. (c) Manila, Philippines. The city is situated between the three-lobed Laguna de Bay that is over 50 km (32 mi.) long and the coast of Luzon. The city has iron and steel, shipyard, wood and paper product, cement, leather, pharmaceutical, fishing, and tourism industries. Manila's harbor needs constant dredging. Taal Volcano on Volcano Island erupted last in 1968 and 1977.

regional headquarters of many multinational corporations. Having a largely Chinese population and many links to other Chinese people throughout Eastern Asia, Singapore offered relocation facilities to Hong Kong businesses that feared the Chinese takeover in 1997. In the late 1990s, falling growth rates in the rest of Eastern Asia slowed growth in Singapore. Its major thrust at the turn of the century is as a regional business center, promoting investments. New centers in Indonesia, China (Suzhou), India (Bangalore), and Vietnam are exports of the Singapore business environment.

## Farming Changes

Most of the people in Southeastern Asia still live in rural areas and gain their living from farming. Many such areas changed from local subsistence to a cash-based economy linked to growing urban markets and overseas demands. These demands led to new crops, new strains of traditional crops, and changing patterns of land ownership. The highest productivity occurs in the major river valleys and island coastal areas (see Figures 6.16 and 6.31).

**FIGURE 6.32 Indonesia: rubber plantation.** A rubber plantation farmer in Sumatra taps a tree to extract the white sap from just beneath the bark.



As in Southern Asia, the Green Revolution combined new varieties of rice with the use of fertilizer, pesticides, mechanization, and irrigation to raise production. The **new rice technology** spread from agricultural research in the Philippines. Although the wealthier owners of larger farms benefited first, increasing numbers of smallholders use it.

The commercialization of farm production changed the use of land and labor. Former communal land was sold to a rich minority of landowners, while poorer people were forced to rent land or work as landless laborers. Landlessness increased further with population growth and the expansion of urban areas across former farmland. For example, the amount of rice land per head in Java halved from 1940 to 1980. Communal rights in the rice harvest—vital in supporting many poor families—were replaced by the sale of the standing crop to middlemen, while mechanization displaced wage labor. In Malaysia, for instance, the introduction of combine harvesters halved the numbers of workers involved in the rice harvest. As the traditional farming practices of supportive and informal labor exchange during planting and harvest gave way to wage labor and mechanization, social impacts multiplied. The growing numbers of landless people in the subregion supported communist groups, as in some of the islands of the Philippines. Such agitation came to a head in the financial crisis of the late 1990s in Indonesia.

While rice growing is the dominant occupier of farmland in all Southeastern Asian countries, many countries in Southeastern Asia also produce tree-crop raw materials in plantations that were often established under colonial regimes (Figure 6.32). Malaysia still produces a large proportion of the world's rubber crop, but a shortage of cheap labor led to falling production as Thailand, Myanmar, and Indonesia increased their output. Malaysia became the world's largest producer of palm oil, aided by the crop being less labor-intensive and the Chinese demand for cooking oil that keeps world prices high. Malaysian policies are, however, diverting investment away from agricultural production and into manufacturing. Indonesian plantation crops include coffee, spices, tea, rubber, and palm oil products, constituting 13 percent of its exports. The expansion of oil palm plantations in Indonesia in the late 1990s by burning rain forest caused widespread smog across the subregion. In the Philippines, coconut products, sugarcane, pineapples, bananas, and coffee make up 16 percent of exports. In Thailand, such crops form 10 percent of exports. Myanmar remains, with Afghanistan (see Chapter 5), the world's leading producer of poppies for opium and heroin. It supplies illegal drug traffic through Thailand and other routes.

### Forest Products

Timber products remain of great significance to Southeastern Asian countries. Tropical rain forest covers much of this subregion, and its products from Malaysia and Indonesia dominate world production—being greater than those from Africa and Latin America combined. Wood-using industries became more important in Sumatra and Kalimantan (Indonesian Borneo) as Indonesia made efforts to increase the income from exploiting its forests. In the 1990s, prices for tropical timber on world markets rose by 50 percent at a time when prices for other commodities fell or remained the same, and this stimulated further cutting. Much of Myanmar remains forest-covered. It has around 250 commercially useful trees, of which teak is a major export.

The limits of cutting without replanting are being reached throughout the subregion. In the Philippines, forest cover was reduced from 60 percent of the country's area in 1950 to 10 percent in the 1990s. Western objections to unmanaged cutting on environmental grounds are rejected by governments in Southeastern Asia, which cite the historic European and American forest removal and their own controls over illegal logging.

### Mining

Mining adds to the primary commodity exports from Southeastern Asia. Indonesia is a major oil producer and much of the world's tin production comes from the subregion. Oil accounts for 20 percent of Indonesian exports, down from 80 percent in the 1980s because of the development of Indonesia's manufacturing base and a fall in new oil discoveries. Growing energy demands with increasing industrialization will soon convert Indonesia into an oil importer unless new discoveries are forthcoming. Malaya was for long the world's largest tin producer and Malaysia still exports some, together with oil and natural gas from Sabah and Sarawak on Borneo.

### Market-Led Industrialization

Although the switch to commercial farming was gradual in the ASEAN countries, industrialization was more rapid. By the time the countries of Southeastern Asia became independent, most had factory-based industries. These were import-substitution industries protected by tariffs and quotas. At first, these industries grew rapidly, but by the 1970s, the small size of domestic markets restricted the scope for further growth. The combination of small markets, limits of local skills, and close involvement of government offices in licensing made the products expensive. In the Philippines, the government licensing role led to corruption and the term "crony capitalism" was applied to describe a linkage between bureaucrats and entrepreneurs that was common throughout Eastern Asia.

Beginning in the 1960s, the success of the Asian Tigers in manufacturing goods for export and producing high rates of economic growth stimulated the countries of Southeastern Asia to switch to such policies. Many retained some of their less-efficient domestic manufacturing, protected by selective tariffs. The promotion of internationally competitive industries forced local manufacturers to cut costs.

Multinational corporations invested in new factories, employed cheap labor to produce goods for export, and used labor-intensive manufacturing processes. In the 1980s, countries such as Thailand and Malaysia moved from labor-intensive industries to high-tech industries. Multinational corporations from the United States, Japan, and South Korea were slow in diffusing innovations to these growing countries and preferred to use them as a cheap source of labor and gain access to their domestic markets. Malaysia and Thailand gained most from this process, although Indonesia and the Philippines also experienced related economic growth. By 1995, the Philippines received major manufacturing plant investments from Japanese auto and electronics corporations.

Much of the export-oriented industrialization relies on imported capital, skills, and management, causing it to be labeled "ersatz capitalism." That suggests it is an inferior substitute for development generated by local entrepreneurs based on local savings and investment. The education gap between these countries and core countries makes it difficult for the former to generate their own innovations. Economic growth of this type brings problems. Growing contrasts between moderate-to-high income urban industrialized areas and low-income rural areas produce internal imbalances.

### Communist Economic Stagnation

Economic growth in Cambodia and Laos remains slow. For most people, only bare survival is possible. Vietnam has larger areas of fertile soil, more mineral resources, and a larger population than the other two countries, but its national income grew more slowly than population, causing per capita incomes to fall in the 1980s. The causes were both external to Vietnam and internal in origin. Much of Vietnam's poverty stems from the effects of the war that lasted from the early 1950s until 1975: the conflict disrupted all aspects of the economy. Later, Vietnam continued to spend heavily on military excursions into Cambodia (1978–1989) and on a border war with China. The economic blockade by ASEAN and the United States prevented Vietnam and its neighbors from receiving overseas aid until 1991. Only the Soviet Union contributed to Vietnam's economy (approximately $1.5 billion per year in aid until the late 1980s), but that ceased after 1989. Internal policies were poorly conceived and failed in large measure. For example, five-year plans set for the late 1970s and early 1980s directed a shift into heavy industry and out of agriculture. The centralized bureaucracy that was needed to carry out such central planning was unwieldy, unresponsive to demands, and failed to gain the support and further sacrifices of the war-weary population.

After 1985, however, Vietnam's leaders realized the problems they faced and allowed crops to be sold on private markets rather than through the communes and cooperatives they had established. The Vietnamese currency was devalued and foreign investment encouraged. By the late 1980s, Vietnam exported rice again, becoming the world's third exporter by 1994—although bad weather in 1995 caused it to be displaced by India. Vietnam established economic leadership over the two smaller countries of Laos and Cambodia to the west. The shift to greater involvement in the world economy was easier in southern Vietnam—a capitalist country until 1975—than in the northern sector, which had been under communist control for longer. In the mid-1990s, corporations from Japan and South Korea, in particular, were investing in Vietnam. In 1995, the United States "normalized" relations with the country as it became the first communist member of ASEAN.

### Myanmar

Myanmar's political isolation is enhanced by an unrealistic official exchange rate, 20 times the black market rate. This makes its own products too expensive to sell on world markets but brings in cheap imports. Such subsidies will be removed as the country seeks to join the world economic system. In the 1990s, Myanmar introduced Chinese contractors to build bridges, railroads, airports, and hotels, and entered joint agreements with Japanese and European corporations to develop its economy.

### A Growth Pattern

In the 1980s and 1990s, all the countries in Southeastern Asia grew economically by having governments that encouraged an openness to foreign investment in export-oriented industries. The initial advantages of low labor costs and government encouragement led to infrastructure developments to produce sufficient power and better transportation facilities, financed by foreign capital. The personal income generated allowed workers to buy consumer goods and obtain access to better education and health care (see "Living in Malaysia," pages 274–275). As labor costs rose, the next step was to develop products within these countries, which needed an increase in technical, research, and management skills that was often more than these countries could fulfill. The experience of assembling vehicles, producing vehicle components, and working on electronics equipment built up technical expertise locally, while the Chinese community often provided a source of finance and management. In the late 1990s, the processes of economic growth were halted by financial disasters that could hold back further growth for some years.

**RECAP 6D: Southeastern Asia**

Southeastern Asia comprises a group of countries at different levels of economic development. Singapore is a wealthy but small country, whereas Laos, Cambodia, and Vietnam remain extremely poor following years of warfare. Myanmar's exit from isolation is slow. All Southeastern Asian countries are switching to the manufacture of goods for export and are attracting foreign investments. They mostly have governments that take an involved part in industrial and trading policy.

6D.1 Which are the ASEAN countries? Why was the organization set up and what is its likely future?

6D.2 Comparing their locations and past and present economies, what are the future prospects of Hong Kong and Singapore?

6D.3 What common issues do North Korea, Vietnam, Laos, Cambodia, and Myanmar face?

6D.4 Which factors led to economic growth in the Asian Tiger and Little Tiger countries? How are they related to the present problems these countries face?

**Key Terms:**

Association of Southeast Asian Nations (ASEAN)
primate city
deurbanization
new rice technology

**Key Places:**

Southeastern Asia
Brunei
Cambodia
*Phnom Penh*
Indonesia
*Jakarta*
*Surabaya*
*Bandung*
*Medan*
*Ujung Padang*
Laos
Malaysia
*Kuala Lumpur*
Myanmar
*Yangon*
Philippines
*Manila*
Singapore
Thailand
*Bangkok*
Vietnam
*Hanoi*
*Haiphong*
*Ho Chi Minh City*

## LANDSCAPES IN EASTERN ASIA

The landscapes of Eastern Asia reflect historic events that range from the large numbers of Buddhist temples built hundreds of years ago to the war-scarred features of Vietnam and the new industrial cities. Both urban and rural landscapes contain many relics of the traditional past. There are many distinctive features associated with each country, but the urban section focuses on China and Japan.

### Urban Landscapes

#### Chinese Cities

The older Chinese cities still retain relics of their imperial past. Former palaces (Figure 6.33) occupy large sections of central Beijing and elsewhere old city walls and cramped housing are only gradually being replaced by modern roads and buildings. Coastal cities often show evidence of colo-

**FIGURE 6.33 Eastern Asia: urban landscapes.** Large areas occupied by former palaces and historic buildings are a feature of Beijing, China. Tiananmen (Gate of Heavenly Peace) is one of these.

© J. Sohm/The Image Works

nial occupation, as in Hong Kong, Macao, and Shanghai (see Figure 6.23). Although China is still largely rural, with one-third of its population living in cities, present urbanization is rapid. The industrial area in southern China inland of Hong Kong is one example of this process. Factories and huge apartment blocks have sprung up in the 1990s to form cities of several million people around Shenzhen (see Figure 6.27). Many people moved into the new cities from rural areas in southern China and some return there after working for a time in the factories.

New areas of Chinese cities built since 1950 are often utilitarian and monolithic—the marks of communist architecture. The periods of urban neglect and the post-1978 renewal of urban and industrial construction can be identified in city zones. Close control on living accommodation, however, reduced the incidence of shantytowns.

A competition for urban investments remains between the production sector and the urban infrastructure, while the rise of free enterprise within a socialist state raises major ideological questions about urbanization for the Chinese. The result of these tensions and the changes of the last 50 years show up in the urban landscapes of large Chinese towns where historic buildings jostle with basic utilitarian apartment blocks, factories, monolithic communist public buildings, new office and apartment blocks, and modern freeways.

### Japanese Cities

Modern Japanese cities are given a unique character by being built around a diverse set of castle towns, ports, and centers on main routes to religious shrines (see Figure 6.11). The older parts of Japanese towns were often built of wood and other nondurable materials that were susceptible to fire and earthquake. The modern cities contain sectors for commerce, industry, and residences but are often interrupted by areas where earthquakes or World War II bombing caused destruction and rebuilding. Many modern Japanese towns retain older street patterns, but the buildings are mostly post-1945.

**FIGURE 6.34 Eastern Asia: urban landscapes.** Seoul, capital of South Korea, where modern buildings overshadow old ones as a result of rapid urbanization. The huge volume of road traffic is accommodated by multilane roads and reflects the growth in economic activity.



Japanese cities often appear to be formless sprawls but in fact reflect irregular historic growth and rapid modern change. All Japanese cities exhibit the crowding of buildings that results from the shortage of buildable land in the country.

The central business districts of the largest Japanese cities contain specific areas devoted to hotels, department stores, financial businesses, and national or local government offices. High-rise offices, hotels, and apartment blocks are a response to high land costs, as are the underground developments in central Tokyo that include restaurants and car parking lots. Most downtown areas of Japanese cities were redeveloped after 1970 but often retain an older sector that acts as a tourist and specialist shopping area. Smaller business districts occur in suburban areas of the major cities. Out-of-town shopping only began to take hold in the 1990s.

Japanese industrial development is strongly localized, with heavy industry in coastal locations on reclaimed land because of the needs for space and proximity to ocean transportation links to raw material suppliers. Such zones shut off city centers from the waterfront. Light industry is more widely distributed and intermixed with housing, with older firms retaining inner-city sites. Location in the inner city reflects the availability of labor, linkages with other firms, and markets. Local governments encourage the establishment of light industrial estates on the city margins, including science parks for high-tech industries. There is thus a clear trend toward the decentralization of industry.

Housing often consists of postwar apartment blocks near the city center, giving way to suburban bedroom areas. Areas of poor housing in downtown areas are being refurbished to maintain the labor force in those areas. Transport routes determine the main expansion of suburban housing.

### Other Cities

In other countries of Eastern Asia, the close location of old and new buildings is also a feature. Seoul, the capital of South Korea (Figure 6.34), demonstrates this.

## Rural Landscapes

The most characteristic feature of rural landscapes in Eastern Asia are the paddy fields of wet rice growing, either in flat lowlands (see Figure 2.38*a*) or on terraced hill slopes. These occur in Japan, southeastern China, the Koreas, Taiwan, and many countries of Southeastern Asia. In Malaysia and Indonesia, they are less obvious, but commercial farming based on tree crops such as rubber and palm oil becomes significant in the landscapes. The remoter islands of Indonesia, such as Kalimantan (Borneo) and Sumatra, have dominantly wooded landscapes under tropical rain forest, although lumbering is felling the trees on a large scale and palm oil plantations are expanding, changing landscapes to agriculture or eroded hillsides.

In inland China and Mongolia, the landscapes change northwestward to increasingly drier farming and eventually to grasslands with nomadic herding (see Figure 6.21*a* and *c*). Population densities, which are high in the rice-growing areas, dwindle until landscapes have few indications of human presence.

# FUTURE PROSPECTS FOR EASTERN ASIA

The modernization of Eastern Asia was a major event of the 1900s, transforming the region's geography and increasing its links to the rest of the world. In the early part of the next century it is possible that Eastern Asia will refocus world economic growth and political power on the Pacific Ocean and the countries situated around it. The future of Eastern Asia may have a greater impact on the world economy than any other major region may have. As with all futurology, however, there are many uncertainties. Predictions can only be made on the grounds of projecting recent trends forward.

## Former Poverty and Recent Economic Growth

Up to 1960, much world misery was concentrated in Eastern Asia. Poverty was the rule. Even after World War II, wars continued to disrupt Korea and the former Indochina. Racial violence and guerrilla wars held back development in Indonesia, Malaysia, Thailand, and the Philippines, while the Great Leap Forward and Cultural Revolution countered any Chinese advances. In 1960, Japan after 100 years of modernization still had a GDP per head that was one-eighth of that of the United States, while other countries had GDPs per head equal to or lower than those of African countries at the time.

The export-led growth that began in Japan in the 1960s led to a quadrupling of its real income between 1960 and 1985 as it became one of the world's richest countries. Some 10 years behind Japan, the newly industrializing Asian Tiger countries—South Korea, Taiwan, Hong Kong, and Singapore—also began to close on the world's rich countries. From the late 1970s, Malaysia, Thailand, Indonesia, and China started to experience similar growth, although progress has not been quite so rapid. Overall, despite some of these countries still having considerable numbers of poor people, Eastern Asian countries experienced the fastest rise in incomes over the shortest period of time, affecting the greatest number of people on Earth—the "East Asian Miracle."

At the same time, countries such as Laos, Myanmar, North Korea, Cambodia, and Vietnam sank to be among the world's poorest in 1990. All of these countries, except for North Korea, which continues to see itself as self-sufficient, changed their orientation to try to follow the others in the late 1980s.

The late 1900s' economic growth in the countries of Eastern Asia was built on such factors as the availability of cheap labor, the accumulation of capital through savings, the quality of labor, entrepreneurial management, and growing domestic markets. Within the context of very large populations, the countries of Eastern Asia experienced swift falls in birth rates, allowing higher household savings and the ability to spend on consumer goods. Furthermore, the countries of Eastern Asia encouraged overseas investment for both manufacturing and the provision of infrastructure. Education programs often spent no more per head than in other poorer countries but focused on primary and secondary education for all rather than on prestigious higher education for the few. The early start to giving girls the same level of education as boys helped in the swift decline of fertility and the availability of cheap semiskilled labor. In South Korea and Japan, however, where married women were largely excluded from the work force so that they could be at home with their families, the following generation had a particularly high education performance. These two countries are at the forefront of progress in manufacturing output and technology.

## Interventionist Governments

In addition to the economic and social factors that gave Eastern Asian countries an edge in economic development, governments encouraged some free market processes while also taking an interventionist role. The free market pricing of labor, capital, and goods existed alongside government maintenance of low inflation, stable currency exchange, small changes in interest rates, and the encouragement of growth industries. The openness to world markets paralleled government involvement in assessing new products and technologies of production and resulted in the countries of Eastern Asia joining the leaders in these areas. The emphasis was always on competing in world markets, while protecting home markets from too much external competition. Asian countries trumpeted the virtues of "Asian values" including strong government, strong family ties, strong basic education, and strong domestic savings.

## Consumerism

During the 1980s and 1990s, the consumer boom expanded the new groups of people between the very rich and the very poor that characterize many Eastern Asian countries today. Trade increased among the countries of Eastern Asia instead of between this region and the rich countries of the world.

To date, the cities contain the largest consumer markets. They form the major marketing focus and have seen huge increases of sales since the 1980s. For example, the Nanjing Road in Shanghai has a huge range of shops in which 1.5 million people spend over $50 million each day. As incomes rise, families reach thresholds above which there is a conspicuous increase in spending: a 40 percent rise in Malaysian incomes led to a 300 percent rise in new car sales.

Economic growth spreads the ability to purchase beyond the rich elites, so that new economic classes of people emerge.

- The "superhaves" are the old elite together with 1980s and 1990s additions from successful manufacturing entrepreneurs. Families with annual household incomes above $30,000 now number at least 12 million in non-Japanese Asia.
- There is a rapidly growing group of "have somes" with annual household incomes over $18,000 and typical middle-class tastes in houses, cars, and consumer durables. It is calculated that there were 14 million to 15 million households of this type in the mid-1990s and that the numbers will increase to 75 million early in the new century.
- The group of "near haves" will soon number around 150 million and have expectations of moving upward, beginning with their purchases of a television and smaller consumer items.

## Infrastructure Needs

Alongside rising consumer activity in growing economies is an increasing need for better roads, airports, ports, telecommunications, and power plants—all forms of infrastructure needed to support economic growth. The ability to supply this infrastructure will affect the speed at which growth continues. In 1990, some 5 percent of total GDP was spent on such facilities in Eastern Asia compared to 2 percent in the United States and Europe. A feature of major construction projects in Eastern Asia is that only one-third of the financing comes from public sources. Increasing reliance is being placed on foreign inward investments. Multinational corporations involved in such projects expected over 50 percent of the world total to be built in Asia in the 1990s, with a doubling of telephones, a huge expansion of electricity-generating equipment, 20 new international airports, and the purchase of 1,000 new passenger aircraft. Overseas investments, however, often introduce tensions between the foreign corporations wishing to see a good return and the country's government that wants to retain control over the new systems.

## Financing Growth

Financing the desired economic developments will require new institutions in Asian countries. To date, the pattern of high levels of savings within the countries has provided investment capital, sometimes controlled in its use by government policy. Further increases in development require major foreign investment. Foreign governments and banks were anxious to supply this capital in the 1990s, when Tokyo financial markets became a world leader. In 1984, stock market capital in Asia outside of China, India, and Japan was $130 billion. By the mid-1990s, it topped $1,000 billion, doubling the area's world share. In the late 1990s, however, the fragility of the Tokyo financial system was exposed after many banks allowed huge debts to accumulate at home and abroad while the yen continued to rise in value.

## Trading Groups

Countries in Eastern Asia were slow to enter trading agreements with each other, certainly on the scale of the European Union (EU, see Chapter 7). The political suspicions of other countries, fears of dominance by China or Japan, and competition among similar products made countries wary of such agreements. In the 1990s, the ASEAN group shifted from a political and defense-oriented group to one with increasing trading objectives.

The Asia-Pacific Economic Cooperation Forum was established to link the growing economies around the Pacific margins. Along with the countries of Eastern Asia (apart from North Korea, Vietnam, Laos, and Cambodia), it includes Australia, New Zealand, and Papua New Guinea (see Chapter 11), Canada, the United States, Mexico, and Chile (see Chapters 9 and 10). In the late 1990s, hopes of agreeing an action agenda for APEC to be implemented by A.D. 2010 faded following the economic recession in Eastern Asia and a failure to agree on whether farm products should be subject to wider trade agreements. Food exporters such as Australia and the United States wanted to include all economic sectors, but countries with uncompetitive farmers, such as Japan, China, and South Korea, wanted to exclude farm products.

## Will Growth Continue?

Seen from the late-1990s, the events that could slow the expected economic growth in Eastern Asia included further wars—such as the two world wars that destroyed the economic leadership of western Europe in the early 1900s—an increase of protectionism in world trade that would exclude Asians from markets in rich countries, and a reduction of

**RECAP 6E: Landscapes and Futures**

In Chinese cities, ancient buildings and communist-era apartment blocks, factories, and monolithic government buildings jostle for space with newer types of architecture that signal the changes since 1978. In Japanese cities, there is a great premium on space and historic buildings may be replaced by newer.

The countries of Eastern Asia have experienced economic growth, and some have risen within the world economic system, going from periphery to semiperiphery or core. Their growth is based on exporting manufactured goods that compete in world markets because of lower labor costs, multinational organization links, and, for some, high product quality.

The future of Eastern Asia depends on whether local and foreign investment can combine to sustain economic growth, the benefits of that growth can be diffused through the whole population, and political tensions can be avoided and peaceful conditions maintained.

6E.1 If you lived in Eastern Asia, what advice would you give to (a) Americans and (b) Africans concerning economic development? What could you learn from them?

6E.2 How has U.S. foreign policy influenced economic development in Eastern Asia in the later 1900s?

6E.3 Why have communities of U.S. retirees established themselves in Southeastern Asia rather than in Japan or South Korea?

6E.4 How would you define the differences between suburbanization, counterurbanization, and deurbanization?

**WEBSITES: Eastern Asia**

Further sites are available through the Dorling Kindersley CD-ROM Atlas online facility.

Use the following daily newspaper source to get up-to-date information on political, economic, cultural, or environmental events in the Koreas and more widely in the region:

*http://www.koreaherald.co.kr*—the *Korea Herald* site

The *Japan Information Network* site provides a wide range of information, including statistics, about Japan. Try accessing the "Atlas" section (at the bottom of the home page) and comparing "Nature" (e.g., Mount Fuji) or "Community" (e.g., Tokyo Waterfront) across Japan: *http://www.jinjapan.org/stat/index.html*

*http://www.athena.ivv.nasa.gov/curric/land/kobe.html*

This site provides an analysis, with photos and maps, of the 1995 Kobe earthquake.

*http://www.insidechina.com*

A site that gathers a wide range of newspaper and other accounts concerning the People's Republic of China and other Asian countries, compiled in Europe (an outsider's view). Investigate accounts of natural events and economic planning within China.

*http://www.soros.org/burma.html*

One of the George Soros organization's sites: the Burma Project of the Open Society Institute collects information about the isolated country of Myanmar. Investigate the role of dissidents and items directed to dissuading tourists from visiting Myanmar.

government abilities and interests in steering economic growth. As in other parts of the world, ethnic sensitivities might erupt, and the subsequent civil disorder would affect economic expansion.

In the late 1990s, governments' attempts to steer economic growth led to a major slowdown and interruption of growth in Eastern Asian countries. Faced with the need to restructure financial institutions, Thailand closed finance companies bankrupted by overfinancing of construction projects but faced political upheaval as so many lost savings, jobs, or merely election kickbacks. Malaysia's overinvestment in construction projects also brought financial disasters and devaluation. In 1997, this had a positive effect in higher income gains for palm oil planters, whose costs were in local currency but sales on world markets in U.S. dollars—hence the scramble to extend areas under the crop leading to great smoke palls. Oil palms now constitute the world's biggest vegetable oil crop, despite containing more "unhealthy" fats than midlatitude vegetable oils. Malaysia contributed 58 percent and Indonesia 28 percent of the world total in the late 1990s.

By 1998, Eastern Asian economies were in major trouble. Economic output in Thailand, South Korea, and Indonesia was expected to fall between 6 and 25 percent in 1998 after years of growth. Indonesian imports were down 70 percent from the previous year, 20,000 South Korean firms were bankrupt, and Thai car production was down by 75 percent. Japan was also in recession and China had to reduce its prices on world markets. Singapore, Taiwan, the Philippines, and the isolated poor countries were affected less, but the whole region, as well as its resource-base and market countries around the world, faced difficulties that were likely to continue into the first decade of the new century.

Governments in South Korea and Thailand fell, while President Suharto of Indonesia was forced to resign. Although individual countries had experienced recession in the last 20 years, this was the first time that all the Asian Tigers were affected at once. External advice was for the countries of Eastern Asia to reform their financial institutions to reduce debts and avoid future overlending. The countries took this advice with reluctance. The "Asian values" and "East Asian Miracle" appeared to have fostered a complacency that caused the growing problems to be overlooked until disaster struck. The rest of the world awaited the outcome, fearing that the problems might spread.

## CHAPTER REVIEW QUESTIONS

Use the questions below to prepare for exams based on this chapter.

1. China's capital city is (a) Shanghai (b) Beijing (c) Hong Kong (d) Lhasa
2. The region of China in which the official language system originated is (a) North Plain (b) Northeast (Manchuria) (c) Sichuan Basin (d) Southeast
3. China's relative size and latitude is ________ to the United States. (a) similar (b) identical (c) opposite (d) uncomparable
4. The last dynasty of Chinese emperors was the (a) Han (b) Manchu (c) Ming (d) Zhou
5. Mao declared the People's Republic of China in (a) 1898 (b) 1933 (c) 1949 (d) 1971
6. Which Japanese island is not adjacent to the Inland Sea? (a) Honshu (b) Hokkaido (c) Kyushu (d) Shikoku
7. The body of water separating Japan and Korea is the (a) Sea of Japan (b) Sea of Siberia (c) East China Sea (d) South China Sea
8. Angkor Wat is located in which country? (a) Bangladesh (b) Cambodia (c) Laos (d) Thailand
9. Malaysia is a federation of states, two of which are located on the island of (a) Borneo (b) New Guinea (c) Sulawesi (d) Taiwan
10. The Indonesian island of ________ is known for its high population density. (a) Bali (b) Jawa (c) Sulawesi (d) Timor
11. Tibetan culture is being destroyed by Chinese policy.
    True/False
12. Tiananmen Square was the site of the 1989 prodemocracy movement's protest.
    True/False
13. Honshu is the largest of the four main Japanese islands.
    True/False
14. The Japanese want Russian forces on Okinawa to leave.
    True/False
15. The Korean Peninsula has been divided since 1910, when the Japanese annexed it.
    True/False
16. Singapore has a Chinese majority.
    True/False
17. Confucius was China's most influential teacher and philosopher.
    True/False
18. Indonesia is a country of hundreds of islands.
    True/False
19. ASEAN includes Hong Kong and Taiwan as members.
    True/False
20. The island archipelagoes of insular Eastern Asia are the result of tectonic activity.
    True/False
21. Which autonomous region in China was considered an independent country from 1911 to 1950?
22. Characterize the China-Japan relationship prior to 1945.
23. What are the two countries on the divided Korean Peninsula?
24. What part of China will take part in the emerging Pacific Rim?
25. What is the largest metropolitan area in Japan (and the world)?
26. List the four largest cities in southeastern Asia.
27. Mongolia is often thought of as a buffer state between ________ and ________.
28. Which countries in Eastern Asia were not incorporated into a European empire?
29. What crop's area is expanding in Malaysia and Indonesia?
30. Which countries in the region have superpower qualities or will soon be a superpower?

# EUROPE

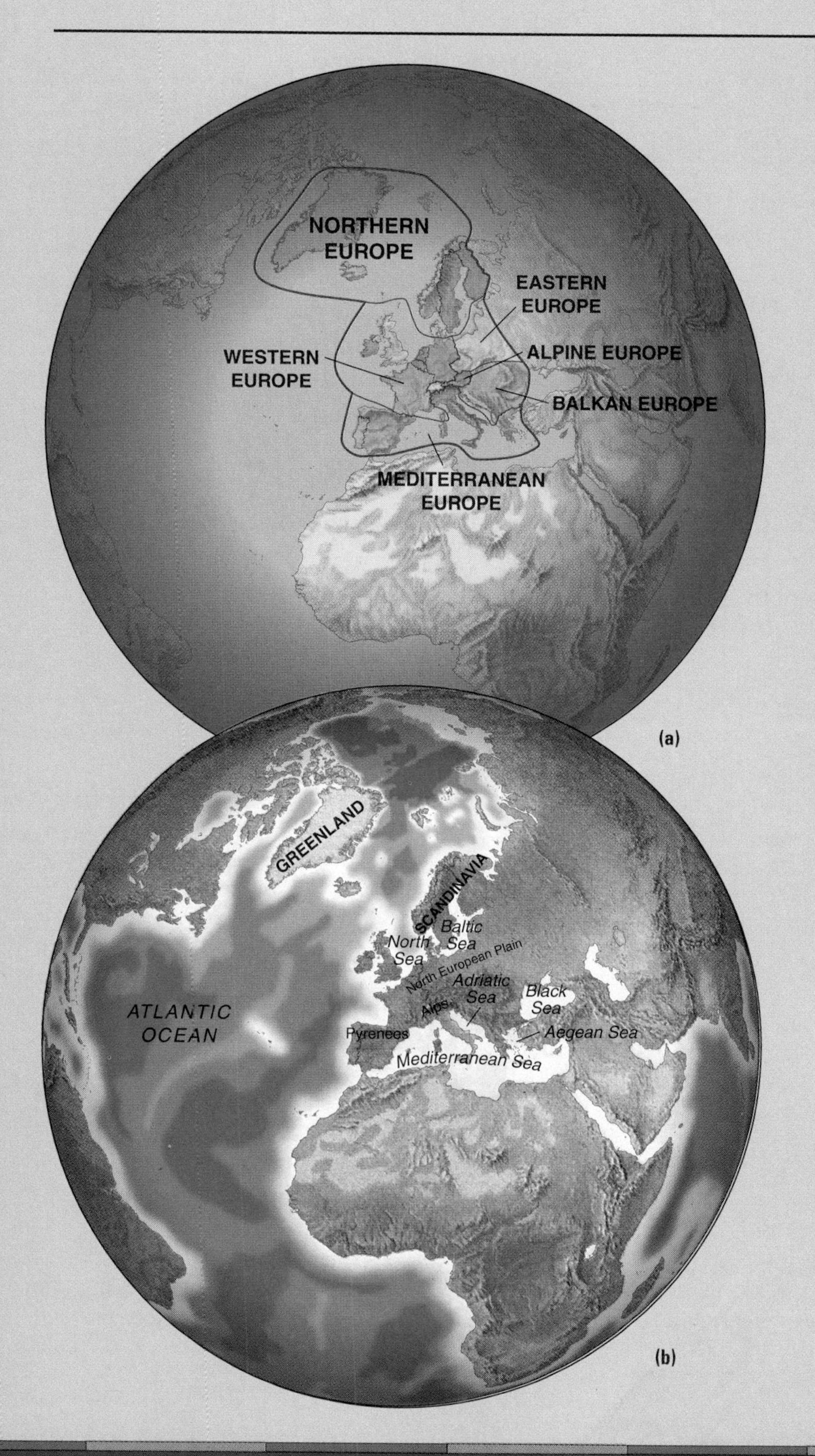


Chapter 7

**THIS CHAPTER IS ABOUT:**

**Europe's leading role in the new global order**

**Historic background to Europe's present geography**

**Midlatitude natural environments with easy access to coasts**

**Six subregions:**

- Western Europe: Europe's industrial heartland
- Northern Europe: high living quality in harsh environments
- Alpine Europe: affluence in the mountains
- Mediterranean Europe: poorer countries catching up
- Eastern Europe: former Soviet bloc countries likely to join the EU
- Balkan Europe: old hatreds die hard

**Urban and rural landscapes that record past and present geographies**

**Future Prospects for Europe**

**World issue: European Union**

**Living in Germany**

**Living in Croatia**

**FIGURE 7.1 Europe in a world context.** (a) The countries and subregions. (b) Natural environments and major physical features.

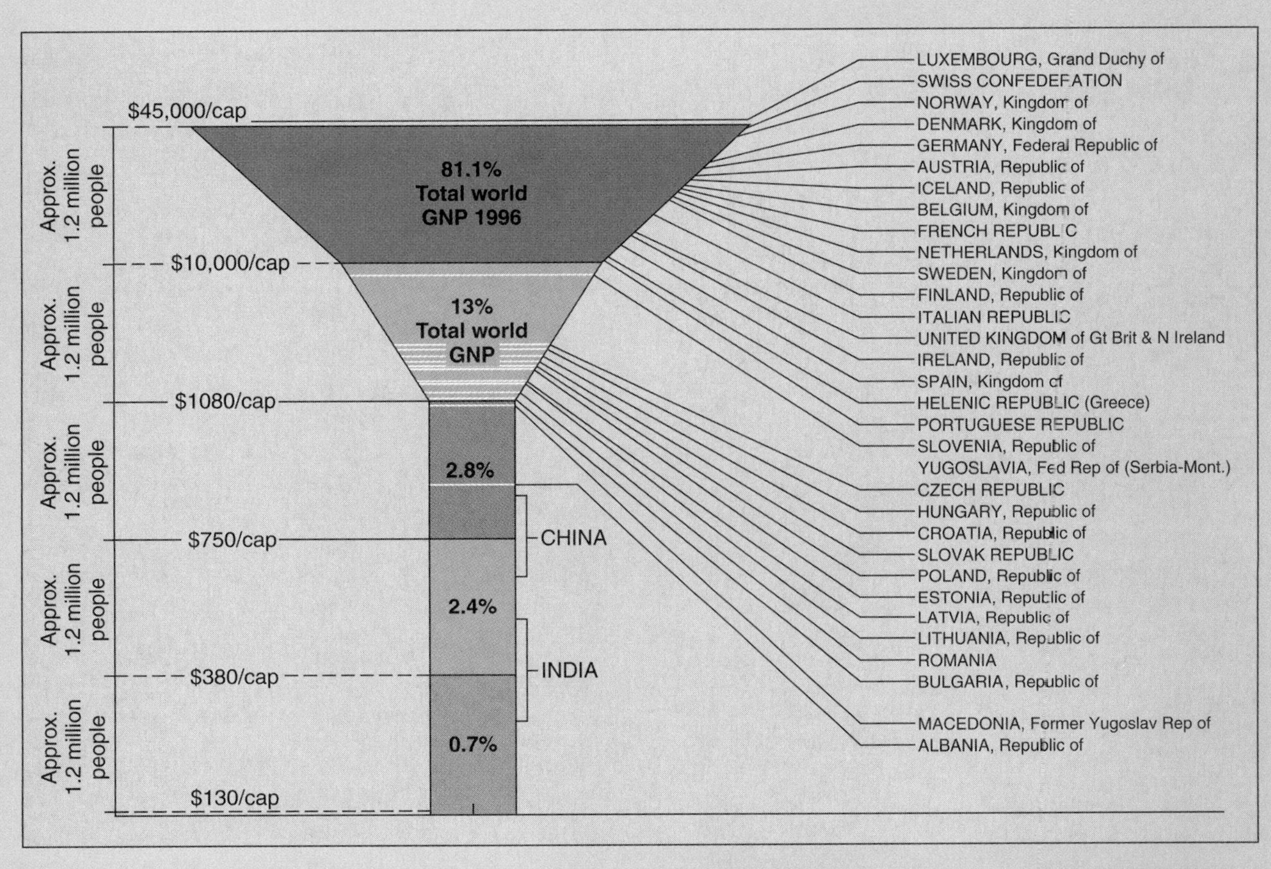


**FIGURE 7.2 Europe: national incomes compared.** Compare the positions of the countries in the two top one-fifth divisions of world countries. Uncertainty over the true position of countries in Eastern and Balkan Europe remains while the countries there are in transition from being controlled by the Soviet Union to becoming oriented to world—and especially European—markets.

Source: Data (for 1996) from *World Development Indicators,* World Bank, 1998.

## FIRST WORLD CORE

The modern world economic system was established in Europe (Figure 7.1), spread around the globe, and then modified and strengthened in Anglo America and Eastern Asia. From the late 1400s, mercantile capitalism spread out of Europe by trade and conquest. Industrial capitalism began in northwest Europe in the late 1700s. Although individual countries in the region lost their economic and political preeminence during the 1900s to the United States, the former Soviet Union, and Japan, they remain supreme in the world economic system as a group (Figure 7.2). In the late 1990s, their combined GDP was 33 percent of the world total, up from 25 percent in 1965. Over the same period, the region moved from 75 percent to 126 percent of the total U.S. economic output.

European influences on the world scale began with the trading and colonial ventures that took Spanish and Portuguese adventurers to the Americas, Africa, and Asia in the 1500s. In the next 300 years, the French, Dutch, British, and Germans followed. Spanish and Portuguese became the main languages in "Latin" America (Chapter 10), French in parts of Africa and Southeastern Asia (Chapters 3 and 6). English is the main language spoken in Anglo America (Chapter 9) and much of Africa, Southern Asia, and the South Pacific (Chapters 3, 5, and 11), and is used more widely in politics and commerce. Roman Catholic and Protestant varieties of Christianity were quickly established on a global basis, diffused through the trading and colonization process.

Geographic differentiation within Europe has a complex history. It reflects the diffusion of modern urban-industrial technical and social cultures, overlaid on older patterns of human activity, and a constantly changing political control. Britain led the way in the industrial revolution of the late 1700s and early 1800s, followed by Germany, France, Belgium, and the Netherlands. By the late 1800s, modernization reached parts of Alpine Europe, Northern Europe, Spain, Italy, and Eastern Europe.

The countries of Europe emerged from centuries of cultural, economic, and political events that sparked wars, political amalgamations and splits. Industrial growth in the 1800s led to competition among Britain, France, Germany, Austria, and Russia. The growing political and military dominance of Germany triggered two wars in the first half of the 1900s that initially focused on European issues but eventually involved most of the world. Countries between the main powers, including Belgium and the Netherlands west of Germany and the Eastern European and Balkan countries between Germany and Russia, became additional battlefields. The world involvement arose from Europe's global colonial expansion, together with the rise of the United States, Japan, and the Soviet Union to world power status.

After World War I, a host of new countries was created in 1919, but this did not allay the tensions that came to a head again in World War II. Following World War II, the countries of Europe were determined to prevent further world wars issuing from conflicts among them. The emergence of the Soviet Union to the east and the perceived threat of its military presence in eastern Europe resulted in not only the signing of the North Atlantic Treaty and the formation of the North Atlantic Treaty Organization (NATO) in 1949—which continued the presence of U.S. forces in Europe—but also the beginnings of economic union. After 1945, the Soviet Union imposed its own economic program, the Council for Mutual Economic Assistance (CMEA, or COMECON), on the countries that it occupied in Eastern Europe and most of the Balkans. The Soviet Union and its occupied countries signed the Warsaw Pact defense treaty in 1955 as a response to NATO admitting West Germany. The Cold War was established. The United States encouraged western European countries to bolster their defenses both by military means and by building strong economies. They reconstructed their productive capacities, creating the world's largest affluent market. The previous history of local conflicts caused moves toward greater political unity to occur slowly and incrementally through the rest of the 1900s (see "World Issue: European Union," pages 292–293).

The subregions of Europe reflect the geographic imprint of the major historical events that molded modern Europe. Within the context of the EU, these subregions and their constituent countries retain distinctive characters. Some could break up into smaller political units. Their constituent countries are:

- **Western Europe**, the industrial heartland: Belgium, France, Germany, Luxembourg, Netherlands, Republic of Ireland, and United Kingdom.
- **Northern Europe:** Denmark, Faeroes, Finland, Greenland, Iceland, Norway, and Sweden.
- **Alpine Europe:** Austria and Switzerland.
- **Mediterranean Europe:** Greece, Italy, Portugal, and Spain.
- **Eastern Europe:** Czech Republic, Estonia, Latvia, Lithuania, Poland, and Slovak Republic.
- **Balkan Europe:** Albania, Bosnia-Herzegovina, Bulgaria, Croatia, Hungary, Romania, Slovenia, and Yugoslavia (Serbia-Montenegro).

The chapter is supported by data in the Reference Section, pages 586–610.

## CHALLENGE AND CHANGE

Although the history of Europe is complex in terms of the economic, political, cultural, and environmental challenges that faced its peoples and produced its current human geography, several major strands can be identified. In each major period of history was a growth of ideas, technology, and facilities that was interrupted by destruction and stagnation but led to increased economic output and political influence for the region. Migrations of peoples into the region brought ideas and technologies that were used, modified, and incorporated into local systems. Many of the forms of organizing the utilization of Earth's surface generated within Europe later diffused to other parts of the world.

New forms of land use replaced old as natural forest vegetation gave way to farmed fields, and then urban areas and factories replaced many of the fields. The landscapes of Europe today reflect this history of continuous changes and the ways in which peoples met the challenges posed.

### Divisions and Integration

As the ice sheets retreated 15,000 years ago, Europe was occupied by small groups of hunters, moving northward. Some 5,000 years ago, world climates reached their warmest point since the end of the last advance of ice sheets and most of Europe was covered by dense deciduous forest with conifers in colder and higher areas. From that time up to around A.D. 1000, Europe had no established political frontiers. Invasions and migrations alternately brought peoples together and separated them into warring groups.

#### Celts, Greeks, and Romans

The first migrations and invasions came mainly from within the region. The **Celts**, with their metal skills in bronze and iron, moved out of the Alps around 1000 B.C. and overran the areas now occupied by modern France, Spain, and Britain

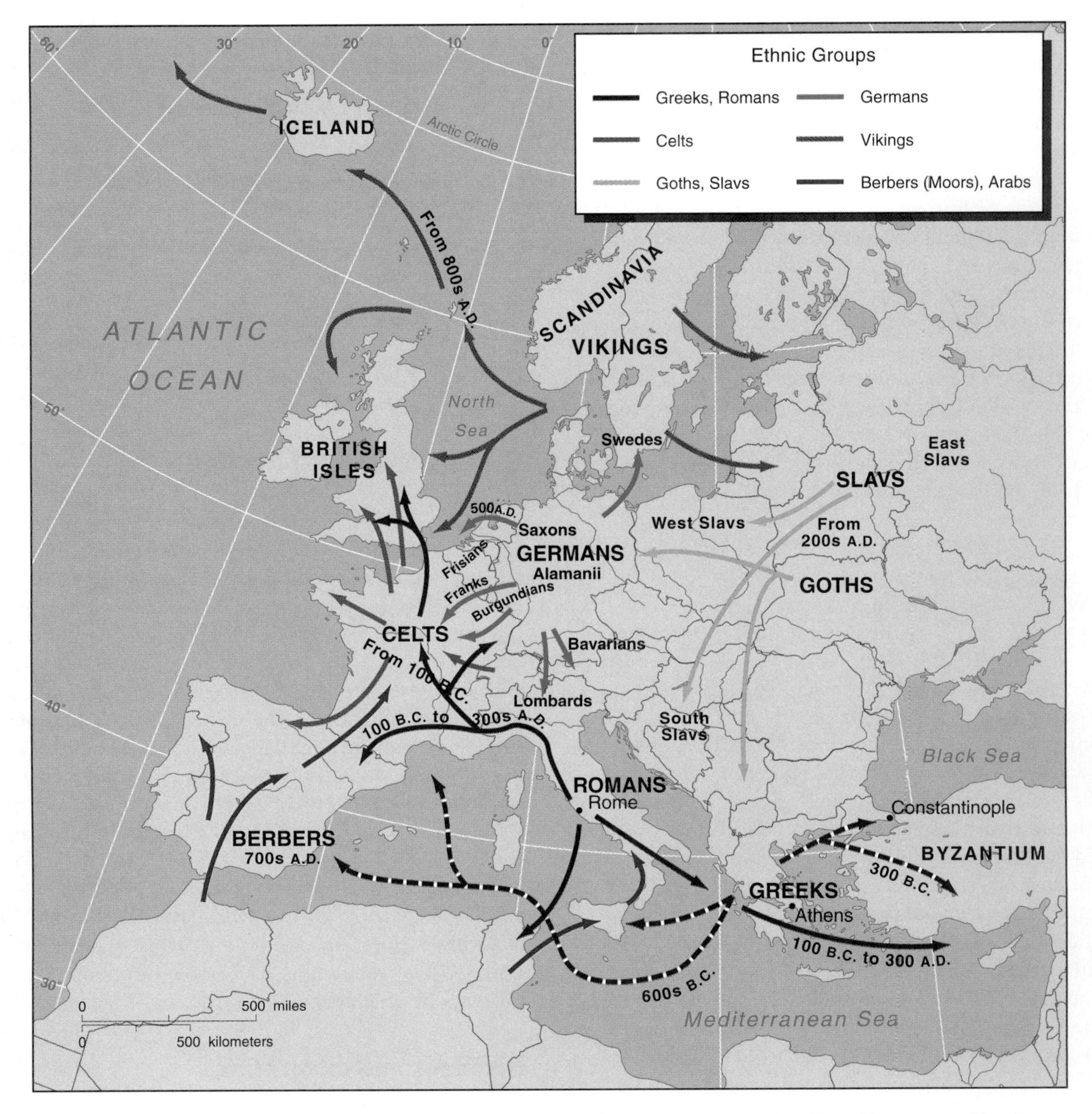


**FIGURE 7.3 Europe: movements and migrations of people.** Some of the early movements that affected later geographic units, from approximately the 300s B.C. to the A.D. 900s. Note how Germanic groups spread outward, displacing the Celts and Romans in the west, and how Slavs moved into eastern Europe, displacing some German groups.

(Figure 7.3). At this time, the **Greeks** were a major force in the eastern Mediterranean, but although they established trading ports around the western parts, they did not venture far inland. From the first century B.C., **Roman** armies took over these Greek ports and expanded their empire inland into Spain (Hispania), across France (Gaul), and eventually into the British Isles. The iron-based technology of the time not only supported powerful armies but also led to extensive woodland clearance in the Roman Empire to increase crop production. The Romans made little headway, however, against the Germanic tribes of central Europe. The Roman Empire ended along the Rhine River from the North Sea lowlands to Helvetia (modern Switzerland) in the Alps and along the Danube River to the east.

## Germans Advance

From the A.D. 300s—almost as soon as Rome became Christianized—**Germanic tribes** extended their influence and the Roman Empire crumbled. While the Alamanii remained in control of central and southern modern Germany, the Burgundians moved into eastern France, the Saxons into the British Isles, the Swedes into Scandinavia, and the Lombards into northern Italy. Most successful were the Franks, who established control over northern France and Germany. The advancing Germanic tribes pushed Celtic peoples into the hilly areas of the far west of Europe including Galicia (northwest Spain), Brittany (northwest France), Cornwall (southwest England), Wales, Ireland, and Scotland. At the same time, Huns, Goths, and Slavs from farther east penetrated Europe for shorter or longer periods. They were the forerunners of other groups from outside the region.

### Moors and Franks

In the early A.D. 700s, Berber and Arab African Muslim groups, known as the **Moors**, invaded southern Spain and Italy, holding territory there for several centuries and advancing rapidly into central France. They were repulsed by Christian Frank armies. In A.D. 800, the Frankish king, Charlemagne, forged links with the popes in Rome and was crowned as holy Roman emperor. At Charlemagne's death, the laws of the Franks decreed that the empire be split among his sons, leading to one of the fundamental divisions of Europe between the West Franks (approximately modern France), the East Franks (Germany), and Lotharingia in between (modern Netherlands, Belgium, Luxembourg, Lorraine in easternmost France, and Switzerland). This division became the basis of claims, counterclaims, and many wars in the succeeding centuries. The power of the Roman Catholic Church and national kings grew after A.D. 1000. People saw themselves as citizens of specific countries and not merely as cogs in the feudal system.

## Europe Goes Global

After absorbing such internal and external influences, the peoples of Europe looked farther afield for conquest and trade. Beginning with tribal groups expanding their horizons, mostly within Europe, the later stages extended worldwide.

### Vikings

In the A.D. 800s, northern European peoples who struggled to exist in villages with meager natural resources became the **Vikings.** Living in the modern area of Denmark, Norway, and Sweden, they developed ocean transport mobility and well-organized fighting abilities at a time of improving climate and overpopulation in their home areas. They sailed northward, colonizing the Faeroe Islands, Iceland, and southwestern Greenland, and probably reached North America by the A.D. 900s. To the west, they invaded the British Isles and northern France. To the south and east, they had major impacts on the lands around the Baltic Sea and as far south as Kiev by the late 800s, where they played a formative part in founding Kievian Rus', which became one basis for the future Russia.

Although they are often caricatured as violent pirates, Vikings mostly settled alongside the local inhabitants of areas that they invaded, organizing wide-ranging trade across Russia. Their influence ended after the 1200s, when cooling climate, volcanic eruptions in Iceland, and political factions caused the Greenland colonies to die out and halved the population of Iceland. Plague illnesses reduced the numbers of people in their homelands. By that time, the mixture of French and Viking peoples in the northern French area of Normandy had invaded and conquered England.

### From the Mediterranean to China

Around the Mediterranean Sea, further stimulus to wider geographic links came through cities such as Genoa and Venice, developing extensive trade with the Arab world and northern Europe. The southern part of Europe was gradually freed from the Muslim invasions, and after the 1200s, only a small area of southern Spain remained in the hands of the Moors until 1492. In the late 1200s, Marco Polo traveled across Asia to China and was employed for a time by the Mongol emperor in Beijing. His account of those travels stimulated others from Italy, Portugal, and Spain to seek routes to Southern and Eastern Asia that avoided the Muslim countries.

### Rise and Fall of Feudalism

In western and central Europe, the period from around A.D. 1000 to 1450 was dominated by inward-looking feudalism in which peasants came under the rule and protection of nobles holding tracts of land for kings and emperors. Feudal estates were largely subsistence farms and sources of manpower in war. They emerged from the unsettled early medieval period of warfare and invasion that demanded a means of cooperative defense. The Roman Catholic Church attempted to overlay local concerns with spiritual and political control by popes and bishops, and worked well with feudalism. The combination produced a major flowering of cultural achievement in Renaissance art and in the construction of massive cathedrals that were built across Europe during the period of warmer climate from the 900s to the early 1300s.

Feudalism in northwestern Europe, however, contained its own seeds of destruction in disputes among nobles and monarchs and among the growing numbers of town-based merchants and the nobles. As climate and technology improved, the production of food surpluses and wool had resulted in craft industries and trade, bringing wealth and increasing power to the merchants. Poorer climatic conditions and plague illnesses in the 1300s cut the populations of France and England by one-third, devaluing the land ownership and servile support on which the feudal nobles depended. The influence of feudal systems declined as people abandoned villages and migrated into towns that were often controlled by merchant groups. Monarchs began to rely on funding in the form of taxes from merchants, establishing more strongly their own position alongside that of the rising merchant class.

### Overseas Exploration

In the 1300s and 1400s, the collapse of feudalism and end of dominance by a single church allowed the inward-looking society to seek wealth farther afield, based on the growing strength of monarchs and the rising number of merchants in

# EUROPEAN UNION

Some of the issues that arise from a study of the European Union (EU) include:

What caused the EU to be set up and has it fulfilled the original aims?

What advantages does the EU bring to its constituent countries?

When will the EU be complete?

The EU is now a major factor in changing the human geography of Europe. Virtually all the countries of this region either belong to it or have applied for membership, as the map (see Figure 7.7b) shows. Yet many issues cause continuing debate, especially as political pressures change.

The EU arose from attempts to ensure that the future of war-torn Europe after 1945 would be peaceful. Two world wars in 30 years resulted in a determination by political leaders to interlock the economies of European countries so closely that further wars would be difficult, if not impossible. Some anticipated that once economic links were established, political union in a federation like that of the United States might follow. With American aid and protected from the expansionist ambitions of the Soviet Union by the North Atlantic Treaty Organization (NATO), the first step was the formation of the Benelux customs union between Belgium, the Netherlands, and Luxembourg in 1949. The *European Coal and Steel Community (ECSC)* in 1952 established common discussions among France, West Germany, and the Benelux countries for what were regarded as basic industries. Governments were directly involved in their management following their nationalization.

In 1957, the Treaty of Rome, signed by the five ECSC countries plus Italy, established the *European Economic Community (EEC)*. The EEC was expected to create a common market in which goods, capital, people, and services moved freely. The *European Commission* became the executive arm of the community and was based in Brussels. After the establishment of a customs union in 1968, further progress was slowed by the addition of new countries, including Denmark, the Republic of Ireland, and the U.K. (1973), Greece (1981), and Portugal and Spain (1986), each of which took time and money to absorb. The European Parliament, with elected members from all these countries, sits in Strasbourg, France.

The Single European Act, signed in 1986 when Spain and Portugal joined, set out the steps needed to implement a single market. These were agreed on by the end of 1992, but not all were adopted by member countries. In 1986 the title was changed to *European Community (EC)* to emphasize the move from economic toward political goals. The Treaty of Maastricht, agreed to by member countries in 1990 but only implemented at the end of 1991 after further discussions, attempted to set a timetable for monetary and political union, and in 1993 the group of countries agreed to call itself the *European Union*. This raised questions of national sovereignty that had not been fully faced by many countries. The ideas of a common currency and common foreign and defence policies, in particular, caused the introduction of the principle of subsidiarity to allow member states to opt out of certain clauses. For example, because it felt it would affect its competitiveness in the world economic system, the U.K. at first opted out of the social chapter that set guarantees on a number of workers' rights, including a minimum wage policy. The new U.K. government elected in 1997 opted into the social chapter but remained wary of deeper commitment to European political integration.

The EU has many achievements. In particular, it provided an expanding internal market that stimulated economic growth in the member countries and investment from other countries that now regard Europe as one of the world's major markets for goods. The EU still struggles with its Common Agricultural Policy (CAP), which uses up over half the annual budget in subsidies to farmers even although such subsidies have been reduced since the mid-1980s. The EU is not yet a convincing political force with a single mind, as its thwarted attempts over several years to engineer reconciliation in Bosnia showed. Nevertheless, countries in Europe that are not members see it as desirable to become so, even those like Sweden and Switzerland, which did very well outside the EU until the 1990s and were previously opposed to formal international involvements. Austria, Sweden, and Finland joined in 1995. The Swiss and Norwegian referendum votes about joining were close—although rejecting membership at the time. Many of the countries of Eastern and Balkan Europe, formerly part of the Soviet bloc, also hope for a future in the EU.

The processes involved in the formation of the EU—bringing countries together in a large grouping—go against trends for splintering and regional pressure groups that are also common within Europe. It appears that the momentum gained by the EU so far will carry it forward because of the economic gains within the world economic system. The extent of political union may not, however, be so great or so rapidly achieved as was once and is still anticipated by some politicians.

### TOPICS FOR DEBATE

1. The European Union's main aims have changed from those contained in the Treaty of Rome.
2. The more countries that wish to join the EU, the less likely it is to achieve fuller integration or union.
3. The success of the European Union has stimulated similar international agreements elsewhere in the world.
4. The EU will never be another United States.

(a)

Palais de l'Europe, Strasbourg, France. The European Parliment building is the rectangular building in the left midground. The members who meet there once a month advise the European Union. The European Court of Human Rights is in the foreground, and medieval Strasbourg Cathedral is in the right background.



(b)

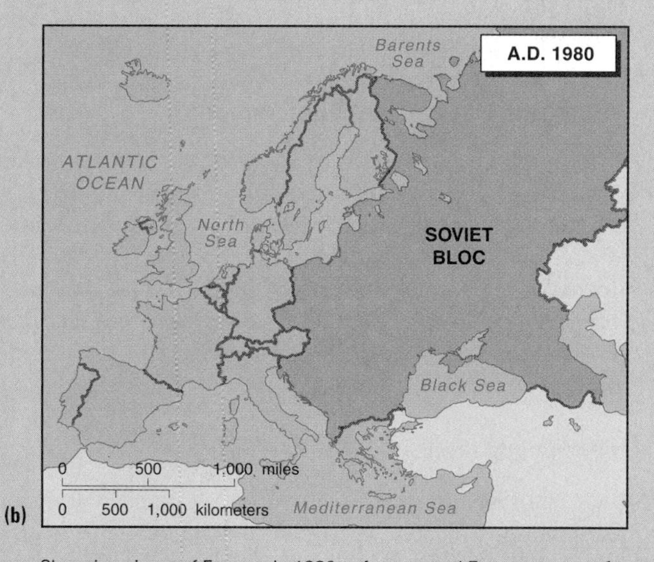



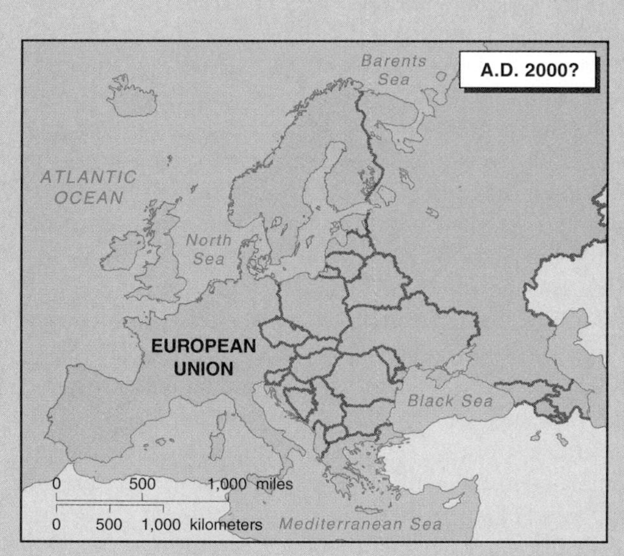



Changing views of Europe. In 1980, a fragmented Europe was confronted by the monolithic Soviet bloc to the east; now the increasingly united Europe confronts the fragmented former Soviet bloc, part of which is included in this chapter and part of which is the subject of Chapter 8.

expanding towns and cities throughout Europe. Merchant wealth and political opportunism combined with improvements in shipbuilding and navigational technology to encourage overseas exploration.

The first voyages had the objective of getting to Asia around the southern tip of Africa. The Portuguese Vasco da Gama took the final stages of this route in the 1490s. Also in that decade, the Genoese Christopher Columbus, funded by the Spanish crown, sailed westward in 1492 in the belief he would get to India by a quicker route. He did not know that the Americas (named after another Italian sailor, Amerigo Vespucci) stood in his path.

Portugal and Spain both gained wealth from trading with and colonizing the Americas, Africa, and parts of Asia. The French, Dutch, and British followed them in the 1500s and 1600s. Wars in Europe and declining home economic bases reduced the roles of Spain and Portugal, which spent their New World wealth on maintaining their positions in Europe. Home-based merchant wealth shifted power toward the northwestern European countries. The Dutch emerged as another maritime power in the 1600s, establishing colonies in the East and West Indies but losing major battles and territory in North America to Britain in the 1660s. Over the following century, Britain won its competition with France for supremacy in North America and India, setting the basis for its worldwide empire in the 1800s.

## Industrial Revolution

The growing overseas trade and merchant wealth of the countries of northwestern Europe led to increasing demands for manufactured goods, stimulating a series of technologic innovations and organizational changes. The resources of cottage weavers and blacksmith forges were hard-pressed to fill the expanding markets for cloth and metal goods. By the late 1700s, machinery, at first driven by waterpower increased productivity in the metal and textile industries. The concentration of machines in factories that were soon powered by steam from coal-burning furnaces required growing amounts of capital and numbers of workers. It led to the expansion of urban-industrial centers on or near coalfields and to systems of banking and investment. Canals were the initial forms of transportation used to assemble raw materials for the new industries and to distribute their products, but railroads replaced the canals during the 1800s.

Western Europe took a central role in the technological innovations that resulted in fresh expansions of industrial activity in new products and new places through the 1800s and 1900s. The initial factory-based technologies of the industrial revolutions in the 1700s and 1800s—metal-smelting and fashioning, textile machinery, steam engines, and chemical refining—all developed first in Europe based on energy from plentiful coal resources. Although European countries did not transfer their new manufacturing technologies to their colonies, the United States and Japan in particular took them over and developed their own production and research centers. Subsequent advances in road vehicles, aircraft, and electronics occurred in Europe and Anglo America in the early 1900s. World War II was responsible for the spin-off of further European research, providing radar technology, jet engines, rocket propulsion, and code breaking advances that led to modern computers. American and Japanese corporations locating facilities in Europe in the 1980s and 1990s included research as well as production units to make use of the continuing stream of innovations from the specialized science graduates in this region.

The innovations emanating from Europe left their marks on the geographic patterns of urban and rural areas. Towns that grew as industrial centers in the 1800s were dominated by coal mines, large factory buildings, and the surrounding low-quality worker housing and white-collar suburbs. Textile, steel, heavy engineering, and chemical industries concentrated on coalfield areas in Britain, northern France, Belgium, and the Ruhr area of Germany (Figure 7.4). Such industries had devastating effects on local environments as they emitted poisonous fumes into the air, poured toxic wastes into the rivers, and left towering waste heaps. The worldwide trading and colonial expansion of European influences resulted in the development of huge port facilities in the estuaries of major European rivers. The largest of these were along the River Thames below London and along the lower River Rhine in the Netherlands around Rotterdam. Other large industrial ports grew near Glasgow and Liverpool (Britain), Marseilles (France), Antwerp (Belgium), and Bremen and Hamburg (Germany).

## World Wars and Global Economic Changes

In the 1900s, two world wars reduced the economies of the European countries that had been world leaders. Between the wars, (p. 478) the economic depression of the late 1920s and 1930s gave the first warnings of future problems to the established areas of heavy industry. Despite the massive changes imposed by these events, Europe regained a major place within the world economic system by the later 1900s and was able to respond to changes in the wider world.

### Increased Government Intervention

The traumas of the 1930s and World War II led to an increasing level of government intervention, ensuring the provision of education, health care, unemployment benefits and pensions, and the restoration of productive capacity. Governments took an increased part in economic planning, the ownership of major industries, and the promotion of welfare legislation. Strong labor unions prospered, and wage levels and social benefits improved.

### Europe and the Wider World

In the 1950s, as the United States assumed world political leadership, it helped to reconstruct the destroyed European cities and transportation links and revive the economy through the Marshall Plan. The growing multinational cor-

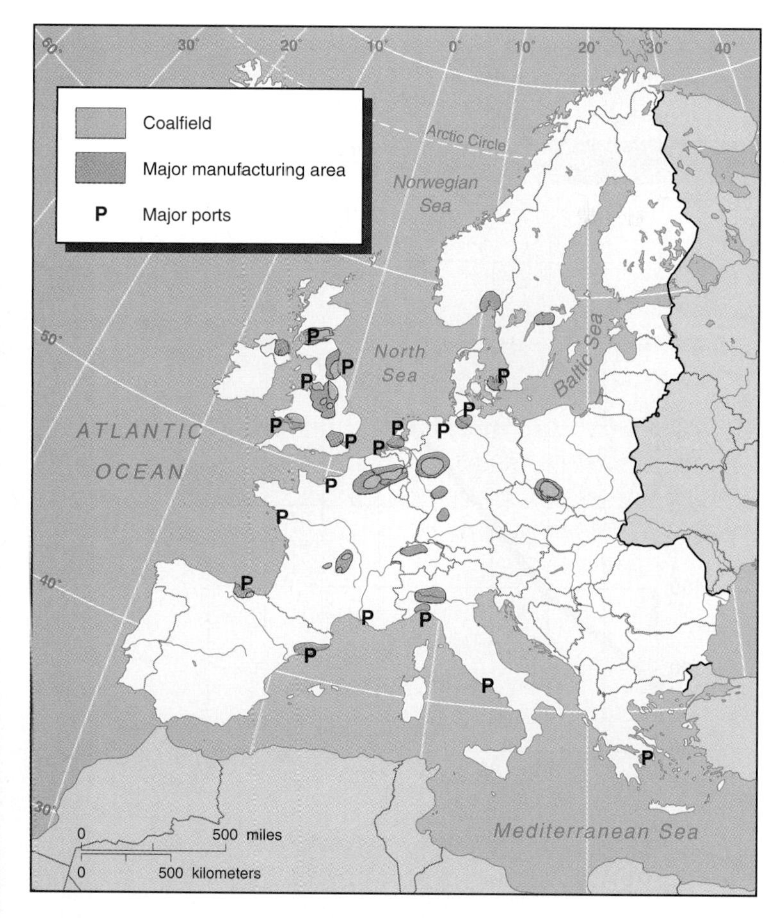


**FIGURE 7.4 Europe: major manufacturing areas at the start of the 1900s.** Compare their distribution with the late 1900s distribution of population (see Figure 7.14).

porations opened Europe to investment, at first largely of American origins. By the 1980s, Japan also gained a position of world economic leadership, targeting Britain within Europe for the largest part of its overseas investments (Figure 7.5). It did so partly because its company executives had learned English for their American expansion and partly because of British government subsidies.

## Changes in Economic and Geographic Structure

During the mid-1900s, the older, "heavy" (producer goods) industries, such as steel-making, heavy engineering, and chemicals located on coalfields in Europe, were replaced in value of output and employment by motor vehicles, consumer goods, and light engineering products. The new products could be manufactured wherever there was an electricity supply and plentiful semiskilled labor. The wealth created and numbers employed in primary and secondary industries gave way to service industries—based in offices rather than factories. This further increased urbanization. New locations for manufacturing became significant around the expanding consumer markets of major cities and in new regions. The polluted old industrial centers with major investments in factories based on obsolete technologies—often kept going by state subsidies and staffed by members of trade unions that resisted changes—declined and unemployment in them rose.

In the second half of the 1900s, two major features marked the economic geography of Western Europe.

- First, the new industrial areas grew at places more suited to the needs of developing technologies and industries, but long-established production continued in the older areas. The major investments in factories, housing, human skills, and infrastructure made it too costly in financial and human terms to undertake sudden locational shifts of production. The advantages of producing goods in an area where there is a trained labor force, assembly and distribution systems of transportation, and support by financial and other services, build up and reinforce the original locational advantages as **agglomeration economies.** Keeping production in an area although its costs may be higher than those in possible competing areas creates **geographic inertia.** It is only when a substantial change in costs occurs, new products emerge, or the demand for the original product is reduced, that new areas develop and older manufacturing centers decline.
- Second, a process known as **deindustrialization** occurred when the numbers of jobs in manufacturing fell rapidly and old factories became derelict (Figure 7.6) as old industrial areas declined. A fall of 20 percent in European manufacturing jobs between 1970 and 1985 was more than balanced by a rise of 40 percent in tertiary sector jobs. The skills of blue-collar miners and factory production-line workers were seldom convertible into the new white-collar office, hospital, or classroom jobs that were taken by younger and better-educated people. In countries where state-run industries had maintained high levels of employees, privatization—with its emphasis on increased productivity—caused rapid increases in unemployment. Long-term unemployment faced those unable to retrain for the new jobs. Unemployment across Europe rose to 12 percent of the labor force throughout the 1990s in France, Italy, Greece, and Finland, with Ireland rising to 16 percent and Spain to 18 percent. In the United Kingdom, such high levels at first fell to 5 percent by the late 1990s. Older production workers, female workers, and young people with a poor education entering the labor force were particularly at risk. In 1996, youth unemployment (ages 15 to 25 years) reached 42 percent in Spain, 34 percent in Italy, and 27 percent in France but fell to 16 percent in the U.K and 8 percent in Germany. In Eastern Europe and the Balkans, unemployment rose sharply when former state-owned factories were exposed to market competition.

## Planning and Privatization

In Europe after 1950, the problems of declining old manufacturing areas in particular became political issues and led to

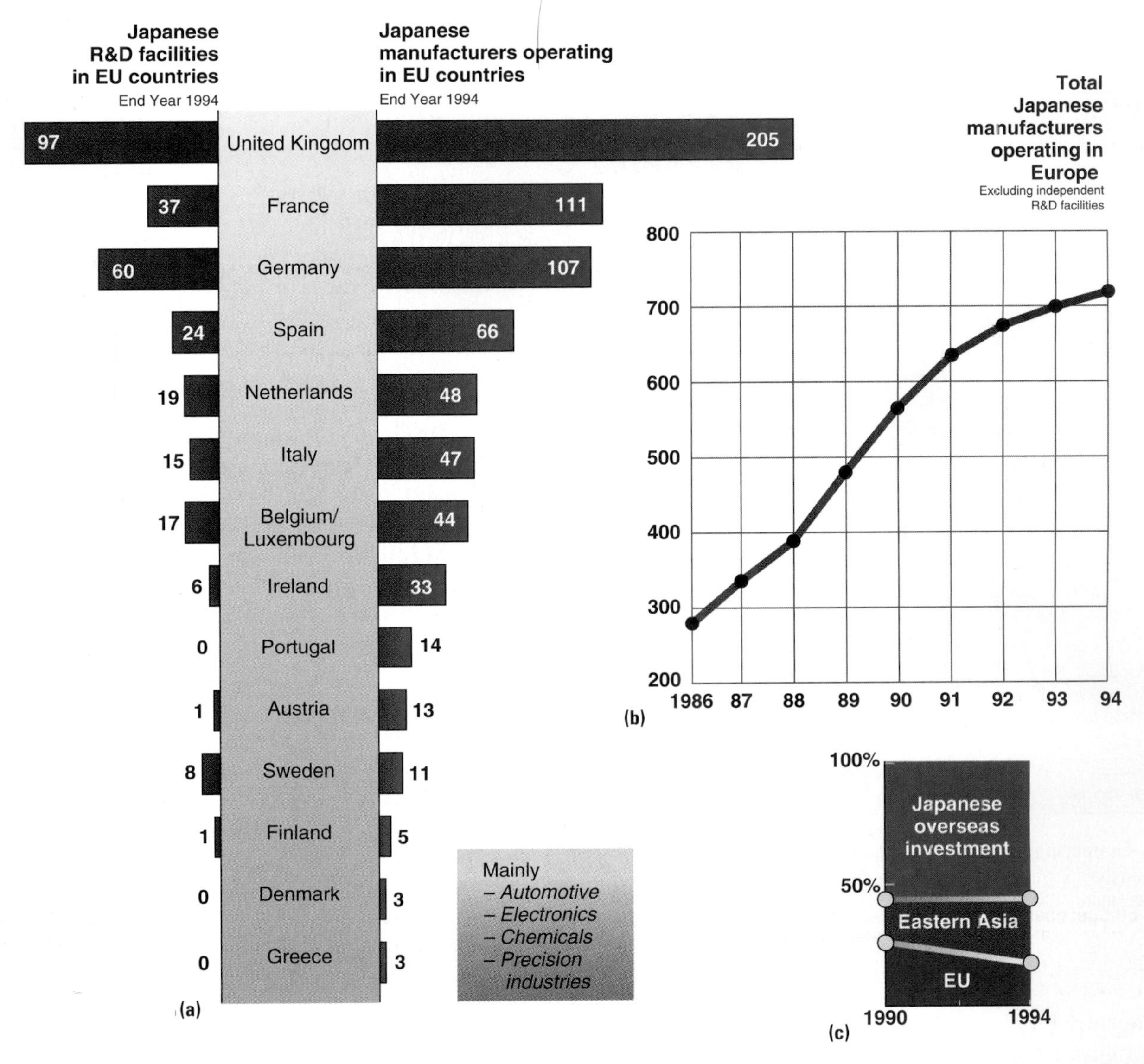


**FIGURE 7.5 Europe: Japanese investments.** (a) The leading role of the U.K., with 30 percent. (b) In the early 1990s, the increase of such investments began to slow. (c) The slowdown of Japanese investments in Europe was related to the Japanese pouring new investments into Eastern Asia.

additional programs of social welfare, regional policies for sitting new industries in the older industrial areas, and retraining programs. Governments attempted to redress the contrasts in unemployment between new and old industrial areas by forcing manufacturers to locate new factories in peripheral, rather than core, areas of their country. State industries were located in such areas despite higher operating costs. Italy, France, and Britain introduced such policies. Investment in southern Italy—the Mezzogiorno—is a major example of government-directed industrial location. By the late 1970s, however, such policies had produced a modest effect at great cost, and by focusing on internal issues, often made the countries uncompetitive in world markets.

During the 1980s, Britain led the way in privatizing state enterprises, many of which made financial losses and were a drain on tax income. The privatized concerns, including steel-making and electricity, gas, and water utilities, commonly made considerable profits after this change, especially by cutting their labor forces.

A major shift in western Europe in the 1980s and early 1990s moved these countries away from a relatively stable economic and social order in which the losers were looked after by government provision of industrial protection and welfare. The other side of this stability had been that the countries of western Europe began to price their products too high for world markets, which drove much of their labor force out of work. Moreover, the increasing welfare costs put country budgets into deficit and raised internal tax demands. To combat this economic problem, many western European countries, led by the United Kingdom, moved

(a)

(b)

**FIGURE 7.6 Europe: changing industrial landscapes.** (a) Some long-established steelworks continue to function, such as the Thyssen works at Duisburg, Germany, where the Ruhr River flows into the Rhine. Such surviving steelworks invested heavily in modern equipment that required fewer workers. This is an example of geographic inertia. (b) Cleared land that was once occupied by major steelworks in Sheffield, England. Two cooling towers remain, due for demolition. Part of the former steelworks is crossed by a new highway. This is an example of deindustrialization.

(a) © Ulrike Welsch (b) © Bill Stephenson/Panos Pictures

toward practices followed in the United States, where private enterprise is encouraged, but potential high rewards are balanced by high levels of risk, making for a less stable human environment.

In the United States in the 1970s, an underclass emerged that was composed of mainly black, unemployed, poorly educated, welfare-dependent people, with disrupted families (see Chapter 9). By the 1990s, a similar underclass, including families of guest workers and former industrial workers, grew in western Europe. Long-term unemployment rather than class was the main segregating factor as a result of deindustrialization, mechanization, and a lack of emigration opportunities. Although the living standards of the poorest groups in western Europe continued to rise—whereas those in the United States fell—the common perception of widening gaps in society became part of the problem as the conspicuous consumption of the majority of affluent groups was clearly evident.

## Expanding Europe

The formation of the **European Union** (see "World Issue: European Union," page 292–293) reflected the desire of European countries to work together in economic—and, for some of them, eventually political—activities. Alongside the EU, NATO strengthened European forces and resolve against possible aggression from the Soviet Union.

In 1997, the EU began expanding alliances into Eastern and Balkan Europe, mainly at the request of the countries that had been satellites of the Soviet Union during the Cold War from 1945 to 1991. Both expansions faced difficulties. Russia resisted moves that threatened to isolate it in its current weak position economically and militarily (see Chapter 8). Internally, NATO and EU countries imposed major hurdles for those wishing to join the established group of countries.

After 1991, NATO had to redefine its purpose, since it had been organized to meet military threat from the Soviet Union

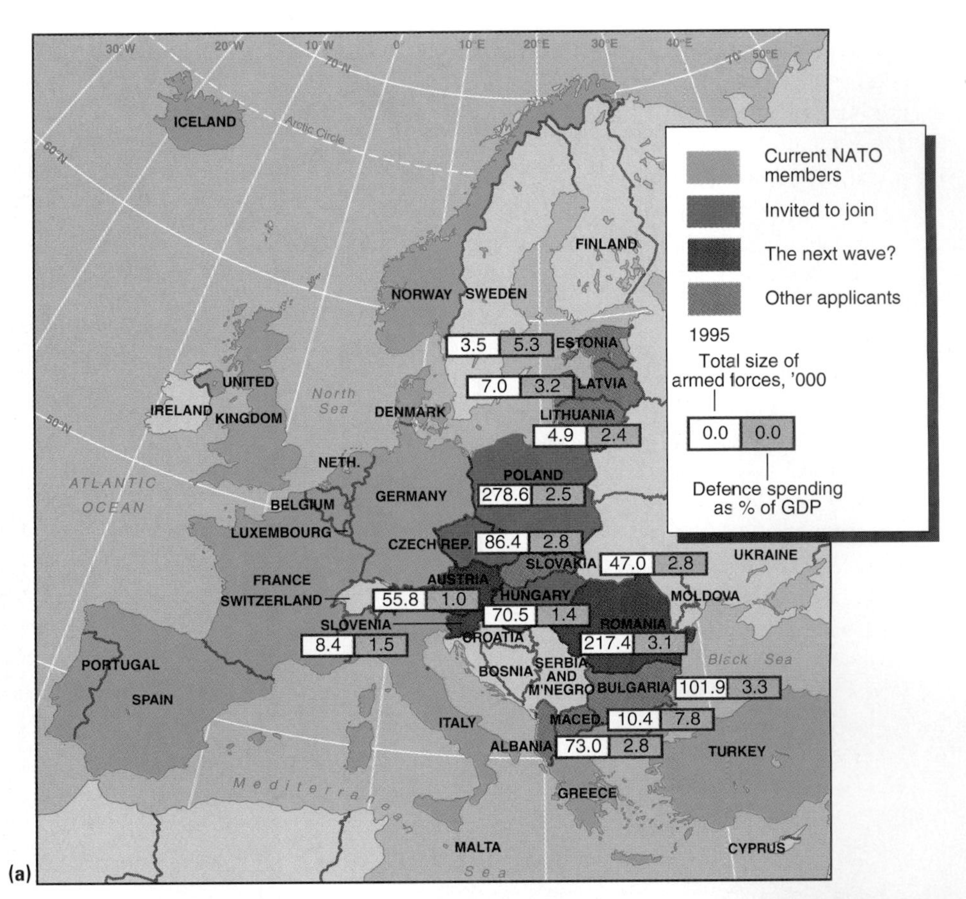


**FIGURE 7.7 Europe: NATO and European Union.** What two roles do these two organizations play? (a) North Atlantic Treaty Organization, showing the countries that are members and those that are applying for membership. Why does Russia disapprove of NATO's expansion? (b) European Union, its growth to date, and possible further expansion.

Source: (b) Reprinted by permission from *Financial Times*, February 2, 1995.

along the eastern borders of West Germany. The breakup of the Soviet Union, consolidation of West and East Germany, and gradual withdrawal of Russian forces left a major buffer zone between western Europe and Russia (Figure 7.7a). The future of NATO is more likely to be in resolving or policing disputes within the expanded Europe and its immediate neighbors—as in Bosnia and Kosovo. In 1997, NATO invited Poland, the Czech Republic, and Hungary to apply for membership in time for the fiftieth anniversary of NATO in 1999. At that stage, if the first group is successful, invitations may be sent to other nonmembers, such as Austria, Slovenia, and Romania. Problems of getting members to agree to the expansion center on the costs. Those include upgrading present forces (to assist new members, funded by current members), modernizing forces in eastern European countries (to be funded by the new countries), and enabling forces to operate together with adequate infrastructure (to be funded by all NATO members).

The problems of expanding NATO, however, are smaller than those in increasing the membership of the EU. While most countries feel more secure militarily, membership of NATO is seen as of less consequence than economic security in the impoverished countries of Eastern and Balkan Europe. The EU offers a large affluent market and regional subsidies, and the prospect of membership provides a stimulus to internal reform, democracy, and market economy procedures. Countries wishing to join the EU have to pass several tests, including political criteria (democratic elections, respect for human rights and minorities) and economic criteria (expanding scope for competition and market regulations). These take time to achieve. The admission of Spain and Portugal in 1986 followed nine years of bargaining over when and how these criteria should apply. The deep current poverty of eastern Europe could involve massive farm and regional aid spending, and applicants with long-sheltered economies need to be ready to compete Europewide.

EU expansion is likely to occur incrementally in groups of five or six countries. The first group, agreed to in December 1997, will be Estonia, Poland, the Czech Republic, Hungary, Slovenia, and Cyprus, possibly joining by 2002 (Figure 7.7b). Five further applicants—Bulgaria, Latvia, Lithuania, Romania, and Slovakia—will be offered financial aid and annual reviews to see if they meet conditions. Turkey's right to join the EU was agreed to at the same time, but German and Greek opposition kept it out of the second group of applicants and led to bad feelings in Turkey and among members of the EU.

## Devolution within European Countries

As Europe moves toward expanded economic and possibly political integration, many local national groups still exert pressure for autonomy. The split of the former Yugoslavia into five separate countries and further tensions within the remnant Serbia-Montenegro concerning the wishes of the Albanian peoples in Kosovo to become independent are examples of this process outside the EU. Within the EU, bloody campaigns have been fought by the Basque peoples straddling the Spain-France border and the Catholic minority in Northern Ireland. In Belgium, the almost evenly balanced Flemish (Dutch-speaking) and Walloon (French-speaking) peoples continue the long-term tension between them. The Welsh and Scots people within the United Kingdom debate whether they would be better off as separate countries.

Such tensions reflect past events and continuing opposition between Catholic and Protestant, Catholic and Orthodox, and Christian and Muslim groups. Originally based on theological differences, the conflicts became politicized and ingrained in the culture of groups. Such conflicts often appear to be insoluble. For example, in Northern Ireland, Protestant groups were introduced in the 1600s as a measure to assist England in gaining control of Ireland. When the Republic of Ireland was created in 1922, Northern Ireland was excluded on the grounds of its Protestant majority and guaranteed continuing union with Great Britain. Within Northern Ireland, the Catholic minority was often discriminated against and campaigned to be part of the Irish Republic. Lack of political action led to guerrilla warfare but, despite political compromises to end the bloodshed in the late 1990s,

**RECAP 7A: Europe and Political and Economic Changes**

Europe was the first core in the world economic system. Although individual countries are now less powerful than they were early in the 1900s, the region as a whole retains a leading role in the world's core of wealthy countries.

The current human geography of Europe grew out of varied influences, including the movements of Mediterranean, Germanic, and Asian peoples and the invasions of the southern parts by Muslims. Europe developed global trading links from the 1400s and dominated the industrial revolution of the 1700s and 1800s.

After World War II, these countries experienced major changes in human geography, including new locations for manufacturing, urban expansion, shifting trading links, and the formation of the European Union. They now face the prospect of adding countries in Eastern and Balkan Europe to their organization.

7A.1 What were Europe's historic contributions to the development of the present world order? How did they affect Europe itself?

7A.2 Which factors promoted European migrations and colonizations from the 1600s to the 1800s?

7A.3 How did the primary, secondary, and tertiary sectors of Europe's economies change from World War II to the 1990s?

7A.4 What are the strengths and weaknesses of the EU?

**Key Terms:**

| | | |
|---|---|---|
| Celts | Moors | geographic inertia |
| Greeks | Vikings | deindustrialization |
| Romans | agglomeration economies | European Union (EU) |
| Germanic tribes | | |

**Key Places:**

| | | |
|---|---|---|
| Western Europe | Alpine Europe | Eastern Europe |
| Northern Europe | Mediterranean Europe | Balkan Europe |

tensions remain between the seemingly irreconcilable views of those who want Northern Ireland to remain part of the United Kingdom and those who want it to become part of the Republic of Ireland.

## NATURAL ENVIRONMENT

The geography of Europe is marked by rapid changes of landscape over short distances and closeness to the ocean, factors that influenced human actions over time. The natural environment, particularly the climatic variety, the small scale of geologic provinces, and the long indented coastline contribute to these abrupt geographic differences from place to place.

### Midlatitude West Coast Climates

Europe has a north-south extent that is slightly larger than its east-west extent. This is reflected in its range of climatic environments (Figure 7.8). Most are **midlatitude west coast climates** ranging from the icy northern Arctic Circle lands of the summer midnight sun to the southern Mediterranean warmth. No part of Europe is more than 500 km (320 mi.) from the coast, allowing mild and humid oceanic atmospheric influences to affect the whole region.

The far north of Norway, Sweden, and Finland have polar climates, with long, very cold winters and months of snow cover. The coldest weather is somewhat ameliorated along the coasts of these northern lands by air warmed over North Atlantic Drift waters. Most of the western coastal countries from Norway to Portugal have mild winters and warm summers with precipitation throughout the year. Much of their warmth and humidity comes from the winds that bring North Atlantic maritime air.

In the far south along the Mediterranean coastlands, there is a seasonal contrast in which the midlatitude belt of cyclones moves southward in winter, bringing rain and wind, while the hot, dry air from the Sahara to the south dominates the summer. This southern variant is often known as a **mediterranean climate**.

In the midlatitude continental interior climate of central and eastern Europe, winters can be severe as cold winds bring freezing temperatures from Russia and the northern countries. Summers are warmer than on the coasts and are marked by thunderstorms.

Climate changes affected Europe during the period of human occupation from the later part of the Pleistocene Ice

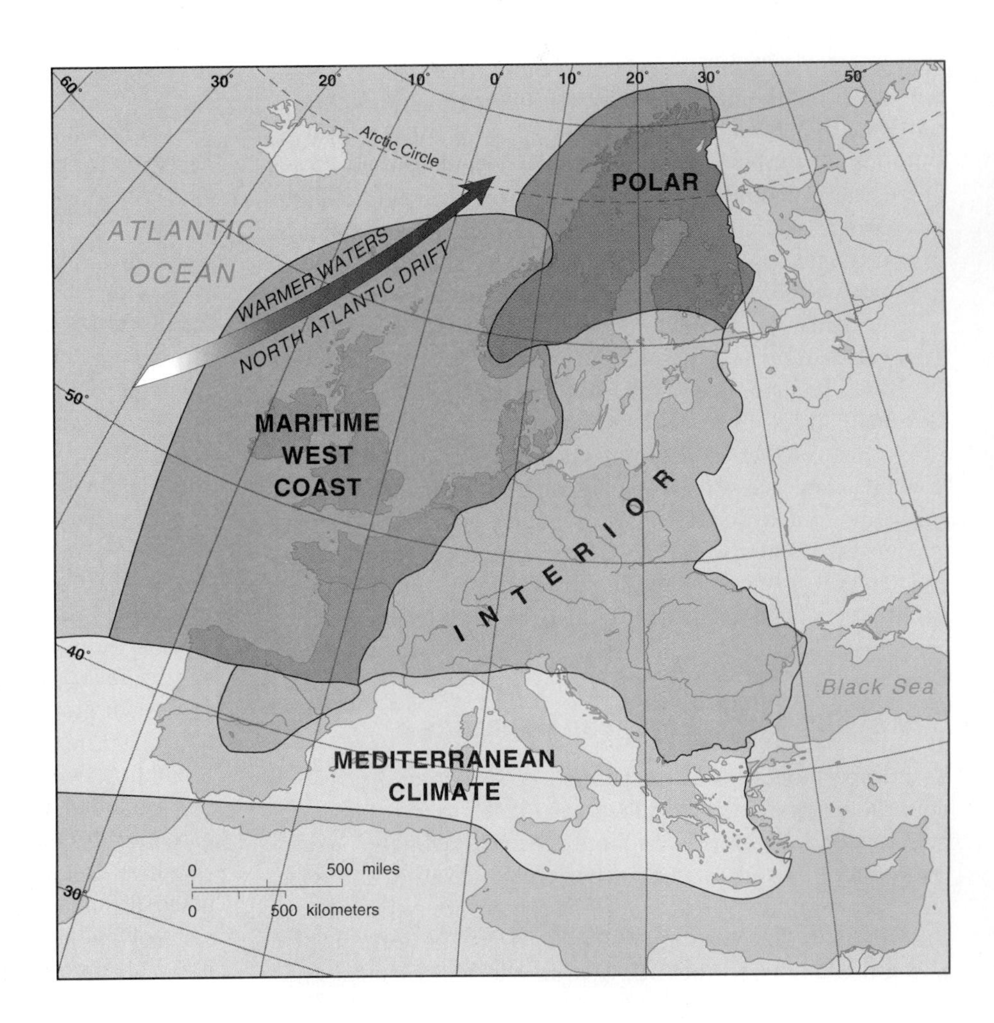


**FIGURE 7.8 Europe: climates.** Europe is marked by the closeness of all parts of the region to oceanic influences. The North Atlantic Drift brings warmer ocean waters across the North Atlantic Ocean from the Gulf Stream, raising the temperature of the air above and increasing its humidity. In the far north, inland areas become polar in climate, and in the far south, the Mediterranean Sea area has alternating dry, hot summers and mild, wet winters. Toward the east, greater continental heating gives higher summer temperatures, but cold winds blowing from the north and east make winters very cold.

Age. As climatic conditions warmed, the ice sheets, which had extended as far south as the Thames River valley in England and the plains of northern Germany and Poland, melted and retreated northward. Climates throughout Europe got warmer until some 5,000 years ago, when they reached their warmest point. After that time, conditions fluctuated as warmer periods in Roman times, from A.D. 900 to 1200, and since around 1850, were punctuated by colder periods. Historians have not fully evaluated the impacts of such climatic changes on economic and political events, but the cooling in the 1300s had much to do with the end of Greenland colonies and with forcing farmers in Europe to lower the upper level of cultivation on hills. Since the mid-1800s, the retreating glaciers in alpine valleys (Figure 7.9) influenced the tourist industry and the generation of hydroelectricity. The retreating glaciers and decreasing extent of summer snow cover reduced the winter sports season but increased the summer holiday warmth. Inlets to hydroelectricity projects are now sited higher up the valleys, increasing the elevation (and energy) of water falling on the electricity-generating turbines.

Although lands around the Baltic Sea continue to rise above sea-level in response to the postglacial melting of the ice sheet burden that caused that area to subside, most of Europe is concerned about the potential impacts of global warming. Rising sea levels would affect the extensive low-lying parts of Europe, including the many areas of former coastal wetland that have been reclaimed and intensively populated. The Netherlands has particular worries in this area, but attention has also focused on the plight of Venice in northern Italy. Venice, a medieval trading city, is one of Europe's greatest architectural treasures and tourist attractions, but it is gradually subsiding. Increasingly frequent high tides weaken its foundations. In 1966, a south wind raised the tide level by two meters (6 ft.), covering St. Mark's Square with oil-polluted water that left its stains on all the buildings. As Venice's buildings became less secure, the city's population fell from 175,000 in 1950 to 80,000 in 1990. The industrial development of the lagoon that backs Venice is a major part of the problem, since its industries and oil refineries pollute the waters and make protection of Venice difficult in the face of demands for deepwater channel access.

## Geologic Variety

Within its relatively small area, Europe includes almost the world's entire range of geologic features. These include the ancient shield areas around the Baltic Sea, the uplands of central Europe, the young folded mountains of the Alps, and the extensive plains in countries around the North and Baltic Seas (Figure 7.10a). Volcanoes erupt and earthquakes occur along the Mediterranean Sea, but much of Europe is almost free of such hazards. The geologic variety in this region, combined with the early development of mining and industrial growth, generated the beginnings of modern geologic science.

The **Mediterranean Sea** is the remnant of a larger ocean that occupied the area between Africa and Europe but closed as the two continents clashed along a convergent tectonic plate margin (Figure 7.10b). Within this zone, the young folded mountains of the **Alps** form the highest ranges, with many peaks rising above the 4,000 m (13,000 ft.) level and the highest point at Mount Blanc (4,807 m, 15,600 ft.) on the French-Italian border. Farther east, the ranges are not so high in Austria, where few peaks exceed 3,000 m (10,000 ft.).

The high Alps are part of a series of ranges that include the **Sierra Nevada** in southern Spain, the **Pyrenees** between France and Spain, the **Apennines** that form the Italian peninsula, the coastal ranges of the **Dinaric Alps** in Croatia, Bosnia, Macedonia, and Albania, and the **Pindus** in the Greek peninsula. The curve of the **Carpathian Mountains** in Slovakia and northern Romania, continued in the **Balkan Mountains** of Bulgaria, forms a further extension. Such ranges create the dominantly mountainous environment of Alpine Europe and the hilly environment of Mediterranean Europe and the southern Balkans. Areas of lowland in these mountainous areas are restricted to river plains and deltas formed by the deposition of rock fragments and particles worn by erosion from the uplands.

To the north of the young folded mountains, most of Europe is lowland, with extensive areas less than 300 m (1,000 ft.) above sea level. The lowlands dominate the north of Germany, Poland, and the Baltic states—part of the **North European Plain** that continues eastward into Russia. They are formed of more recent layers of rock, covered in the north by glacial deposits.

The lowlands are framed by hilly areas of older rocks that once formed mountain ranges but were worn down by erosion and then raised again by faulting. These include much of Spain and Portugal (the **Meseta**), France's **Massif Central** and **Brittany** areas, the **Rhine Highlands** of southern Germany, the **Bohemian Massif** of the Czech Republic, the uplands of western and northern Britain, and those of Norway. As the Atlantic Ocean opened along a divergent plate margin, uplift occurred along continental margins, together with volcanic activity. Rifting extended through the North Sea area, providing a downfaulted block of rocks that became oil reservoirs.

## Long Coastlines and Navigable Rivers

Europe is marked by peninsulas such as **Scandinavia** (Norway and Sweden), Jutland (Denmark), and Brittany (France) in the north, and **Iberia** (Spain and Portugal), Italy, and Greece in the south. The **British Isles** formed another peninsula until waves cut the Strait of Dover a few thousand years ago. Arms of the ocean, such as the **Baltic, North**, Mediterranean, **Adriatic**, and **Aegean Seas**, reach far inland. Smaller islands in the Baltic and Mediterranean Seas add to the length of coastline that encouraged many groups to engage in trade and develop ship technology at a time in history when water transportation was easier than land transportation. Skills and technology of that phase were available to expand mercantile capitalism between A.D. 1450 and 1750.

(a)

(b)

**FIGURE 7.9 Europe: changing climate.** A hundred years that made a difference in Zermatt, Switzerland. (a) The glacier shown on the left side of the 1880 painting disappeared from the modern photo (b), having retreated over a kilometer up the valley, leaving bare rock and icemelt deposits exposed on the valley floor. Locate the church in both views and compare other evidence of the 100-year differences between the two pictures. How might the changing climate have affected tourism?

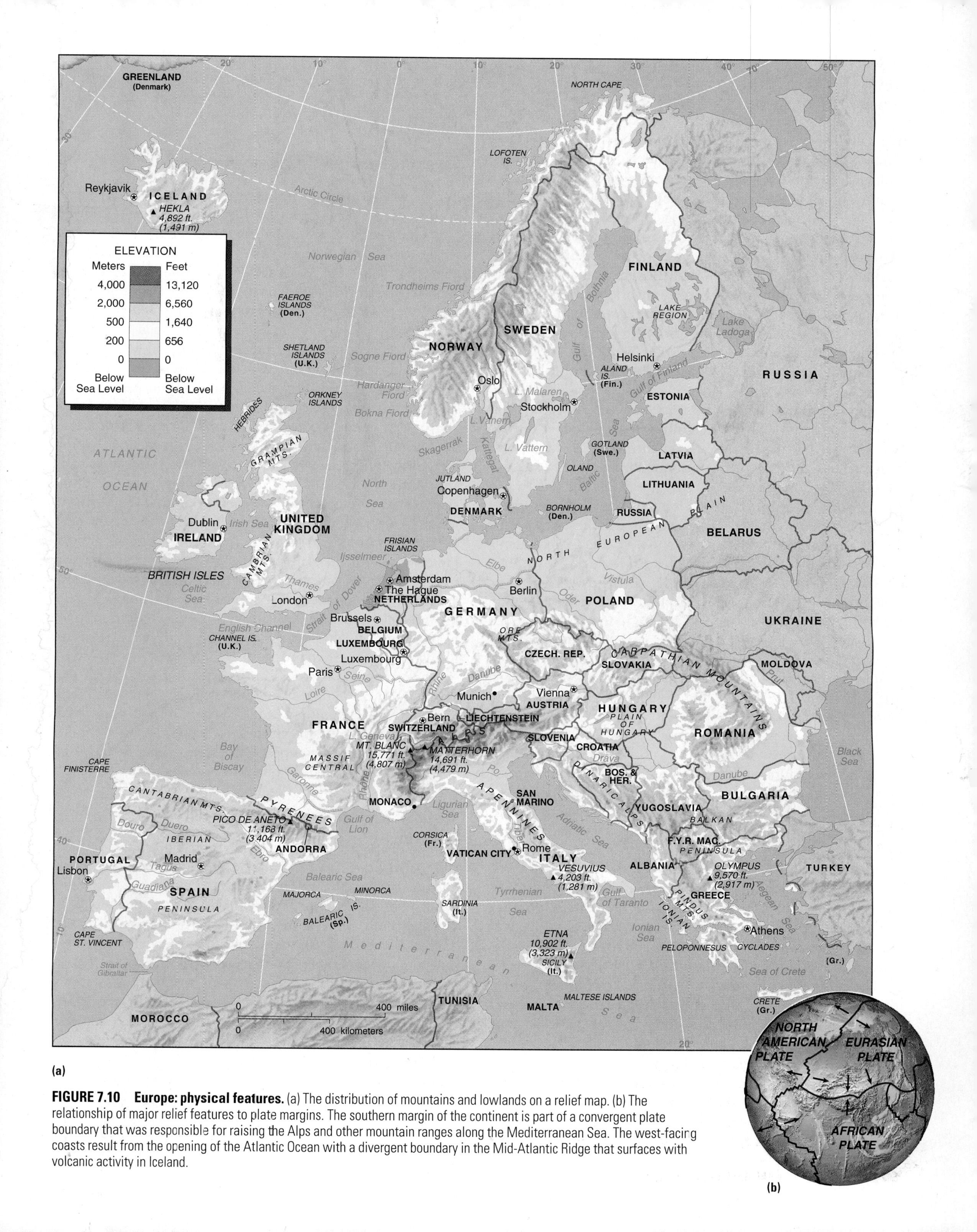


**FIGURE 7.10 Europe: physical features.** (a) The distribution of mountains and lowlands on a relief map. (b) The relationship of major relief features to plate margins. The southern margin of the continent is part of a convergent plate boundary that was responsible for raising the Alps and other mountain ranges along the Mediterranean Sea. The west-facing coasts result from the opening of the Atlantic Ocean with a divergent boundary in the Mid-Atlantic Ridge that surfaces with volcanic activity in Iceland.

On land, connections between places and world trade routes were eased by the existence of major river valleys. The **Rhine** and **Elbe Rivers** in Germany and the **Danube River** flowing from Germany through Austria and the Balkans were particularly important. The **Rhône**, **Seine**, and **Loire Rivers** in France, the **Thames River** in England, the **Vistula River** in Poland, and the **Po River** in Italy were also significant in movements of people and goods from early times. River valleys large and small also provided good soils. Sites for towns at river crossings attracted and concentrated the settlement of growing populations. Modern investments in the Euroports at the mouths of the Rhine and Rhône Rivers are important for maintaing Europe's continuing role in world trade. Europe is marine in economic outlook and culture as well as in climate.

The Danube River is the longest in this region but is less used for transportation than the Rhine because it flows into less economically developed countries. In the last 50 years, efforts to coordinate the management of Danube waters for transportation and hydroelectricity proved difficult. The Danube flows through more countries than any other major world river, and the Cold War limited traffic eastward from Austria. The widening of the Danube-Rhine canal in southern Germany aimed to stimulate greater use of this waterway.

The Rhine is Europe's second longest river and the world's busiest waterway (Figure 7.11), being used annually by some 10,000 ships that carry 250 million tons of cargo. The watershed includes significant parts of four countries (Switzerland, Germany, France, and the Netherlands). The river waters come from melting snow and ice in Switzerland and from tributaries (Aar, Neckar, Main, and Mosel). The Rhine has major roles in transportation, industrial water supply, and, in the Swiss sector, the generation of hydroelectricity. Human management of the river increased in intensity since the 1700s, but the flooding and the high levels of pollution remain potential problems.

Some of the main efforts at managing the Rhine waterway were devoted to making navigation possible for large barges from the international port of Rotterdam at its mouth up to Basel, Switzerland. The river was canalized to that point. As canalization proceeded, rivers were straightened, producing faster flow, channel bed erosion in some sections, and silt deposition in others. Channel deposition partly filled some sectors and caused flooding. Higher levees were constructed to protect from flooding the lands on either side that were used more and more intensively. At the Rhine mouth in the Netherlands, greater efforts were put into keeping out the sea than to maintaining the levees, which were breached in places during the high river levels of early 1995.

The transportation uses of the Rhine River are linked to the growing industrialization of its watershed since the late 1800s. The coalfield and steelmaking areas of the Ruhr (see Figure 7.6a) and Saar are now less productive and polluting, but some 20 percent of the world's chemical industry output occurs along the river with major centers around Basel, Mannheim, and the Ruhr area. Pollution was at its worst in the 1970s, since when there has been an international agreement to reduce the problem, helped by improved water treatment technology. Concerns remain over nondegradeable chemicals and metals that are still at high levels in the river.

## Forests , Fertile Soils, and Marine Resources

Europe was recolonized by forest as the ice sheets retreated northward between 15,000 and 10,000 years ago. Human occupation began in a forested environment in which the circulation of nutrients through an annual leaf fall raised the quality of soils. The most fertile soils developed on river deposits and on the windblown loess covering the common limestone rocks in the northern plains. Fir and pine trees growing on sandy soils and on the thin soils of uplands tended to lower the quality of their soils by acidification.

Because clearance of light woodland was easier, initial settlement favored thinly vegetated uplands on limestone, sandstone, and some granite rocks during the period of warming climate to around 5,000 years ago. Also favored were the easily cultivated loess soils that mantled the southern edge of the North European Plain from Poland to northern France and southern Britain. They provided an easy route of diffusion for farming technology originating around the Mediterranean Sea, via the Danube River valley.

After the introduction of Iron Age tools, human groups rapidly cut into the denser forest on the lowlands. By the time of the Roman occupation of France and lowland Britain, a large proportion of the forest on the lighter lime-rich soils had been cut and some inroads made to the denser forest on heavier soils. Further expansion of the cultivated area on to the clay soils occurred during the Middle Ages.

The marine emphasis of Europe extends to its exploitation of fish resources in the Mediterranean, Baltic, and North Seas and the North Atlantic Ocean. Fishing, related ports, and ships grew in significance in the later medieval period. In the 1900s, the decreasing supply made sea fishing a source of contention and conflict among European countries.

## Natural and Human Resources

Europe contained the natural mineral resources that formed the basis of technological revolutions from the Bronze Age (tin and copper) to modern industrial revolutions (especially coal and iron). Its well-watered lands and moderate temperatures fostered a tree cover and a large proportion of fertile soils that provided the basis for supporting relatively high densities of people at each successive stage of history.

Human ingenuity interacted with the natural environment. By late medieval times, water channels were used widely to generate power in mills and furnaces. Later, the prevalence of upland areas with high levels of precipitation and snow or ice storage later made it possible for engineers to develop hydroelectricity resources there. As populations grew, the development of technologies to support economic

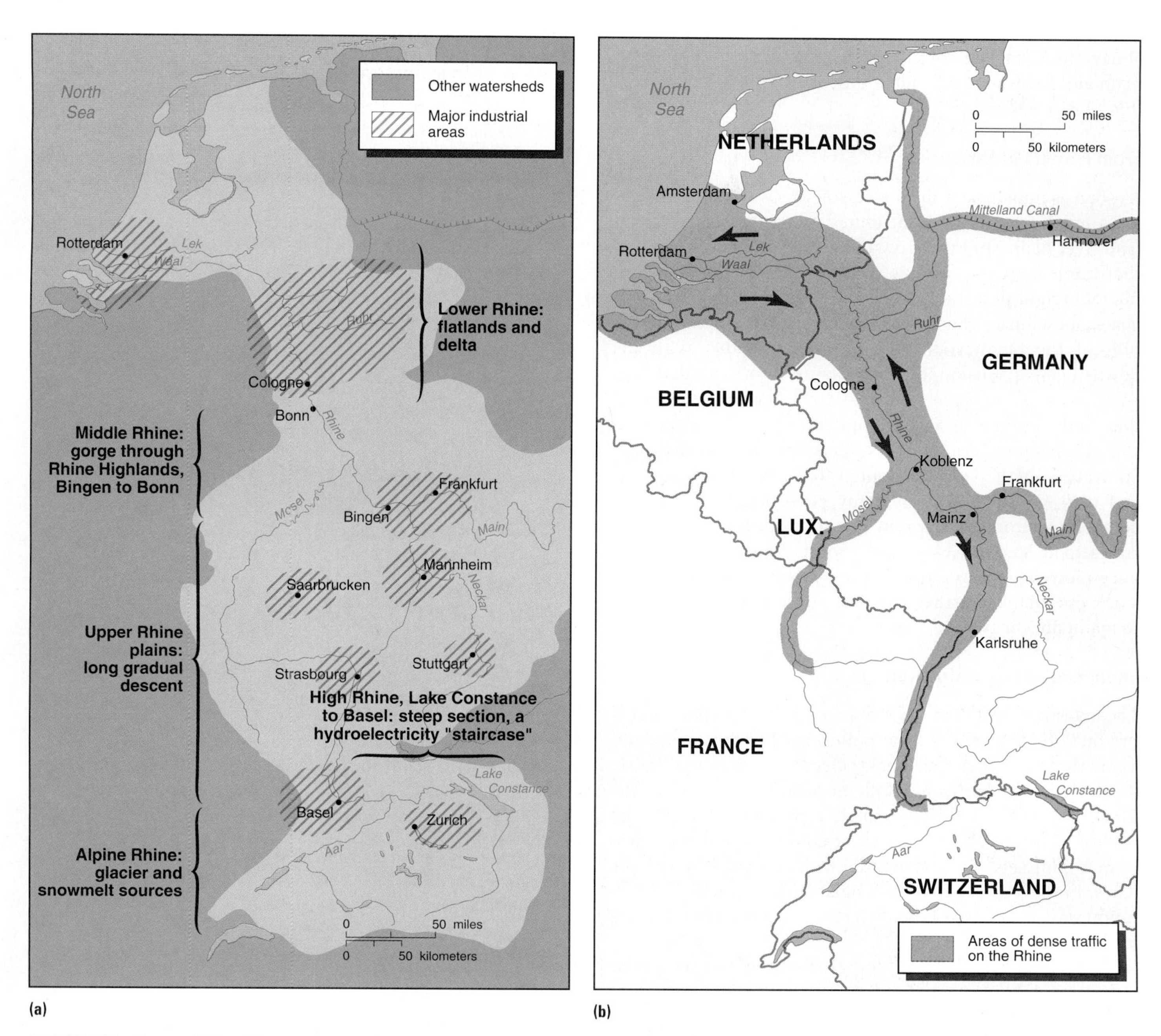


**FIGURE 7.11 Europe: Rhine River waterway.** (a) The major features of the Rhine watershed, including the local relief and locations of industrial areas. (b) Traffic density on the river. Compare the impacts of the industrial areas along the river's length. Rotterdam is Europe's number one port because of its location at the point where maximum Rhine River traffic meets the ocean, requiring change from barge to containers or bulk cargo ship. New canals increased the amount of trade passing through Rotterdam.

Source: (b) Reprinted by permission from *Financial Times*, February 2, 1995.

growth expanded human resources, making it more likely that the new ideas would spawn further changes.

The rivers and extensive coastline encouraged trade and the exchange of ideas across the region. In the 1800s and 1900s, the human landscapes and historic buildings became resources that made Europe the world's major center of tourism. At the same time, some industrial resources that had been important declined in significance. Coal was replaced by oil and natural gas because of high costs of underground mining and the environmental impacts of burning coal.

## Environmental Issues

The occupation of Europe by large numbers of people and their development of technologies for exploiting environmental resources had major impacts on the natural environment.

Today, the advanced societies are particularly concerned to maintain environmental quality, pouring billions of dollars into such measures.

### From Forests to Farms

As soon as humans cut the forest, soil was washed down hillsides more rapidly and contributed to the building of lowland river plains and the growth of deltas at their mouths. In the Middle Ages, for example, a combination of growing populations, rigid political systems, close grazing by sheep and goats, and climate change caused intense soil erosion on the hills of the Mediterranean peninsulas, together with the downstream extension of river plains into coastal deltas. Hilly areas of southern Italy and Greece were denuded of their soils, leaving rocky outcrops. In much of Europe outside of the alpine area, however, the slopes are less steep and cultivation methods were adopted that maintained the soils and their productivity over many centuries. In the 1800s, competition from cheap grain imported from newly opened and settled lands in North and South America resulted in once-plowed lands in Europe being sowed with grass for livestock production, further reducing soil erosion and helping to maintain soil quality.

### Impacts of Industrialization

The industrial revolution and the spread of factory-concentrated production led to widespread pollution of the rivers and air. Many rivers lost their fish stocks. Occasional major pollution incidents still result in fish kills in major rivers such as the Rhine. Great efforts are being made at present to improve the quality of river water in European countries, and the European Union sets rising standards to be attained by specific dates. In the Thames River of England, the reduction of pollution was so successful that fish stocks revived after decades of absence.

The winter smoke fogs (smog) that blighted the major industrial and urban areas of Europe in the 1950s were controlled by legislated reductions in coal burning in some countries. More recent air pollution occurs where high densities of road traffic pour sulfur and nitrogen oxides and carbon particles into the air. Such emissions react with sunlight to lower air quality. Under meteorological conditions of slow-moving air, pollutants may accumulate to dangerous levels in broad valleys. Emissions of sulfur compounds from thermal power stations, especially those burning coal, create **acid deposition**—of dry particles near the source or of wet "acid rain" farther downwind. Acid deposition is common around all the main industrial areas of Europe. It is thought to affect susceptible areas of coniferous forest, thin soils, and shallow lakes away from the sources from the Alps to Scandinavia.

Awareness of environmental problems led to legislation and institutions to monitor and fight pollution. Although road congestion increases with more cars and trucks, new cars in the 1990s emitted 93 percent less carbon dioxide and 85 percent less hydrocarbons and nitrogen gases than in 1970. Lighter vehicles, electronic engine management, more economical engines, and the retiring of older cars will continue to reduce atmospheric pollution. Controlling pollution, however, often increases the costs of industrial production and urban living and causes governments to raise taxes.

**RECAP 7B: Europe's Natural Environments**

Europe has a mainly coastal environment with midlatitude climates that include cold polar varieties in the north, all-year mild and humid conditions in the west, more continental extreme conditions toward the east, and summer droughts around the Mediterranean Sea.

The physical landscapes of Europe include alpine, lowland, hilly plateau, glaciated, and river plain areas. Lowland and better soils predominate between the northern and alpine mountains. The rocks contain energy and metallic mineral resources. Easy access to ocean transportation was an important factor in Europe's economic growth.

The post glacial forest vegetation that covered the region was largely removed from the areas of better soil and replaced by cultivation and urban-industrial areas. Soil erosion, water pollution, and air pollution resulted from intensive use of the natural environment.

7B.1 Do Europe's natural environments give its people advantages over other parts of the world? Attempt to answer this question for the present time and for the period from 1500 to 1900.

**Key Terms:**

midlatitude west coast climate | mediterranean climate | acid deposition

**Key Places:**

| | | |
|---|---|---|
| Mediterranean Sea | Meseta | Aegean Sea |
| Alps | Massif Central | Rhine River |
| Sierra Nevada | Brittany | Elbe River |
| Pyrenees Mountains | Rhine Highlands | Danube River |
| Apennine Mountains | Bohemian Massif | Rhône River |
| Dinaric Alps | Scandinavia | Seine River |
| Pindus Mountains | Iberia | Loire River |
| Carpathian Mountains | British Isles | Thames River |
| Balkan Mountains | Baltic Sea | Vistula River |
| North European Plain | North Sea | Po River |
| | Adriatic Sea | |

### Eastern and Balkan Europe

Environmental degradation was a legacy of communist governments in Eastern and Balkan Europe. Manufacturing industries did not have to adopt procedures to reduce air and water pollution. The industrialized Czech, Slovak, southern Polish, and Romanian areas were badly affected by the burning of low-quality coals that emitted high proportions of sulfur and particles into the atmosphere. Chemical works emissions polluted the rivers. The closure of many antiquated and uneconomic factories in the 1990s reduced the polluting sources.

### Mediterranean Sea

The 1900s growth of population, combined with crowding into coastal locations, industrialization, and the great increase in tourism, led to excessive pollution of the Mediterranean Sea—an important issue for many prospective tourists (Figure 7.12). The Mediterranean Sea is an almost closed sea with hardly any connections to oceans where water mixing dilutes pollution. It is also one of the world's major shipping lanes and has the world's highest level of oil pollution. The most polluted European areas of the sea are those near urban-industrial areas, such as Barcelona, Marseilles, Genoa, Naples, and Athens.

In Spain, for example, the coastal population increased from 12 percent of the national total in 1900 to 35 percent in the 1990s, when the annual influx of tourists reached several million visitors. Although the Mediterranean countries of Europe have populations that are not increasing in total, the movement to the coastal areas and tourism continue to increase. Tourists arrive mainly in July and August, when water is short and local sewerage systems find it hard to cope.

The North African and Asian Mediterranean countries—which also pollute the largely enclosed sea—are experiencing major population increases (see Chapter 4). For example, the coastal cities along the African coast are likely to double their populations by 2025. Tourism around the Mediterranean Sea is expected to grow from around 100 million local and foreign visitors in the early 1990s to between 170 and 340 million.

Following a 1975 UN conference, the governments around the Mediterranean Sea tackled their pollution problem through the Mediterranean Action Plan. Sewerage treatment improved in the wealthier countries such as France, but the poorer countries of North Africa cannot afford the necessary investment. Even if pollution is reduced as countries get richer, the coastline and its delicate ecosystem are changed irrevocably when coastal wetlands are reclaimed and built over.

### Waste Management

In the 1990s, one of the major environmental concerns in Europe was the disposal of the rising quantities of trash and industrial wastes. Since 1950, the amount of wastes per head doubled, although it is still less than half that in the United States. Europe, however, has more people on less land and all

**FIGURE 7.12 Europe: tourism.** Visitors from northern Europe crowd the sunny resorts of Spain, as here at Cadiz. What are the attractions of this place for so many people? The buildings in the background reflect the occupation by the Muslim Moors up to A.D.1492.

its most convenient landfill sites are used up. Further, the nature of trash changed from a majority of ashes and loose dry waste in the 1950s, when coal was burned for domestic heating, to a majority of paper, glass, plastic, and organic waste. Industries produce increasing quantities of hazardous wastes, ranging from poisons to waste oils, heavy metals, and radioactive substances, some of which remain toxic for long periods.

Waste management is becoming a growth industry, requiring major inputs of capital and high technology. As environmental legislation increases, large-scale management corporations will be required. At present, most trash is still buried in landfill sites, the management of which is being improved to exclude gases that might cause combustion or to tap them for use in power generation. Incineration is increasing to reduce the bulk of the large quantities of waste, but it still leaves ashes and emits polluting gases. Recycling is increasingly popular, but the processes are often costly. Recent legislation in Germany obliged companies to take back the packaging used in their products, but the outcome was to overload the German recycling system, creating large "mountains" of paper, glass, and plastic, together with exports of waste to other countries, using up their own capacity.

## WESTERN EUROPE

The countries of Western Europe have been most significant in the development and maintenance of Europe's core position in the world economic system (Figure 7.13). In 1998, they contained 45 percent of the population of Europe's 516

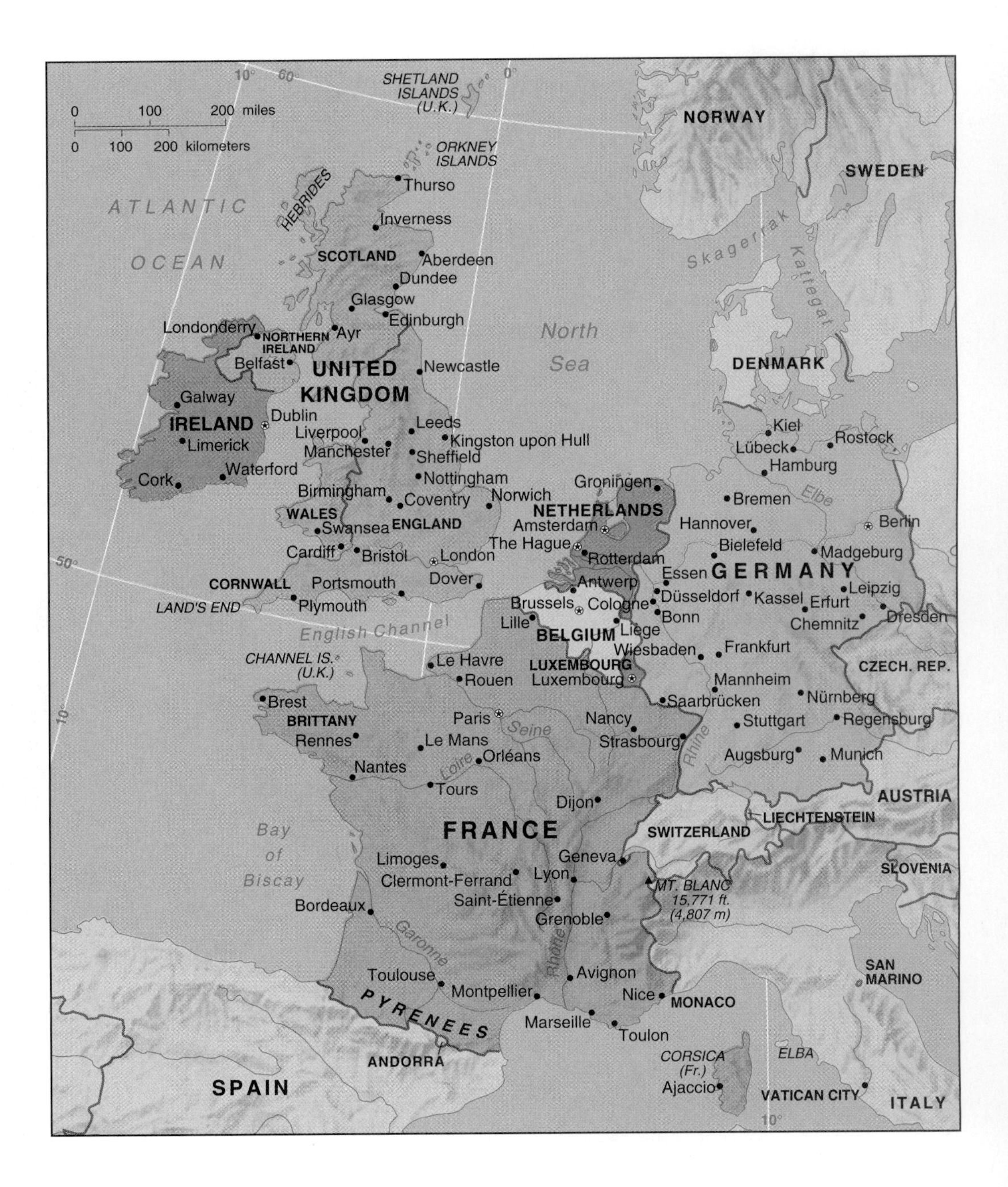


**FIGURE 7.13 Western Europe: countries of the subregion.** The western margins include Scotland, Ireland, Wales, southwest England, and Brittany.

million people, but produced over 60 percent of the region's economic output. Other European countries all have vital links to those of the region's industrial core.

Mercantile and industrial capitalism began here. The United Kingdom was first to establish dominance by a single country until overtaken by Germany in the late 1800s. Despite defeat in two world wars in the 1900s, Germany regained its leading place. In the late 1990s, Germany, France, and the U.K. had the third, fourth, and fifth largest economies, respectively, in the world.

At the end of the 1900s, Western Europe maintained its leading role within the region. Although its natural resource base declined, making imports of raw materials necessary—especially from Africa—it remains a global energy power. As coal declined in importance for cost and environmental reasons, North Sea oil and natural gas and nuclear power replaced it as major energy sources. A net importer of food at the outset of the 1900s, this subregion increased farm productivity to the point where it became a net food exporter. Its high-tech manufacturing and service industries lag somewhat behind those of the United States and Japan, but the subregion continues to lie in the center of the world's largest affluent market and most active trading group of countries. The human environment ranks high in health, education, and income standards.

## Countries

The countries of the industrial core of Western Europe include **Belgium**, **France**, **Germany**, **Luxembourg**, the **Netherlands**, and the **United Kingdom** (U.K.) (England, Wales, Scotland, North Ireland). The **Republic of Ireland** is included in this group because it forms part of the British Isles and is in the western fringe of the subregion that includes Brittany, Cornwall (England), Wales, and Scotland.

The countries of the subregion contain three of the four largest populations in Europe: Germany (see "Living in Germany," pages 316–317) had 82 million people in 1998, while France and the United Kingdom had 59 million each. The Netherlands had 16 million, Belgium 10 million, the Republic of Ireland almost 4 million, and Luxembourg 0.4 million. Most of these countries formed the original European Economic Community in 1957. The United Kingdom and the Republic of Ireland joined in 1973.

## People

Western Europe has high densities of population, except for its marginal highlands (Figure 7.14). The highest are in the urbanized industrial belt from central Britain, through northern France, Belgium, and the Netherlands, and into Germany. These countries are notable for their zero population growth. Their total population is expected to rise slightly from 230 million people in 1998 to 235 million in 2025. The overall change comprises a decrease in Germany and small increases in other countries. Rates of natural increase fell from 0.5 percent in the 1970s to around 0.3 percent in the 1990s. Total fertility rates declined from around 3 births per female in 1965 to around 1.7 in the 1990s. In demographic transition, these countries are into the fourth stage (Figure 7.15), where birth rates and death rates are almost equal: after West and East Germany joined, their death rate was higher than their birth rate through the 1990s.

Following a baby boom throughout the subregion from around 1950 to 1964, the number of children per family declined as later marriages and family planning methods spread. The greater number of divorces, later marriages, and increased numbers of widows resulting from the longer life expectancy for women (nearly 80 years), created more and smaller households so that more housing units were required although the population increased little. Some of the impacts of such demographic changes are illustrated by the age-sex graphs (Figure 7.16).

### Immigrant Workforce

The slowing population growth rate and aging of the populations of these countries at a time of economic growth after 1970 led to a shortage of people in the younger working age groups and to demands for more and cheaper labor. In Britain, the additional labor came from former colonies in the Caribbean, Pakistan, and India. In France, it came from Algeria and other former French colonies. In the Netherlands, it came from Indonesia. In Germany, it came from southern Europe and Turkey (Figure 7.17). Such guest workers often had few rights at first and, only after a lengthy period of residence and political pressure, did they gain full citizenship. Even then, their integration into local communities was slow, and social tensions produced increasing numbers of ethnic incidents in most countries of Western Europe.

In France, there are some 4 million foreign-born workers, with just over 1 million having French citizenship. These people have over 5 million children between them, of which 4.2 million have citizenship. Nearly half of the foreign workers in France come from the Mahgreb region of Algeria, bringing with them a Muslim culture that, when flaunted in political demonstrations, irritates French people.

The United Kingdom is an increasingly multiracial and multicultural country following influxes of Caribbean and Indian subcontinent peoples since the 1950s. These groups now form around 6 percent of the population, mainly in England. The proportion is nearly as high as that in France and Germany but so far has produced fewer social and political problems. Most people in these groups are full citizens of the United Kingdom, which exercises tight controls on further immigration.

### Aging and Declining Populations

Germany's population structure (see Figure 7.16) illustrates an issue that is common in all the countries of Western Europe. Not only will its total population decrease over the next 30 years, even if further immigration is encouraged, but

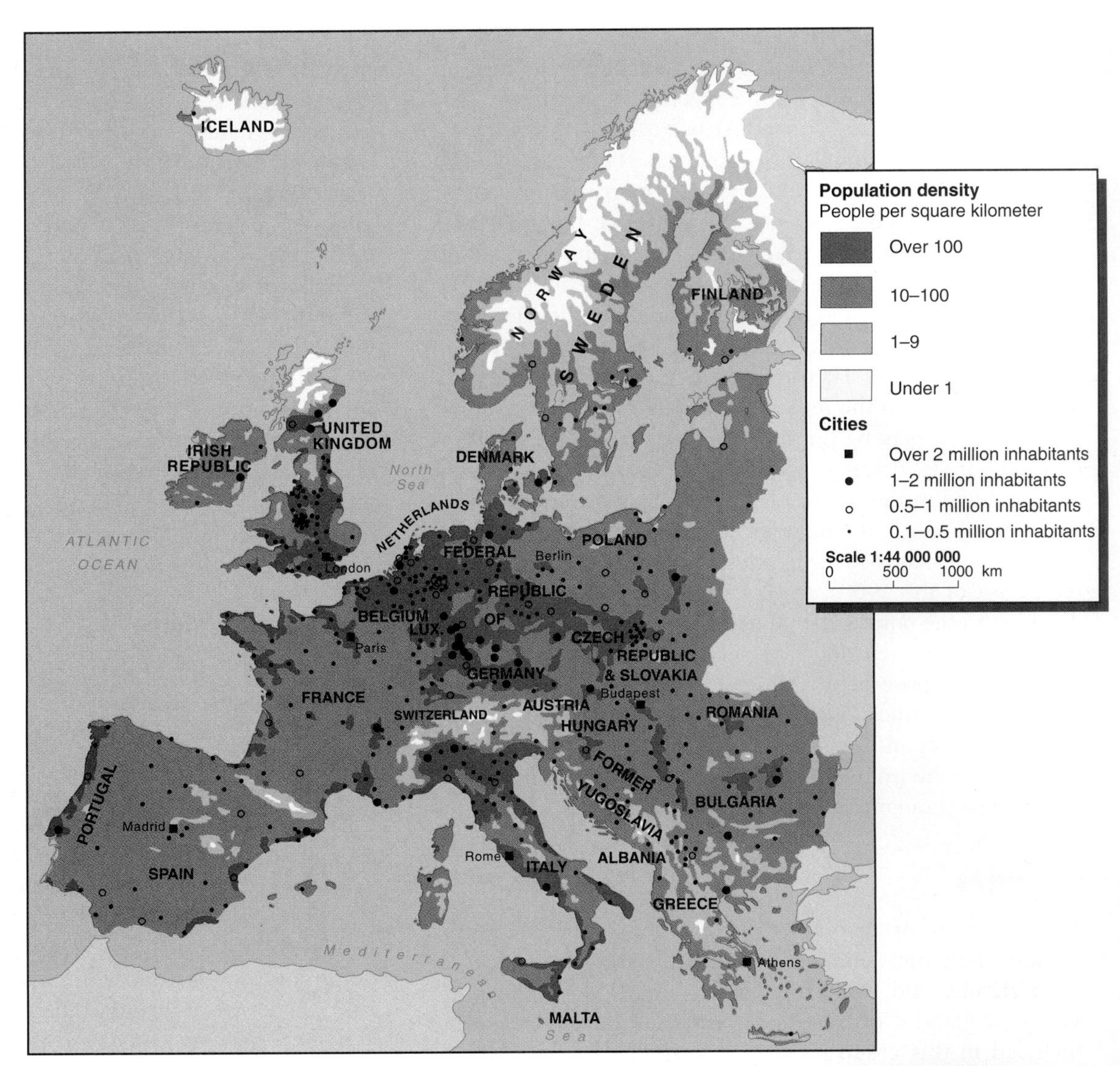


**FIGURE 7.14 Europe: population distribution.** Explain the heaviest and lightest concentrations of people in each subregion.

Souce: Data from *New Oxford School Atlas*, p. 58, Oxford University Press, UK, 1990.

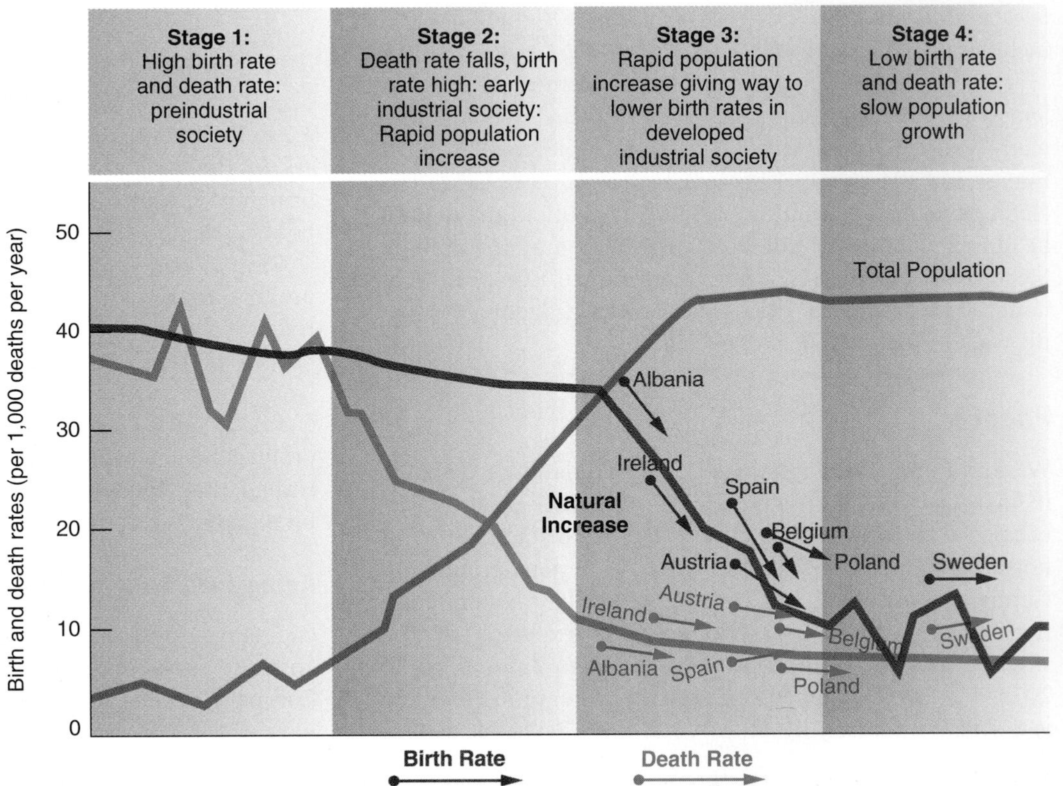


**FIGURE 7.15 Europe: demographic transitions.** Birth rates and death rates are almost equal in most countries. What does that signify?

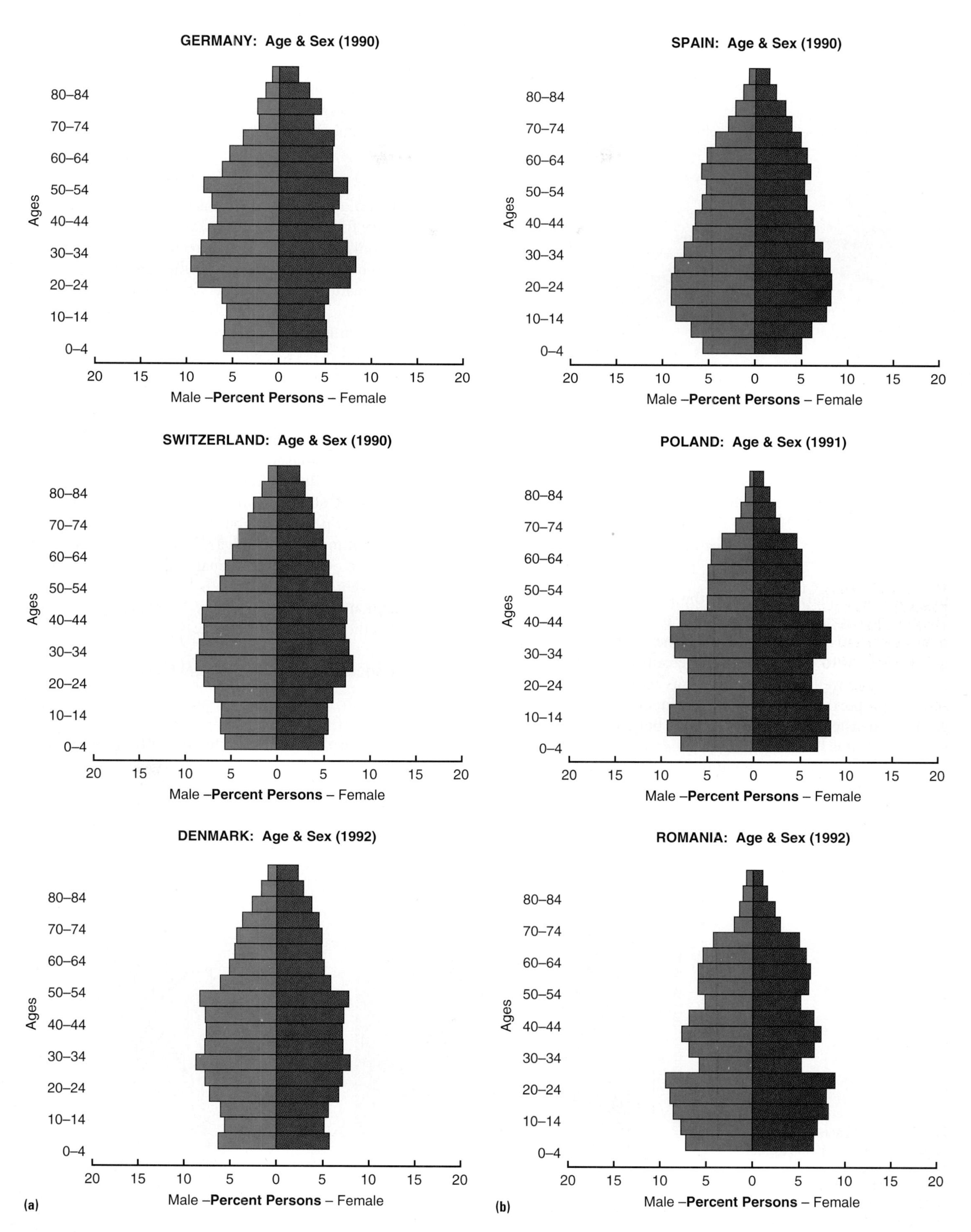


**FIGURE 7.16 Europe: age-sex graphs for countries.** Notice the impacts of the 1950–64 baby boom (ages 24–40 in 1990), the subsequent "baby bust," and the impact of 1900s warfare on the older male populations. Compare the graph for Poland with those for other European countries.

Source: Data from *Demographic Yearbook*, U.N., 1994.

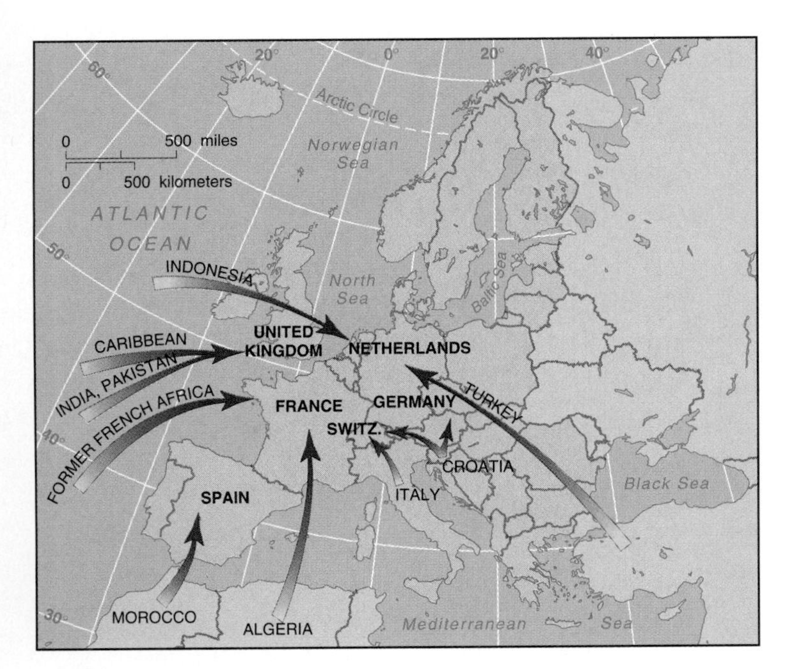


**FIGURE 7.17 Europe: sources of guest workers, 1970 to 1990s.** The sources vary from the former colonies of the United Kingdom, France, and the Netherlands to the countries around the Mediterranean Sea. What contributions have such workers made to the sending and receiving countries?

there will be a marked aging of the population. In the late 1990s, 15 percent of Germans were over age 60; by 2025, that proportion will almost double. The economically active group aged 20 to 60 made up 60 percent of the population but will fall to well below 50 percent as aging progresses. The smaller proportion involved in production and the larger group on pension incomes will impose burdens on the country's budget. It is likely that more women will take jobs.

Germany's current population structure is, however, greatly affected by the situation it inherited in 1990 from former East Germany, where the end of communist rule was followed by several years of population collapse. After 1989, birth rates halved, while death rates increased, especially among teenagers and young adults (Figure 7.18). Birth rates in eastern Germany in 1993 were half the low rates in western Germany. Marriage levels in eastern Germany dropped to the lowest levels of any country. The steep rise in mortality among younger groups appeared to be the result of increases in alcoholism, suicide, and heart ailments.

## Urbanization

The population of Western Europe is highly urbanized. Most countries have 85 to 90 percent of their populations living in towns, with only France (73%) and Ireland (55%) having less in 1998. The high proportions reflect the economic focus on urban-based manufacturing and service industries. There are few extremely large cities, since functions are often spread among several cities. **Paris** (France), with around 10 million people, and **London** (U.K.), with over 7 million, are by far the largest urban centers, combining the government and financial facilities of capital cities with diversified manufacturing. Other major cities in France include **Lyon** and **Marseilles.** Those in the United Kingdom include **Birmingham, Manchester, Leeds,** and **Glasgow.** In Germany, **Berlin** is the reinstated capital. The Ruhr industrial complex around **Essen, Düsseldorf,** and **Cologne** forms the largest concentration of people. The major port of **Hamburg,** the regional manufacturing centers of **Munich** and **Stuttgart,** and the financial center of **Frankfurt** are other major German cities. The largest centers in the low countries are **Brussels** (Belgium) and the ports of **Rotterdam** and **Amsterdam** in the Netherlands. **Dublin** is the capital of the Republic of Ireland.

## Culture of Social Welfare and Global Links

European culture is increasingly difficult to define in the late 1900s because of changes affecting it. The former religious contrast between the Roman Catholic south and Protestant north still features in the statistics and underlies many social institutions and habits, but it has declining relevance. Growing affluence and materialism, reliance on welfare-state provisions, and the coming of immigrant groups from different cultures reduced the impact of traditional religious ideas. The strong emphasis on social welfare and government

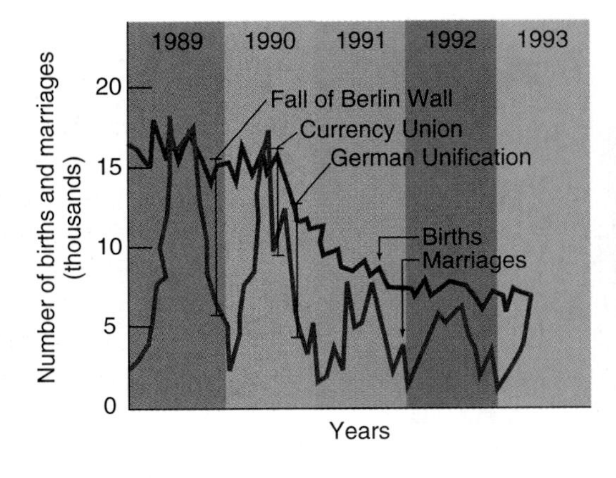


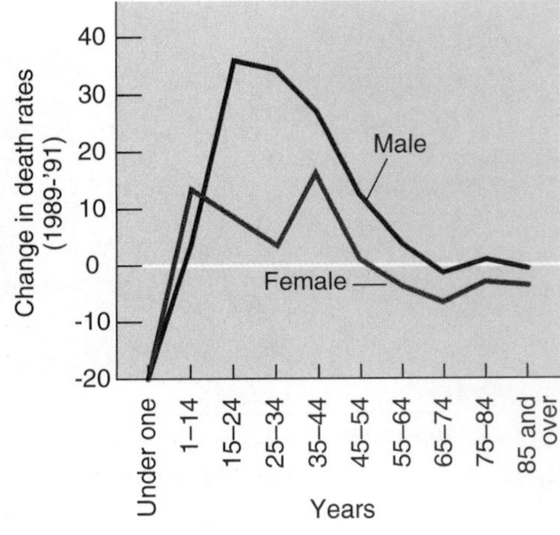


**FIGURE 7.18 Western Europe: population change in the former East Germany.** What happened to birth rates after reunification, and in what age groups did death rates increase?

intervention in economic affairs that characterized these countries from the end of World War II until the early 1980s is now being replaced by market economy processes and the privatization of government-run enterprises.

English, French, and German remain the major languages, comprise three of the six United Nations languages, and have worldwide usage in economic transactions—although they compete increasingly with Chinese and Japanese. Europeans, with their dependence on world trade and internal moves toward unity, are becoming the essential global citizens. They build on past colonial links and cultural understandings, and make their strong diversified and multinational economic base the foundation of competitiveness in trading links with the countries of Africa, Asia, Anglo-America, and Latin America.

## Economic Development

Western Europe remains the economic heart of the region, with 60 percent of all manufacturing jobs in the region and 75 percent of the research and development that devises and applies new technology. From 1950, its economy and the geographic distribution of economic activity changed in response to shifting economic sector structures, the use of new energy sources, increasing involvement in the global economy, and the development of the European Union. The continuing high levels of affluence are reflected in the widespread ownership of consumer goods (Figure 7.19).

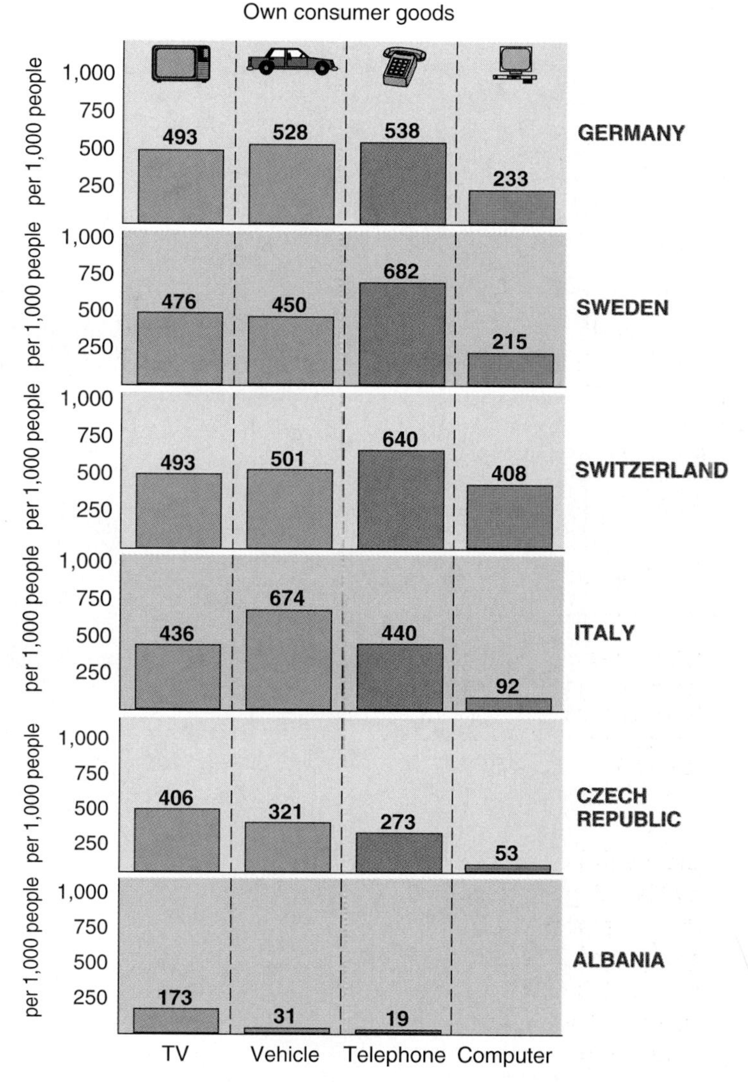


**FIGURE 7.19 Europe: ownership of consumer goods.** Compare these levels with those of the United States (Chapter 9), Japan (Chapter 6), and countries in other regions of the world.

Source: Data (for 1996) from *World Development Indicators*, World Bank, 1998.

### Agriculture

Agriculture still occupies large areas in the countries of Western Europe (Figure 7.20), but provides only 2 or 3 percent of total employment. The subregion's lowland areas contain Europe's main concentration of commercial arable farms, producing grain and other high-value crops from vegetables to sugar beets and oil seeds. It also has the largest expanse of intensive mixed farming, combining livestock and crop production.

After the end of World War II, the changes leading to increased agricultural productivity were as great as those in the so-called "agricultural revolution" before A.D. 1800. New technology, changing world markets for agricultural products, and government policy were important in causing these changes. The results were greater **intensification** (increased output per hectare), **concentration** (in fewer and larger farms and in a smaller group of farming regions), and **specialization** (fewer products).

The technological developments that brought major increases of output included the use of more sophisticated machinery to carry out farm jobs from sowing to harvesting, improved transportation and storage facilities, higher-yielding crop varieties, and the greater availability of fertilizers and pesticides. Farms became links in the chain of **agribusiness** that includes the inputs of mechanization, seeds, fertilizers and pesticides, farm management skills, the outputs of crops and animal products that are fed into food processing factories, and the marketing facilities for the combined range of products. More capital was required to become part of this chain, so farms mechanized and grew in size, while farmers and farm laborers became fewer. Numbers of farms in European Union countries declined by 13 percent from 1980 to 1990 and by 25 percent in Belgium and Ireland, although the area of land in production remained stable. The EU Common Agricultural Policy (CAP) at first provided a set of price supports that made farming almost risk-free at a time of advancing technology. Output increased rapidly in the 1960s and 1970s, although costs to consumers and taxpayers rose.

The most productive farming areas of southeastern England, northeastern France, northern Belgium, and the Netherlands were most favored by EU agricultural policy (Figure 7.21). Farmers in these regions had larger farms, access to more capital, and adapted easily to new methods. "Mountains" of grain and butter, and "lakes" of milk and wine,

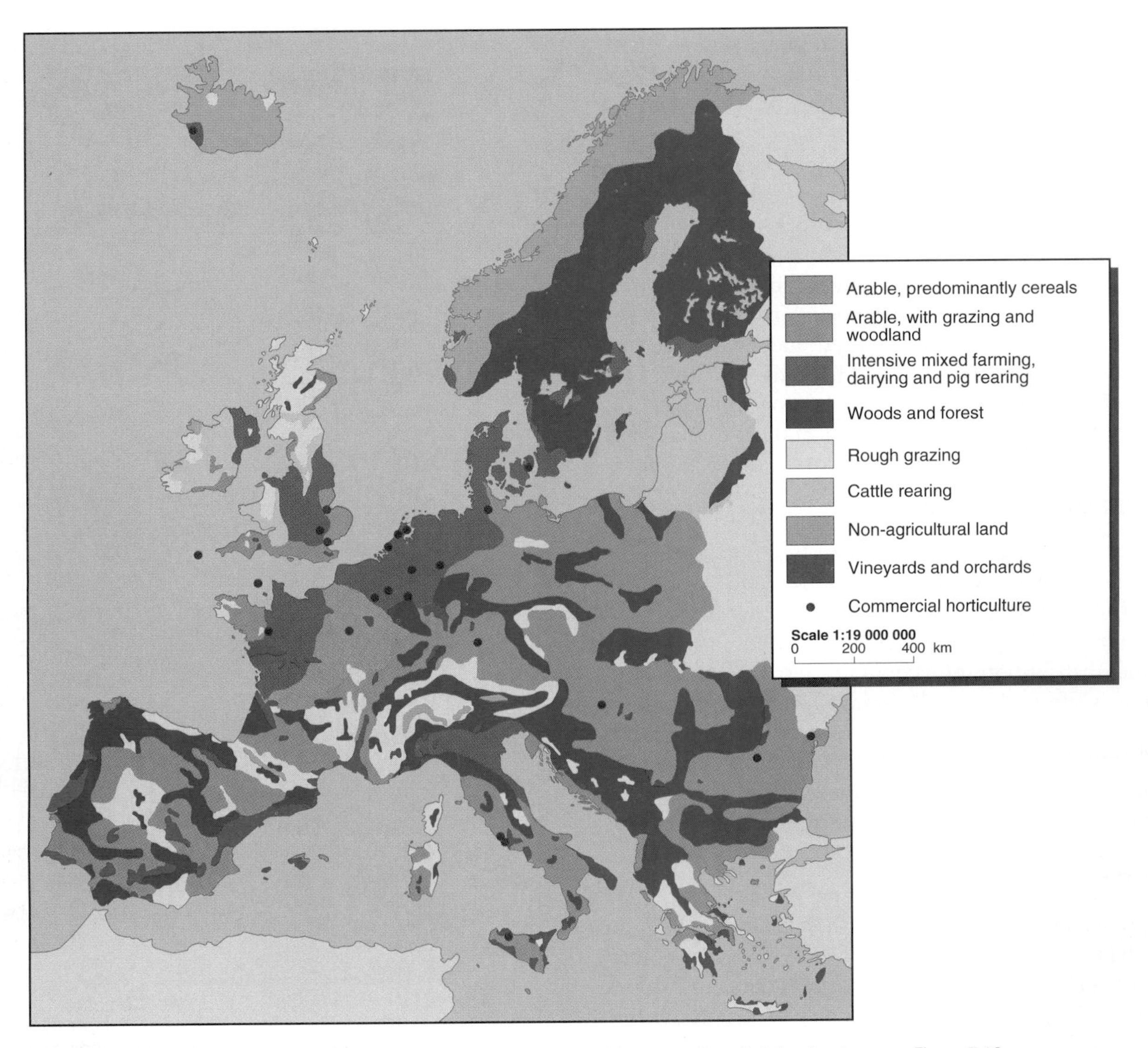


**FIGURE 7.20 Europe: farming land uses.** Relate the categories on this map to the relief details shown on Figure 7.10.

Souce: Data from *New Oxford School Atlas*, p. 55, Oxford University Press, UK, 1990.

resulted. For example, the EC cereal crop rose by 40 percent between 1975 and 1985, generating a surplus and huge storage costs. The main fears of farmers in this subregion in the 1990s concerned the possible admission to the EU of countries in eastern Europe with their lower costs, but this change will take years and at present those farmers are much less efficient than their counterparts in Western Europe.

In the hilly regions where more marginal livestock farming replaced crops, farmers received support for social and environmental reasons rather than for economic ones. Governments feared rural depopulation and soil erosion if the farms in these areas were abandoned.

During the 1980s, the EC agriculture policy changed to reduce the "mountains" and "lakes" of produce in storage as a result of price guarantees to farmers. It was decided that agriculture should be linked more closely to market prices, but that an immediate lowering of guaranteed prices would be devastating to farmers and rural areas alike. More gradual reductions were engineered through downward revisions of price thresholds and milk quotas, and by forcing overproducers to pay part of the storage costs. There was also a greater recognition of environmental issues.

In 1988, further measures encouraged major cutbacks in production. Older farmers were assisted to retire early from farming by being given pensions in exchange for ceasing production of surplus capacity crops. Arable land was retired from production in exchange for grants to farmers who would continue to look after the land but not produce certain crops on it. Grassed fallow, woodland, and nonfarming uses were possible. Livestock farmers were encouraged to produce less from the same area, a process known as **extensification** in which productivity from a particular area is reduced.

By the mid-1990s, the results of the attempts to reduce farm output in the countries of Western Europe were not clear. Setting aside land does not always reduce productivity in proportion to the area set aside. The poorer land is often retired first. Retired land may continue to be used intensively for nonsurplus crops. The grants are often applied to situa-

**FIGURE 7.21 Western Europe: intensive commercial farming.** A grainfield in Berkshire, southern England. Landscapes have changed, increasing field sizes for easier mechanization to raise yields and crop quality.



tions where farmers would have retired or placed land under woodland in any case. The geographic impacts are variable, since France in particular resists any move that reduces its agricultural sector. In other countries, there is debate as to whether the poorer hilly lands should be retired to a greater extent than the productive lowlands. Reducing milk quotas and increasing storage charges appear to be the most effective policies for controlling production levels.

As European farmers faced the loss of long-term subsidies and increased competition from new EU members in the late 1990s, they were hit by other events that reduced their incomes and led to reductions in farm acreage. Urban expansion continued, and planning restrictions on the conversion of farmland were relaxed. The feeding of inappropriate animal matter to cattle in the late 1980s led to the expansion of bovine spongiform encephalopathy (BSE)—a disease similar to Downes cow syndrome in the United States—in Britain and scares over its transmission to humans through ingestion of the infected beef. The EU prevented the U.K. from exporting its beef, but farmers in other countries were affected by a loss of confidence among meat buyers.

## Manufacturing

After World War II, the pattern and geographic distribution of manufacturing industries in Western Europe shifted in response to technological change and world market conditions. Domination by industries such as steel and other metal manufacturing, heavy engineering, chemicals, and textiles gave way to vehicle and consumer goods manufacture and then to high-tech industries such as electronics, aerospace, and pharmaceuticals. New growth areas around major cities and ports, and along new highway corridors, replaced the concentrations of older industries on coalfield areas (Figure 7.22). The area around and west of London, central Scotland, south Wales,

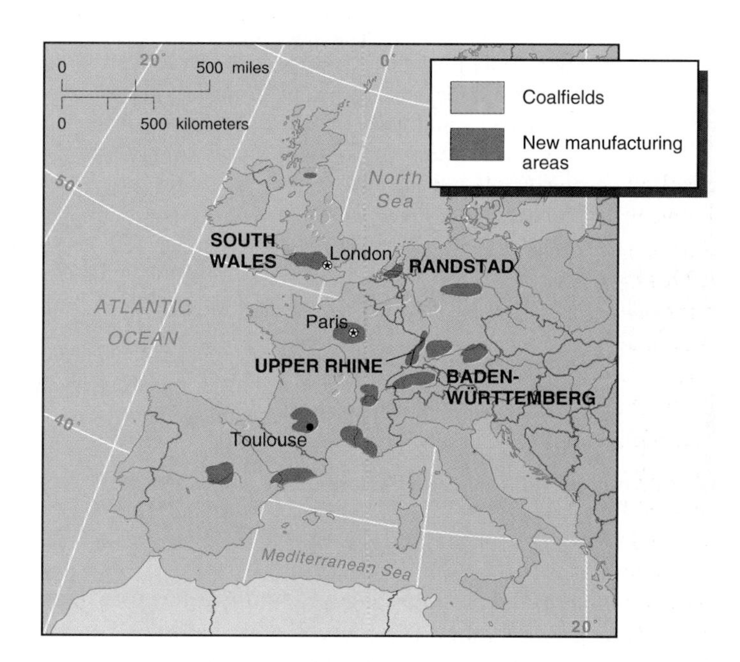


**FIGURE 7.22 Western Europe: the distribution of manufacturing industries in the late 1900s.** Compare this map with Figure 7.4. Although coalfield sites retained some manufacturing, new sites around major cities and along major routeways became more significant.

LIVING IN . . .

## GERMANY

The broad Rhine River valley floor just west of Heidelberg, Germany lies at the crossroads of the wealthiest part of Europe. It has old universities alongside high-tech industries such as Mercedes and IBM. People live in the ancient towns or in new communities built around old villages where the factories and houses are landscaped with a backdrop of wooded hills. The area is central to major European highways that lead to northern Germany and the low countries, south to Switzerland and the Alps, west to Strasbourg and Paris, and east to Frankfurt and Munich. The farming is very commercial and intensive; vine crops and asparagus are specialties of the area. Such crops reflect the very warm summers.

Families where the parents both work in well-paid jobs have affluent lifestyles. As family life progresses, rented apartments give way to owned apartments or houses that are often built to personal specifications. Some houses have swimming pools and children's rooms may have their own bathrooms. Germans are fascinated by technology and acquire the latest home gadgets and entertainment systems as well as the latest Porsche, Mercedes, or BMW car styles if they can afford them.

Nearly all children attend public schools. Private schools exist, but it is unusual to send children there unless they need special instruction. Much education still has a humanistic base, including Latin. Children regularly go abroad, for example spending several weeks of the summer in the United States and a couple of weeks over Easter in the south of France. Student exchanges with other countries are very popular. Germans often speak several languages. Higher education is based on national test ratings, and there is a definite order of preference for local students—beginning with Heidelberg, Freiburg, and Tübingen, and putting Berlin last. Most students in the Heidelberg area avoid Berlin schools if they can for reasons that include their perceptions of lower educational standards and quality of life in Berlin, and feelings of superiority from living in a more affluent part of the country.

Healthcare presents alternatives. There is a government insurance program that is best for those who have large families, lower incomes, or are unemployed. Families may opt out of this scheme and take out private insurance, a preference where both partners are earning and there are two or fewer children. Each family goes to the same specialists for treatment, whether private or not. The government-funded program is giving cause for concern over its expense, especially since it has been extended to the former East Germany.

**(a)**

Map of Germany. The area described is around Heidelberg in the southwest.

Germans are very mobile, and those living near Heidelberg think nothing of driving to Strasbourg, France, for its hypermarkets, to a hairdresser in Frankfurt, or to Zürich, Switzerland, to shop for clothes. They take holidays in all parts of the world and may spend winter weekends skiing in Switzerland or Austria and summer weekends in Paris. All types of leisure activities and a wide range of restaurants are available locally. Outdoor exercise is important to Germans, and there are many tennis courts, while in summer the lakeside areas are packed.

The political issues that were of concern to the Germans of this area in the late 1990s included the problems of integrating the former East Germany, uncertainties over being able to keep up the present standard of living, and environmental concerns. Germans who do not have relatives in the former East Germany have a resentment of "the east" and tend to vote for one of

the Paris region, the Toulouse region (southwest France), the lower Rhône River valley, the Munich area, Baden-Württemburg (southwestern Germany), and Randstad (the Netherlands) were among the new growth areas.

After decades of declining productivity, major state-owned corporations in older industries were privatized. New production technologies were adopted, and greater output came from fewer workers. The major sectors of manufacturing in the 1990s were electrical engineering, mechanical engineering, metal manufacture, food and drink products (together making up 42 percent of the total), chemicals, motor vehicles, clothing, footwear, paper, and printing (together making up a

(b)

Heidelberg, a bridging point on the Neckar River just before it joins the Rhine River. Much of the castle was built in the 1500s and destroyed at the end of the 1600s. The region has become increasingly prosperous since the reconstruction of Germany after World War II.



several centrist or somewhat conservative political parties in reaction to their former fears of communist-inspired aggression or the resumption of Nazi influences. The right wing former Nazi parties have no power in modern Germany despite attracting publicity and winning some local elections in the east. The memories of World War II and the devastating outcome of the Hitler years (1933 to 1945) still form a major barrier.

Many Germans vote for the Green Party, especially in local elections. This is often seen as a way of keeping the main parties on their toes. But people also expect savings in materials and encourage recycling. German supermarkets charge for plastic bags, and there are deposits on all glass and plastic bottles; it is even possible to take washed bottles and get them refilled. Many shoppers unwrap goods they have paid for and leave the wrappings behind in the shops, which are required to recycle them.

Germany has become a more cosmopolitan society. The presence of multinational corporations leads to exchanges of personnel with other countries. The European Union brought increased economic exchange and travel among the countries of Europe. This greater awareness of other European cultures is significant in the wider range of foods that German families eat. Germans particularly favor American ways and contacts. Foreign, or "guest," workers are common in Germany and most—from the Italians, Spaniards, and Portuguese, to the Thais and Vietnamese—have mostly integrated well in the community, although they cannot become citizens. The Turks find it most difficult to be part of the German community, particularly in the main towns, where they have been targets of violence from extremist political groups.

further 27 percent of the total). Food and drink processing dominates manufacturing in the Netherlands, Belgium, France, and Ireland; electrical engineering is most important in Germany, and mechanical engineering in the UK. In the 1980s, all sectors reduced their employees, but electrical and mechanical engineering, food and drink, and paper and printing industries lost under 10 percent, while textiles, footwear, clothing, and nonmetal mineral products all lost over 20 percent and metal manufacturing around 30 percent. The unemployment created by such changes caused particular hardships in the areas that had been dominated by long-established textile and metal manufacturing.

## Energy Sources

Home-produced coal remained the main source of energy until the 1950s: as late as 1952, it provided 90 percent in the U.K. and 92 percent in Germany. After 1955, however, home-produced coal declined rapidly as environmental legislation made burning it more costly, and as cheaper imported oil, mainly from Southwest Asia, took over in residential, industrial, petrochemicals, and transportation uses.

In 1955, Western Europe still used coal for 70 percent of its energy needs, supplemented by 115 million tons (coal equivalent) of oil. By 1972, oil usage rose to over 700 million tons (coal equivalent), amounting to 65 percent of the subregion's energy needs, while coal was reduced to 22 percent. Coal's uses became more restricted so that three-fourths of a much reduced production in the 1990s (135 million tons, plus 200 tons imported) was used to generate electricity. In the late 1990s, that came under threat from natural gas as new technologies using North Sea gas produced electricity more cheaply and with less environmental damage. At the same time, the use of energy in homes and industries increased. Electricity was made available to the rural areas after 1950 and, combined with cheaper road transportation, made it possible for the new "light" industries to be sited away from coalfields.

Projections in the 1970s suggested oil demand in Western Europe would rise to over 1,000 million tons by 1985 and 2,000 million tons (coal equivalent) by 2000. If realized, such expansion would require the building of extra port terminals, refineries, and distribution facilities. It would exhaust the world's oil reserves more rapidly and shorten the length of time over which each oilfield would be productive. Oil demand, however, stabilized at early 1970s levels, while new reserves were discovered around the world and inside Europe. The oil-price shocks of 1973 and 1979–80 raised the oil price ten times above its 1970 level. The growth in oil use slowed, energy conservation became important, and alternative fuels were evaluated. Further, countries instituted policies that reduced their dependence on Southwest Asian oil by developing the North Sea fields and later importing from Russia and Central Asia.

The rising oil prices of the 1970s gave a boost to exploration and production in the North Sea basin, extending northward from the discoveries of natural gas in the Netherlands in 1963. The U.K. and Norway were the main beneficiaries. Early estimates of reserves and production were raised again and again: the Netherlands gas fields have already extracted twice the original reserve estimate and nearly as much again remains available in the rocks. In the North Sea oil and gas fields, production continued to rise in the 1990s and exploration shifted west of the northern British Isles.

In the mid-1980s, oil prices fell in response to conservation by users and new discoveries. Natural gas became the cheapest fuel for electricity generators. By discovering its own oil and gas fields, reducing the growth in its demand, and adopting new technologies, Western Europe obtained good prices and a greater security of delivery from competing producers of oil and gas, particularly those in Northern Africa and Southwestern Asia. The vast natural gas resources of Russia were brought by pipeline into Germany, France, and Italy. In the future, this source of supply is likely to expand since Russia needs to export vast quantities to pay for the consumer, capital, and producer goods that it imports (see Chapter 8).

Of other sources of energy, nuclear-powered electricity generation is the most important, especially in France and Belgium (70 and 60 percent of the national totals respectively). It would have been more important elsewhere if early promises of low cost and safety had materialized and the Chernobyl nuclear power plant explosion in Ukraine had not occurred. Hydroelectricity is important in Northern and Alpine Europe. There is some development of wind power in western Britain and of tidal power in northern France, but these and other alternative energy sources provide a tiny fraction of current needs and are often subject to local opposition—against, for example, the visual pollution of wind farms in the landscape.

## World Economy, Trade, External Investment, and Competition

A greater involvement with the world economy and attempts to develop a common internal market also influenced Western Europe's human geography after 1950. At this time, the United States produced over 50 percent of world manufactures. With the Marshall Plan and other U.S. investments, together with internally generated growth, Western Europe recovered and experienced major economic advances from the 1950s. By the 1970s, Japan formed a third center of economic growth (Figure 7.23). In this triad of core nations, Western Europe formed a large and affluent market with 230 million people in 1998. The rest of Europe adds about 300 million more people. This affluent market contains over 500 million people and has more people than the United States and Japan together (400 million). Europe is the world's largest trading region, dominated by Germany, France, and the United Kingdom. In the 1990s, however, trade increased among the countries of Eastern Asia (See Chapter 6) and between them and Anglo-America. This growth threatens the leading role in world trade taken by the countries of Western Europe, which are themselves trying to develop more trade with countries in Eastern Asia.

Western Europe did not keep up with the other two world core areas in applying advanced technology to the production of manufactured goods. This lag resulted partly from the dual dominance of state-owned firms and firms owned by multinational corporations in this subregion. The emphasis on state-owned industries in some European countries made them subject to internal political pressures, such as the subsidized production of older industries, and reluctance to reduce employment. They continued as moderate-sized, single-focus enterprises catering to the needs of national markets. In steelmaking, for example, the privatization of British Steel in the 1980s made it one of only three private

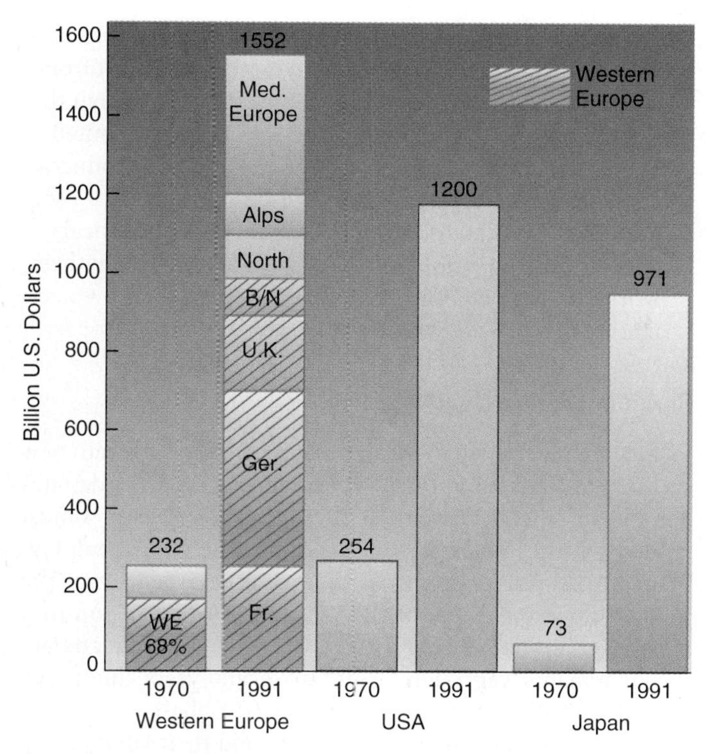


**FIGURE 7.23 Western Europe: changing world balance of manufacturing.** The value added in manufacturing identifies three major world centers in 1991 instead of the two in 1970 and demonstrates an expansion to other areas within Europe. The region of western Europe as a whole (including Northern, Alpine, and Mediterranean Europe) led the world in 1991, while Western Europe was equivalent to Japan. When related to the number of people in each region or country, the value added per head was $4,100 in western Europe, $4,600 in the United States, and $7,800 in Japan.

steel companies in Western Europe. When the world overproduced steel, many European producers attempted to shift into high quality markets and to invest in new technology but did little to cut production or employment. EU directives to reduce output were not always obeyed, and the imposition of quotas resulted in civil disturbances in northeastern France when steel mills were closed in the 1980s.

## Major Manufacturing Industries

Examples of major industries in Western Europe are those that produce automobiles and trucks, airplanes, electronic goods, and pharmaceuticals.

- The *automobile industry* finds it difficult to expand further. It is partly nationalized and partly owned by foreign companies. Renault in France and part of Volkswagen are state corporations, as was the former British Motor Corporation. General Motors and Ford of the United States opened their European plants in the 1950s and built a complex set of interrelationships throughout Europe among assembly plants, independent suppliers of components, and specialist factories producing engines and transmission systems (see Figure 2.14). The Japanese carmakers, Nissan, Toyota, and Honda, all built assembly plants in the U.K. to gain a foothold within the EU market. There are still locally owned private firms, such as Peugeot-Citroen (France) and BMW (Germany).

  Germany produces around 4 million cars per year, France around 3 million, and the United Kingdom and Belgium around 1 million each. The total is well above the numbers that can be sold. In the 1980s, only Germany expanded the number of cars produced as overproduction became common—a situation made worse by the opening of Japanese-owned factories in the U.K. Automation was introduced to cut costs, causing tens of thousands of layoffs in the assembly and component factories. Other cost-cutting moves included increased links between the assemblers and component manufacturers so that inventories could be kept low. This placed a greater emphasis on the close geographic links—including location and transportation—between the carmakers and their suppliers. In 1998, the merger of Daimler-Benz (Mercedes) and Chrysler suggested a new approach toward world markets that involves some of the largest automobile makers in different countries working together.
- The *aerospace industry* is another key sector. Each of the major countries wishes to retain control over its own portion of this industry for defense-related security, and so it remains fragmented with protected, relatively small markets. For example, while American manufacturers think in terms of producing over 2,000 units of a combat aircraft or 700 of an airliner, European firms think in terms of 700 or 150, respectively, because of this fragmentation. Cooperative projects, such as the Airbus (Figure 7.24), the Ariane rockets, and the European fighter project, made Western Europe the world's second largest plane maker but required complex coordination, leading to additional costs and extended time schedules that attracted government subsidies to make sales abroad competitive.

  Airbus was established across several European countries, assisted by government funding in the late 1960s. It was a response to worries that U.S. plane makers could scoop world markets by selling cheaply abroad based on a large home market producing low unit prices. In 1974, American plane makers (Boeing, Lockheed, and McDonnell Douglas) dominated the world market for jetliners, making 4,000 planes. In 1996, Boeing dominated the U.S. market to a greater extent, making 15,000 of the 19,000 planes sold, while European firms—dominated by Airbus—made 9,000.
- Multinational corporations are particularly significant in growth industries such as electronics and pharmaceuticals. The *electronics industry* includes a variety of sectors from the making of microprocessors and integrated circuits to computers and consumer goods such as televisions and stereo sound systems. This spectrum of industries grew since the 1950s, but many sectors were marked by alternate gluts and shortages. The United States and Japan took the lead in various fields of what were the first truly global industries. By the 1990s,

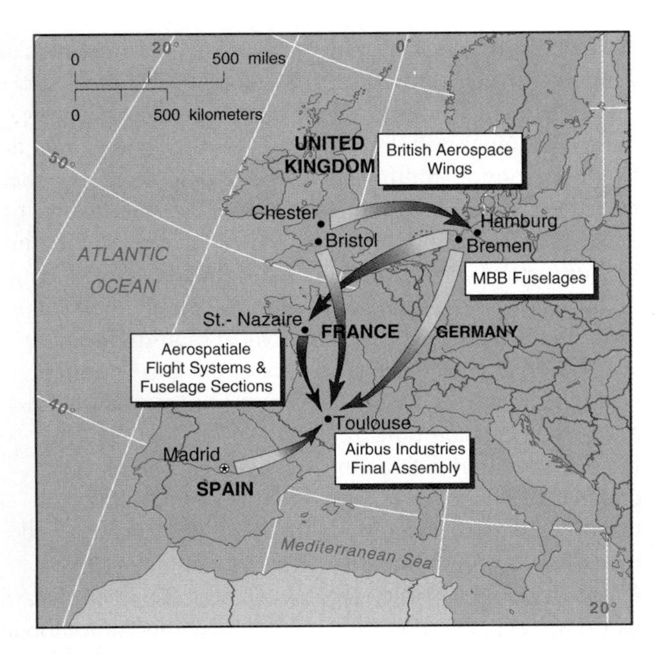


**FIGURE 7.24 Western Europe: aerospace industries.** Cooperative European manufacture of Airbus aircraft. The high levels of capital inputs and technological complexity required for constructing passenger aircraft make it impossible for the aerospace industry in one country of Europe to support the whole process. What are the political and economic implications of these movements?

Western Europe produced 20 percent of the total, the United states 30 percent, and Japan and the newly industrialized countries of Eastern and Southern Asia and Latin America around 50 percent. Philips (the Netherlands) and Thomson (France) are the only world-scale electronics companies with headquarters and origins in Western Europe. In the 1970s, U.S. corporations competed successfully to snatch the integrated circuit market from Western Europe, and in the 1980s, the Japanese became dominant in many sectors, particularly semiconductor production. Both American and Japanese corporations invested in Western Europe, especially in Germany and the U.K., to gain access to EU markets and localized defense industry customers and to utilize good technical and scientific labor. The rise of the Japanese electronics industry and its penetration of markets in Western Europe began with government support at home, leading to rapid increases in exports. It was slowed by voluntary restraint agreements to reduce the impact on Western Europe's factories and ended with the Japanese building their own factories within Western Europe. In the 1990s, American and Western European firms began linking their resources to combat what they perceived as the Japanese threat.

- Western Europe is much stronger in *pharmaceuticals,* another global industry. The production of new drugs requires high levels of research and development and is subject to strong regulatory activities within each country. State health authorities, often with national requirements, dominate major customers. Western European companies are led by those based in Germany, with some input from others in the U.K. The subregion produces nearly half the world total and consumes under 30 percent (the United States produces 30% and consumes 34%; Japan produces 23% and consumes 24%). Western European firms provide more than two-thirds of world pharmaceuticals exports. The Japanese have hardly begun to develop their world market potential in this industry, and their own market remains isolated.

## Service Industries

While manufacturing in Western Europe developed into new sectors but fell behind the world leaders in the United States and Japan, the development of the service sector almost rivaled that in the United States. In the late 1990s, employment in the service sector made up over 65 percent of the total workforce in most of the Western European countries. It is lower, around 50 percent, in Germany, where manufacturing remains very significant. In recent years, industrial countries in Western Europe experienced a phase of deindustrialization as older industries closed or cut their labor forces. By default, this gave greater prominence to the growing service industries.

Service jobs grew in importance as the countries increased their populations, became richer, and instituted strong social welfare programs. Jobs in retailing, wholesaling, education, healthcare, and government employment relate closely to numbers of people and population distribution. The providers range from private corporations, such as food and drink retailing, to state-controlled institutions in health care and education. Such services support and depend on the manufacturing sectors of the economy but play an indirect role in production. For example, better education and health care may provide a more skilled and adaptable work force.

The main growth areas in the service sector are in *producer services* and *tourism*.

- **Producer services** are involved in the output of goods and services, including market research, advertising, accountancy, legal, banking, and insurance. They rank among other key sectors, such as manufacturing, in generating economic development. Producer service jobs increased by a quarter or a third in the countries of Western Europe since the 1970s and employ around 10 percent of the labor force despite productivity gains from the use of information technology. Producer services concentrate in the centers of major cities, where agglomeration economies are significant: such services group near sources of highly qualified labor, national institutions, government departments, and headquarters of major corporations. London, Paris, Amsterdam, Frankfurt (Figure 7.25), and Munich have major shares of this industry. While some functions, such as back-office routine processes, are decentralized in suburban and small-town "paper factories"—often in landscaped office

**FIGURE 7.25 Western Europe: financial services.** One of Europe's major financial centers, Frankfurt, Germany. Medieval buildings are preserved in the heart of the city, surrounded by other buildings limited in height, although some were rebuilt after World War II. Beyond them are high-rise office and apartment blocks.



parks—major city centers retain the high-order functions in which face-to-face personal contact is important.

Mergers and takeovers after the mid-1980s led to a smaller number of very large producer service firms, each broadening the range of services. Banking, insurance, property services, and accountancy were combined. Centrally located specialist firms took over provincial firms to extend their influence nation-wide and internationally. Although financial firms dealing with huge sums of money are liable to occasional scandals, the level of control in Western Europe is high enough to attract many transactions of world financial significance.

- Tourism is another growth sector among service industries. Western Europe dominates the international market with over 320 million foreign visitors each year. In 1997, European countries held 8 of the 11 top places in terms of international tourist receipts: Italy, Spain, France, Britain, Germany, and Austria (2nd to 7th) had total receipts of $134 billion (compared to the United States, which was 1st with $75 billion). The internal and international numbers of tourists, both incoming and outgoing, doubled from 1970 to the 1990s. The growth resulted from increased incomes and leisure time, together with new lifestyle expectations. The expansion of package tourism made foreign travel accessible to greater numbers of people. More people take second holidays or have short weekend breaks. Germany and the U.K. provided the main source of tourists in other European countries. Although the Alpine and Mediterranean countries were the main receiving areas, the rural, coastal, and historic urban areas within Western Europe also increased their tourist trade. From the 1980s, international tourism began to differentiate a less demanding public and a quality market for more affluent people. It is likely that the tourist industries of Western Europe will continue to grow.

  Tourism is encouraged by the governments of each country, EU regional investment in infrastructure (roads, airports), and agreements on the sharing of health facilities, currency regulations, and customs. Individual countries promote their facilities through tourist boards. In the mid-1990s, tourism generated one job in eight in the whole European Union, making it the largest industry in Europe. Yet the industry is fragmented with few large international corporations. Tourism is often promoted to help in regional development, since many popular rural and coastal areas are located away from urban-industrial economic growth areas and need increased local economic activity to prevent further population loss. Tourist developments provide useful additions to local household incomes, often improve the facilities available in an area, and add to amenities that attract other forms of employment. Much of tourism employment, however, is seasonal, poorly paid, and female-dominated. The unskilled and low-paid nature of many jobs lead to them being taken up by migrants and thus having less impact on the local economy than anticipated.

## Regional Policies

Regional differences within Western Europe's countries make significant geographic contrasts in economic well-being between more wealthy and poorer areas and particularly between urban-industrial centers and rural areas (Figure 7.26). Contrasts in well-being increased with the growth of privatization and the emergence of new manufacturing industries and producer services.

Areas of older industry and rural upland experienced similar problems of economic lag compared to growth in the central spine of economic growth extending from central England southeastward along the Rhine axis. Old industrial areas associated with the coalfields of northern France and Britain, central Belgium, and the German Ruhr were affected by the decline of coal and steel industries and the siting of auto and electronics factories in new areas.

After 1945, many countries adopted regional policies. They aimed to reduce spatial differences in standards of living and employment opportunities. Capital and jobs were

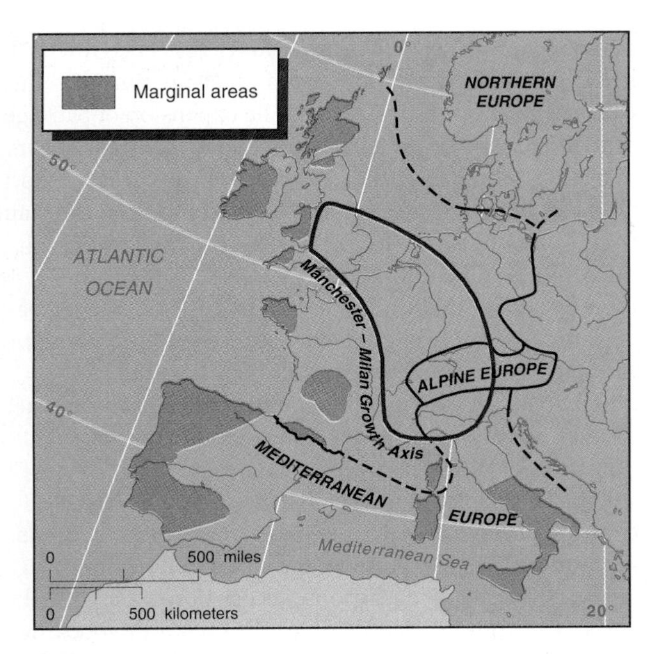


**FIGURE 7.26 Western Europe: regional contrasts.** Variations in economic growth occur within Europe. The wealthiest and most economically active areas in Europe are contained within the Manchester-Milan Growth Axis. This main growth area includes much of Western Europe and continues southward through Switzerland into northern Italy. The poorest marginal areas are peripheral upland areas. The intermediate areas between the growth axis and the marginal areas suffer some of the problems of the peripheries.

redirected from the main growth areas to poorer regions, partly as a means of reducing transportation congestion and labor shortages in the growth areas. For example, manufacturers in southeastern England wishing to expand had to do so in the poorer areas of the north and west. The government relocated several of its major department offices to such areas. In France, policies encouraged decentralization from Paris. Such regional policies were implemented enthusiastically until the 1970s recession, when the poorer regions suffered high unemployment once again but even the former growth areas were also affected. Reductions in government funding, targeting on the most needy areas, and a shift to placing an emphasis on local areas attracting new business replaced the former regional policies in the 1980s.

From the 1970s, EU regional policy took over from national policies aimed at redistributing employment opportunities, and the EU invested large sums in infrastructure construction in the lagging regions. The European Regional Development Fund was established in 1975. There was a net flow of funds to the margins of the countries of Western Europe, to their older industrial areas, and especially to the newer (and poorer) member countries in Mediterranean Europe. The main emphasis of EU regional policy was the granting of loans to create jobs directly or indirectly. The loans were used mainly for infrastructure projects, especially roads, telecommunications, water supplies, and waste disposal. Job creation outcomes had modest success, with a few thousand new jobs created, from the millions of unemployed. As the regional programs were scaled down, many job opportunities moved back to growth areas

Within Western Europe, each country faces specific problems that affect not only its own future and the rest of this subregion but also are significant for the world economy.

## Germany

Continuing economic growth in Germany occurs within a stable economic and political system in which enterprise, worker participation, and government intervention are all significant. Germany has a federal government in which the states (länder) have a strong role (Figure 7.27). Centers of influence, such as Berlin and Bonn (government), Frankfurt and Munich (finance), and Hamburg (media), are spread throughout the country—a contrast to the central roles of Paris in France or London in the U.K.

In the 1990s, Germany appeared able to adapt to new situations and maintain its position as one of the world's richest countries. It contended with tensions arising from reunification, industrial restructuring, and high public spending. Productivity increased and public spending was reduced from the very high levels immediately following the reunification of East and West Germany in 1990, although unemployment remained high.

Germany needs to maintain its competitive position in world markets. Although it exports a higher proportion of its products than any other country in the world, its high wages, expensive social-welfare systems, and aging population place it at a disadvantage compared not only to Eastern Asia but to other countries in Europe. Even in machine tools—one of Germany's major products—cheaper Japanese products took over large sectors of the American market. Japanese firms in this sector are larger, more productive, and keep up with new technologies.

In the early 1990s, the huge costs of reunification were about half Germany's total tax revenues. Extra spending focused on upgrading infrastructure and welfare costs in the former East Germany and servicing the debts incurred. Reunification, however, was not just a problem of initial financial aid. The economic gap between east and west was greater than anticipated before reunification.

- The former East Germany, now six länder within the reunified Germany, suffered economic collapse after 1990: industrial output fell by two-thirds in the first year, with manufacturing jobs reduced from 3.2 million in 1989 to 1.2 million in 1993 and farming jobs from 1 million to 0.25 million. Money from the western länder prevented starvation in the east, where both marriage and birth rates fell by two-thirds. Death rates increased in the active middle years—a demographic shock similar to that caused by a war. People living in the eastern länder retain many of the attitudes and expectations generated by 45 years of communist paternalism. By the late 1990s, the

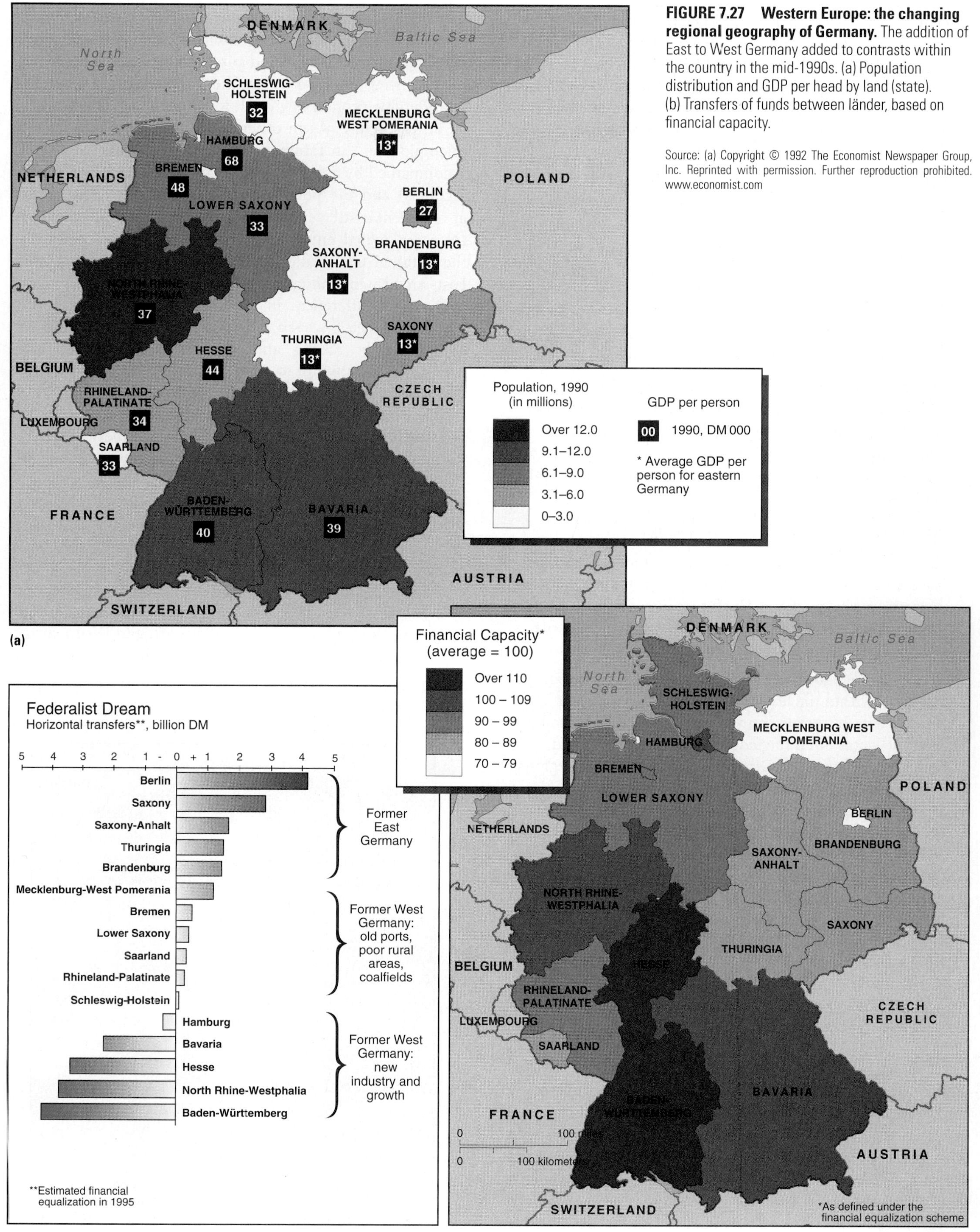


**FIGURE 7.27 Western Europe: the changing regional geography of Germany.** The addition of East to West Germany added to contrasts within the country in the mid-1990s. (a) Population distribution and GDP per head by land (state). (b) Transfers of funds between länder, based on financial capacity.

Source: (a) Copyright © 1992 The Economist Newspaper Group, Inc. Reprinted with permission. Further reproduction prohibited. www.economist.com

Source: (b) Copyright © 1994 The Economist Newspaper Group, Inc. Reprinted with permission. Further reproduction prohibited. www.economist.com

economy of the eastern länder of Germany began to grow as a result of major public and private investments. The central city areas of Dresden and Leipzig, together with a large proportion of the housing, were restored. Over 1,000 foreign firms invested in new factories where workers were more flexible over hours and wages than those in the western länder.

- Reunification made Germany responsible for the environmental reclamation of major military areas and weapons storage sites left by the former Soviet Union and of the areas stripped for the brown coal that was the former East Germany's main energy source. The program to close and reclaim the old brown coal areas is expected to be completed by 2002. Contaminated and stripped land will be returned to farming and recreational uses, and the project provides jobs for unemployed miners.
- Further government expense is involved in the redevelopment and restoration of Berlin as the capital of reunified Germany. The restoration of the Reichstag parliament buildings and the redevelopment of huge tracts of land at Potsdamer Platz and Friedrichstrasse, the main thoroughfares of former East Berlin, will provide office space that is needed for government and service industry expansion. Germany's Top 500 companies were slow to relocate headquarters in Berlin, but their number is expected to increase when more government departments move there. The future of economic development in Berlin remains uncertain in view of the established centers elsewhere in former West Germany.
- Reunification added new regions to Germany. The länder are states within the federal system and have greater influence both internally and over the federal government than the states do in the federal system of the United States (see Chapter 9). Direct influences from the regions are greater than in other large European countries. Bavaria, the northern Rhinelands, Lower Saxony, Hamburg, and the eastern länder are in some ways like different countries. Regional political and economic differences often make overall agreements and consensus difficult. In particular, tax revenue sharing, or financial equalization, shifts funds from the richer to poorer länder (Figure 7.27b). Before reunification, the main shifts were from southern Germany and the Ruhr to Saarland and the north; now it is almost all to the six eastern länder and involves greater sums of money.

## France

France has a more centralized government than Germany and a vision of itself as a distinctive entity that makes a major contribution to the wider world outside Europe, especially in Africa (see Chapters 3 and 4). France's commitment to the EU, however, was accompanied by a decline in its nationalistic assertiveness. Its currency was tied to the German mark in 1982. Decentralization of government in the 1980s gave more control to regional bodies and less to Paris. France bowed to further EU challenges when directives ruled against the protection of its steel and auto factories and farmers.

Its central presence within the EU framework resulted in considerable economic growth for France and investment in its infrastructure. As in other countries in Western Europe, unemployment in France reached 12 percent by the 1990s, accompanied by a growing gap between the rich (with rising incomes) and the poor (with static incomes). The French government is still very involved in the economy of the country. State-owned enterprises produce around 40 percent of the national GDP. Like Germany, the rising social security costs are proving difficult to control.

Regional differences in France are reflected in 22 units established in 1982, each with elected councils and increased autonomy for raising and spending tax revenues (Figure 7.28). The attempt to divert economic development from Paris led to major construction projects in each region and to competition between some of the major regional centers. The Montpellier-Nîmes-Arles grouping at the southern end of the Rhône River valley, close to the major Europort of Marseille (Figure 7.29), markets itself as "Eurocity" with its fiber-optic wired "technopolis" and tourist attractions. Just to the north, the Lyon and Saint Etienne urban areas in the Rhône-Alpes region advertise business links with near-neighbors in Switzerland, Lombardy (northwest Italy), Catalonia (northeast Spain), and Baden-Württemburg (southwest Germany). The old industrial town of Saint Etienne now boasts a high-tech university, a museum of modern art, and leisure attractions. Such decentralizing policies had less impact in the

**FIGURE 7.28 Western Europe: regions of France.** The regions of France and associated major cities. How central is the position of Paris?

**FIGURE 7.29 Western Europe: Marseilles, France.** France's major port on the Mediterranean Sea is in the southeast corner of this space shuttle photo, hemmed in by hills. Port expansion has been around Lake Berre to the west. Roads and railroads tunnel westward to the Rhône River valley routeway connecting to northern France.

NASA

marginal parts of France such as Brittany in the far northwest, the Pyrenees in the south, and the hilly Massif Central —the areas they were supposed to help—and reactions against the policy are growing in these areas.

## United Kingdom

The U.K., consisting of England, Wales, Scotland (together known as Great Britain), and Northern Ireland, did not experience the economic growth that characterized Germany and France in the 1970s and 1980s. The British continued to invest abroad, but at home improvement of Britain's infrastructure was slow, and loss-making nationalized industries held back economic growth. Annual economic growth averaged only 2.3 percent from 1960 to 1980. By comparison, in this period, the other countries of Western Europe averaged 3.5 to 4.5 percent, the United States 3.5 percent, and Japan 7.7 percent.

Policies designed to bring the U.K. into a more competitive position with the rest of the world in the 1980s had some of the desired effects, but engendered growth that was too rapid, leading to inflation and a loss of economic confidence at the end of the decade. The measures taken in the 1980s included deregulation and the privatization of state enterprises, a curbing of trade union powers, and tax cuts. Demand for goods rose, and people bought houses, cars, holidays, school fees, and consumer gadgets beyond the capacity of their incomes and the reduced construction and manufacturing capability to supply these needs. Imports rose, bringing a balance of payments crisis, high interest rates, and shrinking GDP. Manufacturing declined from 27 percent of GDP in 1979 to around 20 percent in the mid-1990s, and this loss in production was not compensated by the rises in income that took place in financial services exports and tourism.

From the early 1980s, manufacturing productivity grew as a result of shedding a third of the labor force. Although it still reached only half of the United States' value added per worker, the U.K. approached that of its other competitors in Europe and Asia. Investment necessary to increase U.K. productivity rose during the boom period of the late 1980s, but then fell in recession. In the early 1990s, the U.K. was one of the first European countries to move out of recession as a result of the stringent measures taken in the 1980s. By the late 1990s, it was the most competitive European country in the world economy.

The U.K. economy remains distinctive and dynamic largely because of the high level of external investment in its industries. Access to EU markets, good local services, lower labor costs, and the speed at which new enterprises are accepted and set up attract overseas investment in the U.K. Good road communications and access to Europe's largest airport at Heathrow, London, are other important factors. The Channel Tunnel, linking England and France, opened in 1994, improving transport links to other EU countries. Over 40 percent of United States' and just under 40 percent of Japanese investment in the EU is in the U.K., while German firms place some 22 percent of their overseas investment there. Foreign firms control around 20 percent of British manufacturing industries. The last British car maker, Rover, was taken over by the German BMW in 1994. In Scotland, the proportion of foreign control is nearly one-third, including the production of semiconductors and desktop computers. South Wales and northeastern England—two other formerly declining industrial areas—also attracted major investments from foreign corporations.

Although London's Heathrow airport continued to dominate European air traffic, it came under threat as the public opposed the proposal to build a fifth terminal that would increase the airport's potential from 50 million to 80 million passengers per year. Although airplane sizes increased while noise from them decreased—0.5 million people living near the airport are now affected instead of 1.5 million in 1978—there is public concern about noise pollution and the potential increase in ground traffic congestion around the airport.

The future prosperity of the U.K. will depend on its continuing attractiveness to new industries and foreign corporations. Further pressures from Scotland, Wales, and especially Northern Ireland for home rule, and the linked political uncertainties, may adversely affect that prosperity. In the late 1990s, the long-term ethnic conflict in North Ireland appeared to be reaching a compromise resolution, while Scotland and Wales voted for a degree of government to be devolved to them from London.

### Netherlands, Belgium, and Luxembourg

These three countries are often called the "Low Countries" because of the low relief of much of their terrain along the coast and at the mouth of the Rhine and Meuse (Maas) Rivers. They linked together economically as the Benelux economic union soon after World War II and stand at the center of the Western European industrial core.

The Netherlands includes large areas of polders—land reclaimed from the sea and river delta. The country continues to export dairy products, meat, vegetables, and horticultural products but also has important manufacturing industries, concentrated in the urbanized Randstad coastal belt from Rotterdam to Amsterdam. Both Rotterdam and Amsterdam are major ocean ports, with Rotterdam at the mouth of the main Rhine, being Europe's busiest. A new railroad line—from Rotterdam to link with the German rail system—will increase Rotterdam's access to the interior of Europe. Amsterdam has many light industries and is one of Europe's busiest airports. The Hague is the Dutch capital. The Dutch possess energy resources in the form of large natural gas fields, the exploitation of which led to the closure of coal mines in the south of the country.

Belgium has a shorter and narrower coastal zone of intensive economic growth from Antwerp to Bruges but also has old industrial areas along the Meuse River valley between Liège and Charleroi in which steel, chemicals, and textiles struggle to adapt to changing markets. Southern Belgium is an upland where few people live. The main tensions within Belgium remain between the more prosperous Flemish (Dutch-speaking) north and the poorer Walloon (French-speaking) old industrial and upland areas of the south. Brussels stands between these zones and is the capital city of the European Commission and NATO as well as of Belgium. Many multinational corporations have offices in Brussels for this reason.

Luxembourg, the smallest country in this subregion, has farming and metal industries as an economic base. Its modern prosperity, however, is based on the growth of service industries and the choice of Luxembourg City as a site for international corporation offices.

**RECAP 7C: Western Europe**

Western Europe includes three of the four largest countries by population in the region (France, Germany, United Kingdom) and the Low Countries (Belgium, the Netherlands, Luxembourg). It initiated the 1700s industrial revolution and dominated the 1800s world, and in combination, its countries are still the wealthiest and largest affluent market in the world. The Republic of Ireland is part of the less affluent western margins of this subregion.

The population of Western Europe is static in overall numbers and aging. Decreasing numbers of people in the working age groups support increasing numbers of retired people. In the 1960s and 1970s, immigration was encouraged to bring in guest workers.

Agriculture was modernized so that the subregion that was a net importer is now a net exporter of food. The manufacturing industries modernized in type and levels of productivity. Service industries, especially financial and tourism aspects, are the main employers, contributing increasingly to the continuing affluence of the subregion.

7C.1 Which five features of each give France, Germany, and the United Kingdom their geographic distinctiveness?

7C.2 What level of success do you think was achieved by regional policies in the countries of Western Europe?

7C.3 Why do so many difficulties confront the movement toward peace in Northern Ireland?

**Key Terms:**

| | | |
|---|---|---|
| intensification (agriculture) | specialization (agriculture) | extensification (agriculture) |
| concentration (agriculture) | agribusiness | producer services |

**Key Places:** *(cities in italics)*

| | | |
|---|---|---|
| Belgium | *Düsseldorf* | United Kingdom |
| *Brussels* | *Cologne* | *London* |
| France | *Hamburg* | *Birmingham* |
| *Paris* | *Munich* | *Manchester* |
| *Lyon* | *Stuttgart* | *Leeds* |
| *Marseilles* | *Frankfurt* | *Glasgow* |
| Germany | Luxembourg | Ireland, Republic of |
| *Berlin* | The Netherlands | *Dublin* |
| *Essen* | *Rotterdam* | |
| | *Amsterdam* | |

### Republic of Ireland

The Republic of Ireland's economy is still mainly devoted to agriculture, which makes up around 40 percent of GDP and 20 percent of exports. The country is a major beneficiary of the EU agriculture policy, which provides 30 percent of Irish farmers' incomes. Over 80 percent of agricultural produce is exported, mainly to the U.K., in a competitive market. Irish farmers specialize, using the deepwater facilities at Dublin to import cheap animal feed at a time when European grain increased in price and raised costs to EU livestock producers.

The Irish livestock herds have increased rapidly since the late 1980s. This country also encourages foreign investment in manufacturing, much of which is centered on Shannon airport in the center of the country, or around Dublin. Ireland is one of the few European countries with a positive trade balance and surplus budget.

## NORTHERN EUROPE

Northern Europe extends from the North European Plain to the Arctic Circle. It is unified by a challenging physical environment and its history. The Viking explorers expanded their domain from this area to the British Isles, northern France, Iceland, Greenland, much of the Baltic area, and Ukraine. The subregion later combined in a single country—but with less overseas activity—for long periods. Today the separate countries have relatively small populations but enjoy considerable prosperity and some of the world's highest GDP per capita figures (see Figure 7.2).

The four countries of Northern Europe are tied to Western Europe in many ways but have jealously guarded their independence. Although Denmark joined the EC in 1973, it remained a critic of too much involvement, while the other three countries remained outside until the 1990s, when Finland and Sweden joined only after considerable debate and close referendum votes. Norway rejected the prospect after a close vote.

### Countries

Northern Europe is sometimes called "Norden" and consists of the Scandinavian and Jutland peninsulas, Finland, and several islands in the Baltic Sea and North Atlantic Ocean. It includes the countries of **Denmark**, **Finland**, **Norway**, and **Sweden**, together with the former colonies of Denmark in **Greenland**, the **Faeroe Islands**, and **Iceland** (Figure 7.30). Iceland became independent in 1944, while the Faeroes (1948) and Greenland (1979) are self-governing associated territories of Denmark, which is responsible for defense and foreign relations.

Denmark, an extension of the North European Plain, has only around one-tenth of the area of Finland, Norway, or Sweden but does not have their large areas of mountainous terrain and arctic climate. Norway has a long coastline giving access to the North Atlantic Ocean, whereas Finland and Sweden look mainly toward the Baltic Sea. This focus and their nearness to Russia affected Sweden, and Finland, from the late 1700s. They remained neutral and avoided close links with NATO and the EU until the breakup of the Soviet Union in 1991. They still remain outside NATO. Since 1991, Finland established close links with the Baltic countries, especially Estonia.

The economies of the countries of Northern Europe relied on primary products until a more recent development of manufacturing and service industries in the 1900s. Denmark is a major agricultural country; Sweden has agriculture in the south, while the north has significant timber and iron-mining industries; Finland is another major wood-producing country; and Norway has fishing and shipping industries. The discovery of major oil and gas reserves beneath the North Sea brought new wealth to Norway from the 1970s. All countries modernized and industrialized in the 1900s, partly based on their natural resources and partly on the production of high-value metal and textile goods, vehicles (Sweden), and electronics (Denmark).

### People

The population of Northern Europe is small, only 24 million people in total in 1998 and likely to rise by less than another million by 2025. Most people live toward the southern part of the subregion. Although (without Greenland and Iceland) they cover an area that is almost as large as Western Europe, these countries, apart from Denmark, have low densities of population because of difficult climates and terrains. compare Figure 7.8 and 7.10a with Figure 7.14).

Rates of population increase were between 0 and 0.3 percent in 1998, and total fertility rates dropped from around 2.5 in 1965 to under 2. Those born in Northern Europe in the 1990s expect to live for 80 years. In all countries, birth rates almost equal the low death rate levels (see Figure 7.15). The age-sex graph for Denmark (see Figure 7.16) shows maximum numbers in the age groups around 30 to 40 years, reflecting the baby boom from the later 1940s into the 1960s and the subsequent fall in births. Those over 65 years make up 15 to 17 percent of the total population, a growing burden on the welfare systems of these countries that is already being felt in increased taxes and the reduction of some welfare programs after years of wide-ranging coverage.

The dearth of farming land and cold northern climates are linked to high percentages of urbanization like those in Western Europe. Finland (65%) and Norway (74%) were the only two countries under 85 percent urbanized. Even so, the major cities are not very large. In each country the capital city is by far the largest, but only **Copenhagen**, (Denmark, Figure 7.31), **Stockholm** (Sweden), and **Helsinki** (Finland) had over 1 million people in 1996. **Oslo** is Norway's capital.

Culturally, the countries of Northern Europe share an almost total dominance of Evangelical Lutheran Christianity as the major religion. Officially it has 90 percent or more of the population as adherents in the four major countries and Iceland. This Protestant religion influenced the lives of the people, inducing very serious and community-conscious attitudes to work and social life. In recent years, the combination of affluence and materialism broke many of these strong cultural links and loosened the control exercised by the churches. The languages of Denmark, Norway, and Sweden are similar, having a Germanic origin, but that of Finland is unique in the region, being akin to Hungarian and some languages in central Asia.

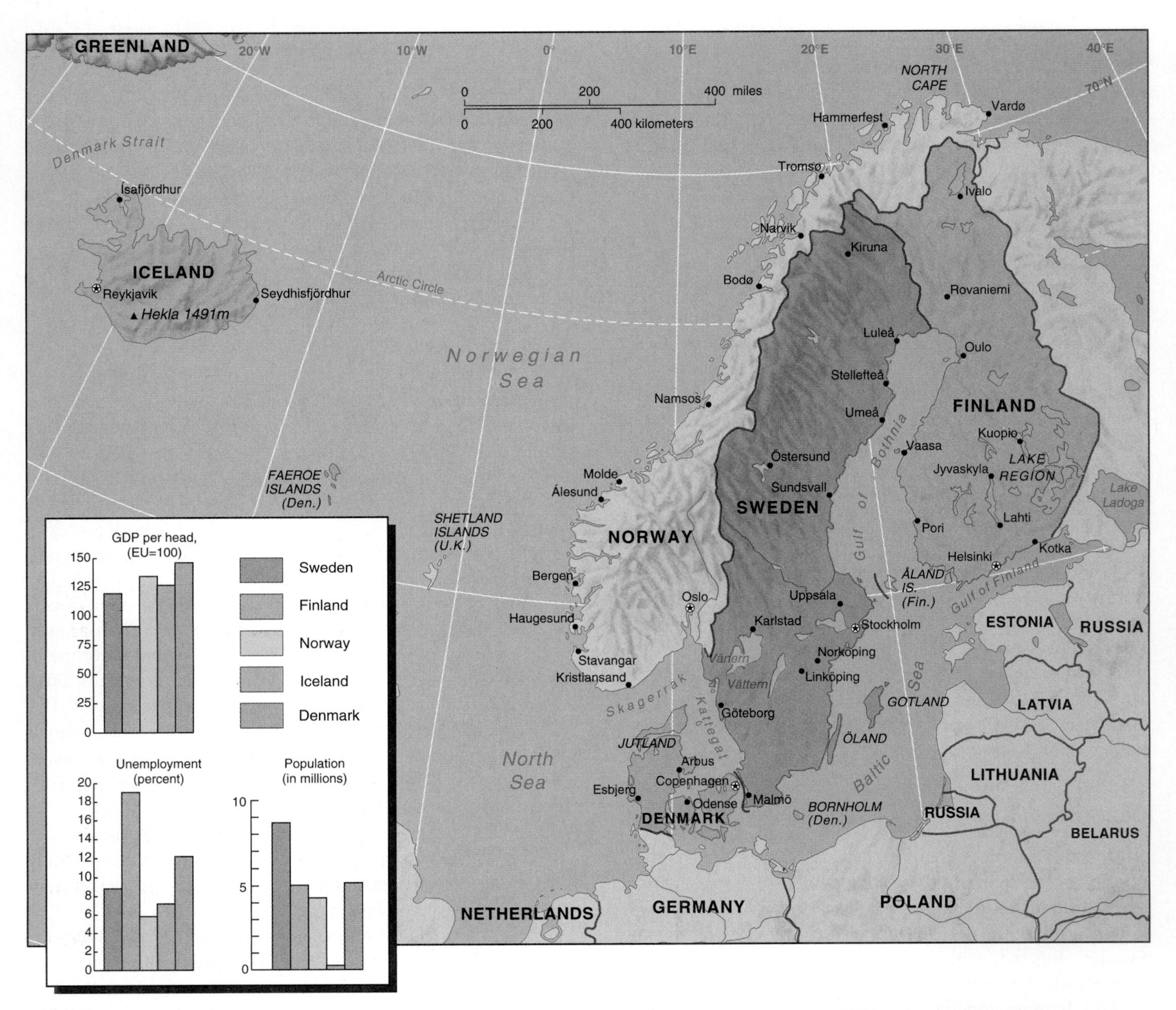


**FIGURE 7.30 Northern Europe: the countries, cities, and physical features.** Compare the countries in terms of GDP per head, unemployment rate, and population. Norway and Sweden comprise the Scandinavian Peninsula; the Jutland Peninsula is part of Denmark.

## Economic Development

The four largest Scandinavian countries have some of the highest GDP per head figures in the world. In recent decades, Sweden maintained its income level, while Finland (new industries), Norway (oil and natural gas), and Denmark (high-tech industries) increased their affluence. The small, stable populations made it possible to develop societies in which poverty is restricted, as reflected in the ownership of consumer goods (see Figure 7.19).

- Denmark's economy is based on agriculture and high-tech manufacturing. Seventy-five percent of its lowland area on the Jutland peninsula and on the islands between the peninsula and Sweden is farmed with an emphasis on dairy and livestock products (see Figure 7.20). Copenhagen, on the largest island, grew as a port that controlled the entrance to the Baltic Sea and acted as an entrepôt for the transhipment of goods. It is now a major service and manufacturing center with government offices and financial services.
- Norway is a mountainous country with limited farmland around Oslo and Trondheim. Until the 1970s, its economy was based on fishing and the smelting of metals using hydroelectricity. The discovery and exploitation of North Sea oil and natural gas then gave Norway a new resource that now makes up half of its income. Norway

**FIGURE 7.31 Northern Europe: Copenhagen, Denmark.** The history of this port city is demonstrated by the 1700s buildings on the waterfront. For many years, Copenhagen controlled entry to the Baltic Sea, and its merchants grew wealthy through trade with surrounding countries.

© Philip Gould/Corbis

became the world's second largest oil exporter after Saudi Arabia in 1995, but production fell after 1997. The funds generated were invested in improving infrastructure throughout the country, encouraging increased tourism (with 2.7 million visitors in 1996). Norway also invested in fish farming for salmon, halibut, and scallops, an industry that is partially replacing the declining offshore fishing.

- Sweden is the largest and most industrialized Scandinavian country. Farming became important in the southern half of the country, with dairy products and some grain forming the main output, but external competition reduced this sector of the economy. Sweden has large deposits of iron ore and smaller reserves of other metallic minerals (copper, lead, zinc, and manganese) in the northern part of the country, where the managed forests provide a continuing source of wood. Much of the iron and wood pulp is exported, but a diversity of manufacturing enterprises in central Sweden in the zone between Stockholm and Göteborg include stainless steel products, ships, automobiles, electronics, furniture, and glassware. Sweden has a long history of manufacturing and managed to develop and maintain a strong social support system, the high cost of which is now questioned.
- Finland is a northern country, having no major farming area like Denmark or Sweden. It has few mineral resources and for long was inhibited in economic development by its position next to Russia and the Soviet Union. It was forced to trade with that country and avoid external political ties. The forests provide Finland's greatest natural resource, with timber and wood pulp forming one-half of its annual export value (Figure 7.32). Most people live along the southern coast, where Helsinki and Turku are centers of metal, machinery, and shipbuilding industries. Tampere has a textile industry. Oulu became the center of high-tech manufacturing, replacing employment in the pulp and paper and chemicals industries, and based around the Nokia mobile telephone factory.
- Iceland and Greenland have total populations of 260,000 and 60,000 respectively. With the Faeroe Islands, they were linked to Denmark until the1900s. Iceland gained some independence in 1918, but retained strong links with Denmark until World War II, when U.S. and U.K. forces occupied it after Germany took Denmark. Iceland declared full independence in 1944. In 1948 the Faeroes, and Greenland in 1979, gained the right to administer their internal affairs in association with Denmark. All are inhibited by physical difficulties of climate and land and rely on fishing the surrounding waters. In the late 1990s, reduction in fish stocks occurred as other countries increased their take. Competition for use of the surrounding waters led Iceland and the Faeroes to extend and defend their 200-mile limits against European fishers until international agreement was obtained. Greenland

FIGURE 7.32 **Northern Europe: Finland manufacturing.** Wood industries play a major part in the Finish economy. The lumber is floating downstream to pulp mills.



has some mining that is carried out by international companies, and the minerals are largely shipped direct to the United States. Iceland has a growing tourist industry.

## ALPINE EUROPE

In its broadest sense, Alpine Europe includes parts of France, Italy, Slovenia, and Croatia, as well as the three small countries of Austria, Liechtenstein, and Switzerland that rule the central parts of the mountain ranges (Figure 7.33). In this subregion, only the two countries of Austria and Switzerland will be considered.

Both Austria and Switzerland are dominated by the Alps and are landlocked. They have populations of 8 and 7 million people, respectively. Their position in the center of Europe, and the energy and skills of their people combined to give them important strategic and economic significance. Both countries have high standards of living. Switzerland, despite the prevalence of high mountains within its borders, is currently the world's wealthiest country in terms of GDP per capita.

### Countries

Although somewhat similar in size, natural features, and current outlook, the main countries of Alpine Europe—**Austria** and **Switzerland**—have very different histories. Austria is the remnant of a large empire that extended over much of southern and eastern Europe until World War I. At one stage in the later Middle Ages, Austria controlled much of Switzerland until some Swiss cantons gained their independence in the 1400s. Others were added later. Switzerland remained traditionally neutral in conflicts affecting the rest of Europe and its financial business grew out of foreigners being able to deposit their wealth in safe, unidentifiable bank accounts. Austria was much involved in the two world wars, losing its former empire after the first and linking with Germany in the second.

Like the Scandinavian countries of Northern Europe, both Austria and Switzerland joined the European Free Trade Association, but stood apart from the EU until the 1990s. Austria joined the EU in 1995. Even Switzerland—long wedded to its neutrality—is split between businesspeople who want to be part of the enlarged EU and more conservative groups who wish to retain their traditional independence.

Both Austria and Switzerland have substantial proportions of their land at lower levels outside the Alps. That is where most of their populations live. Both lowland areas are in the north with good communications into Germany. Switzerland

FIGURE 7.33 **Alpine Europe: the countries, cities, and physical features.** Two small but influential and wealthy countries.

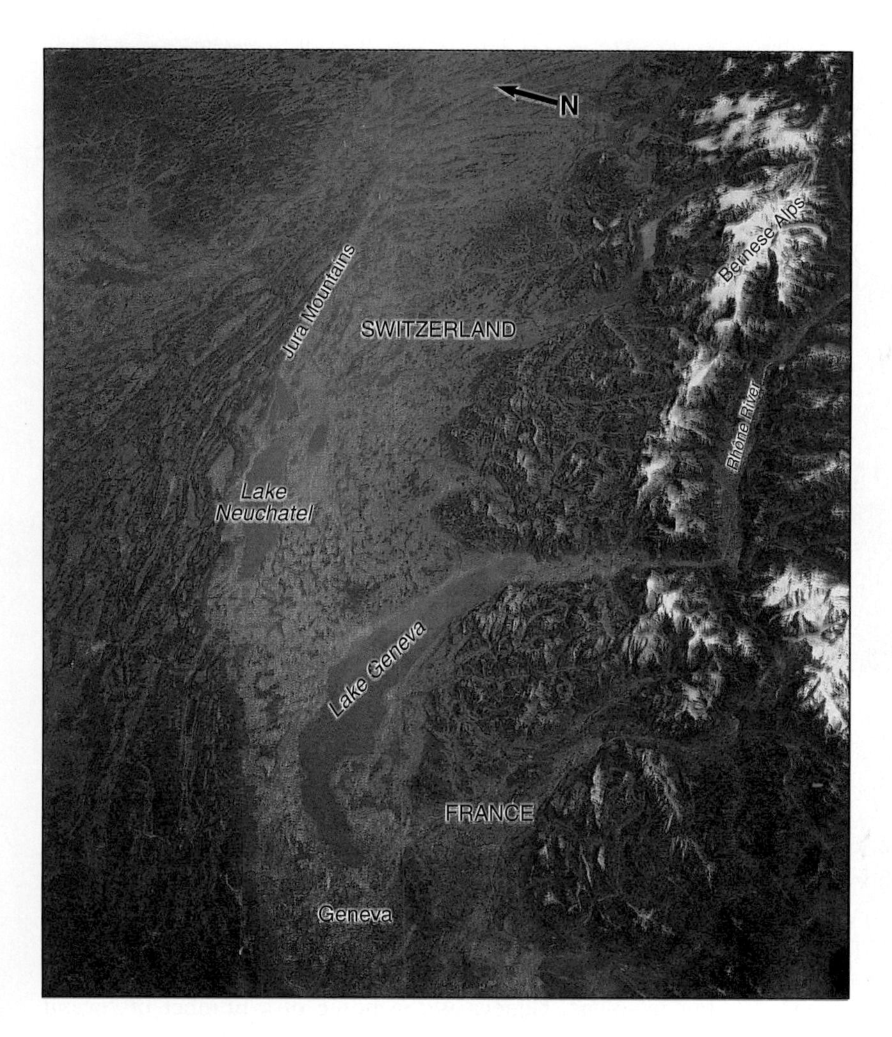


**FIGURE 7.34 Alpine Europe: Switzerland.** A space shuttle view, looking northeast. Lake Geneva in the foreground lies between the forested lower ridges of the Jura Mountains to the north (left) and the snowcapped peaks of the Alps. The Rhône River flows through Lake Geneva after rising in the Alps and following a deep trough excavated by glaciers. Lake Geneva lies in the lower central part of Switzerland that contains most of the country's people.

NASA

also connects into France along the Rhône (Figure 7.34), while Austria's easiest communications are eastward along the Danube River valley. Up to 1990, the Austrian route eastward was largely blocked by Cold War divisions. Although access was gradually eased, it did not provide links to such affluent areas as those adjoining Switzerland. Austria also had to recover from years of World War II liaison with Germany and subsequent occupation by the victorious allied powers until 1955, whereas Switzerland's wartime neutrality enabled it to grow economically without interruption.

## People

Austria and Switzerland show a contrast between the densely populated lower lands and the more thinly populated mountains (see Figure 7.14). Both countries have annual rates of natural increase of under 0.3 percent and total fertility rates around 1.5. Their combined population of 15 million people in 1998 is unlikely to rise beyond 16 million by 2025 (see Figure 7.15). The age-sex graph for Switzerland (see Figure 7.16) records both the reduction in births since 1970 and the increasing numbers of older people, especially females.

Both Austria and Switzerland are urbanized, with almost 65 percent of their populations living in towns. **Vienna**, Austria's capital, is the only city of a million people in Alpine Europe, and is by far the largest city in Austria. In Switzerland, the largest city of **Zürich** has closer rivals in **Geneva**, **Basel**, and the capital **Bern**.

Switzerland has a dominant German-speaking group in the north and east, making up two-thirds of the total population, a smaller French-speaking group in the west, and some Italian-speaking areas in the south. A tiny group of Romansch speakers in the upper Rhine River valley is linked to the ancient Roman occupation and Latin language. Switzerland's affluence and mountain environment attract many immigrants from other parts of the world who are wealthy enough to afford the high cost of living. Austria is a dominantly German-speaking country that also contains smaller groups of peoples from its former empire territories in southeastern Europe.

## Economic Development

Austria and Switzerland are two small countries, but increased their share of world GDP from 1.1 percent in 1965 to 1.9 percent in the late 1990s. Despite their lack of natural resources, the negative potential of the high mountain environment that covers much of both countries, and their landlocked dependence on transportation through

**FIGURE 7.35 Alpine Europe: farming in Switzerland.** Swiss farming is based on using limited alpine resources. The Swiss government subsidizes farmers to maintain the long-established system of summer grazing on high pastures (known as alps). This area, near Mount Blanc, also attracts many summer hikers.



surrounding countries, these two nations became highly prosperous, as their consumer goods ownership shows (see Figure 7.19).

Of the two, Switzerland's economy produces 30 percent more than that of Austria, although the latter was catching up in the 1990s. With a continuous history of neutralism and, hence, little interruption in time of war, the Swiss developed strong manufacturing industries, significant financial institutions, and a major tourist industry. Some of the Swiss manufactures, such as cheese and chocolate, are based on the dairy farming that is maintained with the highest level of government grants in the world (Figure 7.35). The generation of hydroelectricity is so well developed that it meets 60 percent of the country's power needs. Switzerland has a railroad network that is famous for its time-keeping and the feats of engineering that permitted the construction of major routes through the mountains. In the last 20 years, road routes have been improved through the Alps and further capacity is planned, although, for environmental reasons, the federal government is threatening to control the numbers of trucks passing through the country. Good transportation and telecommunications are very important to these mountainous countries.

The Swiss watch-making industry, located in the Jura Mountains in the northwest of the country, was one of the early industrial developments that secured a world wide market. Competition with Japanese digital watches from the 1960s led to the Swiss developing the cheaper "swatches" as well as higher-value watches. Such innovations kept them in the forefront of the industry—one of the few cases where manufacturers in Europe coped with such competition. Other Swiss manufacturing products come from the lower hilly area of central Switzerland, where the main towns are situated, and include pharmaceuticals, precision machinery, and instruments. Zürich and Basel are the main industrial cities, and Zürich is also a major banking and finance center. Bern, the capital, is the center of a number of small manufacturing towns. Geneva is the home of a number of several United Nations agencies and other international organizations that selected Switzerland as a neutral country with high standards of education and international links.

The Swiss tourist industry is based on winter skiing and summer sightseeing and walking holidays in the Alps that provide a year-round demand for accommodations. Tourists are encouraged to travel in the country and visit the central cities by cheap rail passes. The relatively high prices to foreigners, however, result from Switzerland's low inflation and rising currency exchange rate. Larger numbers of tourists in the 1980s declined to 13 million in 1990 and 10 million in 1996. Winter sports remain very important to the economies of many alpine resorts, and there has been investment in new accommodations and ski runs (see Figure 7.9). The need to provide ski runs for competitions, however, led to the construction of features that are unsightly after snowmelt in summer.

The Austrian economy has a growing tourist sector based in the mountains and also in the towns such as Salzburg and Vienna that are associated with classical music. By the mid-1990s, its income from tourism exceeded that of Switzerland, with over 17 million visitors in 1996. Vienna was the center of the Austro-Hungarian Empire in the 1800s. Now the capital of a much smaller country, it is a center of government functions and the main farming region (see Figure 7.20). It has a growing Danube trade assisted by a new canal link through Germany between the Rhine and Danube Rivers, and the opening of trade eastward since the end of the Cold War. Manufacturing industries occur in Vienna and other parts of Austria near small deposits of metal ores, especially in towns along the southern flank of the Alps, such as Graz and Klagenfurt.

## MEDITERRANEAN EUROPE

The countries of Mediterranean Europe are the poorest in the formerly noncommunist Europe. Italy, one of the six original members of the EC, has an overall GNP per head equaling that of the U.K. The country has two parts—the richer north that is an extension of industrial Western Europe and the poorer south that is akin to the rest of Mediterranean Europe. Spain's GNP per capita is almost two-thirds, and Greece's and Portugal's are less than one-half, of Italy's. Portugal, Spain, and Greece are now also EU members, and their economies are improving with massive assistance from the European Union and access to EU markets.

The four main countries of Mediterranean Europe remained as peasant-farming countries with feudal or fragmented types of social organization while northern Europe industrialized in the 1800s. Industrialization began in the late 1800s in northern Italy and in the Catalonian region around Barcelona in northeastern Spain. Most modernization occurred after World War II and the incorporation of the Mediterranean countries in the EU.

### Countries

Mediterranean Europe consists of four large countries—**Portugal, Spain, Italy,** and **Greece**—together with the small independent countries of Andorra, Monaco, Vatican City, San Marino, and Malta (Figure 7.36). Gibralter remains a British colony, although discussions between the U.K. and Spain continue over its future. Portugal is not strictly a "Mediterranean" country, since its coast faces the Atlantic Ocean, but it is part of the Iberian Peninsula with Spain and its history is closely linked to that of its neighbor. The southern coast of France also fronts the Mediterranean Sea and shares many of the physical and locational characteristics common to Spain and Italy. Its position within France, however, resulted in greater well-being for its inhabitants compared with other Mediterranean countries.

All the major Mediterranean countries played important but different roles in the early generation of the cultural and economic foundation that made Europe prominent in the

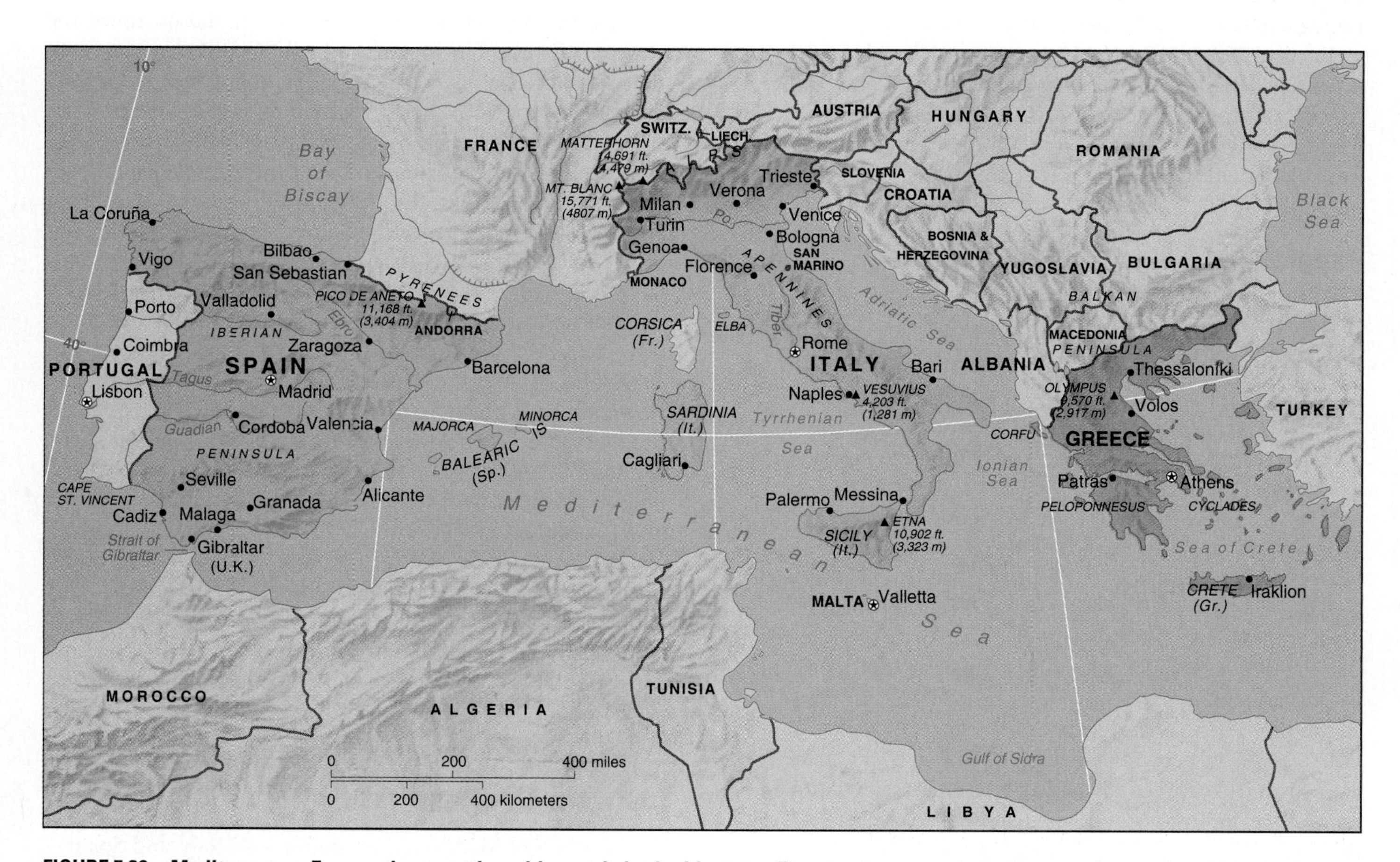


**FIGURE 7.36 Mediterranean Europe: the countries, cities, and physical features.** The countries occupy peninsulas extending southward into the Mediterranean Sea.

expansion of the modern world economic system. Greek history looks back nearly 3,000 years to its short-lived leadership of the Western world under Alexander the Great. Greek ideas from that time provided a basis for many later developments in Europe, including the medieval Renaissance. Italians look back not only to the Roman Empire and its dominance of the whole Mediterranean region in succession to the Greeks but also to its medieval roles in spreading the influence of Roman Catholicism, generating trade with Asia, and sponsoring the intellectual Renaissance.

Portugal and Spain, once freed from occupation by Muslim Moors in the 1400s, led the rest of Europe in explorations of Africa, Asia, and the New Worlds of the Americas. The pope of the time divided the world between them by the Treaty of Tordesillas (1494) that designated a line of longitude 370 leagues (approximately 1,500 km, 1,000 mi.) west of the Azores as the eastern boundary of Spanish influence and the western boundary of Portuguese influence. Subsequently, Portugal colonized the coast of what became modern Brazil and parts of Africa and Asia, while Spain established its rule in the rest of Latin America and some Pacific Ocean islands. Both countries lost their Latin American colonies in the early 1800s, although they maintained African colonies into the 1970s (see Chapter 3). The Portuguese still fear that their larger neighbor Spain—which has four times as many people—will dominate them again through close EU relations. Spanish investments in Portugal rose from 400 firms in 1989 to 3,000 by 1997.

In the 1900s, differences grew among the Mediterranean countries. Italy became the most industrialized and Spain followed, while Portugal and Greece still rely on agriculture, fishing, merchant marines, and tourism for much of their overseas income.

## People

The distribution of population in the Mediterranean countries of Europe reflects the hilly and mountainous nature of the terrain (see Figure 7.14). Most people live in the lower parts of major river valleys and along the coasts. The Po River valley of northern Italy is the largest area of high population density. The Alps, Greek mountains, and Pyrenees have very low densities.

Populations in the Mediterranean countries changed from a period of rapid increase and relatively high fertility and birth rates up to the mid-1900s to those with the world's slowest levels of natural increase and fertility in the 1990s. Total fertility rates dropped from nearly 3 in 1965 to nil in the late 1990s. Projections suggest that the total population of this subregion, some 117 million in 1998, will fall to 113 million by 2025. With levels of life expectancy rising to 80 years, aging is becoming a major issue (see Figure 7.16). Greece and Portugal are likely to retain populations of around 10 million, while Italy and Spain are expected to decrease in population total from 58 and 40 million, respectively, to 55 and 39 million by 2025. Although three countries—Italy, Portugal, and Spain—are predominantly Roman Catholic in religion, that church's opposition to birth control is clearly having little effect. By 1998, Italy and Spain had the lowest total fertility (1.2) in the world. Population growth was only by immigration. In Italy, few babies are born outside marriage, but the access of women to careers, young people continuing to live with parents, and the end of pressures to have children tend to defer marriage and reduce the numbers of children.

Apart from Spain, the Mediterranean countries are less urbanized than those of other parts of western Europe. In 1998, Portugal had only under 50 percent of its population living in towns. Greece and Italy had two-thirds of their populations in urban centers, and Spain 78 percent. **Athens**, Greece, and **Lisbon**, Portugal, are the predominant cities in government, port, and industrial activities in their countries. Italy's capital city of **Rome** (Figure 7.37) is now exceeded in population by the **Milan** metropolitan area, **Naples**, and **Turin**. In Spain, **Madrid** and **Barcelona** are the largest cities, while **Seville** and **Valencia** are regional centers.

Although the major Mediterranean countries have long histories of links with each other, the peninsular nature of Mediterranean land areas encouraged distinctiveness. Spain and Portugal occupy the Iberian Peninsula. The first Iberian peoples came from North Africa but were overrun by the Celts and then influenced by various Mediterranean groups, including Greeks, Phoenicians, and Romans. In A.D. 711, Muslim Moors from North Africa invaded and occupied the whole peninsula. They were gradually pushed back, leaving southern Portugal in 1279 and Spain in 1492. In Spain, minority groups with their own languages include the Catalans of the northeast (17% of the Spanish population), the Basques at the western end of the Pyrenees (and partly in southwest France), and the Galicians of the northwest. These groups often oppose the Madrid government and the Basques have fought for a separate country.

In Italy, there are only tiny areas in the northeast (German) and northwest (French) where Italian is not the dominant language. The modern differences between northern and southern Italy grew out of the medieval occupation of the region south of Rome by Muslims and its later domination by feudal systems under French and Spanish kings. The northern city states maintained trading links and developed industrial output at an earlier date. It was not until late in the 1800s that Italy achieved unification.

Greece lay outside the medieval developments in western Europe and was then submerged within the Ottoman Empire. Nationalistic efforts brought it freedom in the late 1800s. During World War II, it suffered from German occupation, and this was followed by a civil war that further delayed its modernization.

## Economic Development

The two largest Mediterranean countries—Italy and Spain—experienced major economic growth between 1965 and the 1990s. Greece and Portugal remained the poorest EU countries.

## Italy

Italy's GDP is greater than the total of the other Mediterranean Europe countries. The country's economic growth resulted mainly from the revival of its northern-based industries after World War II and the development of high-tech industries. The Po River valley between the Alps and the Apennines is the largest center of manufacturing in Mediterranean Europe. It is based around a series of medieval towns, from Turin and Milan in the west to Venice in the east, that grew up along the margins of what became a major area of commercial agriculture (see Figure 7.20). Hydroelectricity from the Alps partly makes up for a lack of coal and oil, although Italy is also a major customer for Algerian, Libyan, and Russian oil and natural gas. The government-supported Fiat automobile factory is still the main employer in Turin, although automation reduced the labor required. Within Europe, Fiat of Italy has been particularly active in manufacturing its vehicles in the countries of the former Soviet bloc (see Chapter 8). It was responsible for building the Soviet Union's largest truck plant at Minsk (now in Belarus) and its largest car plant in Togliatti, beside the Volga River. Genoa on the Mediterranean coast is a major port and had steel and shipbuilding industries, although these lost their market to Eastern Asia in the 1970s. Milan is the largest city in northern Italy, a major center of financial and other service industries as well as a producer of a diverse group of manufactured goods including tractors, domestic electronic goods, china, shoes, and pharmaceuticals. Milan and its surrounding towns produce nearly one-third of Italian GDP and form one of the major growth areas of Europe. Venice, at the mouth of the Po River, is the center of industries based on imported oil.

Southern Italy still provides a contrast with the north. In the late 1990s, southern Italy's unemployment was over 20 percent, while the north's was 7 percent (compared to the EU average of 10%). While incomes north of Rome averaged over $16,000 in the mid-1990s, those to the south of Rome averaged under $14,000. To overcome the north-south contrasts, Italy instigated a regional policy that poured funds into the industrialization of the southern part of the country. Despite all the Italian and EU investment, the gap between the regions widened. Better roads eased transportation and new industries provided employment in some parts of southern Italy but still could not pull the area forward as fast as the growth in the north. Not only do the southern Italians compare their lot unfavorably to that of the northerners, but the latter resent the taxation that provides funds for building infrastructure in southern Italy fueling political parties that want the north to secede from the rest of Italy. Stories of government corruption and the Mafia organization of crime in the south fan the flames of this issue. Italy continues to suffer from changing governments, but this has not prevented rapid and effective economic growth in the north. Movement of people northward for work, however, almost ceased by the late 1990s because of high living costs in the north and grant programs that supported those living in the south. The shortage of workers in both the north and the south stimulated an increase in immigration from developing countries, leading to a total of nearly 2 million foreign workers (legal and illegal) in the country.

**FIGURE 7.37 Mediterranean Europe: Rome, Italy.** A space shuttle view with Rome at the center and the Tiber River flowing to the Tyrrhenian Sea with a delta flanked by sandbars. The international airport is near the river mouth. Circular lakes are old volcanic features. Rome is Italy's capital, includes the Vatican City, and is a center of tourism, film-making, and manufacturing.

NASA

In 1996, 33 million tourists visited Italy. Most went to the historic centers of Rome, Florence, Venice, and Pompeii (near Naples) (Figure 7.38). The east coast beaches, Sicily, many smaller towns with their art treasures, and the winter sports and lake resorts in the Alps attracted many others.

## Spain

The economic contrasts in Spain are between the poorer rural areas and the wealthier centers of Madrid, the national capital, the industrial Catalonia region of the northeast, and the Basque region inland of Bilbao (Figure 7.39). Most Spanish farmers still work as tenants on large estates; their crop yields are low. The summer sunshine that brings tourists to the coasts also brings water shortages, increasing from the north to the south of Spain as the temperatures rise and rainfall totals decrease.

(a)

(b)

(c)

(d)

**FIGURE 7.38 Mediterranean Europe: Italian tourist attractions.** (a) Pompeii, Naples, is a Roman city that was buried by an eruption from Mount Vesuvius and recently excavated by archaeologists. (b) Central Rome preserves ruins of Roman times, including the Colosseum, the Forum, and temples. (c) Florence has its cathedrals and art galleries by the Arno River. (d) Venice is a medieval port city that accumulated great wealth on its unique site in a lagoon, but is liable to flooding.



The main industrial area is around Barcelona, where established textile industries have been diversified by automobile production and high-tech industries (Figure 7.40). The Bilbao area on the northern coast has metal-based and machine-making industries; farther inland, Pamplona has a Volkswagen car plant. Spain benefited from the influx of foreign investment following its admission to the EU in 1986, taking advantage of the low-cost labor. The flow of funds slowed in the 1990s as labor costs rose, leaving Spain with high unemployment rates. The Barcelona and Bilbao areas gained most from the initial investment, but large areas of the northwest and the interior south remain with little recent economic development.

Spain attracted over 40 million tourists in 1996, mainly to the beaches of the southern coasts and the Balearic Islands. Although their numbers were many, most came on low-cost package holidays. The construction of new tourist facilities along the Mediterranean coast led to the urbanization of long stretches.

## Portugal

Portugal experienced less industrialization than Italy or Spain, and over 60 percent of its population is still rural. Lis-

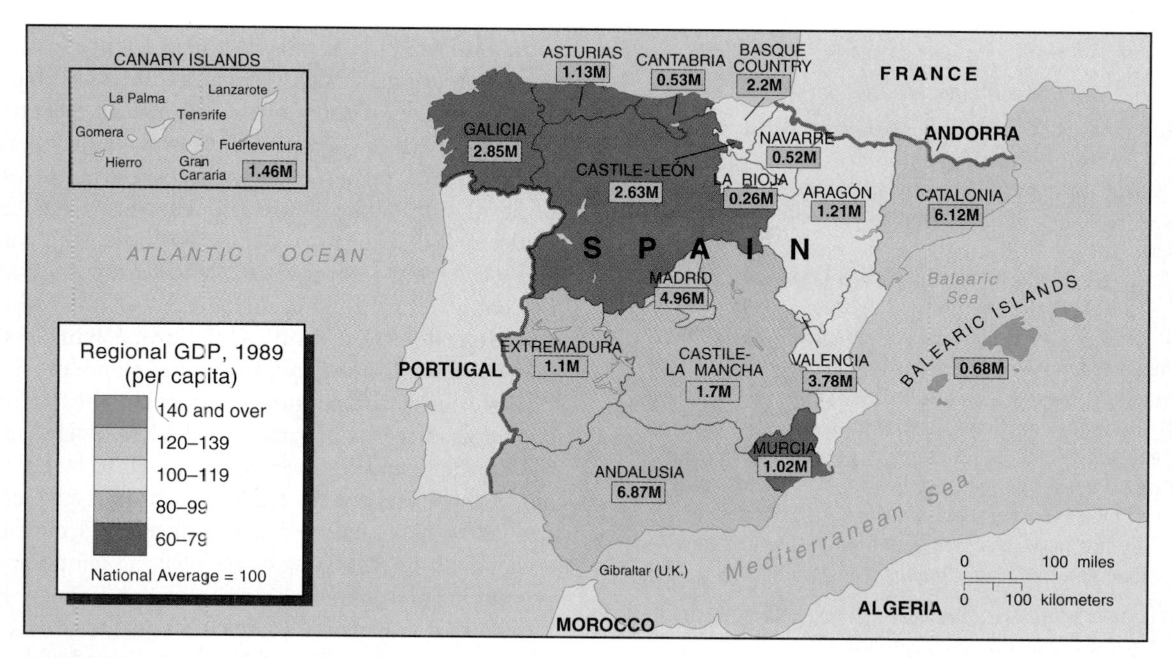


**FIGURE 7.39 Mediterranean Europe: the regional geography of Spain.** The regional GDP per capita and numbers of people in millions (M).

bon remains the main center of population, government, and service jobs. In the late 1990s, the largest industrial development was the joint Ford-Volkswagen project to manufacture multipurpose vehicles at Palmela near Lisbon. When full production is achieved, it will add 15 percent to Portugal's export income. Tourism is a major industry, with over 9 million visitors in 1996.

## Greece

Greece consists of a hilly peninsula and a series of islands including Crete, but only small areas have soils or slopes that support commercial agriculture. Agriculture, however, provides income for most people, who grow olives, grapes, citrus fruits, and figs for export together with cereal crops and

**FIGURE 7.40 Mediterranean Europe: Barcelona, Spain.** The waterfront and city center, including major banks.

**RECAP 7D: Northern, Alpine, and Mediterranean Europe**

Northern Europe includes Denmark, Finland, Iceland, Norway, and Sweden. These are affluent countries with small populations. Their economies were based on primary products from farm, mine, forest, and ocean, but now income is from manufacturing and services.

Alpine Europe includes two countries, Austria and Switzerland, that have most of their territory in the Alps. They are small but affluent countries with high-quality manufactured products and major involvements in financial and tourist industries.

Mediterranean Europe comprises the countries of three peninsulas—Portugal and Spain, Italy, and Greece. Apart from northern Italy, which is as prosperous as parts of Western Europe, these countries lagged far behind their more affluent neighbors until their incorporation in the European Community in the 1980s. They are now developing manufacturing and service industries, and all rely heavily on income from tourism.

7D.1 What factors account for the economic well-being of the Northern and Alpine European countries?

7D.2 Can you suggest solutions to the environmental problems facing the European countries that border on the Mediterranean Sea?

7D.3 Are there geographic reasons that might explain the settlement of Scandinavian migrants in the Great Lakes region of the United States?

**Key Places:**

| | | |
|---|---|---|
| Denmark | Austria | Italy |
| *Copenhagen* | *Vienna* | *Rome* |
| Finland | Switzerland | *Milan* |
| *Helsinki* | *Zürich* | *Naples* |
| Norway | *Geneva* | *Turin* |
| *Oslo* | *Basel* | Spain |
| Sweden | *Bern* | *Madrid* |
| *Stockholm* | Greece | *Barcelona* |
| Greenland | *Athens* | *Seville* |
| Faeroe Islands | Portugal | *Valencia* |
| Iceland | *Lisbon* | |

industrial crops such as tobacco and cotton. Athens has over one-third of Greece's population and remains its main industrial center, capital city, and tourist attraction despite its air pollution. The international airport and marine links to the many Greek islands make Athens the major transportation hub within Greece. The many historic relics, islands, and sunny climate brought 9 million tourists to Greece in 1996.

## EASTERN EUROPE

Eastern Europe—Poland, the Czech and Slovak Republics, and the Baltic States (Figure 7.41)—comprises countries that were part of the Soviet bloc during the Cold War years, 1945–1991, but are now moving along the transition path from state to market capitalism. This subregion has good links to Germany and Scandinavia and its countries eagerly took the 1991 opportunity of independence from Soviet military occupation and involvement in the Soviet bloc economic system. Although some of the countries prefer to be labeled as being part of "Central Europe," rather than "Eastern Europe," the former term includes Germany, Austria, and some of the Balkan countries. The term "Eastern Europe" could include Belarus, Ukraine, Moldova, and part of Russia and may give the impression of eastward links at a time when the countries in the subregion are looking westward. It is used here for the countries of Europe lying between Germany and the Commonwealth of Independent States.

Historically, the countries occupying the North European Plain and adjacent hilly areas, particularly Poland and the three small Baltic countries, were greatly affected by tensions arising out of the expansionary tendencies of Prussia, Germany, Austria, and Russia. For centuries, invading armies crossed the lowlands from east or west, establishing temporary control of this subregion. Following a period when Sweden, Poland, and Lithuania dominated the area, this territory was partitioned among Prussia, Austria, and Russia at the end of the 1700s.

After World War I, independent states were carved out of the German, Russian, and Austrian empires. They were only independent for two decades of hesitant democracy before being invaded by Nazi Germany and the Soviet Union. Germany invaded Czechoslovakia in 1938 and Poland in 1939. Soviet forces invaded Estonia, Latvia, and Lithuania in 1939, incorporating them into the Soviet Union. The German armies combined with the Soviet forces to invade Poland, and then turned on them and invaded the Soviet Union. By 1941, they occupied the whole subregion, but then the Soviet Union forces recaptured their own territory and liberated Eastern Europe. In the resultant peace agreement, Poland's territory was shifted westward and many Germans were replaced by Poles moved from the eastern lands taken over by the Soviet Union. That area is now part of Belarus. The Soviet armed forces stayed in these countries, which were soon ruled by puppet communist governments as part of the Soviet bloc, both militarily and economically.

Since the end of Soviet control in 1991, the countries of Eastern Europe experienced economic crises as they reoriented themselves from incorporation in the Soviet economic system to the world economic system (see Chapter 8). They returned to GDP growth in the mid-1990s, although they lag behind the rest of Europe, and only Poland exceeded its 1989 GDP by 1997. In the late 1990s, Poland, Estonia, the Czech Republic, and Slovakia agreed to forms of association with the European Union, and steps were being taken to admit them. These countries are close to "functioning market economies," having lower government subsidies and better balanced budgets than some EU members. To reduce the differences in income, however, the countries of Eastern Europe need to have more rapid economic growth than the EU countries. Investment in infrastructure and export industries from outside sources is slow. Although exports to the European Union countries rose from 25 to 50 percent of the total exports from Eastern Europe, that total fell. The EU, and particularly Ger-

**FIGURE 7.41 Eastern Europe: countries, cities, and relief features.** This subregion has been a buffer zone between the major powers of Germany and Russia for centuries, and the countries are in transition from a period of domination by the Soviet Union that ended in 1991 to subsequent orientation westward to the rest of Europe.

many, is very conscious of the hopes of peoples in this subregion. There is also a perceived need to control the smuggling of people, drugs, and weapons through Eastern Europe into Western Europe. For example, some 40,000 people from 74 countries, mainly Southern Asians, who were seeking a better life, were intercepted at the Czech-German border in 1994.

## Countries

Eastern Europe consists of the **Czech** and **Slovak Republics**, **Poland**, the three Baltic states of **Estonia**, **Latvia**, **Lithuania**, and **Kaliningrad**, which is still part of the Russian Federation.

Although "voluntarily" incorporated in the Soviet Union in 1940 as republics within the Union of Soviet Socialist Republics, the Baltic countries regarded that as a temporary military occupation of their land in the same way that the peoples of Poland and Czechoslovakia saw their occupation. Deportations of tens of thousands of Baltic people and other enforced policies resulted in armed resistance that was put down and the nationalistic symbols were removed from public view. The Baltic countries' responses to their renewed independence since 1991 vary, but initial threats of Russian intrusion soon receded.

Between Lithuania and Poland, the Russian enclave of Kaliningrad was developed as a naval port, and the German population expelled. This enclave, once claimed by Lithuania, now has a 90 percent Russian population and prospects of being developed as a free trade zone.

Poland and the Czech Republic are among the most advanced along the transition from Soviet domination to independence and involvement in the world economy. In 1993, the Slovak Republic separated from the Czech Republic on ethnic grounds. It has been slower to move away from its command economy of the 1980s and is likely to be a late entrant to the EU.

## People

The distribution of population in Eastern Europe is more even than in other European subregions (see Figure 7.14). The main concentrations of people are in the urban-industrial areas of the Czech and Slovak Republics and southern Poland. Poland has more people (39 million) than the rest of the subregion together, with the Czech Republic having 10 million, Slovak Republic 5 million, and the Baltic states between 1.4 and 3.7 million each in 1998.

The populations of the countries in Eastern Europe are scarcely growing at all. It is estimated that the total population, which was 62 million in 1998, will be only 63 million by 2025. This slow growth is related to very low fertility rates of under 1.6 children born to each woman. Birth rates and death rates both fell to low levels, and some countries have death rates that are above birth rates, giving natural decrease (see Figure 7.15). In 1998, most countries had negative natural population change rates of –0.1 to –0.4 percent; only Poland and the Slovak Republic experienced a little natural growth. The age-sex graph (see Figure 7.16) shows that within overall low rates of increase in the later 1900s, Poland experienced alternate periods of more and fewer births.

Poland has the lowest level of urbanization (62%); other countries have 68 to 75 percent of their populations living in towns. The post–World War II industrialization within the Soviet bloc led to rapid urbanization of previously rural countries. No cities are of great size. In 1996, **Prague** (Czech Republic), **Warsaw** (Poland), **Riga** (Latvia) and **Katowice** (Poland), were the only cities of a million or more people. Tallinn (Estonia), Vilnius (Lithuania), and Bratislava (Slovak Republic) are the other capital cities. The lack of major world cities in Eastern Europe arises from the small size of the countries, the history of warfare and destruction, and the focus on manufacturing growth linked to medium-sized town expansion within the Soviet bloc.

The peoples of Eastern Europe and their languages are all of the Slavic group except for Estonia, which has a language belonging to the Finnish group. Major ethnic issues continue to make governing difficult in the Baltic countries, where the Russians resettled there appealed to Russia for political support until conditions improved more rapidly in the Baltic countries than in Russia. Slovakia still has ethnic tensions as the 600,000 Hungarians living in the southern part of the country campaign for a degree of autonomy.

## Economic Development

The traditional precommunist economies of the Eastern European countries were based on agricultural production. Arable farming on the North European Plain and forests on the sandier soils of the north and uplands to the south were the maintypes of land use (see Figure 7.20). The Soviet control of the economies of these countries after 1945 led to state investment being concentrated in the industrialization of areas linked to coalfields. The exchanges of food, raw materials, and finished or partly finished manufactures with the other countries in the Soviet bloc up to 1991 formed a major part of their economic activities, often subsidized by cheap oil and natural gas from the Soviet Union.

After 1991, such exchanges were greatly reduced, and countries outside Russia had to seek other markets. Countries in Eastern Europe experienced a major drop in income. Their low incomes are reflected in the low levels of consumer goods ownership compared to the rest of Europe (see Figure 7.19). Growing trade with Germany in particular caused incomes to rise gradually. By the late 1990s, the private sector produced over half the GDP of these countries. The process of transition from communist state capitalism is considered more fully in Chapter 8.

Little data is available about economic change during the communist era. In most countries, the HDI rank for the mid-1990s was higher (i.e., better) than the GNP rank, while GDI ranks were also relatively high. This reflects the impact of state-supplied health and education services in the Soviet bloc on improving the quality of life. It did not, however, satisfy people's demands for better incomes and consumer goods. In the late 1990s, most of the countries had GDPs per head that were less than one-third of those in other parts of Europe.

Few Eastern European countries have a developed service sector, although by 1995 all had over 50 percent of GDP from that source. The industrial sector had produced over 50 percent of GDP, but by 1995 this was reduced to 30 to 40 percent. Farming, fishing, mining, and forestry made up around 10 percent of GDP in the Baltic countries, but only 6 percent in Poland and the Czech and Slovak Republics.

### Poland

Poland's most productive farming area is south of Warsaw, where the poor glacial soils give way to loess and wheat and grain crops yield well (Figure 7.42). North of this the soils are stony or dense clays, much of the land is woodland or pasture, and the main crops grown are rye and potatoes. Although Polish farms were not collectivized, investment in modern machinery and other inputs remained low as the communist governments made the development of manufacturing industries their first priority. Productivity is poor.

At the end of World War II, the new borders of Poland included the former German industrial area of Silesia based on a coalfield with other local minerals. This area formed the basis of Polish industrial expansion in and around the towns of Wroclaw, Katowice, and Kraków and their steel and other metal industries (Figure 7.43). Although previously protected from external competition, and despite reducing its late-1990s labor force to half of what it was in the mid-1980s,

**FIGURE 7.42 Eastern Europe: farming in Poland.** The region around Pabianice, southwest of Warsaw, has some of the most fertile soils in Europe, but productivity is low. Mechanization is limited and traditional farming methods are still widely used.



fears remained that the Polish steel industry could not compete in world markets. Closures occurred, and there were few orders for ships from the port of Gdansk-Gdynia. The other main port of Szczecin, however, restructured its shipbuilding industry to specialize in smaller container ships, for which it had full order books in the late 1990s.

## Czech Republic

The Czech Republic's economy centers on the Prague area, which remains a focus of economic activity and technology, closely linked to its western neighbor, Germany. Several small industrial towns in the Bohemian region around Prague produce precision goods. Moravia in the northeast is

**FIGURE 7.43 Eastern Europe: Polish manufacturing.** The Huta Katowice steelmill was the last to be constructed under the communist government, yet it struggles to compete with steelmakers in western Europe.

close to Polish Silesia and was industrialized by the communist government, producing steel, metal goods, and chemicals. In the mid-1990s, the Czech Republic underwent privatization, but its intentions to scale down weapons manufacture were delayed to ease the impacts of change. A major change was the take-over of the Skoda car manufacturer by Volkswagen of Germany, seeking low-cost labor for its cheaper models and bringing better quality control.

### Slovak Republic

Slovakia was always the poorest province of Czechoslovakia, and since declaring independence its government slowed economic reform. Like Moravia, Slovakia was industrialized by the communist government, producing semifinished products for the Soviet bloc and foreign purchasers. These facilities are obsolete and polluting, and their markets have been largely lost. By the late 1990s, the Slovak Republic's arms industry sales fell to less than 10 percent of its 1980s level—when Czechoslovakia was the world's fifth largest arms exporter. The Slovak Republic also faces tensions with its southern neighbor, Hungary, that are not only over its minority Hungarian population but also over the Gaborkovo Dam on the Danube. The dam diverts water across Slovak territory to be used for generating electricity. Hungary petitioned for a realignment of the border so that it could share the electricity.

### Baltic States

The economies of the Baltic countries are limited by their small size and meager natural resources. They now seek new overseas markets after losing access to Russian raw materials and the markets for which many of their production facilities were designed. Estonia's industries center on the mining of oil shale for electricity and the production of cement for export through the reopened port of Kunde. By the late 1990s, Estonia had privatized 80 percent of its former state property and enterprises and had developed economic links with Finland and Sweden. An example of the impact of ending the Soviet bloc arrangements was the 1988 building of a factory in Tallinn to make a million basketballs a year. After the necessary machinery had been imported from Japan, the breakup of the Soviet bloc deprived the factory of its guaranteed market and caused it to close and be sold.

Latvia has a greater range of industries than the other Baltic countries. They were also developed for Soviet bloc markets using Soviet raw materials and include scientific equipment, machinery, transport equipment (electric rail cars), and consumer goods. Lithuania now faces major economic problems since the land is poor, there is a small industrial base, and it is difficult to find new markets for its products.

The Russian enclave around Kaliningrad remains a military fortress largely populated by Russians. Its attempts to become an entrepôt trade center and free trade zone were unsuccessful. After 50 years of being closed to external influences, it is proving difficult to mix the military and commercial functions. The control of major developments in Kaliningrad is still referred to Moscow, and the enclave has much out-of-date and poorly maintained infrastructure, including its transportation links to other parts of Eastern Europe. Although it is a warm-water port on the Baltic Sea, cargo turnover fell in the early 1990s, and potential investors await the withdrawal of the Russian military.

## BALKAN EUROPE

The Balkan countries occupy a broad peninsula in southeastern Europe between the Adriatic and the Black Seas (Figure 7.44). This area is a long-standing zone of international tension on the human fault-line between Christian and Muslim, and between Catholic and Orthodox, influences. Strong nationalist movements survived the disruptions of two world wars and the Cold War occupation by the Soviet Union. Only partly subdued during the communist era from 1945 to 1990, many local tensions surfaced again. Strife began with the early 1990s war in Bosnia-Herzegovina and continued in the late 1990s in Kosovo, part of Serbia-Montenegro.

Before World War I (1914–1918), the Austro-Hungarian Empire governed the north and the Turkish Ottoman Empire the south. There were some incursions from Russia to the east. The new countries carved from the remains of these empires in 1919 were submerged in the German expansion of its territory from 1938. After World War II, the countries had a short independent existence before most became part of the Soviet bloc, or had separate experiences of communist rule (Albania, Yugoslavia, and, to an extent, Romania).

### Countries

The Balkan countries are **Albania**, **Bulgaria**, **Hungary**, **Romania**, and the group of countries that once comprised Yugoslavia (**Bosnia-Herzegovina**, **Croatia**, **Macedonia**, **Slovenia** and the remant Yugoslavia of **Serbia-Montenegro**). They adapted to the Soviet hegemony in different ways. Yugoslavia opted out of Soviet dominance in the 1950s, developing economic links with Germany and Italy. Albania adopted the Chinese form of communism and remained isolated and poor under a repressive dictatorship until the 1990s. Hungary's abortive revolution against Soviet tanks in 1956 was followed by a more subtle but real adjustment to free market principles during the 1980s. Bulgaria's loyalty as the most faithful ally of the Soviet Union was based on continuing gratitude for the Russian expulsion of the Turks in the late 1800s. During the 1970s and 1980s, Romania appeared to be a maverick within the Soviet bloc, frequently expressing its independence of Moscow and gaining sympathy from the West, but it adopted this pose to deflect attention from the repressive rule of Nicolae Ceaucescu.

None of the countries of Balkan Europe had more than just over 11 million people in 1998, with the exception of

**FIGURE 7.44 Balkan Europe: countries, cities, and relief features.** The countries of this subregion, particularly Albania, Romania, and the former Yugoslavia, had less control by Moscow during the Cold War but now have major problems maintaining civil peace that are involving other European countries. In Yugoslavia, K = Kosovo, M = Montenegro, S = Serbia, V = Vojvodina.

Romania, which had 23 million. The former Yugoslavia had a similar total population as Romania, but in 1991 it broke into five separate states each with between 2 and 10 million people. All Balkan countries contain significant minorities and are subject to internal ethnic tensions.

## People

The Danube River valley has the main population centers in Balkan Europe (see Figure 7.14), with the cities of Belgrade (Serbia), Budapest (Hungary), and Bucharest (Romania) along its length. Farming areas outside these cities give way to the lowest densities of people in the highlands of Bosnia, Romania, and Bulgaria.

### Population Decline

Population in Balkan Europe is declining. It is expected that the 1998 total population of 67 million people will be down to 66 million by 2025. Population change rates in the late 1990s ranged from small increases (0.1 to 1.2 percent per year) to decreases in Hungary, Romania, and Bulgaria. Both birth rates and death rates are low (see Figure 7.15). Low birth rates reflect a decline since 1970 in total fertility from over 2 to around 1.5 per woman, but from around 5 to 2 in Albania and Macedonia. The age-sex graph for Romania (see Figure 7.16) demonstrates the fall in birth rates during World War II and the declining effects of Romanian efforts to encourage births within a repressive society.

### Increasing Urbanization

The degree of urbanization was low in Balkan Europe until the 1980s and the early 1990s. Rapid urbanization in Croatia, Macedonia, and Slovenia (under 50% to over 60%) between 1980 and 1995 reflect the impact of civil wars and a decline in rural living. In this period, Romania increased its urbanization to 55 percent, Hungary to 65 percent, and Bulgaria to 68 percent. Albania and Bosnia-Herzegovina are under 40 percent. **Sofia** (Bulgaria), **Budapest** (Hungary), **Bucharest** (Romania, Figure 7.45), and **Belgrade** (Serbia-Montenegro) were the only cities with populations of more than a million people in 1996. They contain between one-tenth and one-fifth of their country's total population.

### Ethnic Tensions

The tensions among ethnic groups within the fragmented political environment of the region are a particularly obvious feature of these countries. The early 1990s civil war in Bosnia and the late 1990s Kosovo strife could multiply in this region.

The former Yugoslavia was a mixture of varied cultural groups with Roman Catholic (Slovenes, Croats), Orthodox (Serbs), and Muslim (Bosnia, Kosovo) allegiances that were the basis of historic antagonisms and maltreatment of rivals by political groups operating under each of the religious heads. The Serbs provided first the royalty and later most of the communist leaders who unified and held this mixture together from 1919 to 1991. Other parts of the country resented Serbian control, and in 1991, Yugoslavia broke into five independent countries. Slovenia led the way and was soon followed by Croatia, Bosnia-Herzegovina, and Macedonia. Serbia-Montenegro, a new country with nearly 11 million inhabitants, still calls itself "Yugoslavia." It contains 6 million Serbs, but another 2 million Serbs live in Bosnia. Serbs also formed a minority in Croatia and set up their own area, calling it Krajina. In Bosnia, the Muslim group makes up 40 percent of the population and dominates the government. Within Serbia are minority groups of Hungarians in

## CROATIA

LIVING IN . . .

In the 1990s, Croatia became involved in a war that affected many aspects of everyday life for its people. Following difficult years during and after World War II, Croatia was part of the Yugoslavia that asserted its independence from the Soviet bloc and built economic links with the richer countries to the north and west. Croatia, with its fertile plains and growing industries, benefited from economic growth. Its beautiful coasts and historic towns attracted tourists from northern Europe. During the 1980s, however, stresses built within Yugoslavia that led to Slovenia, Croatia, Bosnia-Herzegovina, and Macedonia declaring independence in 1991.

Fierce fighting followed these declarations of independence as the Yugoslav military forces and Serb militias bombarded some 300 Croatian towns and villages, destroying 60 of them. Towns such as Osijek, Karlovac, Zadar, and Dubrovnik were bombed again and again. Croats were "ethnically cleansed" from about one-third of their country (diagonal shading on map) by their Serb neighbors with the help of paramilitary forces from the remnant of Yugoslavia dominated by the Serbs. Refugee Croats from these areas swelled the population of other parts of Croatia.

The hostile minority of Serbs made up 10 percent of the Croatian population. Their presence was a historic legacy following the flight of Serbs into Croat lands when the Turks invaded their homeland from the late 1300s onward. The refugee Serbs were allowed to settle and were given the task of defending the Austro-Hungarian Empire from further Turkish attacks. In return, they were exempt from taxes. These Serbs remained under direct rule from Vienna from the 1500s until the late 1800s, when their region was rejoined to Croatia.

After Croatian independence from Yugoslavia was established in 1991, the Serbs living in the border areas of Croatia proclaimed their independence from Croatia as the new state of "Krajina," a name derived from a word meaning "military frontier." This region included the railroad line from Zagreb to the Dalmatian coast, which was cut. It was also dangerously close to the main coastal road, which Serb attacks made periodically impassable. Even in peacetime, Croatia had a strategically impossible shape, as the map shows, and civil war made it vulnerable to dismemberment. In mid-1995, however, the Croatian armies regained this lost territory and most of its population of around 200,000 Serbs fled into Serbian Bosnia and then eastward as the Croat-Bosnian forces took territory from them. The impacts of these disturbances included widespread destruction, the heavy financial burdens of military action, large areas made useless by land mines, and the presence of half a million refugees, including internally displaced Croats and people escaping from Bosnia.

Map of Croatia and Bosnia.

the north (Vojvodina) and Albanians in Kosovo near the Albanian border—a place that reminds Serbs of their 1300s defeat by the Turks. Many in Serbia still call their country "Yugoslavia" to imply that their wider role is not over. Serbians continue to use Cyrillic script (like Russian), whereas people in other parts of the former Yugoslavia write their Serbo-Croat language in Roman script.

The term **"ethnic cleansing"** was first used following Serbian attempts to dominate other ethnic groups within Bosnia, in other countries of the former Yugoslavia, and within Serbia-Montenegro (Kosovo and Vojvodina). In the 1990s, military conflict flared in Bosnia-Herzegovina, where Muslims, Serbs, and Croats all live. Along with having religious and written language differences, the Bosnian Serbs, who are mainly peasant farmers, were jealous of the better lifestyles of the Muslim town dwellers. The former Yugoslavia placed most of its armament factories in Bosnian Serb territory and thus gave the Bosnian Serbs easy access to the tools of war. By mid-1995, the attempts to reorganize the human geography of Bosnia—by such means as the United Nation's "safe

The city of Zagreb, capital of Croatia, showing the large central square and the Roman Catholic Cathedral.



Throughout the war in Bosnia and conflict inside Croatia, some parts of Croatia were less affected than others. Tourists returned to the northern Adriatic coastal resorts in Istria and spent foreign currency. In Zagreb, goods were freely available in the shops but at Western prices in a country where personal incomes are a fraction of those in the West. For many people survival remained difficult. The hospitals were overloaded with war casualties. Patients had to contribute to medical fees, making hospital treatment and medicines almost impossible to afford. People maimed physically or psychologically found it difficult to return to normal life. Schools functioned, but the refugee problem stretched them to the limits.

Despite the difficulties imposed by the war, the Croatian people fought to maintain their independence. Elections held since 1991 are agreed to have been fair in expressing the will of the people. All parties in the Croatian parliament are seen to be aware of past mistakes and to be working for the good of the people.

But Croatians were disappointed by the lack of support from the West. In the early 1990s, they looked to the European Union in particular for help and sympathy, hoping eventually to join that group of affluent countries. They felt betrayed by the failure of the rest of Europe and the United Nations to help them in what they saw as a just fight against a well-planned and executed conspiracy by the Serbs against their country and nation. They were sore that, although the Serbian leadership advertised this conspiracy well in advance and acted as an aggressor defying all internationally recognized norms of behavior, the international community did nothing to prevent the wars in Croatia and adjacent Bosnia. Only when the Serbs in Bosnia proved intractable over cease-fire agreements in 1995 was the Croatian army encouraged to take the initiative—a move that reversed the pattern of events and led to Croatian armies working with the Muslim Bosnian government troops to expel Serbs from large areas of Croatia and Bosnia.

In the late 1990s, cease-fire arrangements held and Croatia began to rebuild its economy. Uncertainty remained over the future of Serbs who fled, or were expelled, from the Eastern Slavonia area of Croatia and wished to return.

havens" for Muslims living among a majority of Bosnia Serbs, and the embargo on foreign arms supplies to the Muslims—failed to obtain peace between the aggressive Serbs and the Muslim-Croat alliance that resulted from their common defense against the Serbs.

In Croatia, the Serb groups that live along the Bosnian border provided a source of continuing disaffection to the government in Zagreb (see "Living in Croatia"), which was highlighted when the Serbs established the breakaway Krajina. In 1995, following the failure of the UN, EU, and NATO negotiations to stop the fighting in Bosnia and Croatia, the combination of Croatian and Bosnian Muslim forces swept the Serbs out of Croatia and northern Bosnia. A peace could then be brokered along the Dayton line that restored much land to Croatia and Muslim Bosnia and labelled Serb leaders as war criminals.

Other ethnic issues continue to raise problems in the subregion. The late 1990s efforts by the Serbian military to rid Kosovo of guerrilla forces also looked like ethnic cleansing when ethnic Albanian villagers were expelled from their

**FIGURE 7.45 Balkan Europe: Budapest, Hungary.** The Danube River separates the twin cities of Buda (left, with castle) and Pest (right, with the dome of the Parliament in the distance). Six bridges link the two towns, and the river is a major transportation artery.



homes, fleeing into the hills or across the border into Albania. A show of force by UN military aircraft and the introduction of UN ground observers failed to stem this aggression.

## Economic Development

Virtually no recent economic data exist for the countries of the former Yugoslavia, although it is clear that the civil war and accompanying sanctions reduced economic growth. Destruction of houses and commercial and industrial facilities was immense, and nearly all the lucrative tourist industry was lost. Data for other Balkan countries are still sparse, but better. Like the countries of Eastern Europe, the countries of Balkan Europe relied on agriculture (see Figure 7.20) until after 1945, when the Soviet Union encouraged them to industrialize.

### Former Yugoslavia

Slovenia is the wealthiest of the former Yugoslav countries. Its population is small (2 million), but its total GDP made up 20 percent of the former Yugoslavia's before 1990. Per head, Slovenia's GDP in the late 1990s was twice that of Hungary and up to ten times that of other Balkan countries. Slovenia's alpine environment attracts tourists and a number of sophisticated industries. Bosnia-Herzegovina's economy was destroyed during the fighting. Macedonia is the poorest of these countries, having a small population and being landlocked among unfriendly neighbors.

Serbia-Montenegro is the largest of the former Yugoslav countries and contains almost half their population. Its northern area is in the Danube River plains, which are farmed commercially and contain oil and natural gas reserves. The southern part of Serbia, like most of Bosnia, Croatia, and Macedonia, is hilly or mountainous and is the home of peasant farmers and towns that had growing manufacturing industries before the civil conflict. Serbia's military actions led to UN sanctions that hurt most of its people but did not move its leaders.

### Hungary

Of the other Balkan countries, Hungary's agricultural produce continues to lead the subregion despite widespread destruction in World War II and collectivization since. Food was produced for the Soviet bloc, and Hungary remains self-sufficient in grain crops and other local foods. Hungary has a small steel industry based on local coal and exports bauxite, the ore of aluminum. It began to privatize businesses in the 1980s and so is farther ahead than other Balkan countries in the process of economic transition. Between 1970 and 1992, the contribution of agriculture to the economy fell from 18 to 7 percent of GDP, while service industries rose from 37 to 63 percent—indicating the extent of economic modernization in Hungary.

### Romania and Bulgaria

Romania and Bulgaria are finding the transition toward democracy and free-market economies very difficult and remain poor. Both Romania and Bulgaria continue to have

**RECAP 7E: Eastern and Balkan Europe**

Eastern Europe consists of three small Baltic countries (Estonia, Latvia, and Lithuania), the plains-dominated Poland with its mixture of farming and manufacturing, and the hilly areas of the Czech and Slovak Republics. Apart from the 1993 breakup of the former Czechoslovakia, the countries of Eastern Europe are passing through the transition from communism to capitalism with fewer internal conflicts than in other parts of the former Soviet bloc and are closest to achieving European Union norms.

In the 1990s, the Balkan states of former Yugoslavia (Bosnia-Herzogovina, Croatia, Macedonia, Serbia-Montenegro, and Slovenia), Albania, Bulgaria, Hungary, and Romania passed through varying degrees of civil strife from full-scale war in Bosnia and Croatia to ethnic clashes in the other countries, including that in Kosovo. The formerly protected state-owned manufacturing industries in the Balkan countries struggle to compete in world markets after losing their links to the former Soviet bloc economy.

7E.1 Can you explain the greater economic growth in Poland and the Czech Republic in the 1990s compared to that in the other countries of Eastern and Balkan Europe?

7E.2 "The countries of Western and Northern Europe are resource-poor but industrial and wealthy; the countries of Eastern and Balkan Europe are resource-rich but economically stagnant." Is this true, and can you explain the differences?

7E.3 What are the causes of, and possible cures for, the strife in Bosnia and Kosovo?

**Key Term:**

ethnic cleansing

**Key Places:**

| | | |
|---|---|---|
| Czech Republic | Lithuania | Bosnia-Herzegovina |
| *Prague* | Kaliningrad | Croatia |
| Slovak Republic | Albania | Macedonia |
| Poland | Bulgaria | Slovenia |
| *Warsaw* | *Sofia* | Yugoslavia |
| *Katowice* | Hungary | Serbia-Montenegro |
| Estonia | *Budapest* | *Belgrade* |
| Latvia | Romania | |
| *Riga* | *Bucharest* | |

15–20 percent of their GDP from agriculture and small proportions (40–50%) from service industries.

Romania is potentially the richer country, with good farmland and oilfields on the southern flanks of the Transylvanian Alps. Romania is more than twice the size in land and population of Hungary or Bulgaria, but its GDP is less than that of Hungary. The communist dictatorship squandered the country's wealth. The heritage of poor economic and social conditions presents a massive challenge for the country, including tens of thousands of orphaned children. Its oil reserves are nearly spent.

Bulgaria's central mountainous area separates the towns along the Danube River valley in the north from those along the Moritsa plains in the south. Its farms range from small peasant plots to larger collective groupings and produce grain as well as fruits in the warm summers of southern Europe.

### Albania

Albania is the poorest country of the Balkans, just emerging from total isolation that a repressive dictatorship imposed until 1992. Livestock herding is important and some tobacco is grown for export. Its economy remains uncertain while the effects of losing migrants to Italy and Greece are assessed. Virtually no data exist to provide an indication of Albania's economic condition. In 1997 and 1998, internal strife continued to delay improvements in well-being.

# LANDSCAPES OF EUROPE

## Urban Landscapes

The geographic characteristics of the city landscapes of Europe result from centuries of making and remaking built environments. They record the consequences of change. Many European cities contain relics of historic cultures, from Greek and Roman times onward (Figure 7.46, see Figure 2.38). In medieval times, a network of towns grew with walled defences and market and administrative functions. Internal differentiation by class was poorly developed.

### From 1800 to 1950

The growth of manufacturing industries during the 1800s involved new location factors. Where they were served by railroads or port facilities, or were near coal mines, the older towns became the centers of urban expansion. Land uses differentiated into central business districts of shops and offices, industrial areas of large factories, worker housing, and suburbs in which the growing management classes lived. Other old towns retained some of their former market, cathedral, or military functions but were not industrialized, remaining small. New factory towns grew up on former farmland, sometimes sponsored by large companies.

Coalfields became the main growth areas for industry and urban centers. In older cities, the medieval walls got in the way of increased mobility and were often pulled down and replaced by wide boulevards, while some central cities were subject to more general reconstruction, as in Paris. In the early 1900s, the development of public transport in cities led to further differentiation of land use and the building of suburban housing linked by streetcar or bus to the city centers where retailing, commercial, and manufacturing activities concentrated. These processes continued until World War II, when bombing and ground fighting reduced parts of many European cities to rubble, destroying older areas, including the medieval and the industrial buildings.

**FIGURE 7.46 Mediterranean Europe: historic city landscapes.** The Roman theater at Orange, southern France, surrounded by the medieval city. Built by the Emperor Augustus, it is considered to be the most beautiful antique theater in existence.



## Changes from 1950 in Noncommunist Europe

After World War II, differences arose between the urban changes in cities of the communist-controlled countries in Eastern and Balkan Europe compared to those in Western cities. In western Europe (Western, Northern, Alpine, and Mediterranean Europe), the period from 1945 to 1970 was one of the rehabilitation, expansion, and restructuring of cities. Nearly all cities expanded their functions and populations.

This was, moreover, a period of government intervention and town planning. In the 1930s, the combination of economic recession and worries about unhealthy urban environments produced a climate in which town planning ideas, first mooted in the early 1900s, gained a hold. After 1945, the need to reconstruct Europe's economies and urban areas put these ideas into practice despite tight financial restraints. City centers were rebuilt with utilitarian buildings, while large tracts of public housing catered to lower-income groups (Figure 7.47a). New towns were built at a distance from the largest cities, separated by "green belts" of rural land in which new building was seldom allowed.

In many countries, planning regulations retained the central business district as the main retailing and financial services sector until the 1980s by preventing out-of-town stores and malls being built. Inner-city slum properties, built as 1800s worker housing, fell below new building standards, and were cleared, being replaced by tracts of public housing. Some public housing tracts had populations of 10,000 and more people. Many were built as apartment blocks rising to 10 or 20 floors. At the same time, factories and some retail outlets moved to the growing suburbs that included large tracts of private or public housing facilitated by improved road connections. The overall result was decentralization—a movement away from the previous focus on the central business district and toward more economic activity in the extensive suburbs built since 1945.

From the 1970s, the governments met difficulties over development during periods of recession and inflation. Costs rose, but funds were scarce. Furthermore, there was disillusionment over the outcome of postwar reconstruction and the bleak nature of low-cost public housing. Large public housing tracts, especially where high-rise buildings were common, proved unsuccessful as attempts at social engineering and often became centers of unemployment and crime. There was a significant move toward public and private co-operation, particularly in efforts to make city centers and old dockland waterfronts more liveable (Figure 7.47b). Areas with buildings of historic interest were declared conservation areas, but the finances for pursuing such a policy now came from private investors who saw the potential returns from encouraging tourism and in building new accommodation for those wishing to move back near the city center.

Older couples whose children had become independent and younger professionals moved to housing near city centers, a process known as **gentrification** because of the addition of high-income people to the previously poor and dilapidated inner city. The allowing of only pedestrian traffic in central areas (see Figure 7.47a) and the building of arcades of specialty shops became features of city centers, together

(a)

(b)

**FIGURE 7.47 Europe: urban landscapes in Plymouth, England.** Modern changes in the city center. (a) The main shopping area was totally rebuilt after World War II bombing and remains the largest retail district in the city despite out-of-town supermarket and warehouse shopping developments from the 1980s. Vehicles were excluded from some streets in this shopping area in the early 1990s to encourage pedestrian traffic. (b) The wharves and warehouses around the old harbor—from which Sir Francis Drake (1570–1590) and the *Mayflower* Pilgrims (1620) sailed—are now adapted for tourists with restaurants and gift shops.

© Michael Bradshaw

with the congregation of high-order services requiring support from people living in a wider area—theaters, opera houses, cinemas, and clubs. Areas around the city center that had become derelict, such as old railroad yards, factories, workshops, and worker housing, were used for building the new facilities. Such activity around city centers, however, often left a zone between it and the suburbs in which little altered and old infrastructure and abandoned buildings existed alongside old housing relics of the early 1900s townscape. Services such as education and healthcare in these areas were often poor in quality. The main directions of planned city restructuring thus had the greatest effects on the inner city and city edges, where the economic outcomes of redevelopment are the dominant considerations.

In the 1980s, the relaxation of planning constraints led to a further dispersal of economic activity from city centers to the suburbs and beyond. Manufacturing had already moved out from cramped inner city sites to suburban industrial estates. Now retail and office facilities followed to out-of-town warehouse shopping, hypermarkets, shopping malls, and office parks.

## Rural Landscape Impacts of Urbanization in Noncommunist Europe

The restructuring and growth of urban areas not only changed the landscapes of towns and cities but also had major impacts on the landscapes of surrounding rural areas. After 1945, the economy of rural areas in western Europe changed from being based almost solely on farming to a more diversified base that was increasingly dependent on proximity to urban facilities. Farming continued to be important in the landscape, but farmers and farm laborers made up smaller proportions of the population of rural areas. From being a peasant occupation based around local subsistence in many hilly areas, farming became business-oriented throughout the region. It could be a profitable business, especially when farmers were backed by EU subsidies.

The distribution of electricity to all rural areas was accomplished by 1960 and, together with the improvements in road transportation, encouraged both the commercialization of farming and the siting of factories unconnected with farming in rural areas. Especially around the larger cities, commuting took high-salaried people up to 80 km (50 mi.) from their places of work and led to the purchase and building of new expensive housing in rural areas that had been in danger of depopulation. Other types of employment, including out-of-town shopping facilities that were soon surrounded by new housing, also moved to rural areas, so that more people worked in formerly urban occupations that were now based in rural villages or small towns. This process of **counterurbanization** was most apparent in the early 1980s, but by the 1990s the focus on the major urban areas once again became stronger. New large-scale centers of high technology offices were built in city centers rather than in rural areas and combined with the gentrification factor to enhance the growth of inner-city facilities and populations once again.

### Subregional Variants

This description of city and rural landscapes outside the former communist countries refers mainly to Western Europe, but it also holds for much of the rest of the region.

- The main cities of Northern Europe were little changed by warfare and their centers continue to be dominated by government and commercial buildings built in the 1800s and early 1900s.
- The medieval cores of Swiss cities are retained with few changes, and the Swiss have strict laws concerning owners' responsibilities for keeping their property in good order.
- Mediterranean cities are most distinctive, with their ancient Greek and Roman historic cores surrounded by medieval, 1800s, and modern industrial and residential developments. Athens (Greece), Rome, Naples, Florence, Venice, and Milan (Italy), Grenada and Barcelona (Spain), and Lisbon (Portugal) have distinctive townscapes that include relics of Greek, Roman, Moorish, and medieval townscapes. Although their historic sectors gain much attention from the many tourists drawn to them, Mediterranean cities are also often ports, industrial centers, administrative centers, or magnets for beach-based tourism. Port sectors are of great significance to Lisbon, Barcelona, Genoa, Naples, Venice, and Athens. Manufacturing industries created distinctive townscapes in Barcelona and the northern Italian cities. Lisbon, Madrid, Rome, and Athens have growing sectors of government offices. The huge new tourist cities of southern Spain, the French Riviera, eastern Italy, and southern Greece add new urban forms to the old established features.

### Cities of Eastern and Balkan Europe

The city landscapes of Eastern and Balkan Europe suffered great war damage and were subject to post-1945 industrialization and the construction of monolithic buildings (Figure 7.48), but often retain older townscapes. They lack the high-rise office buildings containing commercial and professional service facilities and apartments that dominate the central parts of large cities in western Europe. Many, such as the cities in Polish Silesia, are dominated by obsolete and rusting industrial facilities built in the 1960s and 1970s.

The largest cities in the Balkans were often built as expressions of the Austro-Hungarian Empire presence. Budapest, the capital of Hungary, combines two cities on opposite banks of the Danube (see Figure 7.45) and is a government center with industrial suburbs. Its central areas include spacious 1800s buildings. Bucharest, the capital of Romania, had much of its central area, once built to resemble Paris, destroyed and rebuilt in the communist monolithic style with featureless office and apartment blocks.

Other cities in the Balkans often contain historic centers, such as the port of Dubrovnik (Croatia) that was subject to destruction by military action in 1992. An earthquake destroyed much of Skopje, the capital of Macedonia, in the 1960s. Cities such as Sarajevo, the capital of Bosnia, expanded after 1950 with central apartment blocks and suburban industries and single-family homes but suffered destruction in the early 1990s civil war.

## Rural Landscapes

Rural landscapes in Europe range from some of the world's most highly cultivated commercial farmland in the west to patterns of collectivized farming in the east, and to alpine

**FIGURE 7.48 Europe: urban landscape in Warsaw, Poland.** The 38-story Palace of Culture, built with funds from the Soviet Union, 1952–1955. The building, which houses several scientific and cultural institutions, theaters, and Congress Hall, was built on the World War II ruins of the city.

(a)

(b)

**FIGURE 7.49 Europe: rural landscapes.** Golf courses take an increasing proportion of scarce land in western Europe. (a) Near the Ben Lomond mountain, northern Scotland, United Kingdom, provides a dramatic natural setting. (b) A course that is within sight of the old (aqueduct) and new buildings of Rome, Italy.



and arctic lands. Large areas remain forested, or have been reforested. Mining areas, now largely inactive, are marked by old workings and open pits.

The essential feature of most rural landscapes in Europe is the evidence they provide of long-term human occupation. The field patterns and buildings reflect historic choices and allocations of land, even though the original causes of such features have been lost. For example, some field patterns reflect old practices of using oxen plow teams or record Roman survey lines. Many landscapes remain enclosed with hedges formed of trees that often give a wooded appearance when viewed from a distance, although commercial farms may remove them to increase field size. In the late 1900s, many rural landscapes in Europe were converted into urban playgrounds with theme parks and golf courses (Figure 7.49).

**RECAP 7F: Landscapes and Futures**

Towns and cities in western Europe present landscapes that contain relic features from their long history together with massive areas of recent industrial and residential expansion that reflect the growth of urban living in the region.

The city landscapes of Eastern and Balkan Europe also reflect their historic legacies but were heavily rebuilt under communist state planning after World War II.

Rural areas contain a record of history in fields and buildings but are undergoing changes with new transportation infrastructure, industrial buildings, and housing for urban workers.

Europe lags behind the United States and Japan in some aspects of economic development at the forefront of new technology, but it is likely to increase its proportion of world GDP if former Soviet bloc countries are able to join the European Union, to which many are attracted.

7F.1 How do features of rural and urban landscapes in Europe differ from those in another world region? What features are distinctively European?

7F.2 Which tourist attractions do you associate with the different subregions of Europe?

7F.3 How would you define the differences between the following pairs of terms: intensification and extensification of agriculture; gentrification and counterurbanization; geographic inertia and deindustrialization?

**Key Terms:** gentrification counterurbanization

**WEBSITES: Europe**

A huge range of information is available from very sophisticated websites originating in Europe. Many of these can be accessed through the Dorling-Kindersley CD-ROM Eyewitness World Atlas online facility. The following newspapers provide good coverages of recent events in the region and wider world:

*http://www.the-times.co.uk* A major British paper site with free access on registration.

*http://www.belfasttelegraph.co.uk* A Northern Ireland paper that gives the latest information and analysis on political, economic, and social news in that province.

*http://www.info-france.org/* This site is about France. Try "Profile of France"—"General Description"—"Population." Download the age-sex graph (click right mouse button and "Save picture as").

The following sites of embassies in Washington, D.C. and London provide information on regions along with economic and demographic data:

*http://www.germany-info.org* Check out the "Comparative Economic Review" that compares information about Germany and the United States.

*http://www.italyemb.org* Print out a political map of Italy ("General Information"—"Cultural Information"—"Geography") and compare unemployment in the northern and southern parts of the country ("Economic and Statistical Data–Regional Data"). Also examine the "Regional Profiles."

*http://www.poland-embassy.org.uk/* The range of information includes "*Economy*" with reports in English on Polish issues.

*http://www.state.gov/www/regions/eur/country.info.html*

*http://www.state.gov/www/issues/economic/trade_reports/index.html* U.S. State Department sites with general information on countries. Using first site, try "Background Notes"—"France." There are country reports on economic policy and trade practices.

*http://www.kosovo-state.org*

*http://www.gov.yu/terrorism/index.html*

These sites (Kosovo and Republic of Yugoslavia) provide two sides of the Kosovo conflict.

## FUTURE PROSPECTS FOR EUROPE

Europe had a major role in the early development and expansion of today's world economy, including the colonization of much of the rest of the world and the generation of the industrial revolution and many of the subsequent technologic innovations. After two devastating world wars in the first half of the 1900s, Europe faced economic ruin, was divided into western and eastern groups of countries during the Cold War, and had to cope with the emergence of rival economic powers in Eastern Asia. It recovered, at first with U.S. help, and rehabilitated its economy to the extent that it now competes with the United States and Japan in leading the world economy. In the 1990s, the countries of Eastern and Balkan Europe emerged from Russian domination and eastward links, and once again attempted to become part of a whole Europe.

The future of Europe in the new century is uncertain. A number of indicators suggest that it may fall behind the United States and Japan, which led the world economy in the 1990s. The growth of restrictions within the European Union, together with intervention by national governments, has a slowing effect on the adoption of new technology and mobility by industry. Some trends keep costs of energy and transportation high within Europe. For example, the high cost of air fares results from subsidies by governments that wish to maintain the prestige of a national airline against competition. The costs of fuels used in vehicles, and of electricity generation, remain high because of taxes. European investment in fiber-optic cable lags behind that in the United States.

Europe's computer industry has not been able to compete with the American, Japanese, and Taiwanese giants, partly because of continuing government subsidies for firms than do not compete on costs. Many of the best computer software writers are lured away by American firms. Furthermore, the tourist industries of Europe are likely to take a decreasing proportion of the world's expanding tourism market as the countries of Eastern Asia become new centers of demand and supply.

The approach of European governments to economic growth remains very different from that of governments in Eastern Asia. While European governments subsidize and protect their industries, the Eastern Asian governments protect some industries but help companies to identify new areas of technology and market growth and encourage them to develop the capability of competing in those markets on a global scale. It has been suggested that Europe is wealthy enough to control its own future within "Fortress Europe" of the EU, surrounded by tariff walls and with internal laws that guarantee its labor force cozy conditions. It is unlikely that this attitude will enable Europe to maintain its place in the forefront of the affluent and growing economies of the world.

The European Union is negotiating, however, with the Mercosul trading group in Southern South America (see Chapter 10) and the NAFTA group (see Chapter 9). Such links suggest a prospect of increasing integration in world trade, rather than a continuation of "Fortress Europe."

While such prospects are considered, countries in Eastern and Balkan Europe that were impoverished under Soviet bloc conditions wait in line to join the European Union, seek opportunities for trade with the rest of Europe, and try to attract finances and expertise into their own economies. Germany is in the forefront of developing economic links with countries to its east that are likely to form areas of economic expansion.

## CHAPTER REVIEW QUESTIONS

1. Which of the following countries was an original member of the European Economic Community in 1957? (a) The United Kingdom (b) Denmark (c) Spain (d) France (e) All of the above were original members of the EEC.
2. In 1997, NATO invited which former communist country(ies) to apply for membership in 1999? (a) Ukraine (b) Poland (c) Czech Republic (d) Hungary (e) b, c, and d
3. Transportation in Europe during the early periods of colonialization and industrialization was eased by (a) an integrated road network (b) an abundance of navigable major rivers and sea ports (c) comprehensive railroad networks (d) early airline links (e) all of the above
4. Which of the following is a reason for the pollution problem in the Mediterranean Sea? (a) Many surrounding countries discharge their sewage into the sea (b) Oil and other pollutants leak from ships passing through the sea (c) The relatively closed basin limits the amount of seawater coming in and out to dilute the pollutants (d) There are many urban-industrial centers on, or close to, the Mediterranean coast (e) All of the above
5. Europe has some of the highest population densities in the world. Which subregion of Europe has the highest population density? (a) Alpine Europe (b) Western Europe (c) Eastern Europe (d) Balkan Europe (e) Mediterranean Europe
6. Which sector of the economy employs the largest proportion of the workforce in most Western European countries? (a) Mining (b) Agriculture (c) Manufacturing (d) Service sector (e) Military
7. Which of the following European countries are monolingual (i.e. have only one official language)? (a) Belgium (b) France (c) Switzerland (d) Serbia-Montenegro (e) All of the above
8. Which of the following European countries were part of the Soviet Union before 1991? (a) Estonia (b) Czech Republic (c) Slovenia (d) Poland (e) Hungary
9. Which of the former Yugoslav republics is the wealthiest and has the closest ties with the stronger European economies to the north and west? (a) Macedonia (b) Serbia-Montenegro (c) Croatia (d) Slovenia (e) Bosnia-Herzegovina
10. Cities with medieval cores and strict protections on building preservation are characteristic of which subregional urban landscape? (a) Northern European cities (b) Swiss cities (c) Balkan cities (d) Eastern European cities (e) All of the above
11. The modern world economic system was established in Europe.

    True / False
12. The European Union (EU) arose from attempts to reclaim lands invaded by the Soviet Union at the end of World War II.

    True / False
13. From 1750 to 1900, northwest Europe took a central role in the technological innovations that led to fresh expansions of industrial activity in products and new places.

    True / False
14. Countries which are invited to apply for EU membership in the near future include Russia, Turkey, Ukraine, and Switzerland.

    True / False
15. The climatic pattern most prevalent in Europe, and extending from northern Norway to northern Italy, is the Mediterranean climate.

    True / False
16. Europe is the world's major center of tourism.

    True / False

17. The United Kingdom is an increasingly multiracial and multicultural country following major influxes of people from Turkey, North Africa, and Indonesia during the 1950s and 1980s.

    True / False

18. Alpine and Eastern Europe remain the economic heart of Europe with 60 percent of all manufacturing jobs in the region.

    True / False

19. The economic disparity between West and East Germany has proved to be much greater than was anticipated before reunification in 1989.

    True / False

20. By 1998, Italy and Spain had the lowest total fertilities in the world.

    True / False

21. The system by which peasants came under the rule and protection of nobles holding large tracts of land for kings and emperors is called ________________.

22. Continuing production in an area, even though the costs may be higher than those in possible competing areas, creates ________________. It is only when a substantial change in costs occurs, new products decline, or the demand for the original product declines that new areas develop.

23. In the late 1990s, employment in the ________________ sector made up over 65 percent of the total work force in most European countries.

24. The economy of ________________ remains distinctive in Europe and is now the most dynamic in the region because of the high level of external investment in its industries.

25. Estonia, in Eastern Europe, has closer cultural ties to ________________ than to Lithuania and Latvia, with which it is usually grouped.

26. Balkan Europe lies on the ________________, between Christian and Muslim, and between Catholic and Orthodox, influences.

27. In Balkan Europe, ________________ was the only country to adopt the Chinese form of communism and remains the poorest and most isolated state in Europe.

28. The term ________________ was first used following Serbian attempts to dominate other ethnic groups within Bosnia and other countries of the former Yugoslavia.

29. Younger and older professionals moving to housing near city centers is a process known as ________________. This term was coined to describe the addition of higher income people to the previously poor inner city.

30. A future Europe surrounding itself with tariff walls and with internal laws that guarantee its labor-cozy conditions has been termed ________________.

# COMMONWEALTH OF INDEPENDENT STATES

Chapter 8

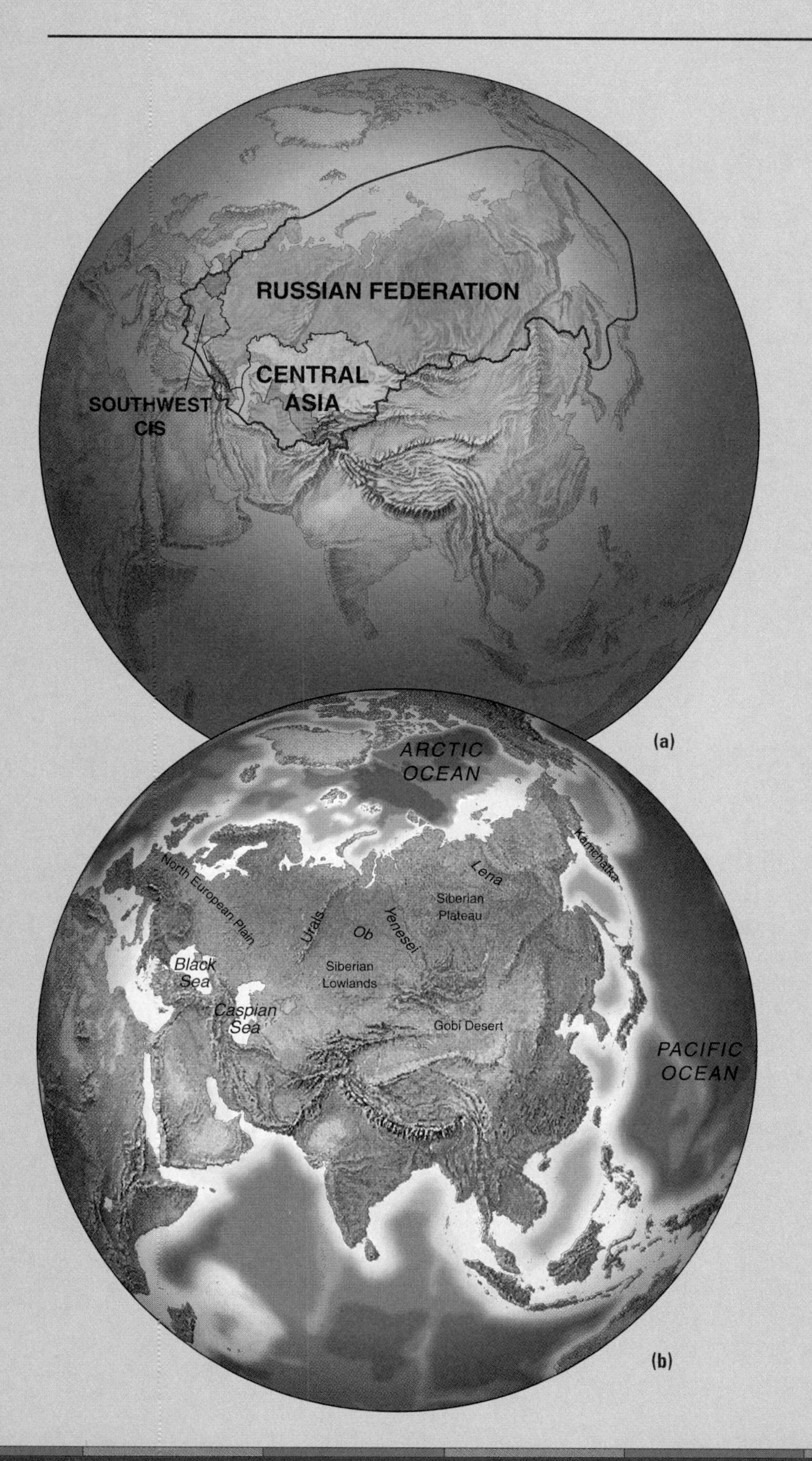


## THIS CHAPTER IS ABOUT:

**Countries in transition from the former Soviet Union to the new global order**

**The cultural and political history of Russian expansion**

**A natural environment of mid- and high-latitude climates and plentiful resources**

**Three subregions:**

Russian Federation: the world's largest country faces major changes

Southwestern CIS: countries of plains and mountains

Central Asia: five Muslim countries with difficult environments

**The urban and rural landscapes that reflect history and geography**

**Future prospects for the Commonwealth of Independent States**

**Living in Russia**

**World issue: Soviet legacy**

**FIGURE 8.1 Commonwealth of Independent States.** (a) The main subregions and country areas within them. (b) Natural environments, showing the mountains, arid (brown), forested (green), and tundra (gray) areas.

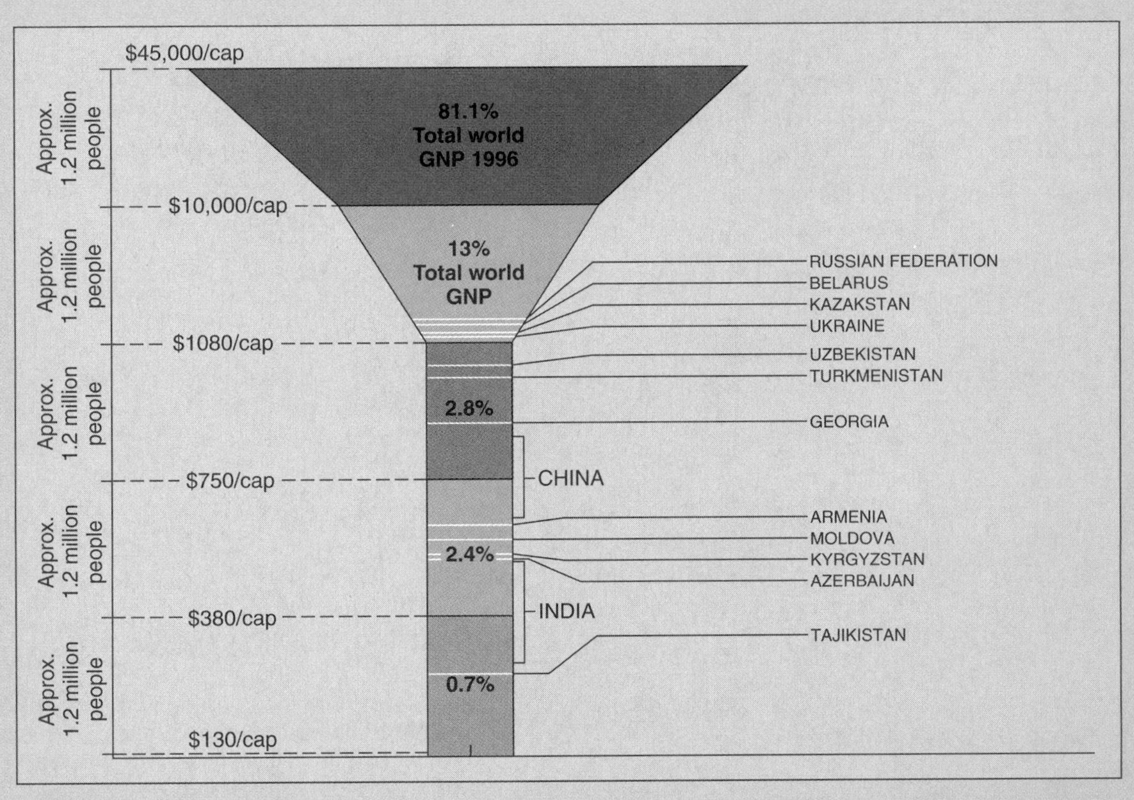


**FIGURE 8.2 Commonwealth of Independent States: global context.** GNP per head and the worldwide share of wealth. The relative positions of CIS countries declined in the 1990s.

Source: Data (for 1996) from World Development Indicators, World Bank, 1998.

## COUNTRIES IN TRANSITION

The collapse of the communist governments in the Soviet bloc of countries between 1989 and 1991 was one of the major turning points in world history. The Soviet bloc of countries had been so powerful that it was termed the "Second World." It was characterized by one-party rule and centrally planned economies. This bloc included Eastern Europe, the Balkans, and the Soviet Union, and extended from central Germany to the eastern tip of Asia, covering one-sixth of the world land surface and containing over 400 million people. It was politically powerful, yet its economic output never matched that of the Western core, "First World" countries. The breakup of the Soviet bloc made the term "Second World" meaningless and resulted in a painful transition from the attempted self-sufficiency of the former Soviet bloc toward involvement in the world economic system. The Commonwealth of Independent States (CIS) is the organization through which Russia hoped to maintain its economic relations with the republics of the former Soviet Union (Figure 8.1) and so retain some of its dissipating economic and political power.

The geographic impacts arising from these changes affected many countries throughout the world with which the Soviet Union and its allies had economic and political links. In Europe, East Germany was soon reunited with West Germany. The countries of Eastern and Balkan Europe, together with the Baltic republics from within the Soviet Union, cut ties with the CIS countries and opted to join NATO and the EU when allowed to (see Chapter 7). Many faced similar problems of transition to the CIS countries. Cuba, North Korea, Iraq, Yemen, Vietnam, and to a lesser extent Angola, Pakistan, and India had strong economic and sometimes political ties with the Soviet Union. The Arab countries of southwestern Asia and African countries had played off Soviet Union support against the United States to gain grants and loans. Such links and support virtually ceased after 1991 as the former parts of the Soviet Union became more concerned with the internal stresses of transition to new ways of government and economic activity. They soon realized that they were relatively poor in global terms (Figure 8.2), and the CIS was an attempt to revive common economic links.

## PROCESSES OF TRANSITION

The transition from one-party rule and centrally planned economies toward democratic institutions and market economics moved with faltering steps through the 1990s and was not completed in that decade. Although some changes occurred and were irreversible, others were resisted, delaying the outcome of this new revolution. The political and economic processes of transition began to have social and geographic impacts.

### Breakup of the Soviet Bloc

The former Soviet Union arose out of the 1917 October Revolution, taking over the Russian Empire that expanded from the 1600s through the 1800s. After World War II, the Soviet Union extended its zone of influence by incorporating into military and economic pacts those countries it had liberated from German occupation in Eastern Europe, including East Germany, and much of the Balkans. This formed the Soviet bloc. Moscow attempted to maintain economic and political control, but countries such as Yugoslavia and Albania became independent (Chapter 7), and others followed separate policies on many issues.

The huge Soviet bloc at its greatest extent contained hundreds of distinctive ethnic groups and the citizens of formerly independent countries who resented and resisted control from Moscow. During the late 1980s, Mikhail Gorbachev's policies of **perestroika** (economic reconstruction) and **glasnost** (information openness) relaxed strict central controls and led to economic as well as political breakup. All the Soviet bloc countries, most of which had been economically interlinked with the Soviet Union, abandoned their basis of communist ideology and socialist economic policies. By 1991, 15 independent countries emerged from the former Soviet Union. Twelve other countries in the former Soviet bloc established new governments.

### Origins of the Commonwealth of Independent States

At the end of 1991, the presidents of Russia, Ukraine, and Belarus, three of the largest and most industrialized republics in the former Soviet Union, agreed to form the Commonwealth of Independent States, since the Soviet Union "was ceasing its existence as a subject of international law and geopolitical reality." Headquarters of CIS were sited in Minsk, Belarus, although there is also a 2,000-person CIS bureaucracy in Moscow. The three countries were soon joined by all the former Soviet republics except for the three small Baltic republics (which opted for total separation from Russia, see Chapter 7) and Georgia (which delayed membership until 1993).

As a political organization, however, CIS had few powers to deal with issues that confronted the region, such as frontier disputes, the future of nuclear weapons, the Russian military installations in what are now separate countries, and the continuing dependence on Russian economic measures. Smaller interest groups within the new countries made greater progress in working with each other for specific objectives, often without a Russian presence. The Central Asian countries were concerned with marketing their oil, the Black Sea countries had common interests, and Belarus wanted to develop a special relationship with Russia. By the late 1990s, Russia continued to see a virtue in keeping the CIS together in order to spread its political influence, but the other members apart from Belarus were not so sure. In 1998, Russia discouraged its officials from referring to the other CIS countries as their "near abroad," which appeared to be patronizing and offensive in indicating that they were somehow still within Russia's sphere of influence. Despite the fact that the CIS is in danger of falling apart, the concept provides a basis for a major regional division, since all the countries have strengths and weaknesses in common that derive from their recent history of incorporation in the Soviet Union and their common geography.

### Facing Transition Problems

When some of the new countries that had been part of the Soviet Union began to distance themselves from Russia, they found geographic, economic, and political difficulties. For example, the landlocked countries of Central Asia had no previous experience of independence to build on, and their transportation links to the rest of the world lay through Russia. They were determined to remain independent, however, and their ethnic identity that grew out of Muslim traditions led them to face up to the Russian inhabitants who held positions of authority in their republics in Soviet days.

The 1990s period of transition from the socialist policies of the Soviet bloc toward the market-based capitalism of the world economic system (Figure 8.3) involved fundamental

**FIGURE 8.3 Russia: symbol of transition.** The first McDonald's fast-food restaurant in Russia opened in Moscow in 1990. It was so popular at the start that people waited in line for 90 minutes. It was followed by many others. The company gains most of its income from its food-processing facility (referred to as "McGulag" because of its barbed-wire fences) that serves hotels and airlines.

© Explorer/Photo Researchers, Inc.

changes in attitudes and institutions. It is not known how long the changes will take or if the peoples or potential ruling factions of each country will have patience to see them through. There is no historic precedent for such a political and economic turnaround, and the understanding gained of economic development in other parts of the world is largely irrelevant in suggesting what might happen. Virtually all aspects of the lives of people living in this region are likely to change within a few years.

The transition processes have many facets, all of which bring problems.

- There is the political breakup of countries and the loosening of ties among groups of countries that were held together by their communist parties and Soviet military power.
- There is the shift from a communist, state-controlled economy based on central planning to market-based economies, together with an unraveling of the economic links and dependencies that had held back the economies of many countries.
- There are economic transitions within agriculture, manufacturing, transport infrastructure, and energy use as countries try to catch up with competing Western technologies and adapt to new trading relationships.
- There is the impact of the émigrés returning from exile abroad and the increasing significance of ethnic populations, while the remnants of the resettlement of Russian peoples in republics outside Russia create continuing tensions. Although tensions among and within these countries in transition caused conflicts, many deaths, and great disruption from local wars, there has been no wave of popular unrest nor mass migration. The worst eruptions were in the small country of Tajikistan and the Russian republic of Chechnya, and in the war between Azerbaijan and Armenia over the Nagorno-Karabakh enclave of Armenians within Azerbaijan.

It is difficult to estimate either the original state or the progress of transition processes in this region. Relative statistical data about conditions in the former Soviet Union is unobtainable apart from population numbers (see Data Tables, Reference Section, page 586-610).

## Political Breakup, New Alliances

The Soviet Union functioned as a single unit for over 70 years. When it broke up in 1991, therefore, the outcomes were of a major order. The Soviet Union governed its huge territory with an intense degree of centralization focused on the Russian heartland around Moscow. The **Council for Mutual Economic Assistance** (CMEA), which became known as COMECON, was established in 1949 to integrate economic activities throughout the Soviet bloc. Based on bartering arrangements, it was somewhat successful in allocating production of agricultural and industrial output to specific locations. Attempts in the late 1980s to revive it as a Common Market for the Soviet bloc failed. CMEA was dissolved in 1991 when Russia demanded that payment in dollars from the other members replace the previous bartering arrangements.

At breakup in 1991, the Russian Federation separated from the other 14 newly independent republics that had been part of the Soviet Union—the three Baltic countries (Chapter 7), and two subregions of the CIS—five Central Asian countries, and six countries in the Southwest CIS. The Russian Federation still has the largest area of any country in the world and maintains a dominating presence in the region. It occupies 77 percent of the CIS land, although its population of 147 million in 1998 was just 52 percent of the region's total. The Russian people are concentrated in the western one-fourth of their country. There are large tracts of low population density extending eastward across Siberia to the Pacific coast. After Russia, Ukraine, with 51 million people, is the next largest single country carved out of the former Soviet Union. Ukraine has a long history of links with Russia and many events in Russian history occurred in its main city, Kiev. Belarus—with a history of shifting control of its lands between Poland and Russia—and Moldova, along the border with Romania, occupy the plains area between Russia and Eastern Europe.

The small, mountainous, southern countries of the Transcaucasus region—Georgia, Armenia, and Azerbaijan—have a history of being affected by conflicts among Turkey, Russia, and Iran. These countries became part of the Russian Empire in the 1800s. When the Soviet Union took over the region in the 1920s, it was at first welcomed by many of the people as a protector from the Turks. Subsequent repressive policies from Moscow led the countries to proclaim independence at the break-up of the Soviet Union. In the early 1990s, after they regained their independence, fighting flared up between Armenia and Muslim Azerbaijan.

The Central Asian republics are Muslim in religion and gained their first recognition as independent political entities in 1991. The CIS nominally includes these new countries within former Soviet territory together and the influence of Russia remains strong. New pulls may draw the countries and distant regions of Russia itself toward closer ties with China, Iran, or Turkey.

## Communism to Capitalism

The change from communist state control of the economy to market-based capitalism is a fundamental and traumatic U-turn. The two systems are incompatible: public versus private property ownership, monopoly of power by communist parties versus multiparty systems, a single bloc versus many countries, state-owned enterprises versus multinational corporations. That is why the Gorbachev reforms of the 1980s could not work and merely succeeded in breaking up the Soviet Union and its linked bloc of countries.

The Soviet Union's economy was based on central planning controlled by Moscow. It organized internal markets for the goods produced. There was a limited, although growing, involvement in selected world markets. A major emphasis was placed on metal-producing industries and their offshoots

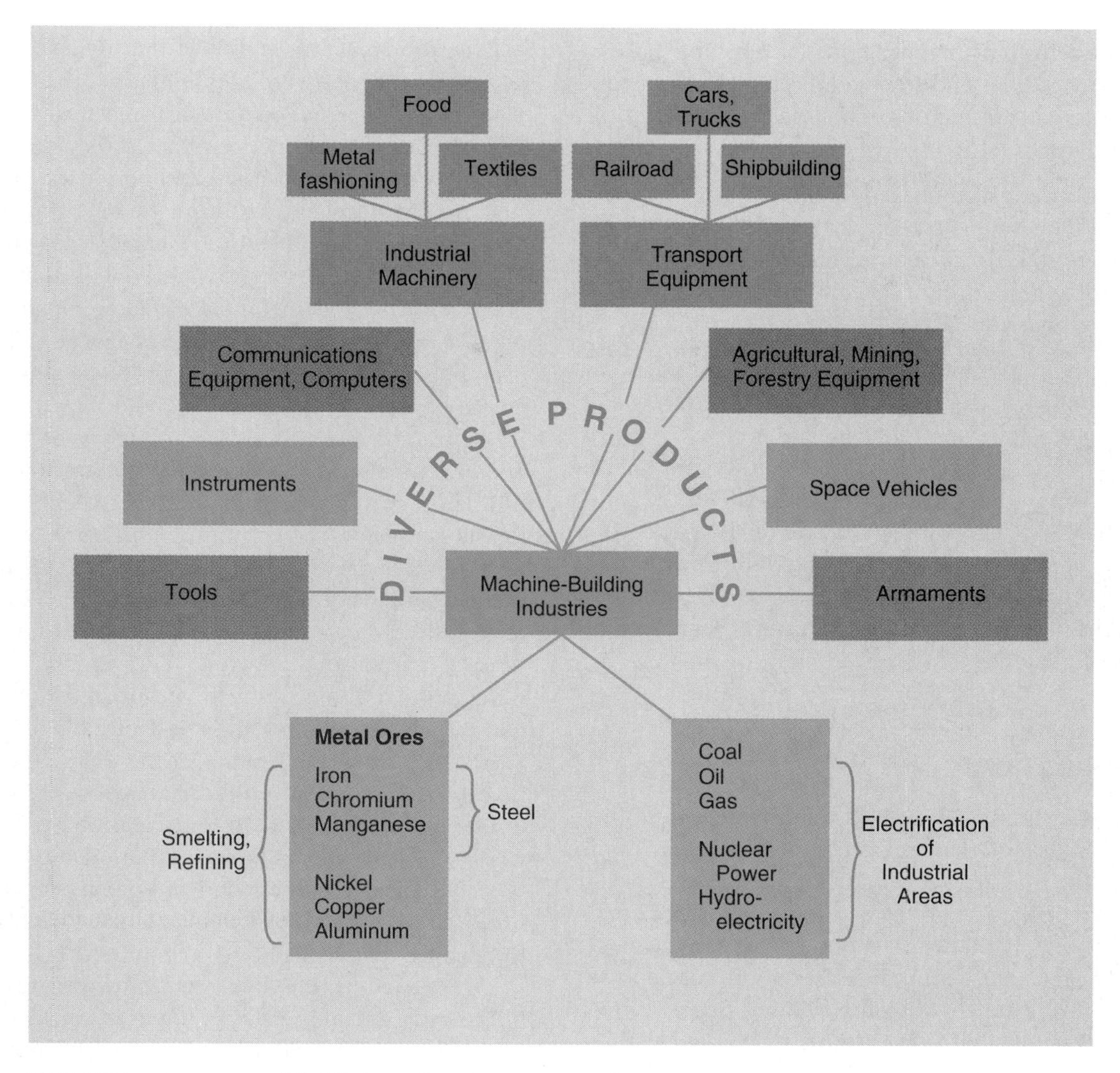


**FIGURE 8.4 Commonwealth of Independent States: emphasis on manufacturing.** The planned Soviet manufacturing links and output. The Soviets overemphasized the production of metals to be used in machine-building industries for a variety of engineering and military purposes and for the products of those industries. Little room existed for the consumer products that comprise so much of the manufacturing output in the rest of the world. Products were functional, not for personal pleasure. The system stressed raw materials and power supplies, which were plentiful.

(Figure 8.4), on economically productive goods instead of consumer goods, on rail and air transportation rather than more individualistic road transportation, and on defense and space industries.

All the factors of production—land, mineral rights, factories, transportation systems, power production, and labor—were owned or directed by the state. Each sector of economic activity was controlled centrally with allocated budgets. The state defined what and how goods were to be produced and the prices at which they would be sold. Little attention was paid to aspects that are significant in market-based economies, such as productivity or market demands. The system was protected by the strict control of external trade and did not allow private enterprise or the development of non-government financial services; they were illegal. Defense, education, and health care were priorities in this system.

Changes in political and financial control were needed to accomplish the move from socialist economics to market capitalism. A range of actions bring about such changes.

- **Price reform** forces prices to respond to conditions of supply and demand. For example, scarce goods attract high prices and stimulate greater production, which pulls down the prices.
- Second, **financial reform** directs the flow of capital to profit-making enterprises. Financial reforms involve **fiscal policy**, in which the government balances its spending with income from taxes, and **monetary control**, in which money supply is managed to prevent inflation. In addition, it is necessary to have a commercial banking sector that is subject to controls, preventing loans to nonviable enterprises.
- Third, **privatization** is the sale of state-owned enterprises to private owners, making managers and workers more accountable for productivity.

- Fourth, **trade reform** dismantles trade barriers and makes currencies convertible. This forces enterprises to compete with foreign producers and domestic prices to move toward world prices.
- Fifth, new institutions are needed to make the free market system work, including corporate ownership laws, rational tax systems, private commercial banks and financial institutions, accounting conventions, and laws to regulate contracts.
- Sixth, and perhaps as important as these basic economic strategies, a change in attitude is required from the socialist point of view that capitalism and private enterprise are morally wrong. Up to 1990, it was a crime in the Soviet Union to trade for profit. After 1991, the absence of banking, legal protection, and other checks and balances of market capitalism allowed the growth of crime and corruption. In the late 1990s, extortion affected the VAZ-LADA auto works at Togliatti on the Volga River, where gangs threatened violence.

## Economic Transitions

The economic transitions since 1991 result largely from the changed political outlook that redirected the focus away from Moscow-based bureaucratic perceptions of the national well-being of the Soviet Union. A policy of internal self-sufficiency outside the workings of the world economic system changed to one of full engagement with it. Features of the communist economy, such as the state ownership of land, mines, factories, transport systems, and power industry, worked within the context of incentives to trade within the Soviet Union and the wider Soviet bloc. Much of the economic geography of Soviet domination related to internal linkages. Those linkages broke down in 1991.

### Shift in Power Sources

Before 1991, the countries of the Soviet bloc depended on the plentiful supplies of oil and natural gas that the Soviet Union delivered at low cost in exchange for products important to the Soviet economy. After 1991, Russia sold as much of its output of oil and natural gas as it could in world markets to obtain hard currency. The local supplies of coal, oil, and natural gas in the other countries of the region were insufficient, partial, or too poorly developed to provide an immediate substitute.

Many countries in the region depend heavily on nuclear power. The fall in economic output in these countries after 1991 caused energy demands to decline—by 20 percent outside Russia and 50 percent in Russia—with recovery to 1980s demand levels unlikely before 2005. Russia finds it is more profitable to continue to use nuclear generators and sell its fossil fuels abroad. Fourteen reactors supply one-fourth of Ukraine's needs. These reactors do not require hard currency to operate and the fuel costs are lower compared to coal-, oil-, or gas-fired generators.

The problems of a policy that relies increasingly on nuclear generators were highlighted by the 1986 Chernobyl (Ukraine) disaster when a nuclear reactor exploded. To shut down other possibly dangerous reactors, however, is not a practical option because the CIS countries cannot afford to stop producing electricity from them. Even nuclear plants that had been shut down in Armenia because they were sited in earthquake-prone territory, as well as some of the Chernobyl units, are being repaired and reopened, while new units of similar design continue to be built. The increasing use of such fallible technology proceeds alongside a debate on the level of Western finance and expertise necessary to regulate and make safe faulty nuclear power plants.

The export sales of oil and natural gas resources to bring in foreign currency remain the main economic lifelines for Russia and some of the other countries. New developments involve working with Western and Arab oil companies, but there was a lengthy transition while the Russian government—which owns most of the reserves (Figure 8.5)—delayed measures that would allow foreign companies to develop them. Projects already negotiated include the oil and natural gas reserves off the coast of Sakhalin Island in the Pacific and along the Arctic Ocean coast. Others in prospect involve the exploitation and transportation of oil and gas resources from Azerbaijan and Central Asia. The Russians wish to retain control of oil exploitation and trade within the CIS, but local pressures for independence within Russian Federation republics close to pipelines, such as Chechnya, lower the confidence of potential investors.

While such negotiations made slow progress, the Russian oil industry declined. Output fell from a peak of 12.5 million barrels per day in 1988 to 6 million in 1997 but then began to expand again. Low production and falling world oil prices in the late 1990s made it difficult to replace plant and machinery. Production wells went out of use earlier than expected. The reduction in domestic demand, however, made it possible to sell as much abroad as before 1990.

### Railroad Infrastructure

During the Soviet era, railroad transport was developed and used for up to two-thirds of total freight compared to less than one-fifth in western Europe. Road transport was regarded as inefficient and oriented to individuals rather than state needs. Roads were left in a poor condition, and there was no freeway network. During the 1980s, however, there was little investment to upgrade the rail track or rolling stock, leading to a declining quality of service. Neither rail nor road facilities could support new economic development.

### Changes in Agriculture and Land Ownership

Agriculture makes a strong contribution to the economies of the CIS countries. Under communism, the land in many Soviet republics was incorporated into the public sector and divided into collective and state farms. Problems of low farmer motivation and mismanagement emerged, and after

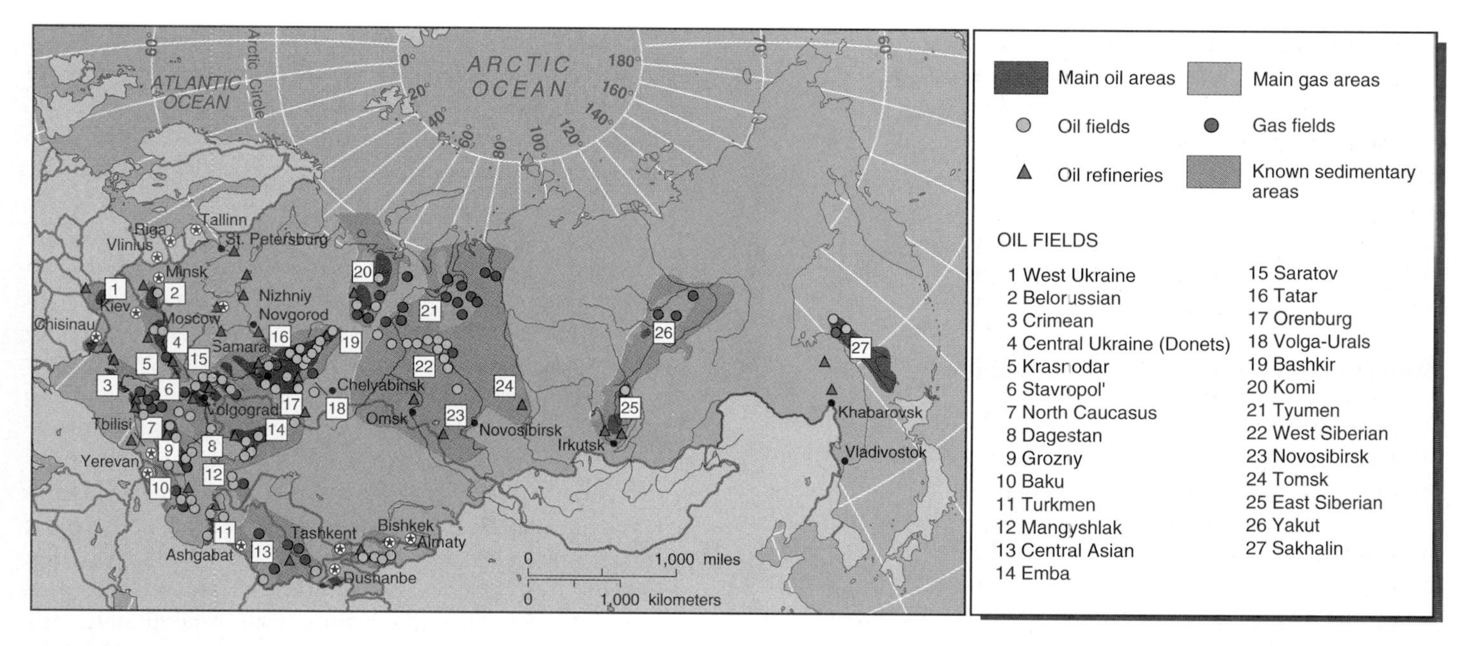


**FIGURE 8.5 Commonwealth of Independent States: oil and gas fields.** Assess the availability of oil and natural gas to Russia, the other CIS countries, and the potential developments in Siberia.

1991, conversion to individually owned plots provided an incentive for increased productivity. It was hoped that a revived commercial farming sector would produce its own capital for modernization and an impetus for other aspects of the economy to move rapidly toward a market economy. Such hopes have not been fulfilled so far. The process of transition in agriculture exposed institutional frictions.

Land ownership is the basis of private agriculture, since farmers need collateral for borrowing to finance modernization. Both the managers of collective and state farms, who had no wish to lose their privileges, and also the farm workers, who were slow to take on additional responsibilities and work, held back the process of land allocation. Bureaucracies became bogged down by millions of claims for land, transferred land in tiny lots, or distributed the poorest land released by collective farms first. Prices for farm products remained low, while privatized food-processing industries increased their profits.

These events began to turn farmers against the prospect of owning their land. In Russia, it was hoped that by 1995 there would be 650,000 private farms with an average of 75 hectares (190 acres) of land each. This target was not reached, although it represented a tiny proportion of the country's farmland compared to the 26,000 state and collective farms that had up to 15,000 hectares (37,500 acres) each and continued to resist government directives to privatize. In 1993, President Boris Yeltsin issued decrees that made the privatization of farming and the redistribution of land easier. These decrees affected the 27 million people working on state and collective farms. An experiment on six farms along the Volga River near Nizhni Novgorod gave all workers convertible land-entitlement certificates, which could be bartered by groups wishing to farm, service machinery, or develop feedlots. This made it possible to assemble viable farm units from tiny plots of individually owned land that could not maintain a family.

**FIGURE 8.6 Russian dachas.** These dachas near Irkutsk, Siberia, reflect the Russian tradition of having a rural retreat and the current need to have some land for growing vegetables.



With so many Russians not receiving salaries or pensions for long periods in the 1990s, former professionals and others owning private plots of land around their rural dachas took to growing their own vegetables (Figure 8.6) to help feed their families.

In the 1990s, farming was a loss-making industry in the CIS. Food prices rose slowly, but the other inputs required for commercial farming, such as machinery, fertilizers, and high-quality seed and stock, increased in price more rapidly. Private farms in Russia had to buy old machinery from the state farms, not being able to afford new. Prices for grain rose more rapidly than those for livestock products. The high tariffs of the United States and EU made it impossible to export surpluses to the world's wealthiest countries, and the United States and EU made matters worse by dumping their surpluses in other world markets at low prices with which the CIS could not compete. Many farmers in the CIS looked back with affection to the times when their prices, although low, were guaranteed by the central government.

### Manufacturing and Services in Transition

Manufacturing industries also faced the difficulties inherent in the transition from state-controlled economic conditions to those of the free market. Privatization of large-scale state-run concerns is a slow process. The large concerns that perpetuated the emphasis on heavy industrial producer goods, rather than consumer goods, remain under state control, struggling to become more efficient and find markets for their output. Many needed massive investment to update machinery and shed large numbers of workers. New private businesses were established, but nearly all were small in size.

Service industries are too poorly developed in the CIS to support an expansion of privatized industries. The 1990s saw a gradual reduction in state-controlled financial and distribution sectors as retail outlets, banks, and other services were privatized. The lack of experienced and trusted financial institutions was a major problem faced by all new ventures. In many cases, control passed from slow-moving bureaucracies to get-rich-quick, short-lived facilities.

## Émigrés and Minorities

The exodus of people from this region increased through the 1800s as many Jews left Russia. The upheavals of the Russian revolution in 1917, the Bolshevik takeover of Muslim Central Asia in the 1920s, and defections from the Soviet Union by artists and athletes led to further groups of émigrés. Many moved to New York, Paris, or London, or, after 1948, to Israel. Central Asian peoples moved into Muslim Iran, Afghanistan, and the northwest parts of China. In the 1990s, some of these émigrés returned home but received different degrees of welcome.

After 1991, few of its many exiles returned to the Russian Federation. They were often better off outside Russia or had lost touch with people inside the country after decades of refused visas, bugged telephone calls, and opened letters. There was never a major émigré political force outside the Soviet Union. The exiled Soviet group consisted mainly of intellectuals, scientists, and musicians. The only business-people to show an interest in returning are Russian bankers living abroad, who are now involved in advising Western businesses about investment inside Russia. While few émigrés return to Russia, many Russians wish to move to Western countries, particularly the United States, because of the poverty and crime in Russia and the new freedom to leave their country. The number able to move is restricted by the receiving countries, although Israel welcomed many Russian Jews in the early 1990s (see Chapter 4).

Ukrainian émigrés included those who left for a better life elsewhere and those who left for political reasons, including the Soviet purges. Those living abroad made much of the differences between Ukraine and Russia rather than the many historic links. Many returned to the country after 1991 and contributed to establishing a national identity. Building on their experiences, they translated American schoolbooks into Ukrainian. While this was welcomed at first, it did not take long for the large Russian-speaking group in Ukraine to object. The current Ukraine government emphasizes territorial rather than ethnic integrity.

Armenia is a very small country but one where émigrés play a major part, even to the extent of keeping a war going with its neighbor Azerbaijan. Over half the world's 10 million Armenians live outside the country, despite the return of many since 1991. They live mainly in the United States, the rest of the CIS, and the countries of southwestern Asia. Divided between the Turks and Russians in the 1700s, Armenia became independent in 1918 following the 1915 killing of 1.5 million people, mainly by Turkish-paid Kurds, and the forceful removal of 4 million people to other parts of the Ottoman Empire. The borders of Armenia today do not satisfy Armenians. The Nagorno-Karabakh enclave within Azerbaijan is important to Armenians as a place where their language and rural culture survives. The ancient center of Armenia, Tbilisi, is now in Georgia. At present, the land-locked country of Armenia is subject to blockades by Azerbaijan and Turkey but continues to place a greater emphasis on establishing the country's identity than on its economy.

Minority groups were largely kept in submission in the former Soviet Union up to 1990. After 1991, however, many of the concerns of such groups surfaced. The most notorious of the disputes occurred in the Transcaucasian area, but other issues continue to feature in the transition phase. The mid-1990s Russian military activity in Chechnya, a tiny republic within the Russian Federation, was one of many confrontations between a Muslim minority group and the central Russian authorities. Feelings rose so high in these clashes that economy and infrastructure were destroyed in Georgia, Armenia, and Tajikistan. There are fears that such actions could spread to other parts of the former Soviet Union.

Perhaps the most significant minority problem waiting to be solved concerns the 26 million Russians stranded by the collapse of the Soviet Union in what are now independent countries outside the Russian Federation. Of these, 16 million are in countries west of Russia. Eleven million are in Ukraine, where the main concentration is in the industrial east in and around the cities of Donetsk, Luhansk, and Kharkov, and in the Crimean peninsula. Russian remains the language of business in Ukraine, where the industrial workers and miners of the eastern sector form a militant political force concerned about mismanagement of the economy by Ukraine's leaders but unwilling to see their uneconomic industries restructured. Moldova is a country that has a mainly Romanian population, but a Russian army unit took control of the Russian-occupied eastern lands in 1992 and does not wish to leave.

## CULTURAL AND POLITICAL HISTORY

Along with the changes of the transitional postcommunist era in the CIS, its modern geography bears the marks of the region's cultural and political history. Ethnic and political groupings had territorial associations that developed through historic events.

### Historic Ethnic Groups

As in Europe (see Chapter 7), the early history of this region was dominated by tribal movements, sometimes involving invasions of people from thousands of kilometers away. The open plains of Ukraine, southern Russia, and Central Asia provided avenues of access to the riches of southern and western Europe. Groups moved westward out of this region, followed by others who replaced them. Each wave of invaders conquered, settled, and either absorbed the local culture or imposed its own—including language and religion. Goths and Huns took the route in the A.D. 300s, Bulgars from east of the Black Sea in the 400s, Avars from Asia in the 500s, Slavs from north of the present Moscow area in the 500s and 600s, and Magyars from Asia in the 800s. The Slavs dominated three areas. While the West and South Slavs moved into areas now part of Poland, former Yugoslavia, and the Balkan countries, the East Slavs formed the basis of the Russian peoples—occupying a forested area bypassed by most Asian tribal movements into Europe.

### Christians, Mongols, and Turks

The region continued to experience movements of people and conquests by different cultures. In the first 10 centuries A.D., much of the region was Christianized. Armenia claims to be the first Christian country, having adopted the religion in the early A.D. 300s, some 90 years before the Roman Empire did so. The expansion of the Byzantine Empire that was based on Constantinople (modern Istanbul) brought Orthodox Christianity to the new Russian Muscovite kingdom in A.D. 988.

The Mongol (often called Tatar by Russians) invasions from eastern Asia in the 1200s had major impacts on the Asian areas east of the Ural Mountains and on the emerging Russian Muscovite kingdom, which had to endure 200 years of tribute payments from the 1200s to the 1400s. Large numbers of people fled westward. The Russian economy stood still while the countries of western Europe moved through the Renaissance, the breakdown of feudalism, and the development of merchant capitalism. The Mongol regime supported the Orthodox Church, which grew stronger. At the end of the Mongol era, after long opposition to Islam, the rulers adopted that religion. Timur Lang, ruler of Samarkand (in modern Uzbekistan), carried out final invasions westward and southward into India in the late 1300s. After the Mongol decline, Samarkand continued as a major trading city and center of Islamic cultures (Figure 8.7).

**FIGURE 8.7 Central Asia: Samarkand, Uzbekistan.** Historic Muslim buildings highlight the importance of Islam in Central Asia. Twin madrasah (theological seminary) buildings of the 1500s and 1600s border Registan Square. Typical features of Islamic art include arches, tilework, domes, and minarets.



Timur Lang was half Turkish, and the future of the southern parts of this region was in the hands of the Turks—another group of Asiatic peoples who moved westward, taking their languages and adopting Islam. The expansion of the Turkish Ottoman Empire occurred from the 1300s, and its conquests rivaled those of the Arabs (see Chapter 4) and Mongols. The Turks took over much of southwestern Asia, northern Africa, and their invasions reached the Balkan area, southern Poland, and parts of Central Asia. Many of the people living in the Ottoman Empire converted to Islam.

### Russian Empire

Amid these changes, Russia emerged. The focal country in this region began as a loose federation of Slavic tribal princedoms around Novgorod and Kiev in the A.D. 800s. The latter city was the center of an area known as "the land of Rus" during the Viking presence. Early primacy fluctuated between the two main cities but moved to Moscow after the 1300s as the Russians retreated into the northern forests during the Mongol invasions through the southern steppes. The Russians also had to defend their lands against the German Order of Teutonic Knights that dominated the Baltic area in the later Middle Ages and against invasions by Sweden and the Lithuanian-Polish kingdom across the North European Plain. Survival after these attacks provided Russian Muscovy with military strength and

**RECAP 8A: Transition and History**

The collapse of the communist government of the Soviet Union in 1991 flung the countries that were its republics into a transitional phase between the socialist economy and market capitalism that will affect their political, economic, and geographic futures. The Soviet Union, centered on the Moscow region, had dependent peripheries, but these geographic relationships were disrupted by the transition to new circumstances. The new freedoms stimulated ethnic conflicts that had been kept down during the Soviet regime.

The present distribution of languages and religions in the region can be traced back to migrations of ethnic groups and the dissemination of religious beliefs up to around A.D. 1000. The Orthodox form of Christianity, the invasions of Mongol hordes and Ottoman Turks, and the conversions to Islam along the region's southern borders had major influences on subregional cultures. Much of the region remained under feudal and agricultural conditions until the early 1900s.

The Russian Empire expanded its territories from the 1600s and began to modernize its productive capacity along Western lines. The Bolshevik Revolution of 1917 led to communist control of the former Russian Empire as the Soviet Union. After World War II, Soviet influence extended over countries in Eastern Europe and the Balkans to form the Soviet bloc with economic and defensive ties.

8A.1 What are the main features of the transition that faces the countries of the former Soviet Union? What progress has been made?

8A.2 How far do the political divisions on the present CIS map reflect historic movements of peoples?

**Key Terms:**

perestroika
glasnost
Council for Mutual Economic Assistance (CMEA)
price reform
financial reform
fiscal policy
monetary control
privatization
trade reform

**Key Places:**

Commonwealth of Independent States (CIS)
Union of Soviet Socialist Republics (USSR)

national memories. By the mid-1600s, the Russian isolation from the rest of Europe that had been imposed by these wars ended as the Romanov dynasty became emperors (or tsars). They introduced Western developments and expanded their territory eastward. As they did so, they incorporated many different ethnic groups under Russian control, sowing the seeds of future multicultural pressures.

The fears generated in Russians by centuries of invasion and subjugation resulted in a determination to use the effective means of increasing wealth and military power that had been developed in western Europe. By the end of the 1700s, Russia consolidated power in its lands west of the Urals, including the defeat and partition of Poland on its western frontier and the incorporation of Ukraine, Belarus, and other areas as provinces within its empire. In the 1800s, Russia pushed its frontiers eastward, across the lands of its former invaders in Central Asia, to the Pacific coast of Siberia. Russia then took over lands in the Caucasus from Persia and Turkey. Between the 1830s and 1881, the lands that today form the five Central Asian countries came under Russian rule, sometimes after bloody warfare. The combination of Slavic settlers, changes in land ownership to accommodate irrigation farming, and forced recruitment into the Russian army led to revolts and declarations of independence in these provinces during World War I.

### Revolution

After the Russian tsar was deposed in early 1917, the Bolshevik group of communists took the opportunity of the new government's weakness to seize power in October 1917. Some provinces within the Russian Empire bowed to nationalistic agitation and declared their independence. The United Kingdom and United States occupied parts of Russia, while Poland invaded the west. Any such independence or alternative control was short-lived, however, for the new communist government in Moscow and its Red Army fought a civil war, subduing opposition.

By 1922, the regime incorporated the former provinces of the empire as soviet socialist republics in the **Union of Soviet Socialist Republics** (USSR). Soviets were the workers' and soldiers' councils that became basic political units. The USSR is often called the Soviet Union: the official title of the main party was the Communist Party of the Soviet Union. Repression of local groups, transportation of dissenting groups to Siberia, and the resettling of large numbers of Slavic Russians in other provinces occurred. New policies of collectivized agriculture and centralized industry were forcefully extended.

## NATURAL ENVIRONMENT

The natural environments of the CIS countries reflect their huge land area. The continental interior climates, often with long and harsh winters, the vast plains, and the massive areas covered by a variety of natural vegetation and soils provide a distinctive stage on which the development of human geographies occurred.

### Midlatitude Continental Interior Climates

The countries of the CIS nearly all lie north of 40° N and most of their area is north of 50° N. Their **midlatitude continental interior climates** are thus more like those of central and eastern Canada than those of the interior United States. The warmest climates are around the Black and Caspian Seas, and in the Central Asian countries (Figure 8.8).

This region contains the world's greatest extent of interior continental conditions. The largest proportion of these lands is more than 500 km (320 mi.) from the ocean, and many

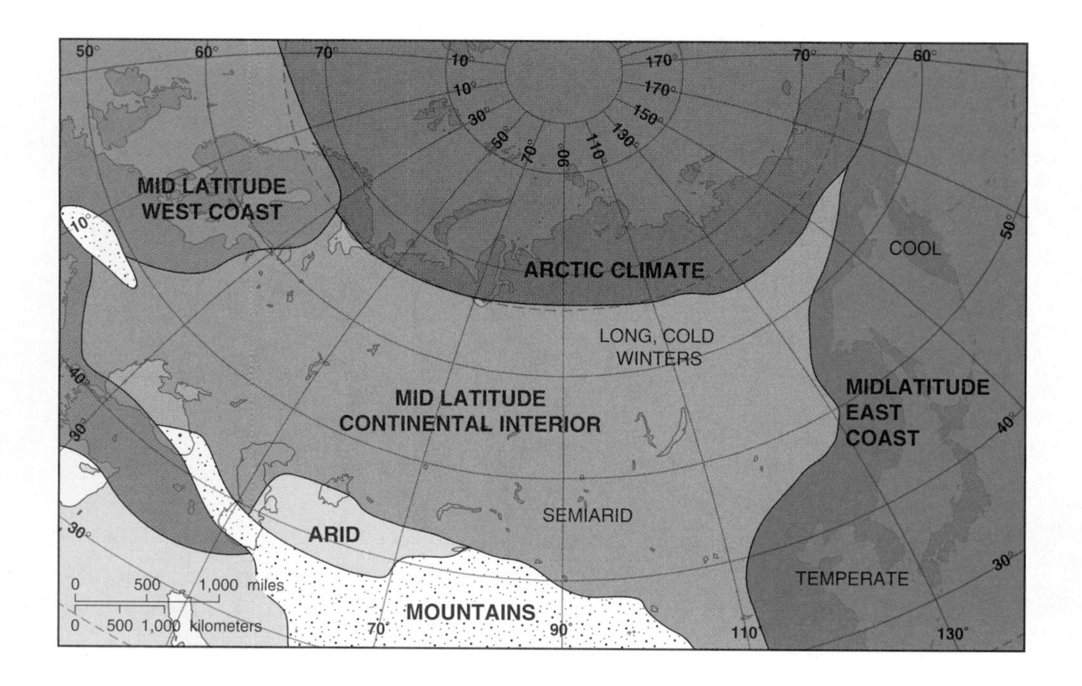


**FIGURE 8.8 Commonwealth of Independent States: climates.** The dominant midlatitude continental interior climatic environments are characterized by harsh winter conditions, aridity in the south, and arctic conditions in the north.

parts are over 2,000 km (1,250 mi.) from the ocean. With greater distance inland, the extremes of summer and winter temperatures increase. The greatest contrast is in eastern Siberia, where large areas have January temperatures below to 30° C (–22° F) and July averages of 12° to 16° C (56° to 60° F; difference of 45° C, or 80° F). In western Russia, winter temperatures average –5° to –10° C (15° to 25° F), and summer temperatures are 15° to 20° C (59° to 68° F; difference of 25° C, or 50° F). With greater distance northward, the climate gets colder and the winters are longer; much of northernmost Russia lies inside the Arctic Circle. In the far east of Siberia, nearness to the Pacific Ocean results in more humid conditions, although the winters remain long and very cold as arctic winds sweep out from the continent's interior.

Greater distance inland reduces precipitation. Few parts of the region have over 80 cm (32 in.) a year. Much has moderate precipitation (40–80 cm, 16–22 in.) falling mainly in summer but producing a long-lasting snow and ice cover during the winter. Parts of southern Central Asia are arid because of the high evaporation rates in warmer latitudes and distance from rain-bearing air masses. Parts of the far east are subhumid. Low precipitation and low temperatures combine to make eastern Siberia inhospitable.

## Southern Mountain Wall

The southern boundary of the region (Figure 8.9a) is marked by mountain systems. They were thrown up along the convergence of tectonic plates carrying the continent of Eurasia on the north and the land masses of Africa, Arabia, and India to the south (Figure 8.9b). The **Caucasus Mountains,** between the Black and Caspian Seas, are part of this line of mountain ranges that extends through the Elburz Mountains of northern Iran to the **Tian Shan** and the **Pamir Mountains** along the southern borders of the Central Asian countries. These mountain ranges rise to over 7,400 m (24,000 ft.) and include many peaks over 3,000 m (10,000 ft.). They are snowcapped and in spring provide considerable meltwater for streams flowing through the dry areas of southern Russia and Central Asia. In the far east, the **East Siberian Uplands** and the volcanic peaks on the **Kamchatka Peninsula** parallel the Pacific coast in areas that have few people.

The earth movements not only raised high mountain ranges but also produced deep basins such as those filled by the **Black Sea, Caspian Sea,** and Lake Baikal. The Caspian Sea has drainage into it that does not flow into the main oceans. Its sea level fluctuates according to the inflow of river water. Between 1922 and 1977, the Caspian Sea level fell to 30 m (65 ft.) below the world ocean level, its lowest level for 500 years. The Gulf of Kara-Bugaz on the eastern shores acted as a natural evaporating pan until the Soviet government blocked the entrance (later reopening it). Since 1980, the Caspian waters rose by just over 2 m (7 ft.) and are likely to rise over 5 m (16 ft.) by A.D. 2010, threatening reclaimed lands, tourist centers, and oil installations around the sea. The rising sea level covered 32,000 km$^2$ (12,360 mi.$^2$) of former land and now threatens over 100 villages and their farmland. The blocking of Kara-Bugaz accounted for only one-fifth of the rise; the rest is due to increased inflows from the Volga River or decreased evaporation from the Caspian Sea.

## Plateaus, Plains, and Major River Valleys

Plains and low plateaus dominate the landscapes of this region. The North European Plain widens eastward from Poland into Belarus, Ukraine, and European Russia until it ends against the **Ural Mountains** that mark the line between Europe and Asia. Plains around the northern shores of the

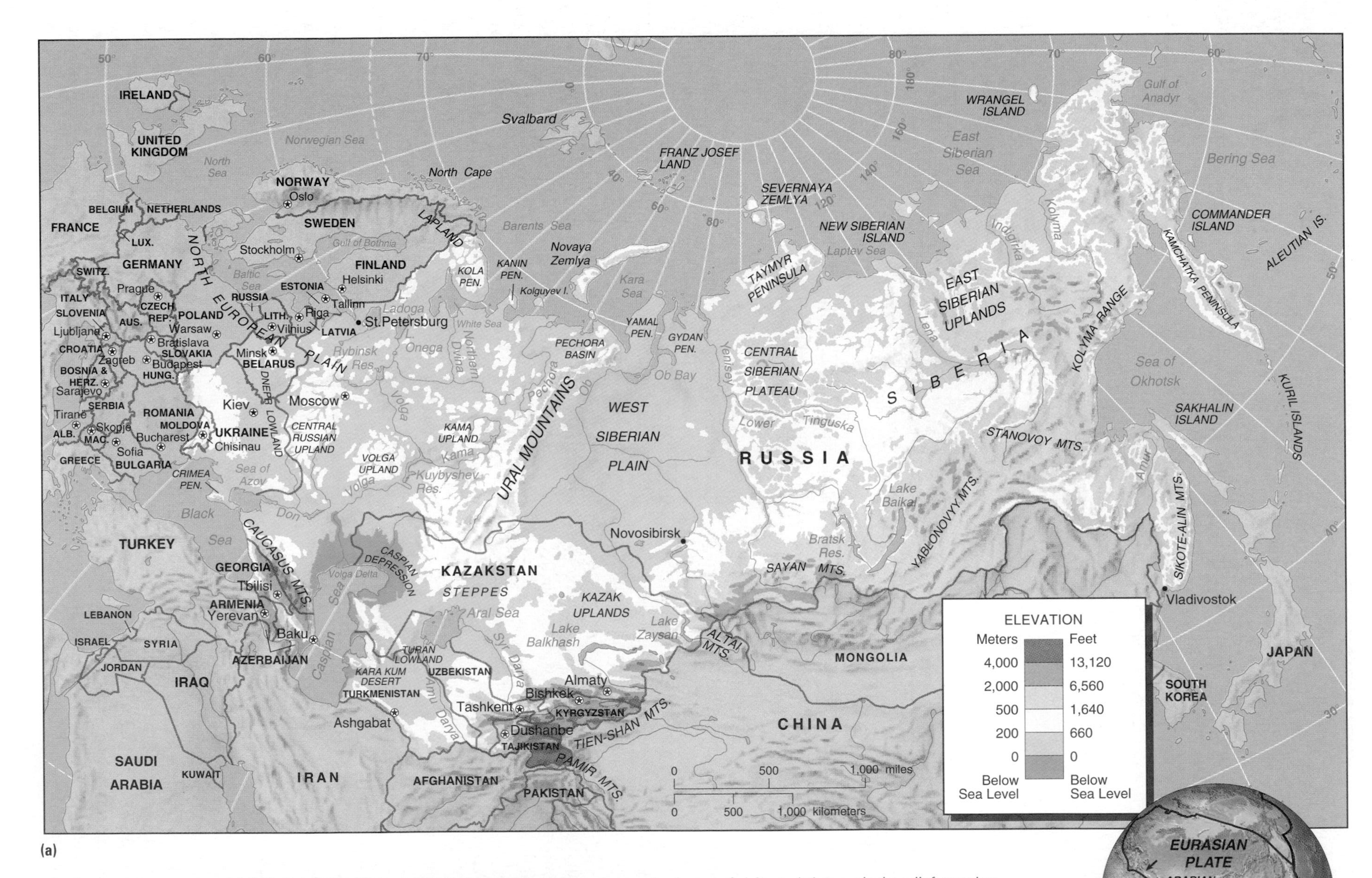


**FIGURE 8.9 Commonwealth of Independent States: physical features.** (a) There is a predominance of plains and plateaus in the relief, together with a southern rim of mountains that resulted from clashing tectonic plates. (b) The map of tectonic plates and their margins shows that the Eurasian plate clashes with other continental plates from the Caucasus to Central Asia; it clashes with the oceanic Pacific plate in the east.

**FIGURE 8.10 Ukraine and Russia: Black Sea and Crimea.** The Crimean Peninsula almost cuts off the Sea of Azov from contact with the Black Sea. This area was ceded to Ukraine in 1954, on the anniversary of Russia-Ukraine links and while the Soviet Union controlled the entire area. Since 1991, much of the former Soviet navy remains in Crimean ports.

NASA

Black Sea (Figure 8.10) and along the rivers leading to the Caspian and Aral Seas extend into the nearly level, vast **West Siberian Plains.** Farther east, the relief again becomes more hilly in the **Central Siberian Plateau** on ancient shield rocks. The existence of such extensive areas of plains, interrupted by low hills, was a major factor in easing the invasions from eastern Asia and central Europe into Russia in medieval times.

Across these mainly low-lying landscapes flow some of the world's longest rivers. In the west, the **Don River** system flows into the Black Sea and the **Volga River,** draining most of the area between Moscow and the Urals, into the Caspian Sea. The **Amu Darya** and **Syr Darya Rivers** flow into the **Aral Sea**—an interior drainage system. The longest rivers of all, however, are those that rise in the southern mountains of Central Asia and eastern Siberia and flow northward to the Arctic Ocean—the **Ob, Yenisey,** and **Lena.** The Ob is the longest of these and drains the largest watershed, but the Yenisey discharges the most water into the Arctic Ocean.

While the rivers in western Russia are intensively used for transportation, generating hydroelectricity, and as sources of industrial and domestic water, those farther east are less exploited. This is partly because the areas through which they flow are sparsely settled but also partly because the rivers flow northward, are frozen for much of the year, and even in midsummer have frozen mouths that cause the backup of meltwater to flood vast areas of wetland on either bank.

## Desert, Grassland, Forest, and Tundra

The natural vegetation, much altered by human occupation, was marked by transitions from hot desert in the south through steppe grassland to deciduous and coniferous forest, and to tundra on the shores of the Arctic Ocean (Figure 8.11).

- The deserts of the area east of the Caspian Sea contained few patches of grass and oasis vegetation and were occupied by nomadic peoples. Trade routes across the region provided long-established links between Europe and China through towns sited along the route where water was available.
- North of the desert, the **steppe grasslands,** and the fertile **black earth soils** on which they grew, extended from southern Poland eastward through Ukraine into Kazakstan, becoming one of the world's major arable regions. In the Middle Ages, the steppe grasslands provided the grazing lands through which wave after wave of invaders moved westward into Europe. They were plowed for centuries in Ukraine, and more recently arable farming was extended eastward into Kazakstan by the Soviet virgin lands program, but droughts and poor management caused the program to fail to make the Soviet Union self-sufficient in grain.
- Farther north, where evaporation rates decrease, trees can grow and a zone of wooded steppe gave way to deciduous forests, linked to fertile **brown earth soils.** The trees were largely cut for the extension of farmland except on slopes that were too steep for plowing.
- The deciduous forest gave way northward and eastward—where temperatures and precipitation declined—to forests dominated by hardy birch trees and evergreen pine, fir, and spruce. The underlying poor **podzol soils** have a gray sandy layer under a surface of decaying leaves and above a layer where iron minerals accumulate, are cemented, and impede drainage. This **northern coniferous forest,** or taiga, dominates vast areas of land from Moscow northward and across most of Siberia (Figure 8.12). Pine and fir trees in the west give way eastward to larch. The taiga forest occupies areas with very long winters and where glacial action deposited poor material for soils.
- Around the shores of the **Arctic Ocean** and extending southward into the plateaus farther east is an area where trees will not grow and that is covered by grasses and low-growing shrubs. Such **tundra** vegetation is underlain by permanently frozen ground, or **permafrost,** that also extends southward beneath the taiga forest. Much of the permafrost, which is 3,000 m (10,000 ft.) thick in parts of eastern Siberia, was formed during the glacial phases of the Pleistocene Ice Age. The continuous and broken permafrost areas cover most of central and eastern Siberia. Warmer conditions over the last 10,000 years melted the

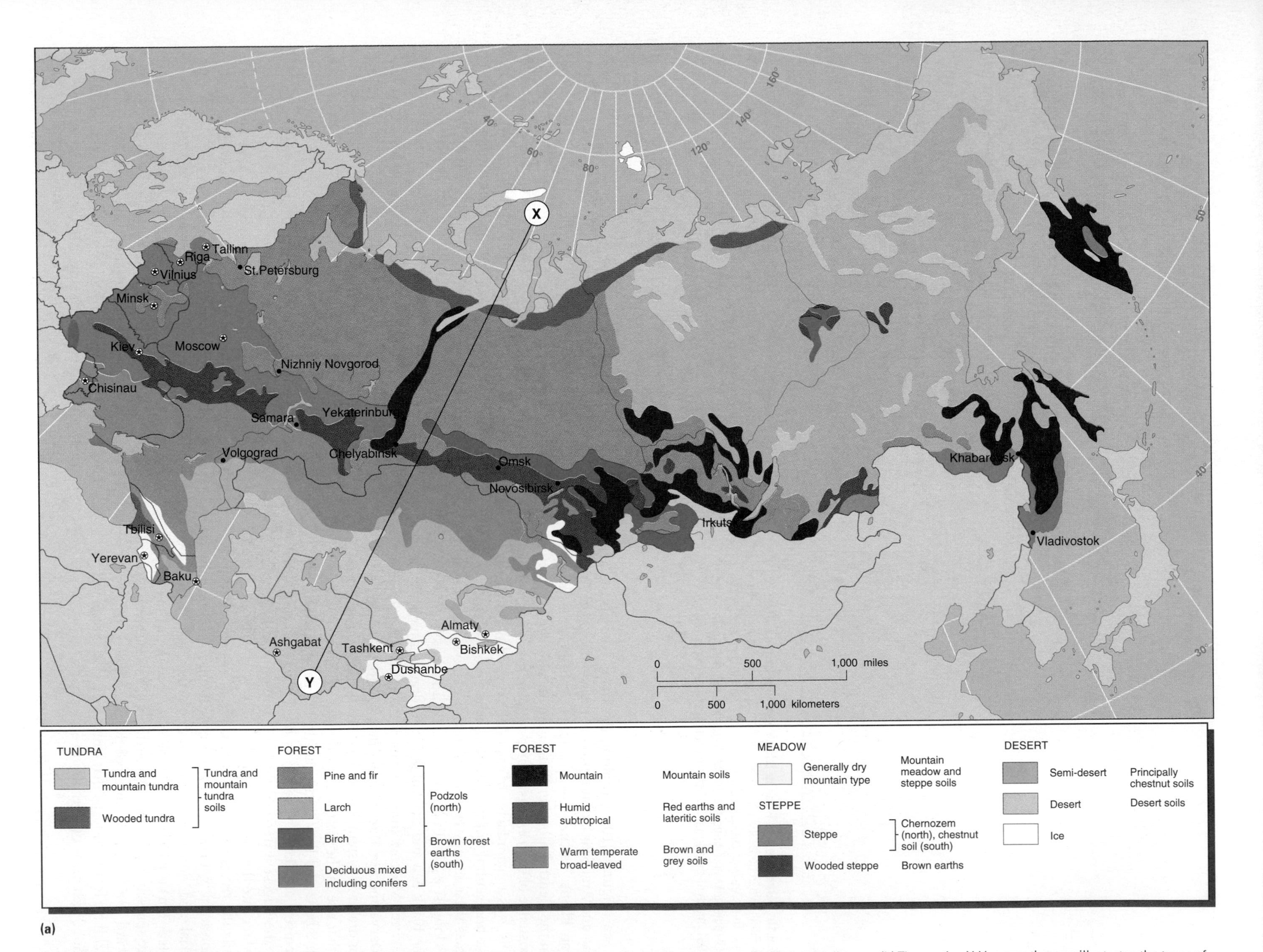


(a)

**FIGURE 8.11 Commonwealth of Independent States: natural vegetation and soils.** (a) Map of natural vegetation and its links to soil types. (b) The section X-Y across the map illustrates the types of vegetation and soils in a north-south succession.

Source: Data from Archie Brown, et al. (eds.) *The Cambridge Encyclopedia of Russia and the Former Soviet Union,* 1994, Cambridge University Press. Reprinted with permission of Cambridge University Press.

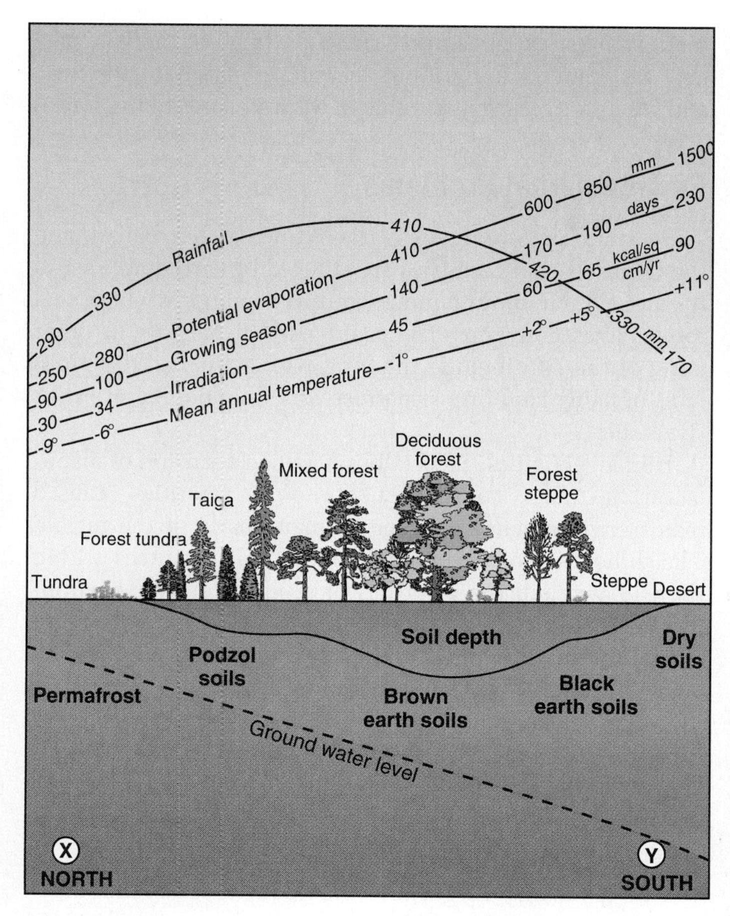


(b)

margins of this area, leaving irregular small-scale subsidence hollows that make farming difficult. Removal of the surface vegetation cover causes melting of the permafrost and collapse of the underlying soil layers.

## Natural Resources

Russian and Soviet political, military, and economic expansions were driven by the use of natural resources in the metal-based manufacturing industries that dominated world core economies in the late 1800s and early 1900s. The rocks of the region contain huge quantities of a wide variety of mineral resources. Fuels include coal, oil, natural gas, and uranium. The Siberian lowland may contain as much oil as the Persian Gulf area. Metallic minerals range from iron to gold, and there are also diamonds and many minerals that are in short supply elsewhere in the world. Many of these resources occur in the ancient shield rocks or have been washed by erosion into the intervening areas of sedimentary rocks often in the northern, sparsely inhabited areas. The present Russian Federation lacks only two major strategic minerals—tin and vanadium. At a time when political power was linked to steel and fuel output, Russia became a superpower.

The CIS continues to be a world leader in producing fuels and other minerals. The first Russian manufacturing industry built up around the iron ore and coal deposits of Ukraine and southern Russia and extended eastward into Kazakstan as new resources were discovered. Other sources of minerals and energy occur in and around the southern mountain ranges of the Transcaucasus and Central Asia. The flanks of the Caspian Sea are particularly rich in oil and natural gas. The main problem is to build the requisite pipelines to a world oil port, preferably on the Mediterranean Sea. This poses many political, economic, and environmental problems at present (Figure 8.13). By the late

(a)

(b)

**FIGURE 8.12 Russia: Siberia.** (a) The birch and larch forests of eastern Siberia, near Irkutsk. The margins provide recreation land and timber poles for an informal volleyball game. (b) Lake Baikal. Students return from a picnic by the lake in summer.

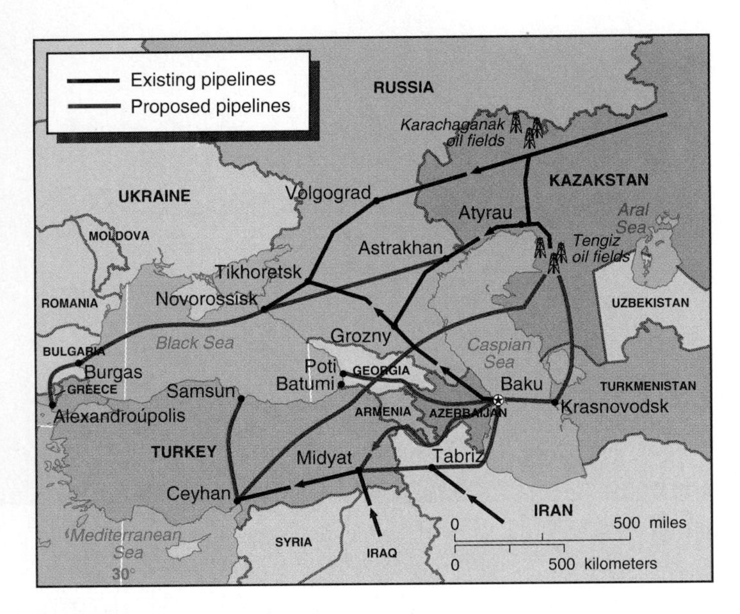


**FIGURE 8.13 Commonwealth of Independent States: potential pipeline outlets for oil and gas.** Russia wants oil and gas exports from the countries of Transcaucasus and Central Asia to be channeled through its Black Sea port of Novorissisk via the Grozny (Chechnya) pipelines. The newly independent countries and their international oil company partners are examining the routes through Turkey to Ceyhan on the Mediterranean Sea or through northern Iran. The Turkish authorities wish to reduce the number of oil tankers passing through the straits at the Black Sea exit, leading to the prospect of a further pipeline linking ports in Bulgaria and northern Greece. How does this example illustrate the interdependence of countries in the world economic system and the importance of individual countries and their actions?

Reprinted by permission from Financial Times, February 2, 1995

1990s, it appeared that a combination of the routes through southern Russia and to the coast in Georgia would be used after the Russians had cleared the way by pacifying Chechnya. Exports of natural resources assumed great significance in the face of transitions in manufacturing and other economic activities.

The taiga across northern Russia forms the world's largest forest area, covering 7.5 million km$^2$ (2.9 million mi.$^2$, compared to 5.5 million km$^2$, 2.12 million mi.$^2$, in Brazil and 4.5 million km$^2$, 1.7 million mi.$^2$, in Canada). It provides a huge reserve of biodiversity and is important in world climate change, taking up one-third as much carbon dioxide as the tropical rain forest absorbs. Owned by the state, this resource has been well regulated and little reduced by logging or fires. State planning located all the Soviet Union paper and pulp mills west of the Urals, making it costly to transport the wood from Siberia. Although large corporations began to cut more of these forests in the 1990s and potential tax revenues from such cutting make it an attractive prospect for Russia, the central bureaucracy continues to slow the process.

Despite its huge area, the extreme climates with extensive subhumid areas and the large areas with arid, podzol, or permafrost soils leave relatively small portions of the CIS suitable for commercial farming. These are mainly in the west and where irrigation is possible in the arid lands of the south.

## Environmental Problems

The natural environments of the Armenia, Azerbaijan, and the Central Asian countries are affected by earthquake activity along their mountainous southern margins where tectonic plates converge. The main natural hazards in other parts of the CIS include the long winter freezes, the problems of using land on permafrost, and the flooding of major river valleys.

Human activities, especially since the expansion of manufacturing under communist economic priorities, caused major environmental damage in many parts of the region. The difficulties of developing the natural resources within the CIS area and the priorities of production over environmental protection led to increasing environmental degradation of the tundra, forest, and desert areas. The emphasis on heavy industry—especially chemicals and steel—the nuclear industry, the faulty storage of toxic rocket fuel and oil, the testing of weapons, and the excavation of metallic minerals laid waste to huge areas, many of which remain unreported and unreclaimed. A few examples illustrate the range of environmental problems facing the CIS countries.

- One of the major legacies is the frequent occurrence of oil pipeline breaks and leakages. In the summer of 1994, one occurred on the Pechora River in northern Russia near Usinsk. An earth dam built in 1988 to contain oil leaking from a pipe collapsed in heavy rain, emptying the oil into the river that drains into the Arctic Ocean. Russian officials played down the scale of the event, but some observers estimated it as being one of the world's largest oil spills, several times larger than the Exxon *Valdez* disaster in Alaska. Earlier in 1994, a pipeline carrying oil to Europe burst some 400 km (260 mi.) south of Moscow, creating a 17-acre petroleum lake on frozen ground.
- The city of Norilsk in western Siberia has the most polluted environment in Russia. Its 250,000 people gain their livings by working in the local nickel and copper works and coal mines that employ 110,000 and account for 80 percent of local tax revenues. They produce nearly 20 percent of the world's nickel and cobalt in a region that has 35 percent of the world's nickel reserves, 10 percent of copper, and 40 percent of platinum, plus substantial cobalt and palladium. Sulfur dioxide emitted by the factory chimneys results in a stinking acidic haze that blackens snow, kills trees for miles around, and poisons the river. The main smelter emits seven times more sulfur dioxide than the entire metals industry of the United States. Male life expectancy is 50 years in Norilsk.

  The city was built in the 1930s Stalinist era, based on the close occurrence of copper and nickel ores and coal

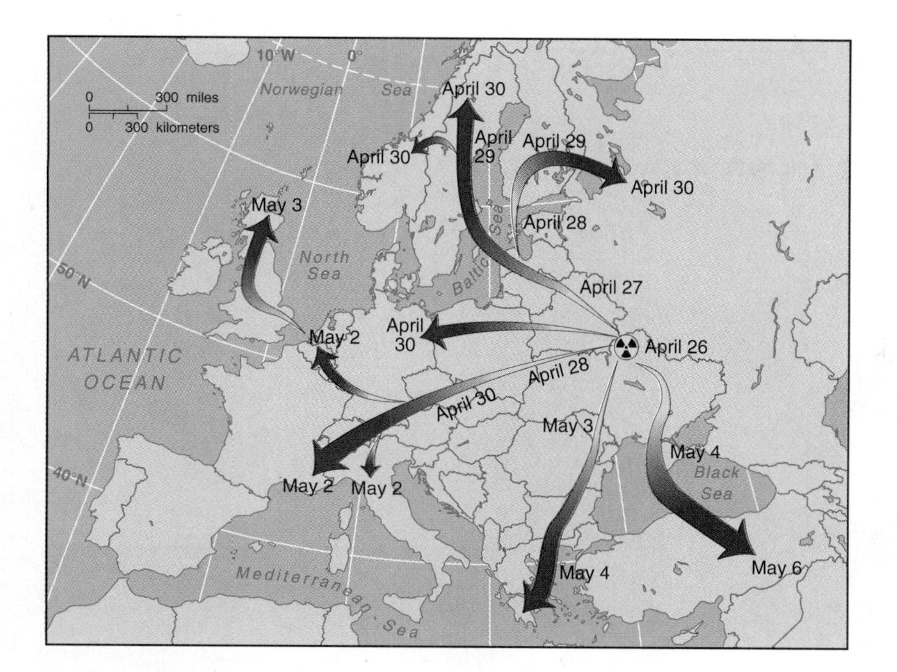


**FIGURE 8.14 Commonwealth of Independent States: impact of a nuclear disaster.** The areas affected by the Chernobyl nuclear power plant explosion that released clouds of radioactive gases in 1986. In Ukraine, over 1.5 million people were affected, 10 percent of whom were severely radiated. All 600,000 members of the cleanup team suffered minor illnesses. Local death rates will increase, since the evacuation of the inhabitants after the explosion was delayed. Farther afield, the radiation was detected over Scandinavia and even in the U.K.

deposits, which were worked at first by political prisoners. The city is not connected by road or rail to other parts of Russia, and life for its people is based in functional apartments, with freezing winters and summer mosquito plagues. Little money earned by the company gets back to the community.

Since privatization, the managing company wishes the government to take responsibility for the social welfare programs that employ 18,000 people in housing, health or sports clubs, holiday homes, day care, and dairy farms. These take any profits, and there is a need to invest in new environmentally friendly facilities and safer working conditions. People want to leave the area but cannot because they lost their savings in the inflation that followed the 1991 breakup of the Soviet Union.

- Environmental damage was also extensive in industrialized Ukraine. The 1986 Chernobyl nuclear reactor explosion, which became a symbol for such disasters, affected most of the area immediately around it north of Kiev, but the fallout was worst to the north in Belarus (Figure 8.14). There are other examples of nuclear reactors failing and exploding in the Urals.
- One of the greatest environmental disasters in the world occurred in the lands east of the Aral Sea (Figure 8.15a). Although arid, Central Asia naturally experiences a major internal circulation of water through the atmosphere and then across the surface in rivers. Water vapor evaporated from the large lakes or inland seas such as the Aral Sea condenses and is precipitated as snow on the high mountains to the east. As the snow melts in spring, the water fills such rivers as the Amu Darya and Syr Darya that flow into the Aral Sea. This water circulation provided the basis of local irrigation farming and small urban settlements through history, but the Soviet Union adopted ambitious plans to use the water on irrigated cotton farms inside and outside the main river basin. Supported by growth-oriented, unbending bureaucrats who often acted against the advice of government scientists, the project was taken too far and so much water was extracted that the rivers stopped flowing into the Aral Sea. The sea is now less than half its previous size (Figure 8.15b), all transportation on it has ceased, and the supply of moisture for the mountain snows has dried up. Glaciers in the mountains are in retreat, groundwater levels have fallen accompanied by ground subsidence, and dust storms now affect areas that seldom experienced them (Figure 8.15c).
- Sumgait, north of Baku in oil-rich Azerbaijan, is another of the world's most polluted places. Once one of the main suppliers of petrochemicals to the Soviet market, its 30 or so industrial plants are now a huge wasteland of broken concrete, rusting factories, and abandoned railroad tracks. The air smells of chlorine and sulfur. Most factories work at 20 to 30 percent capacity, insufficient to maintain corroding pipework. The town is another legacy of environmental costs resulting from Soviet central planning.
- The Black Sea, which is surrounded by six countries, is also under threat from environmental degradation. Concerns were voiced in late 1994 over a proposal by the Ukraine government to build a large oil terminal and refinery near Odessa to enable the country to reduce its dependence on expensive Russian supplies. The proposal threatened local holiday resort beaches

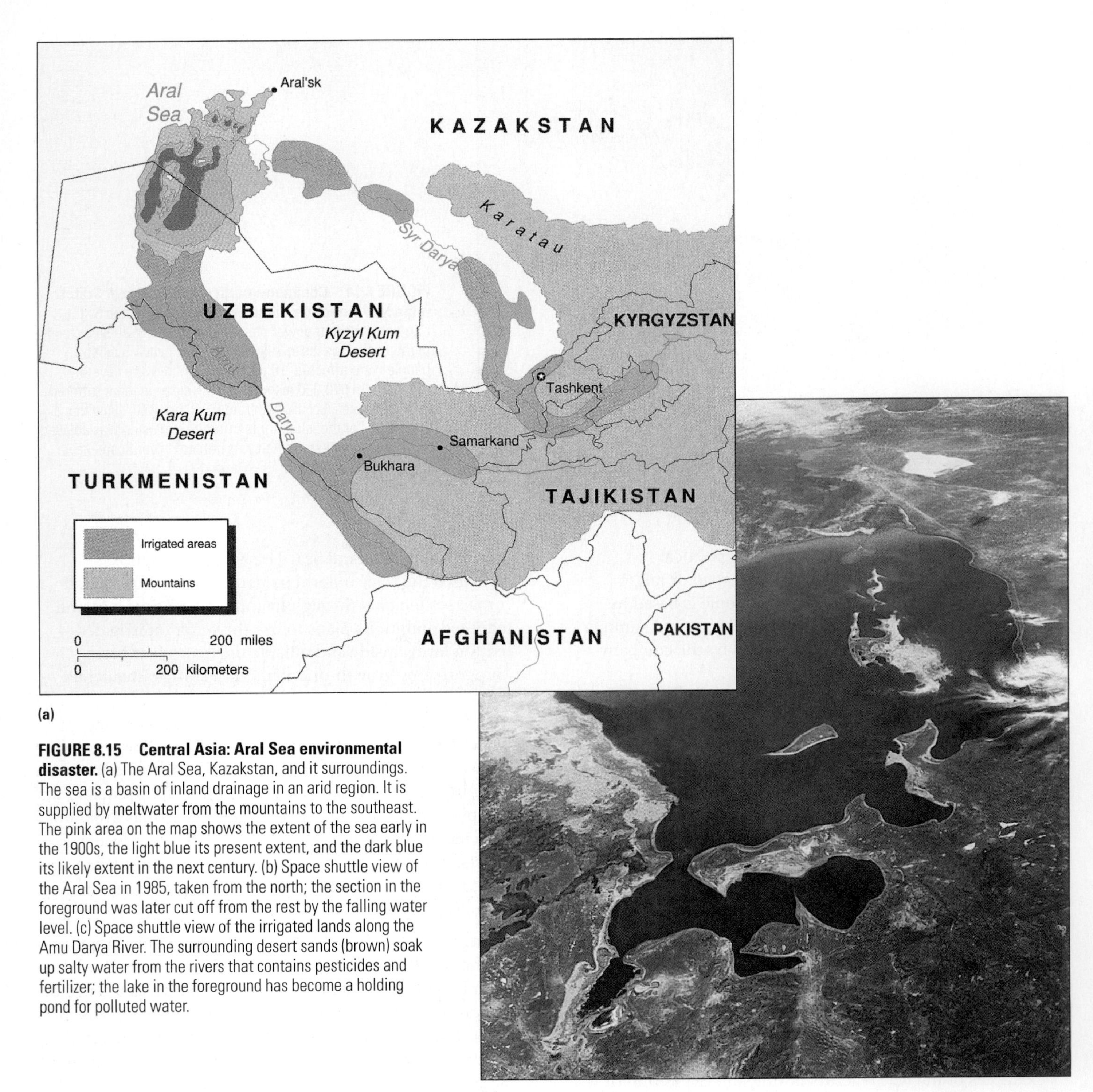


(a)

(b) NASA

**FIGURE 8.15 Central Asia: Aral Sea environmental disaster.** (a) The Aral Sea, Kazakstan, and it surroundings. The sea is a basin of inland drainage in an arid region. It is supplied by meltwater from the mountains to the southeast. The pink area on the map shows the extent of the sea early in the 1900s, the light blue its present extent, and the dark blue its likely extent in the next century. (b) Space shuttle view of the Aral Sea in 1985, taken from the north; the section in the foreground was later cut off from the rest by the falling water level. (c) Space shuttle view of the irrigated lands along the Amu Darya River. The surrounding desert sands (brown) soak up salty water from the rivers that contains pesticides and fertilizer; the lake in the foreground has become a holding pond for polluted water.

and offshore locations that harbor up to 70 percent of the sea's fish. Putting industry before environment occurs in the other five countries bordering the Black Sea (Russia, Georgia, Turkey, Bulgaria, and Romania) and in the 17 countries that drain into the sea. Fish stocks and plant and animal life in the Black Sea are killed by oil, toxic wastes, ships' ballast water, and nutrients from fertilized fields. Only five of the 26 fish species caught in the 1960s were still found in 1992, when the total fish catch of the six rim countries dropped to 10 percent of their 1986 haul. Offshore drilling for oil and gas takes place near the Crimea peninsula, further reducing the prospects of the region's tourist industry.

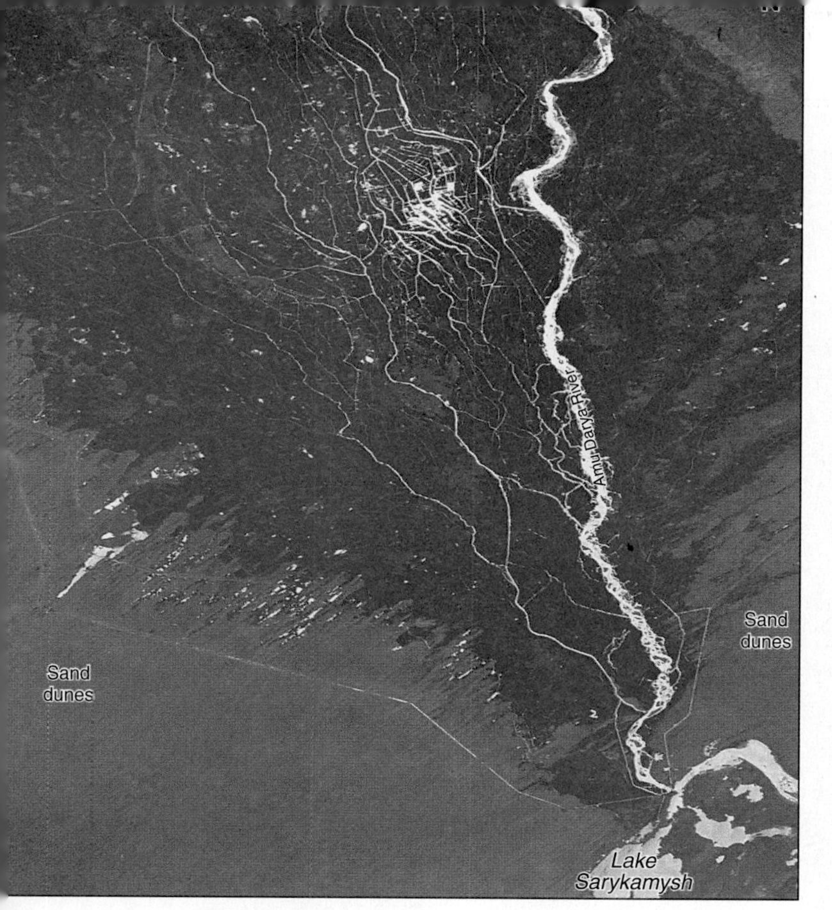


(c) NASA

## RECAP 8B: Natural Environment

The region has midlatitude continental interior climates ranging from arid through humid to arctic conditions and marked by extreme seasonal fluctuations of temperature. Although mainly a region of plain, low plateau, and wide river valleys, its southern margins are marked by high mountain ranges along major junctions of tectonic plates.

The natural vegetation and soils follow the climatic pattern with a south-to-north sequence of desert, steppe grassland, deciduous forest, northern coniferous forest, and tundra.

Natural resources are of vital importance to CIS countries. The rocks of the region contain some the world's most significant reserves of fuels and metallic ores. Human activities, particularly in the communist era, led to major environmental problems.

8B.1 How are the climate and natural vegetation of this region related to each other?

8B.2 Can you explain the susceptibility of countries in this region to environmental problems?

**Key Terms:**

midlatitude continental interior climate
steppe grasslands
black earth soils
brown earth soils
podzol soils
northern coniferous forest (taiga)
tundra
permafrost

**Key Places:**

Caucasus Mountains
Tian Shan Mountains
Pamir Mountains
East Siberian Uplands
Kamchatka Peninsula
Black Sea
Caspian Sea
Ural Mountains
West Siberian Plains
Central Siberian Plateau
Don River
Volga River
Amu Darya Syr Darya River
Aral Sea
Ob River
Yenisey River
Lena River
Arctic Ocean

# RUSSIAN FEDERATION

The **Russian Federation** is the world's largest country, having nearly twice the area of Canada, the United States, or China (Figure 8.16). In 1998 it had 77 percent of the CIS area and 52 percent of its population. The Russian Federation continues to provide the political and economic focus of much of this region (see "Living in Russia," pages 376–378).

From the 1400s to the late 1900s, Russia was governed by feudal, centralizing tsars, or communist dictators. During the 1800s and 1900s, there were only four short periods when wider democratic powers and a freer economy were available to the Russian people. The first was when Alexander II freed the serfs in the early 1860s, the second when land reforms occurred under Peter Stolypin from 1906 to 1911, the third when Lenin introduced economic liberalization in the early 1920s, and the fourth since 1987 following Mikhail Gorbachev's perestroika policies. The processes of centralization that characterized Imperial Russia and the Soviet Union put an end to the first three short phases. Its military power, demonstrated in defeating the German armies in World War II, gained the Soviet Union status as a world power alongside the United States. Economic contacts with the outside world were restricted, leaving behind a lack of experience in the world economic system and of institutions related to it.

Following the breakup of its political and economic empire in 1991, Russia struggled to maintain its position as a leading world power. Its armies slowly returned home from the newly independent countries around it, and the Russian Federation began to embrace the world economic system.

At first, the economy declined and Russia's GDP dropped by 60 percent in the early 1990s. Democratic processes developed through the 1990s. Public institutions and welfare support systems were the main losers as a result of the political and economic changes. As a response to perceived threats to its integrity from outside, many Russians—as reflected in new political parties—wished to resist a Western takeover and to reassert their leadership role in a reestablished "empire," though (for most Russians) without the trappings of communism.

## Regions of the World's Largest Country

The Russian Federation is a huge country. Internal geographic variations are based on political, ethnic, and economic factors, and changed over time.

### Political Divisions

The Russian Federation includes a mixture of political units. There are 21 republics (of which Russia is the largest, with 117 million of the 147 million people in 1998), 6 territories,

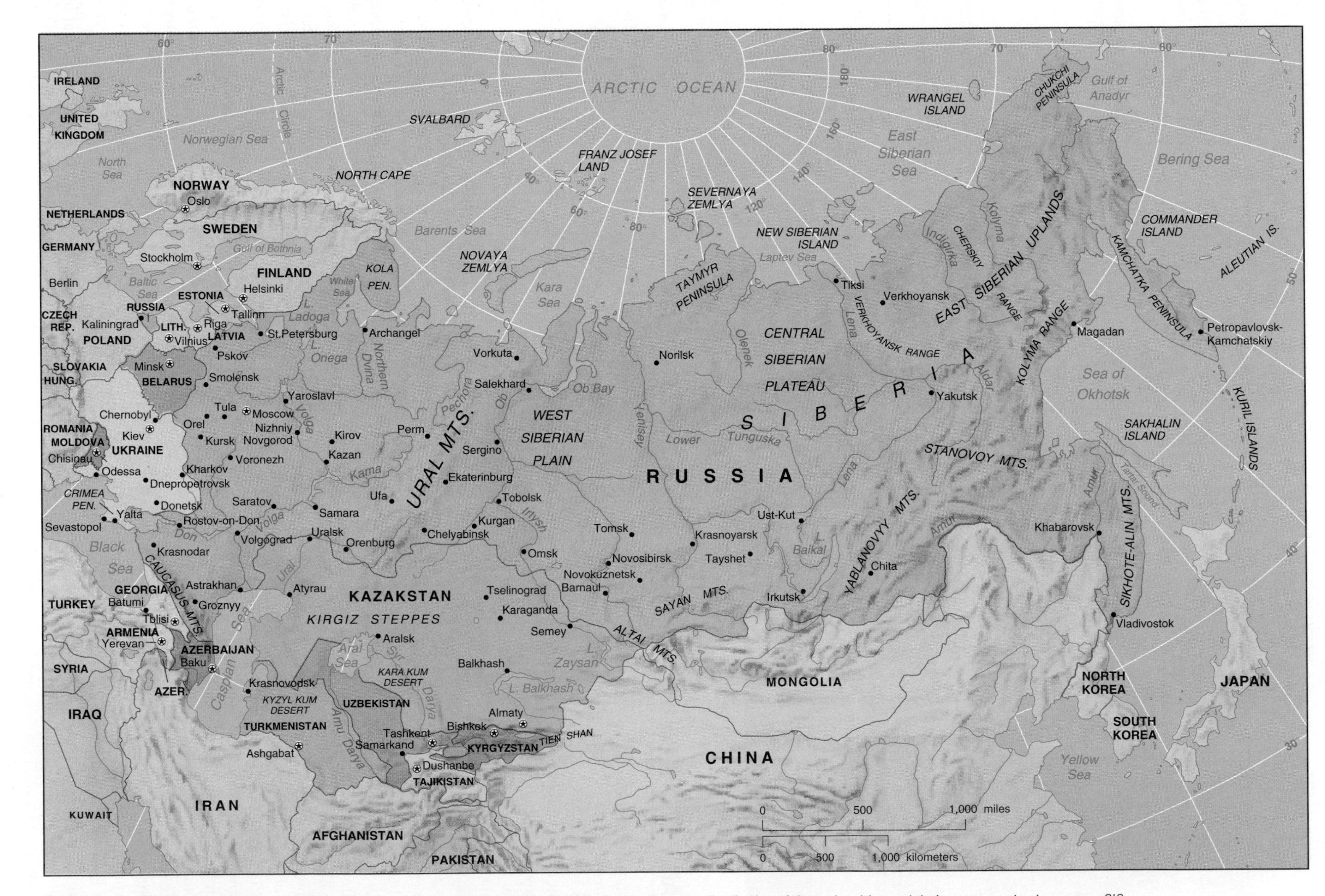


**FIGURE 8.16 Russian Federation and CIS countries: major cities and physical features.** Note the distribution of the major cities and their concentration in western CIS.

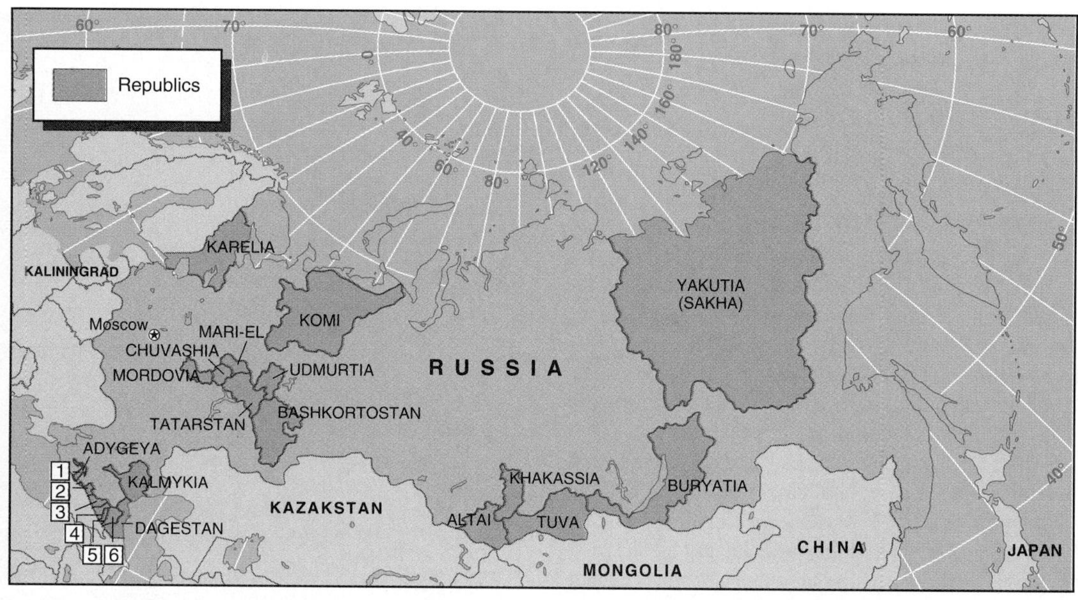


| Ethnic trouble spots | Income per person $, Nov 1994 | Population '000 Jan 1995 | % of Russians 1989 |
|---|---|---|---|
| 6-Chechnya | NA | 1,006 | 22.0 |
| Chuvashia | 36.3 | 1,361 | 26.7 |
| Dagestan | 49.1 | 2,009 | 9.2 |
| 5-Ingushetia | NA | 228 | 23.0 |
| 3-Kabardino-Balkaria | 35.7 | 787 | 32.0 |
| 2-Karachevo-Cherkassia | 35.7 | 435 | 42.4 |
| 4-North Ossetia | 40.5 | 664 | 29.9 |
| Tuva | 57.4 | 308 | 32.0 |

| Volga republics | Income per person $, Nov 1994 | Population '000 Jan 1995 | % of Russians 1989 |
|---|---|---|---|
| Bashkortostan | 51.8 | 4,077 | 39.3 |
| Kalmykia | 36.4 | 320 | 37.7 |
| Mari-El | 41.7 | 767 | 47.5 |
| Mordovia | 37.0 | 959 | 60.8 |
| Tatarstan | 48.7 | 3,754 | 43.3 |
| Udmurtia | 47.9 | 1,641 | 58.9 |
| **All Russia** | **89.9** | **148,200** | **81.5** |

| Resource-rich republics | Income per person $, Nov 1994 | Population '000 Jan 1995 | % of Russians 1989 |
|---|---|---|---|
| Karelia | 116.2 | 789 | 73.6 |
| Komi | 98.7 | 1,203 | 57.7 |
| Yakutia (Sakha) | 142.6 | 1,035 | 50.3 |
| **Others** | | | |
| 1-Adygeya | 36.2 | 450 | 68.0 |
| Altai | 70.6 | 200 | 60.4 |
| Buryatia | 45.2 | 1,052 | 70.0 |
| Khakassia | 56.9 | 583 | 79.5 |

**FIGURE 8.17 Russian Federation and its republics.** Note the special features of republics highlighted in the table, particularly the proportions of ethnic Russians in each republic.

52 oblasts (regions), and 10 autonomous districts (Figure 8.17). The total of 89 units represents different levels of size, resources, and political power. The republics and autonomous districts were established by the Soviet Union to reflect the presence of ethnic minorities, such as the Tatars and Yakuts.

- In some republics, such as Chechnya, Russians are a minority, and ethnic conflicts disrupt life. Their neighbors regard the Chechens as one of the most violent groups in southern Russia.
- A second group includes three large resource-rich republics in the north—Karelia, Komi, and Yakutia—which are closely supported, and intensively exploited, by Russia.
- A third group is formed by republics along the Volga River that have their own oil and straddle railroads and pipelines connecting Moscow to Siberia; most of Russia's 18 million Muslims, including Tatars, live there, and the group is using its strategic position to negotiate more concessions from Moscow.
- A small fourth group of just over 2 million people lives along the sparsely populated Mongolian border.

The 52 oblasts and six territories had governors appointed by the Russian president until 1997, when they began to elect their own leaders, as in the republics. This moved political action from support for the Russian president in Moscow to more local issues, eroding the Kremlin's power to govern centrally. The regional and republic governments increasingly keep tax revenues to spend within their borders, and they compete with each other for foreign investments and loans. The regions of Moscow, St. Petersburg, and Nizhni Novgorod have been most successful at attracting foreign investment. The decision by Saratov, a city on the Volga River, to freely sell surrounding farmland was followed by 12 other regions and led to federationwide progress in a process that had been very slow. The regions that lag behind the rest are those in which ethnic feuding and poverty threaten internal order—such as the Caucasus republics (Chechnya, Kalmykia) and where communist diehards in Russia's southern "red belt" resist changes.

Their variety brings uncertainty over the powers of each political unit, and they have no history of internal government. They are used only to carry out the will of bureaucrats in Moscow. Instead of making decisions in Moscow and transmitting orders throughout the empire or Soviet Union, the politicians of the Russian Federation now have to listen to and persuade the other republics within the federation to agree on common policies. Despite an initial commitment to the decentralization of policy-making, the division of

Irkutsk is a rather special city in Siberia. It lies five time zones and 88 hours by Trans-Siberian Railway east of Moscow and another three time zones and 72 hours on the train from Vladivostok on the Pacific Ocean. It was one of the earliest Russian settlements in Siberia, dating from 1652, when trappers and traders established a wooden fort at a crossing of the Angara River just downstream from Lake Baikal. Today it is home to over 700,000 people and has a much greater variety of buildings than other cities along the railway, ranging from the wooden houses with carved window frames and painted shutters that give it a villagelike appearance to the monolithic former Communist Party headquarters. Its people have a strong local pride in their city and its cultural history, having produced many poets and artists. The expulsion of Europeanized groups from St. Petersburg to Siberia early in the 1900s added to the ethnic range that includes Slavic Russians and Asian groups, such as the Buddhist Buryats living on the Russian side of the Mongolian border and the Yakuts from northern Siberia who speak a Turkic-derived language.

A major stop and junction on the Trans-Siberian Railway, Irkutsk connects southward into Mongolia and China by rail and northward into Siberia by bus and boat in the summer. It also has good air links to Moscow and other Russian cities, as well as places in northern China, Mongolia, and Japan. Chinese traders regularly appear selling a range of cheap goods. Telephone communications are good, with calls abroad often being clearer and cheaper than local or internal Russian calls. Fax, e-mail, and other telecommunications facilities are available, although many people still use telegrams. The post office remains under bureaucratic control with letters having to be sent separately from parcels and many forms needing to be filled out.

Irkutsk is an industrial and commercial city. Its domestic electricity comes from a local hydroelectric dam. Such power is also used in a nearby aluminum factory, although the future of this and other manufacturing facilities based on metal and engineering products is in doubt. Irkutsk is also a center for coal mining and oil production in the surrounding area.

The city-center shops increased their range of wares rapidly in the mid- and late-1990s: a bakery may also sell cameras, while a pharmacy may sell shoes. The department stores are more like indoor markets with different stalls competing to sell the same products. Fruit and vegetables are commonly sold in street or sidewalk booths. Some sports shops sell Western goods, but the prices are too high for ordinary Russians.

The Siberian climate brings varied hardships to the people of Irkutsk. Its winters are long and very cold. The first

(a)

(b)

Irkutsk, Siberia. (a) General view of Irkutsk in winter, with the Angara River and city center left of downtown. (b) The Russian Orthodox cathedral spires, typical of older Russian cities. (c) A typical street market stall selling vegetables and fruit, even in bitter winter temperatures. (d) The buildings on this street were erected by Japanese prisoners after the 1905 war; there is a mixture of trams and trolley buses, Russian-made and foreign autos. (e) The former Communist Party headquarters, typical of many Soviet Union buildings.

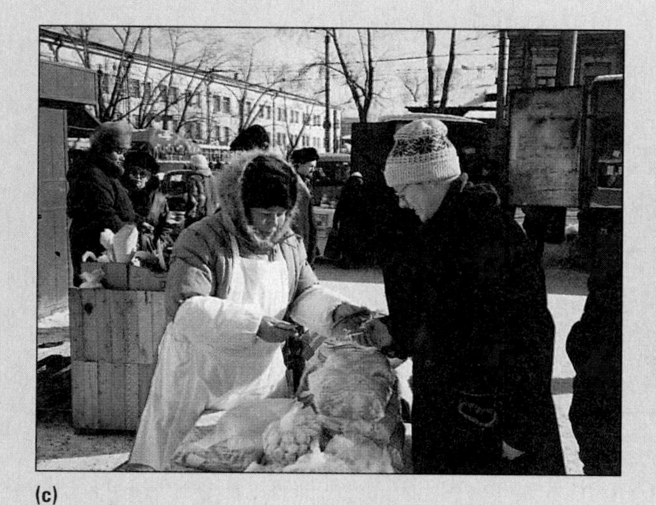

(c)

(e)

(d)

snow flurries may come in September, but by October, the ground is covered by snow that lasts until late March, although it is only 15 cm (6 in.) deep and little more falls until spring. Temperatures at their worst plunge to –40° C (–40° F) for a few days and never rise above freezing point during winter: the January mean temperature is –21° C (–5° F). In summer, daytime temperatures reach 20° to 25° C (70°–75° F) and the thick winter ice on Lake Baikal melts. There is little rain, but dry, dust-laden winds and mosquitoes make life uncomfortable. Although they are used to these conditions, local people need to wear thick fur coats and hats in winter and prefer to live in the centrally heated apartment blocks with running water rather than in the older wooden houses.

Although the city-edge apartment blocks are favored for living, they are often difficult to locate because of irregular numbering. The city authorities and large factories built them for their workers, but during the 1990s, the occupiers assumed ownership. Utilities are monopolies and own the appliances as part of a contract to supply unmetered electricity or gas. Without meters, there is no check on the amounts used, and people tend to be wasteful until supply cuts occur. Water is supplied centrally, as is central heating by water that is pumped around the whole city: warm water is turned on in October and off in April, and there is no other control of this process.

Irkutsk draws its university students mainly from the local area. There are residence halls, but most students live at home. Science, engineering, medical, and language courses, often linked to teacher training, are most popular, but there are also cultural, theater, musical, and historical studies courses. Students get into courses on the basis of their pre-university qualifications, and they have to pass annual tests to progress and qualify for the small monthly maintenance grant. Tuition is free. Personal achievement measured in tests or longer examinations is a problem since Russian students are used to working together and may talk to each other across the exam room.

Students have fewer clubs and societies than their counterparts in Western universities, but they enjoy hockey and basketball in the winter and soccer in the summer. Irkutsk won a national hockey competition in 1998 and has an open-air stadium in the middle of the city (where those watching sometimes endure temperatures of –30° C). Russian hockey is played on larger, more open ice-covered fields and has less personal contact than the hockey played in the United States and Canada. Volleyball is popular as an informal sport that requires little equipment. Theater performances, especially of music or the ballet, are low in cost—a dollar or so for a ticket. But Russians generally prefer entertaining at home rather than going out to be entertained, and there are few night clubs or discos. Birthday parties especially are major social events.

*(continued)*

## RUSSIA (*continued*)

LIVING IN . . .

The university halls provide a diet of basic food, often of low quality based on soups and meat snacks. Men living in apartments struggle to feed themselves, since this is regarded as a woman's job, as is clothes washing and housecleaning. Some students will work part-time to gain income, but this is not common. Resources in the university are poor because of low funding, and staff and students occasionally strike over pay and conditions. Professors are state employees whose pay is often delayed or withheld; when they are retired, their pensions may not be paid for many months.

The issue of low or no pay is important in Irkutsk. People manage to survive, however, by growing their own produce at out-of-town dachas. These are small wooden houses, often near Lake Baikal, on the grounds of which potatoes and other vegetables are grown; they offer some escape from the polluted air of Irkutsk.

Many Russian students entertain strong hopes of moving to a Western country. Their idealistic notions of living elsewhere are fueled by preparations that involve immersing themselves in Western clothes, music, competitive values, and language (English, German, and Japanese courses are popular). Factors such as continuing military service—which all young people have to enter for two years, usually after completing their studies—and the lack of local opportunities also make them want to emigrate. Even if university graduates obtain a good job in Russia, the pay is poor and they will not be able to live independently with their own apartment. Many continue to live with parents, even after they marry.

People living in Irkutsk, Russia, saw many changes in the 1990s. Shops contain a greater range of foods and consumer goods, although prices for the most desirable items remain too high for the majority of people. The job outlook is not hopeful, and housing is generally of poor quality compared to Western norms. Easy access to the surrounding countryside, cheap musical events, and the great strength of deep personal relationships—as opposed to the mere acquaintanceships so common in the individualistic West—remain the strengths of Russian society.

responsibilities and funding sources are still not clear. Such confusion is accompanied by different attitudes to change. Privatization, for example, is rapid in the major cities but extremely slow in the distant parts of the federation and the outlying republics, where small populations are spread over huge areas.

### Core and Periphery in Russia

Another basis of regional differences in Russia is core and periphery. The greatest concentration of Russian people and economic activities occurs in the western part of the country around Moscow. From that base, Russians extended their **hegemony**, or control, eastward across Siberia, then southward to incorporate Ukraine, the Crimea, and Caucasian areas in the 1600s and 1700s (Figure 8.18). Further areas, such as Finland, the Baltic countries, much of Poland, Central Asia, and the area inland of Vladivostok, were added in the 1700s and 1800s. All the added areas were treated as dependent colonial peripheries. Some of this territory, including Finland, was lost in 1919 when the map of Europe was redrawn after World War I. The Baltic states were incorporated in 1940. Much more was lost as republics within the Soviet Union became independent after 1991.

Today the Russian Federation consists of its Moscow-St. Petersburg area core, its immediate farming and industrial areas in the Volga River valley west of the Urals, and a series of peripheries. The peripheries include the Arctic regions around Murmansk, the mining areas east of the Urals, the almost uninhabited but resource-rich far east of Siberia, the new industrial areas along the Trans-Siberian Railway, and the Pacific region inland from Vladivostok. Large expanses of the country are virtually empty of people and economic activity and remain inaccessible to development.

Regional differences and local characteristic lifestyles became more significant within the Russian Federation in the 1990s—following the abandonment of centralized planning that often ignored geographic resource variations and sited production facilities for a variety of political, rather than economic, reasons. Realities of such factors as closeness to consumer markets or ports with world trading connections will cause geographic shifts in regional production patterns during the process of transition and will uproot many people.

## People

The distribution of population in the Russian Federation (Figure 8.19) follows the core-periphery pattern. The greatest concentration is in western Russia. Higher population densities continue eastward along the Trans-Siberian Railway to the southern end of Lake Baikal and Vladivostok on the

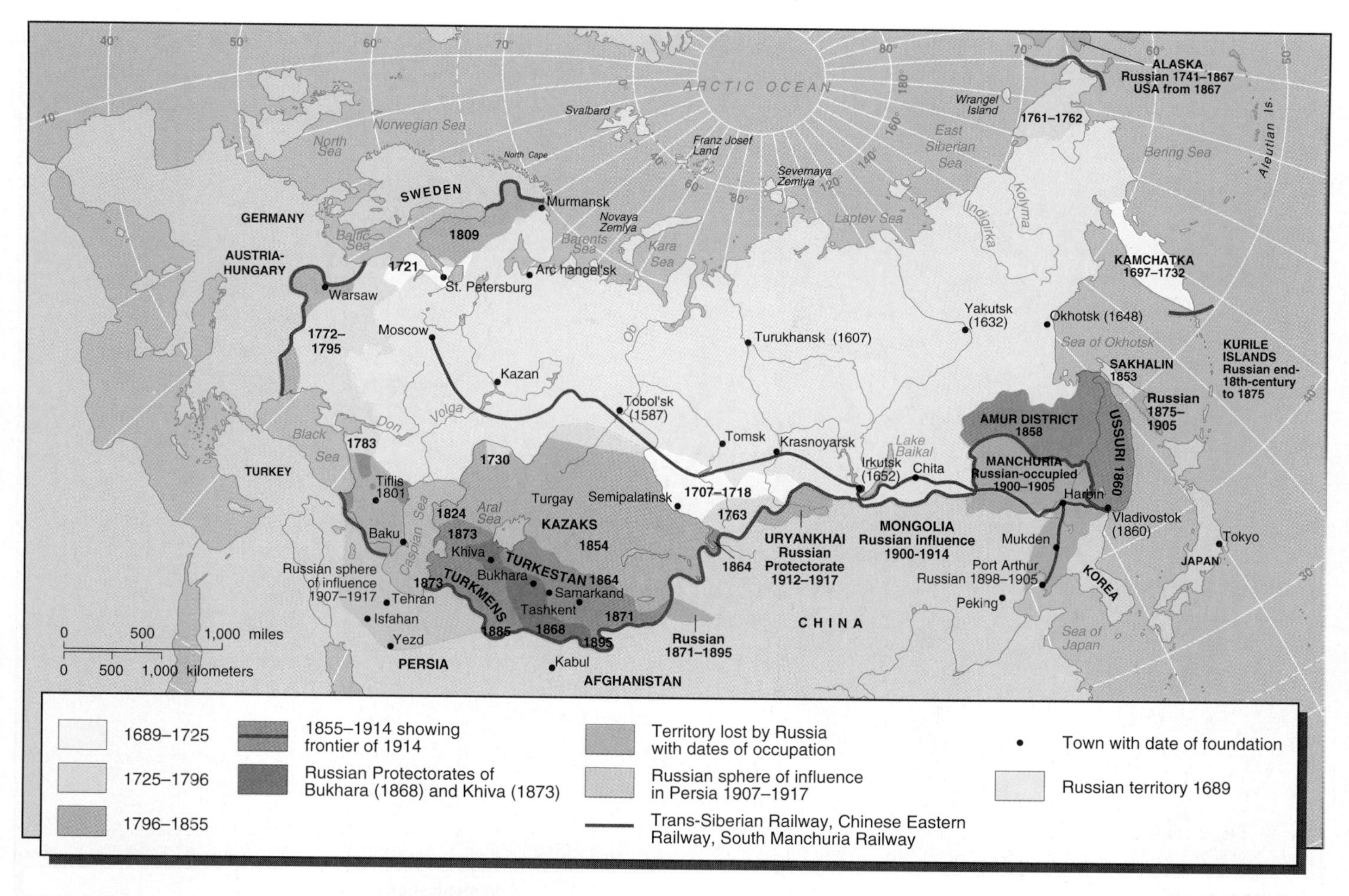


**FIGURE 8.18 Russian Federation: history of growth.** The expansion of the Russian Empire from the 1600s to the early 1900s. The Trans-Siberian Railway was built through territories that were regarded as safely Russian.

Pacific coast. The extensive areas of mountains and desert in the south and of permafrost-ridden lands in the north and east contain few people. The lack of people in eastern Siberia makes it susceptible to Chinese expansion.

## Declining Population

The population of the Russian Federation is declining. Russia's growth rate in 1990 was only 0.1 percent, and its fertility rate was 1.7. The age-sex graph (Figure 8.20) shows an uneven pattern of age groups as a consequence of a baby boom from 1950 to 1970, when large families were encouraged by the state. Women dominate the over-60 age groups to a greater extent than in other parts of the world as a result of World War II military deaths and Stalinist repression. In 1998, the birth rate was 10.9 per 1,000 of the population and falling, while the death rate was 14 and rising (Figure 8.21), giving one of the fastest rates of natural decrease in the world that year (–0.5%).

Such figures hide variations within this huge country. In some parts, deaths outnumber births by two to one. In others, populations increase because of the immigration of Russians leaving other parts of the former Soviet Union. The most significant geographic variation is between the western regions of the Russian Federation and the newly settled regions of Siberia. In the former, the population is stable or declining and aging fast; in the latter, there is some growth. The labor force is not growing in the former, but it is growing farther east.

A major problem that contributed to rising death rates was the decline in health care in the 1990s. The newly rich have access to the excellent facilities formerly available to party leaders and their families, but the rest of the population has access only to poorly paid doctors and underequipped hospitals. In 1994, a cholera epidemic spread northward from the Caucasus.

## Urban Population

The population of the Russian Federation is almost three-fourths urbanized, but the distribution is uneven. **Moscow** (Figure 8.22) and **St. Petersburg** in the Russian core are by far the largest cities. The next largest cities, including **Nizhni Novgorod**, **Novosibirsk**, **Yekaterinburg**, **Samara**, **Omsk**, **Chelyabinsk**, **Volgograd**, **Ufa**, **Rostov**, **Perm**, and **Kazan** all had between 1 and 1.5 million inhabitants in 1996.

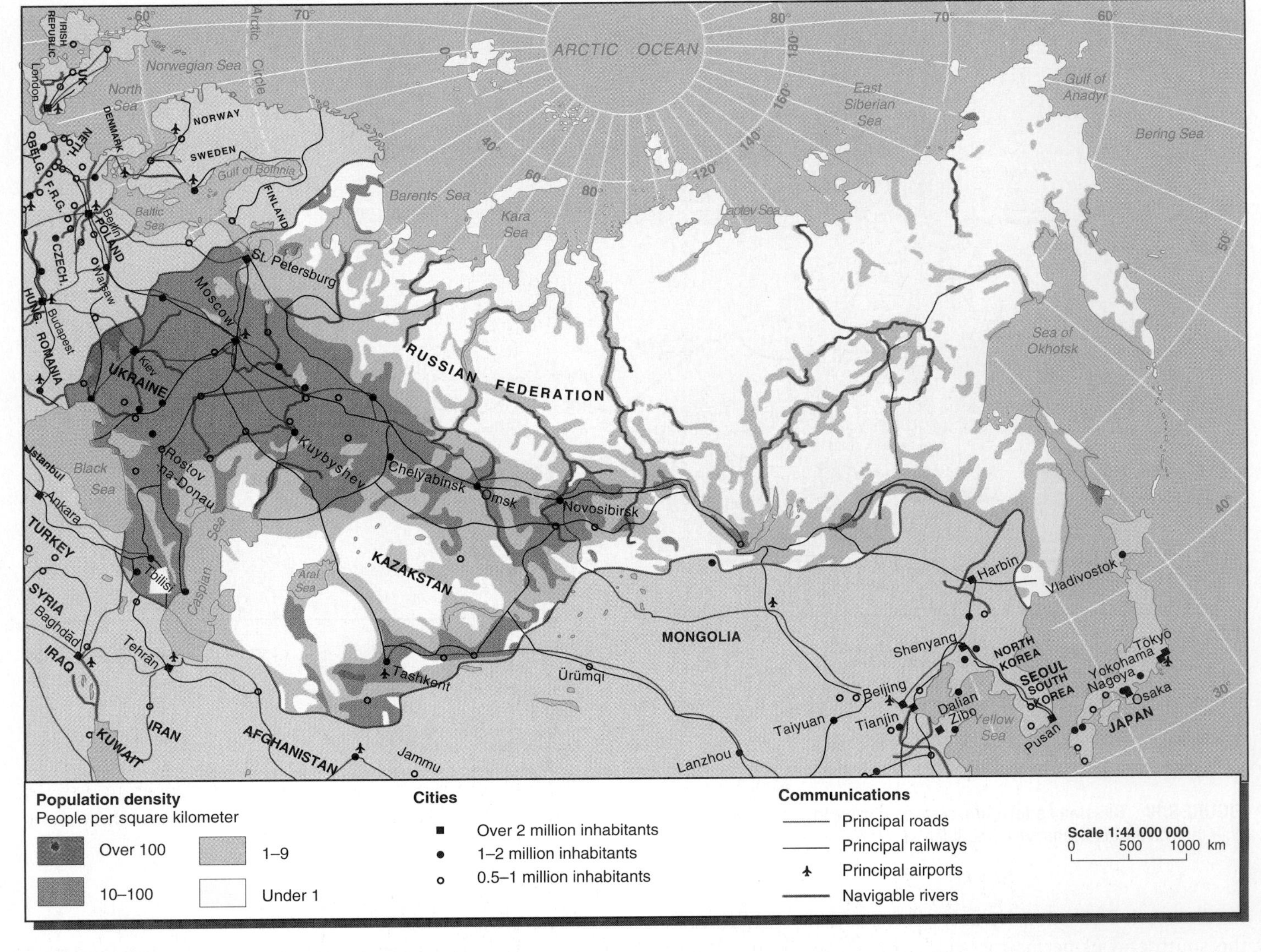


**FIGURE 8.19 Commonwealth of Independent States: distribution of population.** For Russia, comment on the location of the highest and lowest densities of population. How do these reflect climatic conditions, initial historic superiority, the line of the Trans-Siberian Railway, and other factors?

Source: Data from *New Oxford School Atlas*, p.75, Oxford University Press, UK, 1990.

In 1917, at the time of the Bolshevik revolution, only 17 percent of Russians lived in mainly small and provincial towns and cities. By 1959, 48 percent of the population was urban. Such growth reflected the state policies of industrialization and moving people into towns. Up to 1990, city growth in Russia was controlled by central state edicts. There is now a confused transfer of powers to local governments and no overall planning in the period of transition.

## Non-Russian Ethnic Groups

The Russian Federation contains nearly 30 million people who are not ethnic Russians. European Russians dominate the compact core area west of the Urals and a narrow corridor stretching east between the other republics and autonomous regions that were incorporated into the former Russian Empire in the 1800s. Other peoples were incorporated in the former Russian Empire in the 1800s. They retain cultural characteristics of language, religion, and traditional ways that set them apart, despite many years of "russification" under the Russian Empire and Soviet Union aimed at reducing their distinctive character. The Soviet Union's policies toward them were ambiguous. Distinctive language and other cultural characteristics were protected in the republics, although Muslims were often persecuted. Natural assimilation of ethnic groups was advocated but not pressed as hard as in the United States or Canada. These groups persisted and took advantage of the more liberal political environment of the late 1980s and early 1990s. Most live in the republics and

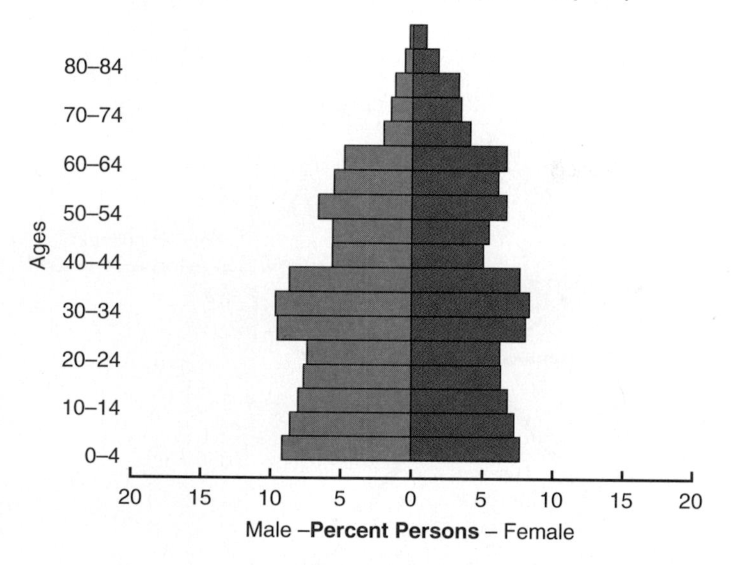


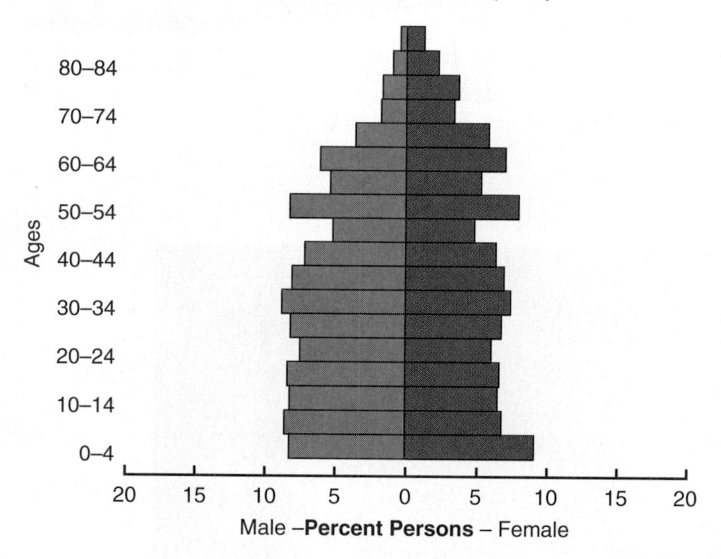


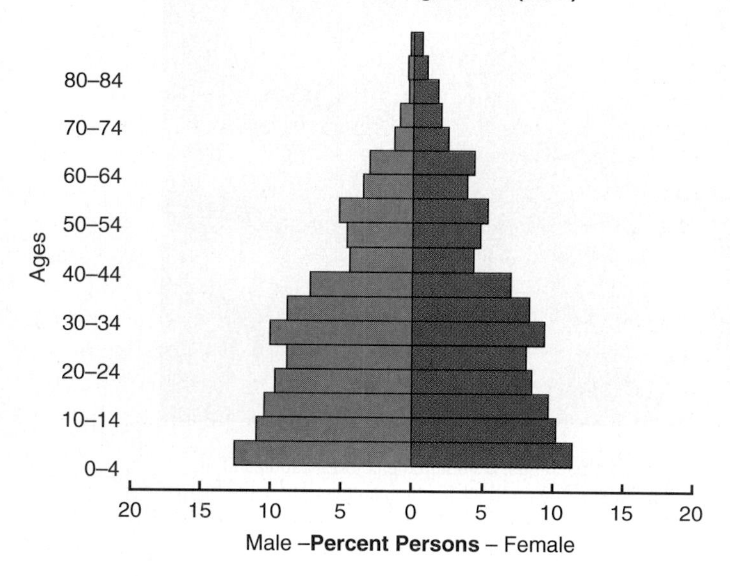


**FIGURE 8.20 Commonwealth of Independent States: age-sex graphs.** Attempt to account for the variety of irregular shapes shown in these three graphs.

autonomous regions, but often they live in political units that have a higher proportion of Russians, and so independence is unlikely.

A group of republics southeast of Moscow contains the remnants of Tatar and Hungarian-related groups. Along the northern edge of the Caucasus Mountains, a series of tiny republics contain a mosaic of Muslim, Christian, and Buddhist cultures and fight fiercely for their independence. Among these, Kalmykia illustrates some of the complex factors that produced the current human geography of the Russian Federation. The Kalmyks were Buddhists pushed out of their ancestral lands in western China by Han expansion in the A.D. 1500s. They reached the steppes west of the Volga River mouth in 1608, where Peter the Great later gave them a kingdom, or khanate. From the mid-1700s, Russians and Germans took some of this land, causing a large group of Kalmyks to attempt to return eastward in 1771. The remnant lost their towns and were resettled on collective farms under Stalin, who accused them of siding with Germany in World War II and scattered them across Siberia. Allowed back in 1957, the Kalmyks are not a majority in their own republic but claim political power. They rebuilt Buddhist temples, and their old Mongol script is being revived in schools. Soviet agricultural policy, however, turned over half of the land into desert through overgrazing and poor irrigation projects and left the republic in poverty.

The other republics and autonomous regions are large territories with few people in the northern lands that stretch from Karelia on the Finnish border to the far eastern tip of Siberia. These lands are the homes for Finns, Yakuts, Buryats (Figure 8.23), and Tuvas.

## Economic Development

The present transitional economic geography of the Russian Federation has to contend with the legacy of policies stemming from the Stalinist regime that ruled the Soviet Union from the mid-1920s to 1953. Only small changes occurred between 1953 and 1985, when Mikhail Gorbachev gained power. The Gorbachev changes conflicted with the essence of the Stalinist regime's goals and led directly to the demise of the Soviet Union.

Observers outside Russia soon realized that previous estimates of economic output in the Soviet Union and its satellite partners had been much too high. Economic output dropped rapidly. By 1995, Russia produced only 1.2 percent of the world GDP. The ownership of consumer goods reflects this modest income (Figure 8.24). TV set ownership is high since it provided mass communication from government to people during the Soviet era.

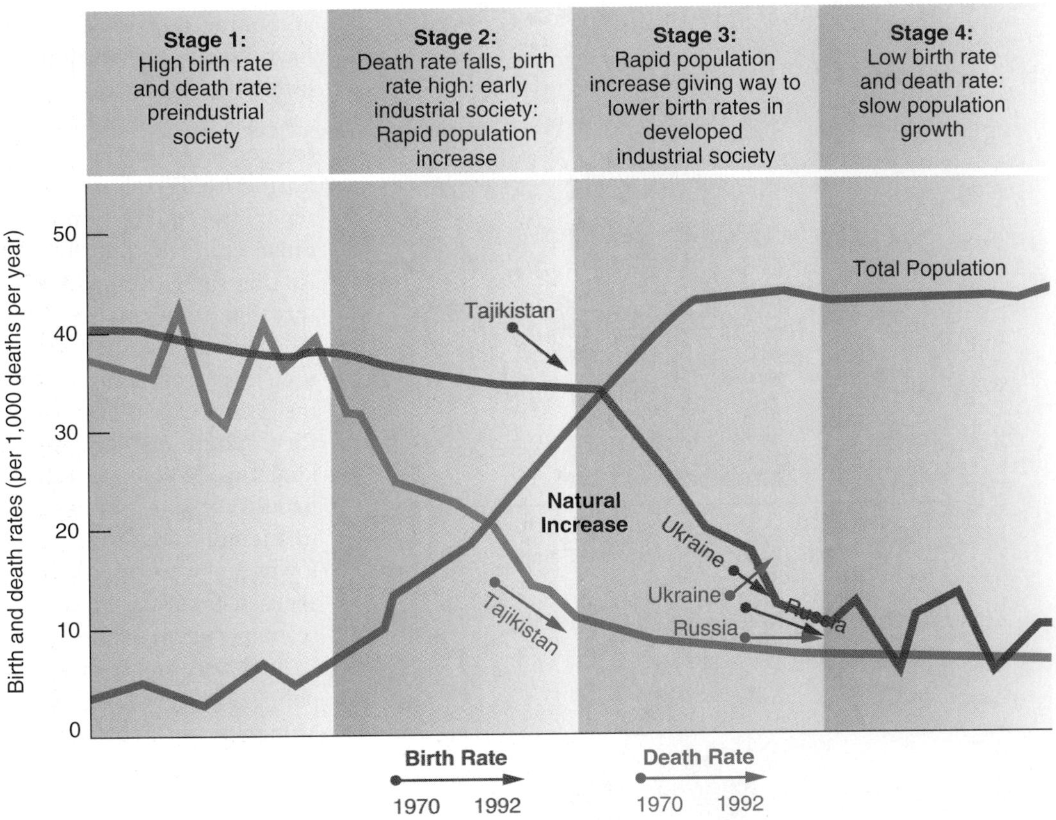

Stage 1:
High birth rate and death rate: preindustrial society

Stage 2:
Death rate falls, birth rate high: early industrial society: Rapid population increase

Stage 3:
Rapid population increase giving way to lower birth rates in developed industrial society

Stage 4:
Low birth rate and death rate: slow population growth

Birth and death rates (per 1,000 deaths per year)

50
40
30
20
10
0

Total Population
Tajikistan
Natural Increase
Ukraine
Tajikistan
Ukraine
Russia
Russia

Birth Rate
1970 1992

Death Rate
1970 1992

**FIGURE 8.21 Commonwealth of Independent States: demographic transition.** Some death rates in the mid-1990s were higher than the birth rates; some countries resembled the world's poor countries.

**FIGURE 8.22 Russia: Moscow urban landscape.** The city is marked by older buildings, including the colorful cupolas of St. Basil's Cathedral in front of the Kremlin's Spasky Tower and the dome of the Senate. The red brick building on the right is the History Museum. Red Square is in the foreground.

**FIGURE 8.23 Russian ethnic groups.** Buryat children in Irkutsk, Siberia, with their distinctive Asian features.



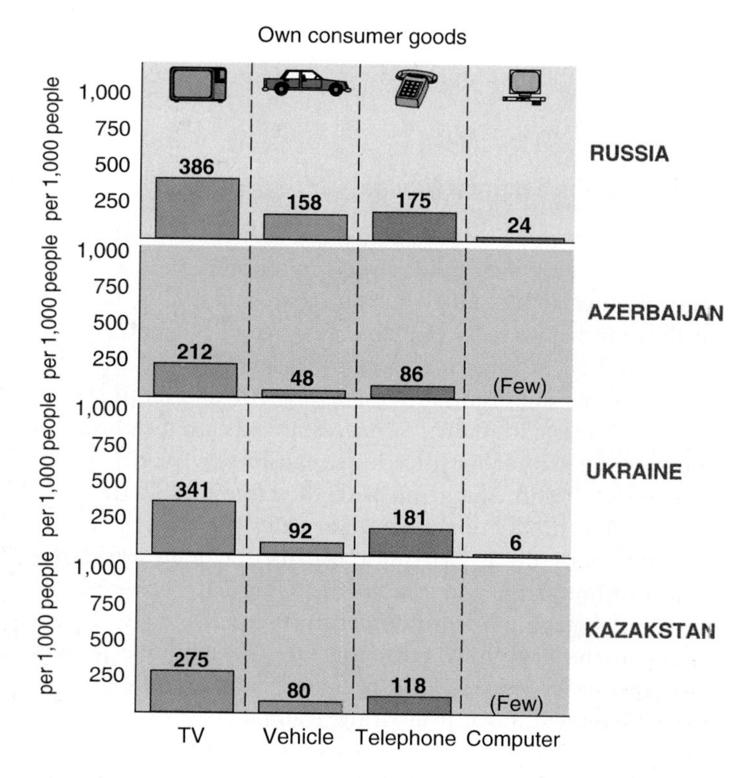


**FIGURE 8.24 Commonwealth of Independent States: ownership of consumer goods.** What is the significance of relatively high television ownership?

Source: Data (for 1996) from *World Development Indicators*, World Bank, 1998.

The balance of the Russian economic sectors remains weighted in favor of manufacturing, with farming declining and services increasing slowly. By 1998, large fortunes were being made in retail distribution and storage through cash-and-carry wholesale businesses, often bringing frozen foods and consumer goods from outside Russia.

## Basis of the Stalinist Centralized Economy

The main features of the Stalinist regime were centralized control, a focus on industrialization, what has been termed "extensive" economics, and resource self-sufficiency.

- **Centralization** involved planning concentrated in the Russian heartland. It prioritized production within sectors of the economy rather than by regional strengths. Such priorities had geographic outcomes.
- The focus on industrialization initially favored growth in the established industrial areas of the western Soviet Union. As the resource frontier was pushed increasingly eastward, new areas were industrialized. For example, the initial industrialization was based on coal and iron mined in the Don River valley but later shifted eastward past the Urals to the Kuzbas area with its newly discovered coalfields and mineral deposits. It was easier to build new factories and towns near the bulky raw materials than to transport coal and iron ore westward. After World War II, coal gave way to oil and then natural gas. Major energy sources and supplies from the Caucasus (Baku) were augmented by new sources in the Volga and Urals regions in the 1960s and by others in western Siberia and Central Asia in the 1980s. The oil and gas were transported westward to the Russian heartland and farther west into Europe by pipeline. New sources of minerals came increasingly from the east after 1960, while the main manufacturing centers remained in the west of the Soviet Union. Transport by pipeline and improved railroads made this possible.
- The **"extensive" economic system** assumed that increases of inputs would produce equivalent increases in output without raising productivity per unit. This created a continuing need for new resources and more labor as production increased—a contrast to the more intensive system of production adopted in Western countries in which greater productivity led to declines in both resource use and employed labor—and hence in manufacturing costs. The extensive system worked as long as the Soviet Union's resource wealth supported it at affordable costs. It was wasteful of resources but maintained high levels of employment.
- Self-sufficiency was a policy designed to demonstrate the superiority of communism over capitalism. It required the production of all goods within the Soviet Union and linked countries, and might have been possible with the natural resources and land areas available. It was not

achieved because of the failure to become self-sufficient in food and the poor quality of the manufactured goods for domestic use.

### Failure of the Stalinist Model

The Stalinist policies of resource self-sufficiency and industrialization were successful at first in a poorly industrialized but resource-rich country. In the 1920s and 1930s, the specialization in products at places designated by central planners resulted in huge increases in the output of goods that were deemed important in Western societies at that time. The state encouraged families to have many children in order to provide the extra labor force for expansion. By World War II, the Soviet Union caught up with the largest Western countries in steel production and military hardware.

After 1945, the Soviet Union became the main world producer of both steel and armaments. Gradually, however, the costs of bringing the resources from increasingly distant areas in the east grew until the rate of economic growth declined in the late 1970s. It became clear that, although the Soviet Union suffered little of the trauma faced by the world economic system in the 1970s as a result of the OPEC oil crisis, it failed to appreciate the changes induced by that crisis. During the 1970s and 1980s, the Soviet Union continued to place its main emphasis on manufacturing the established range of products. This contrasted with the transition from a manufacturing focus in the economy—often termed **Fordist organization** after one of its developers, Henry Ford—to post industrial or post-Fordist organization, where service industries become significant and now dominate the United States, western Europe, and Japan.

The centralization and extensive basis of economic planning caused greater economic crisis in the Soviet Union as increasing production demanded more and more labor and more natural resources from farther afield. In the 1980s, the additional costs caused the economy to stagnate. The Soviet Union remained the world's leading steel producer but its products were no longer needed on such a scale. The Soviet income from sales of oil and natural gas merely paid for grain purchases to increase the amount of food in storage. The central controls were not conducive to innovating the sort of quality consumer products that could compete with those pouring out of Western and Asian countries.

The processes of central control affected both the structure of the economy and the regional distribution of economic activities. In 1991, Russia still had 13 percent of its GDP derived from agriculture, almost half from manufacturing industry, and under 40 percent from services; by 1995, the proportions were 7 percent, 38 percent, and 55 percent, respectively. Despite the rapid shift—which reflected the falling production from farm and factory—these proportions still contrasted with the figures for western Europe and Anglo America—3 percent from agriculture, 30 percent from industry, and 66 percent from services.

### Farming Regions

The core-periphery concept holds for agriculture as well as for industrial activities (Figure 8.25). The most productive farming areas are in the west of Russia in the former steppe and deciduous forest areas. Toward the north, around Moscow, arable gives way to livestock farming. East of the Urals, farming is restricted to areas that have sufficient water and length of summer growing season. When the Soviet Union broke up, Russia lost large areas of productive farmland to Ukraine and of irrigated desert land to Central Asia.

### Russian Core

The Russian economic core region lies between its western border with the Baltic countries, Belarus and Ukraine, and the Urals to the east. The Moscow region, approximately 400 km (250 mi.) square, is home to 50 million people and was the focus of Soviet central planning and transportation routes linking the entire country. Local manufacturing includes vehicle, textile, and metallurgical industries. St. Petersburg, a major Baltic port north of Moscow, is a smaller manufacturing center but still produces around 10 percent of the total Russian output, including shipbuilding, metal goods, and textiles.

Southeast of Moscow, the Volga River is lined by a series of industrial cities that use the river transport, linked since the 1950s by a canal outlet to the Black Sea (Figure 8.26). Around 25 million people live in the Volga River region, which was developed for manufacturing during and after World War II—at a distance from advancing German armies and helped by the local discovery of major oil and natural gas fields. Manufactures include specialized engineering and the Togliatti car plant built by Fiat of Italy.

East of the Volga River basin, the Urals contain metal ores and are only a minor barrier to communications. Like the Volga region, the southern Urals were developed during and after World War II, principally as a metals center. Once again, oil and natural gas fields were discovered to the south and east.

Between these industrial areas are rural farming areas in the south, but they give way to forested lands with poor soils north of Moscow. North of St. Petersburg, there is little farming, but ice-free Murmansk has been an important naval port and center of mining towns, while Archangel'sk is a timber-exporting port.

### Siberian Contrasts

East of the Urals, Siberia forms a huge area that can be divided into the more developed southern margins along the line of the Trans-Siberian Railway and the still inaccessible northern lands. Siberia is an essential part of Russia, making up three-fourths of its land and providing a large proportion of its raw materials—a common feature of peripheral regions. In 1990, Siberia produced 73 percent of Russia's oil, 90 percent of its natural gas, 61 percent of its coal, all of its dia-

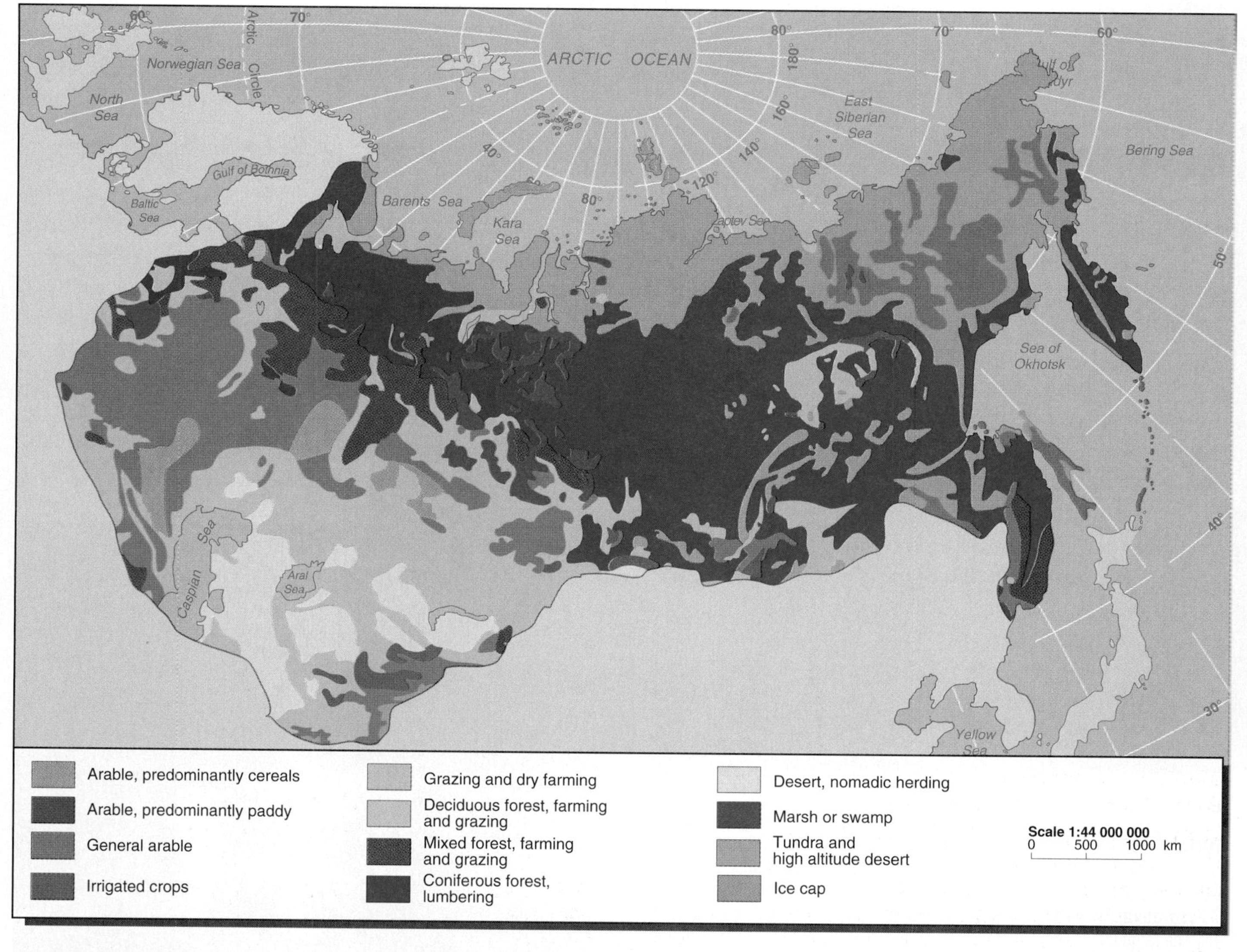


**FIGURE 8.25 Commonwealth of Independent States: major land uses.** Relate the different types of farming to the natural regions as designated by climate, natural vegetation, and soils.

Source: Data from *New Oxford School Atlas,* p.74, Oxford University Press, UK, 1990.

monds, and 30 percent of its timber and electricity. It also has a growing fishing industry on the Pacific coast. Farmed lands are mainly in the west of Siberia (Figure 8.27). Siberia remains a region of promise for enterprising Russians, often based on romantic aspirations such as taming the wilderness or making a fortune.

Along the southern strip of Siberian Russia are a number of centers that were industrialized according to the Soviet planning processes and are separated by large tracts of sparsely settled land. The Kuzbas region around Kuznetz is 2,000 km (1,200 mi.) east of the Urals, where a major coalfield was developed initially to supply coal to the steel manufacturing centers of the Urals. This led to local industrialization, linked to the discovery of iron ore and the return transport of bauxite from the Urals. Aluminum, steel, and products using them are major outputs, but the economics of such facilities after the era of centralized planning are doubtful. The world's largest aluminum plants at Krasnoyarsk, Bratsk, and Sayansk provided metal for the Soviet aircraft and missile industries, but these demands fell away in the 1990s and the output of aluminum fell by 70 percent from 1991 to 1993. Privatization of the aluminum smelters from 1994 involved foreign corporation capital and led to a major increase in exports.

Farther east, centers close to Lake Baikal, including Irkutsk, form a narrow belt of industrialization using hydroelectricity generated in the headwaters of streams flowing to the Yenisey and Lena River systems. Mining and lumbering are also important in isolated localities, while larger cities provide services to extensive areas in northern Siberia where the population and mining activities are scattered.

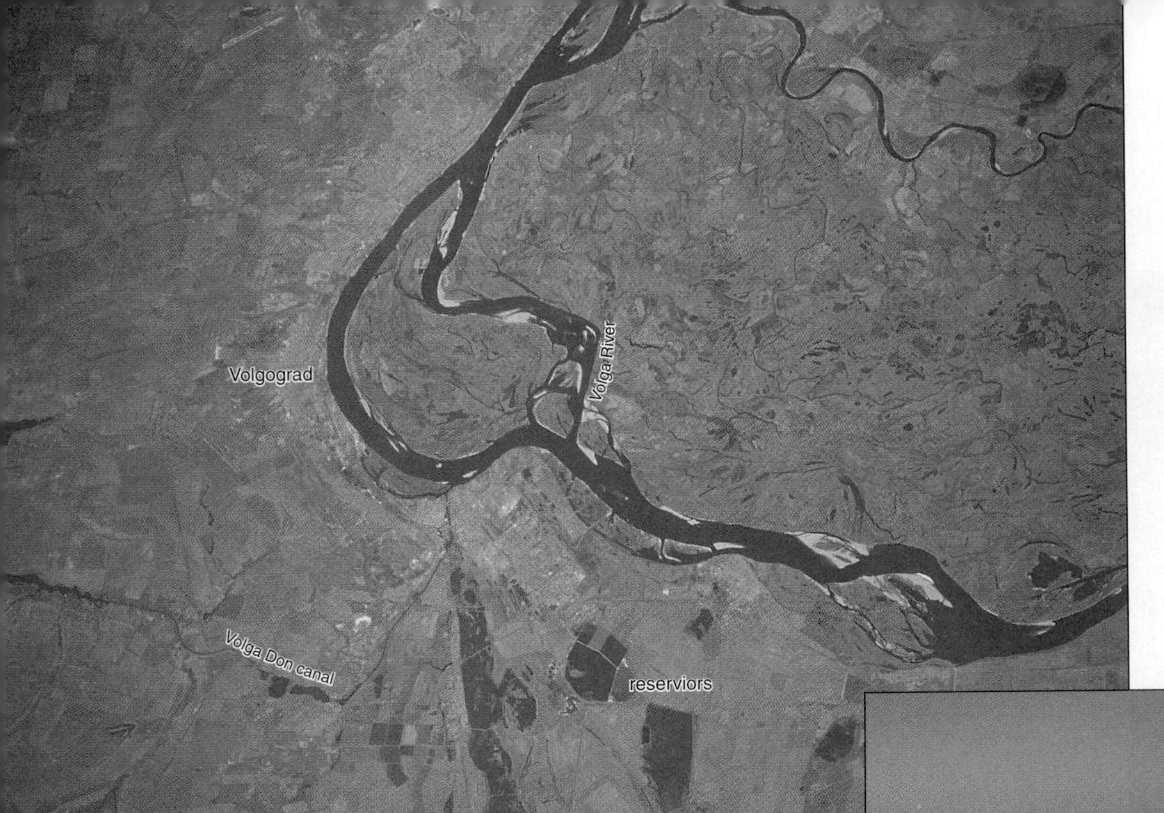


**(a)** NASA

**(b)** © James P. Blair/National Geographic Image Collection

**FIGURE 8.26 Russia: lower Volga River valley.** (a) Space shuttle view of the Volgograd area. Volgograd is a major river port and railroad junction, with two airports and large reservoirs south of the city. The Don-Volga Canal allows waterborne traffic to travel through to the Don River and the Black Sea. The large island in the center of the photo and the area to the east have many former channels of the river and small lakes—an extension of the Caspian Sea depression. (b) Manufacturing industry at Volgograd along its riverbank site. The Khimprom chemical works is one of Europe's major industrial polluters of both air and water.

The far east, flanking the southern Pacific coast of Russia, has ports such as Vladivostok and Nakhodk, and metal industries along the Amur River. Its development as a major area linking Siberian output to the world economy within the Pacific rim (see Chapters 6 and 11) is held up by political differences between Russia and Japan over the possession of former Japanese islands. Japanese markets for Siberian minerals and potential investment in eastern Siberia could be a major factor in the future development of these remote areas if the dispute over the islands can be resolved.

Northern Siberia is larger than the United States or Canada but has only 15 million people living in it. They are the remnants of ethnic groups that formerly lived by herding and hunting and of groups of people forcibly moved to the area to work in mines. The huge distances, intensely cold and long winters, and poor soils hold back development. This is a resource frontier region in which deposits of gold, diamonds, metallic ores, oil, and natural gas are exploited and form a basis for limited, often short-lived settlements.

**FIGURE 8.27 Russia: West Siberian Plain.** The southern edge of the plain on the border between Russia and Kazakstan, looking southwest. This mainly grassland area has desert to the south and woodland to the north. Lakes and dry lakes are interspersed across the flat plains, with part of the Irtysh River system. Different lakewater colors suggest different levels of overenrichment by fertlizer runoff supporting algae growths.

NASA

The possession of huge natural resources and the vast distances from Moscow prompt some Siberians to demand greater autonomy through decolonization. They lobby for greater economic freedom to plan the future and for greater benefits derived from the sale of local products. This feature is typical of peripheral regions in the world economy and affects many parts of Russia. The Tyumen area in western Siberia obtained a right to keep 10 percent of its oil and natural gas export revenues, and now wants powers to grant exploration licenses. In the longer term, Siberia may be a new frontier for Russia that will attract more foreign investment than the older industrial areas to the west. It is unlikely, however, that the Siberian republics and territories will press too strongly for political independence, since the geopolitical realities of Russian support against Chinese expansionism are widely understood.

## Foreign Investment

In the 1990s, part of the economic transition involved foreign corporations joining Russian groups to exploit its vast mineral wealth as a means of balancing the costs of importing new industrial machinery, consumer goods, and food.

The oil and gas reserves are of particular interest to the rest of the world and the main potential source of foreign investment. In early 1995, the Russian government raised oil export quotas, helping to keep internal prices low. In late 1997, the 15 percent ceiling on foreign investment in Russian oil firms was abolished and state shareholdings were offered for sale. The Lukoil company, which controls one-sixth of Russia's oil reserves and is equivalent in size to major world oil producers, now buys equipment from Western companies and enters joint development ventures. Western oil corporations that expected to be placed in charge of oil exploration

**RECAP 8C: Russian Federation**

The Russian Federation is the world's largest country and the most influential member of the former Soviet bloc of countries. It comprises 21 republics, of which Russia is the largest, and 68 other major internal political divisions.

In the 1990s, the population of the Russian Federation began to fall as death rates exceeded low birth rates. The 1998 population of 147 million people included 30 million of non-Russian ethnic groups.

The economy of the Russian Federation is based around metalworking industries, linked engineering, collectivized agriculture, and exports of mineral fuels. Its former centralized organization under the Soviet Union is now proving difficult to adapt to world economic system conditions.

8C.1 How do the terrains and economies of the Russian Federation east and west of the Ural Mountains compare with each other?

8C.2 Why is it difficult for Western corporations to enter partnerships with Russian manufacturing companies?

**Key Words:**

hegemony
centralization
extensive economic system
Fordist organization

**Key Places:**

Russian Federation
*Moscow*
*St. Petersburg*
*Nizhni Novgorod*
*Novosibirsk*
*Yekaterinburg*
*Samara*
*Omsk*
*Chelyabinsk*
*Volgograd*
*Ufa*
*Rostov*
*Perm*
*Kazan*

and marketing at first found that the Russian government expected them to enter joint ventures with companies such as Lukoil. Gazprom is Russia's biggest company with a monopoly of producing and distributing natural gas. Some 34 percent of the world's gas reserves are controlled by Gazprom, which supplies 20 percent of Europe's demands.

As oil and gas became increasingly important to Russia's economy, coal mining declined. Russian production of coal fell from 400 million tons in 1987 to under 300 million in the mid-1990s. In 1997, the home market bought only 127 million tons. Productivity per shift is at best around one-fifth of that in mines in the United States, and there is little hope of competing in world markets or attracting foreign investment. Coal industry restructuring means reducing coal mining jobs by half and closing whole areas of coal mines, leaving local populations without hope of employment. The mines that continue to function find it difficult to get customers who will pay them as the metal-based industries are in decline; moreover, semiprivatized railroads use less coal but charge higher freight costs.

Russia's mineral wealth extends to a wide variety of metal ores. At present, multinational mining corporations are particularly interested in developing gold mining, as in the world's largest gold deposit at Sukhoi Rog in Siberia. Russia also produces one-fourth of the world's diamonds but is suspicious of the De Beers worldwide marketing control of this industry and may lead opposition to it.

One of the most promising areas for foreign investment is in automobile manufacturing. Russia has only 13 million private autos, with their average age being over 10 years. The market is growing and all cars made are sold. In 1998, in one of the biggest foreign investments to date, the Gorky firm in Nizhni Novgorod entered a joint agreement with Fiat of Italy to use idle defense factories for component manufacturers. General Motors is negotiating for a similar arrangement with the largest Russian auto producer, Avtovaz, although criminal gangs besiege its factory.

### Scientific Research

Under the Soviet government, scientific research was a major goal. Scientists were pampered in terms of incomes and equipment and concentrated in new cities established for their work—some of which remained secret because they were dedicated to military purposes. Akademgorodsk, outside Novosibirsk in the center of Russia, has around 30 research institutions. For example, the Budker Institute for Nuclear Physics focused on abstract research into fundamental particle physics. By the late 1990s, its main income source was the commercial sale of spinoff products—such as instruments for irradiating wheat, electric cable insulation, and a process for purifying water—and new products such as low-dose X-ray scanners. Foreign sales are not easy, however, since Russian-made components are often poor in quality. State funds are greatly reduced for such institutes, and many of the best Russian scientists take posts abroad. For many, the move abroad is temporary, and they hope to return to better conditions, bringing knowledge of research in other countries.

### Trade with Europe

A sign of Russia's progress in the transition phase of its economic turnabout was that western Europe became its main trading partner. Over 37 percent of Russia's total trade is with western Europe, compared to 24 percent with other CIS states and 13 percent with former COMECON partners in Eastern and Balkan Europe. The energy exports to countries such as Germany carry high prices. Germany now buys 30 percent of its gas and 12 percent of its crude oil from Russia, and it could be said that Russia is fueling the growth of German industry. The potential for growth in trade between western Europe and Russia is great and may eventually equal EU trade with the United States, at present several times greater than that with Russia.

## SOUTHWESTERN CIS

**Southwestern CIS** includes six former republics that were in the Soviet Union until 1991 but are now independent countries and share the process of transition toward establishing a viable political and economic existence while remaining close to the dominant influence of Russia (Figure 8.28). Georgia did not join the Commonwealth of Independent States until 1993, but is an integral part of the subregion.

Three countries—**Belarus**, **Ukraine**, and **Moldova**—occupy the plains of the western part of the former Soviet Union that stretch from the Baltic countries to the Black Sea. **Georgia**, **Armenia**, and **Azerbaijan** occupy the mountainous Transcaucasus region south of Russia and between the Black and Caspian Seas—a region with the most complex ethnic mix of languages, religions, and historic changes of control in the world.

## Countries

Ukraine is by far the largest in size and population. In 1998 it had 50 million of the 82 million people living in this subregion. Belarus had 10 million people; the others each had between 4 million and 7 million.

Ukraine historically developed alongside Russia and was occupied by the latter for long periods. It was one of the founding republics in the Soviet Union of 1922. In 1954, the Soviet Union transferred the Crimean Peninsula (see Figure 8.10) from Russia to the Ukraine Republic. Nikita Kruschev, who made this inexplicable decision on the anniversary of the 1654 Ukraine-Russian federation, could not have envisaged the future breakup of the Soviet Union, since the port of Odessa is a major trading outlet for the industrial areas of both Ukraine and the Russian areas south of Moscow. Most of the former Soviet naval fleet was also based in Ukrainian ports. The rich agricultural western part of Ukraine has a mainly Ukrainian-speaking population, while the industrial east has a high proportion of Russians.

Belarus is a new independent country with a territory that was controlled by Poland or Russia over several centuries. The Belorussian people (one of three Eastern Slav groups, along with Russians and Ukrainians) and some Russians live on rather poor land.

Moldova is the smallest of the three plains countries and is a part of Romania that was occupied by the Soviet Union in the 1940s and never returned. Its people are Romanians, the majority of whom prefer independence within CIS to being reunited with the even poorer conditions in Romania. Its former Black Sea coast is now part of Ukraine, and so Moldova is landlocked.

The Transcaucasian countries all have mountainous terrain and suffer from internal and external strife. The situation of these countries close to the Turkish and Iranian borders inflicted histories of alternate control by Persian, Turkish, and Russian empires and left sharp antagonisms between Christian and Muslim peoples. Georgia has a Black Sea coast but much is in autonomous regions created by the Soviet Union to protect the rights of Muslim minorities, including those groups it moved into Georgia. Azerbaijan is a Muslim country that has a Caspian Sea coast and major oil fields around Baku. Armenia is a Christian country landlocked between the other two and fighting to retain the enclave of its people of Nagorno-Karabakh within the Azerbaijan borders.

## People

The three plains countries have moderate densities of population (see Figure 8.19), evenly spread, with greater concentrations around industrial cities. The Caucasus Mountains create uneven distributions in Georgia, Armenia, and Azerbaijan, with few people in the mountains and most in the plains.

The populations of the six countries in this region totaled 82 million people in 1998, and a small overall decline to 81 million is expected by 2025. Annual population growth rates by 1998 were around 1 percent in Armenia and Azerbaijan. In the plains countries, death rates exceeded birth rates, and there was –0.5 percent natural decrease in all but Moldova (see Figure 8.21). Apart from Ukraine (see Figure 8.20), the age-sex diagrams show increases in the younger age groups since 1980.

The highest rates of urbanization in 1998 were in Armenia, Belarus, and Ukraine (almost 70%), where there is a lack of good agricultural land (Armenia) or emphases on urban-industrial development. The lowest rate was 46 percent in the largely agricultural Moldova. Major cities with over 1 million people in 1996 included **Yerevan** (Armenia), **Baku** (Azerbaijan), **Minsk** (Belarus), **Tbilisi** (Georgia), **Kiev**, **Kharkov**, **Dnepropetrovsk**, **Donetsk**, and **Odessa** (Ukraine). Moldova had no large cities.

While these countries were republics within the Soviet Union, ethnic tensions were subdued. Since 1991, they came to the fore, including the conflict between Armenia and Azerbaijan over the historic Armenian homeland of Nagorno-Karabakh, the isolation of the Romanian population in Moldova, and the autonomous Muslim regions of Georgia. Muslim groups control major routeways through the mountainous area of the Transcaucasus, making existence difficult for Armenia. The tensions between Russians and ethnic majorities in all these countries affected their governments and policies. The prospects of continuing ethnic cleavages and the likelihood of attempts at ethnic cleansing in Transcaucasia in particular are even greater than those in the European Balkans (Bosnia, Kosovo), although less is heard about them.

## Economic Development

All the countries in this subregion have relatively low ownership levels of consumer goods (see Figure 8.24). In all six countries, agricultural products are important—20 percent of GDP in Ukraine, Belarus, and Azerbaijan but over 40 percent in Armenia, Moldova, and Georgia (67%). Although manufacturing output is a significant proportion of GDP, all these countries lack a strong service sector, mostly around 30

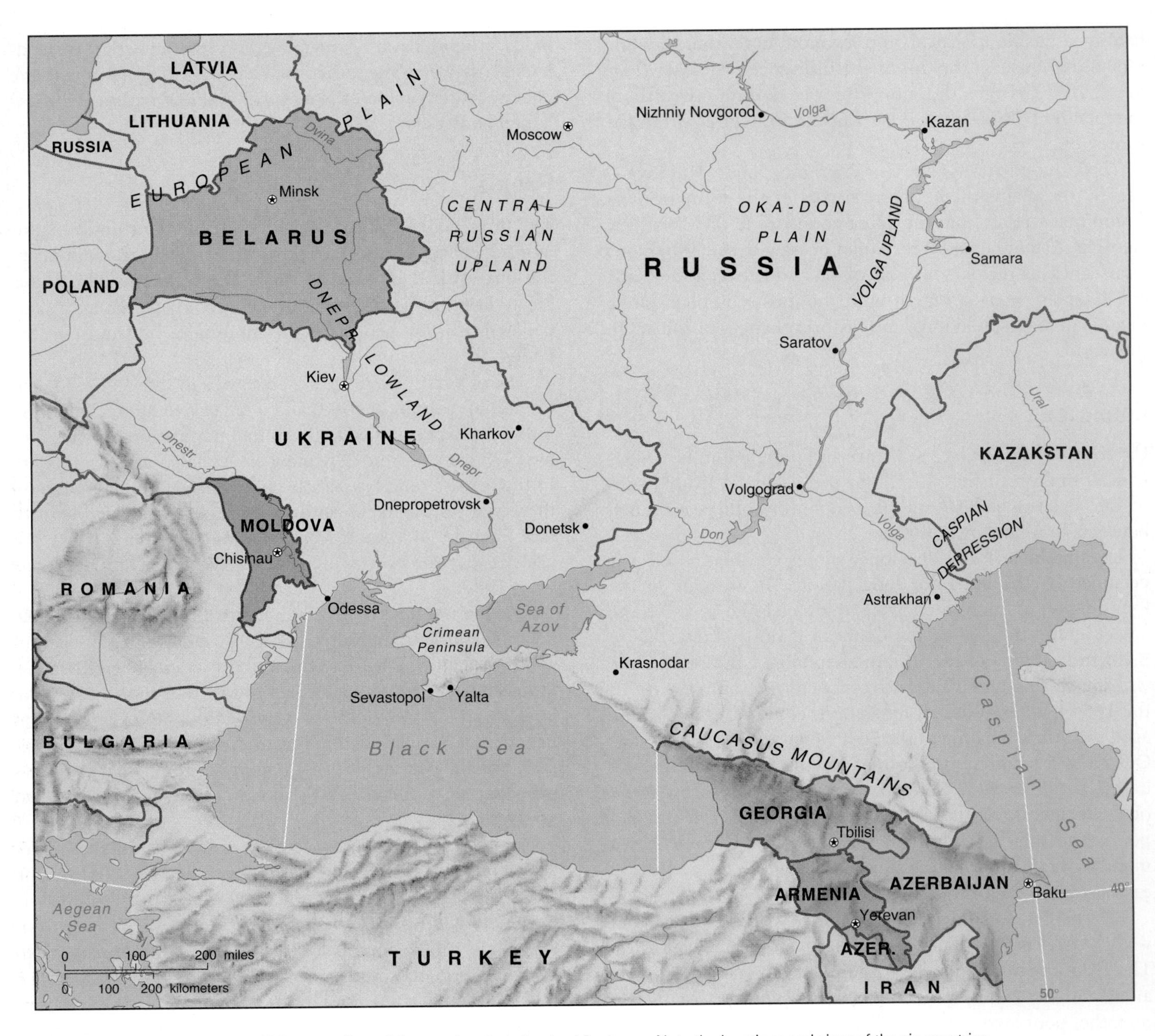


**FIGURE 8.28 Southwestern CIS: countries, cities, and major physical features.** Note the locations and sizes of the six countries.

percent of GDP, with Azerbaijan lowest (18%) and Ukraine highest (42%). That figure reflects a low investment in the financial services and professions during the Soviet era—a gap that needs filling in market economies.

### Transcaucasus

The mountainous Transcaucasian countries, Armenia, Azerbaijan, and Georgia are small, riven by civil strife, and remain very poor. Their lowland areas are more prosperous, being watered from nearby mountains in this most southerly (and warmest) part of the CIS (see Figure 8.25). Commercial farms grow warm temperate crops, with Georgia producing over 90 percent of the former Soviet Union's citrus fruit and tea crops, Armenia producing grapes for brandy, and Azerbaijan producing cotton, fruit, and tobacco.

Azerbaijan has important oil resources around Baku on the Caspian Sea. These oil resources are targeted by foreign investment, although the earthquake-prone physical environment and Russian demands increase the problems of bringing the oil to world markets. Russia tries to make Azerbaijan use the existing oil pipeline via Grozny (Chechnya) to the Russian port of Novorissisk. This is an example of the Russian Federation's continuing attempts to control events in the countries that were once part of the Soviet Union. Alternative routes for exporting oil from Transcaucasian countries and Central Asia (see Figure 8.13) would compete with the Russian outlets and reduce Russian control.

### Moldova and Belarus

Of the plains countries, Moldova, relying almost entirely on farm produce and with its government disrupted by Russian army occupation, is no better off than the Transcaucasian countries. Belarus, despite its relatively poor soils and continuing subservience to Russia, has had high capital investment in a varied industrial base, good transport links westward into Europe, and the highest income in the subregion. Its industries include chemical-based artificial fibers and fertilizers and engineering-based tractors and machine tools. Their output was based on planned former COMECON arrangements and, although they are finding it difficult to attract new destinations for goods that are not always of the quality expected in world markets, they continue to export trucks, tractors, and cheaply assembled pianos. This is helped by the customs union with Russia. Many see the customs union as the first step to a revived Soviet Union. In 1997, the Ford Motor Corporation opened a factory to assemble cars and take advantage of the Belarus customs union with Russia. It was to be followed by Daewoo, Hyundai, and Skoda, but the first two withdrew because of the financial problems in South Korea.

### Ukraine

Ukraine, the country with the greatest potential for development in this subregion, has found transition the most difficult. Ukraine's overall economy is the most diverse in this subregion, although the different aspects are not well integrated, often rely on obsolete equipment and market knowledge, and are still in a stage of reorienting themselves away from a tight integration with the needs of Russia. Ukraine's GDP per head continues to fall by 10 percent a year. Squabbling politicians discourage new investment by changing rules too frequently and by heavily taxing private businesses. U.S. and South Korean potential investors left the country disillusioned. IMF loans remain subject to Ukraine meeting conditions it is unlikely to achieve. There is resistance to privatizing land, to dismantling the bureaucratic restrictions on collective farms, and to restructuring older industries such as coal mining. The fertile soils of western Ukraine produce sugar beets, grain, vegetables, and tobacco, over half of which went to wider Soviet markets under communism. For example, a quarter of the meat produced in the Soviet Union came from this area.

The coal mines and manufacturing industries of eastern Ukraine produce fuels, minerals, steel, heavy engineering goods, chemicals, and consumer goods that were also traded with other parts of the Soviet Union (Figure 8.29). The Donbas area of the Donets Basin produced a quarter of Soviet coal output in the 1980s. Although Ukraine possesses potential oil and natural gas resources, most of these fuels used came from Azerbaijan and Romania. Ukraine has to pay world prices for these fuels while trying to reorientate its agricultural and industrial production to world markets. This reorientation needs investment to upgrade the manufacturing facilities at a time when Ukraine also has to pay the price of closing down or upgrading faulty nuclear power plants such as those at Chernobyl.

**FIGURE 8.29 Ukraine: industrial landscape.** The Donetsk area, with "hills" of coal slag, industrial smoke stacks, and an older residential district.



The lack of economic progress is causing social and political unrest in an area where the miners and factory workers were once considered an elite within the Soviet Union. The cost of producing Ukraine coal is nearly twice the world market price. Large subsidies are voted annually because closure would be a social disaster in eastern Ukraine. Yet this cannot be maintained. Only four of 250 mines were profitable in 1998, and only 50 have long-term futures. Whereas U.S. miners produced 10,000 tons of coal each in 1997, Ukraine miners managed only 200 tons per miner.

Another setback arises because Russian tourists are staying away from their former holiday resorts in the Crimea around Yalta. In the 1980s, 8 million to 10 million tourists came to the warm seashore, but by 1994 there were fewer than 2 million.

## CENTRAL ASIA

The former **Central Asian** republics in the Russian Empire and Soviet Union now form five independent countries—**Kazakstan, Tajikistan, Uzbekistan, Turkmenistan,** and **Kyrgyzstan** (Figure 8.30). These countries have similar landlocked inland situations, arid or semiarid climates, and Muslim faith of many of their peoples. Although apparently isolated from world trade and with little political or economic power in world affairs, this group of countries occupies a strategic geopolitical position with overland access to Russia, Afghanistan, China, Iran, and Pakistan.

The five countries once formed the nation of Turkestan that also included large parts of northwestern China (Xinjiang).

**FIGURE 8.30 Central Asia: countries, cities, and major physical features.** Only Tashkent has a population of over a million people. These countries have a mountainous southern border. Under Russian rule, they were carved out of the former Turkestan area.

They were incorporated into the Russian Empire in the 1800s, partly in response to Chinese expansion toward these lands. Islamic beliefs were suppressed under the Russian Empire and the Soviet Union, with deportations, the breakup of nomadic lifestyles by collectivization, and the immigration and resettlement of Russians on their lands. Since independence in 1991, the Muslim majorities are becoming significant in the political transition of these countries. Some of the countries look to links with Iran (Azerbaijani Shia Muslims) or with other Sunni Muslim countries such as Iraq or Turkey. By the late 1990s, all five countries were moving to reduce the impact of Islamic political groups, holding trials of activists and introducing restrictive laws. In mid-1998, fear of pressure from the victorious Taleban group in Afghanistan caused Tajikistan to agree to 30,000 Russian military entering the country to guard the southern border.

## Countries

In 1998, Kazakstan was the largest of the five countries in area and ninth in the world, but Uzbekistan had the most people (24 million compared to 16 million in Kazakstan). The other countries had small populations of 4 million to 6 million each.

Each country faced specific difficulties in establishing itself as a political entity. Kazakstan straddles the Russian-

Turkestan "border," with Russian people in its northern humid areas and Muslims in the southern desert. Uzbekistan is the heart of former Turkestan, but its people are concentrated at the eastern and western extremities, separated by desert. Turkmenistan has a small population but has oil wealth potential if arrangements can be reached to transport the fuel to world markets. Tajikistan is a mountainous country bordering Afghanistan and China and is the poorest of this group. Since independence, it suffered from civil war between people of Russian origin and those with Islamic beliefs. Kyrgyzstan is a difficult unit to govern because of the central mountain range that separates the lowland communities to the north and south.

Under Soviet rule, Kazakstan became a center for both nuclear testing and space exploration programs. Today Kazakstan is a declared nonnuclear country and hires out the Baikonur space facility to the Russian Federation. In the 1990s, Kazakstan attracted more long-term foreign investment for large projects—mainly oil and gas extraction—than any other country, including Russia, in the region.

## People

### Distribution and Dynamics of Population

The distribution of population in Central Asia is marked by contrasts between areas of few people in the arid and mountainous parts and areas of higher densities in the irrigated lowlands (see Figure 8.19). The main concentrations of people are separated from world economic centers by high mountains and the difficulties of having to travel or send goods through Russia.

While Russia and the Southwestern CIS countries have stable or declining populations, the five countries of Central Asia anticipate growth from the 1995 total of around 55 million people to 83 million by 2025. In the late 1990s, birth rates remained at nearly 2 percent in the countries with large Muslim majorities: Kazakstan (50% non-Muslims) had the lowest increases (0.5% per year), followed by Kyrgyzstan with 70 percent Muslims (1.5% per year). The age-sex graph for Kazakstan (see Figure 8.20) contrasts with those of the other parts of this major world region. Death rates were very low—a recipe for increasing population (see Figure 8.22). The fertility rates fell from around 6 to around 3 between 1970 and 1998. Such rates contrast with the zero growth of the rest of the former Soviet domain, but are set to fall further.

### Urbanization

Urbanization in Central Asia is the lowest in the CIS. Kazakstan, the most industrialized country, had 58 percent living in towns in 1998, but the other countries had around 40 percent. Only Almaty (Kazakstan) and **Tashkent** (Uzbekistan) had a million or more people in 1996. Tashkent was the fourth largest city in the former Soviet Union, and most of the Uzbekistan Russians live there.

### Ethnic Conflicts

Although dominated by Sunni Muslims speaking languages in the Turkish group, the countries of Central Asia contain significant minorities of other ethnic groups. Many of these minorities came from other parts of Central Asia and reflect the often artificial boundaries imposed by the Soviet Union on its former republics. External forces led by ethnic concerns also affect Central Asian policies. For example, Iranian Shiites oppose any alignments with the United States or Russia, while Russians are intolerant of Iranian moves to encourage the expulsion of Russian people.

Kazakstan contains 40 percent of Kazaks and 38 percent Russians, the latter living largely in a northern zone that was settled during the 1800s. Under Soviet industrialization, the northern zone was more integrated with the rest of the Soviet Union than with the arid areas to the south. Since independence in 1991, a former majority of Russians was reduced as Russians left the country and Kazaks from other countries in Central Asia moved in.

Uzbekistan has a more homogeneous population with over 70 percent Uzbeks, but it also has minorities of Russians (declining by outmigration) and other Central Asian groups. Turkmenistan is 73 percent Turkmen and 10 percent Russian and has many small nomadic groups of varied origins in the isolated mountain areas on the southern border. Tajikistan has 65 percent of Tajiks—a people speaking a Persian, not Turkish, language—who share the country with 25 percent Uzbeks and 4 percent Russians in the north and west. Kyrgyzstan has 52 percent of Kyrgyz peoples alongside over 20 percent of Russians and 13 percent Uzbeks. In Uzbekistan and Tajikistan, resistance to poorer immigrants settling the fertile Fergana Valley led to fighting in the early 1990s.

## Economic Development

After a medieval history of trading and conquest that is still reflected in some of the buildings of Tashkent, Samarkand, and Bukhara (see Figure 8.7), the countries of the former Turkestan returned to largely subsistence economies based on nomadic herding. The traditional herding economy with related craft industries had few permanent surface transport routes. Cattle, goats, and sheep, together with yaks on the higher pastures, provided the basis for people's livelihoods.

During the period of the Russian Empire and especially that of Soviet domination, the economic output of these five countries was redirected to supplying the needs of Russia. The Soviet Union imposed its own form of economic development on the subregion, bringing in its technologists and education systems to drastically alter the human geography.

Industrialization was based on the production of iron and steel and tractors, farming focused on the production of irrigated cotton, and the extraction of the rich mineral resources including coal, iron, chromium, oil, and natural gas turned the subregion into a Russian colony producing raw materials for Russian factories.

## Kazakstan

In Kazakstan, the disastrous development of cotton production in the valleys leading to the Aral Sea took priority over food crops, destroyed soil and water resources, and now requires major remedial action (see Figure 8.15). Yet Kazakstan has a huge area and many natural resources with a small population. It has vast agricultural potential despite the past disasters of irrigation and virgin-land plowing. It has gold, uranium, copper, and other minerals, as well as the fuels that bring most income today.

Coal output from the north of Kazakstan and oil from the western sector around the Caspian Sea was sent out of the republic to factories and cities throughout Russia. Nuclear power is used within the country. Kazakstan's oil is concentrated northeast of the Caspian Sea, and several foreign oil companies are negotiating new developments as part of the government's intention to escape Russian domination by opening its economy to world trade. Kazakstan wishes to use exploration and extraction income to fund internal development. In the mid-1990s, however, Russian influences began to intrude, causing some foreign companies to withdraw their interest.

## Uzbekistan

Uzbekistan shares with Kazakstan and Turkmenistan the blight of the Aral Sea disaster that left polluted water and high rates of cancer in cotton-growing areas. Irrigation farming dominates the country's economy but produces insufficient food for home markets. The western sector of Uzbekistan, with its irrigated farming and oil reserves, is separated from the industrialized traditional Uzbek center around Tashkent on the eastern border by desert occupied by nomadic groups. After 1991, the Uzbekistan economy suffered from the disruption of its links with the rest of the former Soviet Union and especially from shortages of fuels.

## Turkmenistan

Turkmenistan has the smallest population of the five countries and still suffers from memories of the draconian measures adopted by the Soviet Union in its attempts to stabilize the nomadic groups that lived in its desert areas. The Kara Kum Canal brings water to the eastern part of the country where cotton, grain, fruits, and vegetables are grown by irrigation. Many herders remain and Astrakhan furs are a famous local product. Natural gas reserves are being discovered in the west, and mineral extraction is more significant than manufacturing.

## Tajikistan

Potentially the poorest of the five Central Asian countries, Tajikistan was industrialized by the Soviet Union following its takeover of the country in the 1920s. Mining, manufacturing, the production of hydroelectricity, and irrigation agriculture were imposed on the country but had little effect on most people's ways of life. After 1991, Tajikistan lost many of the skilled Russian workers and administrators and factories ceased production. It became subject to civil disturbances between the Russian communists and Islamic democrats, which disrupted the economy in such ways as closing the large aluminium factory at Tursunzade. In the late 1990s, Tajikistan, where the Amu Darya and Syr Darya Rivers rise, claimed it had first right to the waters and that downstream countries should pay for their water—as they charge Tajikistan for natural gas.

## Kyrgyz Republic

Kyrgyzstan's economy is linked closely to that of eastern Uzbekistan, but the country is a problematic geographic unit, being divided by the central mountains. In the lowlands, there is irrigation farming and the towns have industries processing the farm output in textiles and food products. On the hills, pastoral farming is the mainstay, producing sheep and cattle with yaks on the high mountains. The country has some oil and natural gas resources but is remote for linking them to world markets. It remains one of the poorest countries in the region. Although a leader among Central Asian countries in economic structural reforms since 1991, Kyr-

### RECAP 8D: Southwestern CIS and Central Asia

Apart from Ukraine, the new countries of Southwestern CIS and Central Asia are small and have only moderate resources with which to establish their economic independence from Russia. Their former ties to centralized communist economics are proving difficult to replace.

Ukraine has some of the world's best quality farmland in the west and major industrial resources and infrastructure in the east, but it still struggles to adjust and make progress in transition.

The Caucasus region is the world's most complex in terms of ethnic differences and conflicts. The Muslim cultures of the Central Asian republics remain isolated economically from the rest of the world. Their world position at the heart of Asia is strategically significant.

8D.1 What are the actual and potential sources of political conflict in the Caucasus and Central Asian countries?

8D.2 What are Ukraine's economic advantages and difficulties?

8D.3 How do the physical and economic geographies of the Central Asian countries differ from each other?

**Key Places:**

| | | |
|---|---|---|
| Southwestern CIS | Moldova | Kazakstan |
| Belarus | Georgia | *Almaty* |
| *Minsk* | *Tbilisi* | Tajikistan |
| Ukraine | Armenia | Uzbekistan |
| *Kiev* | *Yerevan* | Turkmenistan |
| *Kharkov* | Azerbaijan | Kyrgyzstan |
| *Dnepropetrovsk* | *Baku* | *Tashkent* |
| *Donetsk* | Central Asia | |
| *Odessa* | | |

gyzstan's economy continues to stagnate. Its output, however, has dropped less than the other Central Asian countries. For the future, this country has major water resources, a factor that may be more significant than the current scramble for oil and gas may suggest.

### Summary

All these countries have low GDPs, which are reflected in the low levels of consumer goods ownership (see Figure 8.24). Although the irrigation farming, local manufacturing, tourist attractions, and potential oil, natural gas, and metallic mineral production suggest favorable conditions for economic development, most inhabitants remain very poor in monetary terms, and the continuation of traditional ways of life contrasts with efforts to modernize the economies. Agriculture, although restricted to irrigated oasis areas, nomadic herding, and mountain pastures (see Figure 8.25), still provides around 30 percent of GDP in these countries, with service industries around 30 percent. The Soviet development of manufacturing reduced a previous total dependence on agriculture but did not stimulate the range of services needed in a modern economy since there was so much centralized decision making in Moscow. Professional and business services are a major need in these countries during the process of transition.

One of the main problems facing the countries of Central Asia is the need to establish connections with the rest of the world economy through transportation lines. This is illustrated by the difficulties of getting the plentiful oil resources to markets. Russia blocks such trade unless it can control coastal pipeline outlets, and the development of mineral resources, including some of the world's largest gold reserves in Kazakstan, Uzbekistan, and Tajikistan, awaits outlets to world markets. The alternative routes through Russia, China, Pakistan, or Iran are all problematic politically. National airlines have too few resources to maintain a widely based network.

The legacy of post-Soviet industrial collapse, inefficient collectivized agriculture, and environmental disasters pose great problems for Central Asian countries—on top of their isolation and slow progress toward democracy and market economic systems. They now rest on primary-product output that is liable to price and demand fluctuations. Current increasing world demand for oil and gas—especially that produced in areas outside the Persian Gulf—may be affected by political decisions to reduce fossil fuel consumption or by technical breakthroughs, such as the use of hydrogen as a cheaper and more plentiful energy source.

# LANDSCAPES IN THE CIS

The landscapes of other world regions provided a geographic expression of historic changes in political, economic, and cultural conditions within natural environmental settings. The towns and countryside of the CIS countries reflect prerevolution histories and communist, Moscow-centered control.

## Urban Landscapes

Urbanization was a particular feature of Soviet policy. In 1917, at the time of the Bolshevik Revolution, 17 percent of Russians lived in mainly small, provincial towns and cities. By 1959, 48 percent of the population was urban, rising to over 70 percent in the 1990s.

### Prerevolution Russian Urbanization

Prerevolution Russian towns were often more like overgrown villages except for those such as Moscow and St. Petersburg that had grand designs of buildings and roads instigated by the tsars (see Figure 8.22). Although Russia began to industrialize in the 1800s, many developments were in rural areas and the towns that grew were of a frontier, one-industry nature with factories and worker housing.

### Stalinist Urbanization

After the 1917 Revolution, there was an uncertain period during the transition from aristocratic and capitalist control to a new socialist order. By the late 1920s, Josef Stalin demonstrated his way forward in the first of a series of five-year plans in which intensive centralization and rapid industrialization were the basis for achieving national self-sufficiency in resources and products. Such goals overrode considerations of individual living standards and true economic costs. One outcome was rapid city growth in the existing industrial centers and the development of new and specialist resource centers, often in new towns in remote regions. World War II led to massive destruction by the German armies of cities in modern Belarus, Ukraine, and the western parts of Russia. Rebuilding after the war followed many of the priorities established in the 1930s.

Stalinist urbanization produced distinctive but standardized cities throughout the Soviet Union. Rapid building was often little coordinated and of poor quality. Heavy industries and the Moscow-based ministries that developed such industries built new sectors in older towns and completely new towns. **Visual symbolism** was an important part of establishing the new socialist society. The old Tsarist memorials, churches, mosques, and even street and city names were replaced with symbols of Communist Party significance. For example, St. Petersburg became Leningrad and Volgograd became Stalingrad, a city that expanded around a huge tractor plant beside the Volga River. In the 1930s, half the churches in Moscow were demolished. Boulevards and squares were built that could be used for military displays. New statues of Lenin and Stalin, war memorials, and party slogans became a feature of Stalinist cities (Figure 8.31).

After Stalin's death in 1953, the impersonal monolithic architecture style was rejected. Stalingrad was renamed Volgograd. Some attempts were made to humanize the cities Stalin built, including a greater emphasis on housing quality and consumer and welfare needs, but the dominant state pattern of centralization and standardization continued. The quantity of provision caught up with some of the needs, so living standards rose in the 1950s and 1960s before leveling off in the 1970s.

Despite the uniformity of new urban-built environments within the Soviet Union, the stated goals of equal living standards were not achieved; social and spatial inequalities persisted. Inequalities grew among and within the expanding cities, and greater inequalities developed between urban and rural lifestyles. A study of over 200 cities of more than 100,000 people in 1970 identified four general groups of cities in the Soviet Union with distinctive political, economic, and landscape features. The slow speed of changes in the 1980s resulted in the identified features continuing into the 1990s.

- The first group of cities made up 13 percent of the total and included cities with *restricted growth and significant cultural development;* such cities were often regional capitals in which population growth was controlled and there was good infrastructure and a range of services. These were the most "liveable" Soviet cities, benefiting from centrality in the communications structure.
- The second group, making up 19 percent of the total, had both *significant growth and sociocultural development,* including more modern industries and a range of supporting urban services. They occurred mainly in the west and southwest of the Soviet Union near the frontier with the rest of Europe.
- The third group, making up 27 percent of the total, had experienced *significant growth but restricted sociocultural development.* Such cities were often built and run by industrial ministries and were concerned with production rather than worker comforts. They had inadequate social facilities, a narrow range of career opportunities, and transient, unsettled populations. Many were marginal to the main economic and political centers and were located along the southern industrializing belt.
- The fourth and largest group had 41 percent of the cities characterized by both *restricted growth and little sociocultural development.* These cities were the stagnant victims of Stalinist industrialization in the 1930s that produced redundant industries and land use patterns defying new forms of development—the Russian "rust belt." Clusters of such cities occurred in older industrial regions such as the Donbas, Kuzbas, and Urals. By 1970, they were beginning to lose population as their coal mining and steelmaking declined.

Towns smaller than those of 100,000 people also vary in their landscapes. Those close to larger cities are often better off, but those at a distance are little changed from the rural villages out of which they grew, with poor housing and services and a limited range of employment opportunities. Their factories are branch plants with small budgets for local improvements.

**FIGURE 8.31 Soviet urban landscape.** The smokestacks of the Red October steelmill in Volgograd frame the "Mother Russia" statue on Mamayev Hill. The huge statue honors the defenders of the city (then known as Stalingrad) during World War II. Large factories and such statues are common in Soviet urban landscapes.



### New Era in Russian Cities

After 1985, the Gorbachev regime attempted to liberalize the economy. Part of the package gave more planning powers to local groups. Progress on these matters in the late 1980s was slow, because of confusion over the powers of the new decentralized government entities and their sources of funding.

One of the major trends to emerge since the breakup of the Soviet union in 1991 is the building of suburbs, particularly around Moscow. The congestion and pollution of inner-city streets by the growing use of cars and trucks, together with the increase of urban crime, caused richer families to move out. The mayor of Moscow contributed to the movement by ejecting families from apartment blocks designated for renovation and commercial sale. Financial institutions including banks fund the new suburban residential developments and assess the customers' ability to pay. Most new house owners are former members of the communist establishment who transformed their political power into cash. By such means the market economy began to transform Russian landscapes.

### Secret Russian Cities

In addition to the Russian cities marked on official maps, a number of "secret cities" were revealed after 1991. Some have over 100,000 people. These cities were linked to the nuclear industry, biologic warfare research, or missile and weapon design (Figure 8.32). Built in the Stalinist monolithic style, such cities developed as a combination of scientific research institutes and labor camps; each is surrounded by a zone of cleared land and electrified fences up to 30 km (20 mi.) wide. The cities grew out of the early days of secret nuclear bomb

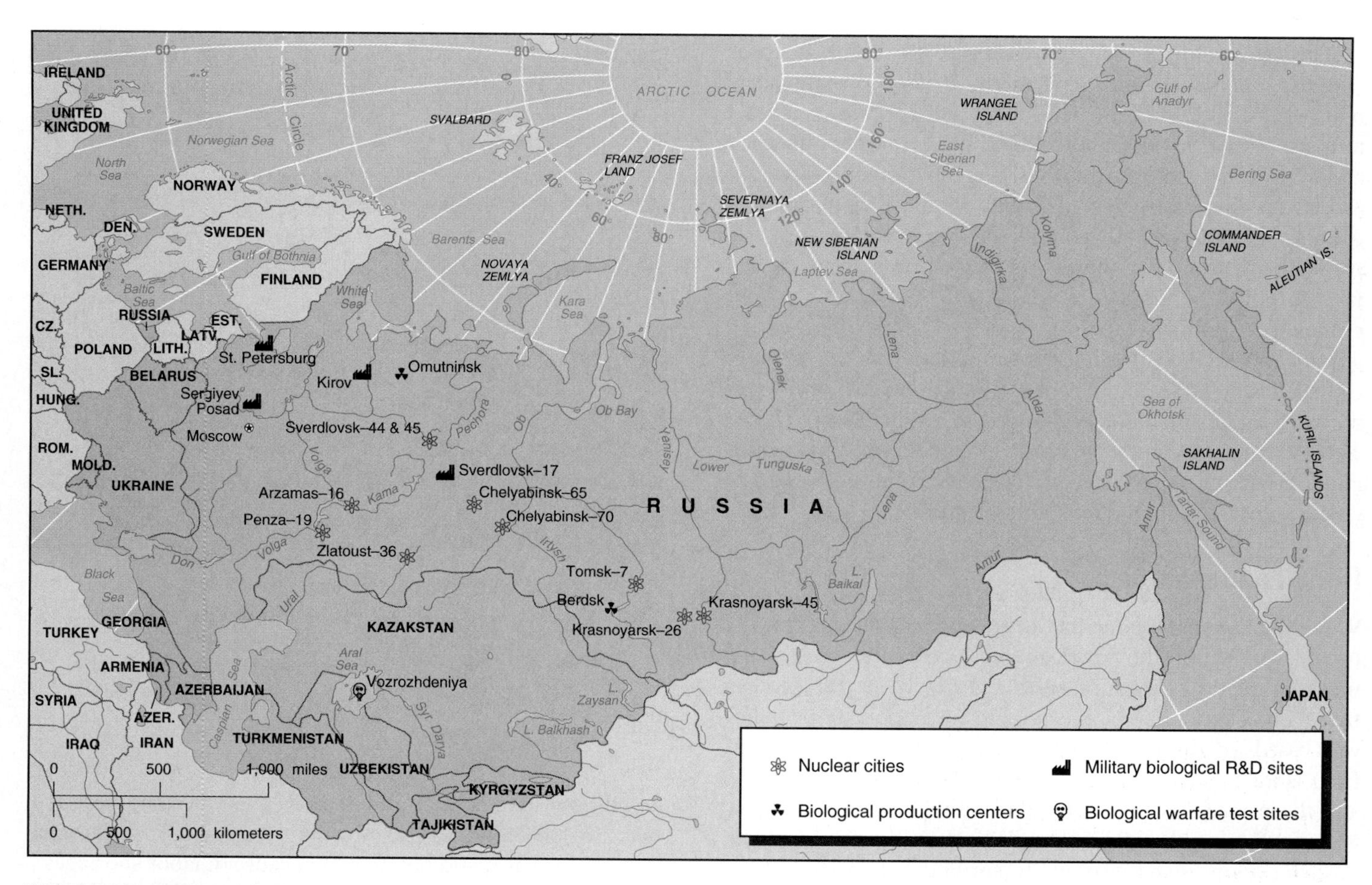


**FIGURE 8.32 Soviet urban landscapes.** The locations of "secret cities" of the former Soviet Union. Built to isolate special types of military research and with access to them limited, they now have little, if any, role.



making. Scientists who were enticed to work in them had better living conditions than other citizens. Many such cities were duplicated at great cost in case the United States struck first in a nuclear war. In some the factories, reactors, and even housing were built underground. After 1991, the Ministry for Medium-Machine Building, which controlled the nuclear program cities, renamed itself Minatom.

The concept of assembling scientists secretly in specialized centers was successful, turning the Soviet Union into a military superpower, albeit at the expense of consumer needs, long-term economic growth, and environmental damage. There is a concern about the level of environmental damage caused by the dumping of wastes and unreported nuclear accidents on a scale greater than the Chernobyl accident. Leaks from the biologic weapons centers gave rise to anthrax outbreaks near Yekaterinburg. The liquid fuel used in Soviet rockets is supertoxic and highly carcinogenic, and the reduction of rocket weapon stocks poses questions about its disposal.

Entry to these cities is still restricted, but life for the people in them has changed. It is difficult for the cities to get new work apart from contracts to carry out the more dangerous types of nuclear experiments. Privileges and status vanished as salaries shrank and were often not paid after 1991. People stay in the cities, however, since they see them as safe havens from the reported crime and instability elsewhere in Russia during the period of transition. New scientists do not volunteer to move there.

## Urban Landscapes in CIS Outside Russia

The cities of Southwestern CIS reflect a combination of historic buildings, rebuilding after World War II destruction, and Soviet-style additions. Kiev, the capital of Ukraine, is a major historic, cultural, and political center. The cities of eastern Ukraine, such as Donetsk, Dnepropetrovsk, and Krivoi Rog, are essentially manufacturing and mining towns that grew with their industries. Minsk, the capital of Belarus, had to be totally rebuilt after World War II and is a major link in the Russian oil and natural gas pipeline network.

In Transcaucasia, Tbilisi is an ancient city that is regarded by Armenians as their historic cultural center. It was taken by the Turks in the 1200s and then by the Soviet Union in 1920. It is now the capital of Georgia. Baku, the oil city and port on the Caspian Sea, is the capital of Azerbaijan. Yerevan, the capital of Armenia, is much impoverished following years of warfare against Azerbaijan.

The major cities of Central Asia are few. Some, such as Bukhara, Samarkand, and Tashkent, have older cores that were built with wealth from medieval trade and continue to attract tourists, but the Soviet Union additions are typically utilitarian and featureless factories, offices, and apartment blocks. In Kazakstan, the government decided to build a new capital at Akmola, in the northern part of the country. It was poorly planned—with utility shortages, dust storms from eroded farmland, little support from private investors or foreign governments in resiting embassies, and few planes calling at the airport. It is unlikely it will attract attention from the previous capital, Tashkent. Even the name, Akmola ("white tomb") was regarded as unsuitable and changed to Astana ("capital") in 1998. Previously, this city was called Tselinograd ("virgin lands city"), reflecting its position on the steppe plains with extreme climate, dust storms, and mosquito plagues.

## Rural Landscapes

While the Stalinist regime had major impacts on the growing urban areas within the Soviet Union, it also influenced settlements in the declining rural areas. For nearly 70 years, collectivization of agriculture was central to Soviet policy. The policy was based on the perceived advantages of joint activities—from labor organization to communal living in rural areas—and the perceived disadvantages of individualism and ethnic identity. During the 1930s, the drive to collectivization and mechanization forced peasants to surrender their land. Rural landscapes changed under large-scale farming. Goals were set including quotas and prices of food and industrial crops that formed a basis for rapid industrialization in local towns. Small private plots of land were retained, however, so that personal food needs could be met. These plots kept many rural families alive in times of famine when the produce of state farms was taken away to feed those living in towns. The policy of collectivization was maintained and post-Stalin reforms only merged collectives into larger state farms (Figure 8.33), while there were campaigns against the retention of private plots.

By 1941, 96 percent of peasant households in the Soviet Union had joined collective farms and the traditional rural society was destroyed. The liquidation of the kulaks—a label for wealthy peasants who resisted collectivization—involved executions or forced moves to other parts of the country. It often included the most skilled, experienced, and energetic peasants. The remaining rural dwellers became workers on large collectives or state farms subject to the commands and investment of central planners. The move to urbanize and industrialize the countryside was reinforced by plans to centralize housing and services in new rural towns with apartments, ceremonial squares, sports facilities, and retail outlets. Few were built and were of low standards, with only a third of apartments having running water or sewage facilities.

In the period from the early 1950s to the late 1980s, many agricultural areas outside the richest lands in Ukraine lost over 50 percent of their population in a stream of migration from the deprived rural areas to towns. Landscapes changed as villages in marginal and remote areas became deserted; over 50,000 ceased to exist during the 1980s. At times, agricultural quotas could not be met because of labor shortages.

**FIGURE 8.33 Soviet rural landscapes.** Collective farm, part of the Kholmogorye model state farm, 80 km (50 mi.) south of Archangel, northern Russia. Dairy products are the main farm output, and milkmaids are still an important part of the labor force.



During the late 1980s, the reforms proposed by Gorbachev established greater freedom for marketing agricultural produce, and the remaining private plots soon supplied nearly a quarter of Soviet food needs through private markets. There were also plans to privatize housing construction in rural areas in the hope that people would invest their future in such areas. Like many of the Gorbachev reforms, however, these changes depended on capital being available, and little was accomplished before the 1991 breakup of the Soviet Union.

Rural landscapes outside the main cultivated areas exhibit great variety because of the different vegetation patterns related to climatic differences set within the dominant plains, plateaus, and wide river valleys. Much of the region is not usable for agriculture because the climate is too arid or cold and remains in a seminatural state modified locally by mining and strategic military facilities.

# FUTURE PROSPECTS FOR THE CIS

The countries of the Commonwealth of Independent States face a future that will be determined by their ability to weather the traumatic series of transitions through which they are passing. Every week there is news of political, economic, or ethnic problems in one or more of these countries, and often several problems occur at once. Strife and self-destruction may result, as in the case of Chechnya and

WORLD ISSUE

## SOVIET LEGACY

Issues involved in this topic include:

What are the problems that affect countries shifting from Soviet control into the world economic system?

Was the Soviet system more egalitarian than the world economic system?

What was the environmental inheritance from the Soviet era?

The shift from control by the Soviet Union and its centrally planned economic system to involvement in the world economic system will take decades. Not only are the two systems opposed and incompatible, but political and ethnic fragmentation within such a large territory hampers progress. Russia, which had a central role within the Soviet Union, planned economic facilities and established a distinctive type of colonial core-periphery system. The other parts of the Soviet bloc, including the eastern parts of Russia itself, were subservient to the identified needs of the country, including its military defenses and worldwide influence. When the Soviet Union broke up, the centralizing influences ceased to affect the newly independent countries and more autonomous Russian regions. Local responsibilities increased but often without the funds to achieve much. There were also liabilities in the form of large production facilities that now had restricted markets. In the late 1990s, Russians continued to seek their empire, holding onto a sphere of influence that was extended during the Russian Empire up to 1917 and then under the Soviet Union until 1991.

The Soviet economic focus differed from the world economic system in three main ways. First, the state assumed ownership of all the factors of production, from land and resources to capital and labor. This contrasted with the emphasis on individual and corporate ownership within the world economic system. Second, the state took responsibility for the planning of production amounts and locations instead of allowing market forces to determine them. Third, the main tenet of such central planning was to increase amounts of products to fulfill increasing demand without increasing the quality or levels of productivity. Quantity and quality are both significant in the world economic system markets for goods.

Such major differences between the Soviet and market focus, combined with political and public unfamiliarity with the new system, made it difficult to switch rapidly and seamlessly from a state-controlled to a market-based economy. Moreover, private enterprise was unlawful under the Soviet state economy and linked to black markets and crooks. After the breakup of the Soviet Union, it was often those who ran the illegal black markets who were the first to set up their own enterprises and make fortunes. Many people still regard private enterprise as illicit.

Other changes that need to be made during this transition include replacing the internal state system of routing, regulation, and ownership and finding other jobs for the bureaucrats who worked to maintain the system of licensing, financing, and financial control. Russia and many of the other countries in this region have the natural resources and many of the skills needed to increase their income, but they lack the capital and managerial skills required to increase productivity.

**(a)** © Dave Bartruff/Corbis

(a) Russian soldiers in Red Square, Moscow. The large former Soviet Army still wields political power and maintains facilities in other CIS countries. (b) Older people are in the majority during a 1997 celebration of the 80th anniversary of the Russian Revolution in Irkutsk.

It is vital for these countries to attract foreign capital and work with foreign experts. By the late 1990s, foreign investment in Russia and Kazakstan was increasing. As Russia went for rapid transition, internal inflation declined and external financiers gambled on the balance between establishing an early stake in Russia's future on the one hand and doubts about the government's stability and intentions on the other. In other countries, particularly in the Caucasus and Central Asia, the balance between foreign investment and political stability was less promising. The continuing Russian influence often works to destabilize the new republics by insisting

WORLD ISSUE

## SOVIET LEGACY (*continued*)

**(b)** © Iain Searle

on continuing involvement in decisions over foreign investment, as in exploiting further the oil fields of Azerbaijan and Kazakstan.

The Soviet system professed to reduce inequalities of well-being and justice under the banner of state control but in fact fostered many inequalities of opportunity. After the 1917 Revolution, the most favored groups changed from those who owned land and other factors of production to Communist Party leaders and the scientists who worked in the secret cities. Geographic inequalities that existed before 1917,

Armenia. These considerations and a lack of statistical information make future predictions more difficult than for other regions of the world.

The rest of the world probably expected too much to happen more rapidly and should not be surprised at the succession of false Russian dawns through the 1990s. The context of the immense changes from communist to market economics involves the total change of many institutions. By the late 1990s, Russia registered its first increase in economic output since the breakup of the Soviet Union, inflation came under control, and a platform was created for economic and political programs. Yet the many possible outcomes still produce conflicting and confusing reports. In 1998, there was a further recession and financial and political crisis.

One of the major considerations is the continuing role of Russia in the internal affairs of other countries, particularly those members of the CIS (see "World Issue: Soviet Legacy," page 399). The independent countries of the CIS continue to feel pressures to reunite with the Russian Federation in economic or political alliances. Many of the CIS countries, for example, depend on Russia for their energy supplies and are running up large debts to Russia. The centripetal forces that would bring these countries closer together once again are, however, matched by centrifugal forces that continue to pull them apart, the latter particularly related to ethnic issues and recent memories of Russian control.

The breakup of the Soviet Union was a signal for many ethnic groups to assert their presence and to work for political independence in republics or autonomous regions within the Russian Federation or in independent countries within the CIS. In the mid-1990s, some ethnic groups were still prepared to press or even fight for further recognition. Many groups within the former Soviet Union have more in com-

focusing the greatest levels of human development in the Moscow region, continued and were enhanced by Soviet policies. The outlying republics and the Siberian parts of Russia continued to be treated as colonies providing resources for the Russian industrial heartland and became dumping grounds for unwanted groups of people.

Since the collapse of the Soviet Union and COMECON in 1991, the peripheral regions, often newly independent countries, experienced great difficulties in adjusting their trade and in developing facilities and systems that do not depend on direction from European Russia. The formerly pampered scientific communities devoted to inventing and producing new military hardware are likely to be abandoned. The attempt to draw newly independent states back into the CIS grouping dominated by Russia did not generate the political and economic links the Russians hoped for. Russia's paternalistic attitude to what is called the "near abroad" had to be revised.

As well as economic, political, and social legacies, the Soviet Union left behind major environmental disasters that require billions of unavailable rubles to manage. The Chernobyl incident in Ukraine is one of several, although world attention was not drawn to others. The disposal of nuclear arsenals, rocket fuel seepages, poorly managed oil installations, irrigation facilities, and the effluent of heavy industries caused a series of environmental disasters across the Soviet Union and the countries occupied by its armies in eastern Europe. For instance, in the former East Germany, the Wünsdorf area contained 250,000 hectares (625,000 acres) of Soviet military camps, airfields, shooting ranges, and waste dumps where sewage, untreated chemicals, oil, unexploded munitions, and abandoned equipment form a formidable task for those clearing it up.

The splintering of the former Soviet Union after 1991 led to groups of Slavic Russian people becoming isolated in new countries that are trying to return to a cultural emphasis suppressed by Russians for many years. The Muslim countries of Central Asia, and the countries and Russian republics of the Caucasus area in particular, provide challenges to the Russian government in Moscow as their Russian populations complain of discrimination. Russia has not yet sent an army into an independent country for this reason, although the difficulties it met in trying to pacify its own republic of Chechnya signaled Russia's restricted military power against guerrilla forces. A new theme of Russian politics—to reestablish a Russian "empire"—could herald further difficulties for the peoples who are beginning to experience a greater level of respect for the integrity of their new countries and deeply-held beliefs.

### TOPICS FOR DEBATE

1. Russia remains a major power within its own region.
2. The other countries of the former Soviet Union face a traumatic future if they do not retain links to Russia.
3. The breakup of the Soviet Union had major effects on other countries throughout the world.

mon with their Asian neighbors to the south, just as the countries of eastern Europe are developing closer ties with western Europe. The countries of the Caucasus area and Central Asia have Turkish and Muslim origins and are adjacent to Turkey, Iran, Afghanistan, and Muslim western China. The peoples of the eastern parts of the Russian Federation have links to Eastern Asian countries, and their future trade might be redirected toward the Pacific if the political estrangement between Russia and Japan could be resolved.

Economic issues are also significant. The period of transition will move faster if external capital can be attracted to fund the necessary changes in infrastructure and to be invested in competitive enterprises. Such capital is urgently needed to help restructure the Russian Federation and its neighboring countries. The CIS members have little capital themselves, and the CIS countries are already deeply in debt as the result of falling internal production and high Russian oil prices. The $24 billion promised by the world's richer countries in the mid-1990s was a tiny sum compared to the needs of 400 million people. By comparison, Germany used over $90 billion in 1991 to restructure the former East Germany with its population of 18 million people.

In the late 1990s, many aspects of the Russian economy looked bleak. Output in 1996 was 50 percent of that in 1989, prices were rising again, and unemployment increased. Over 60 million workers were owed wages, and 32 million people lived below a basic subsistence level. Farm production fell, following the lower demands from poorer people. Schools and hospitals were understaffed and underfunded. Deaths exceeded births and life expectancy fell. The worst aspects of market economies showed up in corruption, mafia-style protection rackets, and abandonment of former state-owned enterprises that could not attract privatization funds. The

**RECAP 8E: Landscapes and Futures**

The urban landscapes of Soviet-dominated eastern Europe and northern Asia contain many buildings and other features that are reminders of the central planning, monolithic architecture, and visual symbolism of the period. In some cities, these features obliterated older townscapes. Rural landscapes range from those of collectivized farmland to underdeveloped wildernesses.

The future prospects for changes of human geography in this region hinge on maintaining internal peace, forging new links with external markets and sources of financial investment, and making the best use of the region's strategic position and natural resources, as well as the will of the Russian people to develop a new area of influence on a shared, rather than imperial, basis.

8E.1 How do features of the urban landscapes generated by the former Soviet Union differ from those of western Europe?

8E.2 To what extent can Russia influence the policies of the former Soviet republics?

8E.3 Why might this world region continue to have an important strategic geopolitical role?

8E.4 Should the rest of the world have the power to insist on the destruction of Russia's nuclear arsenal and the cleanup of its environmental messes?

**Key term:** visual symbolism

legacy of former state control continued bureaucratic activity in many CIS countries, in which rules for private investments kept changing.

Yet much of the stress resulting from the changes to a market economy was past, and massive transfers of property had been made to private hands and improved economic performance started. The main sufferers by the late 1990s were older people who lost their savings in the high initial inflation, the intelligentsia who lost prestige and state income, and those living in small industrial towns where the single industry could not adapt to the changes. Younger people ages 16 to 34 saw their lives getting better, and this group included many of the new entrepreneurs and politicians. The mixed picture reflects one stage in a very difficult transition.

**WEBSITES: *on* The Commonwealth of Independent States**

The following will give you up-to-date and wider information about the countries in this region. Further sites can be accessed through the Dorling-Kindersley CD-ROM Eyewitness World Atlas online facility.

For Russia, try one of the following English-language newspaper sites, where the business news sections feature recent developments in the economy and economic geography of the country. There is an online archive back to 1994. Both papers are produced by an independent group.

*http://www.sptimes.ru/* is the *St. Petersburg Times* site.

*http://www.moscowtimes.ru/* is the *Moscow Times* site.

The following site was set up as a UN project to provide information about Ukraine. Lookup "About Freenet" and "Ukraine Links."

*http://www.freenet.kiev.ua/*

Some sites that refer to other CIS countries include:

*http://www.azer.com.* Azerbaijan international magazine. Try the "Magazine Articles" and Spring 1998 edition with articles on this and other countries in the Caucasus-Iran area.

*http://soros.org/central_eurasia.html.* The Soros organization's Central Eurasia Project provides information on Armenia, Azerbaijan, Georgia, Kazakstan, Kyrgystan, Tajikistan, Turkmenistan, Uzbekistan, and Mongolia.

*http://www.state.gov/www/regions/nis/index.html*

U.S. State Department site on New Independent States (i.e., CIS).

# CHAPTER REVIEW QUESTIONS

1. In the terminology of the cold war era, "Second World" referred to (a) the countries of South America, which were poorer than the First World, but richer than those of the Third World (b) the countries of the Soviet bloc (c) the countries which were not permanent members of the United Nations (d) the countries in the Americas, because Europe and the Middle East were traditionally thought of as the First World (e) China, Japan, and the rest of Eastern Asia
2. Two Russian terms which entered the English language during Mikhail Gorbachev's restructuring of the Soviet Union in the late 1980s were (a) Bolshevik and menshevik (b) Perestroika and zaibatsu (c) Novgorod and glasnost (d) Perestroika and glasnost (e) Chernobyl and hooligan
3. Which of the following is *not* one of the changes needed to accomplish the move from socialist economics to market capitalism? (a) Financial reform (b) Trade reform (c) A change in people's attitudes (d) Privatization (e) Government control over pricing
4. Which country claims to have been the first Christian country in the world? (a) Armenia (b) Israel (c) Italy (d) Egypt (e) Russia
5. The focal point of the earliest Russian culture was centered around Slavic tribal kingdoms in the area of which

two modern cities? (a) Moscow and St. Petersburg (b) Vladivostok and Irkutsk (c) Kiev and Novgorod (d) Berlin and Hamburg (e) Stalingrad and Leningrad

6. Which is the largest lake in the world in terms of both volume and area? It lies at approximately 58 feet below sea level and has major environmental problems. (a) Black Sea (b) Caspian Sea (c) Aral Sea (d) Lake Baikal (e) Gulf of Bothnia

7. The Russian Federation consists of (a) Russia and the Central Asian countries that were formerly called Turkestan (b) Russia and all of the former Soviet countries except Latvia, Lithuania, and Estonia (c) all of the former COMECON countries except East Germany and the Czech Republic (d) 21 republics (of which the largest is Russia), and 68 smaller territorial units with varying degrees of independence (e) the entire former Russian Empire

8. Russia's main trading partner(s) is (are) (a) other CIS countries (b) China (c) Western Europe (d) The United States (e) Japan

9. Which of the following is *not* a Central Asian country in the CIS? (a) Kazakstan (b) Azerbaijan (c) Tajikistan (d) Uzbekistan (e) Kyrgyzstan

10. One of the major trends in urban landscapes to emerge since the breakup of the Soviet Union is (a) upgrading of light rail systems (b) out-of-town shopping areas (c) building of suburbs (d) rapid increase in population leading to new development (e) all of the above are taking place

11. After Russia, Ukraine is the next largest single country (in terms of population) created out of the former Soviet Union.

    True / False

12. The most significant minority problem in the CIS concerns ethnic Mongolians, who have crossed the border into southern Siberia.

    True / False

13. The Russian economy stood still while the countries of Western Europe moved through the Renaissance, the breakdown of feudalism, and the development of mercantile capitalism.

    True / False

14. The southern boundary of the CIS region is marked by mountain systems which were uplifted by the convergence of the Eurasian plate with the African, Arabian, and Indian plates.

    True / False

15. The longest rivers in the CIS are those that flow into the Black Sea and westward into Europe.

    True / False

16. The oil and gas resources in the Caspian Sea Basin are important both to the CIS and the international energy companies.

    True / False

17. The main nickel and cobalt smelter at Norilsk in western Siberia emits seven times more sulfur dioxide than the entire metals industry in the United States.

    True / False

18. The republics of the Caucasus Mountains are entirely Orthodox Christian, except for Armenia.

    True / False

19. Ukraine is considered to have the greatest potential for development in the southwestern CIS region.

    True / False

20. It has been discovered that, since 1991, Russia contained a number of "secret cities" with populations of up to 100,000. They were left off maps for security reasons.

    True / False

21. The change from communist state control of the economy to __________ is a fundamental and traumatic U-turn. The two systems are essentially incompatible.

22. In 1922, the new communist regime incorporated the former provinces of the Russian Empire into the __________ .

23. The __________ climates of most of the CIS countries are more like those of central and eastern Canada than those of the United States.

24. North of the deserts, lying to the east of the Caspian Sea, the __________ and the fertile black earth soils on which they grow have become one of the world's major arable regions.

25. Today, the Russian Federation consists of its area core, __________, and a series of peripheries, including the Arctic regions, the mining areas east of the Urals, and Pacific region.

26. The economic system which was adopted by the USSR assumed that __________ increases of inputs would produce equivalent increases in output without the need for increases in productivity.

27. The __________ region south of Russia, and between the Black and Caspian Seas, is a region with the most complex ethnic mix of languages, religions, and historic changes of control in the world.

28. __________ was an important part of establishing the new socialist society. The old Tsarist memorials, churches, and mosques were removed and replaced with symbols of communist party significance.

29. During the 1930s, the drive to force peasants to surrender their land and to work as a pooled labor resource on socialized land was called __________.

30. In __________, the government decided to build a new capital in the northern part of the country. However, most foreign diplomats and businesspeople stayed in the old capital, Almaty, however, because the new city was poorly designed and not served by international transportation networks.

# ANGLO AMERICA

Chapter 9

## THIS CHAPTER IS ABOUT:

**The new world core: two of the world's largest and most affluent countries**

**Establishing European cultures in new lands**

**The range of natural environments that make the region distinctive**

**Two subregions:**

- United States of America: the world's only superpower
- Canada: a resource-rich country in high latitudes

**Trends in urban and rural landscapes in Anglo America**

**Future prospects for global leaders**

**World Issue: Gentrification**

**Living in the United States of America**

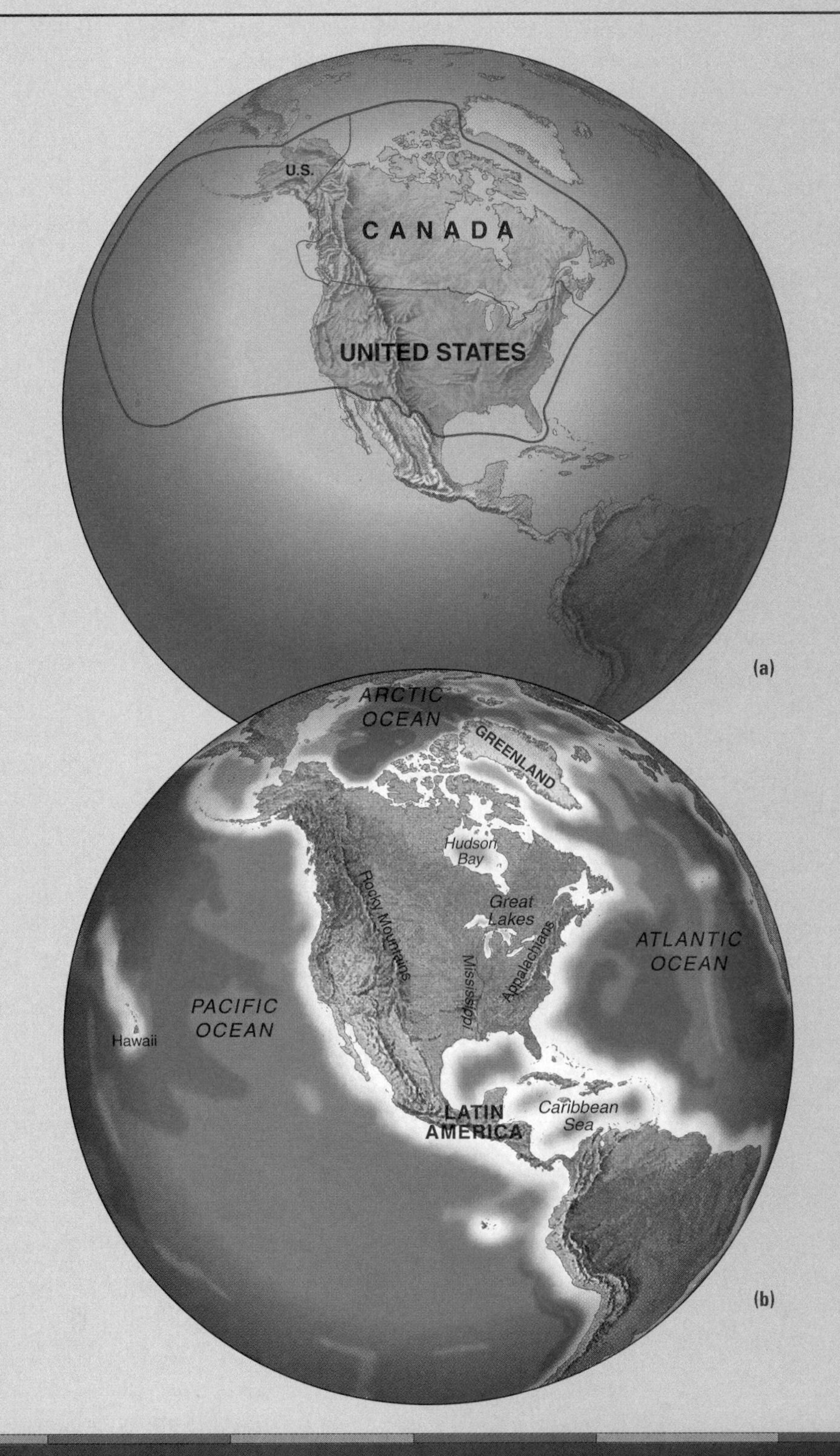


**FIGURE 9.1 Anglo America.** (a) The main subregions. For this region, they comprise two of the world's largest countries, Canada and the United States of America. (b) Natural environments. The main mountain ranges are named; the colors represent the main climatic and natural vegetation environments—desert is brown, grassland light green, woodland dark green, and tundra gray.

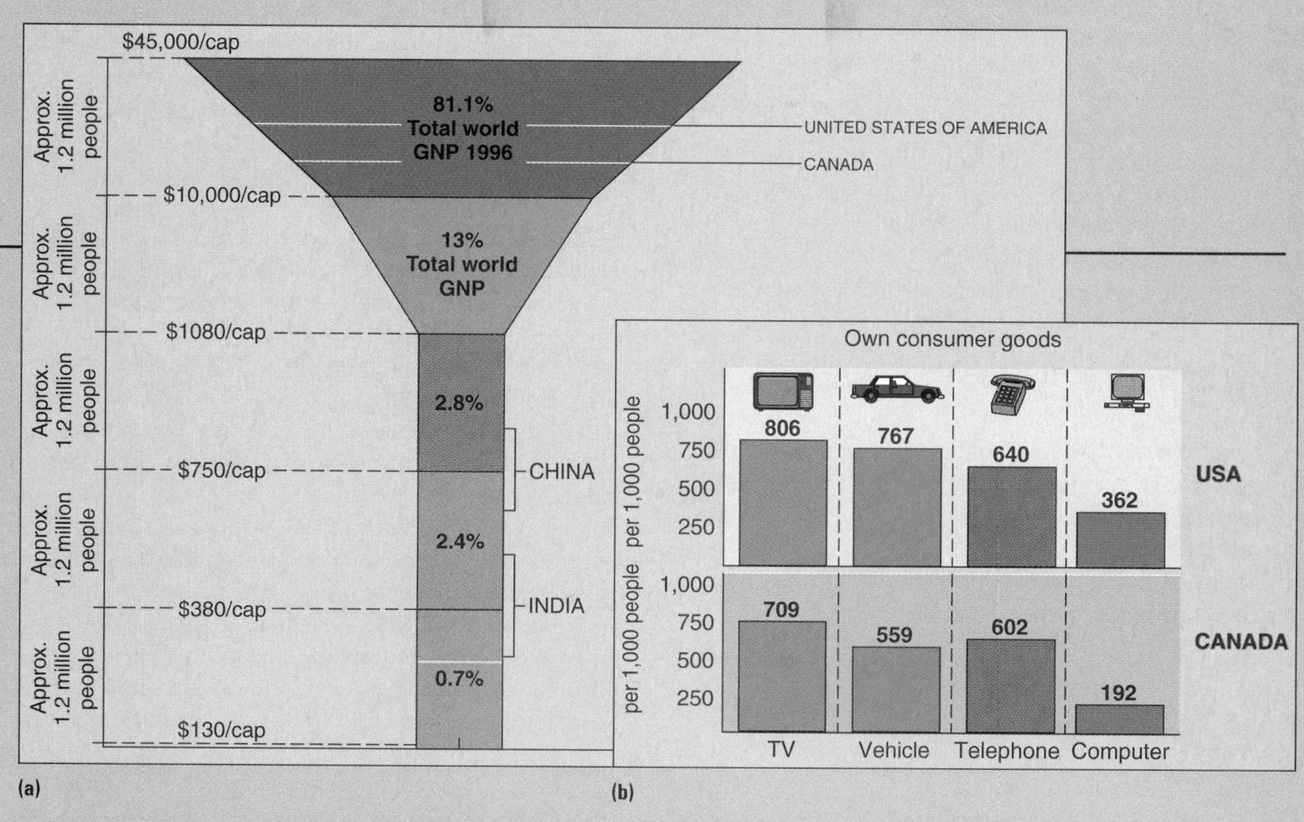


(a) (b)

**FIGURE 9.2 Anglo America: aspects of affluence.** (a) The United States and Canada are in the top group of the world's top countries by income. (b) Ownership of consumer goods is among the highest world levels.

Source: Data (for 1996) from *World Development Indicators*, World Bank, 1998.

## NEW WORLD CORE

Anglo America, composed of Canada and the United States of America (Figure 9.1), is not only a world core in what was once called the "New World." In the early 1900s it became a new core region, joining the established world core region of western Europe. The United States began to make its impression on the world economy in the mid-1800s. By the 1890s, it became the world's most important manufacturing country, outpacing Britain and Germany. In 1950, the United States produced half of the world's goods, and although the proportion fell as Europe and Japan recovered from wartime destruction on their lands, the United States retained its global leadership. In 1995, Canada and the United States produced 27 percent of world GDP (Figure 9.2), Japan 18 percent, and the five main countries of Western Europe (Belgium, France, Germany, Netherlands, and United Kingom) 21 percent.

The term "Anglo America" is used because English is the main language of both countries, setting them apart from Latin America, where Spanish and Portuguese (Brazil) are dominant. Minority languages in Anglo America are, however, significant. The French speakers in Canada constitute only 2 percent of the total population of Anglo America but one-fourth of the Canadian population. In the French-speaking province of Québec, where most Francophones live, there is sufficient political support to consider making it an independent country. In the United States, Spanish is used as a home language by 5 percent of the Anglo American population, located mainly in parts of Florida, Texas, and California. Despite the growing number of other allegiances, "Anglo America" is preferred to "North America" since, although Mexico is geologically part of North America, it has greater cultural affinities with the other countries of Latin America that were colonized by Spain in the 1500s (see Chapter 10).

Canada and the United States of America are two of the world's four largest countries in area. Canada is second after the Russian Federation and the United States is fourth after China. In population, however, the United States comes well behind China and India, having 270 million people in 1998, while Canada had 30 million. The United States had the world's largest economy in 1995. Canada, with a much smaller population, was 7th in 1980 but 9th in 1995 (after the United States, Japan, Germany, France, United Kingdom, Italy, China, and Brazil, but ahead of South Korea and Spain, all of which had much larger populations).

The huge output of Anglo America's industries thus comes from just under 5 percent of the world's population. This makes the majority of the people who live in the region extremely affluent. The high economic output made the United States a major world superpower in politics. Since the end of the Cold War in 1990, it is the only one.

Any explanation of the prominence of Anglo America in the world economic system must take account of both the region's human and physical geography, together with the historic sequence of events that catapulted the United States to world economic leadership. Although the two countries keep their political distance, Canada's geographic proximity to the United States developed increasingly into economic closeness, especially after World War II.

## NEW CULTURES

The modern prominence of Anglo America in the world economy and politics is partly explained by the ways in which people at different stages in the history of Canada and the United States interacted with the natural environment and the developing world economy. They created and recreated the regional patterns of human geography.

Until the end of the A.D. 1400s, the Americas, North and South, were unknown to the rest of the world, and their native peoples lived in societies based mainly on subsistence and the local communal ownership of resources. After Columbus discovered the Americas for Europeans in 1492, the Native Americans were overwhelmed by totally new cultures from Europe based on different technologies and social tenets. It is small wonder that in 1992, Native Americans often opted out of celebrations for the 500th anniversary of Columbus's discovery.

For nearly three centuries after its discovery by Europeans, Anglo America was a set of colonies established by Spanish, French, Dutch, Swedish, and British adventurers. The United States emerged as an independent country in 1783 following the signing of a peace treaty with Britain. At first the United States owned the eastern one-third of its present lands, but added massively to its territory between 1800 and 1867. Canada remained a British colony, achieving a degree of independence through the British North America Act in 1867 but still retaining legal ties to Britain until 1982 and thereafter remaining in the British Commonwealth of Nations as a fully independent country.

### Native Americans

Before A.D. 1500, the area now designated as "Anglo America" was inhabited solely by groups of what are now termed **Native Americans.** The first Americans may have migrated from Siberia to Alaska over 20,000 years ago. By A.D. 1500, they lived in tribal societies adapted in culture and distribution to the physical conditions of terrain and climate (Figure 9.3). The tribes of the eastern forests farmed corn, beans, and squashes, and hunted and fished, living in medium-to-large villages. Those in the warmer lower Mississippi Valley used a surplus from farming to support cities such as Natchez, which had craft industries and many other trappings of civilization.

Farther west, the grassland environment on the plains between the Mississippi River and the Rocky Mountains was inhabited by hunting groups who killed bison to supply their food, clothing, and housing needs. Their tepee tents could be moved to follow the bison herds. This economy grew in significance and in the numbers of people it could support after horses became available in the 1600s, following their introduction by Spanish colonists in the American southwest. Even at its height, such a grassland culture supported a lower density of people and smaller working groups than the eastern forests.

Tribal numbers were even sparser farther west in the more arid parts of the mountains and high plateaus and farther north in the colder parts of northern forests and arctic lands. Some locations in the arid southwest favored the development of primitive irrigation farming and the building of distinctive villages by the pueblo cultures (Figure 9.4). Along the Pacific coast from present-day northern California to southern Alaska, small tribes found plenty of seafood, including salmon and tuna. They carved the tall trees into dugout canoes, plank houses, and totem poles. The Arctic and north Pacific coasts of the continent supported small numbers of Aleuts and Inuits (Eskimos), who hunted and fished the nearby ocean waters.

For many Native Americans, survival was an achievement. Social customs were based on the need to ensure supplies of scarce food and materials for house building, clothing, and crafts. Periods of drought or cold caused groups to starve or migrate to new lands perceived as having better provision for basic needs.

### European Settlers

The regional diversity of the first inhabitants of Anglo America continues today, but the Native Americans played little part in the further evolution of the landscape after people from European countries landed, settled, and developed the natural resources for their own individual and group financial gain. The Native Americans, with their tribal culture that included holding land and resources in common, were killed or swept aside by the advancing settlers and the diseases they introduced. As the settlers' land demands increased, Native Americans were confined into smaller and smaller reservations in areas not wanted by others. This process began in the

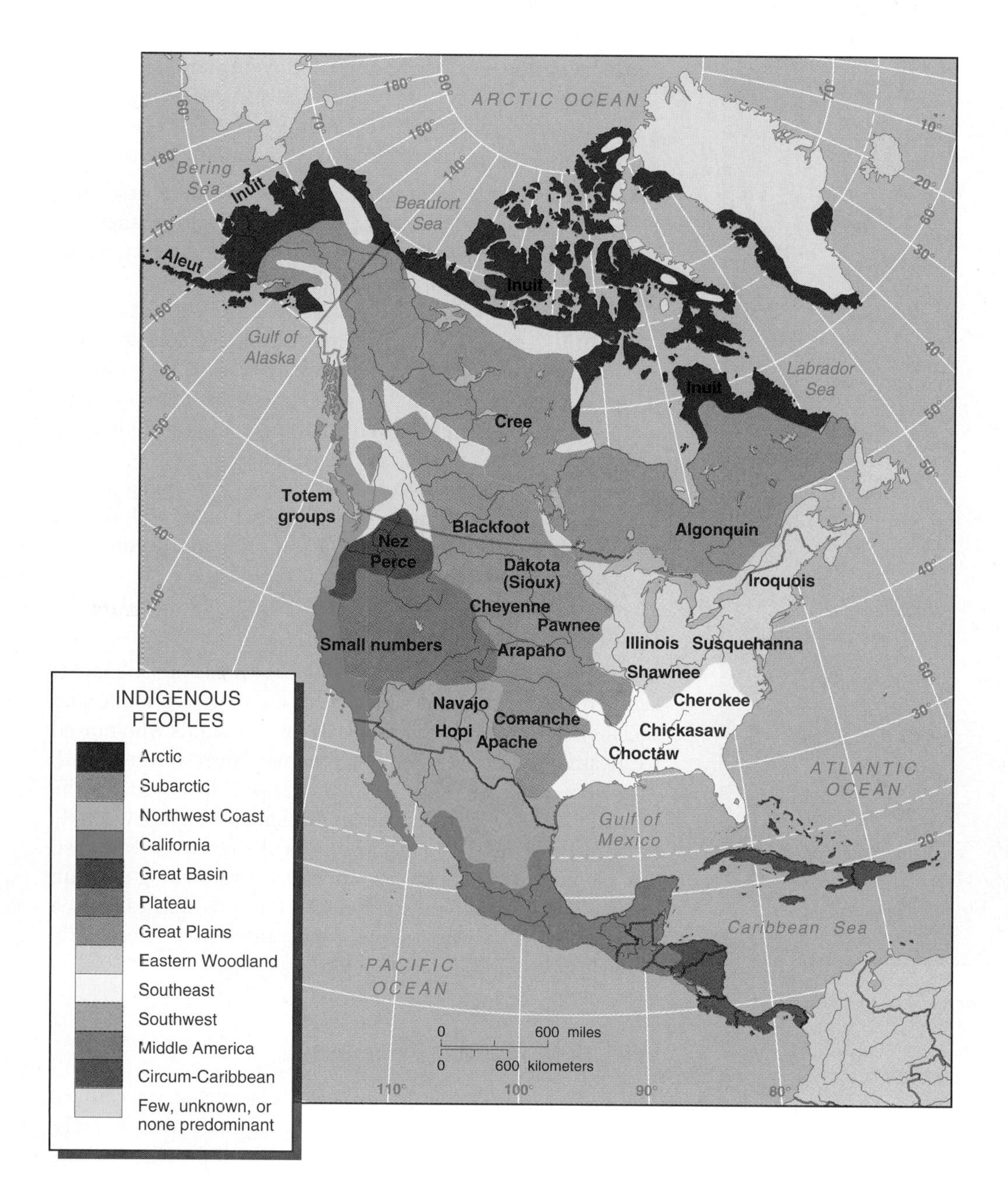


**FIGURE 9.3 Anglo America and Middle America: the distribution of Native Americans.** The eastern groups lived in woodland environments and had well-populated villages; the Great Plains groups had smaller numbers who hunted grassland animals. The southwestern arid areas, the northern forest (subarctic), and the arctic areas had the smallest groups. The woodlands and fisheries of the northwest supported the totem groups. Other distinctive groups of Native Americans lived in Middle America and the Caribbean (see Chapter 10).

east and southwest in the 1600s but lasted into the 1800s in the central and northwestern parts of Anglo America.

Although the "Anglo" element eventually became most prominent in both Canada and the United States, other European countries including France, Spain, the Netherlands, and Sweden were all involved in the early settlement of this region. The geographic pattern of these events had a formative impact on later internal regional distinctions.

- The Spanish settled modern Mexico and parts of modern Colorado, California, Arizona, New Mexico, and Texas from the late 1500s as the colony of New Spain. Not finding gold, they left the lands to Roman Catholic missions and cattle ranches.
- The French settled the mouth of the St. Lawrence River in modern Canada. They established river-related farm settlements along the St. Lawrence River and explored the interior, helped by Native Americans, to hunt for furs, particularly beaver. The French traded in and claimed lands through the Great Lakes and along the Mississippi River, where their long-lot field system still survives.
- Settlers from Britain established permanent settlements along three areas of the **Atlantic coast.** The tidewater areas around Chesapeake Bay provided the sites of the first permanent British settlements. Companies that expected their investments to return profits in the form of precious minerals and stones funded them. Failing that, a

**FIGURE 9.4 Anglo America: Native American settlement.** The Taos pueblo, New Mexico. Such settlements reflected the local climate and materials available. The mud brick houses are typical of arid areas. The beehive ovens are used for cooking. Winters are often cold, but snowfalls are unusual.



commercial crop or hunted animal pelts would do. The first permanent Virginia settlement, first established in A.D.1607, discovered the commercial crop they required in tobacco and adopted Native American crops for subsistence. Sales of tobacco back in England guaranteed the future, attracted more settlers, and led the British government to take over the lands as a Crown colony. This early success led to the production on individual plantations of an increasing range of commercial crops including indigo, rice, sugar cane, and, later, cotton. The plantations and colonies spread southward along the Atlantic coastal plain and in 1619 began importing black slaves from Africa to do the field work—the origins of **African Americans.**

- In 1620, the second wave of British settlers were led by a group of religious emigrants from England on their way to Virginia. They landed at Cape Cod in modern Massachusetts and moved across the bay to Plymouth, beginning the settlement of the area that became known as **New England.** Other groups followed and established a **township**-based culture founded on self-governing villages and a subsistence economy. It contrasted with the southern settlements in being more community-oriented.
- Between New England and Virginia, the **Middle Atlantic** was settled in the 1630s by Dutch settlers around their town of New Amsterdam on the site of modern New York. Swedish settlers took small areas along the lower Delaware River. The Dutch ousted the Swedes, but then the British ousted the Dutch in the 1660s. The Duke of York was presented with their lands and gave his name to the largest city, changing it from New Amsterdam to New York. He paid off his debts by making over most of his remaining lands to William Penn, the founder of Pennsylvania. This Middle Atlantic area attracted a third type of settler, wishing to set up family homes with the least amount of fuss or external control in a new country. Many Scotch-Irish and Germans came to the region in the early 1700s and established a farming system based on growing corn and raising livestock.

After independence in 1783, the three areas of settlements along the Atlantic coast with their different economic and social systems formed springboards for thousands who moved westward within the United States as new lands were acquired. The New England subculture spread into areas around the Great Lakes. The Pennsylvanian system of individual family farms became the basis of farming in the southern Midwest. The plantation system was extended westward along the Gulf lowlands to Texas. The North-South tensions that came to a head in the Civil War of the 1860s arose from these early differences.

## Wealth Of Natural Resources

The early settlers of Anglo America discovered a wealth of natural resources. They included land, good soils, minerals, fur-bearing animals, fish, and timber, which provided a basis for commercial farming and industries.

### Primary Products and New Lands

At first, in the 1600s and 1700s, the wealth was unearthed by an agricultural society. The long, warm summers of the widening coastal plain southward from New York encouraged the commercial farming of subtropical crops such as tobacco, sugar cane, rice, and indigo, and, later, cotton. Investments from London and other European cities could be repaid from the profits of selling such crops in Europe. New England's timber and pitch provided new ships for the British navy, and a third of Admiral Nelson's ships that fought against Napoleon's French fleet in the late 1700s were built there. In Canada, the coastal harvest of fish led to an inland harvest of animal furs, especially beaver.

After establishing full independence in 1783, the United States tripled its area by 1850, buying land from France and Spain and acquiring the western third by negotiation and

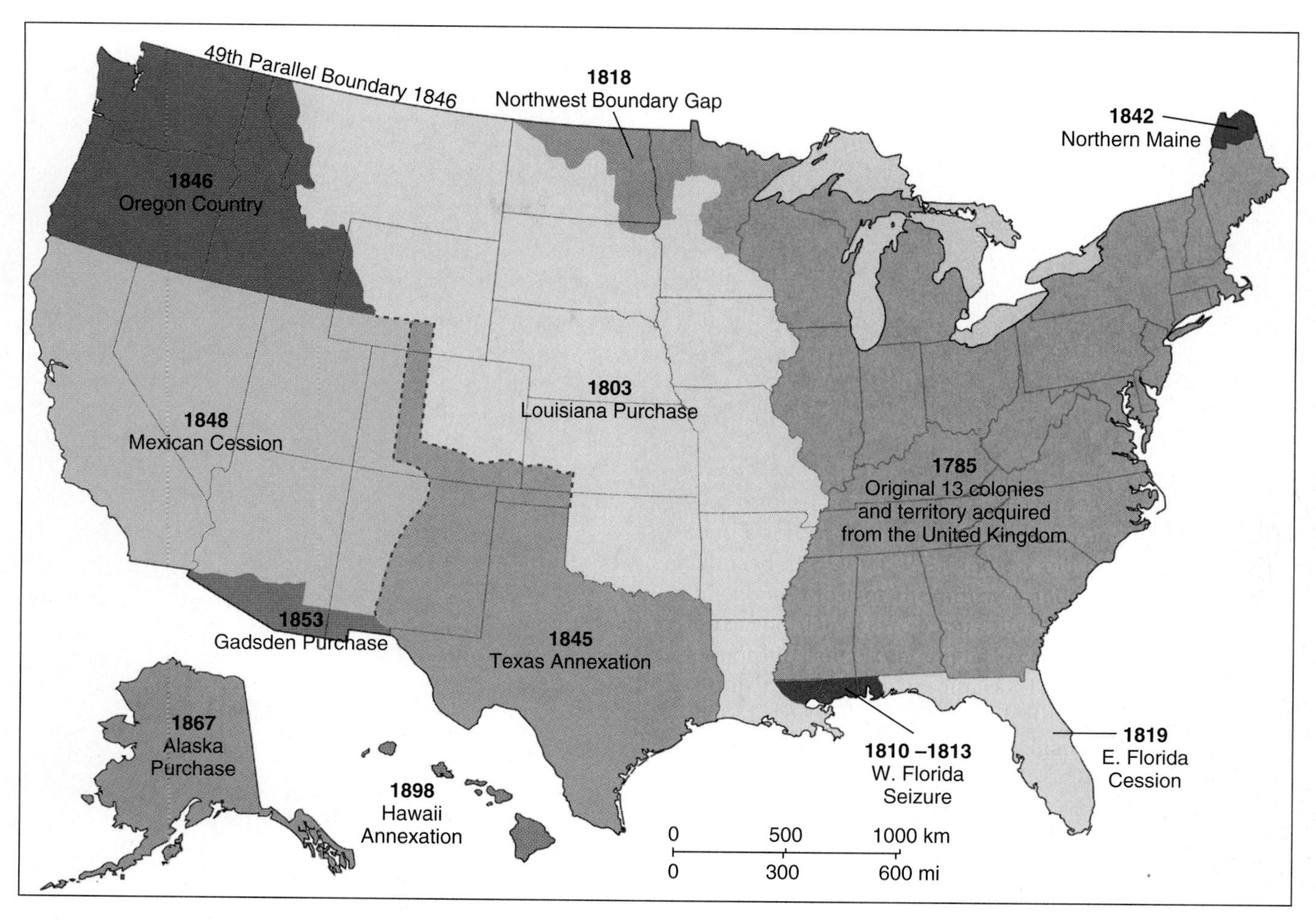


**FIGURE 9.5 The United States: rapid expansion of territory, mainly between 1800 and 1850.** The eastern lands were acquired from the United Kingdom at independence in 1783. The Louisiana Purchase bought lands that France had just won back from Spain. Texas claimed independence after ousting the Mexicans in the early 1830s but then joined the United States in 1845. The western lands of the conterminous United States were added by 1850, including the southwestern lands acquired through the1848 Mexican Cession after a rapid U.S. military victory. Alaska and Hawaii came later.

military conquest (Figure 9.5). To ensure the rapid occupation of these new areas, land was surveyed and sold by the U.S. government at lower and lower prices as time went on until the Homestead Act of 1861 made it possible for families to acquire farmland in the new western areas for almost nothing. Settlers spread to the vast interior plains that had deep, fertile soils and a warm, moist summer climate, a combination that proved ideal for growing corn and wheat and raising cattle and pigs. The United States began to export a wider range of farm produce to Europe, competing directly with European farmers in bulk grain markets as well as continuing to export crops such as cotton and tobacco that could not be grown in northern Europe. Later, the range of climates and area of land within Anglo America made it possible to grow a greater diversity of farm products on a much larger scale than any one country in Europe could manage.

## Slower Canadian Changes

British North America (modern Canada) was a poor and largely hostile remnant left to British authority after the United States declared its independence. British loyalists who left or were forced out of the new country joined the small numbers of Native Americans and mainly French-speaking European settlers living there. Since the French settlers along the lower St. Lawrence resented British rule, the loyalists mostly settled on the east coast or farther inland around the north and west of Lake Ontario. Westward of the Great Lakes there was little settlement in an area still largely administered by the Hudson's Bay Company. The interior remained the realm of Native Americans and isolated groups of French settlers escaping from close British rule. When the Hudson's Bay Company fur trappers began to compete with Americans for land along the lower Columbia River near the Pacific coast, another British-American war appeared possible. Negotiations in 1846, however, set the boundary between the two countries west of Lake Superior at the 49th parallel.

Until the middle of the 1900s, the natural resources of Canada comprised mainly the fish of the Atlantic and Pacific Oceans, the good soils and temperate climate of the Lakes Peninsula in Ontario, the vast prairies that were opened to farming settlement in the early 1900s, and the timber

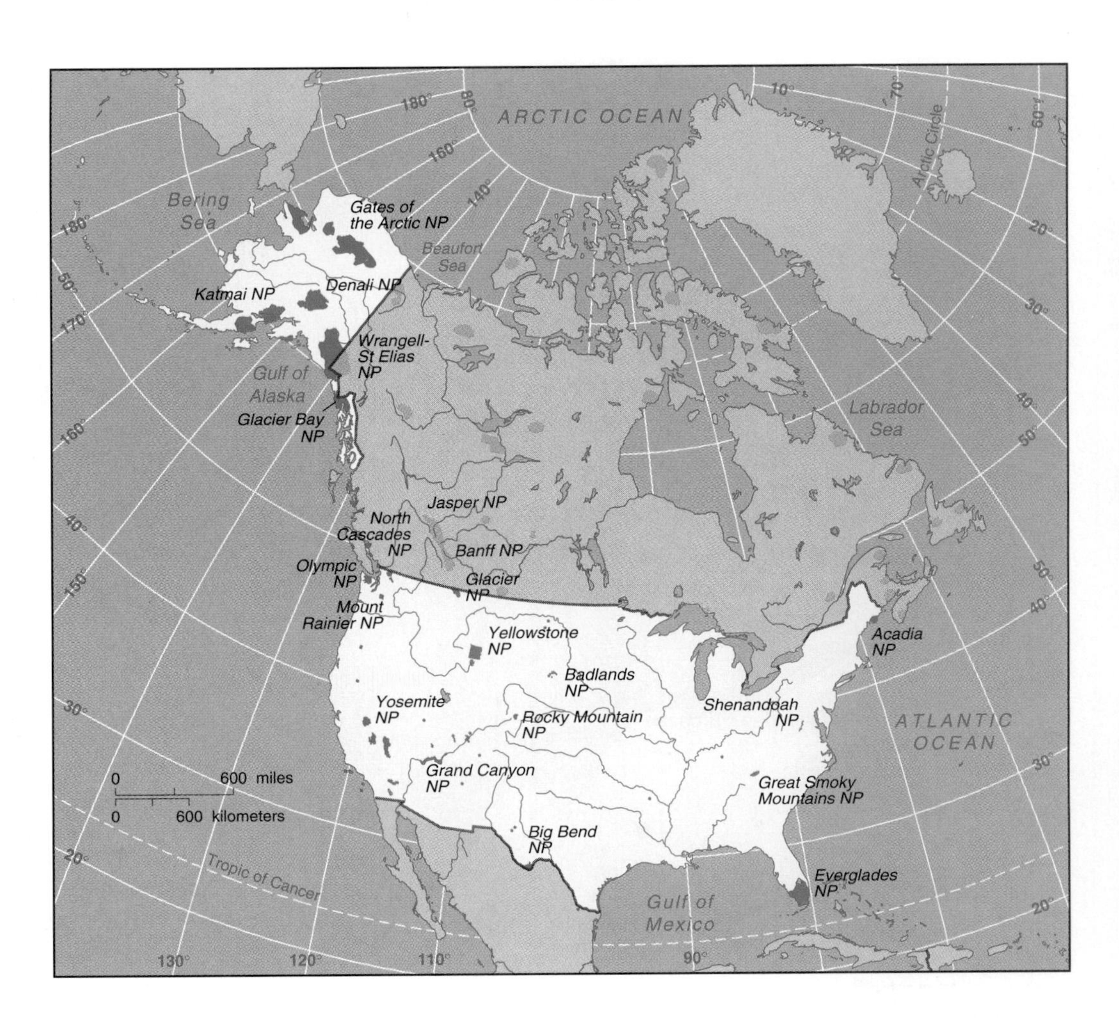


**FIGURE 9.6 Anglo America: national parks.** Some major parks in the United States and the intended expansion of national parks in Canada. In the United States, there are also extensive wild life refuges and national forests, in which varying levels of commercial use are allowed.

resources of the St. Lawrence River valley and British Columbia. From the end of World War II, however, the metal ores of the northern lands, the coal, oil, and natural gas of Alberta, and the hydroelectricity potential of Québec have led to new geographic directions of development and provided the basis for growing Canadian affluence.

## Resources for U.S. Manufacturing

The first industrial revolution began in Western Europe (see Chapter 7), coinciding with the United States obtaining independence. Independence and British hostility motivated Americans to establish their own manufacturing industries, providing goods that were at first not obtainable from Britain. Textiles, metal goods, and leather goods were among the first industries, often powered by water mills in New England, New Jersey, and eastern Pennsylvania. By 1850, the coal in eastern Pennsylvania and around Pittsburgh replaced charcoal in iron-making and later steel industries that set off further industrial revolutions. The northeastern United States soon stood out as the core of the country with most of its people and a growing wealth based on manufacturing industries.

The main symbol of the early steel-based industrial revolution—the railroad—spread across the country. Based initially on local resources of coal, iron, and the agricultural produce of the Midwest, the Manufacturing Belt of the northeastern United States later drew in mineral resources from other parts of the country as railroads linked the mines of the West back to the eastern economic and political core. The mining of gold, silver, and then copper and zinc in the new western lands provided capital for further development and widened the scope of the late-1800s metal industries.

In the 1900s, the availability of such natural resources, the taming of rivers for irrigation, navigation, and hydroelectricity, and the discoveries of oil and natural gas drew more people and manufacturing corporations southward and westward from the national core. Although the emphasis of most Americans with regard to natural resources was on the potential for economic gain, influential groups from the late 1800s persuaded the federal governments to set aside large areas of attractive wilderness as national parks, often before much settlement had occurred. Most of these are in the West, but smaller areas of the eastern lands have also been purchased and designated as national parks (Figure 9.6).

## Human Resources

The variety and richness of the natural resources—such as minerals, climate, and soils—available within such large territories are important in partly explaining the growth of affluence in Anglo America. The human resources, however, are the real key to an understanding of the countries' prominence in the world and the regional patterns of wealth distribution within them.

### New People, New Skills

Immigrants came initially from northwest Europe and were often able individuals or groups of social misfits who wanted freedom from restrictions at home. They brought European cultures, capitalist outlooks, and individual skills and initiative. Many soon established businesses that prospered in the political, economic, and social climate fostered by the independence of the United States after 1783. Tensions among Protestants, Catholics, and Jews were worked out in the context of economic competition and residential congregations of like-minded people. There was space within the United States for many differing groups to establish themselves.

Racial tensions, however, brought greater conflicts than ethnic differences. The reduction of Native American populations and the taking of their lands involved almost 400 years of bloodshed and fear on both sides. The American Civil War in the early 1860s set the North against the South over the latter's wish to secede from the union of states. The South's leaders clung to the right to continue the slavery of the black people who had been imported from Africa to work on their plantations. Southerners saw the North's intervention as going farther than the United States' Constitution warranted. The South's forces were beaten in a hard-fought war, leaving a heritage of hatred and of economic and social backwardness in the South for nearly 100 years.

### Education and Technology

The new United States did not rely on the quality of its immigrants alone but sought to develop conditions in which its whole population could increase their well-being by making the most of their freedoms under the democratic constitution. It established compulsory education much earlier than European countries. Its emphasis on technology transfer began with the late 1800s land grants for establishing engineering and farming universities in each state. Such provision ensured that Americans could not only devise innovative technologies but also widely apply them.

### Management of Manufacturing

Many technical innovations began in the United States, but it was the manufacturing structures and processes in which that country excelled that brought it prominence in world economic activity by the late 1800s. The expanding internal market of the United States provided the demands that stimulated many of the developments.

- First, Americans achieved **economies of scale** by building larger factories for increased output together with access to larger markets for their growing range of products. Such economies cut the cost of individual items by manufacturing large numbers of each item in a single factory.
- Second, **horizontal integration** occurred when financiers bought up several producers of the same product, giving the new owners a large share of the total production of goods and enabling them to set market prices.
- Third, **vertical integration** took this process further by bringing together in a single corporation the producers of inputs to a product and users of that product. As an example of these processes, Andrew Carnegie built larger steel mills to gain economies of scale, then bought out other steel-makers in a time of economic recession in the 1870s to horizontally integrate the steel industry. Then, in a phase of vertical integration, he purchased both the coal and iron mines that provided the raw materials and also the heavy engineering corporations that used the steel. By 1900, Carnegie's U.S. Steel Corporation dominated the industry.
- Fourth, Henry Ford took the concept of the factory **production line** to a new level. He applied it to the assembly of a wide range of components in the production of automobiles. His company produced thousands of a limited number of models for a growing market.
- Fifth, the making of machine tools to produce machines that manufacture and assemble parts was another feature that made the United States the world leader in output of manufactured goods by the mid-1900s.

### Canada Emerges

It was not until the late 1800s that Canada began the process of becoming a major industrial nation, having previously lacked the capital, expertise, and entrepreneurial skills and having been dissuaded by Britain. After the British North America Act of 1867, Canada became a Dominion within the British Empire with a large extent of responsibility for its own affairs. At that stage, Canadian leaders began to integrate their vast country by building transcontinental railroads and encouraging the settlement of the prairie grasslands. Manufacturing industries, including wood processing and steelmaking, were established behind tariff walls to protect them from American competition in particular.

## World Roles

The combination of natural wealth, technological and management leadership, and a large internal market brought the United States world economic leadership. Up to 1950, it dominated the development and output of new products based on automobiles, trucks, airplanes, and consumer goods. Its corporations expanded abroad, taking opportunities to do so after World War II during the economic recoveries in Europe and Eastern Asia, and in the context of Cold War competition in developing countries. After 1950, it established a huge lead in the initial development and use of computers. Although western Europe and especially Japan now challenge this economic and technological superiority, no other single country is close to rivaling the total GDP of the United States or the size of its internal market for products.

Canada lagged behind its neighbor through the early 1900s, partly because of restrictions resulting from its colonial ties to Britain and partly because of the smaller size of its home market for goods and protection of its own industries. It gradually became more closely enmeshed with the United States' economy and dependent on it in many ways. Canada's centers of manufacturing production around Toronto and Vancouver, and to a lesser extent around Montréal and in Alberta, now rival individual centers in the United States in size and diversity of output. Canadians enjoy living standards almost equal to those in the United States, and many who live in Canada strongly support the notion that they enjoy an even better quality of life. Some Americans prefer the Canadian way of life and live across the border from Detroit or Buffalo. The 1997, United Nations human development and gender-related development (HDI and GDI) indices both placed Canada in first position above European countries, the United States, and Japan.

### NAFTA and FTAA

In 1988, the United States and Canada established the **U.S. and Canada Free Trade Agreement**, which led to the **North American Free Trade Agreement** (NAFTA) in 1993 by adding Mexico to the arrangement. The agreement between the United States and Canada continues to thrive under NAFTA, despite a sequence of disputes over individual items. Trade between the two countries rose by 40 percent from 1988 to 1993 and more rapidly than each country's trade with the rest of the world. Long-term investment in Canada rose following the reduction of tariffs into the United States and improved productivity. Canadian car plants expanded in Ontario, the Dutch Philips Corporation moved a light-bulb factory from Mexico to Canada, and Swedish pharmaceuticals and telecom companies chose to locate in Montréal. The Mexican link only began to function in the mid-1990s, and its impact was slowed by the ups and downs of Mexican finances after late 1994 (see Chapter 10).

Other countries in Latin America wish to join NAFTA. Chile is the most likely to be admitted first, since the trading reputation of its rival Colombia is tarnished by its drugs trafficking. Such applications suggested an expansion into a **Free Trade Area of the Americas** (FTAA). All Anglo and Latin American countries except Cuba agreed to this in December 1994. The intention is to complete negotiations by 2005, with the almost redundant Organization of American States coordinating the process. The main problems in the way of such a wider trade grouping are likely to be the small size of Caribbean countries, the U.S. Congress' stances on labor, environmental conditions, and human rights, and the existence of other trading groups such as Mercosul (Argentina, Brazil, Paraguay, and Uruguay, see Chapter 10), the European Union (EU, see Chapter 7), and the Asia-Pacific Economic Community (APEC, see Chapters 6 and 11). Such agreements are increasingly linked to world trade conditions but remain subject to each country's trade laws that from time to time allow them to set new tariffs to resist the dumping of items overproduced in other countries.

## Problems of Affluence

The world's most affluent region has its problems as well as the attractive aspects that cause many to apply each year to these countries for immigrant status. The problems arise from contrasting sets of opportunities, often linked to geographic location.

### Increased Rich-Poor Gap

In the world economic system, increasing wealth in a country often combines greater numbers of very and moderately rich people with continuing large numbers of poor. Gaps increase between the richer and poorer parts of a country—a process known as **uneven development**. Such geographic contrasts occur within cities and also between rural and metropolitan areas.

In the United States, inequalities of wealth narrowed between the early 1900s and 1969, when the top 20 percent of incomes were 7.5 times higher than the bottom 20 percent. By the 1990s, however, the gap widened so that the top 20 percent of incomes were 11 times higher than the bottom 20 percent. The top 20 percent had 45 percent of the national income, while the bottom 20 percent had 4 percent. A large proportion of the increasing inequality is explained by two factors. The first of these is the lightly regulated labor markets that react to world conditions by depressing wages for unskilled jobs in competition with poorer countries and increasing salaries for skilled workers as demand from service industries and professions rises. The second factor is the changes in U.S. households that polarized the increasing incomes in a greater number of two-income homes against the growing numbers of single parents who make up 35 percent of the poorest 20 percent of the population.

As an example of the increasing gap between poor and moderately rich, the contrast in incomes between the affluent Upper East Side of Manhattan, New York, and the poverty in Harlem, just a mile away, increased from 1950 to 1990 (Figure 9.7). In 1950, family incomes in Upper East Side census tracts were around the U.S. average, while those in Harlem were half of that. By 1990, those in Upper East Side were two to four times the U.S. average, while those in Harlem were much less than half the U.S. average. In the period 1950 to 1990, many African Americans earning better incomes moved out of Harlem, while rich white people developed expensive accommodations in the Upper East Side (see "World Issue: Gentrification," pages 414–415). The poorer areas of many inner cities have lower quality buildings and services such as education and health because of the inability of small jurisdictions with predominantly poor populations to support such services. Such effects create a downward spiral of living

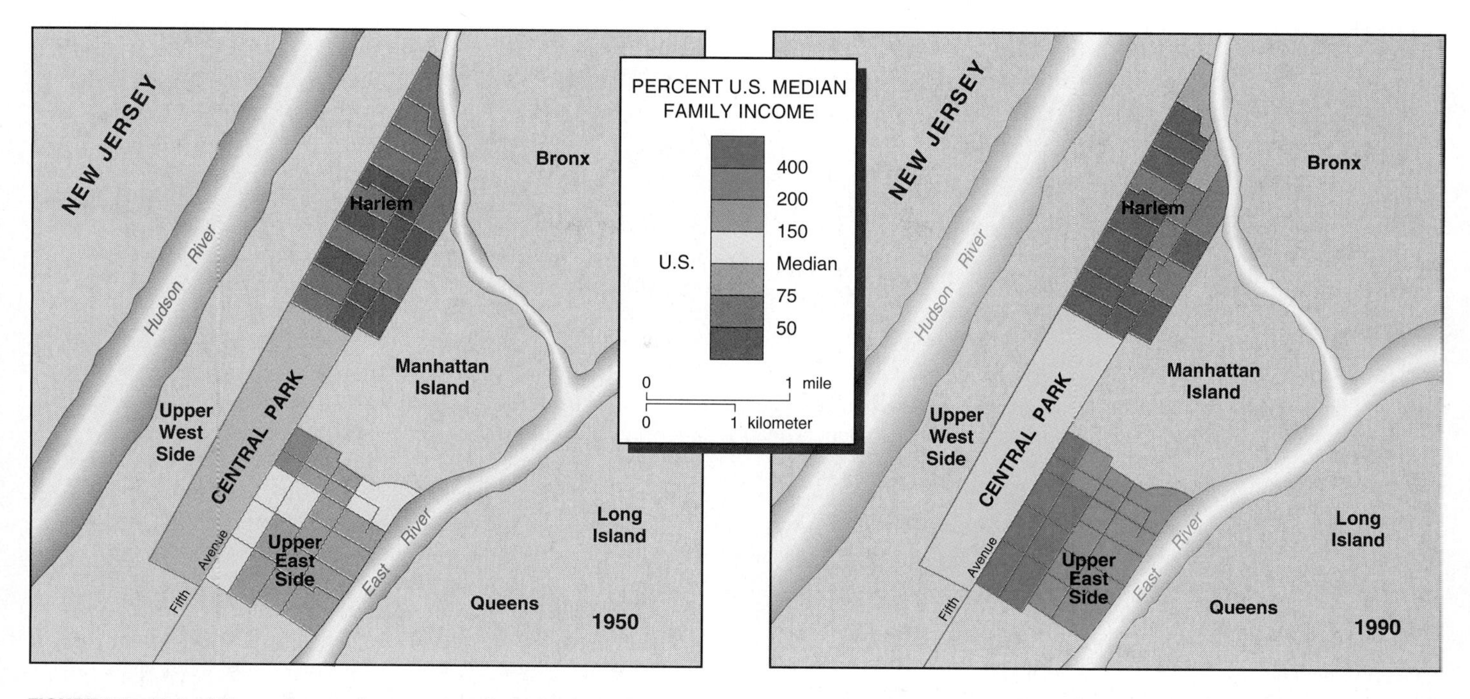


**FIGURE 9.7 United States: increasing contrasts in income.** Family incomes in the groups of census tracts in the Harlem and Upper East Side areas of Manhattan, New York, 1950 and 1990. The Harlem area has been a long-term center for the African American population of New York; the Upper East Side has homes of many wealthy people. The incomes are compared as proportions of the U.S. average. How does the 1990 map indicate increasing differences between the two areas?

conditions. The contrasts produced by uneven development are also illustrated by the fact that some of the best public secondary education in the country is found in the suburbs of Boston but some of the worst in the inner city.

Although the poorer groups in American society gain some sympathy, there is no pressure for taxing the rich further to support the poor. Tax revenues are used for welfare and public health and infrastructure programs, but redistribution of income on a scale that occurred in European countries from 1945 to the 1980s does not occur. This is partly because there is still an unusual level of mobility between economic groups. It is getting harder, however, to climb out of poverty into the next stage of more secure income where the upward mobility mostly begins. And yet, it has been shown that greater inequality often leads to less efficiency in production, since poor people are liable to poor health, social stress, and temptations to be involved in crime.

It appears that the increasingly global scale of technological competition leads to increasing income gaps and uneven development throughout the world and has the potential for setting off social unrest or protectionist policies within countries from the richest to the poorest. Attitudes to poverty in their own countries raise the question of whether American and other core-country peoples are willing to take actions to lessen the differences in their own and worldwide societies.

## Congregation and Segregation

Differences in levels of opportunity for moving from poverty or middle incomes to riches or from powerless positions to those of influence lead to conflicts of interest that affect people's lives and where they live in Anglo America. The gaps between rich and poor are often linked to ethnic and racial differences. People with good incomes and similar cultural and social tastes can live close to others of like mind, choosing to **congregate** in inner city or suburban areas. They may live close to ethnic restaurants, cinemas, places of worship, or recreation facilities—or just with other middle-class groups of varied backgrounds because they appreciate the variety of American ethnicity. Less affluent groups in society have fewer opportunities for choosing where they live and become segregated into communities that are excluded by the more affluent groups. Such **segregated** groups often occupy inner-city areas that contain high proportions of African American or Hispanic people and have poor access to high-quality education and job opportunities. Canadians are less divided by segregation than Americans, but congregation is important among newer immigrant groups in big cities such as Toronto and Vancouver.

## Urban and Rural Contrasts

Another set of contrasts opened up across Anglo America, particularly after 1950, between metropolitan urban and rural communities. The growing affluence arising from the expanding manufacturing and service industries was urban-based. It was enhanced by the U.S. government investment in interstate highways, which provided improved transport among metropolitan centers. In the 1960s, Americans and Canadians became conscious of disparities in well-being between the metropolitan suburbs and the lagging conditions of rural areas such as Appalachia, the Atlantic coastal plain, the Atlantic Provinces of Canada, the rural southwest

WORLD ISSUE

# GENTRIFICATION

Some of the aspects of this issue include:

What factors give rise to gentrification?

What is the extent of gentrification?

Does gentrification have a wider significance?

Gentrification is the movement of higher-income families and individuals into poorer areas of inner cities, leading to the improvement of housing conditions and living environment in the districts affected. It was first noticed in British cities in the early 1970s but is common in the United States and Canada. The interest of geographers in this phenomenon reflects a concern about changes in small areas that may have a wider significance.

The causes of gentrification have been linked, on the one hand, to individual human decisions, and on the other hand, to more general processes at work that cause society to act in this way. Gentrification most commonly involves high-income men and women typically in the age groups of mid-twenties to mid-thirties and mid-fifties to mid-sixties. They are mostly singles or couples with few children. Demographic trends toward later marriage and child-bearing in the younger groups, together with large "empty-nest" homes in the suburbs, rising divorce rates, and widowhood, provide a large affluent group of small households in these age groups. A substantial number is willing and able to take up apartments near the city center. Work in executive offices and the growing financial services sector in the city center, together with access to leisure-time facilities in theaters, symphony halls, opera houses, museums, restaurants, improved parks and waterfront areas, and major sports facilities are attractions that make people wish to live in the vicinity.

With the increasing costs of commuting and suburban housing, many families found it economically worthwhile to move to the inner city, especially in the 1960s and 1970s when inner city properties could be bought cheaply. For the younger groups there were also examples of inner-city areas in which "freer" lifestyles could be pursued. San Francisco and Vancouver are best known for this, but it happens to an extent in all large U.S. and Canadian cities. The main drawbacks of inner-city areas are their crime rates, which can be reduced by security measures, and their poor public school facilities. It remains common for younger affluent families who begin living in the inner city to move out to suburban locations when their children are of school age.

The movements of people with high incomes into city centers were at first heralded as an "urban renaissance," and much was made of the process in the context of the years of declining and abandoned inner-city areas in the 1950s and 1960s. These cries were muted, however, when it was realized that many poorer people were displaced by the process and had nowhere to go. Quality buildings erected at the end of the 1800s, for instance, often became rooming houses before gentrification turned them into more costly accommodations and

(a)

(a) Society Hill, Philadelphia, where row houses built in the 1700s have been renovated and their value risen sharply. (b) San Francisco, California. Renovated Victorian wood-frame homes contrast with the modern downtown buildings in the background.

displaced the previous occupants. Poorer-quality housing was swept aside and replaced by new apartment blocks. In neither case was provision made for the previous inhabitants.

The extent of gentrification is great in terms of the number of cities affected by it but often modest in relation to the total built environment of each inner city—several blocks rather than larger tracts. In Manhattan, New York, for instance, an area of Upper West Side, between Central Park and the Hudson River and near the Lincoln Center and museums, experienced gentrification. Similar areas have been identified in Boston, Philadelphia, Washington, D.C., Atlanta, San Francisco, Vancouver, and Toronto.

Although individual decisions about where to live are important, they are carried out within particular social, political, and economic contexts that impose conditions and limitations. Most people rent their accommodation or purchase it with a loan. The renters and financial institutions thus have a say in guiding peoples' locational decisions. Furthermore, government regulations, tax breaks, and investments in urban infrastructure are other elements that determine financial institution policies. During the movement to the suburbs in the 1950s and 1960s, large sectors of the older parts of American cities were left behind by middle-class white families and taken over by black and other poorer families migrating into the cities. Housing values dropped and many financial institutions drew lines around areas in which they would not support house purchases with their loans—a process known as "redlining." Redlining led to further deterioration of inner-city areas. Once gentrification began to bring higher-income people back to the city center, the process was reversed ("whitelining"). The financial institutions that had contributed to inner-city decline by refusing loans gained much from the revival of inner-city living, backed by government support in renovation grants. They invested heavily in financing the building of new apartments in the 1970s and 1980s.

Opinion is divided as to whether gentrification is a small-scale process of individual choices over short timescales that affects a few people in each city or whether it is symptomatic of large-scale forces at work in remodeling the cities as part of uneven development. It has not involved the numbers of people that took part in the suburbanization of the 1950s and 1960s, but it has been significant in forming a partial reversal of that trend. It may be the outcome of manipulation by financial institutions, but most people who change their housing locations would not see that as a factor they considered. Certainly gentrification is linked to the continuing changes within urban environments in response to economic changes such as the changes in the central business district, demographic changes in family sizes and age structure, social changes in people's living habits and groupings, and perceived improvements in the environmental conditions of city center areas.

### TOPICS FOR DEBATE

1. Gentrification is a beneficial process in cities.
2. Gentrification shows that people's individual choices of where to live interact with attempts by government and financial institutions to manipulate these choices.
3. Gentrification occurs in other parts of the world.

**(b)**

**RECAP 9A: Anglo America's Affluence**

Canada and the United States of America form the most affluent of the world regions. They are two of the world's four largest countries in area, occupying nearly 15 percent of the land. With just over 5 percent of the world's population, they produce nearly 30 percent of the output of goods and services.

Anglo America was historically settled largely by Europeans who displaced native groups from 1500 and especially after 1800. The United States became independent in 1783, while Canada gained the virtual freedom of dominion status within the British Empire in 1867.

Both countries contain a wealth of natural resources and have been settled by people applying the human resources that resulted in high levels of affluence. Both countries have significant world roles. Together with overall affluence, however, go inequalities between rich and poor, the segregation of poor groups—especially in the United States—and environmental impacts.

9A.1 How do the areas, populations, and economic data of Canada and the United States compare with other large countries such as Russia, China, Brazil, and India? (Use data in the Reference Section, page 586–610).

9A.2 Which economic and political factors explain the world leadership position of the United States in the 1900s?

9A.3 What would it take to make the citizens of Canada and the United States more willing to help the poor in their own and other countries?

**Key Terms:**

Native American
African American
township
economies of scale
horizontal integration
vertical integration
production line
U.S. and Canada Free Trade Agreement
North American Free Trade Agreement (NAFTA)
Free Trade Area of the Americas (FTAA)
uneven development
congregate
segregate

**Key Places:**

Atlantic coast
New England
Middle Atlantic

of the United States, and the Upper Great Lakes area. Policy measures passed by governments to help rural areas did not reverse the trend. Farms replaced labor with machines in the most productive farming areas of the Midwest, the high plains and prairies, and central California, and by the 1980s, these areas were losing population.

### Environmental Impacts of Affluence

Other problems of affluence result from the environmental impacts of intensive use and extraction of resources. In 1970, the United States used 40 percent of the world's oil production. Its continuing use of large proportions of the world output of fuels, metal ores, soils, and water led to questions about the impacts of continuing economic growth. From the early 1970, the United States passed stringent environmental legislation to improve its air and water quality, and standards increased in subsequent years. Such regulations, however, sometimes led to polluting industries being resited in poorer countries.

The limits of resource usage occasionally became an issue. In 1972, a scare arose over whether world grain supplies could supply the needs of growing populations. In 1973, the oil-exporting countries raised their prices. To many it seemed that the "American Dream" of access to increasing affluence was coming to an end. The worst, however, did not happen, and new sources of materials, often at cheaper prices, enabled Americans to enjoy increasing affluence and for some other countries elsewhere to enjoy improving conditions.

## NATURAL ENVIRONMENT

Canada and the United States of America contain a wide range of natural environments, from high mountains to plains and from tropical areas to the arctic cold. The distinctive framework of relief that affects other aspects of the natural environments is formed by the north-south mountain systems on the western and eastern sides of the two countries and the intervening wide central lowland. The lowland is drained southward by the Mississippi River and its tributaries and northward to Hudson Bay (Figure 9.8a).

While the natural environments of most of the United States are positive for human settlement, Canada's natural environments have a negative influence on much of its human geography because so much of the country lies near or beyond the margins of productive or habitable land. Continuous settlement in Canada is restricted to a narrow zone just north of the United States' border.

### Tropical to Polar Climates

Climatic environments within Anglo America are mainly midlatitude in type (Figure 9.9). The western coast is dominated by the midlatitude west coast and mediterranean climates that give way eastward to midlatitude interior and midlatitude east coast climates. This is a similar sequence to that found across Europe and northern Asia (see Chapters 7 and 8). The west coast climates of Anglo America are narrowed in extent by the north-south mountain ranges.

The southern United States, northern Canada, and most of Alaska, however, lie outside the midlatitude belt. The northern margin of the tropical arid climatic environment extends from southern California eastward to New Mexico. Southernmost Florida has tropical, humid conditions and the Gulf coast from Texas to Florida and the Atlantic coast as far north as Virginia have subtropical conditions that bring hot, humid summers with many thunderstorms and the annual threat of hurricanes. In northern Canada and Alaska, the climate is deeply arctic in winter and has only short summers.

The north-south lowlands enable bursts of arctic air to move southward in winter, bringing extreme cold especially to the Midwest and northeastern United States but also extending in shorter bursts to Texas and Florida. Warm, dry southwestern air or humid Caribbean air moves northward in spring and summer. The easy confrontation of cold arctic air and humid subtropical air in the central lowlands leads to the formation of active cold fronts with thunderstorms that may develop into

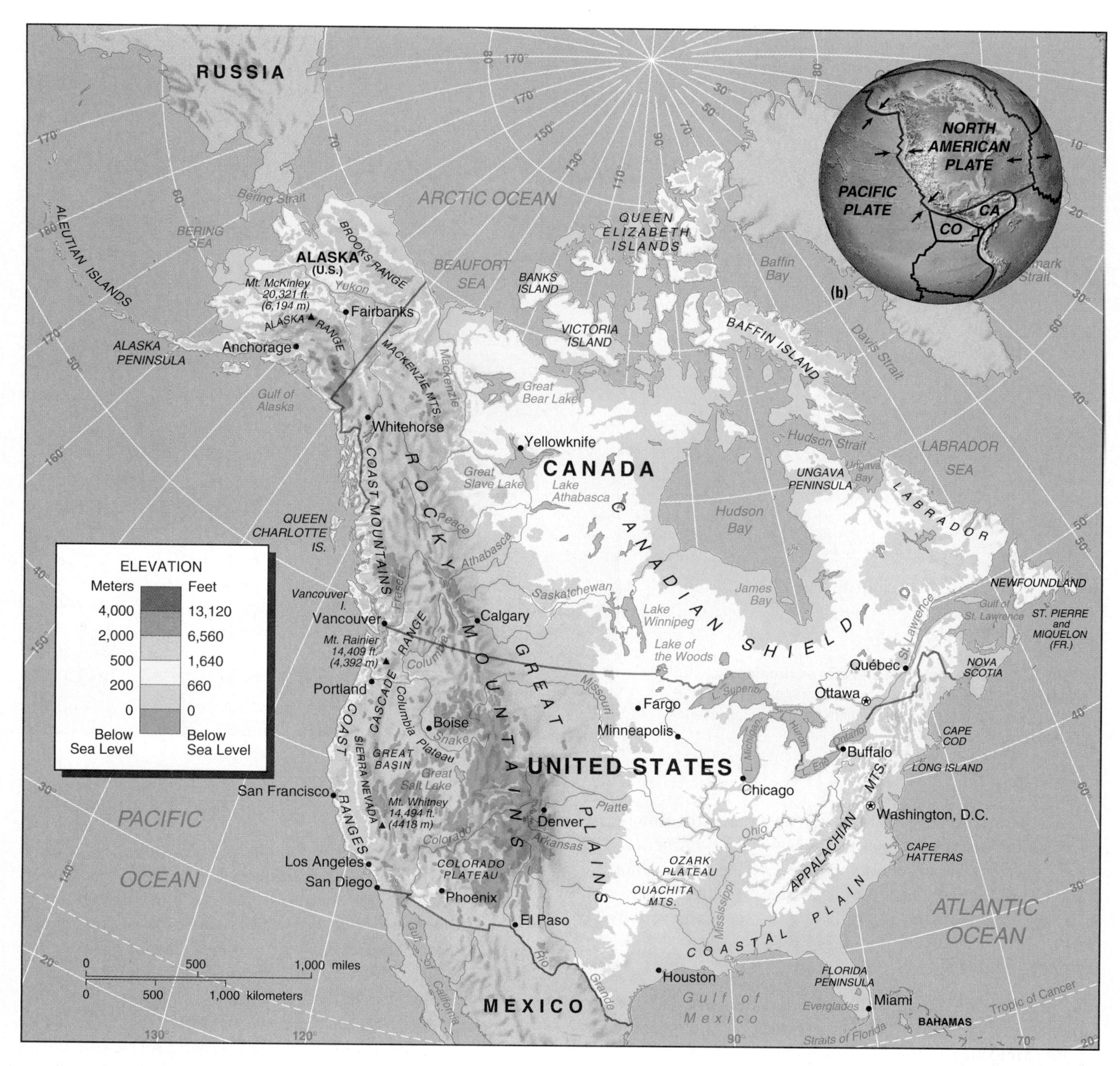


**FIGURE 9.8 Anglo America: physical features.** (a) Map of relief features, major rivers, and lakes. (b) The tectonic plate movements that cause changes in the relief patterns.

local violent tornadoes in spring. Tornado activity moves northward from Texas in February to the Great Lakes in June.

Past climatic changes left their impacts on the present environment. During the Pleistocene Ice Age—the last 2 million years of geologic time—much of Canada and the northern parts of the United States were covered by advancing ice sheets centered on the Hudson Bay area and on the western mountains. The greatest effects occurred in the last half million years, but most of the ice melted beginning about 15,000 years ago, leaving glaciers only in the highest and coldest areas. Distinctive ice-eroded landscapes are relic features in high mountains and in northern Canada, while the melting ice sheets left depositional features mainly in the Midwest. Global warming and rising sea levels in the future may drown low-lying coasts, including wetlands and port areas.

## Mountains and Plains

Anglo America is part of the North American plate (see Figure 9.8b). New plate material forms along the divergent margin at the Mid-Atlantic ridge, forcing the plate to move westward and clash the Pacific plates at the convergent margin along the west

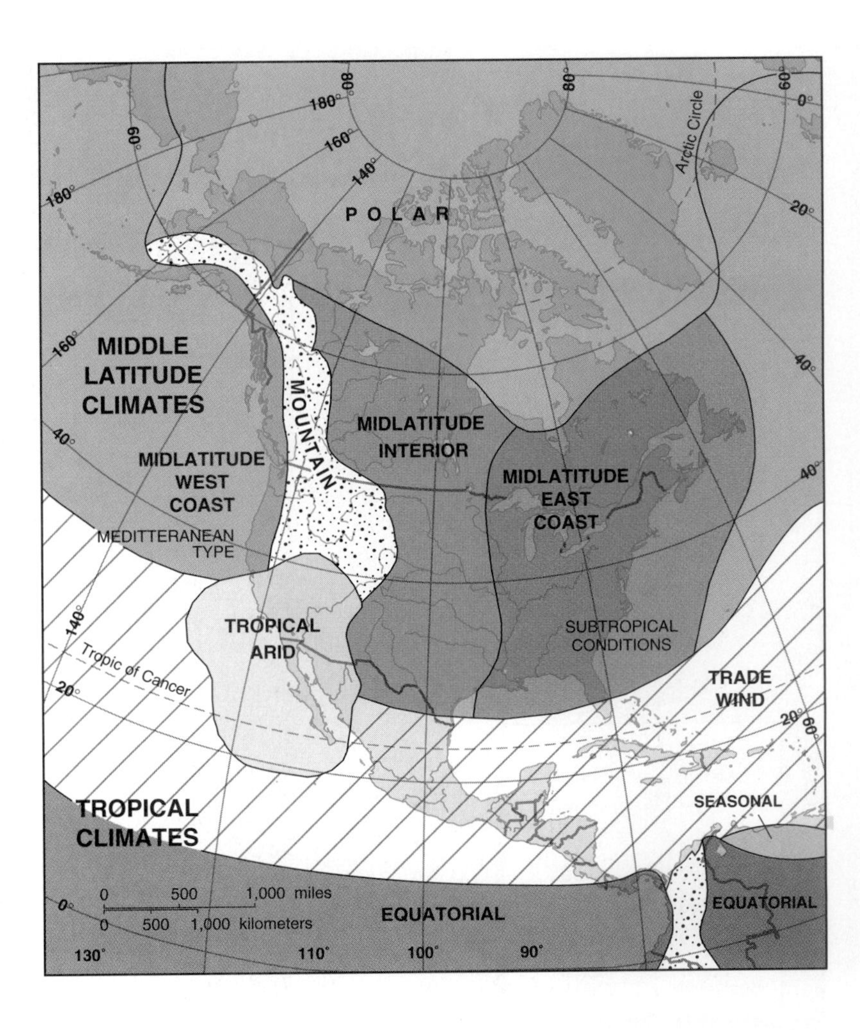


**FIGURE 9.9 Anglo America: climatic environments.** Compare the north-south and west-east distributions of climatic environments with those in the Commonwealth of Independent States (see Figure 8.8).

coast of North America. This pattern explains the differences between the eastern and western coasts of Anglo America. The west coasts are mountainous and are affected by earthquake and volcanic activity; the east coasts are low-lying or hilly with older mountains inland and only rare internal earth activity.

## Western Mountains

The western third of Anglo America is mountainous but made up of several elements. The highest peak in Anglo America is **Mount McKinley** in the Alaskan Range (6,194 m, 20,231 ft.). The **Rocky Mountain** ranges run from Alaska and northwestern Canada to New Mexico. To their west in the United States are extensive high plateaus from the **Colorado Plateau** in the south to the lava layers of the **Columbia Plateau** just south of the Canadian border. The Pacific coastal ranges are backed inland by the Californian Central Valley, Puget Sound lowlands, and the drainage area along the British Columbia and Alaskan panhandle coasts. East of the valleys are the Sierra Nevada Mountains of California and the Cascades farther north. The Cascades have volcanoes such as **Mount St. Helens** (Figure 9.10), and there are others in southwestern Alaska. Movements in Earth's interior every few years make California liable to earthquakes along the **San Andreas Fault** system as the rock on the Pacific side of the fault moves northward (see Figure 2.31).

## Canadian Shield

About half of Canada has the ancient rocks of the **Canadian Shield** at the surface. These are the oldest rocks in Anglo America and contain major deposits of mineral ores. Covered by ice sheets during the last Ice Age, most of the soil was scraped from their surface. Deposits left by the melting ice sheets over the last 15,000 years are scattered and varied in nature and often separated by large lakes that fill hollows gouged out by the ice. A line of very large lakes follows the approximate boundary between the shield rocks and the overlying rocks, from Great Bear Lake in the northwest to the Great Lakes.

## Interior Lowlands

Lowlands with little relief dominate Southern Ontario, the Prairie Provinces, and the central United States. Layers of sedimentary rock cover the shield rocks, forming plateaus and escarpments such as that over which the Niagara Falls plunge. In the U.S. Midwest and Ontario, Canada, most of these rocks are covered by deposits from the melting ice sheets. Similar deposits occur over part of the **Canadian Prairies**, but in areas not covered by the ice sheets, distinctive features such as groups of small mounds and depressions were formed by exposure of the land to an atmosphere of extreme cold.

East of the Rockies, the Mississippi lowlands give the impression of being flat for over a thousand kilometers but rise

(a)

(b)

**FIGURE 9.10 United States: Mount St. Helens eruption, 1980.** (a) Before the eruption. (b) After the eruption, with the mountain top blown away and a new small cone forming in the crater?

U.S. Geological Survey

gradually eastward and westward to the bordering mountains. In the north, they were covered by rock fragments and particles dropped by the ice sheets that covered the area during the Pleistocene Ice Age and by the finer particles blown farther afield by winds and deposited as loess as far south as the state of Mississippi. West of the Mississippi River, the lowlands rise to the **Great Plains** in the United States and the Prairies in Canada.

### Appalachian Mountain System

East of the Mississippi lowlands, the **Appalachian Mountains** form part of a hilly tract extending from northern Georgia into the Adirondacks of New York, the Green and White Mountains of New England, and the Atlantic Provinces of Canada. These mountains are not so high as those on the west coast—few exceeding 2000 m (6500 ft)—and were formed by ancient plate tectonics movements and then worn down. In the northeastern United States and eastern Canada, the glaciers scraped away much of the surface rock and soil, carried it southward, and deposited it along the coast in the low ridges (moraines) that form much of Long Island and Cape Cod (Figure 9.11).

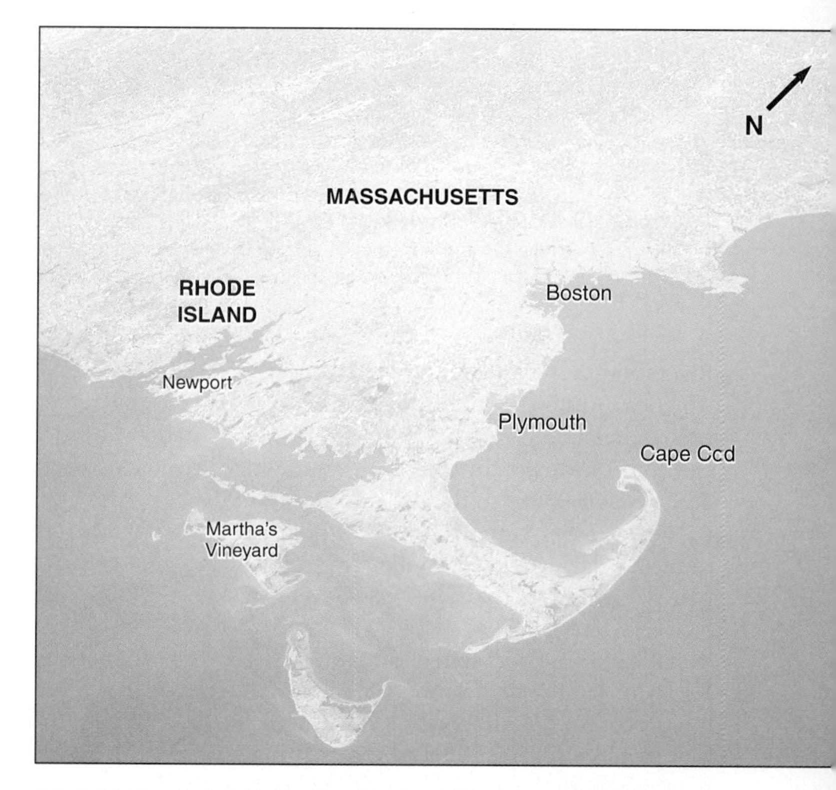


**FIGURE 9.11 United States: New England.** March snows highlight features as seen from a space shuttle. Huge ice sheets crossed the area in the Ice Age, spreading out from Canada. They scraped soil from the mountains and deposited hilly islands in Boston Bay (top right) and the sandy morainic ridges that form the Cape Cod peninsula and the island in the foreground.

NASA

## Major Rivers and the Great Lakes

The physical and human geographies of Anglo America are greatly affected by the surface hydrology. In particular, the major rivers and the Great Lakes provide sources of water together with water and land transportation routes. The **Mississippi River** was the basis of early interior transportation and continues that role for bulk materials, though diminished in significance. Its tributary, the **Ohio River**, was a transportation route at the heart of manufacturing developments in the United States during the late 1800s. The **Colorado River** provides irrigation water for the arid southwest. The **Columbia River** is used to generate hydroelectricity in the northwestern United States. In the 1950s, the **Great Lakes-St. Lawrence Seaway** brought easier trade and associated industries to the heart of Canada and the United States. In northwest Canada, the **Mackenzie River** is a major summer transportation routeway.

The largest area of lowland and fertile farmland occurs in the combined drainage basin of the Mississippi-Missouri-Ohio

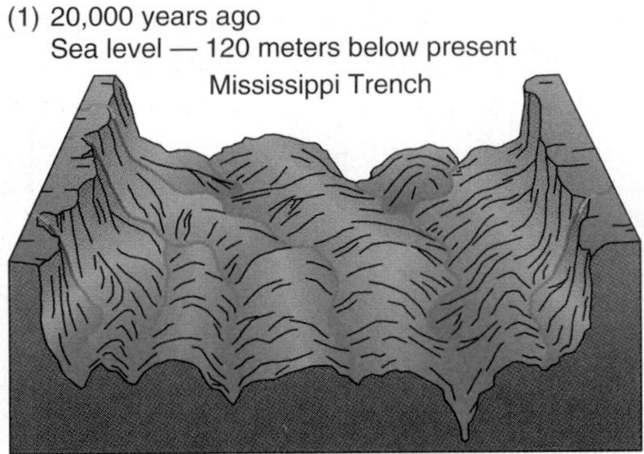


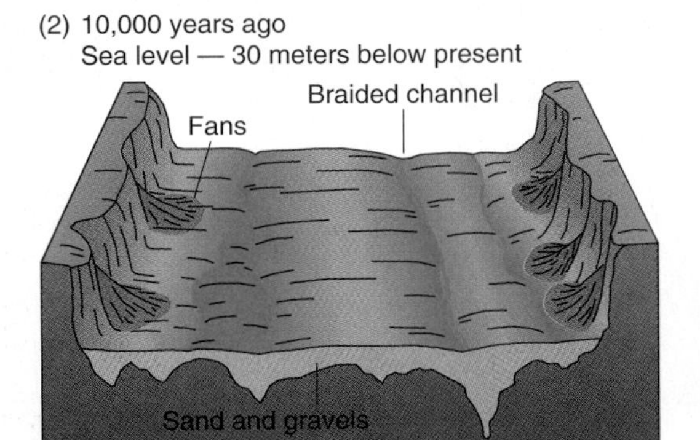


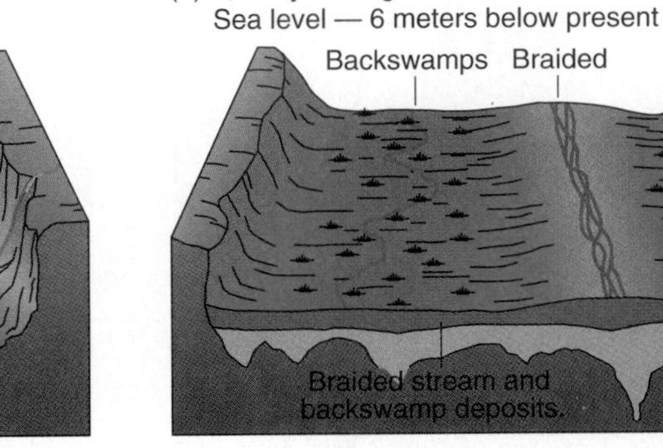


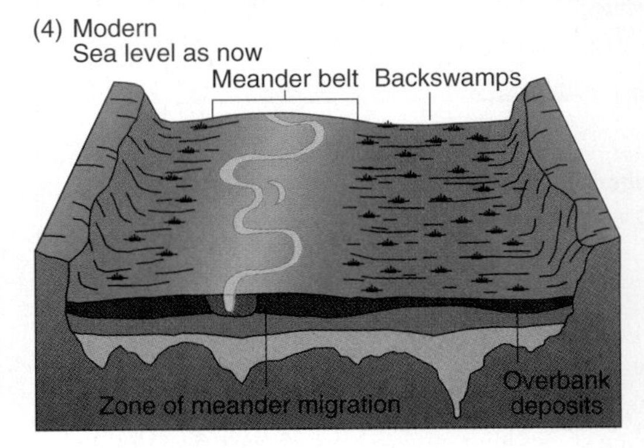


**FIGURE 9.12 Anglo America: impact of ice meltwater on the landscape.** Stages in the evolution of the lower Mississippi River valley. Large flows and low sea level resulted in the Mississippi River carving a deep channel at first. As the sea level rose and the flows lessened, the channel filled with gravel, sand, and mud. Finally, flows from the Great Lakes were diverted eastward, and the Mississippi flow was reduced to its present levels.

Rivers and the Great Lakes, covering nearly one-third of Anglo America. Although a continental interior region, the rivers provide relatively easy outlets to world ocean trade routes to the Gulf of Mexico in the south and along the St. Lawrence and Hudson Rivers to the Atlantic Ocean in the east. Much of the lower relief along the Mississippi Valley was formed by sand and clay deposited from meltwater, as the ice sheets dissipated (Figure 9.12) and the ocean level rose.

The Great Lakes were formed by processes that began as the Pleistocene ice sheets advanced over North America. Lobes of ice gouged depressions between festoons of marginal moraines built from rock fragments dropped at the margins melting ice masses. As the ice sheet retreated northward at the end of the last glacial phase, meltwater accumulated in the depressions, which became the Great Lakes. At first, the meltwater overflowed southward into the Mississippi system, but later the lake waters diverted to flow eastward into the Hudson and St. Lawrence Rivers. Today the Great Lakes function as inland seas with coastlines, ports, and recreation areas. In 1959, the St. Lawrence Seaway was opened as a cooperative project between the Canadian and U.S. governments. Ocean-going ships could then reach Chicago and Duluth. Unfortunately, the locks built then are too small to accommodate modern bulk-carrying ships and are still closed by ice for three months of every year.

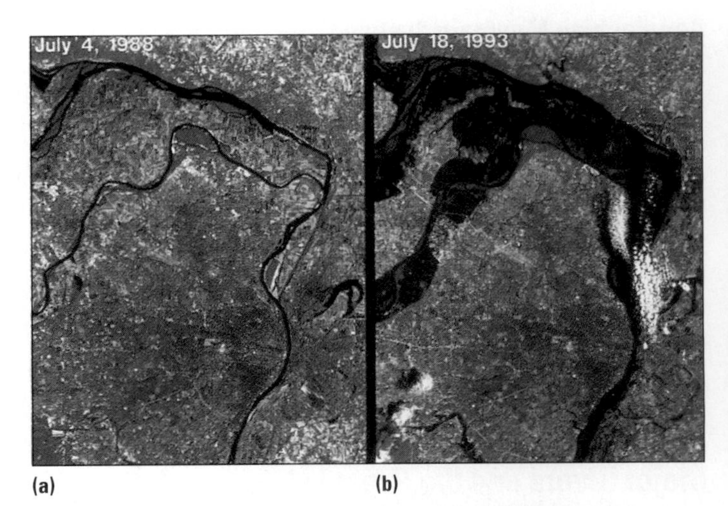


**FIGURE 9.13 United States: flood hazard.** The 1993 floods on the Mississippi River near St. Louis. (a) Before the floods. (b) During the floods. The reddish areas are highways and buildings. The darkened areas on (b) show the extent of flood water cover.

Space Imaging, Thornton CO, USA.

From the early 1900s, the Mississippi-Missouri River system was controlled for flood protection, improved water transportation, and hydroelectricity generation by building a series of dams. The flood protection levels provided prevent all but the severest flooding (Figure 9.13). The dams prevent

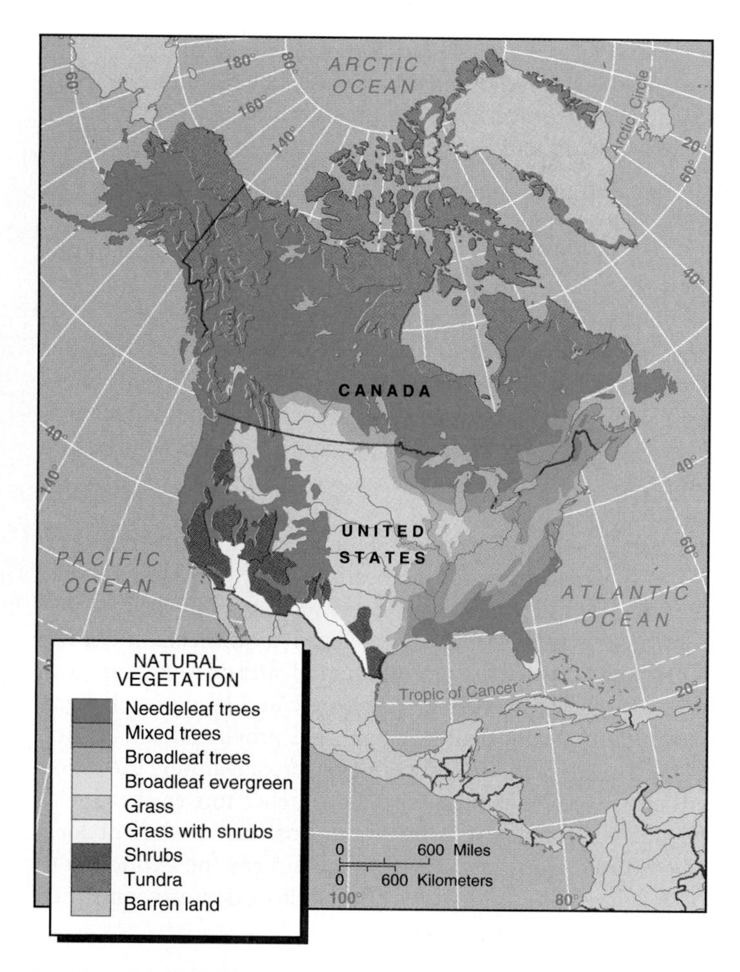


**FIGURE 9.14 Anglo America: natural vegetation.** Relate this distribution to the climatic environments (see Figure 9.9) and land-use patterns (see Figure 9.33).

silt and clay reaching the mouth and so reduce deposition on the delta south of New Orleans. In recent years, the delta surface subsided along its southern margins because it did not receive sediment to build it to sea level.

## Varied Natural Vegetation and Soils

The combinations of terrain and climate produced a range of natural vegetation and soils distributed in a similar north-south pattern to that of the CIS (compare Figure 9.14 with Figure 8.11), although the east-west extent is less in Anglo America. The hot deserts of the southwest support little vegetation, and that mainly of drought-resisting varieties such as cactus (see Figure 2.35b) and low shrubs. The subhumid Great Plains and Prairies of the western Mississippi River basin and south-central Canada had a natural vegetation of prairie grasslands that are now largely plowed to make use of the underlying black earth soils. Eastward, the more humid conditions, from tropical southern Florida to the cool temperate northeast, supported deciduous (broadleaf) forest, which gave rise to brown earth soils of moderate-to-good fertility. North of the deciduous forests and the prairies, a wide band of coniferous (needleleaf) forest has poor podzol soils and gives way to the tundra along the Arctic Ocean shores.

The eastern mountains have thin soils on steep slopes but, before the clear cutting by loggers in the late 1800s and early 1900s, they supported a profusion of trees containing a much greater variety of species than those in western Europe. The best farming soils formed under forest and grassland in the interior plains, where the old glacial and wind-blown materials and the moderate amounts of precipitation produced rich brown (forest) or black (grassland) soils. In the southeastern United States, the frequent presence of sandy soils with low nutrient content caused some areas to be dominated by pine trees. Along the west coast, north of San Francisco, huge firs and cedars, growing over 100 m (300 ft.) tall, formed a massive timber resource.

## Natural Hazards

A greater range of natural hazards than any other country in the world marks the natural environments of the United States. The country is affected hy hurricanes, tornadoes, earthquakes, volcanoes, floods, and ice storms, all of which have major impacts on human activities.

Each hazard has its geographic area of impact. Hurricanes threaten the southern Atlantic and Gulf states in late summer and fall. Tornadoes are most common in the Mississippi Valley, affecting Texas in February and moving to the Great Lakes area by June. River floods occur most commonly in the Mississippi River valley following the spring melting of snow on the surrounding hills or heavy summer rains. Flash floods occur in upland and arid regions in the West after sudden rains fill dry streambeds. Earthquakes and volcanoes occur mainly along the west coast. The meeting of the North America plate and the East Pacific rise causes horizontal displacements along the San Andreas Fault (see Figure 9.10b). Volcanic activity in the Cascades occurs where the Juan de Fuca minor plate plunges beneath part of North America. Canada is less troubled by most of these hazards, apart from earthquakes along the west coast and winter snowfalls across the country.

## Environmental Problems

The emphasis on expansion and economic growth led to many environmental problems in Anglo America before the era of concern that developed through the 1900s. The mining of fuels and metallic ores, the outputs of manufacturing industries, and the plowing of land across Anglo America resulted in spectacular examples of environmental degradation.

- By the early part of the 1900s, many fields in the U.S. South that had been cultivated for decades were destroyed by deep gullies and were abandoned. Heavy rains had washed out the gullies when soil between row crops such as corn, tobacco, or cotton was kept bare by cultivation.
- The plowing of subhumid grasslands in western Oklahoma, Kansas, and Texas resulted in the **Dust Bowl** disaster of the 1930s in which winds blew away the dried

**RECAP 9B: Natural Environment**

Anglo America contains most of the world's natural environmental types except for the most tropical. Its climates range from tropical to arctic and its terrains from young folded mountains to wide river plains and ancient shield areas. Natural vegetation and soils reflect climatic differences, with deserts in the southwest, grasslands and black earths west of the Mississippi River, forests and brown earths in the East, and forests with podzols in the North, giving way to tundra in northern Canada and Alaska.

Natural hazards affecting Anglo America rival the richness and variety of natural resources. They include hurricanes, tornadoes, earthquakes, volcanoes, and floods. Other environmental problems result from the intensive uses made of the soils, forests, and mineral resources and from the pollution of air and water.

9B.1 To what extent have the natural environments of Canada and the United Sates affected their human geography?

9B.2 How do Anglo American environmental problems compare with those in the Commonwealth of Independent States (Chapter 8)?

**Key Terms:** Dust Bowl

**Key Places:**

| | | |
|---|---|---|
| Mount McKinley | Canadian Shield | Colorado River |
| Rocky Mountains | Canadian Prairies | Columbia River |
| Colorado Plateau | Great Plains | Great Lakes-St. Lawrence Seaway |
| Columbia Plateau | Appalachian Mountains | Mackenzie River |
| Mount St. Helens | Mississippi River | |
| San Andreas Fault | Ohio River | |

finer soil particles and left piles of sand behind. Even in the more humid Midwest, the soils thinned as winds and surface runoff removed the topmost layers.

- Farming extended into the arid areas of the southwestern United States, where water for irrigation was available in the underlying rocks or major rivers such as the Colorado, but faced problems from salinization and depletion of groundwater resources.
- Large-scale hydroelectricity projects, often linked to flood control, irrigation, or navigation, as in the Colorado River, Columbia River, and Tennessee Valley Authority projects, had impacts by drowning land and attracting polluting industries (e.g., the Hansford nuclear weapons facility on the Columbia River).
- Mining on a large scale resulted in the digging of huge pits for extracting copper and other metal ores in the West. The strip mining of Appalachian hills for coal after 1950 devastated large areas of eastern Kentucky and West Virginia.
- The increasing scale of industrialization and urbanization in the United States brought problems of storing and getting rid of wastes from factories and homes.
- Cities became plagued by smog generated from the exhaust gases of large numbers of vehicles and thermal power plants.
- Acid rain, derived from power plant emissions, particularly along the Ohio River valley, affected trees and caused rivers and lakes downwind in the northeastern United States and eastern Canada to become more acidic.

As a result of the many laws passed and operated successfully, the impacts of soil erosion, poor water management, and urban air quality reduced in the late 1900s despite the increases of population and intensity of human activities.

Canada, with fewer people in a slightly larger area, has less environmental problems than the United States. It has to cope with environmental pollution exported from the United States in the air flows that bring acid rain and in the industrial effluents fouling the waters of the Great Lakes. Canada reduced polluting outputs from smelters on its territory near the British Columbia border and along the northern shore of Lake Superior. Wherever mining and the refining of minerals occur, there is pressure on the environment, and strict federal laws imposed huge costs on the nickel smelters at Sudbury in Ontario.

The increasing size of hydroelectricity projects in Québec and elsewhere led to environmental and Native American protests over the resulting hydrologic and land-use changes. Uncertainties concerning federal and provincial jurisdiction in these areas led to differences of approach among the different parts of Canada. For example, Québec moves ahead with its huge hydroelectric projects against the wishes of local Native Americans, and Alberta encourages the extraction of coal, oil, and natural gas, while Ontario is more concerned about environmental impacts.

## UNITED STATES OF AMERICA

The **United States of America** is a highly urbanized country with over three-quarters of its population living in urban areas and just over half in metropolitan centers of over 1 million people. Over 90 percent of the U.S. population lives within a two-hour drive of a large city of over 300,000 people. Despite a slight fall in urban population between 1970 and 1980, the rise that began in the early 1800s continued again in the 1980s. Large American cities were the essential feature of the country's geography in the 1990s.

It is in the cities that most of the large manufacturing facilities operate and where the growth in financial and business services is concentrated. Some 85 percent of U.S. real estate value occurs in the 2.5 percent of land occupied by urban areas. Although the contrasts between richer and poorer groups in American society are most obvious in the cities, rural areas struggle to maintain their populations as machines take over for people on farms, land is taken out of agriculture, and low-pay jobs in farming are replaced by low-pay jobs in manufacturing or tourism. Rural areas near large cities show less evidence of such stresses only when they become commuter or second-home hinterlands of the cities.

The United States is the largest market economy in the world, and it has a major influence in spreading this type of economy to the rest of the world. The United States remains

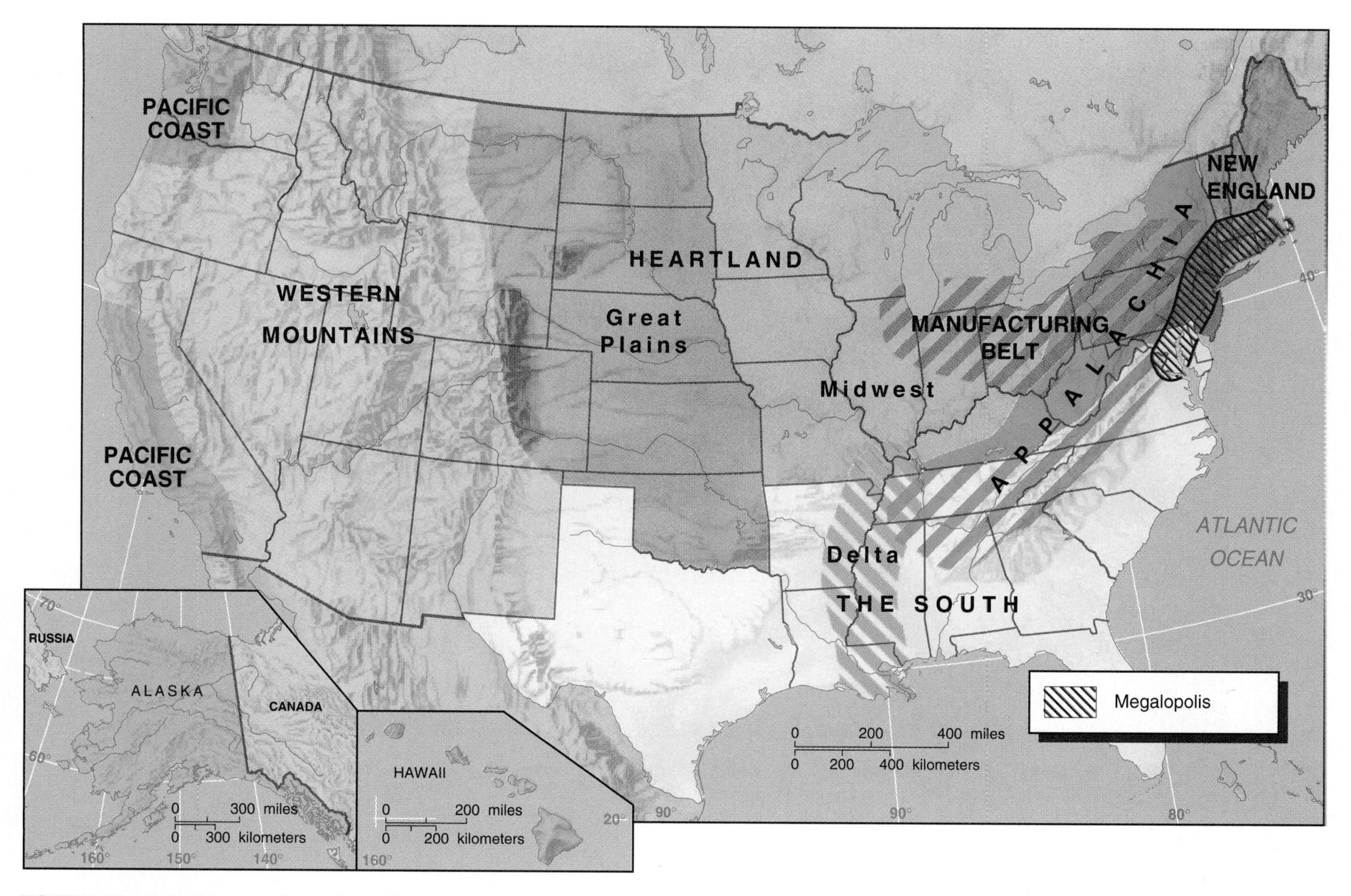


**FIGURE 9.15 United States: the regions.** They are based on differences in natural environments and on patterns of human occupation. The overlap of regions in the northeast reflects the early concentration of people and changing economic and social emphases.

the world's entrepreneurial leader, based on flexible labor markets, low taxes, light regulation, and an easier lot for failures than in other countries. In 1996, the United States raised $10 billion in venture capital, of which $3.7 billion were in California and over $1 billion in Massachusetts. Both states were ahead of the U.K., France, and Germany—the next three—and 10 U.S. states were in the top twenty world "countries."

The growth of market-based activities within the United States and the increasing geographic complexity and mobility of people and jobs during the 1900s led to the greater political involvement of government in regulatory activities and infrastructure provision. The United States has a **federal system of government.** The 50 states have their own constitutions and are linked by the federal government based in Washington, D.C. The states agree to share some functions and to delegate them to a central government. The United States federal government is responsible for financial strategy, defense, international relations, and other matters that involve more than one state, such as building major highways and controlling air traffic. With the growth of government funding through income tax, the expanding role of defense, and increased personal mobility in the 1900s, the federal government became increasingly significant in the country's changing human geography. It grew in importance as an employer and source of grant aid for infrastructure projects that often opened up poorer areas to economic development. The state governments retain responsibilities for local physical planning, education, and other services, many of which are delegated to local authorities at the city or county level.

## Regions of the United States

The large area of the United States makes it a country of many internal geographic regions that developed from the interaction of people with the natural environments and resources. The historic pattern began on the east coast and spread westward. The geographic distribution of its varied natural resources and human conditions resulted in internal regional units being recognized by groups with widely differing interests, such as geographers, novelists, and government and business administrators.

Many different regional divisions of the United States have been proposed, but the one adopted here is based on the human geographic regions that emerged in the 1800s and early 1900s. Current geographic processes are changing their characters and external linkages, but the regional entities established at that time still form an essential basis for understanding the human geography of the country. The overlapping set of regions (Figure 9.15) reflects the outcomes of the

**FIGURE 9.16 New England: Acadia National Park, Maine.** The rocky coast, fir trees, and fall colors provide a distinctive landscape that contrasts with the urban-industrial landscapes of southern New England.



initial geographic concentration of European settlements, and subsequent industrialization and urbanization. Because the regions overlap, it is possible for a city or metropolitan area to be included in more than one region.

## New England

**New England** comprises six states in the northeastern corner of the United States. After the 1620 arrival of the *Mayflower* pilgrims, streams of immigrants first settled the southern parts in townships founded for religious purposes. Secular politics soon took over, and New Englanders played an important part in the revolution that brought U.S. independence in 1783. The southern part of the region then became the first part of the United States to be industrialized in the early 1800s, based on water-mill power, local labor, merchant capital, and access to markets by water transport. The early metal, textile, and leather industries grew in scope and factory size through the 1800s. Possessing no coal or oil, the region was at a disadvantage when new industries based on steam and electrical power developed later in the 1800s and early 1900s.

By the mid-1900s, New England's established manufacturing industries were declining. Cities such as Fall River and Lowell had many derelict textile mills, high unemployment, and out-migration. During and after World War II, many of the traditional industries were replaced by high-tech computer, airplane engine, and other industries that were favored again by high levels of defense spending in the 1980s. At this time, the large number of higher education institutions in the region combined research innovations with the local availability of entrepeneurial capital as a basis for a fresh boom of economic development. This is sometimes referred to as the "Massachusetts Miracle" because of the numbers of new jobs created in a declining area.

In the mid-1990s, some of the computer and defense-based corporations generated in the 1970s and 1980s along the growth corridor of Route 128 west of Boston struggled and many of the offices were "to let." Another industrial boom may be passing, but many parts of New England retain evidence of wealth as its financial and professional services continue to grow. New England's environment of rocky and sandy coasts, wooded hills, and winter sports facilities, together with the sophisticated provision for entertainment in the cities such as Boston, appeals to many Americans as a living environment (Figure 9.16).

## Megalopolis

The **megalopolis** area stretching from Boston in the east to Washington, D.C., in the south includes the huge metropolitan areas of New York, Philadelphia, and Baltimore as well as many smaller urban centers (Figure 9.17). Around 50 million people—nearly one-fifth of the total population of the United States—live in this region, which includes the southern part

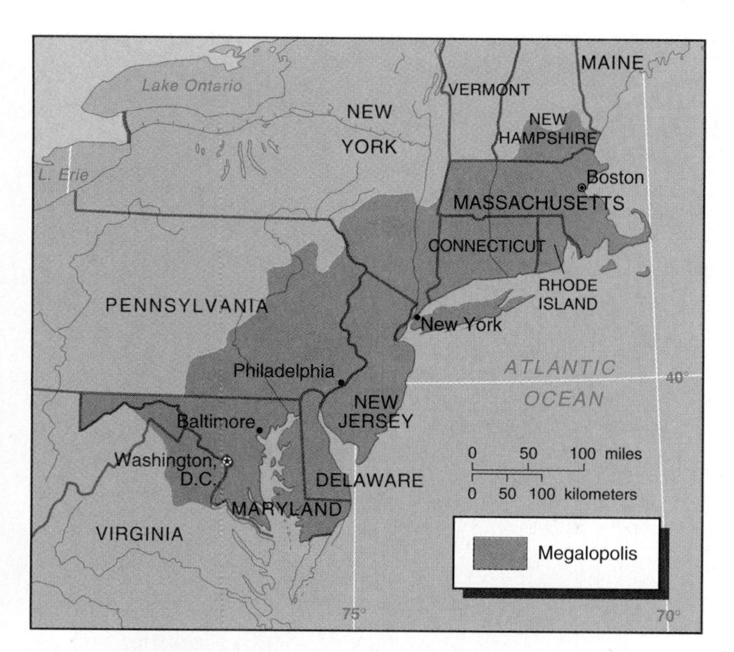


**FIGURE 9.17 Megalopolis: a definition.** The Megalopolis region is defined by largely built-over land that has close-spaced metropolitan cities and a predominance of urban and urban-linked activities. Some areas within this definition are rural but linked closely to the urban functions of the region (commuters, truck farming, dairying).

of New England and the Middle Atlantic coastlands. In places, the roads and railroads are followed by urban land uses from one city to another without a break. Even the rural areas in this belt are subsidiary to the urban activities in the cities. Farmers produce milk and vegetables for the local urban markets and special crops, such as tobacco, for industrial processing, but much of the land was abandoned for farming and covered by trees. Many rural homes set in wooded lots house commuters, while the coasts and hills provide recreation centers for the urban population.

The Megalopolis phenomenon was first identified in the United States in the late 1950s. Such a close-spaced set of cities reflected the historic competition among the closely spaced preindependence colonies and the newly independent states in New England and the Middle Atlantic. Although New York always retained its early advantage as the largest port city, Baltimore grew rapidly in the early 1800s as an exporter of wheat. From 1830, the city fathers of Philadelphia built railroads westward to compete with the New York State Barge (Erie) Canal and New York railroads for access to the newly settled lands in the Midwest. Washington, D.C., emerged as one of the largest U.S. cities when the functions of federal government multiplied in the 1900s. The suburban counties around Washington, D.C., contain some of the most affluent communities in the country.

Despite the migration of some corporation headquarters out of the Megalopolis cities to other parts of the United States, this region still dominates the American economy, federal politics, and the modern media industry (see "Living in the United States," pages 426–427). It is the internal core of the United States. Other groups of cities within the United States have been identified as new versions of the megalopolis phenomenon, but none matches this region in its numbers of people or its economic, political, or social leadership.

## Manufacturing Belt

Between the 1800s and mid-1900s, an area stretching from Boston on the east coast through New York and Philadelphia westward to Pittsburgh, Cleveland, Detroit, and Chicago became so dominated by manufacturing industries that it was termed the **Manufacturing Belt.** The textile and metal goods industries continued to develop in the east as the Pittsburgh-Cleveland area became important in steel and heavy engineering, Detroit became the center of the automobile and truck industries, and the Chicago area produced a range of engineering and farm machinery. The largest cities became the centers of complexes of linked mining and mill towns with high densities of population, high proportions of blue-collar workers, and close-spaced networks of railroad and road links. From Ohio westward, intensive farming produced raw materials for food-processing industries, while other factories made the chemicals and machines needed on the farms. This region produced over two-thirds of U.S. manufactured products until challenged by the west coast and southern states during and after World War II.

The region remains dominant in manufacturing output within the United States, but older industries are less concentrated here and many have been replaced by modern consumer-product manufacturing industries taking up sites close to the country's largest markets for such goods. Although the region lost population to newer urbanizing regions across the United States, some 60 percent of the country's population still live there.

As manufacturing growth slowed from the 1970s, service industries grew in significance in this region. Chicago became a major center for financial services arising out of its futures trading in agricultural products. Some of the smaller towns became higher education centers, where facilities expanded to accommodate 30,000 and more students on each college campus.

After two decades of slower industrial growth, the western part of the Manufacturing Belt experienced a recovery of its manufacturing industries in the 1990s. Steel output along the southern shores of Lake Michigan rose rapidly to supply demands for sheet steel from expanding auto and consumer goods industries. Unemployment dipped below the national average. Since the late 1970s, the economy of the region diversified within each sector (farming, manufacturing, and services) with rising productivity. In the 1990s, manufacturing employment grew faster there than in the rest of the United States. Autos remained central to the region's factory output, and some Japanese manufacturers sited their factories in the region. The problems of the region changed from high unemployment to a potential shortage of skilled labor.

# THE UNITED STATES OF AMERICA

LIVING IN . . .

The United States of America is a huge country, so it is necessary to focus on a quite small but important area to give some idea of what it is like to live there. The Mid-Atlantic seaboard of the United States is the birthplace of the country, since Philadelphia was the first capital of the newly independent nation after 1783. Today is has a great range of industries from agricultural communities producing vegetables, fruits, poultry, and dairy products, to banking, shipping, and chemical companies. The lifestyles of the people in this area are varied. Huge cities are surrounded by sprawling suburbs and those by rural communities. As in other parts of the Megalopolis, economic and social life in these areas are closely integrated.

Americans have enjoyed a high standard of living for many years. Salaries are, on the whole, higher than those in other countries, while food, clothing, and many other items, such as gasoline, electrical goods, and cars, are cheaper. Probably because of its size, and also because this country between Alaska and Florida has practically every type of climate, many Americans prefer to spend their vacations somewhere in the United States. Alternatively, they have easy access through the main air terminals to a variety of exotic lands abroad.

The Mid-Atlantic area offers a wide variety of entertainment for leisure hours. There are casinos at the shore and theaters, first-rate restaurants, museums, and art galleries in the cities. Air-conditioned shopping malls in the suburbs offer movies and special attractions in addition to the wide range of shops. Before the stores open in the morning, those wishing to walk indoors to escape the heat and humidity outside in the summer months can be seen monitoring their speed and distance. Later they might enjoy breakfast in the food courts before beginning their shopping or leaving to go to work. There are probably more varieties of foods available in this area than in any other part of the world because of the kaleidoscope of ethnic backgrounds and the preference of so many people for eating out.

Many families are extremely mobile. Since there is easy access to the shore or the mountains from the Mid-Atlantic cities, many families own two homes. They use one during the week and the other over the weekends or during their vacation periods. It is common to see cars carrying bicycles, canoes, or skis on the highways as families head for the recreational areas. Many retired people have two homes, one in the cooler northern states and one in Florida—or some equally warm climate—which they use in the winter months; like birds, they flock southward at the end of October and migrate north at the end of March, spending time with their families in the Philadelphia area in between. Since Americans frequently relocate with their jobs, or may be filled with the frontier spirit and move around the country, many living in this area do not have roots there. The separation of family members leads to massive migrations on the roads and through the airports for every family festival such as Thanksgiving and the various religious holidays.

(a)

(a) The center of Philadelphia, including City Hall. (b) A house in the Philadelphia suburb of Paoli.



The large family home in the suburbs is an ideal for many. First-time visitors to the United States often comment on the large homes that are not enclosed by the fences, hedges, walls, and gates so beloved in European countries. Instead, expanses of grass surround each home, with shrubs around the house walls to beautify and protect it from the hot summer sun. Flowers are not cultivated in such abundance for the same reason: the extremely hot summers are not conducive to lush gardens. Grass is not the short, velvety European grass, which would die very quickly, but rather a sharper, tougher grass that is kept at a height of three or four inches to protect the roots from the sun.

Americans highly value education. There are many large universities on the Mid-Atlantic seaboard, both in the cities and in the rural areas surrounding them. Most students leaving high school attend either a university or a community col-

**(b)**

© Michael Bradshaw

lege. Freedom of choice is very dear to Americans, and there are many public and private schools, as well as the choice of home-schooling by qualified parents. In recent years, in order to desegregate schools, busing students has been practiced. This has now been stopped in some urban areas, and desegregation by other means, such as magnet or special interest schools, has replaced it.

As the United States is a very health-conscious nation, many large companies offer exercise programs for their employees. Keeping fit is very much a way of life for many suburban Americans. Because most families have a car for every adult member of their households, there is not much need to walk to shops or to work. Few highways or roads have sidewalks; therefore, it is dangerous, and sometimes illegal, to walk on the heavily trafficked roads. Americans have always loved their cars, so public transport is not used as much as in other countries. Sharing rides and the use of buses and trains are encouraged since environmental concerns became much more of an issue and many laws are now in operation to protect the air, water, and land from pollution. Recycling programs are in use in every town and suburb.

People living in this area have continuing concerns. The increasing numbers of senior citizens are particularly concerned about health care issues. The government provides Medicare and Social Security benefits for all seniors during their retirement years if they and their employers paid into the system, but recently the U.S. Congress reduced Medicare benefits and cut cost-of-living adjustments. The issue is not resolved, and there are many debates about it.

The Mid-Atlantic area is part of a very large country, and its population is drawn from every region of the world. Yet life is not chaotic. Violence and lawlessness exist, just as in other countries, but they are not as widespread as the media would have the rest of the world believe. Churches and synagogues attract large congregations, and volunteers are readily available to promote and support worthy causes. As America enters its third century as an independent country, people living around Philadelphia, where the Declaration of Independence was signed, feel proud of its achievements and are increasingly a major part of the global scene.

**FIGURE 9.18 Appalachia: coal mine in West Virginia.** Pickets were posted at the entrance to this mine, near Morgantown, during a miners' strike. Many deep mines in the area have been closed as surface mining with its lower costs has increased.



## Appalachia

The hilly region of the Appalachian Mountains spans the middle part of the Manufacturing Belt and extends into the U.S. South. Each part of **Appalachia** has distinctive economic problems.

In the north, the Pittsburgh region became the national center of iron and steel production in the late 1800s. Many small towns along the Ohio River valleys in Pennsylvania, Ohio, and West Virginia were part of this industrial area, mining coal and iron and manufacturing steel and chemicals. From the mid-1900s, however, the big steel corporations declined, and much industry moved elsewhere, leaving parts of this region to be called the "rustbelt" because of the abandoned and decaying steel mills. Pittsburgh experienced a major replacement of jobs and has many new service functions of a regional metropolis. Smaller towns often had less success in replacing their former economic base.

Eastern Kentucky, West Virginia, western Virginia, and northern central Tennessee form central Appalachia and remain one of the poorest parts of the United States. Coal mining continues to be important but only in areas where the coal has a low sulfur content, making some people rich from this industry (Figure 9.18). After 1950, the lack of other employment opportunities in central Appalachia caused many younger families to leave it for the prospect of jobs elsewhere. They often found difficulty in getting jobs because of their poor education and limited skills. The early 1980s boom in coal output when energy prices were high, together with improved infrastructure, health, and education provision, brought people back to central Appalachia, but conditions deteriorated again as coal prices slumped in the mid-1980s.

Although one of the poorest parts of the United States in the early 1900s, southern Appalachia's economy became diversified with the availability of cheap electrical power through the Tennessee Valley Authority from the late 1930s. Major aluminum companies and federal facilities such as the Oak Ridge atomic laboratories and the Huntsville rocket center were joined by many other manufacturing concerns. Cities such as Asheville, Knoxville, Chattanooga, and Huntsville expanded, developing a wide range of service industries. Contrasts of well-being remain between the urban centers and the surrounding isolated rural mountains.

## U.S. Heartland: Midwest and Great Plains

The lowland area between the Appalachians and Rockies that is north of the Ohio River and the Ozark Mountains is an essentially agricultural region, including eight of the top 10 U.S. farming states and is sometimes called the "breadbasket" of the country because of its production of grain crops. During the period from 1970, this region produced sufficient food to make up for shortages elsewhere in the world, from India to the Soviet Union and parts of Africa. When these markets needed less, land was idled to prevent overproduction.

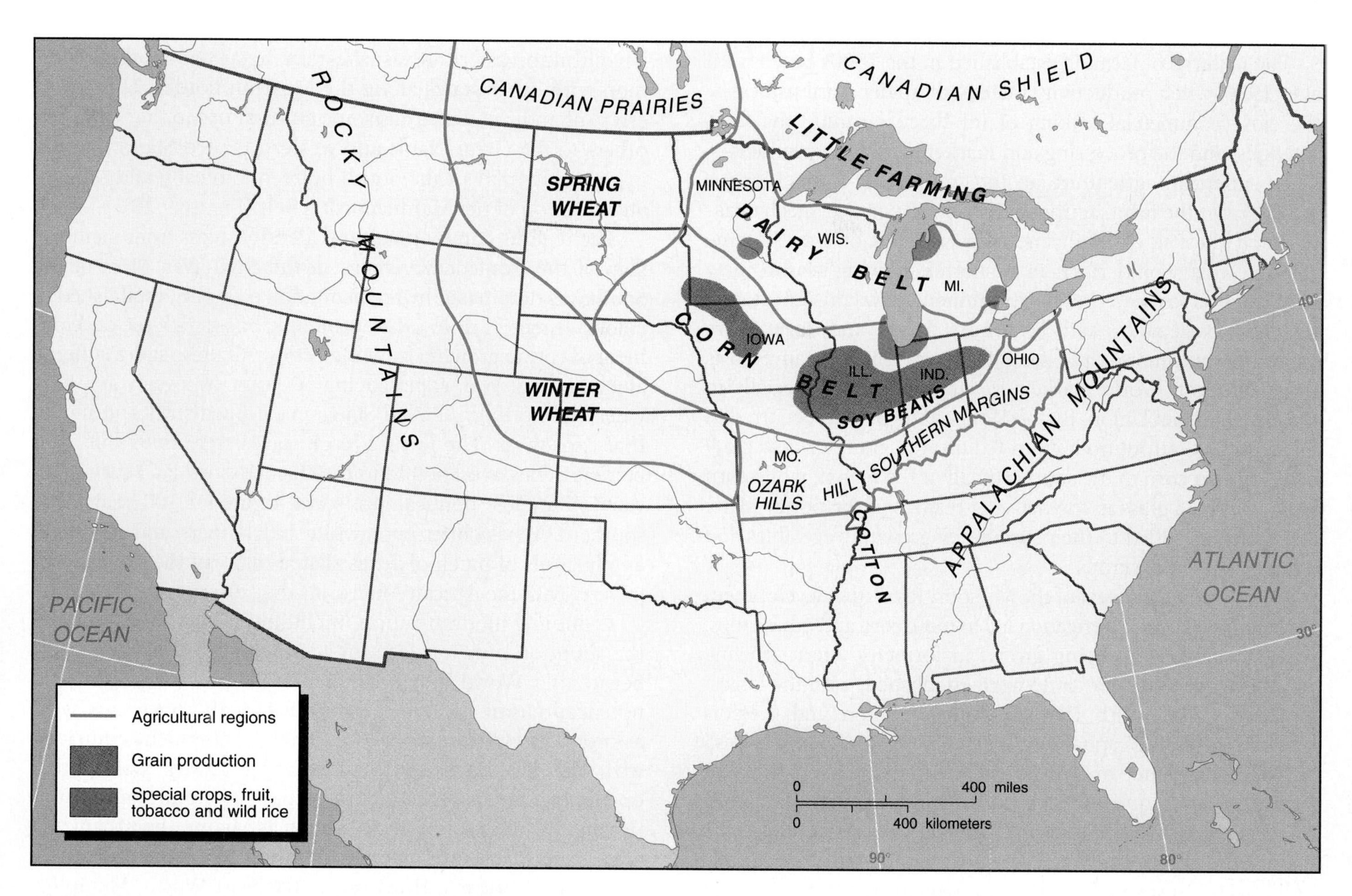


**FIGURE 9.19 United States Heartland: agricultural regions.** The Corn Belt is the most productive and versatile sector, where corn and soybeans ("grain production" areas on map) are grown for direct sale or for feeding livestock (beef in west, pigs in east). In the cooler north, dairying is important, and in the drier west, wheat and other small grains are dominant, while cattle feedlots are toward the south. Farther west, the Great Plains have cattle feedlots and produce wheat in drier conditions. Poor soils, short growing seasons, and mountains mark the boundaries of this productive farming region. High-value crops are grown on costly land close to the main urban areas. Special conditions favor fruit growing on the eastern shores of the lakes where spring comes late, delaying the opening of buds until after killing frosts.

The term "**Midwest**" is mostly applied to the eight states of the Corn Belt and Great Lakes area. The eastern part overlaps with the Manufacturing Belt, so that cities such as Chicago, Detroit, Cleveland, and Cincinnati stand amidst the world's most productive farming region. The Great (or High) Plains include North and South Dakota, Nebraska, Kansas, Oklahoma, and parts of Montana, Wyoming, Colorado, and northern Texas. They have a subhumid climate and rise in elevation to over 1,500 m (5,000 ft.) above sea level at the foot of the Rockies.

The settlement of the **Heartland** area in the early and mid-1800s rapidly established the suitability of its environments for different combinations of crop and livestock products. Growing season declines in length toward the north and precipitation toward the west. The brown soils in the eastern states of Ohio and Indiana mostly developed beneath deciduous forest, while the black soils farther west developed beneath grassland; both are fertile and workable.

Commercial farming grew more important in the mid- and late 1800s, as railroad lines linked the region with markets on the east cost and abroad, replacing the longer distance via the Ohio and Mississippi Rivers and coastal cargo boat. Distance from the main centers such as Chicago and climatic conditions were major factors in the types of farming adopted. A series of agricultural regions emerged, based on a combination of natural advantage and distance costs (Figure 9.19).

- The largest was the Corn Belt, which itself included a mosaic of farming types having varied emphases on corn as a cash crop or—mainly—as feed grain for the fattening of pigs or beef cattle.
- In the cooler conditions north of the Corn Belt, dairy cattle were kept and most of the land left in pasture.
- In the drier Great Plains to the west, wheat became the main grain crop, sown before winter in the south and in spring in the north.
- In favored locations, especially on the eastern lake shores, fruit-growing became a specialty.
- On the hillier southern margins facing the Ohio River and in the Missouri hills, less land was suitable for farming, which was a less profitable occupation.

The patterns of farming established in the 1800s lasted until after 1950, when productivity increased rapidly. Agribusiness—the close commercial linking of inputs to farming, the farm activities, and the processing and marketing of farm products—made American agriculture an integral part of a much larger industry. On the farm, agribusiness brought greater mechanization and the use of fertilizers. New varieties of corn became available that would ripen in a shorter growing season, new crops such as soybeans became prominent, new consumer tastes for more vegetables and less dairy foods and meat were expressed, and more rapid road transport made distance costs less significant. World markets for grain opened, especially in the former Soviet Union. By the 1980s, a major shift occurred in the Corn Belt output from grain-fed meat to grain for cash. Soybeans rivaled corn in the southern half of the region, while corn took over from grass in the dairying regions of the north. Many dairy farmers added to their income by growing vegetables and other valuable cash crops.

In the drier areas west of the Missouri River, the development of center-pivot "circle" irrigation led to more corn and other crops, such as sugar beets, being grown in formerly wheat-growing areas. Huge feedlots for cattle were established, with the largest at Sioux Falls, South Dakota. Both irrigation and feedlots required increasing capital inputs at a time of fluctuating prices.

The dependence on groundwater for the expanded use of irrigation raised questions about the limits of water availability. Some streams dried up and the High Plains aquifer—the main source of groundwater—was drawn down. Water levels in western Kansas wells fell by 62 percent between 1940 and 1991, raising pumping costs and the specter of using up all the groundwater. Better methods of water use later slowed rates of depletion, and the High Plains aquifer still contains a substantial proportion of the water it contained 50 years ago. While some parts of this area are short of water, others have plenty for the foreseeable future.

The Great Plains area, particularly the zone where it meets the Rocky Mountains in Colorado, became a major center of high-tech industries, initially based around the federal facilities in Denver and the University of Colorado at Boulder. The area's population grew rapidly in the early 1990s but slowed late in the decade as air pollution, water shortages, and crowded schools deterred more people moving in.

Despite the high and increasing farm productivity of this region, major problems arose. Fewer people were able to depend totally for their income on farming and obtained off-farm jobs. Many farmers incurred levels of debt that were too high to sustain so they rented land to other farmers or sold out to banks and other financial businesses. Much land that had resided in family farms became corporately owned. Problems of declining population and few alternative occupations that had plagued other rural regions for decades extended to America's richest farming region during the 1980s.

### The South

The American **South** extends from the southern Atlantic coastal plain westward to the southern Mississippi River and beyond into eastern Texas. Western Texas has more in common with other states along the Mexican border. The northern panhandle of Texas has some cultural ties to the South but others to the Great Plains and western United States. Florida is physically part of the South but economically and socially an extension of the Manufacturing Belt (Figure 9.20).

The region's common cultural identity stems from membership of the Confederacy states in the Civil War. Most of the South was dominated by the plantation economy, established in colonial times. It received an economic boost with the development of cotton growing from the 1790s. All these states suffered after the Civil War, experiencing 50 years of poverty at a time when the northern United States made huge strides in industrial expansion. The former black slaves were freed but soon became debt slaves, bound to the sharecropped land they cultivated. They lost political and social rights for 100 years. The southern Appalachian poor white hill farmers and the small family farms of much of Texas shared many of the problems of poverty with the African Americans they discriminated against.

Economic modernization, including industrialization and the adoption of better farming technology and management, began after World War I. The real changes in the South did not occur until the civil rights of African Americans were acknowledged from the 1960s. This development coincided with the building of the interstate highway system that opened up much of the South to new economic opportunities and the expansion of cities such as Miami, the urbanizing area from Charlotte, North Carolina, to Atlanta (Figure 9.21), New Orleans, Houston, Dallas-Fort Worth, Memphis, and Louisville. Service industries, as well as manufacturing, moved into the area, and its changing fortunes were highlighted by local boosters who took up the term "Sun Belt" to refer to the U.S. South with its mild winters and sunnier climate—in contrast to the "Frost Belt" of the U.S. northeast. Although this term was also applied farther west and much of the economic growth with which the Sun Belt became associated was geographically patchy—more a set of "sun spots" than an evenly prosperous belt—the changes in outlook in the region were real and lasting.

The economic advantages enjoyed by the South since the 1960s include an improved and more focused agriculture, using the better lands for cotton, corn, and peanuts and leaving the poorer lands to woodland. After a period of difficulty, demand for cotton surged upward again in the mid-1990s. Cotton products fought back against artificial fibers while bad weather and insect infestations affected other world producers. The United States' 1994 record cotton harvest was one-fourth of the world total, reflecting increasing yields. While South Carolina's previous highest crop in 1877 was grown on 2.8 million acres, the 1994 crop came from 0.23 million acres. Exports took half of the U.S. crop. Poultry raising in factory units is another major capital-intensive farm activity in the region.

Rural areas, however, remain the poorest parts of the South, with the Mississippi Delta lowlands forming one of the most difficult areas in the United States in which to generate economic growth. These lowlands extend up the Mississippi

**FIGURE 9.20 United States South: Florida peninsula.** Space shuttle view from the north. Sea breezes give rise to clouds over the land of the southern part of the state, including Miami and the Everglades. Lake Okeechobee is the cloud-free lake. Tampa Bay is on the right, backed by the white markings of phosphate pits.

NASA

River valley to southern Missouri (see Figure 9.15) and were a major area of cotton production until the mid-1900s.

Most Southerners now work in manufacturing and service industries. In the 1920s, textile mills sprang up in a belt northeast of Atlanta on the basis of cheap local land and labor as the cotton farms were mechanized or abandoned. By the 1990s, many of the textile mills had closed as textile industries moved to Latin America and Asia. Around the city of Greenville, South Carolina, for example, such jobs were replaced by those in a variety of new industries including Michelin tires (from France), BMW cars (from Germany), Lucas automotive electronics (from Britain), and Hitachi electronics (from Japan). Such investments were based on the perception of this region as having skilled labor, often without labor unions, good local support for industry, and pleasant environmental conditions for management.

In Texas, the accumulation of oil wealth was paralleled by manufacturing and services development around Houston and Dallas-Fort Worth, partly related to the National Aeronautics and Space Administration center in Houston and other high-tech advances. The service industries in the South include tourism—spreading northward and westward from

**FIGURE 9.21 United States South: Atlanta, Georgia.** The 1996 Olympic stadium is in the foreground, backed by the baseball stadium and the high-rise skyline of central Atlanta. Atlanta's growth became a symbol of the economic revival of the U.S. South.

©Marvin E. Newman/The Image Bank

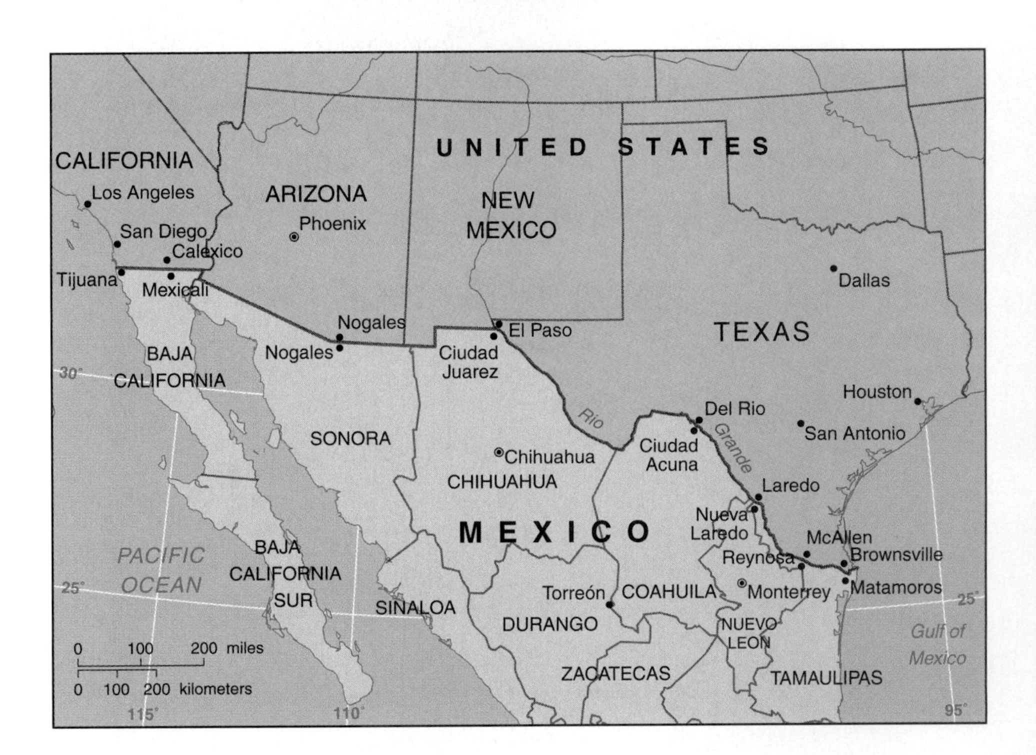


**FIGURE 9.22 Western Mountains: the United States-Mexico borderlands.** The contacts between the two countries led to a series of twinned towns with linked economies. American manufacturers site factories in Mexico to take advantage of cheap labor costs.

Florida—plus health care, education, and financial services. Such industries often pay higher wages than semiskilled jobs in manufacturing.

Even in the Mississippi Delta, one of the poorest parts of the United States, manufacturing brings new jobs. Mississippi County, Arkansas, once the leading cotton county in the country, is now one of the largest steel producers. It has two mills making sheet and I-beam steel using scrap metal brought in by river barge. Other industries also moved into this county, but its unemployment is still twice the Arkansas average.

## Western Mountains

The western one-third of the United States is a mountainous region. The highest ranges are the Rockies on the east and the Sierra Nevada of California and the Cascades on the west. They are separated by broad plateau areas (see Figure 9.8a) and areas of basin and range.

Settlement focuses on irrigation farming and towns that grew up as markets, mining centers, or as nodes on the transcontinental railroads. A large proportion of the land between such centers is still sparsely settled. Much is owned by the federal government and designated as national parks, national forests, or grazing lands. Even these uses in a sparsely populated area generate conflicts when the interests of ranches and farms interact with environmental conservation or tourism.

The **western mountains** of the United States are not only high in altitude, but much of the region is arid. Most of the water available falls on the mountains as snow and runs off in swollen streams in spring. Many federally funded irrigation projects manage this water supply, with the largest groups of projects along the Colorado River in the south and the Columbia River in the north. Irrigated farming and cheap electricity created a series of productive oases that formed the basis for the growth of cities such as Las Vegas, Nevada, Phoenix, Arizona, and Salt Lake City, Utah. These cities are now service and transportation hubs.

In the southern part of the region, Hispanic people make up a large proportion of the population. Defined in part by the Rio Grande, the border with Mexico is largely a mountainous desert where few people live outside the border towns. The border cities are twinned (Figure 9.22) and hold three-fourths of the Americans and Mexicans living within 80 km (50 mi.) of the border—a population that rose from 9 million to over 15 million people since 1980.

From 1965, the Mexican *maquiladora* ("mills") program enabled United States and other foreign companies to assemble imported products tax-free in northern Mexico for immediate reexport without duties (see Chapter 10). In the late 1990s, around 2,000 such factories employed half a million Mexicans and brought in one-third of Mexico's foreign exchange. On the United States side of the border, many Hispanic people find new homes, making up a high proportion of the local population, which is rising so fast that squatter shantytowns form in southern New Mexico. Mexicans like to buy U.S. goods, spending around half of the wages earned at the maquiladoras.

Although the United States-Mexican border areas remain poorer than the United States' average, people living there have more than twice the incomes of people in Mexico. Environmental conditions in this region are also a concern, although both Mexico and the U.S. authorities are dedicated to cleaning up the river water, beginning with waste from industrial facilities. The impact of NAFTA on this region is

**FIGURE 9.23 Pacific Coasts: San Francisco Bay area.** San Francisco occupies the partly cloudy peninsula in the foreground of this space shuttle photo. The Golden Gate inlet is obscured by low clouds forming over the cold water offshore. The line of the San Andreas Fault can be traced from left (north) to right (south) across the foreground. Inland, the urbanized bay shore from Richmond to Berkeley and Oakland, the reclaimed areas of tidal lands, and the complex waterways of the Sacramento-San Joaquin River delta stand out clearly.

NASA

still uncertain, and it will take some years for maquiladora jobs to move to other parts of Mexico.

## Pacific Coast

From the Seattle area around Puget Sound in the north, to San Diego on the Mexican border, the American **Pacific Coast** has become a second national core close in economic importance to that of the region between Boston and Washington, D.C. Settlement increased after the arrival of transcontinental railroads in the 1870s and 1880s, but, until World War II, the main products remained timber in the north and farm products and the growing film industry in the south. National defense needs for the Pacific War then placed manufacturing and military centers on the west coast. Puget Sound in the north and San Diego in the south became major naval centers. Seattle, the home of the Boeing Aircraft Company, grew as the company produced first bombers and then jet airliners. The Columbia River became a staircase of huge hydroelectricity stations that support 11 aluminum smelters and the Hansford nuclear weapons facility among other industries but reduced the salmon catch. Los Angeles in the south also had major aircraft manufacturers, Lockheed and McDonnell-Douglas (now both linked to Boeing), together with a variety of manufacturing and service industries.

After World War II, manufacturing prospered and servicemen returned to live in new communities along the Pacific coast. San Francisco (Figure 9.23) was the only city on the west coast until the early 1900s, when Los Angeles rapidly outstripped its growth. Both cities attracted growing financial service businesses after World War II. Defense cuts in the 1990s led to lost jobs and plant closures, but only a mild recession resulted in California, which was growing again by the mid-1990s as a result of the shift to high-tech jobs, a more diversified range of industries, and increased overseas exports. Over 400,000 new jobs were created in California from mid-1996 to mid-1997, and personal incomes rose. The Los Angeles economy alone is greater than that of South Korea, and the city now contains a reinvigorated film industry, a growing major port, and more computer software jobs

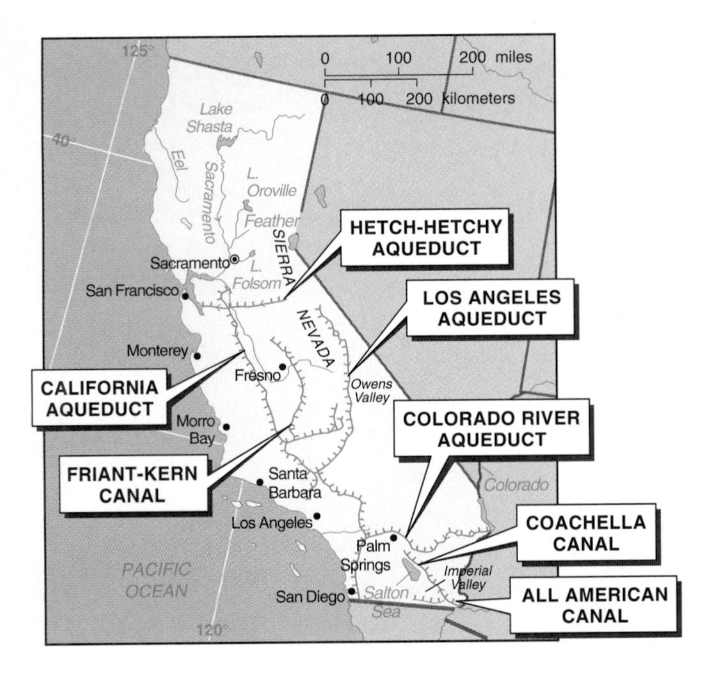


**FIGURE 9.24 Pacific Coasts: water flows in California.** Water flows from the high precipitation areas of the north, where reservoirs store the water, toward the arid south. Water supplies for San Francisco (Hetch-Hetchy) and Los Angeles (Owens Valley) in the early 1900s were later enhanced by federal and state water projects in the Central Valley. The federal Imperial Valley project and Colorado River Aqueduct in the far south took water from the Colorado River to irrigate dry lands and provide water for Los Angeles and San Diego.

than Silicon Valley, the best-known center of high-tech just south of San Francisco. The Seattle area became home to many software firms, including Microsoft.

The basis of economic growth in the southern half of California was the supply of water to its largely arid area (Figure 9.24). Following the early appropriation of water by the cities of Los Angeles (Owens Valley) and San Francisco (Hetch-Hetchy), huge water storage and distribution projects were funded by the federal and state governments. The success of these measures is illustrated by the existence of the world's largest desert city in Los Angeles, the world's eighth largest economy in the State of California, and America's most productive farming state.

In the late 1900s, however, the limits of water availability were approached, highlighted by drought years. In early 1991, after four consecutive drought years, the major state water project reservoirs held less than one-fourth of their capacity and farmers lost their supplies from that source; the federal project rationed farmers to one-third of normal amounts. Cities still receive a water supply but most pay more for it. There are now no further sources of water to be tapped by California. It is likely that farmers, who use 85 percent of the water to produce less than one-tenth of the state's economic output, will have to pay more in the expectation that they will use it more carefully. At present, farmers pay only half the cost of delivering water to them, while city dwellers pay 20 times as much.

In early 1998, two proposals sought different routes to a middle-term solution of the water shortage problem. They may force a closer study of California's limited water resources and their longer-term usage. The state of California, which used up Colorado River water surpluses from the Nevada and Arizona allocations until they were required in those states, is now forced to return to its original allotment—85 percent of current use. The three states can now exchange water, allowing Arizona to sell enough to Nevada to finance the completion of its own distribution system, the Central Arizona Project. San Diego at present depends on Los Angeles sources but proposes to buy its water from the Imperial Valley to the east, which gained larger Colorado River water concessions than other users in the early 1900s.

The west coast developed a new importance through an increasing volume of business and personal movements with Asia and other Pacific countries. Seattle and especially Los Angeles are major world ports. From its development as a front-line base for war operations against Japan during World War II and later in Korea and Vietnam, this region is now America's door to the world's most rapidly growing countries. The west coast attracts most of the growing Asian investment in the United States. Five of California's top 11 banks are Japanese owned; 30 of the smaller ones are backed by Chinese money. Asian capital, particularly Japanese, backs car design centers, media industry corporations, and leisure industries. Half of the 1,400 Taiwanese companies in the United States are based in California, and research-marketing-manufacturing links across the Pacific Ocean are increasing.

The region also has trading links to Europe and Latin America, and close links with British Columbia in Canada to the north and Mexico to the south. In 1990, just under half of California's exports went to Asian countries, around 30 percent went to Europe, and a further 18 percent was shared between Canada and Mexico. Such links attracted Hispanic and Asian peoples to live and work in the region. Southern California has the highest proportion of Hispanic Americans in the country, with Los Angeles having doubled its middle-class Hispanic population since 1980 to half a million and its Hispanic businesses doubling since 1993. There is no suggestion, however, from the rivalry between its Pacific and Atlantic coastal areas that the east- and west-facing parts of the United States are likely to split, for both have worldwide rather than hemispheric visions.

## Alaska and Hawaii

In 1959, Alaska and Hawaii (Figure 9.25) were added to the 48 conterminous states of the mainland United States. They add to the extreme variety of environments and resources within the United States but have their own problems.

**Alaska** is a huge area of northern land that was bought from Russia in 1867. High mountains, icefields, and glaciers mark its southern coast. Inland, the Yukon lowlands have long cold winters, northward of which the bleak Brooks Range and North Slope leading down to the Arctic Ocean get even colder. No commercial farming is possible. The Yukon lowlands widen

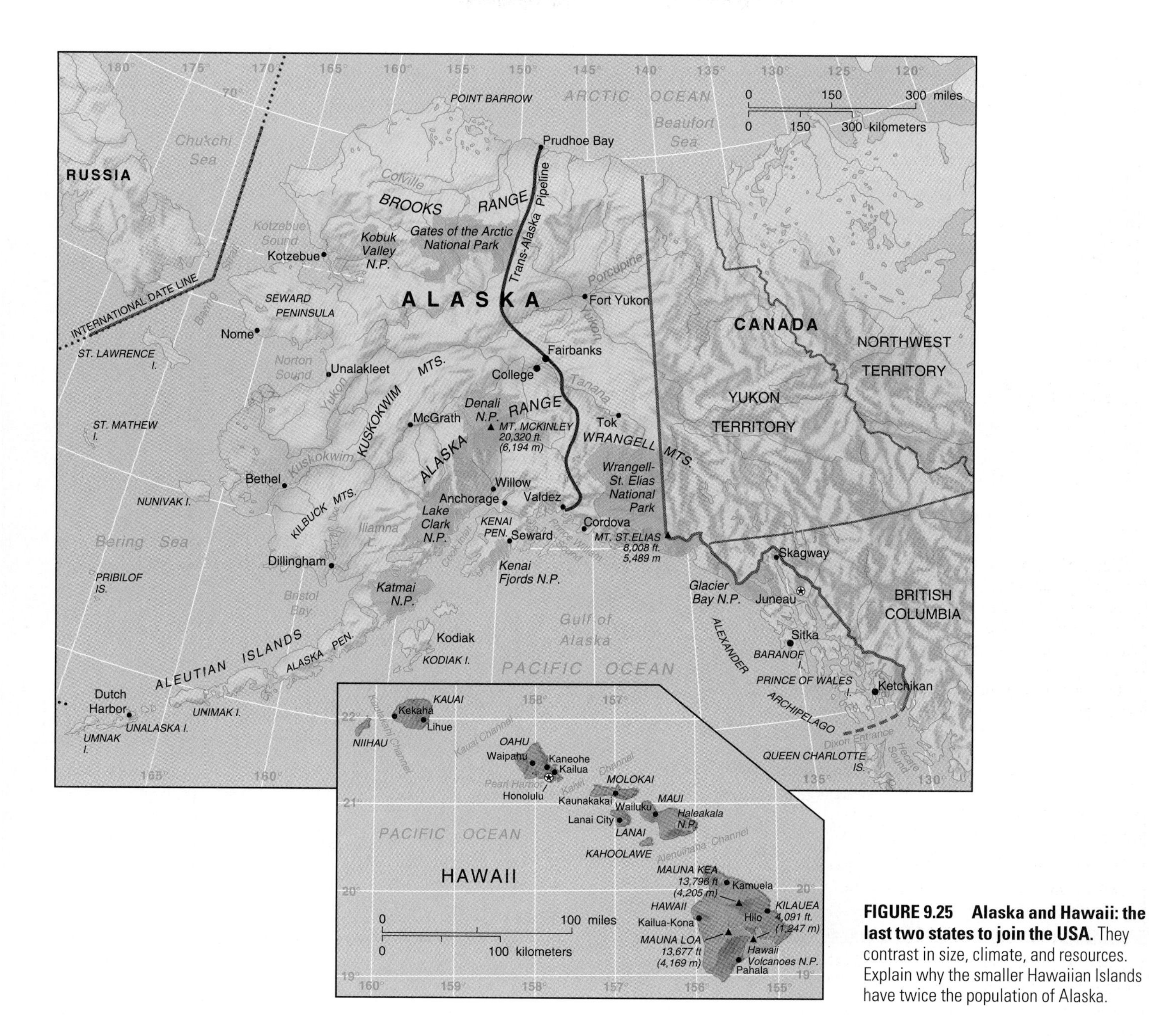


**FIGURE 9.25 Alaska and Hawaii: the last two states to join the USA.** They contrast in size, climate, and resources. Explain why the smaller Hawaiian Islands have twice the population of Alaska.

westward to the Bering Sea, but there is little mining or manufactured output to trade in that direction.

After purchasing Alaska in 1867,the federal government tried to resist attempts to settle and develop it, but gradually salmon fishers and gold miners whittled away this policy. Further exploration showed that Alaska contained resources of fuels and metal ores on a large scale, but only a few thousand people lived in the territory. During World War II, when the Japanese invaded the western islands of the Aleutian chain, the strategic importance of Alaska was recognized, and the military presence in Alaska grew. Military activity increased further through the Cold War because the territory was close to the Soviet Union. Statehood in 1959 gave Alaskans opportunities to select lands for development from the 99 percent owned by the federal government.

The discovery of oil on the North Slope forced the state and federal governments to develop policies for Alaska's native peoples, who live mainly on coastal lands, and for the other settlers, most of whom live in and around Anchorage in the south. The oil from the North Slope made Alaska's 0.5 million people rich in the early 1980s, including state payouts from taxes on oil production. World oil prices fell from the mid-1980s to the late 1990s, with oil production peaking in 1988. The oil discoveries, and especially the *Valdez* oil spill, made Alaska a focus of environmental concern that is now being transferred to other issues in the sparsely settled state. Although oil exploration slowed with the prevention of drilling in the Arctic Wildlife Refuge on the north coast east of Prudhoe Bay, parts of the National Petroleum Reserve to the west of Prudhoe Bay were opened for bids in 1998, looking to a 2010 pumping start.

**Hawaii** was for long a fulcrum of U.S. trade in the Pacific, a naval port, and a source of agricultural produce (pineapples). Involvement in World War II brought its existence

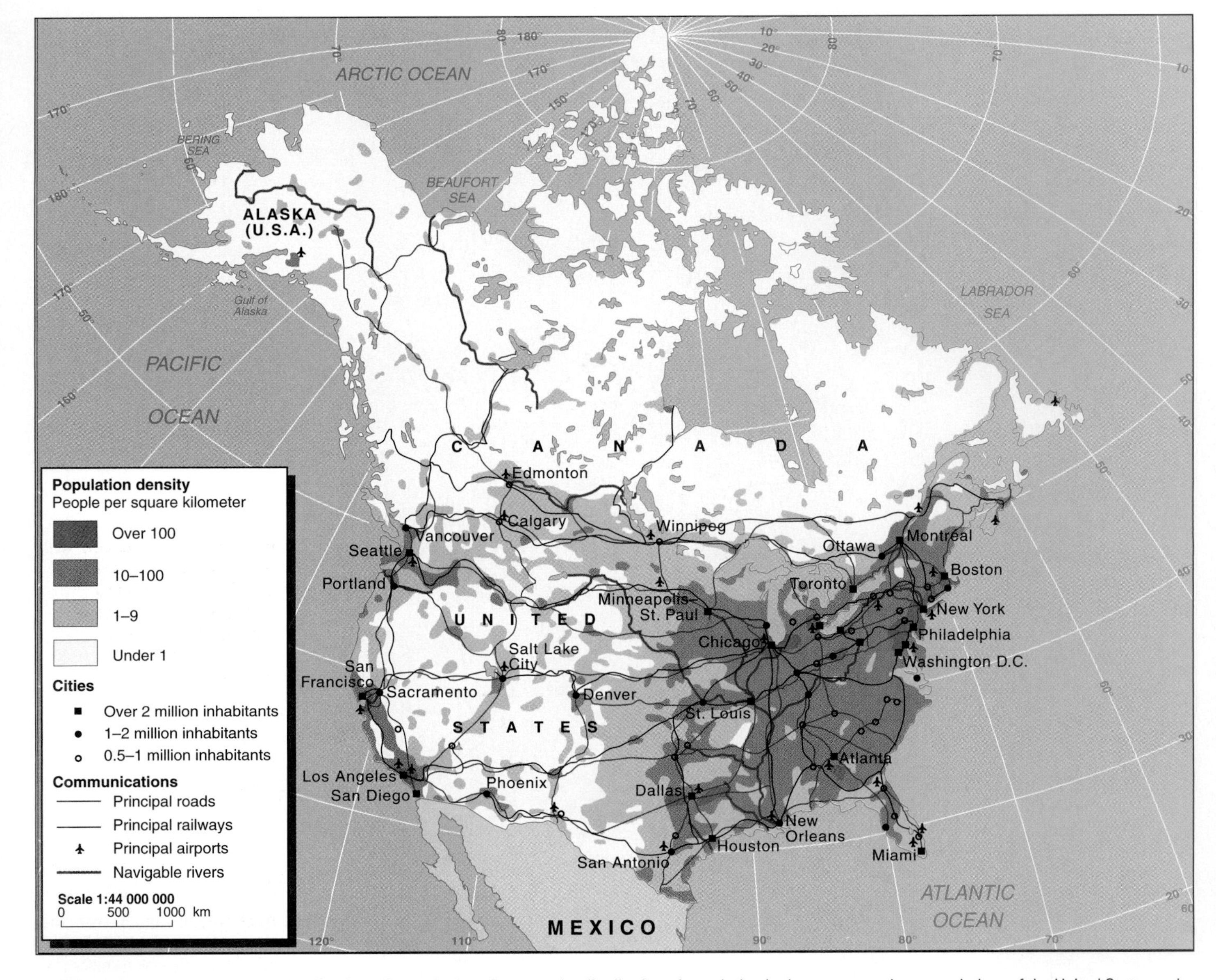


**FIGURE 9.26 Anglo America: distribution of population.** Compare the distribution of population in the eastern and western halves of the United States and note the closeness of most Canadians to the U.S. border.

Source: Data from *New Oxford School Atlas*, p. 99, Oxford University Press, UK, 1990.

home to many more Americans, and it was admitted as the fiftieth state just after Alaska in 1959. Tiny compared to Alaska, but with more people, it has a major international tourist industry. In the late 1990s, Hawaii's economy was in difficulties, with high unemployment. Its tourist industry suffered from cutbacks on travel by Japanese as the yen lost value. Sugarcane plantations closed as Asian competition and changing U.S. consumption habits reduced the market. Hawaii also suffers from a poor business climate due to many regulations, poor schools, and a high union presence.

## People

Distribution of population in the United States divides the country in two (Figure 9.26). The eastern half has more people; the western half has fewer. In the eastern half, highest densities of population occur in the Megalopolis and the Midwest. In the western half, the central Texas and Pacific coastal urban areas and some inland oases (such as Phoenix and Salt Lake City) stand out from areas of very low population.

### Increase by Immigration

The population of the United States is unusual among the world's rich nations because it is still increasing quite rapidly. Its 1998 population of 270 million could rise to 335 million by 2025. The increase is mainly because of continuing immigration, which has been vital to the country's economic growth since its discovery. In 1998, the United States' fertility rate dropped to 2 and the birth rate to 16, while the death rate was 9 (Figure 9.27). With life expectancy over 75 years, the population structure (Figure 9.28) is moving toward

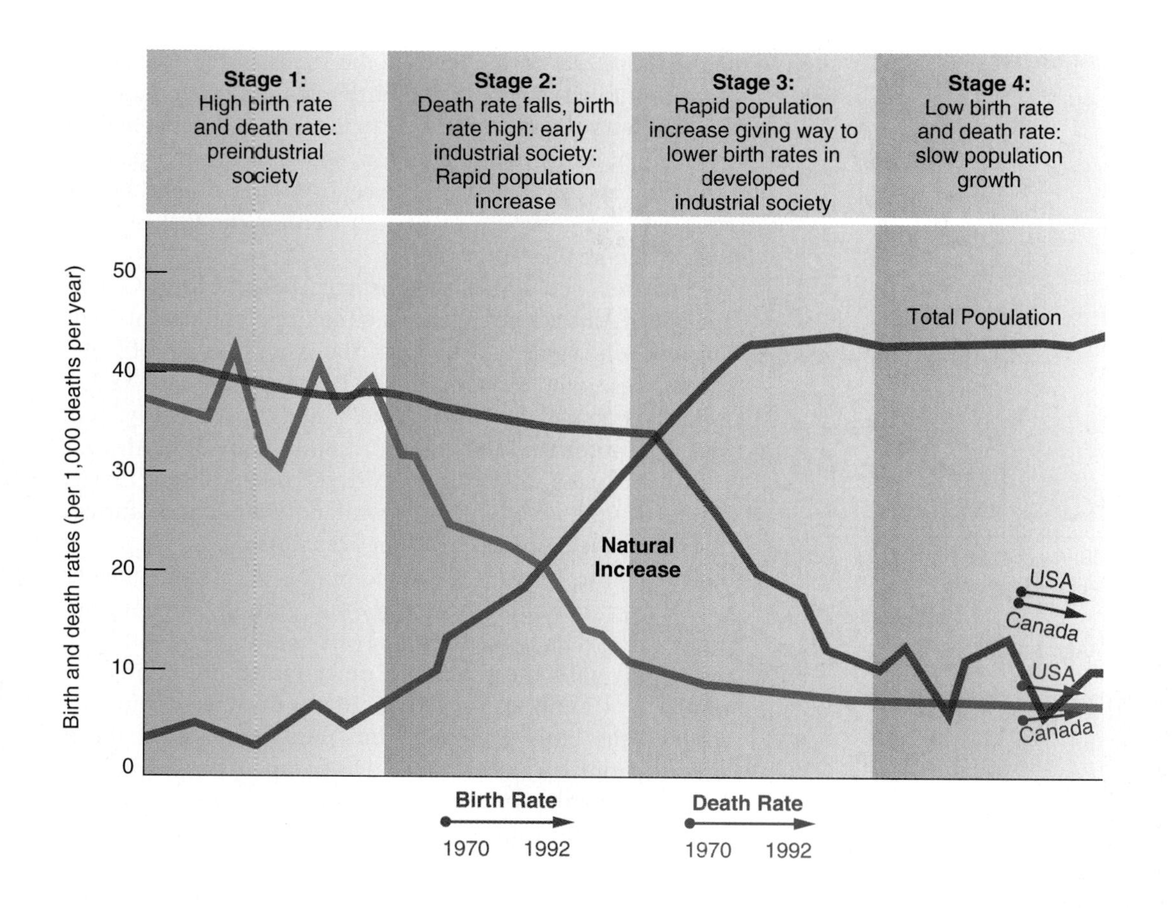


**FIGURE 9.27 Anglo America: demographic transition.** Both Canada and the United States have relatively high birth rates for affluent countries, a sign of the continuing immigration of young families.

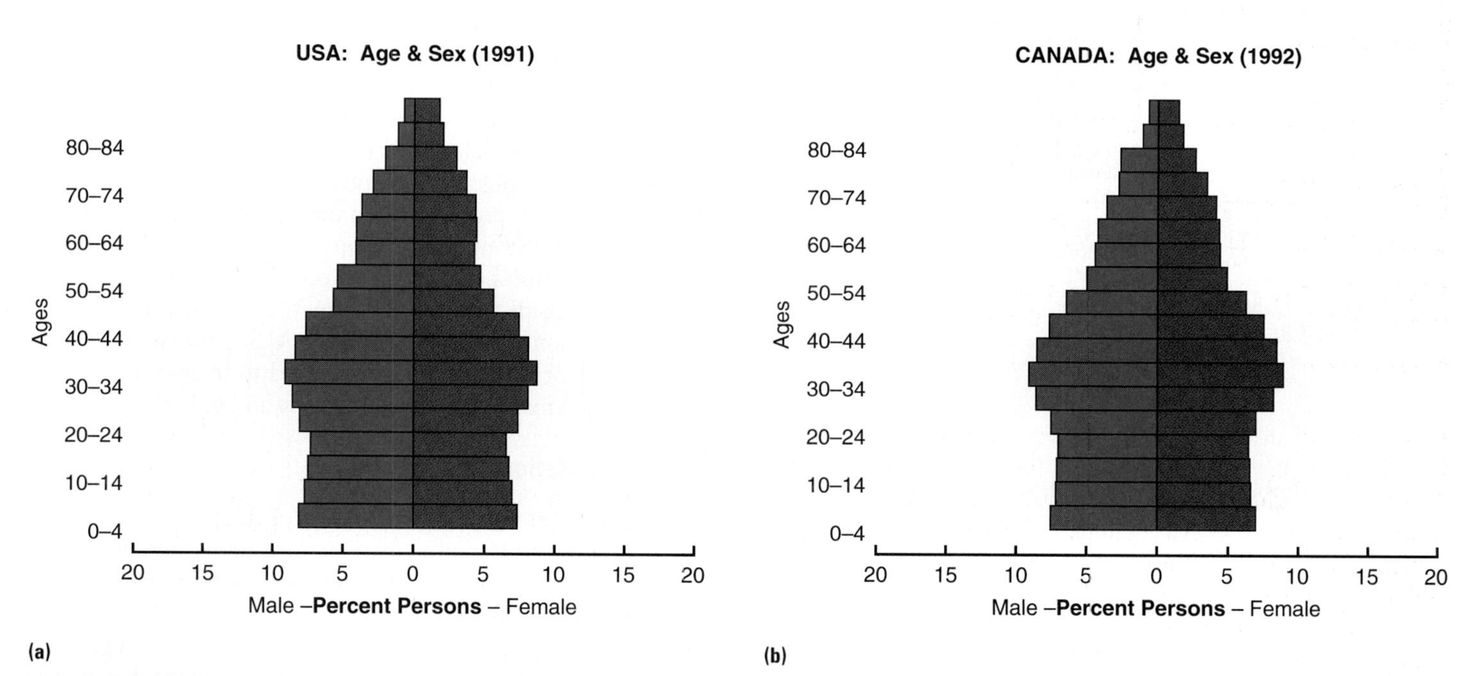


**FIGURE 9.28 (a) & (b) Anglo America: age-sex graphs.** (a), (b) The 1990s graphs show how the baby boom of the 1950s and 1960s produced a bulge in the 20 to 44 age groups. See Figure 9.28c next page.

Source: Data from *Demographic Yearbook,* United Nations, 1994.

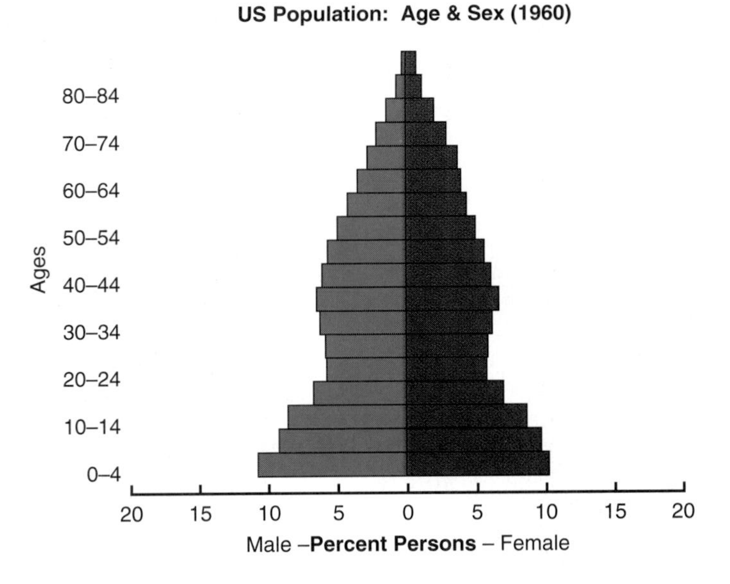

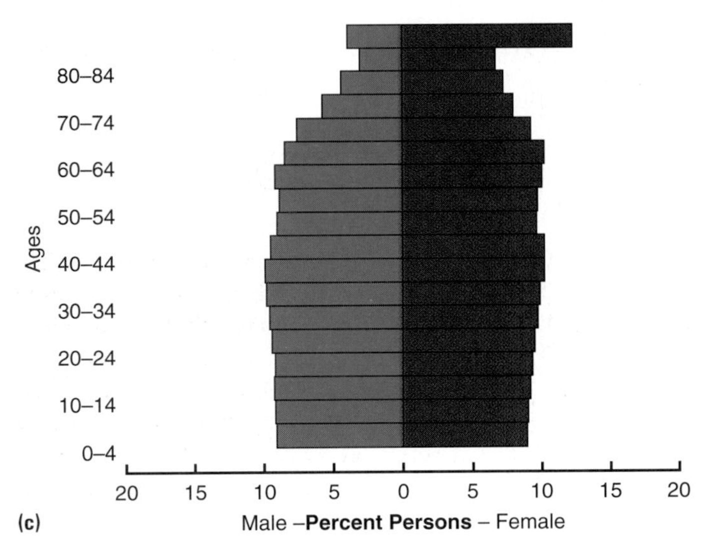


**FIGURE 9.28 (c)** Compare the U.S. graph for 1991 with that for 1960 and the projection for 2050.

increasing numbers of older people, who, with their pension incomes, have become a major economic force in parts of the country such as Florida and Arizona. The immigrant groups maintain a younger age for the working population and have many higher birth rates.

Following the early domination by people of British origin, Irish and German people immigrated in greater numbers during the first half of the 1800s. At its outset, the new American nation debated whether English, German, or both, should be official languages. In 1990, 23 percent of the United States population traced a German ancestry, 15 percent Irish, and 13 percent English. The African American slave population increased by immigration in the 1700s and by natural increase in the 1800s. Slaves were freed in the 1860s, but African Americans continued to have higher birth rates and death rates than the European-origin population. In the later 1800s, the highest numbers of immigrants came from southern Europe, particularly Italy, and at the turn of the century, large numbers came from the Slav countries of Eastern Europe, the Balkans, and Russia (Figure 9.29).

From the mid-1900s, the Hispanic peoples from Central and South America became a major source of immigrants. The Chinese, who came to the United States in the late 1800s, and the Japanese, who had moved in as farmers in the early 1900s, were later joined by other Asian groups, particularly from Vietnam, Korea, and India. Most recently, influxes of Africans and of Russians following the demise of the former Soviet Union and the opening of its frontiers to emigration further diversified the cosmopolitan American population.

During the 1900s, the United States faced a dilemma as to whether it should encourage immigration as a continuing feature of its dynamic society or whether it should discourage it with the perceived objective of preserving the standard of living of its people. Many European countries adopted the latter approach, but they experienced shortages of people in the working age groups (see Chapter 7). Beginning in the 1920s, U.S. immigration laws gave first preference to Europeans, but from the 1960s, they opened access more widely on the basis of family ties or specific skills. There are, however, several million illegal immigrants, or "undocumented aliens," mainly of Hispanic origin, living in the United States.

By 1997, legal immigrants and refugees—who are given special status for a limited period because of adverse political conditions in their own country—topped 0.8 million for the year. Illegal immigrants took numbers well over 1 million. Some 2.7 million illegal immigrants benefited from a 1986 amnesty. New legislation was proposed to insist that sponsorship for relatives should have an income component and to increase patrolling of the Mexican border. Yet immigration responds more to boom or recession inside the U.S. economy than to legislation. Overall, immigration helped to maintain a U.S. population growth of around 1 percent per year, compared to increases of under 0.5 percent in many countries of Europe and in Japan.

## Urban Population

The predominantly urbanized nature of the population of the United States resulted in many very large cities and groups of cities. By 1990, large and growing **metropolitan areas**, often consisting of groups of cities that merged around one major center, were increasingly dominant in the U.S. human geography. In 1996, 104 million people lived in 35 metropolitan areas of more than 1 million people. Of these, 31 million lived in the largest cities of the megalopolis region (mainly **Boston**, **New York**, **Philadelphia**, **Baltimore**, and **Washington**, **D.C.**). A further 23 million people lived in the metropolitan areas of the western Manufacturing Belt and Midwest (**Pittsburgh**, **Cleveland**, **Detroit**, **Chicago**, **St Louis**, and **Minneapolis-St. Paul**,

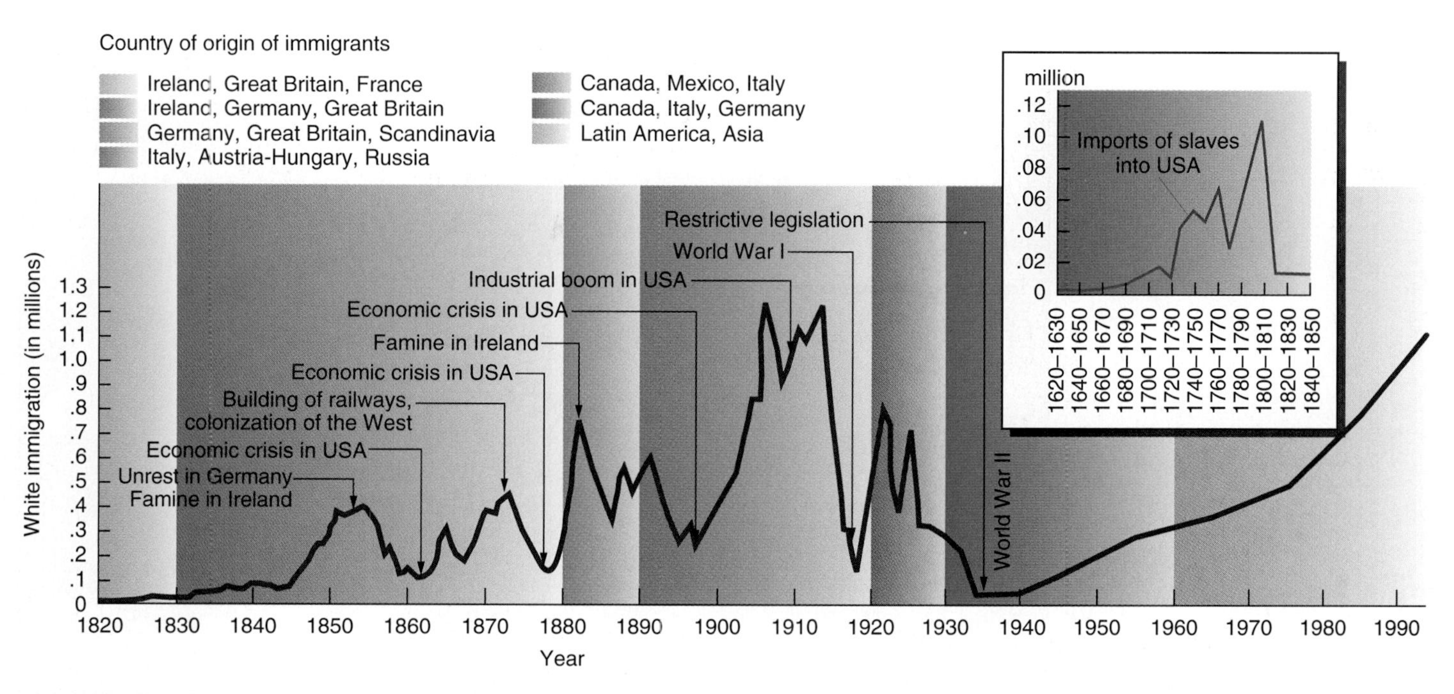


**FIGURE 9.29 United States: pattern of immigration.** The effect of push and pull factors on immigration fluctuations and the changing geographic sources of immigrants. Compare the size of slave inputs to the levels of white immigration up to the mid-1800s.

Source: Data from National Academy of Sciences.

together with the somewhat smaller Kansas City, Milwaukee, Cincinnati, and Columbus). In the South, 16 million people lived in **Miami**, **Tampa**, **Atlanta**, **Houston**, and **Dallas-Fort Worth**, together with Fort Lauderdale, San Antonio, Orlando, and New Orleans. The West had the second highest figure of 31 million people living in metropolitan areas of over 1 million people (**Denver**, **Phoenix**, **San Diego**, **Los Angeles**, **San Francisco**, and **Seattle**, together with San Jose, San Bernardino-Riverside, Sacramento, and Portland).

## Ethnic Variety

The 1990 federal census counted 248 million people in the United States. Of these, almost 200 million were of European origin and 30 million were African origin, 2 million were Native Americans, and just over 7 million were Asians. Some 22 million were Hispanics, an ethnic designation (based mainly on speaking Spanish) that includes several racial types.

There is a regional dimension to population minority groups within the United States. Over half of the African Americans live in the South and the rest mainly in northern and western cities. In the South, many African Americans still live in rural areas. Most of the Hispanic population lives in the West, especially near the Mexican border, and around Denver, and it is diffusing to northern cities. Puerto Ricans live mainly in the New York area. The Cuban population of Miami is wealthier and better educated than the average for Hispanic Americans. Most Asians and Native Americans live in the West.

The growing numbers of Hispanic peoples in the United States will cause them to become the largest minority group early in the new century. At present, over half the local population in southern California, southern Texas, and southern Florida are Hispanic Americans. Pressures to educate their children in their native language were recently rejected by Californian Hispanic Americans, who wish their children to be taught in English so they will be able to compete for jobs.

## Internal Migrations

The dynamics of population geography in the United States involve internal migrations as well as international immigration (Figure 9.30a). During the 1800s, the main internal migrations were from the Atlantic seaboard westward to the interior and then on to the west coast. Within the South, the plantation economy was transferred westward from the Atlantic plain to the Mississippi Delta area, taking with it planters and slaves. The northern (mainly white) and southern (mainly black) streams remained distinct from each other during the 1800s. Even after the Civil War in the 1860s, most African Americans remained working on southern farmland.

From the early to mid-1900s, large numbers of African Americans moved from the rural and urban South to northern cities. Those from the Atlantic coast plain mostly moved to Washington, D.C., Baltimore, Philadelphia, New York, and Boston; those from the Mississippi Delta moved primarily to Cleveland, Cincinnati, Indianapolis, and Chicago. Smaller

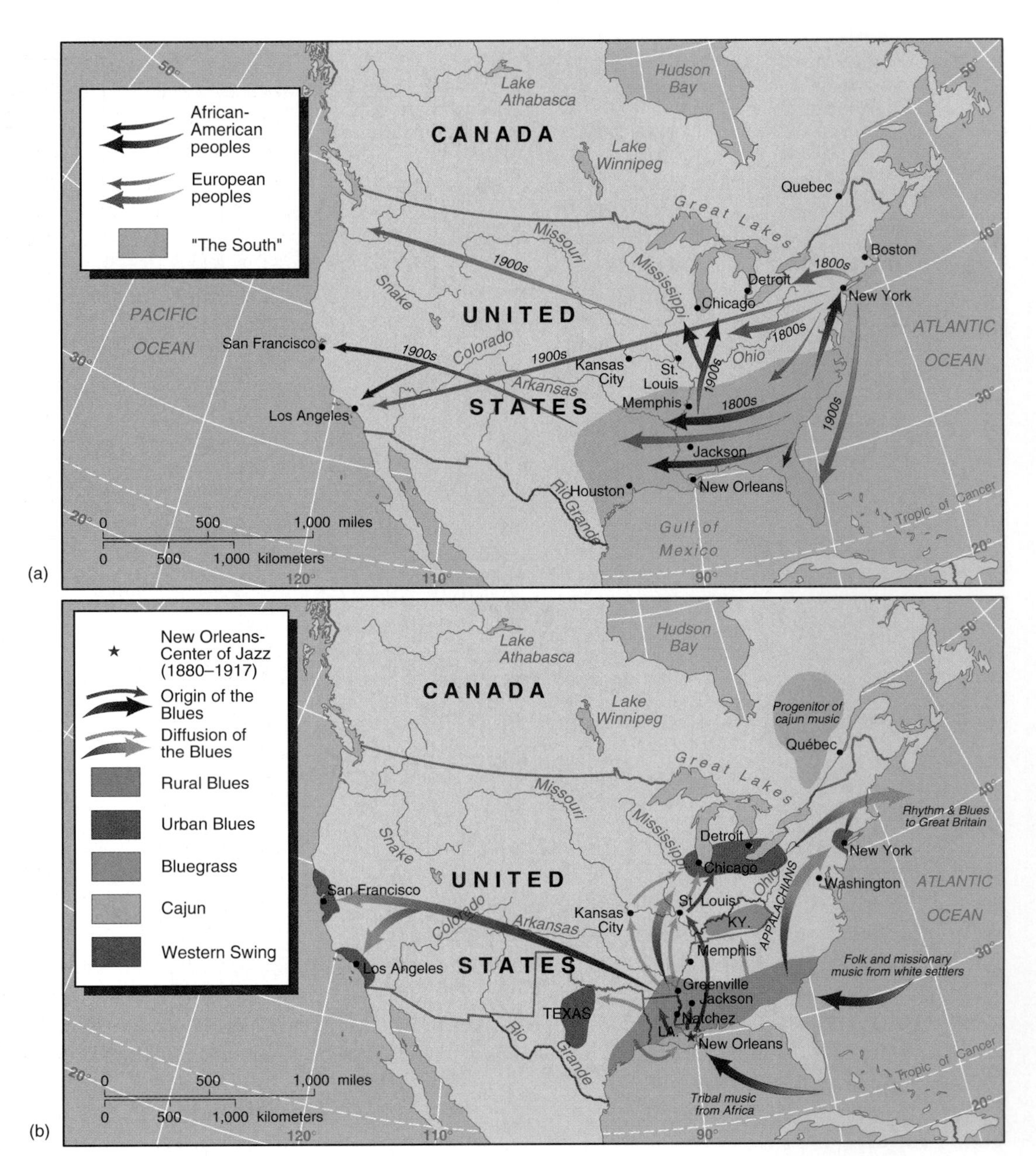


**FIGURE 9.30 United States: internal migrations.** (a) The major movements of African and European Americans in the 1800s and 1900s. (b) Compare the developments in blues music and their diffusion to the movement of African Americans.

numbers of African Americans moved to the west coast. In 1900, nearly 90 percent of African Americans lived in the South, but this proportion declined to just over half by the 1990s. The movements also stimulated the dissemination of cultural features of American society such as the varied types of blues music (Figure 9.30b).

Most of the movements of African Americans out of the South were completed by 1970. After that, smaller return movements, often on retirement, balanced or exceeded the northward flow. Movements of white people from the northeast to both the west coast and the South continued into the 1980s, but recessions in that decade slowed southern economic growth and reduced such migrations.

## Diverse Peoples

The movements of many different groups of people into and around the United States gave rise to two images of the resultant population mixture. Up to around 1950, immigrants were subject to pressures to conform and be anglicised—the "melting pot" concept. In the later 1900s, greater roles were given to African Americans, Native Americans, and recent immigrants from Latin America and Asia, while various European ethnic consciousnesses were revived. The changes led to a new image—the "salad bowl"—in which the variety of contributions rather than pressures to conformity and homogeneity is emphasized.

## Native Americans

In the 1970s, the indigenous groups—called the "First Americans" or "Native Americans"—began to increase their total population for the first time since census records were taken in 1790. Encouraged to develop indigenous lifestyles on their reservations that were administered directly by the federal, rather than state or local, government, they remained poor. By the 1970s, some 50 percent of Native Americans were unemployed and 90 percent on welfare. In the later 1900s, attempts were made to reduce the poverty imposed by living on reservations that were generally placed on land the white people did not want. Helped by the clarification of their rights in Alaska arising from statehood and the selection of state lands from the federal holdings, Native Americans elsewhere in the United States raised questions about the ownership and management of natural resources on their reservations, including minerals and water. They filed suits seeking compensation for being placed on such poor land. Lucrative deals with governments and water projects, however, did little to change the economic structure of the Native American reservations.

New enterprises brought limited satisfaction. Some Native Americans developed tourist facilities on their reservations, but others ignored the opportunities—often to avoid copying white commercial ways. Perhaps the most important source of income to many Native American groups is the series of casino gambling complexes in various states. The main problems are managing the influx of capital and the discouragement of the low-wage jobs that are provided. Such enterprises, however, encourage Indians to engage with the wider U.S. economy and to begin to overcome the isolation and defensiveness they have toward the governments of the United States. To develop their water, minerals, and tourist opportunities Native Americans need to agree on how to administer and develop the lands that are held in trust for them.

# Economic Development

The United States of America is the world's most developed country in economic terms. It has the largest total GDP of any country in the world, although Switzerland, Japan, and some Scandinavian countries exceed its GDP per head. The United States maintains the greatest possible diversity of economic activity and remains at the leading edge of technology in many fields. The U.S. economy is subject to fluctuations; it grew rapidly in the 1980s, but slowed in the early 1990s before growing further.

## Commercial Farming Basis

Commercial farming regions developed as the combination of natural resources (climate, land, soils, water) and economic factors (Figure 9.31) led to areas of specialization. The early farming areas of the east coast suffered soil erosion, and unable to compete with more fertile inland areas, much of the land reverted to woodland. The Midwest grain-livestock area, the Mississippi Delta cotton area, the Great Plains feedlots, and the irrigated lands of California became among the world's most productive areas. Drier and higher parts of the West became low-productivity grazing land.

In 1996, the federal government ended its controls on farming through subsidies to crops and funding the idling of land. Planted acreages surged in response to rising prices and increased exports to Mexico and especially to China. This process is likely to exaggerate the move to increased specialization. The government interventions had helped to moderate the swings of farm income. The farmer's future problem remains the better management of their output.

In the 1990s, many farmers in the U.S. Heartland found that falling or static prices for their output of grain, meat, or milk could not sustain their debts on sophisticated equipment. Some went bankrupt. Others, especially on family-owned farms, returned to lower-cost farming using less equipment and fertilizer and relying on crop rotation to maintain soil fertility. The extra labor previously used on farms before the machinery revolution was not available and so those managing family farms worked harder.

## Manufacturing Becomes Central

Until the 1990s, many economists saw the development of the U.S. economy in the 1800s and 1900s as a model for others to follow. Although it is now realized that the conditions necessary for economic development vary from country to country, and certainly from time to time, it is worth tracking a history of U.S. economic development to help an understanding of the regional differences within the country itself.

- There was a marked break between the economies of the traditional cultures of the Native Americans and the settlers who arrived from Europe after A.D. 1500. From the outset, most settlers followed commercial practices, seeking valuable minerals and producing crops and goods for sale, often back in Europe. Trade flourished as plantation owners in the South and a wealthy merchant class on the Mid-Atlantic coast and in New England emerged among the colonists. Farming and farm produce continued to be the mainstay of the American economy and trade until the mid-1800s.
- Manufacturing developed from local crafts (pottery, smithing, weaving) in the early 1800s in southern New England, New Jersey, and eastern Pennsylvania. It was based mainly on water mill power and produced metal goods, textiles, and leather goods. Mill factories were small, but concentrated the production of goods and required new transportation facilities to take the products to widespread markets. At first, water transportation was also vital. Local capital, accumulated by merchants, was invested in the early factories, often under family ownership.

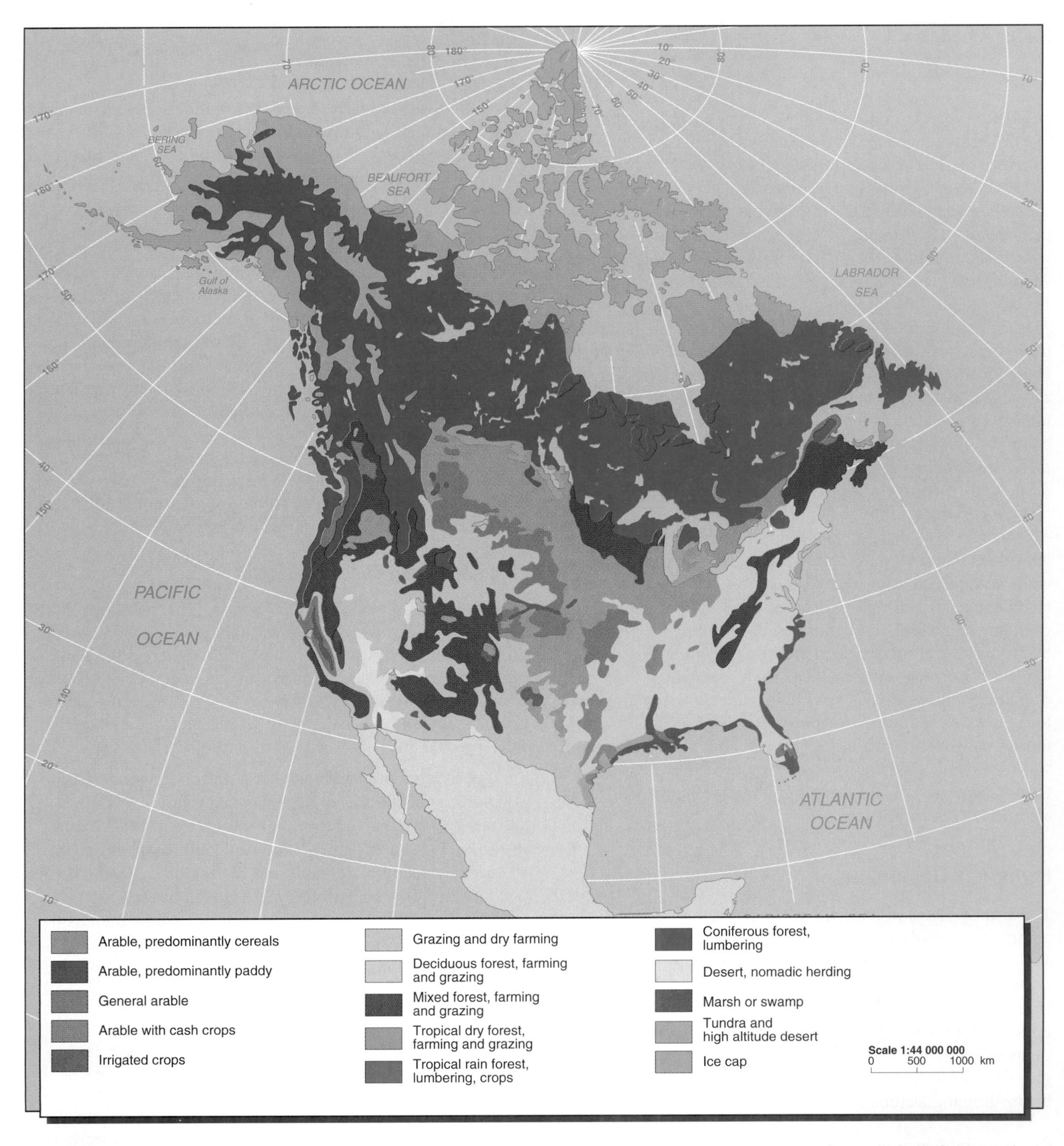


**FIGURE 9.31 Anglo America: dominant land uses.** The most productive farming regions are in the Midwest of the United States, the Prairie Provinces of Canada, and parts of California, Texas, and Florida. What factors make farming difficult in other areas?

Source: Data from *New Oxford School Atlas,* p. 98, Oxford University Press, UK, 1990.

(a)

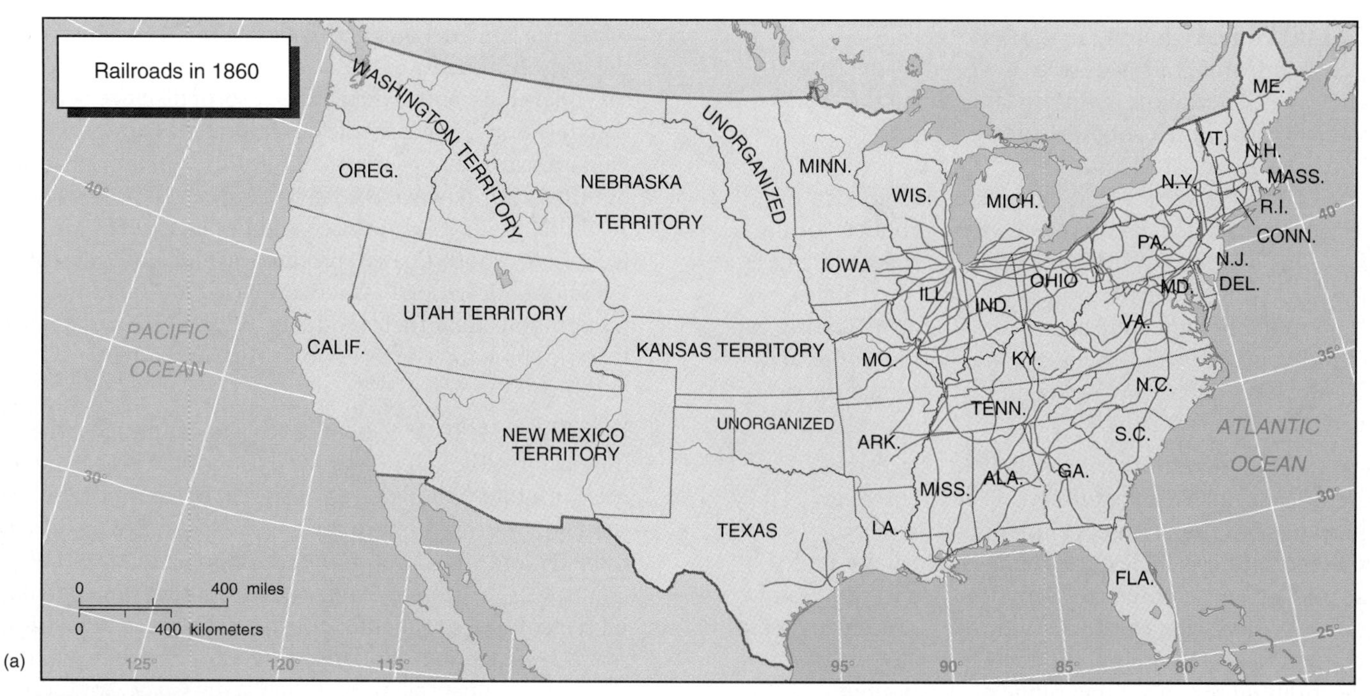


(b)

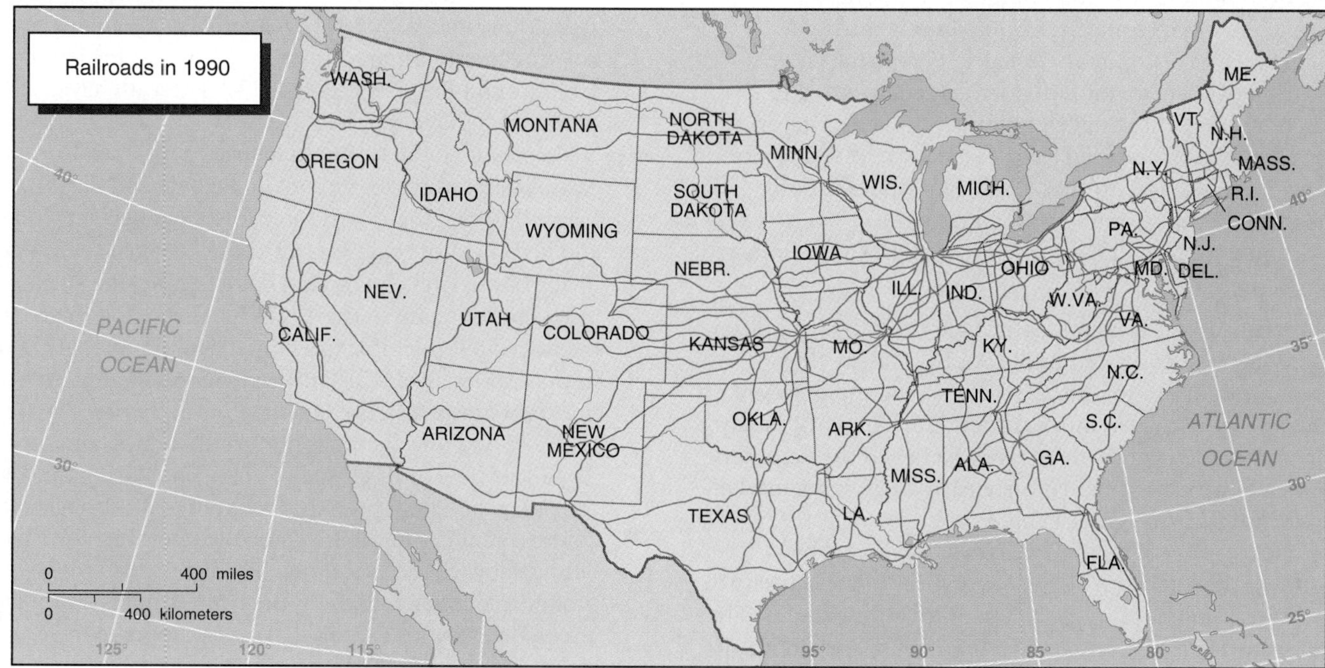


**FIGURE 9.32 United States: railroad networks.** (a) 1860. (b) 1990. The network was virtually complete by 1900: the extension across the continent between the late 1860s and 1900 was linked to the settlement of the U.S. West, although the building of transcontinental routes was held back until after the Civil War of 1861–1864. In the second half of the 1900s, many lines in the east were closed as truck transportation increased on the interstate highway system.

Source: From P. Guinness and M. Bradshaw, *North America*, Copyright © 1985 Maddes & Stoughton Education, London. Reprinted by permission.

By 1850, railroads reached west to Chicago (Figure 9.32a) from several east coast cities, and their construction formed the basis of iron industry expansion. In eastern Pennsylvania the use of charcoal gave way to coal in smelting the iron, spreading to western Pennsylvania around Pittsburgh.

- A major growth in manufacturing occurred after the 1860s with the adoption of steelmaking and the new possibilities it opened in heavy engineering. Many of the manufacturing industries of this phase were tied closely to their raw materials, such as coal and iron ore. New industrial areas sprang up inland from the main markets on the east coast. But new markets also developed as the interior of the United States was settled.

Much of the spread of economic growth was stimulated by major federal investments in transport infrastructure. Beginning in the1860s, the transcontinental railroads were financed by the federal and state governments granting lands along the routes to the railroads, which sold them and encouraged homesteading.

In 1860, the value of U.S. manufactured goods exceeded the value of commercial farm products for the first time. One economist called this the "take-off" phase that led to the modern economy of the United States. Agriculture became industrialized with the increasing use of mechanization and chemicals; markets for crops and livestock products were linked to the growing railroad network (Figure 9.32b).

- Up to 1950, manufacturing remained of greatest significance in the expanding U.S. economy. New products developed in consumer goods and in transportation vehicles, including cars, trucks, and airplanes. These industries were less tied to sources of raw materials than the 1800s industries, and many were market-based or assembly industries where the best locations were central to a range of component producers or large markets. The Manufacturing Belt, however, was both the main market area and the location of basic metal industries and continued to be where most of these developments occurred. Factories were built in established market locations, such as the New York area and around Chicago with its central location on the national railroad network, or in new locations, such as in Detroit, on which the auto industry was centered.
- After World War II, manufacturing industries expanded along the west coast and in the South. The previous concentration in the northeast was spread more widely to site new industries in more appropriate locations, satisfy the new markets growing in what had been the periphery of the country, and relate more clearly to growing markets overseas in Latin America and Asia.

The wider geographic spread was facilitated by building the interstate highways, distributing electricity to rural areas and developing a networkof airline routes. The Interstate Highway System, begun in 1956, created nearly 80,000 km (50,000 mi.) of limited-access highways across the United States, opening many previously isolated parts of the country to economic development. The construction of airport facilities with federal grants increased the amount of traffic possible. In the 1970s, nearly two-thirds of all scheduled passenger flights in the world occurred inside the United States, and air freight increased in importance as a result of the combination of widespread airport facilities and developing aircraft technology. The network then spread more widely around the world, with U.S. airlines having a huge initial advantage because of their flying experience on many internal flights. Air transportation of people and freight inside the United States is still expanding. Combined with the wide availability of high-quality telecommunications, these developments reduced the costs of distance between rival locations for economic development. They provided a vital infrastructure base for continuing economic growth, but their maintenance is costly.

While products diversified and became more technically sophisticated, production and management techniques developed. American-based multinational corporations took their products to the world and opened factories in many countries in Europe and other continents.

- In the later 1900s, U.S. manufacturing continued to be important, although it employed fewer than 20 percent of the work force compared to nearly 40 percent in 1950. Its products were even more diverse, with some of the heavy industry and some producer units moving to lower-cost countries. The United States is a world leader in applying high technology to manufacturing and service industries. An increasing range of high-tech goods are produced in the newer industrial areas of the Pacific Coast, southern Texas, the older South, the Denver area, and many of the growing metropolitan centers, as well as in former textile mills around Boston (Figure 9.33). These all have national as well as regional markets.

As new manufacturing industries developed, the older established industries did not all decline and die out. Many slimmed their workforce, installed new machinery, and became more specialized. Some changed their focus to accommodate new trends. For example, the steel industry, which up to the 1950s had been dominated by the huge and inflexible integrated complexes that resulted from economies of scale and vertical integration, lost out in competition with the Japanese in the 1960s. The Japanese had new facilities and technologies, and the American steel mills closed or changed. During the 1980s, the policy of voluntary export restraints for foreign countries selling to the United States, combined with the falling value of the U.S. dollar—which made foreign products more expensive in the United States—provided an opportunity for the large plants to upgrade their technology. Smaller plants (minimills) established themselves across the United States. For example, the USX Corporation reduced its jobs at Gary, Indiana, from 30,000 to 10,000 and focused its output on sheet steel. Productivity increased, and rising demand in the early 1990s helped much of the U.S. steel industry to meet international competition as the export restraints came to an end.

U.S. economic development in the late 1900s, however, focused increasingly on the growth of information-using occupations, including financial services, publishing, computer services, and design services. Tourism and leisure-based services also became a leading part of many

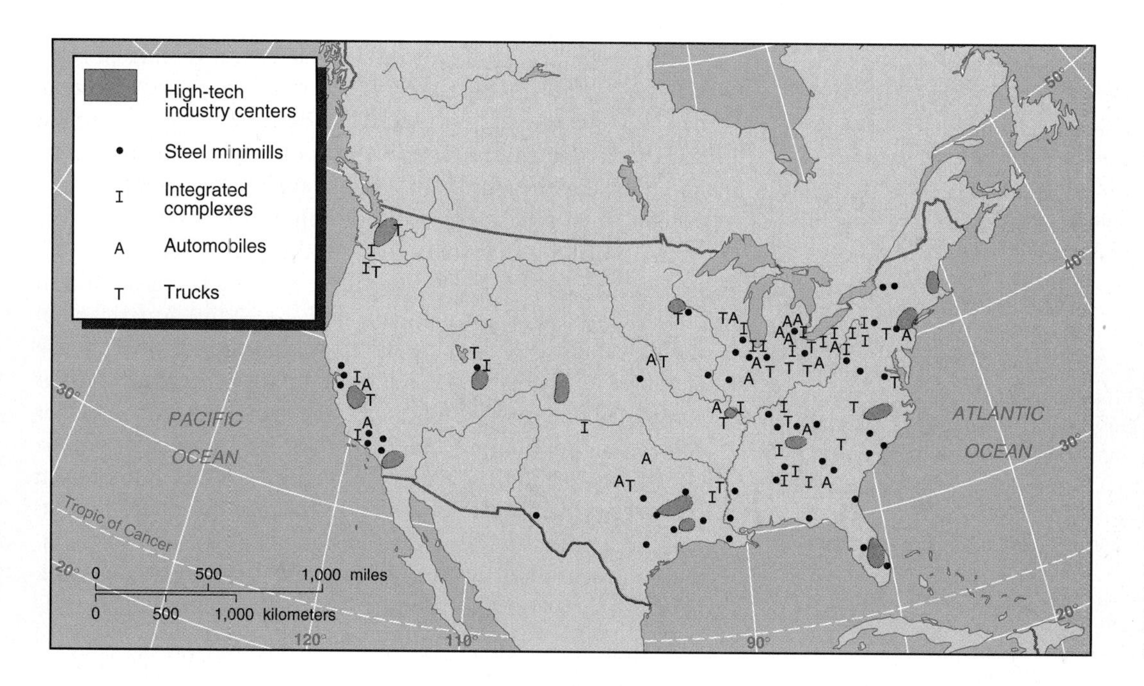


**FIGURE 9.33 United States: distribution of some major manufacturing industries.** The continuing concentration of steelmaking and vehicle manufacturing in the Manufacturing Belt contrasts with the widespread nature of high-tech industry centers. Steel production in minimills has become more widely distributed since 1970.

local economies. In 1996, 46 million tourists from other countries visited the United States, while 51 million went from the United States to other countries and many millions of Americans were tourists in their own country. Such industries locate economic activity away from the workshop and factory.

Integration with the rest of the world became commonplace, with improved communications and transport easing contacts with people and businesses in other countries. The United States, with its economic and technological lead, has so far been able to take greater advantage of this new economic base than any other country and is investing to maintain the lead.

- While comprising a small proportion of U.S. GDP in the 1990s, the primary sector contributed important raw materials and fuels. Mining, forestry, fishing, and agribusiness outputs were often linked to the manufacture of metal goods, energy supplies, paper, fertilizer, textiles, and food products. They were commonly produced by multinational corporations and involved high inputs of capital and technology with declining numbers of employees.

## Rare Regional Policies

During the 1900s, growth in some parts of the United States coincided with slow growth and even poverty in others. Uneven development reflected the concentration of capital investments in some regions and cities rather than in others. The favored and less fortunate regions changed over time. Up to the Civil War, some of the South was prosperous for a few planters, but their wealth came from plantation slavery. After the Civil War, the South was the poor region of the United States for around 100 years. Southern Appalachia was also poor with hill farmers living on small plots of eroded land. Its plight was first attacked after 1933 by the establishment of the **Tennessee Valley Authority** (TVA), one of the few federal programs designed to stimulate economic growth in lagging regions. TVA was initially viewed as a possible model for action in other regions, but the dominant political views in the United States were against government-funded regional aid. The huge investments in hydroelectricity and other forms of cheap electrical power generation in the Tennessee Valley provided a basis for economic growth and population growth—but also brought environmental problems that posed questions. For example, the rising demand for cheap electricity caused TVA to invest in first thermal and then nuclear power plants that became increasingly costly and dangerous to the environment. In the early 1990s, the TVA nuclear program was ended.

By the 1960s, however, it was recognized that a more extensive Appalachian region was the largest of several rural areas that were lagging behind the rest of the country. The main achievement of the **Appalachian Regional Commission** (ARC), established in 1965, was to make available to this region the many federal grant-aid financial packages, particularly in the late 1970s. In the 1980s, the administration of President Ronald Reagan reduced ARC's funding and effectiveness.

The rural problem grew, however, as the metropolitan centers of the United States assumed increasing dominance in the economic and social life of the country. In the 1980s and the early 1990s, the richest farming regions of the U.S. Heartland were affected as drought as well as debt following overinvestment in the 1970s.

**RECAP 9C: United States of America**

The United States of America is a city-based country thriving on high-tech manufacturing and growing financial and business services industries. Its federal government structure led to power moving from the states to Washington, D.C., during the 1900s. The variety of regions within the United States reflects the uneven distribution of natural resources and the historic processes of occupation, industrialization, and developing transportation and telecommunications facilities.

The population of the United States is becoming increasingly multicultural as the Western European groups and African Americans who peopled it until the early 1990s are added to by peoples from Latin America and Asia. Immigration continues to be important in population growth.

The United States' economy has the largest GDP in the world and is market-oriented. Its farming is highly productive and commercially organized, its diversified manufacturing output reacts to changes in technology and world demand, and its service industries are now the dominant employers.

9C.1 What are the comparative advantages of the main regions of the United States?

9C.2 What do you learn about its modern human geography from constructing a flow diagram to summarize the economic development of the United States since 1800?

9C.3 How has the composition of the U.S. population changed since 1900?

**Key Terms:**

federal system of government
Tennessee Valley Authority (TVA)
metropolitan areas
Appalachian Regional Commission (ARC)

**Key Places:**

United States of America
New England
Megalopolis
Manufacturing Belt
Appalachia
Midwest
Heartland
South
western mountains
Pacific Coast
Alaska
Hawaii
*Boston*
*New York*
*Philadelphia*
*Baltimore*
*Washington, D.C.*
*Pittsburgh*
*Cleveland*
*Detroit*
*Chicago*
*St. Louis*
*Minneapolis-St. Paul*
*Miami*
*Tampa*
*Atlanta*
*Houston*
*Dallas-Fort Worth*
*Denver*
*Phoenix*
*San Diego*
*Los Angeles*
*San Francisco*
*Seattle*

# CANADA

That Canada struggles to develop and maintain its national identity is largely a matter of geography (Figure 9.34). It is the second largest country in the world in area but has just over one-tenth of the United States' population. The average population density is low, but there is a strong concentration of Canadians close to the border with the United States—making the rest of Canada virtually empty (see Figure 9.26).

The distance across Canada and the narrow populated band result in a series of regionally individualistic groups of people that have to interact to make Canada function as a country. The distinctive groups are the people of Atlantic Provinces, the French-speaking peoples of Québec, the multicultural population of Ontario and especially of Toronto, the scattered farming populations and industrial cities of the Prairie Provinces, the people of Vancouver and the British Columbia coast with an increasing outlook toward Asia, and the Native Americans of the northlands. In the later 1900s, it appeared likely that the people of Québec might pull Canada apart by insisting on their own independence. In addition, the closeness of the major centers of population and commercial development to the United States and the beaming of media across the border from the United States have powerful influences on modern Canadian life and on the problem of establishing a cohesive Canadian national identity. Perhaps because they choose to work hard at it, however, that identity is recognized at a global level.

The United States continues to be seen as both a friend and an opponent. Canadians enjoy many of the benefits of affluence because of their close integration with the American economy and defenses. Canada is a member of NATO and stationed forces in Europe until 1992. It worked with the United States in maintaining a line of radar warning stations facing the Soviet Union across the Arctic Ocean during the Cold War period. It entered a 1988 trade agreement with the United States that is now being more fully developed in the North American Free Trade Area. Canadians took to U.S. commercial sports such as baseball and football.

Canadians object strongly, however, to being recipients of acid rain from the Ohio River valley power plants. In the late 1990s, there were problems on the west coast over salmon fishing yields in Alaska, British Columbia, and Washington. The smuggling of high-quality marijuana into the United States from its production area around Vancouver was a further source of friction between the two countries. During the Cold War, Canada's external affairs policies were sometimes at variance with those of the United States. It developed relations with some countries in the Soviet camp and some nonaligned poor countries. In the 1990s, Canadian companies took over the development of Cuban mining from the former Soviet groups despite a U.S. embargo. Relations between political leaders in the two countries do not appear to be close. While Canadians keep their distance to preserve their identity, the United States takes Canada too much for granted and gets annoyed when Canadian policies conflict with its own.

## Regions of Canada

**Canada** has a federal government that links the 10 provinces and the Northwest and Yukon Territories—the country's main political regions. This arrangement was decided by Britain in the 1867 Act of Confederation. That act united the different parts of what became modern Canada in the face of a perceived military threat from the United States after its Civil War ended in 1865. In 1982, Canada ended its legal ties to Britain, although it

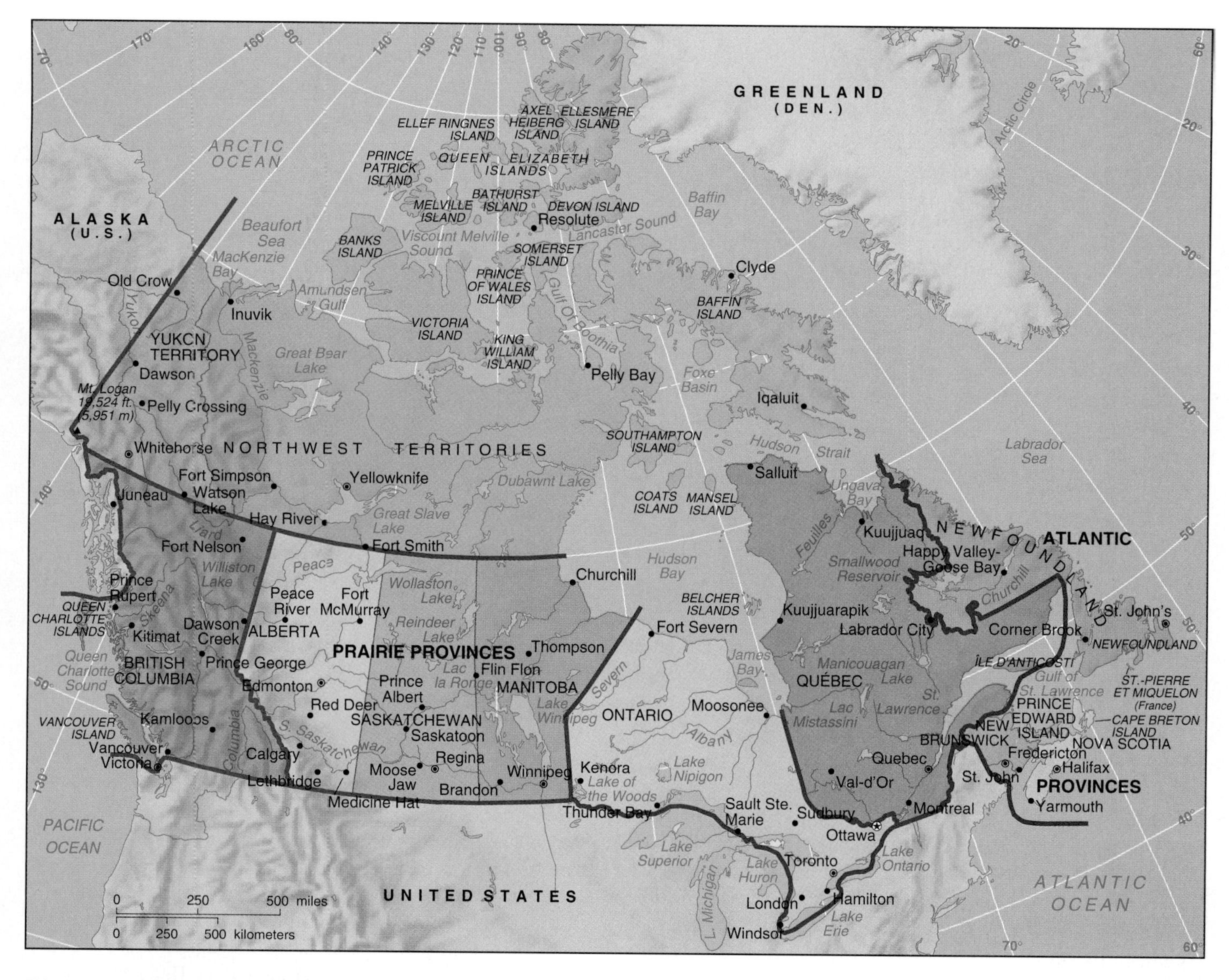


**FIGURE 9.34 Canada: major subregions and cities.** The Canadian provinces have greater powers than the individual states of the United States. Their geographic distribution in a line north of the U.S. border often makes communications difficult among different parts of the country.

remained within the British Commonwealth of Nations. This left it free to determine the constitutional roles of federal and provincial governments. That was not easy. Several attempts to reconcile the wishes of the people of Québec with those of other provinces highlighted the problems of a federal constitution in which the provinces have, in many ways, greater powers than the central government. This situation contrasts with that in the United States, where the constitution was generated internally nearly a century earlier in 1787. The U.S. states are generally smaller in size than Canadian provinces and lost powers to the federal government during the 1900s.

## Atlantic Coast

Canada's political tensions are enhanced by its regional economic and cultural differences. The east coast was settled first and forms the small hilly **Atlantic Provinces** of Newfoundland (which has jurisdiction over the almost uninhabited Labrador), Nova Scotia, Prince Edward Island, and New Brunswick. The small-scale economy of these areas based at first on fishing and farming was augmented locally by mining and manufacturing and by the naval base at Halifax (Figure 9.35). These provinces remain the main recipients of federal regional aid programs.

In the 1980s and early 1990s, the region was hit badly by declining fish stocks on the Grand Banks. Some 30,000 fishers and fish plant workers lost their jobs as cod stocks virtually disappeared. Unemployment in Newfoundland rose to over 20 percent. Few new industries were attracted by government efforts. One source of hope could be to switch some fishing capacity to commercial sealing, although environmental concerns have to be assessed against the economic plight. Other prospects, such as pumping oil from the Hibernia field

**FIGURE 9.35 Canada: Atlantic Provinces.** Halifax, the capital of Nova Scotia, is the largest city in the Atlantic Provinces. It is now a major port and naval center. The huge citadel is a landscape reminder of the city's origin in 1749 as a British fortified town to contain the French presence in nearby Louisburg.



315 km (200 mi.) offshore in the main iceberg lanes, would be very costly. The development of nickel and cobalt mining at Voisey Bay, Labrador, scheduled to start production by A.D. 2000, will bring wealth to the owners and possibly a few local workers but will have little other local impact. It is significant on a world scale, however, that the cobalt from Voisey Bay replaces the falling output from the previously dominant mines in Congo (Zaire).

Although the Atlantic Provinces contain some groups of French speakers, particularly in New Brunswick, sympathy for them is eclipsed by the potential outcomes of Québec leaving Canada. An independent Québec would sever direct ground contacts between the Atlantic Provinces and the rest of Canada. The possibility forces the Atlantic Provinces to consider independence for themselves, a seemingly impossible prospect given the level of federal economic support. If the Québec rift occurred, the Atlantic provinces might look south: New England's shopping opportunities and events in Boston are closer than those in Toronto and already draw many visits across the border.

## Québec

The Province of **Québec** was settled very soon after the first Atlantic coast settlements by French people who developed a distinctive type of **long-lot** land settlement along either side of the St. Lawrence River, at first around Québec City and later upstream to Montréal (Figure 9.36). Québec became part of British Canada following General Wolfe's defeat of the French army under Montcalm before Québec City in 1759. The French settlers and their descendant Québeqois resented living under their conquerors. The economics of their culture were barely above subsistence, but traders bringing furs from far inland supported a merchant class during the 1800s. The combination of French language and Roman Catholic religion forged a strong loyalty that shifted to political activism in the 1900s as a reaction against Anglo-Canadian control. The erosion of French speaking in the rest of Canada as new immigrants settled the prairies formed a resolution that it would not happen in Québec.

As a result of such political activities, French was accepted as an equal national language, and Québec gained other concessions from the rest of Canada despite not acknowledging English as an alternative language within its province. Québec looks beyond Canada to other French-speaking countries and takes a leading role in developing a new global French technical language rather than simply accepting English words.

The province of Québec covers a large area, extending northward to include most of the peninsula east of Hudson Bay (Figure 9.37). Almost the whole population, however, lives along the St. Lawrence River estuary. The two largest cities of Québec and Montréal have a range of manufacturing and service industries that place them among the world's great cities. A series of industrial towns use local hydroelectricity to power timber industries, pulp and paper mills, and aluminum refineries along one of the world's main shipping lanes. With world market production of wood pulp extending to tropical

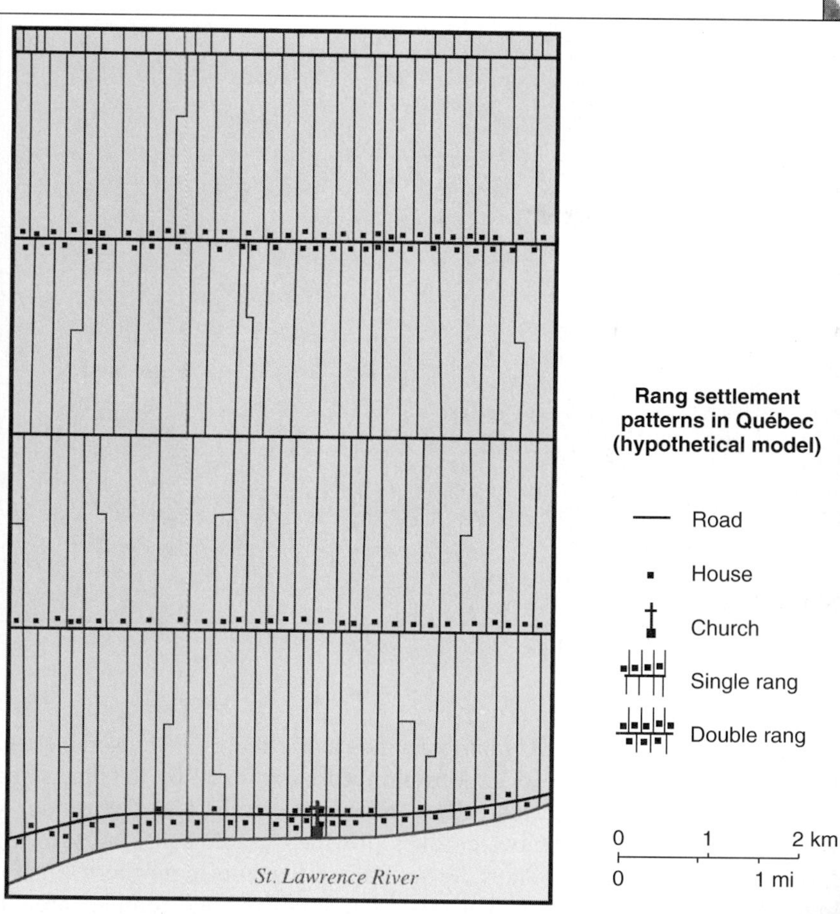


(b)

(a)

**FIGURE 9.36 Canada: Québec land ownership patterns.** (a) Long-lot patterns along the St. Lawrence River. When it was a French colony, New France was divided into seigneuries, each of which was further divided into long lots (ranges) with river or road frontages. (b) An air photo of long lots at L'Isletville, Québec. (c) A painting of the St. Lawrence riverfront in 1787, showing how fishing, farming, and lumbering took place on each long lot.

Source: (a) The excerpt from R. C. Harris and J. Warkenton, *Canada Before Confederation: A Study in Historical Geography,* 1991, is reprinted with permission of the publisher, Carleton University Press, Ottawa, Canada. (b) "This aerial photograph, #A12528-13,© 1950, Her Majesty the Queen in Right of Canada, reproduced from the collection of the National Air Photo Library with permission of Natural Resources Canada." (c) The National Gallery of Canada, Ottawa.

(c)

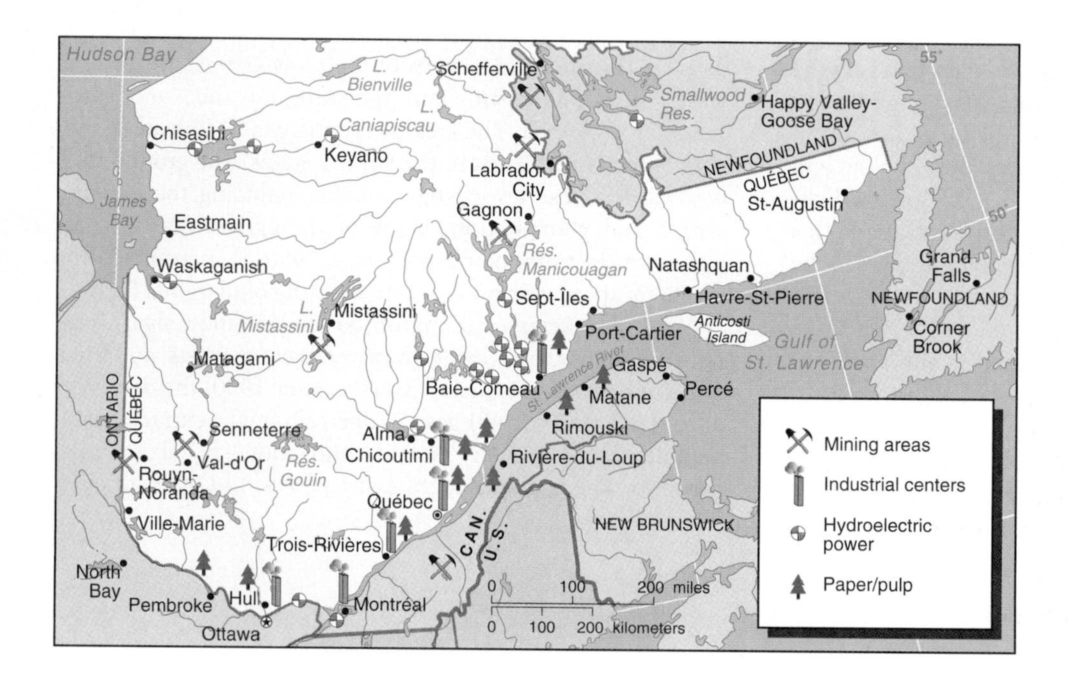


**FIGURE 9.37 Canada: Québec's geography.** The major cities, mining areas, industrial centers, and hydroelectricity sites. Nearly all economic activity crowds into the St. Lawrence River lowlands.

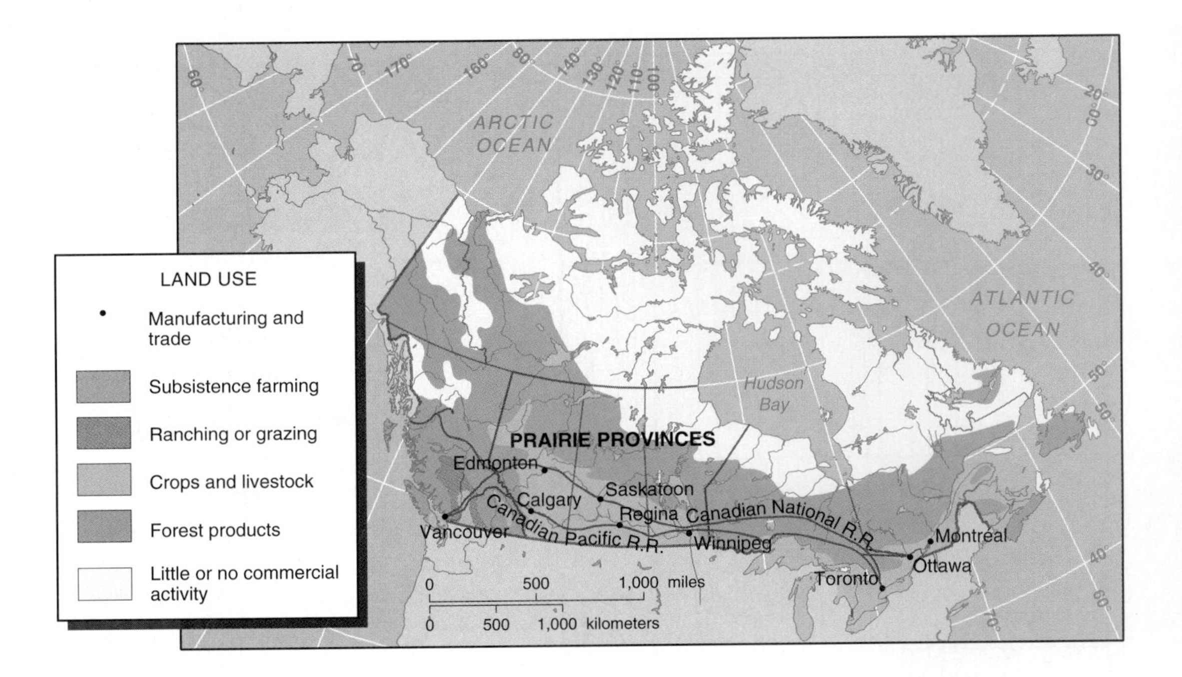


**FIGURE 9.38 Canada: Prairie Provinces in the context of land use patterns.** The building of the transcontinental railways was linked to the farming development of the prairies and the growth of the major cities west of Toronto.

areas, the prices—and hence the well-being of Québec's major industry—fluctuated since the late 1980s. Expansion of world demand, in Asia as well as Europe and Anglo America, does not always balance the production output.

North of the estuary, the land is bleak, eroded by former ice sheets and covered by coniferous forest, lakes, and tundra. The rocks contain large mineral resources that are gradually being exploited as world markets and transport facilities make possible. One of the major potential resources is hydroelectricity, and Québec is investing heavily in new facilities to export power to the United States. In doing this, it came up against problems of Native American land rights that are stalling parts of the plan. Overall, Québec produces 25 percent of Canada's manufacturing output and has a strong economy in its own right.

## Ontario

The province of **Ontario** contrasts with Québec. Although it was settled later than the provinces to the east, Ontario became the center of British rule after 1776 and possessed the best farming land. The Ontario Lakes Peninsula, lying between Lakes Huron, Ontario, and Erie, has a mild climate for central North America due to the lake effects of large water bodies and being farther south than other parts of Canada. Soils and climate suitable for pasture, grain crops, and tobacco made this the most attractive part of Canada for immigrant farmers in the 1800s. The city of Toronto arose at that stage, but its main development occurred from the middle of the 1900s, when the opening of the St. Lawrence Seaway turned it into an ocean port, the French-language policies of Québec drove English-speaking businessmen westward from its former rival in Montréal, and its role as capital of Ontario expanded. Toronto is the largest city of Canada and continues to grow rapidly.

Northern Ontario extends to the shores of Hudson Bay, but its ancient rocks were scraped bare of soil by ice sheets and covered partially by coniferous forest. Local deposition of clays in meltwater lakes provided usable soils, but growing seasons are short. Most of the settlements in northern Ontario are mining settlements, including the huge complex around the nickel mines of Sudbury. Settlements in the areas west of Sudbury and north of Lakes Huron and Superior are few and far between and constitute a major gap in the linear east-west Canadian settlement pattern.

## Prairie Provinces

The **Prairie Provinces** of Manitoba, Saskatchewan, and Alberta were settled following the building of the Canadian Pacific and Canadian National Railroads in the late 1800s (Figure 9.38). In the early 1900s they became major wheat-growing areas. The crop was planted in the spring and ripened in the short growing season, which was as little as 90 days on the northern margin of the plowed area. Toward the west, there was insufficient moisture for this, and cattle ranching took over. The area is essentially a northward extension of the Great Plains in the United States. The ground rises westward in a series of steps, further reducing the growing season and precipitation. Many of the agricultural districts had outmigrations of people since 1950 as a result of the mechanization of farming and the closing of marginal farms.

Winnipeg, Regina, Edmonton, Saskatoon, and Calgary were railroad towns that became grain and meat markets, and three of them became provincial capitals. After 1950, the discovery of coal, oil, and natural gas in Alberta brought extractive and manufacturing industry wealth and more people to that province. The large distances east and west to Canadian ports raised the price of exports and led to some of the main energy markets being southward in the United States.

### West Coast

**British Columbia** faces the Pacific Ocean and includes all the mountainous west of Canada apart from a range shared with southwestern Alberta. It is a province of geographically concentrated economic activity, including pockets of mining, fruit growing, and tourism in small lowland areas that are separated by high mountains. Most of the people of the province live in the Vancouver area, extending to Vancouver Island. Attempts to develop the port of Prince Rupert farther north were not so successful.

The people of British Columbia often distance themselves mentally as well as physically from the rest of Canada. Some of them act more British than the British themselves with over-elaborate rituals of afternoon tea, while others emphasize the Pacific links of Vancouver, which has an increasing Asian population and growing economic links with eastern Asia.

### Northern Canada

The northern parts of all the provinces from Saskatchewan to British Columbia end in the barren rock, trees, and tundra that characterize northern Canada. Here, the **Northwest** and **Yukon Territories** form the largest part of Canada—a federally controlled zone where few people live permanently apart from American Indian and Inuit (Eskimo) groups of Native Americans. Mining settlements produce a range of metallic ores, and there are few communications in this wilderness area apart from those linking mines to their markets. In the late 1990s, a new diamond mine near Yellowknife, Northwest Territories, was expected to produce up to 3 percent of the world's diamonds—if environmental and local community objections based on caribou herd migration paths and threats to hunting and fishing could be overcome. On a global scale, the mine, together with others outside South Africa, could reduce De Beers' control on diamond markets. Native Americans attracted to the mining settlements are often unable to relate to the contrasting lifestyles and in the past have created a social problem that the Canadian government did not tackle well until the 1980s. The future of this region is not bright as mining activities and federal subsidy incomes decline. A possible alternative income is from tourists, both from the United States and Europe.

## People

The concentrated distribution of Canadians along the U.S. border (see Figure 9.26) has some variations. The densest areas are along the Great Lakes and St. Lawrence River valley from Windsor, Ontario, to Québec City, and around the major cities of the Canadian West. The main gap is north of Lake Superior. Moderate densities occur in the Atlantic Provinces and through the Prairie Provinces and the mountains of British Columbia.

### Population Growth

In the century from 1891 to 1998, the Canadian population increased by around 26 million, or six times, from 4.8 million to 30.6 million. Over the same period, the United States' population increased by over 200 million, or more than four times (63 million in 1890 to 270 million in 1998). Both countries received immigrants in the early 1900s, but two world wars separated by a major economic depression in Anglo America deterred further large-scale increases until after World War II. After that, population growth was rapid, doubling from 1941 to 1971 before slowing to a one-fourth increase from 1971 to 1991. At present, Canada has rates of total fertility and natural increase that parallel those of other developed countries. At just under 2 per 1,000 live births, its total fertility rate is less than that of the United States, but its population growth rate is similar. This is partly because of lower death rates than the United States and partly because of continued immigration rates that almost match those of the United States (see Figure 9.27). Canada's age-sex structure (see Figure 9.28) resembles that of the United States.

### Regional Changes

Canada's population growth involved regional shifts (Figure 9.39). In 1891, Ontario and Québec had three-fourths of Canada's people and the Atlantic Provinces just over 18 percent. The rest of Canada held a mere 7 percent. The 1900s growth of the western provinces reduced the dominance of the eastern areas. The Atlantic Provinces declined in their proportion of the Canadian population as the western provinces grew. By 1931, they had less than 10 percent of the total, and although Newfoundland became part of Canada in 1949 and raised the proportion to nearly 12 percent, population growth in these four provinces continued to be slower than in Canada as a whole, and their proportion fell to 8.5 percent in 1991. Ontario and Québec contained over 60 percent of Canada's population between them throughout the 1900s. Québec's proportion stayed around 28 percent until 1971, then fell to 25 percent in 1991; Ontario's rose from around 33 percent in 1931 to almost 37 percent in 1991.

The provinces west of Ontario grew rapidly early in the 1900s, and by 1931, nearly 23 percent of Canada's people lived in the Prairie Provinces and a further 7 percent in British Columbia. At this time, Saskatchewan had a higher population than neighboring Manitoba and Alberta. From 1950, however, Alberta became more industrial and the dominant province of the three. It now has more people than the total of the other two. The proportion of Canadian people living in the Prairie Provinces declined from 23 percent in 1931 to around 17 percent in 1991, reflecting rural outmigration. On

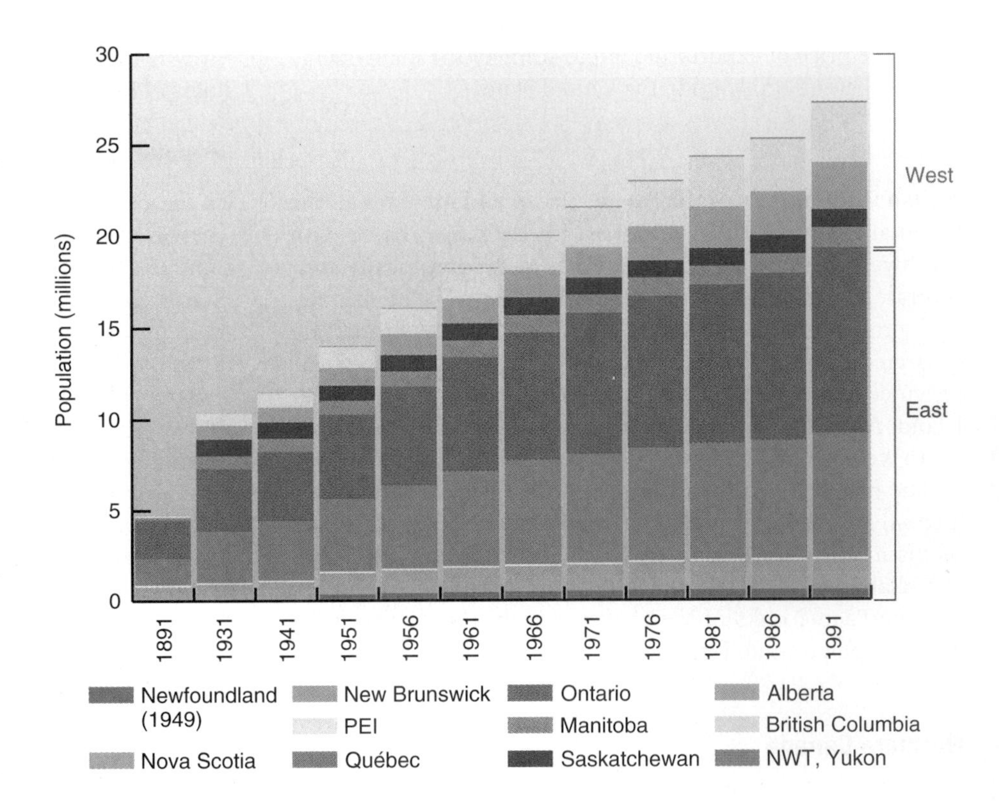


**FIGURE 9.39 Canada: growth of population in Canada by province, 1891 to 1991.** Compare the relative importance of the Atlantic Provinces, Québec, Ontario, the Prairie Provinces, and British Columbia in 1891, 1931, 1961, and 1991.

the Pacific coast, British Columbia had around 7 percent of Canada's people in 1931, rising to over 12 percent in 1991.

Canada is experiencing a trend toward metropolitan expansion. The two largest metropolitan areas, **Toronto** and **Montréal,** rival many U.S. metropolitan areas in size. Together with **Vancouver** on the west coast and **Ottawa,** the capital, these cities contain one-third of Canada's population.

### Ethnic Issues

The major internal political issue for Canadians in the 1980s and 1990s was whether French-speaking Québec would continue to be part of Canada. Within Québec are groups who wish to make their province a separate country. Although only 30 to 40 percent in 1994 opinion polls, support for leaving Canada rose to nearly 50 percent in the late 1995 referendum. Others living in Québec, particularly businesspeople and more recent immigrants from other parts of the world, wish to remain part of Canada. The Québec issue raises questions of whether the varied regions, separated by great overland distances, can continue to provide a complementary unity or will embark on a course that will tear the country apart.

Yet another important issue faced by Canada is the role of the American Indians and Inuits. In 1973, the Canadian government opened the possibility of negotiating land claims with organizations representing native peoples. Up to that date, little had been done in much of Canada to implement treaties negotiated with native peoples in the 1800s. Now several areas are identified for a degree of local government (Figure 9.40). The largest area, Nunavut ("Land of the People" in Inuit language), became a new territory with its own elected government in 1997, although it will remain subject to federal control. Other agreements were reached in northern Québec and with the Inuvialuit people in the northwest Arctic. Further discussions are underway, although many of the smaller claims may take several years to resolve. Some are complex because of overlapping land claims and because bargaining involves the often-opposing interests of native groups, the federal government, and provincial governments.

## Economic Development

Canadian economic development generally followed that of the United States, but often with a lag of several years until the later 1900s. In the 1990s, Canada combined a continuing emphasis on its natural resource base with being an affluent, high-tech society. It had one of the highest proportions of trade per head of its population in the world. Canada's GDP per head in the 1990s rivaled that of many European countries but was not quite equal to that of the United States. Its relative position worsened during the decade.

Until the mid-1900s, Canada's economy depended mainly on primary products such as grain, timber, and minerals. Canada remains a major world producer of newsprint, wood pulp, and timber, and is one of the world's leading exporters of minerals, wheat, and barley. Canada produces 30 percent

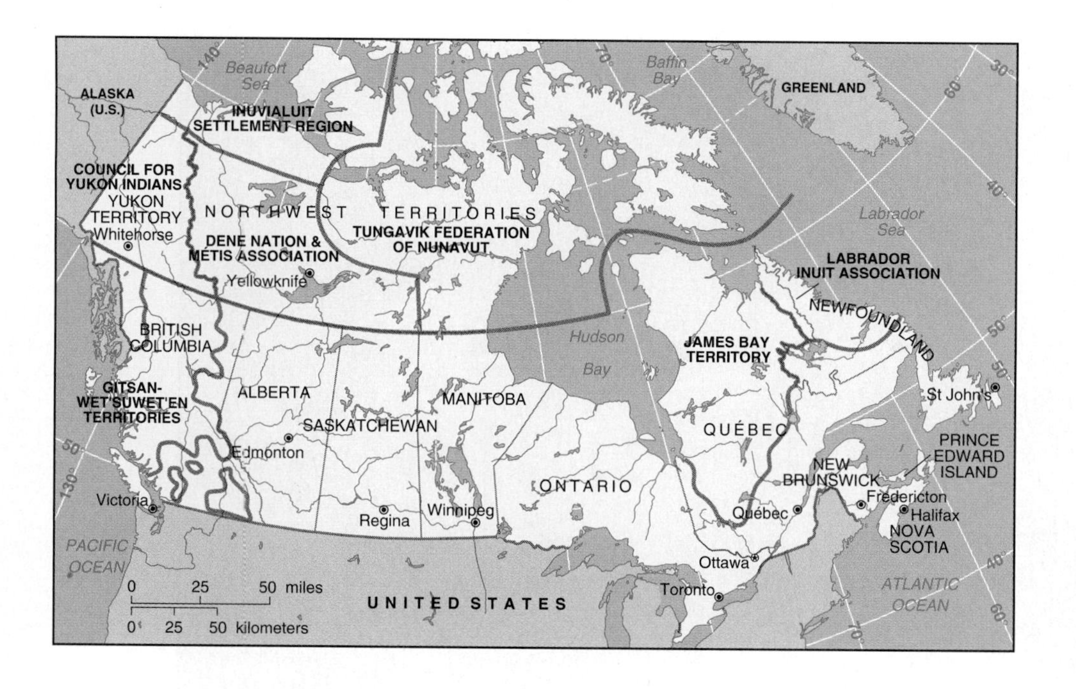


**FIGURE 9.40 Canada: the northlands.** Major Native American land claims in the arctic north, Québec, and British Columbia, late 1990s. Some of these areas may obtain a greater degree of self-government than others.

of the world's newsprint, which, with wood pulp, is manufactured mainly along the lower St. Lawrence River valley. The timber output comes mainly from the west coast forests. The minerals that place Canada in the forefront of world mining countries include coal, oil, and natural gas from Alberta, iron ore from Labrador and Québec, uranium from Ontario and Saskatchewan, nickel from Ontario and Manitoba, and zinc from several places. Agriculture remains significant in the Prairie Provinces, the Ontario Lakes Peninsula, and the specialized fruit-growing districts of British Columbia (see Figure 9.23).

Industrialization in Canada was a much later phenomenon than in the United States. Beginning before World War II and developing rapidly during the 1940s, the production of aluminum, vehicles, and consumer goods became important as Canada developed import-substitution industries to supply its own markets and avoid total U.S. economic domination. Hamilton at the western end of Lake Ontario became a steelmaking center. Montréal, Toronto, and Vancouver became centers of financial services and a wide range of commercial enterprises. Since the signing of the U.S.-Canada Free Trade Agreement, trade between the two countries increased and U.S. investment in Canadian industries rose.

Within Canada, there is a greater political consciousness of regional disparities and a greater will to do something about it than in the United States. Canadian regional policy enables the federal government to stimulate economic growth in the poorer provinces, such as the Atlantic Provinces and the largely agricultural Prairie Provinces, by supporting some services and financial investments in infrastructure. Such policies are popular in the recipient provinces, but their impacts on both source and recipient regions are controversial.

**RECAP 9D: Canada**

Canada is closely tied to the U.S. economy but politically is distinctive—partly because of internal tensions over the future of French-speaking Québec and partly because of its more liberal trading and aid policies. Internal differences are increased by the structure of large provinces that have more powers than states in the United States.

Canada's regions each have an economic heart close to the U.S. border and huge expanses of sparsely settled northern lands. Canada's economy was largely based on agriculture and forestry until around 1950, since then Canada has become high-tech and industrialized, and one of the world's major mining countries.

9D.1 What are the main characteristics of the major Canadian regions?

9D.2 What impacts does the large area of Canadian northlands have on the rest of the country?

**Key Terms:**
long lot

**Key Places:**

| | | |
|---|---|---|
| Canada | Prairie Provinces | *Toronto* |
| Atlantic Provinces | British Columbia | *Montréal* |
| Québec | Northwest Territories | *Vancouver* |
| Ontario | Yukon Territory | *Ottawa* |

# LANDSCAPES OF ANGLO AMERICA

## Urban Landscapes

Urbanization within the United States produced distinctive landscape forms through the types of building, the differentiation of land uses, and the distribution of groups of people—

**FIGURE 9.41 United States: urban landscape.** The ornate former homes of cotton plantation owners in the historic district of Charleston, South Carolina. The large homes in the Battery Street neighborhood were built as summer residences to take advantage of coastal sea breezes.



rich and poor, African Americans, Hispanic, and white—and their neighborhood symbols within the built areas. American cities grew from colonial ports and inland market centers to 1800s industrial and 1900s commercial metropolitan centers. Large cities increasingly dominated the life of the country. Although Americans are quick to knock down the old and build new, most cities retain landscape relics of each historic period that combine with the natural environment to give each city a specific character. Canadian cities followed similar trends but retain some distinctive features of their own.

### Preindustrial U.S. Towns

The colonial ports of New England and the Middle Atlantic were small, with scarcely 10,000 people each and a limited number of functions. Most people lived in the surrounding rural areas. By the mid-1800s, a pattern of market towns and small villages, linked to each other by road, canal, and railroad, covered the farmed plains of Ohio, Indiana, and Illinois. In the northeast, small factory towns huddled around water mills. South of Chesapeake Bay, plantation agriculture in coastal areas produced often lavish individual homes for the planters, surrounded by slave lines and the homes of poor white workers. There were few towns, although places such as Charleston, South Carolina, developed as summer coastal respites for the planters and their families (Figure 9.41). Atlanta, Georgia, became the first inland rail junction in the South and began to grow in size before it was destroyed during the Civil War in the 1860s. Some of the oldest towns grew under the Spanish occupation of the southwestern parts of modern United States, with relics still in Santa Fé and Albuquerque, New Mexico, and San Antonio, Texas. All the early towns comprised relatively small buildings, some in stone or brick, but most in wood. The historic areas composed of such buildings take up smaller areas of modern cities than those in Europe (see Chapter 7).

### Industrial and Commercial Cities

In the later 1800s and early 1900s, many cities in the east and center of the United States expanded into large conurbations of housing, factories, shops, and offices with the rapid growth of manufacturing industry and the railroad network. The larger factories and mills required many workers, who were accommodated in surrounding housing units. Areas of worker housing became differentiated from the better housing areas of the managers and owners. New **central business districts** (CBD), where shops, banks, and other financial services became concentrated, had access by streetcar to and from the industrial areas and leafy suburbs.

By 1920, American cities often evolved into a **concentric pattern of urban zones** around the CBD with poorer housing in inner suburbs and more affluent zones beyond (Figure 9.42a,b). Some of the inner city areas housed distinctive

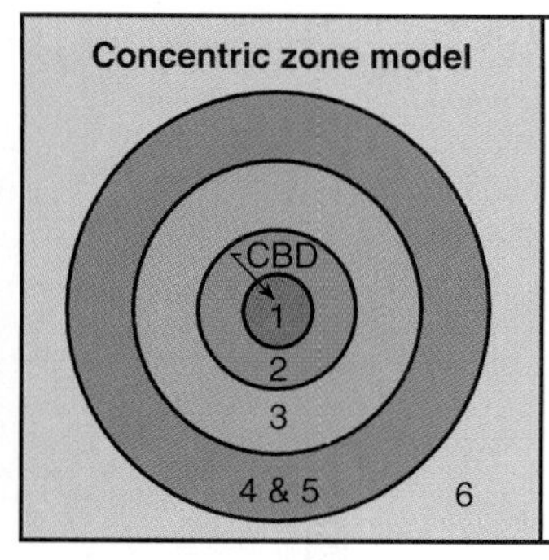


(a)

**FIGURE 9.42 United States: urban landscapes, 1900 to 1950.** (a) The concentric zone model, based on a 1922 study of Chicago (b). At that time, the "ghetto" was not related to African Americans, and other neighborhood names are changed. The right-hand part of (b) is extended over Lake Michigan. (c) The sector model, based on 1930s census information, illustrating the impact of railroads and highways on the location of different land uses. (d) The multiple nuclei model of 1945 still has a single center, but commercial and manufacturing districts are not in close proximity to it.

Source: (a), (c), and (d) From C. D. Harris and E. L. Ullman, "The Nature of Cities" in *The Annals of The American Academy of Political and Social Science,* Vol. 242, 1945. Reprinted by permission of the American Academy of Political & Social Studies.

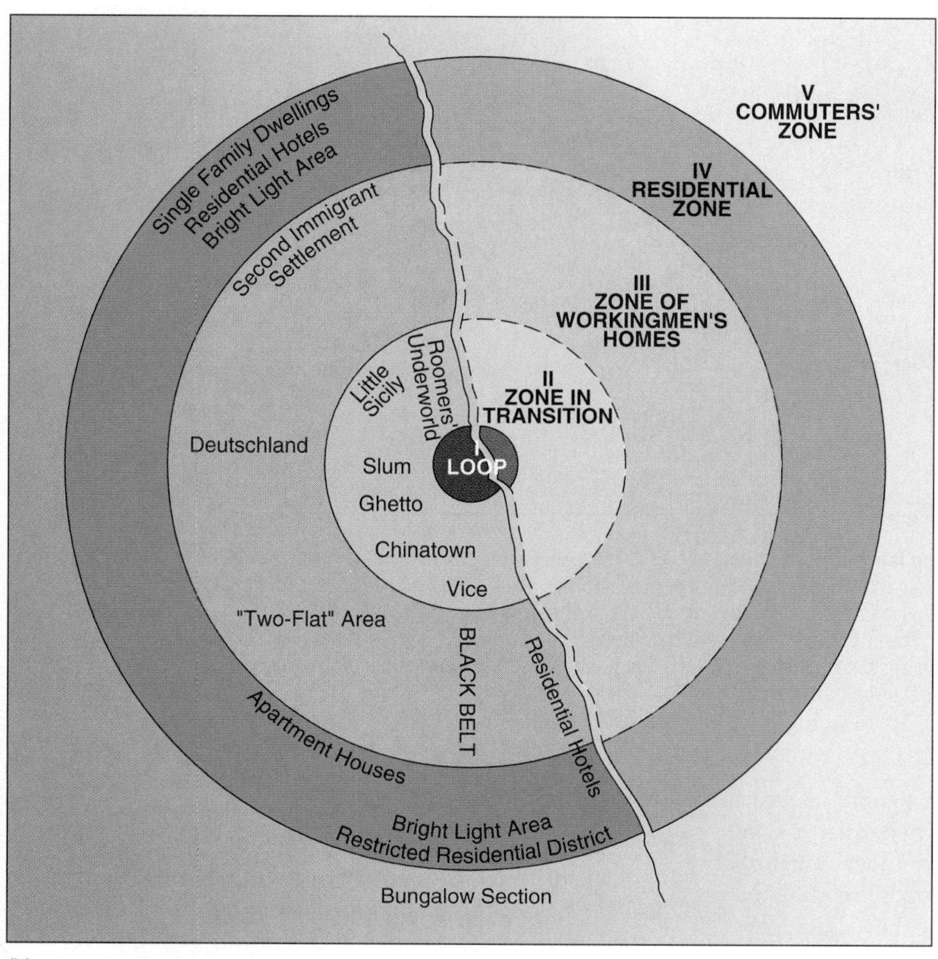


(b)

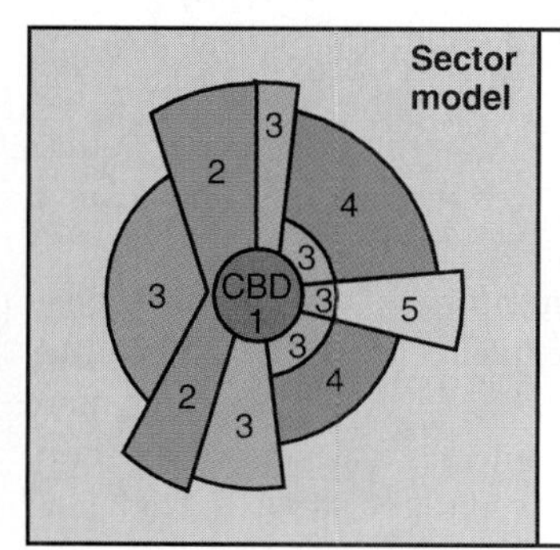


(c)

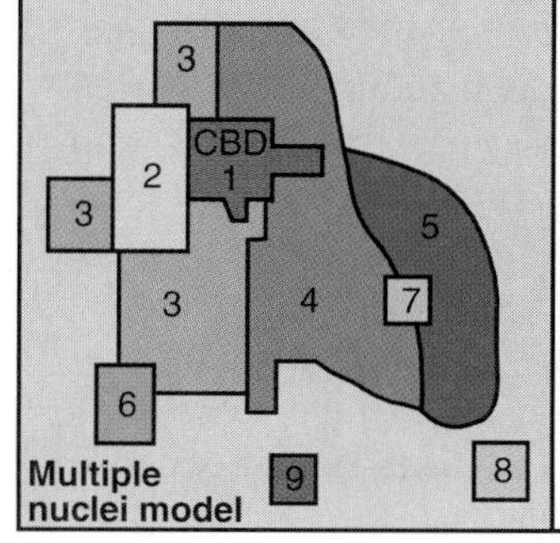


(d)

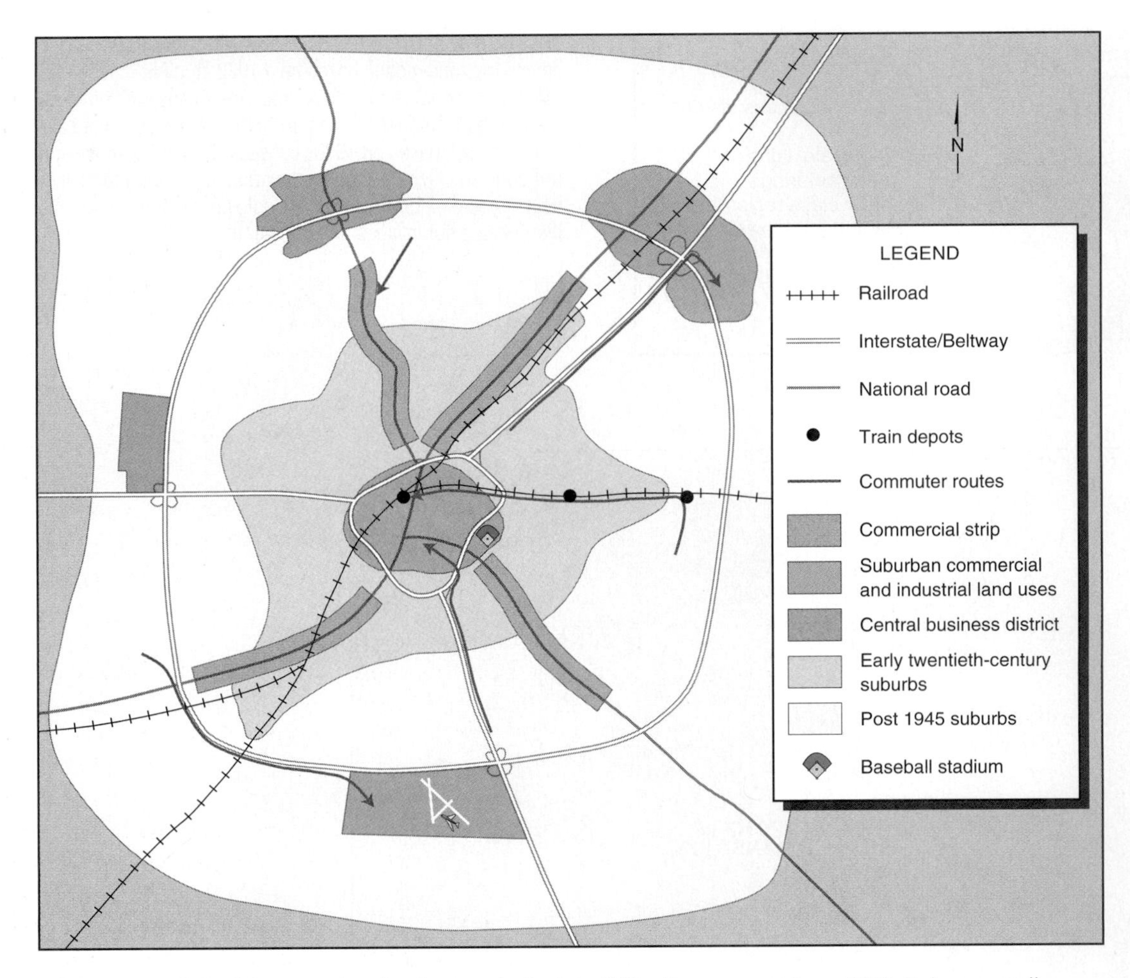


**FIGURE 9.43 United States: urban landscapes in the late 1900s.** The monocentric pre-1940 city is now small compared with the post-1945 suburbs in which interstate and other highway links and junctions provide bases for new centers of retail, office, warehouse, and manufacturing employment—the multicentric American city. Some of these centers produce as much as the CBD. The international airport is the venter of a distinctive suburban commercial zone. Commuter routes can still be from suburbs to CBD, but many go in other directions. The baseball stadium shown may also be a location for other major sports arenas or the symphony hall.

segregated groups such as Slavic or Jewish people in areas that became known as **ghettos** (after Jewish quarters in medieval Italian cities). In cities built across a local relief of hills and valleys the development of railroads, roads, and commercial activities was concentrated along the valleys, giving the city geography a pattern of wedge-shaped **urban sectors** of different land uses (Figure 9.42c). In some cities, the distinction between land uses was clear, but the distribution was not in such regular patterns, establishing **multiple economic nuclei** (Figure 9.42d) in which industrial and commercial activities occurred in several specialized and separated zones instead of around a single CBD.

During the 1930s and 1940s, American cities grew more slowly than they had in the previous 50 years. The economic depression of the 1930s and World War II made capital either in short supply for house building or directed to military purposes. By the mid-1940s, many families shared cramped inner-city housing units.

## Post–1945 Cities

After the war, an explosion of house building and road building opened up huge suburban areas to what had become an ideal of single-family homes for the many with rising incomes as the United States dominated the world economy. New federal roads linking major cities followed by the interstate highway system established in 1956 made greater mobility possible and encouraged suburban housing and commercial development. The outer areas of American cities expanded into formerly rural areas. By the 1990s, for example, metropolitan Chicago occupied 10 times the area its buildings had covered in 1920.

These trends gave rise to a new form of city that developed from the monocentric (single center) before 1940 to the multicentric pattern of today (Figure 9.43). The CBD became more specialized around financial businesses as many of its shopping and industrial functions moved to the suburbs. High land

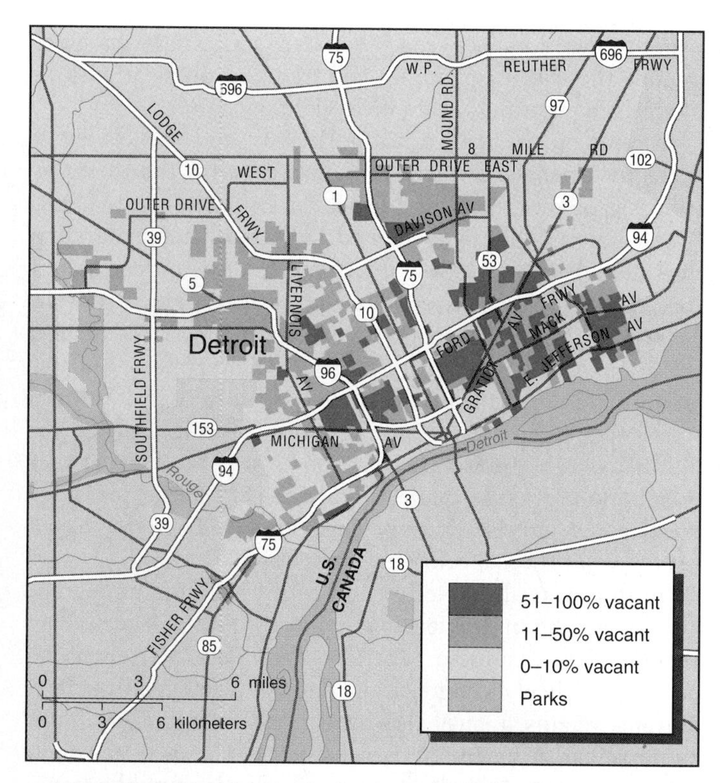


**FIGURE 9.44 United States: urban landscape in Detroit.** The "city of death" shown in the high levels of housing vacancies in the inner city and older industrial areas, compared to the "fat city" of the suburbs.

prices led to the construction of tall office buildings with 50 and more floors to accommodate banks and other commercial corporations (see Figure 9.21). The tall buildings formed the skyline of the central area, which sometimes expanded into surrounding warehouse and poor housing districts. New highways, often elevated or in tunnels, circled the CBD.

Beyond this sector of intense commercial activity, the older pre-1940 suburbs, crossed by railroad yards and older highways with commercial strip developments, were often taken over by the expanding African American, Hispanic, or other ethnic groups. Such groups either did not have the resources to move to suburban homes, faced discrimination by financial institutions or home sellers when they tried to do so, or preferred to stay in an area that was familiar to them. The African American ghettos of northern cities did not appear until the influxes of such people from the South and the suburb-bound movements of white people in the 1950s.

The post-1945 suburbs were quite different from the inner suburbs built before 1930. They were laid out with more space to allow single-family homes and yards. Such areas often occupy the largest amount of land in American cities. They are interrupted by suburban shopping centers, light manufacturing factories, and office buildings that associate together and may include more economic activity than in the CBD. They are linked by interstate highways and beltways. One of these suburban centers in the largest metropolitan complexes often contains the international airport, which itself may be the size of a small city with multiple economic activities from hotels to warehouses and factories.

These major changes in American cities occurred with little or no overall planning apart from designation of land use by zoning. Metropolitan areas in the United States are often a mosaic of local jurisdictions that find it difficult to work together. They take advantage of, or complain about, differences among them instead of looking for possible links. For example, public transit-line construction forced some jurisdictions to work together, but in other cases, one group of jurisdictions excluded others from the planning process. It is, perhaps, surprising that many regular and common patterns of land use emerge in such a situation. Apart from the absence of green belts around U.S. cities, the general built environment patterns are similar to what emerged (with a greater degree of planning) in western Europe (see Chapter 7).

The changing face of U.S. cities since 1950 resulted in many movements of people within the cities. The growth of the suburbs involved people moving out of the older areas to new single-family homes. This was labeled "white flight" since it resulted in African Americans and other minorities being left in the older areas that lost many services and jobs to the growing suburbs. Property values in the older areas declined, sometimes leading to abandonment and dereliction. The social result in Detroit has been described in the graphic terms of a donut—an empty center, the "city of death" with high levels of abandonment (Figure 9.44), surrounded by "fat city" in the suburbs.

### Postindustrial Cities

Since the 1970s, there were some movements of people from the suburbs farther out from the city into surrounding rural areas and other movements back to the city center. The selective movement back to parts of central city areas by more affluent groups is often termed *gentrification* (see "World Issue: Gentrification," pages 414–415). From the late 1970s, expanding American urbanism spread new land uses into former rural areas. This is not a matter of counterurbanization, in which workers stressed by urban living go to live and work in the countryside. The term **"edge city"** was coined for new exurban developments that rely mainly on car and truck transport but at times on airplanes. Edge cities include large developments of shopping malls, offices, warehouses, and factories, located on the edge of suburban areas or on greenfield sites (Figure 9.45).

### Canadian City Landscapes

Of Canadian cities, only Québec has a historic heart with buildings older than the 1800s. Most cities from Toronto westward were built almost entirely in the 1900s. Other

**FIGURE 9.45 An edge city: Bethesda, Maryland.** The characteristics of new blocks of offices and apartments intermingling with parks, highways, and shopping centers, are typical of these new urban developments in the countryside—which is close.

© Jon Feingersh/Stock, Boston

major differences between Canadian and American cities include the level of planning involved and the relationships of government units within metropolitan centers.

Toronto is the largest Canadian city and the center of Canadian financial services and manufacturing industries, as well as capital of the richest province, Ontario (Figure 9.46). When Toronto began to extend suburbs into surrounding jurisdictions in the 1950s, the province of Ontario required that these jurisdictions and the city plan cooperatively in order to make possible the amalgamation of services such as public transit, emergency services, social services, and regional road construction. Further growth in Metro Toronto was encouraged around three hubs outside downtown: North York, the International Airport, and a new center that was built at Scarborough (Figure 9.47). These measures coordinated and concentrated development. There was also a major project to redevelop the waterfront.

Toronto was also able to cope with increasing immigrant groups moving into older inner suburbs: Italian, Greek, Portuguese, and Chinese districts form distinctive ethnic enclaves of lively congregation but not segregated ghettos. Toronto retains a busy downtown and surrounding older suburbs, and it has the lowest homicide rate of large Anglo American cities, uniformly good schools, good public transportation systems, and controlled urban sprawl. Many cities in the United States envy the Toronto situation, since they have problems of multiple jurisdictions but limited overriding planning controls that affect only a few functions. Some U.S. citizens, however, find Toronto too ordered, even monotonous.

**FIGURE 9.46 Canada: urban landscape in central Toronto.** The skydome, CN Tower, and downtown commercial buildings as seen from Lake Ontario. Toronto's lake frontage made it a major port with the opening of the St. Lawrence Seaway in 1959. In the late 1900s, many dock areas became derelict and the waterfront was redeveloped to provide public access to recreation opportunities.

© Wolfgang Kaehler/Corbis

terns around St. Louis and New Orleans before those lands were sold to the United States in 1803: these patterns still provide a framework of land ownership based on parishes that affects city street patterns.

National parks (Figure 9.49) and other preserved areas take up large areas of both countries. Although attempts are made to retain the natural features of these parks, management is essential in allowing fires and maintaining animal populations. There are also federally owned areas that are not claimed by parks or individual ownership and remain part of the federal lands. Some are hired out for grazing, while others are used for military purposes.

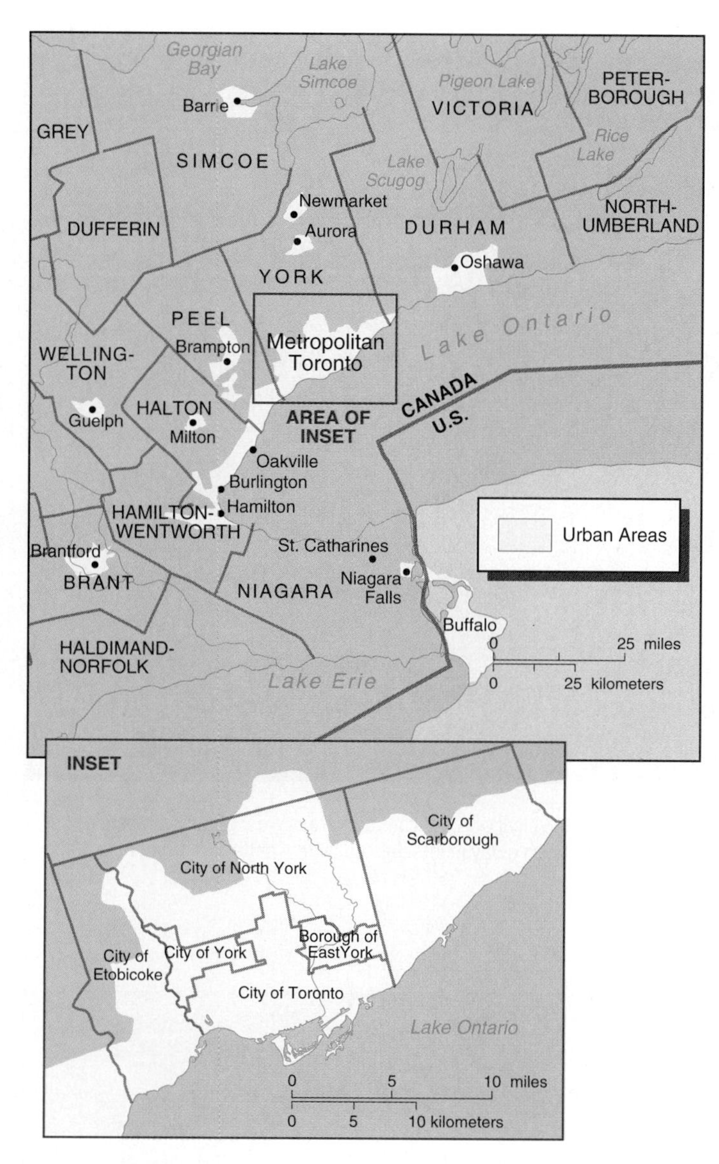


**FIGURE 9.47 Canada: extent of Metro Toronto.** The constituent jurisdictions and surrounding counties in southern Ontario. Toronto's development is controlled by the Province of Ontario and has been planned carefully.

## Rural Landscapes

The rural landscapes of Anglo America include commercially cultivated areas, preserved "wilderness" areas, and areas that are not in demand for commercial use. In both Canada and the United States, the process of converting the land from public to private ownership was haphazard at first, leading to irregular sizes and shapes of plots in the eastern areas. Later settlement was linked to surveys that were based on square plots in the United States (Figure 9.48) from Ohio westward and somewhat skewed squares in southern Canada. The French colonial settlers in Canada favored long lots with river frontages (see Figure 9.36). They instigated similar pat-

# FUTURE PROSPECTS FOR ANGLO AMERICA

To date, the United States has managed to keep ahead of the rest of the world in the generation and take-up of technological innovations and in a worldwide involvement of trade and political links. Such involvements contribute to changes in regional orientations within the United States and to developments such as the edge city phenomenon. Japan and the countries of western Europe will continue to mount challenges, while the future role of China in the world economy may have major consequences for the United States. The more the world economic system develops, trade opens up, and financial resources are free to move around the world, the greater will be the extent and pace of change.

The United States and Canada will continue to have a close association. If Québec becomes independent, there will be a series of new orientations affecting other places inside Canada and across the border in the United States. The future of both countries is tied up with their involvement in the world economy. Anglo America is still the leader among the core countries of the world economy, and its geographic position between the main areas of growth in the world is advantageous. Europe, Eastern Asia, and Latin America all form important and growing trading areas (Figure 9.50). Whether the transatlantic links to Europe will survive or keep pace with transpacific links to Asia is a major debate within Anglo America. The United States and Canada are still very much committed to events in Europe (see Chapter 7), but both countries have also developed close trade and political links with countries in Eastern Asia (see Chapter 6). The recent signing of NAFTA suggests that there may be a wider FTAA to incorporate more countries in Latin America by early in the 2000s (see Chapter 10). These trading zones could increase the roles of Canada and the United States, but they could also result in reassessments of industries and employment within those two countries.

**FIGURE 9.48 Anglo America: rural landscapes.** The rectangular grid pattern of field boundaries typical of American rural areas west of the Appalachian Mountains, here seen in North Dakota. Lands acquired in the early 1800s were surveyed and sold or given away in lots defined by this pattern. Natural boundaries, such as rivers and hills, were overriden by the patterns.

© James P. Blair/National Geographic Image Collection

**FIGURE 9.49 Anglo America: conserved rural landscapes.** Mount Rainier National Park, Washington, USA. The Paradise Lodge and parking lot provide access to climbs and hikes on Mount Rainier and its glaciers. Other commercial developments in the park are forbidden.

© Michael Bradshaw

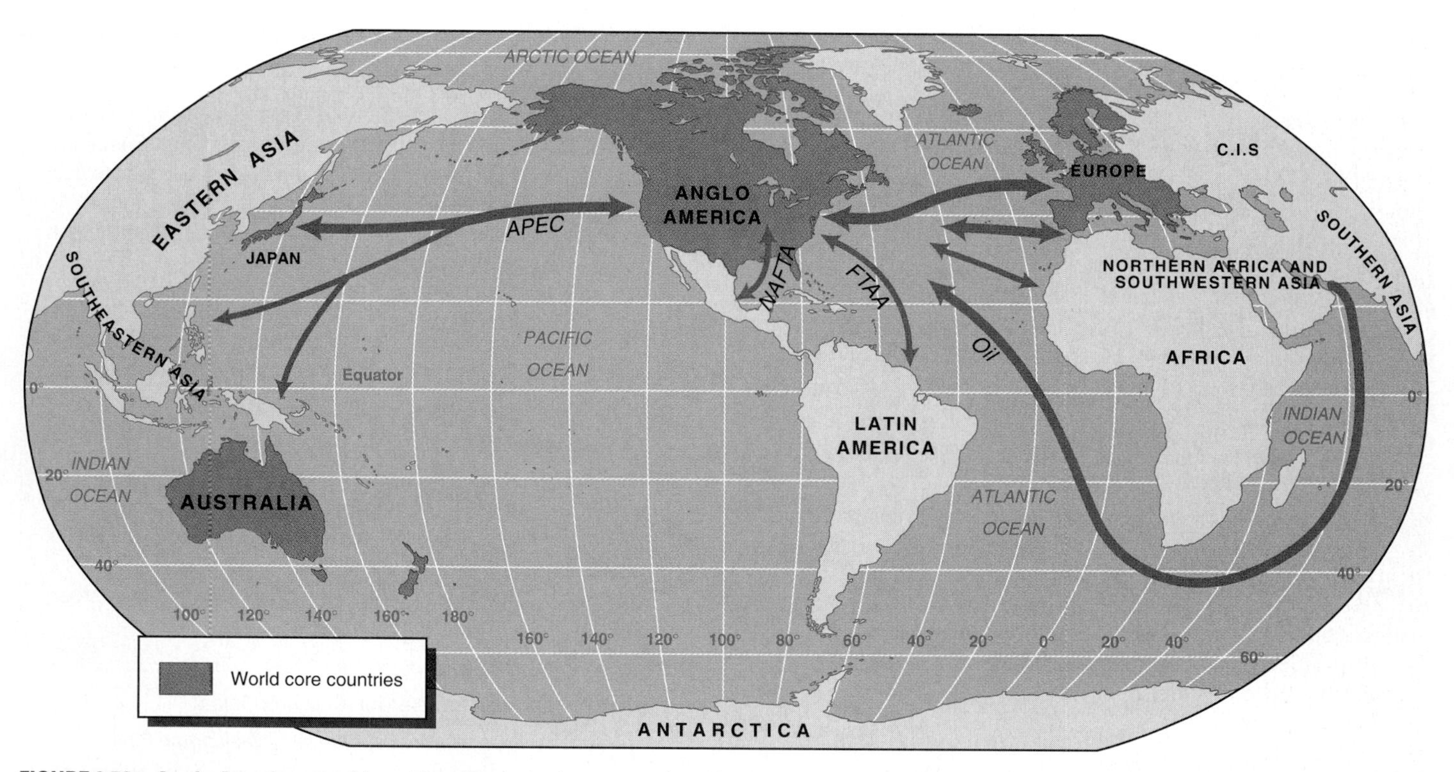


**FIGURE 9.50 Anglo America: world position.** The region has a central position in an era of easy transportation and communication between the Pacific and Atlantic worlds and with its close links to Latin America.

A 1998 study of likely global geopolitical leadership in the early part of the 2000s based its conclusions on answers to four questions about potential leading countries:

1. Is a country rich enough to buy sufficient military forces to deploy around the globe?
2. Does the country have a government able to run a vigorous foreign policy?
3. Do the country's people want an interventionist foreign policy?
4. Has the country practical reasons for getting involved outside its own borders?

African and Latin American countries are unlikely to produce major powers in the medium-term future. If the United States and Europe maintain current levels of cooperation, they will continue to be the dominant powers; if they do not, both would be hurt, but Europe more so. Other likely contenders—Japan, China, and Russia—each have factors that may hold them back, but for China that may be for only a decade or so. Islamic countries are unlikely to be part of the leading group, since their weapons are mostly medium-range, they have little or no agreement on foreign policy, and their worldwide links are limited.

Changes in world linkages will affect the internal human geography of Canada and the United States, both of which are highly mobile societies. A possible further move toward the west coast is, however, likely to be constrained by shortages of water in the southwestern United States and of land in British Columbia. A major concern that remains for the future is the ability of these affluent societies to deal with poverty, reduce extreme contrasts in social and economic well-being among racial and ethnic groups, and lessen the deteriorating quality of life facing both countries. That will require the implementation of improved and acceptable environmental management policies.

Since the United States was in the forefront of computer developments in the late 1900s, the global implications of computer use and telecommunications will have impacts on Anglo America and on its relations with the rest of the world in the 2000s. To date, the conflicting impressions of a few visionaries in the field of technological futurology do not give clear indications of what the future might hold.

## RECAP 9E: Landscapes and Futures

The urban landscapes of the United States are distinguished by their inner-city high-rise buildings that are a response to high land values. U.S. cities have experienced little planning in comparison to Canadian cities. Anglo American rural landscapes include large areas of grid-pattern fields, national parks, and expanses where there is little prospect of economic activity.

The future of Anglo America as world leader is linked to the ability of people in Canada and the United States to maintain their lead in the world economy by encouraging increasingly wide trading exchanges linked to political and technological developments.

9E.1 What factors explain the differences in U.S. cities in the early 1900s, 1940s, and 1990s?

9E.2 How does metropolitan Toronto and its growth compare to processes in major cities in the United States?

**Key Terms:**

central business district (CBD)
concentric pattern of urban zones
ghetto`
urban sectors
multiple economic nuclei
edge city

## WEBSITES: Anglo America

There are thousands! On the Dorling Kindersley CD-ROM Eyewitness World Atlas online facility, sites for the United States are listed under the individual states. Try some of the following to gain an idea of what is available.

For Canada, daily news and views can be accessed through the *Toronto Globe and Mail* site:

*http://www.theglobeandmail.com*

The Canadian government has a wide-ranging site that contains details of how Canada is governed together with informational materials:

*http://canada.gc.ca/*

Select "About Canada"-"Canada Factsheets" "The Intuit" to access information about this group of people living in Arctic Canada. Then "Provinces and Territories"-"Northwest Territories" and "Cool Facts" to find out more about the northern parts of Canada.

For the United States, there are many media news sources, including:

*http://www.latimes.com* The *Los Angeles Times*

*http://www.nytimes.com* The *New York Times*

*http://www.washingtonpost.com* The *Washington Post*

*http://www.pbs.org* Public Broadcasting Service with current issues pages.

As examples of state-based information sources, the following provide information about history and geography of their states:

*http://www.ca.gov/s/* California.

*http://www.state.ma.us/* Massachusetts.

# CHAPTER REVIEW QUESTIONS

1. The percent of the French-speakers in Canada's total population is: (a) 15%. (b) 20%. (c) 25%. (d) 29%.
2. In terms of area, the United States is: (a) only slightly smaller than China. (b) only slightly larger than China. (c) about the same size as China. (d) much larger than China.
3. The area of the United States that was first industrialized in the early 1800s is located in: (a) New England. (b) the Great Plains. (c) the Midwest. (d) the west coast.
4. Which of the following was not considered a factor that brought the United States prominence in world manufacturing by the end of the 1800s? (a) vertical integration (b) horizontal integration (b) economies of scale (c) inflow of immigrants (d) production line.
5. Which part of the United States is sometimes called the country's "breadbasket?" (a) The Midwest and the Great Plains (b) The South (c) The New England (d) California.
6. In which of the U.S. states is 99% of the land owned by the federal government? (a) Oregon (b) Alaska (c) Hawaii (d) Colorado
7. Which statement about the United States is not correct? (a) The U.S. population is aging (i.e., with an increasing number of older people). (b) The majority of the people live in urban areas. (c) The makeup of immigrants to the United States has changed significantly in the last century or so. (d) The dominant direction of population migration is from south to north.
8. Early in the 21st century, which ethnic group is projected to become the largest minority group in the United States? (a) Asians (b) Hispanics (c) African Americans (d) Mexicans
9. When did Canada end its legal ties to Britain? (a) 1865 (b) 1900 (c) 1945 (d) 1982 (e) 1990
10. Which of the following statements about Anglo American cities is not correct? (a) In general, American cities are better-planned and not as old as Canadian cities.

(b) American cities evolved from a pattern of concentric urban zones in the 1920s to the multicentric pattern of today. (c) The African American ghettos in many northern American cities were formed after the influxes of blacks from the South and the suburb-bound movements of white people in the 1950s. (d) Since the 1970s, American cities have experienced movements of people from the suburbs to surrounding rural areas or back to the city center.

11. The importing of black slaves from Africa to the American South started shortly after 1492.
True / False

12. The plate tectonic movements along the San Andreas Fault System make California liable to earthquakes.
True / False

13. The abundance of lakes in Canada and Northern United States. Is attributed to glacial activities during the Pleistocene Ice Age.
True / False

14. One problem of affluence in the United States has been the increasing gap between rich and poor.
True / False

15. The confrontation of cold arctic air and humid subtropical air in the central lowlands of Anglo America leads to the formation of violent tornadoes in spring and summer.
True / False

16. Due to the barrier posed by the Pacific Ocean, the west coast of the United States has not had much business contact with Asian and other Pacific countries.
True / False

17. The manufacturing sector currently employs about 50% of the work force in the United States.
True / False

18. Though comprising only a small proportion of their U.S. economy, the primary sector such as mining, forestry and agribusiness continues to play an important role by contributing raw materials and fuels.
True / False

19. The Tennessee Valley Authority (TVA), established in the 1930s, was one of the few federal programs designed to stimulate economic growth in U.S. regions.
True / False

20. Canadian population shows a strong concentration along the border with the United States.
True / False

21. The closely-spaced set of cities stretching from Boston to Washington, D.C., is often identified in geography as the ________________.

22. NAFTA stands for ________________.

23. The best known center of high technology just south of San Francisco is ________________.

24. The fiftieth state admitted in the U.S., which took place in 1959, is ________________.

25. The French-speaking people in Canada are mainly located in the province of ________________.

26. The two largest metropolitan areas in Canada are ________________ and ________________.

27. The name of the new territory in Canada that was established in 1997 is ________________.

28. The interstate highway system that made rapid suburban development in the United States was established in ________________.

29. The movement back to parts of central city areas of more affluent groups in the United States is termed ________________.

30. "White Flight" in the United States refers to ________________.

# LATIN AMERICA

Chapter

# 10

**THIS CHAPTER IS ABOUT:**

**Upward economic mobility of some Latin American countries**

**Spanish and Portuguese colonial impacts on human geography**

**Tropical and Southern Hemisphere natural environments**

**Six subregions:**

- Mexico: major country and U.S. neighbor
- Central America: small, poor countries
- West Indies: island countries, large to very small
- Northern Andes: mountain environments
- Brazil: second largest economy in the Americas
- Southern South America: midlatitude Latin America

**Changing urban and rural landscapes**

**Future Prospects for Latin America**

**World Issue: Tropical Rain Forest**

**Living in Bolivia**

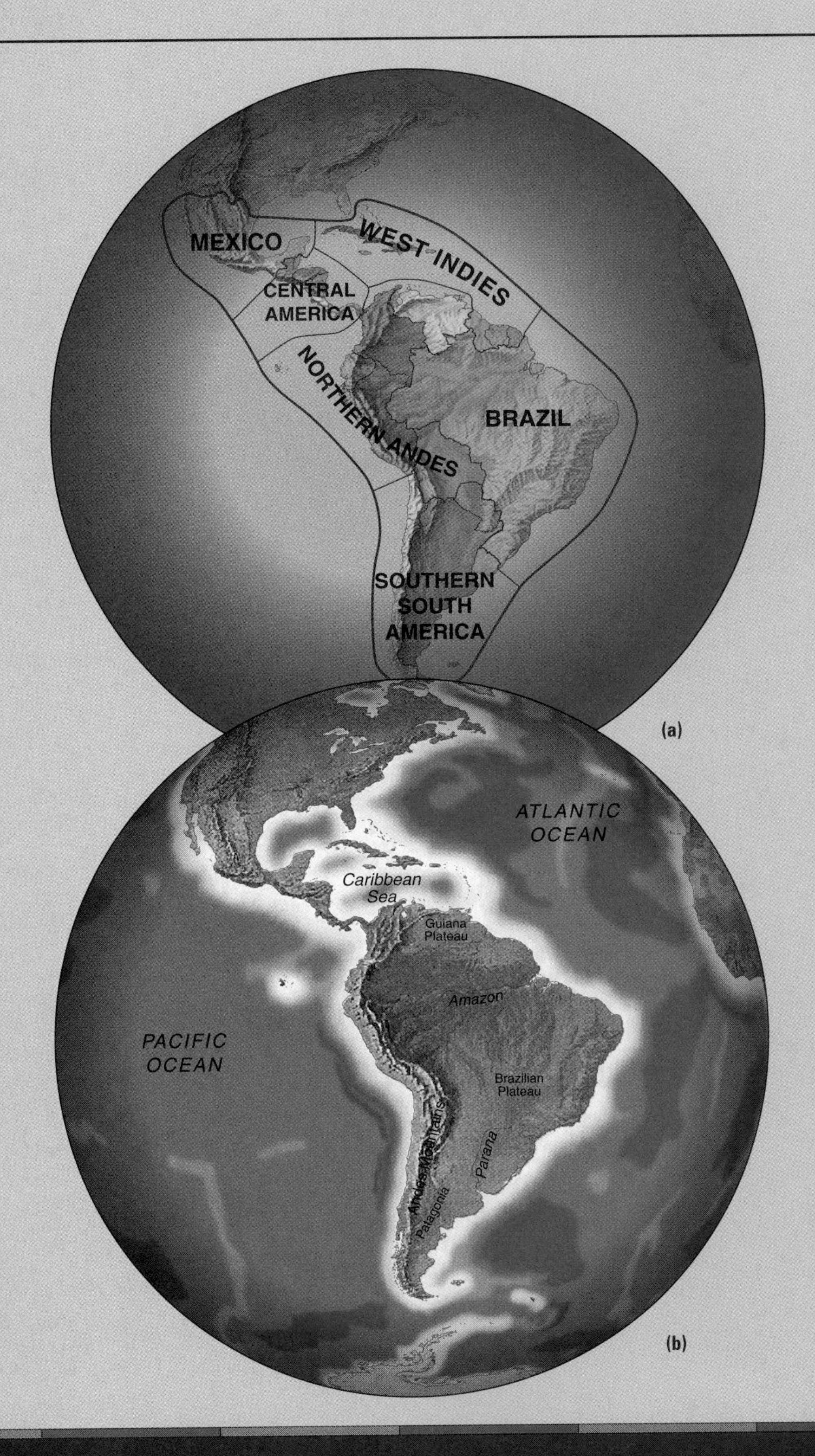


**FIGURE 10.1 Latin America: major geographic features.** (a) Subregions and country boundaries. (b) Natural environments, showing the Andes Mountains along the west coast, the dark green of the tropical rain forest in the Amazon basin, and dry areas in northern Mexico, Peru, northern Chile, and southern Argentina.

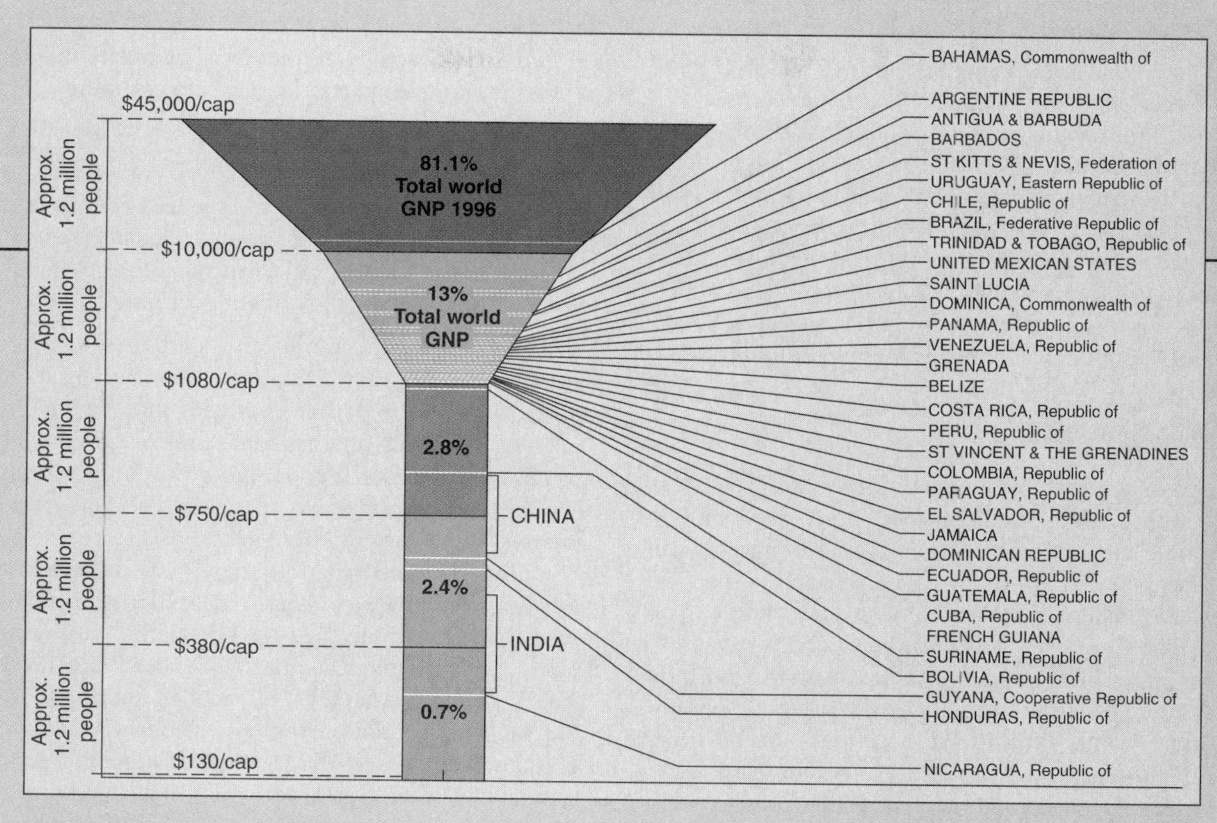


**FIGURE 10.2 Latin America: economic criteria.** The countries of Latin America range from some with deep poverty to many with conditions that place them close to the world's richest countries. In the 1990s, many Latin American countries experienced economic growth, albeit gradual in most cases.

Source: Data (for 1996) from *World Development Indicators,* World Bank, 1998.

## FROM PERIPHERY TO SEMIPERIPHERY

Latin America includes Mexico, Central America from Guatemala to Panama, the West Indian islands, and the continent of South America (Figure 10.1). In this group of countries, Latin languages—especially Spanish but also Portuguese (Brazil) and some French—and the linked social and political cultures provide a degree of unity. Mexico, Central America, and the West Indies are physically part of North America, but their cultures and economic performance link them more closely to South America.

Latin America is part of the developing world, but in the late 1900s, some of the countries were finding their way out of the poverty trap and into greater levels of economic activity (Figure 10.2). The total GDP of the region, however, rises slowly. In 1995, only Brazil, Argentina, and Mexico produced 1 percent or more of world GDP, although Brazil had the world's 8th largest economy. Only these three countries plus Chile were included in the World Bank's Upper-Middle group of countries (see Figure 2.18). Most countries in Latin America still belong to the periphery of the world economic system, while a small number—notably Brazil and Mexico, a few West Indian islands, and possibly Chile and Argentina—are now part of the semiperiphery. Countries in the semiperiphery still have a dependent relationship with core countries, but peripheral countries are dependent on countries in the semiperiphery.

Latin America does not enjoy the same living quality as Anglo America because of its cultural and political history and its natural environments. The history of largely Spanish and Portuguese colonialism in Latin America focused on the extraction of valuable mineral wealth, accompanied by a feudal system of land ownership that lasted into the 1900s. The pre-European populations were dispossessed and further social tensions were stored up within the hierarchical pattern of society imposed by the Europeans. Latin America's natural environment is mainly in tropical latitudes, which proved difficult for European peoples to settle and develop. The combination of physical and human factors slowed the economic progress of Latin American countries long after they gained independence.

It is unlikely that Latin American countries will rival those in Eastern Asia as new world leaders in economic growth. In the late 1990s, both groups of countries faced the problems of making the transition from the periphery toward the core of the world economic system, including unbalanced trading, lack of sufficient capital, poorly developed financial services, and living in the shadow of the core countries. The Latin American countries, however, have very different histories and face different contemporary issues than the countries of Eastern Asia. Their future economic development is likely to be different.

## Geographic Contrasts

Latin America is a region of human and physical geographic contrasts. Countries range in size from Brazil (162 million people in 1998) and Mexico (98 million) to many small countries with populations under 10 million, and to the West Indian islands, where many countries have fewer than 100,000 people. Countries range economically from high to low incomes, from little to great product diversity, and from involvement in to isolation from the world economic system.

Contrasts continue within the countries—inside cities and between the growing cities and hinterland rural areas. Mexico, Brazil, Argentina, Peru, and Chile contain large urban-industrial areas around their major cities, such as Mexico City, the São Paulo and Rio de Janeiro areas, Buenos Aires, Lima-Callao, and Santiago-Valparaiso. The São Paulo and Mexico City metropolitan areas each have nearly 20 million people and are two of the world's largest urban areas. Such cities have many jobs in manufacturing industry and growing professional services. Parts of them appear similar to cities in southern European countries until their huge extents and shantytowns are taken into account. Even within the richest parts of Latin American cities, there are extreme contrasts between the rich and poor. In addition to extensive areas of wealthy and middle-class suburbs, poverty-stricken shantytowns occur throughout their built-up areas.

Some rural areas participate in the world economic system, as the agribusiness equals that in other parts of the world. Brazil's coffee crops are now exceeded by its exports of crops such as soybeans and citrus fruit; Argentina and Uruguay produce high-quality meat and wool from efficient commercial farms; and Chilean farms exploit seasonal markets in Northern Hemisphere countries for fruit and vegetable products. Other rural areas, however, have static or declining economies, with farmers focusing on local markets and making little attempt to become more commercial. Rural depopulation occurs as people migrate to towns. Beyond the fully settled areas are large tracts of Latin America, especially in the tropical rain forest, that are largely undeveloped and still effectively frontier areas.

## Geographic Cooperation

Such contrasts might suggest prospects for cooperation in trading groups of different types, as are being established in other parts of the world. Such groupings are of three types.

- A free-trade area is a grouping of countries that lower the tariffs and trading restrictions between them.
- A **customs union** takes a further step by blending economic policies and business laws.
- A **common market** combines the features of a customs union with establishing a common business environment through the existence of a central authority that may override national government decisions.

For example, the European Union (see Chapter 7) is a common market that encourages trade among its members, provides help for poorer countries and regions within its member countries, and has some control over the production of items such as steel, grain, dairy produce, and wine. Latin American countries have tried all types of grouping, but progress was slow until the 1990s.

Within the potential groupings of countries in Latin America, Brazil and Mexico stand out as countries that are large enough to determine their own economic futures and world links. Central America, the West Indies, and the Northern Andes formed groupings of smaller countries. Argentina, Brazil, Paraguay, and Uruguay established Mercosul. Each group is geographically adjacent and has similar natural environments, histories, and economies that suggest the countries work more closely together in relation to the world economic system. Most groupings make their own attempts at closer economic cooperation. They provide a basis for the subregional divisions in this chapter (see Figure 10.1a), in which further details of such cooperation are discussed.

The groupings of countries face difficulties that contributed to the lack of success achieved by attempts at economic cooperation up to the 1990s. Virtually all the countries of Latin America continue to have boundary disputes with their neighbors. Conflicts between and within the countries of this region are so frequent that the United States will not sell high-tech weapons to Latin American countries since they are likely to be used against each other or to suppress internal dissent.

The boundary dispute between Ecuador and Peru concerned their territories in the western Amazon River basin and lasted into the 1900s, including an outbreak of fighting between the two countries in early 1995. Disputes still exist between Venezuela and Guyana over the Essequibo Valley, while Guatemala took many years to recognize Belize. Memories and impacts of past conflicts continue to sour relations among Chile, Bolivia, and Peru (the Pacific War) and between Paraguay and its neighbors.

In addition, the larger Latin American countries often had visions of grandeur, sometimes termed **geopolitical ambitions**, to extend their territories. Argentina's expansionist stance threatened Paraguay and Uruguay at one stage and later extended to a dispute with Chile over the ownership of the Beagle Channel in the far south. In 1981, Argentina occupied the Falkland (Malvinas) Islands in an attempt to displace the United Kingdom. Brazilian leaders talk about "living frontiers" to indicate that they do not see them as fixed. Since the 1950s, its military expanded the road network across the Amazon River basin, partly for internal security reasons. Surrounding countries, however, see the network as a threatening gesture by Brazil. Mexico and Venezuela used their oil wealth at times of high oil prices to influence other countries. Such actions were perceived as threats by their smaller neighbors and made them see hidden agendas in subsequent proposals for increased economic cooperation.

## HISPANIC CULTURES

The cultural history of Latin America is a vital backdrop to an understanding of its modern geography. Many fluctuations followed a colonial history that decimated the Native American peoples and consigned them to an underclass, left a legacy of ethnic strife, and held back economic development. Habits and attitudes born under colonialism affected the region's countries for long after independence—mostly achieved in the early 1800s.

### Pre-European Peoples

Before Europeans arrived, Latin America was occupied by groups of people similar in racial characteristics to the indigenous peoples of Anglo America. Most had a tribally based subsistence economy supporting modest numbers of people, but several urban-based civilizations or empires emerged (Figure 10.3). Estimates of the numbers of people living in Latin America before European entry range widely, with 50 million as a midway number.

#### Civilizations

In favored areas, the indigenous peoples developed civilizations with religious and urban functions supported by surplus agricultural produce. Among the earliest were groups who moved to the Peruvian coasts from farther north beginning around 1250 B.C. They used the Andean streams for irrigation, culminating in the Chimú society that left behind the ruins of complex buildings and systems at Chanchan around A.D. 1000. In Middle America, the **Mayan civilization** was based on surpluses of maize. Its center was in the eastern lowlands of modern southern Mexico, Guatemala, and western Honduras from the A.D. 200s to the 900s, leaving the remains of large pyramid and monument structures. In the 1100s, a group of tribes from the north, among which the **Aztecs** became dominant, settled in the area of modern Mexico City and in A.D. 1325 built their

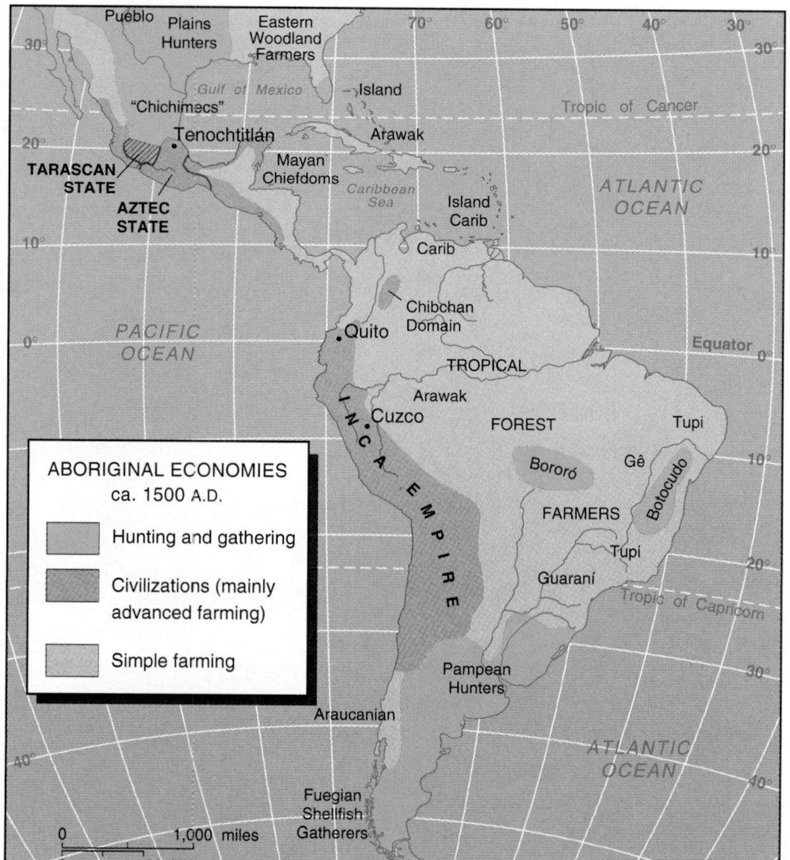

**FIGURE 10.3 Latin America: pre-European economies and empires.** As in Anglo America (see Chapter 9), a variety of Native American groups occupied the region before A.D. 1500. Some lived at subsistence level; others established far-reaching empires.

stronghold capital of Tenochtitlán on a marshland site. They ruled most of Middle America by the 1400s and gathered great tribute riches in their capital. Farther south, at around the same time, the warlike **Incas** were based in southern Peru around the mountain city of Cuzco. They overran their neighbors and controlled the area from Quito (modern Ecuador) in the north to modern Paraguay and Chile in the south. They organized gold and silver mining and a system of foot mail and transportation that kept their empire together.

### Subsistence Culture

The Arawak and Carib groups of the West Indies and northern South America lived more simply by forest hunting and fishing. The Araucarians of modern central Chile and the pampean tribes of modern Argentina had strong economies and organizations that resisted European attempts to occupy their lands for longer than most. The Guarani-speaking tribes of the upper Paraná River basin had a more peaceful way of organizing relationships that enabled their language to become common. All of these groups disappeared or were absorbed into the colonial cultures. Small numbers of people with the most basic cultures penetrated the tropical rain forests, possibly to escape the aggressions of other groups. Ironically, their cultures survived longer in such isolation than the more technically advanced tribes that made use of mineral resources and developed wider handicraft skills but clashed with the European colonizers.

## Spanish Colonization

The Spaniards who conquered most of the region in the 1500s imposed a great degree of control on their colony of New Spain through a close-knit pattern of interconnected urban settlements. The Native American inhabitants could not resist as they were poorly armed or widely scattered and disorganized. They were converted by the Roman Catholic church and taught to provide crops and livestock for the food needs of the foreign urbanites. Mining settlements in the Andes Mountains used various forms of slavery to coerce the local people.

The colonial period in Latin America left a legacy of underdevelopment because it scarcely moved beyond mining and the imposition of large, landed estates with a strict feudal system. The native peoples and many of mixed blood were relegated to being landless laborers. By the end of the Spanish colonial period, antagonisms between privileged and underprivileged groups were ingrained. Subsequent history is largely a record of the antagonisms at work.

### Occupation

Christopher Columbus first crossed the Atlantic Ocean from Spain in 1492 and sailed around the Caribbean area and the coasts of Middle and South America on subsequent voyages. He was soon followed by waves of Spanish military and missionary groups: within 50 years much of the region was conquered and occupied (Figure 10.4). The only exceptions

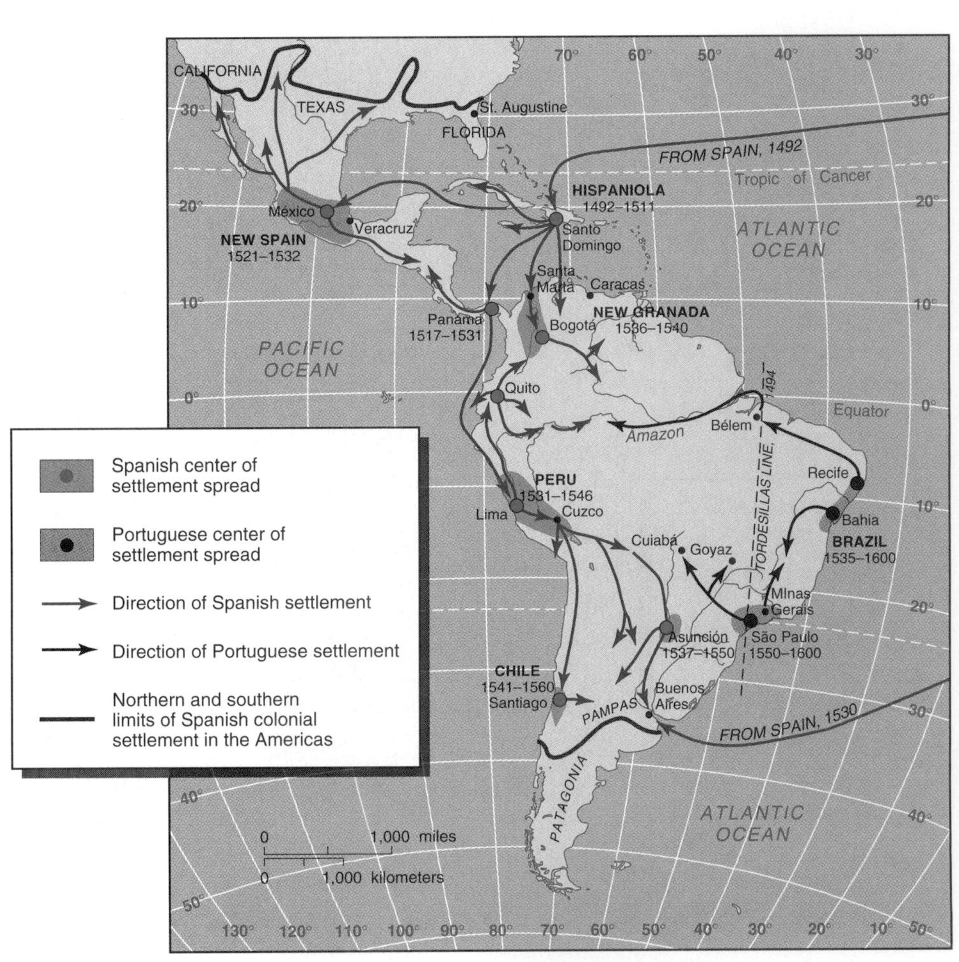


**FIGURE 10.4 Latin America: colonial conquest and settlement.** During the 1500s, Spanish and Portuguese adventurers established bases and conquered huge areas rapidly. The areas of initial conquest became centers of administration and development.

were some of the smaller Caribbean islands, the far southern extremity of South America, and the territory that became Brazil. In 1494, Spain and Portugal signed the Treaty of Tordesillas that established a demarcation line between their global spheres of interest, giving the eastern corner of South America to Portugal.

The initial Spanish occupation around 1510 was based on the major Caribbean islands of Cuba and Hispaniola. Information about the Aztec Empire riches led to invasions of Mexico in 1519 and 1520. The captured Aztec lands formed the heart of the colony of New Spain governed by a Spanish viceroy from 1535. Spain claimed the rest of Middle America, northern Colombia, northern Venezuela, and further West Indian islands such as Jamaica and Trinidad. Expeditions from New Spain pushed northward into the southern parts of the modern United States from the 1530s but found no gold. Attention turned southward (Figure 10.5).

In 1532, a small Spanish force invaded and captured the Cuzco headquarters of the Inca Empire. It founded new cities to control the gold and silver mining and facilitate the traffic home to Spain across the Panama isthmus. In the 1530s, further expeditions linked the rest of South America, although little was fully occupied until later. Lima, with its nearby port of Callao linking the inland wealth of Peru to Spain, became the capital of the Viceroyalty of Peru that included most of northern and western South America. The Viceroyalty of New Granada, centered on Bogota, was established in 1717 and that of La Plata, centered on Buenos Aires, in 1776.

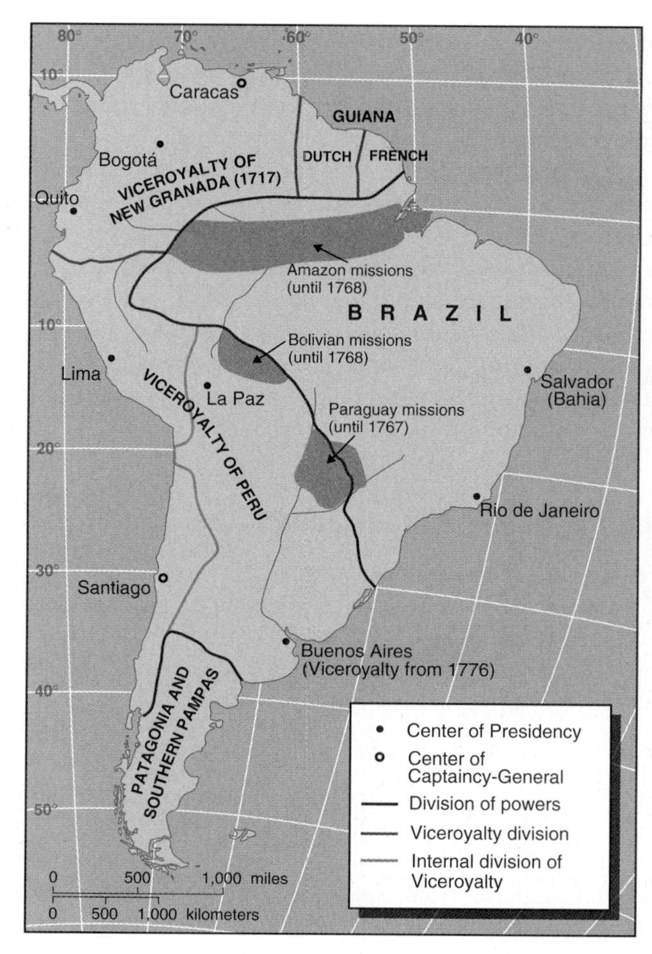


**FIGURE 10.5 South America: colonial divisions.** The jurisdictions established by Spain and Portugal formed the basis of country boundaries at independence but were poorly defined. The red line divides areas of Spanish (to the west) and Portuguese (to the east) influence. The pink shading highlights early attempts to extend Brazilian territory westward. Buenos Aires became the center of the Viceroyalty of Río de la Plata.

## Spanish Control

Having made such rapid conquests of vast tracts of land, Spain granted large areas to nobles, soldiers, and church dignitaries, who were responsible for political control, settling people on the land, and exploiting its wealth. They were given jurisdiction over the native peoples, to use them as laborers. This feudal **encomienda system** applied throughout the Spanish colonies for over 200 years. Although some efforts were made in Spain to pass legislation for the fair treatment of native peoples, such laws were never applied in the colonies.

Alongside an inefficient and corrupt system of colonial administration, this system left a heritage of schisms between groups in society. The native peoples were dispossessed and seldom educated. At the other end of the socioeconomic scale, the *peninsulares*—Spaniards born in Spain—took the highest offices and largest land grants. *Criollos*—Spaniards born in the colonies—and their increasing numbers of **mestizo** (mixtures of European and Native American ancestry) relatives had fewer privileges and became resentful as their links to Spain weakened.

## Portuese Colonization

The Portuguese were slower to settle Brazil, and it was only when the French tried to colonize it in the 1530s that the Portuguese government sent settlers. A governor general was installed in 1549. The first capital was Salvador (Bahia). In the 1550s, São Paulo was founded and French settlers were expelled from Rio de Janeiro Bay, where the city was founded in 1567, becoming the colony's capital by the late 1700s.

The discovery of gold inland of Rio de Janeiro led to increased Portuguese immigration in the 1600s. This in turn set off expeditions to explore and claim the interior. From 1680, there were various attempts to push the Portuguese boundaries southward to the Plate estuary, where Spanish settlement was slow. In 1750, Spain agreed to allow Brazilian interior expansion westward of the Tordesillas line—an expression of its inability to prevent what was already happening (see Figure 10.5).

The Portuguese process of occupying Brazil resembled that of Spain in its colonies, although there were some distinctive features. An elite, who bought African slaves to work on the plantations when the native people ran away, owned large sugar estates along the northeastern coast. Farther south in the São Paulo and Rio de Janeiro areas, there was more opportunity for ordinary Europeans and mestizos to establish their rights to new lands, leading to a greater degree of personal autonomy.

## Other European Colonizations

French, Dutch, and British attempts to colonize Latin America came later and were generally fought off by the Spanish and Portuguese. There were only limited inroads on the dominant colonizers despite the huge extent of their lands and the thin spread of their military forces.

The main area of inroads was in and around the Caribbean. The French took the western one-third of Hispaniola in 1697 but were expelled by an independence movement in 1804, when the former colony became the new country of Haiti. The French also took the rest of Hispaniola (Santo Domingo) from Spain in 1797 until 1809. When Santo Domingo became independent in 1821, Haiti invaded and occupied it until 1844, after which the Dominican Republic was declared. Cuba and Puerto Rico remained Spanish colonies until the United States backed a Cuban uprising in 1898 and won the subsequent Spanish-American War. Cuba became independent. Puerto Rico became a "commonwealth territory" of the United States in 1898, and its people became American citizens in 1917 with a degree of independence and access to markets in the United States.

Some of the smaller West Indian islands that were ignored by Spain had their first European settlement by the French, Dutch, or British. The 1600s was a period of increased wealth creation in these colonies as sugar plantations spread using African slave labor. Many of the islands changed hands, as when Jamaica (1670) and Trinidad (1797) were taken by Britain from Spain.

The Guianas were within the Spanish realm on mainland South America, but little interest was shown in the swampy coasts and forests until the Dutch, British, and French settled them. The Dutch settled first in 1581 and their area (now Suriname) was agreed to following the British capture of New Amsterdam (modern New York) in 1667. The British took the western area in 1814 and established a colony (now Guyana). The French settled the eastern part in the 1600s, where the infamous penal colony of Devil's Island deterred further immigration into French Guiana until the prison was closed in 1938.

## Independence

The initial wealth that attracted Europeans came from exports of high-value minerals such as gold, silver, and gems that their Latin American colonies sent to Europe. When the flows of such wealth slowed in the 1700s, the power of Spain and Portugal weakened, particularly after the Napoleonic conquest of Spain in the 1790s, enabling many Latin American countries to gain independence in the early 1800s. The revolutions often began with uprisings that were suppressed by Spanish soldiers but led to concerted drives centered on Middle America and the Andean states of South America. The new countries emerged from administrative divisions within the Spanish viceroyalties. The boundaries between these divisions were poorly surveyed, if at all, forming a basis for subsequent disputes.

### Independence from Spain

Independence from Spanish rule was often followed by a period when various arrangements were tried before the full identity of the modern countries was achieved. Much of Central America became part of Mexico, then separated as the United Provinces of Central America (1823–1838), before finally breaking into the modern republics. Colombia (including Panama), Ecuador, and Venezuela at first formed Great Colombia but separated in 1830, with Panama gaining its independence in 1903 as a result of United States influence.

Initially, the newly independent countries were poorly prepared and provided for. The 150 years after independence were marked by political instability punctuated by short periods of economic growth. Spain left behind a geographic system based on mineral exploitation, transportation to a limited number of ports, and large estates engaged in livestock raising. It left rivalries between countries and disputes over their boundaries, together with a hierarchical social system that incorporated deep-seated resentments. In most former Spanish colonies in Latin America, government control alternated between the elite conservative land-owning group, often allied to the military, and the more liberal *criollos* who wanted a wider spread of democratic involvement. Many countries had long periods of dictatorship after conservative-liberal tensions made government almost impossible.

### Brazilian Independence

In Brazil, independence also came in the 1820s but with a stronger start that laid the foundations of its modern development. In 1807, the Portuguese royal family evacuated from Europe to Brazil when Napoleon threatened to take their country. Rio de Janeiro became the capital of Portugal for a time. Although the royal family went back to Portugal, the prince regent returned to Brazil in 1816 at a time of flourishing revolutions in the Spanish colonies. In 1822, he proclaimed Brazil's independence from Portugal and himself king as Pedro I. Under King Pedro II (1840–1889), the Brazilian economy grew rapidly with the construction of railroads and ports and the expansion of mining in the east-central parts, commercial farming in the south, and rubber collecting in the Amazon River basin. In 1889, a military coup made Brazil a republic, but falling prices of coffee and rubber led to widespread unrest and a period of dictatorship. Brazil encouraged immigration and received many people from Germany and Italy in the early 1900s. They moved mainly into the southern parts of the country, where environmental conditions were somewhat similar to their homelands.

### Economic Colonialism

Although Spanish and Portuguese rule was broken, Europe remained the main market for exports of mine and farm products from Latin America. Britain in particular established an economic colonialism and dependent core-periphery relationship with several Latin American countries that lasted until the early 1900s. It encouraged the export of raw materials and new products to supply its own industries and sold its own manufactures in the emerging markets. Aimed at developing the main areas that produced minerals, cotton, beef, grain, and coffee, British investments built railroad networks and port installations. Elite families in Latin American countries who worked within the system often sent their children to schools in Europe and Anglo America and banked their wealth in those countries.

## Continuing External Influences

### Import Substitution

The 1900s saw moves toward modernization in many Latin American countries. The disruption of economic activity and trade brought about by the core countries' economic depression of the 1930s and World War II in the 1940s turned Latin American countries toward the idea of becoming more self-sufficient. They would become less dependent on selling the primary sector commodities they produced in exchange for high-priced manufactured goods from industrial countries. Import-substitution manufacturing grew behind tariff walls that protected the new industries against external competition. By the 1970s, this policy resulted in the rapid growth of a few major urban centers in each country. Because of their larger home markets, the most populous countries (Argentina, Brazil, Colombia, Mexico) produced the most under this system. The unevenness of economic growth continued.

### U.S. Influence Grows

From the early 1900s, the decline in the availability of British capital and the growing influence of the United States led to a shift of economic dominance from Europe toward the United States. In 1823, the United States staked its claim on Latin American affairs by formulating the Monroe Doctrine, in which it vowed to resist other nations intervening in Latin American countries. Its own influence was greatest in Middle America and the Caribbean until the mid-1900s. The United States intervened in the affairs of Cuba, Haiti, and the Dominican Republic on several occasions. During the later 1900s, a combination of political and corporate intervention from the United States caused most Latin American countries to look northward for trade, political support, and fashions in clothing, music, and fast foods. Miami, Florida, with its large Hispanic population and its banking and transportation links to all parts of Latin America, was termed "the capital" of Latin America in the absence of coherent links within the region.

### Financial Dependence

In the 1970s, the rising cost of oil first hit countries that did not produce it, but then large quantities of petrodollars were urged on Latin American countries as low-interest loans. Such loans were used to pay for oil imports and major infrastructure projects—and to improve some government officials' overseas bank accounts.

The recession in the richer countries caused by the high oil prices led to cutbacks in markets for products from Latin America and to higher-interest rates on the loans. The combination of debts and falling export income resulted in many Latin American countries defaulting on debt repayments by the mid-1980s. Brazil had to put its large overseas trade balance, generated by import-substitution industries, plus sales of mineral and farm products, toward servicing these debts instead of further investment in production, thereby holding back economic progress. Foreign investment and aid dried up in the late 1980s, but debt liabilities forced countries to emerge from reliance on their internal markets.

### Joining the Global Economy

In the 1990s, Latin American governments changed their approach to the world economy. Elected governments were in power everywhere (except Cuba) and older sources of tension began to be replaced by the need for cooperation. The policies that focused on narrow national markets and protected their local industries from world competition involved high levels of government intervention. These policies gave way to those of structural adjustment that encouraged less government intervention, more foreign investment, and more exporting industries, and opened markets to foreign products and privatized government corporations. Such policies, which had been successful in the growing countries of Eastern Asia, were partly forced on Latin American countries by world bankers as a means of reducing debt burdens or attracting new foreign investment. As with other simplistic approaches to development, this economic restructuring superimposed on the Latin American economic and social

**RECAP 10A: Human Geography of Latin America**

Latin America includes Mexico, Central America, the West Indies, and South America. It is united by its Hispanic heritage, with Spanish and Portuguese remaining the main languages after a colonial history of just over 300 years to A.D. 1820. Latin America contains great contrasts of human wealth and economic development in some of the world's largest cities, some of its richest farming regions, and poverty-stricken and underdeveloped rural areas.

Many Latin American countries are small, especially the West Indian islands, but historic animosities often prevent them from working together for economic growth. The economic and cultural history of Latin America led to slow modernization, internal and international antagonisms, and reliance on one or a few export products.

10A.1 What were the main historic events that imposed a Hispanic culture on Latin America? How did that culture affect postindependence changes and attitudes? Which countries did not have a Hispanic culture, and how did this affect their present human geography?

10A.2 In the past, the United States intervened in Latin American affairs on several occasions. What were those occasions, and should it continue this approach?

10A.3 Why have the countries of Latin America not produced effective international trading groups?

**Key Terms:**
customs union
common market
geopolitical ambitions
Mayan civilization
Aztecs
Incas
encomienda system
mestizo

culture was subject to potential disasters, as Mexico found in 1994–1995, when its economy opened too rapidly, sucking in imports and capital investments and creating a huge trade imbalance.

### Regional Links

At the same time, there was a greater will to seek the cooperation of other countries in the region and beyond. The North American Free Trade Agreement (NAFTA) of 1993 was a major breakthrough for Mexico, opening the prospect of wider trade with Anglo America to other Latin American countries. In the early 1990s, U.S. President George Bush proposed a Free Trade Area of the Americas. By 1994, 34 countries agreed to pursue the idea and bring forward concrete proposals by 2000 that would be implemented by 2005.

Initial enthusiasm was dampened by fears inside and outside the United States about the outcomes. Inside the United States, fears that low-wage jobs would be lost led to Congress denying the president authority for fast-track negotiations. Outside the United States, the other countries, especially the smaller ones, began to think that the costs of adjusting to free trade with the United States might be greater than the benefits. Latin Americans were suspicious of the motives of the United States as it continued to isolate Cuba and to certify or decertify its "friends." For example, while the United States attempted to curb sources of drugs from Latin American countries, it had little success in curbing drug demand at home.

The emphasis shifted from bilateral, "them-us" approaches, with the United States on one side and Latin American countries on the other, to multilateral approaches and a slower rate of achieving cooperation. Tariffs could be cut progressively between 2005 and 2020. The 1990s also saw the rise and regeneration of regional trade groupings. In 1997, Mercosul trade grew by 25 percent, while that in the Central American Common Market and Andean Common Market both grew by 10 percent. The Caribbean area awaited its request for parity with NAFTA, a proposal that did not progress through the U.S. Congress. Mercosul, however, finding that potential FTAA negotiations were slow, opened talks with the European Union, its biggest trading partner and main source of investment since 1996. Outcomes are uncertain. One alternative to FTAA progress could be the incorporation of the northern groups into NAFTA, while the strongest group, Mercosul, negotiates with both the EU and NAFTA countries; another alternative could be each smaller trading group retreating behind its own tariff walls.

## NATURAL ENVIRONMENT

Latin America is a mainly tropical region that includes the world's second highest mountain system, the Andes, and the world's largest river, the Amazon. The region extends through almost 90 degrees of latitude—the greatest north-south distance across any major world region by some 25 degrees—and has a wide variety of climates, natural vegetation types, and soils.

### Tropical and Southern Hemisphere Climates

The range of climatic environments in Latin America (Figure 10.6) provides contrasts in living conditions that affect human development in the region. The climates of upland areas in the tropics, midlatitude humid areas, and coastal areas are linked to the highest densities of population among people of European origin.

#### Middle America and the West Indies

Nearly all Middle America (Mexico with Central America) and the West Indies lie within the tropics. The dominant northeast trade winds that blow across the Gulf of Mexico and Caribbean Sea bring warmth and humidity to these subregions (Figure 10.7). Temperatures remain around 30° C (80° F) through most of the year, although the northern West Indian islands and northern Mexico may be affected by cold air moving down the Mississippi River basin in winter. In Middle America, the rainfall generally increases southward, from the arid region that straddles the Mexican-U.S. border toward the coasts of Nicaragua, Costa Rica, and Panama, which receive over 2,600 mm (100 in.) of rain each year.

Most rain falls in the summer, when the atmosphere allows humid air to rise easily. On West Indian islands with mountains, windward slopes facing into the northeast trades have higher totals than those in the lee of mountains. For example, in Jamaica, the northeast coasts receive over 3,300 mm (130 in.) a year, while the southern coasts require irrigation for farming in areas where the annual rainfall total is under 750 mm (31 in.).

Middle America and most of the West Indies lie in the tracks of summer and fall hurricanes. Hurricanes originate in the eastern tropical Atlantic and move westward, sometimes veering northward on reaching the West Indies and sometimes continuing westward to Middle America. Only the southernmost countries of Middle America, together with Trinidad, the Dutch West Indies, and the Guianas lie outside this zone. Most of the other islands have been affected by hurricane damage to crops, livestock, and property, while loss of life can also be high. Recovery is often slow in small countries with few resources. The 1998 Hurricane Mitch caused damage and loss of life that devastated the already poverty-stricken Honduras and Guatemala.

## Northern South America

All the northern Andean countries lie within the tropics. Temperatures fluctuate little throughout the year. Drier periods and seasons increase in length and severity away from the equator. The amounts of rainfall at a location depend on whether the land faces the direction from which moist air arrives. Humid airflows from the Atlantic Ocean into the Amazon River basin

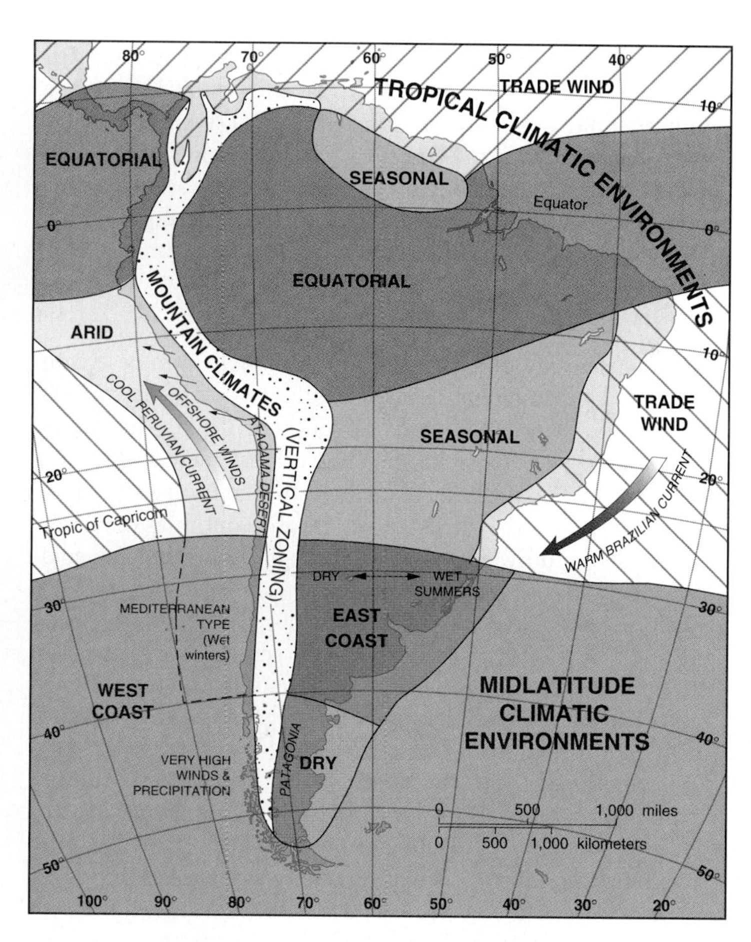


**FIGURE 10.6 Latin America: climatic environments.** Tropical environments dominate all but Southern South America. What effects result from having the widest part of South America in equatorial latitudes and the narrowest in midlatitudes?

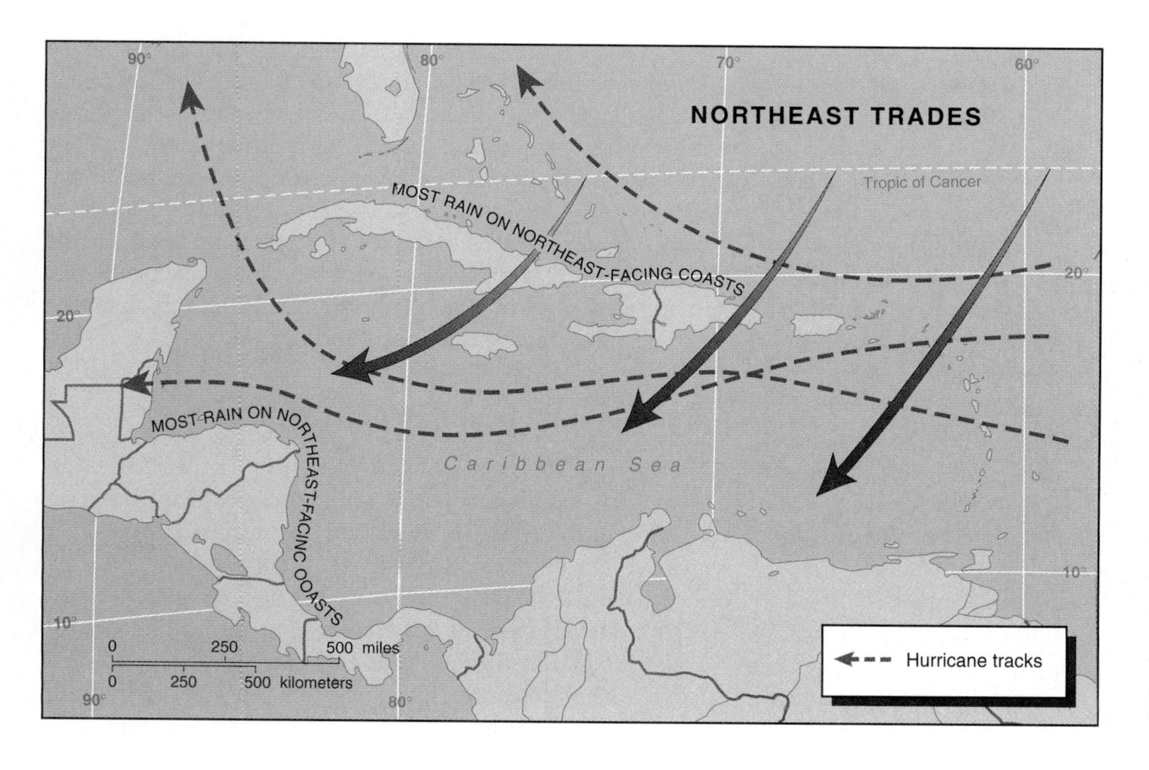


**FIGURE 10.7 Middle America and the West Indies: climatic environments.** The northeast trade winds blow throughout the year, and all but the southernmost part of these subregions lie in the path of hurricane tracks. Three examples of hurricane tracks are shown. Hurricane Mitch took the southernmost route in 1998. The hurricane season lasts from June through November, with most occurrences in August and September. Rainfall is greatest in the summer, with most falling on eastern coasts facing into the winds.

bring heavy rains to the east-facing slopes of the Andes in all these countries. Slopes that face away from the winds, sheltered valleys and basins, and land at high altitude may require irrigation for effective agriculture. On the Pacific coast, rainfall is high on the slopes facing the ocean north of the equator, but southward the offshore winds and cold Peruvian ocean current create an arid climate on the adjacent land.

Northern Brazil experiences an equatorial climate in which the temperatures hover in the low 30s° C (80s° F), humidity is high, and rain falls in all seasons. On the Brazilian and Guiana Plateaus, the rains are more variable and more seasonally concentrated in the high-sun period. The variability is particularly marked in the northeastern corner of Brazil, where severe and prolonged drought occurs every 10 years or so. In southernmost Brazil, the climate becomes midlatitude in type, with cool winters and a reduced length of growing season.

### El Niño

First recognized in the 1970s, the **El Niño phenomenon** is now regarded as a major feature in trying to explain connections among worldwide climatic environments. The major 1997–1998 El Niño event not only brought drought to Middle America and northern South America, and exceptional heavy rains to the deserts of Peru and Chile, but it also was blamed for unusual weather around the world. Droughts in Indonesia and Australia typically balance rains in Peru, but the drought-based fires in Florida and unusually hot weather in southern Europe were also said to be linked.

The basic features of El Niño are understood. At the end of each year, a warm south-flowing current of water replaces the cold water off the Peruvian coast that reinforces the coastal aridity and supports one of the world's major fisheries. The warm current is linked to atmospheric and oceanic movements across the Pacific, the world's largest ocean, and is monitored by satellite. What is not understood is why, every few years, the warm waters stay off the Peruvian coast for more than a year, triggering major weather changes that turn into disasters of flooding and forest fires.

The 1997–1998 El Niño event was the largest since 1982–1983, and its impacts on people's lives in Latin America were diverse. Although some areas had positive outcomes, such as the filling of water reservoirs that had been drawn down after three years of low precipitation and the increased snow depth on the Chilean Andes that prolonged the ski season, many outcomes were negative. Droughts caused lower farm yields in Central America and across northern South America, while floods destroyed crops around the Paraná estuary in Argentina and Uruguay. Forests and savanna grasslands burnt: while the blaze on the Brazil-Venezuela border brought worldwide headlines, fires in Central America were more destructive of trees and wildlife. Transportation was disrupted as the Panama Canal became short of water and as floods destroyed hundreds of roads and bridges in Ecuador and Peru. Shortages of water in some hydroelectricity projects caused power cuts affecting homes and industries. Pollution, increased incidence of diseases, and homelessness affected millions of people, especially the poor.

Although warnings were given, few governments reacted in time. Even in Peru, where funds were spent in advance and relief aid was swift for affected areas, the sheer scale of the damage showed the limits of human action against such natural hazards. While such events have important consequences in richer countries, they are devastating to poor countries that have fewer means of combating them.

### Southern South America

The climates of Southern South America range from deserts in northern Chile and southern Argentina to one of the world's stormiest and wettest regions in southern Chile. The Andes ranges are not so important for human settlement as they are farther north but affect the climates of the lands on either side.

In northern Chile, the **Atacama Desert** continues the Peruvian Desert southward between the high Andes and the cold Peruvian current. Winds blow almost parallel to the coast or offshore, causing cold water to rise along the coast. The cooling effect of the ocean on atmospheric temperatures makes it difficult for air to rise and cause precipitation. The high evaporation rates allow only one river fed by snow meltwater from the high Andes to reach the ocean across this desert.

In southern Chile, by contrast, the midlatitude storms bring precipitation and high winds at all seasons of the year. Between the desert and the stormy southern region of Chile is a transition zone of dry summers, when the arid climate moves south, and wet winters, when the storm tracks move north. This zone is similar in climatic regime to the countries around the Mediterranean Sea in Europe (see Chapter 7) and was familiar to the Spanish settlers in the 1500s and since.

On the eastern side of the Andes is another contrast between northern and southern lands. In the north, the warm Brazilian current allows high rates of evaporation offshore, and trade winds blow humid air into the continent. The highest rainfall totals are near the coast and decline inland, so that the Andean foothills are often arid. In the far south of Argentina, **Patagonia** has a climate that is driest near the coast and becomes somewhat more humid toward the mountains. The westerly airflow descends after crossing the Andes and warms up, becoming drier. Patagonia is often cited as an example of rainshadow conditions in the lee of a mountain range, in contrast to the very wet Pacific slopes of the Andes facing the prevailing westerly winds in southern Chile. The winds blowing across Patagonia are often strong and dry out the surface.

## High Mountains and Island Chains

The major relief features of Latin America (Figure 10.8a) were formed by a combination of clashing tectonic plates along the western and northern margins and of precipitation runoff in major rivers and mountain glaciers. Along the west coast, the South American plate overrides the Nazca plate (Figure 10.8b). The convergent plate margin marks the line of the Andes Mountains that is affected by earthquake and volcanic activity. In Middle America, where the North American and South American plates are separated by the Caribbean plate,

**FIGURE 10.8 Latin America: physical features.** (a) A relief map, showing the mountain ranges, plateaus, and river basins. (b) The tectonic plates and their boundaries.

the pattern is more complex. The eastern Caribbean plate margins are marked by an arc of volcanic islands.

## Middle America and the Caribbean

Middle America is formed of two relief provinces.

- In northern Mexico, an extension of the North American continent is dominated by a high plateau between the eastern and western Sierra Madres. The high land was formed by the North American plate colliding with the Cocos plate. The plateau rises over 2,000 m (6,000 ft.) and contains shallow basins that become more arid northward than the surrounding plateau but are often connected by rivers, forming fertile areas for farming. The western slopes facing the Pacific Ocean are steep, but those on the east are less steep with wide coastal plains. This province terminates at the **Tehuantepec isthmus**—where the land narrows in southern Mexico.

**FIGURE 10.9 Mexico: earthquake damage.** Some buildings collapsed during the 1985 earthquake in Mexico City, causing loss of life and property. Others remained standing but suffered some damage.



- South of the Tehuantepec isthmus, the Caribbean plate collides with the Cocos plate. A single spine of mountains along and parallel to the Pacific coast leaves narrow coastal plains, if any. Eastward of the mountain spine, west-east ranges are separated by deep basins, often occupied by rivers flowing into the Caribbean Sea. A large limestone platform emerged from the seafloor to form the Yucatán Peninsula. The ridges and basins can be traced eastward across the floor of the Caribbean Sea and beneath the island arcs on its eastern edge.

Middle America is subject to the earthquakes and volcanic activity where tectonic plates collide. Mexico City was devastated by a major earthquake centered off the west coast in 1985 (Figure 10.9); earthquakes leveled Managua, the capital city of Nicaragua, twice in the 1900s. Twenty-five volcanoes between the northern Mexican and Colombian borders are regarded as active, erupting lava and ashes.

The island-dominated Caribbean region is also affected by the clashes between tectonic plates. The North American plate

**FIGURE 10.10 West Indies: coral islands.** The Exuma Cays, Bahama Islands, are formed of coral reef limestone and rim the Great Bahama Bank for over 150 km (100 mi.).



drove into the Caribbean plate, producing intense volcanic activity along the plate margin and forming the Lesser Antilles arc. The eruption of lava and ash continued into the 1900s at Martinique (1902), Mount Soufrière on St. Vincent (1979), and the Montserrat eruption of 1995. In the Greater Antilles, the plate movements caused a mass of continental rock to founder, leaving only the highest points above sea level. The Bahamas and an outer group of islands including Anguilla, Barbuda, and Barbados are flat limestone islands constructed of coral reefs on top of subsiding former volcanic peaks (Figure 10.10).

## Andes Mountains

The **Andes Mountains** affect all aspects of the physical environment of western South America. They rise to over 6,500 m (20,000 ft.) in Argentina, Chile, Peru, and Ecuador. In Bolivia, Peru, and Ecuador, the central Andes have two main ranges, the Cordillera Occidental (west) and Cordillera Oriental (east), as shown in Figure 10.11. The collision of the South American and Nazca plates produced a volcanic and earthquake-ridden western mountain range and a folded and faulted eastern range. Between the two ranges, a plateau, the **altiplano**, is widest in Bolivia and narrows northward into Peru. There are also high-level basins within the mountains. In Peru, rivers cut deep gorges as they flow northward and eastward to join the Amazon River tributaries.

In Colombia, the Andean ranges splay out northward into three cordilleras—the Occidental, Central, and Oriental. The Atrato River separates these from a coast range, while the Cauca and Magdalena Rivers separate the three main ranges. The Pacific coast has only narrow coastal plains. In northern Venezuela, the Cordillera Oriental branches into a further series of lower ranges. Along the Caribbean coast, the mouths of the Colombian rivers, the Gulf of Maracaibo, and the Orinoco River delta provide limited areas of lower coastal land between the ranges. Islands such as Trinidad and the Dutch Antilles are extensions of mainland geologic structures.

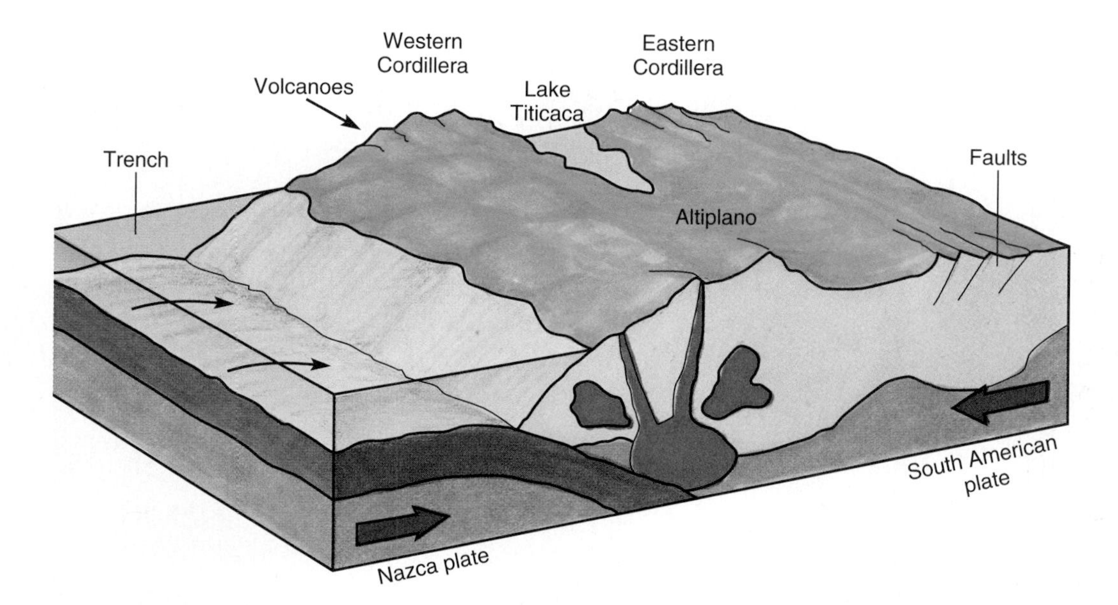


**FIGURE 10.11 South America: Andes Mountains.** The major features of the central Andes Mountains related to their formation along a destructive plate margin. The Nazca plate plunges beneath the South American plate, causing volcanic activity in the western cordillera along the coast and uplift of the eastern cordillera and the high plateau (altiplano) between the two ranges.

**FIGURE 10.12 Southern South America: Andes Mountains.** The Moreno glacier, Glacier National Park, Santa Cruz province, Argentina. The glacier is 5 km (3 mi.) wide, and its front melts on entering Lago Argentino. Such glacial features increase in significance in the southernmost part of the Andes Mountains.



The southern Andes Mountains dominate the landscapes of Chile and the western parts of Argentina, with their highest points constituting the border between the two countries for most of its length. The highest point in the whole range, Cerro Aconcagua (6,958 m, 22,831 ft.), occurs where the central Andean ranges narrow in width southward from about 500 km (300 mi.) to about 150 km (100 mi.). The lowest point for crossing the Andes between the main centers of population in Chile and Argentina is the Uspallata Pass at 3,841 m (12,600 ft.). To the south of this pass, the Andean peaks get lower, continuing for a further 2,700 km (1,800 mi.) to Tierra del Fuego at the tip of the continent. The snow line on the Andes gets lower toward the southern tip of the continent, where glaciers descend to sea level (Figure 10.12). The former more widespread glaciations left behind moraine-dammed lakes and fjords.

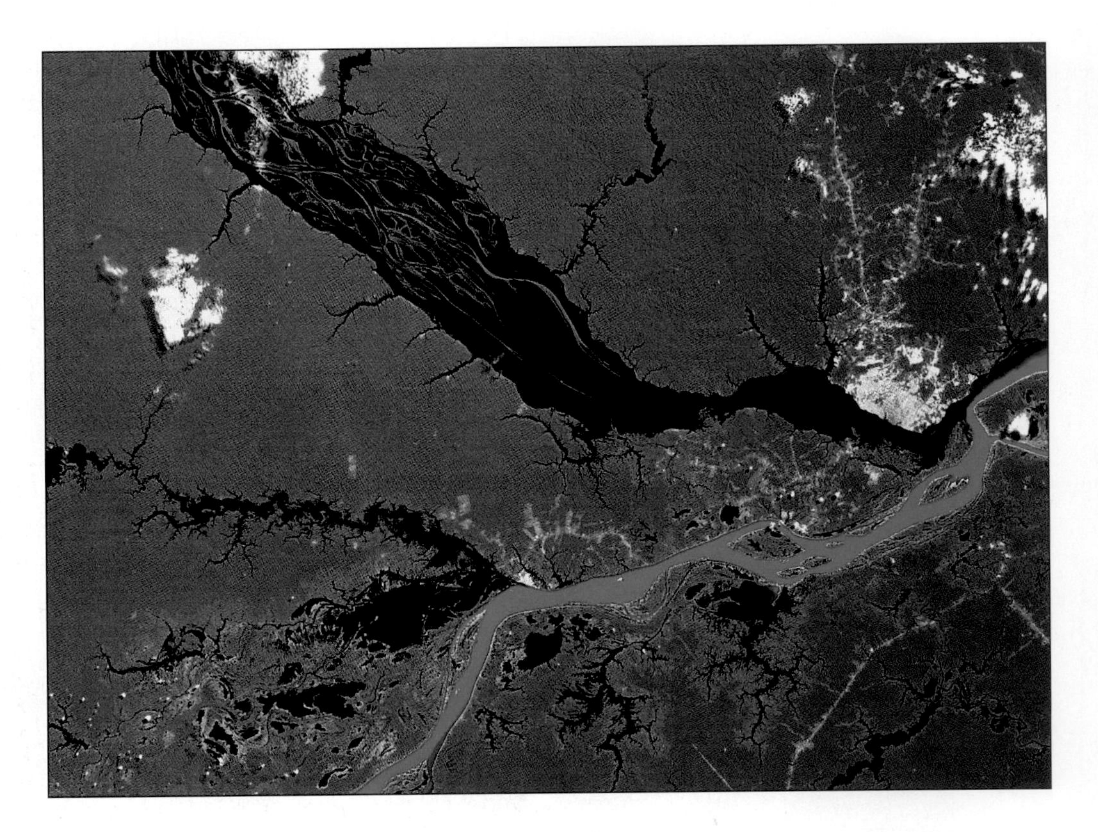

**FIGURE 10.13 Brazil: Amazon rain forest.** A LANDSAT satellite view of the area around Manaus, Brazil, in July 1987 (approximately 150 km, 100 mi. across). Unbroken tropical rain forest is shown as red. The wide, black river is the Río Negro that contains little silt. The blue river is the Solimões branch of the upper Amazon, which brings large quantities of silt from its upper reaches in the Andes Mountains. Manaus is the light-colored area just west of the Negro-Solimões confluence, with radiating roads leading to and from it north and south of the river. Small white areas with shadows to the southwest are clouds.

Image courtesy of Space Imaging, Thornton, CO, USA

On the Chilean side, the Andes come close to the Pacific Ocean in the north. Southward, a coastal range is separated from the main Andes ranges by a series of basins and then a wide continuous valley south of the Chilean capital, Santiago. The coastal range and the valley get lower in height alongside the main Andes range and are drowned by the ocean south of Puerto Montt. On the Argentinean side of the Andes, front ranges in the north are broken by deep river-carved valleys. Their eastern margins have large alluvial fans formed by the deposition of rock material eroded from the mountains. These fans mark both sides of the Andes throughout Chile and Argentina.

## Broad Plateaus

Brazil's physical environment is dominated by broad plateaus and wide river valleys. Locally relief is sharp, as when one travels from the coast to the first plateau level or from one plateau level to the next. More generally, the traveler in Brazil is struck by small differences in the landscape over great distances.

The main relief features of Brazil consist of the ancient rocks of the **Brazilian Highlands**, which are topped in the southeast by layers of lava flows, and the similar ancient rocks of the **Guiana Highlands** on the Venezuelan border to the north. The Guianas have low coastal plains that were formed by the deposition of sediment brought to the Atlantic Ocean by the Amazon River and then moved westward along the coast by offshore currents. Inland, these countries rise to the Guiana Highlands plateau. In southern Argentina, the **Patagonia Plateau** is cut deeply by rivers draining eastward from the Andes.

## Major River Basins

Between the high mountains and lower plateaus, South America is dominated by three major river basins and their depositional features.

- In Colombia and Venezuela, the largest areas of lower land, making up over 50 percent of the countries' area, are drained mainly by tributaries of the **Orinoco River.**
- The **Amazon River** system filled the area between the Guiana and Brazilian Highlands with mud, silt, and sand brought from the plateaus and especially from the Andes Mountains to the west. The rivers flowing from the Andes are muddy "white water" rivers compared to the black rivers, which contain little sediment, flowing from the plateaus. This contrast is demonstrated where the "black" Río Negro joins the muddy Solimões just below Manaus in the center of the Amazon basin (Figure 10.13). The Amazon River is navigable well into Peru; at Manaus, Brazil, it is 2,500 km (1,500 mi.) from the ocean, 15 km (10 mi.) wide, and over 50 m (160 ft.) deep.
- In southern Brazil, rivers drain south to the **Paraná-Paraguay River** system, flowing southward and having massive hydroelectricity potential in the waterfalls at lava plateau breaks. In Argentina and Uruguay, the pampas plains are depositional features formed by the Paraguay-Paraná River system and overlaid by fine windblown loess. These river and wind deposits blanket ancient rocks that occasionally pierce them, forming hilly areas and low ridges in the northern part of Uruguay and on the southern margins of the pampas.

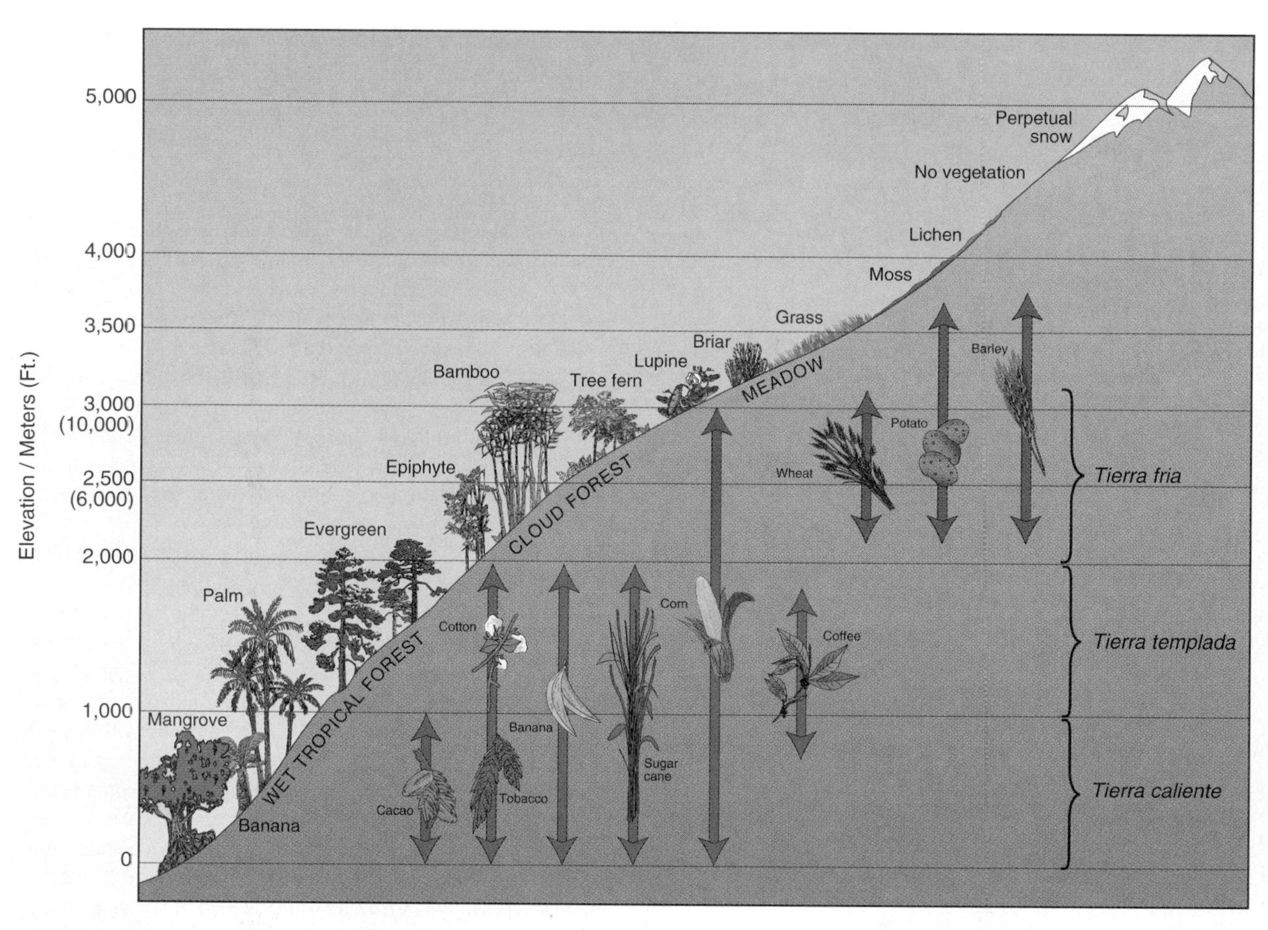


**FIGURE 10.14 Northern Andes: altitudinal zonation of vegetation and crops.** As they cross the equator, the Andes have maximum height and the greatest number of climatic environments from mountain foot to peak, from coastal mangroves and tropical rain forest to perpetual snow. How do the crop zones relate to the vegetation zones?

## Natural Vegetation and Soils

Natural vegetation and soils relate closely to climatic and relief conditions.

- Where tropical rainfall is plentiful and distributed through the year, as in the Amazon River basin, along the northern Pacific and Central American coasts, and on some Caribbean islands, rain forest is the natural vegetation (see "World Issue: Tropical Rain Forest," pages 480–481). Such vegetation spreads several thousand meters up the east-facing slopes of the Andes. The world's largest expanse of tropical rain forest covers most of the Amazon River basin and extends along the eastern coastal lowlands of Brazil to the Tropic of Capricorn. Soils beneath the forest vary, but the areas of good soils are small, apart from the flooded areas close to sediment-carrying rivers.
- Where tropical rainfall is seasonal or lower there are tropical grasslands or shrub vegetation communities that are often thorny. On the Brazilian Highlands, dense deciduous woodland gives way to more open woodland with increasing proportions of shrubs and grasses. Soils there are generally poor beneath the natural vegetation and need treatment for agriculture, but some of the lava flows capping the plateau in southeastern Brazil produce easily worked soils.
- In the southern part of South America, cooler midlatitude conditions bring variations from bare desert, through semiarid bunch grasses and drought-resisting plants, to tall grasses and forest. The Atacama Desert and the eastern margins of the Andes have desert vegetation that ranges from tropical plant species in the north to midlatitude in the south. Before European settlement, central Chile had natural vegetation of trees and shrubs that could grow in a regime of wet winters and dry summers. The plentiful precipitation of southern Chile supports natural vegetation of beech and pine forests. The **pampas** region of central Argentina and Uruguay is named after the tall, lush grasses that grew there before the region was plowed.
- The high altitude of the Andes Mountains results in a series of vertical zones having distinctive vegetation communities, with the greatest number of zones at the equator. These altitudinal zones are linked to the varieties of crops that can be grown (Figure 10.14). The lowest 1,000 m (3,000 ft.) have equatorial climatic conditions and tropical forest in the *Tierra caliente*. The next 1,000 m have warm temperate conditions and deciduous forest in the *Tierra templada*. Between 2,000 m and 3,000 m (6,000

WORLD ISSUE

# TROPICAL RAIN FOREST

Issues connected with tropical rain forest include:

Is tropical rain forest in Latin America being cut too rapidly?

Can areas covered by tropical rain forest be used for other economic activities?

What are the ecologic impacts of cutting rain forest at local and global scales?

Tropical rain forest is the major type of natural vegetation in the Amazon River basin of Brazil and adjacent parts of its neighbors, including eastern Colombia, Ecuador, Peru, Bolivia, southern Venezuela, and the Guianas. Similar vegetation occurs on the Pacific coast of Colombia, along the eastern coasts of the Central American countries, and down the northeast coast of Brazil. Elsewhere in the world, as in West and Central Africa, Malaysia, and Indonesia, much of the tropical rain forest has already been cut. The Latin American areas of tropical rain forest were always the largest in the world, and the Amazon River basin remains the least cut, as the map shows. Although estimates range from 1.5 million to 20 million hectares (4 million to 50 million acres) cut in the Amazon River basin since the entry of Europeans after A.D. 1500, the best estimates are about 8 million hectares (20 million acres), or 7 percent of the total area.

Tropical rain forest is the world's most lavish and diverse ecosystem. It occurs where growing conditions for plants are at their best, with ample rain all year round consistently high temperatures, and intense sunlight. There is a great variety of species in both the plants—commonly 40 to 100 tree species per hectare—and the animals that live in the forest. The soils beneath are not, however, so rich, and although some contain high proportions of nutrients, the heavy rains washed them out of most soils and left behind chemicals that act against soil fertility. The trees on poorer soils maintain their existence by circulating the nutrients without letting them enter the soil: when leaves fall, insects and fungi soon break them down so that roots near and above the surface can capture the chemicals again.

The nature of tropical rain forest and the ways in which people tried to develop areas covered by the forest have not led to convincing alternative land uses after the forest is cut. In many cases, the logging of forest or clearance for agriculture resulted in the impoverishment of land so that it became like a desert—without productive uses apart from settlement sites and mining. In places where better soils coincide with a good local knowledge, and where farmers are not too indebted, there has been some limited success in growing commercial crops of beans and vegetables. In most parts, the cut areas gave way to cattle ranching, but few cattle can be supported per hectare and the carrying capacity soon declines to uneconomic levels. Settlement in the Amazon River basin was nearly always uncontrolled, without proper surveys or understanding of the rain forest ecosystem. The impression is widespread that all settlement attempts must fail.

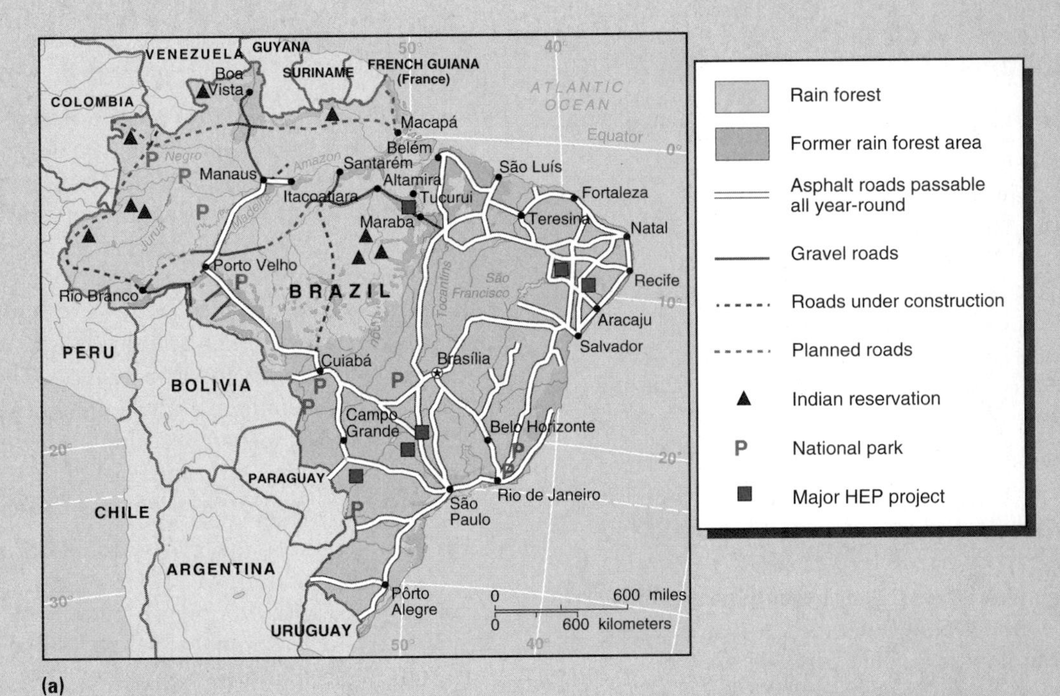


**(a)**

(a) Map of the Amazon River basin in Brazil and the extent of the tropical rain forest. (b), (c) Aerial views of the tropical rain forest—(b) where uncut forest is crossed by the Madeira River and (c) in Rondonia state, where the forest is cut into as farms are established along a road pattern.

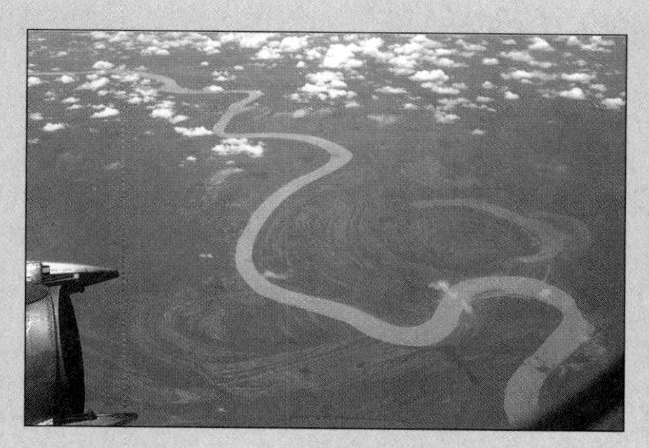

(b) © Michael Bradshaw

(c) © Michael Bradshaw

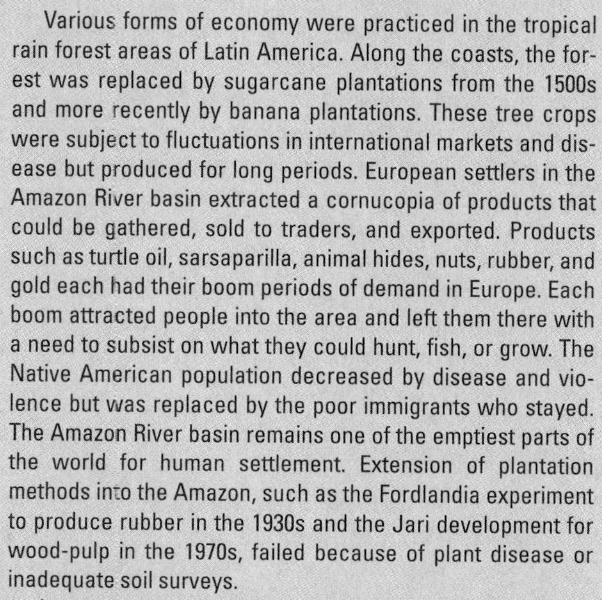

Various forms of economy were practiced in the tropical rain forest areas of Latin America. Along the coasts, the forest was replaced by sugarcane plantations from the 1500s and more recently by banana plantations. These tree crops were subject to fluctuations in international markets and disease but produced for long periods. European settlers in the Amazon River basin extracted a cornucopia of products that could be gathered, sold to traders, and exported. Products such as turtle oil, sarsaparilla, animal hides, nuts, rubber, and gold each had their boom periods of demand in Europe. Each boom attracted people into the area and left them there with a need to subsist on what they could hunt, fish, or grow. The Native American population decreased by disease and violence but was replaced by the poor immigrants who stayed. The Amazon River basin remains one of the emptiest parts of the world for human settlement. Extension of plantation methods into the Amazon, such as the Fordlandia experiment to produce rubber in the 1930s and the Jari development for wood-pulp in the 1970s, failed because of plant disease or inadequate soil surveys.

Larger projects began to change the Amazon River basin after the mid-1900s. Major mining projects and hydroelectric sites resulted in large-scale intrusions and forest cutting, especially along the eastern margins. The building of the Trans-Amazonica Highway in the 1960s and the BR 364 along the Bolivian border in the 1980s made new sections of the rain forest accessible for logging, ranching, and the settlement of landless people on small plots. The designation of Manaus as a free-trade zone brought new people and manufacturing to the heart of the Amazon River basin, but it remains an isolated development without direct road connections to other parts of Brazil. There is still no overall plan for developing the Amazon River basin in the countries that have territory there and no idea of what would be the best policy or even what might succeed there apart from generating hydroelectricity and extracting minerals.

The Brazilian army remains a major influence on Amazonian development, but one that is seldom considered. Arising from the border wars that drained Brazilian manpower earlier in the 1900s, the military encouraged cutting of the rain forest for strategic reasons. When in power, military dictators instigated projects such as the Trans-Amazonica Highway and encouraged families to settle the area as part of the "great march westward." A network of military posts is still being constructed along the western border of Brazil, linked by ground and air transportation.

The increase in cutting and burning of the rain forest in western Brazil in the 1980s led to international ecologic concerns. These included the effect on global warming of adding carbon to the atmosphere and decreases in the diversity of plants from which medicines are made. It was proposed that international loans to Brazil should be conditional on that country taking actions to restrict forest destruction. Cutting and burning decreased in the early 1990s for other reasons, such as families discovering that new plots of land in rain forest areas were not easy to cultivate. The area deforested each year halved but rose again during the 1997–1998 El Niño event that caused drought and fires. As the threat to the forests increased, the Brazilian government passed laws to prevent rapid destruction.

The future of the tropical rain forest in Latin America will be determined by the ability of governments to understand the ecosystem and oversee more considered attempts at its development. There are many renewable resources in the forest, but concentration on those would not employ many people in countries such as Brazil, where many more employment opportunities are needed.

## TOPICS FOR DEBATE

1. Referring to international reactions to the burning of rain forest in Brazil, at what stage do other countries have a right or duty to impose conditions that affect the internal conduct of affairs in a particular country?
2. Is tropical rain forest a development opportunity or inevitable ecologic disaster if not left alone?
3. The needs of poor people come before the preservation of the national environment.

to 10,000 ft.), the *Tierra fria* has cool temperature conditions and pine forests. Above this zone, there is grassland, and at the greatest heights, snow lies all year, even on the equator. Climbers of the Andes near the equator pass through vegetation zones in a few kilometers that would take them from the equator to the pole at sea level. The similarities to sea-level climatic environments and vegetation zones are not complete, however, since temperatures in mountains near the equator vary little from month to month at all altitudes, while winds increase and the air gets thinner at higher altitudes.

## Natural Resources

The natural resources of Latin America include minerals, soils, forests, water, and marine life. The Andes Mountains and the ancient plateau rocks contain considerable resources of metal ores that were among the early attractions for European settlers. They are still not fully exploited where surface transportation is poor. Rivers flowing from the uplands deposited alluvial concentrations of heavy metal-ore minerals in the adjacent lowlands. Sedimentary rock basins between the mountains and in offshore areas became petroleum and natural gas reservoirs, especially in eastern Mexico, northern Venezuela and Colombia, the offshore areas of northeastern Brazil, and along the eastern slopes of the Andes in Ecuador, Peru, and Argentina.

Latin American soils vary in quality, but the alluvial soils of the Amazon River basin, the weathered lava surfaces in southern Brazil, and the pampas of Uruguay and central Argentina form large areas with predominantly good soils for farming. The soils on the eastern plateaus are generally low in plant nutrients, while the steep slopes of the Andes confine potential farming areas to small valleys, high basins, and plateaus.

Forest products are of major significance to many countries in Central America, although some have already depleted their reserves. Only a small proportion of the Amazon rain forest has, however, been cut and it remains the world's largest forest ecosystem.

The water resources of Latin America are huge, including the world's largest river, the Amazon, which carries more than twice the amount of water of the next largest river, the Congo in Africa. Many Latin American countries generate a large proportion of their power as hydroelectricity and have potential for further developments. Although some areas, such as northern Mexico, northeastern Brazil, the Peru–north Chile coastlands, and southern Argentina, are arid, much of the region has plentiful precipitation to support agriculture. For example, rivers flowing from melting snows on the Andes water the arid coast of Peru and the oasis settlements of northern Argentina.

Marine life provides some of the world's richest fisheries along the western coast of Ecuador, Peru, and northern Chile. Other fishing grounds in the Caribbean and off southern Argentina are also being developed. Fish production for Anglo American markets is a growing source of income to many countries in Central America and the Caribbean. Such resources are vulnerable, however, to overfishing and environmental shifts such as the El Niño effect.

## Environmental Problems

Natural hazards such as earthquakes, volcanoes, and hurricanes bring destruction and death to the Andean and Middle American mountain ranges and to the Caribbean islands. Many environmental problems, however, are the outcomes of European settlement and poor management of resources over the last 400 years. Three areas of the environment—the soils, the forests, and the air over major cities—are considered here.

### Soil Erosion

Soil erosion became a major problem in many countries of Middle America, especially where growing populations put increasing pressure on subsistence lands. Such subsistence lands are often on steep slopes, since the major landowners took the best lands for their own farms. The main subsistence crop is maize, which is commonly grown in lines up and down a slope, making it easy for intense rainstorms to wash soil away.

Small Caribbean islands colonized for intensive commercial agriculture became notoriously liable to soil erosion and other forms of environmental degradation. The introduction of sugarcane and banana plantations altered the local environments as well as largely removed the indigenous populations. The natural forests were cleared and the plants and animals replaced by others from Europe. After over 300 years of commercial agriculture, soil fertility decreased to the point where large applications of fertilizer are necessary. The ignorance of impoverished peasants in countries such as Haiti destroys soils through lack of care.

### Air and Water Pollution

Pollution of air and water result from mineral extraction and refining, and from the concentration of human activities in urban areas. Mexico City is the worst offender for air pollution. With the city situated in a bowl-shaped area, over 2,000 m (6,500 ft.) above sea level, the air is often so still that vehicle, industrial, and domestic pollutants drain downward and accumulate in the lowest parts. Pollutant levels exceed maximum tolerances by several times and do not decrease despite stringent government measures. The rate of population increase in the area makes the pollution-reduction efforts of little obvious effect. Brazil has a variety of ecologic problems, ranging from air pollution that reaches high levels around the industries of the lowlands inland of the port of Santos to poisoning of the rivers around mine workings.

### Tropical Rain Forest

Of all Latin America's environmental problems, most publicity is given to the Amazon rain forest. Brazilians are quick to point out that this covers such a vast area that even a continuation of the high rates of cutting attained in the late 1970s and 1980s would take a long time to make a major impression on the

**RECAP 10B: Natural Environments of Latin America**

Latin America's climatic environments are mainly tropical, ranging outward from the equatorial climates of the Amazon basin to seasonal rainy areas and the deserts of northern Mexico and the Peru-Chile coasts. Midlatitude climatic environments affect the southernmost countries.

Tectonic plate movements gave rise to the Andes Mountains and the complex pattern of mountain ranges and islands in Middle America and the West Indies. Brazil is dominated by the Guiana and Brazilian Highlands plateaus, which are separated from each other and the Andes by the Orinoco, Amazon, and Paraná-Paraguay River systems.

Latin America has resources of metal ores, petroleum, water, good soils, and major forests. It is subject to earthquakes, volcanic eruptions, hurricanes in the north, and storms in southern Chile. Human activities have degraded water, air, soil, and forest resources.

10B.1 How do the natural environments of Latin America compare with those of Anglo America?

10B.2 What are the major mineral products of Latin America, and how are they distributed geographically?

**Key Terms:**

El Niño phenomenon

**Key Places:**

Atacama Desert
Patagonia
Tehuantepec isthmus
Andes Mountains
altiplano
Brazilian Highlands
Guiana Highlands
Patagonia Plateau
Orinoco River
Amazon River
Paraná-Paraguay River
pampas

remaining forest. They counter international plans to stop cutting by stating that less than 10 percent of the Amazon forest has been removed, whereas most forest in Anglo America and Europe was cut and replaced by agriculture centuries ago. After 1988, which saw the highest rate of forest cutting in Brazil to date, the amount cut was reduced in the early 1990s. Whether this was because of international pressures, government actions to make forest removal less attractive to settlers and landowners, the lack of success by previous rain forest colonists, or the rival attractions of the Campo Cerrado soybean investments, it is difficult to say (see "World Issue: Tropical Rain Forest," pages 480–481). It appears that greater efforts are being made within Brazil to educate the population in the wisdom of considered resource exploitation and to try to ensure future sustainable development of the Amazon region. Fires on the Venezuelan border in early 1998 resulted in a new annual record of forest acreage destroyed.

Tropical rain forest has already been cut over much of eastern Central America, although some of the more remote regions have been left untouched. The small country of El Salvador has the highest density of population in Middle America and nearly all the forest has been removed. Where cutting is intense, some countries such as Costa Rica and Panama designated national parks and forest reserves. Most areas are now under secondary forest after one or more phases of cutting and regrowth. In Honduras, the secondary growth of pine trees gave rise to sawmill activity that threatens to remove the newer trees.

## MEXICO

Mexico's population was almost 100 million in 1998, and it had a growing and diversifying economy. It has a close relationship with the United States and Canada through NAFTA. **Mexico** is thus considered separately from the other countries occupying the land between the United States of America and South America, and sometimes grouped as **Middle America**.

Mexico was the core cultural area of Middle America in the time of the Mayan and Aztec empires and was the center of New Spain. After the breakup of the Spanish Empire, it remained a more coherent and stable political unit than the countries of Central America. The distinctive regions within Mexico might have been countries themselves if the various parts had opted for independence like the Central American countries in the early 1800s. Despite an outward appearance of stability, Mexico contains many groups that were favored or repressed by successive governments. Tensions among such groups came to a head in the 1990s as economic conditions made single-party central government control more difficult.

Up to 1994, Mexico was regarded as a model of economic restructuring. After serious debt problems in the early 1980s, the Mexican government agreed to allow foreign competitors into its markets by dropping tariff levels, cutting inflation and government deficits, privatizing telecommunications, banks, and agriculture, and reducing a range of central government controls. Mexico put forward the idea of the North American Free Trade Agreement (NAFTA, see Chapter 9), which was signed with Canada and the United States and implemented at the start of 1994. Many Mexicans thought they were close to becoming part of the world's core of rich countries.

A new approach to land tenure tried to deal with rural poverty. The *ejido* system, instituted in 1917, enabled any peasant to ask for land of his own. It resulted in the fragmentation of landholdings. Peasants trapped on small plots of land farmed them more intensively but often without care, leading to declining soil quality and yields of their food crops, corn and beans. In the 1990s, *ejido* was scrapped and rights to individual property ownership were defined clearly, enabling the sale and concentration of land ownership. Foreign agribusiness corporations showed interest with a view to producing on larger farm units high-value cash crops for export.

It now appears that drastic economic and land-tenure measures were forced through too rapidly. The increased demand for imports—which became cheaper with the lower tariffs—caused a rapid rise in Mexico's overseas trade deficit. Foreign investors lost confidence and began to withdraw their capital. In 1993, Mexico attracted $75 billion of foreign investment, mainly from the United States and Japan. In 1994, it attracted $60 billion, but in 1995, the net flow was just over $1 billion. At the end of 1994, Mexico had to devalue its currency. By the late 1990s, it appeared this was a temporary loss of confidence, since the United States backed further reform measures and the devaluation made Mexican goods more attractive in overseas markets.

Other factors of Mexican social, political, and economic conditions, however, may also repel foreign investments. While a few people increased their riches, the poor gained little from the changes. Larger companies were privatized to the advantage of their new owners but shed labor to become more productive. The many smaller companies in Mexico hardly began to make the changes necessary for them to compete in the world economy.

The 1994 challenge mounted by the Zapatist National Liberation Army in the southern province of Chiapas drew attention to the social costs of economic reform. The Zapatistas made the point that the Native Americans, who comprise one-third of the Chiapas population and were already very poor, were further disadvantaged by the economic restructuring and NAFTA. They called it a "death sentence." The Native Americans in Chiapas are only a part of the problem, since 13.5 million Mexicans live in "extreme poverty" and a further 23.6 million are "poor." At the end of the 1990s, the Chiapas situation remained unsolved. Reforms promised in 1996 bought time for the government but were not implemented. Zapatistas set up their own autonomous municipalities in defiance of government orders. Mexico keeps one-third of its military forces in the region but does little to stimulate dialogue. Meanwhile, "revolutionary tourists" from other countries support and perpetuate the Zapatist revolution.

It was also difficult for many Mexicans to adjust to the new norms after decades of state control, protection, and provision. The three areas where Mexico needed to make rapid changes to compete in world markets—to increase productivity throughout the economy, decrease the budget deficit, and maintain civil peace—all confronted contrary forces that slowed progress and pointed to the long-term nature of Mexico's problems.

## Regions of Mexico

Mexico's variety of natural environments produces internal diversity (Figure 10.15). Its southern parts receive adequate rainfall, but the northern region is increasingly arid toward

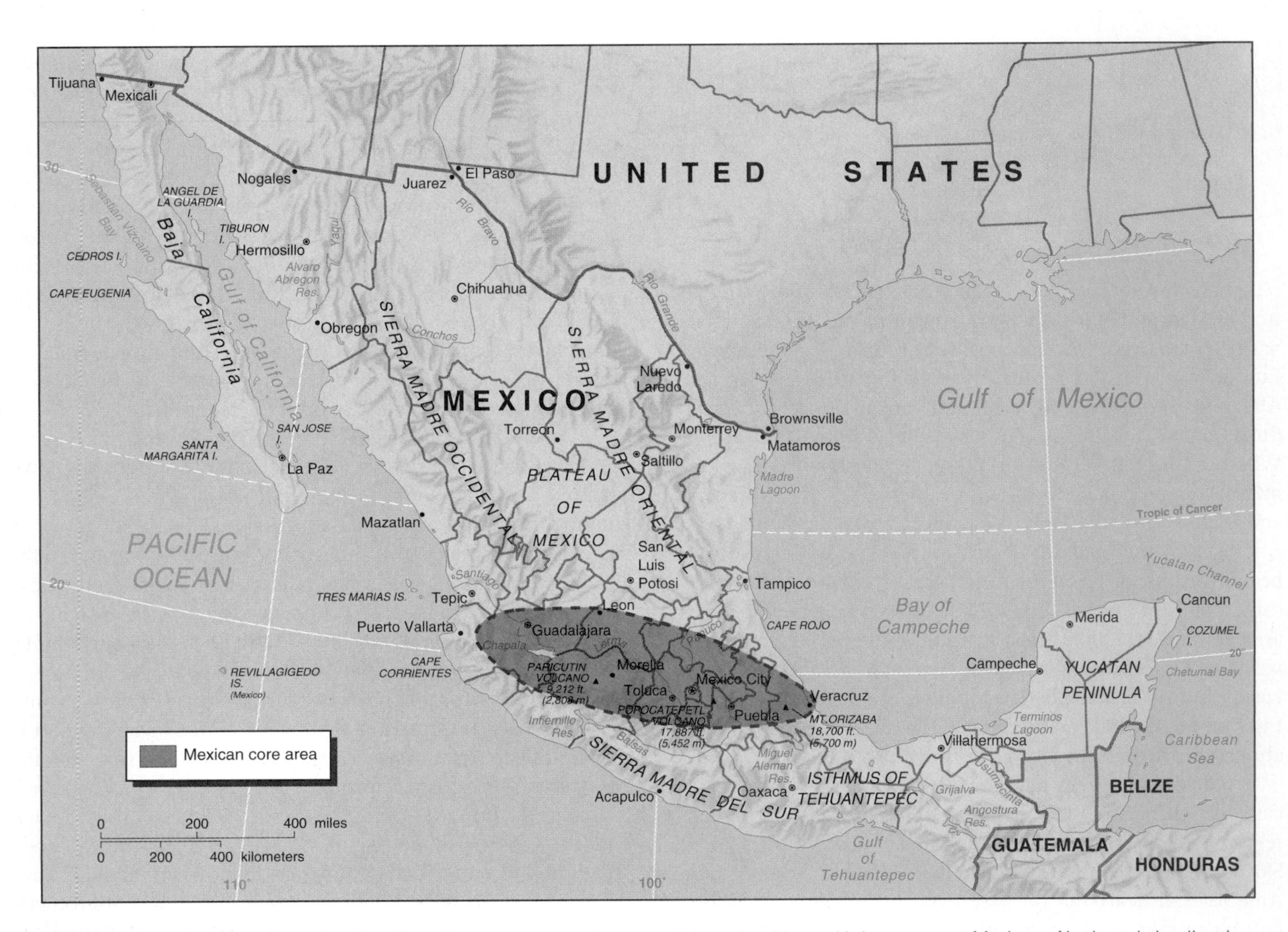


**FIGURE 10.15 Mexico: a diversity of regions.** The core area contains most of the major cities and is home to most Mexicans. Northward, the climatic environments are drier, but economic opportunities increase close to the U.S. border. Southward, the land narrows and localities become increasingly remote from the core.

**FIGURE 10.16 Latin America: distribution of population.** Explain the distribution of heavily and lightly populated parts of this region.

Source: Data from *New Oxford School Atlas*, p. 99 & p. 113, Oxford University Press, UK, 1990.

the U.S. border. It is a country of mountain ranges and basin-shaped lowlands.

The north was settled by ranchers and miners as part of the colonial-period expansion of empire that took Spanish soldiers and Roman Catholic priests from the Mexico City area into California, New Mexico, and Texas from the 1500s to the early 1800s. More recently, northern Mexico became the site for border town industries (see Figure 9.22), while the western coastlands have fishing and commercial agriculture.

Central Mexico is the core region for population and economic activity, centered on Mexico City. Urbanization and industrialization extend westward to Guadalajara and eastward to Veracruz. The east coast around the Gulf of Mexico, for long poorly settled, became a region of oil and natural gas production and tourist opportunity.

The southern mountains and the border regions of Chiapas and Yucatán are the remotest parts of the country. They experienced slow economic development until the Zapatist uprising.

## People

The Mexican population is concentrated in the central core region and around some northern towns—including those along the U.S. border (Figure 10.16). The areas with fewest people include the arid northwest and much of the Yucatán peninsula.

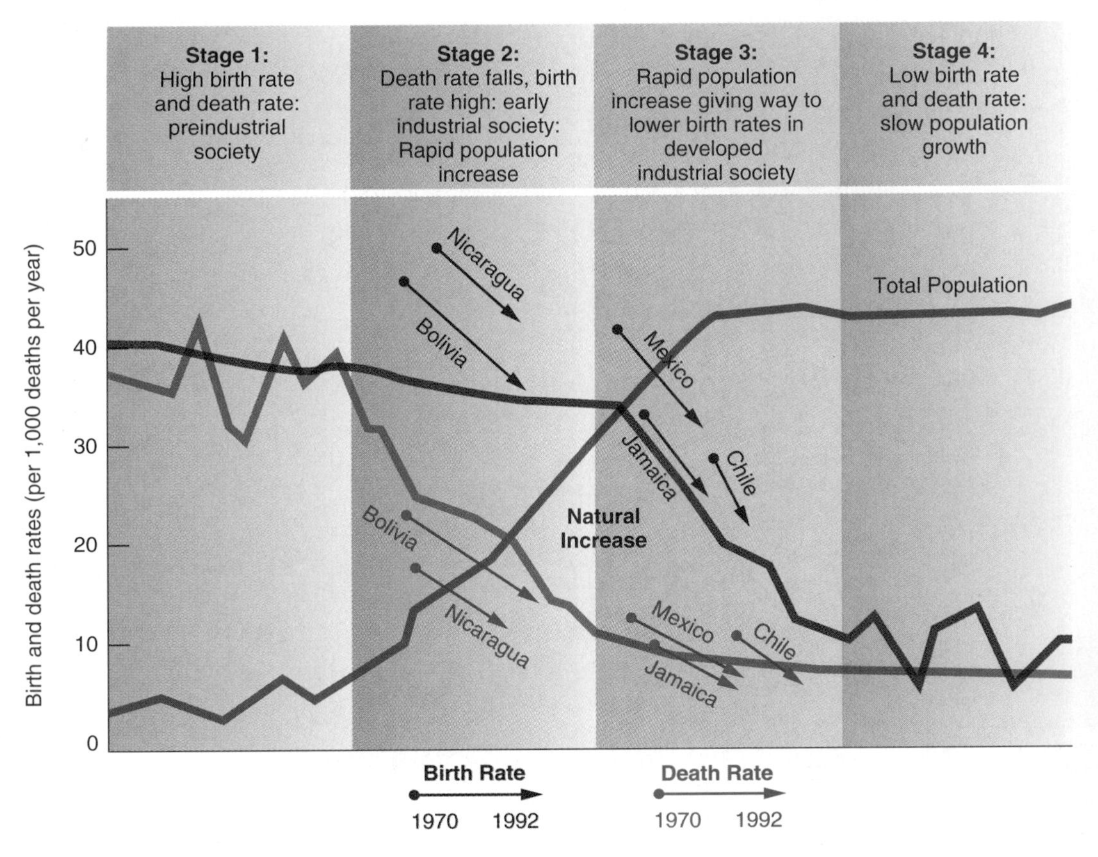


**FIGURE 10.17 Latin America: demographic transition.** How does the process vary among Latin American countries?

Mexico's population grew rapidly to around 70 million in 1980 and 98 million in 1998. Although its fertility rate and birth rate are declining (Figure 10.17), Mexico's population is projected to rise to nearly 140 million by 2025. The age-sex graph (Figure 10.18) shows a decline of births that was achieved through a well-developed family planning program, halving total fertility rates from about 6 in 1976 to about 3 in 1998. The very low death rate and increasing life expectancy, however, maintain a gap between births and deaths that keeps the population totals increasing.

Mexico's population is highly urban. From 1970 to 1998, the proportion of the Mexican population in towns increased from 59 to 75 percent. **Mexico City** grew from a population of 500,000 in 1900 (2.5% of Mexico's population) to over 17 million by 1996 (18% of the total, making it a primate city). Other "million cities" include **Guadalajara**, **Monterrey** (Figure 10.19), and **Puebla**. In Mexico, the stimulus for movements of people to cities is the increasing growth of urban-based manufacturing and service jobs—although the official rates of unemployment remain high and many of the jobs are in the informal sector.

Mexico City is by far the largest urban conglomeration in Middle America. Its site was a major population center before the arrival of the Spaniards, having been the Aztec administrative headquarters. It continued to be the focus of road and rail networks within New Spain and independent Mexico. It is the national capital and media center, and has three-fourths of Mexican manufacturing industry and nearly all of its commercial and financial establishments. From around 1950, some 700,000 people per year moved from rural areas in Mexico to the Mexico City area. The housing supply by government and private constructors could not provide accommodation for almost one-third of the rapidly growing city's population. Squatter settlements mushroomed, including the satellite city of Nezahualcoyatl that now has over a million people living on a former dried lake bed that floods after heavy rain. As job markets also failed to cope with this influx, the informal sector of the economy, in common with other developing countries, grew from around 4 percent to 26 percent of people of working age from 1980 to the 1990s.

Mexico's population is dominated by Native Americans (around 30%), Europeans (9%), and people of mixed Native American and European blood known as **mestizos** (60%). The Native Americans are particularly numerous in the southern province of Chiapas, where language dialects of Mayan origin are commonly spoken in preference to Spanish, the official language. Almost 90 percent of Mexicans are nominally Roman Catholic, although many personal links to

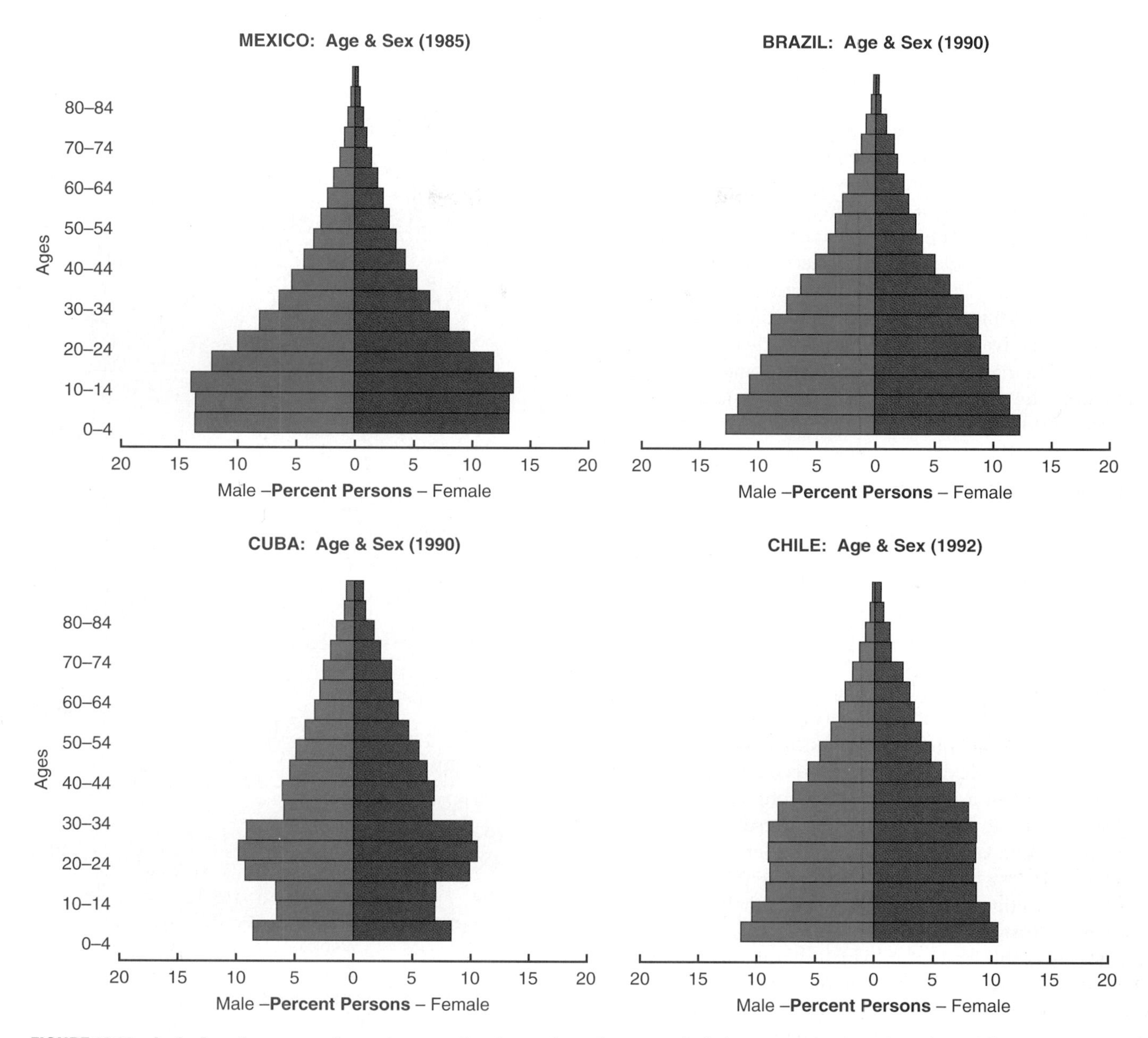


**FIGURE 10.18 Latin America: age and sex structure.** How far are these diagrams typical of poor countries throughout the world?

**FIGURE 10.19 Mexico: urban landscape.** The Gran Plaza in the city center of Monterrey, northern Mexico. The older government and cathedral buildings on the left contrast with the newer ones on the right and the multilevel development in the foreground.

that church are becoming broken as established lifestyles are disrupted by moves to urban centers.

The combination of Mayan or Aztec ancestry, Spanish influences, and, more recently, U.S. commercial and technological links led to a distinctive Mexican culture that diffused into surrounding areas in the southwestern United States and Central America. Mexican cities are built in architectural styles that reflect ancient pyramid shapes and Spanish church architecture, and include murals. Other nationalistic trends are expressed in music, films, and literature. Mexican food, based on corn tortillas, local vegetables, and chilies, is one of the most distinctive national cuisines. Mexican politics still reflect the dominating colonial influences and subsequent dictatorships in the virtually single-party system that was maintained into the 1990s.

## Economic Development

Mexico is by far the most economically developed of the countries of Middle America and continues to outpace the other countries. In the late 1990s, Mexico had a total GDP that was over 80 percent of the total for Middle America. Its GDP per head is one of the highest in Latin America and near the top of the World Bank's Upper-Middle income group. Ownership of consumer goods was well above that in the countries of Central America apart from Costa Rica (Figure 10.20). Mexico's HDI and GDI are much higher in rank than Brazil but not as high as Argentina's. Mexico's size and its economic and human development give it a leadership role among Latin American countries, similar to Brazil's, in foreign investors' perceptions of the region. The late 1994 financial disaster in Mexico had the effect of slowing investments elsewhere in the region.

Mexico has a diversified economy. Before the Mexican Revolution in the early 1900s, most of the country was covered by large estates, or haciendas, on which wheat was grown and livestock were raised for sale, while the peasants grew maize. During most of the 1900s, there was massive and rapid urbanization, but also some redistribution of land and irrigation of the farmland. Farming became commercial with the growing urban market at hand, while foreign companies built packing plants for canned and frozen vegetables. It remains important, employing one-fourth of the labor force. There is a contrast between the more arid north with its livestock grazing, dry farming of grain crops, and some irrigated areas, and the humid south with rainfed crops (Figure 10.21).

Manufacturing became a major source of employment together with increasing numbers of jobs in services and government. Tourism is a major and growing source of income for Mexico, with 21 million visitors in 1996. Mexico City, the Pacific coastal resorts such as Acapulco, and the Mayan temple and coastal areas of the east are major attractions. Variations in the development of manufacturing define economic regions within Mexico.

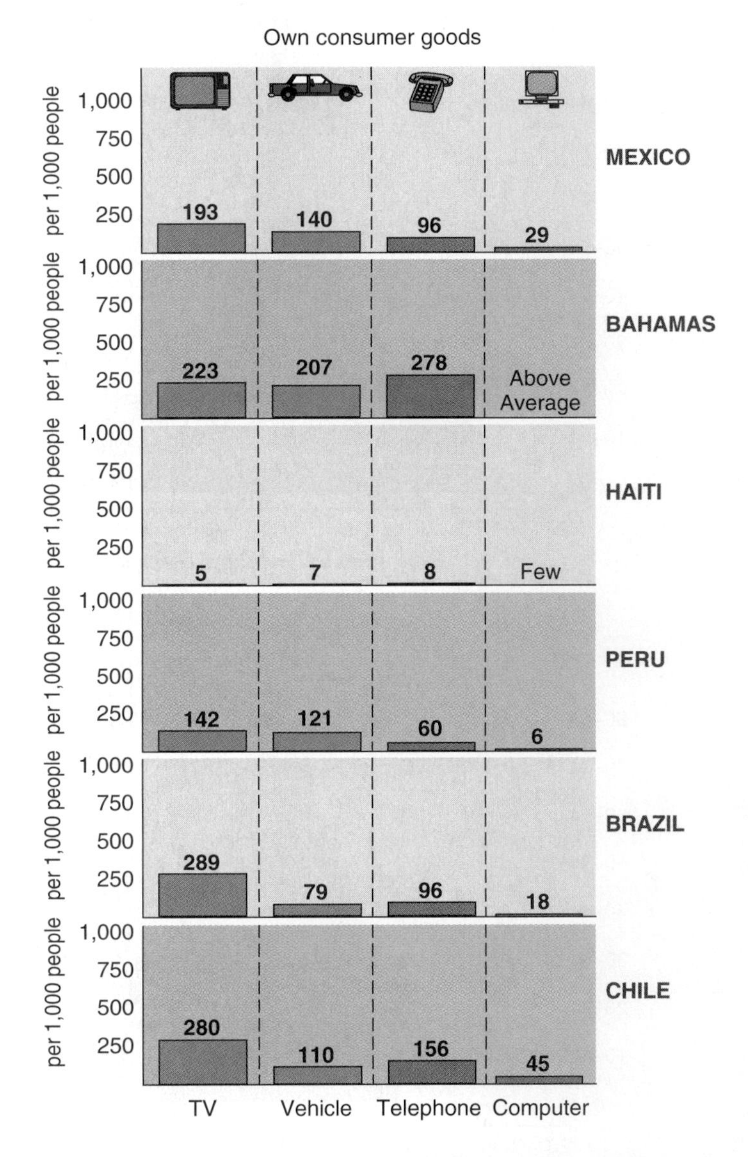


**FIGURE 10.20 Latin America: ownership of consumer goods.** Contrast conditions in rich, moderately rich, and very poor countries.

Source: Data (for 1996) from *World Development Indicators,* World Bank, 1998.

- Nearly two-thirds of the Mexican population live in the urban region focused on Mexico City that includes Guadalajara, Veracruz, and León. This core of the country, where most of its economic activity is centered, is a megalopolis type of urban development like that in the northeastern United States, Japan, and parts of western Europe.
- Northern Mexico became the main growth area in Mexico's economy for new types of industrial development. It was the second area of expansion after the area around Mexico City for the Spanish Empire in the 1600s, an area inhabited by few Native Americans. Large estates and mining towns such as Chihuahua and Monterrey developed along routes following the edges of the eastern and western Sierras. Such towns began to industrialize after 1900, with Monterrey establishing an iron and steel industry.

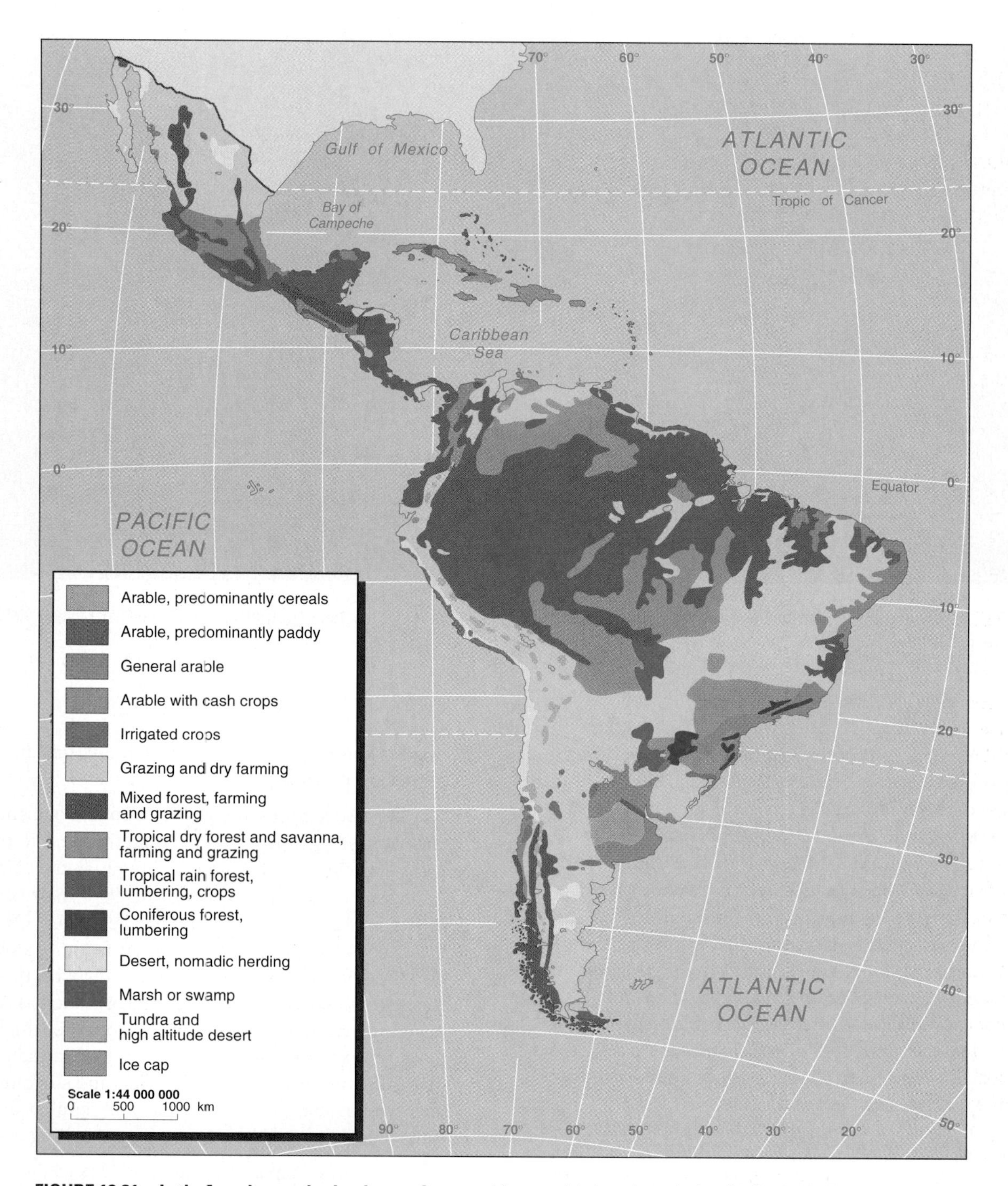


**FIGURE 10.21 Latin America: major land uses.** Compare this map with that of population distribution in Figure 10.16. Are there any links? What factors are missing to explain the distribution of population? Compare it also with the climate map (Figure 10.6).

Source: Data from *New Oxford School Atlas,* p. 98 & p. 112, Oxford University Press, UK, 1990.

More recently, the towns along the U.S. border, from Tijuana in the west to Matamorros in the east, experienced very rapid growth. After 1965, the **maquiladora program** made it possible for factories sited in Mexico to import components without customs duties. Such factories used cheap local labor to assemble goods that could be exported to the United States. U.S. corporations involved in textiles, apparel, electronics, and wood products industries built their factories in these towns, where there was rapid growth in the late 1980s. Asian and European corporations joined them. The concentration of so much economic activity, however, brought increased pollution of air and especially of water. Both Mexico and the United States are trying to implement programs to reduce industrial pollution of the Rio Grande along the border between the two countries.

**FIGURE 10.22 Mexico: rural landscape.** Recently cut clear of the enveloping tropical forest is the former Mayan temple at Palenque, Chiapas state. The large pyramid on the right is the Temple of Inscriptions; the building with the tower was a palace.

© Paul Bradshaw

The maquiladora program is now available in centers farther south in Mexico, and there has been a move toward complete deregulation of foreign involvement in manufacturing industries. Auto assembly and engine manufacturing by Ford, GM, and Nissan, together with IBM, Whirlpool, Kodak, and Caterpillar factories, are located in the growing high-tech manufacturing region of northern Mexico that is developing even stronger links with California under NAFTA. Furthermore, the irrigated northern coastal areas became the main center of modern agriculture, producing winter vegetables for American markets, together with cotton, sugarcane, and rice. Fishing off the Pacific coast makes Mexico the third fish producer in Latin America after Peru and Chile.

- While diversified urban-industrial economies develop in Mexico's center and north, the east coasts facing the Gulf of Mexico supply much of Mexico's wealth from large oil and natural gas fields. Although discovered early in the 1900s, this wealth was not effectively tapped until the 1970s, when oil went from 2 to 80 percent of Mexico's foreign exchange earnings. During the high oil-price period of the late 1970s, Mexico borrowed heavily, but falling prices and demand in the 1980s left it deeply in debt. Oil production had a major impact on the economy of the coastal area between Tampico and Campeche, and recent finds extended this affected area inland. During the 1980s, new refineries and petrochemical industries were built. In the late 1990s, oil formed under 40 percent of exports as more was used by growing industries within Mexico and the products of those industries made up a higher proportion of exports.
- Southern Mexico resembles the countries of Central America, being less developed than the areas to the north. Most of the region is remote and populated by Native American subsistence farmers who grow maize on the hillsides and some wheat on the valley floors. Overgrazing by sheep is common. Poor transport facilities slow development in this region. Contrasting with this poverty, the government-planned tourist resorts such as Acapulco bring American tourists to the coast and new highway links connect the resorts to airports. Unfortunately, sewage disposal from the hastily built hotels and surrounding shantytowns has polluted the bay at Acapulco, and a 1997 hurricane damaged the city's poorest dwelling areas.
- The states of Chiapas, Yucatan, and Oaxaca lie along the southernmost borders of Mexico. Traditional Native American culture is least disturbed in these states, which scarcely feel a part of Mexico. Over a million Native Americans in Chiapas do not speak Spanish. The Yucatán peninsula is seeing development along its northern coast, where tourism facilities attract those interested in the archaeology of former Mayan settlements, and there is pressure for much of the area to be declared a national park (Figure 10.22). The far south remains very remote, with few people or commercial activities. In recent years, government investment in Chiapas resulted in the irrigation of farmland from the Angostura Dam waters that occupy 90 km (60 mi.) of a once-farmed valley. Chiapas

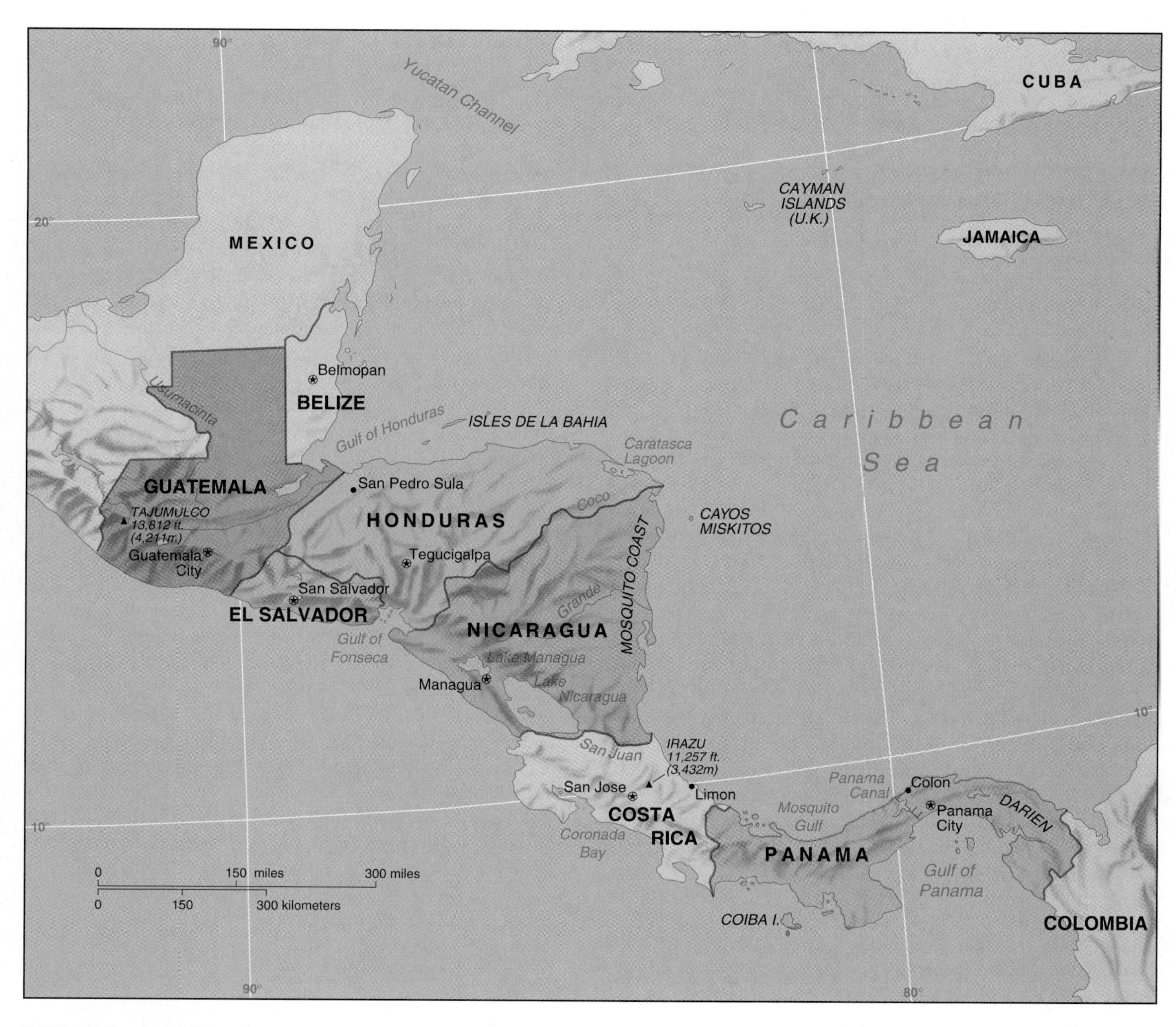


**FIGURE 10.23 Central America: major geographic features, the countries and towns.**

also houses thousands of refugees who fled from civil disturbances in Guatemala and El Salvador. The combination of dispossessed refugees and local disillusionment with the Mexican treatment of these states led to a guerrilla war in the region in 1994 that continued through the 1990s.

## CENTRAL AMERICA

**Central America** is the land area that stretches between Mexico and South America, narrowing southward. It comprises the countries of **Belize**, **Costa Rica**, **El Salvador**, **Guatemala**, **Honduras**, **Nicaragua**, and **Panama** (Figure 10.23). They have a common Hispanic culture and geographic proximity. Although Belize was a British colony, linked to Jamaica from the late 1800s until 1981, and its people include a higher proportion of African Americans, the mainland location, Spanish language, and Roman Catholicism link Belize physically and culturally with Central America rather than with the West Indies.

These seven countries had a total population of 35 million people in 1998 and are among the poorest countries in the Americas, being dependent on external markets for a few, mainly agricultural, products. They were termed **banana republics** because of their economic dependence on one or two export crops, among which bananas figure strongly. Such dependence was linked to the drawbacks of continuing dictatorships and political disruption that most of these countries experience, slowing economic growth. Central America, although at first sight a land bridge between North and South America, acts as a barrier to communications because of the human poverty and physical environment, particularly the narrowing land and almost impenetrable conditions between Panama City and the Colombian border. In the late 1990s, further setbacks to these countries included widespread forest fires linked to an extended period of drought that destroyed crops and the 1998 devastation following Hurricane Mitch.

In 1960, the Central American countries attempted to promote common interests by establishing the **Central American Common Market** (CACM). Almost as soon as it began to boost intraregional trade, however, political instability and civil wars caused the arrangement to decline in significance. It was relaunched in 1993 on a new basis, but the three northern countries (Guatemala, Honduras, and El Salvador) had most to gain and alternative developments involving all the Latin American countries may mute its influence again.

After the failure of other forms of cooperation, Central American countries agreed in 1997 to initiate new forms of common policies for small-scale reforms that boost economic productivity. Helped by internal peace after Guatemalan rebels ended their feud with the government in 1996, economies began to improve. Intraregional trade, previously hampered by many problems, grew following El Salvador's liberalization of its own trade in 1995. Progress is held back by the slow pace of reform, increasing crime among ex-guerrilla fighters, returned refugees, rising numbers of unemployed young people, and natural hazard disasters.

## Countries

The countries of Central America have always been peripheral to a Mexican core. The area was affected little by the Aztec or Mayan civilizations and only partially settled by Spaniards. While the Mexican states voted to stay together at independence from Spain, initial alliances among the Central American countries broke down by 1840.

The four largest countries to the south of Mexico—Guatemala, El Salvador, Honduras, and Nicaragua—had 1998 populations ranging from 5 million (Nicaragua) to 12 million (Guatemala). They all relied on agricultural exports, especially coffee (49 percent of Guatemalan exports, 72 percent of El Salvadoran, 39 percent of Honduran, the poorest country, and 50 percent of Nicaraguan). Apart from Honduras, where land hunger is a growing threat to political stability, the other three of these larger countries were devastated by civil unrest in recent years. The causes of this unrest stem from the unequal division of land by which a few percent of the population own three-fourths or more of the land.

In Guatemala, the Native Americans, who make up half the population and continue to agitate for better conditions, were repressed fiercely by the government. In 1980 to 1981, 150,000 fled into Mexico as 440 of their villages were burned down. El Salvador is ravaged by frequent earthquakes, but greater devastation was caused in the 1980s by attempts to redistribute land. It led to civil war when the former landowners challenged changes imposed by a reformist government. Many people moved into the capital, San Salvador, disrupting the growing economy of that city. In Nicaragua, the Sandinista revolution took over government in 1979, leading to a U.S. trade embargo that caused economic disruption. The 1990s saw a calming of these varied types of civil strife, although it remains difficult for the countries to attract external investment because of their recent histories.

Of the three smaller countries of Central America, Belize has just over 200,000 people and became independent in 1981 after being a British colony. Until 1991, its land was claimed by Guatemala because of possible offshore oil deposits. Costa Rica, with 3 million people, is the only country in the region that has had a long-term democratic government. This may be related to the legacy of its initial colonial settlement, when there were few Native Americans to be dominated and the people of European origin took up medium-sized farms. As a stable democracy, Costa Rica attracted manufacturing and free trade zones financed by U.S. and Taiwanese corporations. Panama became a separate country when its territory was carved out of Colombia in 1903, preparatory to the cutting of the Panama Canal. The construction of the canal brought in people from the Caribbean islands, Asia, and southern Europe. Panama City retains a cosmopolitan population and an international trading and financial role. The United States virtually ran the country until 1979 and then invaded again in 1989, but this caused less disruption than that experienced by civil wars elsewhere in Central America. Outside Panama City, the landscapes and economy of Panama are similar to those of other Central American countries.

## People

In Central America, the main concentrations of people are in and around the largest cities—often in the highlands and closer to the west coast (see Figure 10.16). Eastern coastal areas and parts of the narrow Isthmus of Panama have fewer people where almost impenetrable forest and wetland remain.

The population of Central America is projected to grow by 60 percent in the next 30 years. By 2025, the 1998 population of 35 million may become 56 million people. Such growth will place increasing stress on the already pressured natural environment and the political, economic, and social systems of these countries.

Growth is built into the population by very high fertility rates of the 1960s and 1970s—over 6 children per woman apart from Costa Rica (fewer than 5). Although by 1998 the fertility rates had dropped below 4 except in Guatemala, Honduras, and Nicaragua, while those in Costa Rica and Panama were under 3, the rates were still high compared to

**FIGURE 10.24 Central America: the people.** Native Americans, with their distinctive clothes, make up a large proportion of the people at this market in western Guatemala. These people are descendants of the Mayans. The women wear the traditional "huipil" dress. The designs signify their home villages.



the countries of Europe or Anglo America. In 1998, rates of natural increase remained around 3 percent; only Costa Rica and Panama were below 2 percent (see Figure 10.17). The age-sex patterns for Central American countries are similar to those of other poor countries throughout the world with a growing young population and few in the oldest groups.

Urbanization is lower in Central America than in Mexico, remaining under 50 percent between 1970 and 1995, but Nicaragua went from 47 to 63 percent. In many Central American countries, the civil strife was mainly a rural matter and people moved to towns from fear of living in the countryside. Most countries have a single major, or primate, city with populations that are several times that of the second city. The largest cities are **San Salvador** (El Salvador), **Guatemala City** (Guatemala), **Tegucigalpa** (Honduras), **Managua** (Nicaragua), and **Panama City** (Panama).

The populations of Central American countries comprise racial and ethnic groups that became social classes. The Native Americans, who remain in large numbers in western Guatemala (Figure 10.24) and in smaller numbers in other countries, have been regarded by other inhabitants as a group that should remain outside the processes of economic modernization. They maintained subsistence practices in farming and have their own network of market towns that are often unconnected to the world economic system.

The Europeans and mestizos provide the political and economic leadership. For many years, economic leadership and political power were based on holding large areas of land in cattle ranches, but during the later 1900s they passed to the owners of manufacturing industries and financial and legal services in the modern economy—and sometimes to the military. Smaller groups of people within Central American countries include some of African origin who came after escaping from slavery in the Caribbean, central and southern Europeans who migrated to Central America between or after the two world wars, and Asians who came to work on construction projects such as the Panama Canal or, more recently, to establish factories in Costa Rica or Panama.

## Economic Development

While Mexico has a strongly diversifying economy, the economies of Central America have a narrow base, resulting in slower economic and human development. In Central America, the major difference in economic geography is between the more populous centers of Native American and Hispanic culture in the uplands on the one hand and, on the other, the less populated Caribbean coastal areas with their American-owned plantations growing bananas, sugarcane, and other fruits.

**FIGURE 10.25 Central America: Pacific coast.** An eastward-looking space shuttle view of Guatemala and southern Mexico. The cities are in the highlands, above the narrow coastlands.

NASA

- *Guatemala* is a country of three parts. First, in the uplands northwest of Guatemala City, the largest Central American concentration of Native American peoples live in subsistence conditions, often landless and in extreme poverty. An uprising in the early 1980s resulted in government repression and destruction of many of the poor facilities that existed. Second, there are the people in the southern part of the country who adopted European ways. This area's commercial agriculture is based on coffee on the Pacific slopes and bananas and sugarcane near the Caribbean (Figure 10.25). Guatemala City has manufacturing industries, but recent influxes of people swamped the job markets and half of the working-age population is unemployed. Third, the lowland Peten area of the north is forested and sparsely inhabited, although the Mayan archeologic relics attract tourists. Since the 1960s, up to 40 percent of the forest was logged or burned. The area is known to have a major oil field at depth, but guerrilla warfare prevented its exploitation.
- *El Salvador* has the densest population in the subregion. Most of the people are concentrated in the western highlands, where the dominant crop, coffee, is grown. Light manufacturing industries were established from the 1960s. Civil war over land redistribution since the 1980s disrupted both rural and urban economies.
- *Honduras* was for long the poorest country, but Nicaragua now has that unenviable role. Barely 25 percent of its land can be used for farming. The ranching economy established in colonial times still persists in the west, with

**FIGURE 10.26 Central America: Panama Canal.** The Miraflores locks with a cruise ship entering to be raised to the level for traveling through the canal. Note the land uses on either side of the canal. These locks are now too narrow for the world's largest ships.



banana plantations in the eastern valleys. Diseases affected the banana plantations from the 1930s, and coffee became the main export in the 1960s. Shrimp farming on the Pacific Ocean coast is a new export specialty that is finding markets in the United States. Much of the former tropical rain forest cover was logged, and even the secondary growth of pines is being removed rapidly. Although manufacturing was encouraged after 1950, factories cater only for local needs and are concentrated around San Pedro Sula.

- *Nicaragua* developed a mixture of agriculture and modern manufacturing industries. The trade embargo that followed the Sandinista revolution in 1979 devastated the economy, giving it the lowest GDP in the subregion in the late 1990s. Coffee, cotton, bananas, and sugarcane are important export crops, but the country suffered food shortages from the 1980s. The population is concentrated around the lakes of Nicaragua and Managua in the structural depression that is subject to earthquakes. Other parts of the country have poor transportation links to this core.
- *Costa Rica* has a more prosperous and diversified farming industry than the other countries based mainly on coffee production along the Pacific slopes, bananas and cacao on the eastern coast, and a range of new crops such as tropical fruits, ornamental plants, cut flowers, and tropical nuts. Areas of former rain forest are used for livestock production. Costa Rica has little civil strife. It does not have an army, but the militarized police and better quality of life help to maintain internal peace. The stable government of the country, together with better educational and health care provision than in other Central American countries, made it possible to attract a range of manufacturing industries to free trade zones around the capital, San José. These include textiles and pharmaceuticals for export and cement, tires, and car assembly for the domestic market. Costa Rica attracts tourists to its beaches and rain forest, although the forest resources are much depleted.
- *Panama* has economic contrasts between the Canal Zone, especially Panama City, and the rural areas on either side. The Canal Zone is a cosmopolitan world trading area, where the banking and financial services create twice the GDP of agriculture. In 1991, Taiwanese investors established a free trade zone in Panama City in which some 8,000 people are employed in clothing, furniture, toy, and bike factories. The products are sold in the U.S. market. In the rural districts to the west, coffee is grown on uplands and rice is produced with palm oil along the coasts. To the east of the Canal Zone, the Central American Highway tapers off in impenetrable Darien—the narrow isthmus of land between the Canal and the Colombian border that prevents completion of effective north-south highway links. Panama is dealing with the problem of an aging canal that is now too small for many ships (Figure 10.26). Work began in 1991 to widen and lengthen the main locks, and a pipeline now conveys oil across the isthmus to the west of the Canal Zone. Although the United States' Howard Air Force Base at the Pacific end of the canal is due to be vacated at the end of 1999, it is most likely that its facilities will continue to be used in some fashion to help fight the drug war.

**RECAP 10C: Mexico and Central America**

Mexico is one of the wealthiest countries in Latin America, having a diverse economy, but it has problems adjusting to increased involvement in the world markets and the NAFTA. Mexico City is one of the world's largest metropolitan areas. The contrasts of internal development among regions and groups of people within Mexico cause social tensions and fluctuating patterns of economic growth that threaten to slow the process of human development.

Central America comprises several small countries–Belize, Costa Rica, El Salvador, Guatemala, Honduras, Nicaragua, and Panama. Their peoples, with the exceptions of Costa Rica and Panama, are among the poorest in Latin America. Central America's environment is of mountain ranges and swampy tropical coasts. Its countries produce coffee, bananas, and a few other crops for fluctuating world markets, but many economies have been affected badly by civil wars.

10C.1 How do the main features of the population and economy of Mexico compare with those of the Central American countries?

10C.2 What factors affected the setting of boundaries between countries in Middle America?

10C.3 Do Mexico and the countries of Central America count as the "near abroad" of the United States? (See Chapter 8 for the usage of this term with reference to Russia.)

**Key Terms:**

| | | |
|---|---|---|
| maquiladora program | banana republic | Central American Common Market (CACM) |

**Key Places:**

| | | |
|---|---|---|
| Mexico | Belize | Honduras |
| *Mexico City* | Costa Rica | *Tegucigalpa* |
| *Guadalajara* | El Salvador | Nicaragua |
| *Monterrey* | *San Salvador* | *Managua* |
| *Puebla* | Guatemala | Panama |
| Middle America | *Guatemala City* | *Panama City* |
| Central America | | |

## WEST INDIES

The **West Indies** are a group of former European colonies mainly on islands located around the Caribbean Sea (Figure 10.27). In 1998, almost 35 million people lived in the West Indies, but many of the islands had only a few thousand inhabitants. Over 80 percent of the people lived on the four largest islands of the **Greater Antilles—Cuba, Haiti** and the **Dominican Republic** (both on the island of Hispaniola), **Puerto Rico**, and **Jamaica**.

As well as the Greater Antilles, the West Indies include other groups of countries. The **Lesser Antilles** is an arc of small islands around the eastern edge of the Caribbean Sea, with the Leeward group in the north (Virgin Islands to Guadeloupe) and the Windward group in the south (Dominica to Grenada). It includes several island countries such as Antigua and Barbuda, Barbados, and St. Lucia. **Bermuda** is a group of islands 1,000 km (625 mi.) off the South Atlantic coast of the United States; in 1995, its people voted to remain a British colony. The **Bahamas** include some 700 islands lying east of Florida. Another group of islands, of which the largest form **Trinidad and Tobago**, lies along the northern shore of South America. Also included in this subregion are the three Guianas—**Guyana, Suriname**, and **French Guiana**. Although situated on the South American mainland, their histories, peoples, and present economies are more akin to those of the West Indies.

Located between North and South America, the West Indies had a strategic location from the time of colonization. In the 1500s, they were on the route for Spanish treasure ships, and from the later 1600s, they became a major source of wealth for European colonial countries by growing sugarcane. The opening of the Panama Canal in 1914 placed these countries on a major world routeway. The involvement of the United States in the Cuban crisis of 1962, plus its occupations of the Dominican Republic in 1965, Grenada in 1983, and Haiti in 1994 highlighted the sensitivity of Americans to events in this region close to southern U.S. coasts. The continuing presence of U.S. military bases, including Guantánamo Bay in Cuba, and concerns over drug traffic through the region emphasize the continuing geopolitical significance of the West Indies.

The need for economic integration among the small West Indian countries is clear, but progress has not been easy or impressive. The West Indian countries compete with each other in similar markets and have strong rivalries; there is little internal trade, and they are at different levels of economic development.

The **Caribbean Community and Common Market** (CARICOM) was formed in 1973, taking over from previous attempts at free trade and comprising 13 former British colonies with a total of 5.5 million people. Jamaica and Trinidad are the most powerful members and gain most from the limited degree of integration in the creation of a local food market. The **Caribbean Basin Initiative** of 1984 was set up partly to promote economic growth in the West Indies and partly to protect the United States' interests. It provided special entry to selected U.S. markets but did not assist cooperation among West Indian countries. In 1994, the NAFTA agreement among Canada, Mexico, and the United States fueled fears that future U.S. investment would be directed away from the West Indies and toward Mexico.

In August 1995, the **Association of Caribbean States** was formed by the West Indian countries, Mexico, the Central American countries, Colombia, and Venezuela. It was proposed by the CARICOM group of former British Caribbean island countries and established the fourth largest trading bloc in the world. Its formation was motivated by local issues, and many saw it as a step toward the realization of the Free Trade Area of the Americas. Local issues affecting the new agreement included worries over the impacts of shifting world trade patterns on the narrow economies of the region; the loss of preferential markets in the European Union by

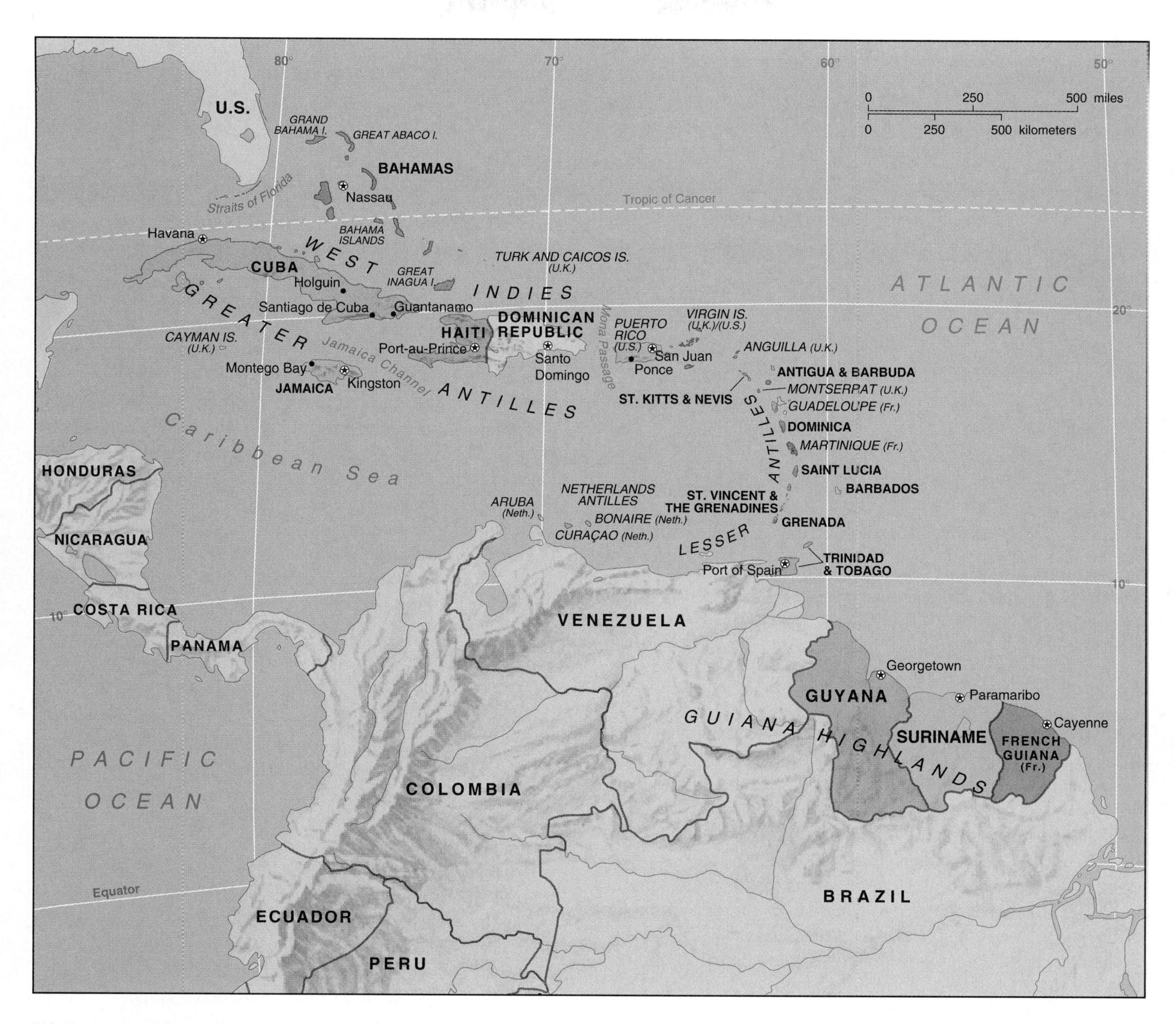


**FIGURE 10.27 West Indies: the island countries.** The many small countries, some of which remain colonies.

A.D. 2000 and the possible loss of entry to U.S. and Canadian markets; Mexico having sole entry to Anglo American markets through NAFTA; and the decline of aid funding from core countries since the end of the Cold War. The FTAA faces major problems, such as the strained relations between the Hispanic countries and the English-speaking West Indies, the differences in country sizes, and the United States' refusal to lift its trade embargo on Cuba.

## Countries

The West Indies contain one of the broadest arrays of political geography in the world. The countries vary in size and population from Cuba, with 11 million people in 1998, to hundreds of tiny island countries with few inhabitants. The ownership of consumer goods (see Figure 10.20) compares poverty-stricken Haiti and the more affluent Bahamas with other Latin American countries.

A major source of variety within the human geography of the region arises out of the different colonial histories of the countries. While Middle America has an almost exclusively Hispanic history, the West Indies were colonized by Spaniards, French, British, Dutch, and Danes. The people have a strong sense of belonging to the places where they live, often expressed as rivalry with their neighbors. For example, sports events among Jamaica, Barbados, Trinidad, and Guyana are prestigious and emotional affairs.

Although the Spanish were the first to discover and claim the largest islands, their main areas of interest shifted to Middle and South America after the invasion of Mexico in 1519.

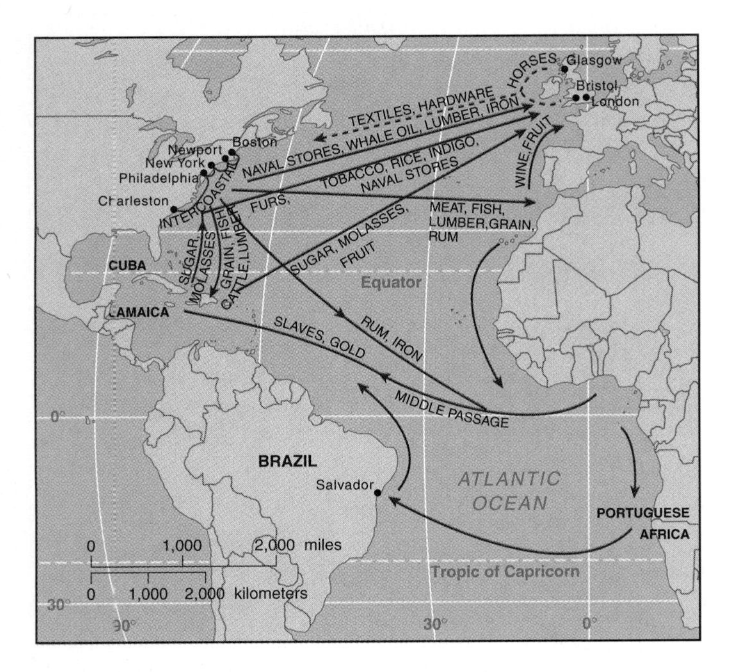


**FIGURE 10.28 Latin America: the slave trade.** The 1700s Atlantic economy, trading colonial commodities with home countries and bringing slaves from Africa to the Americas. Note the different locations of the West Indies and Brazil in this trade.

Spain retained colonies in a few of the West Indian islands to supply meat to the mainland empire of New Spain and sugar and tobacco to the home country. Ports were fortified to protect the trade routes. In the 1600s, the Dutch, British, and French established colonies on Bermuda, the Bahamas, the Lesser Antilles, the Guianas, and the offshore South American islands. At first, some were merely bases for pirates preying on Spanish treasure ships. After 1650, the development of the sugarcane industry made these islands valuable to the home countries. They were linked to the Atlantic economy that involved a pattern of trading products from the West Indies and Americas eastward to Europe, then sailing southward to collect slaves from western Africa, and westward back across the Atlantic (Figure 10.28).

Throughout the 1800s, as the rest of the Latin American countries gained independence and the slave trade was banned, most of the sugar islands remained colonies. They declined in significance as European countries grew their own sugar beet. Independent Haiti and the Dominican Republic failed to develop their economies. The United States took a growing role in the area. Its war with Spain (1898–1902) led to the freeing of Cuba and Puerto Rico from Spanish colonial rule under U.S. oversight. Both new countries found increased economic independence, but Cuba took full political independence while Puerto Rico moved politically closer to the United States. Nearly all the British and Dutch colonies became independent in the mid-1900s, but the remaining French islands—Martinique and Guadeloupe—together with French Guiana remained within the French economic area, sending representatives to the French parliament and officially being regarded as part of the European Union.

## People

Hardly any of the indigenous population of the West Indies survived the early days of European settlement, although a few Caribs live on Dominica and St. Vincent and some Arawaks on Aruba and in French Guiana. Most of the population traces its occupation of the region back less than 400 years.

The former Spanish colonies have relatively high proportions of people with European origins. Eighty percent of the population in Puerto Rico counts itself white, as does over 30 percent in Cuba. In the former British, French, and Dutch colonies, this figure is less than 5 percent and the dominant majority is of African origin. Islands such as Jamaica and Trinidad, together with Guyana and Surinam, also have substantial Asian populations as a result of workers being brought in from India after the abolition of slavery to replace former slaves. Many racial mixtures enhance the cosmopolitan nature of West Indian populations.

### Population Dynamics

The total population of the West Indies doubled from 17 million to 34 million people from 1965 to 1998 and is expected to rise to over 40 million by 2025. Population pressure is already severe in many islands with high population densities of over 600 per km$^2$ (1,500 per mi.$^2$) in Barbados and over 200 in many of the other islands (see Figure 10.16). Only in the Guianas does population density fall below 5 per km$^2$. Jobs available are fewer than the labor force, and there is high unemployment and continuing outmigration.

Population growth rates vary, although they fell in the late 1900s. In 1998, most West Indian countries had growth rates of under 1 percent. Although total fertility remains high, outmigration lowers the overall rate of increase. Jamaica's population trends are compared with other Latin American countries in Figure 10.17.

Recent high growth rates in islands such as St. Lucia and St. Vincent gave rise to large proportions of young people (with 44% under 15 years in these islands), declining fertility rates partly balanced by reduced infant mortality, and increasing life expectancy to over 70 years. Only Haitians had a 1998 life expectancy as low as 50 years; the rest were over 65 years. Cuba's age-sex graph (see Figure 10.18) demonstrates the effects of losing population as refugees fled in 1959, encouraging births in the 1960s, and the more recent fall in births.

### Migrations in the 1900s

The population of the West Indies is a very mobile one, with migrations into the area being dominant up to 1900 but outward migrations becoming significant after 1900. In the early 1900s, many inhabitants of the older British West Indian island colonies moved to new plantations opening in Trinidad and British Guiana, followed by Jamaicans and Haitians moving to Cuba and the Dominican Republic. The cutting of the Panama Canal attracted workers from the West Indies and especially Barbados. Further movements were toward banana plantations established in eastern Central America, U.S. military bases, and new oil-producing centers in Trinidad and Venezuela.

After World War II, West Indians from British, French, and Dutch colonies were attracted to job openings in Europe. The United States and Canada also welcomed people from the West Indies, and over 5 million with such origins now live in the United States—mainly from Cuba and Puerto Rico. The outmigrations relieved local population pressures and provided considerable sums in transfer payments but often took away the most active and able young men, leaving behind dependent populations of old and very young people. Although migration to Europe slowed because of immigration restrictions imposed in the 1980s, West Indians continue to migrate to the United States.

The decline of agriculture as an employer in many of the West Indian countries led to migration to the largest cities in each country. The "push" factor of poor rural job prospects is complemented by the "pull" factors of perceived better employment prospects, social service delivery, and amenities in the towns. Yet most cities were overwhelmed by such rapid growth so that the jobs and housing available were insufficient. Congestion, shantytowns, pollution, and crime resulted.

Even within the context of small total populations in many countries, the size of some of the major cities is impressive. **Havana**, the capital of Cuba, had 20 percent of the country's population in 1996, despite the government policy of focusing economic development in rural areas. In the Dominican Republic, **Santo Domingo** and Santiago do los Caballeros had over a million people each. **San Juan** (Puerto Rico), **Port-au-Prince** (Haiti), and **Kingston** (Jamaica) are other large cities on West Indian islands.

## Economic Development

### Colonial Farming Heritage

The West Indian colonies of Spain, France, Britain, and the Netherlands assumed a greater significance in the world order of the day when the area took over the production of sugarcane from Mediterranean countries in the early 1600s. The heritage from that period still influences island economies. Sugarcane cultivation began in the large islands of the Spanish colonies, but it was the small islands of the Lesser Antilles and the Dutch colonies that made it a major commercial phenomenon. In Barbados—one of the closest islands to Britain—the forest was cleared, holdings expanded, and African slaves brought in. Other islands followed, with Jamaica becoming the richest sugar producer. After cutting and milling, raw sugar was sent to be refined near the markets in Europe and Anglo America.

Island landscapes altered as the plantation system took over. Plantations on the smaller islands contrasted with the Hispanic haciendas—large estates in a society where land ownership brought prestige—on the larger islands. As well as the sugar fields, land was devoted to providing food for the slaves, pasture for livestock, and some woodland for fuel.

From 1500 to the 1800s, some 10 million African slaves were shipped to the Americas, of which half went to the West Indies, 39 percent to Brazil, and under 5 percent to Anglo America. The Africans brought their own culture, handicraft skills, arts, and folkways. At the height of sugar plantation prosperity in the 1700s, the average unit was of 200 acres, used 200 slaves, and produced 200 tons of sugar each year. The British navy stopped the slave trade across the Atlantic after 1807, and slaves were emancipated in British colonies by the 1830s, in French by 1848, and in Spanish by the 1880s. The African American populations increased, but many families left the plantations for their own plots of land. The plantations declined in numbers and output and faced increasing competition from sugar production in other parts of the world.

When new plantations opened in Trinidad and the Guianas, labor was imported from Asia. Cuba increased its sugar production in the 1800s, becoming the largest producer of all, and attracted labor from other West Indian islands. In the 1900s, the family-owned plantations on the sugar islands sold out to major corporations, such as Tate & Lyle (U.K.) in Jamaica and Gulf & Western (United States) in the Dominican Republic. Since the mid-1900s, many island governments took ownership of the sugar industry.

The colonial occupations and economies often took major decisions away from the region and placed them in countries overseas. The interests of external countries and company shareholders often assumed greater importance than local considerations. The colonies were restricted to producing single commercial crops for the needs of overseas markets, and a white elite controlled the imported African- and Asian-origin working populations.

### Economic Strategies Following Independence

After independence, the countries of the West Indies remained small in both area and total population. Their economies depended on exporting a limited range of crops and some minerals, and they imported manufactured goods, food, and fuels. Multinational corporations often marketed the export products. Demand for the main products—bananas, other fruit, sugar, petroleum, and bauxite—fluctuated. The preferential arrangement for West Indian bananas to be sold in the European Union was challenged on the grounds of unfair trading by Central American countries and by the United States on behalf of its own corporations.

To overcome these restrictions, West Indian countries adopted similar economic policies to other developing countries. At first, manufacturing was promoted through import-substitution. Factories were protected from external competition by tariffs, but this often led to high prices and low local profits. Together with the small size of local markets, this strategy had little impact on the economies and employment levels.

In the 1980s, policies switched to making local sites attractive to investment by overseas manufacturers that would take advantage of cheap locations and labor to produce such goods as clothing, sports goods, and electronics for export, principally to the Anglo American market. This policy proved more effective in economic growth but did not reduce external dependency. Manufacturing remains a small part of most island economies.

### Tourism

Tourism has become increasingly important in the West Indies, attracting around 15 million visitors in 1996. Localities within the West Indies cater to a range of tourist interests from the elite to the mass markets. The United States now provides two-thirds of the tourists, while most of the rest come from other world core countries in Europe, as well as Japan. Puerto Rico, the Bahamas, and the Dominican Republic attract the greatest number, while the Virgin Islands, Antigua, Guadeloupe, and Martinique grow in significance. At first, tour operators developed their own hotels and facilities away from local settlements. Many governments now promote the tourist attractions of their countries. While cruise ships also visit the West Indies, they make less impression on the local economies.

Tourism is another economic activity that is dependent on external forces, with demand falling during economic recessions in the wealthier countries or when local terrorism or natural hazards occur in the West Indies. The industry pays low wages, and the beaches and mountains can become polluted. Although tourism brings in foreign exchange, much of it is used to buy the foods and equipment expected by the tourists. Where tourism adds greater diversity to the economy, as in Jamaica, it has a more positive effect than in the small islands, where it dominates the economy.

### Country Specialties

Within this general picture, individual countries have distinctive economies.

- In the early 1900s, Cuba, the largest island, was dominated by U.S. corporation-owned sugar plantations and U.S. corporations. The Fidel Castro-led communist takeover in 1959 severed ties with Anglo America and shifted dependency to the former Soviet bloc countries. Large state farms and cooperative farms replaced confiscated private plantations, but productivity remained low. Some farms diversified to produce citrus fruit for Eastern Europe. Manufacturing included import-substitution units for locally needed goods such as textiles, wood products, and chemicals. Cuba exported sugar, tobacco (cigars), and some strategic minerals to the Soviet bloc in exchange for oil, wheat, fertilizer, and equipment.

  During the 1980s, 85 percent of Cuba's trade was with the Soviet bloc, but this close link was broken after 1991. Cuba suffered as much as other communist countries from the breakup of the Soviet bloc. From 1989 to 1994, Cuban GDP fell by 34 percent as it moved cautiously toward market economy policies. The sugar crop, making up 40 percent of Cuban exports in 1993—down from 60 percent—fell from 8.4 million tons in 1990 to 3.4 million in 1995 because of poor organization, bad weather, and shortages of fuel, fertilizer, and willing labor. Most farm workers found it paid them better to grow fruit and vegetables for the black market as supplies in Havana shops dwindled. The United States maintained its trade embargo through the 1990s. Tourism, however, began to grow as a source of foreign currency, with a million visitors in 1996. Canadian and other non-U.S. mining companies took over Soviet-instigated mining projects, mainly for nickel, cobalt, and gold. In the short term, Cuba faces major economic and social problems.

- Puerto Rico became part of the United States in the early 1900s as a "commonwealth" that retains some independence but is not a full state. Puerto Ricans have open access to migration into the United States, and many took advantage of this possibility since the 1950s. U.S. statehood remains an issue, being rejected again by popular vote in 1993, but most Puerto Ricans prefer a limited degree of autonomy. They still speak only Spanish, which was declared its official language in 1991.

  Puerto Rico has a much higher GDP per head than other West Indian islands. Although its sugar industry expanded, the economy remained narrow until after World War II, when farming shifted to dairying and manufacturing increased rapidly. Machinery, metal products, chemicals, pharmaceuticals, oil refining, rubber, plastics, and garments provide a diverse product mix. Tourism became a leading economic sector with over 3 million visitors in 1996.

- The Dominican Republic, on the larger eastern part of Hispaniola, became independent in 1821 but subsequently suffered from economic stagnation and political instability. As in Cuba, U.S. corporations formed links with the plantation owners, and the U.S. government intervened with military occupations in 1916 to 1924 and in 1965. Spurts of economic development occurred in the mid-1900s, when construction began on irrigation and road infrastructure, and there was industrial investment by U.S. corporations. Agriculture remains the dominant economic activity, based on sugar, while the Standard Fruit Company owns large banana plantations and the Dole Corporation has pineapple plantations. Meat is exported from cattle ranches, and coffee is grown on the uplands. There is some mining and oil reserves are present. Most manufacturing is the assembly of imported components for export. In the 1990s, tourism is a new growth industry, with 2 million visitors in 1996 and an expanding basis of package holidays. Through the gradual diversification of its economy, the Dominican Republic now has one of the highest incomes of the larger West Indian islands.

- Haiti, occupying the smaller western part of Hispaniola, is the poorest country in the Americas. Life expectancy is low, infant mortality high, and 50 percent of its adults are illiterate. Over three-fourths of the people live in rural areas on impoverished land, and there is a land shortage in the overpopulated country resulting from the division of holdings. Sugar plantations provided a major source of income under the Spanish and French colonists, but the value of exports before the 1804 independence was greater than in the 1980s. Independence was followed by the political instability that continues today. The plantations deteriorated, and economic stagnation and decline set in.

  The United States occupied the country from 1915 to 1934 and again in 1994 to overthrow the military government. In between these dates, corrupt and repressive governments produced little economic development, leaving most people dependent on their subsistence plots of land. Some commercial farming for coffee and cacao produced export income, but a 1980 hurricane destroyed many trees and recovery was slow. In the 1970s, tax incentives attracted U.S. corporations to set up factories in Haiti. Exports of clothes, electronics, and sports equipment became more valuable than farm products. Haiti's mineral resources, however, remain undeveloped. Tourism was difficult to establish in a country with so much poverty, political instability, and a high incidence of HIV/AIDS.

  Following the military coup against a democratically elected government in 1992, an international embargo on Haitian products caused the economy to collapse, including the loss of vital aid, default on its public debt, and closure of the new assembly industries. Exports and imports halved. Although a democratic government was reinstated in 1994 after the U.S. intervention, Haiti needed humanitarian assistance and poverty relief before the reconstruction and modernization of the state could proceed. Haitians expect immediate improvements from the new government, but face a bleak future with an economic gap that is being filled by drug-traffic corruption.

- Beginning as a minor Spanish colony, Jamaica gained significance as the largest British colony in the West Indies. Sugar plantations expanded until the abolition of slavery; bananas then became important on the plantations and groups of cooperative farms. Agriculture now provides less than 10 percent of Jamaican GDP, although it still employs one-fourth of the labor force. After World War II, the Jamaican economy diversified. Some 60 percent of its export income now derives from the bauxite mines and alumina-refining plant. Although overseas competition from Australia, Brazil, and Guyana led to a fall in this source of income in the 1980s, the 1990s saw increased output. Jamaican manufacturing is becoming more important, based on the duty-free imports of parts to be assembled for export and textile products. Tourism, with over a million visitors in 1996, contributes income equivalent to half the exports of goods and there are a series of resorts along the north coast.

- Of the Lesser Antilles, Trinidad and Tobago have the largest area and population. Trinidad has an oil-based economy and a unique racial mix. The British took Trinidad from Spain in 1797, and British settlers developed it as a sugar island. Only one-seventh of the land was settled by the time of slave emancipation in 1834–1838, and only one-fifth of that was cultivated. Most plantation development occurred later in the 1800s, when it was necessary to bring in paid labor from Asia. After serving a set number of years on the plantations, these laborers were paid off with land of their own, and today such small farms produce most of the sugar grown. The increasing quality of its product enables Trinidad to expand sugar output as that in other countries declines. Also during the 1800s, Spanish and French plantation owners grew cacao, which covered three times the sugar-growing area by 1905 but later declined in competition with West Africa.

  Oil production began around the La Brea Pitch Lake and then moved offshore. The output of asphalt from the lake declined, but natural gas and oil production, together with the refining of oil from Southwest Asia, is a major industry. In the mid-1990s, oil and natural gas output rose by 10 percent. Manufacturing industries include petrochemicals and steel, and employ 15 percent of the labor force, while government and other service jobs provide 50 percent of employment. Tourism is not well developed as yet.

- The other islands of the Lesser Antilles are small and have restricted economies. On Barbados, sugar production decreased to employing only 5 percent of the work force. Over 80 percent of workers are in service jobs, especially tourism, while Bridgetown has some light industries and the high level of literacy attracts U.S. data processing facilities. The other former British colonies, such as Antigua and Barbuda, St. Kitts and Nevis, Montserrat, Anguilla, Dominica, St. Lucia, St. Vincent, and Grenada, all saw their commercial agriculture decline, although some still grow sugar, cotton, coconuts, and local food crops. Most attempt to promote tourism and have some light industries. Islands such as St. Lucia have spectacular scenery, but all have sun and sea. Some islands have U.S. military bases or oil storage and shipment facilities.

- The French colonies of Martinique and Guadeloupe remain overseas departments of France. Their residents are French citizens and members of the European Union. They were rich sugar islands, but the output declined with a shift to bananas, melons, and pineapples for the

French market in line with European Union policies. The French government is the largest employer, and tourism is increasing. External contacts tend to be with France rather than with other West Indian islands.

- The former Netherlands colonies of Aruba, Bonaire, and Curacao off the northern coast of Venezuela are low-lying and arid, obtaining their water supplies from desalination plants. They are linked administratively to the Netherlands, although Aruba has had a separate status since 1986, voting in 1996 to remain part of the Kingdom of the Netherlands. The people are exceptionally cosmopolitan, including descendants of Africans, Indians, Dutch, Portuguese, Danes, and Jews, all speaking an old trading language, Papiamento, as well as Dutch and English. Oil refineries linked to the Dutch development of Venezuelan oil, along with tourism and offshore banking, are among the main sources of income.
- Guyana (former British Guiana), Suriname (former Dutch Guiana), and French Guiana all have small populations on relatively large areas of land. Made habitable by Dutch drainage engineers in the early 1800s, the coastal plains developed sugar plantations worked by indentured labor, but this industry declined and the drainage works are often in disrepair. In both Guyana and Suriname, living standards declined after independence as the result of civil strife. Guyana and Suriname export bauxite from inland mines, while French Guiana exports timber to the EU, to which it has access as a French territory. The reopening of gold mines by Canadian companies in Guyana caused exports to quadruple in the 1990s, but in 1995, the Omai gold mine was closed for several months following a leakage into the local river of cyanide-rich water from a settling tank.

## NORTHERN ANDES

The Andes Mountains dominate all five countries of the **Northern Andes—Bolivia**, **Colombia**, **Ecuador**, **Peru**, and **Venezuela** (Figure 10.29). The world's second highest mountain range not only looms large in the landscape, but it also creates a multitude of local environments at different heights. The Andes are still in the process of active uplift, which involves frequent earthquakes, landslides, and occasional volcanic eruptions. The mountains isolate large interior sections of the Orinoco and Amazon River basins in these countries from world trade, slowing their potential for development.

The countries of this subregion range from the poorest in South America (Bolivia) to some of the richest (Colombia, Venezuela). Most remain developing countries, although each has some potential in natural resources and economic output. In the later 1990s, only Colombia increased its income per capita; consumer goods ownership remained relatively low (see Figure 10.20), while all aspects of human development have modest scores. Border disputes still erupt, the most notable of which is that between Ecuador and Peru over a short section of their Amazonian lands.

In 1969, five countries (Bolivia, Chile, Colombia, Ecuador, and Peru) formed the Andean Group of trading partners. In the 1970s, Venezuela joined and Chile left. At first, this grouping saw its purpose as creating an internal trading area and keeping out foreign goods. It faltered as members disagreed over which country should produce what goods. In 1988, it was relaunched as the **Andean Common Market** with a greater emphasis on integrating with world trade patterns.

### Countries

The two northern countries, Colombia and Venezuela, have experienced greater development and economic diversification than the southern three. In 1998, Colombia and Venezuela produced over two-thirds of the subregion's total GDP and contained 62 million people between them, while the southern three countries together had 46 million.

Venezuela has the highest income, related to its oil and mineral exports that continue to be in demand. Debts accrued during the expansion of oil production in the late 1970s held back further development when oil prices fell in the 1980s. Colombia has the most diverse mix of economic products, including mineral extraction, agriculture, and manufacturing, but economic growth there is restrained by political instability and civil strife. Ecuador suffers from a small home market. Peru's economy experienced a series of booms and busts as it developed different export products. Bolivia's mining output is less in demand on world markets than it used to be. It lacks an ocean outlet since a war over phosphate deposits in the 1800s resulted in Chile taking over its route to the sea at Arica.

Internally each country has a variety of terrains. The high Andes contrast with the lower lands to the east, much of which are covered naturally by tropical rain forest. There are areas of more seasonal rains in the interior of northern Colombia, central Venezuela, and Bolivia, where tropical grassland is the natural vegetation. The coast of Peru has one of the world's most arid climates.

Although the formal economies of these countries are not thriving, the informal and illegal economy based on the subregion's world leadership in the production of coca and cocaine drugs for the rising demands of the Anglo American and European markets grew rapidly after 1970. Despite efforts to eradicate coca production in Peru and Bolivia, production increased and spread. The former small number of routes from the Colombian processing centers through the West Indies to the United States are now multiplied as coca is processed to cocaine near the growers and exported through Pacific coast ports and Brazil to the wider world. The coca-growing areas spread from the eastern slopes of the Andes into Brazil. The human geography of these Andean countries is increasingly difficult to assess without the illegal drug fac-

**FIGURE 10.29 Northern Andes: main geographic features.** Note the position of the Andes Mountain ranges in each country and how they isolate interior lowlands from the main centers of population and trade.

tor, although precise details of impacts on their economies are difficult to obtain.

## People

The main concentrations of people are in mountain valleys, on plateaus, and on available coastlands (see Figure 10.16). The interior lowlands have few people. This reflects the facts that the Andes act as a barrier and the equator bisects the subregion; in the equatorial climatic environments, life is more comfortable in the highlands.

### Population Growth

In the 1960s and 1970s, the North Andean countries had annual rates of population increase that were among the highest in the world—3 or 4 percent. Since that time, fertility rates fell from over 5 to around 3 by the late 1990s, except in Bolivia, where they are just under 5. These countries lag behind others in South America (see Figure 10.17) in bringing birth rates and death rates closer together and continue to have annual population natural increases of over 2 percent. The total population of 108 million people in 1998 may grow to around 160 million by 2025. The age and sex data confirm the likelihood of continuing population growth, with large numbers of young people who will provide the next generation of parents (see Figure 10.18).

### Urban Growth

The northern Andean countries vary in their levels of urbanization. Venezuela was 86 percent urban in 1998, while Colombia and Peru were over 70 percent. Ecuador and Bolivia, however, were around 50 percent. All these figures show increases of 10 to 20 percent since 1970. In Peru and Bolivia, there is a single primate center, as in the **Lima-Callao** urban complex, where one-third of Peru's population lives. The complex extends along the Rimac Valley from Lima to the port of Callao. **La Paz** has 20 percent of Bolivia's total. Such primate cities are centers of finance and manufacturing as well as government.

In Venezuela, **Caracas**, **Maracaibo**, and **Valencia** are the leading cities. In Ecuador, **Guayaquil**, a center of recent economic growth on the Pacific coast, now surpasses the established capital of **Quito** in the highlands. Colombia has a number of major cities that grew up in once isolated areas and developed in different ways. **Bogotá**, the capital and largest city, has extensive manufacturing industries as well as the national government bureaucracies. **Medellín**, a manufacturing center, grew in the late 1800s, at first making equipment for the local farmers more cheaply than such equipment could be imported and transported from the coast. **Cali** is situated in the Cauca Valley and **Barranquilla** is a port on the north coast.

All these northern Andean cities grew rapidly as rural-to-urban migration led to the building of shantytowns within them. In Lima-Callao, such new growth defeated attempts to control it, and the prevalence of uncontrolled street vendors led to the closing of many established shops, the loss of local taxes, and the deterioration of the whole city infrastructure and built environment. In the 1970s, it was hoped that the shantytowns might provide a transition to more organized urban living, but it was not fulfilled. The reasons for this include the impacts of guerrilla warfare led by the Marxist Shining Path movement that was active in the 1980s and early 1990s, resulting in the deaths of around 90,000 Peruvians, a large-scale abandonment of the Andean provinces south of Lima, and increased migration into the capital's shantytowns.

### Ethnic Contrasts

The Andean countries contain racial and ethnic contrasts that have important economic, social, and political effects. In Colombia and Venezuela, the Native American groups occur mainly in highland and interior areas not taken over by Hispanic or mestizo groups. In Ecuador, Peru, and Bolivia, they tend to be concentrated in the highland areas. Smaller numbers of different Native American tribes occur in the Amazon River lowlands, totally isolated from European intrusions until the 1950s. The Spanish colonizers occupied upland basins and some of the coastal locations that provided overseas links to the home country. Workers from Japan, China, and Africa immigrated to work in the coastal irrigated farmlands of Peru.

The economic and social differences between European-origin and Native American peoples continue to raise tensions within the Andean countries. Despite both formal attempts and unplanned trends toward modernization of the Native Americans, differences remain. The Native Americans often prefer to retain their own culture instead of being modernized. Even in the recent rural-to-urban migrations, the Native Americans from adjacent groups of villages migrate to live close to their kin in the shantytowns of the cities.

## Economic Development

The economic development of the Andean countries during the last two centuries continued to be affected by the Spanish colonization and the network of town-centered areas by which the Spanish overlords controlled the surrounding countryside. After independence in the early 1800s, most countries relied on export crops and minerals to pay for imports that only the rich could buy.

### Export-Led Underdevelopment

The three southern countries in particular, but all five countries to an extent, suffer from what has been called **export-led underdevelopment**. In the late 1800s and early 1900s, the countries relied on a series of export-based booms as specific mining and agricultural products came into demand in Europe and Anglo America. Local people and foreigners invested in these products, while the government taxed the exports as the basis of its own income. Any hard currency earned went to pay for imports, on which tariffs were not charged. In the 1930s, attempts at changing this process by encouraging import-substitution manufacturing at home were not very successful because illegal imports undercut local products and only a restricted range of goods were made in local factories. As a result, the Northern Andean countries found that fuller involvement in the world economic system made them vulnerable to external control. Today's dependence on smuggling coca and cocaine arose from the need to find a high-value product with continuing demand in world markets.

### Patterns of Economic Diversification

Each country evolved distinctive patterns of economic diversification.

- In Peru, the early colonists established mines in the cordilleras and kept farms on the coast to provide food for the people and pack animals that brought the silver and gold to the coast. Later Peruvian export booms from the mid-1800s involved less valuable minerals such as lead and zinc from the cordilleras, guano fertilizer from offshore islands, and crops such as sugarcane and cotton from irrigated coastal lands. Oil from the north, iron from the south, and fishmeal from what became the world's

**FIGURE 10.30 Northern Andes: fishing industry.** Peruvian fishing boats set out to sea at sunset. They fish in the rich grounds of the cold Peruvian current, where nutrients support plankton and so fish and birds. El Niño events bring warm waters that often kill fish and reduce fishing catches.



second largest fishery (Figure 10.30) added new aspects to Peru's income from the 1950s. Although this output appears diverse, each product comes from a particular area and had a limited period of importance. At any one time only a few products dominated Peru's exports, while producing areas became prominent and then faded.

In the 1990s, Peru liberalized its economy, bringing a boom in mining exploration. It already has Latin America's largest gold mine at Yanacocha near Cajamarca, opened in 1992, but other opportunities in copper mining were due to begin production by the late 1990s. Peru's rapid economic growth in the mid-1990s could be slowed by the combination of expanding demand for imports, increasing indebtedness, and continuing poverty among a large sector of the population, so that Peru's economy will experience recession like that of Mexico. The large sums of money entering Peru—the world's largest producer of cocaine—by way of the illegal drug trade may cushion such effects.

Three events in the late 1990s, however, cast doubts on such optimism. First, the 1997–1998 El Niño event brought disaster to coastal farming and fishing communities. Second, in 1998, the Shell Oil Company deferred a final decision on developing the Camisea natural gas field in the Amazon River basin. Potentially the largest field in Latin America, the project includes a gas processing plant, twin pipelines across the Andes, and petrochemical and port complexes. For Peru, it would give entry to higher-level manufacturing skills and products. Third, the economic downturn in Asia reduced demand for Peruvian minerals and caused the postponement of mining projects.

- **Bolivia** for long depended almost entirely on silver and tin for its exports because the altiplano is too high for commercial farming. The tin mines were nationalized in 1952, but the 1980s fall in world prices adversely affected the internal economy (see "Living in Bolivia," page 506–507). Other metallic ores including gold are also mined, and in total they make up 50 percent of Bolivia's exports. The eastern interior plains are now being developed, since the rich soils there and improved transportation make it possible to harvest two crops of soybeans each year. The crops are trucked across the Andes to Peruvian ports for export. Oil from the eastern Andean slopes makes Bolivia self-sufficient in energy.

  Bolivia is the world's second producer of cocaine, and the United States pressures Bolivia to reduce production from the Chaparé Valley east of Cochabamba. The producers, however, oppose such moves by means of political opposition to the government's economic reform program. The Bolivian program of economic reform has done little so far to raise the economic well-being of the people. It is based on the concept of **capitalization**, rather than of privatization: foreign capital is encouraged to invest in state corporations, with the remainder of the investment (and anticipated profits) being spread through Bolivian public institutions such as pension funds.
- **Ecuador** is a small country with limited upland and interior economic potential. Its main economic growth occurs on the coast around Guayaquil, where the expansion of fishing for tuna and white fish, together with shrimp farming, are adding to the base of light manufacturing and commercial farming for sugarcane, bananas, coffee, cacao, and rice. Oil

## BOLIVIA

LIVING IN . . .

Bolivia is one of the poorest countries in Latin America. Although it is five times larger than Britain, it has only one-tenth of the population. It is a mountainous and landlocked country that for long depended on exports of tin and silver for its economy. People living in Bolivia experienced important changes after the 1950s that opened new opportunities to them.

The 1951 revolution led to huge political changes. Up to that time British mining companies virtually ruled Bolivia, but they were dispossessed when the mines were nationalized in 1952. Although the old mining companies were criticized for the slavelike conditions in which they held their laborers, they left an infrastructure heritage of housing, schools, railroads, and medical facilities that were scarcely improved in the next 50 years.

The new government also instituted land reforms, dividing huge landholdings among those who worked them and setting up cooperative arrangements for purchasing and marketing. This policy was an attempt to diversify the economy away from the reliance on mine products by expanding commercial farming. People of Native American origin, who make up around 50 percent of the Bolivian population, were given a vote for the first time and now wield a major influence in government. For them in particular, the revolution brought access to education for a group that was 90 percent illiterate and had no hope of advance. Today many Native Americans are qualifying in the professions, and one has been vice president.

The revolution and its aftermath were not a total success. For example, in the 1980s, a doctor in Catavi, a mining town, found that the government closed mines when local overmanning and other uneconomic measures raised the costs of producing tin to $17 per pound while the world price was only around $4. Numbers of mine workers in the town reduced from 5,000 in 1982 to 500 in 1986. The doctor's salary fell in real value at a time of rapid inflation, but the mining unions forced him to attend to his patients until his contract was up. In the late 1980s and 1990s, many of the former government-owned mines were privatized, often bought by Japanese companies, but the new owners employed few miners.

Bolivia as a whole remains a country of contrasts. There is an affluent class of people who include businesspeople, pro-

Map of Bolivia.

fessional people such as doctors and lawyers, and military officers. Some teachers are in this group if they teach at private schools as well as in the poorly paid, public school jobs. Most of this elite claim Spanish origins, but many are of mixed ancestry. They live in good houses and many own luxury cars bought mainly from Brazilian suppliers. If not travelling by car, they travel in special class trains or well-equipped buses.

There is also a working class of factory, mine, and shop workers. They have regular jobs, but wages are low. They travel in the back of trucks or basic buses. Then there are the landless and jobless people. Many are rural Native Americans who still do not have access to public education. The lack of facilities in rural areas set off a major migration to the urban areas, especially for access to schooling. When such people reach urban areas, they do not have jobs or housing. They move into the shantytowns, or *barrios parifaricos,* often close to people from their home district, and do whatever they can to earn some money. In the barrios, electricity is

production from the interior increased from the 1970s, after a pipeline was built across the Andes to the port of Esmeraldas. Petroleum products make up nearly half of Ecuador's exports. New laws open up mining prospects for the foreign corporations from Canada, South Africa, France, and Belgium that are testing the viability of gold, silver, lead, zinc, and copper deposits.

Shrimp farming produces the second nonoil export, but further expansion is slowed by disputes with environmentalists who claim it is destroying coastal mangroves. The farmers counterclaim that they have to preserve the mangroves in which the shrimps begin life and are providing jobs and services in areas that have great poverty, malaria, no other paying jobs, no electricity, and no health services.

La Paz, the capital of Bolivia, built in a valley in the Andean ranges below Mount Illimani.



available for those who can pay, while water standpipes are shared among some 50 houses. There are stories of Native American families arriving from the country and the parents working in domestic service while the children go to school and become successful professionals. But most Native Americans do not rise so quickly, if at all. Many unemployed peasants obtain a living by begging.

Cutting across these social and economic differences, Bolivia has become a major producer of coca plants for cocaine drugs. When mines closed, many miners moved to the Chaparé area in the eastern lowlands, working on the coca plantations or in the distribution trade. It is said that some school teachers use their vacation periods to earn as much in a month on the coca plantations as in the rest of the year by teaching. Drug sales now underpin the Bolivian economy. The very rich people who control the production of cocaine sometimes invest in other local facilities in an attempt to improve their image: factories are built, stores sell imported goods bought in exchange for the cocaine, and sometimes schools and medical centers are provided. The cocaine trade has made many of those who were very poor quite rich and has altered the existence of many Bolivians. It has not been uncommon for shabbily dressed men to carry thousands of dollar bills to a vehicle dealer to buy a Mercedes truck.

- **Colombia** has a more diversified economy than the other North Andean countries. The Spaniards settled basins high in the Andes where they could raise temperate crops and livestock. The main cities such as Bogotá and Medellín were established there and in the 1800s developed local manufacturing to supply a growing market that was so distant from ports that the costs of transport made it cheaper to manufacture locally than to buy foreign goods. Medellín and the capital, Bogotá, became the main industrial centers for textiles, steel, agricultural equipment, and domestic goods.

In the lower valleys and near the northern coast, tropical cash crops grow on large plantations, including cotton, sugarcane, bananas, cacao, and rice. Colombia is the world's second producer of coffee, after Brazil, based on optimum conditions of well-drained mountain slopes and sufficient rainfall. In the mid-1990s, new varieties of maize and rice were bred in Colombia to grow in the poor soils of the eastern savannas. Trials were successful and suggested a more productive type of farming that combines livestock and locally grown cattle feed for this sparsely settled area extending into southern Venezuela. In the hilly areas, marijuana has become a major crop, together with coca and poppies.

Colombia is a mining country with considerable resources of iron, coal, oil, natural gas, gold, and emeralds. Colombia's overseas trade is important but hampered by the continuing political disturbances. In the 1980s and 1990s, guerrilla warfare, based partly on political repression and partly on strife among the drug syndicates, defeated the Bogotá government's attempts to capitalize on the country's strengths. In 1998, the Occidental Oil Company ended drilling in the Samore field in eastern Colombia when the Native American people threatened mass suicide.

- **Venezuela** had a similar history to Colombia of Hispanic occupation of the upland basins to grow temperate crops. In the 1900s, it became a major oil producer, raising its income to the highest in the subregion. Oil was known in the area around Lake Maracaibo from early in the 1900s, but major boosts in output occurred during World War II and when world oil prices rose in the 1970s. The industry was nationalized, but the economic expansion during the 1970s incurred debts that had to be paid for during the high interest-rate and low oil-price years of the 1980s. Venezuela became deeply in debt. The national debt caused investment in oil exploration to lag in the 1980s. In the 1990s, exploration was opened to private companies, linked to joint production and marketing arrangements with the government if discoveries are proved. The Caracas area is the focus of manufacturing and service industries.

  Southern Venezuela is a source of mineral wealth from the iron and bauxite mines. They provide a basis for manufacturing in Cuidad Guyana on the Orinoco River, powered by hydroelectricity generated in the Guiana Highland valleys. This region has much potential for future economic development but remains isolated by poor transportation facilities and affected by Venezuela's shortage of investment capital.

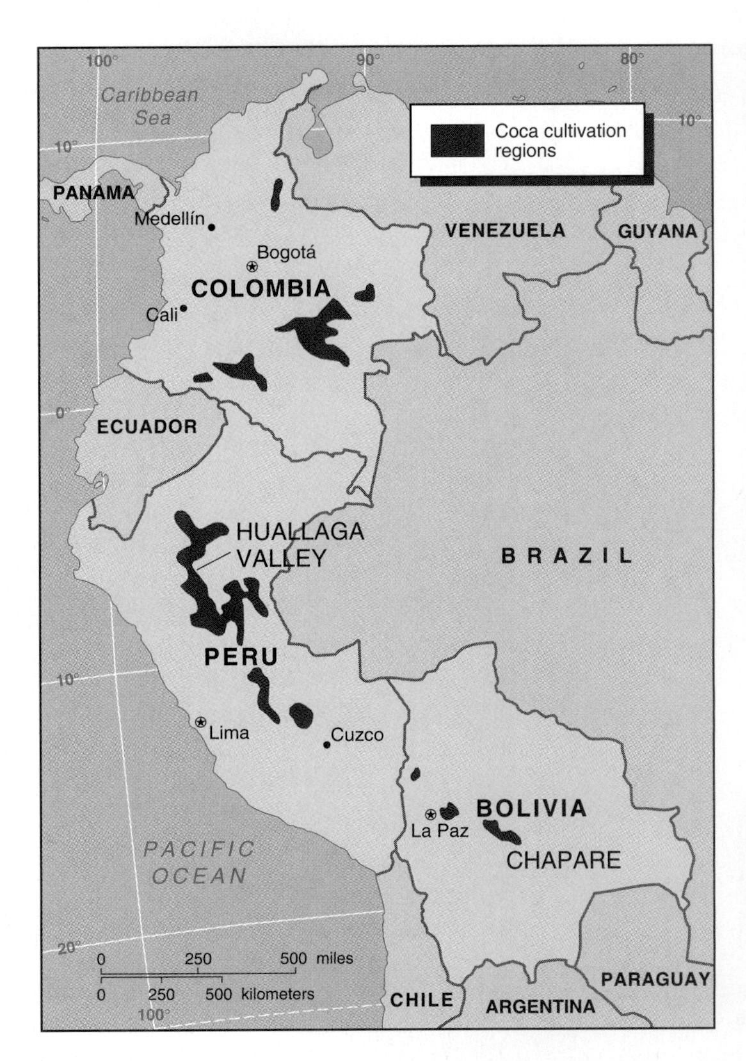


**FIGURE 10.31 Northern Andes: coca-growing areas.** The dominance of Peru and Bolivia in the 1980s gave way to more dispersed production in the 1990s, including Colombia (previously the main gathering and distribution center) and Brazil.

## Drug Trade

The countries of the Northern Andes are one of the world's major drug-producing regions (Figure 10.31). After years of growing the crops mainly in Peru and Bolivia and processing them in Colombia, the Bolivian area stopped expanding in the early 1980s, while the Peruvian area declined in output following a fungus infestation of the crop. Other areas, particularly in Colombia, began to produce coca, together with marijuana and poppies, so that Colombian production is now greater than that of Bolivia.

Although the world market grew, production grew much faster and overcame attempts to restrict output and distribution so well that the price fell by 75 percent. The United States led the way in attempts to prevent drug production in these countries, but Bolivia and Peru reacted half-heartedly to the pressures. Although Bolivian farmers planted other commercially valuable crops on former coca-growing land as a result of U.S. aid, the acreage under coca remained the same. Peruvians fear the resurgence of guerrilla groups among disaffected former coca growers.

In 1995, Colombia attempted a tougher stance and announced its intention of eradicating production within two years. Programs of eradication, however, have to compete with the high prices farmers get for their drug crops and

**RECAP 10D: West Indies and Northern Andes**

The West Indies include four large islands (Cuba, Haiti and Dominican Republic, Jamaica, and Puerto Rico), lots of small islands, and the Guianas on mainland South America. They were mainly settled by Europeans to develop sugar plantations and by imported African slaves.

Most of the former colonies are now independent, and their economies are based on more varied agricultural products, tourism, local mineral products, and some light manufacturing. Some have developed offshore financial services mainly linked to the United States.

The Northern Andean countries of Bolivia, Colombia, Ecuador, Peru, and Venezuela are dominated by the high mountains that isolate internal plains covered by savanna and tropical rain forest from the main centers of population. Their economies are based on mineral production, including increasing oil discoveries, and export crops. These countries include major drug producers, especially of cocaine but increasingly of marijuana and opium.

10D.1 How are the physical and human geographies of the West Indies different from those of Mexico and the Central American countries?

10D.2 What are the impacts of the Andes Mountains on the human geographies of the countries they are part of?

**Key Terms:**

Caribbean Community and Common Market (CARICOM)
Caribbean Basin Initiative
Association of Caribbean States
Andean Common Market
export-led underdevelopment
capitalization

**Key Places:**

West Indies
Greater Antilles
Cuba
*Havana*
Haiti
*Port-au-Prince*
Dominican Republic
*Santo Domingo*
Puerto Rico
*San Juan*
Jamaica
*Kingston*
Lesser Antilles
Bermuda
Bahamas
Trinidad and Tobago
Guyana
Suriname
French Guiana
Northern Andes
Bolivia
*La Paz*
Colombia
*Bogotá*
*Medellín*
*Cali*
*Barranquilla*
Ecuador
*Quito*
*Guayaquil*
Peru
*Lima-Callao*
Venezuela
*Caracas*
*Maracaibo*
*Valencia*

with the guerrilla-backed disorder that such farmers may support, including the strategic occupation of isolated oil-pumping stations or airports. Chemical defoliation of coca-growing areas paralleled a doubling of the output of coca. By 1998, the failure of such programs led the United Nations Drugs Control Program to try to establish a new global drugs control convention. It has to contend with the income from coca-growing that lifts farmers out of poverty and the long-established custom of coca-leaf chewing among Native Americans living at high altitudes in the Andes.

## BRAZIL

**Brazil** is by far the largest country by area and population in Latin America. It has three times the area of Argentina, the second largest. Its 1998 population of 162 million was one-third of the Latin American total, and it had a much higher population than any of the other subregional groupings of countries in the region. In 1995, Brazil had the eighth largest world economy and massive natural resources. Brazil presents an immense market and the greatest diversification of economic development in the region.

Quite clearly the leader in Latin America in its attempts to modernize and develop economically, Brazil's rates of consumer goods ownership (see Figure 10.20) are similar to those of the countries to the south and higher than those of the Northern Andes. Its human development characteristics, however, lie more between those of Argentina and Chile and those of the northern Andean countries.

Despite its rapid economic growth of 7 percent per year between 1940 and 1980, Brazil faced increasing problems through the 1980s that resulted in high inflation, slower economic growth, and increases in the numbers of poor people. Some of these problems arose from the long-term dominance of public involvement in the economy and the high rates of population increase in the 1960s and 1970s. Others arose from the shorter-term impacts of 1985 constitutional changes that enshrined provisions for social programs and job security. They maintained high levels of people in employment but placed a huge burden on the country's finances. Although the federal government was made responsible for administering the new social provisions, it also had to distribute a high proportion of its income to its states. The combination of international debt servicing, the need to cover the finances for constitutional demands, and corruption led to high inflation that rose to 50 percent per month in early 1994. In the mid- and late 1990s, the Brazilian government made changes that reduced the inflation rate and tried to rewrite the constitution to remove many of the 1985 provisions.

Brazil is unique within Latin America because its national language is Portuguese. Portuguese colonial settlement granted large estates to favored men. Others could purchase smaller plots, but the gap between rich and poor was established at the earliest stage by the differences between the large and small landholdings. The import of slaves from Africa added an underclass to the colonial society equivalent to that of the Native Americans in Spanish Latin America. When it became attractive to invest in mines and new coffee estates farther inland, it was the wealthy plantation owners who could afford to do this and so they extended their political power. More recently, the development of manufacturing and commerce resulted in further contrasts among wealthy businesspeople, destitute favela dwellers, and rural peoples (Figure 10.32).

(a)

(b)

**FIGURE 10.32 Brazil: contrasting lifestyles.** (a) Guajara beach, near the port of Santos, southeastern Brazil. The many apartment blocks around the beach house families mainly from prosperous São Paulo, 80 km (50 mi.) inland. (b) Some of the poorest Brazilians live in the dry northeastern region. This farm dwelling in Trapia is in a community with very high infant mortality rates.

**FIGURE 10.33 Brazil: major geographic features.** The states, rivers, and major cities.

## Regions of Brazil

In such a huge country, internal regional contrasts are to be expected (Figure 10.33). Politically, the country is divided into states that are part of a federal government system: the full title of the country is the Federative Republic of Brazil. Further variations in human geography relate to natural environment contrasts, the history of European settlement, and political emphases among the different states.

- Nearly 60 percent of Brazil's territory is drained by the Amazon River network. Much of this area is covered by tropical rain forest, of which some 7 percent has been cut since the first Portuguese settlements. The Amazon River basin is still sparsely settled, despite many efforts to attract people to live there.
- The northeastern coastlands and plateau were the first settled areas of Brazil. Sugarcane plantations prospered until competition with the West Indies began in the later 1600s. Periodic droughts devastate the interior of this region (as in the late 1990s), causing many to emigrate to other parts of Brazil.
- The main economic development in Brazil occurred in the southeast around Rio de Janeiro, the colonial and national capital for 200 years, and São Paulo. Beginning with mining and commercial agriculture based on coffee, this region now dominates manufacturing production and commercial and financial activities in Brazil.
- The three southern states of Paraná, Santa Catarina, and Rio Grande do Sul became centers of growing population and agriculture from the 1930s. After 1950, the coffee crop exhausted the soils of the area west of São Paulo and spread along new railroads into northern Paraná until bursts of cold wintry air from the south demarcated its limits. Cattle and a variety of other temperate agricultural products, including oranges, are produced in these states, settled largely by immigrants encouraged to move from Germany, Italy, and Japan. Hydroelectricity generated on the Paraná River and its tributaries powers manufacturing in the area. This region has the air of southern European countryside and cities.
- Inland of São Paulo, straddling the drainage divide between the Amazon and Paraná-Paraguay River systems, a new area of farming development in Campo Cerrado grew in significance from the 1970s. With improved road communications, this area became one of the world's main soybean producers, with some 6 million hectares growing

12 million tons per year in the 1990s. Farm management, use of machinery, fertilizer inputs, and marketing procedures are highly organized, and yields rival the best in the U.S. Midwest. New towns grew to over a quarter of a million people within 10 years, and Brazil now exports a greater value of soybeans than of coffee.

## People

### Population Distribution and Dynamics

The distribution of population in Brazil reflects the regional developments (see Figure 10.16). The highest densities are in the southeast, around and inland of São Paulo and Rio de Janeiro. Moderate densities occur in a band parallel to the coast from the southeast around Pôrto Alegre to west of Fortaleza in the north. Farther inland, the very low densities of the Amazon rain forest area create a major geographic contrast within the country.

Brazil's population is moving out of a period of rapid growth, which produced the relatively young population of today (see Figure 10.18) and a population total that grew rapidly to the 1998 figure of 162 million people. Total fertility dropped from nearly 6 to 2.5 between 1970 and 1998, while infant mortality went from 116 to 43 between 1960 and 1998. The rate of natural increase fell from 2.4 to 1.4 percent between 1970 and 1998. The fall in birth and fertility rates (see Figure 10.17) is linked to rising education and urbanization levels. Although Brazil is the world's largest Roman Catholic country and that church in Brazil has a conservative hierarchy, its local priests are liberal on birth control matters. Moreover, a high proportion of women choose to have cesarean operation births followed by sterilization, a service available free from the government. Life expectancy in Brazil remains lower than most other South American countries, in the upper 60s.

### Increasing Urbanization

Brazil has two of the three largest metropolitan areas in Latin America. **São Paulo**, with around 17 million people in its built-up area in 1996, was one of the three largest metropolitan areas in the world—after Tokyo (Japan) and approximately equal to Mexico City. It started in 1522 as an inland mission station and was a base for exploration of the interior by paramilitary bands in the 1600s. Later it became the focus of Brazil's world-dominating coffee industry, and railroads were built westward from it to open up new lands. In the 1900s, it developed into the major manufacturing center of Brazil. The state of São Paulo produces nearly half of Brazil's GDP and two-thirds of its manufacturing output. In the late 1990s, São Paulo's population growth slowed (from 5% to 0.5% per year). New industries tended to be sited elsewhere, often farther out in São Paulo State to avoid the high costs of traffic congestion, pollution, and strong union control (high wages, restrictive working conditions). The provision of housing, schools, and health facilities could not keep up with the metropolitan area's growth, and the administration by 40 different cities and a bankrupt state government also slowed economic activities.

**Rio de Janeiro**, with over 10 million people in 1998, first grew as a port for the gold mines that developed in the interior in the early 1900s, becoming Brazil's largest port and capital city from 1763.

Brazil in general is a highly urbanized country, the proportion of the population living in towns having increased from 56 percent in 1970 to 76 percent in 1998. Other major cities include **Belo Horizonte**, the early center of mining and now the focus of a series of growing mining and metal-manufacturing towns, **Recife** and **Salvador** on the northeast coast, **Fortaleza** on the north coast, and **Curitiba** and **Pôrto Alegre** in the south. **Manaus**, in the center of the Amazon River basin, has over half the population of its state, Amazonas. **Brasília**, the new city built since the 1950s and used as the capital since 1960, now has 800,000 people living in the planned central city area but almost 2 million including the surrounding satellite cities (see Figure 2.13).

Within the Brazilian cities, the shantytowns, known as **favelas** in the southern parts of the country, house millions of poor people who cannot be accommodated by the formal patterns of housing construction, infrastructure, and service provision. It is estimated that over 7 million people in São Paulo and up to 6 million people in Rio de Janeiro live in favelas. Favelas vary in their character. In and around the city centers, the favelas occupy gaps in the built environment, in which families build their own minimal accommodations. It is rare for these favelas to have water, electricity, waste disposal, or anything resembling a road. Many such favelas have been in existence for 10 years or so. Old apartment blocks in the city centers, which were abandoned and taken over by poor people, resemble favelas in many ways, with several families dividing each room. On the outskirts of the cities, different types of favelas exist that are pushing the built-up area outward at a rapid pace. These favelas spring up almost overnight and grow by thousands of people within months, often provided with a basic road and plot plan and soon acquiring electricity and other utilities, shops, and schools as they are formalized into suburban townships.

### Ethnic Variety

The large Brazilian population is varied racially and culturally. The original Native American population is much reduced: in the Amazon River basin, it was probably around 3.5 million in A.D. 1500 but is now closer to 200,000. The traditional culture and people's health are increasingly threatened as outsiders invade their territories.

Portuguese and other European peoples make up a major proportion of the present population, being concentrated in the southern states and elite areas of the cities. The descendants of African-origin slaves form a significant proportion of the population along the northeastern coast and have spread into the southern cities. They have had a major impact on Brazilian culture. Immigrants over the last 30 years include

**FIGURE 10.34 Brazil: the people.** Brazilians watch the 1994 World Cup soccer match between their country and the United States on TV. The crowd includes a variety of racial types.



large numbers of Japanese, now accepted as a part of the Brazilian people and taking many of the top jobs in business and politics. A large group in the population are mixtures of the various races (Figure 10.34), including mulattos (African-European mix).

## Economic Development

Brazil has a diversified economy that continues its history as a producer of raw materials for export but now includes manufacturing and service industries. The distribution of economic activity within the country arose from historic events.

### Varied Economic History

Brazil's colonial economic development began with coastal sugarcane plantations along the northeast coast, interior cattle ranches, mining for gold, and trading for Amazon River basin forest products. In the 1800s, the area west of São Paulo became the world's chief coffee producer. This required the recruitment of workers from Europe and the building of railroads with British financing. The Amazon River basin became a rubber producer of world importance in the late 1800s, but competition with Asian producers ended that boom at the outset of the 1900s (Figure 10.35).

By the early 1900s, the southeastern coffee-producing states contributed over 70 percent of the Brazilian GDP. São Paulo and its port Santos grew at the hub of this development, but by the mid-1900s, the soils of the older coffee lands became exhausted and new coffee plantations sprang up westward along the railroads into Paraná state, with growth especially around the city of Londrina. Brazil's development of primary products for export paralleled that in many other Latin American countries. The sugarcane, gold, and rubber exports lasted for a while and then declined; coffee production was maintained only by shifts to new lands.

Economic stagnation in the 1920s and the world economic depression of the 1930s caused a crisis during which the military took over the government of Brazil. From this point, the Brazilian economy diverged from that in many other Latin American countries by basing internal economic growth on a form of state capitalism. The federal government assumed the ownership of major economic activities and established tariffs to protect internal manufacturing. Import-substitution reached its greatest level of activity in Brazil aided by the huge internal market. Durable goods such as automobiles and domestic appliances that were made abroad were subject to an 85 percent import duty, while some, such as computers, were excluded altogether.

In 1937, the new government nationalized the iron mines in the state of Minas Gerais. The huge mines around Itabira became 80 percent government-owned, and production was increased with 70 percent now being sold overseas, delivered by mining company ships. A steel industry grew up along the railroad from Itabira to the port of Tubarão, with the largest integrated mills at Volta Redonda being established in 1940 and extended in 1970 to double their output.

**FIGURE 10.35 Brazil: Amazon River basin.** The city of Manaus has an opera house that was built early in the 1900s to cater to 2,000 or so rubber barons who lived there and controlled trade in the valuable commodity. Now refurbished, the opera house is in the center of a city of over a million people, with high-tech industries and improved communication by river, road, and air.



## Modern Mining

The national iron ore company is developing the iron ore mining Carajás Project in the eastern Amazon River basin. A railroad built in 1985 takes the ore to a port near São Luis on the northern coast for export to Japan, the United States, and Europe, while the Tucurui Dam generates hydroelectricity for mining and industrial needs. The project required the clearance of 3 million hectares of rain forest, and much of that area is now used for ranching and small farms.

Other Brazilian mining developments include the production of manganese (used in hardening steel), tin, and bauxite. One-third of the world's bauxite resources occurs east of Manaus in the Amazon River basin. In the upper reaches of the Amazon River basin, small deposits of gold attract thousands of independent miners, but the use of mercury to separate the gold pollutes the rivers, and this activity forms a major intrusion in the lives of Amazon tribes such as the Yanomami.

Brazil is increasing its production of petroleum. The oil-consumers' crisis of the 1970s hit Brazil hard and was responsible for much of its high debt. An immediate response was the development of ethanol fuel from sugarcane. The 1975 policy for ethanol development guaranteed a price at 65 percent of gasoline prices. The aim was for one-third of all cars to use this fuel. Early technical problems were overcome, although ethanol-using cars still have low resale prices. The ethanol policy was hampered by the lack of rural pumps selling the fuel, the cutback in sugarcane production as a result of lower subsidies, and the development of Brazilian oil products. It is now necessary for Brazil to import sugarcane to make ethanol, and the pump prices of ethanol fuels are higher than those for gasoline. Petrobras, the Brazilian state oil company, was established to import, refine, and distribute oil products but now finds itself an oil producer with offshore wells along the eastern and northern coasts and major reserves in the western Amazon. By the mid-1990s, over 60 percent of Brazil's oil needs were produced from home sources.

## Farming Today

The government continues to encourage agricultural production, but agriculture's proportion of GDP falls as that of manufactured goods grows. Coffee and sugarcane remain important exports, but coffee, once the mainstay of the economy, now makes up less than 10 percent of the agricultural exports. New developments such as the growing of oranges and other citrus in the south and tropical fruits and nuts farther north added to the diversity of commercial farming. The major new development in agriculture is the efficient commercial production of soybeans that shot Brazil to the top of international traders in this crop (**Figure 10.36**). Government-based research, advice to farmers, and marketing facilities contributed to this success. In many areas, however, one commercial crop tends to be dominant, and so the problems of reliance on **monoculture** in the context of shifting world markets are still felt.

The map of farming land uses in Brazil (see Figure 10.21) shows that the most intensive commercial farms are close to the northeastern and eastern coasts, with deeper inland penetration in the temperate climate environments farther south. Much of the grazing land on the drier uplands and rain forest margins has a low intensity of production.

**FIGURE 10.36 Brazil: western Mato Grosso.** A space shuttle photo of new cattle ranches and soybean farms in a 60 $km^2$ (40 $mi.^2$) area. The plateau surface has been cleared and the steep, floodable intervening valleys left in forest. Lands along the Río Sangue (bottom) are flooded up to 10 m (40 ft.) deep for three months in the summer.

NASA

## Manufacturing Expansion

The largest manufacturing sector in Brazil is the production of automobiles and trucks, concentrated in the southern São Paulo suburbs. Autolatina, in which Ford and Volkswagen shared ownership with the government and produced cars specifically for the Brazilian market, made about half the annual output of 1 million cars until 1995, when Ford and Volkswagen split to take advantage of new sales opportunities in Brazil and Argentina. General Motors and Fiat (near Belo Horizonte) have similar but separate manufacturing arrangements. In the 1990s, the lowering of import tariffs led to further competition, falling prices, and expansion of Brazilian car sales and production, especially of small cars. Output of 0.9 million cars in 1992 doubled by the mid-1990s. By 1995, Renault (France) and Hyundai (South Korea) decided to build new factories in Brazil to take advantage of the growing market and the Mercosul free trade agreements.

The state owns several prominent Brazilian manufacturing sectors.

- Embraer is a state-run corporation that became successful in producing airplanes that sell in the United States and many other countries. It produces fighter-bomber aircraft and other military hardware such as missiles and tanks that are sold mainly to Latin American countries.
- Electrobras is a nationalized corporation that produces and distributes electricity. Small hydroelectricity projects

**FIGURE 10.37 Brazil: hydroelectricity.** The spillway at the Itaipu hydroelectricity station on the Paraná River between Brazil and Paraguay. The 18 massive turbines make this Brazil's largest hydroelectricity investment. Although the electricity is shared between the two countries, Paraguay sells nearly all of its quota to Brazil, where there is greater demand.



on the plateaus in the east gave way to huge projects in the interior (Figure 10.37). The world's largest hydroelectricity project is at Itaipu on the Paraná River at the border with Paraguay. It was opened in 1983 and produces one-fifth of Brazil's power (and much of Paraguay's). The Tucurui Dam on the Tocantins was opened in 1985, and there are further plans for 25 large dams, although environmental concerns are increasingly expressed at the drowning of so much land for water storage. The plans for small nuclear plants (Nuclebras) are not going forward at present because of technical and financial problems. Because government investment remains at the same level, the expansion of electricity production in Brazil does not keep pace with demand that quadrupled since the 1970s. In the mid-1990s, a range of projects was opened to private and foreign investment in an effort to increase the power available.

A further example of state enterprise is the **free trade zone** at Manaus in the Amazon River basin. Established in 1966 so that foreign companies could import materials for assembly without tariffs and export the assembled products to other countries, Manaus is now Brazil's fastest-growing city and the second in manufactured goods value after São Paulo. There are 6,000 factories in its huge industrial park with many major foreign companies, including Honda, Sharp, Kodak, Olivetti, Toshiba, Sony, and 3M. Manaus was transformed by the number of visitors from elsewhere in Brazil coming to purchase goods that they are not allowed to import into Brazil. The "electronics bazaar" occupies a maze of streets in the city center. People come to buy the computers and domestic electronic goods that Brazilian companies, protected by high tariffs, make poorly and sell at high prices.

## Amazon Backwater

The main failure of government efforts is in the Amazon River basin. Efforts to settle the lands along the Trans-Amazonica Highway from the 1950s under the regional development ministry, SUDAM (Supertendencia para o Desenvolvimento de Amazonica), saw little progress. Land was not surveyed properly, unsuitable families with little farming experience were recruited, and bureaucratic convenience caused most of the grants to go to large-scale ranches. When a new road, BR 364, was built into the states of Rondõnia and Acre states in the western Amazon, thousands of settlers entered the area on their own initiative and cut plots in the forest without government encouragement. Forest burning increased to the point that international horror was aroused. Many of these settlers failed to grow anything on their plots. The alternative attraction of the rapid development of the soybean farming in the Campo Cerrado in the early 1990s led to a slowing of such settlement along with the linked forest burning and destruction in the western Amazon (see "World Issue: Tropical Rain Forest," pages 480–481). In early 1998, devastating fires reversed this downward trend by destroying rain forest in the north of the Amazon River basin along the Venezuelan border.

Another aspect of government activity in the Amazon is the Indian rights agency, FUNAI (Fundaçao Nactional do Indio), which seeks to provide education, health care, and support for Native Americans. Unfortunately, it is underfunded and, although it has taken some steps to protect its Indian reserves, there are many occasions when ranchers and miners penetrate tribal lands and begin to develop them before FUNAI can act.

### Indebtedness and New Policies

Following its huge infrastructure and state-owned enterprise investments of the 1970s and 1980s, Brazil found itself in deep indebtedness to world banks in the late 1980s, reaching over $120 billion dollars in the 1990s. The cost of oil imports and the huge public works projects to develop iron mining and hydroelectricity, build Brasília, and construct new roads and airports, together with reported political corruption, contributed to this national debt. Despite Brazil's being by far the most prosperous country in Latin America and having a trade balance in which exports were double imports until the early 1990s, most of the surplus income went to the payment of debt interest. Defaults on loan repayments caused wider foreign investments in Brazil to collapse. Hyperinflation is another financial problem that is an obstacle to further development, since it stifles savings. Brazilians became used to living in an environment of high inflation. Large areas of living costs were indexed to this condition. In early 1994, inflation reached 50 percent per month, over 1,000 percent per year, but renewed attempts to bring inflation under control marked government financial policy later in the 1990s.

The 1990s also saw the start of new government policies moving away from the import-substitution and state ownership of manufacturing that had been major factors in economic growth within Brazil. Tariffs on goods not produced in Brazil were lowered. For example, the 85 percent levels on foreign autos and appliances fell to 20 percent. Some products, including computers, remained illegal imports unless purchased in Manaus. Foreign investment was encouraged, and a sign of these changes is that Japanese car manufacturers are establishing outlets in the major cities.

Brazilian economic diversity places it on a par with some of the newly industrialized countries of Eastern Asia. It also hides the gap between rich and poor within Brazil that is reflected in the uneven distribution of land, the high levels of unemployment, and the strength of the informal economy. Brazil has the capability to develop its economy further, but the current economic and social problems and environmental issues make the future difficult to assess.

## SOUTHERN SOUTH AMERICA

**Argentina**, **Chile**, **Paraguay**, and **Uruguay** form the southernmost part of South America and are sometimes called the "Southern Cone" because of their combined shape on a map (Figure 10.38). All four countries have a Hispanic and Roman Catholic cultural base.

Argentina is the dominant country and was home to 36 of the 59 million people who lived in this subregion in 1998. Apart from Paraguay, very high proportions of the people in Southern South America are of European origin, a fact linked to sparse Native American populations, relatively late economic development and settlement by immigrants from Europe in the later 1800s, and the midlatitude climatic environments. The levels of consumer goods ownership and human development are among the highest in Latin America (see Figure 10.20), although Paraguay does not reach the same levels as the other three countries.

In 1991, Argentina, Brazil, Paraguay, and Uruguay formed the trading group of **Mercosul** following discussions between Argentina and Brazil since 1988. The countries agreed to cut internal tariffs as intraregional trade increased, despite the problems of the high inflation rates in Brazil. The grouping produces half of Latin America's GDP, and the European Union—Mercosul's main trading partner—is showing interest in negotiating a closer agreement. Chile and Bolivia are associate members.

Argentina continues to claim sovereignty over the Falkland Islands (Islas Malvinas) despite its expulsion after the 1982 invasion. In 1995, it negotiated a cooperative venture with the United Kingdom to develop the oil potential of the continental shelf area southwest of the islands. It is thought likely that this area may be rich in oil, but none has been recovered as yet. Drilling began in 1998 north of the islands (outside the Argentinean area), but commercial production is many years away.

### Countries

Although **Southern South America**'s distinctive characteristics set it apart from the rest of Latin America, the constituent countries display major differences among themselves. Chile, Argentina, and Uruguay have relatively strong export-based economies, but Paraguay is landlocked and has more restricted economic opportunities. While Argentina had 36 million people in 1998 and Chile just over 15 million, Paraguay had around 5 million, and Uruguay just over 3 million.

The geographic variety among these countries stems partly from physical and especially from historic causes. The Andes Mountains proved an important dividing factor between types of settlement. Under Spanish rule, Chile was governed from Lima, a city that discouraged settlement in Chile as being too far away to control. The lands east of the Andes Mountains became part of the Viceroyalty of La Plata and were first settled with the limited objective of providing food and animals for the Potosi silver mines in Bolivia. The whole subregion lagged behind the Northern Andean countries and Brazil through the 1800s. Despite the small amount of economic development on both sides of the Andes, a series of towns and ports were established early in the 1500s as part of the Spanish control network but grew slowly before independence. The whole subregion lagged behind the Northern Andean countries and Brazil through the 1800s.

Following independence in the early 1800s, Chile became one country, while Argentina at first incorporated modern Paraguay and Uruguay. Paraguay refused to acknowledge

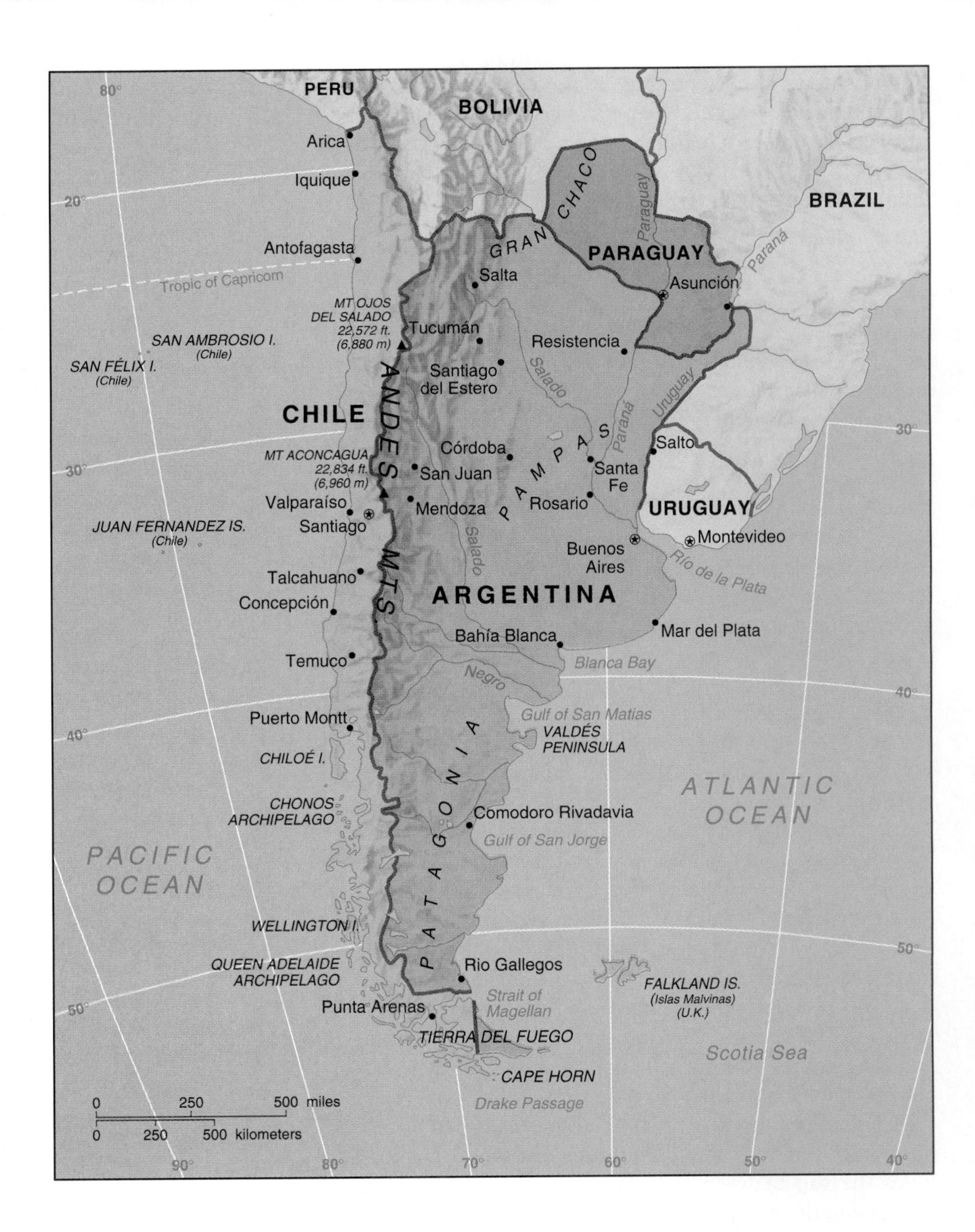


**FIGURE 10.38 Southern South America: major geographic features.** The countries, Andes Mountains, rivers, and major cities.

Argentinean control, however, and declared its own independence in 1811. With British backing, Uruguay also gained its independence from Argentina by the 1820s. Disagreements over borders continue, although Paraguay was subdued after late 1800s border wars with all its neighbors left it with a halved population and with a smaller land area.

Expanding contacts with Europe led to the development of mineral extraction and commercial farming by the end of the 1800s, although Paraguay's political problems and interior location continued to hold it back. Chile's exports were mainly nitrates and copper, Uruguay's were livestock products, and Argentina's livestock and grain. Since the middle of the 1900s, these countries, again with Paraguay lagging, industrialized with movements of people from rural to urban areas.

## People

The main population centers in Southern South America are around the Plate River (Río de la Plata) estuary (Argentina, Uruguay) and in central Chile (see Figure 10.16). Smaller centers occur in the irrigated farming oases of northern Argentina and around Asunción, capital of Paraguay.

The population of Southern South America is growing more slowly than in the other subregions of Latin America. In 1930, this subregion had nearly 20 percent of Latin America's total population, but it now has less than 12 percent. The 59 million people who lived there in 1998 could rise to around 80 million by 2025. Rates of population growth and fertility are low (see Figure 10.17) apart from Paraguay, where the 1990s growth rate was 2.7 percent per year and the total fertility rate was 4.4. Argentina, Chile, and Uruguay had low annual population increase rates, not very different from those in the 1970s. Total fertility rates in 1998 were under 2.5. The age-sex graph for Chile (see Figure 10.18) suggests a rising birth rate in the 1980s.

Apart from Paraguay (52% urban in 1998), all the countries of Southern South America had over 85 percent of their populations living in towns. As in many countries of Latin America, each country has a primate metropolitan center. **Buenos Aires** grew from a population of 170,000 in 1870 to around 12 million people, one-third of the Argentinean total, in 1996. **Montevideo**, Uruguay is home to nearly half the national total. **Santiago** has one-third of the Chilean total. **Asunción** is the largest city in Paraguay with over 20 percent of the country's total. These cities are centers of government, manufacturing industries, and service industries.

Few other cities in any of the four countries approach these national primate cities. In Argentina, the second city is **Córdoba** and the third is the inland port of **Rosario.** Other centers include towns in the northwestern irrigated areas such as Tucumán and Mendoza, and small ports in Patagonia. Chile has a series of port towns along its length that have hinterlands in mining, farming, or forested areas.

In all these countries, immigration from Europe since the late 1800s was of great significance to population growth. Chile grew in population from colonial times, but the growth of commercial farming and mining in the late 1880s led to more rapid increases, including some 200,000 German immigrants between 1881 and 1930. Growth was slower at first in Argentina and Uruguay, but immigration became more important after independence. Between 1821 and 1932, 6.4 million people emigrated to Argentina from southern and eastern European countries, and another million to Uruguay. Paraguay received few immigrants but lost around 60 percent of its population in its wars with Brazil, Argentina, and Uruguay in the 1860s and sustained further losses in the Chaco War against Bolivia in the 1930s. After such losses, Paraguay also encouraged migration—with only moderate success.

The proportions of immigrant population give different emphases to the racial and ethnic mixes of the Southern South American countries. In Chile, around 40 percent of the population claims to be European and the rest mestizo. In Paraguay, virtually all the people are mestizos apart from a few people of European or African origin. Argentina and Uruguay have the largest proportions of Europeans and the smallest proportions of mixed races.

## Economic Development

Argentina, Chile, and Uruguay followed economic development patterns that are somewhat typical of Latin America. Paraguay always lagged behind the others. Up to the mid-1900s, Argentina and Uruguay were the most prosperous, but two disastrous decades set Argentina back among the struggling countries.

### Paraguay

Paraguay remains a poor developing country. The larger portion of the country to the west of the Rio Paraguay, the mainly semiarid Gran Chaco, remains lightly settled apart from military camps and lumber operations. The quebracho ("ax-breaker") tree is its commercial timber product. The tree is very hard and in demand for railroad ties and its tannin extract.

The eastern part of Paraguay is more developed, with forest products being diversified by commercial agriculture. Livestock products and some industrial crops are exported. Paraguay gets its power from hydroelectricity and earns foreign exchange from its share of the Itaipu project on the border with Brazil (see Figure 10.37) by selling most of it to Brazil. Transportation connections with Brazil improved, and that is now the main direction of trade. Paraguayan border towns such as Cuidad del Este sell cheap consumer goods to Brazilians and are involved in smuggling between Brazil and Argentina.

### Colonial Legacy

In the other countries, the European Hispanic culture was imposed on diverse groups of Native Americans, including the grassland nomads east of the Andes and the peoples of southern Chile, who resisted the Spanish arrival. During the 1500s, Spanish settlement in Southern Latin America was hesitant. The initial settlement at Buenos Aires was abandoned in 1641 after five years of occupation. By 1600, however, towns in central Chile and northwestern Argentina were centers of agriculture and roads linked them to more northerly Spanish settlements. The oasis towns in northern Argentina supplied food to the Bolivian silver mines. These towns gradually extended Spanish control over their hinterlands. Montevideo was established to block Portuguese expansion from Brazil toward the La Plata estuary. Following independence, each new government expressed national aspirations to build their capital cities. Economic growth remained slow, however, and a series of wars, mainly focused on Paraguay in the 1860s, also held back development. Argentina took over Patagonia in 1881.

### Growing Engagement with World Economy

In the late 1800s, Britain and other European countries recognized the potential of the pampas grasslands as a source of grain and livestock for the expanding demand in the industrial Northern Hemisphere countries. By then, refrigerated ships could carry meat products to these countries. British capital built railroads around Buenos Aires and Montevideo and encouraged the owners of large estates to move from extensive livestock management to intensive grain farming.

To cope with such economic expansion, the governments of Argentina and Uruguay in particular encouraged immigrants from Europe to come and work on the estates. Many of the immigrants soon moved into the towns. This shift in farming methods led to the final control of the warlike pampas Native Americans that had kept settlement at bay until the 1880s. In similar fashion, German immigrants to the southern part of central Chile in the mid-1800s removed the Araucarian Native Americans who had previously prevented occupation of the land south of the Río Bío Bío. By the early 1900s, the lands around Buenos Aires and Montevideo vied with some European countries in both the quantity and quality of life.

### Economic Isolation

The economic depression of the 1930s and World War II in the 1940s shifted these countries away from dependence on the farm and mine products demanded by Europe. The countries of Southern South America introduced a broader base of manufacturing protected by high tariffs to reduce dependence on Europe and the United States. Import-substitution was at first successful, but its impact was reduced after 1950 as the United States and Europe reestablished their manufacturing markets with better and cheaper goods, while prices for primary products fell because of overproduction.

The economic problems were compounded by a series of governments, both civilian and military, that tried to maintain their protected industries in order to reduce dependence on primary product exports. Their reliance on import-substitution products tied to the needs of their own country's market, however, made it difficult for them to sell these goods abroad in the way Brazil, with its huge internal market and diverse range of products, managed to do in selected sectors. Primary products still constitute a large proportion of the exports of all these countries, although they have also been diversified. Manufacturing constitutes over 25 percent of employment and GDP in Chile, Argentina, and Uruguay but only around 20 percent in Paraguay.

### Modern Chile

In Chile, the dictator Augusto Pinochet's regime that lasted from 1973 to 1989 established an economy based on open trading, increasing exports, good financial management that withstood the fallout of the Mexican collapse in 1995, and low unemployment. After the 1989 democratization of government, the new government followed similar policies with the addition of a greater emphasis on social programs to reduce the numbers of poor people.

At the same time, Chile's economy diversified. Copper, which made up 80 percent of Chile's exports in the 1970s, was down to 44 percent in the late 1980s, although the amounts increased. Agricultural products such as apples, grapes, and wine, together with fish, lumber, and new minerals grew in proportion. The agricultural products and lumber come from central Chile, where a large proportion of Chileans live. Copper mining is important in the Atacama Desert of northern Chile, with the world's two largest mines at Chuquicamata and Escondido along with a series of new large mines being prospected and developed in the late 1990s. After the highest grade copper deposits were mined out by the early 1900s, Chile used U.S. technology to mine lower-grade ores. It now leads the world in refining technology.

**RECAP 10E: Brazil and Southern South America**

Brazil is the largest country in Latin America and contains one-third of its population. Its economy is in the world's top 10, but extremes of riches and poverty exist side by side. Brazil consists of contrasting regions: the sparsely settled tropical rain forest of the Amazon River basin, the industrial cities around São Paulo and Rio de Janeiro, the poverty-stricken northeast, the commercially farmed plateau interior around Brasília and Campo Cerrado, and the southern farming and industrial states. Much of Brazil's economy is controlled by the government through import duties and ownership of corporations that are responsible for utilities, encouraging farming developments, and the high-tech sector.

Southern South America (Argentina, Chile, Paraguay, and Uruguay) has midlatitude climatic environments. The Andes form a distinctive boundary between Chile and Argentina. These countries, apart from Paraguay, have greater proportions of European-origin peoples and more diversified economies than other parts of Latin America outside of Brazil, but they suffered from unstable political conditions and protectionism until the 1990s.

10E.1 What are the main economic uses of the Amazon River basin? How have attempts to settle the area solved the problems faced?

10E.2 How would you summarize the main features of the distribution of agriculture and manufacturing in Brazil?

10E.3 Why did the European settlement of Southern South America occur later than in other parts of Latin America? What effect did this have on those countries' human geography?

**Key Terms:**

favela, free trade zone, Mercosul, monoculture

**Key Places:**

Brazil
- *São Paulo*
- *Rio de Janeiro*
- *Belo Horizonte*
- *Recife*
- *Salvador*
- *Fortaleza*
- *Curitiba*
- *Pôrto Alegre*
- *Manaus*
- *Brasília*

Argentina
- *Buenos Aires*
- *Córdoba*
- *Rosario*

Chile
- *Santiago*

Paraguay
- *Asunción*

Uruguay
- *Montevideo*

Southern South America

### Argentina and Uruguay

Argentina has a larger manufacturing base than Chile. The upturn in world economic activity and economic restructuring in the 1990s resulted in more investment, mainly from foreign sources, in Argentinean manufacturing. Argentina became the fastest-growing economy in Latin America as it established a stable financial system.

Most of the Argentinian factories are based in and around the Buenos Aires metropolitan area. The farm products of the pampas are still important to Argentine trade, but new oil and gas resources fuel new industries, mainly on the coast. Argentinean service industries developed as part of this growth. In 1996, over 4 million foreign tourists visited centers on the coast and in the mountains. Some of the older settlements, such as the northern cities of Tucumán, Córdoba, and Mendoza, where the economy is still based on irrigation agriculture, do not grow as rapidly because of the dominance of Buenos Aires in attracting manufacturing investment. Some of the inland centers, such as Jujuy near the Bolivian border, remain poor and overwhelmed by migrants out of Bolivia. In the late 1990s, the Mexican crisis and fears of open competition with Brazil as Mercosul develops led to a slowing of growth in Argentina, making potential political problems for further economic restructuring.

Uruguay remains a largely agricultural country, with meat and wool exports, but the industrial developments in and around Montevideo, powered by hydroelectricity generated along the Rio Negro, draw workers from the rural areas. Over 80 percent of Uruguayan trade passes through Montevideo, which has over half the country's economic product and population. It developed an offshore banking industry. It attracts tourists to its beaches in an industry that contributes more than either farming or manufacturing to GDP, with over 2 million visitors in 1996.

## LANDSCAPES OF LATIN AMERICA

### Urban Landscapes

Urban landscapes in Latin America combine elements of pre-European, colonial, and more modern buildings. Some of the pre-European elements include the relics of former Mayan (see Figure 10.22) and Inca cities. These were often left to one side—and later abandoned to forest—while the Spaniards built their own cities a short distance away. Spanish and Portuguese colonial architecture is distinctive with towns built in the 1500s and 1600s focusing on the churches that were built as part of a central square. The historic settlements were small and overwhelmed by 1800s and modern urban expansion (see Figure 10.19).

#### Shantytowns

Today the cities of Latin America are pressured by migration from rural areas. People are drawn to cities by expectation of greater opportunities and better education and health services or are driven out of rural areas by civil war or lack of employment. The outcome is similar, for the large-scale and rapid movements of people into towns overwhelmed the planning and financial capabilities of the urban authorities. Informal, often squatter, settlements arose in which people took over unused land and built their shacks on it without water or electricity supplies, sewage disposal, or other services, beginning a new sort of urban expansion. The overall result of such rapid growth is a combination of overcrowding, congestion, and air and water pollution. Once such shantytowns were established, especially on the edges of cities, the authorities began to pave the streets, provide utilities, and make access possible to schooling and health care.

In Brazil, most inner-city favelas (shantytowns) have houses that are built of packing cases and other discarded materials (see Figure 2.21). Disease and infant mortality rates are high, but there is often access to schools and charity-provided preschool education. The city-edge favelas may develop rapidly into more regular suburban facilities. The city authorities often plan a road pattern for new squatter settlements. People build their own housing along the unpaved roads. After a few years, they can tap into electricity supplies, gain a water and sewerage system, and have local access to schools and medical care. Eventually the roads are paved, and a center with shops develops. This process is seen in many city-edge favelas, including San Pedro on the outskirts of San Bernardo do Campo, a major truck-manufacturing town in the São Paulo area (Figure 10.39).

#### Brazilian Cities

The built environment of Brazilian cities is distinctive. Large numbers of middle-class people live in inner-city high-rise apartment blocks with external walls, fences, and guards and internal swimming pools and other services. Others live in single-family homes in the suburbs but guard their possessions with massive steel fences and gates (Figure 10.40).

One of the most remarkable examples of urban growth in Brazil is that around the capital, Brasília—built largely according to the original plans. The central area of government buildings is flanked by two "wings" of apartment accommodation with local shopping facilities (see Figure 2.13). More affluent housing for the diplomatic corps and highly paid bureaucrats was built around the artificial lake. Although commercial facilities have been added centrally and some of the apartment blocks were not built because of unsuitable ground conditions, the center of Brasília reflects the original design. Outside the center, often at a distance of a one-hour bus journey, a series of satellite towns vie with Brasília for housing the growing population. Paranoa, one of these satellite towns, was started when a squatter settlement had to be moved to enable the pope to hold a service on a hillside north of Brasília. The squatters were given land in a new satellite city but erected their own housing. The main street is still like a frontier town, but other signs of town development are there.

#### Buenos Aires, Argentina

The rapid growth of Buenos Aires followed patterns that are common to other cities in Southern Latin America. After slow growth in its early history, Buenos Aires expanded in the late 1800s as commercial farming took hold in its pampas hinterland. The built environment developed a trading and industrial waterfront and a central thoroughfare at right

(a)

(b)

**FIGURE 10.39 Latin America: urban landscapes.** Two views of Curitiba, southern Brazil. (a) The central business district, part of which is designed for exclusively pedestrian access. Curitiba is widely known for its planning of public transportation. (b) View from an affluent suburb, across an area of favelas (dark strip) to the high-rise apartments near the city center on the skyline.

a & b © Michael Bradshaw

(a)

(b)

**FIGURE 10.40 Latin America: urban landscapes.** The San Pedro favela, San Bernadino—part of the urban region centered on São Paulo. (a) The favela becomes established along a grid of unsurfaced roads with houses built of cement blocks and packing cases. There are few services apart from electricity supply. (b) Where the favela joins the main road, the roads are becoming surfaced, and a varied range of services, including shops and schools, are available.

a & b © Michael Bradshaw

angles along the route inland. Rapid growth occurred in both the economy and the immigrant population around 1900 resulted in the middle class of skilled workers and office workers moving out to new suburbs, while the poor and most affluent remained in the inner city. Amenities came slowly to the suburbs. During the later 1900s, increasing rates of population growth produced squatter settlements and rising inner-city population density as apartment blocks replaced mansions. Population growth stagnated in the coastal industrial areas. From the 1960s, much of the commercial and industrial activity moved out of central Buenos Aries to the city edges. There were attempts to divert the overcrowding in the Buenos Aires metropolitan area by placing new projects and development in other centers within the Buenos Aires orbit but farther inland, although these attempts have not been successful.

### Rural Landscapes

Rural landscapes in Latin America include its highly commercial cultivated landscapes such as the Argentine and Uruguay pampas, the Campo Cerrado of Brazil, the West Indian sugar plantations, and the coffee plantations of the Northern Andes and Central America. There are landscapes of subsistence agriculture, as in the Native American areas of Central America and the tribal areas of the Amazon River basin. Poverty-stricken rural areas with partly cultivated fields and straggling villages of mud-brick houses occur in northeastern Brazil and the high Andes. Large areas of mostly untouched natural environments still exist in the high sierras, the remote parts of the Amazon River basin, and on either side of the southern Andes. In places they are interrupted by unreclaimed mining pits and spoilheaps, abandoned plots in cut forest, or huge artificial lakes that are parts of hydroelectric projects.

## FUTURE PROSPECTS FOR LATIN AMERICA

In the 1990s, reports suggested that Latin American countries would advance in economic and human development as they became closely involved with the world economic system. And yet the region remains "the continent of the future" that it has been for many decades. Although there is evidence of economic growth in some Latin American countries, they have often been subject to economic and political fluctuations in the past, and there are still worrisome aspects that could hold them back in the future. The experience of Mexico in the mid-1990s caused any optimistic predictions to be more guarded.

The positive evidence is that countries such as Mexico, Chile, and Argentina experienced rapid economic growth in the early 1990s, while Peru showed rapid growth in the mid-1990s. Brazil has by far the largest economy in Latin America and the greatest potential but is changing more slowly than these other countries. Venezuela, on the other hand, one of the previous leaders because of its oil wealth, lost ground in the 1990s as world oil prices dropped.

The countries showing positive trends are generally those where economic policies reduced inflation, opened the internal markets to foreign firms, and privatized state-run businesses. Improved government financial management, together with exchange rate control, led to spectacular results in Mexico, where a large deficit in the late 1980s was turned into a surplus in 1991, and in Argentina, where a budget surplus was achieved in 1993. The move from import-substitution manufacturing to engagement in world markets led to reductions in tariffs on imports and greater efforts to integrate regional trade. Trade among the largest 11 countries in Latin America doubled between 1989 and the early 1990s. The most active of these agreements is the Mercosul customs union.

The economic openness goes beyond trade in manufactured products to financial sectors and credit systems, so that foreign banks are attracted with their investment management. Privatization of state-owned enterprises was the greatest change, and in many cases, the twin objectives of providing revenue for the government and more efficient working practices were met. Chile, Argentina, and Mexico sold off almost all state concerns apart from the Mexican oil and electricity companies and the largest Chilean copper producer. Peru and Colombia have even wider intentions.

Although a number of Latin American countries show positive signs of economic and human development, others stagnate economically or lose ground. Most of the Central American countries, Haiti, the Guianas, Paraguay, and Bolivia remain poor, often held back by civil wars, lack of resources in world demand, or poor political leadership. They face similar problems to many African countries.

A number of potential dangers faced the growing Latin American countries after the setbacks of the 1980s. Inflows of capital are notoriously linked to world economic conditions and internal political events. A political murder in early 1994 caused capital inflows to Mexico to drop by 20 percent, while the Zapatist guerrilla uprising in Chiapas state was linked to the devaluation of the peso in late 1994 and the financial setbacks of 1995. There is a high reliance on foreign capital and too little on locally saved money: domestic savings in Latin American countries are little more than half the amounts per head in Eastern Asia. Reductions in the role of the state in running industrial and financial businesses are not always escaping the influence of state bureaucrats. In the late 1990s, Latin American countries faced major setbacks from the natural environmental disasters of the 1997–1998 El Niño event, the falling Eastern Asian demands for minerals, and uncertainties as to future international linkages.

Above all, any positive economic changes benefited the wealthier groups in society and left large numbers of poor people behind. Latin American countries had the world's

**RECAP 10F: Landscapes and Futures**

Urban landscapes in Latin America contain relics of pre-European, colonial, and early independence town layouts and buildings, but they have been mostly overwhelmed by modern growth and shantytowns. Rural landscapes in Latin America include commercially farmed areas, subsistence areas, and little-altered natural areas.

The economic future of Latin American countries appeared optimistic until early 1995, when the gains of economic restructuring and potential trading linkages were disrupted by imposing these processes on traditional Latin American society. The Mexican financial problems beginning at the end of 1994 highlighted the insecurities of social, economic, and political conditions in the region.

10F.1 In what ways are cities in Latin America distinctive in comparison, say, to those in Anglo America, Europe, or Asia?

10F.2 What are the main exports of Latin America? What economic problems does such a list reflect?

10F.3 How have Latin American countries responded to involvement in the world economic system in the 1990s?

**WEBSITES: Latin America**

For the smaller Latin American countries, consult the *CIA Factbook* and make comparisons:

*http://www.cia.gov/cia/publications/factbook/index.html*

For Mexico, there is a wealth of sites:

*http://www.presidencia.gob.mx/* The Mexican government site: in Spanish, but go to the "International Page," which is in English (if you need to). Try the "Chiapas Press Room," which has press releases and papers on the Chiapas problems, or the "Briefing Room," where there is access to "Economic and Social Statistics" and other information.

*http://www.qvision.com/palenque* The Virtual Palenque site, giving a tour of the Mayan palace

*http://www.arts-history.mx/* Mexican arts and history with much visual material. Choose "English version."

*http://gorp.com/gorp/location/latamer/mexico.htm Great Outdoor Recreation Pages* site for information on Mexican national parks and other tourist attractions. The GORP home site offers similar information for other parts of the world.

For Brazil:

*http://www.brasil.gov.br* (in Portuguese) Print out the map.

*http://www.amanakaa.org* News on Amazon ethnic issues.

greatest disparities between rich and poor. In Mexico, the wealthiest 20 percent of the population have incomes that are nearly 30 times those of the poorest 20 percent. In Argentina, they are 16 times as great. These rich-poor differences compare with 5–10 times in Eastern Asia and 11 times in the United States. They make social discord more likely.

Furthermore, the processes of privatization and government cutbacks put many people out of work or pushed them into low-wage, casual work. Private industry could not create large numbers of new jobs. Unemployment rose, while infrastructure that might link peripheral parts of countries to trading opportunities was delayed. Levels of education were not high enough to take full advantage of development opportunities. More could be done to lower illiteracy rates in countries such as Brazil, as well as in the poorest countries such as Bolivia and many Central American and West Indian countries.

The future of the countries in Latin America depends on whether they can maintain economic growth and devise and agree on effective policies to overcome the potential dangers inherent in such growth. Increased local savings, a better educated population, a more efficient civil service, and improved facilities for the poor in cities and rural areas are necessary in most countries.

# CHAPTER REVIEW QUESTIONS

1. A group of countries that lowers tariffs and trade restrictions among its members, blends economic policies and business laws, and is ruled by a central authority that may override individual governments is (a) a free trade area (b) a customs (c) a common market (d) the United Nations
2. Which of the following associations is *incorrect?* (a) Aztecs and central Mexico (b) Chimu and Argentina (c) Inca and Peru (d) Maya and southern Mexico, Honduras, and Guatemala
3. Which of the following was *not* a common feature of Spanish colonies in Latin America? (a) Mineral exploitation (b) A hierarchical social structure (c) Transportation to a limited number of ports (d) Firm and well-recognized boundaries
4. In the 1990s, Latin American countries changed their economic policies to include all of the following measures *except* (a) encouraging foreign investment (b) high tariffs on imports (c) privatization of government corporations (d) encouragement of exporting industries

5. The part of Latin America affected by hurricanes includes: (a) Mexico (b) Central America (c) the West Indies (d) all of the above
6. The world's largest expanse of tropical rain forest is in the valley of the (a) Paraná River (b) Paraguay River (c) Amazon River (d) Orinoco River
7. The Central American country with the longest history of democracy and political stability is (a) Honduras (b) Belize (c) Guatemala (d) Costa Rica
8. An industry that is thriving in the Northern Andes is (a) mining (b) bananas (c) textiles (d) illegal drugs
9. Efforts to eliminate the production of illegal drugs in the Northern Andes have been largely unsuccessful because of all of the following *except* (a) Native Americans in the region have a long tradition of chewing coca leaves (b) no other crop can grow in that environment (c) there is an insatiable demand for the product in the United States (d) farmers stand to make a lot of money by growing coca and other drug plants
10. The main mineral produced in the deserts of northern Chile is (a) copper (b) silver (c) petroleum (d) gold
11. The Treaty of Tordesillas provided for the division of the world between the Spanish and the British.
True/False
12. The labor force on Portuguese sugar plantations was Native American people.
True/False
13. Even after political independence, Latin American countries continued to be economically dependent on Europe for markets for their products.
True/False
14. Soils on the Brazilian Plateau are generally fertile.
True/False
15. The soils of tropical rain forests are very fertile.
True/False
16. The West Indies have had the most diverse history and cultural impacts of any part of Latin America.
True/False
17. Brazil's fastest-growing city, based on a free trade zone, is Manaus.
True/False
18. Government programs to settle and develop the Amazon River basin have been a great success.
True/False
19. Development of Argentina and Uruguay in the 1800s was primarily the result of investment of local capital.
True/False
20. Latin America has recently seen economic development and growth, but great gaps remain between the rich and the poor.
True/False
21. People of mixed European and Native American heritage in Latin America are called __________.
22. In 1823, the United States proclaimed its interests in Latin America by the __________.
23. A phenomenon of warmer-than-normal water off the west coast of South America and abnormal weather through most of the world is called __________.
24. Located in a bowl-shaped valley 2,000 m (6,500 ft.) above sea level, which hinders the flow of air, the city with the worst air pollution problem in Latin America is __________.
25. Adopted in 1917 to give Mexican peasants an opportunity to own their own land but scrapped in the 1990s as uneconomical is the __________ system of landholding.
26. Native Americans under the name of Zapatistas have staged guerrilla warfare in protest over the economic reforms of the 1990s in the southern Mexican state of __________.
27. Factories in northern Mexico that import components duty-free, assemble them with cheap labor, and export them to the United States are called __________.
28. In Latin America, repeated cycles of rising demand for mining and agricultural products in return for tax-free imports led to a condition of __________.
29. The Bolivian strategy of inviting foreign capital to invest in state companies and then distribute the proceeds to public institutions is called __________.
30. Shantytowns, frequently built by new migrants to a city, are called __________ in Brazil.

# SOUTH PACIFIC

Chapter 11

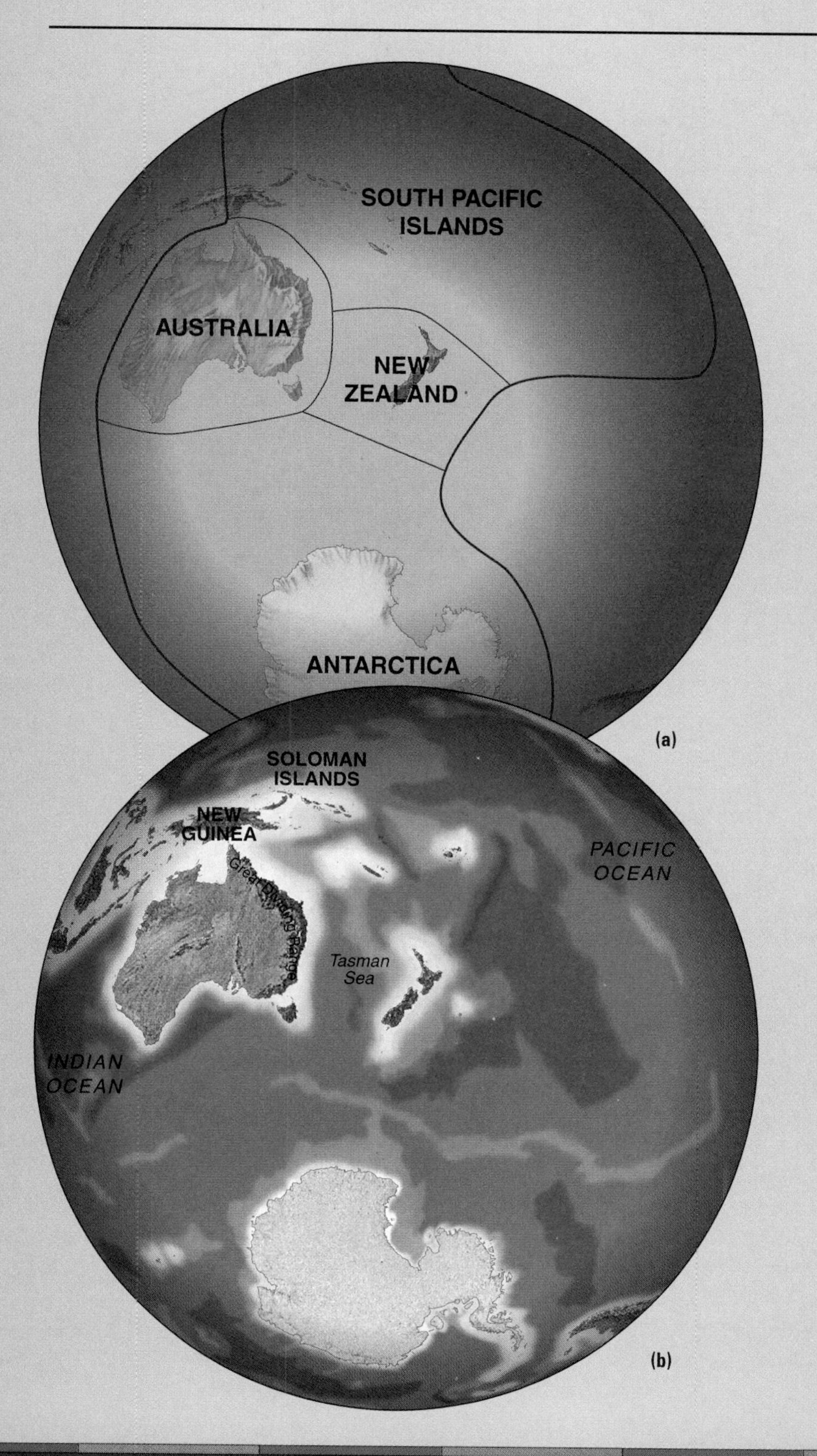


**THIS CHAPTER IS ABOUT:**

**Richer countries among some of the world's poorest in an isolated region**

**European cultures imposed on South Pacific peoples**

**Living in New Zealand**

**Dry and frozen continents plus thousands of small islands**

**Four subregions:**

- Australia: the only single-country continent
- New Zealand: pleasant environment, few people
- South Pacific Islands: the idyllic and the poverty-stricken
- Antarctica: the unclaimed continent

**Trading ports, vast outback, and island landscapes**

**Future prospects for the South Pacific**

**World Issue: Pacific Rim**

**FIGURE 11.1 South Pacific: major geographic features.** (a) The subregions and some countries. (b) Natural environments, showing the great extent of brown arid lands in Australia and the ice desert of Antarctica.

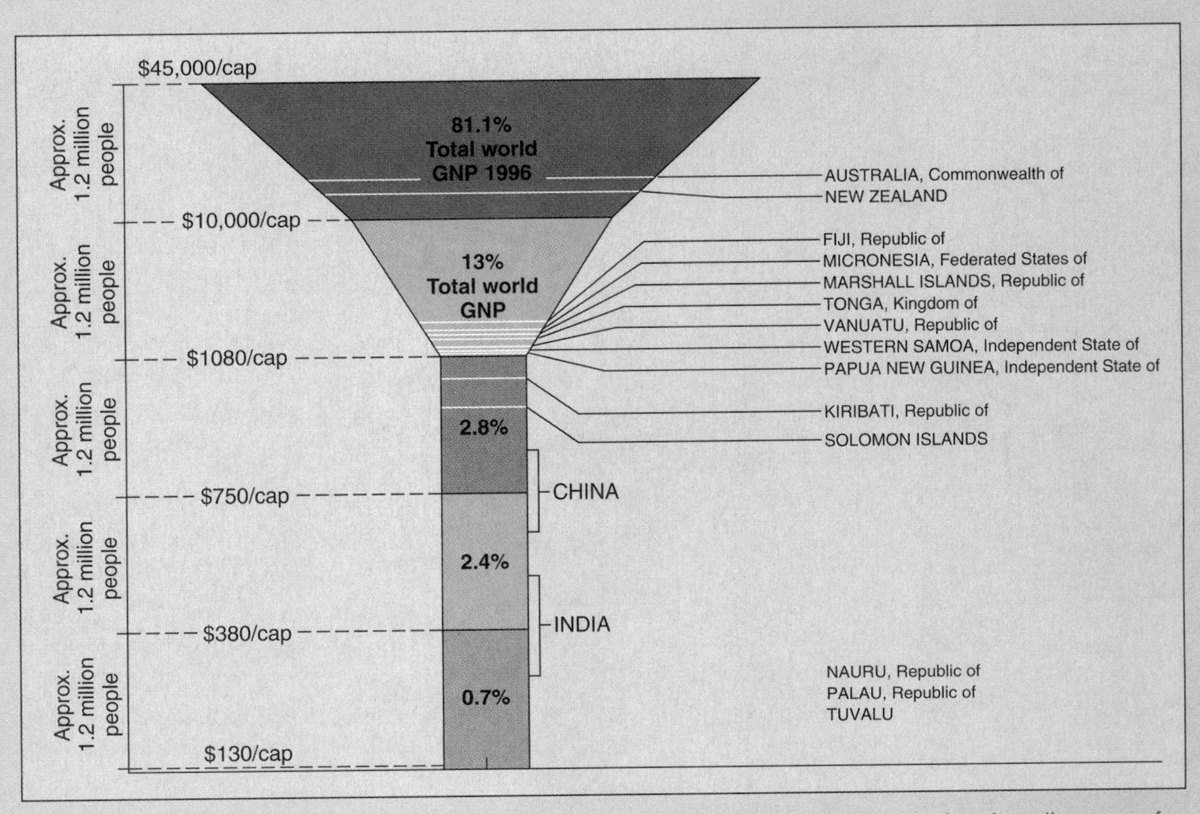


**FIGURE 11.2 South Pacific: comparative economic well-being of the countries.** Note the range from Australia to some of the world's poorest island countries.

Source: Data (for 1996) from *World Development Indicators,* World Bank, 1998.

## CORE IN THE PERIPHERY

The South Pacific region comprises a group of countries in or adjacent to the South Pacific Ocean that have linked experiences of the new global order. They range from some of the world's richest countries to some of the poorest island countries, but all have relatively small populations. The total number of people living in the region in 1998 was 30 million, spread over an area equal to that of the other major world regions that are home to several hundred million people.

Australia and New Zealand, the largest countries in the South Pacific region, are affluent nations—parts of the world core. They are separated by great distances from the other core countries in the northern hemisphere and are set among countries of the periphery (Figure 11.1). Most island countries have low incomes and dependent relationships with Australia, New Zealand, and other core countries (Figure 11.2). The huge ice-covered polar continent of Antarctica occupies the extreme south of the region, with Australia and New Zealand claiming nearly half of it and acting as staging points for access to Antarctica's eastern sectors. Australia and New Zealand are thus an internal core in this region, but they, the other South Pacific countries, and Antarctica also relate closely to the northern hemisphere cores in Europe, Anglo America, and Japan.

Areas and distances within the region are great. Australia is the world's sixth largest country by area, occupying a whole continent. Antarctica is twice as big. From westernmost Australia to Pitcairn Island in the east it is over 12,000 km (8,000 mi.), equal to the distance across Russia, the world's largest country by area and east-west distance. From southernmost New Zealand to the Northern Mariana Islands, it is nearly 10,000 km (6,000 mi.). These distances do not include Antarctica, since it is largely isolated from the rest of the world.

The South Pacific holds only 0.5 percent of the world's population living on 17 percent of the world's land—if Antarctica is included (or 7% of the land if not). The low population totals of these lands relate to the local environmental difficulties, which limited the numbers of pre-European peoples with subsis-

**FIGURE 11.3 South Pacific: the legend and the naked truth.** An idyllic view of Kili, southern Marshall Islands. Kili is just 50 km (35 mi.) from Bikini Atoll, one of the northern Marshall Islands that was used for testing nuclear bombs and suffered radioactive pollution.



tence cultures, and distance from the core countries, which resulted in modest numbers of settlers and financial investments. The population of Australia was nearly 19 million in 1998, no more than medium-sized in world terms. A further 8 million people were shared between the two larger island countries of New Zealand and Papua New Guinea. Under 3 million people were spread through the smaller South Pacific islands. Some of the smaller islands have high population densities but small totals. Fiji, the largest, has 750,000 people—twice as many as the next largest, the Solomon Islands. There is no permanent population on Antarctica.

The contrasts in population totals and size are paralleled by the range of economic well-being among South Pacific countries. In GNP terms, Australia and New Zealand stand out above the various island groups, although the French and United States influences support higher incomes in some islands.

These scattered lands are all distant from the world cores in Europe, Anglo America, and Japan. Australia, New Zealand, and the South Pacific islands were all colonies of the core lands having favored access to the colonial country's home market. After independence, the South Pacific countries had to reorientate their economies to more local and **Pacific Rim** markets and to greater productive efficiency (see "World Issue: Pacific Rim," pages 530–531). Countries that could not readjust still rely on economic support from outside the region. Some islands remain colonies or protectorates of core countries. Now that the costs of distance are falling and the world economic system is becoming ever more pervasive, the transitions to the new orientations impose stresses on Australia, New Zealand, and the South Pacific islands.

The distance factor has positive as well as negative effects. The islands of the South Pacific are perceived by some as the ultimate escape from the rest of the world (Figure 11.3). Australia and New Zealand—and even Antarctica—have a growing range of interests for tourists. Their distance from the urban industrial centers of Europe, Anglo America, and Japan means that they do not have the same potential for mass tourism as Majorca or Miami. They are areas for distant exploration by Europeans and Anglo Americans and closer tourist opportunities for the growing demands in Eastern Asia. Sun, sea, outdoor and outback opportunities, together with golf courses and special developments such as the theme parks of Queensland's Gold Coast in Australia, attract increasing numbers of tourists. In 1996, Australia welcomed over 4 million tourists and New Zealand 1.5 million, while the South Pacific islands attracted a further 1.5 million among them.

The countries and Antarctica have distinctive places in their own region and in the new global order. Elements of world economic interaction and change integrate this region more fully within the worldwide system. It would be a mistake to regard this region as of little significance in world geography: every aspect of our world has its part to play, and regions with smaller populations often illustrate processes

# PACIFIC RIM

The issues facing Australia, New Zealand, and the South Pacific islands include:

Is it necessary to reorient their economies to the needs of Eastern Asian countries?

What changes are occurring in their national economies and attitudes to other peoples?

Can they adjust to the fluctuating economies of Eastern Asia?

Australia, New Zealand, and the South Pacific islands are going through a period of reorientation forced on them by political, economic, and geographic realities. The period in which they were sustained by links mainly to Europe came to an end following two world wars in the first half of the 1900s, after which they gained full independence from the colonial powers. Whereas Britain had relied on imports of cheap food from Australia and New Zealand until the 1970s, its new 1970s ties with the EU led to the exclusion of most goods that competed with European products. Australians continue to assert that this geographic realignment costs British people extra in what they now pay for food. Both European countries and the United States allow limited opportunities for foreign countries to enter their protected markets for agricultural produce. Australia and New Zealand were forced to find new markets in Japan and other countries of the Pacific Rim.

The Pacific Rim includes the countries of the South Pacific, Eastern Asia, Anglo America, and western Latin America. Most of them are members of the Asia-Pacific Economic Cooperation Forum (APEC) that seeks to link the growing economies around the Pacific Ocean margins and provide an alternative major world economic grouping to that of the North Atlantic area. The United States became the major power in the Pacific Ocean realm after World War II, although it is now challenged economically by Japan, the newly industrialized countries of Eastern Asia,

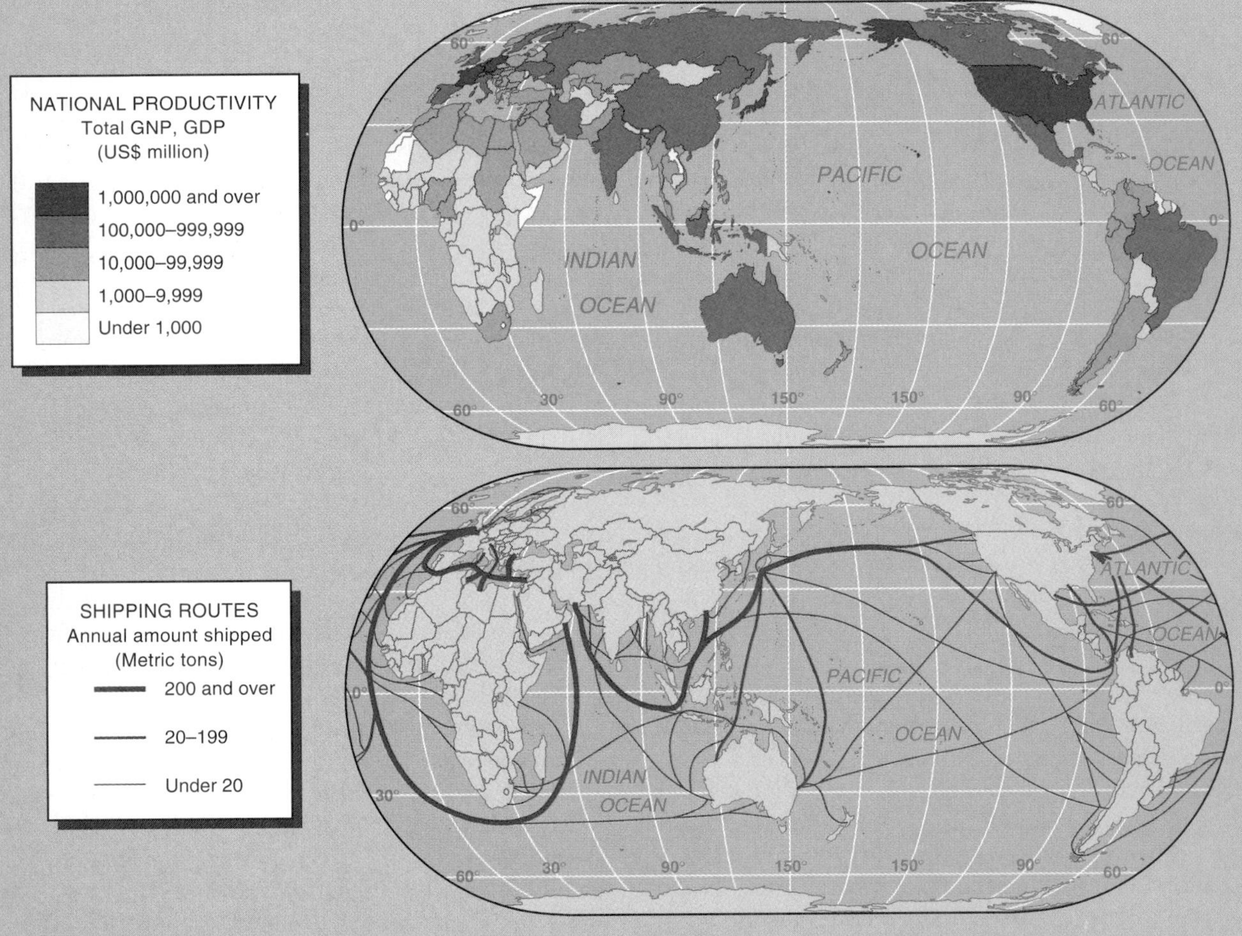


The South Pacific countries in the global context. Compare their productivity and distance from main shipping routes with the positions of other core countries.

WORLD ISSUE

## PACIFIC RIM (*continued*)

and the emergence of China. Both Australia and New Zealand are members of the APEC forum, but other members of this group resist including farm products in trade agreements.

Fortunately for Australia, New Zealand, and some of the South Pacific islands, the countries of Eastern Asia, led by Japan, need the products of this region. Minerals and wood products are particularly in demand, but food tastes are also becoming westernized. It was possible for Australian cattle producers to satisfy the beef preferences of the Japanese public and for wheat farmers to produce higher quality noodle grain. Specialty fruit and nut products became popular in the new markets. Tourism is a major area of economic growth for Australia, New Zealand, and many of the islands, accommodating large increases in the numbers coming from Japan and Asia during the 1980s and early 1990s.

The shift posed difficulties. Australians were very wary of the "yellow peril" in their immigration policies in the early 1900s and were deeply affected by involvement in Asian wars against Japan during World War II, in South Korea in the 1950s, and in Vietnam in the late 1960s. Australians tended to regard Asians as lesser beings until the recent economic successes of Japan and others. The Western attitudes that place a strong emphasis on free democratic activities, allowing influential trade unions a strong voice, and public debate about environmental concerns are often at variance with attitudes in countries of Eastern Asia.

As countries of the South Pacific adjusted to the Asian markets, Japan and Eastern Asian countries maintained their tariffs against processed food and mineral imports, which make it difficult for Australia and New Zealand to make more wealth out of their exports. The Asian countries view Australia and New Zealand as small markets for their own exports of manufactures and pay more attention to the major markets of Europe, Anglo America, and the emerging larger countries of Asia. Closer links to the Asian markets led to problems for the South Pacific countries when those markets went into recession in the late 1990s, affecting exports and numbers of tourists.

Australia and New Zealand found it difficult to penetrate the ASEAN group of Southeast Asian countries to further trading objectives there. They work closely with the broader APEC group to emphasize the fact that they are part of the Pacific Ocean realm and no longer dependent on links outside that realm to Europe. They are also centrally involved in the South Pacific Forum that brings together the many South Pacific Island countries in cooperative concerns. This involvement often leads, however, to new antagonisms with the ASEAN countries that seek to exploit timber or fish in this area but that are also potential trading partners.

Some regard the concept of a Pacific Rim of countries united by common trading interests as unnecessary or an unlikely fiction. They make the point that each group of countries around the Pacific has different cultures, languages, attitudes, demands, and goals, and that the rim idea promotes a false sense of common identity. If that is the case, the future of the South Pacific countries may be merely as dependents of the Eastern Asian countries.

### TOPICS FOR DEBATE

1. Australia, New Zealand, and the South Pacific islands will never fit into the economy of the Pacific Rim.
2. Personal attitudes are more important than economic realities in international relationships.
3. The Pacific Rim does not exist.

more clearly than the greater complexities of other regions allow. Data for this region's countries are found in the Reference Section, pages 586–610.

## EUROPE TRANSPLANTED

The distance from Europe delayed the colonization of the South Pacific region until the 1800s. Much of the region only became known fully to the European colonizers through the late 1700s surveys carried out by Captain James Cook of the British navy.

### South Pacific Cultures

The native groups that inhabited the region before European colonization included different racial and ethnic groups whose origins are still debated. Spread out over vast distances of ocean, the populations of the islands contain many distinct groups, together with mixtures of Melanesian, Polynesian, Micronesian, Asian, and European types. The Aborigines of Australia, the Maoris of New Zealand, and the peoples of the islands have a range of skin colors and body build features, but links between them and other racial types are not clear.

When Europeans first arrived, just over 200 years ago, the population of Australia was between 200,000 and 500,000

**FIGURE 11.4 Australia: the people.** Aboriginal peoples after shopping at a store in Nangalala, Northern Territory. The tension between retaining traditional ways and getting involved in modern Western lifestyles worries many aboriginal people.



indigenous **Aborigines** (Figure 11.4). They were nomadic hunters and gatherers living in groups spread across the continent and speaking 200 different languages. They left a record of rock paintings based on their religious beliefs and social organization. The Europeans took little account of this group, many of whom died from disease and oppression. The numbers of Aborigines were reduced to under 75,000 by 1933 and the Tasmanian Aborigines were wiped out by 1876. In the 1900s, attempts to integrate Aborigines into Australian life partly succeeded in health and education terms, and they now number 300,000. Of these, only 10,000 follow traditional ways of life; the rest live on reservations or in cities. Many are poor and, despite making up only 1.5 percent of Australians, they comprise 29 percent of prison inmates, often being subject to more rigorous judgment for small crimes than people of European heritage.

The **Maoris** of New Zealand came from the wider South Pacific around the A.D. 800s, with the final waves of people arriving from Tahiti in 1350. They replaced an earlier dark-skinned race, the Moriori, most of whom they drove out. Although subject to some segregation practices, the Maoris are more integrated in New Zealand life than the Australian Aborigines in their country.

The inhabitants of the South Pacific oceanic islands are grouped in three geographic categories—the **Melanesian** ("dark-skinned"), **Micronesian** ("small islands"), and **Polynesian** ("many islands") peoples. The Polynesian groups have lighter skins than the Melanesians.

## European Colonization

Europeans colonized little of the region until the 1800s. Spanish, Portuguese, and Dutch ships passed the islands without claiming sovereignty over them.

### Australia

"**Terra Australis**," the Southland, remained a mythical concept as long as the last major continent was kept isolated from passing ships. The southeast trade winds blew early European traders away from Australia and back toward India. The Dutch were the first to discover the continent in the 1600s, when their improved ships and occupation of the Dutch East Indies led them to explore southward. British ships visited the region, but most early assessments were of dismal lands with little economic or settlement potential for Europeans.

Following James Cook's surveys and more encouraging reports in the 1770s, the British claimed the Australian continent, which they first called New South Wales. Part of the colony was used as a penal settlement from 1787, to replace the previous transportation of convicts to the now-independent American colonies. Settlement of New South Wales was stimulated in the early 1800s, but initially there were problems of food supply. Land grants were then handed out freely to encourage sheep farming in the interior, and better government was introduced. Tensions grew between the ex-convicts and other settlers. Most convicts came from the poorer groups within British cities, with a few from better-educated groups. After serving their sentences, some did well in business or government in Australia. A gold mining boom in the 1850s drew many speculators and new settlers. When the last convict ship reached Australia in 1868, voluntary migrants outnumbered the convicts and their descendants by ten to one.

During the 1800s, British groups established new colonies around the Australian coasts, each with its own main city and a competitive pride that generated rivalries among the colonies. The granting of dominion status within the British Empire to Canada in 1867 caused some Australians to think in similar terms, but the internal rivalries held back cooperation. The late 1800s scramble by European powers for colonies in the South Pacific led to worries over common defense, but New South Wales still held aloof from the other colonies' wishing to federate. The federal idea came to fruition in 1901, when the Commonwealth of Australia within the British Empire was created out of the five colonies that then became states.

### New Zealand

In New Zealand, settlement came later. The Maoris resisted British missionaries and whalers in the 1700s and early 1800s. It was not until after 1840 that the British government encouraged systematic immigration of farming settlers. As it took sovereignty, it agreed to respect Maori land ownership. After 1860, there was a short gold rush. The technical advance of refrigerator ships made it possible to export fresh meat to Europe after 1882, and more sheep farmers became established. In 1907, New Zealand gained dominion status within the British Commonwealth and a large degree of autonomy.

### The Islands

Britain was the main colonizer of the islands, including Fiji, Kiribati (Gilbert Islands), Tonga, Tuvalu (Ellice Islands), some of the Solomon Islands, southeastern New Guinea, and parts of modern Vanuatu, which was shared with the French. Guam and the Marianas were Spanish colonies until taken over as United States protectorates just after 1900. The French colonized New Caledonia and the islands around Tahiti, which remain part of France. Decisions are made by the French government in Paris, as recent nuclear tests in French Polynesia emphasized. Like French colonies in the Caribbean (see chapter 10), the French South Pacific islands have access to EU markets.

The Germans were active in the 1880s, when they took the Marshall Islands with Japan, together with Nauru, Western Samoa, northeastern New Guinea, and some of the Solomon Islands. All were lost to the United States (Marshalls), Britain, Australia, or New Zealand in World War I.

**RECAP 11A: South Pacific and Its Cultures**

The South Pacific comprises Australia, New Zealand, hundreds of islands, and Antarctica. Australia and New Zealand act as a local core within part of the world's periphery. Distances within the region and between its countries and the world's major core regions are immense. Following late colonization by mainly Western European countries in the late 1800s, many of the islands remain poor with few products to sell in world markets.

Indigenous groups of people in Australia (Aborigines) and New Zealand (Maoris) fared badly under the European settlers but continue to form the main population groups in many South Pacific islands. Australia and New Zealand, together with a few more prosperous islands, are reorienting their economies to trading with Eastern Asia and the United States as well as Europe.

11A.1 What have been the consequences, good and bad, for the countries of this region of being at a distance from the Northern Hemisphere core countries of the world economy?

**Key Terms:**

| | | |
|---|---|---|
| Pacific Rim | Melanesian people | *Terra Australis* |
| Aborigine | Micronesian people | |
| Maori | Polynesian people | |

## NATURAL ENVIRONMENT

The natural environments of the South Pacific region range from equatorial to polar climatic environments and from volcanic rocks and coral reefs that are forming now to some of the world's oldest landscapes. The long-term geologic isolation of the lands of this region gave them a unique flora and fauna that European intervention changed.

### Oceanic Climates

The climatic environments of the South Pacific (Figure 11.5) are mostly dominated by oceanic influences. The two major exceptions are the interior of Australia and the whole of the Antarctic continent.

The islands of the South Pacific Ocean are mostly in the tropical belt. The tropical oceans, where high temperatures vary little throughout the year, supply moisture and heat to the air above, fueling intense tropical storms and typhoons. The equatorial climatic environment covers up to 10 degrees of latitude on either side of the equator. Air converges and rises, condensing in tall clouds from which rain falls.

North and south of the equatorial belt, the trade winds blowing from the northeast (northern hemisphere) or southeast (southern hemisphere) are constant factors and create their own climatic environment. Annual rainfall totals increase toward the western edges of the ocean. The hilly islands in the path of the trade winds have rainy east-facing flanks. Coral atolls are low-lying and, if they lack hills to

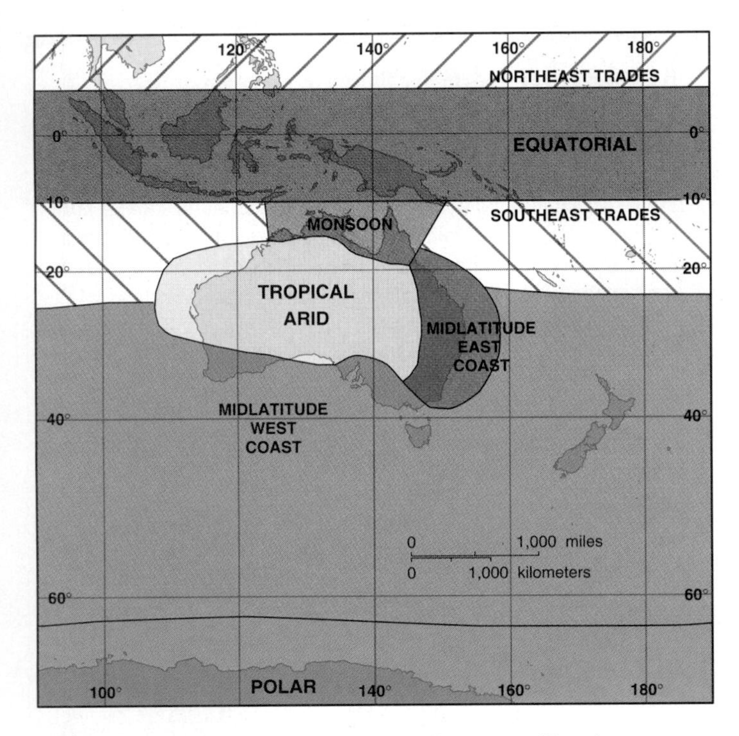


**FIGURE 11.5 South Pacific: climatic environments.** Note the importance of ocean influences in the equatorial, trade wind, and midlatitude west coast climatic environments. Australia is dominated by the tropical arid environment.

attract rain, are often arid with small and uncertain rainfalls. Tropical disturbances and typhoons (equivalent to hurricanes) occur in late summer in the belt between about 10 and 25 degrees north or south of the equator. In early 1993, tropical cyclones hit the Solomon Islands, Papua New Guinea, and Fiji, swamping low-lying land and making over 20,000 people homeless. In 1998, a tsunami caused surges of ocean water that overwhelmed coastal settlements along an isolated part of northern Papua New Guinea. Thousands died as entire villages were swept out to sea, and there were difficulties in getting emergency services to the area.

The tropical part of the region is at the western end of the atmospheric circulations that cause the El Niño fluctuations in Latin America (see Chapter 10). When Peru is dry, this region has plentiful rains. During the major 1997–1998 El Niño event an area stretching from the South Pacific into Indonesia and northern Australia suffered intense drought and forest fires.

Australian climates have a major effect on the human occupation of the different regions. One-third of the continent is arid and another one-third is semiarid. The areas with regular and sufficient rains occur around the margins. Winter rains are characteristic of the midlatitude southwestern corner of the continent and summer rains of the tropical conditions along the northern coasts. The southeastern coastlands, where most Australians live, have a warm midlatitude climate with rains all year, most falling in summer. Water shortages are permanent characteristics of most of the continent, and even the areas that are normally well watered suffer from lengthy droughts that give rise to woodland fires. The aridity of the interior may be broken by storms and flash floods, occasionally filling some of the dried lakes such as Lake Eyre.

New Zealand's climate is humid and similar in many ways to the British climate that many of its settlers left behind. There is no continental interior nearby to provide the coldest winter weather or the hottest summer weather types that are experienced from time to time in Britain because of its closeness to continental Europe. The mountains in South Island create a rain shadow to their east, necessitating the irrigation of some of the farmland.

The southern oceans between Australia and Antarctica have little land to interrupt the strong airflows set in motion by the contrast between Antarctic air and warmer midlatitude air. The strong westerly winds, known as the "**roaring forties**," only touch southernmost New Zealand in winter.

Antarctica forms its own climate. Its extreme cold produces a major contrast between air and water around the continent and that farther north. Perhaps the most significant boundary of Antarctica is the convergence zone between cold and warm ocean water and atmosphere between 50° and 60° S. This contrast affects the overlying air and generates a succession of midlatitude cyclonic storms with high winds that effectively cut off Antarctica from warming influences. Small quantities of chlorine gases, however, penetrate from lower latitudes to produce the ozone hole above the continent at the end of the Antarctic winter. The cold continent controls the weather and ocean movements south of the Antarctic convergence zone. In winter, the ice-covered area of the oceans around Antarctica doubles in size as the sea surface freezes. During the late 1990s, some of Antarctica's ice shelves, where ice flowed off the continents to cover the ocean surface, melted and broke up—seen by many as a sign of global warming.

## Continents and Islands

The surface features of the land areas (Figure 11.6a) were determined by the pattern of tectonic plates and their margins (Figure 11.6b). The collision zone between the Indian and Pacific plates runs east-west through New Guinea and islands to the east. It finishes where a transform plate margin marks the eastern boundary of the Indian plate and then angles southwestward through New Zealand. The divergent plate margin between the Indian and Antarctic plates separates Australia and Antarctica. Most of the earthquakes and volcanic activity in the region take place along these margins.

Many South Pacific islands have a volcanic core, generated by eruptions of lava from beneath the ocean floor as the Pacific plate moved westward. Where the Pacific plate collides with the Indian plate, the line is marked by mountains in Papua New Guinea, and islands along this plate margin continue through the Solomon Islands to Vanuatu and New Caledonia—an island arc. The smaller islands can be divided into those with a hilly volcanic core (Figure 11.7) and those where the volcanic core has sunk beneath the ocean surface,

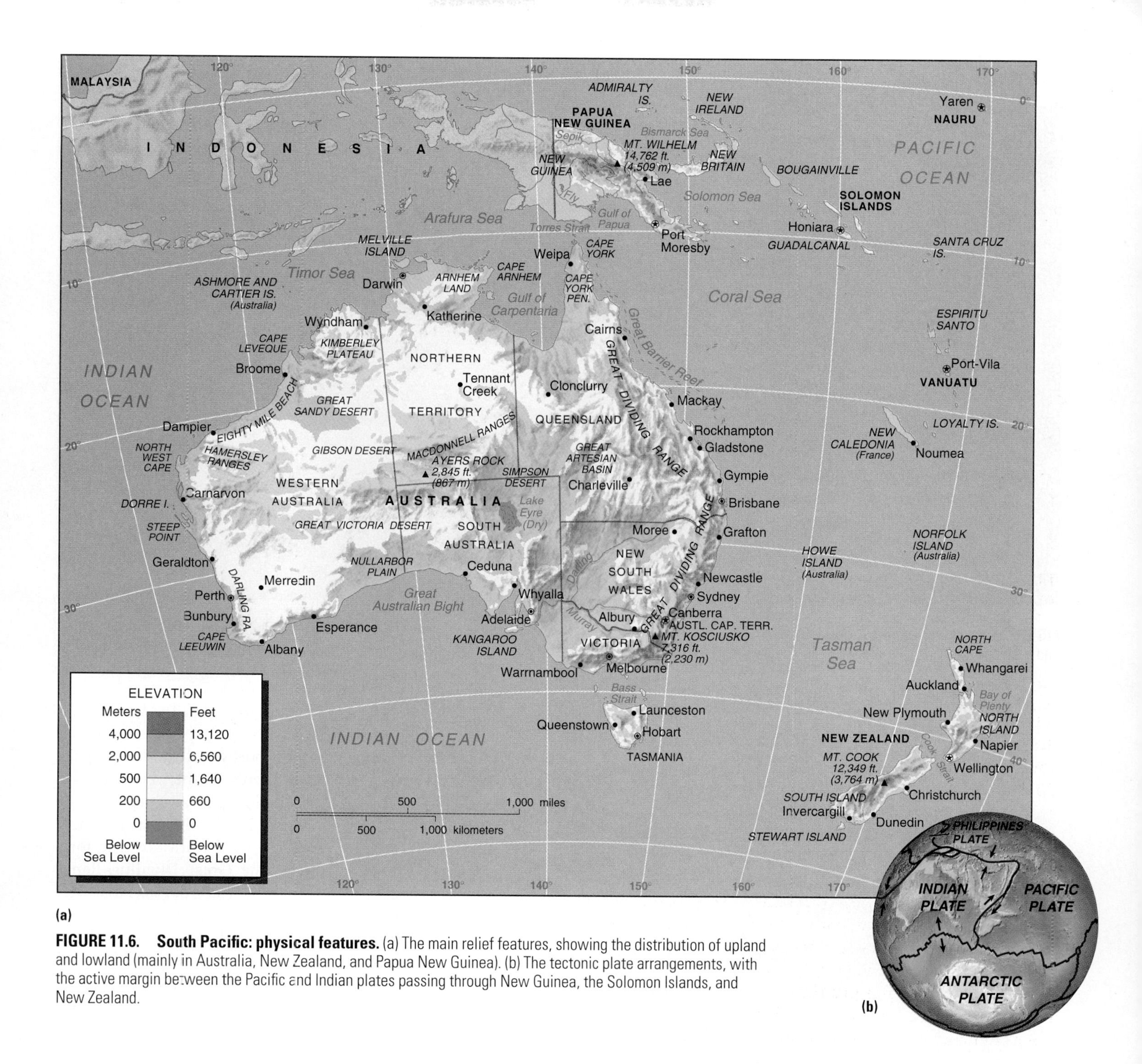


**FIGURE 11.6. South Pacific: physical features.** (a) The main relief features, showing the distribution of upland and lowland (mainly in Australia, New Zealand, and Papua New Guinea). (b) The tectonic plate arrangements, with the active margin between the Pacific and Indian plates passing through New Guinea, the Solomon Islands, and New Zealand.

leaving a coral atoll. Coral reefs form as tiny tropical animals secrete limy structures. The coral animals live near the ocean surface, so that their symbiotic plant algae get access to light and oxygen.

New Zealand is the product of plate margin activity (Figure 11.8). Uplift of the land is as rapid as glaciers and rivers can wear it down. Elsewhere in the world, the surface processes often act more rapidly than the tectonic internal processes. The two islands are part of a fragment of the southern continent that broke away from the main mass of Gondwanaland around 100 million years ago (see Figure 3.7c). Movement along the transform fault between the Indian and Pacific plates led to the formation of the volcanic northern mountains around Mount Egmont and of the Southern Alps that rise to 3,500 m (11,700 ft.) above sea level (Figure 11.9). As the Southern Alps rise, the eastern plains of South Island are uplifted relative to sea level, with the greatest amount near the mountains, thus forming plateaus tilted toward the ocean. Glaciers carved deep valleys in the Southern Alps, while rivers flowing from these mountains cut into the sloping plateaus.

Australia was joined to Africa, Antarctica, South America, and the peninsula of India in the continent of Gondwanaland.

**FIGURE 11.7 South Pacific: coral island.** A typical South Pacific island, with a hilly volcanic core, an outer coral reef, and a shallow lagoon between the reef and mainland. The village is sited where a channel leads from a breach in the reef front.



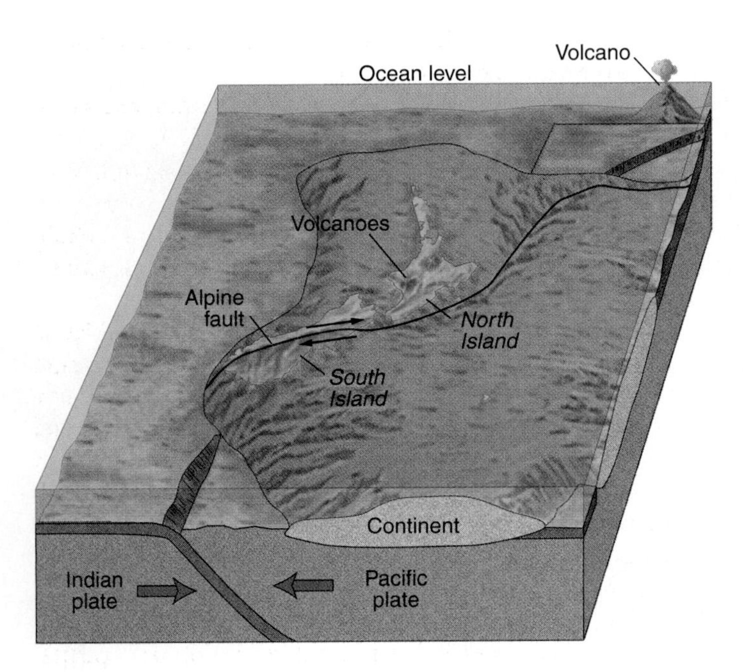


**FIGURE 11.8 New Zealand: major landforms.** The diagram shows the importance of internal Earth processes in causing rapid uplift. New Zealand is on a continental fragment that split from Australia and Antarctica. The two islands lie on a line where the Indian and Pacific plates meet. Volcanic eruptions affected North Island in the mid-1990s. The alpine fault cuts through South Island, and its frequent movements cause earthquakes and uplift.

Australia now lies on the eastern portion of the Indian plate but at a distance from all plate margins. In the western half of the continent, its relief features are low plateaus and plains on the ancient shield rocks (Figure 11.10). Mountain ranges formed over 600 million years ago were worn down to form landscapes of little relief today. The **Great Dividing Range** along the eastern edge of Australia is formed of rocks that were deposited from the erosion of the ancient continent and were then uplifted and broken into blocks by faulting. The lowlands between the shield and eastern highlands have thick layers of water-bearing sediments beneath them. Apart from the steep edges of the eastern plateaus, Australia has fewer major relief contrasts than other continents. The world's largest area of coral reefs forms the Great Barrier Reef off the northeastern cost of Australia.

The Antarctic continent was the core of Gondwanaland. After the combined continent broke apart, Antarctica remained at the South Pole. As Earth's atmosphere cooled some 20 million years ago, ice accumulated on Antarctica, burying the mountain ranges that continue the line of the Andes from South America. A large part of the continent sank under the weight of ice and is now below sea level. If the ice melted, Antarctica would form several land masses. The rocks contain minerals like those found in the Andes (see Chapter 10), together with coal deposits (Figure 11.11).

**FIGURE 11.9 New Zealand: effects of glaciation.** Fiordland National Park and the Southern Alps, South Island. In the distance, Mitre Peak rises from the deep fiord of Milford Sound, carved by glaciers and then drowned by rising sea level at the end of the last glacial phase.



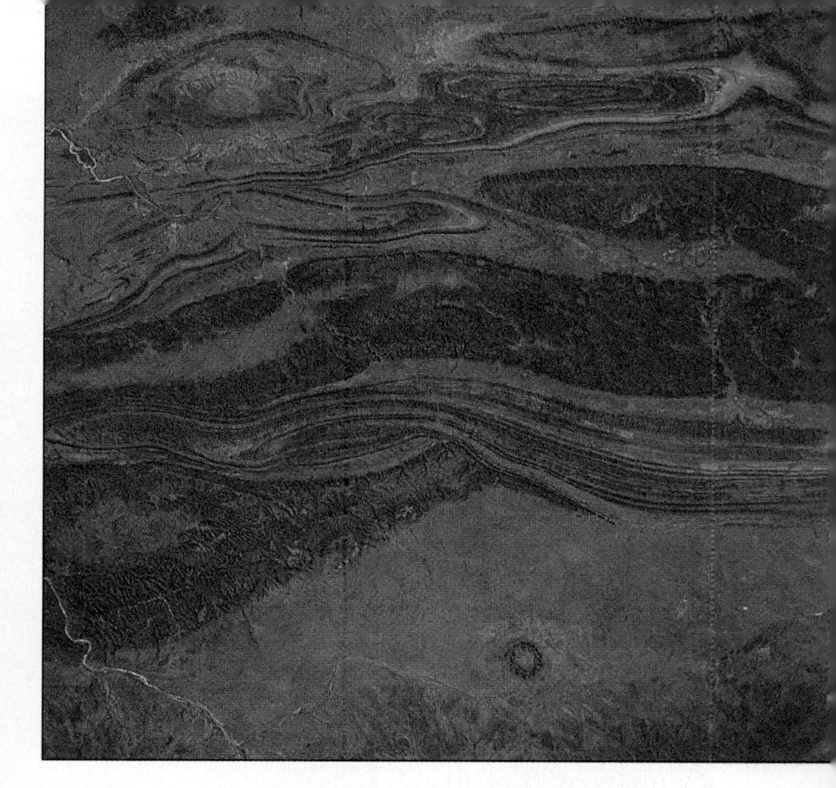

**FIGURE 11.10 Australia: ancient rocks and mountains.** The folded rocks of the Macdonnell Ranges, Northern Territory, the highest parts of Australia's interior. Little vegetation grows except along the river valleys, although the streams are often dry.

NASA

## Unique Biosphere

The isolation of the South Pacific region from the rest of the world dates back several million years, well before humans inhabited it. Australia, New Zealand, and many of the islands have types of plants and animals that are unique.

Separated by plate movements from the other southern continents some 35 million years ago, Australia's mammals were at an early stage in their evolution before the European colonization, when species from this region were taken elsewhere, while external species were introduced. The **marsupials**, such as kangaroos, koalas, wallabies, and possums, raise their young in pouches and compose about half the native animals. They are rare in other parts of the world. The bird life is particularly varied and colorful. Vegetation is dominated by species of eucalyptus and acacia, and there are unique desert species in the dwarf mallee community. **Mallee** is formed of drought-resisting eucalyptus shrubs that grow into almost impenetrable thickets of many close-spaced stems rising to 8 or 9 m (25–30 ft.) high.

Europeans introduced domestic animals, trees, and crops from outside the continent. Predators including wild dogs threatened some of the native groups with extinction. Wild rabbits introduced in 1859 spread across the continent, destroying large areas of grassland. Efforts to control the rabbits continue following their recovery from the myxomatosis virus introduced in the 1930s.

Like Australia, New Zealand had a unique flora and fauna before European settlement, but the immigrants brought domestic animals and crops that replaced many of the native plants and animals. Reindeer damaged trees and shrubs and are now contained. Virtually all the original forest was soon cut but was later reforested by Douglas fir and pines, which are quicker growing for commercial use.

Some of the larger South Pacific islands are forested with species closer to those of Indonesia and Eastern Asia than of Australia. Palms are particularly numerous. Few islands have many animals, although there is a diversity of bird species. The surrounding waters contain a wealth of tropical fish varieties but are too easily overfished.

The oceans surrounding Antarctica are among the world's most profuse life zones. Antarctica's living organisms are dominated by a huge variety of sea birds, including penguins, that rely on the rich ocean life.

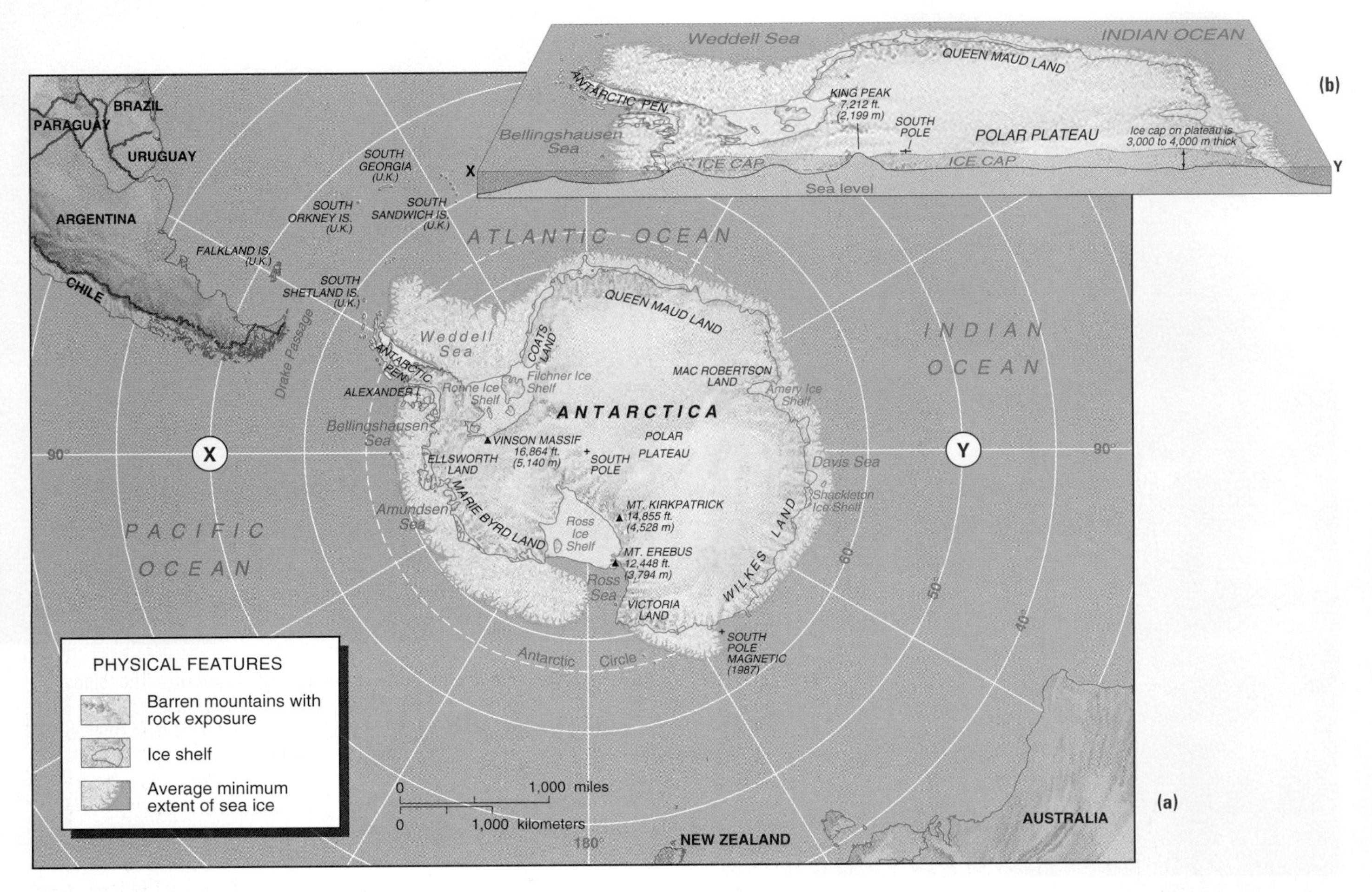


**FIGURE 11.11 Antarctica: the ice-covered continent.** (a) The main features. (b) A section through 90° E and 90° W (across the center of the map, X-Y).

## Natural Resources

The South Pacific region is rich in a number of natural resources that often form a basis for the countries' economies. These resources include minerals, water, forests, and ocean resources from fish to whales. Some of the smaller islands, however, have virtually no natural resources apart from those in the ocean.

The ancient rocks of Western Australia contain many large deposits of iron ore and other metallic ores such as nickel, gold, platinum, uranium, and copper—like the shield rocks of similar geologic age in Africa, northern Canada, and Siberia. The rocks that form the Great Dividing Range in eastern Australia are of more recent origin and contain coal, silver, lead, zinc, and copper ores.

Between the shield rocks and the Great Dividing Range, the lowlands are drained by the **Murray-Darling River system** in the south and form the Great Artesian Basin in the north. Rain falling on the eastern mountains soaks into the rocks and drains westward and downward, accumulating in the sedimentary rocks of the basin (Figure 11.12). Since the rocks in the mountains remain filled with water, the pressure on water in the rocks beneath the lowlands is so great that wells drilled into the rocks cause water to flow out on the surface without a need for pumping. Such a situation is known as **artesian conditions,** and the northern part of the lowlands was named accordingly.

The Pacific islands along the plate collision zone have some mineral resources. The copper deposits on Bougainville, an offshore island that is part of Papua New Guinea, constitute one of the world's largest reserves, which was intensively mined until terrorist activities stopped it. New Caledonia is the world's third largest producer of nickel ore. Most of the larger islands have a covering of rain forest, but some of the drier and flatter islands have only sparse vegetation.

Although the natural resources in its rocks are off-limits to exploitation, Antarctica's surrounding oceans draw fishers from all over the world. The Antarctic oceans are an important basis for wider ocean food chains and are being studied to gain an understanding of the sustainable levels of fishing, sealing, and whaling.

## Environmental Problems

People living in the South Pacific close to plate boundaries have to contend with earthquakes and volcanoes. The 1995 eruption of Mount Ruapehu in the North Island of New Zealand led to the closure of airports and highways and

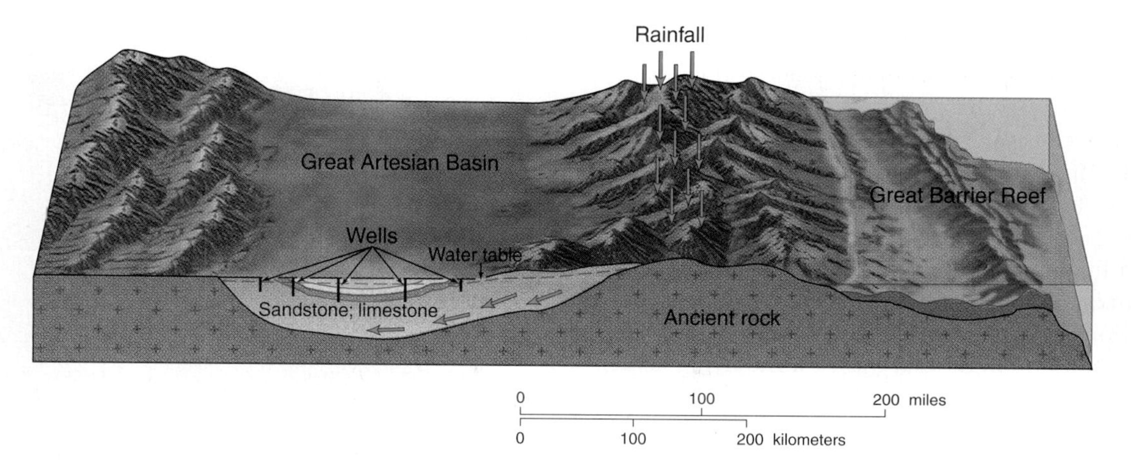


**FIGURE 11.12 Australia: Great Artesian Basin and the Great Barrier Reef.** The artesian basin of northeastern Australia is created by the geologic conditions that produce a major groundwater source as rain falls on the Great Dividing Range, seeps into the rocks, and flows downward to replenish the water in the rocks beneath the basin. The water in the deepest rocks flows out under pressure from the water moving down from the hills. The Great Barrier Reef is one of the world's largest developments of coral reef and forms a feature just off the Queensland coast.

caused worries over the prospect of a larger eruption. Typhoons, floods, and droughts occur in various parts of the region. Human occupation, especially since the European colonization, caused local losses of environmental quality.

Australians are increasingly aware of the environmental damage caused to their fragile landscapes by farmers and miners. Farmers in Western Australia, for instance, cut down the trees that kept down the water table in soil and rock and slowed the wind's force near the ground. The resulting increase of wind erosion and rise in the water table led to surface salinization and made large areas unusable for farming. Replanting and mixing lighter soils with clays from mining wastes combats the dual menace. Overgrazing and overirrigation in the lowlands between the shield area and the Great Dividing Range had similar results.

Such problems came to public consciousness, and the 1990s were declared a "Land Care" decade by the Australian federal government. It is anticipated that the capital being invested in attempts to restore soil quality may have positive results in technologies that will be applicable to arid parts of the world in poorer countries. The Australian forest area was halved following colonial settlement, and millions of new trees were planted in the early 1990s to make up for this loss.

Water management is a major concern of Australians in their dry climate. The largest project completed before environmental concerns were articulated publicly was the Snowy Mountain Scheme that diverted water westward for irrigation and generated hydroelectricity from some 20 dams.

Mining extraction scarred many parts of Australia, causing environmentalists to join their concerns with land claims by Aborigines to delay and impose new conditions on applications for mining licences. British nuclear tests carried out in the interior in the 1950s left scars on the land and on the Aborigines living in those areas. The United Kingdom government still pays compensation for this damage.

**RECAP 11B: Natural Environments**

The climatic environments of the South Pacific are dominated by oceanic influences. They range from equatorial rains through trade wind climates to the stormy seas surrounding Antarctica. Interior Australia and Antarctica remain deserts, with the latter being frozen.

Clashing tectonic plates produce mountains, volcanic islands, and earthquakes from Papua New Guinea to New Zealand and deposit mineral resources such as copper, gold, and nickel. The shield areas of Australia and Antarctica contain other mineral resources such as iron, uranium, and gold.

Environmental problems stem from attempts to establish European farming and tropical plantation crops in these new lands. Soil degradation and mining damage are among the main problems faced.

11B.1 Can the small numbers of people living in this region be related to the natural environments?

**Key Terms:**
"roaring forties" mallee artesian conditions
marsupial

**Key Places:**
Great Dividing Range Murray-Darling River system

In New Zealand, the main forms of environmental degradation include soil erosion and the changes brought about in the fauna and flora by introducing new species. New Zealand takes an increasingly strong stance against potential polluters and was willing to endanger its relationship with the United States to keep nuclear-powered naval ships out of its waters.

The common perception of the South Pacific islands as a pleasant and untouched part of the world is spoiled by the dumping of oil and other materials from ocean transports. Several of the islands are uninhabitable after mining copper, nickel, or phosphate, or after nuclear testing.

For the future, the low-lying parts of islands face environmental disaster if global warming leads to a rising ocean level. The highest parts of some of the islands are only a few meters above that level, and many have a concentration of settlement on low-lying coasts. Many of the coral reefs that form the foundations of the islands, protective barriers, or tourist attractions are under threat from rising sea levels and from predators and pollution that may kill the living corals, leaving the older coral rock to be worn away by the waves.

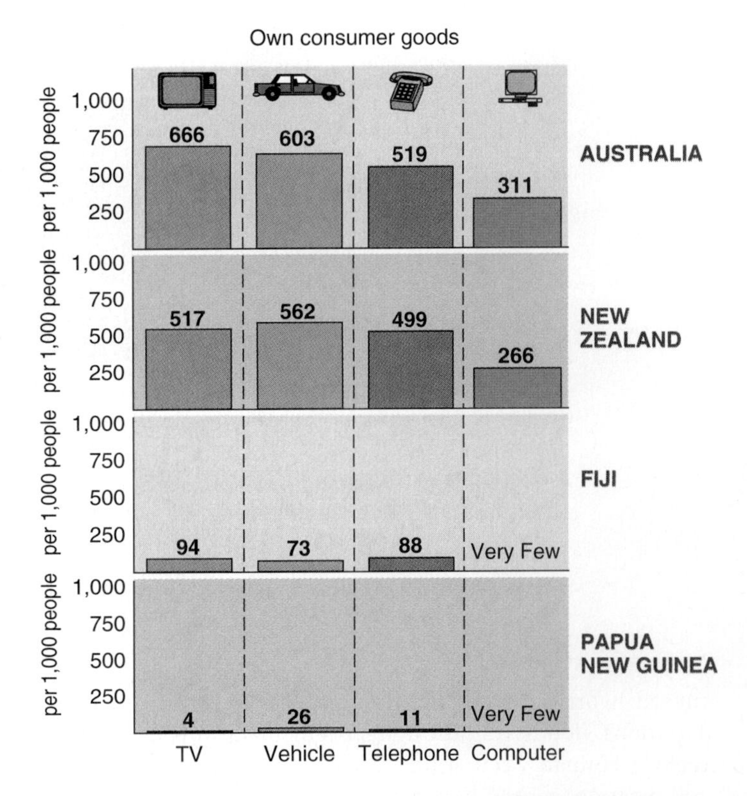


**FIGURE 11.13 South Pacific: consumer goods ownership.** The levels for Australia and New Zealand are similar to those in some countries of western Europe and far above those elsewhere in the South Pacific.

Source: Data (for 1996) from *World Development Indicators*, World Bank, 1998.

## AUSTRALIA

**Australia** becomes a focus of world attention as it hosts the 2000 Olympic Games. It is by far the largest of the settled lands in the South Pacific region and has 60 percent of the total population. It has the most diverse and affluent economy. Although it thus stands apart from the other countries in this region, it faces similar problems. These include great distances from major world markets, a small local market, and the transition from colonial ties with Europe to a closer involvement with Asian countries and the United States.

European colonial settlement from the late 1700s removed almost all trace of the indigenous peoples from Australia. Gradually gaining independence from the United Kingdom during the 1900s, Australia faced major changes in its economic orientation from the U.K. markets toward those of Asian countries. It still relies heavily on the production and sale of primary products from agriculture and mining, and has been called Asia's "farm and quarry." Although this emphasis links it to the economies of many developing countries, Australia has one of the most urbanized populations in the world with affluent lifestyles. Australia's consumer goods ownership levels (Figure 11.13) and human development characteristics relate it closely to other core countries.

### Regions of Australia

Australia is the only country that is also a continent (Figure 11.14). Its size produces internal contrasts of physical and human geography.

#### Regions in the Natural Environment

- Most people live in the temperate and humid southeastern coastal region from Brisbane in the north to Adelaide in the south that includes four of the five largest cities (Sydney, Melbourne, Brisbane, and Adelaide) as well as the federal capital, Canberra.
- For most of its length, that region is bounded by the range of hills and plateaus known as the Great Dividing Range, which formed a barrier to inland movement in the early years of colonization.
- North of Brisbane, the more tropical climate and the offshore Great Barrier Reef—the world's largest continuous coral reef formation—form an area much favored by tourists.
- Inland of the Great Dividing Range, the lowlands have a subhumid climate that supports a major farming region for cattle, sheep, grain, and fruit in western Queensland and New South Wales and northern Victoria.
- Westward of this sparsely populated farming region, the virtually empty Great Australian Desert covers most of the rest of the continent.
- In the far north, the tropical climate brings monsoon rains along the coasts.
- In the southwestern corner, rains come in winter, resembling the conditions around the Mediterranean Sea and providing a basis for agricultural settlements centered on Perth.

#### Political Regions: The States

Separate colonies were established around the ports, giving access to inland areas that they connected back to the mother country (Figure 11.15). The early colonies determined their

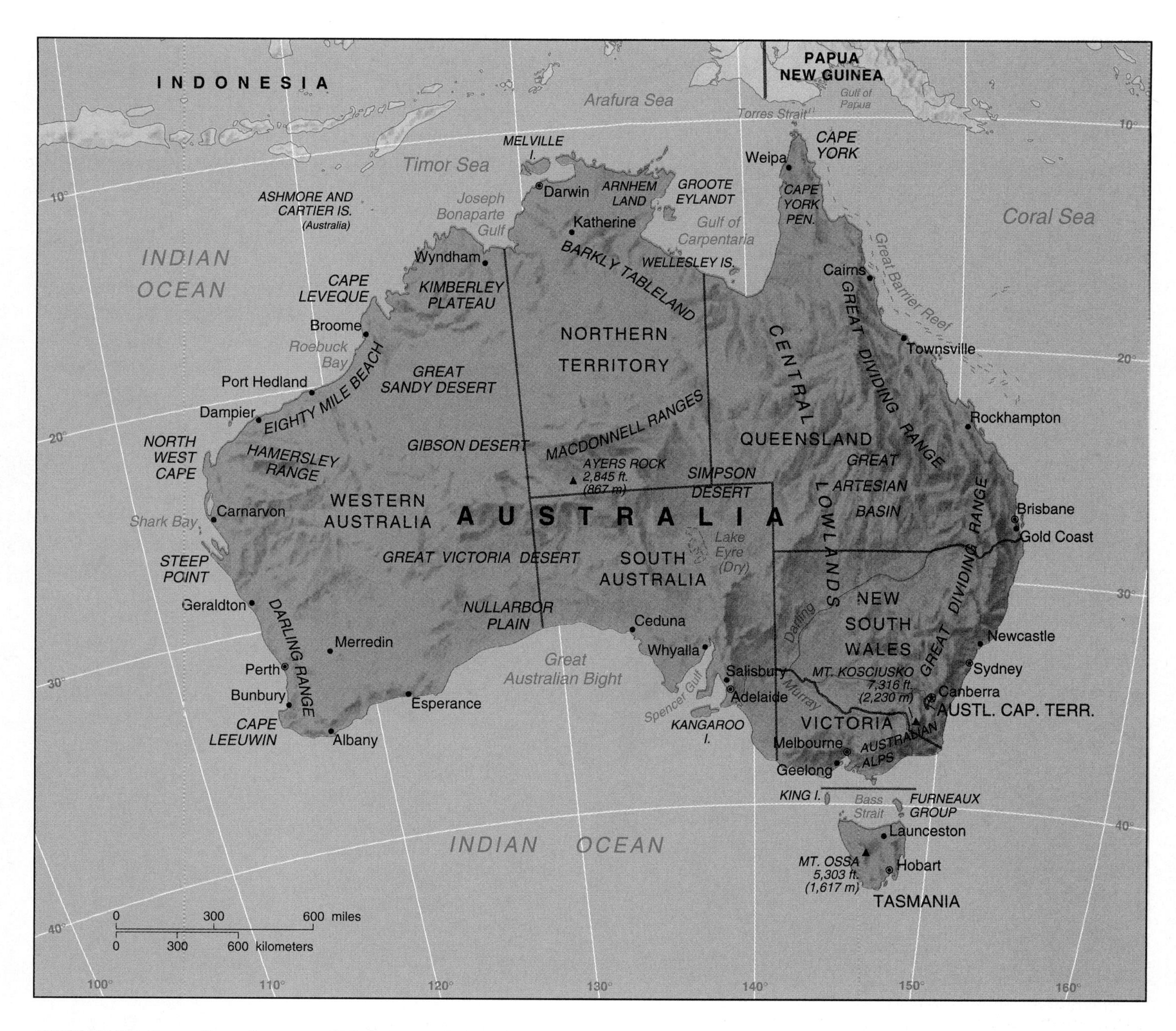


**FIGURE 11.14 Australia: major geographic features.** The states, physical features, and major cities.

own patterns of inland expansion, chose different railroad gauges, and concentrated major developments in their chief cities. In 1901, the Commonwealth of Australia came into being as a federation. The competition between Sydney and Melbourne for the role as national capital was resolved by building a specially planned city in a new Australian Capital Territory centered on Canberra. The federal government moved there in 1928.

Each state retains a specific character as the result of its history, natural resources, and distinctive policies operated by its government.

- Sydney, the largest city, lies at the heart of **New South Wales**, which also contains industrial cities near the coast and the main concentration of Australia's sheep farming on the western side of the Great Dividing Range.
- **Victoria** is focused on Melbourne but is a smaller state, and its inland farming area is mainly concerned with more intensive irrigated farming for grapes, fruit, and grain crops.
- **South Australia** has its main city, Adelaide, in a farmed area at the southern edge. Much of the rest is desert.
- **Queensland** is the second largest state in area with its main city, Brisbane, in the southeast corner. This state has a more tropical climate, is the main cattle state, together with major mining operations, and is developing tourist attractions.

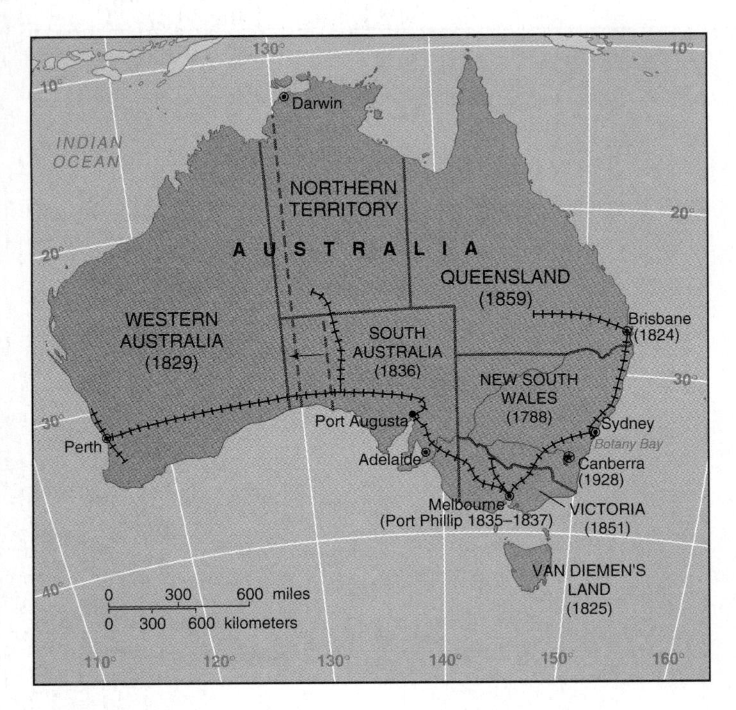


**FIGURE 11.15 Australia: colonization patterns.** Beginning in the late 1700s and continuing through the 1800s to dominion status in 1901, Australia developed political institutions around the separate colonies (later states). The dates in parentheses indicate when the colonies were established.

- **Western Australia** is the largest state in area, but is mostly desert and has its main area of settlement separated by thousands of uninhabited miles from the large cities of southeastern Australia. Perth has farming and mines in its immediate hinterland, and there are large iron ore mines farther north.
- **Northern Territory** is the least populated of the Australian states and territories. Most of the few remaining aboriginal people who live in traditional ways are in this territory, but it has fewer inhabitants than Australian Capital Territory.

## People

The distribution of Australia's population reflects the regional differences (Figure 11.16). Most people live in the southeastern corner. Much of the continent is an almost uninhabited desert wilderness.

Australia had a population of 19 million in 1998, double that in 1950. Like other more wealthy countries, its natural rate of growth is slow, with total fertility rates of less than 2. Immigration continues, resulting in annual population growth rates of over 1 percent, compared to less than 0.5 percent in European countries. Compared to the South Pacific islands, both Australia and New Zealand are well advanced in the demographic transition process (Figure 11.17). Australia's age-sex diagram (Figure 11.18) resembles those of other core countries and reflects a slowing in births after a baby-boom period from the mid-1950s to around 1970. The aging population, particularly of women, contrasts with the age-sex structure of the South Pacific islands such as French Polynesia.

Australia is essentially an urban country, with over 85 percent of its people living in towns in 1998 and 57 percent in the five largest cities of **Sydney, Melbourne, Brisbane, Perth,** and **Adelaide.** This reflects the historic primacy of these cities and the growing significance of service industry employment and output. The national capital, Canberra, has around one-third of a million people.

Up to the mid-1900s, most Australian immigrants came from the British Isles. This was largely because of racial preferences to maintain a **White Australia policy** against the pressures of Asians in particular. Wild claims were made that "empty" Australia could accommodate over 200 million people, although a British geographer in the 1920s doubted it could sustain more than 20 million. To increase its population, Australia encouraged immigrants from Britain by paid passages. After World War II, more immigrants came from other parts of Europe.

Although the white Australia policy ceased in 1972, its legacy still affects some Australian attitudes to new immigrants and the new efforts to focus trade in Asia. It has been difficult for many Australians to realize that countries they once looked down on now have more sophisticated economies and higher wages than Australia. In the 1960s, nearly half of total immigrants came from the United Kingdom. From the early 1970s, the sources of migrants changed. In 1992–1993, only 12 percent came from the United Kingdom, while 43 percent were from Asia (Figure 11.19). Australia is becoming a multiracial, multiethnic, and multicultural country—an environment in which there is also greater sympathy for the Aborigines.

The peak of immigration in 1988 and 1989 added 1 percent to the total population each year and became the main cause of population growth. In 1994, the Australian government began to place upper limits on immigration, halving the immigrant quota to 86,000 per year in the light of continuing unemployment levels of over 10 percent.

## Economic Development

In the late 1980s and early 1990s, Australia's economy boomed because it supplied growing Asian markets. By 1998, worries about its dependence on those markets and weak prices for minerals, wool, and food products caused its currency to drop in value, although it was not as badly hit as the Asian countries (see Chapter 6).

At the end of the 1800s, Australian farmers were major suppliers of wool to British factories. Other products such as fruit, lamb, beef, and dairy products followed the same routes to markets in Britain and Europe. During the 1900s, mining products grew in significance alongside the farm products. The cost of transporting low-value goods from Europe to Australia led to the growth of import-substitution manufacturing in the new country, with foundries, textile mills, breweries, and other food- and drink-processing enterprises. These industries were protected from outside competition by high tariffs. Other industries, such as steel and auto making and

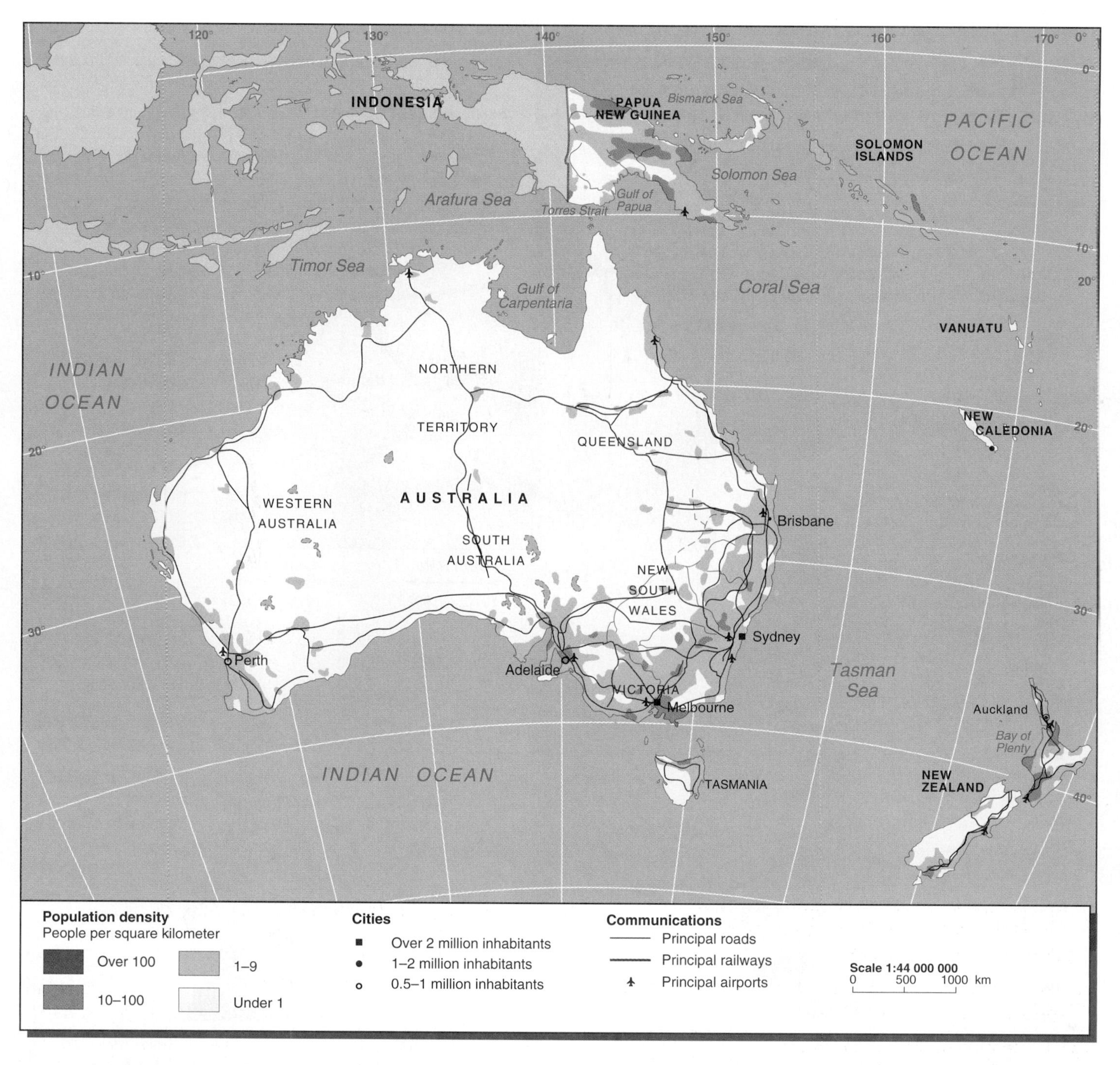


**FIGURE 11.16 South Pacific: population distribution.** Population totals and densities are low everywhere—outside of the major cities.

Source: Data from *New Oxford School Atlas,* p.85, Oxford University Press, UK, 1990.

assembling, were later set up under similar or increased protection levels.

In 1995, Australia's total GDP placed it 13th among world economies. Its GDP per head was just below that of the United Kingdom. As in other richer countries, the GDP was coming less from agriculture (down from 6% in 1970 to 3%) and manufacturing (down from 24% to 15%) and increasingly from service industries, which rose from 55 to 70 percent. A 1995 World Bank study that took account of "natural capital" in terms of land value, water, timber, gold, and other minerals, in addition to human resources, ranked Australia as the top world country, just ahead of Canada. Clearly, mineral resources played a major part in this designation.

## Australia Joins the World Economy

Australia developed as a two-tier economy. The first tier depended on exports from the natural resource base of farm and mine to provide income that paid for imports. The second tier consisted of protected manufacturing and

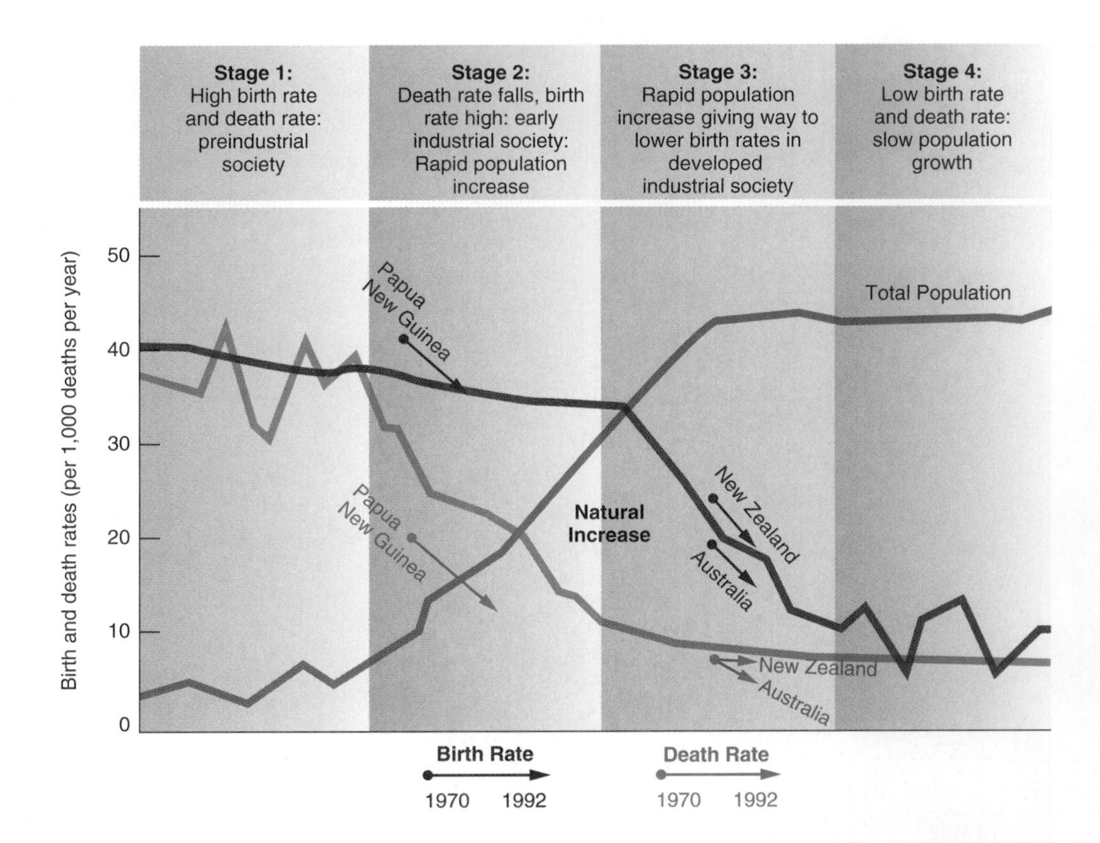


**FIGURE 11.17 South Pacific: demographic transition.** Papua New Guinea is typical of many of the islands. The Australian and New Zealand patterns resemble those of other rich countries.

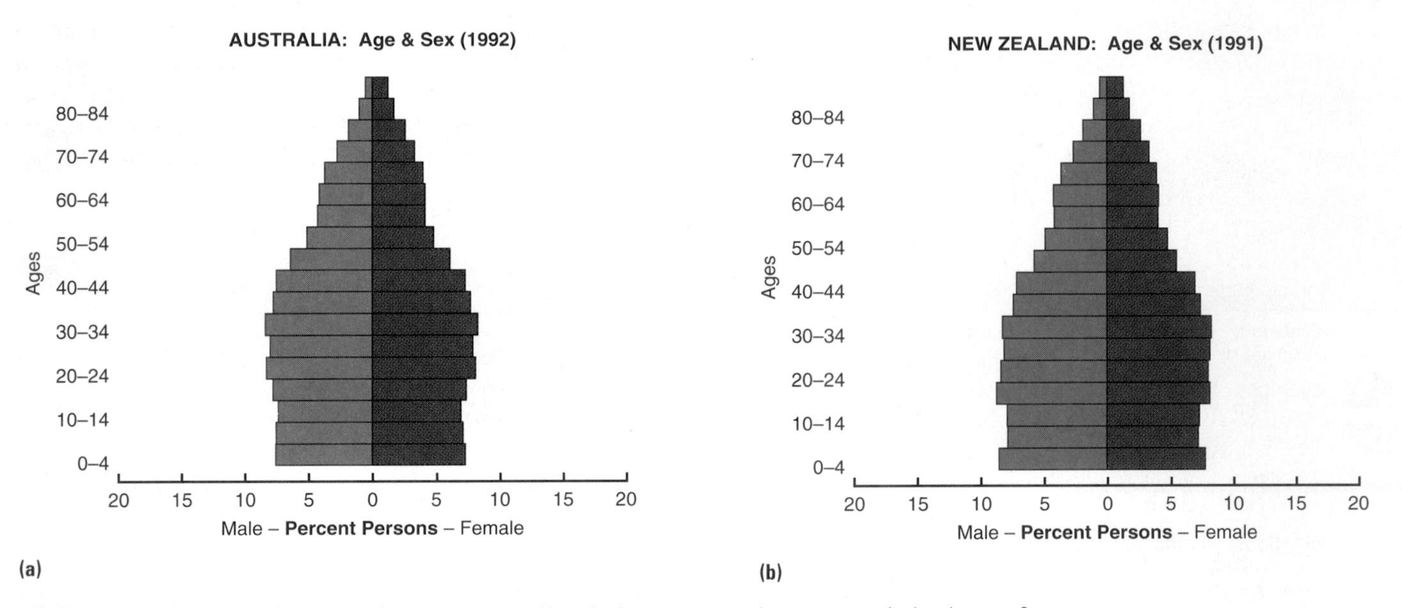


**FIGURE 11.18 South Pacific: age and sex structures.** How do these patterns relate to economic development?

state enterprises such as airlines, railroads, and docks that were often inefficient but had a role of spreading the wealth from the export industries' income over a wider group of people. This two-tier economy was characterized by slow productivity growth. The economic protectionism was linked to the population protectionism of the white Australia policy.

The protectionist policies collapsed in the 1970s and 1980s. The population continued to increase, but exports of agricultural and mineral commodities obtained lower prices on world markets. After Australia's markets in Britain and Europe were lost with the expansion of the European Union, Australia had to shift its export markets to Asian countries and especially Japan. Japan and later South Korea and Taiwan were poor in

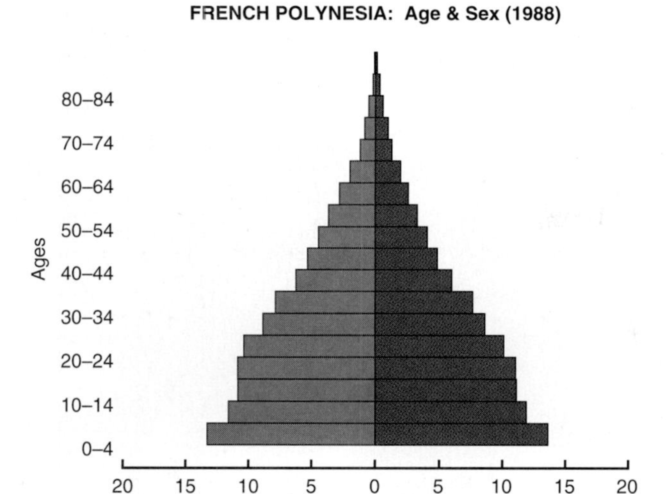


**FIGURE 11.18 (c)**

industrial raw materials. They bought Australian coal, iron ore, and other minerals and some grain and livestock products. Between 1970 and the 1990s, farm exports fell from 44 to 20 percent of total exports, while mine exports rose from 27 to 42 percent. By the mid-1990s, Japan bought more from Australia than from Europe and the United States combined.

Australia could no longer campaign for free markets for its exports while keeping high tariffs for its imports. The federal government lowered tariff rates so that general tariffs of 20 percent on imports common in the early 1980s will be down to 5 percent by 2000, while the 100 percent tariff on cars is scheduled to be 15 percent. One implication of this change is that many Australian manufacturers will not survive, since the manufacturing base was very broad but had little depth in specific areas and there is a relatively small home market. Of the five automakers that sheltered behind protective tariffs, Nissan closed its Australian operation. It is hoped that the remaining manufacturers will begin to compete in export markets as well as at home. This will not be easy.

## Dominant Mining

Natural resources remain of prime importance to Australia and its economy. Mining provides half of the value of goods exported, and large unexploited resources exist.

Australia is the world's largest exporter of coal, and three-fourths of its record production of 190 million tons in the 1990s was exported. Plans for expansion could increase this level of production if world demand for coal remains high. In the late 1990s, demand for coking coal, of which Australia is the largest world supplier, rose as the production of iron and steel increased.

New energy resources are being tapped in natural gas fields along the west coast of the continent, tied to the construction of liquefaction plants that export the gas to Japan. These projects attract investment by oil companies but provide expansion at the same time as in competing producers in nearby Malaysia, Brunei, and Indonesia.

Australia produces a wide variety of other minerals. Expansion of production is made increasingly difficult, however, by the complexities of obtaining new mining licenses, particularly in environmentally sensitive areas and in areas where there might be aboriginal sites. For example, Australia

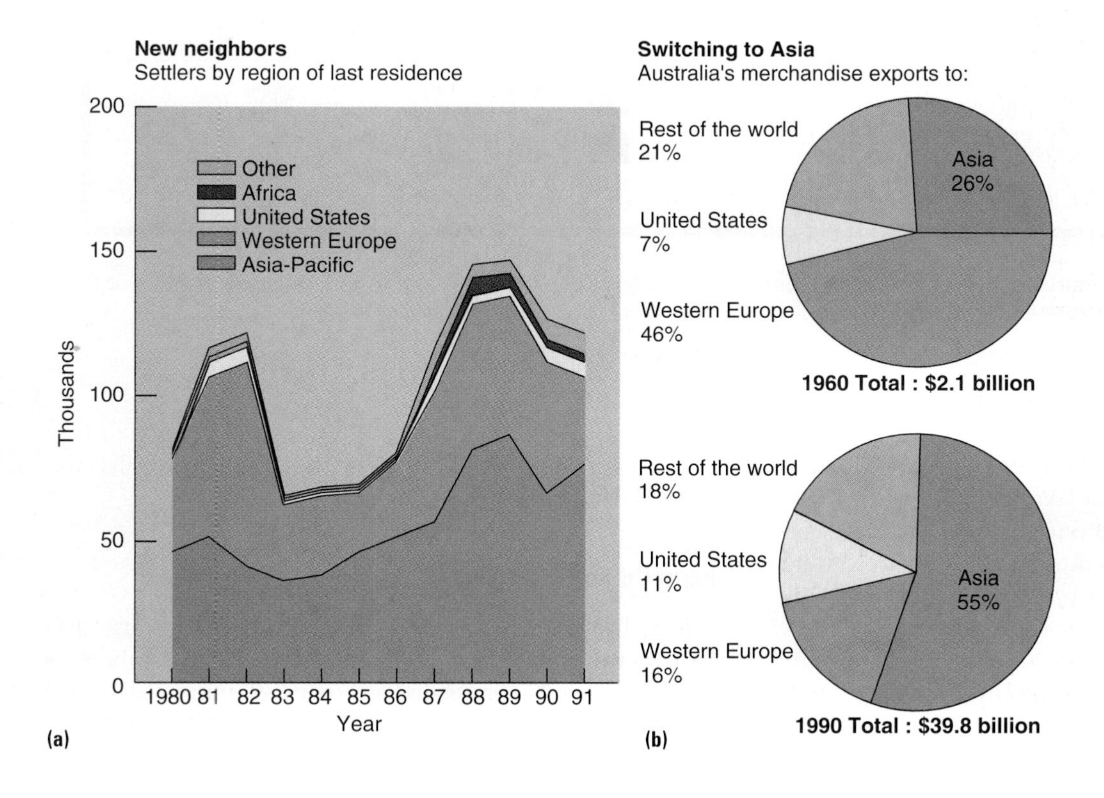


**FIGURE 11.19 Australia: changing immigration and trade patterns.** (a) The increasing numbers of Asian and Pacific immigrants and decreasing numbers from Europe since 1980. (b) Changes in Australian trade partners since 1960.

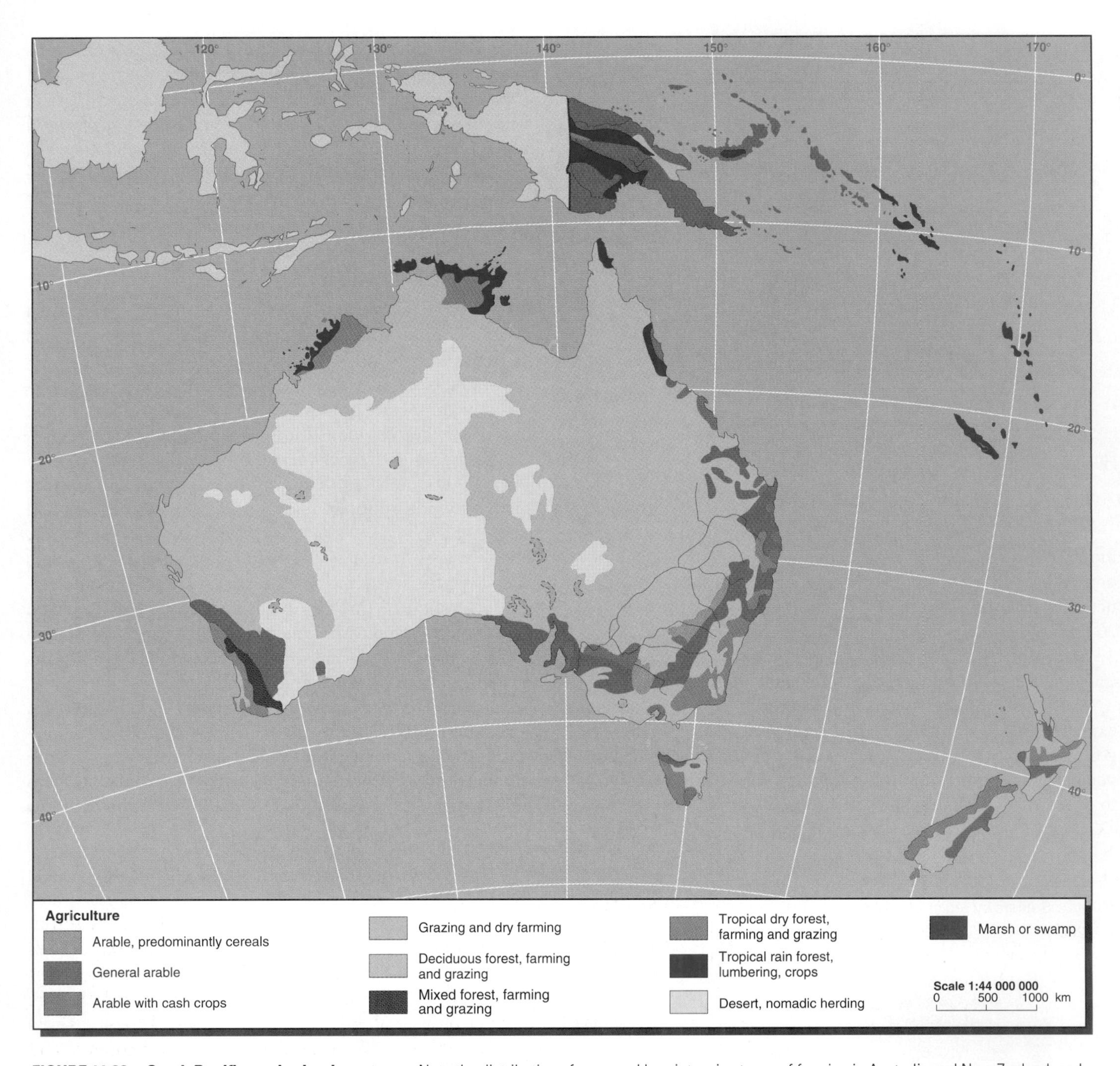


**FIGURE 11.20 South Pacific: major land use types.** Note the distribution of more and less intensive types of farming in Australia and New Zealand, and the contrast of both countries with land uses on Papua New Guinea.

Source: Data from *New Oxford School Atlas,* p.85, Oxford University Press, UK, 1990.

produces 10 percent of the Western world's output of uranium, and has 30 percent of the Western world's reserves. Australian production is restricted to three mines at present, and debate continues between those who wish to expand production and those who wish to curtail it further or see restrictions placed on those buying uranium from Australia. Other minerals of importance to Australia include iron ore, of which it is the 4th largest world producer, bauxite (1st), nickel (4th), and gold (3rd). In the 1990s, Australia also opened the world's largest diamond mine.

## Farm Output

Much of Australia's land is farmed (Figure 11.20). Australia's farmers export 80 percent of their production and sold more in Asia following protectionism in Europe and the United

**FIGURE 11.21 Australia: tourism.** The "Gold Coast" area of Queensland, looking south. The Great Barrier Reef lies offshore. This area attracts many visitors from Asia.



States. In 1994, the United States followed a temporary ban by Japan and halted Australian beef imports on the suspicion that toxic chemicals were entering the food chain. Since Japan and the United States between them take three-fourths of all Australian beef exports, such actions had serious consequences for Australian farmers. Australian farmers suspected that with increasing market freedom and declining tariff protection, unilateral action based on such environmental factors will be used to protect home producers in importing countries. It is difficult to prove either suspicion, but both achieved media attention.

### Underdeveloped North

The northern part of Australia remains the least developed part of the country because of its combination of arid and tropical monsoon climates and distance from other population centers. The 26° S parallel almost cuts the country in half, but only 28 percent of Australia's exports are produced north of this line and only 6 percent of Australia's population (slightly more than 1 million people) live there. A trade development zone in Darwin, the largest city, may focus the increased trade with Asia. It was established in 1985 but initially faced problems, including the world recession and a lack of support from Canberra. It then encouraged medium-sized companies with import or export businesses and has been more successful. The region still lacks good transportation infrastructure, including a rail link south to join up with the country's system at Alice Springs.

### Tourism

Australia's natural resources provide the basis for growth in tourism. Numbers of foreign visitors increased from around 1 million in the early 1980s to over 3 million in 1996 and, it was hoped, before the Asian financial crisis occurred, that the number would double by 2000. Over half of these visitors come from Asia, attracted by the beaches, golf courses, and the theme parks along Queensland's Gold Coast (Figure 11.21); the industry's prosperity depends on growing affluence in Asia. Tourists from North America and Europe come not so much for mass tourist facilities as for the outdoors experience, including the viewing of wildlife and rock formations. Visits to such features form the basis of **ecotourism** for more affluent tourists who have the time and money to take the long journey to Australia.

### Changing Trade Partners

Australia is increasingly tied to Asian markets. As it accepts more imports of manufactured goods from Asian countries, it needs to export more. Australia wishes to add value to its exports by processing its agricultural and mining products in Australia, but Japan and other Asian countries retain import restrictions on processed materials. Certain specialty products created their own growing markets abroad, but Australia can do little more unless the Asian countries lower their tariffs. Furthermore, Australia is a relatively small market for Asian goods.

The six nations of ASEAN (see Chapter 6) have a regional free trade agreement among themselves but are reluctant to admit Australia and are more interested in forging agreements with larger nations. The Asian countries have different attitudes than Australians to the workings of the economy. Asian governments and businesses work closely together, and environmental and civil rights issues assume less importance. Australians remain fixed to Western values and ideas that sometimes upset Asian countries and place political barriers in the way of trade. For example, Australia and Malaysia had major disagreements in the 1990s over Malaysian logging practices in the South Pacific.

Australia lobbies in the **Asia-Pacific Economic Cooperation Forum** (APEC), a wider group of Asian and Pacific countries, including the United States, that is dedicated to trade liberalization by 2000. From Australia's point of view, changes need to come sooner than are likely in this context. The increasing links between the United States and Latin American countries further threaten Australian access to markets in the United States.

Along with its attempts to develop wider trading links with Asian countries in ASEAN and APEC, Australia is involved in links with other countries inside the major region of which it is a part. Attempts to form closer economic links with New Zealand in the 1980s, however, slowed in the 1990s, possibly because the attempts to enter Asian markets were seen as more important or because Australians saw New Zealand as gaining the main benefits from such links. The **South Pacific Forum**, established in 1971, links 13 countries—Australia, New Zealand, Papua New Guinea, the Solomon Islands, the Cook Islands, Fiji, Kiribati, Nauru, Niue, Tonga, Tuvalu, Vanuatu, and Western Samoa. Its aim is to develop regional political cooperation. This agency has had some success in confronting regional problems for the islands in particular. Australia has championed a number of causes, but these often bring it into further confrontations with Asian nations, as when it opposed the exploitive Malaysian logging of Papua New Guinea.

## NEW ZEALAND

**New Zealand** has two main islands—North Island and South Island—but under 4 million people on an area greater than that of the United Kingdom (Figure 11.22). Like Australia, New Zealand was colonized from Britain and is dominated by European culture, economic activities, and attitudes, although the indigenous Maoris form more of a political and social force than the Australian Aborigines. Like Australia, too, New Zealand is going through major changes from being a peripheral colonial supplier of primary products to being an urban-industrial nation related more closely to its Asian neighbors (see "Living in New Zealand," pages 550–551). As a smaller country than Australia, New Zealand has a narrower range of products and a much smaller domestic market.

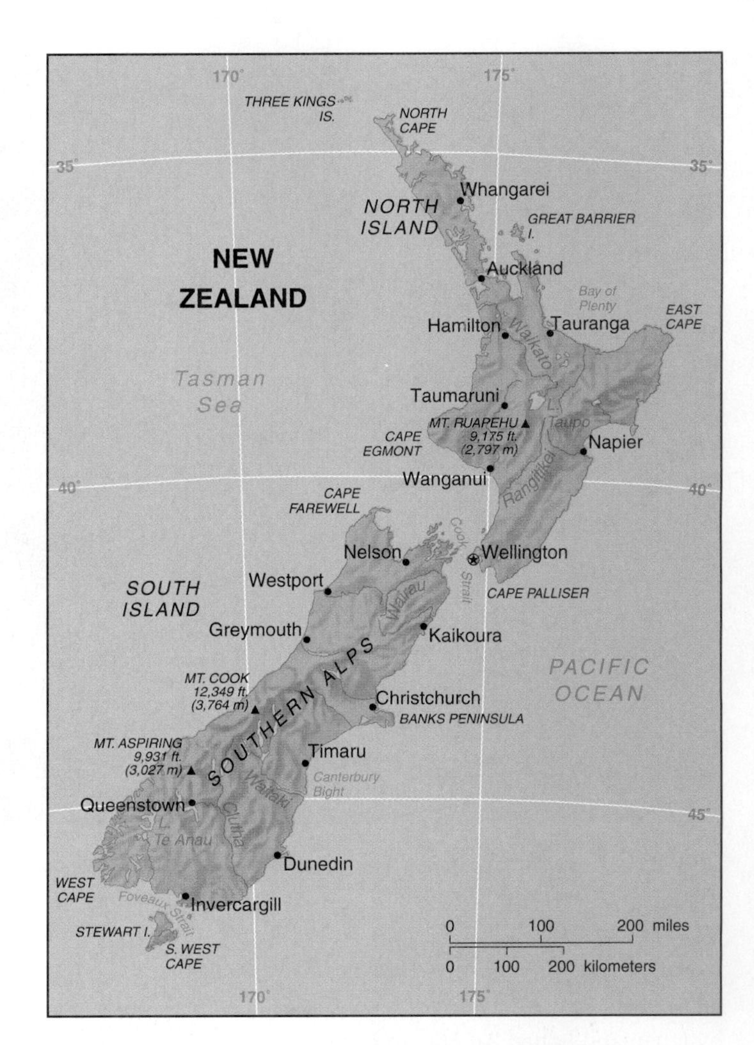


**FIGURE 11.22 New Zealand: major geographic features.** The physical features and major cities. Only Auckland has a population of around 1 million people.

After gaining dominion status, New Zealand had to survive some years of manpower losses from its energetic involvement in two world wars in support of Britain. It also faced economic problems as other countries competed for the markets for their products.

### Regions of New Zealand

Nearly all the Maoris live on North Island, which also contains the largest city, Auckland, and the national capital, Wellington. North Island has a central upland area affected by volcanic activity.

**FIGURE 11.23 New Zealand: urban landscape.** Auckland harbor, North Island. Auckland is the main commercial center of New Zealand. Its waterfront has oil terminals close to the high-rise city-center offices.



South Island has high western mountains, the Southern Alps, that are occasionally rocked by earthquakes. Its eastern plains provide good farming land. Internal rivalries in New Zealand are less marked than those within Australia.

## People

Most of New Zealand's population lives in North Island (see Figure 11.16). On South Island, the east coast is most attractive to people and is where the main towns were built.

New Zealand's total fertility rate remains slightly above that of Australia, but its immigrants keep the annual population rate of increase just over 1 percent. Immigration is less important than in Australia, and the population total remains low. The age-sex diagram (see Figure 11.17) resembles that of Australia apart from an increase in births in the late 1980s.

In the late 1990s, 84 percent of New Zealanders lived in towns and cities. New Zealand's main cities in both islands began as ports that acted as "hinges," giving access to the interior and back to the colonial homeland. Processing industries concentrated in the ports, which became market, commercial, financial, and government centers. **Auckland** is the center of an urbanized area, which stretches along the peninsula to its north (Figure 11.23). **Wellington,** with one-third of Auckland's population, is a major port and the center of government. Christchurch and Dunedin are the main towns of South Island.

New Zealand's population continues to be more British in origin and allegiance than that of Australia. New Zealand receives fewer immigrants, although it does not discourage nonwhites in the way Australia did until 1972. The indigenous people, the Maoris, take a significant part in New Zealand life. They total around half a million people, mostly living in North Island. Auckland has over 100,000 Maoris, many of whom have professional jobs.

## Economic Development

New Zealand's economy, like that of Australia, is based on natural resources, although it has few mineral deposits. In the British colonial era, New Zealand developed an economy based on the exports of mainly farm and forest products to Europe. New Zealand is now adjusting to the switch from European to Asian markets.

New Zealand's GDP per head in 1995 was on a par with the Republic of Ireland. As occurred in Australia, the contributions of agriculture and manufacturing fell as that of services rose to account for around two-thirds of GDP.

### Farming and Forestry

New Zealand's main economic products entering world markets are wool, lamb, and dairy products. Over half of New Zealand is in pasture, and over one-third of its exports is from livestock (see Figure 11.20). Its income from meat products fell in the 1990s, leading to major reductions in herds of beef cattle and sheep, but dairy product demand and deer farming expanded.

In the development of farming, pastureland replaced much forest. The New Zealand government replanted large areas of forest with Douglas firs in the early 1900s. This afforestation policy provides an increasing harvest that finds markets in Asia as those countries cut their natural forest with little replanting. Forestry is now as profitable as farming in New Zealand, and the state-owned forests have nearly all been privatized, albeit with strong conservation laws. Processing industries use the capital accrued from foreign sales of timber.

## NEW ZEALAND

LIVING IN . . .

A person living in New Zealand, especially one who has traveled to other rich world countries, is particularly conscious of the fact that under 4 million people live in a country that has a greater area, for example, than the United Kingdom—in which nearly 58 million people live. There are 12.5 people to the square kilometer in New Zealand compared to 236 in the U.K. The comparison with the United Kingdom is appropriate because New Zealand was settled as a British colony and because the natural environment, particularly the climate, of New Zealand resembles that of Britain.

The low density of population combines with a high percentage of urban living to create large rural areas with little pressure on space. New Zealanders feel at home among their mountains, hills, and coasts. With their pleasant temperate climate, warming in the north to Mediterranean-like conditions, the people of New Zealand experience one of the world's most pleasant living environments. The distance from the former "home country"—it still takes 24 hours by air from London to New Zealand—brought a sense of isolation on the far side of the world, engendering a loneliness and a parochial feeling that makes the inhabitants give a warm welcome to visitors.

The former cozy economic relationship with the home country was shattered in the 1970s when the United Kingdom joined the European Union. People living in New Zealand felt cast adrift from their former economic and cultural partner. The French emphasized this in the U.K. negotiations over entry to the EU: New Zealand lamb meat and fruit competed with French products. At that time and since, New Zealand championed efforts by South Pacific islands and environmental organizations to end French nuclear testing in the region, helping to harden attitudes against it in France.

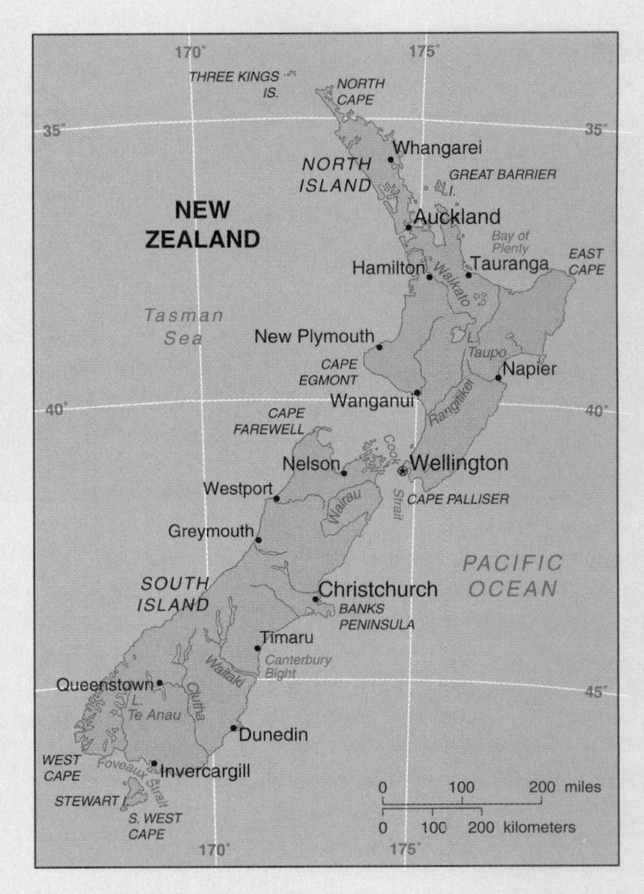


The main features of New Zealand.

Since 1980, people living in New Zealand faced major changes. The main thrust of its economy is still agricultural, but the exports of lamb meat, wool, dairy products, and fruit, mainly to the United Kingdom, had to meet new challenges that affected many aspects of life in New Zealand. Despite such challenges and the costs of transportation to Europe, New Zealand is still able to compete on costs in Britain in such markets as cheese, butter, and lamb. Both Australia and New Zealand can deliver farm products to British homes more cheaply than the EU farmers. Such boasts are based on the internal restructuring of finances and marketing operations within New Zealand and the reduction of restrictive practices among New Zealand dockworkers.

As well as increasing efficiency, New Zealand diversified both its products and its markets. Worldwide concerns over cholesterol led not only to the development of sheep with less body fat but also to a new industry of deer farming. Introduced from Britain as a source of game, deer became a pest that caused the extinction of local herbivores and overgrazed grassland, causing soil erosion. New consumer tastes solved

### Tourism

New Zealand has taken steps to expand its tourist industry, with attractions in the outdoors of both North and South Island. Increasing numbers of tourists now come to this country despite the fact that it lies at a great distance from the main sources of richer tourists. This has been made possible by the deregulation and increase of worldwide air travel. In 1996, New Zealand had 1.5 million tourist visitors.

### Changing Economy

In the 1980s, the New Zealand government instituted economic reforms to revive a rather stagnant economy and produce a trading surplus. Tariffs and restrictive port practices were removed and government spending reduced as a proportion of GDP. By 1995, this resulted in trade surpluses and lower unemployment and inflation. Increases in manufacturing productivity followed as labor costs were kept down and

A New Zealand landscape in Wellington, the national capital. Cricketers practice in the foreground, while the harbor and hilly suburbs provide a backdrop.



this problem by providing a market for deer meat, venison, which is low in fats and increasingly in demand with exports to Australia and Europe. Other deer products such as hides and antlers are developing worldwide markets. Sheep wool remains a major export, mainly for the British carpet industry.

Fruit production, which began to expand in apple growing before the United Kingdom joined the EU, is another facet of product diversification. The best known product is the kiwi fruit, and although other countries now grow their own, New Zealand remains the main southern hemisphere producer and a leader in developing improved varieties. Grape growing and its linked wine industry provided further diversification. The New Zealand vintners selected grape varieties and growing techniques that were suitable and used modern technology to produce high-quality grapes and wines.

New markets, particularly in Japan and other parts of Asia, were developed. Lamb is sold to Muslim countries, and the growing taste for western foods in such countries as Japan provides scope for New Zealand exports of dairy products and fruit.

The pleasant climate, scenic wonders, and open spaces in New Zealand are of increasing interest to tourists, despite the long distances from the world's richest countries. Tourist numbers increase year by year, and so does the range of organized outdoor activities. The people of New Zealand now find signs in tourist resorts written in both English and Japanese.

The ability to make such changes reflects the nature of New Zealand. It is a relatively young country in which cultural traditions imported from Europe were questioned before and after it became a virtually independent dominion within the British Commonwealth. Its political and economic assumptions were challenged again in and since the 1970s, and the responses led to new outlooks for the people.

Alongside the economic changes, New Zealanders are conscious of the people who were in the country before the British—the Maoris. Although these people were subject to repression and reduced to underclass status, like similar groups in other colonized countries, New Zealand made it possible for many people of Maori origins to be integrated as New Zealanders on the sports field and in business life. Their contribution to New Zealand culture is recognized as distinctive, if somewhat marginal in comparison with the imported European features.

New Zealand remains a distinctive country in which to live. Distant from other rich countries, environmentally attractive, and with an integrated pre-European population, it is regarded by many visitors as a pleasant place. Its small population, however, limits the contacts for people in more sophisticated careers, and there is a continuing outmigration to Australia, the United States, and Europe.

transportation and utility costs reduced. Unemployment remains high, however, and there is a steady exodus of people to Australia.

Although New Zealand's attempts to develop an open market with Australia received little encouragement, it successfully developed trade relations with other Pacific countries. New Zealand now sells more goods to Japan than to Australia, the United States, and the U.K. combined. Most of its imports still come from Australia and the United States. Like Australia, New Zealand is a member of the South Pacific Forum, and some of the islands in the South Pacific still have close political and economic ties to New Zealand following New Zealand's role as United Nations agent in those countries before their independence.

**RECAP 11C: Australia and New Zealand**

Australia occupies the world's sixth largest continent but has only 19 million people, largely because half is desert and immigration was restricted for most of the 1900s. Australia's economy ranks it with the world's rich countries and is based on exports of a wide range of minerals and farm products, together with growing manufacturing and tourism industries. Its five largest cities hold nearly 60 percent of the total population.

New Zealand is another rich country. It consists of two main islands. Its 3.5 million people export wool, lamb meat, timber, dairy produce, and fruit.

11C.1 Why does such a high proportion of Australians live in the major coastal cities of the southeastern quadrant of the country?

11C.2 What factors may slow down the further exploitation of Australia's mineral resources?

11C.3 What changes were involved in Australia switching from European to Asian markets?

11C.4 What are some of the environmental impacts faced in Australia and New Zealand?

**Key Terms:**

White Australia policy
ecotourism
Asia-Pacific Cooperation Forum (APEC)
South Pacific Forum

**Key Places:**

Australia
New South Wales
*Sydney*
Victoria
*Melbourne*
South Australia
*Adelaide*
Queensland
*Brisbane*
Western Australia
*Perth*
Northern Territory
Australian Capital Territory
New Zealand
*Auckland*
*Wellington*

# SOUTH PACIFIC ISLANDS

The **South Pacific islands** range from the eastern part of New Guinea—the independent country of Papua New Guinea with 4 million people—to tiny islands with a few hundred people (Figure 11.24). The smaller islands are grouped in independent countries or colonies. Like Australia and New Zealand, many rely on exports of a small range of primary products and growing tourist industries. Their small sizes and markets, however, bring little prestige in arguments with Asian countries or the United States over timber, mineral, agricultural, or fisheries exploitation, and the standards of living of their peoples suffer accordingly. Only occasionally, as in the violence on the French Polynesian Island of Tahiti over the French nuclear testing in mid-1995, do these tensions surface to world news levels.

## Islands

Most of the South Pacific islands gained their independence in the 1970s, although others became independent at varied dates: Western Samoa became independent in 1901 and the Marshall Islands in 1991. The Micronesian group of islands remains a United Nations Trust Territory. New Caledonia and French Polynesia are part of France. Although independence appeared desirable to many of the islands, economic difficulties, internal tensions, and dependence on continuing economic aid and protection keep them linked economically to the United States, France, Britain, Australia, and New Zealand. Kiribati, Tonga, Tuvalu, Western Samoa, and the Solomon Islands remain among the world's poorest countries with few products in world demand. They compete with each other for sales of coconuts and coconut products.

Modern country groupings of the islands cut across ethnic and racial divisions. A more convenient division of the islands than the established Melanesia, Micronesia, and Polynesian distinctions recognizes modern political arrangements by dividing them into the southwestern, northwestern, and central South Pacific Ocean areas.

- The southwestern group approximately coincides with former Melanesia and includes **Papua New Guinea**, the **Solomon Islands**, **Vanuatu**, **Fiji**, and **New Caledonia.** These are the largest and most populated islands. Papua New Guinea includes the eastern part of the largest of the islands; the western part of this island, Irian-Jaya, is in Indonesia (see Chapter 6).
- The northwestern group, roughly coinciding with former Micronesia, includes **Guam**, the **Northern Mariana** and **Marshall Island** groups, **Nauru**, **Palau**, and the **Federated States of Micronesia.**
- The largest of the three groups in terms of ocean area but the poorest in economic development contains the islands of the central Southern Pacific such as **Kiribati**, **Tuvalu**, **Wallis and Futuna**, **Tonga**, **Tokelau**, **Western** and **American Samoa**, **Niue**, **Cook Islands**, **Pitcairn**, and **French Polynesia.**

The islands vary considerably in their well-being, as the consumer goods ownership of Fiji and New Guinea shows (see Figure 11.13). The French colonies of French Polynesia and New Caledonia are heavily subsidized by France and have above-average incomes. Others, such as the Marshall Islands, Palau, Micronesia, Guam, and the Northern Marianas have U.S. support. Nauru has phosphate mines and New Caledonia has nickel mines. Most island countries, however, have few resources that enable them to develop much more than a subsistence economy for their people.

## People

The South Pacific islands have small total populations, with the largest numbers on those islands nearest to Australia and New Zealand. In 1998, the southwestern group had a total population of some 6 million people, 4.3 million of whom lived in Papua New Guinea and 780,000 on Fiji. The total population of the smaller northwestern islands was under half a million, while that of the widespread islands in the central South Pacific was around 2 million.

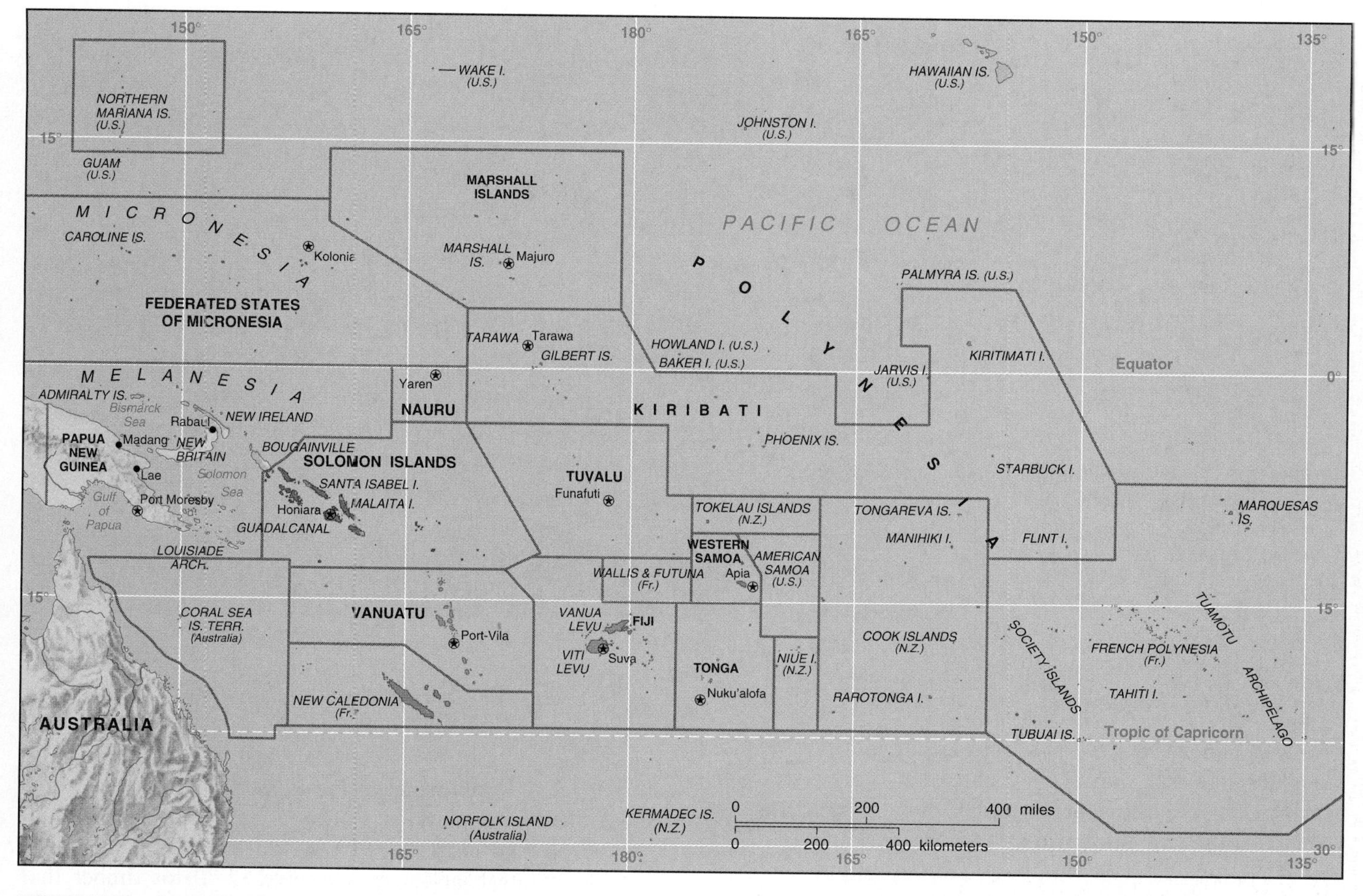


**FIGURE 11.24 South Pacific islands: major geographic features.** The country groups and main towns. These islands are spread across huge distances.

Population growth continues on many of the islands due to high total fertility rates (see Figure 11.17). The age-sex graphs resemble those for poorer countries elsewhere and imply a large proportion of young people in the total population (see Figure 11.18).

Many islands are on the point of **overpopulation**, where the land and its resources cannot support its people on a subsistence basis. Economic development has not occurred to support larger numbers of people. The ease with which basic needs are supplied from fishing and the cultivation of vegetables and fruit results in a high incidence of obesity, heart disease, and diabetes. Only on Fiji, Guam, New Caledonia, and Palau does life expectancy rise above 70 years, and for most islands, it is 55 to 65 years.

Few of the South Pacific islands are large enough to have major cities. Many islands link to the cities of New Zealand and Australia, or to the Hawaiian Islands for access to world air routes. **Suva** on Fiji and **Port Moresby** in Papua New Guinea are the main exceptions. Most island towns began as ports to give access for colonial links and continued as political capitals of the independent countries (Figure 11.25).

## Economic Development

Before colonization, the island communities had complex systems of integrated economies, often linked across the wide ocean expanses. Exchanges took place among the better-watered islands where vegetables and fruit were grown and the arid islands that relied on fishing. Unique maps of direction and distance were constructed, but storms and crop failures brought death and destruction. Although these islands were perceived by early travelers as a "Garden of Eden," day-to-day life was not as idyllic as they wrote.

After colonization, some of the islands were brought into the world economy with sales of coconut palm products such as copra—the dried white meat that lines the inside of the coconut shell and provides an oil used in soaps and candles. The island economies were partly developed by the commercialization of the established inter island trade. Manufactured goods from Europe and the United States and expensive diesel oil for power were imported. Most islands were subsidized by the colonial countries but remained poor because of their limited resources.

**FIGURE 11.25 South Pacific: urban landscape.** Levuka, Fiji, where colonial touches remain in the landscape: a British-style church tower and mainly wooden store buildings on Beach Street.



## Farm and Mine Products

Apart from Papua New Guinea (copper and gold), New Caledonia (nickel), and Nauru (phosphate), few islands have natural resources except for a warm climate, nutrient-rich volcanic soils, and the surrounding ocean (Figure 11.26). The small areas available limit agricultural outputs and few islands diversified beyond coconuts and copra. Fiji produces sugar and Tonga produces bananas and vanilla in addition to coconuts.

Papua New Guinea suffered a major setback when a separatist group on Bougainville forced the closure of copper mining activities. After making protests about the copper mining and its environmental impacts since 1969, the people of Bougainville declared the island's independence in 1990. Papua New Guinea used military force to prevent secession, but the Australian-owned mining operation remains closed because it is located in a separatist-controlled area. The copper made up 30 percent of Papua New Guinea's exports, and the government is trying to replace this income by encouraging the development of gold mines in other parts of its territory.

## Small Islands, Few Resources

In the 1990s, most South Pacific islands faced economic ruin. Islands such as Tuvalu relied on aid for nearly 100 percent of their income. Some see external aid as a compensation for colonial exploitation; others see it as demeaning and sapping self-reliance. With a low economic base, the islands suffer from increased exploitation rates by foreigners of their few natural resources.

World attention was drawn to the cutting of tropical hardwoods on Papua New Guinea and the Solomon Islands by a statement at the South Pacific Forum in Brisbane in mid-1994 that the islands were being "ripped off." The imposition of stricter environmental laws in Malaysia and Indonesia turned logging companies from these countries to the islands that had fewer regulations. Cutting is now at levels that could remove all the timber by 2000, and little replacement is taking place. Local landowners receive \$2.70 for timber that fetches \$350 on world markets.

Similar problems face the local fishers as fishing fleets from Japan, the United States, South Korea, and Taiwan take fish from this region to make up nearly 40 percent of the world catch. Although the island countries of the South Pacific have declared 200-mile exclusion zones and the foreign fishers often pay access fees, it is impossible for the whole area to be patrolled.

## Tourism

Tourism is one of the few ways open to the people of the subregion to increase their incomes. The islands have different levels of access to the rest of the world and of interest shown by tourist companies. Tourism is important on some islands such as Fiji, Guam, the Marshall and Northern Mariana Islands, Toga, Vanuatu, and Western Samoa, but it has little impact on others such as Kiribati, Micronesia, Nauru, New Caledonia, Palau, the Solomon Islands, or Tuvalu. Tourism can, however, destroy the last remnants of traditional culture and often pays low wages to local people. Yet it is often a better industry to work in than Third World shirt factories. Tourism in this region remains dependent on the retention of an attractive environment and political stability and so is likely to encourage good environmental practices. Where tourism has become important, as in Fiji, it has forced a greater openness to foreign visitors' requirements, creating other opportunities for local businesses.

### South Pacific Forum

The South Pacific Forum, formed in 1971 and dominated by Australia and New Zealand, has its headquarters in Suva, Fiji, and is asserting the need for new rules to manage the resources in this region. The growing problems faced by the small islands lead to closer cooperation in fishing and transportation policies. The islands are also cooperating on a political agenda that includes the independence of remaining colonies, increased investment, and trying to stop French nuclear tests.

**FIGURE 11.26 South Pacific: island landscape.** A 1994 space shuttle photo of New Caledonia shows evidence of environmental degradation. Several fires are burning what little remains of the forest cover to provide cattle pasture. Much of the land has been degraded by open-pit mining for nickel. Offshore, the coral reef is being degraded as mud from eroded land fills the lagoon to produce mudflats off the southern coast.

NASA

## ANTARCTICA

For the present, **Antarctica**, which occupies 10 percent of Earth's land surface, is not settled, except for visiting groups of scientists.

### Antarctica's Global Status

International expeditions to Antarctica first concentrated on the personal physical feat of crossing the frozen continent but gave way to international claims on sections of the continent during the mid-1900s (Figure 11.27). In 1961, 39 countries signed the **Antarctic Treaty** as a basis for nonmilitary scientific cooperation, environmental safeguards, and international control. In 1991, the Wellington Agreement banned commercial mining activities and introduced protection regulations. Some countries did not sign this agreement.

Antarctica's natural resources are not exploited, and the international agreements stipulate that they should not be. Only the future will tell how long this protective situation will last or whether Antarctica turns into an open-pit mine for the rest of the world. The main interest in Antarctica in the 1990s concerns its important role as a laboratory for monitoring global climate change through studying the development of the ozone hole above it and the extent of ice on it and the surrounding oceans. Research carried out in Antarctica indicates a decline in world environmental quality. In their small but increasing manner, however, research stations also degrade to local environment by pouring untreated sewage into the ocean and dumping oil drums on sites that seabirds use for nesting.

### Antarctica's Resources

Antarctica is not a country and so does not have an economy of its own. There is no authority to regulate resource usage. Commercial fishers from all over the world exploit the resources of the surrounding oceans. In 1982, it was agreed internationally to regulate such fishing. Some fish stocks, such as cod, together with some groups of whales, were declining because of overfishing. In the mid-1990s, the long line hooks used to catch Patagonian toothfish also snared albatross and petrel birds and caused worries over the future of the total ocean ecology. Some fishing fleets, however, profess ignorance of the quotas that were established or claim that they have fished outside the Antarctic convergence zone that forms the northern boundary of the partially protected area. It is almost impossible to monitor fishing in this extensive area of ocean that has few other ships passing through and is not the responsibility of a particular country.

### Tourism

Tourists, usually based on cruise ships, visit Antarctica to view the scenery, wildlife, and some research stations. In 1994–1995, some 8,500 tourists visited the Antarctic Peninsula, and the numbers are growing, especially with the arrival of sturdy ice-breaking ships that increase landing

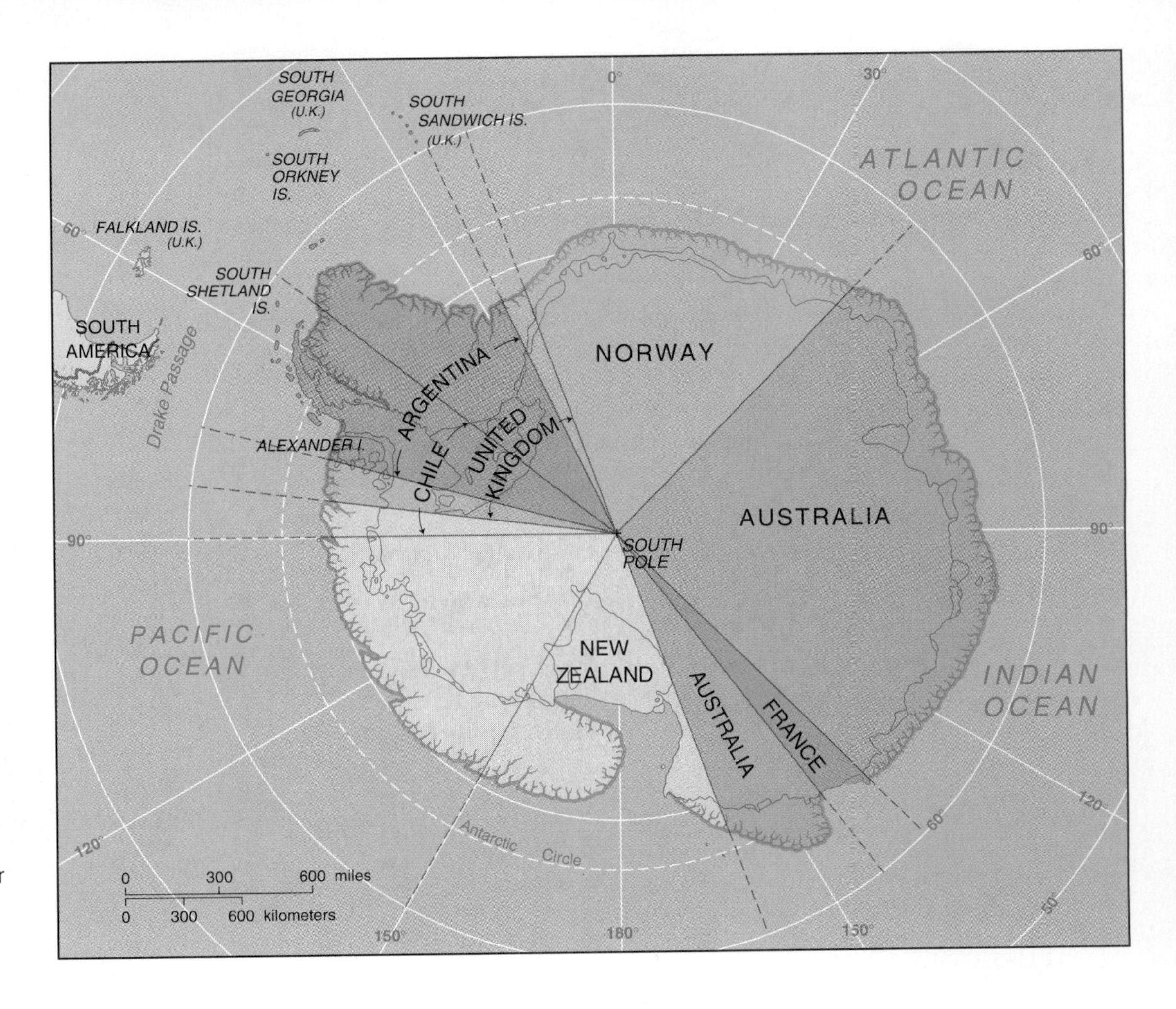


**FIGURE 11.27 Antarctica: political claims.** The United States and Russia, among other countries, refuse to make such claims and do not recognize other countries' claims

possibilities—and as there is a reduction in oceanic ice caused by global warming. There is the prospect that backpackers and mountaineers will visit the continent. As more tourists arrive, however, the dangers of environmental damage increase. At present, the Antarctic Treaty system does not have a code for the tourism industry.

## LANDSCAPES OF THE SOUTH PACIFIC

The landscapes of the South Pacific countries include large areas of seemingly little-altered natural environment, especially in Antarctica and central Australia. Urban landscapes are most fully developed in Australia and New Zealand. Those on the islands are small colonial port towns with only a few large buildings, often added since independence.

Each of the major cities in Australia began as ports of entry for separate colonies and retained the centrality and primacy established at that time (Figure 11.28). They retain their established port functions, although Melbourne grew in importance by specializing in container trade. They have downtowns that combine state government and financial sectors with major facilities such as the Sydney Opera House. Industrial suburbs vie with residential areas for space.

Smaller towns are often linked to specific functions as opposed to the diversified purposes of the largest cities. This is reflected in their urban landscapes. Thus Newcastle, New South Wales, and its neighbor, Port Kembla-Wollongong, are coal and steel towns and ports, while Broken Hill is an inland mining center. Mining towns often have short lives while the mineral resources or markets last, and many have declined in economic activity after a time of boom. Towns along the Queensland coast have tourist functions, while inland towns such as Mildura are farm market centers.

New Zealand towns reflect similar histories (see Figure 11.23), including the most recent growth of manufacturing and service industries. All have port facilities. In the South Pacific islands, the towns have populations of a few thousand people and contain a mixture of colonial facilities, some processing industries, ports, and shantytown areas (see Figure 11.25).

The urban landscapes are set in broader rural landscapes, illustrated by the range mapped in Australia (Figure 11.29). Rural landscapes include the commercial farming areas of

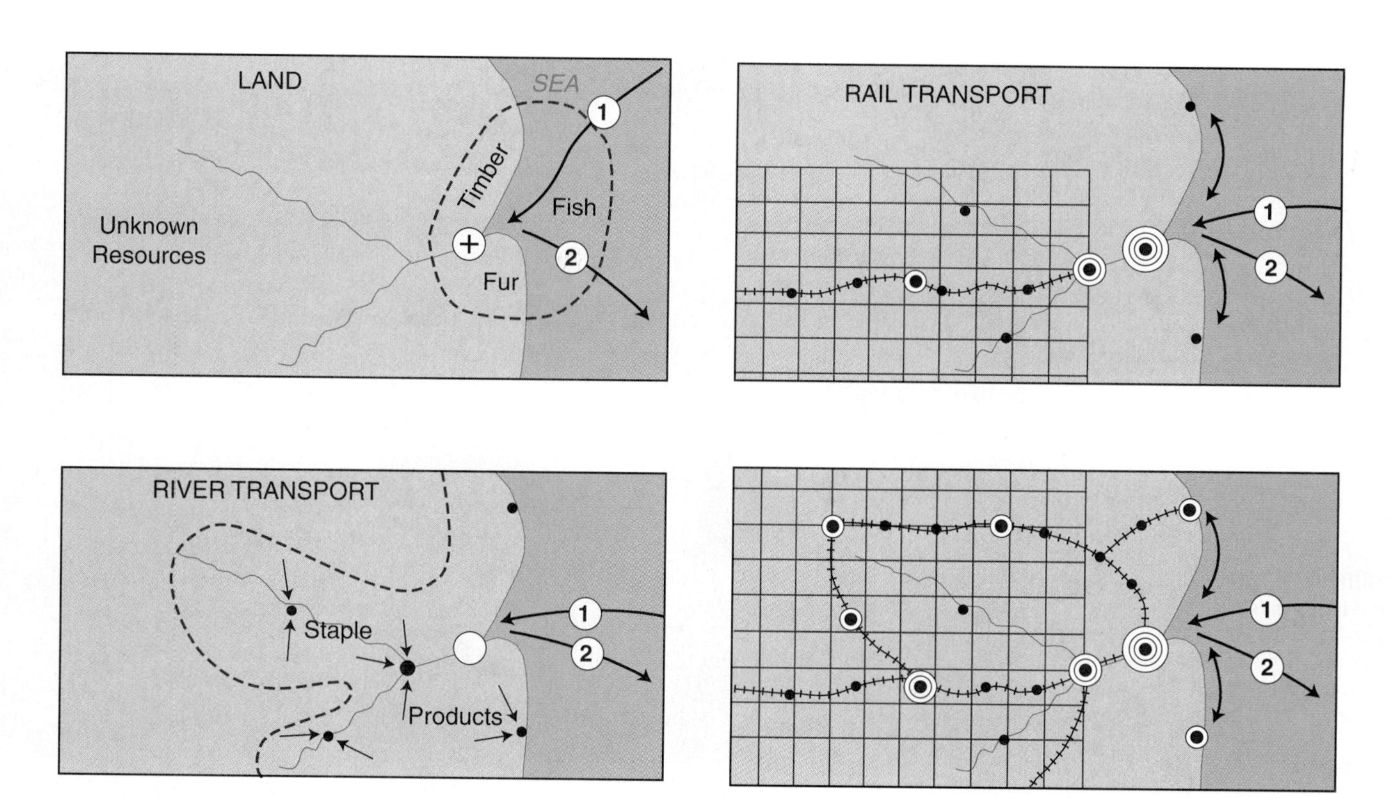


**FIGURE 11.28 South Pacific: mercantile theory of urban patterns in the landscape.** This theory can be applied to Australia and New Zealand, as well as other former colonial lands. The circle size corresponds to the size and range of functions of the urban centers. Imports of manufactures (1) from the home country are paid for by exports of commodities (2).

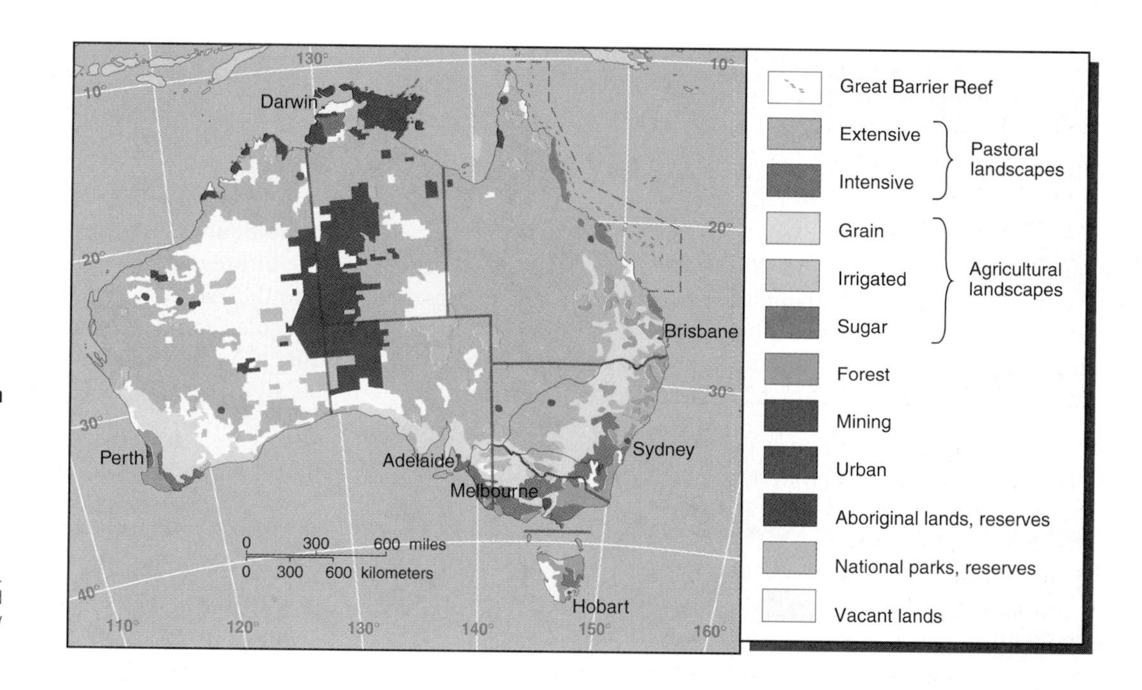


**FIGURE 11.29 Australia: human landscapes.** A variety of human responses to the natural environment, world economic conditions, and the Australian political framework.

**FIGURE 11.30 New Zealand; rural landscape.** Farming area of South Island. The river channels descend from the Southern Alps and form wide valleys across the plateaus of the eastern lowlands.



Australia, New Zealand, and island tree-crop areas (Figure 11.30). Antarctica, interior Australia, and the New Zealand and Papua New Guinea mountains contain landscapes that are little affected by human activities.

## FUTURE PROSPECTS FOR THE SOUTH PACIFIC

The futures of the South Pacific countries clearly lie with that of their neighbors in Eastern Asia. Together they make up part of the western Pacific Rim (see "World Issue: Pacific Rim," pages 530–531). The recent reorientation of the Australian, New Zealand, and many island economies toward Asian, and especially Japanese, customers, are strong indicators of the major shift in geographic alignment and core-periphery relationships in the region. The future of connections to Asia rests on the fragile and changing requirements of the countries of Eastern Asia for the minerals and farm products of the South Pacific and on continuing economic growth in Eastern Asia. Even Australia provides such a small market for the manufactures of Eastern Asia that it is unlikely they will achieve significant sales in this region.

The South Pacific islands continue to be largely dependent on their own internal resources, backed up by external and other forms of support. They need to work together to manage those resources in the light of external attempts to remove them without proper compensation to the local people. They need the South Pacific Forum to voice their problems arising from exploitation by other countries.

The future of Antarctica will depend on the level of adherence to the Wellington Agreement and the prospect of effective regulation of the marine resources surrounding the continent. Some commentators doubt that this will continue to be significant if the world enters a period of natural resource shortages.

## CHAPTER REVIEW QUESTIONS

1. The majority of Australians live in the _________ part of the country. (a) southeastern (b) northeastern (c) southwestern (d) northwestern
2. "Australia" is all of the following except: (a) continent (b) country (c) language (d) all of the above apply to Australia
3. Australia has a(n) _________ percentage of urbanization. (a) high (85%) (b) above average (60%) (c) below average (40%) (d) low (15%)
4. New Zealanders often think of themselves as being on the opposite side of the world to Britain; which is an accurate statement? (a) Britain and New Zealand are in

**RECAP 11D: South Pacific Islands, Antarctica, Landscapes, and Futures**

The South Pacific islands include large numbers of small islands grouped in mainly independent but poor countries. Papua New Guinea is the largest, having over half the total population of all the islands. Many South Pacific islands have no commercial products; others produce coconuts and copra; some produce a particular mineral for export; and some have economies maintained by French colonial support or United States grants.

Antarctica has only a few temporary inhabitants, mostly scientists concerned with monitoring weather processes. International agreements prohibit commercial development of mining opportunities. Despite other agreements, the surrounding seas are fished with little concern for conserving the stocks.

The urban landscapes of the South Pacific range from Western-type urbanization in Australia and New Zealand to small ex-colonial port towns on the islands. Rural landscapes are dominated by palm-covered island landscapes except in the largest land areas of Australia, Antarctica, New Zealand, and Papua New Guinea.

The future prospects of the region rest partly on how Australia and New Zealand meet the challenge of reorienting their economies toward Asian markets and partly on the extent to which the Pacific Rim trading area becomes a reality.

11D.1 What attractions do the countries of this major region offer to tourists? What are the advantages and disadvantages of the tourism industry for the people of this region?

11D.2 What roles does Antarctica play in the physical and human geography of the region?

**Key Terms:**

overpopulation

Antarctic Treaty

**Key Places:**

South Pacific islands
Papua New Guinea
*Port Moresby*
Solomon Islands
Vanuatu
Fiji
*Suva*
New Caledonia
Guam
Northern Mariana Islands
Marshall Islands
Nauru
Palau
Federated States of Micronesia
Kiribati
Tuvalu
Wallis and Futuna Islands
Tonga
Tokelau
Western Samoa
American Samoa
Niue
Cook Islands
Pitcairn
French Polynesia
Antarctica

the Northern Hemisphere. (b) Britain and New Zealand have equatorial climates. (c) Britain and New Zealand have mid-latitude climates. (d) Britain and New Zealand have polar climates.

5. Papuan people are found on the island of New Guinea, and they _________ (a) live divided amongst two countries. (b) live united in one country. (c) live throughout Polynesia. (d) live united with Australian Aborigines.

**WEBSITES: South Pacific**

For Australia, New Zealand, and issues affecting the whole area, the following newspaper sites are important:

*http://www.smh.com.au/* The *Sydney Morning Herald*

*http://www.press.co.nz/* A summary of New Zealand news, including the impacts of the Asian crisis on the New Zealand economy.

*http://www.govt.nz* New Zealand government site giving access to the *New Zealand Official Yearbook* for the current year (published in June) and selected data from its pages.

For other countries in the region, try the CIA *Factbook*:

*http://www.cia.gov/cia/publications/factbook/index.html.*

6. Central areas of Australia (Red Center) can be classified as a _________ environment. (a) desert (b) jungle (c) swamp (d) tundra
7. Which city stands out amongst the following capital city/country associations? (a) Alice Springs / Australia (b) Port Moresby / Papua New Guinea (c) Suva / Fiji (d) Wellington / New Zealand
8. The South Pacific Forum is a(n) _________ (a) military alliance. (b) international political association. (c) tourism promotion board. (d) basketball stadium.
9. The people of one of the following Pacific islands are not automatically American citizens. (a) Guam (b) American Samoa (c) Tonga (d) Wake
10. Antarctica stores a majority of Earth's fresh water as solid ice; this ice covers almost all the land surface which represents _________ percent of the total land surface of the planet. (a) 10 (b) 20 (c) 30 (d) 40
11. Australia and New Zealand are home to unusual numbers of marsupial mammal species.
True / False
12. New Zealand is a net importer of meat products.
True / False
13. The island of New Guinea is divided between two different countries.
True / False
14. Tahiti remains a French territory.
True / False
15. Guam remains a German territory.
True / False
16. South Pacific islanders are mainly Melanesians, Micronesians, or Polynesians.
True / False
17. Some islands in the South Pacific were used for nuclear testing.
True / False

18. The South Pacific islands are united in their ability to resist outside resource exploitation.
    True / False
19. Currently, all South Pacific islands are independent.
    True / False
20. Australia is the only single country continent.
    True / False
21. The Australian Capital Territory encompasses the federal capital of _________ .
22. The Southern Alps are found on the South Island of _________ .
23. Which Australian city is host to the 2000 Summer Olympics?
24. Perth is the largest city in which Australian state?
25. Viticulture is found in which part of Australia?
26. Who is Australia's largest trading partner?
27. Which of the Solomon Islands belongs to Papua New Guinea?
28. Which countries lost direct control of the South Pacific islands as a result of the World Wars?
29. Which two countries do not recognize claims in Antarctica, but maintain their own presence there?
30. Future economic links of the South Pacific will likely be towards where?

# REFERENCE SECTION

# GLOSSARY OF KEY TERMS



Each term is defined and linked to the chapter and RECAP box in which it was a key term.

**Aborigine** (11A). Race of people whose ancestors inhabited Australia before the arrival of Europeans.

**absolute location** (1B). Location of a place on Earth's surface as defined by latitude and longitude or by distance in km (mi.) from another place.

**acid deposition** (7B). Dry or wet deposition of acidic material from the atmosphere, often resulting from sulfur and nitrate gases and particles emitted into the air from coal combustion in power plants.

**African American** (9A). American of African origin.

**agglomeration economies** (7A). The total economies achieved by a production unit because of a large number of related economic activities located in the same area.

**agribusiness** (7C). The large-scale commercialization of agriculture that places farming within the broader context of inputs of seeds, fertilizer, machinery, and so on, and of outputs of processing, marketing, and distribution.

**alluvial fan** (5A). A fan- or cone-shaped river deposit, often formed where a stream issues from mountains into a plain.

**alluvium** (5A). Deposits of sand, clay, and boulders dropped by rivers in the lower parts of their valleys.

**Andean Common Market** (10D). A free trade area among the countries of the Northern Andes—Bolivia, Colombia, Ecuador, Peru, and Venezuela.

**Antarctic Treaty** (11D). A treaty signed by 39 countries in 1961 as a basis for nonmilitary scientific cooperation, environmental safeguards, and international control. Developed further by the 1991 Wellington Agreement with the object of restricting human activities on Antarctica and preventing commercial exploitation of its minerals.

**Appalachian Regional Commission (ARC)** (9C). A U.S. federal government organization established in 1965 with the aim of assisting the poor Appalachian region to catch up with economic growth occurring elsewhere in the country.

**Arabic** (4A). Language spoken and written by Arabs, diffused from Arabia with trade, conquest, and religious use. The only language allowed in the Islamic Qur'an and prayers.

**Arab League** (4A). Organization created in 1945 to encourage the united action of Arab countries for their mutual benefit.

**artesian conditions** (11B). A groundwater situation in which water flows naturally out of springs and wells under pressure that is imposed by the geologic arrangement of water-bearing rocks.

**Asia-Pacific Economic Cooperation Forum** (APEC) (11C). A group of Asian and Pacific countries dedicated to trade liberalization by A.D. 2000.

**Association of Caribbean States** (10D). Formed in 1995 by West Indies countries, Mexico, Central American countries, Colombia, and Venezuela as a response to the formation of NAFTA. A possible step to FTAA.

**Association of Southeast Asian Nations** (ASEAN) (6D). Established in 1967 as a defensive alliance among Indonesia, Malaysia, Philippines, Singapore, and Thailand in response to the advance of communism in Southeastern Asia. Now increasingly a trading group that admitted Vietnam, a communist country, as a member in 1995.

**Aztecs** (10A). A group of Native Americans that moved from northern Mexico and established control by A.D. 1325 over much of the surrounding region from its base at Tenochtitlan, the site of modern Mexico City.

**badlands topography** (5A). Close-spaced networks of deep gullies, often carved by occasional streams in soft sediments unprotected by a vegetation cover. These destroy the usefulness of land.

**banana republic** (10C). A country that relies on a single export crop (such as bananas), is dependent on world core country markets, and has a poor government that is often a dictatorship.

**Bengali** (5C). Language of 98 percent of Bangladesh people; also spoken in eastern India.

**Berber** (4A). Language of African origin spoken, but little written, in North Africa.

**Biharis** (5C). People of non-Bengali origin and Urdu-speaking who migrated to East Pakistan (Bangladesh) at the partition of India in 1947.

**biome** (2D). A world-scale ecosystem type, such as tropical rain forest or savanna grassland.

**birth rate** (2B). The number of live births per 1,000 of the population in a year.

**black earth soils** (8B). Highly fertile soil type in which organic matter accumulates near the surface, commonly beneath midlatitude grassland communities.

**British East India Company** (5A). A British company that traded with and conquered much of the Indian subcontinent. After a 1857 mutiny in India, the British government took over political control.

**British Indian Empire** (5A). Established after 1857 on the Indian subcontinent, including Ceylon and later extended to Burma. Lasted until independence and partition in 1947.

**brown earth soils** (8B). Fertile soil type in which plant matter replenishes nutrients in the upper layers, commonly forming beneath midlatitude deciduous forest.

**Buddhism** (5A). A religion that began in Southern Asia but became the major religion of Eastern Asia. Followers of Buddhism have a greater social openness than do followers of Hinduism and accommodate other philosophies and religions such as Confucianism and Shinto.

**buffer state** (6A). A country that stands between major world powers and helps to reduce direct conflicts between them.

**capitalism** (1C). An economic system in which goods and services are produced and sold by private individuals, corporations, or governments in competitive markets. The means of production are owned by those investing capital, to whom workers sell their labor. Linked to democratic government and increasing trade among places and countries.

**capitalization** (10D). A process adopted by Bolivia in which foreign capital is invested in state corporations instead of privatizing them and in which state institutions benefit from profits.

**Caribbean Basin Initiative** (10D). Formed in 1984 to promote economic growth in the West Indies and to protect U.S. interests.

**Caribbean Community and Common Market** (CARICOM) (10D). Formed in 1973 by 13 former British colonies to provide special entry to U.S. markets.

**caste system** (5A). A social class system associated with Hinduism that is based on the supremacy of Aryan peoples. The Aryan castes include priests (Brahmans), warriors, and other Aryan people; non-Aryan castes include cultivators, craftspeople, and untouchables.

**Celts** (7A). European tribes that diffused skills in metal-making (bronze and iron) as they moved from the Alpine area around 1000 B.C. into Spain, France, and Britain.

**Central American Common Market** (CACM) (10C). Established in 1960 to boost intraregional trade but badly affected by civil wars in member states. Relaunched in 1993.

**central business district** (9E). The area in the center of a city where shops, offices, and financial services are concentrated, normally to the virtual exclusion of residential facilities.

**centralization** (8C). The policy of the Soviet Union in which economic planning was focused on the Russian heartland and economic sectors rather than on regional strengths.

**changing climate** (3A). The process by which the climatic environment of a place changes over time, including, over periods of around 100,000 years, the advance and retreat of ice sheets and the expansion and contraction of arid regions.

**Christianity** (4A). A religion that developed out of Judaism, based on belief that God came to Earth in the form of Jesus Christ. Main religion of the Western world.

**class** (2C). Part of a stratification of society that is based on economic or social criteria.

**climate** (2D). The long-term atmospheric conditions of a place.

**collectivization** (6C). The transformation of rural life in communist countries such as China and the Soviet Union, in which individual farmers were grouped in cooperatives that took ownership of their land and labor.

**common market** (10A). A group of countries that establishes a customs union and a central authority that may override government decisions of individual countries.

**Communauté Financière Africaine** (CFA) (3B). Economic links between France and its former colonies in Africa. From independence until 1994, it involved links to the French franc, but this arrangement ended with the devaluation of the CFA franc.

**commune** (6C). The organization that controls rural life in communist countries, including agriculture, industry, trade, education, local militia, and family life.

**concentration** (of agriculture) (7C). Agricultural production carried out on fewer and larger farms and limited to smaller areas of higher productivity.

**concentric pattern of urban zones** (9E). A pattern of urban geography with a central business district in the middle, surrounded by a hierarchy of residential zones.

**Confucius** (6A). A Chinese administrator who established a system of efficient and humane political and social institutions that became the basis of procedures and ways of life in much of Eastern Asia.

**congregation** (9A). The process by which people choose to live close to others of similar cultural and social character.

**core** (1A). The richer countries in the world economic system that have a favorable balance of trade with poorer countries in the periphery, and a wider range of products, use advanced technology, and pay high wages.

**cottage industry** (5C). Manufacturing based on processes that are carried out in dispersed homes instead of in a central factory.

**Council for Mutual Economic Assistance** (CMEA) (8A). An agreement to link economic planning among Soviet bloc countries, sometimes known as COMECON. Dissolved in December 1991.

**counterurbanization** (6B). The movement of people from metropolitan urban areas to take up residence and employment in small towns or rural areas.

**country** (1B). A self-governing political unit, recognized by other countries.

**cultural hearth** (2C). A small region of the world that acted as a catalyst for developments in technology, religion, or language and as a base for their diffusion.

**Cultural Revolution** (6C). The attempt by Mao Zedong between 1966 and 1976 to change the basis of Chinese society. The disruption held back the country's economic development.

**culture** (1A). The ideas, beliefs, and practices that a group of people hold in common, including language, religion, social activities, and the design of products.

**customs union** (10A). A group of countries that establishes free trade plus a blending of economic policies and business laws among themselves.

**death rate** (2B). The number of deaths per 1,000 of the population in a year.

**deindustrialization** (7A). A rapid fall in manufacturing employment and the abandonment of factories in a once-important industrial region.

**demographic transition** (2B). The process in which the population of a poor country moves from high birth and death rates toward the low birth and death rates of rich countries, through a period of high birth rates and low death rates—when the population increases rapidly.

**demography** (2B). The study of human populations in terms of numbers, density, growth or decline, and migrations from place to place.

**density** (2A). The frequency of a phenomenon within a unit of land area; for example, people per hectare.

**dependency theory** (2C). The theory of development that is a Marxist critique of the world economic system in which the poorer countries (or the periphery) are dependent on and exploited by the core countries.

**deposition** (2D). The dropping of particles of rocks carried by rivers, wind, or glaciers when they stop flowing or blowing, or melt, respectively.

**desalination plant** (4A). A mechanism that extracts fresh water from seawater by evaporation and condensation.

**desert** (3A). The ecosystem type commonly associated with arid climatic environments. The lack of water causes some areas to be without vegetation or animals, but most have scattered drought-resistant shrubs.

**desert biome** (2D). Major biome-type characterized by a discontinuous plant cover or none.

**desertification** (2D). Processes that destroy the productive capacity of an area of land.

**deurbanization** (6D). The process adopted by the communist government of Vietnam after 1975 to reduce the size of Ho Chi Minh City and return people to rural areas.

**development** (2C). The process by which human societies improve their quality of life, including economic, cultural, political, and environmental aspects.

**diffusion** (2A). The movement of people, objects, diseases, or ideas to new areas, as in migration, expansion, or contagion.

**direction** (2A). The position of one place relative to another, measured in degrees from due north or by the points of the compass.

**disease** (3A). A condition in which an animal or plant is prevented from carrying out its normal functions by an infection, which may result in death.

**distance** (2A). The space between places, measured in kilometers (miles), travel time, or travel cost.

**distribution** (2A). The spatial spread of a phenomenon, as in lines, clusters, or at random.

**diversified economy** (4C). An economy in which manufactures are more important than primary products and where there is a variety of manufactured products and a growing service sector.

**dual sectors** (2C). The theory of development that recognizes differences between urban-industrial and rural areas in their potential for economic growth.

**Dust Bowl** (9C). The area of Oklahoma, Kansas, and Texas, USA, where plowing of subhumid grassland resulted in massive wind erosion in the dry years of the 1930s and since.

**economic colonialism** (3D). A modern form of colonialism in which, for example, multinational corporations extract mineral products and produce agricultural raw materials in poorer countries at low prices for their own benefit.

**economics** (2C). The study of the production, distribution, and consumption of goods and resources.

**economic system** (1A). A way in which goods and services are produced, distributed, and consumed.

**economies of scale** (9A). Increased productivity gained by building larger factories and gaining access to larger markets.

**ecosystem** (2D). The total environment of a community of plants and animals, including heat, light, and nutrient supplies.

**ecotourism** (11C). Tourism based on viewing natural wonders such as wildlife and rock formations.

**edge city** (9E). Urban developments at the margins of metropolitan areas in which concentrations of shopping malls, offices, factories, and warehouses establish new economic focuses.

**El Niño phenomenon** (10B). The process by which the warm equatorial waters push back the cold Peruvian current off the western coast of tropical South America, producing local fish kills and having wider world climatic effects.

**encomienda system** (10A). The Spanish colonial system employed in Latin America by which lands were allotted to Spaniards who were responsible for exploiting their wealth and had jurisdiction over native peoples.

**entrepôt** (6C). A port that collects goods from several countries to trade with the wider world and distributes imports to its immediate area or hinterland. Examples include Hong Kong and Singapore.

**equatorial climatic environment** (3A). Climatic environment of areas within 5° to 10° of the equator and characterized by continuous high temperatures and rainfall at all seasons.

**erosion** (2D). The wearing away of rocks at Earth's surface by running water, moving ice, the wind, and the sea to form valleys, cliffs, and other landforms.

**ethnic cleansing** (7E). The process by which a dominant group of people in a country causes another ethnic group to leave a region, often using threats or military force.

**European Union** (EU) (7A). Name adopted by the European Community in 1993, suggesting both an expansion to other European countries following the end of the Cold War and the possibility of a future closer political federation.

**export-led underdevelopment** (10D). The process by which countries export their raw materials while their governments charge taxes on exports to pay for their imports. There is no stimulus to develop manufacturing and so diversify the economy to bring more wealth into it.

**export-oriented industries** (2C). Manufacturing industries that produce goods mainly for competitive export markets.

**extensification** (of agriculture) (7C). The production of fewer livestock or crops from the same area.

**extensive economic system** (8C). The system adopted by the Soviet Union in which manufacturing production was increased by additional inputs of labor and raw materials rather than by greater productivity to reduce costs.

**favela** (10E). A shantytown in a Brazilian city.

**federal system of government** (9C). A system of government in which a group of states agree to share some functions that are delegated to a central government and to retain other functions that they administer in their own way.

**financial reform** (8A). The aspect of transition in former Soviet bloc countries that is linked to fiscal and monetary policy and to redirecting capital to profit-making enterprises.

**First World** (1A). The Western countries, led by the United States.

**fiscal policy** (8A). Government activity that sets taxes and manages a country's budget.

**Fordist organization** (8C). A manufacturing emphasis in the economy, based on complex production-line assembly. Superseded in Western countries by a post-Fordist service industry economy.

**forest biome** (2D). A major biome-type characterized by close-spaced trees.

**formal regions** (1B). Regions defined by internal characteristics, treated as if they are largely self-contained and unique.

**free trade area** (2B). A group of countries that agree to a common set of tariffs on imports.

**Free Trade Area of the Americas (FTAA)** (9A). A free trade area agreed on in 1995 that is intended to include all countries in Anglo America and Latin America by A.D. 2005, with the possible exception of Cuba.

**free trade zone** (10E). An area within a country where components can be imported without tariffs for assembly with a view to exporting the finished goods (see also **maquiladora program**, **special economic zone**).

**friction of distance** (2A). The relative difficulty of moving from one place to another, which increases with kilometers (miles), cost, or travel time.

**functional regions** (1B). Regions described by their interactions with other regions.

**gender** (2C). The cultural implications of being male or female, with particular reference to the inequalities suffered by females in human societies.

**gender-related development index** (GDI) (2C). A measure of gender inequality using differences between male and female values for HDI criteria.

**General Agreement on Tariffs and Trade** (GATT) (2B). The organization, established in 1948, that attempted to persuade countries to reduce tariffs on imports to increase world trade. The final stage of the process, the Uruguay round, ended in 1993.

**gentrification** (7E). The movement of higher-income groups to occupy and improve residences in older and poorer central parts of Western cities.

**geographic inertia** (7A). Once capital investments are made in factories and infrastructure that give a region agglomeration economies, production will continue there for a period of years after other areas emerge with lower production costs.

**geographic information system** (GIS) (2A). The computer-based combination of maps, data, and often satellite images that is a foundation for geographic analysis.

**geography** (1B). The study of how human beings live in varied ways on different parts of Earth's surface; the study of the environment and space in which humans live.

**geopolitical ambitions** (10A). The "vision of grandeur" that some countries have to extend their territory at the expense of others.

**Germanic tribes** (7A). Groups of people that became established within approximately the modern bounds of Germany but moved outward from the A.D. 300s to conquer most of western Europe. Included the Franks, Saxons, Burgundians, and Swedes.

**ghetto** (9E). An urban area in which a particular group of people such as Jews or African Americans is segregated.

**glasnost** (8A). The Soviet Union policy of the late 1980s designed to create greater openness and exchange of information.

**global choke point** (1B). A point at which surface transportation routes converge that is liable to the control by the country in which the choke point occurs; for example, the Suez Canal (Egypt) or the Strait of Hormuz (Iran).

**global warming** (2D). The process by which average temperatures in Earth's atmosphere rise over a period of several decades or centuries, leading to the melting of ice masses and a rising sea level.

**grassland biome** (2D). A major biome type characterized by grasses, with few, scattered trees.

**Great Leap Forward** (6C). The attempt by the Chinese communist government in the late 1950s to increase the pace of industrialization. It failed by ignoring food production at a time when bad weather brought poor harvests and famine.

**Greeks** (7A). Historically, a group of peoples who formed a foundation of European ideas and established a major empire extending to India around 300 B.C

**Green Revolution** (5B). The result of introducing high-yielding strains of wheat and rice. The outputs of commercial farms in Southern and Eastern Asia increased by several times, but the costs of seeds, fertilizers, and pesticides were too high for smaller farmers.

**gross domestic product** (2C). The total value of goods and services produced within a country in a year. Often expressed as GDP per capita, when the total GDP is divided by the country's population.

**Gross National Product** (2C). The total value of goods and services produced within a country in a year, together with income from labor and capital working abroad less deductions for payments to those living abroad.

**Gulf Cooperation Council** (GCC) (4C). Formed in 1981 under the leadership of Saudi Arabia to focus on political problems raised by the Iran-Iraq War.

**Han Chinese** (6C). The largest group (94%) of people in China. An ethnic grouping based on the administrative culture spread by the Chinese empire in the A.D. 200s and 300s.

**hazard** (2D). Difficulties and challenges posed to humans as a result of activities in the natural environment, such as earthquakes and floods.

**Hebrew** (4A). A language related to Arabic that is the official language of Israel.

**hegemony** (8C). An area of influence, or control, such as the Russian realm that was extended to Siberia, the Caucasus, and Central Asia from the A.D. 1500s to 1800s and Eastern Europe in the mid-1900s.

**Hindi** (5B). One of the official languages of India but spoken by only 30 percent of the population, mainly in the north.

**Hinduism** (5A). A religion of Southern Asia, observed mainly in India, that includes the worship of many gods related to varied historic experiences and is associated with the caste system.

**horizontal integration** (9A). The combining of producers of the same product in a single corporation to create economies of scale, including the control of prices.

**household responsibility system** (6C). The replacement for communes in rural China after 1976, returning ownership and decision making to individuals and groups that could sell surpluses in open markets.

**human development index** (HDI) (2C). A measure of human development based on income, life expectancy, adult literacy, and infant mortality.

**human-environment relations** (1B). The study of interactions between human activities and the natural environment.

**human geography** (1B). The study of geographic aspects of human activities, often with a population, economic, cultural, or social focus.

**human poverty index** (HPI) (2C). A measure of human poverty, linked to HDI, that indicates levels of personal deprivation.

**Hutu tribe** (3B). The majority tribe of Burundi and Rwanda with an agricultural culture.

**import substitution** (2C). A development policy in which countries develop manufacturing industries to fulfill internal market demands, often protected by high tariffs to exclude foreign competition.

**Incas** (10A). A group of peoples based in the area of modern southern Peru, South America, that extended their empire northward to modern Ecuador and southward to Chile in the A.D. 1400s.

**Indian National Congress** (5A). Formed by a mainly Hindu elite in 1885, it widened its base to become a major force in the independence movement and then India's main political party.

**industrial capitalism** (1C). The form of the capitalist economic system in which investment in manufacturing industries pays a central part, linked to integrated and world-wide transportation systems.

**industry** (2C). A group of business activities, such as the farming industry, manufacturing industry, or tourism industry.

**infant mortality** (2B). The number of deaths per 1,000 live births in the first year of life.

**informal sector** (2C). Ways in which people make a living outside of waged, salaried, and taxed employment. Informal sector work may lead to and be interlinked with formal sector jobs.

**inselberg** (3A). A hill with bare rocky sides that rise at a sharp angle from the surrounding land. Common in tropical Africa.

**intensification** (of agriculture) (7C). The increased output of crops or livestock per area unit of land.

GLOSSARY

**Islam** (4A). A religion of northern Africa and southwestern Asia, and parts of Southern and Eastern Asia, based on the teachings of Muhammad as recorded in the Qur'an. *Islam* means "reconciliation to God."

**isoline** (2A). A line on a map that joins places of the same value, as in contours (height above sea level) and isobars (atmospheric pressure).

**Judaism** (4A). The religion of the Jewish people who worship Yahweh as the creator and lawgiver.

**kibbutzim** (4C). A communal farming village in Israel, the social and spiritual basis of the new Israeli nation after 1948 but now the home of fewer than 5 percent of the Israeli population.

**Kurdish** (4A). A language akin to Persian spoken by a group of tribes in the hilly lands of eastern Turkey, northern Iraq and Syria, and western Iran. The tribes wish to establish their own country.

**landscape** (2E). The visual appearance of a stretch of countryside or town; a total regional environment; a cultural interpretation of interactions between natural environments and human history.

**language** (2C). The spoken means of communication among people, mostly also capable of being written down. There are 12 dominant language groups but thousands of individual languages in the world.

**latitude** (2A). The distance of a place north or south of the equator, measured in degrees.

**local region** (1B). A region that is a distinctive subdivision of a country, varying in size from an urban neighborhood to an area of distinctive farm products or physical features.

**location** (2A). A place's location is defined by its position on Earth's surface in terms of latitude and longitude (absolute location) or by its level of interaction with other places (relative location).

**longitude** (2A). The distance of a place east or west of 0° longitude, measured in degrees.

**long lot** (9D). The unit of land settlement by French colonists in Anglo America. Each plot of land had a narrow river or road frontage and often extended through different types of land suitable for crops, pasture, or woodland.

**major world region** (1B). The largest regional division of the world, based mainly on distinctive cultural features, with each region including the major part of a continent.

**mallee** (11B). A vegetation community occurring in Australia that is composed of drought-resistant eucalyptus shrubs forming impenetrable thickets.

**Maori** (11A). A people who settled New Zealand from the wider South Pacific around the A.D. 800s and continue to form an important component of the New Zealand population.

**map** (2A). The representation of the features of Earth's surface on paper at varying scales.

**maquiladora program** (10C). Established in 1965 by Mexico to allow factories on Mexican soil to import manufacturing components without duty for assembly and export mainly to the United States.

**marginalization** (2C). The theory of development that recognizes the formation of a group of poor people who do not benefit from economic growth in their country.

**marsupial** (11B). A mammal that raises its young in a pouch instead of a womb. Mainly found in Australia, these include kangaroo and koala.

**Mayan civilization** (10A). A civilization centered on the eastern lowlands of modern southern Mexico, Guatemala, and central Honduras. Main period from A.D. 200s to 900s, with wealth based on a surplus of maize.

**mediterranean climate** (7B). A climatic environment that occurs between midlatitude west coast and tropical arid climatic environments, having wet winters and dry summers.

**megalopolis** (6B). An expanded urbanized area that includes several metropolitan areas with over a million people and dominates the economy of surrounding areas. First identified in the northeastern United States covering the area between Boston and Washington, D.C.

**Melanesian people** (11A). "Dark-skinned" peoples who inhabit the South Pacific islands.

**mercantile capitalism** (1C). The capitalist economic system in which local products are traded with other places and overseas. This form of capitalism came before industrial capitalism in Western Europe.

**Mercosul** (10E). A free trade area in Southern South America, of which Argentina, Brazil, Paraguay, and Uruguay are members.

**meridian of longitude** (2A). A line joining places of the same longitude on Earth's surface.

**mestizo** (10A). A person of mixed Native American and European stock in Latin America.

**metropolitan area** (9C). A large urban area, often consisting of groups of cities that merge around one main center.

**Micronesian people** (11A). Peoples of the South Pacific who inhabit the "small islands" of the northwestern area.

**midlatitude climates** (2D). Climates typical of the zone between the tropics and polar regions.

**midlatitude continental interior climate** (8B). A climatic environment that is marked by very cold and often dry winters and hot summers with thundery rain.

**midlatitude west coast climate** (7B). A climatic environment dominated by cyclonic weather systems linked to oceanic heat and moisture sources, interspersed with anticyclones. Typically with cooler summers and milder winters than continental interiors on the same latitude.

**migration** (2B). The long-term movement of people into (inmigration, immigration) or out of (outmigration, emigration) a place.

**Ministry of International Trade and Industry** (MITI) (6B). The Japanese government ministry charged with promoting Japanese trade and products abroad.

**modernization** (2C). The theory of development in which poorer countries attempt to follow the stages through primary and secondary to tertiary sectors that the countries of Western Europe and Anglo America followed in the A.D. 1800s and early 1900s.

**monetary control** (8A). A government activity that controls the money supply to prevent inflation.

**monoculture** (10E). The reliance on a single commercial crop that may be affected by climatic and world market fluctuations.

**monsoon climatic environment** (3A). A tropical climatic environment in which there are wind shifts between summer and winter, bringing heavy rains from oceanic air in the summer and dry winds of interior continental air in winter.

**Moors** (7A). Muslim groups of people who invaded Spain and southern Italy from North Africa in the A.D. 700s, holding territory in Spain until the 1400s.

**Muhajirs** (5C). Muslim people who migrated to Pakistan at the time of partition in 1947. Many still speak Hindi.

**multinational corporation** (2C). A corporation that makes goods and provides services in several countries but directs operations from headquarters in one country.

**multiple economic nuclei** (9E). The pattern of urban geography in which other economic nuclei as well as the central business district occur among residential sectors.

**Muslim League** (5A). Formed in India in 1906 to uphold the position of Muslims in the British Indian Empire.

**Muslims** (4A). "Those who are reconciled to God." Followers of Islam.

**Native American** (9A). People who inhabited Anglo America before the European arrival, including Amerinds and Inuits (Eskimos).

**natural environment** (1A). The world as it might be without humans, including climate, landforms, plants, animals, soils, and oceans.

**natural population change** (2B). The difference between birth rate and death rate.

**natural resources** (1A). Materials present in the natural environment and recognized by humans as of practical worth (e.g., minerals, soils, water, building stones, timber).

**new international division of labor** (2C). A division of labor adopted by multinational corporations in which high-salaried managerial and research jobs are concentrated in core countries, while low-wage jobs and production units are established in peripheral countries.

**new rice technology** (6D). Improved strains of rice and methods of growing that developed in the Philippines as part of the Green Revolution.

**Nile Waters Agreement** (4A). Agreement between Egypt and Sudan in 1959 to share Nile River waters, with 70 percent allocated to Egypt.

**nonrenewable resource** (2D). A natural resource that is used up once it is all extracted; for example, coal or metallic minerals.

**North American Free Trade Agreement** (NAFTA) (9A). Signed in 1993 by Canada, Mexico, and the United States to lower tariffs and encourage trade between the three countries.

**northern coniferous forest** (taiga) (8B). A forest composed of coniferous trees (firs, pines, cedars) common in the northern parts of midlatitude continental interior climatic environments.

**ocean biome** (2D). The oceans as ecosystems with energy pathways, nutrient cycles, and food chains.

**Organization of Petroleum Exporting Countries** (OPEC) (4A). Established in 1960 to further the interests of oil and gas producers throughout the world, often in peripheral countries, and often to resist the overriding power of multinational oil corporations.

**Organization of the Islamic Conference** (OIC). (4A). Established in 1970 by foreign ministers of Muslim countries throughout the world. It has 45 members but advances individual countries' interests rather than pursuing a common agenda.

**overpopulation** (11D). A situation in which a country's or region's resources cannot support the population on a subsistence basis. The application of more technology and better management may make it possible for such a region to support a larger population.

**Pacific Rim** (11A). The countries that border the Pacific Ocean, including those in the South Pacific, Eastern Asia, Anglo America, and Latin America.

**Palestine Liberation Organization** (PLO) (4A). An organization that promotes the reestablishment of a country of Palestine. It has a secular, left-wing political basis.

**Pan-Arab state** (4A). The idea of Arab countries joining to form a single country. Egypt and Syria were joined for a short while in the 1960s, but few attempts have been made to achieve this end since.

**parallel of latitude** (2A). A circle joining places of the same latitude on Earth's surface.

**perestroika** (8A). The Soviet Union policy of the late 1980s designed to reconstruct the political and economic structure of the country so that it could compete with or in the capitalist world economic system.

**periphery** (1A). The poorer countries of the world economic system. These play a secondary role in world trade, are dependent on core countries, have a narrow range of products, use less advanced technology, and pay low wages.

**permafrost** (8B). Permanently frozen ground extending several hundred meters below the surface in Siberia and northern Canada. In summer, water in the surface active layer, 50 to 100 cm deep, melts.

**Persian** (4A). A language spoken in Iran following its introduction from Eurasia around 3000 B.C. Modern Persian, written in Arabic script, is also known as Farsi.

**physical geography** (1B). The study of geographic aspects of natural environments.

**place** (2A). A point or area on Earth's surface having a geographic character defined by what it looks like, what people do there, and how they feel about it.

**plantation agriculture** (3B). The large-scale and concentrated commercial cultivation of crops often associated with the tropics and colonial attempts to produce such crops for the home countries.

**plateau** (3A). An upland area with flat-topped hills or having a tablelike form.

**podzol soils** (8B). Soils of low fertility in which plant nutrients are removed by water passing through. Commonly develop beneath midlatitude coniferous forest and on sandy soils.

**polar biome** (2D). Major biome-type where plant growth is inhibited by extreme cold and a short, if any, growing season.

**polar climates** (2D). Climates typical of the polar regions.

**political geography** (2B). The study of geographic implications of the ways in which governments control activities within their countries and link with each other.

**Polynesian people** (11A). Lighter-skinned peoples of the South Pacific islands who inhabit mainly the eastern islands.

**poverty** (2C). Low income and deprivation of minimally acceptable material requirements; the absence of basic capabilities that enable people to function fully as human beings.

**price reform** (8A). The aspect of transition in the former Soviet Union bloc that frees prices to respond to conditions of supply and demand instead of setting them centrally without reference to such criteria.

**primary sector** (2C). The sector of an economy that produces output from natural sources, including mining, forestry, fishing, and farming.

**primate city** (6D). A city that contains a large proportion of the urban population of a country, often several times the population of the second city.

**privatization** (8A). The sale of state-owned enterprises to private owners.

**producer services** (7C). Service industries that are involved in the output of goods and services, including market research, advertising, accountancy, legal, banking, and insurance industries.

**production line** (9A). The system of manufacturing in which components are made and assembled into the final product in a sequence of factory-based processes.

**Punjabi** (5C). A language spoken by two-thirds of Pakistan's population.

**quaternary sector** (2C). The sector of an economy that specializes in producer services, including financial services and information services.

**race** (2C). Biological characteristics of a group of people, such as skin color. Such characteristics often have social implications.

**region** (1B). An area of Earth's surface distinguished from others by its physical and human characteristics and interacting with other regions in trade and the exchange of people and ideas.

**regional geography** (1B). The study of different regions at Earth's surface in their country and global contexts.

**relative location** (1B). The location of a place relative to time or cost-distance related to transportation or communications.

**religion** (2C). An organized system of values and practices, including faith in and worship of a divine being or beings.

**remote sensing** (2A). A set of techniques using aerial photographs or satellite images to gather information about Earth's surface.

**renewable resource** (2D). A resource that is replaced by natural processes at a rate that is faster than its usage.

**rift valley** (3A). A deep valley caused by the rocks of Earth's crust cracking along parallel lines to let down a section of crust to form the valley floor.

**"roaring forties"** (11B). Winds that dominate the climatic environment of the stormy, largely oceanic region between approximately 45° S and 60° S.

**Romans** (7A). Historically, a group of people who established an empire around the Mediterranean Sea that extended to northwest Europe from the first century B.C.

**salinization** (4A). The process by which soils become unproductive because of an accumulation of alkaline salts near the surface. Often associated with poorly managed irrigation systems in arid areas.

**savanna grassland** (3A). Tropical grassland containing varied densities of trees and occurring in tropical seasonal climatic environments. In places, the grass content is maintained by annual burning.

**scale** (2A). The relationship of horizontal ground distance to map distance, quoted as a fraction (e.g., 1/10,000) or as a ratio (e.g., 1:10,000) in which one unit on the map represents 10,000 units on the ground.

**Second World** (1A). The communist countries, led by the Soviet Union until 1991. Now redundant.

**secondary sector** (2C). The sector of an economy that changes the raw materials from the primary sector into useful products, thus increasing their value, as in chewing gum or parts for airplanes.

**segregation** (9A). The process by which people exclude others of contrasting ethnic or racial character from their community.

**semiperiphery** (1A). Countries moving between the periphery and core of the world economic system, dependent on core countries but with countries in the periphery dependent on them.

**shantytown** (3D). An unplanned residential sector of urban areas in poorer countries. Housing is often built of any materials that come to hand and does not have electricity, water, or waste disposal.

**Shia Muslims** (4A), also known as Shiites. Muslims who are partisans of the imam Ali (not acceptable to Sunni Muslims) and look to his return. They make up 90 percent of the Iranian population and are associated with radical movements.

**Shinto** (6A). The traditional religion of Japan, built on ancient myths and customs that promote the national interests and identity.

**Sindhi** (5C). A language spoken by 12 percent of Pakistan's population.

**Sinhalese** (5C). A language spoken by the Buddhist majority of people in Sri Lanka.

**soil** (2D). Weathered rock material that develops by the actions of water, animals, and plants into a basis for plant growth.

**Southern Africa Development Conference** (SADC) (3C). Established by countries in Southern Africa opposed to South Africa's apartheid policy to organize alternative trade outlets. Now encouraging trade among the constituent countries, including South Africa.

**South Pacific Forum** (11C). Established in 1971 among 13 countries in the South Pacific, including Australia, New Zealand, and several island countries. Aims to promote regional cooperation and confronts regional problems of external exploitation.

**spatial analysis** (1B). The study of linkages between places at Earth's surface, often in terms of points, lines, and areas, together with statistical associations.

**special economic zones** (6C). Zones established by China in 1979 to encourage foreign investment and export-oriented manufacturing in the southern coastal provinces. Similar zones are found in other countries (see **maquiladora program, free trade area**).

**specialization** (of agriculture ) (7C). The concentration on fewer commercial products within a farming region.

**state capitalism** (1C). The capitalist economic system in which the government, rather than private individuals or corporations, controls production and distribution.

**steppe grasslands** (8B). Midlatitude grasslands typical of the southern parts of midlatitude continental interior climatic environments.

**structural adjustment** (2C). The theory of development in which a country becomes more involved in the world economic system by producing export goods, reducing tariffs, encouraging privatization, and developing good government and a balanced budget.

**suburbanization** (6B). The movement of people and corporations from older central parts of urban areas to new residential and commercial areas on the city outskirts.

**Sunni Muslims** (4A), also known as Sunnites. Traditional, conservative followers of Islam, forming the majority in most Muslim countries.

**sustainable human development** (2C). A level of development in which resources are exploited at a rate that is sustainable for future generations.

**Taoism** (6A). The teachings of Chinese philosopher Lao-zi, who disliked the organized and hierarchical social system of Confucius and advocated a return to local, village-based communities with little outside interference.

**tectonic plate** (2D). A large block of Earth's crust and underlying rocks approximately 100 km thick and up to several thousand kilometers across. Earth's interior heat causes plates to move apart and crash together, forming major relief features including ocean basins, mountain systems, and continental areas.

**Tennessee Valley Authority** (TVA) (9C). Established by the U.S. government in 1933 to stimulate economic growth in southern

Appalachia through the damming of rivers to improve transportation, flood control, and electricity costs.

*Terra Australis* (11A). The "Southland" mythical continent of the South Pacific, now Australia, that was discovered by Dutch and British explorers in the 1700s.

**tertiary sector** (2C). The sector of an economy concerned with the distribution of goods and services, including trade, professions, and government employment.

**Third World** (1A). The poorer countries, mostly in the Southern Hemisphere, not part of the First or Second Worlds.

**total fertility rate** (2B). The number of births per woman in her child-bearing years.

**township** (9A). The initial basis of settlement in New England, Anglo America, based on self-sufficient and self-governing village-sized communities.

**trade reform** (8A). The aspect of economic transition in former Soviet bloc countries in which trade barriers and internal barter agreements (CMEA) are dismantled and currencies become convertible.

**transnational Chinese economy** (6C). Economic and trading links established by Chinese people living outside China, especially in Eastern Asia and Anglo America.

**tribe** (3A). A group of people with similar ethnic characteristics arising from a common cultural heritage and linked to a specific area of land.

**tropical arid climatic environment** (3A). A climatic environment with little rain throughout the year and high evaporation rates related to high temperatures.

**tropical climates** (2D). Climatic environments typical of the tropical zone.

**tropical rain forest** (3A). The ecosystem that occurs in equatorial climatic environments. Complex communities of many tree and shrub species, together with arboreal animals.

**tropical seasonal climatic environment** (3A). A climatic environment in which the high-sun season is wet and the low-sun season is dry. There is a gradual change from rainy equatorial margins to arid margins in which the total rainfall decreases.

**tropical tree crops** (3B). Tree species that produce items in demand in other countries, such as cocoa, coffee, tea, bananas, rubber, palm oil, and coconuts. Often grown on plantations.

**tundra** (8B). Ecosystem type occurring in cold polar environments, consisting of mosses, grasses, and low shrubs.

**Turkish** (4A). A language family spoken by peoples in Central Asia and extended to Turkey. In Turkey, the language is written in Roman script, but Azeris use Arabic script.

**Tutsi tribe** (3B). Originally a cattle-herding tribe that settled in the area of Burundi and Rwanda, assuming a leadership position over the more numerous agricultural Hutu tribe.

**uneven development** (9A). The increase in the gap between poor and wealthy regions in a country and the shifting locations of economic growth and decline over time, seen by Marxists as a natural outcome of capitalism.

**urban sectors** (9E). Wedge-shaped zones of urban land use focused on a central business district, often reflecting local physical environmental contrasts.

**Urdu** (5C). The official language of Pakistan, although spoken by only 7 percent of the population.

**U.S.-Canada Free Trade Agreement** (9A). Signed in 1988 to lower tariffs and encourage trade between the two countries.

**vertical integration** (9A). The combining of producers of raw materials, manufacturers that process the materials, and those that assemble the products in a single corporation to achieve economies of scale.

**Vikings** (7A). A group of peoples who spread out from Scandinavia from the A.D. 700s, conquering much of northwest Europe, Iceland, and Greenland and influencing events in Russia.

**visual symbolism** (8E). The process by which buildings, statues, and other visual features of the built environment promote a dominant social or political message.

**weathering** (2D). The action of atmospheric forces (through water circulation and temperature changes) on rocks at Earth's surface that breaks the rocks into fragments, particles, and dissolved chemicals.

**white Australia policy** (11C). The policy designed to exclude Asian people from Australia. Ended in 1972.

**world system** (1C). The geographic expansion of area subject to wealth accumulation and exchange of goods, beginning with early civilizations that linked with each other.

**World Trade Organization** (2B). Established in 1993 to oversee the liberalizing of world trade and prevent discrimination among trading partners.

**young folded mountains** (4A). Mountain systems produced in the last 50 million years or so by clashing tectonic plates, including the Himalayas, Andes, Alps, and Atlas.

# GLOSSARY OF KEY PLACES

GLOSSARY

Each place listed as a key place in the text is linked to its chapter and RECAP section. Cities are given a description, location, and pronunciation where that is not obvious.

**Abadan** (4C). City in Iran, southwestern Asia, at 30° N, 48° E. (AB a DAN)

**Abidjan** (3B). Port in Côte d'Ivoire, Western Africa, at 5° N, 4° W. (AB ih JAHN)

**Abu Dhabi** (4C). Port in United Arab Emirates, Arab Southwest Asia, at 24° N, 55° E. (AHB OO DAHB ee)

**Accra** (3B). Capital of Ghana, Western Africa, at 6° N, 0°.

**Adana** (4C). City in Turkey, southwestern Asia, at 37° N, 35° E.

**Addis Ababa** (3C). Capital of Ethiopia, Eastern Africa, at 9° N, 39° E. (AD ihs AB uh buh)

**Adelaide** (11C). Capital of the state of Southern Australia, South Pacific, at 35° S, 138° E. (AD uh LAYD)

**Aden** (4C). Port in Yemen, Arab Southwest Asia, at 13° N, 48° E. (AHD un)

**Adriatic Sea** (7B). Arm of Mediterranean Sea between Italy and Croatia, Mediterranean Europe.

**Agean Sea** (7B). Arm of Mediterranean Sea between Greece and Turkey, Mediterranean Europe.

**Afghanistan** (5C). Country in the mountain rim of Southern Asia. Area: 652,090 km$^2$. Population (1998) 24.8 million.

**Africa South of the Sahara** (1B). Major world region, the subject of Chapter 3.

**Ahmadabad** (5B). City in Gujarat, India, Southern Asia, at 23° N, 73° E.

**Alaska** (9C). Forty-ninth state of the United States, Anglo America.

**Albania** (7E). Country in Balkan Europe. Area: 28,750 km$^2$. Population (1998) 3.3 million.

**Aleppo** (4C). City in Syria, Southwest Asia, at 36° N, 37° E.

**Alexandria** (4B). Port in Egypt, Nile River valley, at 31° N, 30° E.

**Algeria** (4B). Country in North Africa. Area: 2,381,740 km$^2$. Population (1998) 30.2 million.

**Algiers** (4B). Capital city of Algeria, North Africa, at 37° N, 3° E. (al JEERS)

**Almaty** (8D). Capital city of Kazakstan, Central Asia, at 43° N, 77° E.

**Alpine Europe** (7A). Subregion of Europe, comprising Austria and Switzerland.

**Alps** (7B). Mountain ranges on the northern margins of Mediterranean Europe.

**altiplano** (10B). High plateau region between the main ranges of the Andes Mountains in Bolivia and southern Peru, Northern Andes, South America.

**Amazon River** (10B). World's largest river, rising in Peru and flowing mainly through Brazil, South America, into the Atlantic Ocean.

**American Samoa** (11D). Island group country in the South Pacific. Area: 300 km$^2$. Population (1998) 0.1 million.

**Amman** (4C). Capital city of Jordan, Arab Southwest Asia, at 32° N, 36° E.

**Amsterdam** (7C). Major city in Netherlands, Western Europe, at 53° N, 5° E.

**Amu Darya Syr Darya River** (8B). Rivers that flow into the Aral Sea, Kazakstan, Uzbekistan, and Turkmenistan, Central Asia.

**Andes Mountains** (10B). World's second highest mountain ranges, situated close to the Pacific coast of South America.

**Anglo America** (1B). Major world region, the subject of Chapter 9.

**Angola** (3C). Country in Southern Africa. Area: 1,246,700 km$^2$. Population (1998) 12 million.

**Ankara** (4C). Capital city of Turkey, southwestern Asia, at 40° N, 33° E.

**Antananarivo** (3C). Capital city of Madagascar, Southern Africa, at 19° S, 47° E. (AHN tuh NAH nuh REE voh)

**Antarctica** (11D). Uninhabited and ice-covered continent that centers on the South Pole.

**Apennine Mountains** (7B). Mountain range that forms the backbone of the Italian peninsula, Mediterranean Europe.

**Appalachia** (9C). Upland region in the eastern United States, Anglo America, that has experienced a lagging economy compared to the rest of the country.

**Appalachian Mountains** (9B). Uplands including plateaus and ridges in the eastern United States, Anglo America.

**Arab Southwest Asia** (4C). Subregion of Northern Africa and Southwestern Asia, comprising Bahrain, Iraq, Jordan, Kuwait, Lebanon, Oman, Qatar, Saudi Arabia, Syria, United Arab Emirates, and Yemen.

**Aral Sea** (8B). Sea in area of inland drainage, supplied by Amu Darya and Syr Darya Rivers. Subject to shrinkage because of overuse of water en route to the sea. Bordered by Uzbekistan and Kazakstan, Central Asia.

**Arctic Ocean** (8B). Ocean that covers the North Pole area. Bordered by Anglo America, Northern Europe, and the Russian Federation.

**Argentina** (10E). Country in Southern South America. Area: 2,791,810 km$^2$. Population (1998) 36.1 million.

**Armenia** (8D). Country in Southwestern CIS. Area: 29,800 km$^2$. Population (1998) 3.8 million.

**Asunción** (10E). Capital city of Paraguay, Southern South America, at 25° S, 57° W. (AH SOON see awn)

**Atacama Desert** (10B). Arid region in northern Chile and southern Peru, South America.

**Athens** (7D). Capital city of Greece, Mediterranean Europe, at 38° N, 24° E.

**Atlanta** (9C). City in the state of Georgia, United States, Anglo America, at 34° N, 84° W.

**Atlantic Provinces** (9D). Easternmost provinces of Canada, Anglo America, including New Brunswick, Newfoundland, Nova Scotia, and Prince Edward Island.

**Atlas Mountains** (4A). Mountain ranges in Morocco, Algeria, and Tunisia, North Africa.

**Auckland** (11C). Port city in North Island, New Zealand, South Pacific, at 37° S, 175° E. (AWK land)

**Australia** (11C). Country in South Pacific. Area: 7,986,850 km$^2$. Population (1998) 18.7 million.

**Australian Capital Territory** (11C). Territory in which Australia's federal government works, based around Canberra at 35° S, 149° E.

**Austria** (7D). Country in Alpine Europe. Area: 83,850 km$^2$. Population (1998) 8.1 million.

**Azerbaijan** (8D). Country in Southwestern CIS. Area: 86,000 km$^2$. Population (1998) 7.7 million.

**Baghdad** (4C). Capital city of Iraq, Arab Southwest Asia, at 33° N, 44° E.

**Bahamas** (10D). Island country in the West Indies. Area: 13,880 km$^2$. Population (1998) 0.3 million.

**Bahrain** (4C). Country on the Persian Gulf in Arab Southwest Asia. Area: 680 km$^2$. Population (1998) 0.6 million. (bach RAIN)

**Baku** (8D). Capital city of Azerbaijan, Southwest CIS, at 40° N, 50° E.

**Balkan Europe** (7A). Subregion of Europe, comprising Albania, Bosnia-Herzegovina, Bulgaria, Croatia, Hungary, Macedonia, Romania, and Serbia-Montenegro.

**Balkan Mountains** (7B). Mountain ranges in Bulgaria, Balkan Europe.

**Baltic Sea** (7B). Sea between Northern and Eastern Europe with outlet to the North Sea.

**Baltimore** (9C). Major port city in state of Maryland, eastern United States, Anglo America, at 39° N, 77° W.

**Bamako** (3B). Capital city of Mali, Western Africa, at 13° N, 8° W.

**Bandung** (6D). City in Indonesia, Southeastern Asia, at 7° S, 109° W.

**Bangalore** (5B). City in southern India, Southern Asia, at 13° N, 78° E.

**Bangkok** (6D). Capital city of Thailand, Southeastern Asia, at 14° N, 100° E.

**Bangladesh** (5C). Country in Southern Asia. Area: 144,000 km$^2$. Population (1998) 129.4 million.

**Barcelona** (7D). City in northeastern Spain, Mediterranean Europe, at 41° N, 2° E. (bahr suh LOH nuh)

**Barranquilla** (10D). City in northern Colombia, Northern Andes, South America, at 11° N, 75° W. (BARR uhn KEE uh)

**Basel** (7D). City in northern Switzerland, Alpine Europe, at 48° N, 8° E. (BAHZ uhl)

**Basra** (4C). Port in Iraq, Arab Southwest Asia, at 31° N, 48° E.

**Bay of Bengal** (5A). Northern part of the Indian Ocean between India and Myanmar.

**Beijing** (6C). Capital city of the People's Republic of China, Eastern Asia, at 40° N, 116° E. (bay JIHNG)

**Beirut** (4C). Capital city of Lebanon, Arab Southwest Asia, at 33° N, 36° E. (bay ROOT)

**Belarus** (8D). Country in Southwestern CIS. Area: 207,600 km$^2$. Population (1998) 10.2 million. (BEHL ah ruhs)

**Belgium** (7C). Country in Western Europe. Area: 33,100 km$^2$. Population (1998) 10.2 million. (BEL jum)

**Belgrade** (7D). Capital city of Serbia-Montenegro, the remnant of Yugoslavia, Balkan Europe, at 45° N, 21° E.

**Belize** (10C). Country in Central America. Area: 22,960 km$^2$. Population (1998) 0.2 million. (buh LEEZ)

**Belo Horizonte** (10E). Manufacturing city in eastern Brazil, South America, at 20° S, 44° W. (BAY low HAWR un ZAHN tee)

**Benghazi** (4B). City of northern Libya, North Africa, at 32° N, 20° E. (ben GAH zee)

**Benin** (3B). Country in Western Africa. Area: 112,020 km$^2$. Population (1998) 6 million.

**Berlin** (7C). Capital of Germany, Western Europe, at 52° N, 13° W.

**Bermuda** (10D). Island in the West Indies, at 32° N, 65° W.

**Bern** (7D). Capital city of Switzerland, Alpine Europe, at 47° N, 7° E.

**Bhutan** (5C). Country in the mountain rim of Southern Asia. Area: 47,000 km$^2$. Population (1998) 0.8 million. (boo THAN)

**Birmingham** (7C). Second city of England, Western Europe, at 53° N, 2° W.

**Black Sea** (8B). Sea bordered by Balkan countries, Southwest CIS, Russia, and Turkey, with outlet through Bosporus and Dardenelles in Turkey to the Aegean Sea.

**Bogotá** (10D). Capital city of Colombia, North Andes, South America, at 4° N, 74° W. (boh guh TAH)

**Bohemian Massif** (7B). Upland area in Czech Republic, Eastern Europe.

**Bolivia** (10D). Country in Northern Andes, South America. Area: 1,098,580 km$^2$. Population (1994) 8 million.

**Bombay (Mumbai)** (5B). Largest city of India, Southern Asia, at 19° N, 73° E.

**Bosnia-Herzegovina** (7D). Country in Balkan Europe, formerly part of Yugoslavia. Area: 51,129 km$^2$. Population (1998) 4 million. (HERT suh goh VEE nuh)

**Bosporus** (4A). Straits between the Black Sea and the Sea of Marmara, Turkey, southwestern Asia. (BAHS pur uhs)

**Boston** (9C). City in state of Massachusetts, United States, Anglo America, at 42° N, 71° W.

**Botswana** (3C). Country in Southern Africa. Area: 1,285,000 km$^2$. Population (1998) 1.4 million.

**Brahmaputra River** (5A). Major river draining eastern Himalayas to a delta with the Ganges River in Bangladesh, Southern Asia. (brah muh POO truh)

**Brasília** (10E). Capital city of Brazil, South America, at 16° S, 48° W. (brah SEEL yuh)

**Brazil** (10E). Largest country in South America. Area: 8,511,970 km$^2$. Population (1998) 162.1 million.

**Brazilian Highlands** (10B). Upland area, mainly of plateau landforms, in eastern Brazil, South America.

**Brazzaville** (3B). Capital city of Congo, Central Africa, at 4° S, 15° E.

**Brisbane** (11C). Capital city of the state of Queensland, Australia, South Pacific, at 27° S, 153° E. (BRIHS bayn)

**British Columbia** (9D). Province on the Pacific coast of Canada, Anglo America.

**British Isles** (7B). Group of islands off northwestern Europe comprising the countries of the Republic of Ireland and the United Kingdom (England, Northern Ireland, Scotland, Wales).

**Brittany** (7B). Northwestern peninsula of France, Western Europe.

**Brunei** (6D). Country on the northern coast of Borneo, Southeastern Asia. Area: 5770 km$^2$. Population (1998) 0.3 million. (broon EYE)

**Brussels** (7C). Capital city of Belgium and center of European Union administration, Western Europe, at 51° N, 4° E.

**Bucharest** (7E). Capital city of Romania, Balkan Europe, at 44° N, 26° E. (BOO cuh REST)

**Budapest** (7E). Capital city of Hungary, Balkan Europe, at 47° N, 19° E. (BOO duh PEST)

**Buenos Aires** (10E). Capital city of Argentina, South America, at 35° S, 58° W. (BWAY nohs EYE rays)

**Bulawayo** (3C). Major city of western Zimbabwe, Southern Africa, at 20° S, 29° E. (boo luh WAH yo)

**Bulgaria** (7E). Country in Balkan Europe. Area: 110,910 km$^2$. Population (1998) 8.3 million.

**Burkina Faso** (3B). Country in Western Africa. Area: 274,200 km$^2$. Population (1998) 11.3 million. (bur KEE nuh FAH soh)

**Bursa** (4C). City in Turkey, southwestern Asia, at 40° N, 29° E.

**Burundi** (3B). Country in Central Africa. Area: 27,830 km$^2$. Population (1998) 5.5 million. (buh RUHN dee)

**Cairo** (4B). Capital city of Egypt, Nile River valley, at 30° N, 31° E. (KY roh)

**Calcutta** (5B). Largest city of the Bengal state of India, Southern Asia, at 23° N, 88° E. (kal KUT uh)

**Cali** (10D). City in central Colombia, Northern Andes, South America, at 3° N, 77° W. (KAH lee)

**Cambodia** (6D). Country in Southeastern Asia. Area: 181,040 km$^2$. Population (1998) 10.8 million.

**Cameroon** (3B). Country in Central Africa. Area: 475,440 km$^2$. Population (1998) 14.3 million. (kam oh ROON)

**Cameroon Highlands** (3A). Upland area in western Cameroon, Central Africa.

**Canada** (9D). Country in Anglo America. Area: 9,976,140 km$^2$. Population (1998) 30.8 million.

**Canadian Prairies** (9B). Natural grassland area between Great Lakes and Rocky Mountains, Anglo America. Now largely plowed.

**Canadian Shield** (9B). Area of northern Canada, Anglo America, formed by ancient rocks, largely scraped bare of soil by ice sheets over the last million years, and containing major mineral resources in the rocks.

**Canberra** (10C). See **Australian Capital Territory**.

**Cape Town** (3C). Major port city on the southwestern tip of the Republic of South Africa, Southern Africa, at 34° S, 18° E.

**Caracas** (10D). Capital city of Venezuela, Northern Andes, South America, at 11° N, 67° W. (kuh RAH kuhs)

**Carpathian Mountains** (7B). Mountain ranges in Slovakia, southwestern Ukraine and Romania, Balkan Europe. (kahr PAY thee un)

**Casablanca** (4B). Major city in Morocco, North Africa, at 34° N, 8° W. (KAS uh BLANG kuh)

**Caspian Sea** (8B). Inland sea bordered by Iran and CIS countries, including Russia, Azerbaijan, Kazakstan, and Turkmenistan.

**Caucasus Mountains** (8B). Mountain ranges in Georgia and Azerbaijan, Southwestern CIS.

**Central Africa** (3B). Subregion of Africa South of the Sahara, comprising Burundi, Central African Republic, Chad, Congo, Equatorial Guinea, Gabon, Rwanda, and Congo (Zaire).

**Central African Republic** (3B). Country in Central Africa. Area: 622,980 km$^2$. Population (1998) 3.4 million.

**Central America** (10C). Subregion in Latin America that comprises Belize, Costa Rica, El Salvador, Guatemala, Honduras, Nicaragua, and Panama.

**Central Asia** (8D). Subregion of CIS, comprising Kazakstan, Kyrgyzstan, Tajikistan, Turkmenistan, and Uzbekistan.

**Central Siberian Plateau** (8B). Upland area in the Russian Federation between Yenisey and Lena Rivers.

**Chad** (3B). Country in Central Africa. Area: 1,284,000 km$^2$. Population (1998) 7.4 million.

**Changchan** (6C). City of China, Eastern Asia, at 44° N, 123° E.

**Chang Jiang** (6A). Major river of southern China (Yangtze), Eastern Asia. (Chung jee AHNG)

**Chao Phraya River** (6A). Major river of Thailand, Southeastern Asia. (CHOW pruh yuh)

**Chelyabinsk** (8C). Major city of the Russian Federation, at 55° N, 61° E.

**Chengdu** (6C). City of China, Eastern Asia, at 30° N, 104° E.

**Chicago** (9C). Major city in the state of Illinois, United States, Anglo America, at 42° N, 88° W.

**Chile** (10E). Country in Southern South America on the Pacific coast. Area: 756,950 km$^2$. Population (1998) 14.8 million.

**China, People's Republic of** (6C). World's most populous country, Eastern Asia. Area: 9,596,960 km$^2$. Population (1998) 1,242.5 million.

**Chittagong** (5C). Major port city in eastern Bangladesh, Southern Asia, at 22° N, 91° E.

**Cleveland** (9C). Major city in the state of Ohio, United States, Anglo America, at 42° N, 82° W.

**Cologne** (7C). City of Germany, Western Europe, at 51° N, 7° E.

**Colombia** (10D). Country in the Northern Andes, South America. Area: 1,138,910 km$^2$. Population (1998) 38.6 million.

**Colombo** (5C). Capital city of Sri Lanka, Southern Asia, at 7° N, 80° E.

**Colorado Plateau** (9B). Upland area in the southwestern United States, Anglo America.

**Colorado River** (9B). Major river draining much of the southwestern United States, Anglo America.

**Columbia Plateau** (9B). Upland area in the northwestern United States, Anglo America.

**Columbia River** (9B). Major river draining part of western Canada and northwestern United States, Anglo America.

**Commonwealth of Independent States** (1B). Major world region that includes the countries of the former Soviet Union, the subject of Chapter 8.

**Conakry** (3B). Port city and capital of Guinea, Western Africa, at 10° N, 14° W.

**Congo (Zaire) River** (3A). World's second largest river in terms of water carried to the sea; in Central Africa.

**Congo, Democratic Republic of the (formerly Zaire)** (3B). Country of Central Africa. Area: 2,344,860 km$^2$. Population (1998) 49 million.

**Congo, Republic of the** (3B). Country of Central Africa. Area: 342,000 km$^2$. Population (1998) 2.7 million.

**Cook Islands** (11D). Island group in South Pacific, colony of New Zealand. Area: 237 km$^2$. Population (1998) under 0.1 million.

**Copenhagen** (7D). Capital city of Denmark, Northern Europe, at 56° N, 13° E.

**Córdoba** (10E). Major inland city of Argentina, Southern South America, at 31° S, 64° W. (KAWRD uh buh)

**Costa Rica** (10C). Country in Central America. Area: 51,100 km$^2$. Population (1998) 3.8 million.

**Côte d'Ivoire** (3B). Country in Western Africa. Area: 322,460 km$^2$. Population (1998) 15.6 million (koht deev WAHR)

**Croatia** (7E). Country in Balkan Europe, formerly part of Yugoslavia. Area: 56,538 km$^2$. Population (1998) 4.2 million. (kro AY shuh)

**Cuba** (10D). Country in the Greater Antilles, West Indies. Area: 114,524 km$^2$. Population (1998) 11.1 million. (KYOO buh)

**Curitiba** (10E). Major city in southern Brazil, South America, at 25° S, 49° W. (KOOR ih TEE buh)

**Czech Republic** (7E). Country in Eastern Europe. Area: 78,864 km$^2$. Population (1998) 10.3 million.

**Dakar** (3B). Capital city of Senegal, Western Africa, at 15° N, 17° W.

**Dallas-Fort Worth** (9C). Major city complex in the state of Texas, United States, Anglo America, at 33° N, 97° W (Dallas).

**Damascus** (4C). Capital city of Syria, Arab Southwest Asia, at 34° N, 36° E.

**Danube River** (7B). Longest river in Europe, rising in southern Germany and flowing through more countries than any other river, reaching the Black Sea in Romania, Balkan Europe.

**Dardenelles** (4A). Narrow strait linking the Aegean Sea to the Sea of Marmara, Turkey, Southwest Asia.

**Dar es Salaam** (3C). Capital city of Tanzania, Eastern Africa, at 7° S, 39° E.

**Dead Sea** (4A). Inland sea into which the Jordan River flows. situated on the border between Israel and Jordan, Arab Southwest Asia.

**Deccan plateau** (5A). Upland in the northwestern part of the Indian peninsula, Southern Asia, formed of layers of volcanic lava erupted as the Indian peninsula cut away from the ancient continent of Gondwanaland.

**Delhi** (5B). Capital city of India, Southern Asia, at 29° N, 77° E. (DEL ee)

**Denmark** (7D). Country in Northern Europe. Area: 43,093 km$^2$. Population (1998) 5.3 million.

**Denver** (9C). Major city in the state of Colorado, United States, Anglo America, at 40° N, 105° W.

**Detroit** (9C). Major city in the state of Michigan, United States, Anglo America, at 42° N, 83° W.

**Dhaka** (5C). Capital city of Bangladesh, Southern Asia, at 24° N, 90° E.

**Djibouti** (3C). City-state, formerly French Somaliland, in Eastern Africa, at 12° N, 43° E. Area: 23,200 km$^2$. Population (1998) 0.7 million. (jih BOO tee)

**Dnepropetrovsk** (8D). Major industrial city in eastern Ukraine, Southwestern CIS, at 48° N, 35° E. (NEHP roh pih TRAWFSK)

**Dominican Republic** (10D). Country occupying the eastern two-thirds of the island of Hispaniola, West Indies. Area: 48,730 km$^2$. Population (1998) 8.3 million.

**Donetsk** (8D). Major industrial city in eastern Ukraine, Southwestern CIS, at 48° N, 38° E.

**Don River** (8B). Major river flowing southward through Russia to the Black Sea.

**Douala** (3B). Port city in Cameroon, Central Africa, at 4° N, 10° E.

**Dublin** (7C). Capital city of the Republic of Ireland, Western Europe, at 53° N, 6° W.

**Durban** (3C). Major port in the Republic of South Africa, Southern Africa, at 30° S, 31° E.

**Düsseldorf** (7C). Industrial city of Germany, Western Europe, at 51° N, 7° E.

**Eastern Africa** (3C). Subregion of Africa South of the Sahara that includes Djibouti, Eritrea, Ethiopia, Kenya, Somalia, Tanzania, and Uganda.

**Eastern Asia** (1B). Major world region, the subject of Chapter 6.

**Eastern Europe** (7A). Subregion of Europe that includes the Baltic states (Estonia, Latvia, Lithuania), Poland, and the Czech and Slovak Republics.

**Eastern Ghats** (5A). Uplands along the eastern coast of the Indian peninsula, Southern Asia.

**East Siberian Uplands** (8B). Group of sparsely inhabited mountain ranges along the Pacific coast of the Russian Federation.

**Ecuador** (10D). Country in the Northern Andes, South America. Area: 283,560 km$^2$. Population (1998) 12.2 million. (EHK wuh dohr)

**Egypt** (4B). Country in the Nile River valley. Area: 1,001,450 km$^2$. Population (1998) 65.5 million.

**Elbe River** (7B). River that flows through northern Germany, Western Europe, into the North Sea.

**El Salvador** (10C). Country in Central America. Area: 21,040 km$^2$. Population (1998) 5.8 million.

**Equatorial Guinea** (3B). Country in Central Africa. Area: 28,050 km$^2$. Population (1998) 0.4 million.

**Eritrea** (3C). Country in Eastern Africa that gained its independence from Ethiopia in 1993. Area: 124,320 km$^2$. Population (1998) 3.8 million. (EH rih TREE yuh)

**Essen** (7C). Industrial city of Germany, Western Europe, at 52° N, 7° E.

**Estonia** (7E). Country on the Baltic Sea in Eastern Europe. Area: 45,274 km$^2$. Population (1998) 1.4 million.

**Ethiopia** (3C). Country in Eastern Africa. Area: 1,221,900 km$^2$. Population (1998) 58.4 million.

**Ethiopian Highlands** (3A). Upland area in central Ethiopia, Eastern Africa.

**Europe** (1B). Major world region, the subject of Chapter 7.

**Faisalabad** (5C). Major city in Pakistan, Southern Asia, at 31° N, 73° E. (FY SAHL uh BAHD)

**Faeroe Islands** (7D). Islands between the British Isles and Iceland, former colony of Denmark, now with substantial responsibility for its own internal affairs.

**Fez** (4B). Major city in Morocco, North Africa, at 34° N, 5° W.

**Fiji** (11D). Group of islands in the South Pacific. Area: 18,270 km$^2$. Population (1998) 0.8 million.

**Finland** (7D). Country in Northern Europe. Area: 338,130 km$^2$. Population (1998) 5.2 million.

**Fortaleza** (10E). Major port city in northeastern Brazil, South America, at 4° S, 39° W.

**France** (7C). Country in Western Europe. Area: 551,500 km$^2$. Population (1998) 58.3 million.

**Frankfurt** (7C). Major city in Germany, Western Europe, at 50° N, 9° E.

**Freetown** (3B). Port and capital city of Sierra Leone, Western Africa, at 8° N, 13° W.

**French Guiana** (10D). Country on the northern coast of South America that is grouped with the West Indies. Area: 91,000 km$^2$. Population (1998) 0.2 million. (gee AH nuh)

**French Polynesia** (11D). Group of islands in the South Pacific. Area: 4000 km$^2$. Population (1998) 0.2 million.

**Gabon** (3B). Country in Central Africa. Area: 267,670 km$^2$. Population (1998) 1.2 million.

**Gambia, The** (3B). Country in Western Africa. Area: 11,300 km$^2$. Population (1998) 1.2 million. (GAM bee uh)

**Ganges River** (5A). Major river that rises in the Himalayas and flows across northern India, Southern Asia.

**Gaza** (4C). Part of Israel, mainly occupied by Palestinians being given more powers, southwestern Asia.

**Geneva** (7D). Major city in western Switzerland, Alpine Europe, at 46° N, 6° E.

**Georgia** (8D). Country with a coast on the Black Sea in Southwestern CIS. Area: 69,700 km$^2$. Population (1998) 5.4 million.

**Germany** (7C). Country in Western Europe. Area: 356,755 km$^2$. Population (1998) 82.3 million.

**Ghana** (3B). Country in Western Africa. Area: 238,540 km$^2$. Population (1998) 18.9 million.

**Glasgow** (7C). Largest city of Scotland, United Kingdom, Western Europe, at 56° N, 4° W. (GLAHZ goh)

**Gobi Desert** (6A). Arid area in Mongolia, Eastern Asia.

**Great Dividing Range** (11B). Upland area that parallels the eastern coast of Australia, South Pacific.

**Greater Antilles** (10D). The major group of West Indian islands, including Cuba, Jamaica, Hispaniola, and Puerto Rico.

**Great Lakes-St. Lawrence Seaway** (9B). Inland waterway that connects the Great Lakes of Anglo America with the Atlantic Ocean, opened in 1959.

**Great Plains** (9B). Area of flat land, rising west of the Mississippi River to the Rocky Mountains, Anglo America. Formerly natural grassland, now mainly plowed.

**Greece** (7D). Country in Mediterranean Europe. Area: 131,990 $km^2$. Population (1998) 10.5 million.

**Greenland** (7D). Country in the North Atlantic, bordering on the Arctic and close to North America. Historic ties to Denmark, Northern Europe, until given responsibility for its own internal affairs in 1979. Area: 2,175,600 $km^2$. Population (1998) 0.1 million.

**Guadalajara** (10C). Major city in central Mexico, Latin America, at 21° N, 103° W. (GWAH duh luh HAHR uh)

**Guam** (11D). Island country in the South Pacific, U.S. territory. Area: 549 $km^2$. Population (1998) 0.2 million.

**Guangzhou** (6C). Major city (formerly Canton) in southern China, Eastern Asia, at 23° N, 113° E. (gwahng joh)

**Guatemala** (10C). Country in Central America. Area: 108,890 $km^2$. Population (1998) 11.6 million.

**Guatemala City** (10C). Capital city of Guatemala, Central America, at 15° N, 91° W.

**Guayaquil** (10D). Major port of Ecuador, Northern Andes, South America, at 2° S, 80° W. (GWAH yah KEEL)

**Guiana Highlands** (10B). Upland area straddling the border between Brazil, Venezuela, and the Guianas, South America. (gee AH nuh hilands)

**Guinea** (3B). Country in Western Africa. Area: 245,860 $km^2$. Population (1998) 7.5 million. (GIHN ee)

**Guinea-Bissau** (3B). Country in Western Africa. Area: 36,120 $km^2$. Population (1998) 1.1 million. (GIHN ee bih SOW)

**Guyana** (10D). Country on the northern coast of South America, part of the West Indies. Area: 214,970 $km^2$. Population (1998) 0.7 million.

**Haifa** (4C). Major city in Israel, southwestern Asia, at 33° N, 35° E. (HYE fah)

**Haiphong** (6D). Major port of northern Vietnam, Southeastern Asia, at 21° N, 107° E. (HY FONG)

**Haiti** (10D). Country occupying western one-third of Hispaniola, West Indies. Area: 27,750 $km^2$. Population (1998) 7.5 million. (HAY tee)

**Hamburg** (7C). Major port of northern Germany, Western Europe, at 54° N, 10° E.

**Hangzhou** (6C). City of China, Eastern Asia, at 30° N, 120° E. (hahng JOH)

**Hanoi** (6D). Capital city of Vietnam, Southeastern Asia, at 21° N, 106° E.

**Harare** (3C). Capital city of Zimbabwe, Southern Africa, at 18° S, 31° E.

**Harbin** (6C). City of China, Eastern Asia, at 46° N, 126° E.

**Havana** (10D). Capital city of Cuba, West Indies, at 23° N, 82° W.

**Hawaii** (9C). Island state of the United States, Anglo America, in the middle of the North Pacific Ocean.

**Heartland** (9C). Northern interior of the United States, Anglo America, a subregion of highly productive farming combined with manufacturing and increasing service industries.

**Helsinki** (7D). Capital city of Finland, Northern Europe, at 60° N, 25° E.

**Himalayan Mountains** (5A). The world's highest mountain system, forming the northern boundary of Southern Asia.

**Ho Chi Minh City** (6D). Major city, formerly Saigon, of southern Vietnam, Southeastern Asia, at 11° N, 107° E.

**Honduras** (10C). Country of Central America. Area: 112,090 $km^2$. Population (1998) 5.9 million.

**Hong Kong** (6B). Port city of China, Eastern Asia, former colony of the United Kingdom, at 22° N, 114° E.

**Houston** (9C). Major city in the state of Texas, United States, Anglo America, at 30° N, 95° W.

**Huang He** (6A). Major river of north-central China, also known as the Yellow River.

**Hungary** (7E). Country of the European Balkans. Area: 93,030 $km^2$. Population (1998) 10.1 million.

**Hyderabad** (5B). Major city in southern India, Southern Asia, at 17° N, 78° E.

**Hyderabad** (5C). Major city of Pakistan, Southern Asia, at 25° N, 68° E.

**Iberia** (7B). Peninsular area of Mediterranean Europe that contains Portugal and Spain.

**Iceland** (7D). Island country in the North Atlantic Ocean that had historic ties to Denmark until its independence. Area: 103,000 $km^2$. Population (1998) 0.3 million.

**Inch'on** (6B). Major city in South Korea, Eastern Asia, at 37° N, 127° E. (IHN CHANN)

**India, Republic of** (5B). Largest country in Southern Asia. Area: 3,287,590 $km^2$. Population (1998) 988.7 million.

**Indian subcontinent** (5A). Land area that forms most of Southern Asia, lying south of the Himalayan Mountains.

**Indonesia** (6D). Country of Southeastern Asia. Area: 1,904,570 $km^2$. Population (1998) 207.4 million.

**Indus River** (5A). Major river that rises in the Himalayan Mountains and flows southwestward through Pakistan.

**Iran** (4C). Country of southwestern Asia. Area: 1,648,000 $km^2$. Population (1998) 64.1 million.

**Iraq** (4C). Country of Arab Southwest Asia. Area: 438,320 $km^2$. Population (1998) 21.8 million.

**Irbid** (4C). City in Jordan, Arab Southwest Asia, at 33° N, 36° E.

**Ireland, Republic of** (7C). Country in Western Europe. Area: 70,280 $km^2$. Population (1998) 3.7 million.

**Irrawaddy River** (6A). Main river of Myanmar, Southeastern Asia.

**Isfahan** (4C). Major city of Iran, Southwest Asia, at 33° N, 52° E.

**Islamabad** (5C). Capital city of Pakistan, Southern Asia, at 34° N, 73° E.

**Israel** (4C). Country in southwestern Asia. Area: 20,770 $km^2$. Population (1998) 6 million.

**Istanbul** (4C). Capital city of Turkey, southwestern Asia (formerly Constantinople), at 41° N, 29° E.

**Italy** (7D). Country of Mediterranean Europe. Area: 301,270 $km^2$. Population (1998) 57.7 million.

**Izmir** (4C). Major city of Turkey, southwestern Asia, at 38° N, 27° E.

**Jakarta** (6D). Capital city of Indonesia, Southeastern Asia, at 6° S, 107° E.

**Jamaica** (10D). Country of the West Indies. Area: 10,990 $km^2$. Population (1998) 2.6 million.

**Japan** (6B). Country of Eastern Asia. Area: 377,800 km$^2$. Population (1998) 126.4 million.

**Jeddah** (4C). Port on the Red Sea in Saudi Arabia, Arab Southwest Asia, at 21° N, 39° E.

**Jerusalem** (4C). City in Israel, in dispute between Israel and the Palestinians, at 32° N, 35° E.

**Johannesburg** (3C). Major industrial city, Republic of South Africa, Southern Africa, at 26° S, 28° E.

**Jordan** (4C). Country in Arab Southwest Asia. Area: 89,210 km$^2$. Population (1998) 4.6 million.

**Jordan River** (4A). River that flows to the Dead Sea, partly along Jordan's border with Israel.

**Kabul** (5C). Capital city of Afghanistan, Southern Asia, at 34° N, 69° E. (KAH bool)

**Kaliningrad** (7E). Small territory on the southern coast of the Baltic Sea, Eastern Europe, that remains part of the Russian Federation.

**Kamchatka Peninsula** (8B). Mountainous and volcanic peninsula in eastern Siberia, Russian Federation.

**Kampala** (3C). Capital city of Uganda, Eastern Africa, at 0°, 33° E.

**Kao-hsiung** (6B). Major port of Taiwan, Eastern Asia, at 23° N, 120° E. (kah oh SYOONG)

**Karachi** (5C). Largest city of Pakistan, Southern Asia, at 25° N, 67° E.

**Kathmandu** (5C). Capital city of Nepal, Southern Asia, at 28° N, 88° E. (kaht man DOO)

**Katowice** (7E). Industrial city of Poland, Eastern Europe, at 50° N, 19 E. (KAT oh we chay)

**Kazakstan** (8D). Country in Central Asia. Area: 2,717,300 km$^2$. Population (1998) 15.6 million.

**Kazan** (8D). City in Russian Federation at 56° N, 49° E.

**Kenya** (3C). Country of Eastern Africa. Area: 580,370 km$^2$. Population (1998) 28.3 million.

**Kharkov** (8C). Major industrial city of Ukraine, CIS, at 50° N, 36° E.

**Khartoum** (4B). Capital city of Sudan, Nile River valley, at 16° N, 33° E. (kahr TOOM)

**Khulna** (5C). Major industrial city of Bangladesh, Southern Asia, at 23° N, 90° E. (KOOL nuh)

**Kiev** (8D). Capital city of Ukraine, Southwestern CIS, at 50° N, 31° E.

**Kingston** (10D). Capital city of Jamaica, West Indies, at 18° N, 77° W.

**Kinshasa** (3B). Capital city of Congo (Zaire), Central Africa, at 4° S, 15° E.

**Kiribati** (11D). Island country in the South Pacific. Area: 728 km$^2$. Population (1998) under 0.1 million. (KIH ruh bahss)

**Kitakyushu** (6B). Main city of Kyushu Island, Japan, Eastern Asia, at 34° N, 130° E.

**Kobe** (6B). Major industrial city of Japan, Eastern Asia, at 35° N, 135° E. (KOH bee)

**Kuala Lumpur** (6D). Capital city of Malaysia, Southeastern Asia, at 3° N, 102° E.

**Kuwait** (4C). Country in Arab Southwest Asia. Area: 17,820 km$^2$. Population (1998) 1.9 million.

**Kyoto** (6B). City in central Japan, its former capital, Eastern Asia, at 35° N, 136° E.

**Kyrgyzstan** (8D). Country in Central Asia. Area: 198,500 km$^2$. Population (1998) 4.7 million.

**Lagos** (3B). Capital city of Nigeria, Western Africa, at 6° N, 3° E. (LAH gohs)

**Lahore** (5C). Major city of Pakistan, Southern Asia, at 32° N, 74° E.

**Laos** (6D). Country in Southeastern Asia. Area: 236,800 km$^2$. Population (1998) 5.3 million. (LAH ohs)

**La Paz** (10D). Capital city of Bolivia, Northern Andes, South America, at 16° S, 68° W.

**Latin America** (1B). Major world region that comprises Mexico, Central America, West Indies, and South America, the subject of Chapter 10.

**Latvia** (7E). Baltic country in Eastern Europe. Area: 66,547 km$^2$. Population (1998) 2.4 million.

**Lebanon** (4C). Country in Arab Southwest Asia. Area: 10,400 km$^2$. Population (1998) 4.1 million.

**Leeds** (7C). Industrial city in England, United Kingdom, Western Europe, at 54° N, 2° W.

**Lena River** (8B). Major river flowing into the Arctic Ocean across Siberia, Russian Federation.

**Lesotho** (3C). Country in Southern Africa. Area: 30,350 km$^2$. Population (1998) 2.1 million. (le so TOO)

**Lesser Antilles** (10D). Group of island countries forming an arc around the eastern margin of the Caribbean Sea, West Indies.

**Liberia** (3B). Country of Western Africa. Area: 111,370 km$^2$. Population (1998) 2.8 million.

**Libreville** (3B). Capital city of Gabon, Central Africa, at 0°, 9° E.

**Libya** (4B). Country in North Africa. Area: 1,759,540 km$^2$. Population (1998) 5.7 million.

**Lima-Callao** (10D). Capital city and main port of Peru, Northern Andes, South America, at 12° S, 77° W.

**Lisbon** (7D). Capital city of Portugal, Mediterranean Europe, at 39° N, 9° W.

**Lithuania** (7E). Baltic country in Eastern Europe. Area: 65,200 km$^2$. Population (1998) 3.7 million.

**Loire River** (7B). Major river of central France, Western Europe.

**London** (7C). Capital city of the United Kingdom, Western Europe, at 52° N, 0°.

**Los Angeles** (9C). Major city of southern California, United States, Anglo America, at 34° N, 118° W.

**Luanda** (3C). Capital city of Angola, Southern Africa, at 9° S, 13° E.

**Lubumbashi** (3B). Major mining city of southern Congo (Zaire), Central Africa, at 12° S, 27° E.

**Lusaka** (3C). Capital city of Zambia, Southern Africa, at 15° S, 28° E.

**Luxembourg** (7C). Country of Western Europe. Area: 2,586 km$^2$. Population (1998) 0.4 million.

**Lyon** (7C). Major city of central France, Western Europe, at 46° N, 5° E. (lyawn)

**Macao** (6C). Port city of China, former colony of Portugal, at 22° N, 114° E.

**Macedonia** (7E). Country in Balkan Europe, formerly part of Yugoslavia. Area: 25,713 km$^2$. Population (1998) 2 million.

**Mackenzie River** (9B). Major river of northwestern Canada, Anglo America.

**Madagascar** (3C). Island country off the eastern coast of Southern Africa. Area: 587,040 km$^2$. Population (1998) 14 million.

**Madras** (5B). Major port city of southern India, Southern Asia, at 13° N, 80° E.

**Madrid** (7D). Capital city of Spain, Mediterranean Europe, at 40° N, 4° W.

**Mahgreb** (4B). Upland area in the Atlas Mountains of North Africa.

**Malawi** (3C). Country in Southern Africa. Areas: 118,480 km$^2$. Population (1998) 9.8 million.

GLOSSARY

**Malaysia** (6D). Country in Southeastern Asia. Area: 329,750 km$^2$. Population (1998) 28.4 million.

**Maldives** (5C). Island country of Southern Asia. Area: 300 km$^2$. Population (1998) 0.3 million.

**Mali** (3B). Country of Western Africa. Area: 1,240,190 km$^2$. Population (1998) 10.1 million.

**Managua** (10C). Capital city of Nicaragua, Central America, at 12° N, 86° W.

**Manaus** (10E). Major city in the heart of the Amazon River basin, Brazil, South America, at 3° S, 60° W.

**Manchester** (7C). Industrial city of England, United Kingdom, Western Europe, at 53° N, 2° W.

**Mandalay** (6D). Capital city of Myanmar (Burma), Southeastern Asia, at 22° N, 96° W.

**Manila** (6D). Capital city of the Philippines, Southeastern Asia, at 14° N, 121° E.

**Manufacturing Belt** (9C). Region of concentrated manufacturing functions in north-central and northeastern United States, Anglo America.

**Maputo** (3C). Capital city of Mozambique, Southern Africa, at 26° S, 33° E.

**Marrakech** (4B). City in Morocco, North Africa, at 32° N, 8° W.

**Marseilles** (7C). Major port on southern coast of France, Western Europe, at 43° N, 5° E. (mahr SAY)

**Marshall Islands** (11D). Island country in the South Pacific. Area: 180 km$^2$. Population (1998) 0.1 million.

**Massif Central** (7B). Upland area in central southern France, Western Europe.

**Matadi** (3B). Major port at mouth of the Congo River, Congo (Zaire), Central Africa, at 6° S, 13° E.

**Mauritania** (3B). Country in Western Africa. Area: 1,025,520 km$^2$. Population (1998) 2.5 million.

**Mecca** (4C). Holy city (Makkah) in Saudi Arabia, Arab Southwest Asia, at 21° N, 40° E.

**Medan** (6D). City in Indonesia, Southeastern Asia, at 4° N, 98° E.

**Medellín** (10D). Major industrial city in Colombia, Northern Andes, South America, at 6° N, 76° W.

**Medina** (4C). Holy city in Saudi Arabia, Arab Southwest Asia, at 24° N, 40° E.

**Mediterranean Europe** (7A). Subregion of Europe comprising Greece, Italy, Portugal, and Spain.

**Mediterranean Sea** (4A). Sea between Europe, southwestern Asia, and northern Africa.

**Megalopolis** (9C). Urbanized region of northeastern United States, Anglo America, extending from Boston, Massachusetts, to Washington, D.C.

**Mekong River** (6A). Major river of Southeastern Asia that rises in southern China and flows through or between Myanmar, Thailand, Laos, Cambodia, and Vietnam.

**Melbourne** (11C). Capital city of the state of Victoria, Australia, South Pacific, at 38° S, 145° E. (MEHL burhn)

**Meseta** (7B). Upland area of central Spain, Mediterranean Europe.

**Meshed** (4C). Major city of Iran, southwestern Asia, at 36° N, 60° E.

**Mexico** (10C). Country in Latin America. Area: 1,958,200 km$^2$. Population (1998) 97.8 million.

**Mexico City** (10C). Capital city of Mexico, Latin America, at 19° N, 99° W.

**Miami** (9C). Major city in the state of Florida, United States, Anglo America, at 26° N, 80° W.

**Micronesia, Federated States of** (11D). Island country in the South Pacific. Area: 700 km$^2$. Population (1998) 0.1 million.

**Middle America** (10C). Part of Latin America comprising Mexico, Central America, and the West Indies.

**Midwest** (9C). Region in north-central area of United States, part of the Heartland, Anglo America.

**Milan** (7D). Major industrial city in northern Italy, Mediterranean Europe, at 45° N, 9° E.

**Minneapolis-St. Paul** (9C). Major urban complex in the state of Minnesota, United States, Anglo America, at 45° N, 93° W.

**Minsk** (8D). Capital city of Belarus, Southwestern CIS, at 54° N, 28° E.

**Mississippi River** (9B). Largest river of Anglo America, draining approximately one-third of United States.

**Moldova** (8D). Country of Southwest CIS. Area: 33,700 km$^2$. Population (1998) 4.2 million.

**Mombasa** (3C). Major port of Kenya, Eastern Africa, at 4° S, 40° E.

**Mongolia** (6C). Country in buffer zone between Russia and China, Eastern Asia. Area: 1,565,000 km$^2$. Population (1998) 2.4 million.

**Monrovia** (3B). Capital city of Liberia, Western Africa, at 6° N, 11° W.

**Monterrey** (10C). City in northern Mexico at 26°N, 100°W.

**Montevideo** (10E). Capital city of Uruguay, Southern South America, at 35° S, 56° W. (MAHN tuh vih DAY oh)

**Montréal** (9D). Major city in Québec, Canada, Anglo America, at 46° N, 73° W.

**Morocco** (with Western Sahara) (4B). Country in North Africa. Area: 446,550 km$^2$. Population (1998) 27.7 million.

**Moscow** (8C). Capital city of the Russian Federation, at 56° N, 38° E.

**Mosul** (4C). City in northern Iraq, southwestern Asia, at 36° N, 43° E.

**Mount Everest** (5A). Highest mountain in the world, part of Himalayan Mountains, Southern Asia.

**Mount Kilimanjaro** (3A). Highest mountain in Africa, in Tanzania, Eastern Africa.

**Mount McKinley** (9B). Highest mountain in Anglo America, located in the state of Alaska, United States.

**Mount St. Helens** (9B). Volcanic mountain that erupted in 1980, located in the state of Washington, United States, Anglo America.

**Mozambique** (3C). Country in Southern Africa. Area: 801,590 km$^2$. Population (1998) 18.8 million. (MOH zuhm BEEK)

**Munich** (7C). Major industrial town in southern Germany, Western Europe, at 48° N, 12° E.

**Murray-Darling River basin** (11B). The best-watered part of Australia west of the Great Dividing Range.

**Myanmar** (6D). Country in Southeastern Asia, called Burma until 1989. Area: 676,550 km$^2$. Population (1998) 47.1 million.

**Nagoya** (6B). Industrial city of central Japan, Eastern Asia, at 35° N, 137° E. (nuh GOY uh)

**Nairobi** (3C). Capital city of Kenya, Eastern Africa, at 1° S, 37° E.

**Namibia** (3C). Country in Southern Africa. Area: 824,290 km$^2$. Population (1998) 1.6 million.

**Naples** (7D). Industrial city in southern Italy, Mediterranean Europe, at 41° N, 14° E.

**Nauru** (11D). Island country in the South Pacific. Area: 21 km$^2$. Population (1998) under 0.1 million.

**N'Djamena** (3B). Capital city of Chad, Central Africa, at 12° N, 15° E. (ehn JAHM un nuh)

**Nepal** (5C). Country in the northern mountain rim of Southern Asia. Area: 140,800 km$^2$. Population (1998) 23.7 million.

**Netherlands** (7C). Country in Western Europe. Area: 37,330 km$^2$. Population (1998) 15.7 million.

**New Caledonia** (11D). Island country in the South Pacific, French colony. Area: 18,600 km$^2$. Population (1998) 0.2 million.

**New England** (9C). Region comprising six states in northeastern United States, Anglo America.

**New South Wales** (11C). State of Australia, South Pacific.

**New York** (9C). Largest city in the United States, Anglo America, at 41° N, 74° W.

**New Zealand** (11C). Country in the South Pacific. Area: 270,028 km$^2$. Population (1998) 3.8 million.

**Nicaragua** (10). Country in Central America. Area: 130,000 km$^2$. Population (1998) 4.8 million.

**Niger** (3B). Country in Western Africa. Area: 1,267,000 km$^2$. Population (1998) 10.1 million. (NY juhr)

**Nigeria** (3B). Country in Western Africa. Area: 923,770 km$^2$. Population (1998) 121.8 million.

**Nile River system** (4A). Largest river system in Africa, fed by waters from Lake Victoria and the Ethiopian Highlands and flowing across Sudan and Egypt.

**Nile River valley** (4B). Subregion of Northern Africa and Southwestern Asia, comprising Egypt and Sudan.

**Niue** (11D). Island country in the South Pacific, dependency of New Zealand. Area: 259 km$^2$. Population (1998) less than 0.1 million.

**Nizhni Novgorod** (8C). Major industrial and historic city in the Russian Federation, at 56° N, 43° E. (NIZH nee NAHV guh RAHD)

**North Africa** (4B). Subregion of Northern Africa and Southwestern Asia comprising Algeria, Libya, Morocco, and Tunisia.

**Northern Africa and Southwestern Asia** (1B). Major world region, the subject of Chapter 4.

**Northern Andes** (10D). Subregion of Latin America comprising Bolivia, Colombia, Ecuador, Peru, and Venezuela.

**Northern Europe** (7A). Subregion of Europe comprising Denmark, Finland, Greenland, Iceland, Norway, and Sweden.

**Northern Mariana Islands** (11D). Island country in the South Pacific. Area: 404 km$^2$. Population (1998) under 0.1 million.

**Northern Territory** (11C). Northern part of Australia, South Pacific.

**North European Plain** (7B). Lowland area extending from northern Germany across Poland and into Russia.

**North Korea** (6C). Country in Eastern Asia; official title, Democratic People's Republic of Korea. Area: 120,540 km$^2$. Population (1998) 22.2 million.

**North Sea** (7B). Sea lying between the British Isles, Scandinavia, and Germany.

**Northwest Territories** (9D). Northwestern part of Canada, Anglo America.

**Norway** (7D). Country of Northern Europe. Area: 323,878 km$^2$. Population (1998) 4.4 million.

**Novosibirsk** (8C). City in the Russian Federation, at 55° N, 83° E. (noo voh suh BEERSK)

**Ob River** (8B). Major river in northern Russia.

**Odessa** (8D). Major port in Ukraine, Southwestern CIS, at 46° N, 31° E. (oh DES uh)

**Ohio River** (9B). Major tributary of the Mississippi River, United States, Anglo America.

**Oman** (4C). Country of Arab Southwest Asia. Area: 212,400 km$^2$. Population (1998) 2.5 million.

**Omsk** (8C). City in the Siberian part of the Russian Federation, at 55° N, 73° E.

**Ontario** (9D). Province of Canada, Anglo America.

**Oran** (4B). Major port of Algeria, North Africa, at 36° N, 1° W.

**Orinoco River** (10B). Major river of Venezuela, Northern Andes, South America.

**Osaka** (6B). Industrial city of central Japan, Eastern Asia, at 35° N, 136° E.

**Oslo** (7D). Capital city of Norway, Northern Europe, at 60° N, 11° E.

**Ottawa** (9D). Capital city of Canada, Anglo America, at 45° N, 76° W.

**Ouagadougou** (3B). Major city of Burkina Faso, Western Africa, at 12° N, 2° W. (wah guh DOO goo)

**Pacific Coast** (9C). Region of the United States, Anglo America, comprising the states of California, Oregon, and Washington.

**Pakistan** (5C). Country of Southern Asia. Area: 766,100 km$^2$. Population (1998) 141.9 million.

**Palau** (11D). Island country of the South Pacific. Area: 494 km$^2$. Population (1998) under 0.1 million. (pah LOW)

**Pamir Mountains** (8B). Mountain ranges on the southern border of Central Asia.

**pampas** (10B). Former grassland area of Argentina and Uruguay, Southern South America.

**Panama** (10C). Country in Central America. Area: 77,080 km$^2$. Population (1998) 2.8 million.

**Panama City** (10C). Capital city of Panama. Central America, at 9° N, 80° W.

**Papua New Guinea** (11D). Country in the South Pacific that includes the eastern part of New Guinea. Area: 462,800 km$^2$. Population (1998) 4.3 million.

**Paraguay** (10E). Country in Southern South America. Area: 406,756 km$^2$. Population (1998) 5.2 million.

**Paraná-Paraguay River system** (10B). River system that drains southeastern Brazil, Paraguay, and parts of Argentina and Uruguay, reaching the ocean in La Plata estuary, Southern South America.

**Paris** (7C). Capital city of France. Western Europe, at 49° N, 2° E.

**Patagonia** (10B). Arid area of southern Argentina, Southern South America.

**Patagonia Plateau** (10B). Upland area that covers the main part of Patagonia.

**Perm** (8C). Major city in the Russian Federation, at 58° N, 57° E.

**Persian Gulf** (4A). Sea connected to the Indian Ocean through the Strait of Hormuz.

**Perth** (11C). Capital of the state of Western Australia, South Pacific, at 31° S, 116° E.

**Peru** (10D). Country in Northern Andes, South America. Area: 1,285,200 km$^2$. Population (1998) 26.1 million.

**Peshawar** (5C). City in northern Pakistan, Southern Asia, at 34° N, 27° E. (puh SHAH wuhr)

**Philadelphia** (9C). Major city in the state of Pennsylvania, United States, Anglo America, at 40° N, 75° W.

**Philippines** (6D). Country in Southeastern Asia. Area: 300,000 km$^2$. Population (1998) 75.3 million.

**Phnom Penh** (6D). Capital city of Cambodia, Southeastern Asia, at 12° N, 105° E. (nawm pehn)

**Phoenix** (9C). Largest city in the state of Arizona, United States, Anglo America, at 33° N, 112° W. (FEE niks)

**Pindus Mountains** (7B). Mountains in Greece, Mediterranean Europe.

GLOSSARY

**Pitcairn** (11D). Island country in the South Pacific. Area: 5 $km^2$. Population (1998) around 100.

**Pittsburgh** (9C). Manufacturing city in western Pennsylvania, United States, Anglo America, at 40° N, 80° W.

**Pointe Noire** (3B). Port city in Congo, Central Africa, at 5° S, 12° E. (PWANT nuh WAHR)

**Poland** (7E). Country in Eastern Europe. Area: 312,680 $km^2$. Population (1998) 38.7 million.

**Po River** (7B). River that drains northern Italy, Mediterranean Europe.

**Port-au-Prince** (10D). Capital city of Haiti, West Indies, at 19° N, 72° W.

**Port Elizabeth** (3C). Port in the Republic of South Africa, Southern Africa, at 34° S, 26° E.

**Port Moresby** (11D). Capital city of Papua New Guinea, South Pacific, at 9° S, 147° E.

**Pôrto Alegre** (10E). Port in southern Brazil, South America, at 30° S, 51° W.

**Portugal** (7D). Country in Mediterranean Europe. Area: 92,390 $km^2$. Population (1998) 10 million.

**Prague** (7E). Capital city of the Czech Republic, Eastern Europe, at 50° N, 14° E.

**Prairie Provinces** (9D). Central part of Canada, Anglo America, comprising the provinces of Alberta, Manitoba, and Saskatchewan.

**Pretoria** (3C). Capital city of the Republic of South Africa, Southern Africa, at 25° S, 28° E.

**Puebla** (10C). City in Mexico, Latin America, at 19° N, 98° W.

**Puerto Rico** (10D). Island country in the West Indies designated as a "commonwealth" within the United States. Area: 8,900 $km^2$. Population (1998) 4 million.

**Pusan** (6B). City in South Korea at 35° N, 129° E.

**Pyongyang** (6C). Capital city of North Korea, Eastern Asia, at 39° N, 126° E.

**Pyrenees Mountains** (7B). Mountain range forming the border between France and Spain, Western Europe.

**Qatar** (4C). Country in Arab Southwest Asia. Area: 11,000 $km^2$. Population (1998) 0.5 million.

**Québec** (9D). Province in Canada, Anglo America.

**Queensland** (11C). Northeastern state of Australia, South Pacific.

**Quito** (10D). Capital city of Ecuador, Northern Andes, South America, at 0°, 78° W. (KEE toh)

**Recife** (10E). Port city in northeastern Brazil, South America, at 8° S, 35° W. (reh SEE fuh)

**Red River** (6A). River flowing through northern Vietnam, Southeastern Asia.

**Red Sea** (4A). Sea between Africa and Arabia with connection to the Indian Ocean.

**Rhine Highlands** (7B). Upland area crossed by the Rhine River between Bingen and Bonn, Germany, Western Europe.

**Rhine River** (7B). River that rises in the Swiss Alps and flows across Germany to the Netherlands and the North Sea, Western Europe.

**Rhône River** (7B). River that rises in the Swiss Alps and flows westward into France and the Mediterranean Sea, Western Europe.

**Riga** (7E). Capital city of Latvia, Eastern Europe, at 58° N, 23° E.

**Rio de Janeiro** (10E). Major city on the eastern coast of Brazil, South America, at 23° S, 43° W. (REE oh day zhuh NAIR oh)

**Riyadh** (4C). Capital city of Saudi Arabia, Arab Southwest Asia, at 25° N, 47° E. (ree AHD)

**Rocky Mountains** (9B). Mountain ranges in western Anglo America.

**Romania** (7E). Country in Balkan Europe. Area: 237,500 $km^2$. Population (1998) 22.5 million.

**Rome** (7D). Capital city of Italy, Mediterranean Europe, at 42° N, 13° E.

**Rosario** (10E). River port city of Argentina, Southern South America, at 33° S, 61° W.

**Rostov** (8C). Russian Black Sea port, CIS, at 47° N, 40° E.

**Rotterdam** (7C). Port at the mouth of the Rhine River, Netherlands, Western Europe, at 52° N, 4° E.

**Russian Federation** (8C). Country that comprises the main heartland of the former Soviet Union. Area: 17,075,400 $km^2$. Population (1998) 146.9 million.

**Rwanda** (3B). Country in Central Africa. Area: 26,340 $km^2$. Population (1998) 8 million. (roo AHN dah)

**Sahara** (3A). World's largest arid area, in northern Africa.

**St. Louis** (9C). City in the state of Missouri, United States, Anglo America, at 39° N, 90° W.

**St. Petersburg** (8C). Port city in northern Russian Federation, formerly known as Leningrad, at 60° N, 30° E.

**Salvador** (10E). Port on the northeastern coast of Brazil, South America, at 13° S, 38° W.

**Samara** (8C). City in the Russian Federation, at 53° N, 50° E.

**Sanaa** (4C). Capital city of Yemen, Arab Southwest Asia, at 15° N, 44° E. (sah NAH)

**San Andreas Fault** (9B). Line of rupture in Earth's crust along which earthquakes occur in the state of California, United States, Anglo America.

**San Diego** (9C). Port city in southernmost California state, United States, Anglo America, at 33° N, 117° W.

**San Francisco** (9C). Major city in central California state, United States, Anglo America, at 38° N, 123° W.

**San Juan** (10D). Capital city of Puerto Rico, West Indies, at 19° N, 66° W. (san HWAHN)

**San Salvador** (10C). Capital city of El Salvador, Central America, at 14° N, 89° W.

**Santiago** (10E). Capital city of Chile, Southern South America, at 34° S, 71° W.

**Santo Domingo** (10D). Capital city of Dominican Republic, West Indies, at 19° N, 70° W.

**São Paulo** (10E). Largest city complex in Brazil, South America, at 24° S, 47° W. (SOWN POW loo)

**Sapporo** (6B). Largest city of Hokkaido Island, Japan, Eastern Asia, at 43° N, 141° E.

**Saudi Arabia** (4C). Country in Arab Southwest Asia. Area: 2,149,690 $km^2$. Population (1998) 20.2 million.

**Scandinavian peninsula** (7B). Land in Northern Europe occupied by Norway and Sweden.

**Sea of Galilee** (4A). Inland sea in northern Israel, also called Lake Tiberias.

**Seattle** (9C). Port city in the state of Washington, United States, Anglo America, at 48° N, 122° W.

**Seine River** (7B). River that drains much of northeastern France, Western Europe.

**Senegal** (3B). Country in Western Africa. Area: 196,720 $km^2$. Population (1998) 9 million.

**Seoul** (6B). Capital city of South Korea, Eastern Asia, at 38° N, 127° E. (sohl)

**Serbia-Montenegro** (7E). The remnant of the former Yugoslavia after Bosnia, Croatia, Macedonia, and Slovenia declared their independence in 1991. Area: 69,780 $km^2$. Population (1998) 10.6 million.

**Seville** (7D). City in southern Spain, Mediterranean Europe, at 37° N, 6° W.

**Shanghai** (6C). Major port city in eastern China, Eastern Asia, at 31° N, 11° E.

**Shenyang** (6C). Major city in northeastern China, Eastern Asia, at 42° N, 123° E.

**Sierra Leone** (3B). Country in Western Africa. Area: 71,740 $km^2$. Population (1998) 4.6 million.

**Sierra Nevada** (7B). Range of mountains in southern Spain, Mediterranean Europe.

**Singapore** (6D). City-state in Southeastern Asia. Area: 620 $km^2$. Population (1998) 3.9 million.

**Slovak Republic** (7E). Country in Eastern Europe. Area: 43,035 $km^2$. Population (1998) 5.4 million.

**Slovenia** (7E). Country in Balkan Europe, formerly part of Yugoslavia. Area: 20,251 $km^2$. Population (1998) 2 million.

**Sofia** (7E). Capital city of Bulgaria, Balkan Europe, at 43° N, 23° E.

**Solomon Islands** (11D). Island country in the South Pacific. Area: 28,900 $km^2$. Population (1998) 0.4 million.

**Somalia** (3C). Country in Eastern Africa composed of former British and Italian Somaliland. Area: 637,660 $km^2$. Population (1998) 10.7 million.

**South, The** (9C). Region of the United States, Anglo America, comprising the southeastern one-fourth of the country.

**South Africa** (3C). Country in Southern Africa. Area: 1,221,040 $km^2$. Population (1998) 38.9 million.

**South Australia** (11C). State of Australia, South Pacific.

**Southeastern Asia** (6D). Subregion of Eastern Asia comprising Cambodia, Indonesia, Laos, Malaysia, Philippines, Singapore, Thailand, and Vietnam.

**Southern Africa** (3C). Subregion of Africa South of the Sahara, comprising Angola, Botswana, Lesotho, Malawi, Mozambique, Namibia, Republic of South Africa, Swaziland, Zambia, and Zimbabwe.

**Southern Asia** (1B). Major world region, the subject of Chapter 5.

**Southern South America** (10E). Subregion of Latin America, comprising Argentina, Chile, Paraguay, and Uruguay.

**South Korea** (6B). Country in Eastern Asia; official name, Republic of Korea. Area: 99,020 $km^2$. Population (1998) 46.4 million.

**South Pacific** (1B). Major world region, the subject of Chapter 11.

**South Pacific Islands** (11D). Subregion of the South Pacific, comprising a number of island groups.

**Southwestern CIS** (8D). Subregion of CIS, comprising Armenia, Azerbaijan, Belarus, Georgia, Moldova, and Ukraine.

**Spain** (7D). Country in Mediterranean Europe. Area: 504,708 $km^2$. Population (1998) 39.4 million.

**Spratly Islands** (GA). Islands in the South China Sea, Eastern Asia, disputed by China, Vietnam, and others.

**Sri Lanka** (5C). Island country in Southern Asia. Area: 65,610 $km^2$. Population (1998) 18.9 million.

**Stockholm** (7D). Capital city of Sweden, Northern Europe, at 59° N, 18° E.

**Strait of Gibraltar** (4A). Narrow connection between the Atlantic Ocean and the Mediterranean Sea.

**Strait of Hormuz** (4A). Narrow entry to the Persian Gulf.

**Stuttgart** (7C). City of southern Germany, Western Europe, at 49° N, 9° E.

**Sudan** (4B). Country in the Nile River valley. Area: 65,610 $km^2$. Population (1998) 28.5 million.

**Suez Canal** (4A). Canal that joins the Red Sea and the Mediterranean Sea in Egypt.

**Surabaya** (6D). City in Indonesia, Eastern Asia, at 8° S, 113° E.

**Suriname** (10D). Country on the northern coast of South America that is part of the West Indies, formerly Dutch Guiana. Area: 163,270 $km^2$. Population (1998) 0.4 million.

**Suva** (11D). Capital city of Fiji, South Pacific, at 20° S, 178° W.

**Swaziland** (3C). Country in Southern Africa. Area: 17,360 $km^2$. Population (1998) 1 million.

**Sweden** (7D). Country in Northern Europe. Area: 449,964 $km^2$. Population (1998) 8.9 million.

**Switzerland** (7D). Country in Alpine Europe. Area: 41,290 $km^2$. Population (1998) 7.1 million.

**Sydney** (11C). Main city in the state of New South Wales, Australia, South Pacific, at 34° S, 151° E.

**Syria** (4C). Country in Arab Southwest Asia. Area: 185,180 $km^2$. Population (1998) 15.6 million.

**Tabriz** (4C). City in Iran, southwestern Asia, at 38° N, 46° E.

**Taegu** (6B). City in South Korea, Eastern Asia, at 36° N, 129° E. (TAG oo)

**Taejon** (6B). City in South Korea, Eastern Asia, at 37° N, 127° E.

**Tai-pei** (6B). Capital city of Taiwan, Eastern Asia, at 25° N, 122° E. (TY PAY)

**Taiwan** (6B). Country in Eastern Asia, claimed by China and not recognized by the United Nations. Area: 35,981 $km^2$. Population (1998) 21.7 million.

**Tajikistan** (8D). Country in Central Asia. Area: 143,100 $km^2$. Population (1998) 6.1 million.

**Tampa** (9C). City on the western coast of the state of Florida, United States, Anglo America at 28° N, 82° W.

**Tanzania** (3C). Country in Eastern Africa. Area: 945,090 $km^2$. Population (1998) 30.8 million.

**Tashkent** (8D). City in Uzbekistan, Central Asia, at 41° N, 69° E.

**Tbilisi** (8D). City in Georgia, Southwest CIS, at 42° N, 45° E. (tuh BIHL us see)

**Tegucigalpa** (10C). Capital city of Honduras, Central America, at 14° N, 87° W. (tuh GOO see GAHL puh)

**Tehran** (4C). Capital city of Iran, Southwest Asia, at 36° N, 51° E.

**Tehuantepec isthmus** (10B). Narrowing of land in southern Mexico. (tuh WAHN tuh PEHK)

**Tel Aviv–Jaffa** (4C). Largest urban complex in Israel, southwestern Asia, at 32° N, 35° E.

**Thailand** (6D). Country in Southeastern Asia. Area: 513,120 $km^2$. Population (1998) 61.1 million.

**Thames River** (7B). River in southern England, Western Europe.

**Thar Desert** (5A). Arid region of Pakistan, Southern Asia.

**Tianjin** (6C). Major industrial city near Beijing, China, Eastern Asia, at 39° N, 117° E.

**Tian Shan Mountains** (8B). Mountain ranges comprising part of the southern border of Central Asia.

**Tigris-Euphrates River system** (4A). River system that rises in Turkey and waters Syria and Iraq.

**Togo** (3B). Country in Western Africa. Area: 56,790 $km^2$. Population (1998) 4.9 million.

**Tokelau** (11D). Island country in the South Pacific, colony of New Zealand. Area: 10 $km^2$. Population (1994) 1,690.

**Tokyo** (6B). Capital city of Japan, world's largest city, Eastern Asia, at 36° N, 140° E.

**Tonga** (11D). Island country in the South Pacific. Area: 699 $km^2$. Population (1998) 0.1 million.

GLOSSARY

GLOSSARY

**Toronto** (9D). Largest city in Canada, Anglo America, at 44° N, 79° W.

**Trinidad and Tobago** (10D). Country in the West Indies. Area: 5,130 km$^2$. Population (1998) 1.3 million.

**Tripoli** (4B). Capital city of Libya, North Africa, at 33° N, 13° E.

**Tunis** (4B). Capital city of Tunisia, North Africa, at 37° N, 10° E. (TOO nihs)

**Tunisia** (4B). Country in North Africa. Area: 163,610 km$^2$. Population (1998) 9.5 million.

**Turin** (7D). Major industrial city in northern Italy, Mediterranean Europe, at 45° N, 8° E. (TOO rin)

**Turkey** (4C). Country in southwestern Asia. Area: 777,450 km$^2$. Population (1998) 64.8 million.

**Turkmenistan** (8D). Country in Central Asia. Area: 448,100 km$^2$. Population (1998) 4.7 million.

**Tuvalu** (11D). Island country in the South Pacific. Area: 2 km$^2$. Population (1998) under 0.1 million.

**Ufa** (8C). City in the Russian Federation, at 55° N, 36° E.

**Uganda** (3C). Country in Eastern Africa. Area: 235,880 km$^2$. Population (1998) 21 million.

**Ujung Pandang** (6D). City in Indonesia, Southeastern Asia, at 5° S, 119° E.

**Ukraine** (8D). Country in Southwestern CIS. Area: 603,700 km$^2$. Population (1998) 50.3 million.

**Ulan Bator** (6C). Capital city of Mongolia, Eastern Asia, at 48° N, 107° E.

**Ulsan** (6B). City in South Korea, Eastern Asia, at 36° N, 129° E. (OOL SAHN)

**Union of Soviet Socialist Republics** (USSR). (8A). Also called the Soviet Union. Country that split into several independent republics in 1991.

**United Arab Emirates** (UAE) (4C). Country in Arab Southwest Asia, comprising several former independent emirates. Area: 83,600 km$^2$. Population (1998) 2.7 million.

**United Kingdom** (UK) (7C). Country in Western Europe, comprising England, Northern Ireland, Scotland, and Wales. Area: 244,880 km$^2$. Population (1998) 59.1 million.

**United States of America** (USA) (9C). Country in Anglo America. Area: 9,372,610 km$^2$. Population (1998) 270.2 million.

**Ural Mountains** (8B). Upland area that forms the boundary between Europe and Asia in the Russian Federation.

**Uruguay** (10E). Country in Southern South America. Area: 177,410 km$^2$. Population (1998) 3.2 million.

**Uzbekistan** (8D). Country in Central Asia. Area: 497,400 km$^2$. Population (1998) 24.1 million.

**Valencia** (7D). City on the southeastern coast of Spain, Mediterranean Europe, at 39° N, 0°.

**Valencia** (10D). City of Venezuela, Northern Andes, Latin America, at 10° N, 68° W.

**Vancouver** (9D). Port city in British Columbia province, western Canada, Anglo America, at 49° N, 123° W.

**Vanuatu** (11D). Island country in the South Pacific. Area: 12,190 km$^2$. Population (1998) 0.2 million.

**Venezuela** (10D). Country in Northern Andes, South America. Area: 912,050 km$^2$. Population (1998) 23.3 million.

**Victoria** (11C). State in southeastern Australia, South Pacific.

**Victoria Falls** (3A). Waterfalls on the Zambezi River on the border of Zambia and Zimbabwe, Southern Africa.

**Vienna** (7D). Capital city of Austria, Alpine Europe, at 48° N, 16° E.

**Vietnam** (6D). Country in Southeastern Asia. Area: 329,560 km$^2$. Population (1998) 78.5 million.

**Vistula River** (7B). River that flows across Poland, Eastern Europe.

**Volga River** (8B). River that flows southward through Russia to the Caspian Sea.

**Volgograd** (8C). Industrial city on the lower Volga River, Russian Federation, at 49° N, 44° E.

**Wallis and Futuna** (11D). Island country in the South Pacific, colony of France. Area: 265 km$^2$. Population (1998) under 0.1 million.

**Warsaw** (7E). Capital city of Poland, Eastern Europe, at 52° N, 21° E.

**Washington, D.C.** (9C). Capital city of the United States, Anglo America, at 39° N, 77° W.

**Wellington** (11C). Capital city of New Zealand, South Pacific, at 41° S, 175° W.

**West Bank** (4C). Area between Jerusalem and the Jordan River that has been occupied by Israel since 1967.

**Western Africa** (3B). Subregion of Africa South of the Sahara, comprising a large number of small countries and Nigeria, the most populous African country.

**Western Australia** (11C). State of Australia, South Pacific.

**Western Europe** (7A). Subregion of Europe, comprising Belgium, France, Germany, Luxembourg, the Netherlands, the Republic of Ireland, and the United Kingdom.

**Western Ghats** (5A). Upland area along the western coast of peninsular India.

**Western Mountains** (9C). Region of the United States, Anglo America, comprising the Rocky Mountains, Colorado and Columbia Plateaus, and the coastal mountain ranges.

**Western Samoa** (11D). Island country of the South Pacific Area: 2,800 km$^2$. Population (1998) 0.2 million.

**West Indies** (10D). Subregion of Latin America comprising the Caribbean islands and the Guianas.

**West Siberian Plains** (8B). Large lowland area east of the Ural Mountains.

**Wuhan** (6C). Inland city of China, Eastern Asia, at 31° N, 114° E.

**Yangon** (6D). Port city in southern Myanmar, also called Rangoon, at 17° N, 96° E.

**Yaoundé** (3B). City in Cameroon, Central Africa, at 4° N, 12° E. (yah oon DAY)

**Yekaterinburg** (8C). Industrial city of the Russian Federation, just east of the Urals, at 57° N, 61° E.

**Yemen** (4C). Country in Arab Southwest Asia. Area: 527,970 km$^2$. Population (1998) 15.8 million.

**Yenisey River** (8B). Major river in Siberia, Russian Federation.

**Yerevan** (8D). Capital city of Armenia, Southwestern CIS, at 40° N, 45° E.

**Yokohama** (6B). Major port city of Japan, Eastern Asia, at 36° N, 140° E.

**Yugoslavia** (7E). See **Serbia-Montenegro**.

**Yukon Territory** (9D). Part of northern Canada, Anglo America.

**Zambia** (3C). Country in Southern Africa. Area: 174,000 km$^2$. Population (1998) 9.5 million.

**Zhu Jiang** (6A). Major river in southern China, Eastern Asia.

**Zimbabwe** (3C). Country in Southern Africa. Area: 390,580 km$^2$. Population (1998) 11 million.

**Zürich** (7D). Major commercial and industrial city in Switzerland, Alpine Europe, at 47° N, 9° E. (ZOO rik).

# JOBS FOR GEOGRAPHERS

Geography incorporates specialist studies such as mapmaking (cartography) and geographic information systems, economic geography, climatology, regional geography, and urban geography. Geographers are trained to use published data and maps, to observe situations and collect their own data, and to analyze data often using computers. This breadth of interests and skills, particularly the world orientation and ability to use maps and computing, opens the doors to many careers. For some of those careers, geography is part of a required specialist training; for others, geography provides a general education that is valued by employers, as shown on the diagram. You may wish to discuss the possibilities further with your geography professors.

Geographers are in demand in a range of specialized jobs. National, state, and local government agencies employ geographers, particularly in mapping and planning. In the United States, the Defense Mapping Agency, Geological Survey, Bureau of the Census, Soil Conservation Survey, Environmental Protection Agency, National Oceanic and Atmospheric Administration, and National Aeronautics and Space Administration all employ geographers in their mapping divisions. The U.S. Department of State has an Office of Geographer involved in international boundary studies. State governments and local city and county governments employ geographers in planning, where the ability to study present and past land-use patterns, transportation flows, and the distribution of economic activity in the jurisdiction provides a basis for decisions by the politicians. In recent years, the State Department, Foreign Service, and Central Intelligence Agency have increased their demand for geographers with expertise in the understanding of other countries.

Private companies such as banks, real estate firms, utilities, retail corporations, airlines, travel agencies, hotel and restaurant chains, and manufacturers employ geographers in locational and market area analysis or to evaluate the potential of location for commercial or residential developments. Geographers are employed by transportation firms to analyze long-distance trucking operations.

Schools and colleges remain significant employers of geographers as teachers, although geographers are not restricted to such careers, as once was thought widely to be the case. People with geography qualifications teach in elementary or secondary schools, or in colleges. At colleges, the teaching is often combined with research and writing.

A geographic education may not necessarily lead to a specific job where geographic knowledge and skills are required. It provides an education in the general understanding of the world and the implications of events that is important in a wide range of jobs. Many geography graduates take further courses and become accountants, lawyers, and business managers.

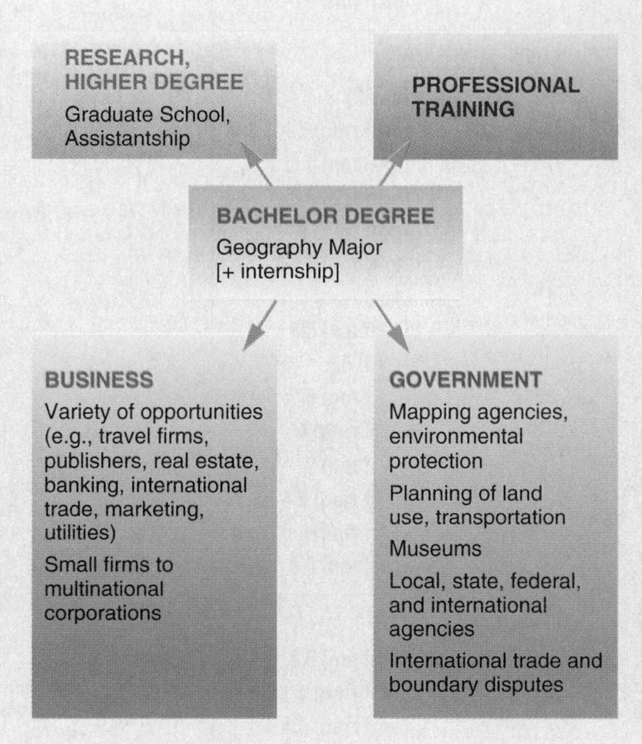


Major types of job opportunities available to people with a geographic education.

# MAP PROJECTIONS

A map should show the spatial relationships of features at Earth's surface as accurately as possible in terms of distance, area, and direction. These requirements can be met when the cartographer is dealing with a relatively small area, such as a county or a small state in the United States. Over this area, Earth's curvature is minimal, and the surfaces can be mapped onto a piece of flat paper with little distortion. However, at a global scale, the problems are acute; the skin of a sphere will simply not lie flat without being distorted in some way. No flat map can portray shape, distance, area, and direction over the spherical globe accurately. The different portraits of Earth shown in the diagrams below are different map projections, each of which attempts to minimize overall distortion or to maximize the accuracy of one measure of space. There are no right or wrong map projections, because they all contain distortion of one sort or another. In virtually all projections, linear scale is particularly not constant across the map, meaning that measurement of long distances will be inaccurate.

The simplest way to imagine constructing a map projection is to think of a transparent globe with a lightbulb inside it and a sheet of paper touching the globe in one of the three ways shown above. The outline of the continents and the lines of latitude and longitude are silhouetted on the paper to form the map. In fact, most projections are not constructed as simply as this. They are calculated mathematically and may be a compromise among a number of methods and designed to preserve the best features of each. Three main properties are desirable in a map, and different projections attempt to preserve each one.

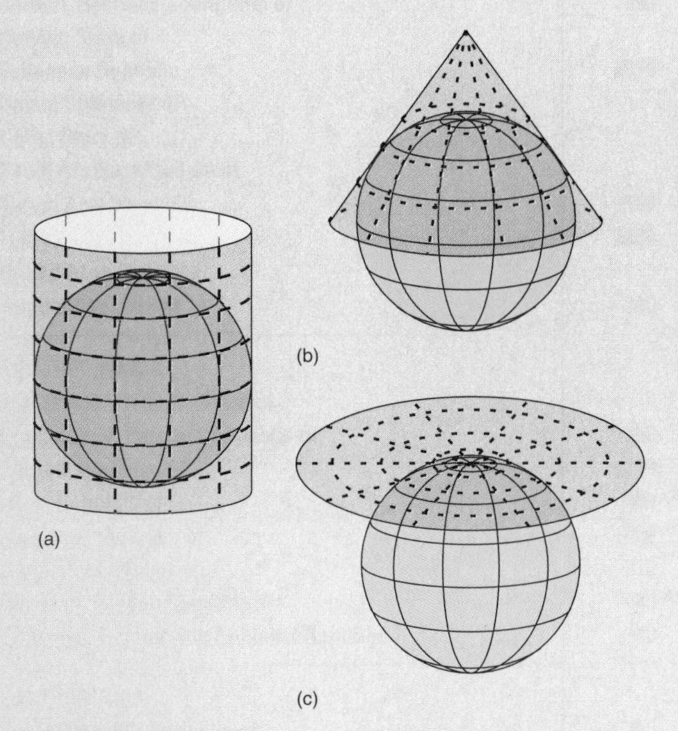


Methods of constructing map projections: (a) cylindrical, (b) conical, and (c) planar. The cylindrical and conical projections are unwrapped from the globe to give a flat map.

*Equal-area projections,* as the name suggests, maintain the areas of the continents in their correct proportions across the globe. They are used to depict distributions such as population, zones of climate, soils, or vegetation. The Robinson projection used for world maps in this book is an equal-area projection. The scale distortion in *conformal maps* is equal in the two main directions from the projection's origin but increases away from the origin. This means that conformal projections maintain the correct shape over small areas, but the outline of continents and oceans mapped over a larger area is distorted. *Azimuthal maps* are constructed around a point or *focus*. Lines of constant bearing or compass direction radiating from the focus are straight on an azimuthal map. Distortion of shape and area is symmetrical around the central point and increases away from it. The azimuthal projection is used most often in geography to portray the polar regions, which are distorted in projections designed for lower latitudes. Azimuthal maps can also be equal-area or conformal.

It is mathematically impossible to combine the properties of equal area and reasonably correct shape on one map. The commonly used Mercator projection is a conformal map and demonstrates graphically how poor this projection is for showing geographical distributions. The Mercator projection was developed in 1569 by a Flemish cartographer, Gerardus Mercator, to help explorers and navigators. It has the important property, for navigation, that a straight line drawn anywhere on the map, in any direction, gives the true compass bearing between these two points. However, the size of land masses in the mid- and high latitudes (toward the poles) is grossly distorted in this projection. Alaska, for example, appears to be the same size as Brazil, although Brazil is actually five times as large. Greenland appears similar in size to the whole of South America. This distortion is partly because the meridians, which actually come together toward the poles, are shown with uniform spacing throughout the Mercator map. Despite its unsuitability for most graphical purposes, the Mercator is still very widely used.

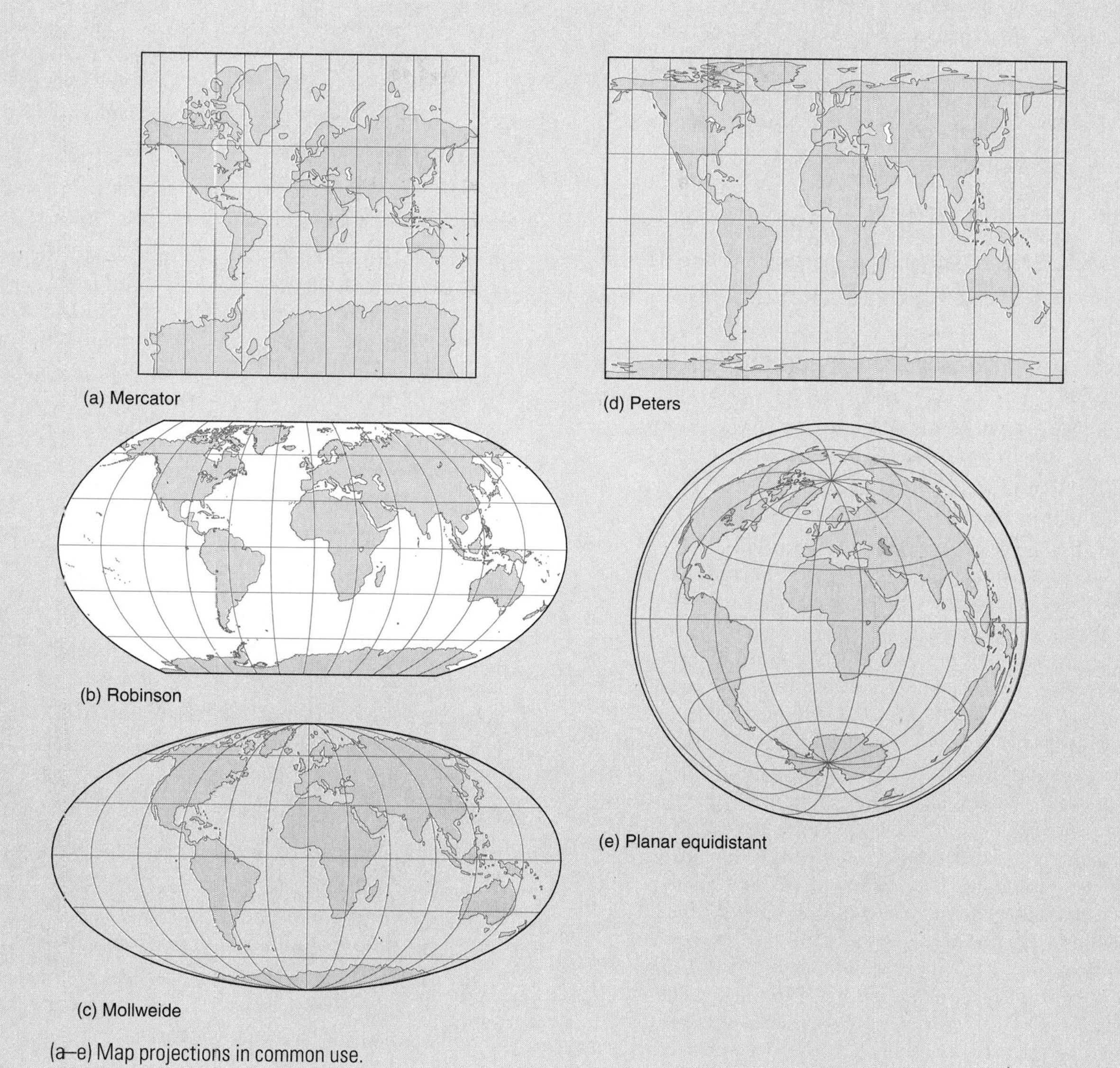


(a–e) Map projections in common use.

# STUDY RESOURCES IN WORLD REGIONAL GEOGRAPHY

STUDY RESOURCES

This section is not a bibliography but a listing of selected study aids in varied media that students might use to obtain more up-to-date information, gain deeper understandings, or explore other aspects related to the subject matter of this book. Websites are listed in Chapter 2 and at the end of Chapters 3 through 11.

**NEWSPAPERS AND PERIODICALS**

These provide current coverage of world events. One of the following should be read regularly: *The Economist* (weekly), *Wall Street Journal, Financial Times, Christian Science Monitor.*

**COMPUTER ATLASES, ENCYCLOPEDIAS, DATABASES, AND VIDEOS**

These provide current and longer-term information. Possible sources include:

*Encarta Virtual Globe* (updated annually), Microsoft Corporation. A CD-ROM encyclopedia that includes descriptions of world countries, cities, and other geographic features, with data, maps, and photos.

*Eyewitness World Atlas,* Dorling-Kindersley Multimedia. Provides additional features compared to most CD-ROM atlas or encyclopedia products. One special feature is an online facility that gives access to websites in each world country under the headings of Profile, People, Tourism, Climate, Media, and Environment. These websites often include *CIA Factbook* pages, local English-language newspapers, locally generated accounts of the country, official country government or embassy sites, travel information, and music. The main features of this CD-ROM include world maps (a three-dimensional globe and a two-dimensional world map) with physical, political, and nighttime views. Both maps are sources for finding a country and information about it. The globe rotates at controllable speeds and can be zoomed. The flat world map can show a range of information selected from a list. All can be printed out. A country index leads to a variety of information on a particular country, including a map, a flag, and 17 categories of data. There are also an index and a glossary of abbreviations, geographic terms (with international name equivalents), and international agreements. A help button tells how to work the programs, which are reasonably intuitive. Nine "amazing journeys" offer satellite-type flybys of different parts of the world with spoken commentary.

*New Millennium World Atlas Deluxe,* Rand McNally. This CD-ROM has excellent maps with facilities for easy movement around the world and zooming. Maps are attached to encyclopedia information on historic, political, and environmental topics that relate to the current view. A Guidebook heading provides country profiles, and a Comparisons heading provides facts, maps, thematic maps, and graphs on a range of topics. There is an online facility for accessing Rand McNally's website. A notebook facility makes it possible to make note of items during a search, and a wide range of customizations is possible for map types.

*World Development Indicators* (annual), World Bank. This is a comprehensive survey of world development statistics covering 1970 to two years before the published date (the 1998 version had data for 1996) and 500 time series for over 200 countries. Topics include population, economy, environment, states, markets, and global links. Data are easy to extract and transfer to other software for making graphs, maps, and so on. This resource is available as an annual CD-ROM or book.

**PRINTED DATA SOURCES**

Some of the data sources used in writing this text include the following, which are attached to written reports:

Buckley, R., ed., *Understanding Global Issues,* Cheltenham, England. Monthly analysis of a world issue, country-based or global in scope. Available in the United States through American Forum for Global Education, New York.

Dorling-Kindersley: London. *World Reference Atlas* (1994).

Instituto del Tercer Mundo, Montevideo, Uruguay (1995). *The World: A Third World Guide 1995/96.* English edition distributed by Oxfam U.K. and Ireland. World geographic information from a Third World point of view.

Population Reference Bureau, Washington, D.C. Monthly. *Population Today,* annual data sheets for the world and the USA, occasional *Population Bulletins.*

United Nations, New York. Annual. *Human Development Report, World Economic and Social Survey, State of the World Population, Demographic Yearbook.*

World Bank, Washington, D.C. Annual. *World Development Report, World Bank Atlas, Global Economic Prospects and the Developing Countries.*

**BOOKS**

The following books provide access to studies in greater depth.

*Cambridge Encyclopedias,* Cambridge University Press, Cambridge, England:

Bowring, R., and P. Kornicki. (1993). *Japan.*

Brown, A., ed. (1994). *Russia and the Former Soviet Union.*

Collier, S., T.E. Skidmore, and H. Blakemore. (1992). *Latin America and the Caribbean.*

Hook, B. (1991). *China.*

Mostyn, T., ed. (1988). *The Middle East and North Africa.*

Robinson, F., ed. (1989). *India, Pakistan, Bangladesh, Sri Lanka, etc.*

Ahuja, V., et al. *Everyone's Miracle? Revisiting Poverty and Inequality in East Asia.* Washington, D.C.: World Bank, 1997.

Aryeetey-Attoh, S. *Geography of Sub-Saharan Africa.* Upper Saddle River, NJ: Prentice-Hall, 1997.

Asian Development Bank. *Asian Development Outlook, 1997 and 1998.* New York: Oxford University Press, 1997.

Asian Development Bank. *Emerging Asia: Changes and Challenges.* Manila: Asian Development Bank, 1997.

Auty, R.M. *Patterns of Development: Resources, Policy and Economic Growth.* London: Edward Arnold, 1995.

Axford, B. *The Global System: Economics, Politics and Culture.* Cambridge, England: Polity Press, 1995.

Binns, T. *Tropical Africa.* New York: Routledge, 1994.

Blouet, B.W., and O.M. Blouet. *Latin America and the Caribbean.* New York: John Wiley, 1993.

Bradshaw, M. *The Appalachian Regional Commission: Twenty-Five Years of Public Policy.* Lexington: University Press of Kentucky, 1992.

Bradshaw, M., and R. Weaver. *Foundations of Physical Geography.* Dubuque, Iowa: Wm. C. Brown, 1995.

Castles, S., and M.J. Miller. *The Age of Migration: International Population Movements in the Modern World.* Basingstoke, England: Macmillan, 1993.

Chapman, G.P., and K.M. Baker, eds. *The Changing Geography of Africa and the Middle East.* New York: Routledge, 1992.

Chapman, G.P., and K.M. Baker, eds. *The Changing Geography of Asia.* New York: Routledge, 1992.

Chinn, J., and R. Kaiser. *Russians as the New Minority.* Boulder, Colo.: Westview (1996).

Cleaver, K.M., and G.A. Schreiber. *Reversing the Spiral: Population, Agriculture, and Environment Nexus in Sub-Saharan Africa.* Washington, D.C.: World Bank, 1994.

Cohen, J.E. *How Many People Can the Earth Support?* New York: W.W. Norton, 1995.

Cole, J., and F. Cole. *A Geography of the European Union.* New York: Routledge, 1996.

Devas, N., and C. Rakodi, eds. *Managing Fast Growing Cities: New Approaches to Management in the Developing World.* New York: Longman/John Wiley & Sons, 1993.

Dawson, A.H. *A Geography of European Integration.* New York: Halstead Press, 1993.

De Almeida, A.L.O., and J.S. Campari. *Sustainable Settlement in the Brazilian Amazon.* Washington, D.C.: World Bank/Oxford University Press, 1995.

Dicken, P. *Global Shift: The Internalizations of Economic Activity.* London: Paul Chapman Publishing, 1992.

Dickenson, J., et al. *A Geography of the Third World.* New York: Routledge, 1996.

Dirlik, A. *What Is in a Rim? Critical Perspectives on the Pacific Region Idea.* Boulder, Colo.: Westview Press, 1993.

Dmitrieva, O. *Regional Development: The USSR and After.* London: UCL Press, 1996.

Edwards, S. *Crisis and Reform in Latin America: From Despair to Hope.* Washington, D.C.: World Bank/Oxford University Press, 1995.

Frank, A.G., and B.K. Gills, eds. *The World System: Five Hundred Years or Five Thousand?* New York: Routledge, 1993.

French, R.A. *Plans, Pragmatism and People: The Legacy of Soviet Planning for Today's Cities.* London: UCL Press, 1995.

Garreau, J. *Edge City: Life on the New Frontier.* New York: Anchor Books, 1991.

Getis, A., and J. Getis, eds. *The United States and Canada.* Dubuque, Iowa: Wm. C. Brown, 1995.

Griffiths, I.L.L. *The African Inheritance.* New York: Routledge, 1995.

Goudie, A., and H. Viles. *The Earth Transformed: An Introduction to Human Impacts on the Environment.* Malden, Mass.: Blackwell, 1997.

Grove, A.T. *The Changing Geography of Africa.* Oxford: Oxford University Press, 1993.

Heathcote, R.L. *Australia.* Harlow, Essex, England: Longman, 1994.

Hodder, R. *The West Pacific Rim.* New York: John Wiley, 1992.

Hoggart, K., H. Buller, and R. Black. *Rural Europe: Identity and Change.* London: Edward Arnold, 1995.

Huntington, S.P. *The Clash of Civilizations and the Remaking of World Order.* New York: Simon and Schuster, 1996.

Johnson, S.P. *World Population—Turning the Tide: Three Decades of Progress.* London: Graham and Trotman, 1994.

Johnston, R.J., P.J. Taylor, and M.J. Watts, eds. *Geographies of Global Change.* Cambridge, Mass.: Blackwell, 1995.

Kofman, E., and G. Youngs, eds. *Globalization, Theory and Practice.* New York: Pinter, 1996.

Klugman, J., ed. *Poverty in Russia.* Washington, D.C.: World Bank, 1997.

Knox, P., and J. Agnew. *The Geography of the World Economy.* London: Edward Arnold, 1994.

Knox, P., and P.J. Taylor, eds. *World Cities in a World System.* Cambridge, England: Cambridge University Press, 1995.

Leeming, F. *The Changing Geography of China.* Cambridge, Mass.: Blackwell, 1993.

McLoud, D.G. *Southeast Asia: Tradition and Modernity in the Contemporary World.* Boulder, Colo.: Westview, 1995.

Myers, N. *Ultimate Security: The Environmental Basis of Political Stability.* New York: W.W. Norton, 1993.

Parker, G., ed. *The Times Atlas of World History.* London: Times Books, 1994.

Paterson, J. *North America.* New York: Oxford University Press, 1994.

Pinder, D., ed. *Western Europe: Challenges and Change.* London: Belhaven Press, 1990.

Preston, D., ed. *Latin American Development.* New York: John Wiley, 1987.

Quibria, M.G., and J.M. Dowling, eds. *Current Issues in Economic Development: An Asian Perspective.* Hong Kong: Oxford University Press, 1996.

Richardson, B.C. *The Caribbean in the Wider World, 1492–1992.* Cambridge, England: Cambridge University Press, 1992.

Shelley, F.M., and A.E. Clarke. *Human and Cultural Geography.* Dubuque, Iowa: Wm. C. Brown, 1994.

Simon, D. *Third World Development: A Reappraisal.* London: Paul Chapman Publishing, 1990.

So, A.Y., and S.W.K. Chiu. *East Asia and the World Economy.* Thousand Oaks, Calif.: Sage Publications, 1995.

Winchester, H. *Contemporary France.* Harlow, England: Longman, 1993.

Witherick, M., and M. Carr. *The Changing Face of Japan.* London: Hodder & Stoughton, 1993.

World Bank. *Adjustment in Africa.* New York: Oxford University Press, 1994.

World Bank. *China 2020: Development Challenges in the New Century.* Washington, D.C.: World Bank, 1997. Main publication in a series covering environment, economy, integration with world economy, and population issues.

World Bank. *The East Asian Miracle.* New York: Oxford University Press, 1993.

World Bank. *Taking Action to Reduce Poverty in Sub-Saharan Africa.* Washington, D.C.: World Bank, 1997.

# DATA BANK

| Country | Capital City (1) | Land Area (1) | | Population (millions) (1) | | | Birth Rate | | Death Rate | |
|---|---|---|---|---|---|---|---|---|---|---|
| | | | | | Projected | | | | | |
| | | (km²) | (mi²) | mid-1998 | 2010 | 2025 | 1970 (2) | 1998 (1) | 1970 (2) | 1998 (1) |
| CENTRAL AFRICA | | | | | | | | | | |
| Burundi, Republic of | Bujumbura | 28,000 | 10,745 | 5.5 | 7.5 | 10.5 | 46 | 43 | 24 | 18 |
| Cameroon, Republic of | Yaoundé | 475,440 | 186,568 | 14.3 | 19.8 | 28.5 | 43 | 41 | 18 | 13 |
| Central African Republic | Bangui | 622,980 | 240,534 | 3.4 | 4.3 | 5.5 | 37 | 38 | 22 | 17 |
| Chad, Republic of | N'Djamena | 1,284,000 | 495,755 | 7.4 | 10.1 | 14.4 | 45 | 50 | 22 | 17 |
| Congo, Republic of the | Brazzaville | 342,000 | 132,047 | 2.7 | 3.4 | 4.2 | 43 | 39 | 16 | 17 |
| Congo, Democratic Republic of the (Zaire) | Kinshasa | 2,344,860 | 905,356 | 49.0 | 70.3 | 105.7 | | 48 | | 16 |
| Equatorial Guinea, Republic of | Malabo | 28,050 | 10,830 | 0.4 | 0.6 | 0.8 | | 44 | | 15 |
| Gabonese Republic | Libreville | 267,670 | 103,348 | 1.2 | 1.6 | 2.1 | 31 | 35 | 21 | 15 |
| Rwanda, Republic of | Kigali | 26,340 | 10,170 | 8.0 | 9.9 | 12.2 | 52 | 39 | 18 | 18 |
| WESTERN AFRICA | | | | | | | | | | |
| Benin, Republic of | Porto-Novo | 112,620 | 43,483 | 6.0 | 8.4 | 12.4 | 50 | 45 | 22 | 14 |
| Burkina Faso, People's Democratic Republic of | Ouagadougou | 274,000 | 105,792 | 11.3 | 15.4 | 21.4 | 48 | 47 | 25 | 18 |
| Côte D'Ivoire, Republic of | Abidjan | 322,460 | 124,503 | 15.6 | 20.3 | 26.1 | 51 | 39 | 20 | 13 |
| Gambia, Republic of the | Banjul | 11,300 | 4,363 | 1.2 | 1.5 | 2.0 | | 43 | | 19 |
| Ghana, Republic of | Accra | 238,540 | 92,101 | 18.9 | 26.0 | 36.3 | 46 | 40 | 16 | 12 |
| Guinea, Republic of | Conakry | 245,860 | 94,927 | 7.5 | 9.4 | 13.1 | 46 | 43 | 28 | 19 |
| Guinea-Bissau, Republic of | Bisau | 36,120 | 13,940 | 1.1 | 1.4 | 1.9 | 41 | 42 | 27 | 21 |
| Liberia, Republic of | Monrovia | 97,750 | 37,741 | 2.8 | 4.3 | 6.5 | | 43 | | 12 |
| Mali, Republic of | Bamako | 1,240,190 | 478,840 | 10.1 | 14.6 | 22.6 | 51 | 51 | 26 | 20 |
| Mauritania, Islamic Republic of | Nouackchott | 1,025,520 | 395,955 | 2.5 | 3.3 | 4.4 | 47 | 40 | 39 | 14 |
| Niger, Republic of | Niamey | 1,267,000 | 489,191 | 10.1 | 14.8 | 22.4 | 50 | 53 | 28 | 19 |
| Nigeria, Federal Republic of | Lagos | 923,770 | 356,670 | 121.8 | 150.3 | 203.4 | 51 | 45 | 21 | 15 |
| Senegal, Republic of | Dakar | 196,720 | 75,954 | 9.0 | 12.3 | 17.0 | 47 | 43 | 22 | 16 |
| Sierra Leone, Republic of | Freetown | 71,740 | 27,699 | 4.6 | 6.1 | 8.2 | 49 | 49 | 30 | 30 |
| Togo, Republic of | Lomé | 56,790 | 21,927 | 4.9 | 7.4 | 11.0 | 50 | 46 | 20 | 11 |
| EASTERN AFRICA | | | | | | | | | | |
| Djibouti, Republic of | Djibouti | 23,200 | 8,958 | 0.7 | 0.9 | 1.1 | | 39 | | 16 |
| Eritrea, Republic of | Asmara | 121,140 | 46,772 | 3.8 | 5.7 | 8.4 | | 43 | | 13 |
| Ethiopia, Federal Democratic Republic of | Addis Ababa | 1,100,760 | 425,000 | 58.4 | 74.8 | 98.8 | 43 | 46 | 20 | 21 |
| Kenya, Republic of | Nairobi | 580,370 | 224,082 | 28.3 | 32.4 | 34.8 | 53 | 33 | 18 | 13 |
| Somali Democratic Republic | Mogadishu | 637,600 | 246,202 | 10.7 | 15.7 | 23.7 | 50 | 50 | 24 | 19 |
| Tanzania, United Republic of | Dar-es-Salaam | 945,090 | 364,901 | 30.6 | 39.4 | 50.7 | 49 | 42 | 22 | 17 |
| Uganda, Republic of | Kampala | 235,880 | 91,074 | 21.0 | 26.4 | 33.5 | 50 | 48 | 17 | 21 |
| SOUTHERN AFRICA | | | | | | | | | | |
| Angola, Republic of | Luanda | 1,246,700 | 481,354 | 12.0 | 17.2 | 25.5 | | 51 | | 19 |
| Botswana, Republic of | Gabarone | 581,730 | 224,607 | 1.4 | 1.6 | 1.6 | 53 | 33 | 17 | 21 |
| Lesotho, Kingdom of | Maseru | 30,350 | 11,718 | 2.1 | 2.4 | 2.7 | 43 | 33 | 20 | 12 |

Sources in column titles: (1) Population Reference Bureau *1998 World Population Data Sheet;* (2) *United Nations Human Development Report 1997;* (3) World Bank *World Development Indicators 1998;* (4) Microsoft *Encarta Virtual Globe* 1998

| % Natural Increase 1998 (1) | Infant Mortality (1) | | Total Fertility Rate 1998 (1) | Life Expectancy at Birth (1) | | % Urban Population (1) | | HDI 1994 Value (rank of 175) (2) | GDI 1994 Value (rank of 146) (2) | HPI (%) 1994 (2) |
|---|---|---|---|---|---|---|---|---|---|---|
| | 1960 | 1998 | | Men | Women | 1980 | 1998 | | | |
| 2.5 | 153 | 105 | 6.6 | 44 | 47 | 4 | 5 | 0.247 (169) | 0.233 (141) | 49.0 |
| 2.8 | 163 | 65 | 5.9 | 53 | 56 | 31 | 44 | 0.468 (133) | 0.444 (115) | 31.4 |
| 2.1 | 163 | 97 | 5.1 | 44 | 48 | 35 | 39 | 0.355 (151) | 0.338 (129) | 41.7 |
| 3.3 | 195 | 110 | 6.6 | 45 | 50 | 19 | 22 | 0.288 (164) | 0.270 (137) | |
| 2.3 | 140 | 107 | 5.1 | 45 | 49 | 41 | 58 | 0.500 (130) | | 29.1 |
| 3.2 | 153 | 106 | 6.6 | 47 | 51 | | 29 | 0.381 (142) | | 41.2 |
| 2.6 | 188 | 117 | 5.9 | 46 | 50 | | 37 | 0.462 (135) | 0.441 (116) | |
| 2.0 | 171 | 94 | 5.0 | 52 | 55 | 36 | 73 | 0.562 (120) | 0.546 (102) | |
| 2.1 | 150 | 114 | 6.0 | 43 | 44 | 5 | 5 | 0.187 (174) | | |
| 3.2 | 179 | 94 | 6.3 | 51 | 56 | 32 | 36 | 0.368 (146) | 0.349 (124) | |
| 2.9 | 186 | 94 | 6.9 | 46 | 47 | 9 | 15 | 0.221 (172) | 0.206 (144) | 58.3 |
| 2.6 | 165 | 89 | 5.7 | 51 | 54 | 35 | 46 | 0.386 (145) | 0.341 (126) | 46.3 |
| 2.4 | 213 | 90 | 5.9 | 43 | 47 | 18 | 37 | 0.281 (165) | 0.283 (138) | |
| 2.9 | 132 | 66 | 5.5 | 54 | 58 | 31 | 35 | 0.468 (132) | 0.459 (111) | 32.6 |
| 2.4 | 203 | 153 | 5.7 | 43 | 47 | 19 | 29 | 0.271 (167) | 0.250 (140) | 50.0 |
| 2.1 | 200 | 141 | 5.8 | 41 | 44 | 17 | 22 | 0.291 (163) | 0.276 (136) | 43.6 |
| 3.1 | | 108 | 3.1 | 56 | 61 | | 45 | | | |
| 3.1 | 209 | 123 | 6.7 | 45 | 47 | 19 | 26 | 0.229 (171) | 0.218 (143) | 54.7 |
| 2.5 | 188 | 101 | 5.4 | 50 | 53 | 29 | 54 | 0.355 (150) | 0.341 (127) | 47.1 |
| 3.4 | 191 | 123 | 7.4 | 45 | 48 | 13 | 15 | 0.206 (173) | 0.193 (145) | 66.0 |
| 3.0 | 189 | 84 | 6.5 | 49 | 52 | 27 | 39 | 0.393 (141) | 0.372 (121) | 41.6 |
| 2.7 | 172 | 68 | 5.7 | 48 | 50 | 36 | 42 | 0.326 (160) | 0.309 (134) | 48.7 |
| 1.9 | 219 | 195 | 6.5 | 33 | 36 | 25 | 36 | 0.176 (175) | 0.155 (146) | 59.2 |
| 3.6 | 182 | 84 | 6.8 | 56 | 60 | 23 | 31 | 0.361 (147) | 0.342 (125) | 39.3 |
| 2.3 | 186 | 115 | 5.8 | 47 | 50 | | 81 | 0.319 (162) | | |
| 3.0 | 166 | 82 | 6.1 | 52 | 57 | | 16 | 0.269 (168) | | |
| 2.5 | 187 | 128 | 7.0 | 41 | 42 | 11 | 16 | 0.244 (170) | 0.233 (142) | 56.2 |
| 2.0 | 124 | 62 | 4.5 | 48 | 49 | 16 | 27 | 0.463 (134) | 0.458 (112) | 26.1 |
| 3.2 | | 122 | 7.0 | 45 | 49 | | 24 | | | |
| 2.5 | 147 | 100 | 5.7 | 45 | 49 | 15 | 21 | 0.357 (149) | 0.352 (123) | 39.7 |
| 2.7 | 133 | 81 | 6.9 | 40 | 41 | 9 | 14 | 0.328 (159) | 0.318 (132) | 41.3 |
| 3.2 | 208 | 124 | 7.2 | 45 | 48 | 21 | 42 | 0.335 (157) | | |
| 1.2 | 116 | 60 | 4.3 | 40 | 42 | 15 | 48 | 0.670 (97) | 0.652 (79) | 22.9 |
| 2.1 | 149 | 80 | 4.3 | 54 | 58 | 13 | 16 | 0.457 (137) | 0.446 (113) | 27.5 |

| Country | Capital City (1) | Land Area (1) | | Population (millions) (1) | | | Birth Rate | | Death Rate | |
|---|---|---|---|---|---|---|---|---|---|---|
| | | | | | Projected | | | | | |
| | | (km²) | (mi²) | mid-1998 | 2010 | 2025 | 1970 (2) | 1998 (1) | 1970 (2) | 1998 (1) |
| SOUTHERN AFRICA *(continued)* | | | | | | | | | | |
| Madagascar, Republic of | Antananarivo | 587,040 | 226,657 | 14.0 | 19.4 | 28.4 | 46 | 44 | 20 | 14 |
| Malawi, Republic of | Lilongwe | 118,480 | 45,745 | 9.8 | 10.7 | 10.9 | 56 | 42 | 24 | 24 |
| Mozambique, Republic of | Maputo | 801,590 | 309,496 | 18.6 | 24.8 | 33.3 | 48 | 41 | 24 | 19 |
| Namibia, Republic of | Windhoek | 824,290 | 318,260 | 1.6 | 1.9 | 2.3 | 44 | 36 | 18 | 20 |
| South Africa, Republic of | Pretoria | 1,221,040 | 471,446 | 38.9 | 43.1 | 45.3 | 37 | 27 | 14 | 11 |
| Swaziland, Kingdom of | Mbabene | 17,360 | 6,703 | 1.0 | 1.2 | 1.6 | | 43 | | 10 |
| Zambia, Republic of | Lusaka | 752,610 | 290,584 | 9.5 | 12.2 | 16.2 | 49 | 42 | 19 | 23 |
| Zimbabwe, Republic of | Harare | 390,760 | 150,873 | 11.0 | 12.0 | 12.4 | 53 | 35 | 16 | 20 |
| NORTH AFRICA | | | | | | | | | | |
| Algeria, Democratic and Popular Republic of | Algiers | 2,381,740 | 919,595 | 30.2 | 38.6 | 47.3 | 49 | 31 | 16 | 7 |
| Socialist People's Libyan Arab Jamahirya | Tripoli | 1,759,540 | 679,362 | 5.7 | 8.7 | 14.2 | | 45 | | 8 |
| Morocco, Kingdom of | Rabat | 446,550 | 172,414 | 27.7 | 34.0 | 41.2 | 47 | 24 | 16 | 7 |
| Tunisia, Republic of | Tunis | 163,610 | 63,170 | 9.5 | 11.4 | 13.5 | 39 | 26 | 14 | 6 |
| NILE RIVER VALLEY | | | | | | | | | | |
| Egypt, Arab Republic of | Cairo | 1,001,450 | 386,662 | 65.5 | 80.3 | 95.8 | 40 | 28 | 17 | 6 |
| Sudan, Republic of | Khartoum | 2,505,810 | 967,499 | 28.5 | 36.9 | 46.9 | 47 | 35 | 22 | 14 |
| SOUTHWESTERN ASIA | | | | | | | | | | |
| Bahrain, State of | Manama | 680 | 263 | 0.6 | 0.8 | 0.9 | | 23 | | 3 |
| Iran | Tehran | 1,648,000 | 636,246 | 64.1 | 75.8 | 92.5 | 45 | 24 | 16 | 6 |
| Iraq, Republic of | Baghdad | 438,320 | 169,236 | 21.8 | 30.4 | 41.6 | | 38 | | 10 |
| Israel | Jerusalem | 21,060 | 8,131 | 6.0 | 7.1 | 8.1 | 26 | 21 | 7 | 7 |
| Jordan, Hashemite Kingdom of | Amman | 89,210 | 34,444 | 4.6 | 7.0 | 10.0 | | 30 | | 5 |
| Kuwait, State of | Kuwait City | 17,820 | 6,880 | 1.9 | 2.5 | 3.0 | | 25 | | 2 |
| Lebanese Republic | Beirut | 10,400 | 4,015 | 4.1 | 4.8 | 5.6 | | 23 | | 7 |
| Oman, Sultanate of | Masqat | 212,460 | 82,031 | 2.5 | 4.0 | 6.5 | 50 | 44 | 21 | 5 |
| Qatar, State of | Doha | 11,000 | 4,247 | 0.5 | 0.6 | 0.7 | | 19 | | 2 |
| Saudi Arabia, Kingdom of | Riyadh | 2,149,690 | 830,000 | 20.2 | 29.2 | 42.6 | 48 | 35 | 18 | 5 |
| Syrian Arab Republic | Damascus | 185,180 | 71,498 | 15.6 | 20.5 | 26.3 | 47 | 33 | 17 | 6 |
| Turkey | Ankara | 779,450 | 300,947 | 64.8 | 76.3 | 88.0 | 36 | 22 | 12 | 7 |
| United Arab Emirates | Abu Dhabi | 83,600 | 32,278 | 2.7 | 3.3 | 3.8 | 35 | 24 | 11 | 2 |
| Yemen, Republic of | Sanaa | 527,970 | 203,850 | 15.8 | 23.9 | 39.0 | 53 | 44 | 23 | 11 |
| SOUTHERN ASIA | | | | | | | | | | |
| Afghanistan, Islamic State of | Kabul | 652,090 | 251,773 | 24.8 | 34.1 | 48.0 | | 43 | | 18 |
| Bangladesh, People's Republic of | Dhaka | 144,000 | 85,599 | 123.4 | 147.5 | 165.6 | 48 | 27 | 21 | 8 |
| Bhutan, Kingdom of | Thimphu | 47,000 | 18,147 | 0.8 | 1.1 | 1.5 | 41 | 40 | 22 | 9 |
| India, Republic of | New Delhi | 3,287,590 | 1,269,346 | 988.7 | 1196.5 | 1441.2 | 41 | 27 | 18 | 9 |
| Maldives, Republic of | Malé | 600 | 116 | 0.3 | 0.4 | 0.6 | | 42 | | 9 |
| Nepal, Kingdom of | Kathmandu | 140,800 | 54,363 | 23.7 | 31.0 | 39.5 | 46 | 33 | 22 | 11 |
| Pakistan, Islamic Republic of | Islamabad | 796,100 | 307,376 | 141.9 | 192.6 | 258.1 | 48 | 39 | 19 | 11 |
| Sri Lanka, Democratic Socialist Republic of | Colombo | 65,610 | 25,332 | 18.9 | 21.5 | 24.1 | 29 | 19 | 8 | 6 |

| % Natural Increase 1998 (1) | Infant Mortality (1) | | Total Fertility Rate 1988 (1) | Life Expectancy at Birth (1) | | % Urban Population (1) | | HDI 1994 Value (rank of 175) (2) | GDI 1994 Value (rank of 146) (2) | HPI (%) 1994 (2) |
|---|---|---|---|---|---|---|---|---|---|---|
| | 1960 | 1998 | | Men | Women | 1980 | 1998 | | | |
| 3.0 | 178 | 96 | 6.0 | 51 | 53 | 18 | 22 | 0.350 (152) | | 49.5 |
| 1.7 | 206 | 140 | 5.9 | 36 | 36 | 9 | 20 | 0.320 (161) | 0.310 (133) | 45.8 |
| 2.2 | 190 | 134 | 5.6 | 43 | 46 | 13 | 28 | 0.281 (166) | 0.262 (139) | 50.1 |
| 1.7 | 146 | 68 | 5.1 | 42 | 42 | 23 | 27 | 0.570 (118) | | 45.1 |
| 1.6 | 89 | 52 | 3.3 | 55 | 60 | 48 | 57 | 0.716 (90) | 0.681 (71) | |
| 3.3 | 157 | 72 | 5.6 | 38 | 41 | | 22 | 0.582 (114) | 0.563 (98) | |
| 1.9 | 135 | 109 | 6.1 | 37 | 38 | 87 | 71 | 0.369 (143) | 0.362 (122) | 35.1 |
| 1.5 | 109 | 53 | 4.4 | 40 | 40 | 64 | 74 | 0.513 (129) | 0.503 (109) | 17.3 |
| | | | | | | | | | | |
| 2.4 | 168 | 44 | 4.4 | 66 | 68 | 43 | 56 | 0.737 (82) | 0.614 (92) | 28.6 |
| 3.7 | 160 | 60 | 6.3 | 62 | 67 | | 86 | (64) | 0.655 (77) | 18.8 |
| 1.8 | 163 | 62 | 3.3 | 69 | 74 | 41 | 52 | 0.566 (119) | 0.515 (105) | 41.7 |
| 1.9 | 159 | 35 | 3.2 | 67 | 69 | 51 | 61 | 0.748 (81) | 0.668 (74) | 24.4 |
| | | | | | | | | | | |
| 2.2 | 179 | 63 | 3.6 | 65 | 69 | 44 | 43 | 0.614 (109) | 0.555 (100) | 34.8 |
| 2.1 | 160 | 70 | 5.0 | 50 | 52 | | 27 | 0.333 (158) | 0.306 (135) | 42.2 |
| | | | | | | | | | | |
| 2.0 | 130 | 14 | 3.2 | 68 | 71 | | 88 | 0.870 (43) | 0.742 (56) | |
| 1.8 | 169 | 35 | 3.0 | 67 | 68 | | 61 | 0.780 (70) | | 22.6 |
| 2.8 | 139 | 127 | 5.7 | 58 | 60 | | 70 | 0.531 (126) | 0.433 (117) | 30.7 |
| 1.5 | | 7 | 2.9 | 76 | 80 | 89 | 90 | 0.913 (23) | 0.872 (22) | |
| 2.5 | 135 | 34 | 4.4 | 66 | 70 | 60 | 78 | 0.730 (84) | | 10.9 |
| 2.3 | 89 | 10 | 3.2 | 72 | 73 | 90 | 100 | 0.844 (53) | 0.769 (51) | |
| 1.6 | 68 | 34 | 2.3 | 68 | 73 | 73 | 87 | 0.794 (65) | 0.708 (66) | |
| 3.9 | 214 | 27 | 7.1 | 68 | 72 | | 72 | 0.718 (88) | | |
| 1.7 | 145 | 12 | 4.1 | 69 | 74 | | 91 | 0.840 (55) | 0.713 (64) | |
| 3.1 | 170 | 29 | 6.4 | 68 | 71 | 67 | 80 | 0.774 (73) | 0.581 (95) | |
| 2.8 | 135 | 35 | 4.6 | 67 | 68 | 47 | 51 | 0.755 (78) | 0.646 (84) | 21.7 |
| 1.6 | 190 | 42 | 2.6 | 66 | 71 | 44 | 64 | 0.772 (74) | 0.737 (58) | |
| 2.2 | 145 | 11 | 4.9 | 73 | 75 | 72 | 82 | 0.866 (44) | 0.727 (61) | 14.9 |
| 3.3 | 224 | 77 | 7.3 | 57 | 60 | 20 | 25 | 0.361 (148) | | 47.6 |
| | | | | | | | | | | |
| 2.5 | | 150 | 6.1 | 46 | 45 | | 18 | | | |
| 1.8 | 156 | 82 | 3.3 | 59 | 58 | 11 | 16 | 0.368 (144) | 0.339 (128) | 48.3 |
| 3.1 | 203 | 71 | 5.6 | | | | 15 | 0.388 (155) | | |
| 1.9 | 165 | 72 | 3.4 | 59 | 59 | 23 | 26 | 0.446 (138) | 0.419 (118) | 36.7 |
| 3.3 | 160 | 30 | 6.4 | 63 | 61 | | 25 | 0.611 (111) | 0.600 (94) | |
| 2.2 | 195 | 79 | 4.6 | 55 | 54 | 7 | 10 | 0.347 (154) | 0.321 (131) | |
| 2.8 | 163 | 91 | 5.6 | 58 | 59 | 28 | 35 | 0.445 (139) | 0.392 (120) | 46.8 |
| | | | | | | | | | | |
| 1.3 | 71 | 17 | 2.2 | 71 | 74 | 22 | 22 | 0.711 (91) | 0.694 (70) | 20.7 |

| Country | Capital City (1) | Land Area (1) | | Population (millions) (1) | | | Birth Rate | | Death Rate | |
|---|---|---|---|---|---|---|---|---|---|---|
| | | | | | Projected | | | | | |
| | | (km²) | (mi²) | mid-1998 | 2010 | 2025 | 1970 (2) | 1998 (1) | 1970 (2) | 1998 (1) |
| **EASTERN ASIA** | | | | | | | | | | |
| China, People's Republic of | Beijing | 9,596,960 | 3,705,407 | 1242.5 | 1394.3 | 1561.4 | 33 | 17 | 8 | 7 |
| Hong Kong (to China, 1997) | | 1,040 | 382 | 6.7 | 7.7 | 7.8 | 21 | 9 | 5 | 5 |
| Japan | Tokyo | 377,800 | 145,569 | 126.4 | 127.6 | 120.9 | 19 | 10 | 7 | 7 |
| Korea, People's Democratic Republic of (North) | Pyongyang | 120,540 | 46,541 | 22.2 | 24.1 | 26.1 | | 18 | | 9 |
| Korea, Republic of (South) | Seoul | 99,020 | 32,232 | 46.4 | 50.6 | 52.7 | 30 | 16 | 9 | 6 |
| Mongolia | Ulan Bator | 1,566,500 | 604,829 | 2.4 | 2.8 | 3.3 | 42 | 24 | 14 | 7 |
| Taiwan | Taipei | 35,760 | 13,807 | 21.7 | 23.8 | 25.4 | | 15 | | 6 |
| Singapore, Republic of | Singapore City | 620 | 239 | 3.9 | 3.9 | 4.2 | 23 | 15 | 5 | 5 |
| **SOUTHEASTERN ASIA** | | | | | | | | | | |
| Cambodia, Kingdom of | Phnom Penh | 181,040 | 69,000 | 10.8 | 13.4 | 17.0 | | 38 | | 14 |
| Indonesia, Republic of | Jakarta | 1,904,570 | 735,359 | 207.4 | 239.4 | 275.4 | 42 | 24 | 18 | 8 |
| Laos, People's Democratic Republic of | Vientiane | 236,800 | 91,429 | 5.3 | 7.2 | 9.8 | 44 | 42 | 23 | 14 |
| Malaysia, Federation of | Kuala Lumpur | 329,750 | 127,317 | 22.2 | 28.4 | 37.0 | 36 | 26 | 10 | 5 |
| Myanmar, Union of (Burma) | Yangon (Rangoon) | 676,580 | 261,229 | 47.1 | 56.3 | 67.8 | | 30 | | 10 |
| Philippines, Republic of | Manilla | 300,000 | 115,831 | 75.3 | 94.1 | 116.8 | 38 | 30 | 11 | 7 |
| Thailand, Kingdom of | Bangkok | 513,120 | 198,117 | 61.1 | 67.3 | 71.6 | 39 | 17 | 9 | 7 |
| Vietnam, Socialist Republic of | Hanoi | 331,690 | 128,066 | 78.5 | 91.9 | 109.5 | | 19 | | 7 |
| **WESTERN EUROPE** | | | | | | | | | | |
| Belgium, Kingdom of | Brussels | 33,100 | 12,780 | 10.2 | 10.3 | 10.3 | 15 | 12 | 12 | 10 |
| French Republic | Paris | 551,500 | 212,935 | 58.8 | 61.7 | 64.2 | 17 | 12 | 11 | 9 |
| Germany, Federal Republic of | Berlin | 356,910 | 137,304 | 82.3 | 81.7 | 76.1 | 14 | 10 | 13 | 10 |
| Ireland, Republic of | Dublin | 70,280 | 27,135 | 3.7 | 3.7 | 3.8 | 22 | 14 | 11 | 9 |
| Luxembourg, Grand Duchy of | Luxembourg City | 2,586 | 998 | 0.4 | 0.5 | 0.5 | | 14 | | 9 |
| Netherlands, Kingdom of | The Hague | 37,330 | 14,413 | 15.7 | 16.7 | 17.3 | 18 | 12 | 8 | 9 |
| United Kingdom of Great Britain and Northern Ireland | London | 244,880 | 94,549 | 59.1 | 60.8 | 62.6 | 16 | 13 | 12 | 11 |
| **NORTHERN EUROPE** | | | | | | | | | | |
| Denmark, Kingdom of | Copenhagen | 43,090 | 16,637 | 5.3 | 5.5 | 5.6 | 14 | 13 | 10 | 11 |
| Finland, Republic of | Helsinki | 338,100 | 130,553 | 5.2 | 5.2 | 5.2 | 14 | 12 | 10 | 10 |
| Iceland, Republic of | Reykjavik | 103,000 | 39,789 | 0.3 | 0.3 | 0.3 | | 15 | | 7 |
| Norway, Kingdom of | Oslo | 323,000 | 125,058 | 4.4 | 4.6 | 4.9 | 17 | 14 | 10 | 10 |
| Sweden, Kingdom of | Stockholm | 449,960 | 173,731 | 8.9 | 9.0 | 9.3 | 14 | 10 | 12 | 11 |
| **ALPINE EUROPE** | | | | | | | | | | |
| Austria, Republic of | Vienna | 83,850 | 32,375 | 8.1 | 8.3 | 8.3 | 15 | 11 | 13 | 12 |
| Swiss Confederation | Bern | 41,290 | 15,942 | 7.1 | 7.6 | 7.5 | 16 | 12 | 9 | 9 |
| **MEDITERRANEAN EUROPE** | | | | | | | | | | |
| Helenic Republic (Greece) | Athens | 131,990 | 50,982 | 10.5 | 10.6 | 10.2 | 17 | 10 | 8 | 10 |
| Italian Republic | Rome | 301,270 | 116,321 | 57.7 | 57.5 | 54.8 | 17 | 9 | 10 | 9 |
| Portuguese Republic | Lisbon | 92,390 | 35,672 | 10.0 | 9.9 | 9.4 | 20 | 11 | 10 | 11 |
| Spain, Kingdom of | Madrid | 504,780 | 194,897 | 39.4 | 39.8 | 39.0 | 20 | 9 | 8 | 9 |

| % Natural Increase 1998 (1) | Infant Mortality (1) | | Total Fertility Rate 1988 (1) | Life Expectancy at Birth (1) | | % Urban Population (1) | | HDI 1994 Value (rank of 175) (2) | GDI 1994 Value (rank of 146) (2) | HPI (%) 1994 (2) |
|---|---|---|---|---|---|---|---|---|---|---|
| | 1960 | 1998 | | Men | Women | 1980 | 1998 | | | |
| 1.0 | 150 | 31 | 1.8 | 69 | 73 | 19 | 30 | 0.626 (108) | 0.617 (90) | 17.5 |
| 0.4 | 43 | 4 | 1.1 | 76 | 82 | 92 | 95 | 0.914 (22) | 0.852 (28) | |
| 0.2 | 10 | 4 | 1.4 | 77 | 84 | 76 | 78 | 0.940 (7) | 0.901 (12) | |
| 0.9 | | 39 | 1.9 | 63 | 69 | | 59 | | | |
| 1.0 | 85 | 11 | 1.7 | 71 | 77 | 57 | 79 | 0.890 (32) | 0.826 (35) | |
| 1.6 | 128 | 49 | 3.1 | 66 | 63 | 52 | 57 | 0.661 (101) | 0.650 (80) | 15.7 |
| 1.0 | | 7 | 1.7 | 72 | 78 | | 75 | | | |
| 1.1 | 36 | 4 | 1.7 | 74 | 80 | 100 | 100 | 0.900 (26) | 0.853 (27) | 6.6 |
| 2.4 | 146 | 116 | 5.2 | 50 | 53 | 12 | 14 | 0.348 (153) | | |
| 1.5 | 139 | 66 | 2.7 | 60 | 64 | 22 | 37 | 0.668 (99) | 0.642 (86) | 20.8 |
| 2.8 | 146 | 97 | 5.9 | 52 | 55 | 12 | 19 | 0.459 (136) | 0.444 (114) | |
| 2.1 | 72 | 10 | 3.2 | 70 | 75 | 42 | 57 | 0.832 (60) | 0.782 (45) | |
| 2.0 | 158 | 83 | 3.8 | 60 | 62 | | 25 | 0.475 (131) | 0.469 (110) | 31.2 |
| 2.3 | 79 | 34 | 3.7 | 63 | 69 | 38 | 47 | 0.672 (98) | 0.650 (81) | 17.7 |
| 1.1 | 103 | 25 | 2.0 | 67 | 72 | 17 | 31 | 0.833 (59) | 0.812 (39) | 11.7 |
| 1.2 | 147 | 38 | 2.3 | 65 | 69 | 19 | 20 | 0.557 (121) | 0.552 (101) | 23.2 |
| 0.1 | | 6 | 1.6 | 74 | 81 | 95 | 97 | 0.932 (13) | 0.891 (14) | |
| 0.3 | | 5 | 1.7 | 74 | 82 | 73 | 74 | 0.946 (2) | 0.926 (6) | |
| –0.1 | | 5 | 1.3 | 73 | 80 | 83 | 85 | 0.924 (19) | 0.886 (16) | |
| 0.5 | | 6 | 1.9 | 72 | 78 | 55 | 57 | 0.929 (17) | 0.851 (29) | |
| 0.4 | | 5 | 1.8 | 73 | 79 | | 86 | 0.899 (27) | 0.813 (38) | |
| 0.3 | | 6 | 1.5 | 75 | 80 | 88 | 89 | 0.940 (6) | 0.901 (11) | |
| 0.2 | | 6 | 1.7 | 74 | 80 | 89 | 90 | 0.931 (15) | 0.896 (13) | |
| 0.2 | | 6 | 1.8 | 73 | 78 | 84 | 85 | 0.927 (18) | 0.916 (10) | |
| 0.2 | | 4 | 1.7 | 73 | 81 | 60 | 65 | 0.940 (8) | 0.925 (7) | |
| 0.9 | | 6 | 3.0 | 76 | 81 | | 92 | 0.942 (5) | 0.932 (4) | |
| 0.3 | | 4 | 1.8 | 75 | 81 | 71 | 74 | 0.943 (3) | 0.934 (2) | |
| 0.0 | | 4 | 1.6 | 77 | 82 | 83 | 83 | 0.936 (10) | 0.932 (3) | |
| 0.1 | | 5 | 1.4 | 74 | 80 | 55 | 65 | 0.932 (12) | 0.890 (15) | |
| 0.3 | | 5 | 1.5 | 76 | 82 | 57 | 68 | 0.930 (16) | 0.874 (20) | |
| 0.0 | | 8 | 1.3 | 75 | 80 | 58 | 59 | 0.923 (20) | 0.873 (21) | |
| 0.0 | | 6 | 1.2 | 75 | 81 | 67 | 67 | 0.921 (21) | 0.867 (23) | |
| 0.0 | | 7 | 1.4 | 71 | 79 | 29 | 48 | 0.890 (31) | 0.850 (30) | |
| 0.0 | | 5 | 1.2 | 73 | 81 | 73 | 77 | 0.934 (11) | 0.874 (19) | |

| Country | Capital City (1) | Land Area (1) | | Population (millions) (1) | | | Birth Rate | | Death Rate | |
|---|---|---|---|---|---|---|---|---|---|---|
| | | | | | Projected | | | | | |
| | | (km²) | (mi²) | mid-1998 | 2010 | 2025 | 1970 (2) | 1998 (1) | 1970 (2) | 1998 (1) |
| EASTERN AND BALKAN EUROPE | | | | | | | | | | |
| Albania, Republic of | Tirana | 28,750 | 11,100 | 3.3 | 4.0 | 4.6 | 33 | 17 | 9 | 5 |
| Bosnia-Herzegovina, Republic of | Sarajevo | 51,130 | 19,741 | 4.0 | 4.4 | 4.3 | | 13 | | 7 |
| Bulgaria, Republic of | Sofia | 110,910 | 42,823 | 8.3 | 8.1 | 7.9 | 16 | 9 | 9 | 14 |
| Croatia, Republic of | Zagreb | 56,540 | 21,830 | 4.2 | 4.4 | 4.2 | | 12 | | 11 |
| Czech Republic | Prague | 78,860 | 30,448 | 10.3 | 10.3 | 10.2 | 16 | 9 | 12 | 11 |
| Estonia, Republic of | Tallinn | 45,100 | 17,413 | 1.4 | 1.3 | 1.2 | 15 | 9 | 11 | 13 |
| Hungary, Republic of | Budapest | 93,030 | 35,919 | 10.1 | 9.7 | 9.3 | 15 | 10 | 12 | 14 |
| Latvia, Republic of | Riga | 64,500 | 24,904 | 2.4 | 2.2 | 2.0 | | 8 | | 14 |
| Lithuania, Republic of | Vilnius | 65,200 | 24,174 | 3.7 | 3.6 | 3.5 | | 11 | | 12 |
| Macedonia, Former Yugoslav Republic of | Slopjec | 25,710 | 9,927 | 2.0 | 2.3 | 2.3 | | 16 | | 8 |
| Poland, Republic of | Warsaw | 312,680 | 120,726 | 38.7 | 40.4 | 40.8 | 17 | 11 | 8 | 10 |
| Romania | Bucharest | 237,500 | 91,699 | 22.5 | 21.2 | 19.7 | 21 | 10 | 10 | 12 |
| Slovak Republic | Bratislava | 49,010 | 18,923 | 5.4 | 5.4 | 5.3 | 19 | 11 | 9 | 10 |
| Slovenia, Republic of | Ljubljiana | 20,050 | 7,819 | 2.0 | 2.0 | 2.0 | | 9 | | 9 |
| Yugoslavia, Federal Republic of (Serbia-Montenegro) | Belgrade | 102,170 | 39,448 | 10.6 | 11.2 | 11.4 | | 13 | | 11 |
| COMMONWEALTH OF INDEPENDENT STATES | | | | | | | | | | |
| Armenia | Yerevan | 29,800 | 11,500 | 3.8 | 3.7 | 4.1 | | 13 | | 7 |
| Azerbaijan | Baku | 86,600 | 33,436 | 7.7 | 8.6 | 9.7 | | 17 | | 6 |
| Belarus | Minsk | 207,600 | 80,155 | 10.2 | 10.1 | 9.8 | | 9 | | 13 |
| Georgia | Tbilisi | 69,700 | 26,911 | 5.4 | 5.2 | 5.1 | | 11 | | 7 |
| Kazakstan | Akmola | 2,717,300 | 1,049,155 | 15.6 | 17.6 | 18.7 | | 15 | | 10 |
| Kyrgyzstan | Bishkek | 198,500 | 76,641 | 4.7 | 5.6 | 7.0 | | 22 | | 8 |
| Moldova | Chisinau | 33,700 | 13,012 | 4.2 | 4.6 | 4.9 | | 12 | | 12 |
| Russian Federation | Moscow | 17,075,400 | 6,592,849 | 146.9 | 141.9 | 134.6 | | 9 | | 14 |
| Tajikistan | Dushanbe | 143,100 | 85,251 | 6.1 | 7.7 | 9.3 | | 22 | | 5 |
| Turkmenistan | Ashkhabad | 488,100 | 188,456 | 4.7 | 5.2 | 6.1 | | 24 | | 7 |
| Ukraine | Kiev | 603,700 | 233,090 | 50.3 | 49.4 | 47.0 | | 9 | | 15 |
| Uzbekistan | Tashkent | 447,400 | 172,742 | 24.1 | 31.9 | 42.3 | | 26 | | 6 |
| ANGLO AMERICA | | | | | | | | | | |
| Canada | Ottawa | 9,976,140 | 3,851,809 | 30.6 | 35.0 | 40.3 | 17 | 12 | 7 | 7 |
| United States of America | Washington, D.C. | 9,809,460 | 3,787,442 | 270.2 | 297.7 | 335.1 | 17 | 15 | 15 | 9 |
| MEXICO AND CENTRAL AMERICA | | | | | | | | | | |
| United Mexican States | Mexico City | 1,956,200 | 756,065 | 97.5 | 117.5 | 140.0 | 43 | 27 | 10 | 5 |
| Belize | Belmopan | 22,960 | 8,865 | 0.2 | 0.3 | 0.4 | | 33 | | 6 |
| Costa Rica, Republic of | San Jose | 51,000 | 19,730 | 3.5 | 4.5 | 5.6 | 34 | 23 | 6 | 4 |
| El Salvador, Republic of | San Salvador | 21,040 | 8,124 | 5.8 | 7.2 | 8.8 | 44 | 29 | 12 | 4 |
| Guatemala, Republic of | Guatemala City | 108,890 | 42,043 | 11.6 | 14.6 | 19.8 | 45 | 38 | 14 | 7 |
| Honduras, Republic of | Tegucigalpa | 112,090 | 43,278 | 5.9 | 7.6 | 9.7 | 48 | 33 | 14 | 6 |
| Nicaragua, Republic of | Managua | 130,000 | 50,193 | 4.8 | 6.5 | 8.5 | 48 | 38 | 14 | 6 |
| Panama, Republic of | Panama City | 75,520 | 29,158 | 2.8 | 3.3 | 3.7 | 37 | 23 | 8 | 5 |
| WEST INDIES | | | | | | | | | | |
| Antigua and Barbuda | St. John's | 440 | 170 | 0.1 | 0.1 | 0.1 | | 17 | | 5 |
| Bahamas, Commonwealth of | Nassau | 13,880 | 5,359 | 0.3 | 0.3 | 0.4 | | 23 | | 6 |

| % Natural Increase 1998 (1) | Infant Mortality (1) | | Total Fertility Rate 1988 (1) | Life Expectancy at Birth (1) | | % Urban Population (1) | | HDI 1994 Value (rank of 175) (2) | GDI 1994 Value (rank of 146) (2) | HPI (%) 1994 (2) |
|---|---|---|---|---|---|---|---|---|---|---|
| | 1960 | 1998 | | Men | Women | 1980 | 1998 | | | |
| 1.2 | | 20 | 2.0 | 70 | 76 | 34 | 37 | 0.655 (102) | 0.643 (85) | |
| 0.6 | | | 1.5 | 70 | 75 | | 40 | | | |
| -0.5 | | 16 | 1.2 | 67 | 75 | 49 | 68 | 0.780 (69) | 0.772 (49) | |
| 0.1 | | 8 | 1.6 | 69 | 76 | 50 | 54 | 0.760 (77) | 0.741 (57) | |
| -0.2 | | 6 | 1.2 | 71 | 77 | 64 | 77 | 0.882 (39) | 0.859 (25) | |
| -0.4 | | 10 | 1.3 | 62 | 74 | 70 | 70 | 0.776 (71) | 0.764 (52) | |
| -0.4 | | 10 | 1.4 | 66 | 75 | 57 | 63 | 0.857 (48) | 0.837 (34) | No Values |
| -0.6 | | 16 | 1.2 | 64 | 76 | 68 | 69 | 0.711 (92) | 0.702 (67) | |
| -0.1 | | 10 | 1.4 | 65 | 76 | 61 | 68 | 0.762 (76) | 0.750 (55) | |
| 0.8 | | 16 | 2.1 | 69 | 73 | 46 | 60 | 0.748 (80) | 0.726 (62) | |
| 0.1 | | 12 | 1.6 | 68 | 77 | 58 | 62 | 0.834 (58) | 0.818 (37) | |
| -0.2 | | 23 | 1.3 | 65 | 73 | 49 | 55 | 0.748 (79) | 0.733 (59) | |
| 0.2 | | 14 | 1.3 | 62 | 73 | 52 | 68 | 0.873 (42) | 0.859 (26) | |
| 0.0 | | 5 | 1.3 | 71 | 78 | 48 | 50 | 0.886 (35) | 0.866 (24) | |
| 0.2 | | 14 | 1.8 | 70 | 75 | | 51 | | | |
| | | | | | | | | | | |
| 0.6 | | 14 | 1.6 | 69 | 76 | | 67 | 0.651 (103) | 0.647 (83) | |
| 1.1 | | 19 | 2.1 | 67 | 74 | | 52 | 0.636 (106) | 0.628 (89) | |
| -0.4 | | 12 | 1.3 | 62 | 74 | | 69 | 0.806 (62) | 0.792 (42) | |
| 0.4 | | 17 | 1.6 | 69 | 76 | | 56 | 0.637 (105) | 0.630 (87) | |
| 0.5 | | 25 | 1.9 | 60 | 70 | | 56 | 0.709 (93) | 0.698 (69) | |
| 1.5 | | 28 | 2.8 | 63 | 71 | | 34 | 0.635 (107) | 0.628 (88) | No Values |
| 0.1 | | 20 | 1.8 | 62 | 70 | | 46 | 0.612 (110) | 0.608 (93) | |
| -0.5 | | 17 | 1.2 | 61 | 73 | | 73 | 0.792 (67) | 0.778 (46) | |
| 1.7 | | 32 | 2.9 | 65 | 71 | | 28 | 0.580 (115) | 0.575 (96) | |
| | | 42 | 2.9 | 62 | 69 | | 45 | 0.723 (85) | 0.712 (65) | |
| -0.6 | | 14 | 1.3 | 62 | 73 | | 68 | 0.689 (95) | 0.681 (72) | |
| 2.0 | | 26 | 3.2 | 66 | 72 | | 38 | 0.662 (100) | 0.655 (78) | |
| | | | | | | | | | | |
| 0.5 | | 6 | 1.6 | 75 | 81 | 76 | 77 | 0.960 (1) | 0.939 (1) | No Values |
| 0.6 | | 7 | 2.0 | 73 | 79 | 76 | 77 | 0.942 (4) | 0.928 (5) | |
| | | | | | | | | | | |
| 2.2 | 95 | 28 | 3.1 | 69 | 75 | 66 | 74 | 0.853 (50) | 0.770 (50) | 10.9 |
| 2.7 | 74 | 34 | 4.1 | 70 | 74 | | 51 | 0.806 (63) | | |
| 1.9 | 85 | 12 | 2.8 | 73 | 78 | 43 | 44 | 0.889 (33) | 0.825 (36) | 6.6 |
| 2.5 | 130 | 41 | 3.9 | 65 | 72 | 42 | 50 | 0.592 (112) | 0.563 (97) | 28.0 |
| 3.1 | 125 | 51 | 5.1 | 63 | 68 | 37 | 38 | 0.572 (117) | 0.510 (107) | 35.5 |
| 2.8 | 145 | 42 | 4.4 | 66 | 71 | 36 | 44 | 0.575 (116) | 0.544 (103) | 22.0 |
| 3.2 | 141 | 46 | 4.6 | 63 | 68 | 53 | 63 | 0.530 (127) | 0.515 (106) | 27.2 |
| 1.8 | 69 | 22 | 2.7 | 71 | 77 | 50 | 55 | 0.864 (45) | 0.802 (41) | 11.2 |
| | | | | | | | | | | |
| 1.2 | | 18 | 1.7 | 71 | 76 | | 36 | 0.892 (29) | | |
| 1.7 | 50 | 19 | 2.0 | 68 | 75 | | 86 | 0.894 (28) | 0.880 (18) | |

| Country | Capital City (1) | Land Area (1) | | Population (millions) (1) | | | Birth Rate | | Death Rate | |
|---|---|---|---|---|---|---|---|---|---|---|
| | | | | | Projected | | | | | |
| | | (km²) | (mi²) | mid-1998 | 2010 | 2025 | 1970 (2) | 1998 (1) | 1970 (2) | 1998 (1) |
| WEST INDIES *(continued)* | | | | | | | | | | |
| Barbados | Bridgetown | 430 | 166 | 0.3 | 0.3 | 0.3 | | 14 | | 9 |
| Cuba, Republic of | Havana | 110,860 | 42,803 | 11.1 | 11.5 | 11.8 | | 14 | | 7 |
| Dominica, Commonwealth of | Roseau | 750 | 290 | 0.1 | 0.1 | 0.1 | | 19 | | 8 |
| Dominican Republic | Santo Domingo | 48,730 | 18,815 | 8.3 | 9.8 | 11.3 | 41 | 27 | 11 | 6 |
| French Guiana | Cayenne | 91,000 | 35,135 | 0.2 | 0.2 | 0.3 | | 30 | | 4 |
| Grenada | St. George's | 340 | 131 | 0.1 | 0.1 | 0.2 | | 29 | | 6 |
| Guyana, Cooperative Republic of | Georgetown | 214,970 | 83,000 | 0.7 | 0.7 | 0.7 | | 24 | | 7 |
| Haiti, Republic of | Port-au-Prince | 27,750 | 10,714 | 7.5 | 9.4 | 12.5 | | 34 | | 13 |
| Jamaica | Kingston | 10,990 | 4,243 | 2.6 | 2.9 | 3.2 | 34 | 23 | 8 | 6 |
| St. Kitts and Nevis, Federation of | Basseterre | 360 | 139 | 0.0 | 0.1 | 0.1 | | 19 | | 9 |
| St. Lucia | Castries | 620 | 239 | 0.1 | 0.2 | 0.2 | | 25 | | 6 |
| St. Vincent and the Grenadines | Kingstown | 390 | 151 | 0.1 | 0.1 | 0.2 | | 22 | | 7 |
| Suriname, Republic of | Paramaribo | 163,270 | 63,039 | 0.4 | 0.5 | 0.5 | | 24 | | 6 |
| Trinidad and Tobago, Republic of | Port-of-Spain | 5,130 | 1,981 | 1.3 | 1.4 | 1.5 | 28 | 15 | 8 | 7 |
| NORTHERN SOUTH AMERICA | | | | | | | | | | |
| Bolivia, Republic of | La Paz | 1,098,580 | 424,164 | 8.0 | 10.3 | 13.2 | 45 | 36 | 20 | 10 |
| Brazil, Federative Republic of | Brasília | 8,511,970 | 3,286,490 | 162.1 | 184.2 | 208.2 | 35 | 22 | 10 | 8 |
| Colombia, Republic of | Bogotá | 1,138,910 | 429,485 | 38.6 | 47.3 | 58.3 | 36 | 27 | 9 | 6 |
| Ecuador, Republic of | Quito | 283,560 | 109,483 | 12.2 | 14.9 | 17.8 | 41 | 28 | 12 | 6 |
| Peru, Republic of | Lima | 1,285,000 | 496,226 | 26.1 | 32.1 | 39.2 | 42 | 28 | 14 | 6 |
| Venezuela, Republic of | Caracas | 912,050 | 352,144 | 23.3 | 28.7 | 34.8 | 37 | 26 | 7 | 5 |
| SOUTHERN SOUTH AMERICA | | | | | | | | | | |
| Argentine Republic | Buenos Aires | 2,766,890 | 1,068,302 | 36.1 | 41.5 | 47.2 | 23 | 19 | 9 | 8 |
| Chile, Republic of | Santiago | 756,950 | 292,260 | 14.8 | 17.0 | 19.5 | 29 | 19 | 10 | 6 |
| Paraguay, Republic of | Asunción | 406,750 | 157,047 | 5.2 | 7.0 | 9.4 | 38 | 32 | 7 | 6 |
| Uruguay, Eastern Republic of | Montevideo | 177,410 | 68,496 | 3.2 | 3.5 | 3.8 | 21 | 18 | 10 | 10 |
| SOUTH PACIFIC | | | | | | | | | | |
| Australia, Commonwealth of | Canberra | 7,713,360 | 2,978,145 | 18.7 | 21.0 | 23.5 | 20 | 14 | 9 | 7 |
| Fiji, Republic of | Suva | 18,270 | 7,054 | 0.8 | 0.9 | 1.6 | | 24 | | 6 |
| Kiribati, Republic of | Tarawa | 730 | 282 | 0.1 | | | | | | |
| Marshall Islands, Republic of | Majuro | 179 | 69 | 0.1 | 0.1 | 0.2 | | 43 | | 7 |
| Micronesia, Federated States of | Palikir | 702 | 271 | 0.1 | 0.1 | 0.2 | | 33 | | 8 |
| Nauru, Republic of | Yaren | 20 | 8 | 0.1 | | | | | | |
| New Zealand | Wellington | 270,990 | 104,630 | 3.8 | 4.0 | 4.3 | 22 | 15 | 9 | 7 |
| Palau, Republic of | Koror | 458 | 177 | 0.1 | 0.1 | 0.1 | | 23 | | 7 |
| Papua New Guinea, Independent State of | Port Moresby | 462,840 | 178,704 | 4.3 | 5.6 | 7.7 | 42 | 34 | 18 | 10 |
| Solomon Islands | Honiara | 28,900 | 11,158 | 0.4 | 0.6 | 0.9 | | 37 | | 4 |
| Tonga, Kingdom of | Nukualofa | 750 | 290 | 0.1 | | | | | | |
| Tuvalu | Funafuti | 30 | 12 | 0.1 | | | | | | |
| Vanuatu, Republic of | Port Vila | 12,190 | 4,707 | 0.2 | 0.2 | 0.3 | | 35 | | 7 |
| Western Samoa, Independent State of | Apia | 2,840 | 1,097 | 0.2 | 0.2 | 0.3 | | 29 | | 5 |

| % Natural Increase 1998 (1) | Infant Mortality (1) | | Total Fertility Rate 1988 (1) | Life Expectancy at Birth (1) | | % Urban Population (1) | | HDI 1994 Value (rank of 175) (2) | GDI 1994 Value (rank of 146) (2) | HPI (%) 1994 (2) |
|---|---|---|---|---|---|---|---|---|---|---|
| | 1960 | 1998 | | Men | Women | 1980 | 1998 | | | |
| 0.5 | 74 | 14 | 1.7 | 73 | 77 | | 38 | 0.907 (25) | 0.885(17) | |
| 0.6 | 65 | 7 | 1.4 | 74 | 77 | | 74 | 0.723 (86) | 0.699 (68) | 5.1 |
| 1.1 | | 16 | 2.0 | | | | | 0.873 (41) | | |
| 2.1 | 125 | 47 | 3.2 | 68 | 72 | 51 | 62 | 0.718 (87) | 0.658 (75) | 18.3 |
| 2.5 | 82 | 14 | 3.7 | 71 | 77 | | | | | |
| 2.3 | | 12 | 3.8 | 68 | 73 | | 32 | 0.843 (54) | | |
| 1.7 | 100 | 63 | 2.7 | 63 | 69 | | 36 | 0.649 (104) | 0.615 (91) | |
| 2.1 | 182 | 74 | 4.8 | 49 | 53 | 24 | 33 | 0.338 (156) | 0.332 (130) | 46.2 |
| 1.8 | 63 | 16 | 3.0 | 70 | 73 | 47 | 50 | 0.736 (83) | 0.726 (63) | 12.1 |
| 1.0 | | 25 | 2.6 | 64 | 70 | | 43 | 0.853 (49) | | |
| 1.9 | | 18 | 2.7 | 68 | 75 | | 48 | 0.838 (56) | | |
| 1.6 | | 19 | 2.4 | 71 | 74 | | 25 | 0.836 (57) | | |
| 1.8 | 70 | 29 | 2.6 | 68 | 73 | | 70 | 0.792 (66) | | |
| 0.8 | 56 | 17 | 1.9 | 68 | 73 | 63 | 65 | 0.880 (40) | 0.841 (32) | 4.1 |
| 2.6 | 167 | 75 | 4.8 | 57 | 63 | 46 | 58 | 0.589 (113) | 0.557 (99) | 22.5 |
| 1.4 | 116 | 43 | 2.5 | 64 | 71 | 66 | 76 | 0.783 (68) | 0.728 (60) | |
| 2.1 | 99 | 28 | 3.0 | 65 | 73 | 64 | 71 | 0.848 (51) | 0.811 (40) | 10.7 |
| 2.2 | 124 | 40 | 3.6 | 66 | 71 | 47 | 61 | 0.775 (72) | 0.675 (73) | 15.2 |
| 2.2 | 142 | 43 | 3.5 | 67 | 71 | 65 | 71 | 0.717 (89) | 0.656 (76) | 22.8 |
| 2.1 | 81 | 21 | 3.1 | 70 | 75 | 83 | 86 | 0.861 (47) | 0.792 (43) | |
| 1.1 | 60 | 22 | 2.5 | 69 | 76 | 83 | 89 | 0.884 (36) | 0.777 (47) | |
| 1.4 | 114 | 11 | 2.4 | 72 | 78 | 81 | 85 | 0.891 (30) | 0.785 (44) | 5.4 |
| 2.7 | 66 | 27 | 4.4 | 66 | 71 | 42 | 52 | 0.706 (94) | 0.649 (82) | 23.2 |
| 0.8 | 50 | 20 | 2.4 | 72 | 78 | 85 | 90 | 0.883 (37) | 0.842 (31) | 11.7 |
| 0.7 | | 5 | 1.8 | 75 | 81 | 86 | 85 | 0.931 (14) | 0.917 (9) | |
| 1.8 | 71 | 17 | 2.8 | 61 | 65 | | 46 | 0.863 (46) | 0.763 (53) | |
| 3.9 | | | | | | | | | | |
| 3.6 | | 26 | 6.7 | 60 | 63 | | 65 | | | |
| 2.6 | 46 | 4.7 | 65 | 67 | 27 | | | | | |
| 1.3 | | | | | | | | | | |
| 0.8 | | 7 | 2 | 69 | 75 | 83 | 85 | 0.937 (9) | 0.918 (8) | |
| 1.6 | | 25 | 2.5 | | | | 69 | | | |
| 2.4 | 165 | 77 | 4.8 | 56 | 57 | 13 | 15 | 0.525 (128) | 0.508 (108) | 32.0 |
| 3.2 | 120 | 28 | 5.4 | 68 | 73 | 13 | | 0.556 (122) | | |
| 0.8 | | | | | | | | | | |
| 1.6 | | | | | | | | | | |
| 2.8 | 141 | 41 | 4.7 | | | | 18 | 0.547 (124) | | |
| 2.4 | 134 | 21 | 4.2 | | | | 21 | 0.684 (96) | | |

| Country | GNP capita 1996 (1) | GDP/cap PPP (1987 constant) (3) 1980 | 1990 | 1996 | Adult Illiteracy (2) %Men 1995 | %Women 1995 |
|---|---|---|---|---|---|---|
| CENTRAL AFRICA | | | | | | |
| Burundi, Republic of | 170 | 591 | 658 | 448 | 51 | 78 |
| Cameroon, Republic of | 610 | 1765 | 1757 | 1461 | 25 | 48 |
| Central African Republic | 310 | 1413 | 1241 | 1103 | 32 | 48 |
| Chad, Republic of | 160 | 546 | 699 | 686 | 38 | 65 |
| Congo, Republic of the | 640 | 1272 | 1522 | 1363 | 17 | 33 |
| Congo, Democratic Republic of the (Zaire) | 130 | 1619 | 1203 | 682 | 13 | 32 |
| Equatorial Guinea, Republic of | 530 | | 907 | 2391 | 10 | 32 |
| Gabonese Republic | 3950 | 7444 | 5615 | 5633 | 26 | 47 |
| Rwanda, Republic of | 190 | 788 | 679 | 481 | 30 | 48 |
| WESTERN AFRICA | | | | | | |
| Benin, Republic of | 350 | 905 | 854 | 948 | 51 | 74 |
| Burkina Faso, People's Democratic Republic of | 230 | 652 | 691 | 719 | 71 | 91 |
| Côte D'Ivoire, Republic of | 660 | 1963 | 1367 | 1306 | 50 | 70 |
| Gambia, Republic of the | | 1103 | 1065 | 979 | 47 | 75 |
| Ghana, Republic of | 360 | 1451 | 1250 | 1390 | 24 | 47 |
| Guinea, Republic of | 560 | | 1262 | 1350 | 50 | 78 |
| Guinea-Bissau, Republic of | 250 | 552 | 722 | 796 | 32 | 57 |
| Liberia, Republic of | | | | | 46 | 78 |
| Mali, Republic of | 240 | 599 | 546 | 550 | 60 | 77 |
| Mauritania, Islamic Republic of | 470 | 1542 | 1328 | 1441 | 50 | 74 |
| Niger, Republic of | 200 | 1263 | 805 | 710 | 79 | 93 |
| Nigeria, Federal Republic of | 240 | 788 | 665 | 679 | 33 | 53 |
| Senegal, Republic of | 570 | 1336 | 1314 | 1284 | 57 | 77 |
| Sierra Leone, Republic of | 200 | 659 | 559 | 395 | 55 | 82 |
| Togo, Republic of | 300 | 1938 | 1531 | 1276 | 33 | 63 |
| EASTERN AFRICA | | | | | | |
| Djibouti, Republic of | | | | | 40 | 67 |
| Eritrea, Republic of | | | | | | |
| Ethiopia, Federal Democratic Republic of | 100 | | 355 | 381 | 55 | 75 |
| Kenya, Republic of | 320 | 904 | 909 | 878 | 14 | 30 |
| Somali Democratic Republic | | | | | 64 | 86 |
| Tanzania, United Republic of | 170 | | | | 21 | 43 |
| Uganda, Republic of | 300 | | 625 | 787 | 26 | 50 |
| SOUTHERN AFRICA | | | | | | |
| Angola, Republic of | 270 | | 1321 | 1514 | 44 | 72 |
| Botswana, Republic of | | 3045 | 5379 | 5793 | 20 | 40 |
| Lesotho, Kingdom of | 660 | 998 | 996 | 1298 | 19 | 38 |
| Madagascar, Republic of | 250 | 1114 | 813 | 706 | 12 | 27 |
| Malawi, Republic of | 180 | 576 | 500 | 534 | 28 | 58 |
| Mozambique, Republic of | 80 | | 360 | 415 | 42 | 77 |
| Namibia, Republic of | 2250 | 4218 | 3625 | 3960 | | |
| South Africa, Republic of | 3520 | 6779 | 5956 | 5763 | 18 | 18 |
| Swaziland, Kingdom of | 1210 | 1928 | 2559 | 2522 | 22 | 24 |

| | Consumer Goods (3) | | | | | | | | |
|---|---|---|---|---|---|---|---|---|---|
| | TVs/1000 | | Vehicles/1000 | | Telephone lines/1000 | | Computers/1000 | | |
| | 1980 | 1995–96 | 1980 | 1995–96 | 1980 | 1995–96 | 1990 | 1995–96 | Political Freedom (4) |
| | — | 2 | — | 6 | — | 3 | — | — | 6.5 (not free) |
| | | 23 | 8 | 12 | 2 | 5 | | | 6 (not free) |
| | | 5 | 8 | 0 | 1 | 3 | | | 3.5 (part free) |
| | | 2 | | 4 | | 1 | | | 5.5 (not free) |
| | 2 | 8 | | 20 | 5 | 8 | | | 4 (part free) |
| | | 41 | | 31 | 1 | 1 | | | 6.5 (not free) |
| | 5 | 98 | | 5 | | 9 | | | 7 (not free) |
| | 11 | 76 | | 38 | | 32 | | 6 | 4.5 (part free) |
| | | | 2 | 4 | 1 | 3 | | | 6.5 (not free) |
| | 1 | 73 | | 8 | 2 | 6 | | | 2 (free) |
| | 3 | 6 | | 5 | | 3 | | | 4.5 (part free) |
| | 37 | 60 | 24 | 32 | 5 | 9 | | 1 | 5.5 (not free) |
| | | | | 15 | 3 | 19 | | | 6.5 (not free) |
| | 5 | 41 | | 8 | 4 | 4 | | 1 | 4 (part free) |
| | 1 | 8 | | 5 | 1 | 2 | | | 5.5 (not free) |
| | | | | 12 | 1 | 7 | | | 3.5 (part free) |
| | 11 | 19 | 1 | 15 | | 2 | | | 6.5 (not free) |
| | | 11 | | 4 | 1 | 2 | | | 2.5 (free) |
| | | 82 | | 13 | 2 | 4 | | 5 | 6 (not free) |
| | 1 | 11 | 6 | 6 | 1 | 2 | | | 4 (part free) |
| | 6 | 55 | 4 | 12 | | 4 | | | 7 (not free) |
| | 1 | 38 | 19 | 14 | 3 | 11 | 2 | 7 | 4.5 (part free) |
| | 6 | 17 | | 6 | 4 | 4 | | | 6.5 (part free) |
| | 4 | 14 | | 27 | 2 | 6 | | | 5.5 (not free) |
| | 18 | 73 | | 17 | 9 | 13 | | 2 | 5.5 (not free) |
| | | 7 | | 2 | | 5 | | | 5 (part free) |
| | 1 | 4 | 2 | 1 | 2 | 3 | | | 4.5 (part free) |
| | 5 | 19 | 8 | 13 | 4 | 8 | | 2 | 6.5 (not free) |
| | | 14 | | 1 | 1 | 2 | | | 7 (not free) |
| | | 2 | 3 | 5 | 2 | 3 | | | 5 (part free) |
| | 5 | 26 | 1 | 4 | 1 | 2 | | 1 | 4.5 (part free) |
| | 4 | 51 | | 21 | 5 | 5 | | | 6 (not free) |
| | | 27 | 27 | 45 | 8 | 48 | | 7 | 2 (free) |
| | | 13 | 10 | 19 | 3 | 9 | | | 4 (part free) |
| | 5 | 25 | | 6 | 2 | 3 | | | 3 (part free) |
| | | | 5 | 6 | 2 | 4 | | | 2.5 (free) |
| | 3 | 4 | | 1 | 1 | 4 | | 1 | 3.5 (part free) |
| | 5 | 29 | | 84 | | 54 | | 13 | 2.5 (free) |
| | 68 | 123 | 133 | 134 | 55 | 100 | | 38 | 1.5 (free) |
| | 2 | 96 | 52 | 66 | 9 | 22 | | | 5.5 (free) |

| Country | GNP capita 1996 (1) | GDP/cap PPP (1987 constant) (3) | | | Adult Illiteracy (2) | |
|---|---|---|---|---|---|---|
| | | 1980 | 1990 | 1996 | %Men 1995 | %Women 1995 |
| SOUTHERN AFRICA *(continued)* | | | | | | |
| Zambia, Republic of | 360 | 1063 | 824 | 672 | 14 | 29 |
| Zimbabwe, Republic of | 610 | 1862 | 1802 | 1737 | 10 | 20 |
| NORTH AFRICA | | | | | | |
| Algeria, Democratic and Popular Republic of | 1520 | 4307 | 4023 | 3688 | 26 | 51 |
| Socialist People's Libyan Arab Jamahirya | | | | | 12 | 37 |
| Morocco, Kingdom of | 1290 | 2223 | 2477 | 2596 | 43 | 69 |
| Tunisia, Republic of | 1930 | 3111 | 3306 | 3644 | 21 | 45 |
| NILE RIVER VALLEY | | | | | | |
| Egypt, Arab Republic of | 1080 | 1460 | 1977 | 2163 | 36 | 61 |
| Sudan, Republic of | | | | | 42 | 65 |
| SOUTHWESTERN ASIA | | | | | | |
| Bahrain, State of | | 16,367 | 10,315 | 11,930 | 11 | 21 |
| Iran | | 4149 | 3737 | 4129 | 22 | 34 |
| Iraq, Republic of | | | | | 29 | 55 |
| Israel | 15,870 | 10,466 | 11,868 | 13,840 | | |
| Jordan, Hashemite Kingdom of | 1650 | 3086 | 2458 | 2763 | 7 | 21 |
| Kuwait, State of | | 23,092 | 10,072 | 20,704 | 18 | 25 |
| Lebanese Republic | 2970 | | | 4481 | 10 | 20 |
| Oman, Sultanate of | | 4334 | 7357 | 7570 | | |
| Qatar, State of | | 43,622 | 16,285 | 13,365 | 21 | 20 |
| Saudi Arabia, Kingdom of | | 15,755 | 8261 | 7649 | 29 | 50 |
| Syrian Arab Republic | 1160 | 2206 | 1881 | 2424 | 14 | 44 |
| Turkey | 2830 | 3225 | 4057 | 4525 | 8 | 28 |
| United Arab Emirates | | 30,867 | 16,607 | 12,145 | 21 | 20 |
| Yemen, Republic of | 380 | | | 684 | 47 | 74 |
| SOUTHERN ASIA | | | | | | |
| Afghanistan, Islamic State of | | | | | 53 | 85 |
| Bangladesh, People's Republic of | 260 | 544 | 654 | 767 | 51 | 74 |
| Bhutan, Kingdom of | 390 | | | | 44 | 72 |
| India, Republic of | 380 | 725 | 982 | 1210 | 35 | 62 |
| Maldives, Republic of | 1080 | | 2060 | 2561 | 7 | 7 |
| Nepal, Kingdom of | 210 | 617 | 709 | 815 | 59 | 86 |
| Pakistan, Islamic Republic of | 480 | 855 | 1083 | 1217 | 50 | 76 |
| Sri Lanka, Democratic Socialist Republic of | 740 | 1133 | 1457 | 1755 | 7 | 13 |
| EASTERN ASIA | | | | | | |
| China, People's Republic of | 750 | 700 | 1376 | 2546 | 10 | 27 |
| Hong Kong (to China, 1997) | 24,290 | 9304 | 14,796 | 18,350 | 4 | 12 |
| Japan | 40,940 | 11,989 | 16,045 | 17,519 | | |
| Korea, People's Democratic Republic of (North) | | | | | 1 | 3 |
| Korea, Republic of (South) | 10,610 | 3428 | 6918 | 9976 | 1 | 3 |
| Mongolia | 360 | | 1792 | 1404 | | |

| Consumer Goods (3) | | | | | | | | |
|---|---|---|---|---|---|---|---|---|
| TVs/1000 | | Vehicles/1000 | | Telephone lines/1000 | | Computers/1000 | | |
| 1980 | 1995–96 | 1980 | 1995–96 | 1980 | 1995–96 | 1990 | 1995–96 | Political Freedom (4) |
| 10 | 80 | | 26 | 6 | 9 | | | 3.5 (part free) |
| 10 | 29 | | 32 | 13 | 15 | | 7 | 5 (part free) |
| 52 | 68 | | 53 | 17 | 44 | | 3 | 6 (not free) |
| 54 | 143 | | 138 | 10 | 59 | | | 7 (not free) |
| 46 | 80 | | 50 | 9 | 45 | | 2 | 5 (part free) |
| 47 | 156 | 38 | 64 | 18 | 64 | 3 | 7 | 5.5 (not free) |
| 32 | 126 | | 30 | 15 | 50 | | 4 | 6 (not free) |
| 43 | 80 | | 12 | 2 | 4 | | 1 | 7 (not free) |
| 259 | 429 | | 296 | | 241 | | 67 | 6 (not free) |
| 51 | 164 | | 38 | 23 | 95 | | 33 | 6.5 (not free) |
| 50 | 78 | | 14 | 19 | 33 | | | 7 (not free) |
| 233 | 303 | 123 | 263 | 222 | 446 | | 118 | 2 (free) |
| 78 | | 56 | 68 | 28 | 60 | | 7 | 4 (part free) |
| 257 | 373 | 390 | 404 | 114 | 232 | 4 | 74 | 5 (part free) |
| 281 | 355 | | 320 | | 149 | | 24 | 5.5 (part free) |
| 32 | 591 | | 134 | 16 | 86 | 2 | 11 | 6 (not free) |
| 349 | 538 | | 287 | 134 | 239 | | 63 | 6.5 (not free) |
| 219 | 263 | 163 | 149 | 33 | 106 | 24 | 37 | 7 (not free) |
| 44 | 91 | | 28 | 28 | 82 | | 1 | 7 (not free) |
| 78 | 307 | 23 | 70 | 26 | 224 | 4 | 14 | 5 (part free) |
| 92 | 276 | | 99 | 119 | 302 | | 65 | 5.5 (part free) |
| 195 | 278 | | 34 | 2 | 13 | | | 5.5 (part free) |
| 3 | 10 | | 2 | 2 | 1 | | | 7 (not free) |
| 1 | 7 | | 1 | | 3 | | | 3.5 (part free) |
| | 6 | | 15 | | 3 | | | 7 (not free) |
| 3 | 64 | 2 | 7 | 3 | 15 | | 1 | 4 (part free) |
| 7 | 39 | | 9 | | 62 | | 12 | 6 (not free) |
| | 4 | | | | 5 | | | 3.5 (part free) |
| 11 | 24 | 2 | 7 | 4 | 18 | | 1 | 4 (part free) |
| 2 | 82 | | 14 | 4 | 14 | | 3 | 4.5 (part free) |
| 5 | 252 | 2 | 8 | 2 | 45 | | 3 | 7 (not free) |
| 221 | 388 | 54 | 78 | 254 | 547 | | 151 | |
| 539 | 700 | 323 | 552 | 342 | 489 | 59 | 128 | 1.5 (free) |
| 7 | | | | | 49 | | | 7 (not free) |
| 165 | 326 | 14 | 195 | 71 | 430 | 37 | 132 | 2 (free) |
| 3 | 63 | | 26 | | 35 | | | 2.5 (free) |

| Country | GNP capita 1996 (1) | GDP/cap PPP (1987 constant) (3) 1980 | 1990 | 1996 | Adult Illiteracy (2) %Men 1995 | %Women 1995 |
|---|---|---|---|---|---|---|
| EASTERN ASIA (*continued*) | | | | | | |
| Taiwan | | | | | | |
| Singapore, Republic of | 30550 | 8644 | 14023 | 20187 | 4 | 14 |
| SOUTHEASTERN ASIA | | | | | | |
| Cambodia, Kingdom of | 300 | | | | 52 | 78 |
| Indonesia, Republic of | 1080 | 1256 | 1853 | 2612 | 10 | 22 |
| Laos, People's Democratic Republic of | 400 | | 761 | 943 | 31 | 56 |
| Malaysia, Federation of | 4370 | 4465 | 5817 | 8246 | 11 | 22 |
| Myanmar, Union of (Burma) | | | | | 11 | 22 |
| Philippines, Republic of | 1160 | 2890 | 2510 | 2588 | 5 | 6 |
| Thailand, Kingdom of | 2960 | 2073 | 3527 | 5197 | 4 | 8 |
| Vietnam, Socialist Republic of | 290 | | | 1190 | 4 | 9 |
| WESTERN EUROPE | | | | | | |
| Belgium, Kingdom of | 26440 | 13953 | 15857 | 16790 | Very low rates | |
| French Republic | 26270 | 13656 | 15667 | 16326 | | |
| Germany, Federal Republic of | 28870 | | | 16046 | | |
| Ireland, Republic of | 17110 | 7868 | 10429 | 14131 | | |
| Luxembourg, Grand Duchy of | 45360 | 13859 | 19559 | 24371 | | |
| Netherlands, Kingdom of | 25940 | 12558 | 14099 | 15505 | | |
| United Kingdom of Great Britain and Northern Ireland | 19600 | 11694 | 14177 | 15070 | | |
| NORTHERN EUROPE | | | | | | |
| Denmark, Kingdom of | 32100 | 13346 | 15529 | 17172 | Very low rates | |
| Finland, Republic of | 23240 | 11795 | 14689 | 14254 | | |
| Iceland, Republic of | 26580 | 14487 | 16233 | 16797 | | |
| Norway, Kingdom of | 34510 | 12509 | 14599 | 17754 | | |
| Sweden, Kingdom of | 25710 | 13099 | 14829 | 14823 | | |
| ALPINE EUROPE | | | | | Very low rates | |
| Austria, Republic of | 28110 | 13176 | 15462 | 16412 | | |
| Swiss Confederation | 44350 | 18049 | 19951 | 18772 | | |
| MEDITERRANEAN EUROPE | | | | | | |
| Helenic Republic (Greece) | 11460 | 8232 | 8759 | 9436 | Very low rates | |
| Italian Republic | 19880 | 12184 | 14444 | 15320 | | |
| Portuguese Republic | 10160 | 7380 | 9283 | 10236 | | |
| Spain, Kingdom of | 14350 | 8779 | 10849 | 11726 | | |
| EASTERN AND BALKAN EUROPE | | | | | | |
| Albania, Republic of | 820 | | | | | |
| Bosnia-Herzogovina, Republic of | | | | | No data | |
| Bulgaria, Republic of | 1190 | 3314 | 4287 | 3375 | | |
| Croatia, Republic of | 3800 | | | | | |
| Czech Republic | 4740 | | 8289 | 8331 | | |
| Estonia, Republic of | 3080 | | 4649 | 3527 | | |

| Consumer Goods (3) | | | | | | | | |
|---|---|---|---|---|---|---|---|---|
| TVs/1000 | | Vehicles/1000 | | Telephone lines/1000 | | Computers/1000 | | |
| 1980 | 1995–96 | 1980 | 1995–96 | 1980 | 1995–96 | 1990 | 1995–96 | Political Freedom (4) |
| | | | | | | | | 3 (part free) |
| 311 | 361 | | 167 | 222 | 513 | | 217 | 5 (part free) |
| 5 | 9 | | 6 | | 1 | | | 6 (part free) |
| 20 | 232 | 8 | 22 | 2 | 21 | | 5 | 6.5 (not free) |
| | 10 | | 4 | 2 | 6 | | 1 | 6.5 (not free) |
| 87 | 228 | | 152 | 29 | 183 | | 43 | 4.5 (part free) |
| | 7 | | 2 | 1 | 4 | | | 7 (not free) |
| 22 | 125 | | 13 | 9 | 25 | 2 | 9 | 3 (part free) |
| 21 | 167 | 13 | 106 | 8 | 70 | | 17 | 3.5 (part free) |
| | 180 | | | 1 | 16 | | 3 | 7 (not free) |
| 387 | 464 | 349 | 469 | 248 | 465 | 88 | 167 | 1 (free) |
| 370 | 598 | 402 | 524 | 295 | 563 | 71 | 151 | 1.5 (free) |
| 439 | 493 | 399 | 528 | 332 | 538 | 83 | 233 | 1.5 (free) |
| 231 | 469 | 236 | 307 | 142 | 395 | 106 | 145 | 1 (free) |
| 302 | 629 | 377 | 602 | 363 | 593 | | | 1 (free) |
| 399 | 495 | 343 | 400 | 346 | 543 | 94 | 232 | 1 (free) |
| 401 | 611 | 303 | 399 | 322 | 528 | 82 | 193 | 1.5 (free) |
| 498 | 533 | 322 | 390 | 434 | 618 | 115 | 304 | 1 (free) |
| 414 | 605 | 288 | 431 | 364 | 549 | 100 | 183 | 1 (free) |
| 285 | 447 | 417 | 524 | 372 | 573 | 39 | 205 | 1 (free) |
| 350 | 569 | 342 | 470 | 293 | 555 | | 273 | 1 (free) |
| 461 | 476 | 370 | 450 | 580 | 682 | 115 | 215 | 1 (free) |
| 391 | 493 | 330 | 494 | 290 | 466 | 62 | 148 | 1 (free) |
| 360 | 493 | 383 | 501 | 445 | 640 | | 408 | 2 (free) |
| 171 | 442 | 134 | 312 | 235 | 509 | 17 | 33 | 2 (free) |
| 390 | 436 | 334 | 674 | 231 | 440 | 36 | 92 | 1.5 (free) |
| 166 | 367 | 145 | 370 | 107 | 375 | 26 | 60 | 1 (free) |
| 255 | 509 | 239 | 455 | 193 | 392 | 28 | 94 | 1.5 (free) |
| 36 | 173 | | 31 | 10 | 19 | | | 3.5 (part free) |
| | 55 | | 24 | | 90 | | | 6 (not free) |
| 243 | 361 | | 234 | 102 | 313 | | 29 | 2 (free) |
| | 251 | | 165 | 83 | 309 | | 21 | 4 (part free) |
| | 406 | | 321 | 114 | 273 | | 53 | 1.5 (free) |
| | 449 | | 329 | 135 | 299 | | 7 | 2 (free) |

| Country | GNP capita 1996 (1) | GDP/cap PPP (1987 constant) (3) | | | Adult Illiteracy (2) | |
|---|---|---|---|---|---|---|
| | | 1980 | 1990 | 1996 | %Men 1995 | %Women 1995 |
| EASTERN AND BALKAN EUROPE *(continued)* | | | | | | |
| Hungary, Republic of | 4340 | 4963 | 5716 | 5259 | | |
| Latvia, Republic of | 2300 | 3993 | 5039 | 2759 | | |
| Lithuania, Republic of | 2280 | | 4612 | 3356 | | |
| Macedonia, Former Yugoslav Republic of | 990 | | | | No data | |
| Poland, Republic of | 3230 | 4439 | 4079 | 4554 | | |
| Romania | 1600 | 3909 | 3593 | 3507 | | |
| Slovak Republic | 3410 | | 6282 | 5652 | | |
| Slovenia, Republic of | 9240 | | | 9087 | | |
| Yugoslavia, Federal Republic of (Serbia-Montenegro) | | | | | | |
| COMMONWEALTH OF INDEPENDENT STATES | | | | | | |
| Armenia | 630 | 4080 | 4483 | 1545 | | |
| Azerbaijan | 480 | | 3314 | 1150 | | |
| Belarus | 2070 | | 5167 | 3322 | | |
| Georgia | 850 | 5076 | 4331 | 1269 | | |
| Kazakstan | 1350 | | 4347 | 2449 | | |
| Kyrgyzstan | 550 | | 3040 | 1560 | No data | |
| Moldova | 590 | | 3213 | 1106 | | |
| Russian Federation | 2410 | 4786 | 5437 | 3231 | | |
| Tajikistan | 340 | | 2149 | 678 | | |
| Turkmenistan | 940 | | 3227 | 1509 | | |
| Ukraine | 1200 | | 3761 | 1713 | | |
| Uzbekistan | 1010 | | 2536 | 1959 | | |
| ANGLO AMERICA | | | | | | |
| Canada | 19000 | 14648 | 16474 | 16725 | Very low rates | |
| United States of America | 28020 | 17743 | 19882 | 21203 | | |
| MEXICO AND CENTRAL AMERICA | | | | | | |
| United Mexican States | 3670 | 6488 | 5934 | 6041 | 8 | 13 |
| Belize | 2700 | 2579 | 3079 | 3290 | | |
| Costa Rica, Republic of | 2640 | 4912 | 4455 | 4903 | 5 | 5 |
| El Salvador, Republic of | 1700 | 2171 | 1798 | 2126 | 27 | 30 |
| Guatemala, Republic of | 1470 | 3506 | 2725 | 2937 | 38 | 51 |
| Honduras, Republic of | 660 | 1891 | 1683 | 1614 | 27 | 27 |
| Nicaragua, Republic of | 380 | 2506 | 1411 | 1571 | 35 | 33 |
| Panama, Republic of | 3080 | 5114 | 4547 | 5452 | 9 | 10 |
| WEST INDIES | | | | | | |
| Antigua and Barbuda | 7330 | 3802 | 6453 | 7017 | | |
| Bahamas, Commonwealth of | | 9033 | 9311 | 8416 | 2 | 2 |
| Barbados | | 7009 | 8258 | 8168 | 2 | 3 |
| Cuba, Republic of | | | | | 4 | 5 |
| Dominica, Commonwealth of | 3090 | 1803 | 2934 | | | |

DATA BANK

| Consumer Goods (3) | | | | | | | | |
|---|---|---|---|---|---|---|---|---|
| TVs/1000 | | Vehicles/1000 | | Telephone lines/1000 | | Computers/1000 | | |
| 1980 | 1995–96 | 1980 | 1995–96 | 1980 | 1995–96 | 1990 | 1995–96 | Political Freedom (4) |
| 310 | 444 | 108 | 273 | 58 | 281 | | 44 | 1.5 (free) |
| | 598 | | 189 | 162 | 296 | | 8 | 2 (free) |
| | 376 | | 237 | 115 | 268 | | 6 | 1.5 (free) |
| | 170 | | 142 | | 170 | | | 3.5 (part free) |
| 246 | 418 | 86 | 248 | 55 | 169 | 7 | 37 | 1.5 (free) |
| 184 | 226 | | 124 | 73 | 140 | | 5 | 3.5 (part free) |
| | 384 | | 216 | 94 | 232 | | 48 | 2.5 (part free) |
| | 374 | | 387 | | 333 | | 48 | 1.5 (free) |
| 446 | 185 | 118 | 163 | 71 | 197 | | | 6 (not free) |
| | 216 | | 2 | 98 | 154 | | | 4 (part free) |
| | 212 | | 48 | 55 | 86 | | | 6 (not free) |
| 218 | 292 | | 101 | 75 | 208 | | | 5 (part free) |
| | 473 | | 87 | 64 | 105 | | | 4.5 (part free) |
| | 275 | | 80 | 44 | 118 | | | 5.5 (not free) |
| | 230 | | 32 | 40 | 75 | | | 4 (part free) |
| | 307 | | 54 | 56 | 140 | | 3 | 4 (part free) |
| | 386 | | 158 | 70 | 175 | | 24 | 3.5 (part free) |
| | 279 | | 1 | 30 | 42 | | | 7 (not free) |
| | 162 | | | 38 | 74 | | | 7 (not free) |
| 255 | 341 | | 92 | 76 | 181 | | 6 | 3.5 (part free) |
| | 190 | | | 36 | 76 | | | 7 (not free) |
| 432 | 709 | 548 | 559 | 406 | 602 | 97 | 192 | 1 (free) |
| 562 | 806 | | 767 | 414 | 640 | 217 | 362 | 1 (free) |
| 57 | 193 | | 140 | 40 | 95 | | 29 | 4 (part free) |
| | 180 | | 88 | 27 | 133 | | 28 | 1 (free) |
| 68 | 220 | | 115 | 69 | 155 | | | 1.5 (free) |
| 66 | 250 | | 77 | 15 | 56 | | | 3 (part free) |
| 25 | 122 | | 18 | 12 | 31 | | 3 | 4.5 (part free) |
| 18 | 80 | | 33 | 8 | 31 | | | 3 (part free) |
| 57 | 170 | | 30 | 11 | 26 | | | 4 (part free) |
| 115 | 229 | | 99 | 65 | 122 | | | 2.5 (part free) |
| 262 | 414 | | | | 429 | | | 3.5 (part free) |
| 148 | 233 | | 207 | 148 | 278 | | | 1.5 (free) |
| 201 | 287 | | 164 | 139 | 370 | | 58 | 1 (free) |
| 131 | 200 | | 5 | 25 | 32 | | | 7 (not free) |
| | 183 | | | 16 | 264 | | | 1 (free) |

| Country | GNP capita 1996 (1) | GDP/cap PPP (1987 constant) (3) | | | Adult Illiteracy (2) | |
|---|---|---|---|---|---|---|
| | | | | | %Men | %Women |
| | | 1980 | 1990 | 1996 | 1995 | 1995 |
| WEST INDIES *(continued)* | | | | | | |
| Dominican Republic | 1600 | 2959 | 2905 | 3426 | 18 | 18 |
| French Guiana | | | | | | |
| Grenada | 2880 | 2204 | 3295 | 3382 | | |
| Guyana, Cooperative Republic of | 690 | 2316 | 1558 | 1870 | 1 | 3 |
| Haiti, Republic of | 310 | 1680 | 1268 | 854 | 52 | 58 |
| Jamaica | 1600 | 2657 | 2809 | 2699 | 19 | 11 |
| St. Kitts and Nevis, Federation of | 5870 | 2758 | 4637 | 5893 | | |
| St. Lucia | 3500 | | 3416 | 4017 | | |
| St. Vincent and the Grenadines | 2370 | 1729 | 2848 | 3261 | | |
| Suriname, Republic of | 1000 | 1297 | | | 5 | 9 |
| Trinidad and Tobago, Republic of | 3870 | 6343 | 4877 | 5048 | 1 | 3 |
| NORTHERN SOUTH AMERICA | | | | | | |
| Bolivia, Republic of | 830 | 2506 | 2038 | 2250 | 10 | 24 |
| Brazil, Federative Republic of | 4400 | 5060 | 4758 | 4911 | 17 | 17 |
| Colombia, Republic of | 2140 | 4158 | 4616 | 5299 | 9 | 9 |
| Ecuador, Republic of | 1500 | 3906 | 3616 | 3860 | 8 | 12 |
| Peru, Republic of | 2420 | 3815 | 2550 | 3422 | 6 | 17 |
| Venezuela, Republic of | 3020 | 7899 | 6237 | 6292 | 8 | 10 |
| SOUTHERN SOUTH AMERICA | | | | | | |
| Argentine Republic | 8380 | 7843 | 5934 | 7299 | 4 | 4 |
| Chile, Republic of | 4850 | 5521 | 6403 | 9089 | 5 | 5 |
| Paraguay, Republic of | 1850 | 3131 | 2923 | 2663 | 7 | 9 |
| Uruguay, Eastern Republic of | 5760 | 6571 | 4813 | 5932 | 3 | 2 |
| SOUTH PACIFIC | | | | | | |
| Australia, Commonwealth of | 20090 | 12432 | 13626 | 15578 | | |
| Fiji, Republic of | 2470 | 3038 | 3049 | 3215 | 6 | 11 |
| Kiribati, Republic of | | | | | | |
| Marshall Islands, Republic of | 1890 | | | | | |
| Micronesia, Federated States of | 2070 | | | | | |
| Nauru, Republic of | | | | | | |
| New Zealand | 15720 | 11789 | 12355 | 13435 | | |
| Palau, Republic of | | | | | | |
| Papua New Guinea, Independent State of | 1150 | 1919 | 1717 | 2308 | 19 | 37 |
| Solomon Islands | 900 | 1255 | 1659 | 1732 | | |
| Tonga, Kingdom of | | | | | | |
| Tuvalu | | | | | | |
| Vanuatu, Republic of | 1290 | | 2377 | 2153 | | |
| Western Samoa, Independent State of | 1170 | | | | | |

| | Consumer Goods (3) | | | | | | | | |
|---|---|---|---|---|---|---|---|---|---|
| | TVs/1000 | | Vehicles/1000 | | Telephone lines/1000 | | Computers/1000 | | |
| | 1980 | 1995–96 | 1980 | 1995–96 | 1980 | 1995–96 | 1990 | 1995–96 | Political Freedom (4) |
| | 70 | 84 | 36 | 47 | 19 | 83 | | | 3.5 (part free) |
| | 199 | 170 | | | 135 | 289 | | | |
| | | 82 | | | 34 | 243 | | | 1.5 (free) |
| | | 42 | | | 12 | 60 | | | 2 (free) |
| | 3 | 5 | | 7 | | 8 | | | 5 (not free) |
| | 78 | 326 | | 50 | 25 | 142 | | 5 | 2.5 (free) |
| | 91 | 244 | | | | 382 | | | 1.5 (free) |
| | 16 | 301 | | 98 | 42 | 235 | | | 1.5 (free) |
| | 53 | 234 | | | 36 | 171 | | | 1.5 (free) |
| | 113 | 208 | | 143 | 43 | 132 | | | 3 (part free) |
| | 195 | 318 | | 113 | 40 | 168 | | 19 | 1.5 (free) |
| | 56 | 202 | 19 | 48 | 25 | 47 | | | 3 (part free) |
| | 124 | 289 | 85 | 79 | 41 | 96 | | 18 | 3 (part free) |
| | 85 | 188 | | 39 | 41 | 118 | | 23 | 4 (part free) |
| | 63 | 148 | | 46 | 29 | 73 | | 4 | 2.5 (free) |
| | 52 | 142 | | 121 | 17 | 60 | | 6 | 4.5 (part free) |
| | 113 | 181 | 112 | 88 | 53 | 117 | | 21 | 3 (part free) |
| | 183 | 347 | 155 | 154 | 67 | 174 | | 25 | 2.5 (free) |
| | 110 | 280 | 61 | 110 | 33 | 156 | 11 | 45 | 2 (free) |
| | 22 | 144 | | 24 | 16 | 36 | | | 3.5 (part free) |
| | 125 | 305 | | 166 | 76 | 209 | | 22 | 2 (free) |
| | 381 | 666 | 502 | 603 | 323 | 519 | 150 | 311 | 1 (free) |
| | | 94 | | 73 | 38 | 88 | | | 3.5 (part free) |
| | | 21 | | | | 26 | | | 1 (free) |
| | | | | | | | | | 1 (free) |
| | 1 | 18 | | | | 65 | | | 1 (free) |
| | | | | | | | | | 2 (free) |
| | 329 | 517 | 492 | 562 | 226 | 499 | | 266 | 1 (free) |
| | | | | | | | | | 1.5 (free) |
| | | 4 | | 26 | 8 | 11 | | | 3 (part free) |
| | | 7 | | | | 18 | | | 1.5 (free) |
| | | 51 | | 18 | 15 | 67 | | | 4 (part free) |
| | | | | | | | | | 1 (free) |
| | | 13 | | 35 | 53 | 117 | | | 2 (free) |
| | | | | | | | | | 2 (free) |

| Country | Ethnic Groups (4) | Languages (4) [Official (O), National (N)] | Religions (4) |
|---|---|---|---|
| CENTRAL AFRICA | | | |
| Burundi, Republic of | Hutu 85%, Tutsi 14% | Kirundi (O), French (O) | Christian 67%, local 32%, Muslim 1% |
| Cameroon, Republic of | 200 groups: Fang, Barnileke, Fulani | English (O), French (O), 24 African | Christian 53%, local 25%, Muslim 22% |
| Central African Republic | Baya 34%, Banada 27%, Mandjia 21% | French (O), Sango (N) | Christian 50%, local 24%, Muslim 15% |
| Chad, Republic of | Muslim groups in N, non-Muslim in S | French (O), Arabic (O), Sara in S | Christian 25%, local 25%, Muslim 50% |
| Congo, Republic of the | Kongo 48%, Sangha 20%, Teke 17% | French (O), Kikongo, Lingala, Teke | Christian 50%, local 48%, Muslim 2% |
| Congo, Democratic Republic of the (Zaire) | Over 200 African groups: Mongo, Luba, Kongo | French (O), Lingala (N), others | Christian 70%, local 20%, Muslim 10% |
| Equatorial Guinea, Republic of | Fang 80%, Bubi 20% | Spanish (O), Fang | Officially Roman Catholic; local practices |
| Gabonese Republic | Fang 30%, Eshira 20%, M'bele 15%, Kota 13% | French (O), Fang, others | Christian 70%, local, Muslim 20% |
| Rwanda, Republic of | Hutu 90%, Tutsi 9% | Kinyarwanda (O), French (O), Kiswahili | Christian 74%, local 25%, Muslim 1% |
| WESTERN AFRICA | | | |
| Benin, Republic of | 42 African groups: Fon, Adju, Yoruba | French (O), Fon, Yoruba | Christian 15%, local 70%, Muslim 15% |
| Burkina Faso, People's Democratic Republic of | Mossi 25%, Gurumsi, Senufo | French (O), Sudanic tribal languages | Christian 10%, local 40%, Muslim 50% |
| Côte D'Ivoire, Republic of | 60 African groups, including foreign: Akan, Kru, Mande | French (O), Akan, Dioula | Christian 12%, local 65%, Muslim 23% |
| Gambia, Republic of the | Mandinke 42%, Fulani, Wolof | English (O), Mandinke, Wolof, Fula | Christian 9%, Muslim 90% |
| Ghana, Republic of | Fanti, Ashanti, Gadangbe, Ewe | English (O), various African | Christian 24%, local 38%, Muslim 30% |
| Guinea, Republic of | Fulani 35%, Malinde 30%, Sousson 20% | French (O), African languages | Christian 8%, local 7%, Muslim 85% |
| Guinea-Bissau, Republic of | Balanta 27%, Fula 23%, many others | Portuguese (O), Kriolu, French, local | Christian 5%, local 65%, Muslim 30% |
| Liberia, Republic of | African groups 95%, U.S. Liberians 5% | English (O), local Niger-Congo groups | Christian 10%, local 70%, Muslim 20% |
| Mali, Republic of | Mande 50%, Peul, Voltaic, Tuareg/Moor, Songhai | French (O), Bambara, others | Local 9%, Muslim 90% |
| Mauritania, Islamic Republic of | Maure (Arab-Berber) 80%, Fulani, Wolof | Arabic (O), Wolof, French | Muslim 100% |
| Niger, Republic of | Hausa 56%, Djena 22% | French (O), 10 other official | Christian/local 20%, Muslim 80% |
| Nigeria, Federal Republic of | Hausa 65%, Fulani, Yoruba, Ibo | English (O), Hausa, others | Christian 40%, local 10%, Muslim 50% |
| Senegal, Republic of | Wolof 36%, Fulani 17%, Serer 17% | French (O), Wolof, Serer, others | Christian 2%, local 6%, Muslim 92% |
| Sierra Leone, Republic of | Mended, Temeden, Krioles, others | English (O), Krio, others | Christian 10%, local 30%, Muslim 60% |
| Togo, Republic of | 37 African groups: Ewe, Kabre | French (O), Ewe, Kabre, others | Christian 35%, local 50%, Muslim 15% |
| EASTERN AFRICA | | | |
| Djibouti, Republic of | Somali 60%, Ethiopian 35% | French (O), Arabic (O) | Christian 6%, Muslim 94% |
| Eritrea, Republic of | Tigray 50%, Tigre-Kunama 40% | Tigre, Afar, others | Christian, Muslim |
| Ethiopia, Federal Democratic Republic of | Oromo 40%, Amharan 32%, Tigre | Amharic (O) | Christian 45%, local 12%, Muslim 40% |
| Kenya, Republic of | Kikuyu 21%, Luo 15%, Luhya 14% | English (O), Swahili (O) | Christian 70%, local 10%, Muslim 6% |
| Somali Democratic Republic | Somali 85% | Somali (O) | Muslim 99% |
| Tanzania, United Republic of | 120 African culture groups | Swahili (O), English (O) | Christian 43%, Muslim 35% |
| Uganda, Republic of | Ganda, Kamora, Teso, others | English (O), Luganda, Swahili, others | Christian 66%, local 18%, Muslim 16% |
| SOUTHERN AFRICA | | | |
| Angola, Republic of | Ovimbundu 37%, Mbundu 27% | Portuguese (O), Bantu languages | Christian 85%, local 10% |
| Botswana, Republic of | Tswana 75% | English (O), Setswana | Christian 50%, local 50% |
| Lesotho, Kingdom of | Basutho 79%, Nguni 20% | Sesotho (O), English (O) | Christian 79%, local 20% |
| Madagascar, Republic of | Malaysian, Indonesian, coastal groups | French (O), Malagasy (O) | Christian 41%, local 52%, Muslim 7% |
| Malawi, Republic of | Chewa, Nyanja, others | English (O), Chichewa (O) | Christian 70%, Muslim 20% |
| Mozambique, Republic of | Makua-Lomwe, Yao, others | Portuguese (O) | Christian 30%, local 60%, Muslim 10% |
| Namibia, Republic of | Orambo 50%, European 6%, mixed 7% | English (O), local, Afrikaans | Christian 90% |
| South Africa, Republic of | African 75%, European 14%, mixed 8% | Afrikaans (O), English (O), Zulu, Xhosa | Christian, Hindu, Muslim, Jewish |
| Swaziland, Kingdom of | African 97%, European 3% | English (O), Siswati (O) | Christian 60%, local 40% |

| Country | Ethnic Groups (4) | Languages (4) [Official (O), National (N)] | Religions (4) |
|---|---|---|---|
| SOUTHERN AFRICA *(continued)* | | | |
| Zambia, Republic of | African 98.7%, European 1.1% | English (O), Ichbula, others | Christian 60%, Muslim/Hindu 40% |
| Zimbabwe, Republic of | Shona 71%, Ndebele 16%, European 2% | English (O), Chishona, Sindebele | Christian 20%, syncretic Christian/local 50% |
| NORTH AFRICA | | | |
| Algeria, Democratic and Popular Republic of | Arab 83%, Berber 16% | Arabic (O), Berber dialects | Muslim 99% |
| Socialist People's Libyan Arab Jamahirya | Arab-Berber 97% | Arabic (O), Italian, English | Muslim 97% |
| Morocco, Kingdom of | Arab-Berber 99% | Arabic, local dialects | Muslim 98% |
| Tunisia, Republic of | Arab-Berber 98% | Arabic (O), French, Berber | Muslim 98% |
| NILE RIVER VALLEY | | | |
| Egypt, Arab Republic of | Egyptian-Bedouin-Berber 99% | Arabic (O), English, French | Christian 6%, Muslim 94% |
| Sudan, Republic of | African 49%, Arab 39%, Nubian 8% | Arabic (O), Nubian, Dinka, others | Christian 5%, local 25%, Muslim 70% |
| SOUTHWESTERN ASIA | | | |
| Bahrain, State of | Bahraini 63%, Asian 13%, 10% other Arab, 8% Iran | Arabic, English, Persian, Urdu | Christian 9%, Muslim 85% |
| Iran | Persian 51%, Azerbaijani 24%, Kurd 7% | Persian (Farsi), Turkic, Kurdish | Muslim 99% |
| Iraq, Republic of | Arab 75%, Kurd 15%, Assyrian, Turkana | Arabic, Kurdish, Assyrian, Armenian | Christian 3%, Muslim 97% |
| Israel | Jewish 82%, non-Jewish (mainly Arab) 18% | Hebrew (O), Arabic, English | Jewish 82%, Muslim 14% |
| Jordan, Hashemite Kingdom of | Arab 98% | Arabic (O), English | Christian 8%, Muslim 92% |
| Kuwait, State of | Kuwaiti 48%, other Arab 35%, 9% Asian, 4% Iran | Arabic (O), English | Muslim 85% |
| Lebanese Republic | Arab 95%, Armenian 4% | Arabic (O), English | Christian 30%, Muslim 70% |
| Oman, Sultanate of | Omani Arab 75%, Pakistani 21% | Arabic (O), English, Urdu | Muslim 98% |
| Qatar, State of | Arab 40%, Pakistani 18%, Indian 18%, Iranian 10% | Arabic (O), English | Muslim 95% |
| Saudi Arabia, Kingdom of | Saudi 73% (90% Arab), foreign workers 27% | Arabic (O), English | Muslim 100% |
| Syrian Arab Republic | Arab 90%, Kurd 7%, Armenian | Arabic (O), Kurdish, French | Christian 10%, Muslim 89% |
| Turkey | Turkish 80%, Kurd 17% | Turkish (O), Kurdish, Arabic | Muslim 99% |
| United Arab Emirates | Emiri 19%, other Arab 23%, Asian 50% | Arabic (O), Persian, English | Muslim 96% |
| Yemen, Republic of | Mainly Arab, with African-Arab and Indian | Arabic (O) | Muslim dominant |
| SOUTHERN ASIA | | | |
| Afghanistan, Islamic State of | Pashtun 38%, Tajik 25%, Hazara 19% | Afghan-Persian 50%, Pashtu 35% | Muslim 99% |
| Bangladesh, People's Republic of | Bengali 98% | Bangla (O), Urdu, English | Hindu 16%, Muslim 83% |
| Bhutan, Kingdom of | Bhote 50%, Nepalese 35% | Dzongkha (O), Tibetan/Nepali dialects | Lamaistic Buddhist 75%, Hindu 25% |
| India, Republic of | Indo-Aryan 72%, Dravidian 25% | Hindi 30%, 14 official regional, English | Christian 2%, Hindu 80%, Muslim 14% |
| Maldives, Republic of | Sinhalese, Dravidian, Arab, African | Divehi (Sinhala), English | Muslim dominant |
| Nepal, Kingdom of | Nawar, Indian, Tibetan, Gurungi, Sherpa | Nepali (O), 20 others, English | Buddhist 5%, Hindu 90%, Muslim 3% |
| Pakistan, Islamic Republic of | Punjabi, Sindhi, Pashtun, Muhajir | Urdu (O), English (O), others | Muslim 97% |
| Sri Lanka, Democratic Socialist Republic of | Sinhalese 74%, Tamil 18%, Moor 7% | Sinhala, Tamil, English | Buddhist 69%, Christian 8%, Hindu 15% |
| EASTERN ASIA | | | |
| China, People's Republic of | Han Chinese 92%, Shuang, Mongolian, Tibetan | Mandarin (N), Yue (S), dialects | (Atheist) Buddhist, Confucian, Taoist |
| Hong Kong (to China, 1997) | Chinese, Europeans | English, Chinese | Buddhist, Christian, other |
| Japan | Japanese 99%, some Korean, indigenous | Japanese | Shinto/Buddhist 84% |
| Korea, People's Democratic Republic of (North) | Korean | Korean | Buddhist, Confucian |
| Korea, Republic of (South) | Korean | Korean (O), English | Buddhist 47%, Christian 49%, Confucian |

DATA BANK

| Country | Ethnic Groups (4) | Languages (4) [Official (O), National (N)] | Religions (4) |
|---|---|---|---|
| EASTERN ASIA *(continued)* | | | |
| Mongolia | Mongolian 90%, Kazak, Chinese, Russian | Khalka Mongol 90%, Turkic, Russian | Tibetan Buddhist, some Muslim |
| Taiwan | Taiwanese 84%, mainland Chinese 14% | Mandarin Chinese, local dialects | Buddhist 50%, Christian 5%, Confucian/Taoist 25% |
| Singapore, Republic of | Chinese 76%, Malay 15%, Indian 6% | Chinese (O), Malay (O), English (O) | Buddhist, Christian, Muslim |
| SOUTHEASTERN ASIA | | | |
| Cambodia, Kingdom of | Khmer 90%, Vietnamese 5% | Khmer (O), French | Buddhist 95% |
| Indonesia, Republic of | Javanese 45%, Sundanese 15%, Madurese, Malay | Bahasa (Malay) (O), English, Dutch | Christian 9%, Muslim 87%, Buddhist, Hindu |
| Laos, People's Democratic Republic of | Lao 50%, Thai 14%, many tribes | Lao (O), Thai, French, local | Buddhist 60%, local 40% |
| Malaysia, Federation of | Malay 59%, Chinese 32%, Indian 9% | Malay (O), Chinese, English, Tamil | Buddhist, Christian, Hindu, Muslim (main) |
| Myanmar, Union of (Burma) | Burman 68%, Shan, Karen, other tribes | Burmese, local languages | Buddhist 89%, Christian 4%, Muslim 4% |
| Philippines, Republic of | Malay 95%, Chinese 2% | Filipino (O), English (O), local | Christian 92%, Muslim 5% |
| Thailand, Kingdom of | Thai 75%, Chinese 14%, Malay 3% | Thai (O), English, Chinese, Malay | Buddhist 95%, Muslim 4% |
| Vietnam, Socialist Republic of | Vietnamese 88%, Chinese 2%, tribal groups | Vietnamese (O), French, English | Buddhist 55%, Christian 12% |
| WESTERN EUROPE | | | |
| Belgium, Kingdom of | Flemish 55%, Walloon 33% | Flemish (Dutch) 56%, French 32% | Protestant 25%, Roman Catholic 75% |
| French Republic | Celtic/Latin with Teutonic, Slavic, Nordic, African | French, with dialects | Roman Catholic 90% |
| Germany, Federal Republic of | German 95%, Turkish 2% | German, English | Protestant 45%, Roman Catholic 37% |
| Ireland, Republic of | Celtic, English | Irish (Gaelic), English | Roman Catholic 93% |
| Luxembourg, Grand Duchy of | Celtic/French/German,some Portuguese, Italian | Luxemburgisch (O), German, French | Roman Catholic 97% |
| Netherlands, Kingdom of | Dutch 96%, Moroccan, Indonesian, Turkish | Dutch (O), Frisian, English, German | Protestant 25%, Roman Catholic 34% |
| United Kingdom of Great Britain and Northern Ireland | English 82%, Scottish (10%), Irish, Welsh | English (O), Welsh, Scottish Gaelic | Protestant 49%, Roman Catholic 16% |
| NORTHERN EUROPE | | | |
| Denmark, Kingdom of | Danish, Inuit (Eskimo), Faroese, Greenlander | Danish, Faroese | Protestant 91% |
| Finland, Republic of | Finnish 90%, Swedish 9% | Finnish (O), Swedish | Protestant 89% |
| Iceland, Republic of | Norwegian/Celtic descendants | Icelandic (O), Danish, English | Protestant 96% |
| Norway, Kingdom of | Germanic (Nordic, alpine, Baltic) | Norwegian (O), English | Protestant 88% |
| Sweden, Kingdom of | Swedish, Finnish, Danish, Norwegian, Greek, Turkish | Swedish, English | Protestant 94% |
| ALPINE EUROPE | | | |
| Austria, Republic of | German | German (O) | Protestant 6%, Roman Catholic 85% |
| Swiss Confederation | German 85%, French 18%, Italian 10% | German (O), French (O), Italian (O) | Protestant 44%, Roman Catholic 48% |
| MEDITERRANEAN EUROPE | | | |
| Helenic Republic (Greece) | Greek 98% | Greek (O), Turkish | Greek Orthodox 98% |
| Italian Republic | Italian, Sicilian, Sardinian, German, French | Italian (O), German, French, Slovenian | Roman Catholic 98% |
| Portuguese Republic | Mediterranean | Portuguese (O) | Roman Catholic 97% |
| Spain, Kingdom of | Mediterranean and Nordic | Castilian Spanish 74%, Catalan 17% | Roman Catholic 99% |
| EASTERN AND BALKAN EUROPE | | | |
| Albania, Republic of | Albanian 95%, Greek 3% | Albanian (O) | Christian 30%, Muslim 70% |
| Bosnia-Herzogovina, Republic of | Muslim 44%, Croat 17%, Serbian 31% | Serbo-Croat 99% | Christian 46%, Muslim 40% |
| Bulgaria, Republic of | Bulgarian 85%, Turkish 9% | Bulgarian | Bulgarian Orthodox 85%, Muslim 13% |
| Croatia, Republic of | Croatian 78%, Serbian 12% | Serbo-Croat 99% | Orthodox 11%, Roman Catholic 77% |
| Czech Republic | Czech 81%, Moravian 13%, Slovak 3% | Czech, Slovak, German | Protestant 5%, Roman Catholic 39%, none 40% |

| Country | Ethnic Groups (4) | Languages (4) [Official (O), National (N)] | Religions (4) |
|---|---|---|---|
| EASTERN AND BALKAN EUROPE *(continued)* | | | |
| Estonia, Republic of | Estonian 62%, Russian 30% | Estonian (O), Latvian, Russian | Protestant, Russian Orthodox |
| Hungary, Republic of | Hungarian (Magyar) 90% | Hungarian 98% | Protestant 25%, Roman Catholic 68% |
| Latvia, Republic of | Latvian 52%, Russian 34% | Latvian (O), Russian | Protestant, Roman Catholic, Orthodox |
| Lithuania, Republic of | Lithuanian 80%, Russian 9%, Polish 8% | Lithuanian (O), Russian | Protestant, Roman Catholic (most) |
| Macedonia, Former Yugoslav Republic of | Macedonian 65%, Albanian 21% | Macedonian, Albanian | Macedonian Orthodox 67%, Muslim 30% |
| Poland, Republic of | Polish 98% | Polish (O), English, German | Roman Catholic 75% |
| Romania | Romanian 89%, Hungarian 9% | Romanian (O), Hungarian | Romanian Orthodox 70%, other Christian 12% |
| Slovak Republic | Slovak 86%, Hungarian 11% | Slovak (O), Hungarian | Protestant 8%, Roman Catholic 60% |
| Slovenia, Republic of | Slovene 91%, Croat 3% | Slovenian, Serbo-Croatian | Roman Catholic 96% |
| Yugoslavia, Federal Republic of (Serbia-Montenegro) | Serbian 63%, Albanian 14%, Montenegrin 6% | Serbo-Croatian 99% | Orthodox 65%, Muslim 19% |
| COMMONWEALTH OF INDEPENDENT STATES | | | |
| Armenia | Armenian 93%, Azeri 3%, Russian 2% | Armenian 96% | Armenian Orthodox 94% |
| Azerbaijan | Azeri 90% | Azeri 89%, Russian | Nominal Muslim 96% |
| Belarus | Belarussian 78%, Russian 13%, Polish 4% | Belarussian, Russian | Eastern Orthodox |
| Georgia | Georgian 70%, Armenian 8%, Russian 6% | Georgian, Russian | Georgian Orthodox 65%, Muslim 11% |
| Kazakstan | Kazak 42%, Russian 37%, Ukraine 5%, German 5% | Kazak, Russian | Muslim 47%, Russian Orthodox 44% |
| Kyrgyzstan | Kirghiz 52%, Russian 21%, Uzbek 13% | Kirghiz, Russian | Muslim 70%, Russian Orthodox, other 30% |
| Moldova | Moldovan/Romanian 65%, Russian 13%, Ukraine 14% | Romanian, Russian | Eastern Orthodox 98% |
| Russian Federation | Russian 81%, Tatar 4%, Ukraine 3% | Russian, Tatar, others | Russian Orthodox, Muslim |
| Tajikistan | Tajik 65%, Uzbek 25%, Russian 4% | Tajik, Russian, Uzbek | Sunni Muslim 80%, other Muslim 7% |
| Turkmenistan | Turkmen 73%, Russian 10%, Uzbek 9% | Turkmen, Russian, Uzbek | Muslim 87%, Eastern Orthodox 11% |
| Ukraine | Ukrainian 73%, Russian 22% | Ukrainian, Russian | Ukraine Orthodox, Catholic |
| Uzbekistan | Uzbek 71%, Russian 8%, Tajik 5% | Uzbek 74%, Russian 14% | Muslim (mostly Sunni) 88% |
| ANGLO AMERICA | | | |
| Canada | European (British 40%, French 27%, other 20%) | English (O), French (O) | Protestant 26%, Roman Catholic 46% |
| United States of America | European 83%, African 12%, Asian 3%, native 1% | English, Spanish | Protestant 56%, Roman Catholic 28% |
| MEXICO AND CENTRAL AMERICA | | | |
| United Mexican States | Mestizo 60%, Native American 30%, European 9% | Spanish (O), local, English | Roman Catholic (nominal) 89% |
| Belize | Mestizo 44%, Creole 30%, Mayan 11% | English (O), Spanish, Mayan | Protestant 30%, Roman Catholic 62% |
| Costa Rica, Republic of | Mestizo and European descendants 95% | Spanish (O), English | Roman Catholic 95% |
| El Salvador, Republic of | Mestizo 94%, Native American 5% | Spanish, native tongues, English | Protestant 25%, Roman Catholic 75% |
| Guatemala, Republic of | Mestizo 56%, Native American 44% | Spanish (O), 20 Native American dialects | Roman Catholic, some Protestant, local |
| Honduras, Republic of | Mestizo 90%, Native American 7% | Spanish, local dialects | Roman Catholic 97% |
| Nicaragua, Republic of | Mestizo 69%, European 17%, African 9% | Spanish (O), local | Protestant 5%, Roman Catholic 95% |
| Panama, Republic of | Mestizo 70%, Caribbean 14%, European 10% | Spanish (O), English, Creole, local | Protestant 15%, Roman Catholic 85% |
| WEST INDIES | | | |
| Antigua and Barbuda | African 96%, European 3% | English (O) | Protestant (main), Roman Catholic, Muslim |
| Bahamas, Commonwealth of | African 75%, European 15% | English, Creole | Protestant 75%, Roman Catholic 19% |
| Barbados | African 80%, mixed 16%, European 4% | English | Protestant 57%, Roman Catholic 4% |
| Cuba, Republic of | African 11%, European 37%, mixed 51% | Spanish | Roman Catholic 40%, none 55% |

| Country | Ethnic Groups (4) | Languages (4) [Official (O), National (N)] | Religions (4) |
|---|---|---|---|
| WEST INDIES *(continued)* | | | |
| Dominica, Commonwealth of | African, Carib native | English (O), French patois | Protestant 15%, Roman Catholic 77% |
| Dominican Republic | African 11%, European 16%, mixed 73% | Spanish | Roman Catholic 95% |
| French Guiana | African, European, Arawak native | French (O), Creole | Roman Catholic, Hindu |
| Grenada | African | English (O), French patois | Protestant 35%, Roman Catholic 60% |
| Guyana, Cooperative Republic of | African/mixed 43%, Asian Indian 51% | English, Hindi, Urdu | Christian 57%, Hindu 33%, Muslim 9% |
| Haiti, Republic of | African 95% | French (O), Creole (O) | Protestant 16%, Roman Catholic/scale voodoo 80% |
| Jamaica | African 76%, mixed 15%, Asian Indian 3% | English (O), Creole | Protestant 56%, Hindu, Muslim, Jewish 39% |
| St. Kitts and Nevis, Federation of | African | English | Protestant, Roman Catholic |
| St. Lucia | African 90%, mixed 6% | English (O), French patois | Protestant 7%, Roman Catholic 90% |
| St. Vincent and the Grenadines | African, European, Asian Indian, Carib native | English, French patois | Protestant, Roman Catholic |
| Suriname, Republic of | Creole 31%, Asian Indian 37%, Indonesian 15% | Dutch (O), English, others | Christian 58%, Hindu 27%, Muslim 20% |
| Trinidad and Tobago, Republic of | African 43%, mixed 14%, Asian Indian 40% | English (O), Hindi, French, Spanish | Christian 60%, Hindu 24%, Muslim 6% |
| NORTHERN SOUTH AMERICA | | | |
| Bolivia, Republic of | Mestizo 30%, Quechua 30%, Aymura 25%, European 10% | Spanish (O), Quechua (O), Aymura (O) | Roman Catholic 95% |
| Brazil, Federative Republic of | African 11%, European 55%, mixed 32% | Portuguese (O), Spanish, English | Roman Catholic 73%, other Christian 20% |
| Colombia, Republic of | Mestizo 58%, European 20% | Spanish (O), Quechua (O), Aymura (O) | Roman Catholic 95% |
| Ecuador, Republic of | Mestizo 55%, Native American 25%, European 10% | Spanish (O), Native American | Roman Catholic 95% |
| Peru, Republic of | Mestizo 37%, Native American 45%, European 15% | Spanish (O), Quechua (O) | Roman Catholic 95% |
| Venezuela, Republic of | Mestizo 67%, European 21%, African 10% | Spanish (O), English, local | Roman Catholic 96% |
| SOUTHERN SOUTH AMERICA | | | |
| Argentine Republic | Mestizo/Native American 15%, European 85% | Spanish (O), English, Italian | Roman Catholic 90% |
| Chile, Republic of | Native American 3%, European 95% | Spanish | Protestant 11%, Roman Catholic 89% |
| Paraguay, Republic of | Mestizo 95% | Spanish (O), Guarani, Portuguese | Protestant 10%, Roman Catholic 90% |
| Uruguay, Eastern Republic of | Mestizo 8%, European 88% | Spanish | Protestant 4%, Roman Catholic 66% |
| SOUTH PACIFIC | | | |
| Australia, Commonwealth of | European 95%, Asian 4%, Aborigine 1% | English | Protestant 50%, Roman Catholic 26% |
| Fiji, Republic of | Fijian 49%, Asian Indian 46% | English (O), Fijian, Hindustani | Christian 52%, Hindu 38%, Muslim 8% |
| Kiribati, Republic of | Micronesian with Tuvaluan minority | English (O), Gilbertese | Protestant 40%, Roman Catholic 52% |
| Marshall Islands, Republic of | Micronesian, German, Japanese, U.S., Filipino | English (O), local Malay dialects | Christian (mainly Protestant) |
| Micronesia, Federated States of | Chuukare 41%, Pohnpeian 26%, other groups | English (O), local | Christian 97% |
| Nauru, Republic of | Naurean 58%, other Pacific 26%, Chinese 8% | Naurean (O), English | Christian (mainly Protestant) |
| New Zealand | Maori 9%, European 88% | English (O), Maori (O) | Christian 67% |
| Palau, Republic of | Malay, Melanesian, Filipino, Polynesian | English (O), state languages | Christian, mainly Roman Catholic, local |
| Papua New Guinea, Independent State of | Melanesian 98% | Melanesian and Papuan languages | Protestant 44%, Roman Catholic 22%, local |
| Solomon Islands | Melanesian 93% | Melanesian, pidgin, English, local | Christian 95% |
| Tonga, Kingdom of | Tongan 98% | Tongan (O), English (O) | Christian |
| Tuvalu | Polynesian 96% | Tuvaluan, English | Church of Tuvalu 97% |
| Vanuatu, Republic of | Melanesian 94% | English (O), French (O) | Christian, mostly Protestant 84% |
| Western Samoa, Independent State of | Samoan 93%, Euronesian mixtures 7% | Samoan, English | Christian 99% |

# INDEX

Note: Page numbers in *italics* indicate illustrations.